Springer Series in Statistics

Springer Series in Statistics (SSS) is a series of monographs of general interest that discuss statistical theory and applications.

The series editors are currently Peter Bühlmann, Peter Diggle, Ursula Gather.

Scott Zeger, Peter Bickel, Ingram Olkin, and Stephen Fienberg were editors of the series for many years.

Yury A. Kutoyants

Hidden Markov Processes and Adaptive Filtering

Springer

Yury A. Kutoyants
Le Mans University
Le Mans, France

ISSN 0172-7397 ISSN 2197-568X (electronic)
Springer Series in Statistics
ISBN 978-3-032-00051-4 ISBN 978-3-032-00052-1 (eBook)
https://doi.org/10.1007/978-3-032-00052-1

This Springer imprint is published by the registered company Springer Nature Switzerland AG
The registered company address is: Gewerbestrasse 11, 6330 Cham, Switzerland

Preface

This work proposes the construction of adaptive, recursive, and asymptotically efficient filtering algorithms for several hidden Markov processes (HMP). There is no need to talk about the importance of this topic for a wide class of applied problems. There is an extensive literature on adaptive filtering, mainly engineering and for discrete-time models. Some works are also devoted to constructing recursive algorithms in these problems. Sometimes the issue of asymptotic efficiency of algorithms using maximum likelihood estimate (MLE) is also discussed. This work considers continuous-time HMPs (except Chap. 6, where hidden AR is studied) for several reasons. First, these are the models I usually work with. Second, I assume that many real dynamic systems "live in real (continuous) time." Third, if such systems are observed at discrete moments, then the results of the analysis of estimates obtained for continuous-time models are interesting as fundamental limits for certain characteristics of estimates in discrete-time models. In all problems, the properties of estimates and filters are described in an asymptotic setting. The asymptotics of either small noise or large samples are considered. One might think that this is a purely mathematical theory, but I believe that such a formulation is still reasonable for applications by the following reason. As a rule, the most interesting situations are those in which it is possible to make decisions (estimates, test hypotheses) with a small error, but a small error implicitly assumes that there is a parameter that ensures this smallness. Most often, this corresponds to either weak noise in observations, or a large volume of observations (large samples), or, as in the case of estimating the volatility of a diffusion process, there is a large number of discrete observations on a fixed interval. In all these cases, the asymptotic approach seems to be reasonable in the applied problems too.

The proposed algorithms for constructing adaptive filters across all models follow a unified approach. First, a consistent estimate of the unknown parameter is obtained over a small learning interval, albeit with a "poor convergence rate" (though not excessively poor). This estimate is then used to build a One-step MLE-process that is asymptotically equivalent to the MLE. Finally, the adaptive filter is derived by substituting this estimate-process into the filtering equations.

All elements of this algorithm are well-known. For instance, method of moments estimates are chosen as preliminary consistent estimates, and these estimates are often used in adaptive filtering. One-step MLE, originally proposed by Fisher and later studied by Le Cam, are widely applied in statistical problems. The novelty here lies in a modification of the One-step MLE, which treats it as a random process, where the role of time is taken by the upper limit of the integral (or sum) in the Score-function. Additionally, the lower bound on the quadratic risk of filter estimation has not, to my knowledge, been addressed before.

The models examined in this work are relatively simple: one-dimensional observation and state processes, uncorrelated noises, mainly linear models, etc., though the proofs can sometimes be nontrivial. Five different models related to the problem of adaptive filtering (Chaps. 2–6), as well as a problem of source localization on a plane and BSDE approximation in the case of hidden forward equation (Chap. 7), are studied. In particular, an adaptive estimate of the source trajectory is constructed. Future research could focus on increasing the complexity of models.

Several chapters of this book were used in my lecture course given in the educational program of the Vega Institute Foundation (Moscow) in 2023.

In working on these models, I greatly benefited from discussions with all my co-authors and colleagues, to whom I am thankful. Particularly I am very grateful to P. Chigansky, S. Dachian, R. Khasminskii, and N. Yoshida for fruitful discussions. R. Liptser introduced me to Kalman filtering in the early 1970s. Notably, he shared with me the martingale proof of Theorem 1.3, which appears in the Chap. 1. I am deeply grateful to him for this and many other contacts and regret that he did not live to see this work.

Thanks are due as well to the Springer staff, especially Eva Hiripi for her kind help and continuous support. My thanks are also due to Le Mans University, under grant ANR-21-CE40-0021, for support during the preparation of this text.

Le Mans, France Yury A. Kutoyants
April 2025

Declarations

Competing Interests The author has no competing interests to declare that are relevant to the content of this manuscript.

Introduction

Hidden Markov models are widely used for the description of many real systems. This is why such systems are extensively studied by engineers and mathematicians. We are interested in the problems of estimating the unobserved (hidden) component. This class of problems, called filtration, has seen spectacular development following the success of the applications of the so-called Kalman filtration.

The filter is named after Rudolf E. Kálmán, one of the primary developers of its theory [83] (discrete-time models) and, together with R.S. Bucy, [84] (continuous-time models). This filter is sometimes called the *Stratonovich-Kalman-Bucy* filter because it is a special case of a more general, nonlinear filter developed somewhat earlier by Ruslan L. Stratonovich [181–184]. We can cite here some details from Kailath et al. [80], p. 651: *Stratonovich recognized that ... for Gaussian Markov processes, the equation for the conditional covariance is a deterministic Riccati differential equation, whose solution can be used to completely define the evolution equation for the conditional means, see, e.g. Stratonovich [183], eq. (59), which gives (of course in different notation) exactly the continuous-time Kalman-Bucy equations!* More about contribution of Stratonovich can be found in [178].

We do not present here a detailed list of references for the works on Kalman filtering, but note that by 1974, there were already about 400 publications [79]. See as well the references in Ephraim and Mehrav [53] and Haykin [67]. There are many books on hidden Markov models and filtering theory, we can mention here [3, 7, 10, 26, 39, 50, 77, 81, 102, 141, 142, 170, 196]. The adaptive filtering as well has a well-developed theory and applications which can be found in the following monographs: [43, 63, 143, 145, 150, 172–174, 194].

Kalman filtering has numerous applications. We do not provide a long list of applied problems here, but note that different versions of discrete and continuous-time models, where the technique of Kalman filters and some of its generalizations to nonlinear models were fruitfully studied and applied. We just mention several references devoted to applications in Financial mathematics [148, 149, 196], Signals processing [65, 155], Biology [1, 25], Econometrics [66, 175], GPS/INS navigation [191], and Vehicle navigation [71].

This work is devoted to different statistical problems related to partially observed systems called *hidden Markov processes*. It is always assumed that these systems depend on some unknown finite-dimensional parameters, and the main problem is to answer the following question: how to construct a good approximation of the conditional expectation of the *hidden* component? The considered models of observations are mainly linear, but similar heuristic solutions are discussed for some nonlinear models as well. In particular, we study (continuous-time) partially observed diffusion processes (Chaps. 2–4), telegraph processes observed in white Gaussian noise (WGN) (Chap. 5), and autoregressive processes observed in WGN (discrete-time, Chap. 6). At the end (Chap. 7), we discuss problems of source localization by observations of partially observed systems and some other problems. Except Chaps. 5 and 6 we study different versions of the equations of well-known Kalman-Bucy filters (continuous-time models) [84]. In the case of the telegraph process (Chap. 5) the basic equation is given by the Wonham filter [197]. In Chap. 6 we use Kalman filter [83].

As the observed systems depend on unknown parameters, the construction of the adaptive Kalman or Wonham filters is realized following the well-known approach in such situations, i.e., in two steps: first, we estimate the unknown parameter, and then this estimator is substituted into the equations of the filter. The solutions of the obtained equations provide the approximation of the conditional expectation, which is called the *adaptive filter*. Due to the importance of Kalman filtering in many applied problems, similar procedures of adaptation have been realized many times, and there exists a vast literature on adaptive filters, especially for discrete-time models. Nevertheless, we suppose that the approach proposed in this work can be interesting as well.

We do not discuss the other methods of estimation like *Expectation-Maximization* (see Dembo and Zeitouni [42] or Campillo and Le Gland [22]) or *Markov chain Monte Carlo* (MCMC) algorithms (Särkä and Svensson [173]), *Particle filter* (Storvik [180]). Following [198], where further references can be found, we just mention here the names of several approaches to the construction of adaptive filters: *innovation-based adaptive filter*, *interacting multiple model filter*, *Sage-Husa adaptive filter*, *variational Bayesian approach*. Remark that it is possible to construct and to study the recursive MLE [137, 185]. For general hidden Markov models the conditions of consistency of MLE were proposed by Leroux [138], see as well Genon-Catalot and Laredo [58] and Douc et al. [45]. The asymptotic normality of estimators was descried by Bickel et al. [12].

To explain the main lines of the approach developed in this work, we illustrate it below with the help of the following *symbolic model* of observations. We do not specify the asymptotic (small noise or large samples) and describe the construction in a heuristic and formal way. The review of such parameter estimation problems can be found here [116].

We are given a couple of equations:

$$\mathrm{d}X_t = f\left(\vartheta, t\right) Y_t \,\mathrm{d}t + \sigma\left(t\right) \mathrm{d}W_t, \qquad X_0, \quad 0 \le t \le T, \tag{1}$$

$$\mathrm{d}Y_t = a(\vartheta, t)\, Y_t\, \mathrm{d}t + b(\vartheta, t)\, \mathrm{d}V_t, \qquad Y_0, \quad 0 \le t \le T, \tag{2}$$

where $W_t, 0 \le t \le T$ and $V_t, 0 \le t \le T$ are independent Wiener processes and the observations are $X^T = (X_t, 0 \le t \le T)$. The Gaussian process $Y^T = (Y_t, 0 \le t \le T)$ is *hidden*. The functions $f(\vartheta, t), \sigma(t), a(\vartheta, t), b(\vartheta, t), t \in [0, T]$ are assumed to be known, sufficiently smooth with respect to ϑ, and the parameter $\vartheta \in \Theta \subset \mathbb{R}^d$ is unknown.

The conditional expectation $m(\vartheta, t) = \mathbf{E}_\vartheta \left(Y_t | \mathfrak{F}_t^X\right)$ and the mean squared error $\gamma(\vartheta, t) = \mathbf{E}_\vartheta (Y_t - m(\vartheta, t))^2$ are solutions of the Kalman-Bucy (K-B) filtration equations [84]. Here $\mathfrak{F}_t^X$ is σ-algebra generated by the observations $X^t = (X_s, 0 \le s \le t)$. The K-B equations are

$$\mathrm{d}m(\vartheta, t) = \left[a(\vartheta, t) - \frac{\gamma(\vartheta, t)\, f(\vartheta, t)^2}{\sigma(t)^2}\right] m(\vartheta, t)\, \mathrm{d}t + \frac{\gamma(\vartheta, t)\, f(\vartheta, t)}{\sigma(t)^2} \mathrm{d}X_t, \tag{3}$$

$$\frac{\partial \gamma(\vartheta, t)}{\partial t} = 2a(\vartheta, t)\, \gamma(\vartheta, t) - \frac{\gamma(\vartheta, t)^2 f(\vartheta, t)^2}{\sigma(t)^2} + b(\vartheta, t)^2, \tag{4}$$

with initial values $m_0 = \mathbf{E}_\vartheta (Y_0 | X_0)$ and $\gamma_0 = \mathbf{E}_\vartheta (Y_0 - m(\vartheta, 0))^2$. Recall that $m(\vartheta, t)$ is optimal in different senses estimator of Y_t.

The main problem is: *To obtain a good recurrent approximation $m_t^\star, 0 < t \le T$ of the process $m(\vartheta, t), 0 < t \le T$.*

Therefore, we need a *good estimator* of the unknown parameter. “Good” in this situation can mean the following:

A. *Easy to calculate for each $t \in (0, T]$.*
B. *At the moment t, it depends on observations $X_s, 0 \le s \le t$.*
C. *The error of estimation of ϑ is small.*

This estimator will be used for the construction of $m_t^\star, 0 < t \le T$ by simple substitution in the Eqs. (3)–(4).

The first candidates are, of course, the maximum likelihood estimator (MLE) $\hat{\vartheta}_t, 0 < t \le T$ and the Bayesian estimator (BE) $\tilde{\vartheta}_t, 0 < t \le T$. Recall that the likelihood ratio function for this model of observations is:

$$L\left(\vartheta, X^t\right) = \exp\left\{\int_0^t \frac{f(\vartheta, s)\, m(\vartheta, s)}{\sigma(s)^2} \mathrm{d}X_s - \int_0^t \frac{f(\vartheta, s)^2\, m(\vartheta, s)^2}{2\sigma(s)^2} \mathrm{d}s\right\}, \quad \vartheta \in \Theta.$$

The MLE and BE are defined by the relations

$$L(\hat{\vartheta}_t, X^t) = \sup_{\vartheta \in \Theta} L\left(\vartheta, X^t\right), \qquad \tilde{\vartheta}_t = \frac{\int_\Theta \vartheta p(\vartheta)\, L\left(\vartheta, X^t\right) \mathrm{d}\vartheta}{\int_\Theta p(\vartheta)\, L\left(\vartheta, X^t\right) \mathrm{d}\vartheta}.$$

In the case of BE, we suppose that the loss function is quadratic and $p(\vartheta), \vartheta \in \Theta$ is a known density of the random variable ϑ.

These estimators are asymptotically efficient (in regular case). Therefore, the conditions **B** and **C** hold, but condition **A** does not. This is because to construct these estimators, we have to calculate the solutions $m(\vartheta, s), \gamma(\vartheta, s), 0 \leq s \leq t$ of the Eqs. (3)–(4) for all or many $\vartheta \in \Theta$ and all $t \in (0, T]$. Of course, these computations are difficult to realize.

To approximate $m(\vartheta, \cdot)$ we propose the following program:

1. *Calculate a preliminary estimator $\bar\vartheta_\tau$ on a relatively small interval of observations $[0, \tau]$ (learning interval) such that $\tau/T \to 0$.*
2. *Using $\bar\vartheta_\tau$, realize the One-step MLE-process $\vartheta_t^\star, \tau < t \leq T$:*

$$\vartheta_t^\star = \bar\vartheta_\tau + \mathbf{I}_\tau^t(\bar\vartheta_\tau)^{-1} \int_\tau^t \frac{\dot{\mathbf{M}}(\bar\vartheta_\tau, s)}{\sigma(s)^2}\left[\mathrm{d}X_s - M(\bar\vartheta_\tau, s)\,\mathrm{d}s\right], \tag{5}$$

where $\mathbf{I}_\tau^t(\vartheta), \tau \leq t \leq T$ is the family of Fisher information matrices, $M(\vartheta, s) = f(\vartheta, s)\,m(\vartheta, s)$ and $\dot{\mathbf{M}}(\vartheta, s) = \partial M(\vartheta, s)/\partial\vartheta$.
3. *As an approximation of $m(\vartheta, t), \tau < t \leq T$ we propose $m_t^\star, \tau < t \leq T$ obtained with the help of Eqs.* (3) *and* (4)*, where ϑ is replaced by $\vartheta_t^\star, \tau < t \leq T$:*

$$\mathrm{d}m_t^\star = \left[a(\vartheta_t^\star, t) - \frac{\gamma_t^\star f(\vartheta_t^\star, t)^2}{\sigma(t)^2}\right] m_t^\star \mathrm{d}t + \frac{\gamma_t^\star f(\vartheta_t^\star, t)}{\sigma(t)^2}\mathrm{d}X_t, \qquad \tau < t \leq T,$$

$$\frac{\partial \gamma_t^\star}{\partial t} = 2a(\vartheta_t^\star, t)\,\gamma_t^\star - \frac{\gamma_t^{\star 2} f(\vartheta_t^\star, t)^2}{\sigma(t)^2} + b(\vartheta_t^\star, t)^2.$$

4. *The last step is to describe the error $m_t^\star - m(\vartheta, t)$ and to discuss the asymptotic optimality of $\mathbf{E}_\vartheta\left(m_t^\star - m(\vartheta, t)\right)^2$. We propose the following minimax lower bound for all estimators $\bar m_t$ of $m(\vartheta, t)$: there exists a function $\mathcal{S}_t^\star(\vartheta)^2, \vartheta \in \Theta$ such that*

$$\lim_{\nu \to 0} \varliminf_{\varphi \to 0} \sup_{\|\vartheta - \vartheta_0\| \leq \nu} \varphi^{-2} \mathbf{E}_\vartheta (\bar m_t - m(\vartheta, t))^2 \geq \mathcal{S}_t^\star(\vartheta_0)^2.$$

Here $\varphi \to 0$ is the normalizing rate. The adaptive filter $m_t^\star, \tau \leq t \leq T$ is called asymptotically efficient if for all $\vartheta_0 \in \Theta$ and $t \in (\tau, T]$:

$$\lim_{\nu \to 0} \lim_{\varphi \to 0} \sup_{\|\vartheta - \vartheta_0\| \leq \nu} \varphi^{-2} \mathbf{E}_\vartheta \left(m_t^\star - m(\vartheta, t)\right)^2 = \mathcal{S}_t^\star(\vartheta_0)^2.$$

This program was realized for all mentioned models up to step 3. Step 4 for some models needs further study.

Therefore, we have no approximation on the learning interval $[0, \tau]$, but the length of this interval is asymptotically negligible. The estimator-process $\vartheta_t^\star$ satisfies all three conditions **A–C** and is asymptotically equivalent to MLE $\hat\vartheta_t$, i.e., it is asymptotically normal:

$$\varphi^{-1}\left(\vartheta_t^\star - \vartheta\right) \Longrightarrow \mathcal{N}\left(0, \mathbf{I}_0^t\left(\vartheta\right)^{-1}\right)$$

and asymptotically efficient.

The important advantage of this estimator-process is the simplicity of its calculation. It requires the solutions of Eqs. (3) and (4) and the equation for the derivative $\dot{\mathrm{M}}\left(\vartheta, s\right)$ for just one value $\vartheta = \bar\vartheta_\tau$. Note that the One-step MLE-process (5) is a particular case of the well-known Le Cam's One-step MLE. Indeed, the One-step MLE for the model (1)–(2) is

$$\vartheta_T^\star = \bar\vartheta_\tau + \mathbf{I}(\bar\vartheta_\tau)^{-1}\int_0^T \frac{\dot{\mathrm{M}}(\bar\vartheta_\tau, s)}{\sigma\left(s\right)^2}\left[\mathrm{d}X_s - M(\bar\vartheta_\tau, s)\mathrm{d}s\right].$$

The difference between this estimator and (5) is in the limits of the integral.

The One-step MLE-process can be written in the recursive form too. Denote $\mathbf{L}\left(\vartheta, t\right) = \dot{\mathrm{M}}\left(\bar\vartheta_\tau, t\right)\dot{\mathrm{M}}\left(\bar\vartheta_\tau, t\right)^\top \sigma\left(t\right)^{-2}$. Then

$$\begin{aligned}\mathrm{d}\vartheta_t^\star &= \mathbf{I}_\tau^t\left(\bar\vartheta_\tau\right)^{-1}\mathbf{L}\left(\bar\vartheta_\tau, t\right)\left[\bar\vartheta_\tau - \vartheta_t^\star\right]\mathrm{d}t \\ &\quad + \mathbf{I}_\tau^t\left(\bar\vartheta_\tau\right)^{-1}\frac{\dot{\mathrm{M}}\left(\bar\vartheta_\tau, t\right)}{\sigma\left(t\right)^2}\left[\mathrm{d}X_t - M\left(\bar\vartheta_\tau, t\right)\mathrm{d}t\right], \quad \vartheta_\tau^\star = \bar\vartheta_\tau, \quad \tau < t \le T.\end{aligned} \tag{6}$$

Therefore, the estimator $\vartheta_t^\star, \tau \le t \le T$ and the adaptive filter $m_t^\star, \tau \le t \le T$ have a recursive structure.

Remark that the recursive methods are well known in statistical estimation theory. We can mention here the monographs by Ljung [145], Nevelson and Khasminskii [157], and Duflo [46]. There is the following essential difference between all these methods and One-step approach in large samples case. In One-step case we need a preliminary estimator whose rate of convergence depends on the final time of observations T. For example, the One-step MLE-process constructed for a given T is not consistent if we use it on the time interval $\left[T^\gamma, T^2\right]$, where $\gamma \in (0, 1)$ (after $[0, T]$). In all other mentioned works there is no preliminary estimators and therefore there is no restriction on the time of observations. For partially observed system an approach similar to considered in this work was applied by Skorokhod and Khasminskii [177] who proposed two-steps procedure of parameter estimation for one nonlinear system.

We apply this construction of an adaptive filter to five different models of observations. The chosen models are, in some sense, the simplest. The observation and state equations are one-dimensional, the noises are uncorrelated, observations are in continuous time, etc. We suppose that this approach can be extended to other models too. There can be problems in constructing preliminary estimators in each particular model, but if the estimators are found, the rest of the approach can be realized as well. For example, if the system is multidimensional or the noises are correlated, then the equations of the filter have to be modified, but the program can remain the same. The discrete-time approximation of the Eqs. (5)–(6) can be studied too.

The goal of Chap. 1 is to introduce the notations and remind the well-known properties of the stochastic integral, equations of Kalman-Bucy filtering, several methods of parameter estimation, etc. It is assumed that the reader is already familiar with basic stochastic calculus and mathematical statistics.

In Chap. 2, we study the models like (1)–(2) where the coefficients $\sigma(t)$ and $b(\vartheta, t)$ are replaced by $\varepsilon\sigma(t)$ and $\varepsilon b(\vartheta, t)$. The first model is

$$\mathrm{d}X_t = f(\vartheta, t)\, Y_t\, \mathrm{d}t + \varepsilon\sigma(t)\, \mathrm{d}W_t, \qquad X_0, \qquad 0 \le t \le T, \tag{7}$$

$$\mathrm{d}Y_t = a(\vartheta, t)\, Y_t\, \mathrm{d}t + \varepsilon b(\vartheta, t)\, \mathrm{d}V_t, \qquad Y_0 = y_0 \neq 0. \tag{8}$$

The time of observation T is fixed, y_0 is known, and the asymptotic corresponds to the limit $\varepsilon \to 0$, i.e., we have *small noise* of the same level in both equations. The construction of the adaptive filter following the algorithm described above is realized with the learning interval $[0, \tau_\varepsilon]$, where $\tau_\varepsilon = \varepsilon^{2/3}$. Here we realize the program: preliminary estimator + One-step MLE-process + adaptive filter. Then in the Sect. 2.2 we consider a similar problem for the model (7)–(8) but with unknown initial value, $\vartheta = y_0$.

The conditionally Gaussian partially observed system, similar to (7)–(8) but with the coefficients $f(\vartheta, t, X_t)$, $\varepsilon\sigma(t, X_t)$, $a(\vartheta, t, X_t)$, $\varepsilon b(\vartheta, t, X_t)$ is studied in Sect. 2.3. The same construction of an adaptive filter is realized for this model too. For nonlinear (with respect to Y_t) partially observed systems where $f(\vartheta, t)\, Y_t$ and $a(\vartheta, t)\, Y_t$ are replaced by $f(\vartheta, t, Y_t)$ and $a(\vartheta, t, Y_t)$, respectively, the similar problem was considered in Sect. 2.4. There we study two extended Kalman filters in the same asymptotic of small noise in both equations. But due to the difficulties related with the calculation of the conditional expectation we do not realize the same program and propose two heuristic approaches which provide consistent adaptive filters.

In the last section we consider once more a partially observed system like (7) and (8) but the noise in the second equation has coefficient $\psi_\varepsilon b(\vartheta, t)$ instead of $\varepsilon b(\vartheta, t)$. Here $\psi_\varepsilon = \varepsilon^\delta$, $\delta \in (0, 1)$. Note that if $\delta = 1$ we have the model (7) and (8) and if $\delta = 0$ then the process $Y_t, 0 \le t \le T$ does not depend on ε and such models are studied in the next chapter. Here we have three different statistical problems depending on the values of δ: $0 < \delta < 1/3$, $\delta = 1/3$, $1/3 < \delta < 1$. It is interesting to note that in the case $0 < \delta < 1/3$ the higher "noise" yields the smaller error of estimation, i.e., $\varepsilon^{\frac{\delta-1}{2}}(\hat{\vartheta}_\varepsilon - \vartheta_0) \Longrightarrow \mathcal{N}$. In all cases we describe the properties of estimators and discuss the construction of adaptive filters.

Chapter 3 is devoted to a version of the system (1)–(2), where the small noise is in the observations (1) only:

$$\mathrm{d}X_t = f(\vartheta, t)\, Y_t\, \mathrm{d}t + \varepsilon\sigma(t)\, \mathrm{d}W_t, \quad X_0, \quad 0 \le t \le T,$$

$$\mathrm{d}Y_t = a(\vartheta, t)\, Y_t\, \mathrm{d}t + b(\vartheta, t)\, \mathrm{d}V_t, \quad Y_0.$$

Therefore, the limit ($\varepsilon = 0$) system corresponds to observations of the Gaussian process. Once more, the mentioned algorithm: preliminary estimator + One-step MLE-process + adaptive filter is realized. The preliminary estimator is related with estimation of the quadratic variation of the derivative of the observations limit. Then, as in Chap. 2, similar problems are studied for conditionally Gaussian and nonlinear partially observed systems.

In Chap. 4, the studied partially observed system is homogeneous:

$$\mathrm{d}X_t = f(\vartheta)\, Y_t\, \mathrm{d}t + \sigma \mathrm{d}W_t, \quad X_0, \quad 0 \le t \le T,$$

$$\mathrm{d}Y_t = -a(\vartheta)\, Y_t\, \mathrm{d}t + b(\vartheta)\, \mathrm{d}V_t, \quad Y_0.$$

The asymptotic corresponds to the limit $T \to \infty$ (*large samples*). The learning interval is $[0, \tau_T]$, where $\tau_T = T^\delta$, $\delta \in (1/2, 1)$. Therefore, $\tau_T/T \to 0$. The preliminary estimators are obtained by the method of moments. The One-step MLE-process has a similar form as in (5), and its substitution in the Kalman-Bucy filter provides the adaptive filter.

In Chap. 5, the observations are

$$\mathrm{d}X_t = Y_t\, \mathrm{d}t + \mathrm{d}W_t, \quad X_0, \quad 0 \le t \le T,$$

where the hidden process $Y_t, 0 \le t \le T$ is a two-state Markov process whose infinitesimal matrix depends on two unknown parameters. We have to estimate these parameters and then approximate the conditional expectation of the trajectory of Y_t. The conditional expectation $m(\vartheta, t) = \mathbf{E}_\vartheta(Y_t | X_s, 0 \le s \le t)$ is calculated with the help of the Wonham equation, which for this model plays the same role as the Kalman-Bucy equations (3) and (4) for the model (1) and (2). The learning interval is the same as in Chap. 4, and preliminary estimators are obtained by the method of moments. This equation allows us to construct, once more, the One-step MLE-process, and then the substitution of the estimator-process in the Wonham equation provides us with the adaptive filter $m_t^\star$, which is an approximation of the conditional expectation $m(\vartheta, t)$. The last step is to introduce the estimator $Y_t^\star$ with the same two states and describe the asymptotic of the error $\mathbf{E}_\vartheta \left(Y_t^\star - Y_t\right)^2$.

The model of observations studied in Chap. 6 is the only one in this work where the time is discrete, and we have the partially observed time series:

$$X_t = f\, Y_{t-1} + \sigma\, w_t, \quad X_0,$$

$$Y_t = a\, Y_{t-1} + b\, v_t, \quad Y_0.$$

Here, $w_t, v_t, t = 1, \ldots, T$ are independent standard Gaussian random variables, and the parameters f, σ, a, b can be unknown. The learning interval is $[1, \tau_T]$, where $\tau_T = \left[T^\delta\right]$, $\delta \in (1/2, 1)$, and $[A]$ here denotes the integer part of A. The preliminary estimators are constructed with the help of the method of moments. The Kalman filter [83] for the conditional expectation $m(\vartheta, t) = \mathbf{E}_\vartheta\left(Y_t | X_s, s = 1, \ldots, t\right)$ allows us to define the One-step MLE-process for the estimation of the unknown parameters and then to write the adaptive filter. The properties of the estimators and the error of approximation are described in the asymptotics of large samples.

In all considered models, we also study the asymptotic behavior of the MLE and BE of the unknown parameter. It is always useful to compare the asymptotics of the One-step MLE-process and the asymptotics of the MLE. The BE are used for the construction of the lower minimax bound on the mean squared risk of all possible estimators (approximations) of the conditional expectations.

Chapter 7 is devoted to several problems related with the partially observed systems. We start (Sect. 7.1) with the problems of localization of the source on the plane by observations of the signals emitted by the source and registered by K detectors. The model of observations in the first problem is described by equations similar to (1)–(2) with the special form of the function $f(\vartheta, t)$ and the small noise in both equations. We describe the properties of the MLE and BE of the position of the source in three cases: smooth function $f(\vartheta, \cdot)$, function $f(\vartheta, \cdot)$ having cusp-type singularity, and discontinuous function $f(\vartheta, \cdot)$. The properties of the least squares estimators of the position of the source are described too. In the second problem, the source is moving along a line, and the problem is tracking this source. The estimation of the current position (trajectory) is done with the help of the One-step MLE-process. In these two problems, the asymptotic corresponds to $\varepsilon \to 0$. Then we discuss several other problems of source localization that can be solved with the help of the developed approach.

In Sect. 7.2 we consider the model of observations

$$
\begin{aligned}
\mathrm{d}X_t &= f\, Y_t\, \mathrm{d}t + \varepsilon\, \sigma\, \mathrm{d}W_t, \qquad X_0 = 0, \qquad 0 \le t \le T,\\
\mathrm{d}Y_t &= S(\vartheta, t)\, \mathrm{d}t + \varepsilon\, b\, \mathrm{d}V_t, \qquad Y_0 = \theta_1.
\end{aligned}
$$

Here $\vartheta = (\theta_1, \ldots, \theta_d)$ and the signal $S(\vartheta, t) = S(\theta_2, \ldots, \theta_d, t)$. The asymptotic properties of the MDE, MLE, and One-step MLE are described under regularity conditions.

Section 7.3 discusses the security price process

$$
\mathrm{d}S_t = \mu_t S_t \mathrm{d}t + \sigma S_t \mathrm{d}W_t, \quad S_0, \quad 0 \le t \le T,
$$

where $\sigma > 0$ is a known constant, but $\mu_t, 0 \le t \le T$ (hidden) is a mean-reverting process satisfying the equation

$$
\mathrm{d}\mu_t = \alpha\left(\upsilon - \mu_t\right) \mathrm{d}t + \beta \mathrm{d}V_t, \qquad \mu_0.
$$

We study the estimators of the parameters of this system in the asymptotic $T \to \infty$.

Three models of change-point type are discussed in the Sect. 7.4. In the first model with the change-point in the observation equation ($f(\vartheta, t) = g_t \mathbb{I}_{\{t<\vartheta\}} + h_t \mathbb{I}_{\{t\geq\vartheta\}}$), the statistical properties of the estimators (MLE, BE) correspond to the singular statistical model, i.e., the rate of convergence is ε^2. In the second model with change-point in the hidden (state) equation ($a(\vartheta, t) = g_t \mathbb{I}_{\{t<\vartheta\}} + h_t \mathbb{I}_{\{t\geq\vartheta\}}$), the properties of the same estimators correspond to the properties of estimators in the regular statistical models with the rate ε. The last model with change-points in both equations yields the combination of the properties of estimators mentioned above, respectively. We discuss the construction of offline adaptive filter but the construction of on-line adaptive filter rest an open problem.

The last Sect. 7.5 is devoted to the problem of approximation of the solution of the backward stochastic differential equation (BSDE) in the situation, where the forward equation

$$\mathrm{d}Y_t = a(t)\, Y_t \mathrm{d}t + b(\vartheta, t)\, \mathrm{d}V_t$$

cannot be observed directly and the observations are

$$\mathrm{d}X_t = f(t)\, Y_t\, \mathrm{d}t + \varepsilon\, \sigma(t)\, \mathrm{d}W_t, \qquad X_0 = 0, \qquad 0 \leq t \leq T.$$

We are given two functions $F(t, y, z, s)$ and $\Phi(y)$ and have to construct an approximation of the couple processes $(Z_t, s_t, 0 \leq t \leq T)$ such that the random process

$$\mathrm{d}Z_t = -F(t, Y_t, Z_t, s_t)\, \mathrm{d}t + s_t\, \mathrm{d}V_t,$$

has the final value $Z_T = \Phi(Y_T)$. As before ϑ is an unknown parameter and the approximation is realized with the help of the corresponding One-step MLE-process.

Throughout the proof, C denotes "generic" positive constants, which can vary from formula to formula (and even in the same formula). The vectors and matrices are denoted using roman $\dot{\mathrm{M}}$ and bold $\mathbf{I}(\vartheta)$ fonts, respectively. The observed process is denoted (as usually in statistics of processes) $X^T = (X_t, 0 \leq t \leq T)$ and the hidden process is $Y^T = (Y_t, 0 \leq t \leq T)$.

Contents

1 Auxiliary Results 1
1.1 Kalman–Bucy Filter and Related Questions 1
1.1.1 Stochastic Integral 1
1.1.2 Kalman–Bucy Filter 8
1.1.3 Extrapolation 23
1.1.4 Limit Theorems 24
1.2 Methods of Estimation 33
1.2.1 Method of Moments 33
1.2.2 Minimum Distance 37
1.2.3 Trajectory Fitting 38
1.2.4 Maximum Likelihood 39
1.2.5 Bayesian 45
1.2.6 Multi-Step 48
1.3 Adaptive Filtration and Extrapolation 53
1.3.1 Adaptive Filtration 53
1.3.2 Adaptive Extrapolation 54
1.4 Asymptotic Efficiency 55

2 Small Noise in Both Equations 69
2.1 Linear System 69
2.1.1 Model of Observations 69
2.1.2 Estimators 72
2.1.3 Adaptive Filter 116
2.2 Unknown Initial Value 127
2.2.1 Deterministic Initial Value 127
2.2.2 Random Initial Value 142
2.3 Conditionally Gaussian Processes 151
2.3.1 Model of Observations 151
2.3.2 Estimators 153
2.3.3 Adaptive Filter 159

2.4 Extended Kalman Filters 164
2.4.1 EKF-1 164
2.4.2 EKF-2 181
2.5 Different Levels of Noises 190
2.5.1 Model of Observations 190
2.5.2 Parameter Estimation 192
2.5.3 One-Step MLE Processes 220
2.5.4 Adaptive Filters 231

3 Low Noise Observations 237
3.1 Linear Model 237
3.1.1 Model of Observations 237
3.1.2 Semiparametric Estimation 239
3.1.3 Parameter Estimation 248
3.1.4 Adaptive Filter 281
3.1.5 Asymptotic Efficiency 288
3.2 Conditionally Gaussian System 293
3.2.1 Model of Observations 293
3.2.2 Parameter Estimator 296
3.2.3 One-Step MLE 308
3.2.4 Adaptive Filter 309
3.3 Extended Kalman Filter 309
3.3.1 Model of Observations 309
3.3.2 Empirical Quadratic Variation 328
3.3.3 MME 336
3.3.4 Adaptive EKF 341

4 Hidden Ergodic O-U Process 343
4.1 Model of Observations 343
4.2 Parameter Estimation 348
4.2.1 MME 348
4.2.2 MLE and BE 361
4.2.3 One-Step MLE-Process 377
4.3 Adaptive Filter 387
4.3.1 Filter 387
4.3.2 Asymptotic Efficiency 397

5 Hidden Telegraph Process 403
5.1 Model of Observations 403
5.2 Parameter Estimation 411
5.2.1 MME 411
5.2.2 MLE and BE 419
5.2.3 One-Step MLE-Process 426
5.3 Adaptive Filter 430

6 Hidden AR Process 439
6.1 Model of Observations 439
6.2 Parameter Estimation 444
6.2.1 MME 444
6.2.2 Fisher Information 451
6.2.3 One-Step MLE-Process 465
6.2.4 MLE and BE 472
6.3 Adaptive Filter 479
6.3.1 Recursive Filter 479
6.3.2 Asymptotic Efficiency 484

7 Miscellaneous 489
7.1 Source Localization 490
7.1.1 Fixed Source 490
7.1.2 Moving Source 523
7.1.3 Some Other Problems 539
7.2 Indirect Observation of a Signal 549
7.2.1 Model of Observations 549
7.2.2 Parameter Estimation 551
7.2.3 One-Step MLE-Process 556
7.3 Security Prices Process 560
7.3.1 Model of Observations 560
7.3.2 Estimators 563
7.3.3 One-Step MLE-Process 573
7.3.4 Adaptive Filter and Extrapolation 575
7.4 Change Point Problems 577
7.4.1 Change Point in the Observation Equation 578
7.4.2 Change Point in the State Equation 596
7.4.3 Change Points in Both Equations 603
7.5 BSDE Approximation 614
7.5.1 BSDE 614
7.5.2 Observable Forward Equation 616
7.5.3 Hidden Forward Equation with Unknown Volatility 626

Bibliographical Notes 639

Bibliography 645

Index 653

Chapter 1
Auxiliary Results

In this chapter, for the convenience of exposition, we remind some basic notions of stochastic calculus and mathematical statistics. It is supposed that the reader is familiar with stochastic calculus; therefore, the properties of the stochastic integral, the likelihood ratio formula, and limit theorems are given without proofs. The definitions and properties of some estimators of the parameters are given in more detail because they are less well-known. The Kalman–Bucy filter is provided with the proof due to the importance of this result for the problems considered in this work. At the end, some lower bounds on the mean-squared risks are presented and the possibility of the construction of a lower bound in the problem of adaptive filtering is discussed.

1.1 Kalman–Bucy Filter and Related Questions

1.1.1 Stochastic Integral

Definition and Properties

It is assumed that the reader is familiar with Itô stochastic calculus. For the convenience of reading, we recall some notations, definitions, and properties following the exposition in [141]. We chose this work because, in the sequel, we will frequently use results presented there without detailed explanations of where these results were first obtained. In any case, the reader can easily find further references in that work.

We are given a probability space with filtration $\left(\Omega, \mathfrak{F}, \{\mathfrak{F}_t\}_{0\leq t\leq T}, \mathbb{P}\right)$. Here, $\Omega = \{\omega\}$ is the space of elementary events ω, $\mathfrak{F}$ is some σ-algebra of subsets of Ω, and $\mathbb{P}$ is a probability measure defined on the sets in $\mathfrak{F}$. A filtration $\{\mathfrak{F}_t\}_{0\leq t\leq T}$ is a collection of sub-σ-algebras of $\mathfrak{F}$ satisfying the *usual properties*: $\mathfrak{F}_s \subseteq \mathfrak{F}_t$ if $s \leq t$. Usually, $\mathfrak{F}_t$ represents the set of events observable by time t. The probability space

Y. A. Kutoyants, *Hidden Markov Processes and Adaptive Filtering*, Springer Series in Statistics, https://doi.org/10.1007/978-3-032-00052-1_1

$(\Omega, \mathfrak{F}, \mathbb{P})$ is assumed to be complete, i.e., the σ-algebra $\mathfrak{F}$ contains all subsets of zero-probability elements of $\mathfrak{F}$.

The family $X = (X_t, t \in \mathfrak{T})$ of random variables $X_t = X_t(\omega)$ is called a stochastic process with continuous time if $t \in \mathfrak{T} = [0, \infty)$, $t \in \mathfrak{T} = [a, b]$, or $t \in \mathfrak{T} = \mathbb{R}$, etc. In the case where the time parameter t is confined to the set $\mathbb{N} = \{0, 1, \ldots\}$, the family $X = (X_t, t \in \mathbb{N})$ is called a random sequence or a stochastic process with discrete time. For a fixed $\omega \in \Omega$, the time function $X_t(\omega)$ (where $t \in \mathfrak{T}$ or $t \in \mathbb{N}$) is called a *realization* or *sample function*.

A stochastic process $X^T = (X_t, 0 \leq t \leq T)$ is *adapted* if X_t is $\mathfrak{F}_t$-measurable for each $t \in [0, T]$. The filtration generated by a process X^T is the smallest filtration with respect to which it is adapted. It is denoted $\mathfrak{F}_t^X, 0 \leq t \leq T$ and is called the *natural filtration*.

A stochastic process X^T is progressively measurable if for each $t \in [0, T]$, the map $(s, \omega) \mapsto X_s$ is $\mathfrak{B}_{[0,t]} \otimes \mathfrak{F}_t$-measurable. Here, $\mathfrak{B}_{[0,t]}$ is the Borel σ-algebra on $[0, t]$.

The adapted random process $W = (W_t, t \geq 0)$ is called a Wiener process if

1. The trajectories $W_t, t \geq 0$ are continuous $\mathbb{P}$-almost surely over t;
2. $W_t, \mathfrak{F}_t, 0 \leq t \leq T$ satisfies the relations

$$\mathbb{E}\,(W_t \mid \mathfrak{F}_s) = W_s,$$

$$\mathbb{E}\left((W_t - W_s)^2 \mid \mathfrak{F}_s\right) = t - s, \quad t \geq s.$$

The measurable function $f(t, \omega), t \geq 0, \omega \in \Omega$ is called *nonanticipative* with respect to the family $(\mathfrak{F}_t, t \geq 0)$ if, for each t, it is $\mathfrak{F}_t$-measurable. The nonanticipative function $f = f(t, \omega)$ is said to be of class $\mathcal{P}_T$ if

$$\mathbb{P}\left(\int_0^T f(t, \omega)^2\, \mathrm{d}t < \infty\right) = 1.$$

If such a function satisfies

$$\int_0^T \mathbb{E}\,|f(t, \omega)|^{2p}\, \mathrm{d}t < \infty,$$

we say that f belongs to the class $\mathcal{M}_T(2p)$.

The stochastic Itô integral

$$I_T\,(f) = \int_0^T f\,(t, \omega)\, \mathrm{d}W_t$$

is defined for nonanticipative functions $f \in \mathcal{P}_T$. It has the following properties:

1. If $f \in \mathcal{M}_T(2)$, then

$$\mathbb{E}\left[\int_0^T f(t,\omega)\,\mathrm{d}W_t\right] = 0,$$

$$\mathbb{E}\left[\left(\int_0^T f(t,\omega)\,\mathrm{d}W_t\right)^2\right] = \int_0^T \mathbb{E}\left[f(t,\omega)^2\right]\mathrm{d}t.$$

2. If $f, g \in \mathcal{M}_T(2)$, then

$$\mathbb{E}\left[\int_0^T f(t,\omega)\,\mathrm{d}W_t \int_0^T g(t,\omega)\,\mathrm{d}W_t\right] = \int_0^T \mathbb{E}\left[f(t,\omega)\,g(t,\omega)\right]\mathrm{d}t.$$

3. If $f \in \mathcal{M}_T(2p)$ where $p \geq 1$, then

$$\mathbb{E}\left[\left(\int_0^T f(t,\omega)\,\mathrm{d}W_t\right)^{2p}\right] \leq [p(2p-1)]^p\, T^{p-1}\int_0^T \mathbb{E}\left|f(t,\omega)\right|^{2p}\mathrm{d}t. \tag{1.1}$$

Furthermore, there exists a constant $C_p > 0$ such that

$$\mathbb{E}\left[\left(\sup_{0\leq t\leq T}\left|\int_0^t f(s,\omega)\,\mathrm{d}W_s\right|\right)^{2p}\right] \leq C_p\mathbb{E}\left[\left(\int_0^T f(t,\omega)^2\,\mathrm{d}t\right)^p\right]. \tag{1.2}$$

For proofs of items 1 and 2 and inequality (1.1), see [141, Section 4.2]. The estimate (1.2) is known as the Burkholder–Davis–Gundy inequality [19].

The continuous random process $X^T = (X_t, \mathfrak{F}_t, 0 \leq t \leq T)$ is called an *Itô process* (relative to the Wiener process $W^T = (W_t, \mathfrak{F}_t, 0 \leq t \leq T)$) if there exist two nonanticipative processes $(A_t, \mathfrak{F}_t, 0 \leq t \leq T)$ and $(B_t, \mathfrak{F}_t, 0 \leq t \leq T)$ such that

$$\mathbb{P}\left(\int_0^T |A_t|\,\mathrm{d}t < \infty\right) = 1, \qquad B_t \in \mathcal{P}_T$$

and, with probability 1 for $0 \leq t \leq T$,

$$X_t = X_0 + \int_0^t A_s\,\mathrm{d}s + \int_0^t B_s\,\mathrm{d}W_s. \tag{1.3}$$

For brevity, it is said that the process X_t has the *stochastic differential*

$$\mathrm{d}X_t = A_t\,\mathrm{d}t + B_t\,\mathrm{d}W_t, \qquad X_0 \tag{1.4}$$

with (1.4) understood as shorthand for the representation given by (1.3).

The Itô process $X^T = (X_t, \mathfrak{F}_t, 0 \le t \le T)$ is called a *process of diffusion type* (relative to the Wiener process $W = (W_t, \mathfrak{F}_t, 0 \le t \le T)$) if the functionals A_s and B_s in (1.3) are $\mathfrak{F}_s^X$-measurable. Here the σ-algebra $\mathfrak{F}_s^X = \sigma\{\omega : X_r, r \le s\}$.

Let $X^T = (X_t, \mathfrak{F}_t, 0 \le t \le T)$ be a random process having the stochastic differential (1.4) and let $F(t, x)$ be a continuous function having partial derivatives $F_t'(t, x)$, $F_x'(t, x)$, and $F_{xx}''(t, x)$. Then the process $(F(t, X_t), \mathfrak{F}_t, 0 \le t \le T)$ also has a stochastic differential

$$\mathrm{d}F(t, X_t) = \left[F_t'(t, X_t) + F_x'(t, X_t) A_t + \frac{1}{2} F_{xx}''(t, X_t) B_t^2\right] \mathrm{d}t + F_x'(t, X_t) B_t \,\mathrm{d}W_t.$$

This formula was obtained by Itô and is called the *Itô formula*.

Let $(\Omega, \mathfrak{F}, \mathbb{P})$ be a probability space, $\mathfrak{F}_t$, $0 \le t \le T$ be a non-decreasing family of sub-σ-algebras of $\mathfrak{F}$, and $(W_t, \mathfrak{F}_t, 0 \le t \le T)$ be a Wiener process. Denote by $(\mathcal{C}_T, \mathfrak{B}_T)$ the measurable space of continuous functions $x = (x_t, 0 \le t \le T)$ with the σ-algebra $\mathfrak{B}_T = \sigma(x : x_s, 0 \le s \le T)$. Also, set $\mathfrak{B}_t = \sigma(x : x_s, 0 \le s \le t)$.

Consider the stochastic differential of diffusion-type process

$$\mathrm{d}X_t = A_t(X)\,\mathrm{d}t + B_t(X)\,\mathrm{d}W_t, \qquad X_0, \quad 0 \le t \le T. \tag{1.5}$$

Here, $A_t(x)$ and $B_t(x)$ are measurable nonanticipative (i.e., $\mathfrak{B}_t$-measurable for each t, $0 \le t \le T$) functionals. The initial value X_0 is $\mathfrak{F}_0$-measurable and $\mathbb{P}(|X_0| < \infty) = 1$.

Introduce the conditions

$$|A_t(x) - A_t(y)| + |B_t(x) - B_t(y)| \le L_1 \int_0^t |x_s - y_s|\,\mathrm{d}s + L_2 |x_t - y_t|, \tag{1.6}$$

$$|A_t(x)| + |B_t(x)| \le L_1 \int_0^t (1 + |x_s|)\,\mathrm{d}s + L_2 (1 + |x_t|), \tag{1.7}$$

where L_1, L_2 are constants and $x, y \in \mathcal{C}_T$.

Theorem 1.1 *Let the nonanticipative functionals $A_t(x)$, $B_t(x)$, $0 \le t \le T, x \in \mathcal{C}_T$ satisfy the Lipschitz* (1.6) *and linear growth* (1.7) *conditions. Then the Eq.* (1.5) *has a unique (strong) solution $X^T = (X_t, \mathfrak{F}_t, 0 \le t \le T)$. If $\mathbb{E}X_0^{2p} < \infty$ for some $p \ge 1$, then there exists a constant c_p such that*

$$\mathbb{E}X_t^{2p} \le \left(1 + \mathbb{E}X_0^{2p}\right) e^{c_p t} - 1. \tag{1.8}$$

For the proof, see Theorem 4.6 in [141].

If $A_t(X) = A(X_t, t)$ and $B_t(X) = B(X_t, t)$, then the process X^T is a solution of the *stochastic differential equation* (SDE)

$$\mathrm{d}X_t = A(X_t, t)\,\mathrm{d}t + B(X_t, t)\,\mathrm{d}W_t, \quad X_0$$

and is called a *diffusion process*.

Let $X^T = (X_t, \mathfrak{F}_t, 0 \le t \le T)$ and $Z^T = (Z_t, \mathfrak{F}_t, 0 \le t \le T)$ be two diffusion-type processes defined by the equations

$$\begin{aligned} \mathrm{d}X_t &= A_t(X)\,\mathrm{d}t + B_t(X)\,\mathrm{d}W_t, \qquad X_0 = Z_0, \\ \mathrm{d}Z_t &= a_t(Z)\,\mathrm{d}t + B_t(Z)\,\mathrm{d}W_t. \end{aligned} \tag{1.9}$$

Let us denote by $\mathbb{P}_X$ and $\mathbb{P}_Z$ the measures induced in the measurable space $(\mathcal{C}_T, \mathfrak{B}_T)$ by the processes $X^T = (X_t, 0 \le t \le T)$ and $Z^T = (Z_t, 0 \le t \le T)$.

Theorem 1.2 *Let the conditions* (1.6) *and* (1.7) *hold, and assume that* $B_t(x)^2 \ge c$, *where* $c > 0$. *Then the measures* $\mathbb{P}_X$ *and* $\mathbb{P}_Z$ *are equivalent, and*

$$\frac{\mathrm{d}\mathbb{P}_X}{\mathrm{d}\mathbb{P}_Z} = \exp\left(\int_0^T \frac{A_t(X) - a_t(X)}{B_t(X)^2}\,\mathrm{d}X_t - \int_0^T \frac{A_t(X)^2 - a_t(X)^2}{2B_t(X)^2}\,\mathrm{d}t\right).$$

For the proof, see [141], Theorem 7.10.

In parameter estimation problems, the processes are indexed by the unknown parameter $\vartheta \in \Theta \subset \mathbb{R}^d$, $d \ge 1$. Suppose we have a family of processes

$$\mathrm{d}X_t = A_t(\vartheta, X)\,\mathrm{d}t + B_t(X)\,\mathrm{d}W_t, \qquad X_0, \qquad 0 \le t \le T, \qquad \vartheta \in \Theta.$$

The likelihood ratio function is denoted as $L\left(\vartheta, X^T\right)$, where

$$L\left(\vartheta, X^T\right) = \exp\left(\int_0^T \frac{A_t(\vartheta, X)}{B(t)^2}\,\mathrm{d}X_t - \int_0^T \frac{A_t(\vartheta, X)^2}{2B(t)^2}\,\mathrm{d}t\right),$$

i.e., in the second process (1.9) we set $a_t(Z) \equiv 0$.

Suppose that there are three values ϑ_0, ϑ_1, and ϑ_2, and denote

$$Z_1 = \frac{L\left(\vartheta_1, X^T\right)}{L\left(\vartheta_0, X^T\right)},$$

$$Z_2 = \frac{L\left(\vartheta_2, X^T\right)}{L\left(\vartheta_0, X^T\right)},$$

$$F_t(\vartheta_2, \vartheta_1) = \frac{A_t(\vartheta_2, X) - A_t(\vartheta_1, X)}{B_t(X)}.$$

Note that under conditions (1.6) and (1.7) and with $B_t(x)^2 \geq c > 0$, all polynomial moments of the functions $A_t(\vartheta, X)\, B_t(X)^{-1}$ are bounded by Theorem 1.1.

In the study of maximum likelihood and Bayesian estimators, we will often use the following lemma, where

$$C_{1,m} = 2^{2m-1}\left(\frac{2m-1}{8m^2}\right)^{2m},$$

$$C_{2,m} = 2^{2m-1}\frac{[m\,(2m-1)]^m}{(2m)^{2m}}.$$

Lemma 1.1 *For any $m \geq 1$,*

$$\mathbb{E}_{\vartheta_0}\left|Z_2^{1/(2m)} - Z_1^{1/(2m)}\right|^{2m} \leq C_{1,m}T^{2m-1}\int_0^T \mathbb{E}_{\vartheta_2}\left[F_t(\vartheta_2,\vartheta_1)\right]^{4m}\,\mathrm{d}t + C_{2,m}T^{m-1/2}\int_0^T \mathbb{E}_{\vartheta_2}\left[F_t(\vartheta_2,\vartheta_1)\right]^{2m}\,\mathrm{d}t. \tag{1.10}$$

Proof We have

$$\mathbb{E}_{\vartheta_0}\left|Z_2^{1/(2m)} - Z_1^{1/(2m)}\right|^{2m} = \mathbb{E}_{\vartheta_0} Z_1\left|\left(\frac{Z_2}{Z_1}\right)^{1/(2m)} - 1\right|^{2m} = \mathbb{E}_{\vartheta_1}\left|\left(\frac{Z_2}{Z_1}\right)^{1/(2m)} - 1\right|^{2m},$$

where we changed the measure $\mathbb{E}_{\vartheta_0}$ to $\mathbb{E}_{\vartheta_1}$.

Set $P_T = \left(\frac{Z_2}{Z_1}\right)^{1/(2m)}$. Then, for the process

$$P_t = \exp\left(\int_0^t \frac{F_s(\vartheta_2,\vartheta_1)}{2m}\,\mathrm{d}W_s - \int_0^t \frac{F_s(\vartheta_2,\vartheta_1)^2}{4m}\,\mathrm{d}s\right), \qquad 0 \leq t \leq T,$$

by Itô's formula, we obtain the representation

$$P_T = 1 + \frac{1-2m}{8m^2}\int_0^T F_t(\vartheta_2,\vartheta_1)^2 P_t\,\mathrm{d}t + \frac{1}{2m}\int_0^T F_t(\vartheta_2,\vartheta_1)\,P_t\,\mathrm{d}W_t.$$

Further

$$\mathbf{E}_{\vartheta_1}\left|P_T - 1\right|^{2m} \leq 2^{2m-1}\left(\frac{(2m-1)}{8m^2}\right)^{2m}\mathbf{E}_{\vartheta_1}\left(\int_0^T F_t(\vartheta_2,\vartheta_1)^2 P_t\,\mathrm{d}t\right)^{2m}$$

$$
\begin{aligned}
&+\frac{2^{2m-1}}{(2m)^{2m}}\mathbf{E}_{\vartheta_1}\left(\int_0^T F_t\left(\vartheta_2,\vartheta_1\right)P_t\,\mathrm{d}W_t\right)^{2m}\\
&\leq 2^{2m-1}\left(\frac{(2m-1)}{8m^2}\right)^{2m}T^{2m-1}\mathbf{E}_{\vartheta_1}\int_0^T F_t\left(\vartheta_2,\vartheta_1\right)^{4m}P_t^{2m}\,\mathrm{d}t\\
&+\left[m\left(2m-1\right)\right]^m T^{m-1}\frac{2^{2m-1}}{(2m)^{2m}}\mathbf{E}_{\vartheta_1}\int_0^T F_t\left(u_2,u_1\right)^{2m}P_t^{2m}\mathrm{d}t\\
&\leq 2^{2m-1}\left(\frac{(2m-1)}{8m^2}\right)^{2m}T^{2m-1}\int_0^T \mathbf{E}_{\vartheta_2}F_t\left(\vartheta_2,\vartheta_1\right)^{4m}\,\mathrm{d}t\\
&+2^{2m-1}\frac{\left[m\left(2m-1\right)\right]^m}{(2m)^{2m}}T^{m-1}\int_0^T \mathbf{E}_{\vartheta_2}F_t\left(\vartheta_2,\vartheta_1\right)^{2m}\mathrm{d}t\\
&=C_{1,m}T^{2m-1}\int_0^T \mathbf{E}_{\vartheta_2}F_t\left(\vartheta_2,\vartheta_1\right)^{4m}\,\mathrm{d}t\\
&+C_{2,m}T^{m-1}\int_0^T \mathbf{E}_{\vartheta_2}F_t\left(\vartheta_2,\vartheta_1\right)^{2m}\mathrm{d}t.
\end{aligned}
$$

Here we used the inequality $(a+b)^{2m} \leq 2^{2m-1}(a^{2m}+b^{2m})$, the property (1.1) of the stochastic integral, and changed the measure from $\mathbb{E}_{\vartheta_1}$ to $\mathbb{E}_{\vartheta_2}$. □

Corollary 1.1 *In the proof of this lemma, we obtained the following inequality:*

$$
\begin{aligned}
\mathbb{E}_{\vartheta_0}\left|Z_2^{1/(2m)}-Z_1^{1/(2m)}\right|^{2m} &\leq C_{1,m}\mathbb{E}_{\vartheta_1}\left(\int_0^T F_t\left(\vartheta_2,\vartheta_1\right)^2 P_t\,\mathrm{d}t\right)^{2m}\\
&+C_{3,m}\mathbb{E}_{\vartheta_1}\left(\int_0^T F_t\left(\vartheta_2,\vartheta_1\right)P_t\,\mathrm{d}W_t\right)^{2m},
\end{aligned}
$$

where

$$
C_{3,m}=\frac{2^{2m-1}}{(2m)^{2m}}.
$$

In some problems, we will use this inequality as well.

Remark 1.1 In the case $m = 1$, the corresponding estimate is more simple. We have

$$
P_T = 1-\frac{1}{8}\int_0^T P_t F_t\left(\vartheta_2,\vartheta_1\right)^2\mathrm{d}t+\frac{1}{2}\int_0^T P_t F_t\left(\vartheta_2,\vartheta_1\right)\mathrm{d}W_t.
$$

Thus,

$$
\begin{aligned}
\mathbf{E}_{\vartheta_0}\left(Z_2^{1/2}-Z_1^{1/2}\right)^2 &= 2-2\mathbf{E}_{\vartheta_1}P_T=\frac{1}{4}\int_0^T \mathbf{E}_{\vartheta_1}P_tF_t\left(\vartheta_2,\vartheta_1\right)^2\mathrm{d}t\\
&\leq \frac{1}{2}\int_0^T \mathbf{E}_{\vartheta_1}F_t\left(\vartheta_2,\vartheta_1\right)^2\mathrm{d}t+\frac{1}{2}\int_0^T \mathbf{E}_{\vartheta_1}P_t^2F_t\left(\vartheta_2,\vartheta_1\right)^2\mathrm{d}t\\
&\leq \frac{1}{2}\int_0^T \mathbf{E}_{\vartheta_1}F_t\left(\vartheta_2,\vartheta_1\right)^2\mathrm{d}t+\frac{1}{2}\int_0^T \mathbf{E}_{\vartheta_2}F_t\left(\vartheta_2,\vartheta_1\right)^2\mathrm{d}t.
\end{aligned}
\tag{1.11}
$$

Here we used the inequality $P_tF_t\left(\vartheta_2,\vartheta_1\right)^2\leq 2F_t\left(\vartheta_2,\vartheta_1\right)^2+2P_tF_t\left(\vartheta_2,\vartheta_1\right)^2$ and changed the measure $\mathbf{E}_{\vartheta_1}P_t^2=\mathbf{E}_{\vartheta_2}$.

1.1.2 Kalman–Bucy Filter

We are given a probability space $(\Omega,\mathfrak{F},\mathbf{P})$ with a family of σ-algebras $(\mathfrak{F}_t, 0\leq t\leq T)$

Consider partially observed linear system

$$
\mathrm{d}X_t=f(t)Y_t\,\mathrm{d}t+\sigma(t)\mathrm{d}W_t,\qquad X_0,\quad 0\leq t\leq T, \tag{1.12}
$$

$$
\mathrm{d}Y_t=a(t)Y_t\,\mathrm{d}t+b(t)\,\mathrm{d}V_t,\qquad Y_0,\quad 0\leq t\leq T. \tag{1.13}
$$

Here $W^T=(W_t,\mathfrak{F}_t,0\leq t\leq T)$ and $V^T=(V_t,\mathfrak{F}_t,0\leq t\leq T)$ are two independent Wiener processes and the Gaussian variables $X_0\sim\mathcal{N}\left(x_0,q^2\right)$ and $Y_0\sim\mathcal{N}\left(y_0,r^2\right)$ are $\mathfrak{F}_0$ measurable and independent of W^T and V^T.

If the functions $f\left(\cdot\right)$ and $b\left(\cdot\right)$ are constant, i.e., $a\left(t\right)=a$ and $b\left(\vartheta,t\right)=b$, then $Y_t, 0\leq t\leq T$ is called *Ornstein–Uhlenbeck* (O-U) process, but we continue to call it O-U even in the case of non-constant coefficients. Introduce the conditions

$$
\int_0^T f(t)^2\,\mathrm{d}t<\infty,\qquad \int_0^T b(t)^2\,\mathrm{d}t<\infty,\qquad \int_0^T |a(t)|\,\mathrm{d}t<\infty, \tag{1.14}
$$

$$
\int_0^T \sigma(t)^2\,\mathrm{d}t<\infty,\qquad \sigma\left(t\right)^2\geq c>0,\qquad 0\leq t\leq T. \tag{1.15}
$$

Below we suppose that these conditions are fulfilled. The Eqs. (1.13) and (1.12) are called *state equation* and *observation equation* respectively.

We have to estimate Y_t for each $t\in[0,T]$ by observations $X^t=(X_s,0\leq s\leq t)$. Recall that the σ-algebra generated by observations X^t, we denote as $\mathfrak{F}_t^X$.

Let us define mean square risk of an arbitrary estimator $\bar{Y}_t=\bar{Y}_t(X^t)$ as follows:

$$R\left(\bar{Y}_t\right) = \mathbf{E}\left|Y_t - \bar{Y}_t\right|^2.$$

Note that

$$Y_t = Y_0\,\phi\,(0,t) + \int_0^t \phi\,(s,t)\, b\,(s)\ \mathrm{d}V_s, \qquad \phi\,(s,t) = \exp\left(\int_s^t a(v)\mathrm{d}v\right).$$

Hence $Y_t \sim \mathcal{N}\left(y_t, d_t^2\right)$, i.e., is Gaussian r.v. with parameters

$$y_t = y_0\phi\,(0,t)\,, \qquad d_t^2 = r^2\phi\,(0,t)^2 + \int_0^t \phi\,(s,t)^2\, b\,(s)^2\ \mathrm{d}s.$$

Bayesian estimator $\tilde{Y}_t$ of Y_t is defined by the relation

$$\mathbf{E}|Y_t - \tilde{Y}_t|^2 = \inf_{\bar{Y}_t} \mathbf{E}\left|Y_t - \bar{Y}_t\right|^2.$$

Proposition 1.1 *The mean square optimal estimator of Y_t is a Bayesian estimator $m_t = \mathbf{E}\left(Y_t|\mathfrak{F}_t^X\right)$, i.e., conditional mathematical expectation.*

Proof We can write

$$\begin{aligned}\mathbf{E}\left|Y_t - \bar{Y}_t\right|^2 &= \mathbf{E}\left|Y_t - m_t + m_t - \bar{Y}_t\right|^2 = \mathbf{E}\left|Y_t - m_t\right|^2 + \mathbf{E}\left|m_t - \bar{Y}_t\right|^2 \\ &\quad + \mathbf{E}\left(Y_t - m_t\right)\left(m_t - \bar{Y}_t\right) = \mathbf{E}\left|Y_t - m_t\right|^2 + \mathbf{E}\left|m_t - \bar{Y}_t\right|^2.\end{aligned} \tag{1.16}$$

because

$$\begin{aligned}\mathbf{E}\left(Y_t - m_t\right)\left(m_t - \bar{Y}_t\right) &= \mathbf{E}\left(\mathbf{E}(Y_t - m_t)(m_t - \bar{Y}_t)|\mathfrak{F}_t^X\right) \\ &= \mathbf{E}\left[\left(m_t - \bar{Y}_t\right)\left(\mathbf{E}(Y_t - m_t)|\mathfrak{F}_t^X\right)\right] = \mathbf{E}\left[\left(m_t - \bar{Y}_t\right)\left(\mathbf{E}(Y_t|\mathfrak{F}_t^X\right) - m_t)\right] = 0.\end{aligned}$$

Minimizing the r.h.s. of (1.16) by $\bar{Y}_t$, we obtain for $\bar{Y}_t = m_t$ the equality

$$\inf_{\bar{Y}_t} \mathbf{E}\left|Y_t - \bar{Y}_t\right|^2 = \mathbf{E}\left|Y_t - m_t\right|^2 + \inf_{\bar{Y}_t}\mathbf{E}\left|m_t - \bar{Y}_t\right|^2 = \mathbf{E}\left|Y_t - m_t\right|^2.$$

Therefore, the estimator m_t is optimal in the mean square sense. □

The main result of this section is the following theorem.

Theorem 1.3 (Kalman–Bucy [84]) *Let $(X_t, Y_t, 0 \le t \le T)$ be a two-dimensional Gaussian process satisfying the system of Eqs.* (1.12) *and* (1.13) *and the conditions* (1.14), (1.15) *be satisfied. Then the process $m_t, 0 \le t \le T$ and the mean square error $\gamma_t = \mathbf{E}\left(Y_t - m_t\right)^2$ satisfy the system of equations*

$$dm_t = a(t)m_t\,dt + \frac{f(t)\gamma_t}{\sigma(t)^2}\left[dX_t - f(t)m_t\,dt\right], \qquad m_0, \quad 0 \le t \le T, \tag{1.17}$$

$$\frac{d\gamma_t}{dt} = 2a(t)\gamma_t - \frac{f(t)^2\gamma_t^2}{\sigma(t)^2} + b(t)^2, \qquad \gamma_0, \quad 0 \le t \le T, \tag{1.18}$$

with initial values $m_0 = \mathbf{E}(Y_0|X_0)$ *and* $\gamma_0 = \mathbf{E}(Y_0 - m_0)^2$. *The system* (1.17) *and* (1.18) *has a unique continuous solution (for* γ_t*, in the class of nonnegative functions).*

Equations (1.17), (1.18) are called *Kalman–Bucy filter* and (1.18) is *Riccati equation*. Before the start of the proof, we remind some auxiliary results.

Introduce a Gaussian vector (Y, X), where the components are $\mathrm{Y} = (Y_1, \ldots, Y_K)^\top, \mathrm{X} = (X_1, \ldots, X_L)^\top$. The vectors of means and covariance matrices are denoted $\mathrm{m}_y = \mathbf{E}Y$, $\mathrm{m}_x = \mathbf{E}\mathrm{X}$, $\mathbf{D}_{yy} = \mathrm{Cov}\,(\mathrm{Y}, \mathrm{Y})$,

$$\mathbf{D}_{yx} = \mathrm{Cov}\,(\mathrm{Y}, \mathrm{X}) = \mathbf{E}\left[\left(\mathrm{Y} - \mathrm{m}_y\right)(\mathrm{X} - \mathrm{m}_x)\right], \qquad \mathbf{D}_{xx} = \mathrm{Cov}\,(\mathrm{X}, \mathrm{X}).$$

We suppose for simplicity that the matrix $\mathbf{D}_{xx}$ is nondegenerate.

Theorem 1.4 (Theorem on Normal Correlation) *The conditional mathematical expectation* $\mathbf{E}\,(\mathrm{Y}|\mathrm{X})$ *and the conditional covariance*

$$\mathrm{Cov}\,(\mathrm{Y}, \mathrm{Y}|\mathrm{X}) = \mathbf{E}\,(\mathrm{Y} - \mathbf{E}\,(\mathrm{Y}|\mathrm{X}))\,(\mathrm{Y} - \mathbf{E}\,(\mathrm{Y}|\mathrm{X}))^\top$$

are given by the formulae

$$\mathbf{E}\,(\mathrm{Y}|\mathrm{X}) = \mathrm{m}_y + \mathbf{D}_{yx}\mathbf{D}_{xx}^{-1}\,(\mathrm{X} - m_x), \qquad \mathrm{Cov}\,(\mathrm{Y}, \mathrm{Y}|\mathrm{X}) = \mathbf{D}_{yy} - \mathbf{D}_{yx}\mathbf{D}_{xx}^{-1}\mathbf{D}_{yx}^\top.$$

The proof of this theorem and the examples below can be found in [141, Theorem 13.1].

Example 1.1 If $K = L = 1$ and $D_{xx} > 0$, then

$$\mathbf{E}\,(Y|X) = \mathbf{E}Y + \frac{\mathrm{Cov}\,(Y, X)}{D_{xx}}\,(X - \mathbf{E}X), \quad D\,(Y|X) = D_{yy} - \frac{[\mathrm{Cov}\,(Y, X)]^2}{D_{xx}}.$$

Here $D\,(Y|X) = \mathbf{E}\left([Y - \mathbf{E}\,(Y|X)]^2\,|X\right)$.

Example 1.2 If $Y = a\epsilon_1 + b\epsilon_2$ and $X = f\epsilon_1 + \sigma\epsilon_2$, where $\epsilon_1 \sim \mathcal{N}(0, 1)$ and $\epsilon_2 \sim \mathcal{N}(0, 1)$ are independent and $f^2 + \sigma^2 > 0$, then

$$\mathbf{E}\,(Y|X) = \frac{af + b\sigma}{f^2 + \sigma^2}\,X, \qquad D\,(Y|X) = \frac{(fb - a\sigma)^2}{f^2 + \sigma^2}.$$

Example 1.3 Let the random variables $(Y, X_1, \ldots, X_L)^\top$ form a Gaussian vector, where $X_1, \ldots, X_L$ are independent and $DX_l > 0$. Then

$$\mathbf{E}\left(Y|\mathrm{X}^L\right) = \mathbf{E}Y + \sum_{l=1}^{L} \frac{\mathrm{Cov}\,(Y, X_l)}{DX_l}\,(X_l - \mathbf{E}X_l)\,.$$

Here $\mathrm{X}^L = (X_1, \ldots, X_L)$. In particular, if $\mathbf{E}Y = \mathbf{E}X_l = 0$, then

$$\mathbf{E}\left(Y|\mathrm{X}^L\right) = \sum_{l=1}^{L} \frac{\mathrm{Cov}\,(Y, X_l)}{DX_l}\,X_l.$$

Example 1.4 Let $W^T = (W_t, \mathfrak{F}_t, 0 \le t \le T)$, be a Wiener process and $\mu = \mu\left(W^t\right)$ be a Gaussian random variable $\mathfrak{F}_t^W$-measurable, then there exists a deterministic function $F(t, s)$ such that

$$\mu = \int_0^t F(t, s)\,\mathrm{d}W_s \qquad \text{and} \qquad \int_0^t F(t, s)^2\,\mathrm{d}s < \infty. \tag{1.19}$$

Example 1.5 Let the conditions (1.14) and (1.15) be satisfied and $m_0 = 0$. Then, for each $t \in [0, T]$, there exists a function $F(t, s)\,, 0 \le s \le t$, such that

$$\int_0^t F(t, s)^2\,\mathrm{d}s < \infty, \qquad \int_0^t F(t, s)^2\,\sigma(s)^2\,\mathrm{d}s < \infty,$$

$$\int_0^t \int_0^t F(t, u)\,F(t, v)\,f(u)\,f(v)\,\mathbf{E}\,(Y_u Y_v)\,\mathrm{d}u\,\mathrm{d}v < \infty,$$

and $m_t = \mathbf{E}\left(Y_t|X^t\right)$ is given by

$$m_t = \int_0^t F(t, s)\,\mathrm{d}X_s. \tag{1.20}$$

For the proofs of (1.19) and (1.20), see [141], Lemma 10.1 and Corollaries 1 and 2 of this lemma, respectively.

If $q^2 > 0$ and $\mathbf{E}\,(X_0 Y_0) = \varrho \neq 0$, then according to the Example 1.12 $m_0 = \frac{\varrho}{q^2} X_0$ and $\gamma_0 = r^2 - \frac{\varrho^2}{q^2}$. If $X_0 = 0 (q^2 = 0)$, then $m_0 = \mathbf{E}\,(Y_0|X_0) = \mathbf{E}Y_0 = 0$ and $\gamma_0 = \mathbf{E}Y_0^2 = r^2$.

Notice thar the process $m_t, 0 \le t \le T$ admits another representation as well

$$m_t = \int_0^t G(t, s)\,\mathrm{d}\bar{W}_s, \qquad \int_0^t G(t, s)^2\,\mathrm{d}s < \infty.$$

Here the process $\bar{W}_t, 0 \le t \le T$ is defined by the relation

$$\bar{W}_t = \int_0^t \frac{1}{\sigma(s)} \, \mathrm{d}X_s - \int_0^t \frac{f(s) m_s}{\sigma(s)} \, \mathrm{d}s, \qquad 0 \le t \le T. \tag{1.21}$$

and is called *innovation Wiener process*. For the proof, see Theorems 7.12, 7.16, and 7.17 in [141].

Lemma 1.2 (Innovation) *$\bar{W}_t = \left(\bar{W}_t, \mathfrak{F}_t^X, 0 \le t \le T\right)$ is Wiener process.*

Proof By Itô formula, we can write (below $s < t$ and $\mathrm{i} = \sqrt{-1}$)

$$\begin{aligned} e^{\mathrm{i}\lambda(\bar{W}_t - \bar{W}_s)} &= \exp\left(\mathrm{i}\lambda\,(W_t - W_s) + \int_s^t \frac{f(r)\,(Y_r - m_r)}{\sigma(r)}\, \mathrm{d}r\right) \\ &= 1 + \mathrm{i}\lambda \int_s^t e^{\mathrm{i}\lambda(\bar{W}_t - \bar{W}_r)} \mathrm{d}W_r \\ &\quad + \int_s^t e^{\mathrm{i}\lambda(\bar{W}_t - \bar{W}_r)} \left[\frac{f(r)\,(Y_r - m_r)}{\sigma(r)} - \frac{\lambda^2}{2}\right] \mathrm{d}r. \end{aligned}$$

Hence, for the characteristic function of $\bar{W}_t - \bar{W}_s$ we obtain the equation

$$\mathbf{E}e^{\mathrm{i}\lambda(\bar{W}_t - \bar{W}_s)} = 1 - \frac{\lambda^2}{2} \int_s^t \mathbf{E}e^{\mathrm{i}\lambda(\bar{W}_t - \bar{W}_r)} \mathrm{d}r \tag{1.22}$$

because

$$\mathbf{E}\left(\int_s^t e^{\mathrm{i}\lambda(\bar{W}_t - \bar{W}_r)} \mathrm{d}W_r \middle| \mathfrak{F}_s^X\right) = 0,$$

$$\mathbf{E}\left(e^{\mathrm{i}\lambda(\bar{W}_t - \bar{W}_r)} \frac{f(r)\mathbf{E}\left[(Y_r - m_r) | \mathfrak{F}_r^X\right]}{\sigma(r)} \middle| \mathfrak{F}_s^X\right) = 0.$$

The solution of the Eq. (1.22) is

$$\mathbf{E}e^{\mathrm{i}\lambda(\bar{W}_t - \bar{W}_s)} = e^{-\frac{\lambda^2}{2}(t-s)}.$$

□

Now we can start the proof of the Theorem 1.3.

Proof The processes $\left(X^T, Y^T\right)$ are jointly Gaussian and by Theorem 1.4 the conditional expectation m_t can be presented as the integral (1.20)

$$m_t = \int_0^t G(t, s)\, \mathrm{d}\bar{W}_s$$

(we assume that $m_0 = 0$), where the Wiener process $\bar{W}_t, \mathfrak{F}_s^X, 0 \leq t \leq T$ is determined by the equality (1.21).

Let us fix $t \in [0, T]$ and let $g(t, s)$ be a bounded function. Then

$$\mathbf{E}(Y_t - m_t) \int_0^t g(t, s)\, \mathrm{d}\bar{W}_s = 0$$

or

$$\mathbf{E}Y_t \int_0^t g(t, s)\, \mathrm{d}\bar{W}_s = \mathbf{E}m_t \int_0^t g(t, s)\, \mathrm{d}\bar{W}_s = \int_0^t G(t, s)\, g(t, s)\, \mathrm{d}s.$$

By definition (1.21) of innovation process, we have

$$\bar{W}_t = W_t + \int_0^t \frac{f(s)}{\sigma(s)} (Y_s - m_s)\, \mathrm{d}s,$$

and therefore,

$$\begin{aligned}\int_0^t G(t, s)\, g(t, s)\, \mathrm{d}s &= \mathbf{E}\left[Y_t \int_0^t G(t, s)\, \mathrm{d}W_s\right] \\ &\quad + \mathbf{E}\left[Y_t \int_0^t \frac{f(s) g(t, s)}{\sigma(s)} (Y_s - m_s)\, \mathrm{d}s\right] \\ &= \int_0^t \frac{f(s) g(t, s)}{\sigma(s)} \mathbf{E}\left[Y_t (Y_s - m_s)\right] \mathrm{d}s.\end{aligned}$$

Further, almost surely

$$\mathbf{E}(Y_t | \mathfrak{F}_s) = Y_s \exp\left(\int_s^t a(r)\, \mathrm{d}r\right) = Y_s\, \phi(s, t), \qquad s < t.$$

Hence,

$$\begin{aligned}\mathbf{E}Y_t (Y_s - m_s) &= \mathbf{E}\left[\mathbf{E}(Y_t | \mathfrak{F}_s)(Y_s - m_s)\right] = \phi(s, t)\, \mathbf{E}Y_s (Y_s - m_s) \\ &= \phi(s, t)\, \mathbf{E}(Y_s - m_s)^2 = \phi(s, t)\, \gamma_s,\end{aligned}$$

and therefore,

$$\int_0^t G(t, s)\, g(t, s)\, \mathrm{d}s = \int_0^t g(t, s) \frac{f(s)}{\sigma(s)} \phi(s, t)\, \gamma_s\, \mathrm{d}s.$$

As this equality is valid for any function $g(t, s)$, we obtain

$$G(t,s) = \phi(s,t)\,\frac{f(s)}{\sigma(s)}\gamma_s.$$

Thus,

$$\begin{aligned} m_t &= \int_0^t G(t,s)\,\mathrm{d}\bar{W}_s = \int_0^t \phi(s,t)\,\frac{f(s)\,\gamma_s}{\sigma(s)}\,\mathrm{d}\bar{W}_s \\ &= \phi(0,t)\int_0^t \phi(0,s)^{-1}\,\frac{f(s)\,\gamma_s}{\sigma(s)^2}\,[\mathrm{d}X_s - f(s)m_s\,\mathrm{d}s]\,. \end{aligned}$$

From this by the Itô formula for m_t, we obtain the stochastic differential (1.17).

To obtain the Eq. (1.18) for $\gamma_t = \mathbb{E}\,(Y_t - m_t)^2$, we denote $\delta_t = Y_t - m_t$. Then from (1.13) and (1.17), we have

$$\mathrm{d}\delta_t = a(t)\delta_t\,\mathrm{d}t + b(t)\,\mathrm{d}V_t - \frac{f(t)^2\gamma_t}{\sigma(t)^2}\delta_t\,\mathrm{d}t - \frac{f(t)\gamma_t}{\sigma(t)}\,\mathrm{d}W_t.$$

Hence, by the Itô formula, we find that

$$\begin{aligned} \delta_t^2 = \delta_0^2 &+ 2\int_0^t \left[a(s) - \frac{f(s)^2\gamma_s}{\sigma(s)^2}\right]\delta(s)^2\,\mathrm{d}s + \int_0^t \left[b(s)^2 + \frac{f(s)^2\gamma_s^2}{\sigma(s)^2}\right]\mathrm{d}s \\ &+ 2\int_0^t \delta_s b(s)\,\mathrm{d}V_s - 2\int_0^t \delta_s\frac{f(s)\gamma_s}{\sigma(s)}\,\mathrm{d}W_s. \end{aligned}$$

From this equation, we obtain the relation

$$\gamma_t = \gamma_0 + 2\int_0^t \left[a(s) - \frac{f(s)^2\gamma_s}{\sigma(s)^2}\right]\gamma_s\,\mathrm{d}s + \int_0^t \left[b(s)^2 + \frac{f(s)^2\gamma_s^2}{\sigma(s)^2}\right]\mathrm{d}s$$

which is equivalent to (1.18). □

Let us consider several examples of Kalman–Bucy filters, where the Riccati equation (1.18) has explicite solutions.

Example 1.6 Suppose that in (1.12)–(1.13) we have $Y_t \equiv Y_0, 0 \le t \le T$, where $Y_0 \sim \mathcal{N}\left(0, r^2\right)$. The observation equation is

$$\mathrm{d}X_t = f(t)Y_t\mathrm{d}t + \sigma(t)\,\mathrm{d}W_t, \qquad X_0 = 0, \qquad 0 \le t \le T,$$

and therefore, the Riccati equation is

$$\frac{\mathrm{d}\gamma_t}{\mathrm{d}t} = -\frac{f(t)^2\gamma_t^2}{\sigma(t)^2}, \qquad \gamma_0 = r^2.$$

The solution of this equation is

$$\gamma_t = \frac{r^2}{1 + r^2 F_t}, \qquad F_t = \int_0^t \frac{f(s)^2}{\sigma(s)^2}\, ds, \qquad 0 \le t \le T.$$

Therefore,

$$dm_t = -\frac{r^2 f(t)^2}{\sigma(t)^2 \left(1 + r^2 F_t\right)}\, m_t\, dt + \frac{r^2 f(t)}{\sigma(t)^2 \left(1 + r^2 F_t\right)}\, dX_t, \qquad m_0 = 0.$$

Suppose that the functions $f(t) \equiv f \neq 0$ and $\sigma(t) \equiv \sigma \neq 0$. Then it is easy to verify that the random process

$$m_t = \frac{r^2 f}{\sigma^2 + r^2 f^2 t}\, X_t, \qquad 0 \le t \le T,$$

has the stochastic differential

$$dm_t = -\frac{r^2 f^2}{\left(\sigma^2 + r^2 f^2 t\right)}\, m_t\, dt + \frac{r^2 f}{\left(\sigma^2 + r^2 f^2 t\right)}\, dX_t, \qquad m_0 = 0.$$

Hence,

$$m_T = \frac{r^2 f}{\sigma^2 + r^2 f^2 T}\, X_T \longrightarrow Y_0, \qquad \sqrt{T}\,(m_T - Y_0) \Longrightarrow \mathcal{N}\left(0, \sigma^2\right)$$

as $T \to \infty$.

Example 1.7 Suppose that the coefficients are $f(t) = f > 0, \sigma(t) = \sigma > 0, a(t) = 0, b(t) = b > 0$ and $Y_0 \sim \mathcal{N}\left(0, r^2\right)$. Then, the Riccati equation is

$$\frac{d\gamma_t}{dt} = -\frac{f^2 \gamma_t^2}{\sigma^2} + b^2, \qquad \gamma_0 = r^2.$$

It can be written as

$$\frac{d\gamma_t}{\gamma_*^2 - \gamma_t^2} = \frac{f^2\, dt}{\sigma^2}, \qquad \frac{f^2 t}{\sigma^2} = \int_{r^2}^{\gamma_t} \frac{1}{\gamma_*^2 - \gamma^2}\, d\gamma, \qquad \gamma_* = \frac{b\sigma}{f}.$$

This integral can be easy calculated, and we obtain

$$\gamma_t = \frac{\left(\gamma_* + r^2\right) e^{2\frac{fb}{\sigma} t} - \gamma_* + r^2}{\left(\gamma_* + r^2\right) e^{2\frac{fb}{\sigma} t} + \gamma_* - r^2}\, \gamma_* \tag{1.23}$$

Therefore,

$$\gamma_t \longrightarrow \frac{b\,\sigma}{f} = \gamma_* \tag{1.24}$$

with exponential rate.

For the large values of T, the calculation of the conditional expectation can be simplified too

$$\begin{aligned} m_T &= m_0 \exp\left(-\frac{f^2}{\sigma^2}\int_0^T \gamma_s \mathrm{d}s\right) + \frac{f}{\sigma^2}\int_0^T \exp\left(-\frac{f^2}{\sigma^2}\int_s^T \gamma_u \mathrm{d}u\right)\gamma_s\, \mathrm{d}X_s \\ &= \frac{b}{\sigma}\int_0^T e^{-\frac{fb}{\sigma}(T-s)}\mathrm{d}X_s + o\left(1\right), \end{aligned}$$

and

$$\mathbf{E}\left(m_T - bV_T\right)^2 = \frac{\sigma b}{f} + o\left(1\right).$$

Example 1.8 Suppose that we have partially obsevred system with the constant coefficients

$$\begin{aligned} \mathrm{d}X_t &= f Y_t\, \mathrm{d}t + \sigma\, \mathrm{d}W_t, \qquad X_0, \qquad 0 \le t \le T, \\ \mathrm{d}Y_t &= -aY_t\, \mathrm{d}t + b\, \mathrm{d}V_t, \qquad Y_0, \end{aligned}$$

where $\sigma > 0$. Then, once more the solution of the Riccati equation

$$\frac{\mathrm{d}\gamma_t}{\mathrm{d}t} = -2a\gamma_t - \frac{f^2\gamma_t^2}{\sigma^2} + b^2, \qquad \gamma_0 = r^2$$

has an explicite expression. It is sufficient to write it as

$$\frac{\mathrm{d}\gamma_t}{\mathrm{d}t} = -\frac{f^2}{\sigma^2}\left(\gamma_t + \frac{a\sigma^2}{f^2}\right)^2 + \frac{a^2\sigma^2}{f^2} + b^2, \qquad \gamma_0 = r^2.$$

Then put $\hat{\gamma}_t = \gamma_t + a\sigma^2 f^{-2}$, $\hat{b}^2 = a^2\sigma^2 f^{-2} + b^2$ and the equation for $\hat{\gamma}_t$ is reduced to the Riccati equation of Example 1.7 with the initial value $\hat{\gamma}_0 = r^2 + a\sigma^2 f^{-2}$.

Denote

$$\rho = \left(a^2 + \frac{b^2 f^2}{\sigma^2}\right)^{1/2}, \qquad \gamma_* = \frac{\sigma^2}{f^2}\left(\rho - a\right). \tag{1.25}$$

If $\gamma_0 = \gamma_*$, then $\gamma_t = \gamma_*$. If $\gamma_0 \neq \gamma_*$, then

$$\gamma_t = e^{-2\rho t}\left[\frac{1}{\gamma_0-\gamma_*}+\frac{f^2}{2\rho\sigma^2}\left(1-e^{-2\rho t}\right)\right]^{-1}+\gamma_*. \tag{1.26}$$

It is easy to see that if $a>0$ and $\gamma_0\neq\gamma_*$, then we have exponential convergence

$$|\gamma_t-\gamma_*|\leq Ce^{-2\rho t}, \qquad \text{as}\quad t\rightarrow\infty.$$

Of course, if $a=0$, then (see (1.24))

$$\gamma_*=\frac{\sigma b}{f}.$$

Consider the partially observed system (1.12) and (1.13) where the coefficients of the system depend on finite-dimensional parameter $\vartheta\in\Theta\subset\mathcal{R}^d, d\geq 1$, i.e.,

$$\mathrm{d}X_t = f(\vartheta,t)Y_t\,\mathrm{d}t+\sigma(t)\,\mathrm{d}W_t, \qquad X_0, \quad 0\leq t\leq T, \tag{1.27}$$

$$\mathrm{d}Y_t = a(\vartheta,t)Y_t\,\mathrm{d}t+b(\vartheta,t)\,\mathrm{d}V_t, \qquad Y_0, \quad 0\leq t\leq T. \tag{1.28}$$

Here $W^T=(W_t,\mathfrak{F}_t,0\leq t\leq T)$ and $V^T=(V_t,\mathfrak{F}_t,0\leq t\leq T)$ as before are two independent Wiener processes and the Gaussian variables $X_0\sim\mathcal{N}\left(x_0,q^2\right)$ and $Y_0\sim\mathcal{N}\left(y_0,r^2\right)$ are $\mathfrak{F}_0$ measurable and independent of W^T and V^T. The set Θ is open and bounded.

The conditions (1.14) and (1.15) are replaced by the conditions

$$\sup_{\vartheta\in\Theta}\int_0^T f(\vartheta,t)^2\,\mathrm{d}t<\infty, \qquad \sup_{\vartheta\in\Theta}\int_0^T b(\vartheta,t)^2\,\mathrm{d}t<\infty, \qquad \sigma(t)^2\geq c>0, \tag{1.29}$$

$$\sup_{\vartheta\in\Theta}\int_0^T |a(\vartheta,t)|\,\mathrm{d}t<\infty, \qquad \int_0^T\sigma(t)^2\,\mathrm{d}t<\infty \tag{1.30}$$

and the corresponding equations of Kalman–Bucy filter (1.17) and (1.18) are

$$\mathrm{d}m(\vartheta,t)=a(\vartheta,t)\,m(\vartheta,t)\,\mathrm{d}t+\frac{f(\vartheta,t)\,\gamma(\vartheta,t)}{\sigma(t)^2}\left[\mathrm{d}X_t-f(\vartheta,t)\,m(\vartheta,t)\,\mathrm{d}t\right], \tag{1.31}$$

$$\frac{\partial\gamma(\vartheta,t)}{\partial t}=2a(\vartheta,t)\,\gamma(\vartheta,t)-\frac{f(\vartheta,t)^2\,\gamma(\vartheta,t)^2}{\sigma(t)^2}+b(\vartheta,t)^2, \qquad 0\leq t\leq T, \tag{1.32}$$

Here $m(\vartheta,t)=\mathbf{E}_\vartheta\left(Y_t|\mathfrak{F}_t^X\right)$ and $\gamma(\vartheta,t)=\mathbf{E}_\vartheta\left(Y_t-m(\vartheta,t)\right)^2$ and for simplicity, we suppose that the initial values $m(\vartheta,0), \gamma(\vartheta,0)$ do not depend of ϑ, i.e., $m(\vartheta,0)=m(0)$ and $\gamma(\vartheta,0)=\gamma(0)$.

Under conditions (1.29) and (1.30) the system (1.31) and (1.32) for each $\vartheta \in \Theta$ by Theorem 1.3 has a unique solution $m(\vartheta, t), \gamma(\vartheta, t), 0 \leq t \leq T$.

Let us denote

$$A(\vartheta, t) = a(\vartheta, t) - \frac{f(\vartheta, t)^2 \gamma(\vartheta, t)}{\sigma(t)^2}, \qquad B(\vartheta, t) = \frac{f(\vartheta, t)\gamma(\vartheta, t)}{\sigma(t)^2},$$

$$\Phi(\vartheta, s, t) = \exp\left(\int_s^t A(\vartheta, r)\,\mathrm{d}r\right).$$

Sometimes, we will need to substitute in $m(\vartheta, \cdot)$ an estimator $\bar\vartheta = \bar\vartheta(X^\tau)$ constructed by the observations $X^\tau = (X_s, 0 \leq s \leq \tau)$. To avoid the problem with definition of the Itô integral in such situations, we will use the following lemma.

Lemma 1.3 *Suppose that the functions $a(\vartheta, t), f(\vartheta, t), \sigma(t), t \in [0, T], \vartheta \in \Theta$ have continuous derivatives w.r.t. t. Then the solution $m(\vartheta, \cdot)$ of* (1.31) *admits the representation*

$$m(\vartheta, t) = m(0)\,\Phi(\vartheta, 0, t) + X_t B(\vartheta, t) - \int_0^t X_s \Phi(\vartheta, s, t)\, B_s'(\vartheta, s)\,\mathrm{d}s$$
$$+ \int_0^t X_s \Phi(\vartheta, s, t)\, A(\vartheta, s)\, B(\vartheta, s)\,\mathrm{d}s. \tag{1.33}$$

Proof The solution $m(\vartheta, \cdot)$ of the linear equation (1.31) can be written in the integral form

$$m(\vartheta, t) = m(0)\,\Phi(\vartheta, 0, t) + \int_0^t \Phi(\vartheta, s, t)\, B(\vartheta, s)\,\mathrm{d}X_s$$
$$= \Phi(\vartheta, 0, t)\left[m(0) + \int_0^t \Phi(\vartheta, 0, s)^{-1} B(\vartheta, s)\,\mathrm{d}X_s\right].$$

For continuously differentiable function $H(\cdot)$, we have the relation

$$\int_a^b H(s)\,\mathrm{d}X_s = H(b)\,X_b - H(a)\,X_a - \int_a^b X_s H_s'(s)\,\mathrm{d}s.$$

Therefore, we can write

$$m(\vartheta, t) = \Phi(\vartheta, 0, t)\Big[m(0) + X_t \Phi(\vartheta, 0, t)^{-1} B(\vartheta, t)$$
$$- \int_0^t X_s \left[\Phi(\vartheta, 0, s)^{-1} B(\vartheta, s)\right]_s' \mathrm{d}s\Big]$$
$$= m(0)\,\Phi(\vartheta, 0, t) + X_t B(\vartheta, t) - \int_0^t X_s \Phi(\vartheta, s, t)\, B_s'(\vartheta, s)\,\mathrm{d}s$$

$$+\int_0^t X_s \Phi(\vartheta, s, t) A(\vartheta, s) B(\vartheta, s) \mathrm{d}s.$$

Note that the derivative $\gamma_s'(\vartheta, s)$ is given by the Eq. (1.32) □

Conditions $\mathscr{C}^\dagger$.

$\breve{\mathscr{C}}_1^\dagger$. *The functions $a(\vartheta, t), b(\vartheta, t), f(\vartheta, t), \sigma(t); \vartheta \in \bar{\Theta}, t \in [0, \tau]$ have continuous bounded derivatives w.r.t. t.*

$\breve{\mathscr{C}}_2^\dagger$. *The functions $a(\vartheta, t), b(\vartheta, t), f(\vartheta, t), b_t'(\vartheta, t), f_t'(\vartheta, t); \vartheta \in \bar{\Theta}, t \in [0, \tau]$ have continuous bounded derivatives w.r.t. ϑ.*

$\breve{\mathscr{C}}_3^\dagger$. *The estimator $\bar{\vartheta}_{\tau,\varepsilon}$ converges to ϑ_0 and $\varphi_\varepsilon^{-1}(\bar{\vartheta}_{\tau,\varepsilon} - \vartheta_0) \Longrightarrow \zeta(\vartheta_0)$, where $\varphi_\varepsilon \to 0$.*

Below $\dot{A}(\vartheta_0, r)$ and $\dot{B}(\vartheta_0, r)$ are derivatives w.r.t. ϑ.

Corollary 1.2 *Let conditions $\mathscr{C}^\dagger$ hold and*

$$\begin{aligned} m\left(\bar{\vartheta}_{\tau,\varepsilon}, \tau\right) = m(0)\, \Phi\left(\bar{\vartheta}_{\tau,\varepsilon}, 0, \tau\right) + X_t B\left(\bar{\vartheta}_{\tau,\varepsilon}, \tau\right) \\ - \int_0^\tau X_s \Phi\left(\bar{\vartheta}_{\tau,\varepsilon}, s, \tau\right) B_s'\left(\bar{\vartheta}_{\tau,\varepsilon}, s\right) \mathrm{d}s \\ + \int_0^\tau X_s \Phi\left(\bar{\vartheta}_{\tau,\varepsilon}, s, \tau\right) A\left(\bar{\vartheta}_{\tau,\varepsilon}, s\right) B\left(\bar{\vartheta}_{\tau,\varepsilon}, s\right) \mathrm{d}s. \end{aligned}$$

Then $m\left(\bar{\vartheta}_{\tau,\varepsilon}, \tau\right)$ admits the representation

$$\begin{aligned} m\left(\bar{\vartheta}_{\tau,\varepsilon}, \tau\right) &= m(0)\, \Phi(\vartheta_0, 0, \tau) + \int_0^\tau \Phi(\vartheta_0, s, \tau)\, B(\vartheta_0, s)\, \mathrm{d}X_s \\ &\quad + \left(\bar{\vartheta}_{\tau,\varepsilon} - \vartheta_0\right)\Big[m(0)\, \Phi(\vartheta_0, 0, \tau) \int_0^\tau \dot{A}(\vartheta_0, s)\, \mathrm{d}s \\ &\quad + \int_0^\tau \Phi(\vartheta_0, s, \tau) \int_s^\tau \dot{A}(\vartheta_0, r)\, \mathrm{d}r\ B(\vartheta_0, s)\, \mathrm{d}X_s \\ &\quad + \int_0^\tau \Phi(\vartheta_0, s, \tau)\ \dot{B}(\vartheta_0, s)\, \mathrm{d}X_s\Big] + o(\varphi_\varepsilon) \\ &= m(\vartheta_0, \tau) + \left(\bar{\vartheta}_{\tau,\varepsilon} - \vartheta_0\right)\Big[m(0)\, \Phi(\vartheta_0, 0, \tau) \int_0^\tau \dot{A}(\vartheta_0, s)\, \mathrm{d}s \\ &\quad + \int_0^\tau \Phi(\vartheta_0, s, \tau) \int_s^\tau \dot{A}(\vartheta_0, r)\, \mathrm{d}r\ B(\vartheta_0, s)\, \mathrm{d}X_s \\ &\quad + \int_0^\tau \Phi(\vartheta_0, s, \tau)\ \dot{B}(\vartheta_0, s)\, \mathrm{d}X_s\Big] + o(\varphi_\varepsilon) \end{aligned}$$

The proof follows directly from the Taylor expansion at the point ϑ_0.

We have as well the convergence

$$\varphi_\varepsilon^{-1}\left(m\left(\bar\vartheta_{\tau,\varepsilon},\tau\right)-m\left(\vartheta_0,\tau\right)\right)\Longrightarrow\zeta\left(\vartheta_0\right)\Big[m\left(0\right)\Phi\left(\vartheta_0,0,\tau\right)\int_0^\tau\dot A\left(\vartheta_0,s\right){\rm d}s$$
$$+\int_0^\tau\Phi\left(\vartheta_0,s,\tau\right)\int_s^\tau\dot A\left(\vartheta_0,r\right){\rm d}r\;B\left(\vartheta_0,s\right){\rm d}X_s$$
$$+\int_0^\tau\Phi\left(\vartheta_0,s,\tau\right)\;\dot B\left(\vartheta_0,s\right){\rm d}X_s\Big].$$

Introduce the condition

$\mathscr{C}^\diamond$. *The functions* $a\left(\vartheta,t\right),b\left(\vartheta,t\right),f\left(\vartheta,t\right),\vartheta\in\bar\Theta,t\in[0,T]$ *have two continuous bounded derivatives w.r.t.* ϑ.

Here $\bar\Theta$ is a closure of Θ. The derivatives on ϑ and the derivatives on t are denoted $\dot{\rm a}\left(\vartheta,t\right)$ and $a'\left(\vartheta,t\right)$, respectively.

Lemma 1.4 *Let the conditions* (1.29), (1.30) *and* $\mathscr{C}^\diamond$ *be satisfied. Then the function* $\gamma\left(\vartheta,t\right),t\in[0,T]$ *has continuous derivative* $\dot\gamma\left(\vartheta,t\right),t\in[0,T]$, *and the vector* $\dot\gamma\left(\vartheta,t\right),t\in[0,T]$ *is solution of the system of equations*

$$\frac{\partial\dot\gamma\left(\vartheta,t\right)}{\partial t}=2A\left(\vartheta,t\right)\dot\gamma\left(\vartheta,t\right)+2\dot{\rm a}\left(\vartheta,t\right)\gamma\left(\vartheta,t\right)-\frac{2f\left(\vartheta,t\right)\dot{\rm f}\left(\vartheta,t\right)}{\sigma\left(t\right)^2}\gamma\left(\vartheta,t\right)^2$$
$$+2b\left(\vartheta,t\right)\dot{\rm b}\left(\vartheta,t\right),\qquad\dot\gamma\left(\vartheta,0\right)=0.\tag{1.34}$$

The random process $m\left(\vartheta,\cdot\right)$ *has mean square continuous derivative* $\dot{\rm m}\left(\vartheta,t\right)$, $t\in[0,T]$ *and the vector* $\dot{\rm m}\left(\vartheta,t\right),t\in[0,T]$ *satisfies the equations*

$${\rm d}\dot{\rm m}\left(\vartheta,t\right)=A\left(\vartheta,t\right)\dot{\rm m}\left(\vartheta,t\right){\rm d}t+\dot{\rm A}\left(\vartheta,t\right)m\left(\vartheta,t\right){\rm d}t+\dot{\rm B}\left(\vartheta,t\right){\rm d}X_t,\;\;\dot{\rm m}\left(\vartheta,0\right)=0.\tag{1.35}$$

Proof We have to show that for any vector h satisfying $\|{\rm h}\|\rightarrow0$ we have the relations

$$\Delta_\gamma\left({\rm h},t\right)=\gamma\left(\vartheta+{\rm h},t\right)-\gamma\left(\vartheta,t\right)-\langle{\rm h},\dot\gamma\left(\vartheta,t\right)\rangle=O\left(\|h\|^2\right),\tag{1.36}$$

$$\Delta_m\left({\rm h},t\right)={\bf E}_\vartheta\big[m\left(\vartheta+{\rm h},t\right)-m\left(\vartheta,t\right)-\langle{\rm h},\dot{\rm m}\left(\vartheta,t\right)\rangle\big]^2=o\left(\|{\rm h}\|^2\right).\tag{1.37}$$

Here $\langle\cdot,\cdot\rangle$ and $\|\cdot\|$ are the inner product and norm in $\mathcal{R}^d$, respectively. To prove (1.36), we first note that the function $\gamma\left(\vartheta,t\right),\vartheta\in\Theta,t\in[0,T]$ is bounded. Then using (1.34) for $\delta_\gamma\left({\rm h},t\right)=\gamma\left(\vartheta+{\rm h},t\right)-\gamma\left(\vartheta,t\right)$ we write

$$\frac{\partial}{\partial t}\delta_\gamma\left({\rm h},t\right)=\left[2a\left(\vartheta_h,t\right)-\frac{f\left(\vartheta_h,t\right)^2\left(\gamma\left(\vartheta_h,t\right)+\gamma\left(\vartheta,t\right)\right)}{\sigma\left(t\right)^2}\right]\delta_\gamma\left({\rm h},t\right)$$

$$+ 2\left[a\left(\vartheta_h, t\right) - a\left(\vartheta, t\right)\right]\gamma\left(\vartheta, t\right) + \left[b\left(\vartheta_h, t\right)^2 - b\left(\vartheta, t\right)^2\right]$$
$$+ \left[f\left(\vartheta, t\right) - f\left(\vartheta_h, t\right)\right]\frac{\left[f\left(\vartheta, t\right) + f\left(\vartheta_h, t\right)\right]\gamma\left(\vartheta, t\right)^2}{\sigma\left(t\right)^2}.$$

The solution of this equation can be written explicitly. Then using estimates like $\left|a\left(\vartheta_h, t\right) - a\left(\vartheta, t\right)\right| \leq C\left\|\mathrm{h}\right\|$, we obtain the similar estimate for $\delta_\gamma\left(\mathrm{h}, t\right)$

$$\left|\gamma\left(\vartheta_h, t\right) - \gamma\left(\vartheta, t\right)\right| \leq C\left\|\mathrm{h}\right\|.$$

The same procedure for $\Delta_\gamma\left(h, t\right)$ gives us the differential equation

$$\mathrm{d}\Delta_\gamma\left(\mathrm{h}, t\right) = 2\left[a\left(\vartheta, t\right) - \frac{f\left(\vartheta, t\right)^2\gamma\left(\vartheta, t\right)}{\sigma\left(t\right)^2}\right]\Delta_\gamma\left(\mathrm{h}, t\right)\mathrm{d}t$$
$$- \frac{f\left(\vartheta, t\right)^2}{\sigma\left(t\right)^2}\left[\gamma\left(\vartheta_h, t\right)^2 - \gamma\left(\vartheta, t\right)^2 - 2\gamma\left(\vartheta, t\right)\langle\mathrm{h}, \dot{\gamma}\left(\vartheta, t\right)\right]$$
$$+ 2\left[a\left(\vartheta_h, t\right) - a\left(\vartheta, t\right) - \langle\mathrm{h}, \dot{\mathrm{a}}\left(\vartheta, t\right)\rangle\right]\gamma\left(\vartheta, t\right)$$
$$+ 2\left[a\left(\vartheta_h, t\right) - a\left(\vartheta, t\right)\right]\left[\gamma\left(\vartheta_h, t\right) - \gamma\left(\vartheta, t\right)\right]$$
$$+ \left[b\left(\vartheta_h, t\right)^2 - b\left(\vartheta, t\right)^2 - 2b\left(\vartheta, t\right)\langle\mathrm{h}, \dot{\mathrm{b}}\left(\vartheta, t\right)\rangle\right]$$
$$- \frac{\gamma\left(\vartheta, t\right)^2}{\sigma\left(t\right)^2}\left[f\left(\vartheta_h, t\right)^2 - f\left(\vartheta, t\right)^2 - 2f\left(\vartheta, t\right)\langle\mathrm{h}, \dot{\mathrm{f}}\left(\vartheta, t\right)\right]$$
$$+ \frac{1}{\sigma\left(t\right)^2}\left[f\left(\vartheta_h, t\right)^2 - f\left(\vartheta, t\right)^2\right]\left[\gamma\left(\vartheta, t\right)^2 - \gamma\left(\vartheta_h, t\right)^2\right].$$

We have

$$\gamma\left(\vartheta_h, t\right)^2 - \gamma\left(\vartheta, t\right)^2 - 2\gamma\left(\vartheta, t\right)\langle\mathrm{h}, \dot{\gamma}\left(\vartheta, t\right)\rangle$$
$$= \left(\gamma\left(\vartheta_h, t\right) - \gamma\left(\vartheta, t\right)\right)^2 + 2\gamma\left(\vartheta, t\right)\Delta_\gamma\left(\mathrm{h}, t\right).$$

Once more we obtained a linear equation, which has explicite solution. Using the estimates like

$$\left|b\left(\vartheta_h, t\right)^2 - b\left(\vartheta, t\right)^2 - 2b\left(\vartheta, t\right)\langle\mathrm{h}, \dot{\mathrm{b}}\left(\vartheta, t\right)\rangle\right| \leq C\left\|\mathrm{h}\right\|^2,$$
$$\left|\left[a\left(\vartheta_h, t\right) - a\left(\vartheta, t\right)\right]\left[\gamma\left(\vartheta_h, t\right) - \gamma\left(\vartheta, t\right)\right]\right| \leq C\left\|\mathrm{h}\right\|^2$$

we obtain (1.36).

To verify (1.37), we write the Eq. (1.31) for ϑ, ϑ_h and the Eq. (1.35) for the scalar product $\langle\mathrm{h}, \dot{\mathrm{m}}\left(\vartheta, t\right)\rangle$. This leads to the following equation for $\delta_m\left(\mathrm{h}, t\right) = m\left(\vartheta + \mathrm{h}, t\right) - m\left(\vartheta, t\right) - \langle\mathrm{h}, \dot{\mathrm{m}}\left(\vartheta, t\right)\rangle$

$$
\begin{aligned}
\mathrm{d}\delta_m(\mathrm{h},t) &= A(\vartheta,t)\,\delta_m(\mathrm{h},t)\,\mathrm{d}t + \left[A(\vartheta_h,t) - A(\vartheta,t) - \langle \mathrm{h}, \dot{\mathrm{A}}(\vartheta,t)\rangle\right] m(\vartheta,t)\,\mathrm{d}t \\
&\quad + \left[A(\vartheta_h,t) - A(\vartheta,t)\right]\left[m(\vartheta_h,t) - m(\vartheta,t)\right]\mathrm{d}t \\
&\quad + \left[B(\vartheta_h,t) - B(\vartheta,t) - \langle \mathrm{h}, \dot{\mathrm{B}}(\vartheta,t)\rangle\right]\mathrm{d}X_t .
\end{aligned}
$$

This linear equation has the solution

$$
\begin{aligned}
\delta_m(h,t) &= \int_0^t \phi(\vartheta,s,t)\left[A(\vartheta_h,s) - A(\vartheta,s) - \langle \mathrm{h}, \dot{\mathrm{A}}(\vartheta,s)\rangle\right] m(\vartheta,s)\,\mathrm{d}s \\
&\quad + \int_0^t \phi(\vartheta,s,t)\left[A(\vartheta_h,s) - A(\vartheta,s)\right]\left[m(\vartheta_h,s) - m(\vartheta,s)\right]\mathrm{d}s \\
&\quad + \int_0^t \phi(\vartheta,s,t)\left[B(\vartheta_h,s) - B(\vartheta,s) - \langle \mathrm{h}, \dot{\mathrm{B}}(\vartheta,s)\rangle\right] f(\vartheta,s)\, m(\vartheta,s)\,\mathrm{d}s \\
&\quad + \int_0^t \phi(\vartheta,s,t)\left[B(\vartheta_h,s) - B(\vartheta,s) - \langle \mathrm{h}, \dot{\mathrm{B}}(\vartheta,s)\rangle\right]\sigma(s)\,\mathrm{d}\bar{W}_s,
\end{aligned}
\tag{1.38}
$$

where

$$
\phi(\vartheta,s,t) = \exp\left(\int_s^t A(\vartheta,v)\,\mathrm{d}v\right).
$$

Now the estimate (1.36) for $\mathbf{E}_\vartheta \delta(m,t)^2$ can be obtained from elementary calculations. □

Remark 1.2 From the representation (1.38) and boundness of all underlying functions, it is possible to obtain the estimate: for any integer $p \geq 1$

$$
\sup_{\vartheta\in\Theta} \mathbf{E}_\vartheta \left|m(\vartheta+\mathrm{h},t) - m(\vartheta,t)\right|^{2p} \leq C\,\|\mathrm{h}\|^{2p}. \tag{1.39}
$$

Remark 1.3 Using Eq. (1.35) for $\dot{m}(\vartheta+\mathrm{h},t)$ and $\dot{\mathrm{m}}(\vartheta,t)$, we can write the equation for $\delta_{\dot{\mathrm{m}}}(h,t)$. Then once more the boundness of all underlying functions and their derivatives allow us to obtain another estimate: for any integer $p \geq 1$

$$
\sup_{\vartheta\in\Theta} \mathbf{E}_\vartheta \left|\dot{\mathrm{m}}(\vartheta+\mathrm{h},t) - \dot{\mathrm{m}}(\vartheta,t)\right|^{2p} \leq C\,\|\mathrm{h}\|^{2p}. \tag{1.40}
$$

Hence, the random processes $m(\vartheta,t)$ and $\dot{\mathrm{m}}(\vartheta,t)$ are continuous with probability 1 functions of ϑ because we can always take $2p > d$.

Remark 1.4 This work is mainly devoted to the statistical problems related with the systems like (1.27) and (1.28). Sometimes, it is interesting to identify a slightly more general models with trends in the both equations containing deterministic functions

$$
\begin{aligned}
\mathrm{d}X_t &= [h(\vartheta,t)+f(\vartheta,t)Y_t]\,\mathrm{d}t+\sigma(t)\,\mathrm{d}W_t, \qquad X_0, \quad 0\le t\le T,\\
\mathrm{d}Y_t &= [c(\vartheta,t)+a(\vartheta,t)Y_t]\,\mathrm{d}t+b(\vartheta,t)\,\mathrm{d}V_t, \qquad Y_0, \quad 0\le t\le T.
\end{aligned}
$$

Suppose that the functions $f(\cdot), h(\cdot), a(\cdot), c(\cdot), \sigma(\cdot)$ and $b(\cdot)$ satisfy the conditions (1.29) and (1.30). Then, the Kalman–Bucy filter for $m(\vartheta,t)=\mathbf{E}_\vartheta\left(Y_t|\mathfrak{F}_t^X\right)$ is given by the equations

$$
\begin{aligned}
\mathrm{d}m(\vartheta,t) &= [c(\vartheta,t)+a(\vartheta,t)\,m(\vartheta,t)]\,\mathrm{d}t\\
&\quad+\frac{f(\vartheta,t)\,\gamma(\vartheta,t)}{\sigma(t)^2}\left[\mathrm{d}X_t-[h(\vartheta,t)+f(\vartheta,t)\,m(\vartheta,t)]\,\mathrm{d}t\right],
\end{aligned}
\tag{1.41}
$$

$$
\frac{\partial\gamma(\vartheta,t)}{\partial t}=2a(\vartheta,t)\,\gamma(\vartheta,t)-\frac{f(\vartheta,t)^2\,\gamma(\vartheta,t)^2}{\sigma(t)^2}+b(\vartheta,t)^2, \qquad 0\le t\le T,
\tag{1.42}
$$

subject to the corresponding initial values (see Theorem 10.3 in [141]).

1.1.3 Extrapolation

Suppose that the partially observed system $\left(X^T, Y^T\right)$ is (1.27) and (1.28) and the conditions (1.29) and (1.30) are fulfilled. Therefore, the equations of Kalman–Bucy filter are (1.31) and (1.32). Consider the problem of extrapolations of the values Y_t and X_t, where $t>T$ by observations $X^T=(X_s, 0\le s\le T)$. The mean-squared optimal extrapolations are given by the conditional expectations

$$
m(\vartheta,T,t)=\mathbf{E}_\vartheta\left(Y_t|\mathfrak{F}_T^X\right), \qquad n(\vartheta,T,t)=\mathbf{E}_\vartheta\left(X_t|\mathfrak{F}_T^X\right)
$$

respectively.

Consider the joint extrapolation of $Y_t, t>T$ and $X_t, t>T$.

Theorem 1.5 *Under the assumptions made, $m(\vartheta,T,t)$ and $n(\vartheta,T,t)$ (for each $t>T$) are solutions of the system of equations*

$$
\begin{aligned}
\frac{\mathrm{d}m(\vartheta,T,t)}{\mathrm{d}t} &= a(\vartheta,t)\,m(\vartheta,T,t),\\
\frac{\mathrm{d}n(\vartheta,T,t)}{\mathrm{d}t} &= f(\vartheta,t)\,m(\vartheta,T,t)
\end{aligned}
$$

with the initial conditions $m(\vartheta,T,T)=m(\vartheta,T)$, $n(\vartheta,T,T)=X_T$.

For the proof, see Theorem 12.13 in [142].

Hence,

$$m(\vartheta, T, t) = m(\vartheta, T)\exp\left(\int_T^t a(\vartheta, s)\,\mathrm{d}s\right), \tag{1.43}$$

$$n(\vartheta, T, t) = X_T + \int_T^t f(\vartheta, s)\, m(\vartheta, T, s)\,\mathrm{d}s \tag{1.44}$$

$$= X_T + m(\vartheta, T)\int_T^t f(\vartheta, s)\exp\left(\int_T^s a(\vartheta, r)\,\mathrm{d}r\right)\mathrm{d}s. \tag{1.45}$$

Therefore, if we have good estimators of ϑ and adaptive filter for $m(\vartheta, T)$, then the adaptive extrapolation of the processes Y_t and X_t can be written too.

1.1.4 Limit Theorems

We consider two types of limits. The first one can be called "small noise" ($\varepsilon \to 0$), and the second, "large samples" ($T \to \infty$).

We start with the example of the first type of limit.

Small Noise

Let $X^T = (X_t,\ 0 \le t \le T)$ be the solution of the SDE

$$\mathrm{d}X_t = A(t, X_t)\,\mathrm{d}t + \varepsilon\, B(t, X_t)\,\mathrm{d}W_t, \qquad X_0 = x_0, \qquad 0 \le t \le T, \tag{1.46}$$

and let $x^T = (x_t,\ 0 \le t \le T)$ be the solution of the ordinary differential equation (ODE)

$$\frac{\mathrm{d}x_t}{\mathrm{d}t} = A(t, x_t), \qquad x_0, \qquad 0 \le t \le T. \tag{1.47}$$

Here $\varepsilon \in (0, 1]$, and we are interested in the behavior of X^T as $\varepsilon \to 0$.

Theorem 1.6 *Suppose that the functions $A(t, x)$ and $B(t, x)$ satisfy the Lipschitz condition*

$$|A(t, x) - A(t, y)| + |B(t, x) - B(t, y)| \le L\,|x - y|, \tag{1.48}$$

and the linear growth condition

$$|A(t, x)| + |B(t, x)| \le L(1 + |x|). \tag{1.49}$$

Then there exist constants $C_* = C_*(T, L) > 0$ and $C_{**} = C_{**}(p, T, L)$ such that

$$\sup_{0\le t\le T} |X_t - x_t| \le C_* \,\varepsilon\, \eta_T, \qquad a.s., \tag{1.50}$$

and for any $p \ge 2$

$$\sup_{0\le t\le T} \mathbb{E}\,|X_t - x_t|^p \le C_{**}\,\varepsilon^p, \tag{1.51}$$

where

$$\eta_T = \sup_{0\le t\le T} \left|\int_0^t B(s, X_s)\, dW_s\right|.$$

Proof The integral form of the Eqs. (1.46) and (1.47) are

$$X_t = x_0 + \int_0^t A(s, X_s)\, ds + \varepsilon \int_0^t B(s, X_s)\, dW_s,$$

$$x_t = x_0 + \int_0^t A(s, x_s)\, ds.$$

Hence,

$$\begin{aligned}|X_t - x_t| &\le \int_0^t |A(s, X_s) - A(s, x_s)|\, ds + \varepsilon \left|\int_0^t B(s, X_s)\, dW_s\right| \\ &\le L \int_0^t |X_s - x_s|\, ds + \varepsilon\eta_T. \end{aligned} \tag{1.52}$$

Further, using elementary inequalities, we obtain

$$\begin{aligned}\mathbf{E}\,|X_t - x_t|^p &\le 2^{p-1}L^p t^{p-1} \int_0^t \mathbf{E}\,|X_s - x_s|^p\, ds + \varepsilon^p\, 2^{p-1}\mathbf{E}\eta_T^p \\ &\le 2^{p-1}L^p T^{p-1} \int_0^t \mathbf{E}\,|X_s - x_s|^p\, ds + \varepsilon^p\, 2^{p-1}\mathbf{E}\eta_T^p. \end{aligned} \tag{1.53}$$

Note that by the Burkholder–Davis–Gundy inequality (1.2), for any $p > 2$, we have

$$\begin{aligned}\mathbf{E}\,|\eta_T|^p &\le C_p \mathbf{E}\left(\int_0^T B(t, X_t)^2\, dt\right)^{p/2} \le T^{\frac{p}{2}-1} \int_0^T \mathbf{E}\,|B(t, X_t)|^p\, dt \\ &\le T^{\frac{p}{2}-1} L^p \int_0^T \mathbf{E}\,(1 + |X_t|)^p\, dt \le C(p, T, L), \end{aligned}$$

with the corresponding constant $C(p, T, L)$. Note that $\sup_{0\le t\le T} \mathbf{E}|X_t|^p < \infty$ (see (1.8)). □

Lemma 1.5 (Grönwall's Lemma) *Let $u(t)$, $v(t)$, and $r(t)$ be nonnegative continuous functions for $0 \le t \le T$ such that*

$$u(t) \le r(t) + \int_0^t v(s)\, u(s)\, \mathrm{d}s. \tag{1.54}$$

Then

$$u(t) \le r(t) + \int_0^t r(s)\, v(s) \exp\left(\int_s^t v(q)\, \mathrm{d}q\right) \mathrm{d}s. \tag{1.55}$$

If the function $r(\cdot)$ is non-decreasing, then

$$u(t) \le r(t) \exp\left(\int_0^t v(s)\, \mathrm{d}s\right).$$

Proof Introduce the function

$$p(t) = \exp\left(-\int_0^t v(s)\, \mathrm{d}s\right) \int_0^t v(s)\, u(s)\, \mathrm{d}s.$$

For the derivative $p'(t)$, we have

$$\begin{aligned} p'(t) &= \left(u(t) - \int_0^t v(s)\, u(s)\, \mathrm{d}s\right) v(t) \exp\left(-\int_0^t v(s)\, \mathrm{d}s\right) \\ &\le r(t)\, v(t) \exp\left(-\int_0^t v(s)\, \mathrm{d}s\right). \end{aligned}$$

Since $p(0) = 0$, integration gives us

$$p(t) \le \int_0^t r(s)\, v(s) \exp\left(-\int_0^s v(q)\, \mathrm{d}q\right) \mathrm{d}s.$$

Further,

$$\begin{aligned} \int_0^t v(s)\, u(s)\, \mathrm{d}s &= \exp\left(\int_0^t v(s)\, \mathrm{d}s\right) p(t) \\ &\le \int_0^t r(s)\, v(s) \exp\left(\int_s^t v(q)\, \mathrm{d}q\right) \mathrm{d}s. \end{aligned}$$

Substituting this result into (1.54) gives us (1.55).

If the function $r(\cdot)$ is non-decreasing, then using $r(s) \leq r(t)$, we obtain from (1.55) the relation

$$u(t) \leq r(t) \exp\left(\int_0^t v(s)\, ds\right).$$

□

By this lemma and the estimates (1.52) and (1.53), we obtain (1.50) and (1.51).

Corollary 1.3 *If the conditions* (1.48) *and* (1.49) *are satisfied and* $\varepsilon \to 0$, *then the process* X_t, $0 \leq t \leq T$ *converges a.s. to* x_t, $0 \leq t \leq T$, *and the p-th moments of the difference* $|X_t - x_t|$ *converge as well:*

$$\sup_{0\leq t\leq T} |X_t - x_t| \longrightarrow 0,$$

$$\sup_{0\leq t\leq T} \mathbf{E}\, |X_t - x_t|^p \longrightarrow 0.$$

We also need the central limit theorem (CLT) for stochastic integrals in the two mentioned situations. Both limits can be combined in the following statement.

We are given a family of Wiener processes $\left(W_\varepsilon(t), \mathfrak{F}_t^\varepsilon,\ 0 \leq t \leq T_\varepsilon\right)$, $\varepsilon \in (0, 1]$, and a family of random vector-functions $\mathrm{f}_\varepsilon(t) = \left(f_{1,\varepsilon}(t,\omega), \ldots, f_{K,\varepsilon}(t,\omega)\right)^\top$, $(\mathrm{f}_\varepsilon(t), \mathfrak{F}_t^\varepsilon, 0 \leq t \leq T_\varepsilon)$, $\varepsilon \in (0, 1]$. Suppose that for all $\varepsilon \in (0, 1]$

$$\mathbb{P}\left(\int_0^{T_\varepsilon} \|\mathrm{f}_\varepsilon(t)\|^2\, dt < \infty\right) = 1.$$

Then the vector of stochastic integrals

$$\mathrm{J}_\varepsilon = \int_0^{T_\varepsilon} \mathrm{f}_\varepsilon(t)\, dW_\varepsilon(t)$$

is well defined.

Theorem 1.7 (CLT) *Suppose that*

$$\int_0^{T_\varepsilon} \mathrm{f}_\varepsilon(t)\, \mathrm{f}_\varepsilon(t)^\top\, dt \longrightarrow \mathbf{D}, \qquad as \qquad \varepsilon \to 0,$$

where $\mathbf{D}$ *is some non-degenerate matrix. Then,*

$$\int_0^{T_\varepsilon} \mathrm{f}_\varepsilon(t)\, dW_\varepsilon(t) \Longrightarrow \mathcal{N}(0, \mathbf{D}).$$

This is a particular case of the central limit theorem for continuous martingales [75]. In this form, it was published in [105].

Consider two particular cases.

Small noise asymptotics. Let us put $T_\varepsilon = T$, $W_\varepsilon(t) = W_t$, and $\mathrm{f}_\varepsilon(t) = h(t, X_t)$. Here, X_t satisfies the Eq. (1.46), and the function $h(t, x)$ is Lipschitz:

$$|h(t, x) - h(t, y)| \leq C|x - y|.$$

Suppose that the conditions (1.48) and (1.49) are fulfilled. Then,

$$\sup_{0\leq t\leq T} \mathbf{E}|h(t, X_t) - h(t, x_t)|^2 \leq C\varepsilon^2.$$

Let us denote

$$D = \int_0^T h(t, x_t)^2 \, \mathrm{d}t.$$

Then we have

$$\int_0^T h(t, X_t)^2 \, \mathrm{d}t \longrightarrow D$$

and by Theorem 1.7,

$$\int_0^T h(t, X_t) \, \mathrm{d}W_t \Longrightarrow \mathcal{N}(0, D).$$

Moreover, we have the asymptotic normality "in probability" too:

$$\int_0^T h(t, X_t) \, \mathrm{d}W_t \longrightarrow \int_0^T h(t, x_t) \, \mathrm{d}W_t$$

because the limit is defined on the same probability space, and we can write: for any $\nu > 0$

$$\begin{aligned}
&\mathbb{P}\left(\left|\int_0^T h(t, X_t) \, \mathrm{d}W_t - \int_0^T h(t, x_t) \, \mathrm{d}W_t\right| > \nu\right)\\
&\leq \frac{1}{\nu^2}\int_0^T \mathbf{E}|h(t, X_t) - h(t, x_t)|^2 \, \mathrm{d}t\\
&\leq \frac{C}{\nu^2}\int_0^T \mathbf{E}|X_t - x_t|^2 \, \mathrm{d}t \leq \frac{C\varepsilon^2}{\nu^2}.
\end{aligned}$$

Consider the convergence of ordinary integrals too. Suppose that there is a family of integrals indexed by ε:

$$J_\varepsilon(\vartheta) = \int_0^T f_\varepsilon(\vartheta, t)\,\mathrm{d}t, \qquad \varepsilon \in (0, 1].$$

Here, $f_\varepsilon(\cdot), \varepsilon \in (0, 1]$ are measurable random functions depending on the parameter $\vartheta \in \Theta$.

Theorem 1.8 *Assume that the following conditions are satisfied:*

- $$\sup_{\varepsilon\in(0,1],\vartheta\in\Theta} \mathbf{E}_\vartheta \int_0^T |f_\varepsilon(\vartheta, t)|\,\mathrm{d}t < \infty.$$
- *There exists a number $\alpha > 0$ such that*

$$\sup_{\varepsilon\in(0,1],\vartheta\in\Theta} \mathbf{E}_\vartheta\, |f_\varepsilon(\vartheta, t) - f_\varepsilon(\vartheta, s)| \le H\, |t - s|^\alpha$$

  with $H > 0$ being a constant.
- *The marginal distributions of $f_\varepsilon(\vartheta, \cdot)$ converge to the marginal distributions of $f(\vartheta, \cdot)$ uniformly in Θ.*

Then the distribution of the integrals $J_\varepsilon(\vartheta)$ converges to the distribution of the integral

$$J(\vartheta) = \int_0^T f(\vartheta, t)\,\mathrm{d}t$$

uniformly in Θ.

For the proof, see [73, Theorem A.22].

Large Samples

Consider the SDE

$$\mathrm{d}X_t = A(X_t)\,\mathrm{d}t + B(X_t)\,\mathrm{d}W_t, \qquad X_0, \qquad 0 \le t \le T,$$

where the functions $A(x)$ and $B(x)$ satisfy the following conditions:

$$G = \int_{\mathcal{R}} B(x)^{-2} \exp\left(2\int_0^x \frac{A(y)}{B(y)^2}\,\mathrm{d}y\right) \mathrm{d}x < \infty,$$

$$\lim_{x\to\pm\infty} \int_0^x \exp\left(-2\int_0^y \frac{A(z)}{B(z)^2}\,\mathrm{d}z\right) \mathrm{d}y \to \pm\infty.$$

Under these conditions, the process $X_t, t \geq 0$ has ergodic properties, i.e., there exists an invariant distribution with the density function

$$f(x) = G^{-1} B(x)^{-2} \exp\left(2 \int_0^x \frac{A(y)}{B(y)^2}\, \mathrm{d}y\right),$$

and for any absolutely integrable with respect to this density function $H(\cdot)$, the law of large numbers

$$\frac{1}{T} \int_0^T H(X_t)\, \mathrm{d}t \longrightarrow \int_{\mathcal{R}} H(x) f(x)\, \mathrm{d}x$$

holds (see [91]).

Define the stochastic integral

$$J_T = \frac{1}{\sqrt{T}} \int_0^T h(X_t)\, \mathrm{d}W_t.$$

Let us change the variable $t = sT,\ s \in [0, 1]$, $T = \varepsilon^{-1}$, put $W_\varepsilon(s) = T^{-1/2} W_{sT}$, and $f_\varepsilon(s) = h(X_{sT})$. Then

$$J_T = \int_0^1 f_\varepsilon(s)\, \mathrm{d}W_\varepsilon(s)$$

and

$$\int_0^1 f_\varepsilon(s)^2\, \mathrm{d}s = \frac{1}{T} \int_0^T h(X_t)^2\, \mathrm{d}t \longrightarrow \int_{\mathcal{R}} h(x)^2 f(x)\, \mathrm{d}x \equiv D.$$

Hence, by Theorem 1.7, we have the CLT for the stochastic integral in this case too:

$$\frac{1}{\sqrt{T}} \int_0^T h(X_t)\, \mathrm{d}W_t \Longrightarrow \mathcal{N}(0, D).$$

The following Lemma will be used several times.

Lemma 1.6 *Let $\{Y_t, t > 0\}$ be a stochastic process with zero mean, satisfying*

$$\mathbf{E}\,|Y_t|^{m(2k-1)} < C_1, \qquad \int_0^\infty t^{k-1} [\alpha(t)]^{(m-2)/m}\, \mathrm{d}t < C_2,$$

for some $m > 2$ and $k \geq 1$, where $\alpha(t)$ is the strong mixing coefficient. Then

$$\mathbf{E}\left|\int_0^T Y_t\, \mathrm{d}t\right|^{2k} \leq C_3\, T^k.$$

For the proof, see Lemma 2.1 in [90].

We will use several times the following elementary technical lemma.

Lemma 1.7 *Suppose that the functions $F(t)$, $G(t)$, $H(t)$, $t \in [0, T]$ are continuously differentiable on $[0, T]$, that $F(0) = 0$ and that $F(t) > 0$ for all $t \in (0, T]$. Then, for any $t \in (0, T]$, we have, as $\varepsilon \to 0$, the relations*

$$N_\varepsilon(t) = \frac{1}{\varepsilon}\int_0^t \exp\left\{-\frac{1}{\varepsilon}\int_s^t F(v)\,\mathrm{d}v\right\} G(s)\,\mathrm{d}s = \frac{G(t)}{F(t)}\{1 + O(\varepsilon)\} \tag{1.56}$$

$$M_\varepsilon(t) = \frac{1}{\sqrt{\varepsilon}}\int_0^t \exp\left\{-\frac{1}{\varepsilon}\int_s^t F(v)\,\mathrm{d}v\right\} H(s)\,\mathrm{d}W_s \Longrightarrow \frac{H(t)}{\sqrt{2F(t)}}\,\xi_t. \tag{1.57}$$

Here, the variables $\xi_t \sim \mathcal{N}(0, 1)$ are mutually independent for all $t \in (0, T]$.

Proof Let us consider $t_0 > 0$ (small), $t_1 \in (t_0, t)$ and denote

$$\inf_{t_0 \le s \le T} F(s) = c_1 > 0, \qquad \sup_{0 \le s \le T} F(s) = C_1 < \infty, \qquad \sup_{0 \le s \le T} |G(s)| = C_2 < \infty.$$

We have the following estimates:

$$\begin{aligned}
J_0 = \int_0^{t_0} \exp\left\{-\frac{1}{\varepsilon}\int_s^t F(v)\,\mathrm{d}v\right\} |G(s)|\,\mathrm{d}s &\le C_2 t_0 \exp\left\{-\frac{1}{\varepsilon}\int_{t_0}^t F(v)\,\mathrm{d}v\right\}\\
&\le C_2 t_0 \exp\left\{-\frac{c_1(t - t_0)}{\varepsilon}\right\}
\end{aligned}$$

and

$$\begin{aligned}
J_1 = \int_{t_0}^{t_1} \exp\left\{-\frac{1}{\varepsilon}\int_s^t F(v)\,\mathrm{d}v\right\} |G(s)|\,\mathrm{d}s &\le \int_{t_0}^{t_1} \exp\left\{-\frac{c_1(t - s)}{\varepsilon}\right\} |G(s)|\,\mathrm{d}s\\
&\le \frac{C_2 \varepsilon}{c_1} \exp\left\{-\frac{c_1(t - t_1)}{\varepsilon}\right\}.
\end{aligned}$$

Then, by setting $s = t - u\varepsilon$, $v = t - q\varepsilon$ and using Taylor's expansion, we obtain

$$\begin{aligned}
J_2 &= \int_{t_1}^t \exp\left\{-\frac{1}{\varepsilon}\int_s^t F(v)\,\mathrm{d}v\right\} G(s)\,\mathrm{d}s\\
&= -\varepsilon \int_{(t-t_1)/\varepsilon}^0 \exp\left\{-\frac{1}{\varepsilon}\int_{t-u\varepsilon}^t F(v)\,\mathrm{d}v\right\} G(t - u\varepsilon)\,\mathrm{d}u\\
&= \varepsilon\, G(t) \int_0^{(t-t_1)/\varepsilon} \exp\left[-\frac{1}{\varepsilon}\int_{t-u\varepsilon}^t \left[F(t) + (t - v)\,F'(\tilde{t})\right]\mathrm{d}v\right](1 + O(\varepsilon))
\end{aligned}$$

$$= \varepsilon\, G(t) \int_0^{(t-t_1)/\varepsilon} \exp\{-uF(t) + O(\varepsilon)\}\, du\, (1 + O(\varepsilon))$$

$$= \varepsilon \frac{G(t)}{F(t)} (1 + O(\varepsilon)) .$$

Therefore, (1.56) is verified.

To show (1.57), we make a similar change of variables and obtain the representation

$$\begin{aligned} M_\varepsilon(t) &= H(t) \int_0^{(t-t_1)/\varepsilon} e^{-uF(t)}\, d\tilde{W}_{t,\varepsilon}(u)\, (1 + O_p(\varepsilon)) \\ &= \frac{H(t)}{\sqrt{2F(t)}} \sqrt{2} \int_0^{(t-t_1)F(t)/\varepsilon} e^{-r}\, dw_{t,\varepsilon}(r)\, (1 + O_p(\varepsilon)) \\ &= \frac{H(t)}{\sqrt{2F(t)}}\, \xi_{t,\varepsilon}(1 + O_p(\varepsilon)), \end{aligned}$$

where we introduced the Wiener processes $\tilde{W}_{t,\varepsilon}(u) = (W_{t-u\varepsilon} - W_t)/\sqrt{\varepsilon}$ and $w_{t,\varepsilon}(r) = \sqrt{F(t)}\, \tilde{W}_{t,\varepsilon}\{r/F(t)\}$. Here and in the sequel $O_p(\varepsilon)$ means $O(\varepsilon)$ in probability. Furthermore, we have

$$\xi_{t,\varepsilon} = \sqrt{2} \int_0^{(t-t_1)F(t)/\varepsilon} e^{-r}\, dw_{t,\varepsilon}(r) \Longrightarrow \xi_t = \sqrt{2} \int_0^\infty e^{-r} dW(r) \sim \mathcal{N}(0,1).$$

From this representation of $\xi_{t,\varepsilon}$, it follows that the limits ξ_{t_1}, ξ_{t_2} of two random variables $\xi_{t_1,\varepsilon}, \xi_{t_2,\varepsilon}$, where $t_1 \neq t_2$ are independent. □

Suppose that

$$Y_t = \int_0^t B(r)\, dW_r,$$

where the function $B(t), 0 \le t \le T$ is continuously differentiable and the conditions of Lemma 1.7 hold. Then,

$$Q_\varepsilon(t) = \frac{1}{\varepsilon} \int_0^t \exp\left\{-\frac{1}{\varepsilon} \int_s^t F(v)\, dv\right\} G(s)\; Y_s\, ds = \frac{G(t)\, Y_t}{F(t)} + O\left(\varepsilon^{1/2}\right). \tag{1.58}$$

This is almost consequence of the relation (1.56) because

$$Q_\varepsilon(t) = Y_t\, \frac{1}{\varepsilon} \int_0^t \exp\left\{-\frac{1}{\varepsilon} \int_s^t F(v)\, dv\right\} G(s)\, ds$$

$$
\begin{aligned}
&-\frac{1}{\varepsilon}\int_0^t \exp\left\{-\frac{1}{\varepsilon}\int_s^t F\left(v\right)\mathrm{d}v\right\} G\left(s\right)\int_s^t B\left(r\right)\mathrm{d}W_r\,\mathrm{d}s\\
&=\frac{G\left(t\right)Y_t}{F\left(t\right)}-B\left(t\right)G\left(t\right)\frac{1}{\varepsilon}\int_0^t \exp\left\{-\frac{1}{\varepsilon}\int_s^t F\left(v\right)\mathrm{d}v\right\}\left(W_t-W_s\right)\,\mathrm{d}s+O\left(\varepsilon\right)\\
&=\frac{G\left(t\right)Y_t}{F\left(t\right)}+O\left(\varepsilon^{1/2}\right).
\end{aligned}
$$

1.2 Methods of Estimation

Below, we introduce and discuss several methods of parameter estimation applied in this work to different models of hidden Markov processes. We chose model+estimator because we use different estimators for different models.

1.2.1 Method of Moments

We start with the hidden ergodic Ornstein–Uhlenbeck process (asymptotic $T \to \infty$). Suppose that the observed process $X^T = (X_t, 0 \le t \le T)$ is

$$
\mathrm{d}X_t = f\left(\vartheta\right)Y_t\,\mathrm{d}t + \sigma\,\mathrm{d}W_t, \qquad X_0, \qquad 0 \le t \le T, \tag{1.59}
$$

where the hidden Markov process $Y^T = (Y_t, 0 \le t \le T)$ satisfies the stochastic differential equation

$$
\mathrm{d}Y_t = -a\left(\vartheta\right)Y_t\,\mathrm{d}t + b\left(\vartheta\right)\mathrm{d}V_t, \qquad Y_0, \qquad 0 \le t \le T. \tag{1.60}
$$

Here $f\left(\cdot\right)$, $a\left(\cdot\right)$, and $b\left(\cdot\right)$ are known smooth functions of $\vartheta \in \bar{\Theta}$, where Θ is an open, convex, bounded subset of $\mathcal{R}^d$, $d \ge 1$. The constant σ is known as well. The Wiener processes $W^T = (W_t, 0 \le t \le T)$ and $V^T = (V_t, 0 \le t \le T)$ are independent, and the initial value X_0 is independent of W^T, while Y_0 is independent of V^T. The Gaussian random variables X_0 and Y_0 can be correlated. For the conditional expectation $m\left(\vartheta, t\right) = \mathbf{E}_\vartheta\left(Y_t \mid X^t\right)$, the Kalman–Bucy filter is described by the Eqs. (1.31) and (1.32)

$$
\mathrm{d}m\left(\vartheta, t\right) = -a\left(\vartheta\right)m\left(\vartheta, t\right)\,\mathrm{d}t + \frac{f\left(\vartheta\right)\gamma\left(\vartheta, t\right)}{\sigma^2}\left[\mathrm{d}X_t - f\left(\vartheta\right)m\left(\vartheta, t\right)\,\mathrm{d}t\right], \tag{1.61}
$$

$$
\frac{\partial\gamma\left(\vartheta, t\right)}{\partial t} = -2a\left(\vartheta\right)\gamma\left(\vartheta, t\right) - \frac{f\left(\vartheta\right)^2\gamma\left(\vartheta, t\right)^2}{\sigma^2} + b\left(\vartheta\right)^2, \qquad 0 \le t \le T. \tag{1.62}
$$

Recall that $\gamma(\vartheta, t) = \mathbf{E}_\vartheta (Y_t - m(\vartheta, t))^2$, and the initial values are $m(\vartheta, 0)$ and $\gamma(\vartheta, 0)$.

As the functions $f(\cdot)$, $a(\cdot)$, and $b(\cdot)$ are bounded, the conditions (1.29) and (1.30) are fulfilled, and the system (1.61) and (1.62) has a unique solution $m(\vartheta, t)$ and $\gamma(\vartheta, t)$ for each $\vartheta \in \Theta$, as guaranteed by Theorem 1.3.

We always assume that the following conditions hold:

$$\inf_{\vartheta\in\Theta} f(\vartheta)^2 > 0, \qquad \inf_{\vartheta\in\Theta} a(\vartheta) > 0, \qquad \inf_{\vartheta\in\Theta} b(\vartheta)^2 > 0, \qquad \sigma^2 > 0.$$

Note that if for some ϑ_* we have $f(\vartheta_*) = 0$, this implies that there is no Y^T in the observations X^T and hence, $m(\vartheta_*, t) \equiv \mathbf{E}_{\vartheta_*} Y_t = 0$. If $a(\vartheta_*) \leq 0$, then the process $Y_t, t \geq 0$, is not ergodic, and the construction proposed below does not work. It is possible to find other estimators for this model, but we do not consider that problem here. If $b(\vartheta_*) = 0$, then the process $Y_t, t \geq 0$, is almost deterministic, and once again, the model requires a different construction of estimators. The case $\sigma = 0$ allows for the exact calculation of ϑ without error, which is discussed in Sect. 3.1.

Below, we discuss how the parameter ϑ can be estimated from the observations (1.59) and present the properties of the estimators without proofs. The proofs can be found in Chap. 4.

We denote the true value by ϑ_0.

Remark 1.5 We do not consider the problem of estimating ϑ for models where the coefficient $\sigma(\cdot)$ in (1.59) depends on this parameter. Recall that in such cases, the estimation problem is singular, and ϑ can be calculated exactly. For example, if $\sigma(\vartheta, t) = \vartheta g(t)$, where ϑ and $g(\cdot)$ are positive, then by Itô's formula, we have the equality

$$\mathrm{d}X_t^2 = 2X_t \mathrm{d}X_t + \vartheta_0^2 g(t)^2 \,\mathrm{d}t, \qquad X_0.$$

For any $t \in (0, T]$, we then have

$$\vartheta_0^2 = \left(\int_0^t g(s)^2 \,\mathrm{d}s\right)^{-1} \left[X_t^2 - X_0^2 - 2\int_0^t X_s \,\mathrm{d}X_s\right].$$

The process $X_t, t \geq 0$, of the partially observed system (1.59)–(1.60) does not have ergodic properties, and the construction of method of moments estimators proposed in Section 2.4 of [109] does not apply here. We explain the difference between the ergodic case (see [109]) and our case (see (1.59)) in the following example.

Example 1.9 We have the following limits in distribution:

$$\frac{1}{T^{3/2}} \int_0^T X_t \,\mathrm{d}t \Longrightarrow \int_0^1 \left[\frac{f(\vartheta_0) b(\vartheta_0)}{a(\vartheta_0)} v_r + \sigma w_r\right] \mathrm{d}r \sim \mathcal{N}\left(0, D(\vartheta_0)^2\right), \tag{1.63}$$

$$\frac{1}{T^2}\int_0^T X_t^2 \, \mathrm{d}t \Longrightarrow \int_0^1 \left[\frac{f(\vartheta_0)\, b(\vartheta_0)}{a(\vartheta_0)} v_r + \sigma w_r\right]^2 \mathrm{d}r. \tag{1.64}$$

Here, v_r and w_r, for $0 \leq r \leq 1$, are independent Wiener processes, and

$$D(\vartheta_0)^2 = \frac{2f(\vartheta_0)^2\, b(\vartheta_0)^2 + 2a(\vartheta_0)^2\,\sigma^2}{3a(\vartheta_0)^2}.$$

These limits follow from the representation

$$\int_0^T X_t \, \mathrm{d}t = f(\vartheta_0)\int_0^T\int_0^t Y_s \, \mathrm{d}s \, \mathrm{d}t + \sigma\int_0^T W_t \, \mathrm{d}t$$

and the equality (with $Y_0 = 0$)

$$\begin{aligned}\int_0^t Y_s \, \mathrm{d}s &= b(\vartheta_0)\int_0^t\int_0^s e^{-a(\vartheta_0)(s-z)} \, \mathrm{d}V_z \, \mathrm{d}s \\ &= b(\vartheta_0)\int_0^t e^{a(\vartheta_0)z}\int_z^t e^{-a(\vartheta_0)s} \, \mathrm{d}s \, \mathrm{d}V_z \\ &= \frac{b(\vartheta_0)}{a(\vartheta_0)}\left[V_t + O(1)\right].\end{aligned}$$

Therefore,

$$\begin{aligned}T^{-3/2}\int_0^T X_t \, \mathrm{d}t &= \frac{f(\vartheta_0)\, b(\vartheta_0)}{a(\vartheta_0)} T^{-3/2}\int_0^T V_t \, \mathrm{d}t + \sigma T^{-3/2}\int_0^T W_t \, \mathrm{d}t + O\left(T^{-1/2}\right) \\ &= \frac{f(\vartheta_0)\, b(\vartheta_0)}{a(\vartheta_0)}\int_0^1 v_T(r) \, \mathrm{d}r + \sigma\int_0^1 w_T(r) \, \mathrm{d}r + O\left(T^{-1/2}\right),\end{aligned}$$

where $v_T(r) = T^{-1/2} V_{rT}$ and $w_T(r) = T^{-1/2} W_{rT}$, for $0 \leq r \leq 1$, are independent Wiener processes, and $v_T(r) \Rightarrow v_r$, $w_T(r) \Rightarrow w_r$. The convergences (1.63) and (1.64) follow from this representation.

Suppose that T is an integer and the parameter ϑ is one-dimensional, $\Theta = (\alpha, \beta)$. Introduce the empirical variance (empirical second moment) of the increments of the observations:

$$R_{1,T}\left(X^T\right) = \frac{1}{T}\sum_{t=1}^{T}\left[X_t - X_{t-1}\right]^2$$

and the function

$$\Phi_1(\vartheta) = \frac{f(\vartheta)^2 b(\vartheta)^2}{a(\vartheta)^3}\left[e^{a(\vartheta)} - 1 + a(\vartheta)\right] + \sigma^2, \qquad \vartheta \in \Theta.$$

It is shown (see Proposition 4.1) that

$$R_{1,T}\left(X^T\right) \longrightarrow \Phi_1(\vartheta_0), \qquad \sqrt{T}\left(R_{1,T}\left(X^T\right) - \Phi_1(\vartheta_0)\right) \Longrightarrow \mathcal{N}(0, K(\vartheta_0)). \tag{1.65}$$

These convergences allow us to define the *method of moments estimator* (MME) ϑ_T^* as the solution of the equation

$$\Phi_1\left(\vartheta_T^*\right) = R_{1,T}\left(X^T\right). \tag{1.66}$$

Of course, we must assume that the function $\Phi(\vartheta)$, $\vartheta \in \bar{\Theta}$, is strictly monotone (say, increasing) and define ϑ_T^* in the situations where

$$\inf_{\vartheta\in\Theta}\left|R_{1,T}\left(X^T\right) - \Phi_1(\vartheta)\right| > 0. \tag{1.67}$$

We can set (where $R_{1,T} = R_{1,T}\left(X^T\right)$)

$$\vartheta_T^* = \alpha \mathbf{1}_{\{R_{1,T}<\Phi_1(\alpha)\}} + \bar{\vartheta}_T \mathbf{1}_{\{\Phi_1(\alpha)<R_{1,T}<\Phi_1(\beta)\}} + \beta \mathbf{1}_{\{R_{1,T}>\Phi_1(\beta)\}}.$$

The probabilities of the events $\left\{R_{1,T} < \Phi_1(\alpha)\right\}$ and $\left\{R_{1,T} > \Phi_1(\beta)\right\}$ are asymptotically negligible, and we will often define the MME by the Eq. (1.66) without paying attention to the situations with (1.67).

The consistency $\vartheta_T^* \to \vartheta_0$, asymptotic normality $\sqrt{T}\left(\vartheta_T^* - \vartheta_0\right) \Rightarrow \mathcal{N}$, and convergence of polynomial moments of these estimators are obtained directly from (1.65).

For example, if $f(\vartheta) = \vartheta$, $a(\vartheta) = a$, and $b(\vartheta) = b$, then

$$\Phi_1(\vartheta) = \vartheta^2 b^2 a^{-3}\left(e^{-a} + 1 - a\right),$$

the MME is

$$\vartheta_T^* = f_T^* = \frac{a^{3/2}\sqrt{R_{1,T} - \sigma^2}}{b\sqrt{e^{-a} + 1 - a}} \longrightarrow \vartheta_0,$$

and

$$\sqrt{T}\left(f_T^* - \vartheta_0\right) \Longrightarrow \mathcal{N}\left(0, K(\vartheta_0)\left(\frac{d\Phi_1}{d\vartheta}(\vartheta_0)\right)^{-2}\right).$$

Of course, for other one-dimensional parameters $\vartheta = a$ and $\vartheta = b$, the results are similar.

If the parameter ϑ is two or three-dimensional, then we need two or three statistics respectively; see details in Sect. 4.2.1.

Similar MMEs are used in the cases of the hidden telegraph process (Chap. 5) and the hidden AR process (Chap. 6).

In Chap. 3, where we consider the model with small noise in the observations only, we use the MME constructed on the increments of the observed process once more, but the basic statistic is essentially different.

1.2.2 Minimum Distance

The minimum distance estimators we introduce for the model of a partially observed system with small noise in both equations given by:

$$\mathrm{d}X_t = f\left(\vartheta, t\right) Y_t\, \mathrm{d}t + \varepsilon\sigma\left(t\right) \mathrm{d}W_t, \qquad x_0, \qquad 0 \le t \le T, \tag{1.68}$$

$$\mathrm{d}Y_t = a\left(\vartheta, t\right) Y_t\, \mathrm{d}t + \varepsilon b\left(t\right) \mathrm{d}V_t, \qquad y_0. \tag{1.69}$$

Here, as before, the Wiener processes W_t and V_t for $0 \le t \le T$ are independent, and the functions $f\left(\cdot,\cdot\right)$, $\sigma\left(\cdot\right)$, $a\left(\cdot,\cdot\right)$, and $b\left(\cdot\right)$ are known and smooth with respect to $\vartheta \in \Theta \subset \mathbb{R}^d$, $d \ge 1$. The initial values are real numbers x_0 and $y_0 \ne 0$. The noise level $\varepsilon \in (0, 1]$ is assumed to be small, meaning we consider the behavior of estimators in the asymptotic limit $\varepsilon \to 0$.

Let us denote the limit solutions of this system as $x^T\left(\vartheta\right) = \left(x_t\left(\vartheta\right), 0 \le t \le T\right)$ and $y^T\left(\vartheta\right) = \left(y_t\left(\vartheta\right), 0 \le t \le T\right)$, i.e., we set $\varepsilon = 0$ in the Eqs. (1.68) and (1.69):

$$\frac{\partial x_t\left(\vartheta\right)}{\partial t} = f\left(\vartheta, t\right) y_t\left(\vartheta\right), \qquad x_0\left(\vartheta\right) = x_0, \quad 0 \le t \le T,$$

$$\frac{\partial y_t\left(\vartheta\right)}{\partial t} = a\left(\vartheta, t\right) y_t\left(\vartheta\right), \qquad y_0\left(\vartheta\right) = y_0.$$

Clearly,

$$y_t\left(\vartheta\right) = y_0 \exp\left(\int_0^t a\left(\vartheta, s\right) \mathrm{d}s\right),$$

$$x_t\left(\vartheta\right) = x_0 + \int_0^t f\left(\vartheta, s\right) y_s\left(\vartheta\right) \mathrm{d}s.$$

It is easy to see that the Gaussian process $x_t^{(1)}\left(\vartheta_0\right) = \varepsilon^{-1}\left(X_t - x_t\left(\vartheta_0\right)\right)$ is a solution of the system:

$$\mathrm{d}x_t^{(1)}\left(\vartheta_0\right) = f\left(\vartheta_0, t\right) y_t^{(1)}\, \mathrm{d}t + \sigma\left(t\right) \mathrm{d}W_t, \qquad x_0^{(1)}\left(\vartheta_0\right) = 0,$$

$$\mathrm{d}y_t^{(1)}(\vartheta_0) = a(\vartheta_0, t)\, y_t^{(1)}\,\mathrm{d}t + b(t)\,\mathrm{d}V_t, \qquad y_0^{(1)}(\vartheta_0) = 0.$$

The minimum distance estimator (MDE) $\check{\vartheta}_\varepsilon$ is defined by the relation:

$$\check{\vartheta}_\varepsilon = \arg\inf_{\vartheta\in\Theta} \int_0^T [X_t - x_t(\vartheta)]^2\,\mathrm{d}t. \tag{1.70}$$

The identifiability condition is: for any $\nu > 0$ and any $\vartheta_0 \in \Theta$,

$$\inf_{|\vartheta-\vartheta_0|\geq\nu} \int_0^T [x_t(\vartheta) - x_t(\vartheta_0)]^2\,\mathrm{d}t > 0.$$

If this condition is fulfilled, then the MDE is consistent. Suppose that the matrix

$$\mathbf{T}(\vartheta_0) = \int_0^T \dot{x}_t(\vartheta_0)\,\dot{x}_t(\vartheta_0)^\top\,\mathrm{d}t$$

is uniformly non-degenerate on $\vartheta_0 \in \Theta$. Then, the MDE is asymptotically normal:

$$\varepsilon^{-1}\left(\check{\vartheta}_\varepsilon - \vartheta_0\right) \Longrightarrow \mathbf{T}(\vartheta_0)^{-1} \int_0^T \dot{x}_t(\vartheta_0)\, x_t^{(1)}(\vartheta_0)\,\mathrm{d}t.$$

For the proofs, see Sect. 2.1.2.

1.2.3 Trajectory Fitting

Consider the same model of observations (1.68)–(1.69). The definition of the trajectory fitting estimator (TFE) ϑ_T^* is in some sense similar to the definition of the MDE (1.70). To write the Kalman–Bucy filter for the system (1.68)–(1.69), we use a slightly different notation for the error of estimation: $\gamma_*(\vartheta, t) = \varepsilon^{-2}\mathbf{E}_\vartheta\left[(m(\vartheta, t) - Y_t)^2\right]$. Then we obtain the system:

$$\mathrm{d}m(\vartheta, t) = -a(\vartheta, t)\, m(\vartheta, t)\,\mathrm{d}t + \frac{f(\vartheta, t)\,\gamma_*(\vartheta, t)}{\sigma(t)^2}\left[\mathrm{d}X_t - f(\vartheta, t)\, m(\vartheta, t)\,\mathrm{d}t\right], \tag{1.71}$$

$$\frac{\partial\gamma_*(\vartheta, t)}{\partial t} = -2a(\vartheta, t)\,\gamma_*(\vartheta, t) - \frac{f(\vartheta, t)^2\,\gamma_*(\vartheta, t)^2}{\sigma(t)^2} + b(t)^2, \qquad 0 \leq t \leq T,$$

with the same initial values and which do not depend explicitly on ε. Of course, X_t for $t \in [0, T]$ depends on ε.

Introduce the family of random processes ($X_0 = 0$ for simplicity):

$$\breve{X}_t(\vartheta) = \int_0^t f(\vartheta, s)\, m(\vartheta, s)\, \mathrm{d}s, \qquad 0 \le t \le T, \qquad \vartheta \in \Theta,$$

and define the TFE $\breve{\vartheta}_\varepsilon$ by the relation:

$$\breve{\vartheta}_\varepsilon = \arg\inf_{\vartheta\in\Theta} \int_0^T \left[X_t - \breve{X}_t(\vartheta)\right]^2 \mathrm{d}t.$$

It can be shown that:

$$\breve{X}_t(\vartheta_0) \longrightarrow x_t(\vartheta_0) = \int_0^t f(\vartheta_0, s)\, y_s(\vartheta_0)\, \mathrm{d}s,$$

$$\breve{X}_t(\vartheta) \longrightarrow x_t(\vartheta, \vartheta_0) = \int_0^t f(\vartheta, s)\, y_s(\vartheta, \vartheta_0)\, \mathrm{d}s.$$

Here, $y_s(\vartheta, \vartheta_0)$ is the solution of the Eq. (1.71) for $\varepsilon = 0$.

The identifiability condition is: for any $\nu > 0$

$$\inf_{|\vartheta-\vartheta_0|>\nu} \int_0^T \left[x_t(\vartheta_0) - x_t(\vartheta, \vartheta_0)\right]^2 \mathrm{d}t > 0.$$

If this condition is fulfilled, then the TFE $\breve{\vartheta}_\varepsilon$ is consistent.

Introduce the matrix:

$$\breve{\mathbf{I}}(\vartheta_0) = \int_0^T \dot{x}_t(\vartheta_0, \vartheta_0)\, \dot{x}_t(\vartheta_0, \vartheta_0)^\top \mathrm{d}t$$

and suppose that this matrix is non-degenerate. Then, we obtain the asymptotic normality (in probability) of this estimator:

$$\varepsilon^{-1}\left(\breve{\vartheta}_\varepsilon - \vartheta_0\right) \longrightarrow \breve{\mathbf{I}}(\vartheta_0)^{-1} \int_0^T \sigma(t) \int_t^T \dot{x}_t(\vartheta_0, \vartheta_0)\, \mathrm{d}s\, \mathrm{d}W_t.$$

For the proofs, see Sect. 2.1.2.

Of course, TFE is the same time MDE, and we introduce two different names just to distuingish the cases of deterministic ($x_t(\vartheta)$) and random ($\breve{X}_t(\vartheta)$) fitting trajectories.

1.2.4 Maximum Likelihood

Remark that in this work we consider several models of partially observed Markov processes and describe the properties of estimators in different asymptotics. In

Chaps. 2 and 3, we have the asymptotic of *small noise* ($\varepsilon \to 0$), in Chaps. 4 and 5, the asymptotic of *large samples* for continuous time observations ($T \to \infty$), and in Chap. 6, the asymptotic of *large samples* ($T \to \infty$) for the discrete time model. In all these chapters, the properties of the maximum likelihood and Bayesian estimators of the parameters of the corresponding models are studied with the help of two fundamental theorems of Ibragimov and Khasminskii [73]. These theorems are formulated in a somewhat universal form. Roughly speaking, they can be described as follows.

Consider the problem of estimation of $\vartheta \in \Theta \subset \mathcal{R}^d$ by observations X^ε in the *scheme of series*, i.e., we have a series of problems indexed by some parameter $\varepsilon \in (0, 1]$ and the asymptotic corresponds to $\varepsilon \to 0$. If we have a small noise case, then ε can be the order of noise; in the case of large samples, we can put $\varepsilon = T^{-1}$. Therefore, this approach covers all mentioned limits. We denote the likelihood ratio functions as $L\left(\vartheta, X^{(\varepsilon)}\right), \vartheta \in \Theta$. Suppose that the re-normalized LR function, denoted $Z_\varepsilon(\cdot)$, has some non-degenerate limit $Z_\varepsilon(u) \Rightarrow Z(u)$ as $\varepsilon \to 0$. Then the conditions of these theorems and properties of the MLE and BE are given in terms of the properties of the random functions $Z_\varepsilon(\cdot)$ and $Z(\cdot)$.

The maximum likelihood estimator (MLE) $\hat\vartheta_\varepsilon$ is defined by the equation

$$L(\hat\vartheta_\varepsilon, X^{(\varepsilon)}) = \sup_{\vartheta\in\Theta} L\left(\vartheta, X^{(\varepsilon)}\right).$$

If this equation has several solutions, then any one can be called MLE. The asymptotic properties of the MLE and the Bayesian estimator, discussed in the next section, are studied with the help of several general results by Ibragimov and Khasminskii [73]. Below, we give their Theorems 1.10.1 and 3.1.1, which describe the behavior of the MLE in the general case (Theorem 1.9) and in the regular case (Theorem 1.59), respectively. A similar result concerning the BE will be given later too.

Introduce the *normalized likelihood ratio* ($\varphi_\varepsilon \to 0$)

$$Z_\varepsilon(\vartheta_0, u) = \frac{L\left(\vartheta_0 + \varphi_\varepsilon u, X^{(\varepsilon)}\right)}{L\left(\vartheta_0, X^{(\varepsilon)}\right)}, \qquad \mathbb{U}_\varepsilon = \left\{u : \vartheta_0 + \varphi_\varepsilon u \in \Theta\right\},$$

where the parameter set Θ is an open, convex subset of $\mathcal{R}^d$.

The *Conditions* $\mathscr{IK}_d$ *are: For any compact* $\mathbb{K} \subset \Theta$, *there correspond numbers* $a(\mathbb{K}) = a$, $B(\mathbb{K}) = B$, $C(\mathbb{K}) = C$, *such that*

- *There exist numbers* $k > d$ *and* $m \geq k$ *such that*

$$\sup_{\vartheta_0\in\mathbb{K}} \sup_{|u_1|\leq R, |u_2|\leq R} \|u_2 - u_1\|^{-k} \mathbf{E}_{\vartheta_0} \left|Z_\varepsilon(\vartheta_0, u_2)^{1/m} - Z_\varepsilon(\vartheta_0, u_1)^{1/m}\right|^m \leq B\left(1 + R^a\right).$$

- *For any* $N > 0$

$$\sup_{\vartheta_0 \in \mathbb{K}} \sup_{\varepsilon} \sup_{u \in \mathcal{U}_\varepsilon} \|u\|^N \mathbf{E}_{\vartheta_0} Z_\varepsilon(\vartheta_0, u)^{1/2} < \infty. \tag{1.72}$$

Introduce as well the condition $\mathscr{IK}_0$ which coincides with $\mathscr{IK}_d$ except for the condition $k > d$, which is replaced by $k > 0$.

Denote $\mathcal{C}_0(\mathcal{R}^d)$ as the space of continuous functions on $\mathcal{R}^d$ vanishing at infinity, i.e., $\lim_{\|u\| \to \infty} f(u) = 0$ with the norm $\|f\|_\infty = \sup_u |f(u)|$ and denote $\mathfrak{B}$ the corresponding Borelian σ-algebra.

The following theorem will be used in this work each time we describe the properties of MLEs in singular cases.

Theorem 1.9 (Ibragimov, Khasminskii [73], Theorem 1.10.1) *Suppose that the functions $Z_\varepsilon(\vartheta_0, u)$, $u \in \mathbb{U}_\varepsilon$, are continuous with probability 1 and possess the following properties:*

1. *The condition $\mathscr{IK}_d$ is fulfilled.*
2. *Uniformly in $\vartheta \in \mathbb{K}$, the marginal (finite-dimensional) distributions of the random functions $Z_\varepsilon(\vartheta_0, \cdot)$ converge to the marginal distributions of the random function $Z(\vartheta_0, u)$, $u \in \mathcal{R}^d$, $Z(\vartheta_0, \cdot) \in \mathcal{C}_0(\mathcal{R}^d)$.*
3. *The limit function $Z(\cdot)$ attains the maximum at a unique point $\hat{u}(\vartheta_0)$ with probability 1.*

Then, uniformly in $\vartheta_0 \in \mathbb{K}$, the distribution of the random variables $\varphi_\varepsilon^{-1}(\hat{\vartheta}_\varepsilon - \vartheta_0)$ converges to the distribution of $\hat{u}(\vartheta_0)$, and for any $p > 0$, we have uniformly in $\vartheta_0 \in \mathbb{K}$ the convergence of moments

$$\varphi_\varepsilon^{-p} \mathbf{E}_{\vartheta_0} \|\hat{\vartheta}_\varepsilon - \vartheta_0\|^p \longrightarrow \mathbf{E}_{\vartheta_0} \left\|\hat{u}(\vartheta_0)\right\|^p.$$

This is a slightly simplified version of Theorem 1.10.1 in [73]. The simplification concerns the normalizing function φ_ε, which in [73] is a matrix function admitting the models where different components of the parameter ϑ can be estimated with different rates, and the condition (1.72) is in a less general form.

Let us recall some steps of this proof. For simplicity, we consider a one-dimensional parameter $\vartheta \in \Theta = (\alpha, \beta)$ $(d = 1)$.

The random process

$$Z_\varepsilon(\vartheta_0, u), u \in \mathbb{U}_\varepsilon, \qquad \mathbb{U}_\varepsilon = \left(\varphi_\varepsilon^{-1}(\alpha - \vartheta_0), \varphi_\varepsilon^{-1}(\beta - \vartheta_0)\right) \equiv (u_{m,\varepsilon}, u_{M,\varepsilon})$$

we define linearly increasing from zero to $Z_\varepsilon\left(\vartheta_0, u_{m,\varepsilon}\right)$ on the interval $\left[u_{m,\varepsilon} - 1, u_{m,\varepsilon}\right]$ and linearly decreasing from $Z_\varepsilon\left(\vartheta_0, u_{M,\varepsilon}\right)$ to zero on the interval $\left[u_{M,\varepsilon}, u_{M,\varepsilon} + 1\right]$. Outside of $\left[u_{m,\varepsilon} - 1, u_{M,\varepsilon} + 1\right]$, we put $Z_\varepsilon(\vartheta_0, u) = 0$. Note that $u_{m,\varepsilon} \to -\infty$ and $u_{M,\varepsilon} \to +\infty$. Now, $Z_\varepsilon(\vartheta_0, \cdot)$ is defined on $\mathcal{R}$ and belongs to $\mathcal{C}_0(\mathcal{R})$.

Let us denote $\mathbf{P}_{\vartheta_0}^{(\varepsilon)}$ the measure induced by $Z_\varepsilon(\vartheta_0, \cdot)$ on the measurable space $(\mathcal{C}_0(\mathcal{R}), \mathfrak{B})$, and $\mathbf{P}_{\vartheta_0}$ the measure induced on this space by the limit process

$Z(\vartheta_0, \cdot)$. It is shown that under the condition of this theorem, we have the uniform on compacts weak convergence

$$\mathbf{P}_{\vartheta_0}^{(\varepsilon)} \quad \Longrightarrow \quad \mathbf{P}_{\vartheta_0}.$$

This means that for any continuous bounded functional $\Phi(z)$, $z \in \mathcal{C}_0(\mathcal{R})$, and any compact $\mathbb{K} \subset \Theta$

$$\sup_{\vartheta_0 \in \mathbb{K}} \left| \mathbf{E}_{\vartheta_0} \Phi(Z_\varepsilon) - \mathbf{E}_{\vartheta_0} \Phi(Z) \right| \longrightarrow 0$$

and as a consequence, the distributions of such functionals converge too. The asymptotic behavior of the MLE is described with the help of the convergence of distributions of the functionals $\Phi(Z_\varepsilon) = \sup_{u<x} Z_\varepsilon(\vartheta_0, u) - \sup_{u \geq x} Z_\varepsilon(\vartheta_0, u)$ as follows (below $\vartheta_u = \vartheta_0 + \varphi_\varepsilon u$).

$$\begin{aligned}
&\mathbf{P}_{\vartheta_0}^{(\varepsilon)}\left(\varphi_\varepsilon^{-1}(\hat{\vartheta}_\varepsilon - \vartheta_0) < x\right) \\
&\quad = \mathbf{P}_{\vartheta_0}^{(\varepsilon)}\left(\hat{\vartheta}_\varepsilon < \vartheta_0 + \varphi_\varepsilon x\right) \\
&\quad = \mathbf{P}_{\vartheta_0}^{(\varepsilon)}\left(\sup_{\vartheta < \vartheta_0 + \varphi_\varepsilon x} L\left(\vartheta, X^{(\varepsilon)}\right) > \sup_{\vartheta \geq \vartheta_0 + \varphi_\varepsilon x} L\left(\vartheta, X^{(\varepsilon)}\right)\right) \\
&\quad = \mathbf{P}_{\vartheta_0}^{(\varepsilon)}\left(\sup_{\vartheta_u < \vartheta_0 + \varphi_\varepsilon x} \frac{L\left(\vartheta_u, X^{(\varepsilon)}\right)}{L\left(\vartheta_0, X^{(\varepsilon)}\right)} > \sup_{\vartheta_u \geq \vartheta_0 + \varphi_\varepsilon x} \frac{L\left(\vartheta_u, X^{(\varepsilon)}\right)}{L\left(\vartheta_0, X^{(\varepsilon)}\right)}\right) \\
&\quad = \mathbf{P}_{\vartheta_0}^{(\varepsilon)}\left(\sup_{u<x} Z_\varepsilon(\vartheta_0, u) > \sup_{u \geq x} Z_\varepsilon(\vartheta_0, u)\right) \\
&\quad = \mathbf{P}_{\vartheta_0}^{(\varepsilon)}\left(\sup_{u<x} Z_\varepsilon(\vartheta_0, u) - \sup_{u \geq x} Z_\varepsilon(\vartheta_0, u) > 0\right) \\
&\quad \longrightarrow \mathbf{P}_{\vartheta_0}\left(\sup_{u<x} Z(\vartheta_0, u) - \sup_{u \geq x} Z(\vartheta_0, u) > 0\right) \\
&\quad = \mathbf{P}_{\vartheta_0}\left(\sup_{u<x} Z(\vartheta_0, u) > \sup_{u \geq x} Z(\vartheta_0, u)\right) \\
&\quad = \mathbf{P}_{\vartheta_0}\left(\hat{u}(\vartheta_0) < x\right).
\end{aligned} \tag{1.73}$$

Therefore,

$$\varphi_\varepsilon^{-1}(\hat{\vartheta}_\varepsilon - \vartheta_0) \Longrightarrow \hat{u}(\vartheta_0). \tag{1.74}$$

In the regular cases, the limit likelihood ratio function $Z(\vartheta_0, u)$, $u \in \mathcal{R}^d$, is well known, and it will be convenient to cite the corresponding version of Theorem 1.9, which is a slightly simplified version of Theorems 3.1.1 and 3.1.2 in [73].

Theorem 1.10 *Suppose that the functions $Z_\varepsilon(\vartheta_0, u)$, $u \in \mathbb{U}_\varepsilon$, are continuous with probability 1 and possess the following properties:*

1. *The condition $\mathscr{IK}_d$ is fulfilled.*
2. *For an arbitrary sequence $\vartheta_\varepsilon = \vartheta_0 + \varphi_\varepsilon u_\varepsilon \in \mathbb{K}$, $u_\varepsilon \to u$, the random function $Z_\varepsilon(\vartheta_\varepsilon, u_\varepsilon)$ admits the representation*

$$Z_\varepsilon(\vartheta_\varepsilon, u_\varepsilon) = \exp\left(u_\varepsilon^\top \Delta_\varepsilon\left(\vartheta_\varepsilon, X^{(\varepsilon)}\right) - \frac{1}{2} u_\varepsilon^\top \mathbf{I}(\vartheta_\varepsilon) u_\varepsilon + r_\varepsilon\left(u_\varepsilon, \vartheta_\varepsilon, X^{(\varepsilon)}\right)\right)$$

where uniformly in $\vartheta_0 \in \mathbb{K}$ the $d \times d$ matrix $\mathbf{I}(\vartheta_0)$ (Fisher information) is non-degenerate and the following convergences hold:

$$\Delta_\varepsilon\left(\vartheta_\varepsilon, X^{(\varepsilon)}\right) \Longrightarrow \Delta(\vartheta_0) \sim \mathcal{N}(0, \mathbf{I}(\vartheta_0)), \qquad r_\varepsilon\left(u_\varepsilon, \vartheta_\varepsilon, X^{(\varepsilon)}\right) \longrightarrow 0.$$

Then, uniformly in $\vartheta_0 \in \mathbb{K}$,

- *the MLE is consistent,*
- *it is asymptotically normal:*

$$\varphi_\varepsilon^{-1}(\hat{\vartheta}_\varepsilon - \vartheta_0) \Longrightarrow \hat{u}(\vartheta_0) = \mathbf{I}(\vartheta_0)^{-1} \Delta(\vartheta_0) \sim \mathcal{N}\left(0, \mathbf{I}(\vartheta_0)^{-1}\right), \tag{1.75}$$

- *its polynomial moments converge: for any $p > 0$*

$$\varphi_\varepsilon^{-p} \mathbb{E}_{\vartheta_0} \|\hat{\vartheta}_\varepsilon - \vartheta_0\|^p \longrightarrow \mathbb{E}_{\vartheta_0} \left\|\hat{u}(\vartheta_0)\right\|^p,$$

- *for any $\delta > 0$*

$$\lim_{T\to\infty} \sup_{\vartheta_0 \in \mathbb{K}} \mathbb{P}_{\vartheta_0}\left(\left\|\varphi_\varepsilon^{-1}\left(\hat{\vartheta}_\varepsilon - \vartheta_0\right) - \mathbf{I}(\vartheta_0)^{-1} \Delta_\varepsilon\left(\vartheta_0, X^{(\varepsilon)}\right)\right\| > \delta\right) = 0.$$

The limit process is

$$Z(\vartheta_0, u) = \exp\left(u^\top \Delta(\vartheta_0) - \frac{1}{2} u^\top \mathbf{I}(\vartheta_0) u\right), \qquad u \in \mathbb{R}^d.$$

The point of maximum of $Z(\vartheta_0, \cdot)$ is $\hat{u}(\vartheta_0) = \mathbf{I}(\vartheta_0)^{-1} \Delta(\vartheta_0)$ because

$$\frac{\partial \ln Z(\vartheta_0, u)}{\partial u} = \Delta(\vartheta_0) - \mathbf{I}(\vartheta_0)\hat{u}(\vartheta_0) = 0.$$

This asymptotic normality, together with an estimate of the tails of the process $Z_\varepsilon(\vartheta_0,\cdot)$, allows us to prove the convergence of all polynomial moments too.

Example 1.10 As an example, consider the model (1.59)–(1.60). Under the regularity conditions presented in Sect. 4.2.2, it is shown that the normalized LR $Z_T(\vartheta_0,u)$ with $\varphi_\varepsilon = T^{-1/2}$ satisfies the conditions of Theorem 1.10. Therefore, the MLE $\hat\vartheta_T$ is consistent, asymptotically normal:

$$\sqrt{T}\left(\hat\vartheta_\varepsilon-\vartheta_0\right)\Longrightarrow \mathcal{N}\left(0,\mathbf{I}\left(\vartheta_0\right)^{-1}\right)$$

and the moments converge.

The Fisher information (for $d=1$) is

$$\mathbf{I}(\vartheta)=\frac{\dot a(\vartheta)^2}{2a(\vartheta)}-\frac{2\dot a(\vartheta)\dot r(\vartheta)}{a(\vartheta)+r(\vartheta)}+\frac{\dot r(\vartheta)^2}{2r(\vartheta)},\quad r(\vartheta)=\left(a(\vartheta)^2+\frac{f(\vartheta)^2b(\vartheta)^2}{\sigma^2}\right)^{1/2}. \tag{1.76}$$

Remark 1.6 There exists a formal link between MLE and one MDE. Consider the model (1.68)–(1.69) with the LR function

$$L\left(\vartheta,X^T\right)=\exp\left(\int_0^T\frac{f(\vartheta,t)\,m(\vartheta,t)}{\varepsilon^2\sigma(t)^2}\mathrm{d}X_t-\int_0^T\frac{f(\vartheta,t)^2\,m(\vartheta,t)^2}{2\varepsilon^2\sigma(t)^2}\mathrm{d}t\right),\quad\vartheta\in\Theta.$$

Suppose that all functions are smooth enough w.r.t. ϑ. Then, the MLE $\hat\vartheta_\varepsilon$ is one of the solutions of *maximum likelihood equation* (MLEq), which we obtain by differentiating $\ln L\left(\vartheta,X^T\right)$ on ϑ:

$$\int_0^T\frac{\left[\dot f(\vartheta,t)\,m(\vartheta,t)+f(\vartheta,t)\,\dot m(\vartheta,t)\right]}{\sigma(t)^2}\left[\mathrm{d}X_t-f(\vartheta,t)\,m(\vartheta,t)\,\mathrm{d}t\right]=0.$$

Suppose for the moment that the observed process $X_t, 0\le t\le T$ has derivative X_t' on t what is of course wrong. Introduce formally the MDE

$$\check\vartheta_\varepsilon=\arg\inf_{\vartheta\in\Theta}\int_0^T\left[X_t'-f(\vartheta,t)\,m(\vartheta,t)\right]^2\sigma(t)^{-2}\,\mathrm{d}t.$$

Differentiating once more on ϑ this integral, we obtain the *minimum distance equation* (MDEq)

$$\begin{aligned}0&=\int_0^T\left[\dot f(\vartheta,t)\,m(\vartheta,t)+f(\vartheta,t)\,\dot m(\vartheta,t)\right]\left[X_t'-f(\vartheta,t)\,m(\vartheta,t)\right]\sigma(t)^{-2}\,\mathrm{d}t\\&=\int_0^T\frac{\left[\dot f(\vartheta,t)\,m(\vartheta,t)+f(\vartheta,t)\,\dot m(\vartheta,t)\right]}{\sigma(t)^2}\left[\mathrm{d}X_t-f(\vartheta,t)\,m(\vartheta,t)\,\mathrm{d}t\right].\end{aligned}$$

The last integral is already well defined and the both equations coincide. If we replace X_t' in the definition of MDE by $\left(X_{t+h_\varepsilon} - X_t\right)/h_\varepsilon$, where $h_\varepsilon \to 0$, then we obtain one more MDE.

1.2.5 Bayesian

Suppose that the unknown parameter ϑ is random with the probability density $p(\vartheta)$, $\vartheta \in \Theta$. The function $p(\cdot)$ is supposed to be continuous and positive on Θ. The mathematical expectation with respect to this density is denoted by $\mathbf{E}_p$. The Bayesian estimator (BE) $\tilde\vartheta_\varepsilon$ is defined by the relation: for any estimator $\bar\vartheta_\varepsilon$,

$$\mathbb{E}\left\|\bar\vartheta_\varepsilon - \vartheta\right\|^2 \geq \mathbb{E}\|\tilde\vartheta_\varepsilon - \vartheta\|^2. \tag{1.77}$$

Here, we denote $\mathbb{E}$ as the double expectation: for a function $h(\vartheta, X^{(\varepsilon)})$,

$$\mathbb{E}\, h(\vartheta, X^{(\varepsilon)}) = \mathbf{E}_p \mathbf{E}_\vartheta h(\vartheta, X^{(\varepsilon)}) = \int_\Theta \mathbf{E}_\vartheta h(\vartheta, X^{(\varepsilon)})\, p(\vartheta)\, \mathrm{d}\vartheta. \tag{1.78}$$

Let us show that the conditional expectation

$$\tilde\vartheta_\varepsilon = \mathbf{E}_p\left(\vartheta \mid X^{(\varepsilon)}\right) = \int_\Theta \vartheta\, p(\vartheta \mid X^{(\varepsilon)})\, \mathrm{d}\vartheta = \frac{\displaystyle\int_\Theta \vartheta\, p(\vartheta)\, L(\vartheta, X^{(\varepsilon)})\, \mathrm{d}\vartheta}{\displaystyle\int_\Theta p(\vartheta)\, L(\vartheta, X^{(\varepsilon)})\, \mathrm{d}\vartheta}$$

is the BE. We have

$$\begin{aligned}\mathbb{E}\left\|\bar\vartheta_\varepsilon - \vartheta\right\|^2 &= \mathbb{E}\|\bar\vartheta_\varepsilon - \tilde\vartheta_\varepsilon + \tilde\vartheta_\varepsilon - \vartheta\|^2\\ &= \mathbb{E}\|\bar\vartheta_\varepsilon - \tilde\vartheta_\varepsilon\|^2 + \mathbb{E}\|\tilde\vartheta_\varepsilon - \vartheta\|^2 + 2\mathbb{E}\left[(\bar\vartheta_\varepsilon - \tilde\vartheta_\varepsilon)^\top(\tilde\vartheta_\varepsilon - \vartheta)\right]\\ &= \mathbb{E}\|\bar\vartheta_\varepsilon - \tilde\vartheta_\varepsilon\|^2 + \mathbb{E}\|\tilde\vartheta_\varepsilon - \vartheta\|^2 \end{aligned} \tag{1.79}$$

because

$$\begin{aligned}\mathbf{E}_p\left[(\bar\vartheta_\varepsilon - \tilde\vartheta_\varepsilon)^\top(\tilde\vartheta_\varepsilon - \vartheta) \mid X^{(\varepsilon)}\right) &= \mathbf{E}_p\left[(\bar\vartheta_\varepsilon - \tilde\vartheta_\varepsilon)^\top \mathbf{E}_p\left(\tilde\vartheta_\varepsilon - \vartheta \mid X^{(\varepsilon)}\right)\right]\\ &= \mathbf{E}_p\left[(\bar\vartheta_\varepsilon - \tilde\vartheta_\varepsilon)^\top\left(\tilde\vartheta_\varepsilon - \mathbf{E}_p\left(\vartheta \mid X^{(\varepsilon)}\right)\right)\right] = 0.\end{aligned}$$

Therefore, from (1.79), it follows that for all estimators $\bar\vartheta_\varepsilon$,

$$\mathbb{E}\left\|\bar\vartheta_\varepsilon - \vartheta\right\|^2 \geq \mathbb{E}\|\tilde\vartheta_\varepsilon - \vartheta\|^2$$

with equality iff $\bar{\vartheta}_\varepsilon = \tilde{\vartheta}_\varepsilon$ with probability 1. Hence, according to the definition (1.77), $\tilde{\vartheta}_\varepsilon$ is the BE.

The asymptotic behavior of the BE is studied with the help of other theorems from [73]. As in the case of the MLE above, we consider once more a general statistical model with the likelihood ratio $L(\vartheta, X^{(\varepsilon)}), \vartheta \in \Theta$. The normalizing function φ_ε and the processes $Z_\varepsilon(\vartheta_0, u), u \in \mathbb{U}_\varepsilon$ and $Z(\vartheta_0, u), u \in \mathcal{R}^d$ are the same too. Define

$$\tilde{u}(\vartheta_0) = \frac{\int_{\mathcal{R}^d} u\, Z(\vartheta_0, u)\, du}{\int_{\mathcal{R}^d} Z(\vartheta_0, u)\, du}.$$

The following theorem will be used in this work each time when we describe the properties of BEs in singular cases. It is a particular case that corresponds to the quadratic loss function of Theorem 1.10.2 in [73].

Theorem 1.11 *Let $\tilde{\vartheta}_\varepsilon$ be a Bayesian estimator. Assume that the normalized likelihood ratio $Z_\varepsilon(\vartheta_0, u), u \in \mathbb{U}_\varepsilon$ possesses the following properties:*

1. *The condition $\mathscr{I}\mathscr{K}_0$ is fulfilled.*
2. *Uniformly in $\vartheta \in \mathcal{K}$, the marginal distributions of the random functions $Z_\varepsilon(\vartheta_0, \cdot)$ converge to the marginal distributions of the random function $Z(\vartheta_0, u), u \in \mathcal{R}^d$.*

Then, uniformly in $\vartheta_0 \in \mathbb{K}$, the distribution of the random variables $\varphi_\varepsilon^{-1}(\tilde{\vartheta}_\varepsilon - \vartheta_0)$ converges to the distribution of $\tilde{u}(\vartheta_0)$, and for any $p > 0$, we have uniformly in $\vartheta_0 \in \mathbb{K}$,

$$\varphi_\varepsilon^{-p} \mathbf{E}_{\vartheta_0} \|\tilde{\vartheta}_\varepsilon - \vartheta_0\|^p \longrightarrow \mathbf{E}_{\vartheta_0} \|\tilde{u}(\vartheta_0)\|^p .$$

The asymptotic behavior of the BE can be formally explained as follows. We can write (below $\vartheta_u = \vartheta_0 + \varphi_\varepsilon u$)

$$\tilde{\vartheta}_\varepsilon = \frac{\int_\Theta \vartheta\, p(\vartheta) L(\vartheta, X^{(\varepsilon)})\, d\vartheta}{\int_\Theta p(\vartheta) L(\vartheta, X^{(\varepsilon)})\, d\vartheta} = \frac{\int_{\mathbb{U}_\varepsilon} (\vartheta_0 + \varphi_\varepsilon u)\, p(\vartheta_u) L(\vartheta_u, X^{(\varepsilon)})\, du}{\int_{\mathbb{U}_\varepsilon} p(\vartheta_u) L(\vartheta_u, X^{(\varepsilon)})\, du}$$

$$= \vartheta_0 + \varphi_\varepsilon \frac{\int_{\mathbb{U}_\varepsilon} u\, p(\vartheta_u) \frac{L(\vartheta_u, X^{(\varepsilon)})}{L(\vartheta_0, X^{(\varepsilon)})}\, du}{\int_{\mathbb{U}_\varepsilon} p(\vartheta_u) \frac{L(\vartheta_u, X^{(\varepsilon)})}{L(\vartheta_0, X^{(\varepsilon)})}\, du} = \vartheta_0 + \varphi_\varepsilon \frac{\int_{\mathbb{U}_\varepsilon} u\, p(\vartheta_u) Z_\varepsilon(\vartheta_0, u)\, du}{\int_{\mathbb{U}_\varepsilon} p(\vartheta_u) Z_\varepsilon(\vartheta_0, u)\, du}.$$

Hence,

$$\varphi_\varepsilon^{-1}(\tilde\vartheta_\varepsilon-\vartheta_0)=\frac{\displaystyle\int_{\mathbb{U}_\varepsilon}u\,p(\vartheta_u)Z_\varepsilon(\vartheta_0,u)\,{\rm d}u}{\displaystyle\int_{\mathbb{U}_\varepsilon}p(\vartheta_u)Z_\varepsilon(\vartheta_0,u)\,{\rm d}u}\Longrightarrow\frac{\displaystyle\int_{\mathcal{R}^d}u\,Z(\vartheta_0,u)\,{\rm d}u}{\displaystyle\int_{\mathcal{R}^d}Z(\vartheta_0,u)\,{\rm d}u}=\tilde u(\vartheta_0).$$

Here, we used the convergence $p(\vartheta_u)\to p(\vartheta_0)>0$.

Once more, for the convenience of citation, we formulate the version of this theorem in the regular case as it was done for the MLE above.

Theorem 1.12 (Ibragimov, Khasminskii [73]) *Suppose that the functions* $Z_\varepsilon(\vartheta_0,u),\ u\in\mathbb{U}_\varepsilon$ *possess the following properties:*

1. *The condition* $\mathscr{IK}_0$ *is fulfilled.*
2. *For an arbitrary sequence* $\vartheta_\varepsilon=\vartheta_0+\varphi_\varepsilon u_\varepsilon\in\mathbb{K}$, $u_\varepsilon\to u$, *the random function* $Z_\varepsilon(\vartheta_\varepsilon,u_\varepsilon)$ *admits the representation*

$$Z_\varepsilon(\vartheta_\varepsilon,u_\varepsilon)=\exp\left(u_\varepsilon^\top\Delta_\varepsilon(\vartheta_\varepsilon,X^{(\varepsilon)})-\frac{1}{2}u_\varepsilon^\top\mathbf{I}(\vartheta_\varepsilon)u_\varepsilon+r_\varepsilon(u_\varepsilon,\vartheta_\varepsilon,X^{(\varepsilon)})\right)$$

where, uniformly in $\vartheta_0\in\mathbb{K}$, *the* $d\times d$ *matrix* $\mathbf{I}(\vartheta_0)$ *is non-degenerate and the following convergences hold*

$$\Delta_\varepsilon(\vartheta_\varepsilon,X^{(\varepsilon)})\Longrightarrow\Delta(\vartheta_0)\sim\mathcal{N}\left(0,\mathbf{I}(\vartheta_0)\right),\qquad r_\varepsilon(u_\varepsilon,\vartheta_\varepsilon,X^{(\varepsilon)})\longrightarrow 0.$$

Then uniformly in $\vartheta_0\in\mathbb{K}$:

- *The BE is consistent.*
- *It is asymptotically normal*

$$\varphi_\varepsilon^{-1}(\tilde\vartheta_\varepsilon-\vartheta_0)\Longrightarrow\tilde u(\vartheta_0)=\mathbf{I}(\vartheta_0)^{-1}\Delta(\vartheta_0)\sim\mathcal{N}\left(0,\mathbf{I}(\vartheta_0)^{-1}\right),\tag{1.80}$$

- *Its polynomial moments converge: for any* $p>0$

$$\varphi_\varepsilon^{-p}\mathbf{E}_{\vartheta_0}\|\tilde\vartheta_\varepsilon-\vartheta_0\|^p\longrightarrow\mathbf{E}_{\vartheta_0}\|\tilde u(\vartheta_0)\|^p,$$

- *For any* $\delta>0$

$$\lim_{T\to\infty}\sup_{\vartheta_0\in\mathbb{K}}\mathbf{P}_{\vartheta_0}\left(\|\varphi_\varepsilon^{-1}(\tilde\vartheta_\varepsilon-\vartheta_0)-\mathbf{I}(\vartheta_0)^{-1}\Delta_\varepsilon(\vartheta_0,X^{(\varepsilon)})\|>\delta\right)=0.\tag{1.81}$$

For the proof, see Theorems 3.2.1 and 3.2.2 in [73].

Consider the case $d=1$ and show that $\tilde u(\vartheta_0)=\mathbf{I}(\vartheta_0)^{-1}\Delta(\vartheta_0)$. We have:

$$u\Delta(\vartheta_0) - \frac{u^2}{2}\mathrm{I}(\vartheta_0) = -\frac{1}{2}\mathrm{I}(\vartheta_0)\left[u^2 - 2u\frac{\Delta(\vartheta_0)}{\mathrm{I}(\vartheta_0)} - \frac{\Delta(\vartheta_0)^2}{\mathrm{I}(\vartheta_0)^2}\right] - \frac{\Delta(\vartheta_0)^2}{2\mathrm{I}(\vartheta_0)}$$

$$= -\frac{1}{2}\mathrm{I}(\vartheta_0)\left[u - \frac{\Delta(\vartheta_0)}{\mathrm{I}(\vartheta_0)}\right]^2 - \frac{\Delta(\vartheta_0)^2}{2\mathrm{I}(\vartheta_0)}.$$

Therefore,

$$\tilde{u}(\vartheta_0) = \frac{\int_{\mathcal{R}} u\, Z(\vartheta_0, u)\, \mathrm{d}u}{\int_{\mathcal{R}} Z(\vartheta_0, u)\, \mathrm{d}u} = \frac{\int_{\mathcal{R}} \frac{\Delta(\vartheta_0)}{\mathrm{I}(\vartheta_0)} \exp\left(-\frac{1}{2}\mathrm{I}(\vartheta_0)\left[u - \frac{\Delta(\vartheta_0)}{\mathrm{I}(\vartheta_0)}\right]^2\right) \mathrm{d}u}{\int_{\mathcal{R}} \exp\left(-\frac{1}{2}\mathrm{I}(\vartheta_0)\left[u - \frac{\Delta(\vartheta_0)}{\mathrm{I}(\vartheta_0)}\right]^2\right) \mathrm{d}u}$$

$$+ \frac{\int_{\mathcal{R}} \left[u - \frac{\Delta(\vartheta_0)}{\mathrm{I}(\vartheta_0)}\right] \exp\left(-\frac{1}{2}\mathrm{I}(\vartheta_0)\left[u - \frac{\Delta(\vartheta_0)}{\mathrm{I}(\vartheta_0)}\right]^2\right) \mathrm{d}u}{\int_{\mathcal{R}} \exp\left(-\frac{1}{2}\mathrm{I}(\vartheta_0)\left[u - \frac{\Delta(\vartheta_0)}{\mathrm{I}(\vartheta_0)}\right]^2\right) \mathrm{d}u}$$

$$= \frac{\Delta(\vartheta_0)}{\mathrm{I}(\vartheta_0)} \sim \mathcal{N}\left(0, \mathrm{I}(\vartheta_0)^{-1}\right).$$

Here we used the equality

$$\int_{\mathcal{R}} (u - a) \exp\left(-b(u - a)^2\right) \mathrm{d}u = 0, \tag{1.82}$$

which is valid for any a and $b > 0$.

1.2.6 Multi-Step

We illustrate the construction of One-step and Two-step MLE processes with the help of the model (1.59)–(1.60), which we repeat here:

$$\mathrm{d}X_t = f(\vartheta)\, Y_t\, \mathrm{d}t + \sigma\, \mathrm{d}W_t, \qquad X_0, \quad 0 \le t \le T, \tag{1.83}$$

$$\mathrm{d}Y_t = -a(\vartheta)\, Y_t\, \mathrm{d}t + b(\vartheta)\, \mathrm{d}V_t, \qquad Y_0, \tag{1.84}$$

together with the Kalman–Bucy filter (1.61)–(1.62):

$$\mathrm{d}m(\vartheta, t) = -a(\vartheta)\, m(\vartheta, t)\, \mathrm{d}t + \frac{f(\vartheta)\, \gamma(\vartheta, t)}{\sigma^2}\left[\mathrm{d}X_t - f(\vartheta)\, m(\vartheta, t)\, \mathrm{d}t\right], \tag{1.85}$$

$$\frac{\partial \gamma(\vartheta,t)}{\partial t}=-2a(\vartheta)\gamma(\vartheta,t)-\frac{f(\vartheta)^2\gamma(\vartheta,t)^2}{\sigma^2}+b(\vartheta)^2,\qquad 0\le t\le T \tag{1.86}$$

with the initial values $m(\vartheta,0)=y_0$ and $\gamma(\vartheta,0)=0$.

Recall that

$$\mathrm{d}X_t=M(\vartheta,t)\,\mathrm{d}t+\sigma\mathrm{d}\bar W_t,\qquad X_0,\quad 0\le t\le T,$$

where $M(\vartheta,t)=f(\vartheta)m(\vartheta,t)$. It is shown in Chap. 4 that the solution $\gamma(\vartheta,t)$ of the Riccati equation converges to $\gamma_*(\vartheta)=\sigma^2 f(\vartheta)^{-2}[r(\vartheta)-a(\vartheta)]$ with an exponential rate. Therefore, we replace below $\gamma(\vartheta,t)$ by $\gamma_*(\vartheta)$, and the two equations (1.85) and (1.86) we replace by one:

$$\mathrm{d}m(\vartheta,t)=-a(\vartheta)m(\vartheta,t)\,dt+\frac{f(\vartheta)\gamma_*(\vartheta)}{\sigma^2}[\mathrm{d}X_t-f(\vartheta)m(\vartheta,t)\,\mathrm{d}t]$$

with the initial value $m(\vartheta,0)\sim\mathcal{N}(0,\gamma_*(\vartheta))$.

One-Step MLE Process

Let us remind how the one-step MLE $\vartheta_T^\star$ can be defined (for simplicity, $d=1$). The maximum likelihood equation (MLEq) is (formally)

$$\int_0^T\dot M(\hat\vartheta_T,t)\left[\mathrm{d}X_t-M(\hat\vartheta_T,t)\mathrm{d}t\right]=0.$$

The function $M(\hat\vartheta_T,t)$ is expanded in the vicinity of the true value ϑ_0, i.e., we write $M(\hat\vartheta_T,t)=M(\vartheta_0,t)+(\hat\vartheta_T-\vartheta_0)\dot M(\bar\vartheta_T,t)$. Then we obtain the equation

$$\int_0^T\dot M(\hat\vartheta_T,t)\left[\mathrm{d}X_t-M(\vartheta_0,t)\mathrm{d}t\right]-(\hat\vartheta_T-\vartheta_0)\int_0^T\dot M(\hat\vartheta_T,t)\dot M(\bar\vartheta_T,t)\mathrm{d}t=0,$$

and

$$\hat\vartheta_T=\vartheta_0+\left(\int_0^T\dot M(\hat\vartheta_T,t)\dot M(\bar\vartheta_T,t)\mathrm{d}t\right)^{-1}\int_0^T\dot M(\hat\vartheta_T,t)\left[\mathrm{d}X_t-M(\vartheta_0,t)\mathrm{d}t\right]$$

$$=\vartheta_0+\frac{1}{\sqrt{T}}\left(\frac{1}{T}\int_0^T\frac{\dot M(\vartheta_0,t)^2}{\sigma^2}\mathrm{d}t\right)^{-1}\frac{1}{\sqrt{T}}\int_0^T\frac{\dot M(\vartheta_0,t)}{\sigma}\mathrm{d}\bar W_t\,(1+o(1)).$$

This last expression is used to define the one-step MLE as follows. First, recall that by the law of large numbers,

$$\frac{1}{\sigma^2 T}\int_0^T \dot M\left(\vartheta_0,t\right)^2\,\mathrm{d}t \longrightarrow \mathrm{I}\left(\vartheta_0\right), \tag{1.87}$$

and the CLT,

$$\frac{1}{\sigma\sqrt{T}}\int_0^T \dot M\left(\vartheta_0,t\right)\,\mathrm{d}\bar W_t \Longrightarrow \Delta\left(\vartheta_0\right)\sim\mathcal{N}\left(0,\mathrm{I}\left(\vartheta_0\right)\right). \tag{1.88}$$

For the Fisher information $\mathrm{I}\left(\vartheta\right)$, see (1.76).

Suppose that we already have some consistent estimator $\bar\vartheta_T$. Then, the Le Cam's one-step MLE $\vartheta_T^\star$ of the parameter ϑ is defined by the equality

$$\begin{aligned}\vartheta_T^\star &= \bar\vartheta_T + \frac{1}{\mathrm{I}\left(\bar\vartheta_T\right)T}\left.\frac{\partial \ln L\left(\vartheta,X^T\right)}{\partial\vartheta}\right|_{\vartheta=\bar\vartheta_T}\\ &= \bar\vartheta_T + \frac{1}{\sigma^2\mathrm{I}\left(\bar\vartheta_T\right)T}\int_0^T \dot M\left(\bar\vartheta_T,t\right)\left[\mathrm{d}X_t - M\left(\bar\vartheta_T,t\right)\mathrm{d}t\right].\end{aligned} \tag{1.89}$$

Forget, for an instant, that the stochastic integral is not defined because $\bar\vartheta_T$ depends on the whole trajectory X^T and study this estimator formally.

We have

$$\begin{aligned}\sqrt{T}\left(\vartheta_T^\star-\vartheta_0\right) &= \sqrt{T}\left(\bar\vartheta_T-\vartheta_0\right) + \frac{1}{\sigma\mathrm{I}\left(\bar\vartheta_T\right)\sqrt{T}}\int_0^T \dot M\left(\bar\vartheta_T,t\right)\mathrm{d}\bar W_t\\ &\quad + \frac{1}{\sigma^2\mathrm{I}\left(\bar\vartheta_T\right)\sqrt{T}}\int_0^T \dot M\left(\bar\vartheta_T,t\right)\left[M\left(\vartheta_0,t\right)-M\left(\bar\vartheta_T,t\right)\right]\mathrm{d}t\\ &= \sqrt{T}\left(\bar\vartheta_T-\vartheta_0\right) + \frac{1}{\sigma\mathrm{I}\left(\vartheta_0\right)\sqrt{T}}\int_0^T \dot M\left(\vartheta_0,t\right)\mathrm{d}\bar W_t\left(1+o\left(1\right)\right)\\ &\quad - \frac{\sqrt{T}\left(\bar\vartheta_T-\vartheta_0\right)}{\sigma^2\mathrm{I}\left(\vartheta_0\right)T}\int_0^T \dot M\left(\vartheta_0,t\right)^2\mathrm{d}t\left(1+o\left(1\right)\right)\\ &= \frac{\sqrt{T}\left(\bar\vartheta_T-\vartheta_0\right)}{\mathrm{I}\left(\vartheta_0\right)}\left[\mathrm{I}\left(\vartheta_0\right)-\frac{1}{T}\int_0^T\frac{\dot M\left(\vartheta_0,t\right)^2}{\sigma^2}\mathrm{d}t\right]\left(1+o\left(1\right)\right)\\ &\quad + \frac{1}{\sigma\mathrm{I}\left(\vartheta_0\right)\sqrt{T}}\int_0^T \dot M\left(\vartheta_0,t\right)\mathrm{d}\bar W_t\left(1+o\left(1\right)\right)\\ &= \frac{1}{\sigma\mathrm{I}\left(\vartheta_0\right)\sqrt{T}}\int_0^T \dot M\left(\vartheta_0,t\right)\mathrm{d}\bar W_t\left(1+o\left(1\right)\right)+o\left(1\right)\\ &\Longrightarrow \frac{\Delta\left(\vartheta_0\right)}{\mathrm{I}\left(\vartheta_0\right)}\sim\mathcal{N}\left(0,\mathrm{I}\left(\vartheta_0\right)^{-1}\right).\end{aligned} \tag{1.90}$$

Here we used the LLN (1.87) and the CLT (1.88). Therefore, this one-step MLE is (formally) asymptotically equivalent to the MLE (1.75) and BE (1.80). Our main goal is to construct an on-line adaptive filter, and we seek an estimator that can be substituted into the Eqs. (1.85)–(1.86). Of course, none of the estimators considered so far satisfy this on-line condition because in $m(\vartheta, t)$ and $\gamma(\vartheta, t)$ in (1.85), we can only put an estimator that depends on the observations up to time t only, i.e., based on the observations $X^t = (X_s, 0 \leq s \leq t)$. We cannot use the estimators $\breve{\vartheta}_t$, $\hat{\vartheta}_t$ or $\tilde{\vartheta}_t$ because their calculation requires solutions of the Eqs. (1.85) for different values of ϑ. For example, to calculate the MLE, we need the trajectories $m(\vartheta, s), 0 \leq s \leq t$ for many values of ϑ.

To avoid all these difficulties and to have a well-defined corresponding stochastic integral, we propose a one-step MLE process, which is a version of the one-step MLE, where we use a preliminary estimator $\bar{\vartheta}_{\tau_T}$ constructed by a small part of the initial observations $X^{\tau_T} = \left(X_t, 0 \leq t \leq \tau_T\right)$. Here $\tau_T = T^\delta$, $\delta \in (1/2, 1)$.

Suppose that $\bar{\vartheta}_{\tau_T}$ is such that $\sqrt{\tau_T}\left(\bar{\vartheta}_{\tau_T} - \vartheta_0\right)$ is bounded in probability. The one-step MLE-process $\vartheta^\star_{t,T}$, $\tau_T < t \leq T$, is defined by the formula

$$\vartheta^\star_{t,T} = \bar{\vartheta}_{\tau_T} + \frac{1}{\sigma^2 \mathrm{I}\left(\bar{\vartheta}_{\tau_T}\right)\left(t - \tau_T\right)} \int_{\tau_T}^{t} \dot{M}\left(\bar{\vartheta}_{\tau_T}, s\right)\left[\mathrm{d}X_s - M\left(\bar{\vartheta}_{\tau_T}, s\right)\mathrm{d}s\right]. \tag{1.91}$$

Now, the integrand $\dot{M}\left(\bar{\vartheta}_{\tau_T}, X_{\tau_T}\right)$ is measurable with respect to $\mathfrak{F}^X_{\tau_T}$, and therefore the stochastic integral is well-defined. The calculations like (1.90) for $\vartheta^\star_{t,T}$ can be valid too if we note two important differences. First, we note that if we put $t = vT$, $v \in [\epsilon_0, 1)$, $\epsilon_0 \in (0, 1)$,

$$\frac{1}{\sqrt{t}} \int_{\tau_T}^{t} \left[\frac{\dot{M}(\vartheta_0, t)^2}{\sigma^2} - \mathrm{I}(\vartheta_0)\right] \mathrm{d}t \Longrightarrow \mathcal{N}\left(0, D^2\right)$$

with some $D^2 < \infty$ (see Proposition 1.23 in [109]). Therefore, the term

$$\frac{\sqrt{T}\left(\bar{\vartheta}_{\tau_T} - \vartheta_0\right)}{\mathrm{I}(\vartheta_0)} \left[\mathrm{I}(\vartheta_0) - \frac{1}{T}\int_{\tau_T}^{T} \frac{\dot{M}(\vartheta_0, t)^2}{\sigma^2} \mathrm{d}t\right] \longrightarrow 0.$$

Further,

$$\begin{aligned}
&\frac{1}{\sigma^2 \mathrm{I}\left(\bar{\vartheta}_{\tau_T}\right)\sqrt{T}} \int_{\tau_T}^{T} \dot{M}\left(\bar{\vartheta}_{\tau_T}, t\right)\left[M(\vartheta_0, t) - M\left(\bar{\vartheta}_{\tau_T}, t\right)\right]\mathrm{d}t \\
&\quad = \frac{\sqrt{T}\left(\bar{\vartheta}_{\tau_T} - \vartheta_0\right)}{\sigma^2 \mathrm{I}\left(\bar{\vartheta}_{\tau_T}\right) T} \int_{\tau_T}^{T} \dot{M}\left(\bar{\vartheta}_{\tau_T}, t\right)\dot{M}\left(\bar{\bar{\vartheta}}_{\tau_T}, t\right)\mathrm{d}t \\
&\quad = \frac{\sqrt{T}\left(\bar{\vartheta}_{\tau_T} - \vartheta_0\right)}{\sigma^2 \mathrm{I}(\vartheta_0) T} \int_{\tau_T}^{T} \dot{M}(\vartheta_0, t)^2 \mathrm{d}t + \sqrt{T}\left(\bar{\vartheta}_{\tau_T} - \vartheta_0\right)^2 O(1).
\end{aligned}$$

Now the convergence

$$\sqrt{T}\left(\bar{\vartheta}_{\tau_T}-\vartheta_0\right)^2=\frac{\sqrt{T}}{\tau_T}\left[\sqrt{\tau_T}\left(\bar{\vartheta}_{\tau_T}-\vartheta_0\right)\right]^2\longrightarrow 0$$

defines the condition $\delta \in (\frac{1}{2},1)$ because $\tau_T = T^\delta$ and we supposed that $\sqrt{\tau_T}\left(\bar{\vartheta}_{\tau_T}-\vartheta_0\right)$ is bounded in probability.

Therefore,

$$\begin{aligned}\sqrt{t}\left(\vartheta^\star_{t,T}-\vartheta_0\right)&=\frac{\sqrt{t}}{\sigma \mathrm{I}\left(\vartheta_0\right)\left(t-\tau_T\right)}\int_{\tau_T}^t \dot M\left(\vartheta_0,s\right)\mathrm{d}\bar W_s\left(1+o\left(1\right)\right)+o\left(1\right)\\&\Longrightarrow \mathcal{N}\left(0,\mathrm{I}(\vartheta_0)^{-1}\right).\end{aligned}$$

Remark that the estimator (1.91) can be written in the recurrent form as follows

$$\mathrm{d}\vartheta^\star_{t,T}=\frac{\bar{\vartheta}_{\tau_T}-\vartheta^\star_{t,T}}{t-\tau_T}\mathrm{d}t+\frac{\dot M\left(\bar{\vartheta}_{\tau_T},t\right)\left[\mathrm{d}X_t-M\left(\bar{\vartheta}_{\tau_T},t\right)\mathrm{d}t\right]}{\sigma^2\mathrm{I}\left(\bar{\vartheta}_{\tau_T}\right)\left(t-\tau_T\right)},\qquad \tau_T<t\leq T. \tag{1.92}$$

Two-Steps MLE-Process

The one-step MLE-process $\vartheta^\star_{t,T}$, $\tau_T<t\leq T$ does not provide an estimator for $t<\tau_T$. If the learning interval $\left[0,\tau_T\right]$ with $\tau_T=T^\delta,\delta\in\left(\frac{1}{2},1\right)$, is considered too long, there is a possibility to use a shorter interval $\left[0,\tau_T\right]$ with $\tau_T=T^\delta,\ \delta\in\left(\frac{1}{3},\frac{1}{2}\right)$. In this case, the construction of the estimator will be modified. First, we obtain a preliminary estimator $\bar{\vartheta}_{\tau_T}$, such that $\sqrt{\tau_T}\left(\bar{\vartheta}_{\tau_T}-\vartheta_0\right)$ is bounded in probability. The second step is the calculation of the following estimator process:

$$\mathrm{d}\tilde{\vartheta}_{t,T}=\frac{\bar{\vartheta}_{\tau_T}-\tilde{\vartheta}_{t,T}}{t-\tau_T}\mathrm{d}t+\frac{\dot M\left(\bar{\vartheta}_{\tau_T},t\right)}{\sigma^2\mathrm{I}\left(\bar{\vartheta}_{\tau_T}\right)\left(t-\tau_T\right)}\left[\mathrm{d}X_t-M\left(\bar{\vartheta}_{\tau_T},t\right)\mathrm{d}t\right],\ \tau_T<t\leq T.$$

The last step is the calculation of the two-step MLE process:

$$\vartheta^{\star\star}_{t,T}=\tilde{\vartheta}_{t,T}+\frac{1}{\sigma^2\mathrm{I}(\tilde{\vartheta}_{t,T})\left(t-\tau_T\right)}\int_{\tau_T}^t \dot M(\tilde{\vartheta}_{s,T},s)\left[\mathrm{d}X_s-M(\tilde{\vartheta}_{t,T},s)\mathrm{d}s\right],$$

$$\tau_T\quad <t\leq T.$$

Here, $\dot M(\tilde{\vartheta}_{s,T},s)=\dot M_s$ is the solution of the equation

$$\mathrm{d}\dot M_s=-r(\tilde{\vartheta}_{s,T})\dot M_s\mathrm{d}s+\dot\Gamma(\tilde{\vartheta}_{s,T})\mathrm{d}X_s,$$

and $\dot{M}(\tilde{\vartheta}_{t,T}, s)$ is the solution of the equation

$$\mathrm{d}\dot{M}_t = -r(\tilde{\vartheta}_{t,T})\dot{M}_t\mathrm{d}t - \dot{r}(\tilde{\vartheta}_{t,T})M_t\mathrm{d}t + \dot{\Gamma}(\tilde{\vartheta}_{t,T})\mathrm{d}X_t,$$

with corresponding initial values M_{τ_T} and $\dot{M}_{\tau_T}$. The Two-step MLE-process $\vartheta^{\star\star}_{t,T}, \tau_T < t \leq T$ is consistent and asymptotically normal with optimal rate and limit variance:

$$\sqrt{t}\left(\vartheta^{\star\star}_{t,T} - \vartheta_0\right) \Longrightarrow \mathcal{N}\left(0, \mathrm{I}(\vartheta_0)^{-1}\right).$$

The proof of a similar results but for the ergodic diffusion and inhomogeneous Poisson processes of observations can be found in [112] and [118], respectively.

If we compare one-step and two-step procedures, we must remark that the advantage of the two-step MLE process is to have a smaller learning interval, which can be considered as not too interesting because it requires essentially more calculations.

1.3 Adaptive Filtration and Extrapolation

1.3.1 Adaptive Filtration

Suppose that the model of observations is given by (1.83)–(1.84), and therefore, the Kalman–Bucy filter is given by (1.85)–(1.86). The adaptive filter is constructed in several steps:

1. Choose the learning interval $[0, \tau_T]$, where $\tau_T = [T^\delta]$, $[A]$ is the integer part of A, and $\delta \in \left(\frac{1}{2}, 1\right)$.
2. Calculate a consistent EMM $\vartheta^*_{\tau_T}$ using observations $X^{\tau_T} = \left(X_s, 0 \leq s \leq \tau_T\right)$ (see (1.66)).
3. Calculate the recurrent One-step MLE-process $\vartheta^\star_{t,T}, \tau_T < t \leq T$ given by (1.92):

$$\mathrm{d}\vartheta^\star_{t,T} = \frac{\vartheta^*_{\tau_T} - \vartheta^\star_{t,T}}{t - \tau_T}\mathrm{d}t + \frac{\dot{M}(\vartheta^*_{\tau_T}, t)}{\sigma^2 \mathrm{I}(\vartheta^*_{\tau_T})(t - \tau_T)}\left[\mathrm{d}X_t - M(\vartheta^*_{\tau_T}, t)\mathrm{d}t\right].$$

4. Then the adaptive filter for the estimate $m^\star_{t,T}$ of the conditional expectation $m\left(\vartheta_0, t\right)$ is defined as the solution of the equation

$$\mathrm{d}m^\star_{t,T} = -r(\vartheta^\star_{t,T})m^\star_{t,T}\mathrm{d}t + B(\vartheta^\star_{t,T})\mathrm{d}X_t, \qquad \tau_T < t \leq T.$$

Here $m^\star_{\tau_T,T} = m(\vartheta^*_{\tau_T}, \tau_T)$, $r\left(\vartheta\right) = \left(a\left(\vartheta\right)^2 + \sigma^{-2} f\left(\vartheta\right)^2 b\left(\vartheta\right)^2\right)^{1/2}$, and $B\left(\vartheta\right) = f\left(\vartheta\right)^{-1}\left(r\left(\vartheta\right) - a\left(\vartheta\right)\right)$.

Remark that in such problems, special attention has to be paid to the calculation of the initial value $m^\star_{\tau_T,T}$, because the Eq. (1.85) cannot be used directly for the calculation of $m(\vartheta^*_{\tau_T},\tau_T)$, and this calculation can be done with the help of Corollary 1.2. More details will be given in Chap. 4. The solution of (1.85) is

$$m\left(\vartheta_0,t\right)=y_0\Phi\left(\vartheta,0,t\right)+\frac{f\left(\vartheta\right)\gamma_*\left(\vartheta\right)}{\sigma^2}\int_0^t\Phi\left(\vartheta,s,t\right){\rm d}X_s,$$

where

$$\Phi\left(\vartheta,s,t\right)=\exp\Big(-A\left(\vartheta\right)\left(t-s\right)\Big),\qquad A\left(\vartheta\right)=a\left(\vartheta\right)+\frac{f\left(\vartheta\right)^2\gamma_*\left(\vartheta\right)}{\sigma^2}.$$

It can be written without the stochastic integral with the help of (1.33) as follows:

$$\begin{aligned}m(\vartheta^*_{\tau_T},\tau_T)&=y_0\Phi(\vartheta^*_{\tau_T},0,\tau_T)+X^*_{\tau_T}\frac{f(\vartheta^*_{\tau_T})\gamma_*(\vartheta^*_{\tau_T})}{\sigma^2}\\&\quad-\frac{A(\vartheta^*_{\tau_T})\gamma_*(\vartheta^*_{\tau_T})f(\vartheta^*_{\tau_T})}{\sigma^2}\int_0^{\tau_T}X_s\Phi(\vartheta^*_{\tau_T},s,\tau_T){\rm d}s.\end{aligned}\tag{1.93}$$

More details and the large sample asymptotic of the error $m^\star_{\tau_T,T}-m\left(\vartheta_0,t\right)$ are given in Chap. 4.

1.3.2 Adaptive Extrapolation

Suppose once more that the partially observed system is given by (1.83), (1.84), and the Kalman–Bucy filter is

$${\rm d}m\left(\vartheta,s\right)=-r\left(\vartheta\right)m\left(\vartheta,s\right){\rm d}s+B\left(\vartheta\right){\rm d}X_s,\qquad m\left(\vartheta,0\right),\quad 0\leq s\leq T.$$

The unknown parameter is $\vartheta\in\Theta=(\alpha,\beta)$. Consider the problem of adaptive extrapolation of $Y_t,t>T$ and $X_t,t>T$.

Recall that if ϑ is known and we have to calculate $m\left(\vartheta,T,t\right)=\mathbf{E}_\vartheta\left(Y_t|\mathfrak{F}^X_T\right)$ and $n\left(\vartheta,T,t\right)=\mathbf{E}_\vartheta\left(X_t|\mathfrak{F}^X_T\right)$, then the solution is given by the equalities (1.43) and (1.45), respectively. For adaptive extrapolation, we do not need a one-step MLE process but just a one-step MLE (1.87). Therefore, we take the preliminary estimator $\vartheta^*_{\tau_T}$ constructed using the observations $X^{\tau_T}=\left(X_s,0\leq s\leq\tau_T\right)$. Then, the one-step MLE can be:

$$\vartheta^\star_T=\vartheta^*_{\tau_T}+\frac{1}{\sigma^2{\rm I}(\vartheta^*_{\tau_T})(T-\tau_T)}\int_{\tau_T}^T\dot M(\vartheta^*_{\tau_T},s)\left[{\rm d}X_s-M(\vartheta^*_{\tau_T},s){\rm d}s\right].$$

Here $M(\vartheta, s) = f(\vartheta) m(\vartheta, s)$ and the Fisher information $\mathrm{I}(\vartheta)$ can be found in (1.76).

According to (1.43), (1.45) and Corollary 1.2, the adaptive extrapolations $m_{t,T}^{\star}$ and $n_{t,T}^{\star}$ can be

$$\begin{aligned}
m_{t,T}^{\star} &= m\left(\vartheta_T^{\star}, T\right) \exp\left(-a\left(\vartheta_T^{\star}\right)(t-T)\right),\\
n_{t,T}^{\star} &= X_T + m\left(\vartheta_T^{\star}, T\right) \int_T^t f\left(\vartheta_T^{\star}\right) \exp\left(-a\left(\vartheta_T^{\star}\right)(s-T)\right) \mathrm{d}s\\
&= X_T + \frac{m\left(\vartheta_T^{\star}, T\right) f\left(\vartheta_T^{\star}\right)}{a\left(\vartheta_T^{\star}\right)}\left[1-\exp\left(-a\left(\vartheta_T^{\star}\right)(t-T)\right)\right].
\end{aligned}$$

Here for calculation $m\left(\vartheta_T^{\star}, T\right)$, we use the equality (1.93), where $\vartheta_{\tau_T}^{*}$ is replaced by $\vartheta_T^{\star}$.

The errors of extrapolations $\sqrt{T}\left(m_{t,T}^{\star} - m\left(\vartheta_0, T, t\right)\right)$ and $\sqrt{T}\left(n_{t,T}^{\star} - n\left(\vartheta_0, T, t\right)\right)$ can be easily described too.

1.4 Asymptotic Efficiency

Let us recall the well-known lower bounds on the mean-squared errors of all estimators of parameters in the case of a regular statistical experiment and then propose a lower bound on the mean squared risk for the estimators of the random process $m(\vartheta_0, \cdot)$. As the model of observations, we take the partially observed system

$$\begin{aligned}
\mathrm{d}X_t &= f(\vartheta, t) Y_t \mathrm{d}t + \sigma(t) \mathrm{d}W_t, \qquad X_0 = 0, \quad 0 \le t \le T, \qquad (1.94)\\
\mathrm{d}Y_t &= a(\vartheta, t) Y_t \mathrm{d}t + b(\vartheta, t) \mathrm{d}V_t, \qquad y.
\end{aligned}$$

We suppose that the functions $f(\vartheta, t), a(\vartheta, t), b(\vartheta, t)$ for $\vartheta \in \Theta, t \in [0, T]$, $\Theta = (\alpha, \beta)$, have continuous derivatives with respect to ϑ, and $\sigma(t)^2 > 0$.

The corresponding Kalman–Bucy filter is given by the equations

$$\mathrm{d}m(\vartheta, t) = a(\vartheta, t) m(\vartheta, t) \mathrm{d}t + \frac{f(\vartheta, t)\gamma(\vartheta, t)}{\sigma(t)^2}\left[\mathrm{d}X_t - f(\vartheta, t) m(\vartheta, t) \mathrm{d}t\right],$$

$$\frac{\partial \gamma(\vartheta, t)}{\partial t} = 2a(\vartheta, t)\gamma(\vartheta, t) - \frac{f(\vartheta, t)^2 \gamma(\vartheta, t)^2}{\sigma(t)^2} + b(\vartheta)^2, \qquad 0 \le t \le T.$$

The initial values are $m(\vartheta, 0) = y$, $\gamma(\vartheta, 0) = 0$.

The likelihood ratio formula is

$$L\left(\vartheta, X^T\right) = \exp\left(\int_0^T \frac{M(\vartheta,t)}{\sigma(t)^2}\mathrm{d}X_t - \int_0^T \frac{M(\vartheta,t)^2}{2\sigma(t)^2}\mathrm{d}t\right),$$

where $M(\vartheta,t) = f(\vartheta,t)\, m(\vartheta,t)$.

Note that $L\left(\vartheta, X^T\right)$ is the Radon–Nikodym derivative of the measure $\mathbf{P}_\vartheta$ generated by the observations (1.94) with respect to the measure $\mathbf{P}_*$ induced in the space of observations by a stochastic process (1.94) with $f(\vartheta, 0) \equiv 0$. The mathematical expectation with respect to this last measure we denote by $\mathbf{E}_*$.

Cramér–Rao Lower Bound

The first one is the *Cramér–Rao* lower bound, , which can be obtained as follows. Let $\bar\vartheta_T$ be some estimator of ϑ and denote $N(\vartheta) = \mathbf{E}_\vartheta \bar\vartheta_T$ its mean value. Then, for the derivative of the mean, we can write

$$\begin{aligned}
\dot N(\vartheta) &= \frac{\partial}{\partial\vartheta}\mathbf{E}_\vartheta\bar\vartheta_T = \frac{\partial}{\partial\vartheta}\mathbf{E}_*\left(\bar\vartheta_T L\left(\vartheta, X^T\right)\right) = \mathbf{E}_*\left(\bar\vartheta_T \dot L\left(\vartheta, X^T\right)\right)\\
&= \mathbf{E}_*\left(L\left(\vartheta, X^T\right)\bar\vartheta_T \frac{\partial}{\partial\vartheta}\ln L\left(\vartheta, X^T\right)\right)\\
&= \mathbf{E}_\vartheta\left(\bar\vartheta_T \int_0^T \frac{\dot M(\vartheta,t)}{\sigma(t)^2}\left[\mathrm{d}X_t - M(\vartheta,t)\,\mathrm{d}t\right]\right)\\
&= \mathbf{E}_\vartheta\left(\left[\bar\vartheta_T - N(\vartheta)\right]\int_0^T \frac{\dot M(\vartheta,t)}{\sigma(t)^2}\left[\mathrm{d}X_t - M(\vartheta,t)\,\mathrm{d}t\right]\right)\\
&\le \left[\mathbf{E}_\vartheta\left[\bar\vartheta_T - N(\vartheta)\right]^2 \mathbf{E}_\vartheta\left(\int_0^T \frac{\dot M(\vartheta,t)}{\sigma(t)^2}\left[\mathrm{d}X_t - M(\vartheta,t)\,\mathrm{d}t\right]\right)^2\right]^{1/2}\\
&= \left[\mathbf{E}_\vartheta\left[\bar\vartheta_T - N(\vartheta)\right]^2 \mathrm{I}_T(\vartheta)\right]^{1/2}.
\end{aligned}$$

Here we changed the measure $\mathbf{E}_* L\left(\vartheta, X^T\right) = \mathbf{E}_\vartheta$ and used the Cauchy–Shwarz inequality. This allow us to write

$$\begin{aligned}
\mathbf{E}_\vartheta\left(\int_0^T \frac{\dot M(\vartheta,t)}{\sigma(t)^2}\left[\mathrm{d}X_t - M(\vartheta,t)\,\mathrm{d}t\right]\right)^2 &= \mathbf{E}_\vartheta\left(\int_0^T \frac{\dot M(\vartheta,t)}{\sigma(t)}\,\mathrm{d}W_t\right)^2\\
&= \mathbf{E}_\vartheta\int_0^T \frac{\dot M(\vartheta,t)^2}{\sigma(t)^2}\,\mathrm{d}t = \mathrm{I}_T(\vartheta).
\end{aligned}$$

Therefore,

$$\mathbf{E}_\vartheta\left[\bar\vartheta_T - N(\vartheta)\right]^2 \ge \frac{\dot N(\vartheta)^2}{\mathrm{I}_T(\vartheta)}.$$

If we introduce the bias $B(\vartheta) = N(\vartheta) - \vartheta$ and write $\mathbf{E}_\vartheta\left[\bar\vartheta_T - N(\vartheta)\right]^2 = \mathbf{E}_\vartheta\left[\bar\vartheta_T - \vartheta\right]^2 - B(\vartheta)^2$, then we obtain the final form of the Cramér–Rao inequality

$$\mathbf{E}_\vartheta\left[\bar\vartheta_T - \vartheta\right]^2 \geq \frac{\left[1 + \dot B(\vartheta)\right]^2}{\mathrm{I}_T(\vartheta)} + B(\vartheta)^2 .$$

For the unbiased estimators ($B(\vartheta) \equiv 0$), it has more simple form

$$\mathrm{I}_T(\vartheta)\,\mathbf{E}_\vartheta\left[\bar\vartheta_T - \vartheta\right]^2 \geq 1. \tag{1.95}$$

The unbiased estimator $\bar\vartheta_T$ satisfying the equality

$$\mathrm{I}_T(\vartheta)\,\mathbf{E}_\vartheta\left[\bar\vartheta_T - \vartheta\right]^2 = 1$$

for all $\vartheta \in \Theta$ is called *Cramér–Rao—efficient.*

The inequality (1.95) is sometimes used to define asymptotically efficient estimators as follows. First, it is supposed that for all estimators, the asymptotic lower bound

$$\lim_{T\to\infty} \mathrm{I}_T(\vartheta)\,\mathbf{E}_\vartheta\left[\bar\vartheta_T - \vartheta\right]^2 \geq 1 \tag{1.96}$$

holds. Then, the estimators for which the limit

$$\lim_{T\to\infty} \mathrm{I}_T(\vartheta)\,\mathbf{E}_\vartheta\left[\bar\vartheta_T - \vartheta\right]^2 = 1 \tag{1.97}$$

holds for all $\vartheta \in \Theta$ are called *asymptotically efficient.* The inconvenience of this definition is that there are few unbiased estimators and even if we consider the class of estimators such that $B(\vartheta) \to 0$, nevertheless the bound (1.97) is not correct due to the example of Hodges (see, e.g., [73]). We will return to the question of asymptotic efficiency later in this section.

Note that if the unknown parameter is a vector $\vartheta \in \Theta \subset \mathbb{R}^d$, then for the covariance matrix of the error of estimation, the Cramér–Rao inequality for unbiased estimators is

$$\mathbf{I}_T(\vartheta)\,\mathbf{E}_\vartheta\left(\bar\vartheta_T - \vartheta\right)^\top\left(\bar\vartheta_T - \vartheta\right) \geq \mathrm{I}_{d\times d}$$

where $\mathrm{I}_{d\times d}$ is the unit matrix and the inequality $\mathbf{A} \geq \mathbf{B}$ for matrices is understood as the inequality

$$\lambda^\top \mathbf{A}\lambda \geq \lambda^\top \mathbf{B}\lambda$$

for any non-null vector $\lambda \in \mathbb{R}^d$.

Example 1.11 To construct a Cramér–Rao efficient estimator, we consider a different model of observations

$$\mathrm{d}X_t = \vartheta y(t)\,\mathrm{d}t + \sigma(t)\,\mathrm{d}W_t, \quad X_0 = 0, \quad 0 \le t \le T, \tag{1.98}$$

where $y(\cdot)$ and $\sigma(\cdot)$ are bounded non-identically null functions, $\sigma(t)^2 \ge c > 0$. Then the MLE $\hat{\vartheta}_T$ and Fisher information are

$$\hat{\vartheta}_T = \left(\int_0^T \frac{y(t)^2}{\sigma(t)^2}\,\mathrm{d}t\right)^{-1} \int_0^T \frac{y(t)}{\sigma(t)^2}\,\mathrm{d}X_t,$$

$$\mathrm{I}_T = \int_0^T \frac{y(t)^2}{\sigma(t)^2}\,\mathrm{d}t > 0$$

respectively.

It can be shown that the Cramér–Rao lower bound is: for any unbiased estimator $\bar{\vartheta}_T$, the following inequality

$$\mathbf{E}_\vartheta \left(\bar{\vartheta}_T - \vartheta\right)^2 \ge \left(\int_0^T \frac{y(t)^2}{\sigma(t)^2}\,\mathrm{d}t\right)^{-1}$$

holds for all $\vartheta \in \Theta$.

We have $\mathbf{E}_\vartheta \left(\hat{\vartheta}_T - \vartheta\right)^2 = \mathrm{I}_T^{-1}$. Hence, the estimator $\hat{\vartheta}_T$ is *efficient in the class of unbiased estimators* or *Cramér-Rao-efficient.*

Van Trees Lower Bound

There exists a Bayesian version of the Cramér-Rao lower bound, called the *Van Trees inequality*, the right-hand side of which does not depend on the characteristics of the particular estimator like $B(\vartheta)$, $\dot{B}(\vartheta)$.

Suppose that ϑ is a random variable with the density function $p(\vartheta)$, $\vartheta \in \Theta = (\alpha, \beta)$, where the function $p(\cdot)$ satisfies the conditions $p(\alpha) = 0$, $p(\beta) = 0$, and has the continuous derivative $\dot{p}(\vartheta)$. The Fisher information related to $p(\cdot)$ is finite:

$$\mathrm{I}_p = \int_\alpha^\beta \frac{\dot{p}(\vartheta)^2}{p(\vartheta)}\,\mathrm{d}\vartheta.$$

The mathematical expectation with respect to this density is denoted $\mathbf{E}_p$.

Integrating by parts gives us the relations

$$\int_\alpha^\beta \left(\bar{\vartheta}_T - \theta\right) \frac{\partial}{\partial\theta}\left[L\left(\theta, X^T\right) p(\theta)\right] \mathrm{d}\theta = \left[\left(\bar{\vartheta}_T - \theta\right) L\left(\theta, X^T\right) p(\theta)\right]\Big|_\alpha^\beta$$
$$+ \int_\alpha^\beta L\left(\theta, X^T\right) p(\theta)\,\mathrm{d}\theta = \int_\alpha^\beta L\left(\theta, X^T\right) p(\theta)\,\mathrm{d}\theta$$

and

$$\mathbf{E}_* \int_\alpha^\beta L\left(\theta, X^T\right) p(\theta)\, \mathrm{d}\theta = \int_\alpha^\beta p(\theta)\, \mathrm{d}\theta = 1$$

since $\mathbf{E}_* L\left(\theta, X^T\right) = 1$. Therefore, we can write

$$\begin{aligned}
1 &= \left(\mathbf{E}_* \int_\alpha^\beta \left(\bar\vartheta_T - \theta\right) \frac{\partial}{\partial\theta}\left[L\left(\theta, X^T\right) p(\theta)\right] \mathrm{d}\theta\right)^2 \\
&= \left(\mathbf{E}_* \int_\alpha^\beta \left(\bar\vartheta_T - \theta\right) \frac{\partial}{\partial\theta}\left[\ln\left(L\left(\theta, X^T\right) p(\theta)\right)\right] L\left(\theta, X^T\right) p(\theta)\, \mathrm{d}\theta\right)^2 \\
&\leq \mathbf{E}_* \int_\alpha^\beta \left(\bar\vartheta_T - \theta\right)^2 L\left(\theta, X^T\right) p(\theta)\, \mathrm{d}\theta \\
&\quad \times \mathbf{E}_* \int_\alpha^\beta \left[\frac{\partial}{\partial\theta} \ln L\left(\theta, X^T\right) + \frac{\dot p(\theta)}{p(\theta)}\right]^2 L\left(\theta, X^T\right) p(\theta)\, \mathrm{d}\theta.
\end{aligned}$$

Recall that $\mathbf{E}_*\left[L\left(\theta, X^T\right)\xi\right] = \mathbf{E}_\theta \xi$. Then

$$\begin{aligned}
\mathbf{E}_* \int_\alpha^\beta \left(\bar\vartheta_T - \theta\right)^2 L\left(\theta, X^T\right) p(\theta)\, \mathrm{d}\theta &= \int_\alpha^\beta \mathbf{E}_\theta \left(\bar\vartheta_T - \theta\right)^2 p(\theta)\, \mathrm{d}\theta \\
&= \mathbb{E}\left(\bar\vartheta_T - \vartheta\right)^2.
\end{aligned}$$

Here we have denoted by $\mathbb{E}$ the *double* mathematical expectation defined by the last equality. Recall that $\bar\vartheta_T$ and ϑ are random variables. Further,

$$\begin{aligned}
&\mathbf{E}_* \int_\alpha^\beta \left[\frac{\partial}{\partial\theta} \ln L\left(\theta, X^T\right) + \frac{\dot p(\theta)}{p(\theta)}\right]^2 L\left(\theta, X^T\right) p(\theta)\, \mathrm{d}\theta \\
&\quad = \mathbf{E}_* \int_\alpha^\beta \left[\int_0^T \frac{\dot M(\theta, t)}{\sigma(t)} \left[\mathrm{d}X_t - M(\theta, t)\, \mathrm{d}t\right] + \frac{\dot p(\theta)}{p(\theta)}\right]^2 L\left(\theta, X^T\right) p(\theta)\, \mathrm{d}\theta \\
&\quad = \int_\alpha^\beta \mathbf{E}_\theta \left[\int_0^T \frac{\dot M(\theta, t)}{\sigma(t)} \left[\mathrm{d}X_t - M(\theta, t)\, \mathrm{d}t\right]\right]^2 p(\theta)\, \mathrm{d}\theta + \mathrm{I}_p \\
&\quad = \int_\alpha^\beta \mathrm{I}_T(\theta)\, p(\theta)\, \mathrm{d}\theta + \mathrm{I}_p = \mathbf{E}_p \mathrm{I}_T(\vartheta) + \mathrm{I}_p.
\end{aligned}$$

This gives us the *Van Trees bound* : *For ll estimators $\bar\vartheta_T$ the following inequality holds*:

$$\mathbb{E}\left(\bar\vartheta_T - \vartheta\right)^2 \geq \frac{1}{\mathbf{E}_p \mathrm{I}_T(\vartheta) + \mathrm{I}_p}. \tag{1.99}$$

Observe that the right-hand side of this inequality does not depend on characteristics of the estimator $\bar{\vartheta}_T$ in contrast to the Cramér–Rao inequality.

The estimator $\hat{\vartheta}_T$ satisfing the equality

$$\mathbb{E}\left(\bar{\vartheta}_T-\vartheta\right)^2=\frac{1}{\mathbf{E}_p\mathrm{I}_T\left(\vartheta\right)+\mathrm{I}_p}$$

can be called *Van Trees-efficient.*

Example 1.12 We have the same model of observations (1.98) as in the Example 1.11, but in this case, the unknown parameter ϑ is a Gaussian random variable, $\vartheta\sim\mathcal{N}\left(0,b^2\right)$. Recall that the Fisher information related with the density $p\left(\cdot\right)$ is $\mathrm{I}_p=b^{-2}$.

The BE is

$$\begin{aligned}\tilde{\vartheta}_T&=\frac{\int_{\mathcal{R}}\theta p\left(\theta\right)L\left(\theta,X^T\right)\mathrm{d}\theta}{\int_{\mathcal{R}}p\left(\theta\right)L\left(\theta,X^T\right)\mathrm{d}\theta}=\frac{\int_{\mathcal{R}}\theta\exp\left(-\frac{\theta^2}{2b^2}+\theta J_T\left(X^T\right)-\frac{\theta^2}{2}\mathrm{I}_T\right)\mathrm{d}\theta}{\int_{\mathcal{R}}\exp\left(-\frac{\theta^2}{2b^2}+\theta J_T\left(X^T\right)-\frac{\theta^2}{2}\mathrm{I}_T\right)\mathrm{d}\theta}\\&=\frac{\int_{\mathcal{R}}\theta\exp\left(-\frac{1}{2}\left[\theta^2\left(\mathrm{I}_p+\mathrm{I}_T\right)-2\theta J_T\left(X^T\right)\right]\right)\mathrm{d}\theta}{\int_{\mathcal{R}}\exp\left(-\frac{1}{2}\left[\theta^2\left(\mathrm{I}_p+\mathrm{I}_T\right)-2\theta J_T\left(X^T\right)\right]\right)\mathrm{d}\theta},\end{aligned}$$

where we denoted

$$J_T\left(X^T\right)=\int_0^T\frac{y\left(t\right)}{\sigma\left(t\right)^2}\,\mathrm{d}X_t,\qquad\mathrm{I}_T=\int_0^T\frac{y\left(t\right)^2}{\sigma\left(t\right)^2}\,\mathrm{d}t.$$

Using the elementary relations, we obtain the equalities

$$\theta^2\left(\mathrm{I}_p+\mathrm{I}_T\right)-2\theta J_T\left(X^T\right)=\left[\mathrm{I}_p+\mathrm{I}_T\right]\left[\vartheta-\frac{J_T\left(X^T\right)}{\mathrm{I}_p+\mathrm{I}_T}\right]^2+\frac{J_T\left(X^T\right)^2}{\mathrm{I}_p+\mathrm{I}_T}$$

and

$$\begin{aligned}\tilde{\vartheta}_T&=\frac{\int_{\mathcal{R}}\theta\exp\left(-\frac{\mathrm{I}_p+\mathrm{I}_T}{2}\left[\vartheta-\frac{J_T\left(X^T\right)}{\mathrm{I}_p+\mathrm{I}_T}\right]^2\right)\mathrm{d}\theta}{\int_{\mathcal{R}}\exp\left(-\frac{\mathrm{I}_p+\mathrm{I}_T}{2}\left[\vartheta-\frac{J_T\left(X^T\right)}{\mathrm{I}_p+\mathrm{I}_T}\right]^2\right)\mathrm{d}\theta}=\frac{J_T\left(X^T\right)}{\mathrm{I}_p+\mathrm{I}_T}\\&=\frac{1}{\mathrm{I}_p+\mathrm{I}_T}\int_0^T\frac{y\left(t\right)}{\sigma\left(t\right)^2}\,\mathrm{d}X_t.\end{aligned}$$

Here we used the equality (1.82).

For the mean-squared error by the direct calculations, we obtain

$$\mathbf{E}_{\vartheta_0}\left(\tilde{\vartheta}_T-\vartheta_0\right)^2=\mathbf{E}_{\vartheta_0}\left(\frac{\vartheta_0\,\mathrm{I}_T}{\mathrm{I}_p+\mathrm{I}_T}+\frac{1}{\mathrm{I}_p+\mathrm{I}_T}\int_0^T\frac{y\left(t\right)}{\sigma\left(t\right)}\,\mathrm{d}W_t-\vartheta_0\right)^2$$

$$=\frac{1}{\left(\mathrm{I}_p+\mathrm{I}_T\right)^2}\mathbf{E}_{\vartheta_0}\left(\int_0^T\frac{y\left(t\right)}{\sigma\left(t\right)}\,\mathrm{d}W_t-\vartheta_0\mathrm{I}_p\right)^2=\frac{\mathrm{I}_T+\vartheta_0^2\,\mathrm{I}_p^2}{\left(\mathrm{I}_p+\mathrm{I}_T\right)^2}.$$

Finally,

$$\mathbb{E}\left(\tilde{\vartheta}_T-\vartheta_0\right)^2=\frac{\mathrm{I}_T+\mathrm{I}_p^2\,\mathbf{E}_p\vartheta_0^2}{\left(\mathrm{I}_p+\mathrm{I}_T\right)^2}=\frac{1}{\mathrm{I}_p+\mathrm{I}_T}.$$

Therefore, the BE $\tilde{\vartheta}_T$ in this problem is *Van Trees-efficient.*

This proof of (1.99) and the wide discussion of the different forms of Van Trees inequality can be found in [59].

Hájek's Lower Bound

If we have to define an asymptotic lower bound on the mean-squared risk of all estimators, the direct use of the mentioned above bounds is not evident because the inequality (1.96) was never proved, and the inequality (1.99) depends on the prior distribution $p\left(\vartheta\right), \vartheta\in\Theta$, which in many problems does not exist or can be unknown.

A universal, in some sense, asymptotic lower bound, which we describe below, was constructed in two steps. First, it was L. Le Cam [134] who introduced the notion of *local asymptotic normality* (LAN) for the families of measures corresponding to observations in regular statistical models. Then, J. Hájek [64] proposed an asymptotic lower bound on the risks of all estimators, which is valid for statistical models with LAN families of distributions. Sometimes, it is called the *Hájek–Le Cam lower bound.* To define an LAN family, we follow Chap. 2 in [73], but we present a slightly modified definition, which is sufficient for the models considered in this work.

Suppose that once more we have a series of problems indexed by a parameter $\varepsilon\in(0,1]$. Let $\left(\mathcal{X}^{(\varepsilon)},\mathfrak{U}^{(\varepsilon)},\mathbf{P}_\vartheta^{(\varepsilon)},\vartheta\in\Theta\right)$, $\Theta\subset\mathcal{R}^d$ be a family of statistical experiments and $X^{(\varepsilon)}$ be the corresponding observation. We shall refer to $\mathrm{d}\mathbf{P}_{\vartheta_2}^{(\varepsilon)}/\mathrm{d}\mathbf{P}_{\vartheta_1}^{(\varepsilon)}\left(X^{(\varepsilon)}\right)$ as the derivative of the measure $\mathbf{P}_{\vartheta_2}^{(\varepsilon)}$ with respect to measure $\mathbf{P}_{\vartheta_1}^{(\varepsilon)}$ at the observation $X^{(\varepsilon)}$ (likelihood ratio).

Definition 1.1 A family $\left(\mathbf{P}_\vartheta^{(\varepsilon)},\vartheta\in\Theta\right)$ is called locally asymptotically normal at the point $\vartheta_0\in\Theta$ as $\varepsilon\to 0$ if there exists a nondegenerate $d\times d$ matrix $\mathbf{I}\left(\vartheta_0\right)$, and a vector $\Delta_\varepsilon\left(\vartheta_0,X^{(\varepsilon)}\right)$ such that for some $\varphi_\varepsilon\to 0$ and any $u\in\mathcal{R}^d$ the representation

$$Z_\varepsilon(\vartheta_0, u) = \frac{d\mathbf{P}^{(\varepsilon)}_{\vartheta_0+\varphi_\varepsilon u}}{d\mathbf{P}^{(\varepsilon)}_{\vartheta_0}}\left(X^{(\varepsilon)}\right)$$
$$= \exp\left(u^\top \Delta_\varepsilon\left(\vartheta_0, X^{(\varepsilon)}\right) - \frac{1}{2}u^\top \mathbf{I}(\vartheta_0)\, u + r_\varepsilon(u, \vartheta_0)\right)$$

is valid, where

$$\Delta_\varepsilon\left(\vartheta_0, X^{(\varepsilon)}\right) \Longrightarrow \mathcal{N}(0, \mathbf{I}(\vartheta_0)),$$

and

$$r_\varepsilon(u, \vartheta_0) \longrightarrow 0$$

in probability $\mathbf{P}^{(\varepsilon)}_{\vartheta_0}$.

If the condition of LAN is valid for all $\vartheta_0 \in \Theta$, we say that the family is LAN in Θ.

If we consider the sequences $\varepsilon_n \to 0, \vartheta_n \to \vartheta_0, u_n \longrightarrow u$ and obtain the representation

$$Z_{\varepsilon_n}(\vartheta_n, u_n) = \exp\left(u_n^\top \Delta_{\varepsilon_n}(\vartheta_n, X^{\varepsilon_n}) - \frac{1}{2}u_n^\top \mathbf{I}(\vartheta_n)\, u + r_{\varepsilon_n}(u_n, \vartheta_n)\right),$$

where

$$\Delta_{\varepsilon_n}(\vartheta_n, X^{\varepsilon_n}) \Longrightarrow \mathcal{N}(0, \mathbf{I}(\vartheta_0)), \qquad r_{\varepsilon_n}(u_n, \vartheta_n) \xrightarrow{\mathbf{P}^{(\varepsilon_n)}_{\vartheta_n}} 0,$$

then we say that the family is uniformly LAN (ULAN).

The following theorem was proved by J. Hájek [64].

Theorem 1.13 *Let the family* $\left(\mathbf{P}^{(\varepsilon)}_{\vartheta_0}, \vartheta \in \Theta\right)$ *satisfy the LAN condition at the point* ϑ_0 *and* $\varphi_\varepsilon \to 0$ *as* $\varepsilon \to 0$. *Then, for any estimator* $\bar\vartheta_\varepsilon$, *the inequality*

$$\lim_{\nu\to 0} \underline{\lim}_{\varepsilon\to 0} \sup_{\|\vartheta-\vartheta_0\|<\nu} \varphi_\varepsilon^{-2}\mathbf{E}_\vartheta \left\|\bar\vartheta_\varepsilon - \vartheta\right\|^2 \geq \operatorname{Tr}\left(\mathbf{I}(\vartheta_0)^{-1}\right) \tag{1.100}$$

holds.

If, moreover, $\Theta \subset \mathcal{R}$, *then the equality*

$$\lim_{\nu\to 0} \lim_{\varepsilon\to 0} \sup_{|\vartheta-\vartheta_0|<\nu} \varphi_\varepsilon^{-2}\mathbf{E}_\vartheta \left\|\bar\vartheta_\varepsilon - \vartheta\right\|^2 = \mathbf{I}(\vartheta_0)^{-1}$$

is possible if and only if the difference $\varphi_\varepsilon^{-1}\left[\bar\vartheta_\varepsilon - \vartheta_0\right] - \Delta_\varepsilon\left(\vartheta_0, X^{(\varepsilon)}\right) \to 0$ *in* $\mathbf{P}_{\vartheta_0}^{(\varepsilon)}$*-probability, as* $\varepsilon \to 0$.

For the proof, see Theorem 2.12.1 in [73].

The estimator ϑ_ε^* we call *asymptotically efficient* if the equality

$$\lim_{\nu\to 0}\lim_{\varepsilon\to 0}\sup_{\|\vartheta-\vartheta_0\|<\nu}\varphi_\varepsilon^{-2}\mathbf{E}_\vartheta\left\|\vartheta_\varepsilon^*-\vartheta\right\|^2 = \mathrm{Tr}\left(\mathbf{I}\left(\vartheta_0\right)^{-1}\right)$$

holds for all $\vartheta_0 \in \Theta$. This estimator can be called *Hájek efficient* .

We present here the elementary proof of it in the one-dimensional case ($d = 1$) for the model of observations (1.83) and (1.84), following [59]. We have to prove that for all estimators $\bar\vartheta_T$, the following lower bound holds:

$$\lim_{\nu\to 0}\varliminf_{T\to\infty}\sup_{|\vartheta-\vartheta_0|<\nu} T\mathbf{E}_\vartheta|\bar\vartheta_T-\vartheta|^2 \geq \mathrm{I}\left(\vartheta_0\right). \tag{1.101}$$

Here $\varphi_T = T^{-1/2}$, and for Fisher information, see (1.76). This proof is based on the Van Trees lower bound. The first step is the elementary inequality

$$\sup_{|\vartheta-\vartheta_0|<\nu} T\mathbf{E}_\vartheta|\bar\vartheta_T-\vartheta|^2 \geq T\int_{\vartheta_0-\nu}^{\vartheta_0+\nu}\mathbf{E}_\vartheta|\bar\vartheta_T-\vartheta|^2 p\left(\vartheta\right)\,\mathrm{d}\vartheta = T\mathbb{E}|\bar\vartheta_T-\vartheta|^2.$$

Here we introduced a probability density $p\left(\vartheta\right), \vartheta_0-\nu \leq \vartheta \leq \vartheta_0+\nu$ with finite Fisher information I_p and conditions $p\left(\vartheta_0-\nu\right) = 0, p\left(\vartheta_0+\nu\right) = 0$. As before, $\mathbb{E}$ means double mathematical expectation (see (1.78)). By the Van Trees inequality (1.99)

$$T\mathbb{E}|\bar\vartheta_T-\vartheta|^2 \geq \frac{T}{T\mathbf{E}_p\mathrm{I}\left(\vartheta\right)+\mathrm{I}_p} = \left(\int_{\vartheta_0-\nu}^{\vartheta_0+\nu}\mathrm{I}\left(\vartheta\right)p\left(\vartheta\right)\mathrm{d}\vartheta + T^{-1}\mathrm{I}_p\right)^{-1}.$$

If we take $T \to \infty$ and $\nu \to 0$, then from the continuity of $\mathrm{I}\left(\vartheta\right)$ we obtain the limits

$$\int_{\vartheta_0-\nu}^{\vartheta_0+\nu}\mathrm{I}\left(\vartheta\right)p\left(\vartheta\right)\mathrm{d}\vartheta + T^{-1}\mathrm{I}_p \xrightarrow{T\to\infty} \int_{\vartheta_0-\nu}^{\vartheta_0+\nu}\mathrm{I}\left(\vartheta\right)p\left(\vartheta\right)\mathrm{d}\vartheta \xrightarrow{\nu\to 0} \mathrm{I}\left(\vartheta_0\right).$$

This proves (1.101).

Lower Bound for Adaptive Filter

Here we present a lower bound on the mean-squared risks of all estimators of the conditional expectation for partially observed system, which we will use several times later for different particular models of observations. As we have limits like $\varepsilon \to 0$ and $T \to \infty$, we propose to express this bound in the "scheme of series" indexed by a parameter $\varepsilon \in (0,1]$ as it was done in the Sects. 1.2.4 and 1.2.5 in the cases of presenting the general results for the MLE and BE. Therefore, we suppose that there is a family of statistical experiments $\mathscr{E} = \left(\mathscr{X}^{(\varepsilon)}, \mathfrak{F}^{(\varepsilon)}, \mathbf{P}_\vartheta^{(\varepsilon)} \vartheta \in \Theta\right), \varepsilon \in (0,1]$ generated by observations $\mathscr{X}^{(\varepsilon)}$ of some partially observed system $\left(X^{(\varepsilon)}, Y^{(\varepsilon)}\right)$, where $X^{(\varepsilon)} = \left(X_t^{(\varepsilon)}, 0 \le t \le T^{(\varepsilon)}\right)$ are observations and $Y^{(\varepsilon)} = \left(Y_t^{(\varepsilon)}, 0 \le t \le T^{(\varepsilon)}\right)$ is the *hidden* process. Denote $m^{(\varepsilon)}(\vartheta, t) = \mathbf{E}_\vartheta\left(Y_t^{(\varepsilon)} | \mathfrak{F}_t^X\right)$ the conditional expectation.

For any $t \in (0, T]$, the measures $\mathbf{P}_\vartheta^{(t,\varepsilon)}, \vartheta \in \Theta$ generated by the observations $X^{(t,\varepsilon)} = \left(X_s^{(\varepsilon)}, 0 < s \le t\right)$ are supposed to be equivalent, the likelihood ratio is $L\left(\vartheta, X^{(t,\varepsilon)}\right), \vartheta \in \Theta$ and the normalized likelihood ratio

$$Z^{(t,\varepsilon)}(\vartheta_0, u) = \frac{L\left(\vartheta_0 + \varphi_\varepsilon u, X^{(t,\varepsilon)}\right)}{L\left(\vartheta_0, X^{(t,\varepsilon)}\right)}, u \in \mathbb{U}_\varepsilon, \qquad \mathbb{U}_\varepsilon = \left(u : \vartheta_0 + \varphi_\varepsilon u \in \Theta\right).$$

Introduce the condtions

1. *The conditions of the Theorem 1.11 are fulfilled.*
2. *The following normalized difference converges uniformly on compacts to some limit $S_t^\star(\vartheta)^2 > 0, \vartheta \in \Theta$*

$$\lim_{\varepsilon \to 0} \varphi_\varepsilon^{-2} \mathbf{E}_\vartheta \left[m^{(\varepsilon)}(\tilde{\vartheta}_\varepsilon, t) - m^{(\varepsilon)}(\vartheta, t)\right]^2 = S_t^\star(\vartheta)^2.$$

3. *The function $S_t^\star(\vartheta)^2, \vartheta \in \Theta$ is continuous.*

Theorem 1.14 *Suppose that the conditions 1–3 are fulfilled. Then, for any estimator $\bar{m}_{t,\varepsilon}$ of $m(\vartheta, t)$ the following lower bound holds*

$$\lim_{\nu \to 0} \varliminf_{\varepsilon \to 0} \sup_{\|\vartheta - \vartheta_0\| \le \nu} \varphi_\varepsilon^{-2} \mathbf{E}_\vartheta \left[\bar{m}_{t,\varepsilon} - m^{(\varepsilon)}(\vartheta, t)\right]^2 \ge S_t^\star(\vartheta_0)^2.$$

Proof The first steps are motivated by the proof of Theorems 1.9.1 in [73]

$$\sup_{\|\vartheta - \vartheta_0\| \le \nu} \mathbf{E}_\vartheta \left|\bar{m}_{t,\varepsilon} - m^{(\varepsilon)}(\vartheta, t)\right|^2 \ge \int_{\Theta_\nu} \mathbf{E}_\vartheta \left|\bar{m}_{t,\varepsilon} - m^{(\varepsilon)}(\vartheta, t)\right|^2 p_\nu(\vartheta)\, d\vartheta.$$

Here $\Theta_\nu = \{\vartheta : \|\vartheta - \vartheta_0\| \le \nu\}$ and $p_\nu(\vartheta), \vartheta \in \Theta_\nu$ is a positive continuous density. If we denote by $\tilde{m}_{t,\varepsilon}$ the Bayesian estimator of $m(\vartheta, t)$, which corresponds to this density $p_\nu(\cdot)$, then

$$\tilde m_{t,\varepsilon} = \int_{\Theta_\nu} m^{(\varepsilon)}(\theta, t)\, p_\nu\left(\theta | X^{(t,\varepsilon)}\right) \mathrm{d}\theta,$$

$$p_\nu\left(\theta | X^{(t,\varepsilon)}\right) = \frac{p_\nu(\theta)\, L\left(\theta, X^{(t,\varepsilon)}\right)}{\int_{\Theta_\nu} p_\nu(\theta)\, L\left(\theta, X^{(t,\varepsilon)}\right) \mathrm{d}\theta}$$

and

$$\int_{\Theta_\nu} \mathbf{E}_\vartheta \left|\bar m_{t,\varepsilon} - m^{(\varepsilon)}(\vartheta, t)\right|^2 p_\nu(\vartheta)\, \mathrm{d}\vartheta \geq \int_{\Theta_\nu} \mathbf{E}_\vartheta \left|\tilde m_{t,\varepsilon} - m^{(\varepsilon)}(\vartheta, t)\right|^2 p_\nu(\vartheta)\, \mathrm{d}\vartheta.$$

By condition 2

$$\lim_{\varepsilon \to 0} \int_{\Theta_\nu} \varphi_\varepsilon^{-2} \mathbf{E}_\vartheta \left|\tilde m_{t,\varepsilon} - m^{(\varepsilon)}(\vartheta, t)\right|^2 p_\nu(\vartheta)\, \mathrm{d}\vartheta \longrightarrow \int_{\Theta_\nu} S_t^\star(\vartheta)^2 p_\nu(\vartheta)\, \mathrm{d}\vartheta$$

and by condition 3

$$\lim_{\nu \to 0} \int_{\Theta_\nu} S_t^\star(\vartheta)^2 p_\nu(\vartheta)\, \mathrm{d}\vartheta = S_t^\star(\vartheta_0)^2 .$$

□

Remark 1.7 Let us see how $S_t^\star(\vartheta)^2$ can be calculated. Consider the example with observations (1.83) and (1.84), where $\vartheta \in \Theta = (\alpha, \beta)$. The asymptotic behavior of BE $\tilde m_t$ can be described as follows (below $\theta_u = \vartheta + \varphi_t u, \varphi_t = t^{-1/2}, \mathbb{U}_\nu = \left(\sqrt{t}\,(\vartheta_0 - \nu - \vartheta), \sqrt{t}\,(\vartheta_0 + \nu - \vartheta)\right)$)

$$\tilde m_t = \frac{\int_{\vartheta_0-\nu}^{\vartheta_0+\nu} m(\theta, t)\, p_\nu(\theta)\, L\left(\theta, X^t\right) \mathrm{d}\theta}{\int_{\vartheta_0-\nu}^{\vartheta_0+\nu} p_\nu(\theta)\, L\left(\theta, X^t\right) \mathrm{d}\theta} = \frac{\int_{\mathbb{U}_\nu} m(\theta_u, t)\, p_\nu(\theta_u)\, L\left(\theta_u, X^{(t,\varepsilon)}\right) \mathrm{d}u}{\int_{\mathbb{U}_\nu} p_\nu(\theta_u)\, L\left(\theta_u, X^t\right) \mathrm{d}u}$$

$$= m(\vartheta, t) + \varphi_t \dot m(\vartheta, t) \frac{\int_{\mathbb{U}_\nu} u\, p_\nu(\theta_u) \frac{L\left(\theta_u, X^t\right)}{L(\vartheta, X^t)} \mathrm{d}u}{\int_{\mathbb{U}_\nu} p_\nu(\theta_u) \frac{L\left(\theta_u, X^t\right)}{L(\vartheta, X^t)} \mathrm{d}u} (1 + o(1))$$

$$= m(\vartheta, t) + \varphi_t \dot m(\vartheta, t) \frac{\int_{\mathbb{U}_\nu} u p_\nu(\vartheta)\, Z_t(\vartheta, u)\, \mathrm{d}u}{\int_{\mathbb{U}_\nu} p_\nu(\vartheta)\, Z_t(\vartheta, u)\, \mathrm{d}u} (1 + o(1)) .$$

Hence,

$$\sqrt{t}\left(\tilde{m}_t - m(\vartheta, t)\right) = \dot{m}(\vartheta, t)\frac{\displaystyle\int_{\mathbb{U}_\nu} u\ Z_t(\vartheta, u)\, \mathrm{d}u}{\displaystyle\int_{\mathbb{U}_\nu} Z_t(\vartheta, u)\, \mathrm{d}u}(1 + o(1))$$

$$= \dot{m}(\vartheta, t)\frac{\Delta_t\left(\vartheta, X^t\right)}{\mathrm{I}(\vartheta)}(1 + o(1)).$$

Here

$$\Delta_t\left(\vartheta, X^t\right) = \frac{1}{\sigma\sqrt{t}}\int_0^t \dot{M}(\vartheta, s)\, \mathrm{d}\bar{W}_s \Longrightarrow \Delta(\vartheta) \sim \mathcal{N}\big(0, \mathrm{I}(\vartheta)\big). \tag{1.102}$$

Recall that

$$Z_t(\vartheta, u) \Longrightarrow Z(\vartheta, u) = \exp\left(u\Delta(\vartheta) - \frac{u^2}{2}\mathrm{I}(\vartheta)\right), \quad \frac{\displaystyle\int_{\mathcal{R}} uZ(\vartheta, u)\, \mathrm{d}u}{\displaystyle\int_{\mathcal{R}} Z(\vartheta, u)\, \mathrm{d}u} = \frac{\Delta(\vartheta)}{\mathrm{I}(\vartheta)}.$$

We have to study the convergence of moments

$$t\mathbf{E}_\vartheta\left(\tilde{m}_t - m(\vartheta, t)\right)^2 \approx \lim_{t\to\infty} t\mathbf{E}_\vartheta\left[\dot{m}(\vartheta, t)^2\left(\tilde{\vartheta}_t - \vartheta\right)^2\right]. \tag{1.103}$$

Therefore, we have to obtain the asymptotic representation of the derivative $\dot{m}(\vartheta, t)$. Note that in the Chap. 6 we obtain the relations

$$\dot{m}(\vartheta, t) = -\left(\dot{a}(\vartheta) + \frac{f(\vartheta)\dot{f}(\vartheta)\gamma_*(\vartheta)}{\sigma^2}\right)\int_0^t \Phi(\vartheta, s, t)\, m(\vartheta, s)\, \mathrm{d}s$$
$$+ \frac{\dot{f}(\vartheta)\gamma_*(\vartheta) + f(\vartheta)\dot{\gamma}_*(\vartheta)}{\sigma}\int_0^t \Phi(\vartheta, s, t)\, \mathrm{d}\bar{W}_s + o(1) \tag{1.104}$$

and

$$m(\vartheta, t) = \frac{f(\vartheta)\gamma_*(\vartheta)}{\sigma}\int_0^t \phi(\vartheta, s, t)\, \mathrm{d}\bar{W}_s + o(1) \tag{1.105}$$

Having the representations (1.102), (1.104) and (1.105) and the convergence (1.103) the value of $\mathcal{S}^\star(\vartheta)^2$ can be calculated.

Of course, what is written above is not a real proof but just a sketch of such proof. The detailed proof contains the technical tools developed in [73] in the study of behavior of Bayesian estimators.

Definition 1.2 The estimator $m^{\circ}_{t,T}$, is called *asymptotically efficient* if for all $\vartheta_0 \in \Theta$, $t = vT$ and any $v \in [\varepsilon_0, 1]$, $\varepsilon_0 \in (0, 1)$

$$\lim_{\nu\to 0}\lim_{T\to\infty}\sup_{|\vartheta-\vartheta_0|\le\nu} t\mathbf{E}_{\vartheta}\left|m^{\circ}_{t,T} - m(\vartheta, t)\right|^2 = \mathcal{S}^{\star}(\vartheta_0)^2 .$$

To construct the asymptotically efficient estimator of $m(\vartheta_0, t)$ we have to write the solution $m(\vartheta, t)$ of (1.85) without stochastic integral (using (1.33)) and then to substitute there the one-step MLE-process $\vartheta^{\star}_{t,T}$. Recall that $\vartheta^{\star}_{t,T}$ is asymptotically equivalent to the MLE $\hat{\vartheta}_t$ and to the BE $\tilde{\vartheta}_t$ and has the same asymptotic representation.

Remark 1.8 Consider the partially observed system (1.83)–(1.84) and suppose that we already have the solutions $m(\vartheta, t), 0 \le t \le T, \vartheta \in \Theta$ of the Eq. (1.85) and suppose as well that the BE $\tilde{\vartheta}_t$ is asymptotically efficient:

$$\lim_{\nu\to 0}\lim_{t\to\infty}\sup_{|\vartheta-\vartheta_0|\le\nu} T^{-1}\mathbf{E}_{\vartheta}\left|\tilde{\vartheta}_t - \vartheta_0\right|^2 = \mathrm{I}(\vartheta_0)^{-1} .$$

The substitution of this estimator in $m(\vartheta, t)$ leads to the expression $m(\tilde{\vartheta}_t, t)$, and of course, this leads to the problem with the stochastic integral. The calculation of $m(\tilde{\vartheta}_t, t)$ can be realized if we write

$$m(\tilde{\vartheta}_t, t) = \frac{\int_{\Theta} m(\vartheta, t)\, p(\vartheta)\, L\left(\vartheta, X^t\right) \mathrm{d}\vartheta}{\int_{\Theta} p(\vartheta)\, L\left(\vartheta, X^t\right) \mathrm{d}\vartheta}.$$

This expression can be used to calculate the limit bound, but for construction of the asymptotically efficient adaptive filter, we can use the asymptotic equivalence of the one-step MLE-process, MLE $\hat{\vartheta}_t$ and BE $\tilde{\vartheta}_t$.

Chapter 2
Small Noise in Both Equations

Several models of partially observed diffusion processes with small diffusion coefficients of the same order in the state and observation equations are considered. These models are assumed to depend on some finite-dimensional unknown parameters, and the main problem is the construction of adaptive filters. These constructions are realized in several steps. First, by observations on a relatively small learning interval, the preliminary consistent estimators are proposed, then these estimarors are used for the construction of the One-step MLE-process. In the last step, the adaptive filter is obtained by the substitution of this estimator process in the equations of filtration. In the first section, the initial value is supposed to be known. In the second section, the initial values are estimated in different situations. Then the same program of adaptive filtering is realized in the third section for conditionally Gaussian systems. The nonlinear partially observed systems are studied in the fourth section. Using two approximations of these systems, some heuristic adaptive filters are proposed. In the last section, the partially observed linear system with different order small noises in the observation and state equations is studied. It is supposed that the system depends on some unknown parameters and the adaptive filters are costructed.

2.1 Linear System

2.1.1 Model of Observations

Let us consider the partially observed linear system of equations

$$\mathrm{d}X_t = f(\vartheta, t)\, Y_t\, \mathrm{d}t + \varepsilon\sigma(t)\, \mathrm{d}W_t, \qquad X_0 = 0, \qquad 0 \le t \le T, \tag{2.1}$$

$$\mathrm{d}Y_t = a(\vartheta, t)\, Y_t\, \mathrm{d}t + \varepsilon b(\vartheta, t)\, \mathrm{d}V_t, \qquad Y_0, \qquad 0 \le t \le T. \tag{2.2}$$

Y. A. Kutoyants, *Hidden Markov Processes and Adaptive Filtering*, Springer Series in Statistics, https://doi.org/10.1007/978-3-032-00052-1_2

where W_t and V_t, $0 \leq t \leq T$, are independent Wiener processes. The parameter $\varepsilon \in (0, 1]$ is known, and the parameter $\vartheta \in \Theta$ is supposed to be unknown. The observations are $X^T = (X_t, 0 \leq t \leq T)$, and the process $Y^T = (Y_t, 0 \leq t \leq T)$ is "hidden". The set $\Theta \subset \mathbb{R}^d$, with $d \geq 1$, is open, convex, and bounded. The functions $f(\vartheta, t)$, $\sigma(t)$, $a(\vartheta, t)$, and $b(\vartheta, t)$, for $0 \leq t \leq T$ and $\vartheta \in \Theta \subset \mathbb{R}^d$, are assumed to be known. For simplicity of exposition, we replace the conditions (1.29) and (1.30) with the following condition $\mathscr{A}_*$.

Condition $\mathscr{A}_*$. *The functions* $f(\vartheta, t)$, $a(\vartheta, t)$, $b(\vartheta, t)$, $\sigma(t)$, *for* $t \in [0, T]$ *and* $\vartheta \in \bar{\Theta}$, *are bounded, and* $\sigma(t)^2 \geq c > 0$ *for* $t \in [0, T]$.

The problems to solve are:

- *Estimation of the parameter* ϑ.
- *Construction of an adaptive filter.*
- *Analysis of the corresponding errors in the asymptotics* $\varepsilon \to 0$.

The Kalman–Bucy filter (see (1.17) and (1.18)) is given by the following couple of equations:

$$\mathrm{d}m(\vartheta, t) = a(\vartheta, t)m(\vartheta, t)\,\mathrm{d}t + \frac{f(\vartheta, t)\gamma(\vartheta, t)}{\varepsilon^2\sigma(t)^2}\left[\mathrm{d}X_t - f(\vartheta, t)m(\vartheta, t)\,\mathrm{d}t\right], \tag{2.3}$$

$$\frac{\partial\gamma(\vartheta, t)}{\partial t} = 2a(\vartheta, t)\gamma(\vartheta, t) - \frac{f(\vartheta, t)^2\gamma(\vartheta, t)^2}{\varepsilon^2\sigma(t)^2} + \varepsilon^2 b(\vartheta, t)^2, \quad 0 \leq t \leq T. \tag{2.4}$$

with initial values: $m(\vartheta, 0) = \mathbf{E}_\vartheta(Y_0|X_0)$ and $\gamma(\vartheta, 0) = \mathbf{E}_\vartheta(Y_0 - m(\vartheta, 0))^2$. The initial value Y_0 is non-random, $Y_0 = y_0 \neq 0$, and known. Therefore, $m(\vartheta, 0) = y_0$ and $\gamma(\vartheta, 0) = 0$. The cases of random Y_0 and unknown y_0 will be discussed below in Sect. 2.2.

Set $\gamma_*(\vartheta, t) = \varepsilon^{-2}\gamma(\vartheta, t)$. Then these equations become

$$\mathrm{d}m(\vartheta, t) = a(\vartheta, t)m(\vartheta, t)\,\mathrm{d}t + \frac{f(\vartheta, t)\gamma_*(\vartheta, t)}{\sigma(t)^2}\left[\mathrm{d}X_t - f(\vartheta, t)m(\vartheta, t)\,\mathrm{d}t\right], \tag{2.5}$$

$$\frac{\partial\gamma_*(\vartheta, t)}{\partial t} = 2a(\vartheta, t)\gamma_*(\vartheta, t) - \frac{f(\vartheta, t)^2\gamma_*(\vartheta, t)^2}{\sigma(t)^2} + b(\vartheta, t)^2, \quad 0 \leq t \leq T. \tag{2.6}$$

with the same initial values: $m(\vartheta, 0) = y_0 \neq 0$ and $\gamma_*(\vartheta, 0) = 0$. We see that the Eqs. (2.5) and (2.6) do not depend directly on ε, but the process X^T does. The true value is denoted as ϑ_0.

We need several technical results. Introduce the following notations:

$$\phi(\vartheta, s, t) = \exp\left(\int_s^t a(\vartheta, v)\, dv\right),$$

$$B(\vartheta, t) = \frac{f(\vartheta, t)\, \gamma_*(\vartheta, t)}{\sigma(t)^2},$$

$$\tilde{\eta}_t(\vartheta) = \int_0^t \phi(\vartheta, s, t)\, B(\vartheta, s)\, \sigma(s)\, d\bar{W}_s,$$

$$A(\vartheta, t) = a(\vartheta, t) - \frac{f(\vartheta, t)^2\, \gamma_*(\vartheta, t)}{\sigma(t)^2},$$

$$\Phi(\vartheta, s, t) = \exp\left(\int_s^t A(\vartheta, v)\, dv\right),$$

$$\hat{\eta}_t(\vartheta) = \int_0^t \Phi(\vartheta, s, t)\, B(\vartheta, s)\, \sigma(s)\, d\bar{W}_s,$$

$$y_t(\vartheta, \vartheta_0) = y_0\, \Phi(\vartheta, 0, t) + \int_0^t \Phi(\vartheta, s, t)\, B(\vartheta, s)\, f(\vartheta_0, s)\, y_s(\vartheta_0)\, ds,$$

$$y_t(\vartheta) = y_0\, \phi(\vartheta, 0, t).$$

Recall that by Lemma 1.2

$$\begin{aligned} dm(\vartheta, t) &= a(\vartheta, t) m(\vartheta, t)\, dt + \varepsilon \frac{f(\vartheta, t)\gamma_*(\vartheta, t)}{\sigma(t)}\, d\bar{W}_t \\ &\quad + \frac{f(\vartheta, t)\gamma_*(\vartheta, t)}{\sigma(t)^2} \left[f(\vartheta_0, t) m(\vartheta_0, t) - f(\vartheta, t) m(\vartheta, t)\right] dt \\ &= A(\vartheta, t) m(\vartheta, t)\, dt + B(\vartheta, t) f(\vartheta_0, t) m(\vartheta_0, t)\, dt + \varepsilon B(\vartheta, t)\sigma(t)\, d\bar{W}_t. \end{aligned} \tag{2.7}$$

Lemma 2.1 *Let the conditions $\mathscr{A}_*$ be satisfied. Then*

$$m(\vartheta_0, t)\,|_{\varepsilon=0} = y_t(\vartheta_0), \quad m(\vartheta, t)|_{\varepsilon=0} = y_t(\vartheta, \vartheta_0),$$

$$m(\vartheta_0, t) - y_t(\vartheta_0) = \varepsilon\, \tilde{\eta}_t(\vartheta_0), \tag{2.8}$$

$$\begin{aligned} m(\vartheta_0, t) - Y_t &= -\varepsilon \int_0^t \phi(\vartheta_0, s, t)\, b(\vartheta_0, s)\, dV_s \\ &\quad + \varepsilon \int_0^t \phi(\vartheta_0, s, t)\, B(\vartheta_0, s)\, \sigma(s)\, d\bar{W}_s, \end{aligned}$$

$$m(\vartheta, t) - y_t(\vartheta, \vartheta_0) = \varepsilon\, \hat{\eta}_t(\vartheta) + \varepsilon \int_0^t \Phi(\vartheta, s, t)\, B(\vartheta, s)\, f(\vartheta_0, s)\, \tilde{\eta}_s(\vartheta_0)\, ds. \tag{2.9}$$

Proof The relations (2.8)–(2.9) follow directly from the expression of the solution of the linear equation (2.7)

$$m(\vartheta,t)=y_0\,\Phi(\vartheta,0,t)+\int_0^t\Phi(\vartheta,s,t)\,B(\vartheta,s)\,f(\vartheta_0,s)\,m(\vartheta_0,s)\mathrm{d}s$$
$$+\,\varepsilon\int_0^t\Phi(\vartheta,s,t)\,B(\vartheta,s)\,\sigma(s)\mathrm{d}\bar W_s. \qquad \square$$

2.1.2 Estimators

The model of observations is (2.1) and (2.2). Consider the problem of estimation $\vartheta\in\Theta$ by observations $X^T=(X_t,0\le t\le T)$. The set $\Theta\subset\mathcal{R}^d, d\ge 1$ is open, convex, and bounded. The properties of estimators we describe in the asymptotic of small noise: $\varepsilon\to 0$. We suppose that the condition $\mathscr{A}_*$ is always fulfilled and discuss several methods of estimation.

TFE

Let us study the *trajectory fitting estimator* (TFE) $\check\vartheta_\varepsilon$ for this model of observations. Introduce the further notations:

$$\check X_t(\vartheta)=\int_0^t f(\vartheta,s)\,m(\vartheta,s)\,\mathrm{d}s,\qquad x_t(\vartheta_0)=\int_0^t f(\vartheta_0,s)\,y_s(\vartheta_0)\,\mathrm{d}s,$$

$$\check x_t(\vartheta,\vartheta_0)=\int_0^t f(\vartheta,s)\,y_s(\vartheta,\vartheta_0)\,\mathrm{d}s,\qquad \dot{\check x}_t(\vartheta_0,\vartheta_0)=\left.\frac{\partial\check x_t(\vartheta,\vartheta_0)}{\partial\vartheta}\right|_{\vartheta=\vartheta_0},$$

$$\check g(\vartheta_0,\nu)^2=\inf_{|\vartheta-\vartheta_0|>\nu}\int_0^T\left[\check x_t(\vartheta,\vartheta_0)-x_t(\vartheta_0)\right]^2\mathrm{d}t,\qquad \eta_t=\int_0^t\sigma(s)\,\mathrm{d}\bar W_s.$$

By Innovation Lemma 1.2 the observed process can be written as follows:

$$X_t=\int_0^t f(\vartheta_0,s)\,m(\vartheta_0,s)\,\mathrm{d}s+\varepsilon\int_0^t\sigma(s)\,\mathrm{d}\bar W_s=\check X_t(\vartheta_0)+\varepsilon\eta_t.$$

The TFE $\check\vartheta_\varepsilon$ is introduced by the relation

$$\int_0^T\left[X_t-\check X_t(\check\vartheta_\varepsilon)\right]^2\mathrm{d}t=\inf_{\vartheta\in\Theta}\int_0^T\left[X_t-\check X_t(\vartheta)\right]^2\mathrm{d}t.$$

Note that by Lemma 2.1, we have

$$m(\vartheta_0,t)=y_t(\vartheta_0)+\varepsilon\int_0^t \phi(\vartheta_0,s,t)\,B(\vartheta_0,s)\sigma(s)\mathrm{d}\bar W_s,$$

$$\breve X_t(\vartheta)=\breve x_t(\vartheta,\vartheta_0)+O(\varepsilon),\qquad\qquad m(\vartheta,t)=y_t(\vartheta,\vartheta_0)+O(\varepsilon).$$

The values of $O(\varepsilon)$ here can be written explicitly.

Define the vectors

$$\mathrm{Q}(\vartheta_0,s)=\int_s^T \dot{\breve{\mathrm{x}}}_t(\vartheta_0,\vartheta_0)\,\mathrm{d}t,\qquad \breve\zeta(\vartheta_0)=\breve{\mathbf{T}}(\vartheta_0)^{-1}\int_0^T \dot{\breve{\mathrm{x}}}_t(\vartheta_0,\vartheta_0)\,\eta_t\mathrm{d}t,$$

and two matrices

$$\breve{\mathbf{T}}(\vartheta_0)=\int_0^T \dot{\breve{\mathrm{x}}}_t(\vartheta_0,\vartheta_0)\,\dot{\breve{\mathrm{x}}}_t(\vartheta_0,\vartheta_0)^\top\,\mathrm{d}t,$$

$$\breve{\mathbf{D}}(\vartheta_0)=\breve{\mathbf{T}}(\vartheta_0)^{-1}\int_0^T \mathrm{Q}(\vartheta_0,s)\,\mathrm{Q}(\vartheta_0,s)^\top\sigma(s)^2\,\mathrm{d}s\,\breve{\mathbf{T}}(\vartheta_0)^{-1}.$$

Introduce the *Conditions* $\breve{\mathscr{B}}$.

$\breve{\mathscr{B}}_1$. *The functions* $f(\vartheta,t), a(\vartheta,t), b(\vartheta,t)$ *have two continuous bounded derivatives on* $\bar\Theta$.

$\breve{\mathscr{B}}_2$. The matrix $\breve{\mathbf{T}}(\vartheta_0)$ *is uniformly nondegenerate*

$$\inf_{\vartheta_0\in\Theta}\ \inf_{\|\mathrm{e}\|=1,\mathrm{e}\in\mathcal{R}^d}\ \mathrm{e}^\top\breve{\mathbf{T}}(\vartheta_0)\,\mathrm{e}>0.$$

$\breve{\mathscr{B}}_3$. *For all* $\nu>0$

$$\inf_{\vartheta_0\in\Theta}\breve g(\vartheta_0,\nu)>0.$$

Theorem 2.1 *Let the conditions* $\breve{\mathscr{B}}$ *hold. Then the TFE* ϑ_ε^* *is uniformly on compacts* $\mathbb{K}\subset\Theta$ *consistent, asymptotically normal*

$$\frac{\breve\vartheta_\varepsilon-\vartheta_0}{\varepsilon}\Longrightarrow\breve\zeta(\vartheta_0)\sim\mathcal{N}\left(0,\breve{\mathbf{D}}(\vartheta_0)\right),$$

and the moments converge: for any $p\geq 2$

$$\varepsilon^{-p}\mathbf{E}_{\vartheta_0}\|\breve\vartheta_\varepsilon-\vartheta_0\|^p\to\mathbf{E}_{\vartheta_0}\|\breve\zeta(\vartheta_0)\|^p.$$

Proof Below, $\|h_t\|_2$ and $\langle h_t,q_t\rangle_2$ denote the norm and scalar product in $\mathcal{L}^2(0,T)$.

Consistency. For any $\nu>0$, we can write

$$
\begin{aligned}
&\sup_{\vartheta_0\in\mathbb{K}}\mathbf{P}_{\vartheta_0}\left(\|\check\vartheta_\varepsilon-\vartheta_0\|>\nu\right)\\
&\quad=\sup_{\vartheta_0\in\mathbb{K}}\mathbf{P}_{\vartheta_0}\left(\inf_{\|\vartheta-\vartheta_0\|<\nu}\left\|X_t-\check X_t\left(\vartheta\right)\right\|_2>\inf_{|\vartheta-\vartheta_0|\geq\nu}\left\|X_t-\check X_t\left(\vartheta\right)\right\|_2\right)\\
&\quad\leq\sup_{\vartheta_0\in\mathbb{K}}\mathbf{P}_{\vartheta_0}\left(\inf_{\|\vartheta-\vartheta_0\|<\nu}\left(\left\|X_t-\check X_t\left(\vartheta_0\right)\right\|_2+\left\|\check X_t\left(\vartheta\right)-\check X_t\left(\vartheta_0\right)\right\|_2\right)\right.\\
&\qquad\left.>\inf_{\|\vartheta-\vartheta_0\|\geq\nu}\left(\left\|\check X_t\left(\vartheta\right)-\check X_t\left(\vartheta_0\right)\right\|_2-\left\|X_t-\check X_t\left(\vartheta_0\right)\right\|_2\right)\right)\\
&\quad=\sup_{\vartheta_0\in\mathbb{K}}\mathbf{P}_{\vartheta_0}\left(2\left\|X_t-\check X_t\left(\vartheta_0\right)\right\|_2>\inf_{|\vartheta-\vartheta_0|\geq\nu}\left\|\check X_t\left(\vartheta\right)-\check X_t\left(\vartheta_0\right)\right\|_2\right)\\
&\quad=\sup_{\vartheta_0\in\mathbb{K}}\mathbf{P}_{\vartheta_0}\left(2\varepsilon\left\|\eta_t\right\|_2>\inf_{|\vartheta-\vartheta_0|\geq\nu}\left\|\check x_t\left(\vartheta,\vartheta_0\right)-x_t\left(\vartheta_0\right)\right\|_2+O\left(\varepsilon\right)\right)\\
&\quad\leq\sup_{\vartheta_0\in\mathbb{K}}\mathbf{P}_{\vartheta_0}\bigl(2\varepsilon\left\|\eta_t\right\|_2+O\left(\varepsilon\right)>g\left(\vartheta_0,\nu\right)\bigr)\leq\frac{\varepsilon^N\left(C+2^NK_N\right)}{\inf_{\vartheta_0\in\mathbb{K}}\breve g\left(\vartheta_0,\nu\right)^N}\rightarrow0,
\end{aligned}
\tag{2.10}
$$

where K_N is from the estimate (see (1.1))

$$
\begin{aligned}
\mathbf{E}_{\vartheta_0}\left\|\eta_t\right\|^N&=\mathbf{E}_{\vartheta_0}\left|\int_0^T\left(\int_0^t\sigma\left(s\right)\mathrm{d}\bar W_s\right)^2\mathrm{d}t\right|^M\\
&\leq T^{M-1}\int_0^T\mathbf{E}_{\vartheta_0}\left(\int_0^t\sigma\left(s\right)\mathrm{d}\bar W_s\right)^{2M}\mathrm{d}t\\
&\leq\left[M\left(2M-1\right)\right]^MT^{M-1}\int_0^Tt^{M-1}\left(\int_0^t\sigma\left(s\right)^{2M}\mathrm{d}s\right)\mathrm{d}t=K_N.
\end{aligned}
$$

Here $M=N/2$.

Asymptotic normality. The *trajectory fitting equation* (TFEq) in this problem is the following system of equations:

$$
\int_0^T\left[X_t-\check X_t(\check\vartheta_\varepsilon)\right]\dot{\check X}_t(\check\vartheta_\varepsilon)\mathrm{d}t=0.
$$

Note that

$$
\frac{\partial}{\partial\vartheta}\check X_t\left(\vartheta\right)\Bigr|_{\vartheta=\vartheta_0}=\int_0^t\left[\dot{\mathrm{f}}\left(\vartheta_0,s\right)m\left(\vartheta_0,s\right)+f\left(\vartheta_0,s\right)\dot{\mathrm{m}}\left(\vartheta_0,s\right)\right]\mathrm{d}s
$$

$$= \int_0^t \left[\dot{\mathrm{f}}(\vartheta_0, s)\, y_s(\vartheta_0) + f(\vartheta_0, s)\, \dot{\mathrm{y}}_s(\vartheta_0, \vartheta_0)\right] \mathrm{d}s + O(\varepsilon)$$

$$= \check{\dot{\mathrm{x}}}_t(\vartheta_0, \vartheta_0) + O(\varepsilon).$$

The expansion (below $\vartheta_r = \vartheta_0 + r(\check{\vartheta}_\varepsilon - \vartheta_0)$)

$$\check{X}_t(\check{\vartheta}_\varepsilon) - \check{X}_t(\vartheta_0) = \int_0^1 \check{\dot{\mathrm{X}}}_t(\vartheta_r)^\top \mathrm{d}r(\check{\vartheta}_\varepsilon - \vartheta_0)$$

allows us to write

$$\int_0^T \left[X_t - \check{X}_t(\vartheta_0) + \check{X}_t(\vartheta_0) - \check{X}_t(\check{\vartheta}_\varepsilon)\right] \check{\dot{\mathrm{X}}}_t(\check{\vartheta}_\varepsilon)\mathrm{d}t$$

$$= \varepsilon \int_0^T \eta_t\, \check{\dot{\mathrm{X}}}_t(\check{\vartheta}_\varepsilon)\mathrm{d}t - \int_0^1 \int_0^T \check{\dot{\mathrm{X}}}_t(\vartheta_\varepsilon)\, \check{\dot{\mathrm{X}}}_t(\check{\vartheta}_r)^\top \mathrm{d}t\, \mathrm{d}r(\check{\vartheta}_\varepsilon - \vartheta_0)$$

$$= \varepsilon \int_0^T \eta_t\, \check{\dot{\mathrm{x}}}_t(\vartheta_0)\, \mathrm{d}t - \int_0^T \check{\dot{\mathrm{x}}}_t(\vartheta_0, \vartheta_0)\, \check{\dot{\mathrm{x}}}_t(\vartheta_0, \vartheta_0)^\top \mathrm{d}t(\check{\vartheta}_\varepsilon - \vartheta_0) + o(\varepsilon) = 0.$$

Hence,

$$\varepsilon^{-1}(\check{\vartheta}_\varepsilon - \vartheta_0) = \check{\mathbf{I}}(\vartheta_0)^{-1} \int_0^T \eta_t\, \check{\dot{\mathrm{x}}}_t(\vartheta_0)\, \mathrm{d}t + o(1).$$

The value $\check{\mathbf{D}}(\vartheta_0)$ can be obtained from this formula by changing the order of integration

$$\int_0^T \eta_t\, \check{\dot{\mathrm{x}}}_t(\vartheta_0, \vartheta_0)\, \mathrm{d}t = \int_0^T \left(\int_0^t \sigma(s)\, \mathrm{d}W_s\right) \check{\dot{\mathrm{x}}}_t(\vartheta_0, \vartheta_0)\, \mathrm{d}t$$

$$= \int_0^T \left(\int_s^T \check{\dot{\mathrm{x}}}_t(\vartheta_0, \vartheta_0)\, \mathrm{d}t\right) \sigma(s)\, \mathrm{d}W_s$$

$$\sim \mathcal{N}\left(0, \int_0^T \mathrm{Q}(\vartheta_0, s)\, \mathrm{Q}(\vartheta_0, s)^\top \sigma(s)^2\, \mathrm{d}s\right).$$

Convergence of moments. Consider the function

$$G(\vartheta, \vartheta_0) = \int_0^T [x_t(\vartheta_0) - x_t(\vartheta, \vartheta_0)]^2\, \mathrm{d}t.$$

We can write for $\|\vartheta - \vartheta_0\| \leq \nu$

$$G(\vartheta,\vartheta_0)=(\vartheta-\vartheta_0)^\top\breve{\mathbf{T}}(\vartheta_0)(\vartheta-\vartheta_0)+O(\nu)\geq\frac{1}{2}(\vartheta-\vartheta_0)^\top\breve{\mathbf{T}}(\vartheta_0)(\vartheta-\vartheta_0)$$
$$\geq\frac{1}{2}\|\vartheta-\vartheta_0\|^2\inf_{\vartheta_0\in\Theta}\mathrm{e}^\top\breve{\mathbf{T}}(\vartheta_0)\,\mathrm{e}\geq c_1\|\vartheta-\vartheta_0\|^2.$$

For $\|\vartheta-\vartheta_0\|>\nu$ we have

$$G(\vartheta,\vartheta_0)\geq g(\vartheta_0,\nu)^2\geq g(\vartheta_0,\nu)^2\frac{\|\vartheta-\vartheta_0\|^2}{\sup_{\vartheta,\vartheta_1\in\Theta}\|\vartheta-\vartheta_1\|^2}\geq c_2\|\vartheta-\vartheta_0\|^2.$$

Therefore, for all $\vartheta\in\Theta$

$$\int_0^T[x_t(\vartheta_0)-x_t(\vartheta,\vartheta_0)]^2\,\mathrm{d}t\geq c_*\|\vartheta-\vartheta_0\|^2,$$

where $c_*=c_1\wedge c_2$. From this estimate and (2.10) we obtain: for any $L>0$ and any $N>0$

$$\sup_{\vartheta_0\in\mathbb{K}}\mathbf{P}_{\vartheta_0}\left(\frac{\|\breve{\vartheta}_\varepsilon-\vartheta_0\|}{\varepsilon}>L\right)\leq\frac{\varepsilon^N\left(C_N+2^NK_N\right)}{\inf_{\vartheta_0\in\mathbb{K}}g(\vartheta_0,\varepsilon L)^N}\leq\frac{C}{L^N}. \tag{2.11}$$

Introduce the function

$$F(\vartheta_0,u)=\mathbf{P}_{\vartheta_0}\left(\frac{\|\breve{\vartheta}_\varepsilon-\vartheta_0\|}{\varepsilon}<u\right),\qquad u>0.$$

For any $P>1$ and $N>P+1$ (2.11) implies

$$\varepsilon^{-P}\mathbf{E}_{\vartheta_0}\left|\|\breve{\vartheta}_\varepsilon-\vartheta_0\|\right|^P=\int_0^\infty u^P\mathrm{d}F(\vartheta_0,u)=1-\int_1^\infty u^P\mathrm{d}(1-F(\vartheta_0,u))$$
$$\leq 2+P\int_1^\infty u^{P-1}(1-F(\vartheta_0,u))\,\mathrm{d}u$$
$$\leq 2+CP\int_1^\infty\frac{u^{P-1}}{u^N}\mathrm{d}u\leq C.$$

Therefore, the random variables $\|\breve{u}_\varepsilon\|^P=\varepsilon^{-P}\|\breve{\vartheta}_\varepsilon-\vartheta_0\|^P$ are uniformly integrable, and we have the convergence of moments. □

In the future, to verify the convergence of moments, we will just obtain the estimates like (2.11).

Example 2.1 Consider the simple model

$$
\begin{aligned}
\mathrm{d}X_t &= \theta_1 Y_t\,\mathrm{d}t + \varepsilon\,\mathrm{d}W_t, \qquad X_0 = 0, \qquad 0 \le t \le T,\\
\mathrm{d}Y_t &= \theta_2 Y_t\,\mathrm{d}t + \varepsilon\,\mathrm{d}V_t, \qquad Y_0 = y_0 \neq 0.
\end{aligned}
$$

In this case, we have $\vartheta = (\theta_1, \theta_2)^\top$ and

$$
r(\vartheta) = \|\vartheta\|, \qquad \gamma_*(\vartheta) = \frac{\|\vartheta\| - \theta_2}{\theta_1^2}, \qquad \gamma_*(\vartheta, t) = \gamma_*(\vartheta)\tanh\big(r(\vartheta)\,t\big),
$$

$$
\Phi(\vartheta, s, t) = \exp\left(\int_s^t \Big[\theta_2 - \theta_1^2\gamma_*(\vartheta, v)\Big]\mathrm{d}v\right), \qquad x_t(\vartheta) = \frac{\theta_1 y_0}{\theta_2}\left[e^{\theta_2 t} - 1\right],
$$

$$
y_t(\theta_2) = y_0 e^{\theta_2 t}, \qquad y_t(\vartheta, \vartheta_0) = y_0\Phi(\vartheta, 0, t) + \theta_1\int_0^t \Phi(\vartheta, s, t)\,\gamma_*(\vartheta, s)\,\mathrm{d}s
$$

These expressions are obtained directly from (1.25) and (1.26). The filter is

$$
\mathrm{d}m(\vartheta, t) = \Big[\theta_2 - \theta_1^2\gamma_*(\vartheta, t)\Big]m(\vartheta, t)\,\mathrm{d}t + \theta_1\gamma_*(\vartheta, t)\,\mathrm{d}X_t
$$

and the TFE is

$$
\breve{\vartheta}_\varepsilon = \arg\inf_{\vartheta\in\Theta}\int_0^T\left[X_t - \theta_1\int_0^t m(\vartheta, s)\,\mathrm{d}s\right]^2\mathrm{d}s.
$$

Let us check the consistency of $\breve{\vartheta}_\varepsilon$. The only condition to verify is the identifiability condition $\breve{\mathscr{B}}_3$ of the Theorem 2.1. Suppose that there exists $\vartheta^* = \left(\theta_1^*, \theta_2^*\right)^\top \neq \vartheta_0 = \left(\theta_{0,1}, \theta_{0,2}\right)^\top$ such that

$$
\int_0^T \left[x_t(\vartheta_0) - \breve{x}_t(\vartheta^*, \vartheta_0)\right]^2\mathrm{d}t = 0.
$$

Then $x_t(\vartheta_0) \equiv \breve{x}_t(\vartheta^*, \vartheta_0)$, and this implies $\theta_{0,1}y_t\left(\theta_{0,2}\right) \equiv \theta_1 y_t(\vartheta^*, \vartheta_0)$. Let us denote $z_t^\circ = \theta_{0,1}y_t\left(\theta_{0,2}\right)$ and $z_t = \theta_1 y_t(\vartheta^*, \vartheta_0)$ and show that $z_t^\circ \not\equiv z_t$. The equations for these functions are

$$
\begin{aligned}
\frac{\mathrm{d}z_t^\circ}{\mathrm{d}t} &= \theta_{0,2}z_t^\circ, \qquad z_0^\circ = \theta_{0,1}y_0,\\
\frac{\mathrm{d}z_t}{\mathrm{d}t} &= \theta_2^* z_t - \theta_1^{*2}\gamma_*\left(\vartheta^*, t\right)\left(z_t - z_t^\circ\right), \qquad z_0 = \theta_1^* y_0.
\end{aligned}
$$

From these equations, we obtain the relation

$$
z_t - z_t^\circ = y_0\left(\theta_1^* - \theta_{0,1}\right)\psi\left(\vartheta^*, \theta_{0,2}, 0, t\right) + \left(\theta_2^* - \theta_{0,2}\right)\int_0^t \psi\left(\vartheta^*, s, t\right) z_s\mathrm{d}s,
$$

where

$$\psi\left(\vartheta^*,\theta_{0,2},s,t\right)=\exp\left(\int_s^t\left[\theta_{0,2}-\theta_1^{*2}\gamma_*\left(\vartheta^*,v\right)\mathrm{d}v\right]\right).$$

Therefore, if $z_t\equiv z_t^\circ$, then

$$y_0\left(\theta_1^*-\theta_{0,1}\right)=\left(\theta_{0,2}-\theta_2^*\right)\int_0^t\psi\left(\vartheta^*,\theta_{0,2},0,s\right)^{-1}z_s\mathrm{d}s.$$

If $\vartheta^*\neq\vartheta_0$, then this equality is impossible, and we obtain $\breve g\left(\vartheta_0,\nu\right)>0$.

MDE

Remind that the process $\breve X_t\left(\vartheta\right)$ converges to the function $\breve x_t\left(\vartheta,\vartheta_0\right)$. Direct calculation gives $\breve y_t\left(\vartheta,\vartheta\right)=y_t\left(\vartheta\right)$ and $\breve x_t\left(\vartheta,\vartheta\right)=x_t\left(\vartheta\right)$.

Introduce the MDE $\breve\vartheta_\varepsilon$ by the equation

$$\left\|X_t-x_t\left(\breve\vartheta_\varepsilon\right)\right\|_2=\inf_{\vartheta\in\Theta}\left\|X_t-x_t\left(\vartheta\right)\right\|_2$$

and the notation:

$$\breve g\left(\vartheta_0,\nu\right)=\inf_{|\vartheta-\vartheta_0|\geq\nu}\left\|x_t\left(\vartheta\right)-x_t\left(\vartheta_0\right)\right\|_2,\ \breve\eta_t\left(\vartheta_0\right)=\int_0^t\phi\left(\vartheta_0,s,t\right)b\left(\vartheta_0,s\right)\mathrm{d}V_s,$$

$$X_t-x_t\left(\vartheta_0\right)=\varepsilon\int_0^t f\left(\vartheta_0,s\right)\breve\eta_s\left(\vartheta_0\right)\mathrm{d}s+\varepsilon\,\eta_t,\qquad \eta_t=\int_0^t\sigma\left(s\right)\mathrm{d}W_s,$$

$$\mathrm{N}\left(\vartheta_0,q\right)=\int_q^T\left(\int_q^t f\left(\vartheta_0,s\right)\phi\left(\vartheta_0,q,s\right)\mathrm{d}s\right)\dot{\mathrm{x}}_t\left(\vartheta_0\right)\mathrm{d}t,$$

$$\breve{\mathbf{R}}\left(\vartheta_0\right)=\int_0^T\mathrm{N}\left(\vartheta_0,t\right)\mathrm{N}\left(\vartheta_0,t\right)^\top b\left(\vartheta_0,t\right)^2\mathrm{d}t$$
$$+\int_0^T\int_s^T\dot{\mathrm{x}}_t\left(\vartheta_0\right)\mathrm{d}t\int_s^T\dot{\mathrm{x}}_t\left(\vartheta_0\right)^\top\mathrm{d}t\,\sigma\left(s\right)^2\mathrm{d}s,$$

$$\breve{\mathbf{T}}\left(\vartheta_0\right)=\int_0^T\dot{\mathrm{x}}_t\left(\vartheta_0\right)\dot{\mathrm{x}}_t\left(\vartheta_0\right)^\top\mathrm{d}t,\qquad \breve{\mathbf{D}}\left(\vartheta_0\right)=\breve{\mathbf{T}}\left(\vartheta_0\right)^{-1}\breve{\mathbf{R}}\left(\vartheta_0\right)\breve{\mathbf{T}}\left(\vartheta_0\right)^{\top-1}$$

Introduce the conditions $\breve{\mathscr{B}}$:

$\breve{\mathscr{B}}_1$. *The functions $f(\vartheta,t),a(\vartheta,t),b(\vartheta,t)$ are twice continuously differentiable on $\bar\Theta$.*

$\breve{\mathscr{B}}_2$. *The matrix $\breve{\mathbf{T}}\left(\vartheta_0\right)$ is uniformly nondegenerate*

$$\inf_{\vartheta_0\in\Theta}\ \inf_{\|\mathrm{e}\|=1,\mathrm{e}\in\mathcal{R}^d} \mathrm{e}^\top \check{\mathbf{T}}(\vartheta_0)\,\mathrm{e} > 0.$$

$\check{\mathscr{B}}_3$. *For all $\nu > 0$*

$$\inf_{\vartheta_0\in\Theta} \check{g}(\vartheta_0,\nu) > 0.$$

Theorem 2.2 *Suppose that the conditions $\mathscr{A}_*$, $\check{\mathscr{B}}$ are fulfilled. Then the MDE ϑ_ε^* is uniformly on compacts $\mathbb{K}\subset\Theta$ consistent, asymptotically normal*

$$\frac{\check{\vartheta}_\varepsilon-\vartheta_0}{\varepsilon} \Longrightarrow \check{\zeta}(\vartheta_0) \sim \mathcal{N}\left(0,\check{\mathbf{D}}(\vartheta_0)\right), \tag{2.12}$$

and the moments converge: for any $p\geq 2$

$$\varepsilon^{-p}\mathbf{E}_{\vartheta_0}\|\check{\vartheta}_\varepsilon-\vartheta_0\|^p \rightarrow \mathbf{E}_{\vartheta_0}\left\|\check{\zeta}(\vartheta_0)\right\|^p. \tag{2.13}$$

Proof The proof of consistency is quite close to the given above in the proof of Theorem 2.1. We have to show that for any $\nu>0$

$$\sup_{\vartheta_0\in\mathbb{K}} \mathbf{P}_{\vartheta_0}\left(\left\|\check{\vartheta}_\varepsilon-\vartheta_0\right\|>\nu\right)\rightarrow 0.$$

Following (2.10) we can write

$$\begin{aligned}
&\mathbf{P}_{\vartheta_0}\left(\left\|\check{\vartheta}_\varepsilon-\vartheta_0\right\|>\nu\right)\\
&\quad=\mathbf{P}_{\vartheta_0}\left(\inf_{\|\vartheta-\vartheta_0\|<\nu}\|X_t-x_t(\vartheta)\|>\inf_{\|\vartheta-\vartheta_0\|\geq\nu}\|X_t-x_t(\vartheta)\|\right)\\
&\quad\leq\mathbf{P}_{\vartheta_0}\left(2\,\|X_t-x_t(\vartheta_0)\|>\inf_{\|\vartheta-\vartheta_0\|\geq\nu}\|x_t(\vartheta_0)-x_t(\vartheta)\|\right)\\
&\quad\leq\frac{4}{\check{g}(\vartheta_0,\nu)^2}\mathbf{E}_{\vartheta_0}\int_0^T[X_t-x_t(\vartheta_0)]^2\,\mathrm{d}t\leq C\varepsilon^2.
\end{aligned} \tag{2.14}$$

Therefore, the consistency was proved.

To verify the asymptotic normality, we write the *minimum distance equation* (MDEq), i.e., the following system of equations:

$$\int_0^T\left[X_t-x_t\left(\check{\vartheta}_\varepsilon\right)\right]\dot{\mathrm{x}}_t\left(\check{\vartheta}_\varepsilon\right)\mathrm{d}t=0,$$

where we expand $x_t\left(\check\vartheta_\varepsilon\right)$ at the vicinity of the point ϑ_0.

$$
\begin{aligned}
&\int_0^T \left[X_t - x_t\left(\vartheta_0\right) + x_t\left(\vartheta_0\right) - x_t\left(\check\vartheta_\varepsilon\right)\right] \dot{\mathrm{x}}_t\left(\check\vartheta_\varepsilon\right) \mathrm{d}t \\
&\qquad = \int_0^T \left[X_t - x_t\left(\vartheta_0\right) - \left(\check\vartheta_\varepsilon - \vartheta_0\right)^\top \dot{\mathrm{x}}_t\left(\vartheta_0\right)\right] \dot{\mathrm{x}}_t\left(\vartheta_0\right) \mathrm{d}t\left(1 + o\left(1\right)\right) \\
&\qquad = \int_0^T \left[X_t - x_t\left(\vartheta_0\right)\right] \dot{\mathrm{x}}_t\left(\vartheta_0\right) \mathrm{d}t\left(1 + o\left(1\right)\right) \\
&\qquad\qquad - \int_0^T \dot{\mathrm{x}}_t\left(\vartheta_0\right) \dot{\mathrm{x}}_t\left(\vartheta_0\right)^\top \mathrm{d}t \left(\check\vartheta_\varepsilon - \vartheta_0\right)\left(1 + o\left(1\right)\right).
\end{aligned}
$$

Hence,

$$
\check\vartheta_\varepsilon - \vartheta_0 = \left(\int_0^T \dot{\mathrm{x}}_t\left(\vartheta_0\right) \dot{\mathrm{x}}_t\left(\vartheta_0\right)^\top \mathrm{d}t\right)^{-1} \int_0^T \left[X_t - x_t\left(\vartheta_0\right)\right] \dot{\mathrm{x}}_t\left(\vartheta_0\right) \mathrm{d}t\left(1 + o\left(1\right)\right).
$$

□

Lemma 2.2 *We have*

$$
\begin{aligned}
&\int_0^T \left[X_t - x_t\left(\vartheta_0\right)\right] \dot{\mathrm{x}}_t\left(\vartheta_0\right) \mathrm{d}t \\
&\qquad = \varepsilon \int_0^T \left(\int_s^T \dot{\mathrm{x}}_t\left(\vartheta_0\right) \mathrm{d}t\right) \sigma\left(s\right) \mathrm{d}W_s + \varepsilon \int_0^T \mathrm{N}\left(\vartheta_0, s\right) b\left(\vartheta_0, s\right) \mathrm{d}V_s.
\end{aligned}
\tag{2.15}
$$

Proof The proof consists of the change of order of integration. Recall that

$$
X_t - x_t\left(\vartheta_0\right) = \int_0^t f\left(\vartheta_0, s\right)\left[Y_s - y_s\left(\vartheta_0\right)\right] \mathrm{d}s + \varepsilon \int_0^t \sigma\left(s\right) \mathrm{d}W_s.
$$

The first integral in (2.15) by Fubini theorem can be written as stochastic integral

$$
\int_0^T \left(\int_0^t \sigma\left(s\right) \mathrm{d}W_s\right) \dot{\mathrm{x}}_t\left(\vartheta_0\right) \mathrm{d}t = \int_0^T \left(\int_s^T \dot{\mathrm{x}}_t\left(\vartheta_0\right) \mathrm{d}t\right) \sigma\left(s\right) \mathrm{d}W_s.
$$

To write the second integral, we note that

$$
Y_s - y_s\left(\vartheta_0\right) = \varepsilon \int_0^s \phi\left(\vartheta_0, q, s\right) b\left(\vartheta_0, q\right) \mathrm{d}V_q.
$$

Therefore, using once more the Fubini theorem, we can write

$$\int_0^t f(\vartheta_0, s)\left[Y_s - y_s(\vartheta_0)\right] \mathrm{d}s = \varepsilon \int_0^t f(\vartheta_0, s) \int_0^s \phi(\vartheta_0, q, s)\, b(\vartheta_0, q)\, \mathrm{d}V_q\, \mathrm{d}s$$

$$= \varepsilon \int_0^t \left(\int_q^t f(\vartheta_0, s)\, \phi(\vartheta_0, q, s)\, \mathrm{d}s\right) b(\vartheta_0, q)\, \mathrm{d}V_q.$$

The last step is

$$\int_0^T \int_0^t f(\vartheta_0, s)\left[Y_s - y_s(\vartheta_0)\right] \mathrm{d}s\ \dot{\mathrm{x}}_t(\vartheta_0)\, \mathrm{d}t$$

$$= \varepsilon \int_0^T \int_0^t \left(\int_q^t f(\vartheta_0, s)\, \phi(\vartheta_0, q, s)\, \mathrm{d}s\right) b(\vartheta_0, q)\, \mathrm{d}V_q\ \dot{\mathrm{x}}_t(\vartheta_0)\, \mathrm{d}t$$

$$= \varepsilon \int_0^T \int_q^T \left(\int_q^t f(\vartheta_0, s)\, \phi(\vartheta_0, q, s)\, \mathrm{d}s\right) \dot{\mathrm{x}}_t(\vartheta_0)\, \mathrm{d}t\ b(\vartheta_0, q)\, \mathrm{d}V_q$$

$$= \varepsilon \int_0^T \mathrm{N}(\vartheta_0, q)\ b(\vartheta_0, q)\ \mathrm{d}V_q.$$

□

Therefore, the MDE has the representation

$$\frac{\check{\vartheta}_\varepsilon - \vartheta_0}{\varepsilon} = \check{\mathbf{T}}(\vartheta_0)^{-1}\left[\int_0^T \left(\int_s^T \dot{\mathrm{x}}_t(\vartheta_0)\, \mathrm{d}t\right) \sigma(s)\, \mathrm{d}W_s + \int_0^T \mathrm{N}(\vartheta_0, s)\, b(\vartheta_0, s)\, \mathrm{d}V_s\right] + O(\varepsilon). \tag{2.16}$$

From this representation, it follows that the MDE is asymptotically normal with the given Theorem 2.2 limit covariance.

The convergence of polynomial moments can be proved following the same lines as it was done in the case of TFE above.

Example 2.2 Consider the model (2.1)–(2.2) with $f(\vartheta, t) = \sum_{k=1}^d [\vartheta_k h_k(t) + l_k(t)]$, $a(\vartheta, t) = a(t)$, $b(\vartheta, t) = b(t)$. Then $y_t(\vartheta) = y_t$ and

$$x_t(\vartheta) = \sum_{k=1}^d [\vartheta_k H_k(t) + L_k(t)], \qquad \mathrm{H}(t) = (H_1(t), \ldots, H_d(t))^\top,$$

$$H_k(t) = \int_0^t h_k(s)\, y_s \mathrm{d}s, \quad L_k(t) = \int_0^t l_k(s)\, y_s \mathrm{d}s, \quad L(t) = \sum_{k=1}^d L_k(t).$$

Suppose that the matrix

$$\mathbf{T}=\int_0^T \mathrm{H}\left(t\right)\mathrm{H}\left(t\right)^\top \mathrm{d}t$$

is nondegenerate:

$$\inf_{\mathrm{e}\in\mathcal{R}^d,\|\mathrm{e}\|=1} \mathrm{e}^\top\mathbf{T}\mathrm{e}>0.$$

Then we have

$$\breve{g}\left(\vartheta_0,\nu\right)=\inf_{\|\vartheta-\vartheta_0\|\geq\nu}\left(\vartheta-\vartheta_0\right)^\top\mathbf{T}\left(\vartheta-\vartheta_0\right)\geq\nu^2\inf_{\mathrm{e}\in\mathcal{R}^d,\|\mathrm{e}\|=1}\mathrm{e}^\top\mathbf{T}\mathrm{e}=c_*\nu^2>0.$$

Then MDE can be calculated explicitly

$$\check{\vartheta}_\varepsilon=\left(\int_0^T \mathrm{H}\left(t\right)\mathrm{H}\left(t\right)^\top \mathrm{d}t\right)^{-1}\int_0^T\left[X_t-R\left(t\right)\right]\mathrm{H}\left(t\right)\mathrm{d}t. \tag{2.17}$$

It can be shown that this estimator is unbiased, has Gaussian distribution, and is consistent. The limit variance and convergence of moments can be obtained as well from the representation (2.17).

Remark 2.1 It is known that the MDE is not asymptotically efficient, but it has remarcable property concerning its computational relative simplicity. Its calculation does not involve solution of filtration equations (2.5) and (2.6) and can be numerically preferable. This advantage became more clear in the next section, where the MLE and BE are studied. The calculation of these estimators requires a large number of solutions of Eqs. (2.5) and (2.6) for different values ϑ. To avoid this difficulty and to have computationaly relatively simple estimator, which is asymptotically efficient, we propose later One-step MLE. In that case, the MDE can be used as a preliminary estimator in the construction of One-step MLE.

Remark 2.2 Consider the behavior of MDE $\check{\vartheta}_{\tau_\varepsilon}$ on the small interval of observations $[0,\tau_\varepsilon]$, where $\tau_\varepsilon\rightarrow 0$. We suppose that $d=1$.

Then using elementary expansion technique, we can write

$$T\left(\vartheta\right)=\int_0^{\tau_\varepsilon}\dot{x}_t\left(\vartheta\right)^2\mathrm{d}t=\frac{\tau_\varepsilon^3}{3}\,y_0^2\,\dot{f}\left(\vartheta,0\right)^2\left(1+O\left(\tau_\varepsilon\right)\right),$$

because

$$\dot{x}_t\left(\vartheta\right)=\int_0^t\left(\dot{f}\left(\vartheta,s\right)y_s\left(\vartheta\right)+f\left(\vartheta,s\right)\dot{y}_s\left(\vartheta\right)\right)\mathrm{d}s=t\,y_0\,\dot{f}\left(\vartheta,0\right)\left(1+O\left(t\right)\right)$$

and $\dot{y}_s\left(\vartheta\right)=s\,y_0\,\dot{a}\left(\vartheta,0\right)\left(1+O\left(s\right)\right)$.

Further,

$$X_t - x_t(\vartheta_0) = \varepsilon \int_0^t f(\vartheta_0, s) \int_0^s \phi(\vartheta_0, r, s)\, b(\vartheta_0, r)\, \mathrm{d}V_r\, \mathrm{d}s + \varepsilon \int_0^t \sigma(s)\, \mathrm{d}W_s$$

$$= \varepsilon\, f(\vartheta_0, 0)\, b(\vartheta_0, 0) \int_0^t V_s\, \mathrm{d}s\, (1 + o(1)) + \varepsilon\, \sigma(0)\, W_t\, (1 + o(1))$$

as $t \to 0$ and therefore

$$\int_0^{\tau_\varepsilon} [X_s - x_s(\vartheta_0)]\, \dot{x}_s(\vartheta_0)\, \mathrm{d}s = \varepsilon\, y_0 \dot{f}(\vartheta_0, 0)\, \sigma(0) \int_0^{\tau_\varepsilon} t\, W_t\, \mathrm{d}t$$

$$+ \varepsilon\, y_0\, f(\vartheta_0, 0)\, \dot{f}(\vartheta_0, 0)\, b(\vartheta_0, 0) \int_0^{\tau_\varepsilon} t \int_0^t V_s \mathrm{d}s\, \mathrm{d}t\, (1 + o(1))$$

$$= \varepsilon\, \tau_\varepsilon^{5/2}\, y_0 \dot{f}(\vartheta_0, 0)\, \sigma(0) \int_0^1 r\, w(r)\, \mathrm{d}r\, (1 + o(1))$$

$$+ \varepsilon\, \tau_\varepsilon^{7/2}\, y_0\, f(\vartheta_0, 0)\, \dot{f}(\vartheta_0, 0)\, b(\vartheta_0, 0) \int_0^1 q \int_0^q v(r)\, \mathrm{d}r\, \mathrm{d}q\, (1 + o(1)).$$

The last equality is in distribution and $w(r), v(r), 0 \le r \le 1$ are independent Wiener processes.

Suppose that the following condition is fulfilled:

Condition $\breve{\mathscr{B}}_4$: $\inf_{\vartheta \in \Theta} |\dot{f}(\vartheta, 0)| > 0$.

Then we have the convergence in distribution

$$\frac{\tau_\varepsilon^{1/2}}{\varepsilon} \left(\breve{\vartheta}_{\tau_\varepsilon} - \vartheta_0 \right) = \frac{3\sigma(0)}{y_0\, \dot{f}(\vartheta_0, 0)} \int_0^1 r w_r \mathrm{d}r\, (1 + o(1))$$

$$\Longrightarrow \frac{3\sigma(0)}{y_0\, \dot{f}(\vartheta_0, 0)} \int_0^1 r w(r)\, \mathrm{d}r.$$

This means that

$$\frac{\tau_\varepsilon^{1/2}}{\varepsilon} \left(\breve{\vartheta}_{\tau_\varepsilon} - \vartheta_0 \right) \Longrightarrow \mathcal{N}(0, D(\vartheta_0)), \qquad D(\vartheta_0) = \frac{6}{5} \frac{\sigma(0)^2}{y_0^2\, \dot{f}(\vartheta_0, 0)^2}. \tag{2.18}$$

We have to suppose that $y_0 \neq 0, \sigma(0) \neq 0$ and $\tau_\varepsilon^{1/2} \varepsilon^{-1} \to \infty$. If we put $\tau_\varepsilon = \varepsilon^\delta$, then $\delta \in (0, 2)$. To have a good rate of convergence of $\breve{\vartheta}_{\tau_\varepsilon}$, we need the values of δ close to 0.

One-Step MDE Processes

The MDE is computationally simpler than the MLE or BE but is not asymptotically efficient. There exists one class of problems, where the MDE can be nevereless preferable. To illustrate this, we introduce the diffusion process

$$\mathrm{d}X_t = S(\vartheta_0, X_t)\,\mathrm{d}t + \varepsilon \mathrm{d}W_t, \qquad X_0 = x_0, \quad 0 \leq t \leq T, \tag{2.19}$$

where $S(\vartheta, x)$ is a known smooth function of $\vartheta \in \Theta = (\alpha, \beta)$ and x. Suppose that the observations $X^T = (X_t, 0 \leq t \leq T)$ are *contaminated*:

$$\mathrm{d}X_t = [S(\vartheta_0, X_t) + H(X_t)]\,\mathrm{d}t + \varepsilon \mathrm{d}W_t, \qquad X_0 = x_0, \quad 0 \leq t \leq T, \tag{2.20}$$

where the contamination $H(\cdot)$ (unknown) is small. The MDE is defined by the equation

$$\check{\vartheta}_\varepsilon = \arg \inf_{\vartheta \in \Theta} \int_0^T [X_t - x_t(\vartheta)]^2\,\mathrm{d}t, \tag{2.21}$$

where $x^T(\vartheta) = (x_t(\vartheta), 0 \leq t \leq T)$ is solution of (2.19) for the value $\varepsilon = 0$ and observations $X^T = (X_t, 0 \leq t \leq T)$ are from (2.20). Let us denote $x_t^H(\vartheta_0), 0 \leq t \leq T$ the solution of (2.20) for $\varepsilon = 0$ and suppose that we want to find in the family $\left\{x^T(\vartheta), \vartheta \in \Theta\right\}$ a trajectory $x_t\left(\vartheta^H\right), 0 \leq t \leq T$ closest in $\mathcal{L}_2$-sens to the unknown (true) trajectory $x_t^H(\vartheta_0), 0 \leq t \leq T$ -solution of the equation

$$\frac{\mathrm{d}x_t^H(\vartheta_0)}{\mathrm{d}t} = S\left(\vartheta_0, x_t^H(\vartheta_0)\right) + H\left(x_t^H(\vartheta_0)\right), \qquad x_0^H(\vartheta_0) = x_0, \quad 0 \leq t \leq T.$$

Therefore, we seek ϑ^H such that

$$\inf_{\vartheta \in \Theta} \int_0^T \left[x_t^H(\vartheta_0) - x_t(\vartheta)\right]^2 \mathrm{d}t = \int_0^T \left[x_t^H(\vartheta_0) - x_t\left(\vartheta^H\right)\right]^2 \mathrm{d}t.$$

It is easy to see that under the mild regularity conditions, the estimator $\check{\vartheta}_\varepsilon \to \vartheta^H$ is "consistent." The estimated trajectory is $x_t(\check{\vartheta}_\varepsilon), 0 \leq t \leq T$.

Now if we denote $\mathbb{H}_\nu$ a small ball of radius ν in $\mathcal{L}_2([0, T])$, then the following minimax lower bound on the mean-squared risk of any estimator $\bar{\vartheta}_\varepsilon$ holds

$$\lim_{\nu \to 0} \underline{\lim}_{\varepsilon \to 0} \sup_{H(\cdot) \in \mathbb{H}_\nu} \varepsilon^{-2} \mathbf{E}^H \left(\bar{\vartheta}_\varepsilon - \vartheta^H\right)^2 \geq \mathrm{I}_*(\vartheta_0)^{-1},$$

where expectation $\mathbf{E}^H$ corresponds to the observations (2.20) and the quantity $\mathrm{I}_*(\vartheta_0)$ plays the role of Fisher information.

The estimator that attains this bound for all $\vartheta_0 \in \Theta$ is called *locally asymptotically minimax* (LAM) . It is shown that the MDE $\check\vartheta_\varepsilon$ defined in (2.21) is LAM.

Note that the MLE $\hat\vartheta_\varepsilon$ for the model (2.19)–(2.20) converges to the value $\hat\vartheta^H$, which minimizes the limit of Kulback–Leibner divergency

$$\hat\vartheta_\varepsilon \longrightarrow \hat\vartheta^H = \arg\inf_{\vartheta\in\Theta}\int_0^T \left[S\left(\vartheta_0, x_t^H\left(\vartheta_0\right)\right) + H\left(x_t^H\left(\vartheta_0\right)\right)\right.$$
$$\left. - S\left(\vartheta_0, x_t^H\left(\vartheta_0\right)\right)\right]^2 {\rm d}t,$$

see [108, Section 2.6].

Therefore, in some situations, it can be interesting to have *on-line* MDE, which has a recurrent structure and is similar to the One-step MLE-process (1.91).

Consider the model of observations (2.1) and (2.2) with unknown parameter $\vartheta \in \Theta \subset \mathcal{R}^d$ and recall formally the construction of One-step estimator by observations $X^\tau = (X_s, 0 \le s \le \tau)$. The MDEq is a system of d equations

$$\int_0^\tau \left[X_s - x_s\left(\check\vartheta_{\tau,\varepsilon}\right)\right]\dot{\rm x}_s\left(\check\vartheta_{\tau,\varepsilon}\right){\rm d}s = 0, \tag{2.22}$$

where the vector $\dot{\rm x}_s\left(\vartheta\right) = \partial {\rm x}_s\left(\vartheta\right)/\partial\vartheta$. If we expand the function $x_s\left(\check\vartheta_{\tau,\varepsilon}\right)$ at the vicinity of the point ϑ_0, then we obtain $x_s\left(\check\vartheta_{\tau,\varepsilon}\right) = x_s\left(\vartheta_0\right) + \left(\check\vartheta_{\tau,\varepsilon} - \vartheta_0\right)^\top \dot{\rm x}_s(\vartheta_0)\left(1+o\left(1\right)\right)$. Then we can write

$$\int_0^\tau \left[X_s - x_s\left(\vartheta_0\right) - \left(\check\vartheta_{\tau,\varepsilon} - \vartheta_0\right)^\top \dot{\rm x}_s\left(\vartheta_0\right)\left(1+o\left(1\right)\right)\right]\dot{\rm x}_s\left(\check\vartheta_{\tau,\varepsilon}\right){\rm d}s = 0.$$

Hence,

$$\begin{aligned}\check\vartheta_{\tau,\varepsilon} &= \vartheta_0 + \left(\int_0^\tau \dot{\rm x}_s\left(\vartheta_0\right)\dot{\rm x}_s\left(\check\vartheta_{\tau,\varepsilon}\right)^\top {\rm d}s\right)^{-1}\\ &\quad\times\int_0^\tau \left[X_s - x_s\left(\vartheta_0\right)\right]\dot{\rm x}_s\left(\check\vartheta_{\tau,\varepsilon}\right){\rm d}s\left(1+o\left(1\right)\right)\\ &= \vartheta_0 + \left(\int_0^\tau \dot{\rm x}_s\left(\vartheta_0\right)\dot{\rm x}_s\left(\vartheta_0\right)^\top {\rm d}s\right)^{-1}\int_0^\tau \left[X_s - x_s\left(\vartheta_0\right)\right]\dot{\rm x}_s\left(\vartheta_0\right){\rm d}s\left(1+o\left(1\right)\right).\end{aligned}$$

Introduce the matrices:

$$\check{\mathbf{T}}\left(\vartheta_0, t\right) = \int_0^t \dot{\rm x}_s\left(\vartheta_0\right)\dot{\rm x}_s\left(\vartheta_0\right)^\top {\rm d}s,$$

$$\check{\mathbf{D}}(\vartheta_0,t)=\check{\mathbf{T}}(\vartheta_0,t)^{-1}\check{\mathbf{R}}(\vartheta_0,t)\left(\check{\mathbf{T}}(\vartheta_0,t)^\top\right)^{-1},$$

$$\check{\mathbf{R}}(\vartheta_0,t)=\int_0^t\left[\mathrm{N}(\vartheta_0,s)\,\mathrm{N}(\vartheta_0,s)^\top b(\vartheta_0,s)^2\right.$$

$$\left.+\sigma(s)^2\int_s^t\dot{\mathrm{x}}_r(\vartheta_0)\,\mathrm{d}r\int_s^t\dot{\mathrm{x}}_r(\vartheta_0)^\top\,\mathrm{d}r\right]\mathrm{d}s$$

and suppose that we have a consistent preliminary MDE $\check{\vartheta}_{\tau,\varepsilon}$. Then the One-step MDE-process can be defined by the expression

$$\check{\vartheta}^\star_{t,\varepsilon}=\check{\vartheta}_{\tau,\varepsilon}+\check{\mathbf{T}}\left(\check{\vartheta}_{\tau,\varepsilon},t\right)^{-1}\int_0^t\left[X_s-x_s\left(\check{\vartheta}_{\tau,\varepsilon}\right)\right]\dot{\mathrm{x}}_s\left(\check{\vartheta}_{\tau,\varepsilon}\right)\mathrm{d}s,\qquad \tau\le t\le T. \tag{2.23}$$

The conditions $\check{\mathscr{B}}$, where T is replaced by $\tau\in(0,T)$, we denote $\check{\mathscr{B}}^\tau$ and the condition $\check{\mathscr{B}}_2$, where the matrix $\check{\mathbf{T}}(\vartheta_0)$ is replaced by $\check{\mathbf{T}}(\vartheta_0,t)$ we denote $\check{\mathscr{B}}_2^t$.

Theorem 2.3 *Let the conditions $\check{\mathscr{B}}^\tau$, $\check{\mathscr{B}}_1$ and $\check{\mathscr{B}}_2^t$ hold. Then the One-step MDE process $\check{\vartheta}^\star_{t,\varepsilon}$, $\tau<t\le T$ is uniformly consistent: for any $t_0\in(\tau,T]$ and $\nu>0$*

$$\sup_{\vartheta_0\in\mathbb{K}}\mathbf{P}_{\vartheta_0}\left(\sup_{t_0\le t\le T}\left\|\check{\vartheta}^\star_{t,\varepsilon}-\vartheta_0\right\|\ge\nu\right)\longrightarrow 0, \tag{2.24}$$

is asymptotically normal

$$\frac{\check{\vartheta}^\star_{t,\varepsilon}-\vartheta_0}{\varepsilon}\Longrightarrow\check{\zeta}_t(\vartheta_0)\sim\mathcal{N}\left(0,\check{\mathbf{D}}(\vartheta_0,t)\right) \tag{2.25}$$

and the moments converge: for any $p\ge 2$

$$\varepsilon^{-p}\mathbf{E}_{\vartheta_0}\left\|\check{\vartheta}^\star_{t,\varepsilon}-\vartheta_0\right\|^p\longrightarrow\mathbf{E}_{\vartheta_0}\left\|\check{\zeta}_t(\vartheta_0)\right\|^p. \tag{2.26}$$

Proof We have with probability 1 the estimate

$$\sup_{t_0\le t\le T}\left\|\check{\vartheta}^\star_{t,\varepsilon}-\vartheta_0\right\|\le\left\|\check{\vartheta}_{\tau,\varepsilon}-\vartheta_0\right\|$$

$$+\left\|\check{\mathbf{T}}\left(\check{\vartheta}_{\tau,\varepsilon},t_0\right)\right\|^{-1}\int_0^T\left|X_s-x_s\left(\check{\vartheta}_{\tau,\varepsilon}\right)\right|\left\|\dot{\mathrm{x}}_s\left(\check{\vartheta}_{\tau,\varepsilon}\right)\right\|\mathrm{d}s$$

$$\le\left\|\check{\vartheta}_{\tau,\varepsilon}-\vartheta_0\right\|+C\int_0^T\left|X_s-x_s(\vartheta_0)\right|\,\mathrm{d}s+C\int_0^T\left|x_s(\vartheta_0)-x_s\left(\check{\vartheta}_{\tau,\varepsilon}\right)\right|\,\mathrm{d}s.$$

Now the uniform consistency (2.24) follows from this estimate.

Consider the difference

$$
\begin{aligned}
\frac{\vartheta^{\star}_{t,\varepsilon}-\vartheta_0}{\varepsilon} &= \frac{\check\vartheta_{\tau,\varepsilon}-\vartheta_0}{\varepsilon}+\check{\mathbf{T}}\left(\check\vartheta_{\tau,\varepsilon},t\right)^{-1}\int_0^t \frac{\left[X_s-x_s\left(\vartheta_0\right)\right]}{\varepsilon}\dot{\mathbf{x}}_s\left(\check\vartheta_{\tau,\varepsilon}\right)\,\mathrm{d}s\\
&\quad+\frac{1}{\varepsilon}\check{\mathbf{T}}\left(\check\vartheta_{\tau,\varepsilon},t\right)^{-1}\int_0^t\left[x_s\left(\vartheta_0\right)-x_s\left(\check\vartheta_{\tau,\varepsilon}\right)\right]\dot{\mathbf{x}}_s\left(\check\vartheta_{\tau,\varepsilon}\right)\,\mathrm{d}s\\
&=\frac{\check\vartheta_{\tau,\varepsilon}-\vartheta_0}{\varepsilon}+\check{\mathbf{T}}\left(\vartheta_0,t\right)^{-1}\int_0^t\frac{\left[X_s-x_s\left(\vartheta_0\right)\right]}{\varepsilon}\dot{\mathbf{x}}_s\left(\vartheta_0\right)\,\mathrm{d}s\\
&\quad\times\left(1+O\left(\left\|\check\vartheta_{\tau,\varepsilon}-\vartheta_0\right\|\right)\right)\\
&\quad-\check{\mathbf{T}}\left(\vartheta_0,t\right)^{-1}\int_0^t\dot{\mathbf{x}}_s\left(\vartheta_0\right)\dot{\mathbf{x}}_s\left(\vartheta_0\right)^{\top}\,\mathrm{d}s\frac{\left(\check\vartheta_{\tau,\varepsilon}-\vartheta_0\right)}{\varepsilon}\\
&\quad\times\left(1+O\left(\left\|\check\vartheta_{\tau,\varepsilon}-\vartheta_0\right)\right\|\right)\\
&=\check{\mathbf{T}}\left(\vartheta_0,t\right)^{-1}\int_0^t\frac{\left[X_s-x_s\left(\vartheta_0\right)\right]}{\varepsilon}\dot{\mathbf{x}}_s\left(\vartheta_0\right)\,\mathrm{d}s+O\left(\varepsilon^{-1}\left\|\check\vartheta_{\tau,\varepsilon}-\vartheta_0\right\|^2\right).
\end{aligned}
$$

Note that

$$
\varepsilon^{-1}\left\|\check\vartheta_{\tau,\varepsilon}-\vartheta_0\right\|^2\longrightarrow 0.
$$

Using the representation (2.16) of the MDE, we obtain the asymptotic normality (2.25).

This representation and the conditions of the theorem allow us to verify the convergence of moments (2.26) by direct calculations. □

Remark 2.3 The One-step MDE process (2.23) can be written in recurrent form as solution of ordinary differential equation. We have

$$
\check{\mathbf{T}}\left(\check\vartheta_{\tau,\varepsilon},t\right)\vartheta^{\star}_{t,\varepsilon}=\check{\mathbf{T}}\left(\check\vartheta_{\tau,\varepsilon},t\right)\check\vartheta_{\tau,\varepsilon}+\int_0^t\left[X_s-x_s\left(\check\vartheta_{\tau,\varepsilon}\right)\right]\dot{\mathbf{x}}_s\left(\check\vartheta_{\tau}\right)\,\mathrm{d}s,\quad \tau\leq t\leq T.
$$

Hence,

$$
\begin{aligned}
\frac{\mathrm{d}\vartheta^{\star}_{t,\varepsilon}}{\mathrm{d}t}&=\check{\mathbf{T}}\left(\check\vartheta_{\tau,\varepsilon},t\right)^{-1}\dot{\mathbf{x}}_t(\check\vartheta_{\tau,\varepsilon})\dot{\mathbf{x}}_t(\check\vartheta_{\tau,\varepsilon})^{\top}\left[\check\vartheta_{\tau,\varepsilon}-\vartheta^{\star}_{t,\varepsilon}\right]\\
&\quad+\check{\mathbf{T}}\left(\check\vartheta_{\tau,\varepsilon},t\right)^{-1}\left[X_t-x_t\left(\check\vartheta_{\tau,\varepsilon}\right)\right]\dot{\mathbf{x}}_t\left(\check\vartheta_{\tau,\varepsilon}\right),\qquad \tau\leq t\leq T,
\end{aligned}
$$

with the corresponding initial value $\check\vartheta^\star_{\tau,\varepsilon}$. Note that the other elements of this formula can be written in recurrent form too:

$$\frac{\mathrm{d}x_t(\check\vartheta_{\tau,\varepsilon})}{\mathrm{d}t}=f(\check\vartheta_{\tau,\varepsilon},t)\,y_t\left(\check\vartheta_{\tau,\varepsilon}\right),\qquad x_\tau(\check\vartheta_{\tau,\varepsilon})=\int_0^t f(\check\vartheta_{\tau,\varepsilon},s)\,y_s\left(\check\vartheta_{\tau,\varepsilon}\right)\mathrm{d}s,$$

$$\frac{\mathrm{d}\dot{\mathrm{x}}_t(\check\vartheta_{\tau,\varepsilon})}{\mathrm{d}t}=\dot{\mathrm{f}}(\check\vartheta_{\tau,\varepsilon},t)\,y_t\left(\check\vartheta_{\tau,\varepsilon}\right)+f(\check\vartheta_{\tau,\varepsilon},t)\,\dot{\mathrm{y}}_t\left(\check\vartheta_{\tau,\varepsilon}\right),\qquad \dot{\mathrm{x}}_\tau(\check\vartheta_{\tau,\varepsilon}),$$

$$\frac{\mathrm{d}\check{\mathbf{I}}\left(\check\vartheta_{\tau,\varepsilon},t\right)}{\mathrm{d}t}=\dot{\mathrm{x}}_t(\check\vartheta_{\tau,\varepsilon})\dot{\mathrm{x}}_t(\check\vartheta_{\tau,\varepsilon})^\top,\qquad \check{\mathbf{I}}\left(\check\vartheta_{\tau,\varepsilon},\tau\right),\qquad \tau\le t\le T.$$

Remark 2.4 The MDEq (2.22) can be rewritten as follows. Let us denote

$$\dot\Lambda\,(\vartheta,t)=\int_t^T \dot{\mathrm{x}}_s\,(\vartheta)\,\mathrm{d}s.$$

Then integrating by parts, we obtain

$$\int_0^T [X_t-x_t\,(\vartheta)]\,\dot{x}_t\,(\vartheta)\;\mathrm{d}t=\int_0^T \dot\Lambda\,(\vartheta,s)\,[\mathrm{d}X_s-f\,(\vartheta,s)\,y_s\,(\vartheta)\,\mathrm{d}s]=0.$$

This form of MDEq allows us to use a slightly different One-step MDE process

$$\check\vartheta^\star_{t,\varepsilon}=\check\vartheta_{\tau,\varepsilon}+\check{\mathbf{I}}\left(\check\vartheta_{\tau,\varepsilon},t\right)^{-1}\int_\tau^t \dot\Lambda\left(\check\vartheta_{\tau,\varepsilon},s\right)\left[\mathrm{d}X_s-f\left(\check\vartheta_{\tau,\varepsilon},s\right)y_s\left(\check\vartheta_{\tau,\varepsilon}\right)\mathrm{d}s\right],$$
$$\tau<t\le T.$$

This form is in some sense similar to that of One-step MLE process considered below in this section. Of course, this One-step MDE process has the same asymptotic properties as the One-step MDE process (2.23).

Remark 2.5 Let us introduce one else MDE (see Remark 1.6). Suppose that the process X_t has derivative X_t' with respect to t (what is, of course, wrong) and write formally the (diverging) integral

$$I\,(\vartheta)=\int_0^T\left[X_t'-f\,(\vartheta,t)\,y_t\,(\vartheta)\right]^2\mathrm{d}t.$$

Consider the minimization of this integral on ϑ. After formal derivation of $I\,(\vartheta)$ w.r.t. ϑ, we obtain the system of equations

$$\int_0^T\left[X_t'-f\,(\vartheta,t)\,y_t\,(\vartheta)\right]\left[\dot{\mathrm{f}}\,(\vartheta,t)\,y_t\,(\vartheta)+f\,(\vartheta,t)\,\dot{\mathrm{y}}_t\,(\vartheta)\right]\mathrm{d}t=0,$$

which we rewrite as follows:

$$\int_0^T \left[\dot{\mathrm{f}}(\vartheta,t)\,y_t(\vartheta)+f(\vartheta,t)\,\dot{\mathrm{y}}_t(\vartheta)\right]\left[\mathrm{d}X_t-f(\vartheta,t)\,y_t(\vartheta)\,\mathrm{d}t\right]=0 \qquad (2.27)$$

or as

$$\int_0^T \dot{\mathrm{S}}(\vartheta,t)\left[\mathrm{d}X_t-S(\vartheta,t)\,\mathrm{d}t\right]=0,\qquad \vartheta\in\Theta,$$

where $S(\vartheta,t)=f(\vartheta,t)\,y_t(\vartheta)$.

Now we can forget about wrong preliminary calculations and consider this equation as one, whose solution $\check\vartheta_\varepsilon^*$ defines another MDE.

Introduce the function

$$g(\vartheta,\vartheta_0)=\int_0^T\left[S(\vartheta,t)-S(\vartheta_0,t)\right]^2\mathrm{d}t,\qquad \vartheta\in\Theta$$

and the condition:

$$\textit{The equation}\quad \dot{\mathrm{g}}(\vartheta,\vartheta_0)=0,\quad \vartheta\in\Theta\quad \textit{has a unique solution.} \qquad (2.28)$$

Then it can be shown that this estimator is consistent and under usual conditions is asymptotically normal. Note that the Eq. (2.27) has to be understand in a symbolique way, because we can not write

$$\int_0^T \dot{\mathrm{S}}\left(\check\vartheta_\varepsilon^*,t\right)\left[\mathrm{d}X_t-S\left(\check\vartheta_\varepsilon^*,t\right)\mathrm{d}t\right]=0,$$

and expand $\dot{\mathrm{S}}\left(\check\vartheta_\varepsilon^*,t\right)$ because $\check\vartheta_\varepsilon^*$ depends on all observations X^T and the stochastic integral is not defined. Exception is the linear case $f(\vartheta,t)=\vartheta f_t, a(\vartheta,t)=a_t$. Then the solution of the Eq. (2.27) is

$$\check\vartheta_\varepsilon^*=\left(\int_0^T f_s^2y_s^2\,\mathrm{d}s\right)^{-1}\int_0^T f_sy_s\,\mathrm{d}X_s.$$

If we need an on-line MDE process of such type and there is some preliminary consistent estimator $\bar\vartheta_{\tau,\varepsilon}$, then the corresponding One-step MDE process based on the MDEq (2.27) can be

$$\check\vartheta_{t,\varepsilon}^\star=\check\vartheta_{\tau,\varepsilon}+\mathbf{J}_\tau\left(\check\vartheta_{\tau,\varepsilon},t\right)^{-1}\int_\tau^t\dot{\mathrm{S}}\left(\check\vartheta_{\tau,\varepsilon},s\right)\left[\mathrm{d}X_s-S\left(\check\vartheta_{\tau,\varepsilon},s\right)\mathrm{d}s\right],\qquad \tau<t\le T,$$

under mild regularity conditions is consistent and asymptotically normal. Here the matrix

$$\mathbf{J}_\tau(\vartheta,t)=\int_\tau^t \dot{\mathbb{S}}(\vartheta,s)\,\dot{\mathbb{S}}(\vartheta,s)^\top\,\mathrm{d}s.$$

The proofs use the same arguments as above.

Method of Moments Estimators

We have the model of observations (2.1) and (2.2), where $\vartheta\in\Theta\subset\mathcal{R}^d$. Introduce some functions $H_{\varepsilon,1}(x),\ldots,H_{\varepsilon,d}(x)$, $H_1(x),\ldots,H_d(x)$, $x\in\mathcal{R}$ such that for all $\vartheta_0\in\Theta$ (ϑ_0 is the true value)

$$\int_0^T H_{\varepsilon,k}(X_t)\,\mathrm{d}t\longrightarrow h_k(\vartheta_0)=\int_0^T H_k\big(x_t(\vartheta_0)\big)\,\mathrm{d}t,\qquad k=1,\ldots,d,$$

uniformly in $\vartheta_0\in\Theta$. Denote $\mathrm{h}(\vartheta)=(h_1(\vartheta),\ldots,h_d(\vartheta))^\top$.

The *method of moments estimator* (MME) ϑ_ε^* is defined as solution of the system of equations

$$h_1\left(\vartheta_\varepsilon^*\right)=\int_0^T H_{\varepsilon,1}(X_t)\,\mathrm{d}t,$$

$$\cdots\cdots\cdots\cdots\cdots\cdots\cdots\cdots$$

$$h_d\left(\vartheta_\varepsilon^*\right)=\int_0^T H_{\varepsilon,d}(X_t)\,\mathrm{d}t.$$

The condition of identifiability is: for any $\nu>0$ and any $\vartheta_0\in\Theta$

$$g^*(\vartheta_0,\nu)=\inf_{\|\vartheta-\vartheta_0\|>\nu}\|\mathrm{h}(\vartheta)-\mathrm{h}(\vartheta_0)\|>0. \tag{2.29}$$

Here $\mathrm{h}(\vartheta)=(h_1(\vartheta),\ldots,h_d(\vartheta))^\top$. By this condition, there exists one-to-one correspondence between the sets $\bar\Theta$ and $\mathbb{H}=\big(\mathrm{h}(\vartheta),\quad\vartheta\in\bar\Theta\big)$, i.e., for any $\mathrm{g}\in\mathbb{H}$ the system of equations $\mathrm{h}(\vartheta)=\mathrm{g}$ has a unique solution.

The identifiability condition allows to prove the consistency of the MME, and it is possible to describe the conditions of asymptotic normality of this estimator, but we will give just one example.

Example 2.3 Suppose that the system is

$$\mathrm{d}X_t=f(t)\,Y_t\mathrm{d}t+\varepsilon\mathrm{d}W_t,\qquad X_0=0,\quad 0\le t\le T,$$

$$\mathrm{d}Y_t=\sum_{k=1}^d a_k(t)\,\theta_k\,Y_t\,\mathrm{d}t+\varepsilon\mathrm{d}V_t,\qquad Y_0=y_0>0,$$

where $f(t) > 0$. We have to estimate $\vartheta = (\theta_1, \ldots, \theta_d)^\top$ by observations X^T.

Note that if $\varepsilon = 0$, then

$$x_t(\vartheta) = y_0 \int_0^t f(s) \exp\left(\sum_{k=1}^d A_k(s)\,\theta_k\right) \mathrm{d}s, \qquad A_k(t) = \int_0^t a_k(s)\,\mathrm{d}s.$$

To "estimate" ϑ, we denote

$$z_t = \ln \frac{x_t'(\vartheta)}{y_0 f(t)} = \sum_{k=1}^d A_k(t)\,\theta_k,$$

introduce the functions $\phi_1(t), \ldots, \phi_d(t)$ and calculate d integrals

$$Z_j = \int_0^T z_t\,\phi_j(t)\,\mathrm{d}t = \sum_{k=1}^d \int_0^T A_k(t)\,\phi_j(t)\,\mathrm{d}t\,\theta_k, \qquad j = 1, \ldots, d.$$

Suppose that the functions $A_k(\cdot)$ and $\phi_j(\cdot)$ are such that the $d \times d$ matrix $\mathbf{M} = (\mathrm{M}_{k,j})$

$$\mathrm{M}_{k,j} = \int_0^T A_k(t)\,\phi_j(t)\,\mathrm{d}t, \qquad k, j = 1, \ldots, d,$$

is nondegenerate. Then $\vartheta = \mathbf{M}^{-1}\mathrm{Z}$, where $\mathrm{Z} = (\mathrm{Z}_1, \ldots, \mathrm{Z}_d)^\top$.

For construction of the corresponding MME ϑ_ε^*, we have to estimate the derivative $x_t'(\vartheta)$. This can be done with the help of the statistic

$$\begin{aligned}\bar x_{t,\varepsilon}' &= \frac{X_{t+\delta_\varepsilon} - X_t}{\delta_\varepsilon} = \frac{1}{\delta_\varepsilon}\int_t^{t+\delta_\varepsilon} f(s)\,y_s(\vartheta_0)\,\mathrm{d}s + \frac{\varepsilon}{\delta_\varepsilon}\int_t^{t+\delta_\varepsilon} f(s)\,\check\eta_s(\vartheta_0)\,\mathrm{d}s \\ &\qquad + \frac{\varepsilon}{\delta_\varepsilon}\left[W_{t+\delta_\varepsilon} - W_t\right] \\ &\ge \kappa_1 + r_{t,\varepsilon}.\end{aligned}$$

Here $\delta_\varepsilon = \varepsilon^b$, $b \in (0, 1/2)$, $\sup_{t\in[0,T]} \left|r_{t,\varepsilon}\right| \to 0$ and

$$\kappa_1 = \inf_{t\in[0,T],\vartheta\in\Theta} f(t)\,y_t(\vartheta) > 0.$$

Introduce the statistics

$$z_{t,\varepsilon} = \mathbb{I}_{\{\mathbb{B}_\varepsilon\}} \ln\left(\frac{X_{t+\delta_\varepsilon} - X_t}{y_0 f(t)\,\delta_\varepsilon}\right),$$

where the set

$$\mathbb{B}_\varepsilon = \left(\omega \ : \quad \sup_{0\leq t\leq T} \left|r_{t,\varepsilon}\right| < \kappa_1/2\right)$$

and the MME estimator

$$\vartheta_\varepsilon^* = \mathbf{M}^{-1}\mathrm{Z}_\varepsilon, \qquad\qquad Z_{j,\varepsilon} = \int_0^T z_{t,\varepsilon}\,\phi_j\left(t\right)\mathrm{d}t, \quad j=1,\ldots,d.$$

Remark that for any $p>0$, there exists a constant C_p such that $\mathbf{P}_{\vartheta_0}\left(\mathbb{B}_\varepsilon^c\right) \leq C_p\varepsilon^p$ an on the set $\mathbb{B}_\varepsilon$

$$\frac{X_{t+\delta_\varepsilon}-X_t}{\delta_\varepsilon} \geq \kappa_1/2.$$

It is easy to see that $\vartheta_\varepsilon^* \longrightarrow \vartheta_0$. It is possible to show its asymptotic normality too.

MLE and BE

We have the model of partially observed linear system (2.1) and (2.2), where the unknown parameter is $\vartheta \in \Theta \subset \mathcal{R}^d$, the set Θ is open, bounded, and convex, and the condition $\mathscr{A}_*$ holds.

Introduce the notation: $S_*\left(\vartheta,\vartheta_0,t\right) = f\left(\vartheta,t\right)y_t\left(\vartheta,\vartheta_0\right)$,

$$\dot{\mathrm{S}}_*\left(\vartheta,\vartheta_0,t\right) = \frac{\partial S_*\left(\vartheta,\vartheta_0,t\right)}{\partial\vartheta} = \dot{\mathrm{f}}\left(\vartheta,t\right)y_t\left(\vartheta,\vartheta_0\right) + f\left(\vartheta,t\right)\dot{\mathrm{y}}_t\left(\vartheta,\vartheta_0\right),$$

$$\dot{\mathrm{y}}_t\left(\vartheta,\vartheta_0\right) = \int_0^t \Phi\left(\vartheta,s,t\right)\left[\dot{\mathrm{A}}\left(\vartheta,s\right)y_s\left(\vartheta,\vartheta_0\right) + \dot{\mathrm{B}}\left(\vartheta,s\right)f\left(\vartheta_0,s\right)y_s\left(\vartheta_0\right)\right]\mathrm{d}s,$$

$$\mathbf{I}\left(\vartheta_0\right) = \int_0^T \frac{\dot{\mathrm{S}}_*\left(\vartheta_0,\vartheta_0,t\right)\dot{\mathrm{S}}_*\left(\vartheta_0,\vartheta_0,t\right)^\top}{\sigma\left(t\right)^2}\mathrm{d}t, \tag{2.30}$$

$$g\left(\vartheta_0,\nu\right)^2 = \inf_{\left\|\vartheta-\vartheta_0\right\|>\nu}\int_0^T \frac{\left[S_*\left(\vartheta,\vartheta_0,t\right)-S_*\left(\vartheta_0,\vartheta_0,t\right)\right]^2}{\sigma\left(t\right)^2}\mathrm{d}t.$$

The $d\times d$ matrix $\mathbf{I}\left(\vartheta_0\right)$, $\vartheta_0\in\Theta$ is the Fisher information. Below, we verify that the model of observations is regular with LAN family of measures. Therefore, the *Hájek's lower bound* is valid, and we can write: for all estimators $\bar\vartheta_\varepsilon$ and all $\vartheta_0\in\Theta$

$$\lim_{\nu\to0}\varliminf_{\varepsilon\to0}\sup_{\left|\vartheta-\vartheta_0\right|\leq\nu}\varepsilon^{-2}\mathbf{E}_\vartheta\left\|\bar\vartheta_\varepsilon-\vartheta\right\|^2 \geq \mathbf{E}_{\vartheta_0}\left\|\zeta\left(\vartheta_0\right)\right\|^2, \quad \zeta\left(\vartheta_0\right)\sim\mathcal{N}\left(0,\mathbf{I}\left(\vartheta_0\right)^{-1}\right).$$

We call an estimator ϑ_ε^* *asymptotically efficient* if for all $\vartheta_0\in\Theta$

$$\lim_{\nu\to 0}\lim_{\varepsilon\to 0}\sup_{|\vartheta-\vartheta_0|\leq\nu}\varepsilon^{-2}\mathbf{E}_{\vartheta}\left\|\vartheta_{\varepsilon}^{*}-\vartheta\right\|^2=\mathbf{E}_{\vartheta_0}\left\|\zeta\left(\vartheta_0\right)\right\|^2.$$

The LR formula for this model of observations is the function

$$L\left(\vartheta,X^T\right)=\exp\left(\int_0^T\frac{f\left(\vartheta,t\right)m\left(\vartheta,t\right)}{\varepsilon^2\sigma\left(t\right)^2}\mathrm{d}X_t-\int_0^T\frac{f\left(\vartheta,t\right)^2m\left(\vartheta,t\right)^2}{2\varepsilon^2\sigma\left(t\right)^2}\mathrm{d}t\right),\ \vartheta\in\Theta. \tag{2.31}$$

The process $m\left(\vartheta,t\right),0\leq t\leq T$ is solution of the system of Eqs. (2.5) and (2.6).

Remind that the MLE and BE are defined by the relations

$$L(\hat\vartheta_{\varepsilon},X^T)=\sup_{\vartheta\in\Theta}L\left(\vartheta,X^T\right),\qquad\tilde\vartheta_{\varepsilon}=\frac{\int_{\Theta}\vartheta\, p\left(\vartheta\right)L\left(\vartheta,X^T\right)\mathrm{d}\vartheta}{\int_{\Theta}p\left(\vartheta\right)L\left(\vartheta,X^T\right)\mathrm{d}\vartheta}. \tag{2.32}$$

Here $p\left(\vartheta\right),\vartheta\in\Theta$ is positive continuous function (density a priory).

We have to study the asymptotic properties of these estimators as $\varepsilon\to 0$.

Introduce the *Conditions* $\hat{\mathscr{B}}$:

$\hat{\mathscr{B}}_1$. *The functions* $f(\vartheta,t),a(\vartheta,t),b(\vartheta,t),\vartheta\in\Theta,t\in[0,T]$ *are twice continuously differentiable on* $\bar\Theta$.

$\hat{\mathscr{B}}_2$. *The Fisher information matrix* $\mathbf{I}\left(\vartheta_0\right)$ *is uniformly nondegenerate*

$$\inf_{\vartheta_0\in\Theta}\ \inf_{\|\mathrm{e}\|=1,\mathrm{e}\in\mathcal{R}^d}\mathrm{e}^{\top}\mathbf{I}\left(\vartheta_0\right)\mathrm{e}>0.$$

$\hat{\mathscr{B}}_3$. *For all* $\nu>0$

$$\inf_{\vartheta_0\in\Theta}g\left(\vartheta_0,\nu\right)>0.$$

Theorem 2.4 *Let the conditions* $\hat{\mathscr{B}}$ *hold. Then the MLE and BE are uniformly on compacts* $\mathbb{K}\subset\Theta$ *consistent, asymptotically normal*

$$\frac{\hat\vartheta_{\varepsilon}-\vartheta_0}{\varepsilon}\Longrightarrow\zeta\left(\vartheta_0\right)\sim\mathcal{N}\left(0,\mathbf{I}\left(\vartheta_0\right)^{-1}\right),\qquad\frac{\tilde\vartheta_{\varepsilon}-\vartheta_0}{\varepsilon}\Longrightarrow\zeta\left(\vartheta_0\right). \tag{2.33}$$

The moments converge: for any $p>0$

$$\varepsilon^{-p}\mathbf{E}_{\vartheta_0}\|\hat\vartheta_{\varepsilon}-\vartheta_0\|^p\longrightarrow\mathbf{E}_{\vartheta_0}\left\|\zeta\left(\vartheta_0\right)\right\|^p,\ \varepsilon^{-p}\mathbf{E}_{\vartheta_0}\|\tilde\vartheta_{\varepsilon}-\vartheta_0\|^p\longrightarrow\mathbf{E}_{\vartheta_0}\left\|\zeta\left(\vartheta_0\right)\right\|^p.$$

The both estimators are asymptotically efficient.

Proof The proof is based on the Theorems 1.10 and 1.12, and therefore, we verify the corresponding properties of the normalized LR

$$Z_\varepsilon(\vartheta_0,u)=\frac{L\left(\vartheta_0+\varepsilon u,X^T\right)}{L\left(\vartheta_0,X^T\right)},\qquad u\in\mathbb{U}_\varepsilon=\left(u\ :\ \vartheta_0+\varepsilon u\in\Theta\right).$$

Let us denote $\vartheta_u=\vartheta_0+\varepsilon u$, $M(\vartheta_u,t)=f(\vartheta_u,t)\,m(\vartheta_u,t)$. Then

$$\ln Z_\varepsilon(\vartheta_0,u)=\int_0^T\frac{M(\vartheta_u,t)-M(\vartheta_0,t)}{\varepsilon\sigma(t)}\mathrm{d}\bar W_t-\int_0^T\frac{[M(\vartheta_u,t)-M(\vartheta_0,t)]^2}{2\varepsilon^2\sigma(t)^2}\mathrm{d}t.$$

We have to prove three lemmae describing the properties of $Z_\varepsilon(\vartheta_0,\cdot)$. □

Lemma 2.3 *The normalized LR $Z_\varepsilon(\cdot)$ admits the representation ULAN*

$$Z_\varepsilon(\vartheta_0,u)=\exp\left(u^\top\Delta_\varepsilon\left(\vartheta_0,X^T\right)-\frac{1}{2}u^\top\mathbf{I}(\vartheta_0)\,u+r_\varepsilon(\vartheta_0,u)\right),\qquad u\in\mathbb{U}_\varepsilon, \tag{2.34}$$

where uniformly on compacts $\mathbb{K}\subset\Theta$ we have the convergences

$$\Delta_\varepsilon\left(\vartheta_0,X^T\right)\Longrightarrow\mathcal{N}\left(0,\mathbf{I}(\vartheta_0)\right),\qquad r_\varepsilon(\vartheta_0,u)\longrightarrow 0.$$

Proof By Lemma 1.3, the random process $M(\vartheta,t)$ has continuous derivative $\dot{\mathrm{M}}(\vartheta_0,t)=\dot{\mathrm{f}}(\vartheta_0,t)\,m(\vartheta_0,t)+f(\vartheta_0,t)\,\dot{\mathrm{m}}(\vartheta_0,t)$ and this means, that

$$\mathbf{E}_{\vartheta_0}\left(M(\vartheta_u,t)-M(\vartheta_0,t)-\varepsilon\,u^\top\dot{\mathrm{M}}(\vartheta_0,t)\right)^2=o\left(\varepsilon^2\right).$$

Therefore,

$$\ln Z_\varepsilon(\vartheta_0,u)=u^\top\int_0^T\frac{\dot{\mathrm{M}}(\vartheta_0,t)}{\sigma(t)}\mathrm{d}\bar W_t-u^\top\int_0^T\frac{\dot{\mathrm{M}}(\vartheta_0,t)\,\dot{\mathrm{M}}(\vartheta_0,t)^\top}{2\sigma(t)^2}\mathrm{d}t\,u+o(1).$$

Recall that the equation for derivative $\dot{\mathrm{m}}(\vartheta_0,t)$ (see (1.35)) is

$$\begin{aligned}
\mathrm{d}\dot{\mathrm{m}}(\vartheta_0,t)&=A(\vartheta_0,t)\,\dot{\mathrm{m}}(\vartheta_0,t)\,\mathrm{d}t+\left[\dot{\mathrm{A}}(\vartheta_0,t)+\dot{\mathrm{B}}(\vartheta_0,t)\,f(\vartheta_0,t)\right]m(\vartheta_0,t)\,\mathrm{d}t\\
&\quad+\varepsilon\dot{\mathrm{B}}(\vartheta_0,t)\,\sigma(t)\,\mathrm{d}\bar W_t,\\
&=A(\vartheta_0,t)\,\dot{\mathrm{m}}(\vartheta_0,t)\,\mathrm{d}t+\left[\dot{\mathrm{a}}(\vartheta_0,t)-\mathrm{B}(\vartheta_0,t)\,\dot{f}(\vartheta_0,t)\right]m(\vartheta_0,t)\,\mathrm{d}t\\
&\quad+\varepsilon\dot{\mathrm{B}}(\vartheta_0,t)\,\sigma(t)\,\mathrm{d}\bar W_t,\qquad \dot{\mathrm{m}}(\vartheta_0,0)=0,
\end{aligned}$$

where

$$A(\vartheta,t)=a(\vartheta,t)-\frac{f(\vartheta,t)^2\,\gamma_*(\vartheta,t)}{\sigma(t)^2},\qquad B(\vartheta,t)=\frac{f(\vartheta,t)\,\gamma_*(\vartheta,t)}{\sigma(t)^2}.$$

Therefore,

$$
\begin{aligned}
&\dot{\mathrm{m}}(\vartheta_0, t) - \dot{\mathrm{y}}(\vartheta_0, \vartheta_0, t) \\
&\quad = \varepsilon \int_0^t \Phi(\vartheta_0, s, t)\left[\dot{\mathrm{a}}(\vartheta_0, s) - \mathrm{B}(\vartheta_0, s)\, \dot{f}(\vartheta_0, s)\right] \eta_s(\vartheta_0)\, \mathrm{d}s
\end{aligned}
$$

and we have convergence in probability

$$
\int_0^T \frac{\dot{\mathrm{M}}(\vartheta_0, t)\, \dot{\mathrm{M}}(\vartheta_0, t)^\top}{\sigma(t)^2} \mathrm{d}t \longrightarrow \int_0^T \frac{\dot{\mathrm{S}}_*(\vartheta_0, \vartheta_0, t)\, \dot{\mathrm{S}}_*(\vartheta_0, \vartheta_0, t)^\top}{\sigma(t)^2} \mathrm{d}t = \mathbf{I}(\vartheta_0).
$$

This convergence provides the asymptotic normality (see CLT Theorem 1.7)

$$
\int_0^T \frac{\dot{\mathrm{M}}(\vartheta_0, t)}{\sigma(t)} \mathrm{d}\bar{W}_t \Longrightarrow \mathcal{N}(0, \mathbf{I}(\vartheta_0)).
$$

It is easy to verify that the both convergences are uniform on compacts $\mathbb{K} \subset \Theta$.

Hence, the family of measures is LAN, i.e., the LR $Z_\varepsilon(\cdot)$ admits the representation (2.34), where

$$
\Delta_\varepsilon\left(\vartheta_0, X^T\right) = \int_0^T \frac{\dot{\mathrm{M}}(\vartheta_0, t)}{\varepsilon\sigma(t)^2} \left[\mathrm{d}X_t - f(\vartheta_0, t)\, m(\vartheta_0, t)\, \mathrm{d}t\right].
$$

□

Lemma 2.4 *For any $m > d/2$, there exists a constant $C > 0$ such that for any $R > 0$ and u_1, u_2 satisfying $\|u_1\|, \|u_2\| \leq R$, we have the estimate*

$$
\sup_{\vartheta_0 \in \mathbb{K}} \mathbf{E}_{\vartheta_0} \left| Z_\varepsilon(\vartheta_0, u_2)^{1/2m} - Z_\varepsilon(\vartheta_0, u_1)^{1/2m} \right|^{2m} \leq C\left(1 + R^{2m}\right) \|u_2 - u_1\|^{2m}
$$

Proof By Lemma 1.1

$$
\begin{aligned}
\mathbf{E}_{\vartheta_0} \left| Z_\varepsilon(\vartheta_0, u_2)^{1/2m} - Z_\varepsilon(\vartheta_0, u_1)^{1/2m} \right|^{2m} &\leq CT^{2m-1} \int_0^T \mathbf{E}_{\vartheta_{u_2}} F_t(u_2, u_1)^{4m}\, \mathrm{d}t \\
&\quad + CT^{m-1/2} \int_0^T \mathbf{E}_{\vartheta_{u_2}} F_t(u_2, u_1)^{2m}\, \mathrm{d}t.
\end{aligned}
$$

Now the estimates

$$
\mathbf{E}_{\vartheta_{u_2}} F_t(u_2, u_1)^{4m} \leq C\|u_2 - u_1\|^{4m}, \qquad \mathbf{E}_{\vartheta_{u_2}} F_t(u_2, u_1)^{2m} \leq C\|u_2 - u_1\|^{2m},
$$

follow from (1.39). □

Lemma 2.5 *For any $N > 0$, there exists a constant $C = C(N) > 0$, such that*

$$\sup_{\vartheta_0 \in \mathbb{K}} \mathbf{E}_{\vartheta_0} Z_\varepsilon(\vartheta_0, u)^{1/2} \leq \frac{C}{\|u\|^N}. \tag{2.35}$$

Proof Set $G(u,t) = \varepsilon^{-1}[S_*(\vartheta_u, \vartheta_0, t) - S_*(\vartheta_0, \vartheta_0, t)]\,\sigma(t)^{-1}$ and $G_\varepsilon(u,t) = F_t(u, 0)$. We obtain the estimates: for $\varepsilon\|u\| \leq \nu$ and sufficiently small $\nu > 0$ we have

$$\inf_{\vartheta_0 \in \mathbb{K}} \int_0^T G(u,t)^2 \,\mathrm{d}t = \inf_{\vartheta_0 \in \mathbb{K}} u^\top \mathbf{I}(\vartheta_0)\, u\,(1 + o(1)) \geq \frac{c_1}{2}\|u\|^2.$$

Here we used the condition $\hat{\mathscr{B}}_2$ of Theorem and denoted

$$c_1 = \inf_{\vartheta_0 \in \mathbb{K}} \inf_{\mathrm{e} \in \mathcal{R}^d,\ \|\mathrm{e}\|=1} \mathrm{e}^\top \mathbf{I}(\vartheta_0)\,\mathrm{e}.$$

For $\varepsilon\|u\| > \nu$ using condition $\hat{\mathscr{B}}_3$ of the same theorem, we can write

$$\inf_{\vartheta_0 \in \mathbb{K}} \int_0^T G(u,t)^2 \,\mathrm{d}t \geq \inf_{\vartheta_0 \in \mathbb{K}} \frac{g(\vartheta_0, \nu)^2}{\varepsilon^2} \geq \frac{c_2}{\varepsilon^2} \geq \frac{c_2\|u\|^2}{\sup_{\vartheta,\vartheta_0 \in \Theta}\|\vartheta - \vartheta_0\|} \geq c_3\|u\|^2.$$

Hence,

$$\inf_{\vartheta_0 \in \mathbb{K}} \int_0^T G(u,t)^2 \,\mathrm{d}t \geq c_*\|u\|^2$$

with $c_* = c_1 \wedge c_3$.

Consider the probability (below $c > 0$, $p > 0$)

$$\begin{aligned}
&\mathbf{P}_{\vartheta_0}\left(Z_\varepsilon(\vartheta_0, u) \geq e^{-c\|u\|^2}\right)\\
&\quad= \mathbf{P}_{\vartheta_0}\left(p\int_0^T G_\varepsilon(u,t)\,\mathrm{d}W_t - \frac{p}{2}\int_0^T G_\varepsilon(u,t)^2\,\mathrm{d}t \geq -pc\|u\|^2\right)\\
&\quad\leq \mathbf{P}_{\vartheta_0}\left(p\int_0^T G_\varepsilon(u,t)\,\mathrm{d}W_t - \frac{p^2}{2}\int_0^T G_\varepsilon(u,t)^2\,\mathrm{d}t\right.\\
&\qquad\left.+\frac{p^2-p}{2}\int_0^T G_\varepsilon(u,t)^2\,\mathrm{d}t + q\int_0^T G(u,t)^2\,\mathrm{d}t \geq -pc\|u\|^2 + qc_*\|u\|^2\right)\\
&\quad\leq \mathbf{P}_{\vartheta_0}\left(p\int_0^T G_\varepsilon(u,t)\,\mathrm{d}W_t - \frac{p^2}{2}\int_0^T G_\varepsilon(u,t)^2\,\mathrm{d}t \geq \frac{(qc_* - pc)}{2}\|u\|^2\right)\\
&\qquad+ \mathbf{P}_{\vartheta_0}\left(q\int_0^T G(u,t)^2\,\mathrm{d}t - \frac{p-p^2}{2}\int_0^T G_\varepsilon(u,t)^2\,\mathrm{d}t \geq \frac{(qc_* - pc)}{2}\|u\|^2\right).
\end{aligned} \tag{2.36}$$

Let us put $\mu = (qc_* - pc)/2 > 0$ and $q = \left(p - p^2\right)/2 > 0$, then

$$\mathbf{P}_{\vartheta_0}\left(p\int_0^T G_\varepsilon(u,t)\,\mathrm{d}W_t - \frac{p^2}{2}\int_0^T G_\varepsilon(u,t)^2\,\mathrm{d}t \geq \mu\,\|u\|^2\right) \leq e^{-\mu\|u\|^2}.$$

Consider the difference

$$\begin{aligned}&\varepsilon\sigma(t)\left[G(u,t) - G_\varepsilon(u,t)\right]\\&\quad = S_*(\vartheta_u,\vartheta_0,t) - M(\vartheta_u,t) - S_*(\vartheta_0,\vartheta_0,t) + M(\vartheta_0,t).\end{aligned}$$

Recall that by Lemma 2.1

$$\begin{aligned}M(\vartheta_u,t) - S_*(\vartheta_u,\vartheta_0,t) &= \varepsilon\, f(\vartheta_u,t)\,\hat\eta_t(\vartheta_u)\\&\quad + \varepsilon f(\vartheta_u,t)\int_0^t H(\vartheta_u,\vartheta_0,s,t)\,\tilde\eta_s(\vartheta_0)\,\mathrm{d}s,\\M(\vartheta_0,t) - S_*(\vartheta_0,\vartheta_0,t) &= \varepsilon\, f(\vartheta_0,t)\,\tilde\eta_t(\vartheta_0)\end{aligned}$$

where

$$\begin{aligned}H(\vartheta,\vartheta_0,s,t) &= \Phi(\vartheta,s,t)\,B(\vartheta,s)\,f(\vartheta_0,s),\\\tilde\eta_t(\vartheta_0) &= \int_0^t \phi(\vartheta_0,s,t)\,B(\vartheta_0,s)\,\sigma(s)\,\mathrm{d}\bar W_s,\\\hat\eta_t(\vartheta) &= \int_0^t \Phi(\vartheta,s,t)\,B(\vartheta,s)\,\sigma(s)\,\mathrm{d}\bar W_s.\end{aligned}$$

Therefore,

$$\begin{aligned}\sigma(t)\left[G(u,t) - G_\varepsilon(u,t)\right] &= f(\vartheta_0,t)\,\tilde\eta_t(\vartheta_0) - f(\vartheta_u,t)\,\hat\eta_t(\vartheta_u)\\&\quad - f(\vartheta_u,t)\int_0^t H(\vartheta_u,\vartheta_0,s,t)\,\tilde\eta_s(\vartheta_0)\,\mathrm{d}s.\end{aligned}$$

All (deterministic) functions have bounded derivatives w.r.t. ϑ and the derivative of $\hat\eta_t(\vartheta)$ is a Gaussian process with finite variance. This allows us to write the estimate: for any $N \geq 2$

$$\mathbf{E}_{\vartheta_0}\left|G(u,t) - G_\varepsilon(u,t)\right|^N \leq C\,\varepsilon^N\,\|u\|^N$$

We have as well: for any $N > 2$

$$\varepsilon^N\mathbf{E}_{\vartheta_0}\left|G(u,t) + G_\varepsilon(u,t)\right|^N \leq C.$$

From these two estimates, we obtain

$$\mathbf{E}_{\vartheta_0}\left|G\left(u,t\right)^2-G_\varepsilon\left(u,t\right)^2\right|^N\leq C\ \|u\|^N.$$

For the last probability in (2.36), we can write

$$\begin{aligned}
&\mathbf{P}_{\vartheta_0}\left(\int_0^T\left[G\left(u,t\right)^2-G_\varepsilon\left(u,t\right)^2\right]\mathrm{d}t\geq\frac{\mu}{q}\ \|u\|^2\right)\\
&\qquad\leq\left(\frac{q}{\mu\ \|u\|^2}\right)^N\mathbf{E}_{\vartheta_0}\left(\int_0^T\left[G\left(u,t\right)^2-G_\varepsilon\left(u,t\right)^2\right]\mathrm{d}t\right)^N\leq\frac{C}{\|u\|^N}
\end{aligned}$$

Finally, we obtain: for any $N>2$, there exist constants $c>0$, $L_0>0$ and $C_*>0$ such that for all $\|u\|\geq L_0$

$$\mathbf{P}_{\vartheta_0}\left(Z_\varepsilon\left(u\right)\geq\frac{1}{\|u\|^{4N}}\right)\leq\mathbf{P}_{\vartheta_0}\left(Z_\varepsilon\left(u\right)\geq e^{-c\|u\|^2}\right)\leq\frac{C_*}{\|u\|^{2N}}.$$

As N is arbitrary we can always replace N by $2N$. Therefore,

$$\begin{aligned}
\mathbf{E}_{\vartheta_0}Z_\varepsilon^{1/2}\left(\vartheta_0,u\right)&=\mathbf{E}_{\vartheta_0}Z_\varepsilon^{1/2}\left(\vartheta_0,u\right)\mathbb{I}_{\left\{Z_\varepsilon^{1/2}(\vartheta_0,u)<\frac{1}{\|u\|^{2N}}\right\}}\\
&\quad+\mathbf{E}_{\vartheta_0}Z_\varepsilon^{1/2}\left(\vartheta_0,u\right)\mathbb{I}_{\left\{Z_\varepsilon^{1/2}(\vartheta_0,u)>\frac{1}{\|u\|^{2N}}\right\}}\\
&\leq\|u\|^{-2N}+\left(\mathbf{E}_{\vartheta_0}Z_\varepsilon\left(\vartheta_0,u\right)\right)^{1/2}\mathbf{P}_{\vartheta_0}\left(Z_\varepsilon^{1/2}\left(\vartheta_0,u\right)>\|u\|^{-2N}\right)^{1/2}\\
&\leq\frac{C}{\|u\|^N}.
\end{aligned}$$

□

Now the properties of the MLE and BE follow from the mentioned Theorems 1.10 and 1.12.

Remark 2.6 It is easy to see that if $f\left(\vartheta,t\right)=f\left(t\right)$, $a\left(\vartheta,t\right)=a\left(t\right)$ and $b\left(\vartheta,t\right)$ is the only function in the model, which depends of ϑ, then the consistent estimation is impossible. Let us verify that $g\left(\vartheta_0,\nu\right)=0$ for any $\nu>0$. It is sufficient to show that

$$\int_0^T f\left(t\right)^2\left[y_t\left(\vartheta,\vartheta_0\right)-y_t\left(\vartheta_0\right)\right]^2\mathrm{d}t=0 \tag{2.37}$$

for any $\vartheta\in\Theta$, i.e., $y_t\left(\vartheta,\vartheta_0\right)=y_t\left(\vartheta_0\right),0\leq t\leq T$. We can write

$$
\begin{aligned}
y_t(\vartheta,\vartheta_0) &= y_0\Phi(\vartheta,0,t)+\int_0^t \Phi(\vartheta,s,t)\,\frac{f(s)^2\gamma_*(\vartheta,s)}{\sigma(s)^2}\,y_0(s)\,\mathrm{d}s\\
&= y_0\Phi(\vartheta,0,t)\\
&\quad + y_0\Phi(\vartheta,0,t)\int_0^t \exp\left(\int_0^s \frac{f(v)^2\gamma_*(\vartheta,v)}{\sigma(v)^2}\mathrm{d}v\right)\mathrm{d}\\
&\quad\times\left(\int_0^s \frac{f(v)^2\gamma_*(\vartheta,v)}{\sigma(v)^2}\mathrm{d}v\right)\\
&= y_0\Phi(\vartheta,0,t)+y_0\Phi(\vartheta,0,t)\left[\exp\left(\int_0^t \frac{f(s)^2\gamma_*(\vartheta,s)}{\sigma(s)^2}\mathrm{d}v\right)-1\right]\\
&= y_0\phi(\vartheta,0,t)=y_t(\vartheta_0)\,.
\end{aligned}
$$

Therefore, we proved (2.37) and $g(\vartheta_0,\nu)=0$ for any ν.

The presence of ϑ in the function $b(\vartheta,t)$ is in some sense *noninformative*, but we less it in the description of the model (2.6) because the real system can have $b(\cdot)$ depending of ϑ and the calculation of the filter and estimators depends on the true $b(\vartheta,t)$.

Remark 2.7 Consider the situation similar to that of Remark 2.6 but for the MLE $\hat\vartheta_{\tau_\varepsilon}$, i.e., where the observations are X^{τ_ε} and $\tau_\varepsilon\rightarrow 0$. Without discussing the underlying conditions, we just describe the limit variance. We have

$$
\mathbf{I}_0^{\tau_\varepsilon}(\vartheta_0)=\tau_\varepsilon\frac{y_0^2\dot f(\vartheta_0,0)^2}{\sigma(0)^2}\left(1+o(1)\right).
$$

Hence, it is possible to show that

$$
\frac{\tau_\varepsilon^{1/2}}{\varepsilon}\left(\hat\vartheta_{\tau_\varepsilon}-\vartheta_0\right)\Longrightarrow\mathcal{N}\left(0,D(\vartheta_0)^2\right),\qquad D(\vartheta_0)^2=\frac{\sigma(0)^2}{y_0^2\dot f(\vartheta_0,0)^2}.
$$

We see that the gain of the MLE with respect to MDE (2.18) is just 20%, but the calculation of MLE involves the solution of filtration equations (2.5) and (2.6) for many values of ϑ and therefore is much more complicate. That is why the use of MDE can be more interesting in the problems, where the time of calculation is more important parameter than asymptotic efficiency of the estimator.

One-Step MLE-process

The calculation of the MLE $\hat\vartheta_\varepsilon$ and BE $\tilde\vartheta_\varepsilon$ requires knowledge of the solutions $m(\vartheta,t)$ and $\gamma(\vartheta,t)$ for $0\le t\le T$ of the Eqs. (2.5) and (2.6) for $\vartheta\in\Theta\subset\mathcal{R}^d$.

With these solutions, one needs to compute the likelihood ratio (LR) $L\left(\vartheta, X^T\right)$ using the formula (2.31) and then calculate the estimators according to their definitions in (2.33). Of course, the realization of these procedures will require a significant amount of calculations, and as a result, the obtained estimators will be approximations of $\hat{\vartheta}_\varepsilon$ and $\tilde{\vartheta}_\varepsilon$.

In this section, our goal is to construct a One-step MLE $\vartheta^\star_\varepsilon$ (see (1.89)) and a One-step MLE process $\vartheta^\star_{t,\varepsilon}$ for $\tau_\varepsilon < t \leq T$ (see (1.91)). These estimators have the same asymptotic properties as the MLE and BE but require substantially fewer calculations. Moreover, in the second case, we obtain an *on-line* estimator of the unknown parameter, which can be expressed in a recurrent form. This construction is based on the well-known Le Cam's One-step MLE .

Recall the model (2.1) and (2.2). The observations $X^T = (X_t, 0 \leq t \leq T)$ satisfy the equation

$$\mathrm{d}X_t = f\left(\vartheta, t\right) Y_t \,\mathrm{d}t + \varepsilon\sigma\left(t\right)\mathrm{d}W_t, \quad X_0 = 0, \quad 0 \leq t \leq T, \tag{2.38}$$

where the Gaussian process $Y^T = (Y_t, 0 \leq t \leq T)$ is not observable (hidden) and satisfies the Ornstein–Uhlenbeck (O-U) equation

$$\mathrm{d}Y_t = a\left(\vartheta, t\right) Y_t \,\mathrm{d}t + \varepsilon b\left(\vartheta, t\right)\,\mathrm{d}V_t, \quad Y_0 = y_0. \tag{2.39}$$

The Wiener processes W^T and V^T are independent. Note that the process (2.38) can be represented as

$$\mathrm{d}X_t = f\left(\vartheta, t\right) m(\vartheta, t) \,\mathrm{d}t + \varepsilon\sigma\left(t\right)\mathrm{d}\bar{W}_t, \quad X_0 = 0, \quad 0 \leq t \leq T,$$

where $\bar{W}^T$ is the *innovation* Wiener process (see Lemma 1.12).

We assume that the conditions $\mathscr{A}_*$ are satisfied. Therefore, by Theorem 1.3, the conditional expectation $m(\vartheta, t) = \mathbf{E}_\vartheta\left(Y_t|\mathfrak{F}^X_t\right)$ and the mean-square error of estimation $\gamma_*\left(\vartheta, t\right) = \varepsilon^{-2}\mathbf{E}_\vartheta\left(Y_t - m\left(\vartheta, t\right)\right)^2$ are solutions of the Eqs. (2.5) and (2.6), which are repeated here:

$$\mathrm{d}m(\vartheta, t) = a(\vartheta, t)m(\vartheta, t)\,\mathrm{d}t + \frac{f(\vartheta, t)\gamma_*(\vartheta, t)}{\sigma(t)^2}\left[\mathrm{d}X_t - f(\vartheta, t)m(\vartheta, t)\,\mathrm{d}t\right], \tag{2.40}$$

$$\frac{\partial\gamma_*(\vartheta, t)}{\partial t} = 2a(\vartheta, t)\gamma_*(\vartheta, t) - \frac{f(\vartheta, t)^2\gamma_*(\vartheta, t)^2}{\sigma(t)^2} + b(\vartheta, t)^2, \quad 0 \leq t \leq T. \tag{2.41}$$

with initial values $m(\vartheta, 0) = y_0$ and $\gamma_*(\vartheta, 0) = 0$. The true value, as before, is denoted by ϑ_0.

The construction of the one-step MLE process is proposed in two stages. First, we consider the case of a multidimensional ($d \geq 1$) parameter ϑ and a fixed preliminary interval $[0, \tau]$. Then we study the case of a one-dimensional parameter ϑ with $|\dot{f}(\vartheta, 0)| > c > 0$ but with a vanishing learning interval $[0, \tau_\varepsilon]$, where $\tau_\varepsilon \to 0$.

Fixed Learning Interval

Let us fix an interval $[0, \tau]$, where τ is small and $\tau < T$, and consider the MDE $\check{\vartheta}_{\tau,\varepsilon}$ constructed from the observations $X^\tau = (X_s, 0 \le s \le \tau)$:

$$\check{\vartheta}_{\tau,\varepsilon} = \arg\min_{\vartheta\in\Theta} \int_0^\tau [X_s - x_s(\vartheta)]^2\, ds.$$

Suppose that the conditions of Theorem 2.2 are fulfilled for $T = \tau$. Then this estimator is consistent, asymptotically normal (see (2.12)), and the moments converge (see (2.13)).

The Fisher score function for this model is

$$\Delta_\varepsilon\left(\vartheta, X^T\right) = \int_0^T \frac{\dot{\mathrm{M}}(\vartheta, t)}{\varepsilon^2\sigma(t)^2}\left[\mathrm{d}X_t - M(\vartheta, t)\,\mathrm{d}t\right], \qquad M(\vartheta, t) = f(\vartheta, t)\, m(\vartheta, t).$$

Therefore, the One-step MLE $\vartheta_\varepsilon^\star$ and One-step MLE process $\vartheta_{t,\varepsilon}^\star, \tau < t \le T$ are defined by the relations

$$\vartheta_\varepsilon^\star = \check{\vartheta}_{\tau,\varepsilon} + \mathbf{I}_\tau^T\left(\check{\vartheta}_{\tau,\varepsilon}\right)^{-1} \int_\tau^T \frac{\dot{\mathrm{M}}\left(\check{\vartheta}_{\tau,\varepsilon}, s\right)}{\sigma(s)^2}\left[\mathrm{d}X_s - M\left(\check{\vartheta}_{\tau,\varepsilon}, s\right)\mathrm{d}s\right],$$

$$\vartheta_{t,\varepsilon}^\star = \check{\vartheta}_{\tau,\varepsilon} + \mathbf{I}_\tau^t\left(\check{\vartheta}_{\tau,\varepsilon}\right)^{-1} \int_\tau^t \frac{\dot{\mathrm{M}}\left(\check{\vartheta}_{\tau,\varepsilon}, s\right)}{\sigma(s)^2}\left[\mathrm{d}X_s - M\left(\check{\vartheta}_{\tau,\varepsilon}, s\right)\mathrm{d}s\right], \qquad \tau < t \le T. \tag{2.42}$$

Here

$$\mathbf{I}_\tau^t(\vartheta) = \int_\tau^t \frac{\dot{\mathrm{S}}_*(\vartheta_0, \vartheta_0, s)\, \dot{\mathrm{S}}_*(\vartheta_0, \vartheta_0, s)^\top}{\sigma(s)^2}\mathrm{d}s,$$

$$\dot{\mathrm{S}}_*(\vartheta, \vartheta_0, s) = \dot{\mathrm{f}}(\vartheta, s)\, y_s(\vartheta, \vartheta_0) + f(\vartheta, \vartheta_0)\, \dot{\mathrm{y}}_s(\vartheta, \vartheta_0).$$

The random processes $M\left(\check{\vartheta}_{\tau,\varepsilon}, s\right), \tau \le s \le T$ and $\dot{\mathrm{M}}\left(\check{\vartheta}_{\tau,\varepsilon}, s\right), \tau \le s \le T$ are defined as follows. The process $M\left(\check{\vartheta}_{\tau,\varepsilon}, s\right) = f\left(\check{\vartheta}_{\tau,\varepsilon}, s\right) m\left(\check{\vartheta}_{\tau,\varepsilon}, s\right)$, where $m(\vartheta, s), \tau \le t \le T$ is solution of the Eqs. (2.40) and (2.41) for the value $\vartheta = \check{\vartheta}_{\tau,\varepsilon}$. To obtain the initial value $m\left(\check{\vartheta}_{\tau,\varepsilon}, \tau\right)$, we write the equation for $M(\vartheta, t)$ using (2.40)

$$\mathrm{d}M(\vartheta, t) = \bar{A}(\vartheta, t)\, M(\vartheta, t)\,\mathrm{d}t + \bar{B}(\vartheta, t)\,\mathrm{d}X_t, \tag{2.43}$$

where $M(\vartheta, 0) = f(\vartheta, 0)\, y_0$, and we denoted

$$\bar{A}(\vartheta, t) = \frac{f'(\vartheta, t)}{f(\vartheta, t)} + a(\vartheta, t) - \bar{B}(\vartheta, t), \qquad \bar{B}(\vartheta, t) = \frac{\gamma_*(\vartheta, t)\, f(\vartheta, t)^2}{\sigma(t)^2}.$$

The solution of (2.43) by Lemma 1.3 can be written as follows:

$$\begin{aligned} M(\vartheta, \tau) = M(\vartheta, 0)\, \bar{\Phi}(\vartheta, 0, \tau) &+ \int_0^\tau X_s \bar{\Phi}(\vartheta, s, \tau)\, \bar{A}'_s(\vartheta, s)\, \bar{B}(\vartheta, s)\, \mathrm{d}s \\ &+ X_\tau \bar{B}(\vartheta, \tau) - \int_0^\tau X_s \bar{\Phi}(\vartheta, s, \tau)\, \bar{B}'_s(\vartheta, s)\, \mathrm{d}s. \end{aligned}$$

Here

$$\bar{\Phi}(\vartheta, s, \tau) = \exp\left(\int_s^t \bar{A}(\vartheta, r)\, \mathrm{d}r\right).$$

Therefore, we can put

$$\begin{aligned} M\left(\check{\vartheta}_{\tau,\varepsilon}, \tau\right) &= M\left(\check{\vartheta}_{\tau,\varepsilon}, 0\right) \bar{\Phi}\left(\check{\vartheta}_{\tau,\varepsilon}, 0, \tau\right) \\ &\quad + \int_0^\tau X_s \bar{\Phi}\left(\check{\vartheta}_{\tau,\varepsilon}, s, \tau\right) \bar{A}'_s\left(\check{\vartheta}_{\tau,\varepsilon}, s\right) \bar{B}\left(\check{\vartheta}_{\tau,\varepsilon}, s\right) \mathrm{d}s \\ &\quad + X_\tau \bar{B}\left(\check{\vartheta}_{\tau,\varepsilon}, \tau\right) - \int_0^\tau X_s \bar{\Phi}\left(\check{\vartheta}_{\tau,\varepsilon}, s, \tau\right) \bar{B}'_s\left(\check{\vartheta}_{\tau,\varepsilon}, s\right) \mathrm{d}s. \end{aligned} \tag{2.44}$$

The random process $\hat{M}(t)$, $\tau \le t \le T$ is defined by the Eq. (2.43), where $\vartheta = \check{\vartheta}_\tau$, i.e.,

$$\mathrm{d}M\left(\check{\vartheta}_{\tau,\varepsilon}, t\right) = \bar{A}\left(\check{\vartheta}_{\tau,\varepsilon}, t\right) M\left(\check{\vartheta}_{\tau,\varepsilon}, t\right) \mathrm{d}t + \bar{B}\left(\check{\vartheta}_{\tau,\varepsilon}, t\right) \mathrm{d}X_t, \qquad \tau \le t \le T \tag{2.45}$$

with the initial value $M\left(\check{\vartheta}_{\tau,\varepsilon}, \tau_\varepsilon\right)$.

For the vector process $\dot{\mathrm{M}}\left(\check{\vartheta}_{\tau,\varepsilon}, t\right)$, $\tau \le t \le T$, we have the similar construction. Using (1.37), we obtain the equation for $\dot{\mathrm{M}}\left(\check{\vartheta}_{\tau,\varepsilon}, t\right)$

$$\begin{aligned} \mathrm{d}\dot{\mathrm{M}}\left(\check{\vartheta}_{\tau,\varepsilon}, t\right) &= \bar{A}\left(\check{\vartheta}_{\tau,\varepsilon}, t\right) \dot{\mathrm{M}}\left(\check{\vartheta}_{\tau,\varepsilon}, t\right) \mathrm{d}t + \dot{\bar{\mathrm{A}}}\left(\check{\vartheta}_{\tau,\varepsilon}, t\right) \mathrm{M}\left(\check{\vartheta}_{\tau,\varepsilon}, t\right) \mathrm{d}t \\ &\quad + \dot{\bar{\mathrm{B}}}\left(\check{\vartheta}_{\tau,\varepsilon}, t\right) \mathrm{d}X_t, \qquad \dot{\mathrm{M}}\left(\check{\vartheta}_\tau, \tau\right), \end{aligned} \tag{2.46}$$

where $t \in [\tau, T]$. The initial value $\dot{M}\left(\check{\vartheta}_\tau, \tau\right)$ can be calculated by the same way as it was done above for $M\left(\check{\vartheta}_\tau, \tau\right)$.

Now the processes $M\left(\check{\vartheta}_\tau, t\right), \dot{M}\left(\check{\vartheta}_\tau, t\right), \tau < t \leq T$ are defined, and the One-step estimators (2.45) and (2.46) can be calculated.

Introduce the conditions $\mathscr{B}^\star$:

$\mathscr{B}_1^\star$. *The conditions of the Theorem 2.2 with T replaced by τ are fulfilled.*

$\mathscr{B}_2^\star$. *The functions $f(\vartheta, t), a(\vartheta, t), \dot{f}(\vartheta, t), \dot{a}(\vartheta, t), \sigma(t), t \in [0, T], \vartheta \in \Theta$ are continuously differentiable on t.*

$\mathscr{B}_3^\star$. *For any $\tau \in (0, T]$ and $t \in (\tau, T]$*

$$\inf_{\vartheta_0 \in \Theta} \inf_{e \in \mathcal{R}^d, \|e\|=1} e^\top \mathbf{I}_\tau^t(\vartheta_0) e > 0.$$

Theorem 2.5 *Suppose that the conditions $\mathscr{B}^\star$ are satisfied. Then the One-step MLE process $\vartheta_{t,\varepsilon}^\star, \tau < t \leq T$ is uniformly consistent: for any $\nu > 0$ and any $t_0 \in (\tau, T]$*

$$\mathbf{P}_{\vartheta_0}\left(\sup_{t_0 \leq t \leq T} \left\|\vartheta_{t,\varepsilon}^\star - \vartheta_0\right\| \geq \nu\right) \rightarrow 0, \tag{2.47}$$

asymptotically normal

$$\frac{\vartheta_{t,\varepsilon}^\star - \vartheta_0}{\varepsilon} \Longrightarrow \zeta_t(\vartheta_0) \sim \mathcal{N}\left(0, \mathbf{I}_{t_0}^t(\vartheta_0)^{-1}\right), \tag{2.48}$$

the moments converge: for any $p \geq 0$

$$\varepsilon^{-p}\mathbf{E}_{\vartheta_0}\left\|\vartheta_{t,\varepsilon}^\star - \vartheta_0\right\|^p \longrightarrow \mathbf{E}_{\vartheta_0}\left\|\zeta_t(\vartheta_0)\right\|^p.$$

The random vector process $\zeta_{t,\varepsilon}^\star = \varepsilon^{-1}\left(\vartheta_{t,\varepsilon}^\star - \vartheta_0\right), t_0 \leq t \leq T$ converges in distribution in the measurable space $(\mathcal{C}_d[t_0, T], \mathfrak{B})$ to the Gaussian vector process $\zeta_t(\vartheta_0), t_0 \leq t \leq T$, where

$$\zeta_t(\vartheta_0) = \mathbf{I}_{t_0}^t(\vartheta)^{-1} \int_{t_0}^t \frac{\dot{S}_*(\vartheta_0, \vartheta_0, s)}{\sigma(s)} \mathrm{d}w(s).$$

Proof The proof is similar to the proof of Theorem 2.3. The uniform consistency (2.47) follows from the estimate:

$$\sup_{t_0 \leq t \leq T} \left\|\vartheta_{t,\varepsilon}^\star - \vartheta_0\right\| \leq \left\|\check{\vartheta}_{\tau,\varepsilon} - \vartheta_0\right\| + C\varepsilon \sup_{t_0 \leq t \leq T} \left\|\int_\tau^t \frac{\dot{M}\left(\check{\vartheta}_{\tau,\varepsilon}, s\right)}{\sigma(s)} \mathrm{d}W_s\right\|$$

$$+C\int_{\tau}^{T}\left\|\dot{\mathrm{M}}\left(\check{\vartheta}_{\tau,\varepsilon},s\right)\right\|\left|M\left(\vartheta_0,s\right)-M\left(\check{\vartheta}_{\tau,\varepsilon},s\right)\right|\mathrm{d}s. \tag{2.49}$$

As $\dot{M}\left(\check{\vartheta}_{\tau,\varepsilon},s\right)$ is bounded in probability, the RHS of this inequality tends to zero.

To prove the asymptotic normality (2.48) we write

$$\begin{aligned}\frac{\vartheta_{t,\varepsilon}^{\star}-\vartheta_0}{\varepsilon}&=\frac{\check{\vartheta}_{\tau,\varepsilon}-\vartheta_0}{\varepsilon}+\mathbf{I}_{\tau}^{t}\left(\check{\vartheta}_{\tau,\varepsilon}\right)^{-1}\int_{\tau}^{t}\frac{\dot{\mathrm{M}}\left(\check{\vartheta}_{\tau,\varepsilon},s\right)}{\sigma\left(s\right)}\mathrm{d}\bar{W}_s\\&\quad+\mathbf{I}_{\tau}^{t}\left(\check{\vartheta}_{\tau,\varepsilon}\right)^{-1}\int_{\tau}^{t}\frac{\dot{\mathrm{M}}\left(\check{\vartheta}_{\tau,\varepsilon},s\right)}{\varepsilon\sigma\left(s\right)^2}\left[M\left(\vartheta_0,s\right)-M\left(\check{\vartheta}_{\tau,\varepsilon},s\right)\right]\mathrm{d}s.\end{aligned}$$

The consistency of $\check{\vartheta}_{\tau,\varepsilon}$ allows us to write

$$\begin{aligned}&\mathbf{I}_{\tau}^{t}\left(\check{\vartheta}_{\tau,\varepsilon}\right)^{-1}\int_{\tau}^{t}\frac{\dot{\mathrm{M}}\left(\check{\vartheta}_{\tau,\varepsilon},s\right)}{\sigma\left(s\right)}\mathrm{d}\bar{W}_s\\&\qquad=\mathbf{I}_{\tau}^{t}\left(\vartheta_0\right)^{-1}\int_{\tau}^{t}\frac{\dot{\mathrm{M}}\left(\vartheta_0,s\right)}{\sigma\left(s\right)}\mathrm{d}\bar{W}_s\left(1+o\left(\left\|\check{\vartheta}_{\tau,\varepsilon}-\vartheta_0\right\|\right)\right).\end{aligned}$$

We have as well the convergencies

$$\mathbf{I}_{\tau}^{t}\left(\check{\vartheta}_{\tau,\varepsilon}\right)\longrightarrow\mathbf{I}_{\tau}^{t}\left(\vartheta_0\right),\qquad\qquad\dot{\mathrm{M}}\left(\vartheta_0,s\right)\longrightarrow\dot{\mathrm{S}}\left(\vartheta_0,\vartheta_0,s\right).$$

Therefore,

$$\mathbf{I}_{\tau}^{t}\left(\check{\vartheta}_{\tau,\varepsilon}\right)^{-1}\int_{\tau}^{t}\frac{\dot{\mathrm{M}}\left(\check{\vartheta}_{\tau,\varepsilon},s\right)}{\sigma\left(s\right)}\mathrm{d}\bar{W}_s\Longrightarrow\mathbf{I}_{\tau}^{t}\left(\vartheta_0\right)^{-1}\int_{\tau}^{t}\frac{\dot{\mathrm{S}}\left(\vartheta_0,\vartheta_0,s\right)}{\sigma\left(s\right)}\mathrm{d}w\left(s\right)=\zeta_t\left(\vartheta_0\right),$$

where $w\left(s\right),\tau\leq s\leq t$ is some Wiener process.

The ordinary integral has the following asymptotics:

$$\begin{aligned}&\int_{\tau}^{t}\frac{\dot{\mathrm{M}}\left(\check{\vartheta}_{\tau,\varepsilon},s\right)}{\varepsilon\sigma\left(s\right)^2}\left[M\left(\vartheta_0,s\right)-M\left(\check{\vartheta}_{\tau,\varepsilon},s\right)\right]\mathrm{d}s\\&\quad=\int_{\tau}^{t}\frac{\dot{\mathrm{M}}\left(\check{\vartheta}_{\tau,\varepsilon},s\right)}{\varepsilon\sigma\left(s\right)^2}\int_0^1\left(\vartheta_0-\check{\vartheta}_{\tau,\varepsilon}\right)^{\top}\dot{\mathrm{M}}\left(\check{\vartheta}_{\tau,\varepsilon}+v\left(\vartheta_0-\check{\vartheta}_{\tau,\varepsilon}\right),s\right)\mathrm{d}v\,\mathrm{d}s\end{aligned}$$

$$
= \int_{\tau}^{t} \frac{\dot{\mathrm{M}}\left(\check{\vartheta}_{\tau,\varepsilon}, s\right)\dot{\mathrm{M}}\left(\check{\vartheta}_{\tau,\varepsilon}, s\right)^{\top}}{\varepsilon\sigma\left(s\right)^{2}}\left(\vartheta_0 - \check{\vartheta}_{\tau,\varepsilon}\right)\mathrm{d}s\left(1 + o\left(\left\|\vartheta_0 - \check{\vartheta}_{\tau,\varepsilon}\right\|\right)\right)
$$

$$
= \int_{\tau}^{t} \frac{\dot{\mathrm{S}}\left(\check{\vartheta}_{\tau,\varepsilon}, \vartheta_0, s\right)\dot{\mathrm{S}}\left(\check{\vartheta}_{\tau,\varepsilon}, \vartheta_0, s\right)^{\top}}{\varepsilon\sigma\left(s\right)^{2}}\left(\vartheta_0 - \check{\vartheta}_{\tau,\varepsilon}\right)\mathrm{d}s\left(1 + o\left(\left\|\vartheta_0 - \check{\vartheta}_{\tau,\varepsilon}\right\|\right)\right)
$$

because $\dot{\mathrm{M}}\left(\vartheta, s\right) - \dot{\mathrm{S}}\left(\vartheta, \vartheta_0, s\right) = O\left(\varepsilon\right)$. Therefore

$$
\begin{aligned}
\mathbf{I}_{\tau}^{t}\left(\check{\vartheta}_{\tau,\varepsilon}\right)^{-1}\int_{\tau}^{t}\frac{\dot{\mathrm{M}}\left(\check{\vartheta}_{\tau,\varepsilon}, s\right)}{\varepsilon\sigma\left(s\right)^{2}}\left[M\left(\vartheta_0, s\right) - M\left(\check{\vartheta}_{\tau,\varepsilon}, s\right)\right]\mathrm{d}s \\
= \frac{\left(\vartheta_0 - \check{\vartheta}_{\tau,\varepsilon}\right)}{\varepsilon}\left(1 + o\left(\left\|\vartheta_0 - \check{\vartheta}_{\tau,\varepsilon}\right\|\right)\right)
\end{aligned}
$$

and we obtain the convergence (2.48)

$$
\begin{aligned}
\frac{\vartheta_{t,\varepsilon}^{\star} - \vartheta_0}{\varepsilon} &= \frac{\check{\vartheta}_{\tau,\varepsilon} - \vartheta_0}{\varepsilon} + \frac{\vartheta_0 - \check{\vartheta}_{\tau,\varepsilon}}{\varepsilon}\left(1 + o\left(\left\|\vartheta_0 - \check{\vartheta}_{\tau,\varepsilon}\right\|\right)\right) \\
&\quad + \mathbf{I}_{\tau}^{t}\left(\check{\vartheta}_{\tau,\varepsilon}\right)^{-1}\int_{\tau}^{t}\frac{\dot{\mathrm{M}}\left(\check{\vartheta}_{\tau,\varepsilon}, s\right)}{\sigma\left(s\right)}\mathrm{d}\bar{W}_s \\
&= \mathbf{I}_{\tau}^{t}\left(\check{\vartheta}_{\tau,\varepsilon}\right)^{-1}\int_{\tau}^{t}\frac{\dot{\mathrm{M}}\left(\check{\vartheta}_{\tau,\varepsilon}, s\right)}{\sigma\left(s\right)}\mathrm{d}\bar{W}_s + O\left(\frac{\left\|\vartheta_0 - \check{\vartheta}_{\tau,\varepsilon}\right\|^{2}}{\varepsilon}\right) \qquad (2.50) \\
&\Longrightarrow \mathbf{I}_{0}^{t}\left(\vartheta_0\right)^{-1}\int_{0}^{t}\frac{\dot{\mathrm{S}}_{*}\left(\vartheta_0, \vartheta_0, s\right)}{\sigma\left(s\right)}\mathrm{d}w\left(s\right) = \zeta_t\left(\vartheta_0\right).
\end{aligned}
$$

Here $w\left(s\right), \tau \leq s \leq T$ is some Wiener process. The convergence of moments follows from the estimates (2.26) and (1.40).

Let us denote $\zeta_{t,\varepsilon} = \varepsilon^{-1}(\vartheta_{t,\varepsilon}^{\star} - \vartheta_0)$ and rewrite the representation (2.50) as

$$
\begin{aligned}
\zeta_{t,\varepsilon} &= \mathbf{I}_{\tau}^{t}\left(\check{\vartheta}_{\tau,\varepsilon}\right)^{-1}\int_{\tau}^{t}\frac{\dot{\mathrm{M}}\left(\check{\vartheta}_{\tau,\varepsilon}, s\right)}{\sigma\left(s\right)}\mathrm{d}\bar{W}_s + \frac{\left\|\vartheta_0 - \check{\vartheta}_{\tau,\varepsilon}\right\|^{2}}{\varepsilon}R_{t,\varepsilon} \\
&= \mathbf{I}_{\tau}^{t}\left(\vartheta_0\right)^{-1}\int_{\tau}^{t}\frac{\dot{\mathrm{M}}\left(\vartheta_0, s\right)}{\sigma\left(s\right)}\mathrm{d}\bar{W}_s + \left\|\vartheta_0 - \check{\vartheta}_{\tau,\varepsilon}\right\|Q_{t,\varepsilon} + \frac{\left\|\vartheta_0 - \check{\vartheta}_{\tau,\varepsilon}\right\|^{2}}{\varepsilon}R_{t,\varepsilon}.
\end{aligned}
$$

Here the functions $Q_{t,\varepsilon}, R_{t,\varepsilon}$ are uniformly in $t \in [\tau, T]$ bounded in probability.

Therefore, for any $\tau < t_1 < \ldots < t_k \leq T$, we have the convergence in distribution

$$\left(\zeta_{t_1,\varepsilon}, \ldots, \zeta_{t_k,\varepsilon}\right) \Longrightarrow \left(\zeta_{t_1}(\vartheta_0), \ldots, \zeta_{t_k}(\vartheta_0)\right).$$

Set

$$u_{t,\varepsilon} = \mathbf{I}_\tau^t(\vartheta_0)^{-1} \int_\tau^t \frac{\dot{\mathrm{M}}(\vartheta_0, s)}{\sigma(s)} \mathrm{d}\bar{W}_s.$$

The direct calculations allow us to write

$$\mathbf{E}_{\vartheta_0} \left\| u_{t_2,\varepsilon} - u_{t_1,\varepsilon} \right\|^4 \leq C \left|t_2 - t_1\right|^4 + C \left|t_2 - t_1\right|^2 \leq C \left|t_2 - t_1\right|^2.$$

Therefore, we have the weak convergence of the process $\zeta_{t,\varepsilon}, \tau \leq t \leq T$ to the Gaussian process $\zeta_t, \tau \leq t \leq T$. □

Remark 2.8 The One-step MLE process can be written in recurrent form. Introduce the $d \times d$ matrix

$$\mathbf{L}(\vartheta, \vartheta, t) = \sigma(t)^{-2} \dot{\mathrm{S}}(\vartheta, \vartheta, t) \dot{\mathrm{S}}(\vartheta, \vartheta, t)^\top$$

and note that

$$\mathbf{I}_\tau^t(\vartheta) = \int_\tau^t \mathbf{L}(\vartheta, \vartheta, s)\, \mathrm{d}s, \qquad \text{and} \qquad \frac{\mathrm{d}}{\mathrm{d}t} \mathbf{I}_\tau^t(\vartheta) = \mathbf{L}(\vartheta, \vartheta, t), \quad \tau < t \leq T.$$

We have the equality

$$\mathbf{I}_\tau^t\left(\vartheta_{\tau,\varepsilon}^*\right) \vartheta_{t,\varepsilon}^\star = \mathbf{I}_\tau^t\left(\vartheta_{\tau,\varepsilon}^*\right) \vartheta_{\tau,\varepsilon}^* + \int_\tau^t \frac{\dot{\mathrm{M}}\left(\vartheta_{\tau,\varepsilon}^*, s\right)}{\sigma(s)^2} \left[\mathrm{d}X_s - M\left(\vartheta_{\tau,\varepsilon}^*, s\right) \mathrm{d}s\right], \quad \tau_\varepsilon < t \leq T.$$

Hence,

$$\begin{aligned} \mathrm{d}\vartheta_{t,\varepsilon}^\star &= \mathbf{I}_{\tau,\varepsilon}^t\left(\vartheta_{\tau,\varepsilon}^*\right)^{-1} \mathbf{L}\left(\vartheta_{\tau,\varepsilon}^*, \vartheta_{\tau,\varepsilon}^*, t\right) \left[\vartheta_{\tau,\varepsilon}^* - \vartheta_{t,\varepsilon}^\star\right] \mathrm{d}t \\ &\quad + \mathbf{I}_{\tau,\varepsilon}^t\left(\vartheta_{\tau,\varepsilon}^*\right)^{-1} \dot{\mathrm{M}}\left(\vartheta_{\tau,\varepsilon}^*, t\right) \sigma(t)^{-2} \left[\mathrm{d}X_t - M\left(\vartheta_{\tau,\varepsilon}^*, t\right) \mathrm{d}t\right], \quad \tau < t \leq T. \end{aligned} \tag{2.51}$$

This is already recurrent presentation of the One-step MLE process. Remind as well that the random processes $\dot{\mathrm{M}}\left(\vartheta_{\tau,\varepsilon}^*, t\right)$ and $M\left(\vartheta_{\tau,\varepsilon}^*, t\right)$ are solutions of stochastic differential equations too.

Remark 2.9 The calculation of the One-step MLE process can be simplified as follows. The solution of the Eq. (2.46) converges to the function $\dot{\mathrm{S}}_*(\vartheta_0, \vartheta_0, t)$, i.e.,

$$\dot{\mathrm{M}}\left(\vartheta^*_{\tau,\varepsilon},t\right) \longrightarrow \dot{\mathrm{S}}_*\left(\vartheta_0,\vartheta_0,t\right) = \dot{\mathrm{f}}\left(\vartheta_0,t\right)\mathrm{y}_t\left(\vartheta_0,\vartheta_0\right) + f\left(\vartheta_0,t\right)\dot{\mathrm{y}}_t\left(\vartheta_0,\vartheta_0\right)$$
$$= \dot{\mathrm{f}}\left(\vartheta_0,t\right)\mathrm{y}_t\left(\vartheta_0\right) + f\left(\vartheta_0,t\right)\dot{\mathrm{y}}_t\left(\vartheta_0,\vartheta_0\right).$$

The analysis of the proof of the Theorem 2.5 shows that the $\dot{\mathrm{M}}\left(\vartheta^*_{\tau_\varepsilon},t\right)$ can be replaced by the function $\mathrm{S}_*\left(\vartheta^*_{\tau_\varepsilon},\vartheta^*_{\tau_\varepsilon},t\right)$. The estimators

$$\vartheta^\star_\varepsilon = \vartheta^*_{\tau,\varepsilon} + \mathbf{I}^T_\tau\left(\vartheta^*_{\tau,\varepsilon}\right)^{-1}\int_\tau^T \frac{\dot{\mathrm{S}}_*\left(\vartheta^*_{\tau,\varepsilon},\vartheta^*_{\tau,\varepsilon},s\right)}{\sigma\left(s\right)^2}\left[\mathrm{d}X_s - M\left(\vartheta^*_{\tau,\varepsilon},s\right)\mathrm{d}s\right],$$

$$\vartheta^\star_{t,\varepsilon}=\vartheta^*_{\tau,\varepsilon} + \mathbf{I}^t_\tau\left(\vartheta^*_{\tau,\varepsilon}\right)^{-1}\int_\tau^t \frac{\dot{\mathrm{S}}_*\left(\vartheta^*_{\tau,\varepsilon},\vartheta^*_{\tau,\varepsilon},s\right)}{\sigma\left(s\right)^2}\left[\mathrm{d}X_s-M\left(\vartheta^*_{\tau,\varepsilon},s\right)\mathrm{d}s\right],\quad \tau<t\leq T,$$

and, of course,

$$\mathrm{d}\vartheta^\star_{t,\varepsilon} = \mathbf{I}^t_\tau\left(\vartheta^*_{\tau,\varepsilon}\right)^{-1}\mathbf{L}\left(\vartheta^*_{\tau,\varepsilon},\vartheta^*_{\tau,\varepsilon},t\right)\left[\vartheta^*_{\tau,\varepsilon}-\vartheta^\star_{t,\varepsilon}\right]\mathrm{d}t$$
$$+\,\mathbf{I}^t_\tau\left(\vartheta^*_{\tau,\varepsilon}\right)^{-1}\dot{\mathrm{S}}_*\left(\vartheta^*_{\tau,\varepsilon},\vartheta^*_{\tau,\varepsilon},s\right)\sigma\left(t\right)^{-2}\left[\mathrm{d}X_t - M\left(\vartheta^*_{\tau,\varepsilon},t\right)\mathrm{d}t\right],\quad \tau<t\leq T$$

have the same asymptotic properties.

Remark that $M\left(\vartheta^*_{\tau,\varepsilon},t\right)$ cannot be replaced by $f\left(\vartheta^*_{\tau,\varepsilon},t\right)y_t\left(\vartheta^*_{\tau,\varepsilon}\right)$.

Example 2.4 Linear function $f\left(\vartheta,t\right)$. Suppose that the system is

$$\mathrm{d}X_t = \langle\vartheta,\mathrm{f}\left(t\right)\rangle Y_t\,\mathrm{d}t + \varepsilon\sigma\left(t\right)\mathrm{d}W_t,\qquad X_0=0,\qquad 0\leq t\leq T,$$
$$\mathrm{d}Y_t = a\left(t\right)Y_t\,\mathrm{d}t + \varepsilon b\left(t\right)\mathrm{d}V_t,\qquad Y_0 = y_0\neq 0.$$

Here $\langle\vartheta,f\left(t\right)\rangle = \sum_{k=1}^d \theta_k f_k\left(t\right)$ and the functions $\sigma\left(t\right),a\left(t\right),b\left(t\right),f_k\left(t\right),k=1,\ldots,d$ are known and satisfy the condition $\mathcal{A}_*$. The parameter $\vartheta\in\Theta=\left(\alpha_1,\beta_1\right)\times\left(\alpha_2,\beta_2\right)\times\ldots\times\left(\alpha_d,\beta_d\right),\alpha_k>0$.

The limit of X_t is the function

$$x_t\left(\vartheta\right) = y_0\int_0^t \langle\vartheta_0,f\left(s\right)\rangle\phi\left(0,s\right)\mathrm{d}s = \langle\vartheta_0,\mathrm{F}\left(t\right)\rangle.$$

where the vector $\mathrm{F}\left(t\right) = \left(F_1\left(t\right),\ldots,F_d\left(t\right)\right)^\top$,

$$\phi\left(0,s\right) = \exp\left(\int_0^s a\left(r\right)\mathrm{d}r\right),\qquad F_k\left(t\right) = y_0\int_0^t f_k\left(s\right)\phi\left(0,s\right)\mathrm{d}s.$$

Hence for any $\tau\in(0,T]$ the MDE is

$$\check\vartheta_{\tau,\varepsilon} = \arg\inf_{\vartheta\in\Theta}\int_0^\tau\left[X_t - \langle\vartheta,\mathrm{F}\left(t\right)\rangle\right]^2\mathrm{d}t.$$

Introduce the condition

$\mathscr{B}_4^\star$. *For any* $\tau \in (0, T]$ *the matrix*

$$\mathbf{G}(\tau) = \int_0^\tau \mathrm{F}(t)\,\mathrm{F}(t)^\top\,\mathrm{d}t$$

is nondegenerate.

If the condition $\mathscr{B}_4^\star$ is satisfied, then the MDE is

$$\check{\vartheta}_{\tau,\varepsilon} = \mathbf{G}(\tau)^{-1}\int_0^\tau \mathrm{F}(t)\;X_t\;\mathrm{d}t.$$

It is easy to see that this estimator has Gaussian distribution with the mean

$$\mathbf{E}_{\vartheta_0}\check{\vartheta}_{\tau,\varepsilon} = \mathbf{G}(\tau)^{-1}\int_0^\tau \mathrm{F}(t)\;\mathbf{E}_{\vartheta_0}X_t\;\mathrm{d}t = \vartheta_0$$

and the covariance

$$\begin{aligned}&\mathbf{E}_{\vartheta_0}(\check{\vartheta}_{\tau,\varepsilon}-\vartheta_0)(\check{\vartheta}_{\tau,\varepsilon}-\vartheta_0)^\top\\ &\qquad = \varepsilon^2\mathbf{G}(\tau)^{-1}\int_0^\tau\int_0^\tau \mathrm{F}(t)\,R(\vartheta_0,t,s)\,\mathrm{F}(s)\;\mathrm{d}t\mathrm{d}s\left(\mathbf{G}(\tau)^{-1}\right)^\top,\end{aligned}$$

where

$$R(\vartheta_0,t,s) = \int_0^t\int_0^s \langle\vartheta_0, f(u)\rangle\langle\vartheta_0, f(v)\rangle\int_0^{u\wedge v}\phi(u,r)\,\phi(v,r)\,b(r)^2\,\mathrm{d}r\;\mathrm{d}u\mathrm{d}v.$$

It can be shown that

$$\varepsilon^{-1}\tau^{-1/2}(\check{\vartheta}_{\tau,\varepsilon}-\vartheta_0)(\check{\vartheta}_{\tau,\varepsilon}-\vartheta_0)^\top \Longrightarrow \mathcal{N}(0,\mathbf{K}(\vartheta_0))$$

with some matrix $\mathbf{K}(\vartheta_0)$.

The Kalman–Bucy filter is

$$\mathrm{d}m(\vartheta,t) = \left[a(t) - \frac{\langle\vartheta,\mathrm{f}(t)\rangle^2\gamma_*(\vartheta,t)}{\sigma(t)^2}\right]m(\vartheta,t)\,\mathrm{d}t + \frac{\langle\vartheta,\mathrm{f}(t)\rangle\gamma_*(\vartheta,t)}{\sigma(t)^2}\mathrm{d}X_t,$$

$$\frac{\partial\gamma_*(\vartheta,t)}{\partial t} = 2a(t)\,\gamma_*(\vartheta,t) - \frac{\langle\vartheta,\mathrm{f}(t)\rangle^2\gamma_*(\vartheta,t)^2}{\sigma(t)^2} + b(t)^2,$$

$$\gamma_*(\vartheta,0) = 0,\;\; m(\vartheta,0) = y_0.$$

The limit ($\varepsilon = 0$) solution of the first equation is the function

$$y_t(\vartheta,\vartheta_0)=\int_0^t \Phi_*(\vartheta,s,t)\,\frac{\langle\vartheta,\mathrm{f}(s)\rangle\langle\vartheta_0,\mathrm{f}(s)\rangle y_s(\vartheta_0)\,\gamma_*(\vartheta,s)}{\sigma(s)^2}\,\mathrm{d}s$$

where

$$\Phi_*(\vartheta,s,t)=\exp\left(\int_s^t\left[a(r)-\frac{\langle\vartheta,\mathrm{f}(r)\rangle^2\gamma_*(\vartheta,r)}{\sigma(r)^2}\right]\mathrm{d}r\right)$$

and

$$y_t(\vartheta_0)=y_0\exp\left(\int_0^t a(s)\,\mathrm{d}s\right).$$

Let us denote

$$\mathrm{z}_t(\vartheta_0)=\left.\frac{\partial y_t(\vartheta,\vartheta_0)}{\partial\vartheta}\right|_{\vartheta=\vartheta_0}.$$

The Fisher information matrix is

$$\mathbf{I}_\tau^t(\vartheta_0)=\int_\tau^t\frac{[\mathrm{f}(s)\,y_s(\vartheta_0)+\langle\vartheta_0,\mathrm{f}(s)\rangle\mathrm{z}_s(\vartheta_0)]\,[\mathrm{f}(s)\,y_s(\vartheta_0)+\langle\vartheta_0,\mathrm{f}(s)\rangle\mathrm{z}_s(\vartheta_0)]^\top}{\sigma(s)^2}$$
$$\times\,\mathrm{d}s.$$

The One-step MLE process for this model is

$$\vartheta_{t,\varepsilon}^\star=\check\vartheta_\varepsilon+\mathbf{I}_\tau^t(\check\vartheta_{\tau,\varepsilon})^{-1}\int_\tau^t\frac{\left[\mathrm{f}(s)\,y_s(\check\vartheta_{\tau,\varepsilon})+\langle\check\vartheta_{\tau,\varepsilon},\mathrm{f}(s)\rangle\mathrm{z}_s(\check\vartheta_{\tau,\varepsilon})\right]}{\sigma(s)^2}$$
$$\times\left[\mathrm{d}X_s-\langle\check\vartheta_{\tau,\varepsilon},\mathrm{f}(s)\rangle m(\check\vartheta_{\tau,\varepsilon},s)\mathrm{d}s\right].$$

The asymptotic normality with the rate ε of this estimator can be verified following the same steps as it was done above.

Vanishing Learning Interval

We have the same model of observations (2.38) and (2.39) and Kalman–Bucy filter (2.40), (2.41), but now we suppose that the unknown parameter ϑ is one-dimensional ($d=1$) and that the condition $\dot f(\vartheta,0)\neq 0$ is fulfilled. Then, it is possible to realize the construction of adaptive Kalman filter with the learning interval $[0,\tau_\varepsilon]$, where $\tau_\varepsilon\to 0$ but sufficiently slowly.

The random process X_{τ_ε} admits the Taylor expansion

$$
\begin{aligned}
X_{\tau_\varepsilon} &= y_0 f\left(\vartheta_0,0\right)\tau_\varepsilon+\frac{y_0}{2}\left[f_t'\left(\vartheta_0,0\right)+f\left(\vartheta_0,0\right)a\left(\vartheta_0,0\right)\right]\tau_\varepsilon^2+O\left(\tau_\varepsilon^3\right)\\
&\quad+\sigma\left(0\right)\bar\zeta_{1,\varepsilon}\,\varepsilon\sqrt{\tau_\varepsilon}+f\left(\vartheta_0,0\right)b\left(\vartheta,0\right)\bar\zeta_{2,\varepsilon}\,\varepsilon\tau_\varepsilon^{3/2}+O\left(\varepsilon\tau_\varepsilon\right), \qquad (2.52)
\end{aligned}
$$

where $\bar\zeta_{1,\varepsilon}\sim\mathcal{N}\left(0,1\right)$ and $\bar\zeta_2=\int_0^1 v_\varepsilon\left(q\right)\mathrm{d}q$, $v_\varepsilon\left(\cdot\right)$ is a Wiener process. This expansion allows us to write

$$
\frac{X_{\tau_\varepsilon}}{y_0\tau_\varepsilon}=f\left(\vartheta_0,0\right)+N\left(\vartheta_0\right)\tau_\varepsilon+\frac{\sigma\left(0\right)\bar\zeta_{1,\varepsilon}}{y_0}\frac{\varepsilon}{\sqrt{\tau_\varepsilon}}+O\left(\varepsilon\right)+O\left(\tau_\varepsilon^2\right)
$$

where $N\left(\vartheta_0\right)=2^{-1}\left[f_t'\left(\vartheta_0,0\right)+f\left(\vartheta_0,0\right)a\left(\vartheta_0,0\right)\right]$. To chose the rate $\tau_\varepsilon\to0$, we solve the equation $\tau_\varepsilon=\varepsilon\tau_\varepsilon^{-1/2}$, which gives us the rate $\tau_\varepsilon=\varepsilon^{2/3}$.

The MME $\vartheta_{\tau_\varepsilon}^*$ can be defined as solution of the equation

$$
\frac{X_{\tau_\varepsilon}}{y_0\tau_\varepsilon}=f\left(\vartheta_{\tau_\varepsilon}^*,0\right).
$$

Hence,

$$
f\left(\vartheta_{\tau_\varepsilon}^*,0\right)=f\left(\vartheta_0,0\right)+\left[N\left(\vartheta_0\right)+y_0^{-1}\sigma\left(0\right)\bar\zeta_{1,\varepsilon}\right]\varepsilon^{2/3}+o\left(\varepsilon^{2/3}\right)
$$

and

$$
\begin{aligned}
\varepsilon^{-2/3}\left(\vartheta_{\tau_\varepsilon}^*-\vartheta_0\right)&=\frac{N\left(\vartheta_0\right)+y_0^{-1}\sigma\left(0\right)\bar\zeta_{1,\varepsilon}}{\dot f(\tilde\vartheta,0)}+o\left(1\right)\\
&\Longrightarrow\mathcal{N}\left(N\left(\vartheta_0\right)\dot f\left(\vartheta_0,0\right)^{-1},D\left(\vartheta_0\right)\right).
\end{aligned}
$$

The limit variance is

$$
D\left(\vartheta_0\right)=\frac{\sigma\left(0\right)^2}{y_0^2\,\dot f\left(\vartheta_0,0\right)^2}.
$$

Therefore, we have a preliminary consistent MME $\vartheta_{\tau_\varepsilon}^*$, constructed by observations $X^{\tau_\varepsilon}=\left(X_t,0\le t\le\tau_\varepsilon\right)$, $\tau_\varepsilon\to0$, and satisfying the relation $\varepsilon^{-1}\left(\vartheta_{\tau_\varepsilon}^*-\vartheta_0\right)^2\to0$. This MME allows us to introduce the One-step MLE $\vartheta_\varepsilon^\star$ and One-step MLE-process $\vartheta_{t,\varepsilon}^\star$, $\tau_\varepsilon<t\le T$ by the relations

$$
\vartheta_\varepsilon^\star=\vartheta_{\tau_\varepsilon}^*+\mathrm{I}_{\tau_\varepsilon}^T\left(\vartheta_{\tau_\varepsilon}^*\right)^{-1}\int_{\tau_\varepsilon}^T\frac{\dot M\left(\vartheta_{\tau_\varepsilon}^*,s\right)}{\sigma\left(s\right)^2}\left[\mathrm{d}X_s-M\left(\vartheta_{\tau_\varepsilon}^*,s\right)\mathrm{d}s\right],
$$

$$
\vartheta_{t,\varepsilon}^\star=\vartheta_{\tau_\varepsilon}^*+\mathrm{I}_{\tau_\varepsilon}^t\left(\vartheta_{\tau_\varepsilon}^*\right)^{-1}\int_{\tau_\varepsilon}^t\frac{\dot M\left(\vartheta_{\tau_\varepsilon}^*,s\right)}{\sigma\left(s\right)^2}\left[\mathrm{d}X_s-M\left(\vartheta_{\tau_\varepsilon}^*,s\right)\mathrm{d}s\right],\qquad\tau_\varepsilon<t\le T.
$$

As the One-step MLE is a particular case of the One-step MLE-process, we will study the One-step MLE-process only. Recall that

$$\mathrm{I}_{\tau_\varepsilon}^{t}\left(\vartheta\right)=\int_{\tau_\varepsilon}^{t}\frac{\dot S_*\left(\vartheta,\vartheta,s\right)^2}{\sigma\left(s\right)^2}\mathrm{d}s.$$

The random processes $M\left(\vartheta_{\tau_\varepsilon}^*,s\right)=f\left(\vartheta_{\tau_\varepsilon}^*,s\right)m\left(\vartheta_{\tau_\varepsilon}^*,s\right),\ \tau_\varepsilon\leq t\leq T$ and $\dot M\left(\vartheta_{\tau_\varepsilon}^*,s\right),\ \tau_\varepsilon\leq t\leq T$ are defined as solutions of the Eqs. (2.45) and (2.46), respectively.

Introduce the conditions $\bar{\mathscr{B}}^\star$:

$\bar{\mathscr{B}}_1^\star$. *The functions* $f\left(\vartheta,t\right),a\left(\vartheta,t\right),b\left(\vartheta,t\right),t\in\left[0,T\right],\vartheta\in\bar\Theta$ *are twice continuously differentiable on* $\bar\Theta$.

$\bar{\mathscr{B}}_2^\star$. *The functions* $f\left(\vartheta,t\right),t\in\left[0,T\right],\vartheta\in\bar\Theta$ *and* $\dot f\left(\vartheta,0\right),\vartheta\in\bar\Theta$ *are separated from zero.*

$\bar{\mathscr{B}}_3^\star$. *The functions* $f\left(\vartheta,t\right),a\left(\vartheta,t\right),b\left(\vartheta,t\right),\dot f\left(\vartheta,t\right),\dot a\left(\vartheta,t\right),\sigma\left(t\right)t\in\left[0,T\right]$, $\vartheta\in\bar\Theta$ *are continuously differentiable on* $t\in\left[0,T\right]$.

$\bar{\mathscr{B}}_4^\star$. *For any* $\tau\in\left(0,T\right]$ *and* $t\in\left(\tau,T\right]$ *the Fisher information is positive*

$$\inf_{\vartheta_0\in\Theta}\mathrm{I}_\tau^t\left(\vartheta_0\right)>0.$$

Theorem 2.6 *Suppose that the conditions* $\mathscr{A}_*$, $\bar{\mathscr{B}}^\star$ *are satisfied. Then the One-step MLE-process* $\vartheta_{t,\varepsilon}^\star$, $\tau_\varepsilon<t\leq T$ *is uniformly consistent: for any* $\nu>0$ *and any* $t_0\in\left(0,T\right]$

$$\mathbf{P}_{\vartheta_0}\left(\sup_{t_0\leq t\leq T}\left|\vartheta_{t,\varepsilon}^\star-\vartheta_0\right|\geq\nu\right)\rightarrow 0,$$

asymptotically normal

$$\frac{\vartheta_{t,\varepsilon}^\star-\vartheta_0}{\varepsilon}\Longrightarrow\zeta_t^*\left(\vartheta_0\right)\sim\mathcal{N}\left(0,\mathrm{I}_0^t\left(\vartheta_0\right)^{-1}\right),$$

the moments converge: for any $p\geq 2$

$$\varepsilon^{-p}\mathbf{E}_{\vartheta_0}\left|\vartheta_{t,\varepsilon}^\star-\vartheta_0\right|^p\longrightarrow\mathbf{E}_{\vartheta_0}\left|\zeta_t^*\left(\vartheta_0\right)\right|^p.$$

The random process $\zeta_{t,\varepsilon}^\star=\varepsilon^{-1}\left(\vartheta_{t,\varepsilon}^\star-\vartheta_0\right),t_0\leq t\leq T$ *converges in distribution in the measurable space* $\left(\mathcal{C}\left[t_0,T\right],\mathfrak{B}\right)$ *to the Gaussian process* $\zeta_t^*\left(\vartheta_0\right),t_0\leq t\leq T$, *where*

$$\zeta_t^*\left(\vartheta_0\right)=\mathrm{I}_0^t\left(\vartheta\right)^{-1}\int_0^t\frac{\dot S_*\left(\vartheta_0,\vartheta_0,s\right)}{\sigma\left(s\right)}\mathrm{d}w\left(s\right).$$

Here $w(s), 0 \leq s \leq T$ is some Wiener process.

Proof The proof repeats the proof of the Theorem 2.5. As the estimator $\vartheta^*_{\tau_\varepsilon}$ is consistent and $\varepsilon^{-1}\left(\vartheta^*_{\tau_\varepsilon}-\vartheta_0\right)^2 \rightarrow 0$ the estimator $\vartheta^\star_{t,\varepsilon}$ is consistent and asymptotically normal. The uniform consistency follows from the proof of the Theorem 2.5.

To prove asymptotic normality and convergence of moments, we write the same relations as above with the only difference in the lower limits of integrals.

$$\frac{\vartheta^\star_{t,\varepsilon}-\vartheta_0}{\varepsilon}=\frac{\vartheta^*_{\tau_\varepsilon}-\vartheta_0}{\varepsilon}+\mathrm{I}^t_{\tau_\varepsilon}\left(\vartheta^*_{\tau_\varepsilon}\right)^{-1}\int_{\tau_\varepsilon}^{t}\frac{\dot M\left(\vartheta^*_{\tau_\varepsilon},s\right)}{\sigma(s)}\mathrm{d}\bar W_s$$
$$+\mathrm{I}^t_{\tau_\varepsilon}\left(\vartheta^*_{\tau_\varepsilon}\right)^{-1}\int_{\tau_\varepsilon}^{t}\frac{\dot M\left(\vartheta^*_{\tau_\varepsilon},s\right)}{\varepsilon\sigma(s)^2}\left[M(\vartheta_0,s)-M\left(\vartheta^*_{\tau_\varepsilon},s\right)\right]\mathrm{d}s.$$

From the consistency of $\vartheta^*_{\tau_\varepsilon}$ and $\tau_\varepsilon\rightarrow 0$, we obtain

$$\mathrm{I}^t_{\tau_\varepsilon}\left(\vartheta^*_{\tau_\varepsilon}\right)^{-1}\int_{\tau_\varepsilon}^{t}\frac{\dot M\left(\vartheta^*_{\tau_\varepsilon},s\right)}{\sigma(s)}\mathrm{d}\bar W_s=\mathrm{I}^t_{\tau_\varepsilon}(\vartheta_0)^{-1}\int_{\tau_\varepsilon}^{t}\frac{\dot M(\vartheta_0,s)}{\sigma(s)}\mathrm{d}\bar W_s\left(1+O\left(\left|\vartheta^*_{\tau_\varepsilon}-\vartheta_0\right|\right)\right),$$
$$\mathrm{I}^t_{\tau_\varepsilon}\left(\vartheta^*_{\tau_\varepsilon}\right)\longrightarrow \mathrm{I}^t_0(\vartheta_0),\qquad \dot M(\vartheta_0,s)\longrightarrow \dot S(\vartheta_0,\vartheta_0,s),$$
$$\mathrm{I}^t_{\tau_\varepsilon}\left(\vartheta^*_{\tau_\varepsilon}\right)^{-1}\int_{\tau_\varepsilon}^{t}\frac{\dot M\left(\vartheta^*_{\tau_\varepsilon},s\right)}{\sigma(s)}\mathrm{d}\bar W_s\Longrightarrow \mathrm{I}^t_0(\vartheta_0)^{-1}\int_0^t\frac{\dot S(\vartheta_0,\vartheta_0,s)}{\sigma(s)}\mathrm{d}w(s)=\zeta^*_t(\vartheta_0),$$

where $w(s), 0\leq s\leq t$ is some Wiener process.

The ordinary integral has the following asymptotics

$$\int_{\tau_\varepsilon}^{t}\frac{\dot M\left(\vartheta^*_{\tau_\varepsilon},s\right)}{\varepsilon\sigma(s)^2}\left[M(\vartheta_0,s)-M\left(\vartheta^*_{\tau_\varepsilon},s\right)\right]\mathrm{d}s$$
$$=\int_{\tau_\varepsilon}^{t}\frac{\dot S\left(\vartheta^*_{\tau_\varepsilon},\vartheta_0,s\right)^2}{\varepsilon\sigma(s)^2}\left(\vartheta_0-\vartheta^*_{\tau_\varepsilon}\right)\mathrm{d}s\left(1+O\left(\left|\vartheta_0-\vartheta^*_{\tau_\varepsilon}\right|\right)\right)$$

because $\dot M(\vartheta,s)-\dot S(\vartheta,\vartheta_0,s)=O(\varepsilon)$. Therefore,

$$\mathrm{I}^t_{\tau_\varepsilon}\left(\vartheta^*_{\tau_\varepsilon}\right)^{-1}\int_{\tau_\varepsilon}^{t}\frac{\dot M\left(\vartheta^*_{\tau_\varepsilon},s\right)}{\varepsilon\sigma(s)^2}\left[M(\vartheta_0,s)-M\left(\vartheta^*_{\tau_\varepsilon},s\right)\right]\mathrm{d}s$$
$$=\frac{\left(\vartheta_0-\vartheta^*_{\tau_\varepsilon}\right)}{\varepsilon}\left(1+O\left(\left|\vartheta_0-\vartheta^*_{\tau_\varepsilon}\right|\right)\right)$$

and we obtain the the same convergence (2.48)

$$
\begin{aligned}
\frac{\vartheta^{\star}_{t,\varepsilon}-\vartheta_0}{\varepsilon} &= \frac{\vartheta^*_{\tau_\varepsilon}-\vartheta_0}{\varepsilon}+\frac{\vartheta_0-\vartheta^*_{\tau_\varepsilon}}{\varepsilon}\left(1+O\left(\left|\vartheta_0-\vartheta^*_{\tau_\varepsilon}\right|\right)\right)\\
&\quad +\mathrm{I}^t_{\tau_\varepsilon}\left(\vartheta^*_{\tau_\varepsilon}\right)^{-1}\int_{\tau_\varepsilon}^t\frac{\dot M\left(\vartheta^*_{\tau_\varepsilon},s\right)}{\sigma\left(s\right)}\mathrm{d}\bar W_s\\
&=\mathrm{I}^t_{\tau_\varepsilon}\left(\vartheta^*_{\tau_\varepsilon}\right)^{-1}\int_{\tau_\varepsilon}^t\frac{\dot M\left(\vartheta^*_{\tau_\varepsilon},s\right)}{\sigma\left(s\right)}\mathrm{d}\bar W_s+O\left(\frac{\left|\vartheta_0-\vartheta^*_{\tau_\varepsilon}\right|^2}{\varepsilon}\right)\\
&\Longrightarrow \mathrm{I}^t_0\left(\vartheta_0\right)^{-1}\int_0^t\frac{\dot S_*\left(\vartheta_0,\vartheta_0,s\right)}{\sigma\left(s\right)}\mathrm{d}w\left(s\right).
\end{aligned}
$$

Here $w\left(s\right), 0\leq s\leq T$ is some Wiener process. The convergence of moments follows from the estimates (2.26) and (1.40). □

Remark 2.10 The consistency of the preliminary estimator and therefore the consistency of the One-step MLE-process are provided by the condition $\bar{\mathscr{B}}^{\star}_2$, which plays the role of identifiability condition. This means that if in the observed equation $f\left(\vartheta,t\right)=f\left(t\right)$ and $\dot a\left(\vartheta,t\right)\neq 0$, then the proposed above preliminary estimator cannot be constructed. Consider the problem of adaptive filtration for the model of observations

$$
\begin{aligned}
\mathrm{d}X_t&=f\left(t\right)Y_t\mathrm{d}t+\varepsilon\sigma\left(t\right)\mathrm{d}W_t, &\quad X_0&=0, \quad 0\leq t\leq T,\\
\mathrm{d}Y_t&=a\left(\vartheta,t\right)Y_t\mathrm{d}t+\varepsilon b\left(t\right)\mathrm{d}V_t, &\quad Y_0&=y_0\neq 0.
\end{aligned}
$$

The Taylor expansion (2.52) is

$$
\begin{aligned}
X_{\tau_\varepsilon}&=y_0f\left(0\right)\tau_\varepsilon+\frac{y_0}{2}\left[f'_t\left(0\right)+f\left(0\right)a\left(\vartheta_0,0\right)\right]\tau_\varepsilon^2+O\left(\tau_\varepsilon^3\right)\\
&\quad+\sigma\left(0\right)\bar\zeta_{1,\varepsilon}\,\varepsilon\sqrt{\tau_\varepsilon}+f\left(0\right)b\left(0\right)\bar\zeta_{2,\varepsilon}\,\varepsilon\tau_\varepsilon^{3/2}+O\left(\varepsilon\tau_\varepsilon\right).
\end{aligned}
$$

Therefore,

$$
\begin{aligned}
\left[X_{\tau_\varepsilon}-y_0f\left(0\right)\tau_\varepsilon-y_02^{-1}f'_t\left(0\right)\tau_\varepsilon^2\right]2f\left(0\right)^{-1}\tau_\varepsilon^{-2}&=a\left(\vartheta_0,0\right)+O\left(\tau_\varepsilon\right)\\
&\quad+\frac{2\sigma\left(0\right)\bar\zeta_{1,\varepsilon}}{y_0f\left(0\right)}\,\varepsilon\tau_\varepsilon^{-3/2}+O\left(\varepsilon\tau_\varepsilon^{-1}\right)
\end{aligned}
$$

and the MME $\vartheta^*_{\tau_\varepsilon}$ defined by the equation

$$
\left[X_{\tau_\varepsilon}-y_0f\left(0\right)\tau_\varepsilon-y_02^{-1}f'_t\left(0\right)\tau_\varepsilon^2\right]2f\left(0\right)^{-1}\tau_\varepsilon^{-2}=a\left(\vartheta^*_{\tau_\varepsilon},0\right)
$$

satisfies the equation

$$
a\left(\vartheta^*_{\tau_\varepsilon},0\right)=a\left(\vartheta_0,0\right)+O\left(\tau_\varepsilon\right)+\frac{2\sigma\left(0\right)\bar\zeta_{1,\varepsilon}}{y_0f\left(0\right)}\,\varepsilon\tau_\varepsilon^{-3/2}+O\left(\varepsilon\tau_\varepsilon^{-1}\right).
$$

The equality $\tau_\varepsilon = \varepsilon \tau_\varepsilon^{-3/2}$ gives the rate $\tau_\varepsilon = \varepsilon^{2/5}$ and the asymptotic normality

$$\varepsilon^{2/5}\left(\vartheta^*_{\tau_\varepsilon} - \vartheta_0\right) \Longrightarrow \mathcal{N}\left(n\left(\vartheta_0\right), \check{D}\left(\vartheta_0\right)\right).$$

Note that this estimator cannot be used for the construction of a consistent One-step MLE-process because

$$\frac{\left(\vartheta^*_{\tau_\varepsilon} - \vartheta_0\right)^2}{\varepsilon} \longrightarrow \infty.$$

In such situation, it is possible to use Two-step MLE-process, i.e., first to improve the rate of convergence and then to use the obtained estimator as preliminary for One-step procedure.

The first step is to define the estimator process $\bar{\vartheta}_{t,\varepsilon}, \tau_\varepsilon < t \leq T$

$$\bar{\vartheta}_{t,\varepsilon} = \vartheta^*_{\tau_\varepsilon} + \mathrm{I}^t_{\tau_\varepsilon}\left(\vartheta^*_{\tau_\varepsilon}\right)^{-1} \int_{\tau_\varepsilon}^t \frac{f(s)\,\dot{m}(\vartheta^*_{\tau_\varepsilon}, s)}{\sigma(s)^2}\left[\mathrm{d}X_s - f(s)\,m\left(\vartheta_{\tau_\varepsilon}, s\right)\mathrm{d}s\right]. \tag{2.53}$$

Then consider the difference

$$\begin{aligned}
\frac{\bar{\vartheta}_{t,\varepsilon} - \vartheta_0}{\varepsilon^{3/5}} &= \frac{\vartheta^*_{\tau_\varepsilon} - \vartheta_0}{\varepsilon^{3/5}} + \frac{\varepsilon^{2/5}}{\mathrm{I}^t_{\tau_\varepsilon}\left(\vartheta^*_{\tau_\varepsilon}\right)} \int_{\tau_\varepsilon}^t \frac{f(s)\,\dot{m}(\vartheta^*_{\tau_\varepsilon}, s)}{\sigma(s)^2}\mathrm{d}\bar{W}_s \\
&\quad + \frac{\varepsilon^{-3/5}}{\mathrm{I}^t_{\tau_\varepsilon}\left(\vartheta^*_{\tau_\varepsilon}\right)} \int_{\tau_\varepsilon}^t \frac{f(s)^2\,\dot{m}(\vartheta^*_{\tau_\varepsilon}, s)}{\sigma(s)^2}\left[m\left(\vartheta_0, s\right) - m\left(\vartheta_{\tau_\varepsilon}, s\right)\right]\mathrm{d}s
\end{aligned}$$

We have

$$\frac{\varepsilon^{2/5}}{\mathrm{I}^t_{\tau_\varepsilon}\left(\vartheta^*_{\tau_\varepsilon}\right)} \int_{\tau_\varepsilon}^t \frac{f(s)\,\dot{m}(\vartheta^*_{\tau_\varepsilon}, s)}{\sigma(s)^2}\mathrm{d}\bar{W}_s \longrightarrow 0$$

and for ordinary integral

$$\begin{aligned}
&\frac{\varepsilon^{-3/5}}{\mathrm{I}^t_{\tau_\varepsilon}\left(\vartheta^*_{\tau_\varepsilon}\right)} \int_{\tau_\varepsilon}^t \frac{f(s)^2\,\dot{m}(\vartheta^*_{\tau_\varepsilon}, s)}{\sigma(s)^2}\left[m\left(\vartheta_0, s\right) - m\left(\vartheta^*_{\tau_\varepsilon}, s\right)\right]\mathrm{d}s \\
&\quad = \frac{\vartheta_0 - \vartheta^*_{\tau_\varepsilon}}{\varepsilon^{3/5}} \frac{1}{\mathrm{I}^t_{\tau_\varepsilon}\left(\vartheta^*_{\tau_\varepsilon}\right)} \int_{\tau_\varepsilon}^t \frac{f(s)^2\,\dot{m}(\vartheta^*_{\tau_\varepsilon}, s)\dot{m}(\tilde{\vartheta}^*_{\tau_\varepsilon}, s)}{\sigma(s)^2}\mathrm{d}s \\
&\quad = \frac{\vartheta_0 - \vartheta^*_{\tau_\varepsilon}}{\varepsilon^{3/5}}\left(1 + O\left(\left|\vartheta^*_{\tau_\varepsilon} - \vartheta_0\right|\right)\right).
\end{aligned}$$

Therefore,

$$
\begin{aligned}
\frac{\bar\vartheta_{t,\varepsilon}-\vartheta_0}{\varepsilon^{3/5}} &= \frac{\vartheta^*_{\tau_\varepsilon}-\vartheta_0}{\varepsilon^{3/5}}+\frac{\vartheta_0-\vartheta^*_{\tau_\varepsilon}}{\varepsilon^{3/5}}\left(1+O\left(\left|\vartheta^*_{\tau_\varepsilon}-\vartheta_0\right|\right)\right)+O\left(\varepsilon^{2/5}\right)\\
&= O\left(\frac{\left|\vartheta^*_{\tau_\varepsilon}-\vartheta_0\right|^2}{\varepsilon^{3/5}}\right)+O\left(\varepsilon^{2/5}\right)\longrightarrow 0.
\end{aligned}
$$

The rate $\varepsilon^{-3/5}$ is sufficient for the construction of the Two-step MLE process because

$$
\frac{\left|\bar\vartheta_{t,\varepsilon}-\vartheta_0\right|^2}{\varepsilon}\longrightarrow 0.
$$

Therefore, the Two-step MLE process can be

$$
\begin{aligned}
\vartheta^{\star\star}_{t,\varepsilon} &= \bar\vartheta_{t,\varepsilon}+\mathrm{I}^t_{\tau_\varepsilon}\left(\bar\vartheta_{t,\varepsilon}\right)^{-1}\int_{\tau_\varepsilon}^t\frac{f\left(s\right)\dot m(\bar\vartheta_{s,\varepsilon},s)}{\sigma\left(s\right)^2}\mathrm{d}X_s\\
&\quad -\mathrm{I}^t_{\tau_\varepsilon}\left(\bar\vartheta_{t,\varepsilon}\right)^{-1}\int_{\tau_\varepsilon}^t\frac{f\left(s\right)^2\dot m(\bar\vartheta_{s,\varepsilon},s)m\left(\bar\vartheta_{t,\varepsilon},s\right)}{\sigma\left(s\right)^2}\mathrm{d}s,\qquad \tau_\varepsilon<t\leq T.
\end{aligned}
$$

Here for the calculation of $m\left(\bar\vartheta_{t,\varepsilon},s\right)$ we have to use the representation (1.33). The asymptotic normality of this estimator is obtained as before

$$
\begin{aligned}
\frac{\vartheta^{\star\star}_{t,\varepsilon}-\vartheta_0}{\varepsilon} &= \frac{\bar\vartheta_{t,\varepsilon}-\vartheta_0}{\varepsilon}+\frac{1}{\mathrm{I}^t_{\tau_\varepsilon}\left(\bar\vartheta_{t,\varepsilon}\right)}\int_{\tau_\varepsilon}^t\frac{f\left(s\right)\dot m(\bar\vartheta_{s,\varepsilon},s)}{\sigma\left(s\right)}\mathrm{d}\bar W_s-\frac{1}{\mathrm{I}^t_{\tau_\varepsilon}\left(\bar\vartheta_{t,\varepsilon}\right)}\\
&\quad\times\int_{\tau_\varepsilon}^t\frac{f\left(s\right)^2\dot m(\bar\vartheta_{s,\varepsilon},s)[m\left(\vartheta_0,s\right)-m\left(\bar\vartheta_{t,\varepsilon},s\right)]}{\varepsilon\sigma\left(s\right)^2}\mathrm{d}s,\ \tau_\varepsilon<t\leq T.
\end{aligned}
$$

We have

$$
\frac{1}{\mathrm{I}^t_{\tau_\varepsilon}\left(\bar\vartheta_{t,\varepsilon}\right)}\int_{\tau_\varepsilon}^t\frac{f\left(s\right)\dot m(\bar\vartheta_{s,\varepsilon},s)}{\sigma\left(s\right)}\mathrm{d}\bar W_s\Longrightarrow\mathcal{N}\left(0,\mathrm{I}^t_0\left(\vartheta_0\right)^{-1}\right)
$$

and

$$
\begin{aligned}
&\frac{1}{\mathrm{I}^t_{\tau_\varepsilon}\left(\bar\vartheta_{t,\varepsilon}\right)}\int_{\tau_\varepsilon}^t\frac{f\left(s\right)^2\dot m(\bar\vartheta_{s,\varepsilon},s)[m\left(\vartheta_0,s\right)-m\left(\bar\vartheta_{t,\varepsilon},s\right)]}{\varepsilon\sigma\left(s\right)^2}\mathrm{d}s\\
&\qquad=\frac{\vartheta_0-\bar\vartheta_{t,\varepsilon}}{\varepsilon}\frac{1}{\mathrm{I}^t_{\tau_\varepsilon}\left(\bar\vartheta_{t,\varepsilon}\right)}\int_{\tau_\varepsilon}^t\frac{f\left(s\right)^2\dot m(\bar\vartheta_{s,\varepsilon},s)\dot m(\tilde\vartheta_{\varepsilon},s)}{\varepsilon\sigma\left(s\right)^2}\mathrm{d}s\\
&\qquad=\frac{\vartheta_0-\bar\vartheta_{t,\varepsilon}}{\varepsilon}+O\left(\varepsilon^{-1}\left(\vartheta_0-\bar\vartheta_{t,\varepsilon}\right)^2\right)
\end{aligned}
$$

Therefore,

$$\varepsilon^{-1}\left(\vartheta^{\star\star}_{t,\varepsilon}-\vartheta_0\right)\Longrightarrow\mathcal{N}\left(0,\mathrm{I}_0^t\left(\vartheta_0\right)^{-1}\right).$$

2.1.3 Adaptive Filter

We have the model of observations

$$\begin{aligned}\mathrm{d}X_t&=f\left(\vartheta,t\right)Y_t\,\mathrm{d}t+\varepsilon\sigma\left(t\right)\mathrm{d}W_t, &&X_0=0,\quad 0\le t\le T,\\ \mathrm{d}Y_t&=a\left(\vartheta,t\right)Y_t\,\mathrm{d}t+\varepsilon b\left(\vartheta,t\right)\mathrm{d}V_t, &&Y_0=y_0,\end{aligned}$$

where all functions $f\left(\cdot\right),\sigma\left(\cdot\right),a\left(\cdot\right),b\left(\cdot\right)$ and the constant $y_0\neq 0$ are known and satisfy the condition $\mathscr{A}_*$. The unknown parameter $\vartheta\in\Theta\subset\mathcal{R}^d$. The set Θ is open, convex and bounded. The true value is denoted as ϑ_0. The process $X^T=(X_t,0\le t\le T)$ is observed and Y^T is hidden. Our goal is to construct an adaptive filter, i.e., to approximate the conditional expectation $m\left(\vartheta_0,t\right)=\mathbf{E}_{\vartheta_0}\left(Y_t|\mathfrak{F}^X_t\right)$ by observations $X^t=(X_s,0\le s\le t)$ for all $t\in(0,T]$. We follow the procedure described in the Sect. 1.3.

If ϑ is known, then the solution is well known and is given by the Eqs. (2.5) and (2.6), which we recall here

$$\mathrm{d}m(\vartheta,t)=\left[a(\vartheta,t)-\frac{f(\vartheta,t)^2\gamma_*(\vartheta,t)}{\sigma\left(t\right)^2}\right]m(\vartheta,t)\,\mathrm{d}t+\frac{f(\vartheta,t)\gamma_*(\vartheta,t)}{\sigma\left(t\right)^2}\mathrm{d}X_t, \tag{2.54}$$

$$\frac{\partial\gamma_*(\vartheta,t)}{\partial t}=2a(\vartheta,t)\gamma_*(\vartheta,t)-\frac{f(\vartheta,t)^2\gamma_*(\vartheta,t)^2}{\sigma\left(t\right)^2}+b(\vartheta,t)^2,\quad 0\le t\le T, \tag{2.55}$$

with initial values $m(\vartheta,0)=y_0$ and $\gamma_*\left(\vartheta,0\right)=0$. Remind as well that $m\left(\vartheta_0,t\right)$ is mean square optimal estimator of the hidden process Y_t and the error of estimation is $\mathbf{E}_{\vartheta_0}\left[Y_t-m\left(\vartheta_0,t\right)\right]^2=\varepsilon^2\gamma_*\left(\vartheta_0,t\right)$.

It is quite natural to define this approximation $\hat m_t$ with the help of the Eqs. (2.54) and (2.55), where we replace ϑ by some estimator $\bar\vartheta_\varepsilon$. Note that we cannot put $\bar\vartheta_\varepsilon$ directly in the function

$$m\left(\vartheta,t\right)=y_0\Phi\left(\vartheta,0,t\right)+\int_0^t\Phi\left(\vartheta,s,t\right)\frac{f(\vartheta,s)\gamma_*(\vartheta,s)}{\sigma\left(s\right)^2}\mathrm{d}X_s$$

because there will be a problem with the definition of the stochastic integral.

It is desirable that the estimator has the following properties.

1. *The estimator* $\bar\vartheta_\varepsilon$ *for* $m\left(\vartheta,t\right)$ *depends of the observations* X^t *up to time* t.

2. *It is consistent and has asymptotically minimal mean square error.*
3. *It can be easily calculated for all* $t \in (0, T]$.

Recall that the MLE $\hat\vartheta_{t,\varepsilon}$ and BE $\tilde\vartheta_{t,\varepsilon}$ constructed by observations X^t (see (2.33), where $T = t$) satisfy the conditions 1 and 2, but their calculation is numerically quite cumbersome problem. The One-step MLE process satisfies well these conditions for the approximation of $m(\vartheta_0, t), 0 < t \leq T$, and the proposed below estimation procedure is based on this estimator process.

The calculations of MDE $\check\vartheta_{t,\varepsilon}$ or One-step MDE process $\check\vartheta^\star_{t,\varepsilon}$ are much easier than that of the MLE or BE. Moreover, its calculation is even easier than that of One-step MLE process because it does not involves the stochastic integral. Remark that the One-step MDE process is not asymptotically efficient and therefore do not satisfy the condition 2, but its computational simplicity sometimes can compensate the losses in the errors. Recall that if the model is contaminated, the approximation of $m(\vartheta_0, t)$ with the help of One-step MDE process can be interesting too (see the model (2.20) and the corresponding discussion there).

AF Based on One-Step MLE Process

We consider once more two statements of the problem. One with the fixed learning interval and multidimensional parameter $\vartheta \in \Theta \subset \mathcal{R}^d$ and the second with a vanishing learning interval, one-dimensional parameter ϑ and condition $\dot f(\vartheta, 0) \neq 0$.

Fixed Learning Interval

We propose to use One-step MLE process $\vartheta^\star_{t,\varepsilon}$ because it satisfies abovementioned conditions 1–3.

1. *Fix a learning interval of* $[0, \tau_*]$, *where* $\tau_* \in (0, \tau)$ *with the observations* $X^{\tau_*} = (X_s, 0 \leq s \leq \tau_*)$.
2. *As the preliminary estimator take the MDE*

$$\check\vartheta_{\tau_*,\varepsilon} = \arg\inf_{\vartheta\in\Theta} \int_0^{\tau_*} [X_s - x_s(\vartheta)]^2\,\mathrm{d}s$$

and suppose that conditions of the Theorem 2.2 where T is replaced by τ) are fulfilled.

3. *Calculate the One-step MLE process* (2.42)

$$\vartheta^\star_{t,\varepsilon} = \check\vartheta_{\tau_*,\varepsilon} + \mathbf{I}^t_{\tau_*}\left(\check\vartheta_{\tau_*,\varepsilon}\right)^{-1} \int_{\tau_*}^t \frac{\dot{\mathrm{M}}\left(\check\vartheta_{\tau_*,\varepsilon}, s\right)}{\sigma(s)^2}\left[\mathrm{d}X_s - M\left(\check\vartheta_{\tau_*,\varepsilon}, s\right)\mathrm{d}s\right],$$

$$\tau_* < t \leq T.$$

Here the random processes $M\left(\check\vartheta_{\tau_*,\varepsilon},s\right),\dot{M}\left(\check\vartheta_{\tau_*,\varepsilon},s\right)$ *are given by the Eqs.* (2.45) *and* (2.46). *Note that these equations have to be solved just for one value of parameter* $\vartheta=\check\vartheta_{\tau_*,\varepsilon}$.

4. *The adaptive filter is given by the equations*

$$\mathrm{d}\hat m_t=\left[a(\vartheta^\star_{t,\varepsilon},t)-\frac{f(\vartheta^\star_{t,\varepsilon},t)^2\hat\gamma_t}{\sigma(t)^2}\right]\hat m_t\,\mathrm{d}t+\frac{f(\vartheta^\star_{t,\varepsilon},t)\hat\gamma_t}{\sigma(t)^2}\mathrm{d}X_t,\qquad \tau\le t\le T,\tag{2.56}$$

$$\frac{\partial\hat\gamma_t}{\partial t}=2a(\vartheta^\star_{t,\varepsilon},t)\hat\gamma_t-\frac{f(\vartheta^\star_{t,\varepsilon},t)^2\hat\gamma_t^2}{\sigma(t)^2}+b(\vartheta^\star_{t,\varepsilon},t)^2,\qquad \tau\le t\le T.\tag{2.57}$$

Here $\tau>\tau_*$. *The initial value* $\hat m_{\tau,\varepsilon}$ *is calculated similary to the calculation of* $M\left(\check\vartheta_{\tau,\varepsilon},\tau\right)$ *in* (2.44) *but with the help of the One-step MLE process as follows (see* (1.33))

$$\begin{aligned}m\left(\vartheta^\star_{\tau,\varepsilon},\tau\right)&=y_0\hat\Phi\left(\vartheta^\star_{\tau,\varepsilon},0,\tau\right)-\int_0^\tau X_s\hat\Phi\left(\vartheta^\star_{\tau,\varepsilon},s,\tau\right)\hat B_s'\left(\vartheta^\star_{\tau,\varepsilon},s\right)\mathrm{d}s\\&\quad+\int_0^\tau X_s\hat\Phi\left(\vartheta^\star_{\tau,\varepsilon},s,\tau\right)\hat A_s'\left(\vartheta^\star_{\tau,\varepsilon},s\right)\hat B\left(\vartheta^\star_{\tau,\varepsilon},s\right)\mathrm{d}s+X_\tau\hat B\left(\vartheta^\star_{\tau,\varepsilon},\tau\right).\end{aligned}\tag{2.58}$$

Here $\hat A\left(\vartheta^\star_{\tau,\varepsilon},t\right)=a(\vartheta^\star_{\tau,\varepsilon},t)-f(\vartheta^\star_{\tau,\varepsilon},t)^2\hat\gamma_t\sigma(t)^{-2}$, $\hat B\left(\vartheta^\star_{\tau,\varepsilon},t\right)=f(\vartheta^\star_{\tau,\varepsilon},t)\hat\gamma_t\sigma(t)^{-2}$ *and*

$$\hat\Phi\left(\vartheta^\star_{\tau,\varepsilon},s,t\right)=\exp\left(\int_s^t\hat A\left(\vartheta^\star_{\tau,\varepsilon},r\right)\mathrm{d}r\right).$$

The initial value $\hat\gamma_\tau$ *is the final value of the following equation solution*

$$\frac{\partial\hat\gamma_t}{\partial t}=2a(\vartheta^\star_{\tau,\varepsilon},t)\hat\gamma_t-\frac{f(\vartheta^\star_{\tau,\varepsilon},t)^2\hat\gamma_t^2}{\sigma(t)^2}+b(\vartheta^\star_{\tau,\varepsilon},t)^2,\qquad \hat\gamma_0=0,\quad 0\le t\le\tau.$$

The system (2.56) and (2.57) is given in recurrent form, but the error $\hat\gamma_t-\gamma(\vartheta_0,t)$ in general is greater than $\gamma\left(\vartheta^\star_{t,\varepsilon},t\right)-\gamma(\vartheta_0,t)$. That is why it can be reasonable to replace $\hat\gamma_t$ in the Eq. (2.56) by (nonrecurrent) function $\gamma_*\left(\vartheta^\star_{t,\varepsilon},t\right)$. The corresponding family of functions $\left(\gamma_*(\vartheta,t),0\le t\le T\right),\vartheta\in\Theta$ can be calculated before the experiment, and in this case, the values $\gamma_*\left(\vartheta^\star_{t,\varepsilon},t\right)$ can be easily calculated too. Therefore, we consider the adaptive filter $m^\star_{t,\varepsilon},\tau\le t\le T$ governed by the equation

$$\mathrm{d}m_{t,\varepsilon}^{\star}=\left[a(\vartheta_{t,\varepsilon}^{\star},t)-\frac{f(\vartheta_{t,\varepsilon}^{\star},t)^2\gamma_*\left(\vartheta_{t,\varepsilon}^{\star},t\right)}{\sigma\left(t\right)^2}\right]m_{t,\varepsilon}^{\star}\,\mathrm{d}t+\frac{f(\vartheta_{t,\varepsilon}^{\star},t)\gamma_*\left(\vartheta_{t,\varepsilon}^{\star},t\right)}{\sigma\left(t\right)^2}\mathrm{d}X_t.$$

The initial value $m_{\tau,\varepsilon}^{\star}=m\left(\vartheta_{\tau,\varepsilon}^{\star},\tau\right)$ given in (2.58) with the corresponding modifications of the functions $A\left(\cdot,\cdot\right)$, $B\left(\cdot,\cdot\right)$ and $\Phi\left(\cdot,\cdot,\cdot\right)$:

$$A\left(\vartheta_0,t\right)=a\left(\vartheta_0,t\right)-\frac{f\left(\vartheta_0,t\right)^2\gamma_*\left(\vartheta_0,t\right)}{\sigma\left(t\right)^2},\qquad B\left(\vartheta_0,t\right)=\frac{f\left(\vartheta_0,t\right)\gamma_*\left(\vartheta_0,t\right)}{\sigma\left(t\right)^2},$$

$$\Phi\left(\vartheta_0,s,t\right)=\exp\left(\int_s^t A\left(\vartheta_0,v\right)\mathrm{d}v\right).$$

Introduce the notation:

$$\mathrm{K}\left(\vartheta_0,t\right)=\left[\dot{\mathrm{a}}\left(\vartheta_0,t\right)-B\left(\vartheta_0,t\right)\dot{\mathrm{f}}\left(\vartheta_0,t\right)\right]y_t\left(\vartheta_0\right),$$

$$\zeta_s\left(\vartheta_0\right)=\mathbf{I}_{\tau_*}^{s}\left(\vartheta_0\right)^{-1}\int_{\tau_*}^{s}\frac{\dot{\mathrm{S}}_*\left(\vartheta_0,\vartheta_0,r\right)}{\sigma\left(r\right)}\mathrm{d}w\left(r\right),$$

$$n^{\star}\left(\vartheta_0,t\right)=\int_0^t\Phi\left(\vartheta_0,s,t\right)\mathrm{K}\left(\vartheta_0,s\right)^{\top}\left[\zeta_\tau\left(\vartheta_0\right)\mathbb{I}_{\{s\le\tau\}}+\zeta_s\left(\vartheta_0\right)\mathbb{I}_{\{s>\tau\}}\right]\mathrm{d}s,$$

$$D^{\star}\left(\vartheta_0,t\right)=\mathbf{E}_{\vartheta_0}\left[n^{\star}\left(\vartheta_0,t\right)^2\right].$$

Theorem 2.7 *Let the conditions $\mathscr{B}^{\star}$ hold. Then as $\varepsilon\to0$ the error of approximation $m_{t,\varepsilon}^{\star}-m\left(\vartheta_0,t\right)$ is asymptotically normal*

$$\frac{m_{t,\varepsilon}^{\star}-m\left(\vartheta_0,t\right)}{\varepsilon}\Longrightarrow n^{\star}\left(\vartheta_0,t\right),\qquad\tau\le t\le T$$

and for any $p\ge2$

$$\varepsilon^{-p}\mathbf{E}_{\vartheta_0}\left|m_{t,\varepsilon}^{\star}-m\left(\vartheta_0,t\right)\right|^p\longrightarrow\mathbf{E}_{\vartheta_0}\left|n^{\star}\left(\vartheta_0,t\right)\right|^p.$$

If $p=2$, then $\mathbf{E}_{\vartheta_0}n^{\star}\left(\vartheta_0,t\right)^2=D^{\star}\left(\vartheta_0,t\right)$.

Proof The equations for $m_{t,\varepsilon}^{\star}$ (see (2.56)) and $m\left(\vartheta_0,t\right)=m_t$, we write as follows:

$$\begin{aligned}\mathrm{d}m_{t,\varepsilon}^{\star}&=A\left(\vartheta_{t,\varepsilon}^{\star},t\right)m_{t,\varepsilon}^{\star}\mathrm{d}t+B\left(\vartheta_{t,\varepsilon}^{\star},t\right)\mathrm{d}X_t, && m_{\tau,\varepsilon}^{\star}, && \tau<t\le T,\\ \mathrm{d}m_t&=A\left(\vartheta_0,t\right)m_t\,\mathrm{d}t+B\left(\vartheta_0,t\right)\,\mathrm{d}X_t, && m_0=y_0, && 0\le t\le T.\end{aligned}$$

First, we estimate the rate of $\mathbf{E}_{\vartheta_0}\left[m_{t,\varepsilon}^{\star}-m_t\right]^2$. We have

$$
\begin{aligned}
\mathrm{d}\left(m_{t,\varepsilon}^{\star}-m_t\right) &= A\left(\vartheta_0,t\right)\left(m_{t,\varepsilon}^{\star}-m_t\right)\mathrm{d}t+\left[A\left(\vartheta_{t,\varepsilon}^{\star},t\right)-A\left(\vartheta_0,t\right)\right]m_{t,\varepsilon}^{\star}\,\mathrm{d}t\\
&\quad+\left[B\left(\vartheta_{t,\varepsilon}^{\star},t\right)-B\left(\vartheta_0,t\right)\right]\mathrm{d}X_t\\
&= A\left(\vartheta_0,t\right)\left(m_{t,\varepsilon}^{\star}-m_t\right)\mathrm{d}t+\left[A\left(\vartheta_{t,\varepsilon}^{\star},t\right)-A\left(\vartheta_0,t\right)\right]m_{t,\varepsilon}^{\star}\,\mathrm{d}t\\
&\quad+\left[B\left(\vartheta_{t,\varepsilon}^{\star},t\right)-B\left(\vartheta_0,t\right)\right]f\left(\vartheta_0,t\right)m_t\,\mathrm{d}t\\
&\quad+\varepsilon\left[B\left(\vartheta_{t,\varepsilon}^{\star},t\right)-B\left(\vartheta_0,t\right)\right]\sigma\left(t\right)\mathrm{d}\bar W_t,\qquad \tau<t\leq T,
\end{aligned}
$$

with the initial value $m_{\tau,\varepsilon}^{\star}-m\left(\vartheta_0,\tau\right)$.

Consider now the difference $\delta_m\left(\vartheta,\tau\right)=m\left(\vartheta,\tau\right)-m\left(\vartheta_0,\tau\right)$. To simplify the exposition, we make the calculations formally as if the problem of stochastic integral does not exist. The detailed proof will require more cumbersom expressions like (2.58) but will lead to the same result. We have $\delta_m\left(\vartheta,0\right)=0$ and the equation

$$
\begin{aligned}
\mathrm{d}\delta_m\left(\vartheta,t\right)&=A\left(\vartheta_0,t\right)\delta_m\left(\vartheta,t\right)\mathrm{d}t+\delta_A\left(t\right)m\left(\vartheta,t\right)\mathrm{d}t\\
&\quad+\delta_B\left(t\right)f\left(\vartheta_0,t\right)m\left(\vartheta_0,t\right)\mathrm{d}t+\varepsilon\delta_B\left(t\right)\sigma\left(t\right)\mathrm{d}\bar W_t.
\end{aligned}
$$

Here $\delta_A\left(t\right)=A\left(\vartheta,t\right)-A\left(\vartheta_0,t\right)$ and $\delta_B\left(t\right)=B\left(\vartheta,t\right)-B\left(\vartheta_0,t\right)$.

The solution of this equation is

$$
\begin{aligned}
\delta_m\left(\vartheta,\tau\right)&=\int_0^{\tau}\Phi\left(\vartheta_0,s,\tau\right)\delta_A\left(t\right)m\left(\vartheta,s\right)\mathrm{d}s+\varepsilon\int_0^{\tau}\Phi\left(\vartheta_0,s,\tau\right)\delta_B\left(t\right)\sigma\left(s\right)\mathrm{d}\bar W_s\\
&\quad+\int_0^{\tau}\Phi\left(\vartheta_0,s,\tau\right)\delta_B\left(t\right)f\left(\vartheta_0,s\right)m\left(\vartheta_0,s\right)\mathrm{d}s.
\end{aligned}
$$

For the small values of $\left\|\vartheta-\vartheta_0\right\|$, we obtain the equation

$$
\begin{aligned}
&\delta_m\left(\vartheta,\tau\right)\\
&\quad=\int_0^{\tau}\Phi\left(\vartheta_0,s,\tau\right)\left[\dot{\mathrm{A}}\left(\vartheta_0,s\right)+\dot{\mathrm{B}}\left(\vartheta_0,s\right)f\left(\vartheta_0,s\right)\right]^{\top}m\left(\vartheta_0,s\right)\mathrm{d}s\left(\vartheta-\vartheta_0\right)\\
&\quad\quad+\varepsilon\int_0^{\tau}\Phi\left(\vartheta_0,s,\tau\right)\dot{\mathrm{B}}\left(\vartheta_0,s\right)^{\top}\sigma\left(s\right)\mathrm{d}\bar W_s\left(\vartheta-\vartheta_0\right)+O\left(\left\|\vartheta-\vartheta_0\right\|^2\right).
\end{aligned}
$$

Therefore,

$$
\mathbf{E}_{\vartheta_0}\left|m\left(\vartheta_{\tau,\varepsilon}^{\star},\tau\right)-m\left(\vartheta_0,\tau\right)\right|^2\leq C\,\varepsilon^2
$$

and

$$
\varepsilon^{-1}\left(m\left(\vartheta_{\tau,\varepsilon}^{\star},\tau\right)-m\left(\vartheta_0,\tau\right)\right)\Longrightarrow\int_0^{\tau}\Phi\left(\vartheta_0,s,\tau\right)\mathrm{K}\left(\vartheta_0,s\right)^{\top}\mathrm{d}s\,\zeta_\tau\left(\vartheta_0\right).
$$

Further,

$$
\begin{aligned}
m^{\star}_{t,\varepsilon}-m_t&=\left[m^{\star}_{\tau,\varepsilon}-m_\tau\right]\Phi\left(\vartheta_0,0,t\right)\\
&\quad+\int_\tau^t\Phi\left(\vartheta_0,s,t\right)\left[A\left(\vartheta^{\star}_{s,\varepsilon},s\right)-A\left(\vartheta_0,s\right)\right]m^{\star}_{s,\varepsilon}\,\mathrm{d}s\\
&\quad+\int_\tau^t\Phi\left(\vartheta_0,s,t\right)\left[B\left(\vartheta^{\star}_{s,\varepsilon},s\right)-B\left(\vartheta_0,s\right)\right]f\left(\vartheta_0,s\right)m_s\,\mathrm{d}s\\
&\quad+\varepsilon\int_\tau^t\Phi\left(\vartheta_0,s,t\right)\left[B\left(\vartheta^{\star}_{s,\varepsilon},s\right)-B\left(\vartheta_0,s\right)\right]\sigma\left(s\right)\mathrm{d}\bar W_s.
\end{aligned}
$$

After elementary calculations, we obtain the estimate

$$
\begin{aligned}
\mathbf{E}_{\vartheta_0}\left(m^{\star}_{t,\varepsilon}-m_t\right)^2&\leq\mathbf{E}_{\vartheta_0}\left(m^{\star}_{\tau,\varepsilon}-m_\tau\right)^2+C\int_\tau^t\mathbf{E}_{\vartheta_0}\left\|\vartheta^{\star}_{s,\varepsilon}-\vartheta_0\right\|^2\mathrm{d}s\\
&\qquad+C\varepsilon^2\int_\tau^t\mathbf{E}_{\vartheta_0}\left\|\vartheta^{\star}_{s,\varepsilon}-\vartheta_0\right\|^2\mathrm{d}s\\
&\leq C\varepsilon^2.
\end{aligned}
$$

Therefore,

$$
\begin{aligned}
&\left[A\left(\vartheta^{\star}_{t,\varepsilon},t\right)-A\left(\vartheta_0,t\right)\right]m^{\star}_{t,\varepsilon}+\left[B\left(\vartheta^{\star}_{t,\varepsilon},t\right)-B\left(\vartheta_0,t\right)\right]f\left(\vartheta_0,t\right)m_t\\
&\quad=\left[A\left(\vartheta^{\star}_{t,\varepsilon},t\right)-A\left(\vartheta_0,t\right)+B\left(\vartheta^{\star}_{t,\varepsilon},t\right)-B\left(\vartheta_0,t\right)\right]f\left(\vartheta_0,t\right)m_t+O\left(\varepsilon^2\right)\\
&\quad=\left[a\left(\vartheta^{\star}_{t,\varepsilon},t\right)-a\left(\vartheta_0\right)-B\left(\vartheta^{\star}_{t,\varepsilon},t\right)\left(f\left(\vartheta^{\star}_{t,\varepsilon},t\right)-f\left(\vartheta_0\right)\right)\right]m_t+O\left(\varepsilon^2\right)\\
&\quad=\left[a\left(\vartheta^{\star}_{t,\varepsilon},t\right)-a\left(\vartheta_0\right)-B\left(\vartheta_0,t\right)\left(f\left(\vartheta^{\star}_{t,\varepsilon},t\right)-f\left(\vartheta_0\right)\right)\right]m_t+O\left(\varepsilon^2\right).
\end{aligned}
$$

This allows us to write

$$
\begin{aligned}
m^{\star}_{t,\varepsilon}-m_t&=\left(m^{\star}_{\tau,\varepsilon}-m_\tau\right)\Phi\left(\vartheta_0,\tau,t\right)\\
&\quad+\int_\tau^t\Phi\left(\vartheta_0,s,t\right)\left[\dot{\mathrm{a}}\left(\vartheta_0,s\right)-B\left(\vartheta_0,s\right)\dot{\mathrm{f}}\left(\vartheta_0,s\right)\right]^\top\left(\vartheta^{\star}_{s,\varepsilon}-\vartheta_0\right)m_s\mathrm{d}s+O\left(\varepsilon^2\right),
\end{aligned}
$$

and

$$
\begin{aligned}
\varepsilon^{-1}\left(m^{\star}_{t,\varepsilon}-m_t\right)&=\int_0^\tau\Phi\left(\vartheta_0,s,t\right)\mathrm{K}\left(\vartheta_0,s\right)^\top\mathrm{d}s\;\varepsilon^{-1}\left(\vartheta^{\star}_{\tau,\varepsilon}-\vartheta_0\right)\\
&\quad+\int_\tau^t\Phi\left(\vartheta_0,s,t\right)\mathrm{K}\left(\vartheta_0,s\right)^\top\varepsilon^{-1}\left(\vartheta^{\star}_{s,\varepsilon}-\vartheta_0\right)\mathrm{d}s+O\left(\varepsilon\right)\\
&\Longrightarrow\int_0^t\Phi\left(\vartheta_0,s,t\right)\mathrm{K}\left(\vartheta_0,s\right)^\top\left[\zeta_\tau\left(\vartheta_0\right)\mathbb{I}_{\{s\leq\tau\}}+\zeta_s\left(\vartheta_0\right)\mathbb{I}_{\{s>\tau\}}\right]\mathrm{d}s\\
&=n^\star\left(\vartheta_0,t\right).
\end{aligned}
$$

The proof of the convergence of moments follows the same arguments. □

Vanishing Learning Interval

Suppose that the parameter $\vartheta \in \Theta$ is one-dimensional, $\Theta = (\alpha, \beta)$ and the *learning interval* is $[0, \tau_\varepsilon]$, $\tau_\varepsilon = \varepsilon^\delta, \delta \in \left(\frac{1}{2}, 1\right)$. The preliminary estimator is the MME $\vartheta^*_{\tau_\varepsilon}$ defined by the Eq. (2.53). Then the One-step MLE process (2.42)

$$\vartheta^\star_{t,\varepsilon} = \vartheta^*_{\tau_\varepsilon} + \mathrm{I}^t_{\tau_\varepsilon}\left(\vartheta^*_{\tau_\varepsilon}\right)^{-1} \int_{\tau_\varepsilon}^t \frac{\dot M\left(\vartheta^*_{\tau_\varepsilon}, s\right)}{\sigma\left(s\right)^2}\left[\mathrm{d}X_s - M\left(\vartheta^*_{\tau_\varepsilon}, s\right)\mathrm{d}s\right], \qquad \tau_\varepsilon < t \le T.$$

The Eqs. (2.45) and (2.46) for $M\left(\vartheta^*_{\tau_\varepsilon}, s\right)$ and $\dot M\left(\vartheta^*_{\tau_\varepsilon}, s\right)$ have to be solved just for one value of parameter $\vartheta = \vartheta^*_{\tau_\varepsilon}$.

Of course, it is possible to use the simplified version

$$\vartheta^\star_{t,\varepsilon} = \vartheta^*_{\tau_\varepsilon} + \mathrm{I}^t_{\tau_\varepsilon}\left(\vartheta^*_{\tau_\varepsilon}\right)^{-1} \int_{\tau_\varepsilon}^t \frac{\dot y_s\left(\vartheta^*_{\tau_\varepsilon}, \vartheta^*_{\tau_\varepsilon}\right)}{\sigma\left(s\right)^2}\left[\mathrm{d}X_s - M\left(\vartheta^*_{\tau_\varepsilon}, s\right)\mathrm{d}s\right], \qquad \tau_\varepsilon < t \le T$$

too.

The adaptive filter $m^\star_{t,\varepsilon}$, $\tau_\varepsilon < t \le T$ is almost the same

$$\mathrm{d}m^\star_{t,\varepsilon} = \left[a(\vartheta^\star_{t,\varepsilon}, t) - \frac{f(\vartheta^\star_{t,\varepsilon}, t)^2\gamma_*\left(\vartheta^\star_{t,\varepsilon}, t\right)}{\sigma\left(t\right)^2}\right] m^\star_{t,\varepsilon}\,\mathrm{d}t + \frac{f(\vartheta^\star_{t,\varepsilon}, t)\gamma_*\left(\vartheta^\star_{t,\varepsilon}, t\right)}{\sigma\left(t\right)^2}\mathrm{d}X_t. \tag{2.59}$$

Here $\gamma_*\left(\vartheta^\star_{t,\varepsilon}, t\right)$ is obtained by substitution of $\vartheta^\star_{t,\varepsilon}$ in the solution $\gamma\left(\vartheta, t\right), \vartheta \in \vartheta$ of the Eq. (2.55). The initial value $m^\star_{\tau_\varepsilon,\varepsilon} = m\left(\vartheta^\star_{\tau_\varepsilon}, \tau_\varepsilon\right)$ is like (2.58)

$$\begin{aligned} m\left(\vartheta^\star_{\tau_\varepsilon}, \tau\right) &= m\left(\vartheta^\star_{\tau_\varepsilon}, 0\right)\Phi\left(\vartheta^\star_{\tau_\varepsilon}, 0, \tau_\varepsilon\right) - \int_0^{\tau_\varepsilon} X_s\Phi\left(\vartheta^\star_{\tau_\varepsilon}, s, \tau\right)B'_s\left(\vartheta^\star_{\tau_\varepsilon}, s\right)\mathrm{d}s \\ &\quad + \int_0^\tau X_s\Phi\left(\vartheta^\star_{\tau_\varepsilon}, s, \tau\right)A'_s\left(\vartheta^\star_{\tau_\varepsilon}, s\right)B\left(\vartheta^\star_{\tau_\varepsilon}, s\right)\mathrm{d}s + X_{\tau_\varepsilon}B\left(\vartheta^\star_{\tau_\varepsilon}, \tau_\varepsilon\right). \end{aligned}$$

Introduce almost the same notation:

$$n_\star\left(\vartheta_0, t\right) = \int_0^t \Phi\left(\vartheta_0, s, t\right)K\left(\vartheta_0, s\right)\hat\zeta_s\left(\vartheta_0\right)\,\mathrm{d}s,$$

$$\hat\zeta_s\left(\vartheta_0\right) = \mathrm{I}^t_0\left(\vartheta_0\right)^{-1}\int_0^t \frac{\dot S_*\left(\vartheta_0, \vartheta_0, v\right)}{\sigma\left(v\right)}\mathrm{d}w\left(v\right),$$

$$D_\star(\vartheta_0, t) = \int_0^t \left(\left(\int_s^t \Phi(\vartheta_0, v, t)\, K(\vartheta_0, v)\, \mathrm{I}_0^v(\vartheta_0)^{-1}\, dv \right) \right.$$

$$\left. \times \dot{S}_*(\vartheta_0, \vartheta_0, s)\, \sigma(s)^{-1} \right)^2 \mathrm{d}s.$$

Theorem 2.8 *Let the conditions $\mathscr{B}^\star$ hold. Then the error of approximation of $m(\vartheta_0, t)$ by $m_{t,\varepsilon}^\star$ is asymptotically normal*

$$\frac{m_{t,\varepsilon}^\star - m(\vartheta_0, t)}{\varepsilon} \Longrightarrow n_\star(\vartheta_0, t), \qquad 0 < t \leq T$$

and for any $p \geq 2$

$$\varepsilon^{-p} \mathbf{E}_{\vartheta_0} \left| m_{t,\varepsilon}^\star - m(\vartheta_0, t) \right|^p \longrightarrow \mathbf{E}_{\vartheta_0} \left| n_\star(\vartheta_0, t) \right|^p.$$

If $p = 2$, then $\mathbf{E}_{\vartheta_0}\left[n_\star(\vartheta_0, t)^2 \right] = D_\star(\vartheta_0, t)$.

Proof The proof is a slight modification of the above given proof of the Theorem 2.7. The main difference is in the contribution of the initial values $m_{\tau_\varepsilon,\varepsilon}^\star - m(\vartheta_0, \tau_\varepsilon) = o(\varepsilon)$. □

AF Based on One-Step MDE Process

The adaptive filter (2.56) and (2.57) is computationally much simpler than the filter based on the real MLE $\hat{\vartheta}_{t,\varepsilon}$, but nevertheless its calculation involves the solutions of filtration equations for $M(\vartheta, t)$ and $\dot{M}(\vartheta, t)$ with corresponding Riccati equations. Sometimes it is reasonable to lose a bit in the efficiency of the estimator in profit of essential simplification of the calculations. That is why it can be interesting to use One-step MDE process $\check{\vartheta}_{t,\varepsilon}^\star$, $\tau_\varepsilon \leq t \leq T$ for the construction of adaptive filter.

Let us remind some details of its construction. We have the same model of observations, where the unknown parameter $\vartheta \in \Theta \subset \mathcal{R}^d$. The preliminary estimator $\check{\vartheta}_{\tau,\varepsilon}$ is defined by the equation

$$\int_0^\tau \left[X_t - x_t\left(\check{\vartheta}_{\tau,\varepsilon} \right) \right]^2 \mathrm{d}t = \inf_{\vartheta \in \Theta} \int_0^\tau \left[X_t - x_t(\vartheta) \right]^2 \mathrm{d}t.$$

The correponding One-step MDE process is

$$\check{\vartheta}_{t,\varepsilon}^\star = \check{\vartheta}_{\tau,\varepsilon} + \mathbf{T}_\tau^t\left(\check{\vartheta}_{\tau,\varepsilon} \right)^{-1} \int_\tau^t \left[X_s - x_s\left(\check{\vartheta}_{\tau,\varepsilon} \right) \right] \dot{\mathrm{x}}_s\left(\check{\vartheta}_{\tau,\varepsilon} \right) \mathrm{d}s, \qquad \tau < t \leq T.$$

The adaptive filter is

$$\mathrm{d}\check m^\star_{t,\varepsilon}=\left[a(\check\vartheta^\star_{t,\varepsilon},t)-\frac{f(\check\vartheta^\star_{t,\varepsilon},t)^2\gamma_*\left(\check\vartheta^\star_{t,\varepsilon},t\right)}{\sigma\left(t\right)^2}\right]\check m^\star_{t,\varepsilon}\,\mathrm{d}t+\frac{f(\check\vartheta^\star_{t,\varepsilon},t)\gamma_*\left(\check\vartheta^\star_{t,\varepsilon},t\right)}{\sigma\left(t\right)^2}\mathrm{d}X_t,$$

where the initial value $\check m^\star_{\tau_\varepsilon}=m\left(\check\vartheta_{\tau,\varepsilon},\tau\right)$ is calculated with the replacing the stochastic integral by an ordinary one and $\gamma_*\left(\vartheta,t\right),0\leq t\leq T,\vartheta\in\Theta$ is calculated before the experiment. Hence, the values $\gamma_*\left(\check\vartheta^\star_{t,\varepsilon},t\right)$ are obtained by the simple substitution $\check\vartheta^\star_{t,\varepsilon}$ in $\gamma_*\left(\vartheta,t\right)$. The error can be estimated using the same calculations as above. This leads to the expression

$$\varepsilon^{-1}\left(\check m^\star_{t,\varepsilon}-m_t\right)\Longrightarrow\int_0^t\Phi\left(\vartheta_0,s,t\right)\mathrm{K}\left(\vartheta_0,s\right)^\top\check\zeta_s\left(\vartheta_0\right)\mathrm{d}s=\check n\left(\vartheta_0,t\right)$$

with the corresponding Gaussian process $\check\zeta_s\left(\vartheta_0\right),\tau<s\leq T$.

Asymptotic Efficiency

We consider the one-dimensional case $d=1$. Suppose that the observations are $X^t=\left(X_s,0\leq s\leq t\right)$ and the conditions of Theorem 2.4 are satisfied for any $t\in(0,T]$. Then for BE $\tilde\vartheta_{t,\varepsilon}$, we have the representation (see (1.81))

$$\varepsilon^{-1}\left(\tilde\vartheta_{t,\varepsilon}-\vartheta_0\right)=\mathrm{I}_0^t\left(\vartheta_0\right)^{-1}\int_0^t\frac{\dot y_s\left(\vartheta_0,\vartheta_0\right)}{\sigma\left(s\right)}\mathrm{d}\bar W_s\left(1+o\left(1\right)\right)$$

and the uniform on compacts $\mathbb{K}\subset\Theta$ convergence of moments

$$\varepsilon^{-2}\mathbf{E}_{\vartheta_0}|\tilde\vartheta_{t,\varepsilon}-\vartheta_0|^2\longrightarrow\mathbf{E}_{\vartheta_0}\left|\zeta_t\left(\vartheta_0\right)\right|^2=\mathrm{I}_0^t\left(\vartheta_0\right)^{-1},\quad\zeta_t\left(\vartheta_0\right)\sim\mathcal{N}\left(0,\mathrm{I}_0^t\left(\vartheta_0\right)^{-1}\right).$$

As it was discussed in the Sect. 1.4 for any estimator $\bar m_{t,\varepsilon}$ the lower bound

$$\lim_{\nu\to0}\varliminf_{\varepsilon\to0}\sup_{|\vartheta-\vartheta_0|\leq\nu}\varepsilon^{-2}\mathbf{E}_\vartheta\left|\bar m_{t,\varepsilon}-m\left(\vartheta,t\right)\right|^2\geq S_t^\star\left(\vartheta_0\right)^2,$$

holds. Here

$$S_t^\star\left(\vartheta_0\right)^2=\lim_{\varepsilon\to0}\varepsilon^{-2}\mathbf{E}_{\vartheta_0}\left(\dot m\left(\vartheta_0,t\right)\left(\tilde\vartheta_{t,\varepsilon}-\vartheta_0\right)\right)^2.$$

Recall that we have the convergence

$$\dot m\left(\vartheta_0,t\right)\longrightarrow\dot y_t\left(\vartheta_0,\vartheta_0\right)=\int_0^t\Phi\left(\vartheta_0,s,t\right)K\left(\vartheta_0,s\right)\mathrm{d}s$$

$$= \int_0^t \Phi\left(\vartheta_0, s, t\right)\left[\dot{a}\left(\vartheta_0, s\right) - B\left(\vartheta_0, s\right)\dot{f}\left(\vartheta_0, s\right)\right] y_s\left(\vartheta_0\right) \mathrm{d}s.$$

Therefore,

$$S_t^\star\left(\vartheta_0\right)^2 = \dot{y}_t\left(\vartheta_0, \vartheta_0\right)^2 \mathrm{I}\left(\vartheta_0, t\right)^{-1}.$$

The estimator $m_{t,\varepsilon}^*$ of $m\left(\vartheta_0, t\right)$ we call *asymptotically efficient*, if the equality

$$\lim_{\nu\to 0}\lim_{\varepsilon\to 0}\sup_{|\vartheta-\vartheta_0|\leq\nu} \varepsilon^{-2}\mathbf{E}_\vartheta\left|m_{t,\varepsilon}^* - m\left(\vartheta, t\right)\right|^2 = \dot{y}_t\left(\vartheta_0, \vartheta_0\right)^2 \mathrm{I}\left(\vartheta_0, t\right)^{-1}$$

holds for all $\vartheta_0 \in \Theta$.

Remark that $m_t^\star$ depends on the whole trajectory $\vartheta_{s,\varepsilon}^\star$, $\tau_\varepsilon < s \leq t$ and not on $\vartheta_{t,\varepsilon}^\star$ (see (2.59)). Therefore, $m_t^\star - m\left(\vartheta_0, t\right) \neq \dot{m}\left(\vartheta_0, t\right)\left(\vartheta_{t,\varepsilon}^\star - \vartheta_0\right)\left(1 + o\left(1\right)\right)$

Denote $Q\left(\vartheta, s\right) = \Phi\left(\vartheta, 0, s\right)^{-1} B\left(\vartheta, s\right)$ and consider another adaptive filter based on the stochastic process (see Lemma 1.3)

$$\begin{aligned}
m\left(\vartheta, t\right) &= y_0\Phi\left(\vartheta, 0, t\right) + \Phi\left(\vartheta, 0, t\right)\int_0^t \Phi\left(\vartheta, 0, s\right)^{-1} B\left(\vartheta, s\right)\mathrm{d}X_s\\
&= y_0\Phi\left(\vartheta, 0, t\right) + \Phi\left(\vartheta, 0, t\right)\int_0^t Q\left(\vartheta, s\right)\mathrm{d}X_s\\
&= y_0\Phi\left(\vartheta, 0, t\right) + \Phi\left(\vartheta, 0, t\right)\left[Q\left(\vartheta, t\right)X_t - \int_0^t X_s Q_s'\left(\vartheta, s\right)\mathrm{d}s\right]\\
&= y_0\Phi\left(\vartheta, 0, t\right) + B\left(\vartheta, t\right)X_t\\
&\qquad - \int_0^t X_s\Phi\left(\vartheta, s, t\right)\left[B\left(\vartheta, s\right)f\left(\vartheta, s\right) - a\left(\vartheta, s\right) + B_s'\left(\vartheta, s\right)\right]\mathrm{d}s.
\end{aligned}$$

Let us put

$$\begin{aligned}
\hat{m}_t &= y_0\Phi\left(\vartheta_{t,\varepsilon}^\star, 0, t\right) + B\left(\vartheta_{t,\varepsilon}^\star, t\right)X_t\\
&\quad - \int_0^t X_s\Phi\left(\vartheta_{t,\varepsilon}^\star, s, t\right)\left[B\left(\vartheta_{t,\varepsilon}^\star, s\right)f\left(\vartheta_{t,\varepsilon}^\star, s\right) - a\left(\vartheta_{t,\varepsilon}^\star, s\right) + B_s'\left(\vartheta_{t,\varepsilon}^\star, s\right)\right]\mathrm{d}s.
\end{aligned}\tag{2.60}$$

All functions of $\vartheta_{t,\varepsilon}^\star$ in this expression are sufficiently smooth w.r.t. ϑ and can be expanded using the representation $\vartheta_{t,\varepsilon}^\star = \vartheta_0 + \varepsilon\,\zeta_t\left(\vartheta_0\right) + o\left(\varepsilon\right)$. Then we obtain the expression

$$\hat{m}_t = y_0\Phi\left(\vartheta_0, 0, t\right) + B\left(\vartheta_0, t\right)X_t$$

$$
\begin{aligned}
&-\int_0^t X_s \Phi(\vartheta_0, s, t)\left[B(\vartheta_0, s) f(\vartheta_0, s) - a(\vartheta_0, s) + B_s'(\vartheta_0, s)\right]\mathrm{d}s \\
&+\Big[y_0 \dot\Phi(\vartheta_0, 0, t) + \dot B(\vartheta_0, t) X_t \\
&\quad -\int_0^t X_s \dot\Phi(\vartheta_0, s, t)\left[B(\vartheta_0, s) f(\vartheta_0, s) - a(\vartheta_0, s) + B_s'(\vartheta_0, s)\right]\mathrm{d}s \\
&\quad -\int_0^t X_s \Phi(\vartheta_0, s, t)\left[\dot B(\vartheta_0, s) f(\vartheta_0, s) + B(\vartheta_0, s) \dot f(\vartheta_0, s)\right. \\
&\qquad \left. - \dot a(\vartheta_0, s) + \dot B_s'(\vartheta_0, s)\right]\mathrm{d}s\Big] \left(\vartheta_{t,\varepsilon}^\star - \vartheta_0\right) + o(\varepsilon) \\
&= m(\vartheta_0, t) + \dot m(\vartheta_0, t)\left(\vartheta_{t,\varepsilon}^\star - \vartheta_0\right) + o(\varepsilon).
\end{aligned}
$$

Now

$$
\begin{aligned}
\frac{\hat m_t - m(\vartheta_0, t)}{\varepsilon} &= \dot m(\vartheta_0, t)\frac{\left(\vartheta_{t,\varepsilon}^\star - \vartheta_0\right)}{\varepsilon}(1 + o(1)) \Longrightarrow \dot y_t(\vartheta_0, \vartheta_0)\,\zeta_t(\vartheta_0) \\
&= \hat n(\vartheta_0, t),
\end{aligned}
$$

where convergence is uniform on compacts and for the moments we have as well the uniform convergence

$$
\varepsilon^{-2}\mathbf{E}_{\vartheta_0}\left(\hat m_t - m(\vartheta_0, t)\right)^2 \longrightarrow \dot y_t(\vartheta_0, \vartheta_0)^2\,\mathbf{E}_{\vartheta_0}\zeta_t(\vartheta_0)^2 = S_t^*(\vartheta_0)^2.
$$

Therefore, the adaptive filter defined by the relation (2.60) is asymptotically efficient.

Let us compare $S_t^*(\vartheta_0)^2$ and $\hat D(\vartheta_0)$. We have to compare the limit random variables $n^\star(\vartheta_0, t)$ and $\hat n(\vartheta_0, t)$

$$
\begin{aligned}
n^\star(\vartheta_0, t) &= \int_0^t \Phi(\vartheta_0, s, t) K(\vartheta_0, s)\,\zeta_s(\vartheta_0)\,\mathrm{d}s, \\
\hat n(\vartheta_0, t) &= \int_0^t \Phi(\vartheta_0, s, t) K(\vartheta_0, s)\,\zeta_t(\vartheta_0)\,\mathrm{d}s.
\end{aligned}
$$

Here the variances of the Gaussian variables $\zeta_s(\vartheta_0)$, $0 < s < t$ and $\zeta_t(\vartheta_0)$ are

$$
\mathrm{I}(\vartheta_0, s)^{-1} = \left(\int_0^s \frac{\dot S_r(\vartheta_0, \vartheta_0)^2}{\sigma(v)^2}\mathrm{d}v\right)^{-1} \geq \left(\int_0^t \frac{\dot S_r(\vartheta_0, \vartheta_0)^2}{\sigma(v)^2}\mathrm{d}v\right)^{-1} = \mathrm{I}(\vartheta_0, t)^{-1},
$$

respectively with strict inequality for the small values of s. Therefore, $\hat D(\vartheta_0) > S_t^*(\vartheta_0)^2$.

2.2 Unknown Initial Value

Consider the partially observed linear system of equations

$$\mathrm{d}X_t = f(t)\,Y_t\,\mathrm{d}t + \varepsilon\sigma(t)\,\mathrm{d}W_t, \qquad X_0 = 0, \quad 0 \le t \le T, \tag{2.61}$$

$$\mathrm{d}Y_t = a(t)\,Y_t\,\mathrm{d}t + \varepsilon b(t)\,\mathrm{d}V_t, \qquad Y_0 = \vartheta. \tag{2.62}$$

where $W_t, V_t, 0 \le t \le T$ are independent Wiener processes. The functions $f(t), a(t), \sigma(t), b(t); t \in [0, T]$ and the parameter $\varepsilon \in (0, 1]$ are supposed to be known. The initial value $\vartheta \in \Theta$ is the unknown parameter. The observations are $X^T = (X_t, 0 \le t \le T)$ and the process $Y^T = (Y_t, 0 \le t \le T)$ is "hidden." Below, we consider three problems:

1. The system (2.61) and (2.62) with unknown deterministic value $Y_0 = \vartheta$.
2. The system like (2.61) and (2.62), but with $Y_0 = \theta_1$ and $a(t) = a(\theta_2, t)$. The unknown parameter is $\vartheta = (\theta_1, \theta_2)$.
3. The homogeneous system ($f(t) = f, \sigma(t) = \sigma, a(t) = \vartheta, b(t) = b$), but with $Y_0 \sim \mathcal{N}(0, d^2)$.

In all three models, we discuss the construction of the preliminary estimators, One-step MLE-process, and adaptive filters.

2.2.1 *Deterministic Initial Value*

Unknown Initial Value

Suppose that the partially observed system is

$$\mathrm{d}X_t = f(t)\,Y_t\,\mathrm{d}t + \sigma(t)\,\mathrm{d}W_t, \qquad X_0 = 0, \qquad 0 \le t \le T,$$

$$\mathrm{d}Y_t = a(t)\,Y_t\,\mathrm{d}t + b(t)\,\mathrm{d}V_t, \qquad Y_0 = \vartheta,$$

where the parameter $\varepsilon \in (0, 1]$ and the functions $f(t), \sigma(t), a(t), b(t)$ are known and satisfy the

Conditions $\mathscr{A}_*$. *The functions* $f(t), \sigma(t), a(t), b(t), t \in [0, T]$ *are bounded and the functions* $f(\cdot), \sigma(\cdot)$ *are separated from 0.*

The unknown parameter is the initial value $Y_0 = \vartheta \in \Theta = (\alpha, \beta)$ and we have to estimate ϑ by observations X^T. Moreover, this estimator will be used later for the construction of the adaptive K-B filter. Our goal is to obtain a recurrent adaptive filter and a recurrent presentation of the estimator, because at each moment $t \in [0, T]$ we need an estimator constructed by the observations $X_s, 0 \le s \le t$. Then this estimator will be used in the construction of the filter.

Recall that the conditional expectation $m(\vartheta,t)=\mathbf{E}_\vartheta\left(Y_t|\mathfrak{F}_t^X\right)$ (here $\mathfrak{F}_t^X$ is the σ-algebra generated by the observations $X^t=(X_s,0\leq s\leq t)$) satisfies the equations of Kalman–Bucy filtration

$$\mathrm{d}m(\vartheta,t)=a(t)\,m(\vartheta,t)\,\mathrm{d}t+\frac{f(t)\,\gamma(t)}{\varepsilon^2\sigma(t)^2}\left[\mathrm{d}X_t-f(t)\,m(\vartheta,t)\,\mathrm{d}t\right],\quad m(\vartheta,0)=\vartheta,$$

$$\frac{\partial\gamma(t)}{\partial t}=2a(t)\,\gamma(t)-\frac{f(t)^2\,\gamma(t)^2}{\varepsilon^2\sigma(t)^2}+\varepsilon^2 b(t)^2,\quad \gamma(0)=0,\quad 0\leq t\leq T.$$

These equations can be slightly simplified by introducing the function $\gamma_*(t)=\frac{\gamma(t)}{\varepsilon^2}$. Then we can write

$$\mathrm{d}m(\vartheta,t)=a(t)\,m(\vartheta,t)\,\mathrm{d}t+\frac{f(t)\,\gamma_*(t)}{\sigma(t)^2}\left[\mathrm{d}X_t-f(t)\,m(\vartheta,t)\,\mathrm{d}t\right],\quad m(\vartheta,0)=\vartheta, \tag{2.63}$$

$$\frac{\partial\gamma_*(t)}{\partial t}=2a(t)\,\gamma_*(t)-\frac{f(t)^2\,\gamma_*(t)^2}{\sigma(t)^2}+b(t)^2,\quad \gamma_*(0)=0,\quad 0\leq t\leq T.$$

The solution of the Eq. (2.63) is

$$m(\vartheta,t)=\vartheta\,\Phi(0,t)+\int_0^t\Phi(s,t)\,f(s)\,\gamma_*(s)\sigma(s)^{-2}\mathrm{d}X_s=\vartheta\,\Phi(0,t)+H\left(t,X^t\right) \tag{2.64}$$

where

$$\Phi(s,t)=\exp\left(\int_s^t A(r)\,\mathrm{d}r\right),\qquad A(t)=a(t)-f(t)^2\,\gamma_*(t)\sigma(t)^{-2},$$

$$H\left(t,X^t\right)=\int_0^t\Phi(s,t)\,f(s)\,\gamma_*(s)\sigma(s)^{-2}\mathrm{d}X_s.$$

Therefore, the good approximation of $m(\vartheta,t)$ needs a good estimator of the parameter ϑ.

One way to estimate ϑ is to use the MLE $\hat\vartheta_{t,\varepsilon}$. The measures $\left(\mathbf{P}_\vartheta^{(t,\varepsilon)},\vartheta\in\Theta\right)$ induced by observations X^t in the measurable space $(\mathcal{C}[0,T],\mathfrak{B})$ under condition $\mathscr{A}_*$ are equivalent, and the likelihood ratio (LR) function is

$$L\left(\vartheta,X^t\right)=\exp\left(\int_0^t\frac{f(s)\,m(\vartheta,s)}{\varepsilon^2\sigma(s)^2}\,\mathrm{d}X_s-\int_0^t\frac{f(s)^2\,m(\vartheta,s)^2}{2\varepsilon^2\sigma(s)^2}\,\mathrm{d}s\right),\qquad \vartheta\in\Theta.$$

Hence, the Fisher information is

$$\mathrm{I}^t(\vartheta)=\varepsilon^2\mathbf{E}_\vartheta\left[\frac{\partial\ln L\left(\vartheta,X^t\right)}{\partial\vartheta}\right]^2=\int_0^t\frac{f(s)^2\,\Phi(0,s)^2}{\sigma(s)^2}\mathrm{d}s$$

and the MLE has an explicit form

$$\hat\vartheta_{t,\varepsilon}=\left(\int_0^t\frac{f(s)^2\,\Phi(0,s)^2}{\sigma(s)^2}\mathrm{d}s\right)^{-1}\int_0^t\frac{f(s)\,\Phi(0,s)}{\sigma(s)^2}\left[\mathrm{d}X_s-f(s)\,H\left(s,X^s\right)\mathrm{d}s\right].\tag{2.65}$$

Of course, in this model, the Fisher information does not depend on ϑ and we write $\mathrm{I}^t(\vartheta)$ to mention the estimated parameter.

This is a "toy model," where the MLE has an explicite expression, and we present it for the illustration of the recurrent representation of the MLE and One-step MLE process.

It is easy to see that

$$\begin{aligned}\frac{\hat\vartheta_{t,\varepsilon}-\vartheta_0}{\varepsilon}&=\left(\int_0^t\frac{f(s)^2\,\Phi(0,s)^2}{\sigma(s)^2}\mathrm{d}s\right)^{-1}\int_0^t\frac{f(s)\,\Phi(0,s)}{\sigma(s)}\mathrm{d}\bar W_s\\&\sim\mathcal{N}\left(0,\mathrm{I}^t(\vartheta_0)^{-1}\right).\end{aligned}$$

Here we used the representation of observations

$$\mathrm{d}X_s=f(s)\,m(\vartheta_0,s)\,\mathrm{d}s+\varepsilon\sigma(s)\,\mathrm{d}\bar W_s,\qquad X_0=0,\quad 0\le s\le t$$

where $\bar W_s, 0\le s\le t$ is an innovation Wiener process (Lemma 1.2).

The estimator $\hat\vartheta_{t,\varepsilon}$ can be written in the recurrent form as follows:

$$\hat\vartheta_{t,\varepsilon}=\frac{S(t)}{\mathrm{I}^t}$$

where

$$\begin{aligned}\mathrm{d}S(t)&=\frac{f(t)\,\Phi(0,t)}{\sigma(t)^2}\left[\mathrm{d}X_t-f(t)\,H\left(t,X^t\right)\mathrm{d}t\right],\qquad S(0)=0,\\\mathrm{d}\mathrm{I}^t&=\frac{f(t)^2\,\Phi(0,t)^2}{\sigma(t)^2}\mathrm{d}t,\qquad \mathrm{I}^0=0,\\\mathrm{d}H\left(t,X^t\right)&=A(t)\,H\left(t,X^t\right)\mathrm{d}t+f(t)\,\gamma_*(t)\sigma(t)^{-2}\mathrm{d}X_t.\end{aligned}$$

It is possible to write the stochastic differential of the process $\hat\vartheta_{t,\varepsilon}, 0\le t\le T$. Let us multiply the both sides of (2.65) by $\mathrm{I}^t(\vartheta)$, denote $g(t)=f(t)\,\Phi(0,t)\sigma(t)^{-1}$

and take the differential, then we obtain

$$\hat{\vartheta}_{t,\varepsilon}\frac{f(t)^2\,\Phi(0,t)^2}{\sigma(t)^2}\mathrm{d}t+\mathrm{I}^t(\vartheta)\,\mathrm{d}\hat{\vartheta}_{t,\varepsilon}=\frac{f(t)\,\Phi(0,t)}{\sigma(t)^2}\left[\mathrm{d}X_t-f(t)\,H\left(t,X^t\right)\mathrm{d}t\right]$$

and

$$\mathrm{d}\hat{\vartheta}_{t,\varepsilon}=-\frac{g(t)^2}{\mathrm{I}^t(\vartheta)}\hat{\vartheta}_{t,\varepsilon}\mathrm{d}t+\frac{g(t)}{\mathrm{I}^t(\vartheta)\,\sigma(t)}\left[\mathrm{d}X_t-f(t)\,H\left(t,X^t\right)\mathrm{d}t\right]. \tag{2.66}$$

In these two representations, there is a question of the initial value because $\hat{\vartheta}_{0,\varepsilon}=0/0$. To avoid this problem, we can fix some (small) τ and calculate the preliminary estimator

$$\hat{\vartheta}_{\tau,\varepsilon}=\frac{1}{\mathrm{I}^\tau}\int_0^\tau\frac{g(s)}{\sigma(s)}\left[\mathrm{d}X_s-f(s)\,H\left(s,X^s\right)\mathrm{d}s\right] \tag{2.67}$$

and then to use the Eq. (2.66) for calculation $\hat{\vartheta}_{t,\varepsilon},\ \tau\leq t\leq T$.

Let us verify that (2.66)–(2.67) gives the same estimator, i.e., that the solution $\hat{\vartheta}_{\tau,\varepsilon}\sim\mathcal{N}\left(\vartheta_0,\varepsilon^2\mathrm{I}^t(\vartheta_0)^{-1}\right)$. It is sufficient to calculate $\mathbf{E}_{\vartheta_0}\left(\hat{\vartheta}_{\tau,\varepsilon}-\vartheta_0\right)^2=\varepsilon^2\mathrm{I}^t(\vartheta_0)^{-1}$. We have

$$\hat{\vartheta}_{t,\varepsilon}=\hat{\vartheta}_{\tau,\varepsilon}\tilde{\Phi}(\tau,t)+\int_\tau^t\tilde{\Phi}(s,t)\frac{g(s)}{\mathrm{I}^s\sigma(s)}\left[\mathrm{d}X_s-f(s)\,H\left(s,X^s\right)\mathrm{d}s\right],$$

where

$$\tilde{\Phi}(s,t)=\exp\left(-\int_s^t\frac{g(r)^2}{\mathrm{I}^r}\mathrm{d}r\right).$$

Therefore, the substitution of the observations gives the expression

$$\hat{\vartheta}_{t,\varepsilon}=\hat{\vartheta}_{\tau,\varepsilon}\tilde{\Phi}(\tau,t)+\vartheta_0\int_\tau^t\tilde{\Phi}(s,t)\frac{g(s)^2}{\mathrm{I}^s}\mathrm{d}s+\varepsilon\int_\tau^t\tilde{\Phi}(s,t)\frac{g(s)}{\mathrm{I}^s}\mathrm{d}\bar{W}_s.$$

Remark that

$$\int_s^t\frac{g(r)^2}{\mathrm{I}^r}\mathrm{d}r=\int_s^t\frac{1}{\mathrm{I}^r}\mathrm{d}\left(\int_0^r g(q)^2\,\mathrm{d}q\right)=\int_s^t\frac{1}{\mathrm{I}^r}\mathrm{d}\mathrm{I}^r=\ln\mathrm{I}^t-\ln\mathrm{I}^s$$

and

$$\tilde{\Phi}(s,t)=\frac{\mathrm{I}^s}{\mathrm{I}^t}.$$

Therefore,

$$\int_\tau^t \tilde{\Phi}(s,t)\frac{g(s)^2}{\mathrm{I}^s}\mathrm{d}s = \frac{1}{\mathrm{I}^t}\int_\tau^t g(s)^2\,\mathrm{d}s = \frac{\mathrm{I}^t - \mathrm{I}^\tau}{\mathrm{I}^t}$$

and we obtain the relation

$$\hat{\vartheta}_{t,\varepsilon} = \hat{\vartheta}_{\tau,\varepsilon}\frac{\mathrm{I}^\tau}{\mathrm{I}^t} + \vartheta_0\frac{\mathrm{I}^t - \mathrm{I}^\tau}{\mathrm{I}^t} + \frac{\varepsilon}{\mathrm{I}^t}\int_\tau^t g(s)\,\mathrm{d}\bar{W}_s.$$

Hence,

$$\varepsilon^{-1}\left(\hat{\vartheta}_{t,\varepsilon} - \vartheta_0\right) = \varepsilon^{-1}\left(\hat{\vartheta}_{\tau,\varepsilon} - \vartheta_0\right)\frac{\mathrm{I}^\tau}{\mathrm{I}^t} + \frac{1}{\mathrm{I}^t}\int_\tau^t g(s)\,\mathrm{d}\bar{W}_s$$

and

$$\begin{aligned}\varepsilon^{-2}\mathbf{E}_{\vartheta_0}\left(\hat{\vartheta}_{t,\varepsilon} - \vartheta_0\right)^2 &= \varepsilon^{-2}\mathbf{E}_{\vartheta_0}\left(\hat{\vartheta}_{\tau,\varepsilon} - \vartheta_0\right)^2\left(\frac{\mathrm{I}^\tau}{\mathrm{I}^t}\right)^2 + \frac{1}{(\mathrm{I}^t)^2}\int_\tau^t g(s)^2\,\mathrm{d}s\\ &= \frac{\mathrm{I}^\tau}{(\mathrm{I}^t)^2} + \frac{\mathrm{I}^t - \mathrm{I}^\tau}{(\mathrm{I}^t)^2} = \frac{1}{\mathrm{I}^t}.\end{aligned}$$

Therefore, $\varepsilon^{-1}\left(\hat{\vartheta}_{t,\varepsilon} - \vartheta_0\right) \sim \mathcal{N}\left(0, \mathrm{I}^t(\vartheta_0)^{-1}\right)$.

We have recurrent equations for the estimator of the parameter. The adaptive Kalman filter is defined in two steps. First, we introduce the random process

$$\mathrm{d}n_t^* = \left[a(t) - \frac{f(t)^2\gamma_*(t)}{\sigma(t)^2}\right]n_t^*\mathrm{d}t + \frac{f(t)\gamma_*(t)}{\sigma(t)^2}\mathrm{d}X_t, \qquad n_0^* = 0.$$

The solution of this equation is (see (2.64))

$$n_t^* = \int_0^t \Phi(s,t)\,f(s)\,\gamma_*(s)\,\sigma(s)^{-2}\,\mathrm{d}X_s.$$

The adaptive Kalman filter defined by the equation

$$m_t^\star = \hat{\vartheta}_{t,\varepsilon}\Phi(0,t) + n_t^*.$$

The error of estimation is

$$\varepsilon^{-1}\left(m_t^\star - m(\vartheta_0,t)\right) = \varepsilon^{-1}(\hat{\vartheta}_{t,\varepsilon} - \vartheta_0)\Phi(0,t) = \frac{\Phi(0,t)}{\mathrm{I}^t}\int_0^t \frac{f(s)\gamma_*(s)}{\sigma(s)}\mathrm{d}\bar{W}_s$$

The lower bound in this problem is

$$\lim_{\nu\to 0}\varliminf_{\varepsilon\to 0}\sup_{|\vartheta-\vartheta_0|\leq\nu}\varepsilon^{-2}\mathbf{E}_\vartheta\left[\bar m_{t,\varepsilon}-m\left(\vartheta,t\right)\right]^2\geq\frac{\Phi\left(0,t\right)^2}{\mathrm{I}^t}.$$

As

$$\varepsilon^{-2}\mathbf{E}_\vartheta\left[m_t^\star-m\left(\vartheta,t\right)\right]^2=\frac{\Phi\left(0,t\right)^2}{\mathrm{I}^t}$$

the approximation $m_t^\star$ is asymptotically efficient.

Unknown Initial Value and Parameter

Suppose that the partially observed system is

$$\mathrm{d}X_t=f\left(t\right)Y_t\,\mathrm{d}t+\varepsilon\sigma\left(t\right)\mathrm{d}W_t,\qquad X_0=0,\qquad 0\leq t\leq T,$$
$$\mathrm{d}Y_t=a\left(\theta_2,t\right)Y_t\,\mathrm{d}t+\varepsilon b\left(t\right)\mathrm{d}V_t,\qquad Y_0=\theta_1.$$

The unknown parameter is $\vartheta=(\theta_1,\theta_2)\in\Theta=(\alpha_1,\beta_1)\times(\alpha_2,\beta_2)$, $\alpha_1>0$. The functions $f\left(\cdot\right),\sigma\left(\cdot\right),a\left(\cdot\right),b\left(\cdot\right)$ satisfy the condition $\mathcal{A}_*$. Our goal is to find an approximation of the conditional expectation $m\left(\vartheta,t\right)=\mathbf{E}_\vartheta\left(Y_t|\mathfrak{F}_t^X\right)$ and describe its properties in the asymptotic $\varepsilon\longrightarrow 0$.

Introduce conditions $\mathscr{C}^\circ$.

$\mathscr{C}_1^\circ$. *The function* $a\left(\vartheta,\cdot\right),\vartheta\in\bar\Theta$ *has two continuous derivatives w.r.t.* ϑ.

$\mathscr{C}_2^\circ$. *The functions* $f\left(\cdot\right),\sigma\left(\cdot\right),a\left(\cdot\right),b\left(\cdot\right)$ *have continuous derivatives w.r.t.* $t\in[0,T]$.

The adaptive recurrent K-B filter is constructed in several steps. First, we estimate the parameter θ_2 by the observations on the small interval $[0,\tau_\varepsilon]$, where $\tau_\varepsilon\to 0$. Then we use this estimator in the construction of the parameter θ_1 by observations on the interval $[0,\tau]$. The obtained estimator of ϑ is used to introduce the One-step MLE process. The last step is to substitute this One-step estimator process in the equations of Kalman–Bucy filter.

The limit system $(\varepsilon=0)$ is, of course,

$$\frac{\mathrm{d}x_t\left(\vartheta_0\right)}{\mathrm{d}t}=f\left(t\right)y_t\left(\vartheta_0\right),\qquad x_0\left(\vartheta_0\right)=0,\qquad 0\leq t\leq T,$$
$$\frac{\mathrm{d}y_t\left(\vartheta_0\right)}{\mathrm{d}t}=a\left(\theta_{2,0},t\right)y_t\left(\vartheta_0\right),\qquad y_0\left(\vartheta_0\right)=\theta_{1,0}.$$

Here $\vartheta_0=\left(\theta_{1,0},\theta_{2,0}\right)$ is the true value of ϑ.

Note that the process Y_t converges to the function

$$y_t\left(\vartheta_0\right)=\theta_{1,0}\,\phi\left(\theta_{2,0},0,t\right),\qquad\qquad \phi\left(\theta_{2,0},s,t\right)=\exp\left(\int_s^t a\left(\theta_{2,0},v\right)\mathrm{d}v\right)$$

and

$$Y_t=\theta_{1,0}\,\phi\left(\theta_{2,0},0,t\right)+\varepsilon\int_0^t\phi\left(\theta_{2,0},s,t\right)b\left(s\right)\mathrm{d}V_s.$$

The observations have the following representation as $\tau_\varepsilon\to 0$

$$\begin{aligned}X_{\tau_\varepsilon}&=\theta_{1,0}\int_0^{\tau_\varepsilon}f\left(t\right)\phi\left(\theta_{2,0},0,t\right)\,\mathrm{d}t+\varepsilon\int_0^{\tau_\varepsilon}\sigma\left(s\right)\mathrm{d}W_s\\&\quad+\varepsilon\int_0^{\tau_\varepsilon}f\left(t\right)\int_0^t\phi\left(\theta_{2,0},s,t\right)b\left(s\right)\mathrm{d}V_s\,\mathrm{d}t\\&=\theta_{1,0}f\left(0\right)\tau_\varepsilon+\varepsilon\sigma\left(0\right)W_{\tau_\varepsilon}+\varepsilon\tau_\varepsilon^{3/2}f\left(0\right)b\left(0\right)\int_0^1 r\mathrm{d}v_\varepsilon\left(r\right)+O\left(\varepsilon\tau_\varepsilon^{3/2}\right),\end{aligned}$$

where we used Taylor formula and denoted $v_\varepsilon\left(r\right),0\leq r\leq 1$ some Wiener process independent of W_{τ_ε}. Hence, the MME can be

$$\begin{aligned}\theta^*_{1,\tau_\varepsilon}=\frac{X_{\tau_\varepsilon}}{f\left(0\right)\tau_\varepsilon}&=\theta_{1,0}+\frac{\varepsilon}{\sqrt{\tau_\varepsilon}}\sigma\left(0\right)\xi_{1,\varepsilon}+\tau_\varepsilon O\left(1\right)\\&\quad+\varepsilon\tau_\varepsilon^{1/2}b\left(0\right)\int_0^1 r\mathrm{d}v_\varepsilon\left(r\right)+\varepsilon\tau_\varepsilon O\left(1\right).\end{aligned}$$

The solution of the equation $\varepsilon/\sqrt{\tau_\varepsilon}=\tau_\varepsilon$ gives us the rate $\tau_\varepsilon=\varepsilon^{2/3}$. Therefore, we obtain the asymptotic normality

$$\varepsilon^{-2/3}\left(\theta^*_{1,\tau_\varepsilon}-\theta_{1,0}\right)\Longrightarrow\mathcal{N}\left(q,d^2\right)$$

with some parameters q,d^2. The particular values of these parameters are not interesting. Moreover, we have the convergence of all polynomial moments too.

Let us fix some (small) $\tau>0$ and define the random process

$$x_{t,\varepsilon}\left(\theta_2\right)=\theta^*_{1,\tau_\varepsilon}\int_0^t f\left(s\right)\phi\left(\theta_2,0,s\right)\mathrm{d}s,\qquad\qquad 0\leq t\leq\tau,$$

and define the minimum distance estimator (MDE) $\check\theta_{2,\tau}$ by the equation

$$\int_0^\tau\left[X_t-x_{t,\varepsilon}\left(\check\theta_{2,\tau}\right)\right]^2\mathrm{d}t=\inf_{\alpha_2<\theta_2<\beta_2}\int_0^\tau\left[X_t-x_{t,\varepsilon}\left(\theta_2\right)\right]^2\mathrm{d}t.$$

Introduce the identifiability condition

$\mathscr{C}_3^\circ$. *For a given τ, any $\nu>0$ and any $\theta_{2,0}\in[\alpha_2,\beta_2]$*

$$g\left(\vartheta_0,\nu\right)=\inf_{|\theta_2-\theta_{2,0}|\geq\nu}\int_0^\tau\left[F\left(\theta_2,t\right)-F\left(\theta_{2,0},t\right)\right]^2\mathrm{d}t>0.$$

Here

$$F\left(\theta,t\right)=\int_0^t f\left(s\right)\phi\left(\theta,0,s\right)\mathrm{d}s.$$

We have

$$\begin{aligned}X_t-x_{t,\varepsilon}\left(\check\theta_{2,\tau}\right)&=x_t\left(\vartheta_0\right)-x_{t,\varepsilon}\left(\check\theta_{2,\tau}\right)+\varepsilon\int_0^t\sigma\left(s\right)\mathrm{d}W_s\\&\quad+\varepsilon\int_0^t f\left(s\right)\int_0^s\phi\left(\theta_{2,0},r,s\right)b\left(r\right)\mathrm{d}V_r\,\mathrm{d}s\\&=\left(\theta_{1,0}-\theta^*_{1,\tau_\varepsilon}\right)F\left(\theta_{2,0},t\right)+\theta^*_{1,\tau_\varepsilon}\left[F\left(\theta_{2,0},t\right)-F\left(\check\theta_{2,\tau},t\right)\right]\\&\quad+\varepsilon\,\eta_t,\end{aligned}$$

where $\eta_t,0\leq t\leq\tau$ is a Gaussian process with bounded variance defined by the last equality.

Below, $\|h_t\|_\tau$ is $\mathcal{L}_2\left[0,\tau\right]$-norm. To prove the consistency of the estimator $\check\theta_{2,\tau}$ we write

$$\begin{aligned}&\mathbf{P}_{\vartheta_0}\left(\left|\check\theta_{2,\tau}-\theta_{2,0}\right|\geq\nu\right)\\&\quad=\mathbf{P}_{\vartheta_0}\left(\inf_{|\theta_2-\theta_{2,0}|\leq\nu}\left\|X_t-x_{t,\varepsilon}\left(\theta_2\right)\right\|_\tau>\inf_{|\theta_2-\theta_{2,0}|\geq\nu}\left\|X_t-x_{t,\varepsilon}\left(\theta_2\right)\right\|_\tau\right)\\&\quad\leq\mathbf{P}_{\vartheta_0}\Big(\inf_{|\theta_2-\theta_{2,0}|\leq\nu}\left(\left\|X_t-x_t\left(\vartheta_0\right)\right\|_\tau+\left\|x_t\left(\vartheta_0\right)-x_{t,\varepsilon}\left(\theta_{2,0}\right)\right\|_\tau\right.\\&\qquad\left.+\left\|x_{t,\varepsilon}\left(\theta_2\right)-x_{t,\varepsilon}\left(\theta_{2,0}\right)\right\|_\tau\right)>\inf_{|\theta_2-\theta_{2,0}|\geq\nu}\left\|x_{t,\varepsilon}\left(\theta_2\right)-x_{t,\varepsilon}\left(\theta_{2,0}\right)\right\|_\tau\\&\qquad-\left\|X_t-x_t\left(\vartheta_0\right)\right\|_\tau-\left\|x_t\left(\vartheta_0\right)-x_{t,\varepsilon}\left(\theta_{2,0}\right)\right\|_\tau\Big)\\&\quad\leq\mathbf{P}_{\vartheta_0}\left(2\left\|X_t-x_t\left(\vartheta_0\right)\right\|_\tau+2\left\|x_t\left(\vartheta_0\right)-x_{t,\varepsilon}\left(\theta_{2,0}\right)\right\|_\tau>\alpha_1\,g\left(\theta_{2,0},\nu\right)\right).\end{aligned}$$

By Tchebyshev inequality we have

$$
\mathbf{P}_{\vartheta_0}\left(\left|\check{\theta}_{2,\tau}-\theta_{2,0}\right|\geq\nu\right)\leq\frac{8}{\alpha_1^2 g\left(\vartheta_0,\nu\right)^2}\left(\mathbf{E}_{\vartheta_0}\left\|X_t-x_t\left(\vartheta_0\right)\right\|_\tau^2\right.
$$

$$
\left.+\,\mathbf{E}_{\vartheta_0}\left\|x_t\left(\vartheta_0\right)-x_{t,\varepsilon}\left(\theta_{2,0}\right)\right\|_\tau^2\right)\leq\frac{C\varepsilon^2+C\varepsilon^{4/3}}{\alpha_1^2 g\left(\vartheta_0,\nu\right)^2}\longrightarrow 0.
$$

Therefore, the estimator $\bar{\vartheta}_{\tau,\varepsilon}=\left(\theta_{1,\tau_\varepsilon},\theta_{2,\tau}\right)$ is consistent. It can be shown that the difference $\varepsilon^{-2/3}\left(\bar{\vartheta}_{\tau,\varepsilon}-\vartheta_0\right)$ is asymptotically normal and the polynomial moments $\varepsilon^{-2p/3}\mathbf{E}_{\vartheta_0}\left\|\bar{\vartheta}_{\tau,\varepsilon}-\vartheta_0\right\|^p$ are bounded for any $p>0$. This estimator will be used for the construction of One-step MLE process. We need the equations of Kalman–Bucy filter for this model of observations:

$$
\mathrm{d}m\left(\vartheta,t\right)=A\left(\theta_2,t\right)m\left(\vartheta,t\right)\mathrm{d}t+\frac{f\left(t\right)\gamma_*\left(\theta_2,t\right)}{\sigma\left(t\right)^2}\,\mathrm{d}X_t,\qquad m\left(\vartheta,0\right)=\theta_1,
$$

$$
\frac{\mathrm{d}\gamma_*\left(\theta_2,t\right)}{\mathrm{d}t}=2a\left(\theta_2,t\right)\gamma_*\left(\theta_2,t\right)-\frac{f\left(t\right)^2\gamma_*\left(\theta_2,t\right)^2}{\sigma\left(t\right)^2}+b\left(t\right)^2,\qquad \gamma_*\left(\theta_2,0\right)=0,
$$

where

$$
A\left(\theta_2,t\right)=a\left(\theta_2,t\right)-\frac{f\left(t\right)^2\gamma_*\left(\theta_2,t\right)}{\sigma\left(t\right)^2}.
$$

The derivatives $\dot{m}_1\left(\vartheta,t\right)=\partial m\left(\vartheta,t\right)/\partial\theta_1$ and $\dot{m}_2\left(\vartheta,t\right)=\partial m\left(\vartheta,t\right)/\partial\theta_2$ satisfy the equations

$$
\frac{\mathrm{d}\dot{m}_1\left(\vartheta,t\right)}{\mathrm{d}t}=A\left(\theta_2,t\right)\dot{m}_1\left(\vartheta,t\right),
$$

$$
\mathrm{d}\dot{m}_2\left(\vartheta,t\right)=A\left(\theta_2,t\right)\dot{m}_2\left(\vartheta,t\right)\mathrm{d}t+\dot{A}_2\left(\theta_2,t\right)m\left(\vartheta,t\right)\mathrm{d}t+\frac{f\left(t\right)\dot{\gamma}_*\left(\theta_2,t\right)}{\sigma\left(t\right)^2}\,\mathrm{d}X_t,
$$

subject to initial values $\dot{m}_1\left(\vartheta,t\right)=1$ and $\dot{m}_2\left(\vartheta,0\right)=0$. Therefore,

$$
\dot{m}_1\left(\vartheta,t\right)=\Phi\left(\theta_2,0,t\right),\qquad \Phi\left(\theta_2,s,t\right)=\exp\left(\int_s^t A\left(\theta_2,r\right)\mathrm{d}r\right),
$$

$$
\dot{m}_2\left(\vartheta,t\right)=\int_0^t\Phi\left(\theta_2,s,t\right)\dot{A}_2\left(\theta_2,s\right)m\left(\vartheta,s\right)\mathrm{d}s
$$

$$
+\int_0^t\Phi\left(\theta_2,s,t\right)\frac{f\left(s\right)\dot{\gamma}_*\left(\theta_2,s\right)}{\sigma\left(s\right)^2}\,\mathrm{d}X_s.
$$

The limit ($\varepsilon=0$) values are: for $\dot{m}_1\left(\vartheta,t\right)=\Phi\left(\theta_2,0,t\right)\equiv\dot{y}_1\left(\vartheta,t\right)$ and for $\dot{m}_2\left(\vartheta,t\right)$ we obtain the expression

$$\dot{y}_2(\vartheta,\vartheta_0,t)=\int_0^t \Phi(\theta_2,s,t)\,\dot{A}_2(\theta_2,s)\,y_s(\vartheta,\vartheta_0)\,\mathrm{d}s$$
$$+\int_0^t \Phi(\theta_2,s,t)\,\frac{f(s)^2\,\dot{\gamma}_*(\theta_2,s)}{\sigma(s)^2}\,y_s(\vartheta_0)\;\mathrm{d}s$$

where

$$y_s(\vartheta_0)=\theta_{1,0}\phi\left(\theta_{2,0},0,t\right),\qquad \phi\left(\theta_{2,0},0,t\right)=\exp\left(\int_0^t a\left(\theta_{2,0},s\right)\mathrm{d}s\right),$$

$$y_t(\vartheta,\vartheta_0)=\theta_1\Phi(\theta_2,0,t)+\theta_{1,0}\int_0^t \Phi(\theta_2,s,t)\,\frac{f(s)^2\,\dot{\gamma}_*(\theta_2,s)}{\sigma(s)^2}\,\phi\left(\theta_{2,0},0,s\right)\mathrm{d}s.$$

The function $\dot{y}_2(\vartheta,\vartheta_0,t)$ at the point $\vartheta=\vartheta_0$ takes the value $\dot{y}_2(\vartheta_0,t)$ where

$$\dot{y}_2(\vartheta_0,t)=\int_0^t \Phi\left(\theta_{2,0},s,t\right)\dot{a}\left(\theta_{2,0},s\right)y_s(\vartheta_0)\,\mathrm{d}s$$
$$=\theta_{1,0}\int_0^t \Phi\left(\theta_{2,0},s,t\right)\phi\left(\theta_{2,0},0,s\right)\dot{a}\left(\theta_{2,0},s\right)\mathrm{d}s.$$

The Fisher information matrix is

$$\mathbf{I}_\tau^t(\vartheta)=\begin{pmatrix}\displaystyle\int_\tau^t \frac{f(s)^2\,\dot{y}_1(\vartheta,s)^2}{\sigma(s)^2}\mathrm{d}s & \displaystyle\int_\tau^t \frac{f(s)^2\,\dot{y}_1(\vartheta,s)\,\dot{y}_2(\vartheta,s)}{\sigma(s)^2}\mathrm{d}s\\ \displaystyle\int_\tau^t \frac{f(s)^2\,\dot{y}_1(\vartheta,s)\,\dot{y}_2(\vartheta,s)}{\sigma(s)^2}\mathrm{d}s & \displaystyle\int_\tau^t \frac{f(s)^2\,\dot{y}_2(\vartheta,s)^2}{\sigma(s)^2}\mathrm{d}s\end{pmatrix}.$$

Introduce the condition

$\mathscr{C}_4^\circ$. *For a given τ and any $t\in(\tau,T]$ and any $\vartheta\in\Theta$ the Fisher information is nondegenerate*

$$\inf_{\|\mathrm{e}\|=1,\mathrm{e}\in\mathcal{R}^2}\mathrm{e}^\top\mathbf{I}_\tau^t(\vartheta)\,\mathrm{e}>0.$$

The One-step MLE process is

$$\vartheta_{t,\varepsilon}^\star=\bar{\vartheta}_{\tau,\varepsilon}+\mathbf{I}_\tau^t\left(\bar{\vartheta}_{\tau,\varepsilon}\right)^{-1}\int_\tau^t \frac{f(s)\,\dot{\mathrm{m}}\left(\bar{\vartheta}_{\tau,\varepsilon},s\right)}{\sigma(s)^2}\left[\mathrm{d}X_s-f(s)\,m\left(\bar{\vartheta}_{\tau,\varepsilon},s\right)\mathrm{d}s\right].\tag{2.68}$$

Here the vector $\dot{\mathrm{m}}(\vartheta,s)=\left(\dot{m}_1(\vartheta,t),\dot{m}_2(\vartheta,t)\right)^\top$. Remark that the vector $\dot{\mathrm{m}}(\vartheta,s)$ can be replaced by the vector $\dot{\mathrm{y}}(\vartheta,s)=\left(\dot{y}_1(\vartheta,s),\dot{y}_2(\vartheta,s)\right)^\top$ and the asymptotic properties of the estimator will be the same.

Define the Gaussian process $\eta_t(\vartheta_0) = \left(\eta_{1,t}(\vartheta_0), \eta_{2,t}(\vartheta_0)\right)^\top$

$$\eta_t(\vartheta_0) = \mathbf{I}_\tau^t(\vartheta)^{-1} \int_\tau^t \frac{f(s)\,\dot{\mathbf{y}}_s(\vartheta_0)}{\sigma(s)}\,\mathrm{d}w(s), \qquad \tau \le t \le T,$$

where $\dot{\mathbf{y}}_t(\vartheta_0) = \left(\dot{y}_1(\vartheta_0, t), \dot{y}_2(\vartheta_0, t)\right)^\top$ and $w(s), 0 \le s \le T$ is some Wiener process.

Proposition 2.1 *Let the conditions* $\mathscr{C}_{1-4}^{\circ}$ *hold. Then the One-step MLE process* $\vartheta_{t,\varepsilon}^{\star}, \tau < t \le T$ *is consistent, asymptotically normal*

$$\eta_{t,\varepsilon}(\vartheta_0) = \left(\eta_{1,t,\varepsilon}(\vartheta_0), \eta_{2,t,\varepsilon}(\vartheta_0)\right)^\top \equiv \varepsilon^{-1}\left(\vartheta_{t,\varepsilon}^{\star} - \vartheta_0\right) \Longrightarrow \mathcal{N}\left(0, \mathbf{I}_\tau^t(\vartheta_0)^{-1}\right) \tag{2.69}$$

and the process $\eta_{t,\varepsilon}(\vartheta_0), \tau_* \le t \le T$ *for any* $\tau_* \in (\tau, T]$ *converges in distribution in* $(\mathcal{C}(\tau_*, T), \mathfrak{B})$ *to the Gaussian process*

$$\eta_{t,\varepsilon}(\vartheta_0) \Longrightarrow \eta_t(\vartheta_0), \qquad \tau_* \le t \le T. \tag{2.70}$$

Proof Consider the difference

$$\begin{aligned}\varepsilon^{-1}\left(\vartheta_{t,\varepsilon}^{\star} - \vartheta_0\right) &= \varepsilon^{-1}\left(\bar{\vartheta}_{\tau,\varepsilon} - \vartheta_0\right) + \mathbf{I}_\tau^t\left(\bar{\vartheta}_{\tau,\varepsilon}\right)^{-1} \int_\tau^t \frac{f(s)\,\dot{\mathbf{m}}\left(\bar{\vartheta}_{\tau,\varepsilon}, s\right)}{\sigma(s)}\mathrm{d}\bar{W}_s \\ &\quad + \mathbf{I}_\tau^t\left(\bar{\vartheta}_{\tau,\varepsilon}\right)^{-1} \int_\tau^t \frac{f(s)^2\,\dot{\mathbf{m}}\left(\bar{\vartheta}_{\tau,\varepsilon}, s\right)}{\varepsilon\sigma(s)^2}\left[m(\vartheta_0, s) - m\left(\bar{\vartheta}_{\tau,\varepsilon}, s\right)\right]\mathrm{d}s.\end{aligned} \tag{2.71}$$

The consistency of the estimator $\bar{\vartheta}_{\tau,\varepsilon}$ and the central limit theorem for stochastic integrals allow us to write

$$\pi_{t,\varepsilon} \equiv \mathbf{I}_\tau^t\left(\bar{\vartheta}_{\tau,\varepsilon}\right)^{-1} \int_\tau^t \frac{f(s)\,\dot{\mathbf{m}}\left(\bar{\vartheta}_{\tau,\varepsilon}, s\right)}{\sigma(s)}\mathrm{d}\bar{W}_s \Longrightarrow \eta_t(\vartheta_0) \sim \mathcal{N}\left(0, \mathbf{I}_\tau^t(\vartheta_0)^{-1}\right).$$

If we consider the vector $\pi_{t_1,\varepsilon}, \ldots, \pi_{t_k,\varepsilon}$, where $\tau \le t_1 \le t_2 \le \ldots \le t_k \le T$, then we obtain the convergence

$$\left(\pi_{t_1,\varepsilon}, \ldots, \pi_{t_k,\varepsilon}\right) \Longrightarrow \left(\eta_{t_1}(\vartheta_0), \ldots, \eta_{t_k}(\vartheta_0)\right).$$

For the ordinary integral, we have the relations

$$\mathbf{I}_\tau^t\left(\bar{\vartheta}_{\tau,\varepsilon}\right)^{-1} \int_\tau^t \frac{f(s)^2\,\dot{\mathbf{m}}\left(\bar{\vartheta}_{\tau,\varepsilon}, s\right)}{\varepsilon\sigma(s)^2}\left[m(\vartheta_0, s) - m\left(\bar{\vartheta}_{\tau,\varepsilon}, s\right)\right]\mathrm{d}s$$

$$
\begin{aligned}
&= \mathbf{I}_\tau^t\left(\vartheta_0\right)^{-1} \int_\tau^t \frac{f(s)^2 \dot{\mathrm{m}}\left(\vartheta_0, s\right)}{\varepsilon \sigma(s)^2}\left[m\left(\vartheta_0, s\right)-m\left(\bar{\vartheta}_{\tau, \varepsilon}, s\right)\right] \mathrm{d} s\left(1+O\left(\varepsilon^{2/3}\right)\right) \\
&= \mathbf{I}_\tau^t\left(\vartheta_0\right)^{-1} \int_\tau^t \frac{f(s)^2 \dot{\mathrm{m}}\left(\vartheta_0, s\right)}{\sigma(s)^2} \dot{\mathrm{m}}\left(\vartheta_0, s\right)^{\top} \mathrm{d} s \frac{\left(\vartheta_0-\bar{\vartheta}_{\tau, \varepsilon}\right)}{\varepsilon}\left(1+O\left(\varepsilon^{2/3}\right)\right) \\
&= \frac{\left(\vartheta_0-\bar{\vartheta}_{\tau, \varepsilon}\right)}{\varepsilon}+\frac{\left(\vartheta_0-\bar{\vartheta}_{\tau, \varepsilon}\right)}{\varepsilon} O\left(\varepsilon^{2/3}\right)=\frac{\left(\vartheta_0-\bar{\vartheta}_{\tau, \varepsilon}\right)}{\varepsilon}+O\left(\varepsilon^{1/3}\right)
\end{aligned}
$$

because

$$
\int_\tau^t \frac{f(s)^2 \dot{\mathrm{m}}\left(\vartheta_0, s\right) \dot{\mathrm{m}}\left(\vartheta_0, s\right)^{\top}}{\sigma(s)^2} \mathrm{d} s \longrightarrow \mathbf{I}_\tau^t\left(\vartheta_0\right).
$$

The substitution of these limits in (2.71) proves (2.69).

To prove the weak convergence (2.70), we have to verify the estimate

$$
\mathbf{E}_{\vartheta_0}\left\|\eta_{t_2, \varepsilon}\left(\vartheta_0\right)-\eta_{t_1, \varepsilon}\left(\vartheta_0\right)\right\|^4 \leq C\left|t_2-t_1\right|^2 .
$$

The calculations are direct but cumbersome. It uses the non-deneracy of the Fisher information, boundedness of the derivatives, and the elementary properties of stochastic and ordinary integrals. □

Remark that the equations of the adaptive filter (2.72) and (2.73) are given in recurrent form. It is possible to write the One-step MLE process in the recurrent form too. To simplify its calculation, we replace $\dot{\mathrm{m}}\left(\bar{\vartheta}_{\tau,\varepsilon}, s\right)$ by $\dot{\mathrm{y}}\left(\bar{\vartheta}_{\tau,\varepsilon}, s\right)$ in (2.68). Then we multiply the both sides of obtained equation by $\mathbf{I}_\tau^t\left(\bar{\vartheta}_{\tau,\varepsilon}\right)$, denote the matrix

$$
\mathbf{L}(\vartheta, t)=\frac{f(t)^2 \dot{\mathrm{y}}(\vartheta, t) \dot{\mathrm{y}}(\vartheta, t)^{\top}}{\sigma(t)^2}
$$

and take the differential of the equation

$$
\mathbf{I}_\tau^t\left(\bar{\vartheta}_{\tau, \varepsilon}\right) \vartheta_{t, \varepsilon}^{\star}=\mathbf{I}_\tau^t\left(\bar{\vartheta}_{\tau, \varepsilon}\right) \bar{\vartheta}_{\tau, \varepsilon}+\int_\tau^t \frac{f(s) \dot{\mathrm{y}}\left(\bar{\vartheta}_{\tau, \varepsilon}, s\right)}{\sigma(s)^2}\left[\mathrm{d} X_s-f(s) m\left(\bar{\vartheta}_{\tau, \varepsilon}, s\right) \mathrm{d} s\right] .
$$

Then we obtain

$$
\begin{aligned}
\mathbf{I}_\tau^t\left(\bar{\vartheta}_{\tau, \varepsilon}\right) \mathrm{d} \vartheta_{t, \varepsilon}^{\star}+\mathbf{L}\left(\bar{\vartheta}_{\tau, \varepsilon}, t\right) \vartheta_{t, \varepsilon}^{\star} \mathrm{d} t &= \mathbf{L}\left(\bar{\vartheta}_{\tau, \varepsilon}, t\right) \bar{\vartheta}_{\tau, \varepsilon} \mathrm{d} t \\
&\quad+\frac{f(t) \dot{\mathrm{y}}\left(\bar{\vartheta}_{\tau, \varepsilon}, t\right)}{\sigma(t)^2}\left[\mathrm{d} X_t-f(t) m\left(\bar{\vartheta}_{\tau, \varepsilon}, t\right) \mathrm{d} t\right]
\end{aligned}
$$

and

$$
\begin{aligned}
\mathrm{d}\vartheta^\star_{t,\varepsilon} &= \mathbf{I}^t_\tau\left(\bar\vartheta_{\tau,\varepsilon}\right)^{-1}\mathbf{L}\left(\bar\vartheta_{\tau,\varepsilon},t\right)\left[\bar\vartheta_{\tau,\varepsilon}-\vartheta^\star_{t,\varepsilon}\right]\mathrm{d}t \\
&\quad + \mathbf{I}^t_\tau\left(\bar\vartheta_{\tau,\varepsilon}\right)^{-1}\frac{f(t)\,\dot{y}\left(\bar\vartheta_{\tau,\varepsilon},t\right)}{\sigma(t)^2}\left[\mathrm{d}X_t - f(t)\,m\left(\bar\vartheta_{\tau,\varepsilon},t\right)\mathrm{d}s\right].
\end{aligned}
$$

The Fisher information can be written in the recurrent form.

Adaptive Filter

Suppose that we already have the estimators $\bar\vartheta_{\tau,\varepsilon}$ and $\vartheta^\star_{t,\varepsilon} = \left(\theta^\star_{1,t,\varepsilon},\theta^\star_{2,t,\varepsilon}\right)^\top$, $\tau < t \leq T$. Then the adaptive filter $m^\star_{t,\varepsilon}$, $\tau_* < t \leq T$, where $\tau_* > \tau$ is

$$
\mathrm{d}m^\star_{t,\varepsilon} = a\left(\theta^\star_{2,t,\varepsilon},t\right)m^\star_{t,\varepsilon}\mathrm{d}t + \frac{f(t)\,\gamma^\star_*(t)}{\sigma(t)^2}\left[\mathrm{d}X_t - f(t)\,m^\star_{t,\varepsilon}\mathrm{d}t\right],\quad m^\star_{\tau_*,\varepsilon},
\tag{2.72}
$$

$$
\frac{\mathrm{d}\gamma^\star_*(t)}{\mathrm{d}t} = 2a\left(\theta^\star_{2,t,\varepsilon},t\right)\gamma^\star_*(t) - \frac{f(t)^2\,\gamma^\star_*(t)^2}{\sigma(t)^2} + b(t)^2,\qquad \gamma^\star_*(\tau_*).
\tag{2.73}
$$

The initial value $\gamma^\star_*(\tau_*)$ is the solution of the equation

$$
\frac{\mathrm{d}\gamma^\star_*(s)}{\mathrm{d}s} = 2a\left(\theta^\star_{2,\tau,\varepsilon},s\right)\gamma^\star_*(s) - \frac{f(s)^2\,\gamma^\star_*(s)^2}{\sigma(s)^2} + b(s)^2,\ \gamma^\star_*(0)=0,\ 0\leq s\leq\tau_*
\tag{2.74}
$$

at the point $s=\tau_*$.

To calculate $m^\star_{\tau_*,\varepsilon}$, we denote

$$
R(\theta_2,t) = \Phi(\theta_2,0,t)^{-1} f(t)\,\gamma_*(\theta_2,t)\,\sigma(t)^{-2} = \Phi(\theta_2,0,t)^{-1}B(\theta_2,t)
$$

and remark that

$$
\begin{aligned}
m(\vartheta,t) &= \Phi(\theta_2,0,t)\left[\theta_1 + \int_0^t R(\theta_2,s)\,\mathrm{d}X_s\right] \\
&= \Phi(\theta_2,0,t)\left[\theta_1 + R(\theta_2,t)\,X_t - \int_0^t R'_s(\theta_2,s)\,X_s\mathrm{d}s\right] \\
&= \theta_1\Phi(\theta_2,0,t) + \frac{f(t)\,\gamma_*(\theta_2,t)}{\sigma(t)^2}X_t \\
&\quad + \int_0^t \Phi(\theta_2,s,t)\left[A(\theta_2,s) - B'_s(\theta_2,s)\right]X_s\mathrm{d}s.
\end{aligned}
$$

The initial value is calculated as follows:

$$m^\star_{\tau_*,\varepsilon}=\theta^\star_{1,\tau_*,\varepsilon}\Phi\left(\theta^\star_{2,\tau_*,\varepsilon},0,\tau_*\right)+\frac{f\left(\tau_*\right)\gamma_*\left(\theta^\star_{2,\tau_*,\varepsilon},\tau_*\right)}{\sigma\left(\tau_*\right)^2}X_{\tau_*}$$
$$+\int_0^{\tau_*}\Phi\left(\theta^\star_{2,\tau_*,\varepsilon},s,\tau_*\right)\left[A\left(\theta^\star_{2,\tau_*,\varepsilon},s\right)-B'_s\left(\theta^\star_{2,\tau_*,\varepsilon},s\right)\right]X_s\mathrm{d}s. \qquad (2.75)$$

Theorem 2.9 *Let the conditions* $\mathscr{C}^\circ_{1-4}$ *hold. Then the error of approximation*

$$\varepsilon^{-1}\left(m^\star_{t,\varepsilon}-m\left(\vartheta_0,t\right)\right)\Longrightarrow\xi_m\left(\vartheta_0,t\right),\qquad \tau_*\leq t\leq T$$

where the Gaussian process $\xi_m\left(\vartheta_0,t\right),\ \tau_*\leq t\leq T$ *is defined below in* (2.76).

Proof Denote $\delta_m\left(t\right)=m^\star_{t,\varepsilon}-m\left(\vartheta_0,t\right)$, $\delta_\vartheta\left(t\right)=\vartheta^\star_{t,\varepsilon}-\vartheta_0$, $\delta_{\theta_2}\left(t\right)=\theta^\star_{2,t,\varepsilon}-\theta_{2,0}$, $\delta_\gamma\left(t\right)=\gamma^\star_*\left(t\right)-\gamma_*\left(\vartheta_{2,0},t\right)$. We have already the estimate $\delta_\vartheta\left(t\right)=O\left(\varepsilon\right)$. The equations for $\delta_m\left(t\right)$ and $\delta_\gamma\left(t\right)$ we obtain as difference of equations for corresponding functions

$$\mathrm{d}\delta_m\left(t\right)=A\left(\theta^\star_{2,t,\varepsilon},t\right)\delta_m\left(t\right)\mathrm{d}t+\left[a\left(\theta^\star_{2,t,\varepsilon},t\right)-a\left(\theta_{2,0},t\right)\right]m\left(\vartheta_0\right)\mathrm{d}t$$
$$+\varepsilon\frac{f\left(t\right)\delta_\gamma\left(t\right)}{\sigma\left(t\right)}\mathrm{d}\bar W_t,\qquad \delta_m\left(\tau_*\right),$$
$$\frac{\mathrm{d}\delta_\gamma\left(t\right)}{\mathrm{d}t}=2\left[a\left(\theta^\star_{2,t,\varepsilon},t\right)-a\left(\theta_{2,0},t\right)\right]\gamma_*\left(\theta_{2,0},t\right)$$
$$+\left[2a\left(\theta_{2,0},t\right)-\frac{f\left(t\right)^2\left[\gamma^\star_*\left(t\right)+\gamma_*\left(\vartheta_{2,0},t\right)\right]}{\sigma\left(t\right)^2}\right]\delta_\gamma\left(t\right),\qquad \delta_\gamma\left(\tau_*\right).$$

We have $a\left(\theta^\star_{2,t,\varepsilon},t\right)-a\left(\theta_{2,0},t\right)=\dot a\left(\theta_{2,0},t\right)\delta_{\theta_{2,0}}\left(t\right)\left(1+O\left(\varepsilon\right)\right)$. Using these relations, boundedness of the derivatives, and Grönwall's lemma , it can be shown that $\delta_\gamma\left(t\right)=O\left(\varepsilon\right),\delta_m\left(t\right)=O\left(\varepsilon\right)$.

Therefore,

$$\mathrm{d}\delta_m\left(t\right)=A\left(\theta_{2,0},t\right)\delta_m\left(t\right)\mathrm{d}t\left(1+O\left(\varepsilon\right)\right)+\varepsilon\frac{f\left(t\right)\delta_\gamma\left(t\right)}{\sigma\left(t\right)}\mathrm{d}\bar W_t$$
$$+\dot a\left(\theta_{2,0},t\right)m\left(\vartheta_0,t\right)\delta_\vartheta\left(t\right)\mathrm{d}t\left(1+O\left(\varepsilon\right)\right),$$
$$\frac{\mathrm{d}\delta_\gamma\left(t\right)}{\mathrm{d}t}=2\dot a\left(\theta_{2,0},t\right)\gamma_*\left(\theta_{2,0},t\right)\delta_\vartheta\left(t\right)+2A\left(\theta_{2,0},t\right)\delta_\gamma\left(t\right)+O\left(\varepsilon^2\right).$$

The solutions of these linear equations are

$$\delta_\gamma\left(t\right)=\delta_\gamma\left(\tau_*\right)\Phi\left(\theta_{2,0},\tau_*,t\right)^2$$

$$+2\int_{\tau_*}^{t}\Phi\left(\theta_{2,0},s,t\right)^2\dot a\left(\theta_{2,0},s\right)\gamma_*\left(\theta_{2,0},s\right)\delta_\vartheta\left(s\right)\mathrm{d}s+O\left(\varepsilon^2\right),$$

$$\delta_m\left(t\right)=\delta_m\left(\tau_*\right)\Phi\left(\theta_{2,0},\tau_*,t\right)+\int_{\tau_*}^{t}\Phi\left(\theta_{2,0},s,t\right)\dot a\left(\theta_{2,0},s\right)m\left(\vartheta_0,s\right)\delta_\vartheta\left(s\right)\mathrm{d}s$$

$$+\varepsilon\int_{\tau_*}^{t}\Phi\left(\theta_{2,0},s,t\right)f\left(s\right)\delta_\gamma\left(s\right)\sigma\left(s\right)^{-1}\mathrm{d}\bar W_s+O\left(\varepsilon^2\right).$$

Solution of the equation related with (2.74) gives us the representation

$$\varepsilon^{-1}\delta_\gamma\left(\tau_*\right)=2\varepsilon^{-1}\int_0^{\tau_*}\Phi\left(\theta_{2,0},s,\tau_*\right)\dot a\left(\theta_{2,0},s\right)\gamma_*\left(\theta_{2,0},s\right)\delta_{\theta_2}\left(\tau_*\right)\mathrm{d}s+O\left(\varepsilon\right)$$

$$\Longrightarrow 2\int_0^{\tau_*}\Phi\left(\theta_{2,0},s,\tau_*\right)\dot a\left(\theta_{2,0},s\right)\gamma_*\left(\theta_{2,0},s\right)\eta_{2,\tau_*}\left(\theta_2\right)\mathrm{d}s\equiv\xi_\gamma\left(\vartheta_0\right).$$

Hence,

$$\begin{aligned}\varepsilon^{-1}\delta_\gamma\left(t\right)&\Longrightarrow\xi_\gamma\left(\vartheta_0\right)\Phi\left(\theta_{2,0},\tau_*,t\right)^2\\&\quad+2\int_{\tau_*}^{t}\Phi\left(\theta_{2,0},s,t\right)^2\dot a\left(\theta_{2,0},s\right)\gamma_*\left(\theta_{2,0},s\right)\eta_{2,s}\left(\vartheta_0\right)\mathrm{d}s\\&\equiv\xi_\gamma\left(\vartheta_0,t\right),\qquad\tau_*\le t\le T.\end{aligned}$$

Let us denote $C\left(\theta_2,s\right)=-A\left(\theta_2,s\right)+B_s'\left(\theta_2,s\right)$, then we can write (see (2.75))

$$\begin{aligned}\frac{\delta_m\left(\tau_*\right)}{\varepsilon}&\Longrightarrow\Phi\left(\theta_{2,0},0,\tau_*\right)\eta_{1,\tau_*}\left(\vartheta_0\right)+\theta_{1,0}\dot\Phi\left(\theta_{2,0},0,\tau_*\right)\eta_{2,\tau_*}\left(\vartheta_0\right)\\&+\frac{f\left(\tau_*\right)x_{\tau_*}\left(\vartheta_0\right)}{\sigma\left(\tau_*\right)^2}\xi_\gamma\left(\vartheta_0\right)\\&-\eta_{2,\tau_*}\left(\vartheta_0\right)\int_0^{\tau_*}\left[\dot\Phi\left(\theta_{2,0},0,\tau_*\right)C\left(\theta_{2,0},s\right)+\Phi\left(\theta_{2,0},0,\tau_*\right)\dot C\left(\theta_{2,0},s\right)\right]\\&\times x_s\left(\vartheta_0\right)\mathrm{d}s\\&\equiv\xi_m\left(\vartheta_0\right).\end{aligned}$$

Therefore,

$$\frac{\delta_m(t)}{\varepsilon} \Longrightarrow \xi_m(\vartheta_0)\,\Phi\left(\theta_{2,0},\tau_*,t\right) + \int_{\tau_*}^{t} \Phi\left(\theta_{2,0},s,t\right)\dot a\left(\theta_{2,0},s\right) y_s(\vartheta_0)\,\eta_{2,s}(\vartheta_0)\,\mathrm{d}s$$

$$+ \int_{\tau_*}^{t} \Phi\left(\theta_{2,0},s,t\right) f(s)\sigma(s)^{-1}\xi_\gamma(\vartheta_0,s)\,\mathrm{d}w(s) \equiv \xi_m(\vartheta_0,t), \qquad \tau_* \le t \le T. \tag{2.76}$$

□

2.2.2 Random Initial Value

Let us consider an example of partially observed system with the random initial value of non-observed component:

$$\mathrm{d}X_t = fY_t\,\mathrm{d}t + \varepsilon\sigma\,\mathrm{d}W_t, \qquad X_0 = 0, \qquad 0 \le t \le T,$$

$$\mathrm{d}Y_t = \vartheta Y_t\,\mathrm{d}t + \varepsilon b\,\mathrm{d}V_t, \qquad Y_0 \sim \mathcal{N}\left(0,d^2\right).$$

Here $f > 0, \sigma > 0, b > 0, d^2 > 0$ and $\vartheta \in \Theta = (\alpha,\beta), \alpha > 0$. The random variable Y_0 is independent of the Wiener processes $W_t, V_t; 0 \le t \le T$. The sample value of Y_0 we denote y_0. As usual, we have to estimate the parameter ϑ by observations $X^T = (X_t, 0 \le t \le T)$.

Parameter Estimators

Remark that these processes can be written in explicit form

$$X_t = \frac{fy_0}{\vartheta}\left[e^{\vartheta t} - 1\right] + \varepsilon fb\int_0^t\int_0^s e^{\vartheta(s-r)}\mathrm{d}V_r\,\mathrm{d}s + \varepsilon\sigma W_t, \qquad 0 \le t \le T,$$

$$Y_t = y_0\,e^{\vartheta t} + \varepsilon b\int_0^t e^{\vartheta(t-s)}\mathrm{d}V_s.$$

The limit system (ϑ_0 is the true value)

$$\frac{\mathrm{d}x_t(\vartheta_0)}{\mathrm{d}t} = f\,y_t(\vartheta_0), \qquad x_0(\vartheta_0) = 0, \qquad 0 \le t \le T,$$

$$\frac{\mathrm{d}y_t(\vartheta_0)}{\mathrm{d}t} = \vartheta_0\,y_t(\vartheta_0), \qquad y_0(\vartheta_0) = y_0 \sim \mathcal{N}\left(0,d^2\right)$$

gives us a couple of random processes $x_t(\vartheta_0), y_t(\vartheta_0), 0 \le t \le T$. As $y_t(\vartheta) = y_0e^{\vartheta t}$, we have the equality

$$x_t(\vartheta) = \frac{y_0 f}{\vartheta}\left(e^{\vartheta t} - 1\right), \qquad 0 \le t \le T.$$

There are several ways to construct a preliminary estimator of ϑ_0. Note that we cannot define the MDE by the equality

$$\check{\vartheta}_{\tau,\varepsilon} = \arg\inf_{\vartheta \in \Theta} \int_0^\tau \left[X_t - x_t(\vartheta)\right]^2 \mathrm{d}t$$

because y_0 is unknown and its conditional expectation is always $m(\vartheta, 0) = 0$.

Let us denote $x_t'(\vartheta)$ the derivative of $x_t(\vartheta)$ w.r.t. t and remark that ($t_1 < t_2$)

$$\vartheta_0 = \frac{\ln\left|x_{t_2}'(\vartheta_0)\right| - \ln\left|x_{t_1}'(\vartheta_0)\right|}{|t_2 - t_1|}.$$

Therefore, if we have a consistent estimator $X_{t,\varepsilon}'$ of the derivative, then the MME

$$\vartheta_{t_2,\varepsilon}^* = \frac{\ln\left|X_{t_2,\varepsilon}'\right| - \ln\left|X_{t_1,\varepsilon}'\right|}{|t_2 - t_1|}$$

can be consistent too.

Note that for the estimator of the derivative $X_{t,\varepsilon}' = \left(X_{t+\delta_\varepsilon} - X_t\right)/\delta_\varepsilon$ we have

$$\begin{aligned} X_{t,\varepsilon}' &= \frac{f y_0}{\delta_\varepsilon} \int_t^{t+\delta_\varepsilon} e^{\vartheta_0 s} \mathrm{d}s + \frac{f b \varepsilon}{\delta_\varepsilon} \int_t^{t+\delta_\varepsilon} \int_0^s e^{\vartheta_0 (s-r)} \mathrm{d}V_r \, \mathrm{d}s + \frac{\varepsilon}{\delta_\varepsilon} \sigma \left[W_{t+\delta_\varepsilon} - W_t\right] \\ &= f y_0 e^{\vartheta_0 t} + \frac{f \vartheta_0 y_0}{2} e^{\vartheta_0 t} \delta_\varepsilon + f b \varepsilon \int_0^t e^{\vartheta_0 (t-r)} \mathrm{d}V_r + \frac{\varepsilon}{\sqrt{\delta_\varepsilon}} \sigma \, \xi_{t,\varepsilon} + O\left(\varepsilon\sqrt{\delta_\varepsilon}\right) \end{aligned}$$

where $\xi_{t,\varepsilon} \sim \mathcal{N}(0, 1)$. The correct rate $\delta_\varepsilon = \varepsilon^{2/3}$ we obtain as solution of the equation $\delta_\varepsilon = \varepsilon/\sqrt{\delta_\varepsilon}$. Taylor expansion allows to write

$$\begin{aligned} \ln\left|X_{t,\varepsilon}'\right| &= \ln\left| f y_0 e^{\vartheta_0 t} \left(1 + \left[\frac{\vartheta_0}{2} + \frac{\sigma \, \xi_{t,\varepsilon}}{f y_0} e^{-\vartheta_0 t}\right] \varepsilon^{2/3} (1 + o(1))\right)\right| \\ &= \ln(f\,|y_0|) + \vartheta_0 t + \ln\left|1 + \left[\frac{\vartheta_0}{2} + \frac{\sigma \, \xi_{t,\varepsilon}}{f y_0} e^{-\vartheta_0 t}\right] \varepsilon^{2/3} (1 + o(1))\right| \\ &= \ln(f\,|y_0|) + \vartheta_0 t + \left[\frac{\vartheta_0}{2} + \frac{\sigma \, \xi_{t,\varepsilon}}{f y_0} e^{-\vartheta_0 t}\right] \varepsilon^{2/3} (1 + o(1)), \end{aligned}$$

where the last equality is valid asymptotically as $\varepsilon \to 0$.

Therefore, we proved the following proposition.

Proposition 2.2 *For the MME* $\vartheta_{t_2,\varepsilon}^*$, *the following relation holds:*

$$\varepsilon^{-2/3}\left(\vartheta^*_{t_2,\varepsilon}-\vartheta_0\right)\Longrightarrow \frac{\sigma}{f\left(t_2-t_1\right)}\left[\frac{\xi_{t_2}}{y_0}e^{-\vartheta_0 t_2}-\frac{\xi_{t_1}}{y_0}e^{-\vartheta_0 t_1}\right].$$

Here $\xi_{t_1}\sim\mathcal{N}\left(0,1\right), \xi_{t_2}\sim\mathcal{N}\left(0,1\right)$ and $y_0\sim\mathcal{N}\left(0,d^2\right)$ are independent Gaussian random variables.

Remark that $\xi_{t_1}d/y_0$ and $\xi_{t_2}d/y_0$ are Cauchy variables, and hence, there is no convergence of moments of this estimator.

It is easy to see that $\varepsilon^{-2/3}y_0\left(\vartheta^*_{t_2,\varepsilon}-\vartheta_0\right)$ is asymptotically normal.

Remark 2.11 Introduce $t_{i,\varepsilon}=p_i/\ln\left(1/\varepsilon\right)\rightarrow 0, i=1,2$. Here $p_1>0$, $p_2>p_1$ and $0<t_{i,\varepsilon}<T$. Then the normalizing and limit change

$$\ln\left(1/\varepsilon\right)\varepsilon^{-2/3}\left(\vartheta^*_{t_2,\varepsilon}-\vartheta_0\right)\Longrightarrow \frac{\sigma}{f\left(p_2-p_1\right)}\frac{\left(\xi_{p_2}-\xi_{p_1}\right)}{y_0}.$$

Here $\xi_{p_1}\sim\mathcal{N}\left(0,1\right)$, $\xi_{p_2}\sim\mathcal{N}\left(0,1\right)$ are idependent random ariables.

The Kalman–Bucy filter for conditional expectation $m\left(\vartheta,t\right), 0\leq t\leq T$ is

$$\mathrm{d}m\left(\vartheta,t\right)=\vartheta m\left(\vartheta,t\right)\mathrm{d}t+\frac{f\gamma\left(\vartheta,t\right)}{\varepsilon^2\sigma^2}\left[\mathrm{d}X_t-fm\left(\vartheta,t\right)\mathrm{d}t\right],\quad m\left(\vartheta,0\right)=0, \tag{2.77}$$

$$\frac{\partial\gamma\left(\vartheta,t\right)}{\partial t}=2\vartheta\gamma\left(\vartheta,t\right)-\frac{f^2\gamma\left(\vartheta,t\right)^2}{\varepsilon^2\sigma^2}+\varepsilon^2b^2,\qquad \gamma\left(\vartheta,0\right)=d^2. \tag{2.78}$$

The solution of the Eq. (2.77) can be written explicitly

$$\begin{aligned}m\left(\vartheta,t\right)&=\int_0^t\Phi\left(\vartheta,s,t\right)\frac{f^2\gamma\left(\vartheta,s\right)}{\varepsilon^2\sigma^2}Y_s\,\mathrm{d}s+\int_0^t\Phi\left(\vartheta,s,t\right)\frac{f\gamma\left(\vartheta,s\right)}{\varepsilon\sigma}\mathrm{d}W_s\\&=y_0\int_0^t\Phi\left(\vartheta,s,t\right)\frac{f^2\gamma\left(\vartheta,s\right)}{\varepsilon^2\sigma^2}e^{\vartheta_0 s}\,\mathrm{d}s+\int_0^t\Phi\left(\vartheta,s,t\right)\frac{f\gamma\left(\vartheta,s\right)}{\varepsilon\sigma}\mathrm{d}W_s\\&\quad+\int_0^t\Phi\left(\vartheta,s,t\right)\frac{f^2\gamma\left(\vartheta,s\right)}{\varepsilon\sigma^2}\int_0^s e^{\vartheta_0(s-r)}\mathrm{d}V_r\,\mathrm{d}s\end{aligned}$$

where

$$\Phi\left(\vartheta,s,t\right)=\exp\left(\int_s^t\left[\vartheta-\frac{f^2\gamma\left(\vartheta,r\right)}{\varepsilon^2\sigma^2}\right]\mathrm{d}r\right).$$

Let us see the asymptotic of the main term

$$I\left(\tau,\varepsilon\right)\equiv y_0\int_0^\tau\Phi\left(\vartheta,s,\tau\right)\frac{f^2\gamma\left(\vartheta,s\right)}{\varepsilon^2\sigma^2}e^{\vartheta_0 s}\,\mathrm{d}s$$

for the small values of τ. Integrating by parts, we obtain

$$\begin{aligned} I(\tau,\varepsilon) &= y_0\Phi(\vartheta,0,\tau)\int_0^\tau \exp\left(\int_0^s \frac{f^2\gamma(\vartheta,r)}{\varepsilon^2\sigma^2}\mathrm{d}r\right)\frac{f^2\gamma(\vartheta,s)}{\varepsilon^2\sigma^2}\,e^{(\vartheta_0-\vartheta)s}\,\mathrm{d}s \\ &= y_0\Phi(\vartheta,0,\tau)\int_0^\tau e^{(\vartheta_0-\vartheta)s}\,\mathrm{d}\exp\left(\int_0^s \frac{f^2\gamma(\vartheta,r)}{\varepsilon^2\sigma^2}\mathrm{d}r\right) \\ &= y_0\,e^{\vartheta_0\tau} - y_0\Phi(\vartheta,0,\tau) - y_0(\vartheta_0-\vartheta)\int_0^\tau \Phi(\vartheta,s,\tau)\,e^{\vartheta_0 s}\,\mathrm{d}s \\ &= y_0\,e^{\vartheta_0\tau} + O\left(\varepsilon^2\right). \end{aligned}$$

The function $\gamma_*(\vartheta,t)=\varepsilon^{-2}\gamma(\vartheta,t)$ is solution of the Riccati equation

$$\frac{\partial\gamma_*(\vartheta,t)}{\partial t} = 2\vartheta\gamma_*(\vartheta_*,t) - \frac{f^2\gamma_*(\vartheta,t)^2}{\sigma^2} + b^2, \qquad \gamma_*(\vartheta,0)=\frac{d^2}{\varepsilon^2}.$$

Recall that this equation with constant coefficients has the following explicite solution (1.26):

$$\begin{aligned} \gamma_*(\vartheta,t) &= e^{-2r(\vartheta)t}\left[\frac{1}{\frac{d^2}{\varepsilon^2}-\frac{\sigma^2}{f^2}(\vartheta+r(\vartheta))} + \frac{f^2\left(1-e^{-2r(\vartheta)t}\right)}{2\sigma^2 r(\vartheta)}\right]^{-1} \\ &\quad + \frac{\sigma^2}{f^2}(\vartheta+r(\vartheta)). \end{aligned}$$

Here $r(\vartheta)=\left(\vartheta^2+b^2f^2\sigma^{-2}\right)^{1/2}$. The function $\gamma_*(\vartheta,t)$ for any $t\in(0,T]$ has the asymptotic expansion

$$\gamma_*(\vartheta,t)=\frac{\sigma^2}{f^2}\left[\vartheta+r(\vartheta)+\frac{2r(\vartheta)}{e^{2r(\vartheta)t}-1}\right]\left(1+O\left(\varepsilon^2\right)\right)=\gamma^*(\vartheta,t)\left(1+O\left(\varepsilon^2\right)\right)$$

where the function $\gamma^*(\vartheta,t)$ is defined by the last equality. Moreover, for any $\tau\in(0,T]$ we have the uniform convergence too

$$\sup_{\tau\le t\le T}\left|\gamma_*(\vartheta,t)-\gamma^*(\vartheta,t)\right|\le C\varepsilon^2.$$

Therefore, for any $t\in(0,T)$ we have the approximation

$$\gamma(\vartheta,t)=\varepsilon^2\gamma^*(\vartheta,t)\left(1+O\left(\epsilon^2\right)\right).$$

Omitting the terms of order $O\left(\varepsilon^2\right)$, we obtain the representation

$$
\begin{aligned}
m^*(\vartheta,t) &= m(\vartheta,\tau)\,\Phi^*(\vartheta,\tau,t) + y_0\int_\tau^t \Phi^*(\vartheta,s,t)\,\frac{f^2\gamma^*(\vartheta,s)}{\sigma^2}\,e^{\vartheta_0 s}\,\mathrm{d}s\\
&\quad + \varepsilon\int_\tau^t \Phi^*(\vartheta,s,t)\,\frac{f\gamma(\vartheta,s)}{\sigma}\,\mathrm{d}W_s\\
&\quad + \varepsilon\int_\tau^t \Phi(\vartheta,s,t)\,\frac{f^2\gamma^*(\vartheta,s)}{\sigma^2}\int_0^s e^{\vartheta_0(s-r)}\mathrm{d}V_r\,\mathrm{d}s\\
&= y_0 e^{\vartheta_0\tau}\,\Phi^*(\vartheta,\tau,t) + y_0\int_\tau^t \Phi^*(\vartheta,s,t)\,\frac{f^2\gamma^*(\vartheta,s)}{\sigma^2}\,e^{\vartheta_0 s}\,\mathrm{d}s \;+ o(1),
\end{aligned}
$$

where

$$
\Phi^*(\vartheta,s,t) = \exp\left(\int_s^t\left[\vartheta - \frac{f^2\gamma^*(\vartheta,r)}{\sigma^2}\right]\mathrm{d}r\right).
$$

The initial value $m(\vartheta,\tau)$ we obtain as solution of the Eq. (2.77) and τ takes small value such that $m(\vartheta,\tau)\,\Phi^*(\vartheta,0,\tau) = o(1)$. Recall that $m(\vartheta,0)=0$.

For the derivative $\dot m_t(\vartheta) = \partial m_t(\vartheta)/\partial\vartheta$ we have the equation

$$
\begin{aligned}
\mathrm{d}\dot m_t(\vartheta) &= \left[\vartheta - \frac{f^2\gamma^*(\vartheta,t)}{\sigma^2}\right]\dot m_t(\vartheta)\,\mathrm{d}t + \frac{f\dot\gamma^*(\vartheta,t)}{\sigma^2}\mathrm{d}X_t\\
&\quad + \left[1 - \frac{f^2\dot\gamma^*(\vartheta,t)}{\sigma^2}\right]m_t(\vartheta)\,\mathrm{d}t
\end{aligned}
$$

and obtain the expression

$$
\begin{aligned}
\dot m_t(\vartheta) &= y_0\int_\tau^t \Phi^*(\vartheta,s,t)\left[t-s+\frac{f^2}{\sigma^2}\left(\gamma^*(\vartheta,s) - \int_s^t \dot\gamma^*(\vartheta,s)\,\mathrm{d}r\right)\right]\\
&\quad e^{\vartheta_0 s}\mathrm{d}s + o(1)\\
&= y_0\,H(\vartheta,\vartheta_0,t) + o(1) \longrightarrow y_0\,H(\vartheta,\vartheta_0,t)
\end{aligned}
$$

as $\varepsilon\to 0$.

Therefore, the empirical Fisher information function converges in probability

$$
\begin{aligned}
\mathrm{I}_\tau^t(\vartheta_0) &= \frac{f^2}{\sigma^2}\int_\tau^t \dot m_s(\vartheta_0)^2\,\mathrm{d}s \longrightarrow \frac{y_0^2 f^2}{\sigma^2}\int_\tau^t H(\vartheta_0,\vartheta_0,s)^2\,\mathrm{d}s\\
&= y_0^2 J_\tau^t(\vartheta_0).
\end{aligned}
$$

The family of measures $\left(\mathbf{P}_\vartheta^{(\varepsilon)},\vartheta\in\Theta\right)$ induced in the space of observations is locally asymptotically mixing normal (LAMN). The normalized likelihood ratio function constructed by the observations $X_\tau^T = (X_t,\tau\le t\le T)$

$$Z_\varepsilon(\vartheta_0, u) = \frac{L\left(\vartheta_0 + \varepsilon u, X_\tau^T\right)}{L\left(\vartheta_0, X_\tau^T\right)}, \qquad u \in \mathbb{U}_\varepsilon = \left(\frac{\alpha - \vartheta_0}{\varepsilon}, \frac{\beta - \vartheta_0}{\varepsilon}\right)$$

admits the representation

$$Z_\varepsilon(\vartheta_0, u) = \exp\left(u\, y_0\, \Delta_\varepsilon\left(\vartheta_0, X_\tau^T\right) - \frac{u^2}{2} y_0^2\, J_\tau^t(\vartheta_0) + r_\varepsilon\right).$$

Here $r_\varepsilon \to 0$ and

$$\Delta_\varepsilon\left(\vartheta_0, X^T\right) \Longrightarrow \Delta(\vartheta_0) \sim \mathcal{N}\left(0, J_\tau^t(\vartheta_0)\right).$$

We do not study the MLE but note that it has the limit Cauchy distribution

$$\frac{\hat\vartheta_\varepsilon - \vartheta_0}{\varepsilon} \Longrightarrow \hat u = \frac{\Delta(\vartheta_0)}{y_0\, J_\tau^t(\vartheta_0)}$$

The K-B filter on the time interval $[\tau, T]$ $(\tau > 0)$ we rewrite as follows

$$\mathrm{d}m_*(\vartheta, t) = \left[\vartheta - \frac{f^2\gamma^*(\vartheta, t)}{\sigma^2}\right] m_*(\vartheta, t)\,\mathrm{d}t + \frac{f\gamma^*(\vartheta, t)}{\sigma^2}\mathrm{d}X_t. \tag{2.79}$$

The initial value $m_*(\vartheta, \tau) = m(\vartheta, \tau)$ can be calculated by solving the system (2.77)–(2.78) on the interval $[0, \tau]$.

The equation for the derivative $\dot m_*(\vartheta, t)$ is

$$\begin{aligned}
\mathrm{d}\dot m_*(\vartheta, t) &= \left[1 - f^2\dot\gamma^*(\vartheta, t)\sigma^{-2}\right] m_*(\vartheta, t)\,\mathrm{d}t \\
&\quad + \left[\vartheta - f^2\gamma^*(\vartheta, t)\sigma^{-2}\right] \dot m_*(\vartheta, t)\,\mathrm{d}t \\
&\quad + f\dot\gamma^*(\vartheta, t)\sigma^{-2}\mathrm{d}X_t, \qquad \dot m_*(\vartheta, \tau), \quad \tau < t \le T.
\end{aligned} \tag{2.80}$$

Note that the calculation of the $m_*(\vartheta, t)$ and $\dot m_*(\vartheta, t)$ by these equations does not depend on y_0, which appears in the limits only.

If $\varepsilon \to 0$, then

$$\begin{aligned}
m_*(\vartheta, t) \longrightarrow y_t(\vartheta, \vartheta_0) &= y_0 e^{\vartheta_0\tau}\Phi^*(\vartheta, \tau, t) \\
&\quad + y_0 \int_\tau^t \Phi^*(\vartheta, s, t)\, f^2\gamma^*(\vartheta, s)\sigma^{-2}\, e^{\vartheta_0 s}\,\mathrm{d}s = y_0\, h_t(\vartheta, \vartheta_0),
\end{aligned}$$

where the function $h_t(\vartheta, \vartheta_0)$ is defined by the last equality.

The expression for the limit of the derivative $\dot m(\vartheta, t) \longrightarrow \dot y_t(\vartheta, \vartheta_0)$ is the following random function

$$
\begin{aligned}
\dot{y}_t(\vartheta,\vartheta_0) &= \dot{y}_\tau(\vartheta,\vartheta_0)\,\Phi^*(\vartheta,0,\tau)\\
&\quad + y_0\int_\tau^t \Phi^*(\vartheta,s,t)\left[h_s(\vartheta,\vartheta_0)+\frac{f^2\dot{\gamma}^*(\vartheta,s)}{\sigma^2}\left[e^{\vartheta_0 s}-h_s(\vartheta,\vartheta_0)\right]\right]\mathrm{d}s\\
&= y_0 e^{\vartheta_0\tau}\tau\Phi^*(\vartheta,0,\tau)+y_0\,\bar{H}_t(\vartheta,\vartheta_0)\equiv y_0\,\bar{G}_t(\vartheta,\vartheta_0),
\end{aligned}
$$

where $\bar{H}_t(\vartheta,\vartheta_0)$ and $\bar{G}_t(\vartheta,\vartheta_0)$ are defined by the last two equalities. If $\vartheta=\vartheta_0$, then $h_s(\vartheta_0,\vartheta_0)=e^{\vartheta_0 s}$ and

$$
\begin{aligned}
\dot{y}_t(\vartheta_0,\vartheta_0) &= \dot{y}_\tau(\vartheta_0,\vartheta_0)\,\Phi^*(\vartheta,0,\tau)+y_0e^{\vartheta_0 t}\int_\tau^t \exp\left(-\frac{f^2}{\sigma^2}\int_s^t \gamma^*(\vartheta_0,r)\,\mathrm{d}r\right)\mathrm{d}s\\
&= y_0e^{\vartheta_0\tau}\tau\Phi^*(\vartheta,0,\tau)+y_0\,\bar{H}_t(\vartheta_0,\vartheta_0)=y_0\,\bar{G}_t(\vartheta,\vartheta_0).
\end{aligned}
$$

These limits are proved for the integrals on the intervals $[\tau,T]$, but the value τ can be taken close to 0. It can be verified that the integral on the interval $[0,\tau]$ can be done less than any constant by choosing sufficiently small τ. That is why in the expressions for the limits we use the integrals on the intervals $(0,T]$.

The Fisher information we write in two forms: for $\varepsilon>0$ and $\varepsilon=0$

$$
\begin{aligned}
\mathrm{I}_{\tau,\varepsilon}^t(\vartheta) &= \frac{f^2}{\sigma^2}\int_\tau^t \dot{m}_*(\vartheta,s)^2\,\mathrm{d}s, \qquad \tau<t\leq T,\\
\mathrm{I}_\tau^t(\vartheta_0) &= y_0^2\,J_\tau^t(\vartheta_0),
\end{aligned}
$$

where $J_\tau^t(\vartheta_0)$ is the corresponding deterministic function. We have $\mathrm{I}_{\tau,\varepsilon}^t(\vartheta_0)\longrightarrow \mathrm{I}_\tau^t(\vartheta_0)$ as $\varepsilon\to 0$.

The One-step MLE process $\vartheta_{t,\varepsilon}^\star$, $\tau<t\leq T$ can be defined by the equality

$$
\vartheta_{t,\varepsilon}^\star=\vartheta_{\tau_*,\varepsilon}^*+\mathrm{I}_{\tau_*,\varepsilon}^t\left(\vartheta_{\tau,\varepsilon}^*\right)^{-1}\frac{f}{\sigma^2}\int_\tau^t \dot{m}_*(\vartheta_{\tau_*,\varepsilon}^*,s)\,\left[\mathrm{d}X_s-f m_*(\vartheta_{\tau_*,\varepsilon}^*,s)\mathrm{d}s\right].
$$

Here the MME $\vartheta_{\tau_*,\varepsilon}^*$ is calculated for the values $t_2=\frac{3\tau}{4}$ and $t_1=\frac{\tau}{2}$. The random processes $m_*(\vartheta_{\tau_*,\varepsilon}^*,s)$, $\tau\leq t\leq T$ and $\dot{m}_*(\vartheta_{\tau_*,\varepsilon}^*,s)$, $\tau\leq t\leq T$ are obtained as solutions of (2.79) and (2.80) subject to the initial values calculated as solutions of the same equations but written without stochastic integrals as it was done in (2.75).

We have the representation

$$
\begin{aligned}
\frac{\vartheta_{t,\varepsilon}^\star-\vartheta_0}{\varepsilon} &= \frac{\vartheta_{\tau_*,\varepsilon}^*-\vartheta_0}{\varepsilon}+\mathrm{I}_{\tau_*,\varepsilon}^t\left(\vartheta_{\tau,\varepsilon}^*\right)^{-1}\frac{f}{\sigma}\int_\tau^t \dot{m}_*(\vartheta_{\tau_*,\varepsilon}^*,s)\,\mathrm{d}\bar{W}_s\\
&\quad+\mathrm{I}_{\tau_*,\varepsilon}^t\left(\vartheta_{\tau,\varepsilon}^*\right)^{-1}\frac{f^2}{\varepsilon\sigma^2}\int_\tau^t \dot{m}_*(\vartheta_{\tau_*,\varepsilon}^*,s)\,\left[m(\vartheta_0,s)-m_*(\vartheta_{\tau_*,\varepsilon}^*,s)\right]\mathrm{d}s.
\end{aligned}
$$

Following the main steps of the proof of the Proposition 2.1 it can be shown that this estimator for any $t \in (\tau, T]$ is asymptotically conditionally normal:

$$\hat\eta_{t,\varepsilon}(\vartheta_0) = \frac{\vartheta^\star_{t,\varepsilon} - \vartheta_0}{\varepsilon} \Longrightarrow \hat\eta_t(\vartheta_0) = \mathrm{I}^t_\tau(\vartheta_0)^{-1} \frac{f}{\sigma} \int_\tau^t \dot y(\vartheta_0, \vartheta_0, s)\, \mathrm{d}w(s),$$

$$= y_0^{-1}\, \mathrm{J}^t_\tau(\vartheta_0)^{-1} \frac{f}{\sigma} \int_\tau^t \bar G_s(\vartheta_0, \vartheta_0)\, \mathrm{d}w(s) \equiv y_0^{-1}\, \zeta_\vartheta(\vartheta_0, t),$$

where $\zeta_m(\vartheta_0, t)$, $\tau \le t \le T$ is a Gaussian process.

We have as well the following asymptotic normality

$$\frac{y_0\left(\vartheta^\star_{t,\varepsilon} - \vartheta_0\right)}{\varepsilon} \Longrightarrow \zeta_\vartheta(\vartheta_0, t) \sim \mathcal{N}\left(0, \mathrm{J}^t_\tau(\vartheta_0)^{-1}\right), \qquad \tau < t \le T.$$

Adaptive Filter

The adaptive Kalman filter is

$$\mathrm{d}m^\star_{t,\varepsilon} = \left[\vartheta^\star_{t,\varepsilon} - f^2\gamma^*\left(\vartheta^\star_{t,\varepsilon}, t\right)\sigma^{-2}\right] m^\star_{t,\varepsilon}\mathrm{d}t + f\gamma^*\left(\vartheta^\star_{t,\varepsilon}, t\right)\sigma^{-2}\mathrm{d}X_t, \qquad \tau \le t \le T.$$

The initial value $m^\star_{\tau,\varepsilon} = m\left(\vartheta^\star_{\tau,\varepsilon}, \tau\right)$ is calculated as it was explained above in (2.75).

Proposition 2.3 *The error of approximation $m^\star_{t,\varepsilon} - m(\vartheta_0, t)$ has the following: asymptotics*

$$\varepsilon^{-1}\left(m^\star_{t,\varepsilon} - m(\vartheta_0, t)\right) \Longrightarrow \xi^*_m(\vartheta_0, t), \qquad \tau \le t \le T,$$

where the Gaussian process $\xi^*_m(\vartheta_0, t)$ defined in (2.81) *below does not depend on y_0.*

Proof This equation and the Eq. (2.79) with $\vartheta = \vartheta_0$ allow us to write the equation for the error of estimation $\delta_m(t) = m^\star_{t,\varepsilon} - m(\vartheta_0, t)$

$$\delta_m(t) = \delta_m(\tau)\,\Phi^*(\vartheta_0, 0, \tau) + \int_\tau^t \Phi(\vartheta_0, s, t)\left[A\left(\vartheta^\star_{t,\varepsilon}, s\right) - A(\vartheta_0, s)\right] m^\star_{s,\varepsilon}\mathrm{d}s$$

$$+ f\sigma^{-2}\int_\tau^t \Phi^*(\vartheta_0, s, t)\left[\gamma^*\left(\vartheta^\star_{s,\varepsilon}, s\right) - \gamma^*(\vartheta_0, s)\right]\mathrm{d}X_s.$$

We have

$$\delta_m(t) = \dot m(\vartheta_0, \tau)\,\delta_\vartheta(\tau)\,\Phi^*(\vartheta_0, 0, \tau) + \int_\tau^t \Phi^*(\vartheta_0, s, t)\,\dot A(\vartheta_0, s)\, m^\star_{s,\varepsilon}\delta_\vartheta(s)\,\mathrm{d}s$$

$$+ f\sigma^{-2}\int_{\tau}^{t} \Phi^{*}\left(\vartheta_0, s, t\right)\dot{\gamma}^{*}\left(\vartheta_0, s\right)\delta_{\vartheta}\left(s\right){\rm d}X_s + O\left(\varepsilon^2\right)$$

where $\delta_{\vartheta}\left(s\right) = \vartheta^{\star}_{s,\varepsilon} - \vartheta_0$. Therefore,

$$\begin{aligned}
\varepsilon^{-1}\delta_m\left(t\right) &= \dot{m}\left(\vartheta_0, \tau\right)\Phi^{*}\left(\vartheta_0, 0, \tau\right)\varepsilon^{-1}\delta_{\vartheta}\left(\tau\right)\\
&\quad + \int_{\tau}^{t} \Phi^{*}\left(\vartheta_0, s, t\right)\dot{A}\left(\vartheta_0, s\right)m^{\star}_{s,\varepsilon}\varepsilon^{-1}\delta_{\vartheta}\left(s\right){\rm d}s\\
&\quad + f\sigma^{-2}\int_{\tau}^{t} \Phi^{*}\left(\vartheta_0, s, t\right)\dot{\gamma}^{*}\left(\vartheta_0, s\right)\varepsilon^{-1}\delta_{\vartheta}\left(s\right){\rm d}X_s + O\left(\varepsilon\right)\\
&\Longrightarrow \dot{y}\left(\vartheta_0, \vartheta_0, \tau\right)\Phi^{*}\left(\vartheta_0, 0, \tau\right)\eta_{\tau}\left(\vartheta_0\right)\\
&\quad + \int_{\tau}^{t} \Phi^{*}\left(\vartheta_0, s, t\right)\dot{A}\left(\vartheta_0, s\right)y_s\left(\vartheta_0\right)\eta_s\left(\vartheta_0\right){\rm d}s\\
&\quad + f^2\sigma^{-2}\int_{\tau}^{t} \Phi^{*}\left(\vartheta_0, s, t\right)\dot{\gamma}^{*}\left(\vartheta_0, s\right)y_s\left(\vartheta_0\right)\eta_s\left(\vartheta_0\right){\rm d}s\\
&= G_{\tau}\left(\vartheta_0, \vartheta_0\right)\Phi^{*}\left(\vartheta_0, 0, \tau\right)\zeta\left(\vartheta_0, \tau\right)\\
&\quad + \int_{\tau}^{t} \Phi^{*}\left(\vartheta_0, s, t\right)\dot{A}\left(\vartheta_0, s\right)H_s\left(\vartheta_0, \vartheta_0\right)\zeta\left(\vartheta_0, s\right){\rm d}s\\
&\quad + f^2\sigma^{-2}\int_{\tau}^{t} \Phi^{*}\left(\vartheta_0, s, t\right)\dot{\gamma}^{*}\left(\vartheta_0, s\right)H_s\left(\vartheta_0, \vartheta_0\right)\zeta\left(\vartheta_0, s\right){\rm d}s\\
&= G_{\tau}\left(\vartheta_0, \vartheta_0\right)\Phi^{*}\left(\vartheta_0, 0, \tau\right)\zeta\left(\vartheta_0, \tau\right) + \int_{\tau}^{t} \Phi^{*}\left(\vartheta_0, s, t\right)H_s\left(\vartheta_0, \vartheta_0\right)\zeta\left(\vartheta_0, s\right){\rm d}s\\
&\equiv \xi^{*}_{m}\left(\vartheta_0, t\right).
\end{aligned} \tag{2.81}$$

□

Remark 2.12 Remark that in the third problem (random initial value) the family of measures of this statistical experiment with unknown parameter ϑ is locally asymptotically mixing normal (LAMN). The normalized likelihood ratio function constructed by the observations $X_{\tau}^{T} = \left(X_t, \tau \leq t \leq T\right)$

$$Z_{\varepsilon}\left(\vartheta_0, u\right) = \frac{L\left(\vartheta_0 + \varepsilon u, X_{\tau}^{T}\right)}{L\left(\vartheta_0, X_{\tau}^{T}\right)}, \qquad u \in \mathbb{U}_{\varepsilon} = \left(\frac{\alpha - \vartheta_0}{\varepsilon}, \frac{\beta - \vartheta_0}{\varepsilon}\right)$$

admits the representation

$$Z_{\varepsilon}\left(\vartheta_0, u\right) = \exp\left(u\, y_0\, \Delta_{\varepsilon}\left(\vartheta_0, X_{\tau}^{T}\right) - \frac{u^2}{2}y_0^2\, J_{\tau}^{t}\left(\vartheta_0\right) + r_{\varepsilon}\right).$$

Here $r_\varepsilon \to 0$ and

$$\Delta_\varepsilon\left(\vartheta_0, X^T\right) \Longrightarrow \Delta\left(\vartheta_0\right) \sim \mathcal{N}\left(0, J_\tau^t\left(\vartheta_0\right)\right).$$

We do not study the MLE but note that it has the limit Cauchy distribution

$$\frac{\hat{\vartheta}_\varepsilon - \vartheta_0}{\varepsilon} \Longrightarrow \hat{u} = \frac{\Delta\left(\vartheta_0\right)}{y_0\, J_\tau^t\left(\vartheta_0\right)}$$

(see Remark 2.11) The study of MLE can be an interesting problem but we will not do it here.

Of course, there is a problem of definition of asymptotic efficiency of adaptive filter in the problem of estimation of $m\left(\vartheta_0, t\right)$. Remind that in the construction of the lower bound, we obtain on the right-hand side the expression like

$$\mathbf{E}_\vartheta\left[\dot{m}\left(\vartheta, t\right) \mathrm{I}_0^T\left(\vartheta\right)^{-1} \Delta_\varepsilon\left(\vartheta, X^T\right)\right]^2$$

where the random variable y_0 vanishes in the limit. This probably can be proved, but we do not occupy this problem here. In any case, the proposed here One-step MLE process with fixed τ is not asymptotically efficient.

2.3 Conditionally Gaussian Processes

2.3.1 *Model of Observations*

The Kalman–Bucy filtration equations can be extended on the conditionally Gaussian processes. The general theory can be found in [142, Chapters 11–13]. We consider a simple model of such processes. Suppose that the following partially observed system is given:

$$\begin{aligned} \mathrm{d}X_t &= f\left(\vartheta, t, X_t\right) Y_t\, \mathrm{d}t + \varepsilon\sigma\left(t, X_t\right) \mathrm{d}W_t, \qquad & X_0 &= 0, \quad 0 \le t \le T, \\ \mathrm{d}Y_t &= a\left(\vartheta, t, X_t\right) Y_t\, \mathrm{d}t + \varepsilon b\left(\vartheta, t, X_t\right) \mathrm{d}V_t, \qquad & Y_0 &= y_0 \neq 0. \end{aligned}$$

As before the process $X^T = \left(X_t, 0 \le t \le T\right)$ is observed and the process $Y^T = \left(Y_t, 0 \le t \le T\right)$ is *hidden*. The Wiener processes $W_t, 0 \le t \le T$ and $V_t, 0 \le t \le T$ are independent and the functions $f\left(\cdot\right), \sigma\left(\cdot\right), a\left(\cdot\right),\, b\left(\cdot\right)$ are known. The initial value y_0 and parameter $\varepsilon \in (0, 1]$ are known too. The unknown parameter $\vartheta \in \Theta \subset \mathcal{R}^d$, $d \ge 1$, where Θ is an open, convex, and bounded set.

We write below the optimal filtering equations for $m\left(\vartheta, t\right) = \mathbf{E}_\vartheta\left(Y_t | \mathfrak{F}_t^X\right)$ and for the conditional mean square error $\gamma\left(\vartheta, t, X\right) = \mathbf{E}_\vartheta\left(\left(Y_t - m\left(\vartheta, t\right)\right)^2 | \mathfrak{F}_t^X\right)$.

Conditions $\mathscr{D}$.

$\mathscr{D}_1$. *The functions* $f(\vartheta,t,x), \sigma(t,x), a(\vartheta,t,x), b(\vartheta,t,x); \vartheta \in \Theta, t \in [0,T], x \in \mathcal{R}$ *are bounded and have continuous bounded derivatives w.r.t. x.*
$\mathscr{D}_2$. *There exists a constant* $c > 0$ *which does not depend on t and x such that*

$$\sigma(t,x)^2 > c.$$

Theorem 2.10 *Let the conditions* $\mathscr{D}$ *hold. Then the random process* $X_t, Y_t, 0 \leq t \leq T$ *is conditionally Gaussian, i.e., for any* $t \in (0,T]$, *and any* $0 \leq t_0 \leq t_1 < \ldots < t_n \leq t$, *the conditional distributions*

$$\mathbf{P}_\vartheta\left(Y_{t_0} \leq x_0, \ldots, Y_{t_n} \leq x_n | \mathfrak{F}_t^X\right)$$

are ($\mathbf{P}_\vartheta$*-a.s.) Gaussian.*

This is particular case of the Theorem 11.1 in [142].

If the condition $\mathscr{D}_1$ is fulfilled, then the limit ($\varepsilon = 0$) system is

$$\frac{\mathrm{d}x_t(\vartheta)}{\mathrm{d}t} = f(\vartheta,t,x_t(\vartheta))\, y_t(\vartheta), \qquad x_0(\vartheta) = 0,$$
$$\frac{\mathrm{d}y_t(\vartheta)}{\mathrm{d}t} = a(\vartheta,t,x_t(\vartheta))\, y_t(\vartheta), \qquad y_0(\vartheta) = y_0.$$

Let us denote

$$A(\vartheta,t,X) = a(\vartheta,t,X_t) - \frac{f(\vartheta,t,X_t)^2\gamma(\vartheta,t,X)}{\varepsilon^2\sigma(t,X_t)^2},$$
$$B(\vartheta,t,X) = \frac{f(\vartheta,t,X_t)\gamma(\vartheta,t,X)}{\varepsilon^2\sigma(t,X_t)^2}.$$

Theorem 2.11 *Let the conditions* $\mathscr{D}$ *hold, then* $m(\vartheta,t)$ *and* $\gamma(\vartheta,t,X)$ *are solutions of the equations*

$$\mathrm{d}m(\vartheta,t) = A(\vartheta,t,X)\, m(\vartheta,t)\, \mathrm{d}t + B(\vartheta,t,X)\, \mathrm{d}X_t, \qquad m(\vartheta,0) = y_0, \tag{2.82}$$

$$\frac{\mathrm{d}\gamma(\vartheta,t,X)}{\mathrm{d}t} = 2a(\vartheta,t,X_t)\gamma(\vartheta,t,X) - \frac{f(\vartheta,t,X_t)^2\gamma(\vartheta,t,X)^2}{\varepsilon^2\sigma(t,X_t)^2} + \varepsilon^2 b(\vartheta,t,X_t)^2, \qquad \gamma(\vartheta,0,X) = 0. \tag{2.83}$$

For the proof of more general result see [141], Theorem 12.1.

If we denote $\gamma_*(\vartheta,t,X) = \varepsilon^{-2}\gamma(\vartheta,t,X)$, then the Eqs. (2.82) and (2.83) can be replaced by the equations

$$\mathrm{d}m(\vartheta,t)=A_*(\vartheta,t,X)\,m(\vartheta,t)\,\mathrm{d}t+B_*(\vartheta,t,X)\,\mathrm{d}X_t, \tag{2.84}$$

$$\frac{\mathrm{d}\gamma_*(\vartheta,t,X)}{\mathrm{d}t}=2a(\vartheta,t,X_t)\,\gamma_*(\vartheta,t,X)-\frac{f(\vartheta,t,X_t)^2\,\gamma_*(\vartheta,t,X)^2}{\sigma(t,X_t)^2}+b(\vartheta,t,X_t)^2, \tag{2.85}$$

with the obvious notations for $A_*(\vartheta,t,X)$, $B_*(\vartheta,t,X)$ and the same initial values.

2.3.2 Estimators

MDE

As usual ϑ_0 is the true value. Let us fix the learning interval $[0,\tau]$ and construct the MDE by observations $X^\tau=(X_t, 0\leq t\leq\tau)$, where $\tau\in(0,T]$. Introduce notations:

- Deterministic functions

$$y_t(\vartheta)=y_0\exp\left(\int_0^t a(\vartheta,s,x_s(\vartheta))\,\mathrm{d}s\right),\quad x_t(\vartheta)=\int_0^t f(\vartheta,s,x_s(\vartheta))\,y_s(\vartheta)\,\mathrm{d}s.$$

- The vector function $\dot{\mathrm{x}}_t(\vartheta), t\in[0,T], \vartheta\in\Theta$ is solution of the system of equations

$$\frac{\mathrm{d}\dot{\mathrm{x}}_t(\vartheta)}{\mathrm{d}t}=\dot{\mathrm{f}}(\vartheta,t,x_t(\vartheta))\,y_t(\vartheta)+f_x'(\vartheta,t,x_t(\vartheta))\,\dot{\mathrm{x}}_t(\vartheta)\,y_t(\vartheta)$$
$$+f(\vartheta,t,x_t(\vartheta))\,y_t(\vartheta)\int_0^t\left[\dot{\mathrm{a}}(\vartheta,s,x_s(\vartheta))+a_x'(\vartheta,s,x_s(\vartheta))\,\dot{\mathrm{x}}_s(\vartheta)\right]\mathrm{d}s$$

  with the initial value $\dot{\mathrm{x}}_0(\vartheta)=0$.
- The $d\times d$ the matrix

$$\check{\mathbf{T}}(\vartheta)=\int_0^\tau \dot{\mathrm{x}}_t(\vartheta)\,\dot{\mathrm{x}}_t(\vartheta)^\top\,\mathrm{d}t.$$

- Zero-mean Gaussian process $x_t^{(1)}(\vartheta_0), 0\leq t\leq T$ defined as the solution of the linear equation

$$\mathrm{d}x_t^{(1)}(\vartheta_0)=f_x'(\vartheta_0,t,x_t(\vartheta_0))\,x_t^{(1)}(\vartheta_0)\,y_t(\vartheta_0)\,\mathrm{d}t+\sigma(t,x_t(\vartheta_0))\,\mathrm{d}W_t$$
$$+f(\vartheta_0,t,x_t(\vartheta_0))\int_0^t\phi(\vartheta_0,s,t)\,a_x'(\vartheta_0,s,x_s(\vartheta_0))\,x_s^{(1)}(\vartheta_0)\,y_s(\vartheta_0)\,\mathrm{d}s\,\mathrm{d}t$$

$$+ f(\vartheta_0, t, x_t(\vartheta_0)) \int_0^t \phi(\vartheta_0, s, t)\, b(\vartheta_0, s, x_s(\vartheta_0))\, dV_s\, dt,\ \ x_0^{(1)}(\vartheta_0) = 0,$$

where

$$\phi(\vartheta_0, s, t) = \exp\left(\int_s^t a(\vartheta_0, r, x_r(\vartheta_0))\, dr \right).$$

- Gaussian vector

$$\check{\zeta}_\tau(\vartheta_0) = \check{\mathbf{T}}(\vartheta_0)^{-1} \int_0^\tau x_t^{(1)}(\vartheta_0)\, \dot{\mathrm{x}}_t(\vartheta_0)\, dt.$$

Then we can introduce the MDE

$$\check{\vartheta}_{\tau,\varepsilon} = \arg\inf_{\vartheta\in\Theta} \int_0^\tau [X_t - x_t(\vartheta)]^2\, dt. \tag{2.86}$$

The regularity *Conditions* $\check{\mathscr{D}}$ *are*

$\check{\mathscr{D}}_1$. *The conditions* $\mathscr{D}$ *are fulfilled.*
$\check{\mathscr{D}}_2$. *For any* $\nu > 0$ *we have*

$$\inf_{\vartheta_0\in\Theta} g(\vartheta_0, \nu) > 0, \qquad g(\vartheta_0, \nu) = \inf_{\|\vartheta-\vartheta_0\|>\nu} \int_0^\tau [x_t(\vartheta_0) - x_t(\vartheta)]^2\, dt.$$

$\check{\mathscr{D}}_3$. *The functions* $f(\vartheta, t, x), a(\vartheta, t, x), b(\vartheta, t, x), \vartheta \in \Theta, t \in [0, \tau], x \in \mathcal{R}$ *have two continuous bounded derivatives w.r.t.* ϑ.
$\check{\mathscr{D}}_4$. *The matrix* $\check{\mathbf{T}}(\vartheta)$ *is uniformly nondegenerate*

$$\inf_{\vartheta\in\Theta} \inf_{\mathrm{e}=1,\mathrm{e}\in\mathcal{R}^d} \mathrm{e}^\top \check{\mathbf{T}}(\vartheta)\, \mathrm{e} > 0.$$

Theorem 2.12 *Let the conditions* $\check{\mathscr{D}}$ *hold then the MDE* $\check{\vartheta}_{\tau,\varepsilon}$ *is uniformly on compacts* $\mathbb{K} \subset \Theta$ *consistent, asymptotically normal*

$$\varepsilon^{-1}\left(\check{\vartheta}_{\tau,\varepsilon} - \vartheta_0\right) \Longrightarrow \check{\zeta}_\tau(\vartheta_0)$$

and the polynomial moments converge: for any $p > 0$

$$\lim_{\varepsilon\to 0} \varepsilon^{-p} \mathbf{E}_{\vartheta_0} \left\| \check{\vartheta}_{\tau,\varepsilon} - \vartheta_0 \right\|^p = \mathbf{E}_{\vartheta_0} \left\| \check{\zeta}_\tau(\vartheta_0) \right\|^p.$$

Proof The proof of this theorem almost coincides with the proof of the Theorem 2.2. The only difference is in the description of the asymptotic of the random process $X_t - x_t(\vartheta_0)$.

Below, we omit ϑ_0 and write $f(t, x)$ instead of $f(\vartheta_0, t, x)$ *etc.* and denoted $\delta_x(t) = X_t - x_t$, $\delta_y(t) = Y_t - y_t$. We have

$$\begin{aligned}
\delta_x(t) &= \int_0^t [f(s, X_s) Y_s - f(s, x_s) y_s] \, \mathrm{d}s + \varepsilon \int_0^t \sigma(s, X_s) \, \mathrm{d}W_s \\
&= \int_0^t f(s, X_s) \delta_y(s) \, \mathrm{d}s + \int_0^t [f(s, X_s) - f(s, x_s)] \, y_s \mathrm{d}s \\
&\quad + \varepsilon \int_0^t \sigma(s, X_s) \, \mathrm{d}W_s \\
&= \int_0^t f(s, X_s) \delta_y(s) \, \mathrm{d}s + \int_0^t f_x'(s, \tilde{X}_s) \delta_x(s) \, y_s \, \mathrm{d}s + \varepsilon \int_0^t \sigma(s, X_s) \, \mathrm{d}W_s .
\end{aligned}$$

The difference $Y_t - y_t$ has the following expression:

$$\begin{aligned}
\delta_y(t) &= \int_0^t a(s, X_s) \delta_y(s) \, \mathrm{d}s + \int_0^t [a(s, X_s) - a(s, x_s)] \, y_s \mathrm{d}s + \varepsilon \int_0^t b(s, X_s) \, \mathrm{d}V_s \\
&= \int_0^t a(s, X_s) \delta_y(s) \, \mathrm{d}s + \int_0^t a_x'(s, \tilde{X}_s) \delta_x(s) \, y_s \mathrm{d}s + \varepsilon \int_0^t b(s, X_s) \, \mathrm{d}V_s .
\end{aligned}$$

For the functions $v_x(t) = \mathbf{E}_{\vartheta_0} \delta_x(t)^2$ and $v_y(t) = \mathbf{E}_{\vartheta_0} \delta_y(t)^2$ we obtain the relations

$$\begin{aligned}
v_x(t) &\leq C \int_0^t v_y(s) \, \mathrm{d}s + C \int_0^t v_x(s) \, \mathrm{d}s + C\varepsilon^2, \\
v_y(t) &\leq C \int_0^t v_y(s) \, \mathrm{d}s + C \int_0^t v_x(s) \, \mathrm{d}s + C\varepsilon^2.
\end{aligned}$$

Hence, the function $u(t) = v_x(t) + v_y(t)$ satisfies the iequality

$$u(t) \leq C \int_0^t u(s) \, \mathrm{d}s + C\varepsilon^2.$$

Thus, by Grönwall's lemma 1.5 $u(t) \leq C\varepsilon^2$, and this implies

$$\sup_{0 \leq t \leq T} \mathbf{E}_{\vartheta_0} [X_t - x_t(\vartheta_0)]^2 \leq C \varepsilon^2, \qquad \sup_{0 \leq t \leq T} \mathbf{E}_{\vartheta_0} [Y_t - y_t(\vartheta_0)]^2 \leq C \varepsilon^2,$$

where the constants $C > 0$ do not depend on ϑ_0.

These estimates together with the condition $\check{\mathscr{D}}_2$ allow to prove the consistency of MDE.

Therefore, we can write for $x_{t,\varepsilon}^{(1)} = \varepsilon^{-1}(X_t - x_t)$ the representation

$$
\begin{aligned}
x_{t,\varepsilon}^{(1)} &= \int_0^t f_x'(s, \tilde{X}_s)\, x_{s,\varepsilon}^{(1)}\, Y_s\, \mathrm{d}s + \int_0^t f(s, x_s) \int_0^s \phi(r, s)\, b(r, X_r)\, \mathrm{d}V_r\, \mathrm{d}s \\
&\quad + \int_0^t f(s, x_s) \int_0^s \phi(r, s)\, a_x'(r, \tilde{X}_r)\, x_{r,\varepsilon}^{(1)}\, y_r\, \mathrm{d}r\, \mathrm{d}s + \int_0^t \sigma(s, X_s)\, \mathrm{d}W_s \\
&= x_t^{(1)}(\vartheta_0) + O(\varepsilon).
\end{aligned}
$$

Now the asymptotic normality follows from MDEq as follows:

$$
\begin{aligned}
0 &= \int_0^\tau \left[X_t - x_t(\check{\vartheta}_{\tau,\varepsilon})\right] \dot{\mathrm{x}}_t(\check{\vartheta}_{\tau,\varepsilon})\, \mathrm{d}t \\
&= \int_0^\tau \left[X_t - x_t(\vartheta_0) + x_t(\vartheta_0) - x_t(\check{\vartheta}_{\tau,\varepsilon})\right] \dot{\mathrm{x}}_t(\check{\vartheta}_{\tau,\varepsilon}) \mathrm{d}t \\
&= \int_0^\tau x_{t,\varepsilon}^{(1)}\, \dot{\mathrm{x}}_t(\check{\vartheta}_{\tau,\varepsilon})\, \mathrm{d}t\, (1 + O(\varepsilon)) - \check{\mathbf{T}}(\vartheta_0)\, \varepsilon^{-1} \left(\check{\vartheta}_{\tau,\varepsilon} - \vartheta_0\right)(1 + O(\varepsilon)).
\end{aligned}
$$

Hence,

$$
\begin{aligned}
\varepsilon^{-1}\left(\check{\vartheta}_{\tau,\varepsilon} - \vartheta_0\right) &= \check{\mathbf{T}}(\vartheta_0)^{-1} \int_0^\tau x_{t,\varepsilon}^{(1)}\, \dot{\mathrm{x}}_t(\check{\vartheta}_{\tau,\varepsilon})\, \mathrm{d}t\, (1 + O(\varepsilon)) \\
&\Longrightarrow \check{\mathbf{T}}(\vartheta_0)^{-1} \int_0^\tau x_t^{(1)}(\vartheta_0)\, \dot{\mathrm{x}}_t(\vartheta_0)\, \mathrm{d}t \equiv \check{\zeta}_\tau(\vartheta_0).
\end{aligned}
$$

The rest of the proof is close to the proof of the Theorem 2.2. □

MLE and BE

The LR for this model is

$$
L\left(\vartheta, X^T\right) = \exp\left(\int_0^T \frac{M(\vartheta, t, X)}{\varepsilon^2 \sigma(t, X_t)^2}\, \mathrm{d}X_t - \int_0^T \frac{M(\vartheta, t, X)^2}{2\varepsilon^2 \sigma(t, X_t)^2}\, \mathrm{d}t\right), \qquad \vartheta \in \Theta,
$$

where $M(\vartheta, t, X) = f(\vartheta, t, X_t)\, m(\vartheta, t)$. The MLE $\hat{\vartheta}_\varepsilon$ and BE $\tilde{\vartheta}_\varepsilon$ are defined by the usual relations (2.32).

The random processes $m(\vartheta, t)$ and $\gamma_*(\vartheta, t, X)$ from the Eqs. (2.84) and (2.85), respectively converge to the following limits $m(\vartheta, t) \longrightarrow y_t(\vartheta, \vartheta_0)$ and $\gamma_*(\vartheta, t, X) \longrightarrow \gamma_t^*(\vartheta, \vartheta_0, x)$ which are solutions of the equations

$$
\frac{\partial y_t(\vartheta, \vartheta_0)}{\partial t} = A_*(\vartheta, t, x(\vartheta_0))\, y_t(\vartheta, \vartheta_0) + B_*(\vartheta, t, x(\vartheta_0))\, f(\vartheta_0, t, x_t(\vartheta_0))\, y_t(\vartheta_0),
$$

$$\frac{\mathrm{d}\gamma_t^*(\vartheta,\vartheta_0)}{\mathrm{d}t}=2a(\vartheta,t,x_t(\vartheta_0))\,\gamma_t^*(\vartheta,\vartheta_0)-\frac{f(\vartheta,t,x_t(\vartheta_0))^2\,\gamma_t^*(\vartheta,\vartheta_0)^2}{\sigma(t,x_t(\vartheta_0))^2}$$
$$+\,b(\vartheta,t,x_t(\vartheta_0))^2,\qquad \gamma_0^*(\vartheta,\vartheta_0)=0.$$

Here obviously

$$A_*(\vartheta,t,x(\vartheta_0))=a(\vartheta,t,x_t(\vartheta_0))-\frac{f(\vartheta,t,x_t(\vartheta_0))^2\,\gamma_t^*(\vartheta,\vartheta_0)}{\sigma(t,x_t(\vartheta_0))^2},$$
$$B_*(\vartheta,t,x(\vartheta_0))=\frac{f(\vartheta,t,x_t(\vartheta_0))\,\gamma_t^*(\vartheta,\vartheta_0)}{\sigma(t,x_t(\vartheta_0))^2}.$$

Having all these relations, it is possible to construct and describe the same estimators as in the Sect. 2.1 and these estimators will have the similar asymptotic properties. As the proofs will be quite close to those given above, we describe the estimators without detailed proofs.

Let us set

$$S(\vartheta,\vartheta_0,t)=f(\vartheta,t,x_t(\vartheta_0))\,y_t(\vartheta,\vartheta_0),\quad \dot{\mathrm{f}}(\vartheta,t,x_t(\vartheta_0))=\frac{\partial}{\partial\vartheta}f(\vartheta,t,x_t(\vartheta_0))$$
$$\dot{\mathrm{S}}(\vartheta,\vartheta_0,t)=\dot{\mathrm{f}}(\vartheta,t,x_t(\vartheta_0))\,y_t(\vartheta,\vartheta_0)+f(\vartheta,t,x_t(\vartheta_0))\,\dot{\mathrm{y}}_t(\vartheta,\vartheta_0).$$

The Fisher information $d\times d$ matrix is

$$\mathbf{I}(\vartheta_0)=\int_0^T\frac{\dot{\mathrm{S}}(\vartheta_0,\vartheta_0,t)\,\dot{\mathrm{S}}(\vartheta_0,\vartheta_0,t)^\top}{\sigma(t,x_t(\vartheta_0))^2}\mathrm{d}t.$$

Conditions $\hat{\mathscr{D}}$.

$\hat{\mathscr{D}}_1$. *The conditios* $\mathscr{D}$ *and* $\check{\mathscr{D}}_3$ *are satisfied.*

$\hat{\mathscr{D}}_2$. *The Fisher information matrix is uniformly non degenerate*

$$\inf_{\vartheta_0\in\Theta}\ \inf_{\mathrm{e}\in\mathcal{R}^d,\|\mathrm{e}\|=1}\mathrm{e}^\top\mathbf{I}(\vartheta_0)\,\mathrm{e}>0.$$

$\hat{\mathscr{D}}_3$. *For any* $\nu>0$

$$\inf_{\vartheta_0\in\Theta}\ \inf_{\|\vartheta-\vartheta_0\|>\nu}\int_0^T\left[\frac{S(\vartheta,\vartheta_0,t)-S(\vartheta_0,\vartheta_0,t)}{\sigma(t,x_t(\vartheta_0))}\right]^2\mathrm{d}t\geq 0.$$

Theorem 2.13 *Let the conditions* $\hat{\mathscr{D}}$ *hold. Then the MLE* $\hat\vartheta_\varepsilon$ *and BE* $\tilde\vartheta_\varepsilon$ *are uniformly consistent, asymptotically normal*

$$\varepsilon^{-1}\left(\hat{\vartheta}_\varepsilon - \vartheta_0\right) \Longrightarrow \mathcal{N}\left(0, \mathbf{I}(\vartheta_0)^{-1}\right), \qquad \varepsilon^{-1}\left(\tilde{\vartheta}_\varepsilon - \vartheta_0\right) \Longrightarrow \mathcal{N}\left(0, \mathbf{I}(\vartheta_0)^{-1}\right),$$

the moments converge, and the both estimators are asymptotically efficient.

Proof The proof of this theorem is similar to the proof of the Theorem 2.4. First, we verify that the normalized LR

$$Z_\varepsilon(\vartheta_0, u) = \frac{L\left(\vartheta_0 + \varepsilon u, X^T\right)}{L\left(\vartheta_0, X^T\right)}, \qquad u \in \mathbb{U}_\varepsilon = (u\ :\ \vartheta_0 + \varepsilon u \in \Theta)$$

admits the representation (LAN)

$$Z_\varepsilon(\vartheta_0, u) = \exp\left(u^\top \Delta_\varepsilon\left(\vartheta_0, X^T\right) - \frac{1}{2} u^\top \mathbf{I}(\vartheta_0)\, u + r_\varepsilon\right).$$

Here $r_\varepsilon \to 0$ and

$$\begin{aligned}\Delta_\varepsilon\left(\vartheta_0, X^T\right) &= \int_0^T \frac{\dot{\mathrm{S}}(\vartheta_0, \vartheta_0, t)}{\varepsilon\sigma(t, x_t(\vartheta_0))^2}\left[\mathrm{d}X_t - M(\vartheta_0, t, X)\,\mathrm{d}t\right]\\ &\Longrightarrow \int_0^T \frac{\dot{\mathrm{S}}(\vartheta_0, \vartheta_0, t)}{\sigma(t, x_t(\vartheta_0))}\mathrm{d}w(t) \equiv \Delta(\vartheta_0) \sim \mathcal{N}(0, \mathbf{I}(\vartheta_0)).\end{aligned}$$

Then using the Lemma 1.1, we check the condition

$$\mathbf{E}_{\vartheta_0}\left|Z_\varepsilon(\vartheta_0, u_2)^{1/2m} - Z_\varepsilon(\vartheta_0, u_1)^{1/2m}\right|^{2m} \leq C\left\|u_2 - u_1\right\|^{2m}$$

where $2m > d$.

To verify the last condition (2.35), we use the proof of the similar result in Sect. 2.1 (Lemma 2.5) and the condition $\hat{\mathscr{D}}_3$. □

One-Step MLE-Process

The One-step MLE process is defined by the obvious generalization of (2.42)

$$\vartheta^\star_{t,\varepsilon} = \check{\vartheta}_{\tau,\varepsilon} + \mathbf{I}^t_\tau\left(\check{\vartheta}_{\tau,\varepsilon}\right)^{-1}\int_\tau^t \frac{\dot{\mathrm{S}}\left(\check{\vartheta}_{\tau,\varepsilon}, \check{\vartheta}_{\tau,\varepsilon}, s\right)}{\sigma(s, X_s)^2}\left[\mathrm{d}X_s - M\left(\check{\vartheta}_{\tau,\varepsilon}, s, X\right)\mathrm{d}s\right],$$
$$\tau < t \leq T,$$

where $\check{\vartheta}_{\tau,\varepsilon}$ is the MDE constructed by observations X^τ (see (2.86)) and the Fisher information matrix

$$\mathbf{I}_\tau^t\left(\vartheta_0\right)=\int_\tau^t \frac{\dot{\mathrm{S}}\left(\vartheta_0, \vartheta_0, s\right) \dot{\mathrm{S}}\left(\vartheta_0, \vartheta_0, s\right)^{\top}}{\sigma\left(s, x_s\left(\vartheta_0\right)\right)^2} \mathrm{d} s .$$

Introduce the condition

$\mathscr{D}^\star$. *The matrix* $\mathbf{I}_\tau^t(\vartheta)$ *is uniformly nondegenerate: for all* $t \in(\tau, T]$

$$\inf _{\vartheta \in \Theta} \inf _{\mathrm{e} \in \mathcal{R}^d,\|\mathrm{e}\|=1} \mathrm{e}\, \mathbf{I}_\tau^t(\vartheta)\, \mathrm{e}^{\top}>0 .$$

Introduce the Gaussian process

$$\eta_t\left(\vartheta_0\right)=\mathbf{I}_\tau^t\left(\vartheta_0\right)^{-1} \int_\tau^t \frac{\dot{\mathrm{S}}\left(\vartheta_0, \vartheta_0, s\right)}{\sigma\left(s, x_s\left(\vartheta_0\right)\right)} \mathrm{d} w(s), \quad \tau<t \leq T$$

Here $w(\cdot)$ is some Wiener rocess.

It can be shown that under conditions $\mathscr{D}, \check{\mathscr{D}}, \mathscr{D}^\star$ the estimator process $\vartheta_{t, \varepsilon}^\star$ is consistent and asymptotically normal

$$\varepsilon^{-1}\left(\vartheta_{t, \varepsilon}^\star-\vartheta_0\right) \Longrightarrow \eta_t\left(\vartheta_0\right) \sim \mathcal{N}\left(0, \mathbf{I}_\tau^t\left(\vartheta_0\right)^{-1}\right) .$$

Moreover, if we consider the random vector process $\eta_{t, \varepsilon}=\varepsilon^{-1}\left(\vartheta_{t, \varepsilon}^\star-\vartheta_0\right), \tau+\nu \leq t \leq T$ where $\nu>0$, then it can be shown that for any $K \geq 1$ the random variables $\eta_{t_k, \varepsilon}, k=1, \ldots, K$ converge in distribution

$$\left(\eta_{t_1, \varepsilon}, \ldots, \eta_{t_K, \varepsilon}\right) \Longrightarrow\left(\eta_{t_1}\left(\vartheta_0\right), \ldots, \eta_{t_K}\left(\vartheta_0\right)\right) . \tag{2.87}$$

The condition

$$\mathbf{E}_{\vartheta_0}\left\|\eta_{t_1, \varepsilon}-\eta_{t_2, \varepsilon}\right\|^4 \leq C\left|t_2-t_1\right|^2$$

can be verified too. These two conditions yield the weak convergence of the random process $\eta_{t, \varepsilon}, \tau \leq t \leq T$ to the random process $\eta_t\left(\vartheta_0\right), \tau \leq t \leq T$.

2.3.3 Adaptive Filter

The construction of the adaptive filter is very close to that described in Sect. 2.1.2.

1. *Fix some small* $\tau \in(0, T)$ *and calculate the MDE*

$$\check{\vartheta}_{\tau, \varepsilon}=\arg \inf _{\vartheta \in \Theta} \int_0^\tau\left[X_t-x_t(\vartheta)\right]^2 \mathrm{d} t .$$

2. *Define the One-step MLE process* $\vartheta^\star_{t,\varepsilon},\ \tau < t \leq T$

$$\vartheta^\star_{t,\varepsilon} = \check\vartheta_{\tau,\varepsilon} + \mathbf{I}_\tau^t\left(\check\vartheta_{\tau,\varepsilon}\right)^{-1}\int_\tau^t \frac{\dot{\mathrm{S}}\left(\check\vartheta_{\tau,\varepsilon},\check\vartheta_{\tau,\varepsilon},s\right)}{\sigma\left(s,X_s\right)^2}\left[\mathrm{d}X_s - M\left(\check\vartheta_{\tau,\varepsilon},s,X\right)\mathrm{d}s\right].$$

Here $\dot{\mathrm{S}}\left(\vartheta,\vartheta_0,t\right) = \dot{\mathrm{f}}\left(\vartheta,t,x_t\left(\vartheta_0\right)\right)y_t\left(\vartheta,\vartheta_0\right) + f\left(\vartheta,t,x_t\left(\vartheta_0\right)\right)\dot{\mathrm{y}}_t\left(\vartheta,\vartheta_0\right)$, *Fisher information matrix*

$$\mathbf{I}_\tau^t\left(\vartheta\right) = \int_\tau^t \frac{\dot{\mathrm{S}}\left(\vartheta,\vartheta,s\right)\dot{\mathrm{S}}\left(\vartheta,\vartheta,s\right)^\top}{\sigma\left(s,x_s\left(\vartheta\right)\right)^2}\mathrm{d}s.$$

The process $M\left(\check\vartheta_{\tau,\varepsilon},s,X\right) = f\left(\check\vartheta_{\tau,\varepsilon},s,X\right)m\left(\check\vartheta_{\tau,\varepsilon},s\right)$ *is solution of the equations*

$$\begin{aligned}\mathrm{d}M\left(\check\vartheta_{\tau,\varepsilon},t,X\right) &= A_*\left(\check\vartheta_{\tau,\varepsilon},t,X\right)M\left(\check\vartheta_{\tau,\varepsilon},t,X\right)\mathrm{d}t\\ &\quad + f\left(\check\vartheta_{\tau,\varepsilon},t,X\right)B_*\left(\check\vartheta_{\tau,\varepsilon},t,X\right)\mathrm{d}X_t,\\ \frac{\mathrm{d}\gamma_*\left(\check\vartheta_{\tau,\varepsilon},t,X\right)}{\mathrm{d}t} &= 2a\left(\check\vartheta_{\tau,\varepsilon},t,X_t\right)\gamma_*\left(\check\vartheta_{\tau,\varepsilon},t,X\right)\\ &\quad - \frac{f\left(\check\vartheta_{\tau,\varepsilon},t,X_t\right)^2\gamma_*\left(\check\vartheta_{\tau,\varepsilon},t,X\right)^2}{\sigma\left(t,X_t\right)^2}\\ &\quad + b\left(\check\vartheta_{\tau,\varepsilon},t,X_t\right)^2,\end{aligned}$$

subject to initial values $M\left(\check\vartheta_{\tau,\varepsilon},0,0\right) = f\left(\check\vartheta_{\tau,\varepsilon},0,0\right)y_0$ *and* $\gamma_*\left(\check\vartheta_{\tau,\varepsilon},0,X\right) = 0$.

3. *The adaptive filter we obtain by substitution of* $\vartheta^\star_{t,\varepsilon}$ *in the Eqs.* (2.84) *and* (2.85). *This leads to the equations for* $m^\star_{t,\varepsilon},\ \tau_* \leq t \leq T$ *and* $\gamma^\star_t$ *(here* $\tau_* \in (\tau, T]$*)*

$$\mathrm{d}m^\star_{t,\varepsilon} = A_*\left(\vartheta^\star_{t,\varepsilon},t,X\right)m^\star_{t,\varepsilon}\,\mathrm{d}t + B_*\left(\vartheta^\star_{t,\varepsilon},t,X\right)\mathrm{d}X_t,\qquad m^\star_{\tau_*,\varepsilon} = y_\tau\left(\vartheta^\star_{\tau,\varepsilon}\right), \tag{2.88}$$

$$\frac{\mathrm{d}\gamma^\star_t}{\mathrm{d}t} = 2a\left(\vartheta^\star_{t,\varepsilon},t,X_t\right)\gamma^\star_t - \frac{f\left(\vartheta^\star_{t,\varepsilon},t,X_t\right)^2\left(\gamma^\star_t\right)^2}{\sigma\left(t,X_t\right)^2} + b\left(\vartheta^\star_{t,\varepsilon},t,X_t\right)^2,\quad \gamma^\star_0 = 0$$

with obvious notations.

We suppose that the conditions $\mathscr{D}$, $\check{\mathscr{D}}$ and $\hat{\mathscr{D}}$ are fulfilled and the functions $a\left(\vartheta,t,x\right), f\left(\vartheta,t,x\right), \sigma\left(t,x\right)$ have continuous derivatives on $t \in [0,T]$.

For the difference $\delta_m(t) = m^\star_{t,\varepsilon} - m_t$, where $m_t = m(\vartheta_0, t)$, we have the equation

$$\begin{aligned}\mathrm{d}\delta_m(t) &= \left[A_*\left(\vartheta^\star_{t,\varepsilon}, t, X\right) m^\star_t - A_*(\vartheta_0, t, X)\, m_t\right]\mathrm{d}t\\ &\quad + \left[B_*\left(\vartheta^\star_{t,\varepsilon}, t, X\right) - B_*(\vartheta_0, t, X)\right] f(\vartheta_0, t)\, m_t\,\mathrm{d}t\\ &\quad + \varepsilon\left[B_*\left(\vartheta^\star_{t,\varepsilon}, t, X\right) - B_*(\vartheta_0, t, X)\right]\sigma(t, X_t)\,\mathrm{d}W_t.\end{aligned}$$

Let us denote

$$\begin{aligned}R_t &= \int_{\tau_*}^t \left[B_*\left(\vartheta^\star_{s,\varepsilon}, s, X\right) - B_*(\vartheta_0, s, X)\right]\sigma(s, X_s)\,\mathrm{d}W_s,\\ Z_t &= m^\star_{t,\varepsilon} - m_t - \varepsilon R_t.\end{aligned}$$

Then we can write the linear equation for the process Z_t

$$\begin{aligned}\mathrm{d}Z_t &= A_*\left(\vartheta^\star_{t,\varepsilon}, t, X\right) Z_t\,\mathrm{d}t + \left[A_*\left(\vartheta^\star_{t,\varepsilon}, t, X\right) - A_*(\vartheta_0, t, X)\right] m_t\,\mathrm{d}t\\ &\quad + \left[B_*\left(\vartheta^\star_{t,\varepsilon}, t, X\right) - B_*(\vartheta_0, t, X)\right] f(\vartheta_0, t)\, m_t\,\mathrm{d}t + \varepsilon A_*\left(\vartheta^\star_{t,\varepsilon}, t, X\right) R_t\,\mathrm{d}t.\end{aligned}$$

Therefore, if we denote

$$\Phi\left(\vartheta^\star_\varepsilon, s, t, X\right) = \exp\left(\int_s^t A_*\left(\vartheta^\star_{v,\varepsilon}, v, X\right)\mathrm{d}v\right),$$

then the solution of this linear equation can be written as follows:

$$\begin{aligned}Z_t &= Z_{\tau_*}\Phi\left(\vartheta^\star_\varepsilon, 0, t\right) + \int_{\tau_*}^t \Phi\left(\vartheta^\star_\varepsilon, s, t, X\right)\left[A_*\left(\vartheta^\star_{s,\varepsilon}, s, X\right) - A_*(\vartheta_0, s, X)\right] m_s\,\mathrm{d}s\\ &\quad + \int_{\tau_*}^t \Phi\left(\vartheta^\star_\varepsilon, s, t, X\right)\left[B_*\left(\vartheta^\star_{s,\varepsilon}, s, X\right) - B_*(\vartheta_0, s, X)\right] f(\vartheta_0, s)\, m_s\,\mathrm{d}s\\ &\quad + \varepsilon\int_{\tau_*}^t \Phi\left(\vartheta^\star_\varepsilon, s, t, X\right) A_*\left(\vartheta^\star_{s,\varepsilon}, s, X\right) R_s\,\mathrm{d}s\\ &= Z_{\tau_*}\Phi\left(\vartheta^\star_\varepsilon, 0, t, X\right) + \int_{\tau_*}^t \Phi(\vartheta_0, s, t, X)\,\dot{\mathrm{A}}_*(\vartheta_0, s, X)^\top\left[\vartheta^\star_{s,\varepsilon} - \vartheta_0\right] m_s\,\mathrm{d}s\\ &\quad \times (1 + o(1))\\ &\quad + \int_{\tau_*}^t \Phi(\vartheta_0, s, t, X)\,\dot{\mathrm{B}}_*(\vartheta_0, s, X)^\top\left[\vartheta^\star_{s,\varepsilon} - \vartheta_0\right] f(\vartheta_0, s)\, m_s\,\mathrm{d}s\,(1 + o(1))\\ &\quad + o\left(\varepsilon^2\right).\end{aligned}$$

Recall that $R_t = O(\varepsilon)$. Thus

$$\begin{aligned}\delta_m(t) &= \left[y_{\tau_*}\left(\vartheta^\star_{\tau_*,\varepsilon}\right) - y_{\tau_*}(\vartheta_0) + y_{\tau_*}(\vartheta_0) - m(\vartheta_0,\tau_*)\right]\Phi\left(\vartheta^\star_\varepsilon,0,t,X\right)(1+o(1))\\ &\quad+\int_{\tau_*}^t \Phi(\vartheta_0,s,t,X)\,\dot{\mathbf{A}}_*(\vartheta_0,s,X)^\top\left[\vartheta^\star_{s,\varepsilon}-\vartheta_0\right]m_s\,\mathrm{d}s\,(1+o(1)) + O\left(\varepsilon^2\right)\\ &\quad+\int_{\tau_*}^t \Phi(\vartheta_0,s,t,X)\,\dot{\mathbf{B}}_*(\vartheta_0,s,X)^\top\left[\vartheta^\star_{s,\varepsilon}-\vartheta_0\right]f(\vartheta_0,s)\,m_s\,\mathrm{d}s\,(1+o(1)).\end{aligned}$$

For the inner product we have the convergence

$$\lambda_{t,\varepsilon} = \varepsilon^{-1}\dot{\mathbf{B}}_*(\vartheta_0,s,X)^\top\left[\vartheta^\star_{s,\varepsilon}-\vartheta_0\right] \Longrightarrow \lambda_t = \dot{\mathbf{B}}_*(\vartheta_0,s,x(\vartheta_0))^\top\,\eta_t(\vartheta_0)$$

and the convergence of the vectors

$$\left(\lambda_{t_1,\varepsilon},\ldots,\lambda_{t_K,\varepsilon}\right) \Longrightarrow \left(\lambda_{t_1},\ldots,\lambda_{t_K}\right) \tag{2.89}$$

(see (2.87)). It is possible to verify the estimate

$$\mathbf{E}_{\vartheta_0}\left|\lambda_{t,\varepsilon}-\lambda_{s,\varepsilon}\right|^2 \le C\left|t-s\right|.$$

This estimate and the convergence (2.89) by the Theorem 1.8 provide us the convergence of the integrals

$$\begin{aligned}&\varepsilon^{-1}\int_{\tau_*}^t \Phi(\vartheta_0,s,t,X)\,\dot{\mathbf{B}}_*(\vartheta_0,s,X)^\top\left[\vartheta^\star_{s,\varepsilon}-\vartheta_0\right]f(\vartheta_0,s)\,m_s\,\mathrm{d}s\\ &\qquad\Longrightarrow \int_{\tau_*}^t \Phi(\vartheta_0,s,t,x(\vartheta_0))\,\dot{\mathbf{B}}_*(\vartheta_0,s,x(\vartheta_0))^\top\,\eta_s(\vartheta_0)\,f(\vartheta_0,s)\,y_s(\vartheta_0)\,\mathrm{d}s.\end{aligned}$$

We have the convergence of the second integral too

$$\begin{aligned}&\varepsilon^{-1}\int_{\tau_*}^t \Phi(\vartheta_0,s,t,X)\,\dot{\mathbf{A}}_*(\vartheta_0,s,X)^\top\left[\vartheta^\star_{s,\varepsilon}-\vartheta_0\right]m^\star_s\,\mathrm{d}s\\ &\qquad\Longrightarrow \int_{\tau_*}^t \Phi(\vartheta_0,s,t,x(\vartheta_0))\,\dot{\mathbf{A}}_*(\vartheta_0,s,x(\vartheta_0))^\top\,\eta_s(\vartheta_0)\,y_s(\vartheta_0)\,\mathrm{d}s.\end{aligned}$$

Further,

$$\varepsilon^{-1}\left[y_{\tau_*}\left(\vartheta^\star_{\tau_*,\varepsilon}\right)-y_{\tau_*}(\vartheta_0)\right] \Longrightarrow y_{\tau_*}(\vartheta_0)\int_0^{\tau_*}\dot{a}(\vartheta_0,s,\tau_*)^\top\,\eta_{\tau_*}(\vartheta_0)\,\mathrm{d}s. \tag{2.90}$$

Notice that

$$\begin{aligned}\varepsilon^{-1}\left(m\left(\vartheta_0,t\right)-y_t\left(\vartheta_0\right)\right)&=a\left(\vartheta_0,t,x_t\right)\varepsilon^{-1}\left[m\left(\vartheta_0,t\right)-y_t\left(\vartheta_0\right)\right]\mathrm{d}t\\&\quad+\varepsilon^{-1}\left[a\left(\vartheta_0,t,X_t\right)-a\left(\vartheta_0,t,x_t\right)\right]m\left(\vartheta_0,t\right)\mathrm{d}t\\&\quad+\frac{f\left(\vartheta_0,t,X_t\right)\gamma_*\left(\vartheta_0,t,X\right)}{\sigma\left(t,X_t\right)}\,\mathrm{d}\bar W_t.\end{aligned}$$

Hence,

$$\begin{aligned}\frac{m\left(\vartheta_0,\tau_*\right)-y_{\tau_*}\left(\vartheta_0\right)}{\varepsilon}&=\int_0^{\tau_*}\phi\left(\vartheta_0,s,\tau_*,x\right)a_x'\left(\vartheta_0,s,x_s\right)x_s^{(1)}\left(\vartheta_0\right)\,y_s\left(\vartheta_0\right)\mathrm{d}s\\&\quad+\int_0^{\tau_*}\phi\left(\vartheta_0,s,\tau_*,x\right)\frac{f\left(\vartheta_0,s,x_s\right)\gamma_s^*\left(\vartheta_0,\vartheta_0\right)}{\sigma\left(s,x_s\right)}\,\mathrm{d}\bar W_s+o\left(1\right)\\&\Longrightarrow\int_0^{\tau_*}\phi\left(\vartheta_0,s,\tau_*,x\right)a_x'\left(\vartheta_0,s,x_s\right)x_s^{(1)}\left(\vartheta_0\right)\,y_s\left(\vartheta_0\right)\mathrm{d}s\\&\quad+\int_0^{\tau_*}\phi\left(\vartheta_0,s,\tau_*,x\right)\frac{f\left(\vartheta_0,s,x_s\right)\gamma_s^*\left(\vartheta_0,\vartheta_0\right)}{\sigma\left(s,x_s\right)}\,\mathrm{d}w\left(s\right)\\&\equiv\xi_{\tau_*}\left(\vartheta_0\right)\end{aligned}$$

where

$$\phi\left(\vartheta_0,s,\tau_*,x\right)=\exp\left(\int_s^{\tau_*}a\left(\vartheta_0,v,x_v\left(\vartheta_0\right)\right)\mathrm{d}v\right).$$

Finally, we obtain the error of approximation (equality in distribution)

$$\begin{aligned}\frac{m_{t,\varepsilon}^{\star}-m\left(\vartheta_0,t\right)}{\varepsilon}&=y_{\tau_*}\left(\vartheta_0\right)\int_0^{\tau_*}\dot{\mathrm{a}}\left(\vartheta_0,s,\tau_*\right)^{\top}\eta_{\tau_*}\left(\vartheta_0\right)\mathrm{d}s+\xi_{\tau_*}\left(\vartheta_0\right)\\&\quad+\int_{\tau_*}^{t}\Phi\left(\vartheta_0,s,t,x\left(\vartheta_0\right)\right)\dot{\mathrm{A}}_*\left(\vartheta_0,s,x\left(\vartheta_0\right)\right)^{\top}\eta_s\left(\vartheta_0\right)y_s\left(\vartheta_0\right)\mathrm{d}s\\&\quad+\int_{\tau_*}^{t}\Phi\left(\vartheta_0,s,t,x\left(\vartheta_0\right)\right)\dot{\mathrm{B}}_*\left(\vartheta_0,s,x\left(\vartheta_0\right)\right)^{\top}\eta_s\left(\vartheta_0\right)f\left(\vartheta_0,s\right)y_s\left(\vartheta_0\right)\mathrm{d}s+o\left(1\right)\\&\Longrightarrow\pi_t\left(\vartheta_0\right).\end{aligned}$$

Therefore, the following result was proved.

Proposition 2.4 *Let the conditions $\mathscr{D}$, $\check{\mathscr{D}}$ and $\mathscr{D}^{\star}$ hold. Then the error of approximation of the adaptive filter* (2.88) *is asymptotically normal*

$$\frac{m_t^{\star}-m\left(\vartheta_0,t\right)}{\varepsilon}\Longrightarrow\pi_t\left(\vartheta_0\right),\qquad\tau_*\leq t\leq T.$$

Remark 2.13 If the unknown parameter is one-dimensional, then it is possible to consider the vanishing learning interval and to discuss the construction of the asymptotically efficient adaptive filter. The proofs will follow approximately the same lines, but there will be a problem with the definition of the initial value $m^{\star}_{\tau_{\varepsilon},\varepsilon}$.

2.4 Extended Kalman Filters

2.4.1 EKF-1

Model of Observations

Let us consider a version of the the extended Kalman filter following [168]. The nonlinear partially observed system is

$$\mathrm{d}X_t = f\left(\vartheta, t, Y_t\right)\mathrm{d}t + \varepsilon\sigma\left(t\right)\mathrm{d}W_t, \qquad X_0 = 0, \quad 0 \le t \le T, \tag{2.91}$$

$$\mathrm{d}Y_t = a\left(\vartheta, t, Y_t\right)\mathrm{d}t + \varepsilon b\left(t\right)\mathrm{d}V_t, \qquad Y_0 = y_0, \tag{2.92}$$

where $W_t, V_t, 0 \le t \le T$ are independent Wiener processes, the observations are $X^T = (X_t, 0 \le t \le T)$. The unknown parameter $\vartheta \in \Theta = (\alpha, \beta)$ (α and β are finite). We have to find an approximation of the unobserved component Y_t and to describe its properties in the asymptotic of *small noise* $\varepsilon \to 0$.

Introduce the *Conditions* $\mathscr{E}$.

$\mathscr{E}_1$. *The functions* $f\left(\vartheta, t, y\right), a\left(\vartheta, t, y\right); \vartheta \in \Theta, t \in [0, T], y \in \mathcal{R}$ *have two continuous bounded derivatives w.r.t.* y.

$\mathscr{E}_2$. *The functions* $f\left(\vartheta, t, y\right), a\left(\vartheta, t, y\right); \vartheta \in \Theta, t \in [0, T], y \in \mathcal{R}$ *have two continuous derivatives w.r.t.* ϑ *with polynomial majorants in* y.

$\mathscr{E}_3$. *The functions* $\sigma\left(\cdot\right)$ *and* $b\left(\cdot\right)$ *are bounded on* $[0, T]$ *and* $\sigma\left(t\right)^2 > c$, *where* $c > 0$.

The limit ($\varepsilon = 0$) system is, of course,

$$\frac{\mathrm{d}x_t\left(\vartheta\right)}{\mathrm{d}t} = f\left(\vartheta, t, y_t\left(\vartheta\right)\right), \qquad x_0\left(\vartheta\right) = 0, \qquad 0 \le t \le T,$$

$$\frac{\mathrm{d}y_t\left(\vartheta\right)}{\mathrm{d}t} = a\left(\vartheta, t, y_t\left(\vartheta\right)\right), \qquad y_0\left(\vartheta\right) = y_0.$$

To verify this convergence, we note that for the function $\delta_y\left(t\right) = Y_t - y_t\left(\vartheta\right)$ we have the estimate

$$\begin{aligned}\left|\delta_y\left(t\right)\right| &\le \int_0^t \left|a\left(\vartheta, s, Y_s\right) - a\left(\vartheta, s, y_s\left(\vartheta\right)\right)\right| \mathrm{d}s + \varepsilon\left|\int_0^t b\left(s\right)\mathrm{d}V_s\right| \\ &\le C\int_0^t \left|\delta_y\left(s\right)\right| \mathrm{d}s + \varepsilon \sup_{0\le s\le t}\left|\int_0^s b\left(r\right)\mathrm{d}V_r\right|.\end{aligned}$$

Thus, by Grönwall's lemma 1.5

$$\sup_{0\le t\le T}\left|\delta_y\left(t\right)\right|\le C\,\varepsilon\,\sup_{0\le t\le T}\left|\int_0^t b\left(s\right){\rm d}V_s\right|\longrightarrow 0.$$

Further,

$$\begin{aligned}\left|X_t-x_t\left(\vartheta\right)\right|&\le\int_0^t\left|f\left(\vartheta,s,Y_s\right)-f\left(\vartheta,s,y_s\left(\vartheta\right)\right)\right|{\rm d}s+\varepsilon\left|\int_0^t\sigma\left(s\right){\rm d}W_s\right|\\&\le C\int_0^t\left|\delta_y\left(s\right)\right|{\rm d}s+\varepsilon\left|\int_0^t\sigma\left(s\right){\rm d}W_s\right|\\&\le C\,\varepsilon\,\sup_{0\le t\le T}\left|\int_0^t b\left(s\right){\rm d}V_s\right|+\varepsilon\left|\int_0^t\sigma\left(s\right){\rm d}W_s\right|\longrightarrow 0.\end{aligned}$$

Recall that the conditional expectation $m\left(\vartheta,t\right)=\mathbf{E}_\vartheta\left(Y_t|\mathfrak{F}_t^X\right)$ is the best in the mean-square approximation of Y_t, but we have no explicit expressions for $m\left(\vartheta,t\right)$ and for the corresponding error of estimation $\mathbf{E}_\vartheta\left(Y_t-m\left(\vartheta,t\right)\right)^2$.

Below, we propose heuristic approximations for these quantities with the help of two equations for some random functions $m_t\left(\vartheta\right)=m_t$ and $\gamma\left(\vartheta,t,m^t\right)$, which are considered as approximations of $\mathbf{E}_\vartheta\left(Y_t|\mathfrak{F}_t^X\right)$ and $\mathbf{E}_\vartheta\left(Y_t-m\left(\vartheta,t\right)\right)^2$, respectively, and which coincide with the solutions of the equations of Kalman–Bucy in the linear case ($f\left(\vartheta,t,y\right)=f\left(\vartheta,t\right)y$, $a\left(\vartheta,t,y\right)=a\left(\vartheta,t\right)y$).

Introduce these stochastic processes as solutions of the equations

$${\rm d}m_t=a\left(\vartheta,t,m_t\right){\rm d}t+\frac{f_y'\left(\vartheta,t,m_t\right)\gamma\left(\vartheta,t,m^t\right)}{\varepsilon^2\sigma\left(t\right)^2}\left[{\rm d}X_t-f\left(\vartheta,t,m_t\right){\rm d}t\right],\tag{2.93}$$

$$\frac{{\rm d}\gamma\left(\vartheta,t,m^t\right)}{{\rm d}t}=2a_y'\left(\vartheta,t,m_t\right)\gamma\left(\vartheta,t,m^t\right)-\frac{f_y'\left(\vartheta,t,m_t\right)^2\gamma\left(\vartheta,t,m^t\right)^2}{\varepsilon^2\sigma\left(t\right)^2}+\varepsilon^2b\left(t\right)^2,\tag{2.94}$$

sbject to initial values $m_0=y_0$ and $\gamma\left(\vartheta,0,m_0\left(\vartheta\right)\right)=0$. The solution $\gamma\left(\vartheta,t,m^t\right)$ of the Eq. (2.94) depends on the trajectory $m^t=\left(m_s,0\le s\le t\right)$. This couple of equations with known ϑ is called *extended Kalman filter* (EKF-1). Note that in the linear case: $f\left(\vartheta,t,y\right)=f\left(\vartheta,t\right)y$, $a\left(\vartheta,t,y\right)=a\left(\vartheta,t\right)y$ these equations coincide with the equations of the usual Kalman–Bucy filter (2.3) and (2.4).

Once more we denote $\gamma_*\left(\vartheta,t,m\right)=\varepsilon^{-2}\gamma\left(\vartheta,t,m^t\right)$ and replace the Eqs. (2.93) and (2.94) by

$$\mathrm{d}m_t = a(\vartheta, t, m_t)\,\mathrm{d}t + \frac{f'_y(\vartheta, t, m_t)\,\gamma_*(\vartheta, t, m)}{\sigma(t)^2}\left[\mathrm{d}X_t - f(\vartheta, t, m_t)\,\mathrm{d}t\right], \tag{2.95}$$

$$\frac{\mathrm{d}\gamma_*(\vartheta, t, m)}{\mathrm{d}t} = 2a'_y(\vartheta, t, m_t)\,\gamma_*(\vartheta, t, m) - \frac{f'_y(\vartheta, t, m_t)^2\,\gamma_*(\vartheta, t, m)^2}{\sigma(t)^2} + b(t)^2, \tag{2.96}$$

with the same initial values.

Substitution of the observations in (2.95) with the true value $\vartheta = \vartheta_0$ yields the equation

$$\begin{aligned}\mathrm{d}m_t &= a(\vartheta, t, m_t)\,\mathrm{d}t + \varepsilon\frac{f'_y(\vartheta, t, m_t)\,\gamma_*(\vartheta, t, m)}{\sigma(t)}\mathrm{d}W_t\\ &\quad + f'_y(\vartheta, t, m_t)\,\gamma_*(\vartheta, t, m)\sigma(t)^{-2}\left[f(\vartheta_0, t, Y_t) - f(\vartheta, t, m_t)\right]\mathrm{d}t.\end{aligned}$$

If $\varepsilon = 0$, then the solution of this equation and of the Eq. (2.96) are the deterministic functions $y_t = y_t(\vartheta, \vartheta_0),\ 0 \leq t \leq T$ and $\gamma_t = \gamma_*\left(\vartheta, t, y^t\right),\ 0 \leq t \leq T$

$$\frac{\mathrm{d}y_t}{\mathrm{d}t} = a(\vartheta, t, y_t) + f'_y(\vartheta, t, y_t)\,\gamma_*(\vartheta, t, y)\sigma(t)^{-2}\left[f\left(\vartheta_0, t, y_t^*\right) - f(\vartheta, t, y_t)\right],\quad y_0,$$

$$\frac{\mathrm{d}\gamma_*\left(\vartheta, t, y^t\right)}{\mathrm{d}t} = 2a'_y(\vartheta, t, y_t)\,\gamma_*\left(\vartheta, t, y^t\right) - \frac{f'_y(\vartheta, t, y_t)^2\,\gamma_*\left(\vartheta, t, y^t\right)^2}{\sigma(t)^2} + b(t)^2,$$

Here $y_t^* = y_t(\vartheta_0) = Y_t|_{\varepsilon=0}$. Remark that the function $\gamma_*\left(\vartheta, t, y^t\right)$ depends on ϑ_0.

We need the derivatives $\dot{y}_t = \dot{y}_t(\vartheta, \vartheta_0) = \partial y_t(\vartheta, \vartheta_0)/\partial\vartheta$ and $\dot{\gamma}_t = \partial\gamma_*\left(\vartheta, t, y^t\right)/\partial\vartheta$, which satisfy the equations

$$\begin{aligned}\frac{\mathrm{d}\dot{y}_t}{\mathrm{d}t} &= \left[\dot{f}'_{\vartheta y}(\vartheta, t, y_t)\,\gamma_t + f''_{yy}(\vartheta, t, y_t)\,\gamma_t\dot{y}_t + f'_y(\vartheta, t, y_t)\,\dot{\gamma}_t\right]\\ &\quad\times\frac{\left[f\left(\vartheta_0, t, y_t^*\right) - f(\vartheta, t, y_t)\right]}{\sigma(t)^2}\\ &\quad + \dot{a}_\vartheta(\vartheta, t, y_t) + a'_y(\vartheta, t, y_t)\,\dot{y}_t\\ &\quad - \frac{f'_y(\vartheta, t, y_t)\,\gamma_t}{\sigma(t)^2}\left[\dot{f}_\vartheta(\vartheta, t, y_t) + f'_y(\vartheta, t, y_t)\,\dot{y}_t\right],\\ \frac{\mathrm{d}\dot{\gamma}_t}{\mathrm{d}t} &= 2\dot{a}'_{\vartheta y}(\vartheta, t, y_t)\,\gamma_t + 2a''_{yy}(\vartheta, t, y_t)\,\dot{y}_t\gamma_t + 2a'_y(\vartheta, t, y_t)\,\dot{\gamma}_t\\ &\quad - 2f'_y(\vartheta, t, y_t)^2\,\gamma_t\dot{\gamma}_t\sigma(t)^{-2}\\ &\quad - 2f'_y(\vartheta, t, y_t)\left[\dot{f}'_{\vartheta y}(\vartheta, t, y_t) + f''_{yy}(\vartheta, t, y_t)\,\dot{y}_t\right]\gamma_t^2\sigma(t)^{-2}.\end{aligned} \tag{2.97}$$

These are linear equations w.r.t. $\dot{y}_t$ and $\dot{\gamma}_t$ with initial values $\dot{y}_0(\vartheta,\vartheta_0)=0$, $\dot{\gamma}_0=0$ and the solutions can be written explicitly.

Introduce as well the notation for the EKF-1 with known ϑ_0, i.e. for $m_t^* = m_t(\vartheta_0)$, $\gamma_t^* = \gamma_*(\vartheta_0, t, m^*)$:

$$\mathrm{d}m_t^* = a\left(\vartheta_0, t, m_t^*\right)\mathrm{d}t + \frac{f_y'\left(\vartheta_0, t, m_t^*\right)\gamma_t^*}{\sigma(t)^2}\left[\mathrm{d}X_t - f\left(\vartheta_0, t, m_t^*\right)\mathrm{d}t\right], \qquad m_0^* = y_0,$$

$$\frac{\mathrm{d}\gamma_t^*}{\mathrm{d}t} = 2a_y'\left(\vartheta_0, t, m_t^*\right)\gamma_t^* - \frac{f_y'\left(\vartheta_0, t, m_t^*\right)^2\gamma_t^{*2}}{\sigma(t)^2} + b(t)^2, \qquad \gamma_0^* = 0.$$

Notice that $m_t^*\big|_{\varepsilon=0} = y_t(\vartheta_0,\vartheta_0) = y_t(\vartheta_0) = y_t^*$. Indeed, denote $y_t^\circ = y_t(\vartheta_0,\vartheta_0)$, then for the difference $\delta_y(t) = y_t^\circ - y_t^*$ we have the estimates

$$\begin{aligned}
\left|\delta_y(t)\right| &\le \int_0^t \left|a\left(\vartheta_0, s, y_s^\circ\right) - a\left(\vartheta_0, s, y_s^*\right)\right| \mathrm{d}s \\
&\quad + \int_0^t \left|f_y'\left(\vartheta_0, s, y_s^\circ\right)\gamma^\circ(\vartheta_0, s, y^\circ)\sigma(s)^{-2}\left[f\left(\vartheta_0, s, y_s^*\right) - f\left(\vartheta_0, s, y_s^\circ\right)\right]\right| \mathrm{d}s \\
&= \int_0^t \left|a_y'\left(\vartheta_0, s, \bar{y}_s\right)\right| \left|\delta_y(s)\right| \mathrm{d}s \\
&\quad + \int_0^t \left|f_y'\left(\vartheta_0, s, y_s^\circ\right)\gamma^\circ(\vartheta_0, s, y^\circ)\sigma(s)^{-2} f_y'\left(\vartheta_0, s, \bar{y}_s\right)\right| \left|\delta_y(s)\right| \mathrm{d}s.
\end{aligned}$$

Hence,

$$\left|\delta_y(t)\right| \le C\int_0^t \left|\delta_y(s)\right| \mathrm{d}s$$

and by Grönwall's lemma 1.5, it follows that $y_t(\vartheta_0,\vartheta_0) - y_t^* \equiv 0$.

Using elementary expansions and once more the Grönwall lemma, we obtain the following limits in probability

$$\begin{aligned}
\varepsilon^{-1}\left(X_t - x_t(\vartheta_0)\right) &\longrightarrow \int_0^t \sigma(s)\,\mathrm{d}W_s \\
&\quad + \int_0^t f_y'\left(\vartheta_0, s, y_s(\vartheta_0)\right)\int_0^s \phi(\vartheta_0, r, s)\, b(r)\,\mathrm{d}V_r\,\mathrm{d}s \equiv x_t^{(1)}(\vartheta_0), \\
\varepsilon^{-1}\left(m_t(\vartheta_0) - Y_t\right) &\longrightarrow \int_0^t \Phi(\vartheta_0, s, t)\frac{f_y'\left(\vartheta_0, s, y_s(\vartheta_0)\right)\gamma_s^*}{\sigma(s)}\,\mathrm{d}W_s \\
&\quad - \int_0^t \Phi(\vartheta_0, s, t)\, b(s)\,\mathrm{d}V_s \equiv \mu_t(\vartheta_0),
\end{aligned} \tag{2.98}$$

$$\varepsilon^{-1}\left(Y_t - y_t(\vartheta_0)\right) \longrightarrow \int_0^t \phi(\vartheta_0, s, t)\, b(s)\, \mathrm{d}V_s \equiv y_t^{(1)}(\vartheta_0), \tag{2.99}$$

$$\varepsilon^{-1}\left(m_t(\vartheta_0) - y_t(\vartheta_0)\right) \longrightarrow \mu_t(\vartheta_0) + y_t^{(1)}(\vartheta_0), \tag{2.100}$$

where

$$\phi(\vartheta_0, r, s) = \exp\left(\int_r^s a_y'(\vartheta_0, v, y_v(\vartheta_0))\, \mathrm{d}v\right),$$

$$\Phi(\vartheta_0, s, t) = \exp\left(\int_s^t \left[a_y'(\vartheta_0, v, y_v(\vartheta_0)) - \frac{f_y'(\vartheta_0, v, y_v(\vartheta_0))^2\, \gamma_v^*}{\sigma(v)^2}\right] \mathrm{d}v\right).$$

Remark that $\mu_t^{(1)}(\vartheta_0)$, $y_t^{(1)}(\vartheta_0)$ and $x_t^{(1)}(\vartheta_0)$ are zero mean correlated Gaussian processes, and the convergence (2.100) is obtained directly from (2.98) and (2.99).

The proposed below approximation is motivated by such approximation in the linear case, i.e., we start with the construction of the preliminary estimator by the observations on the learning interval, describe the properties of the pseudo-MLE and One-step pseudo-MLE process and then we use the last one for the construction of the adaptive filter. It is supposed that if the model is close to the linear model, then such construction will provide a good approximation of the optimal filter.

MDE

As preliminary estimator, we take the MDE for this model of observations. The learning interval is $[0, \tau]$, $\tau \in (0, T)$. We have

$$x_t(\vartheta) = \int_0^t f(\vartheta, s, y_s(\vartheta))\, \mathrm{d}s, \qquad 0 \le t \le \tau$$

and recall that for the observed process, we have the equality

$$X_t|_{\varepsilon=0} = x_t(\vartheta_0) = \int_0^t f\left(\vartheta_0, s, y_s^*\right) \mathrm{d}s.$$

The MDE $\check\vartheta_{\tau,\varepsilon}$ can be defined by the relation

$$\check\vartheta_{\tau,\varepsilon} = \arg\inf_{\vartheta\in\Theta} \int_0^\tau \left[X_t - x_t(\vartheta)\right]^2 \mathrm{d}t. \tag{2.101}$$

Introduce the conditions $\mathscr{E}_{4-5}$.

$\mathscr{E}_4$. *For any* $\nu > 0$

$$\inf_{\vartheta_0\in\Theta}\inf_{|\vartheta-\vartheta_0|\geq\nu}\int_0^\tau\left[x_t\left(\vartheta_0\right)-x_t\left(\vartheta\right)\right]^2 \mathrm{d}t>0. \tag{2.102}$$

$\mathscr{E}_5$.

$$\inf_{\vartheta\in\Theta}T\left(\vartheta\right)>0,\qquad\qquad T\left(\vartheta\right)=\int_0^\tau\dot x_t\left(\vartheta\right)^2\mathrm{d}t.$$

Here

$$\dot x_t\left(\vartheta\right)=\int_0^t\left[\dot f\left(\vartheta,s,y_s\left(\vartheta\right)\right)+f_y'\left(\vartheta,s,y_s\left(\vartheta\right)\right)\dot y_s\left(\vartheta\right)\right]\mathrm{d}s.$$

Remark that if the condition (2.102) is not fulfilled then there exists $\vartheta_*\neq\vartheta_0$ such that $x_t\left(\vartheta_0\right)\equiv x_t\left(\vartheta_*\right), 0\leq t\leq\tau$ and this means that $f\left(\vartheta_*,t,y_t\left(\vartheta_*\right)\right)\equiv f\left(\vartheta_0,t,y_t\left(\vartheta_0\right)\right)$.

Proposition 2.5 *Let the conditions $\mathscr{E}_{1-5}$ hold. Then the MDE $\check\vartheta_{\tau,\varepsilon}$ is consistent, asymptotically normal*

$$\varepsilon^{-1}\left(\check\vartheta_{\tau,\varepsilon}-\vartheta_0\right)\Longrightarrow\rho\left(\vartheta_0\right)\equiv T\left(\vartheta_0\right)^{-1}\int_0^\tau\dot x_t\left(\vartheta_0\right)x_t^{(1)}\left(\vartheta_0\right)\mathrm{d}t\sim\mathcal{N}\left(0,D\left(\vartheta_0\right)\right)$$

with the corresponding limit variance $D\left(\vartheta_0\right)$ and the polynomial moments converge.

Proof The proof of the consistency is similar to (2.14) and for the proofs of asymptotic normality and convergence of moments follow the same lines as it was done in the proof of the Theorem 2.2. □

Example 2.5 Suppose that $f\left(\vartheta,t,y\right)=\vartheta f\left(t,y\right)$ and $a\left(\vartheta,t,y\right)=a\left(t,y\right)$. Then

$$\check\vartheta_{\tau,\varepsilon}=\left(\int_0^\tau F\left(t,y\right)^2\mathrm{d}t\right)^{-1}\int_0^\tau X_t\,F\left(t,y\right)\,\mathrm{d}t$$

where

$$F\left(t,y\right)=\int_0^t f\left(s,y_s\right)\mathrm{d}s.$$

The estimator $\check\vartheta_{\tau,\varepsilon}$ under corresponding regularity conditions is consistent and asymptotically normal.

Pseudo-MLE

It is possible to introduce and study the pseudo-MLE $\hat\vartheta_\varepsilon$ with the help of the pseudo-LR (p-LR)

$$L\left(\vartheta, X^T\right) = \exp\left(\int_0^T \frac{f\left(\vartheta, t, m_t\left(\vartheta\right)\right)}{\varepsilon^2\sigma\left(t\right)^2}\mathrm{d}X_t - \int_0^T \frac{f\left(\vartheta, t, m_t\left(\vartheta\right)\right)^2}{2\varepsilon^2\sigma\left(t\right)^2}\mathrm{d}t\right), \quad \vartheta\in\Theta$$

using the definition

$$L(\hat\vartheta_\varepsilon, X^T) = \sup_{\vartheta\in\Theta} L\left(\vartheta, X^T\right).$$

We have the following limit:

$$-2\varepsilon^2 \ln L\left(\vartheta, X^T\right) \longrightarrow \int_0^T \frac{\left[f\left(\vartheta, t, y_t\left(\vartheta, \vartheta_0\right)\right) - f\left(\vartheta_0, t, y_t\left(\vartheta_0\right)\right)\right]^2}{\sigma\left(t\right)^2}\mathrm{d}t + \int_0^T \frac{f\left(\vartheta_0, t, y_t\left(\vartheta_0\right)\right)^2}{\sigma\left(t\right)^2}\mathrm{d}t.$$

It can be shown that this convergence is uniform on $\vartheta\in\Theta$. Therefore,

$$\hat\vartheta_\varepsilon \quad\longrightarrow\quad \hat\vartheta = \arg\inf_{\vartheta\in\Theta}\int_0^T \frac{\left[f\left(\vartheta, t, y_t\left(\vartheta, \vartheta_0\right)\right) - f\left(\vartheta_0, t, y_t\left(\vartheta_0\right)\right)\right]^2}{\sigma\left(t\right)^2}\mathrm{d}t.$$

The identifiability condition is:
$\mathscr{E}_6$. *For any* $\nu>0$

$$\inf_{\vartheta_0\in\bar\Theta}\ \inf_{|\vartheta-\vartheta_0|\geq\nu}\int_0^T \frac{\left[f\left(\vartheta, t, y_t\left(\vartheta, \vartheta_0\right)\right) - f\left(\vartheta_0, t, y_t\left(\vartheta_0\right)\right)\right]^2}{\sigma\left(t\right)^2}\mathrm{d}t > 0.$$

By this condition $\hat\vartheta=\vartheta_0$, i.e., the pseudo-MLE is consistent.
Let us describe the normalized p-LR (below $\vartheta_u=\vartheta_0+\varepsilon u$)

$$Z_\varepsilon\left(\vartheta_0, u\right) = \frac{L\left(\vartheta_u, X^T\right)}{L\left(\vartheta_0, X^T\right)}, \qquad u\in\mathbb{U}_\varepsilon = \left(\frac{\alpha-\vartheta_0}{\varepsilon}, \frac{\beta-\vartheta_0}{\varepsilon}\right).$$

It has the following form:

$$\ln Z_\varepsilon\left(\vartheta_0, u\right) = \int_0^T \frac{f\left(\vartheta_u, t, m_t\left(\vartheta_u\right)\right) - f\left(\vartheta_0, t, m_t\left(\vartheta_0\right)\right)}{\varepsilon\sigma\left(t\right)}\mathrm{d}W_t - \int_0^T \frac{\left[f\left(\vartheta_u, t, m_t\left(\vartheta_u\right)\right) - f\left(\vartheta_0, t, m_t\left(\vartheta_0\right)\right)\right]^2}{2\varepsilon^2\sigma\left(t\right)^2}\mathrm{d}t$$

$$
-\int_0^T \frac{[f(\vartheta_0,t,m_t(\vartheta_0))-f(\vartheta_0,t,Y_t)][f(\vartheta_u,t,m_t(\vartheta_u))-f(\vartheta_0,t,m_t(\vartheta_0))]}{\varepsilon^2\sigma(t)^2}\,dt
$$

$$
=\int_0^T \frac{\delta_f(t,m,u)}{\varepsilon\sigma(t)}\mathrm{d}W_t-\int_0^T \frac{\delta_f(t,m,u)^2}{2\varepsilon^2\sigma(t)^2}\mathrm{d}t-\int_0^T \frac{\delta_f(t,m,u)\,\delta_f(t,Y)}{\varepsilon^2\sigma(t)^2}\mathrm{d}t,
$$

where we denoted $\delta_f(t,m,u) = f(\vartheta_u,t,m_t(\vartheta_u)) - f(\vartheta_0,t,m_t(\vartheta_0))$ and $\delta_f(t,Y)=f(\vartheta_0,t,Y_t)-f(\vartheta_0,t,m_t(\vartheta_0))$. We have the expansions

$$
\begin{aligned}
&f(\vartheta_u,t,m_t(\vartheta_u))-f(\vartheta_0,t,m_t(\vartheta_0))=\varepsilon u\,\dot S_t(\vartheta_0,y)\,(1+o(1)),\\
&[f(\vartheta_u,t,m_t(\vartheta_u))-f(\vartheta_0,t,m_t(\vartheta_0))][f(\vartheta_0,t,m_t(\vartheta_0))-f(\vartheta_0,t,Y_t)]\\
&\qquad=\varepsilon u\,\dot S_t(\vartheta_0,y)\,f_y'(\vartheta_0,t,y)\,(m_t(\vartheta_0)-Y_t)\,(1+o(1))\\
&\qquad=\varepsilon^2 u\,\dot S_t(\vartheta_0,y)\,f_y'(\vartheta_0,t,y)\,\mu_t(\vartheta_0)\,(1+o(1)).
\end{aligned}
$$

Here

$$
\dot S_t(\vartheta_0,y)=\dot f_\vartheta(\vartheta,t,y_t(\vartheta_0,\vartheta_0))\,|_{\vartheta=\vartheta_0}+f_y'(\vartheta_0,t,y_t(\vartheta_0))\,\dot y_t(\vartheta,\vartheta_0)\Big|_{\vartheta=\vartheta_0}.
$$

Therefore,

$$
\begin{aligned}
\ln Z_\varepsilon(\vartheta_0,u)&=u\int_0^T\frac{\dot S_t(\vartheta_0,y)}{\sigma(t)}\,\mathrm{d}W_t-\frac{u^2}{2}\int_0^T\frac{\dot S_t(\vartheta_0,y)^2}{\sigma(t)^2}\,\mathrm{d}t\\
&\quad-u\int_0^T\frac{\dot S_t(\vartheta_0,y)\,f_y'(\vartheta_0,t,y_t(\vartheta_0))\,\mu_t(\vartheta_0)}{\sigma(t)^2}\,\mathrm{d}t+o(1)\\
&=u\Delta(\vartheta_0)-\frac{u^2}{2}\mathrm{I}(\vartheta_0)-uK(\vartheta_0)+o(1),
\end{aligned}
$$

where we used (2.98) and denoted

$$
\Delta(\vartheta_0)=\int_0^T\frac{\dot S_s(\vartheta_0,y)}{\sigma(s)}\,\mathrm{d}W_s,\qquad \mathrm{I}(\vartheta_0)=\int_0^T\frac{\dot S_s(\vartheta_0,y)^2}{\sigma(s)^2}\,\mathrm{d}s,
$$

$$
K(\vartheta_0)=\int_0^T\frac{\dot S_s(\vartheta_0,y)\,f_y'(\vartheta_0,s,y_s(\vartheta_0))\,\mu_s(\vartheta_0)}{\sigma(s)^2}\,\mathrm{d}s.
$$

Introduce the limit p-LR

$$
Z(\vartheta_0,u)=\exp\left(u\Delta(\vartheta_0)-\frac{u^2}{2}\mathrm{I}(\vartheta_0)-uK(\vartheta_0)\right),\qquad u\in\mathcal{R}
$$

and the condition

$\mathscr{E}_7$. *The function* $\mathrm{I}(\vartheta_0)$, $\vartheta_0\in\Theta$ *is strictly positive*

$$\inf_{\vartheta_0\in\Theta}\mathrm{I}(\vartheta_0)>0.$$

Recall the relations (1.73)

$$\begin{aligned}\mathbf{P}_{\vartheta_0}\left(\varepsilon^{-1}(\hat\vartheta_\varepsilon-\vartheta_0)<x\right)&=\mathbf{P}_{\vartheta_0}\left(\sup_{\vartheta-\vartheta_0<\varepsilon x}L\left(\vartheta,X^T\right)>\sup_{\vartheta-\vartheta_0\geq\varepsilon x}L\left(\vartheta,X^T\right)\right)\\&=\mathbf{P}_{\vartheta_0}\left(\sup_{\vartheta-\vartheta_0<\varepsilon x}\frac{L\left(\vartheta,X^T\right)}{L\left(\vartheta_0,X^T\right)}>\sup_{\vartheta-\vartheta_0\geq\varepsilon x}\frac{L\left(\vartheta,X^T\right)}{L\left(\vartheta_0,X^T\right)}\right)\\&=\mathbf{P}_{\vartheta_0}\left(\sup_{u<x}\frac{L\left(\vartheta_0+\varepsilon u,X^T\right)}{L\left(\vartheta_0,X^T\right)}>\sup_{u\geq x}\frac{L\left(\vartheta_0+\varepsilon u,X^T\right)}{L\left(\vartheta_0,X^T\right)}\right)\\&=\mathbf{P}_{\vartheta_0}\left(\sup_{u<x}Z_\varepsilon\left(\vartheta_0,u\right)>\sup_{u\geq x}Z_\varepsilon\left(\vartheta_0,u\right)\right).\end{aligned}$$

If the weak convergence of the process $Z_\varepsilon(\vartheta_0,\cdot)$ to the process $Z(\cdot)$ was proved, then

$$\begin{aligned}\mathbf{P}_{\vartheta_0}\left(\varepsilon^{-1}(\hat\vartheta_\varepsilon-\vartheta_0)<x\right)&=\mathbf{P}_{\vartheta_0}\left(\sup_{u<x}Z_\varepsilon\left(\vartheta_0,u\right)>\sup_{u\geq x}Z_\varepsilon\left(\vartheta_0,u\right)\right)\\&\longrightarrow\mathbf{P}_{\vartheta_0}\left(\sup_{u<x}Z\left(\vartheta_0,u\right)>\sup_{u\geq x}Z\left(\vartheta_0,u\right)\right)=\mathbf{P}_{\vartheta_0}\left(\zeta\left(\vartheta_0\right)<x\right),\end{aligned}$$

where

$$\zeta\left(\vartheta_0\right)=\mathrm{I}\left(\vartheta_0\right)^{-1}\left[\Delta\left(\vartheta_0\right)-K\left(\vartheta_0\right)\right]$$

is the zero mean Gaussian random variable (see (1.74)).

We can suppose that the p-MLE $\hat\vartheta_\varepsilon$ is asymptotically normal

$$\varepsilon^{-1}\left(\hat\vartheta_\varepsilon-\vartheta_0\right)\Longrightarrow\zeta\left(\vartheta_0\right).$$

Remark 2.14 This convergence can be verified as follows. The limit likelihood ratio $Z(\vartheta_0,u)$, $u\in\mathcal{R}$ is typical for the misspecification problems. For example, in general $\mathbf{E}_{\vartheta_0}Z_\varepsilon(\vartheta_0,u)\neq 1$. Even it is possible for some $K(\vartheta_0)$ to have $\mathbf{E}_{\vartheta_0}Z(\vartheta_0,u)\to\infty$ as $|u|\to\infty$. To verify the conditions $\mathscr{IK}_1$ we have to modify the function $Z_\varepsilon(\cdot)$ as it was done, i.e., in the Section 5.1 in [118]. We introduce the function $Z_{\varepsilon,q}(\vartheta_0,u)=Z_\varepsilon(\vartheta_0,u)^q$, where $q>0$, then it is possible to take such sufficiently small $q>0$ that

$$\mathbf{E}_{\vartheta_0} Z_{\varepsilon,q}(\vartheta_0, u) \leq C,$$

where the constant $C > 0$ does not depend on u and ε.

Note that if $Z_\varepsilon(\vartheta_0, \cdot)^q \Longrightarrow Z_0(\vartheta_0, \cdot)^q$ in $(\mathcal{C}_0(\mathcal{R}), \mathfrak{B})$ then

$$\mathbf{P}_{\vartheta_0}\left(\varepsilon^{-1}(\hat{\vartheta}_\varepsilon - \vartheta_0) < x\right) = \mathbf{P}_{\vartheta_0}\left(\sup_{u<x} Z_\varepsilon(\vartheta_0, u)^q > \sup_{u\geq x} Z_\varepsilon(\vartheta_0, u)^q\right)$$

$$\longrightarrow \mathbf{P}_{\vartheta_0}\left(\sup_{u<x} Z(\vartheta_0, u)^q > \sup_{u\geq x} Z(\vartheta_0, u)^q\right) = \mathbf{P}_{\vartheta_0}(\zeta(\vartheta_0) < x).$$

Thus, it is sufficient to verify the conditions $\mathscr{I}\mathscr{K}_1$ for the process $Z_{\varepsilon,q}(\vartheta_0, u), u \in \mathbb{U}_\varepsilon$.

Let us consider the verification of the condition

$$\mathbf{E}_{\vartheta_0}\left|Z_{\varepsilon,q}(\vartheta_0, u_2) - Z_{\varepsilon,q}(\vartheta_0, u_1)\right|^p \leq C\left|u_2 - u_1\right|^p \tag{2.103}$$

for some $p > 1$. Denote

$$\begin{aligned} P_T = \frac{Z_{\varepsilon,q}(\vartheta_0, u_2)}{Z_{\varepsilon,q}(\vartheta_0, u_1)} = \exp\Bigg(& q\int_0^T \frac{\delta_f(s, m, u_2) - \delta_f(s, m, u_1)}{\varepsilon\sigma(s)}\,\mathrm{d}W_s \\ & - q\int_0^T \frac{\left[\delta_f(s, m, u_2) - \delta_f(s, m, u_1)\right]^2}{2\varepsilon^2\sigma(s)^2}\,\mathrm{d}s \\ & - q\int_0^T \frac{\left[\delta_f(s, m, u_2) - \delta_f(s, m, u_1)\right]\delta_f(s, Y)}{\varepsilon^2\sigma(s)^2}\,\mathrm{d}s\Bigg) \end{aligned}$$

Then the random process $P_t, 0 \leq t \leq T$ has the stochastic differential (see the proof of Lemma 1.1)

$$\begin{aligned} \mathrm{d}P_t = {} & P_t\left(q^2 - q\right)\frac{\left[\delta_f(t, m, u_2) - \delta_f(t, m, u_1)\right]^2}{2\varepsilon^2\sigma(t)^2}\,\mathrm{d}t \\ & - P_t q\frac{\left[\delta_f(t, m, u_2) - \delta_f(t, m, u_1)\right]\delta_f(t, Y)}{\varepsilon^2\sigma(t)^2}\,\mathrm{d}t \\ & + P_t q\frac{\left[\delta_f(t, m, u_2) - \delta_f(t, m, u_1)\right]}{\varepsilon\sigma(t)}\,\mathrm{d}W_t, \qquad P_0 = 1. \end{aligned}$$

Thus,

$$
\begin{aligned}
P_T = 1 &+ \left(q^2 - q\right) \int_0^T P_t \frac{\left[\delta_f(t,m,u_2) - \delta_f(t,m,u_1)\right]^2}{2\varepsilon^2\sigma(t)^2}\,\mathrm{d}t \\
&- q \int_0^T P_t \frac{\left[\delta_f(t,m,u_2) - \delta_f(t,m,u_1)\right]\delta_f(t,Y)}{\varepsilon^2\sigma(t)^2}\,\mathrm{d}t \\
&+ q \int_0^T P_t \frac{\left[\delta_f(t,m,u_2) - \delta_f(t,m,u_1)\right]}{\varepsilon\sigma(t)}\,\mathrm{d}W_t.
\end{aligned}
$$

Therefore, we can write

$$
\begin{aligned}
\mathbf{E}_{\vartheta_0}\left|Z_{\varepsilon,q}(\vartheta_0,u_2) - Z_{\varepsilon,q}(\vartheta_0,u_1)\right|^p &= \mathbf{E}_{\vartheta_0} Z_{\varepsilon,q}(\vartheta_0,u_1)^p \left|1 - P_T\right|^p \\
&\leq \left(\mathbf{E}_{\vartheta_0} Z_{\varepsilon,q}(\vartheta_0,u_1)^{2p}\, \mathbf{E}_{\vartheta_0}\left|1 - P_T\right|^{2p}\right)^{1/2} \\
&\leq C\left(\mathbf{E}_{\vartheta_0}\left|1 - P_T\right|^{2p}\right)^{1/2},
\end{aligned}
$$

with some constant $C > 0$, because for any $p > 0$ and safficiently small $q > 0$ we have $\mathbf{E}_{\vartheta_0} Z_{\varepsilon,q}(\vartheta_0,u_1)^{2p} < C$.

Now (2.103) follows from the estimates like

$$
\begin{aligned}
&\mathbf{E}_{\vartheta_0}\left|\int_0^T P_t \frac{\left[\delta_f(t,m,u_2) - \delta_f(t,m,u_1)\right]\delta_f(t,Y)}{\varepsilon^2\sigma(t)^2}\,\mathrm{d}t\right|^p \\
&\leq T^{p-1}\int_0^T \mathbf{E}_{\vartheta_0} P_t^p \frac{\left|\delta_f(t,m,u_2) - \delta_f(t,m,u_1)\right|^p \left|\delta_f(t,Y)\right|^p}{\varepsilon^{2p}\sigma(t)^{2p}}\,\mathrm{d}t \\
&\leq C\frac{T^{p-1}}{\varepsilon^{2p}}\int_0^T \left(\mathbf{E}_{\vartheta_0} P_t^{2p}\, \mathbf{E}_{\vartheta_0}\left|\delta_f(t,m,u_2) - \delta_f(t,m,u_1)\right|^{2p}\left|\delta_f(t,Y)\right|^{2p}\right)^{1/2}\mathrm{d}t \\
&\leq C\frac{T^{p-1}}{\varepsilon^{2p}}\int_0^T \left(\mathbf{E}_{\vartheta_0}\left|\delta_f(t,m,u_2) - \delta_f(t,m,u_1)\right|^{4p}\, \mathbf{E}_{\vartheta_0}\left|\delta_f(t,Y)\right|^{4p}\right)^{1/4}\mathrm{d}t \\
&\leq C\left|u_2 - u_1\right|^p.
\end{aligned}
$$

To prove the estimate

$$
\mathbf{P}_{\vartheta_0}\left(Z_{\varepsilon,q}(\vartheta_0,u) > e^{-cu^2}\right) \leq \frac{C_N}{|u|^N}
$$

on the tails of $Z_{\varepsilon,q}(\cdot)$ we first verify with the help of conditions $\mathscr{E}_6$ and $\mathscr{E}_7$ the estimate

$$
\int_0^T \frac{\left[f(\vartheta_u,t,y_t(\vartheta_u,\vartheta_0)) - f(\vartheta_0,t,y_t(\vartheta_0))\right]^2}{2\varepsilon^2\sigma(t)^2}\,\mathrm{d}t \geq \kappa\, u^2
$$

and then follow the proof of Lemma 2.5.

One-Step MLE Process

Recall the notation $S_t(\vartheta, m) = f(\vartheta, t, m_t(\vartheta))$, $S_t(\vartheta, y) = f(\vartheta, t, y_t(\vartheta))$. Then the One-step pseudo-MLE process can be

$$\vartheta^\star_{t,\varepsilon} = \check{\vartheta}_{\tau,\varepsilon} + \mathrm{I}^t_\tau\left(\check{\vartheta}_{\tau,\varepsilon}\right)^{-1}\int_\tau^t \frac{\dot{S}_s\left(\check{\vartheta}_{\tau,\varepsilon}, y\right)}{\sigma(s)^2}\left[\mathrm{d}X_s - f\left(\check{\vartheta}_{\tau,\varepsilon}, s, m_s\left(\check{\vartheta}_{\tau,\varepsilon}\right)\right)\mathrm{d}s\right], \tag{2.104}$$

where

$$\dot{S}_s(\vartheta, y) = \dot{f}_\vartheta(\vartheta, s, y_s(\vartheta)) + f'_y(\vartheta, s, y_s(\vartheta))\,\dot{y}_s(\vartheta),$$

$$\mathrm{I}^t_\tau(\vartheta_0) = \int_\tau^t \frac{\dot{S}_s(\vartheta_0, y)^2}{\sigma(s)^2}\,\mathrm{d}s.$$

Denote

$$\Delta^t_\tau(\vartheta_0) = \int_\tau^t \frac{\dot{S}_s(\vartheta_0, y)}{\sigma(s)}\,\mathrm{d}W_s,$$

$$K^t_\tau(\vartheta_0) = \int_\tau^t \frac{\dot{S}_s(\vartheta_0, y)\, f'_y(\vartheta_0, s, y_s(\vartheta_0))\,\mu_s(\vartheta_0)}{\sigma(s)^2}\,\mathrm{d}s.$$

Using the limit (2.98), we obtain the following convergence

$$\frac{\vartheta^\star_{t,\varepsilon} - \vartheta_0}{\varepsilon} = \frac{\check{\vartheta}_{\tau,\varepsilon} - \vartheta_0}{\varepsilon} + \mathrm{I}^t_\tau\left(\check{\vartheta}_{\tau,\varepsilon}\right)^{-1}\int_\tau^t \frac{\dot{S}_s\left(\check{\vartheta}_{\tau,\varepsilon}, y\right)}{\sigma(s)}\mathrm{d}W_s$$

$$+\,\mathrm{I}^t_\tau\left(\check{\vartheta}_{\tau,\varepsilon}\right)^{-1}\int_\tau^t \frac{\dot{S}_s\left(\check{\vartheta}_{\tau,\varepsilon}, y\right)}{\varepsilon\sigma(s)^2}\left[f(\vartheta_0, s, Y_s) - f\left(\check{\vartheta}_{\tau,\varepsilon}, s, m_s(\check{\vartheta}_{\tau,\varepsilon})\right)\right]\mathrm{d}s$$

$$\Longrightarrow \mathrm{I}^t_\tau(\vartheta_0)^{-1}\left[\int_\tau^t \frac{\dot{S}_s(\vartheta_0, y)}{\sigma(s)}\mathrm{d}W_s - \int_\tau^t \frac{\dot{S}_s(\vartheta_0, y)\, f'_y(\vartheta_0, s, \vartheta_0)\,\mu_s(\vartheta_0)}{\sigma(s)^2}\mathrm{d}s\right]$$

$$= \mathrm{I}^t_\tau(\vartheta_0)^{-1}\left[\Delta^t_\tau(\vartheta_0) - K^t_\tau(\vartheta_0)\right] \equiv \zeta^t_\tau(\vartheta_0).$$

Of course, the One-step pseudo-MLE process is no more asymptotically efficient, but if the observation time is sufficiently long and the system is close to linear, then it is reasonable to suppose that the precision of approximation of the given below adaptive filter will be improved with time.

Adaptive EKF-1

It is possible to have different adaptive filters. The simplest can be, for example, almost deterministic. As the limit of Y_t is the function $y_t(\vartheta_0)$, it is possible to consider the functions

$$\check{y}_t = y_0 + \int_0^t a\left(\check{\vartheta}_{\tau,\varepsilon}, s, \check{y}_s\right) ds, \qquad \tau \leq t \leq T$$

and

$$y_t^\star = y_\tau\left(\check{\vartheta}_{\tau,\varepsilon}\right) + \int_\tau^t a\left(\vartheta_{s,\varepsilon}^\star, s, y_s^\star\right) ds, \qquad \tau \leq t \leq T$$

as approximations of $Y_t, 0 \leq t \leq T$. The error in the first case is

$$\begin{aligned} y_t\left(\check{\vartheta}_{\tau,\varepsilon}\right) - Y_t &= y_t\left(\check{\vartheta}_{\tau,\varepsilon}\right) - y_t(\vartheta_0) + y_t(\vartheta_0) - Y_t \\ &= \dot{y}_t(\vartheta_0)\left(\check{\vartheta}_{\tau,\varepsilon} - \vartheta_0\right) - \varepsilon y_t^{(1)}(\vartheta_0) + o(\varepsilon) \\ &= \varepsilon\left[\dot{y}_t(\vartheta_0)\,\rho_\tau(\vartheta_0) - y_t^{(1)}(\vartheta_0)\right](1 + o(1)). \end{aligned}$$

Here

$$\begin{aligned} \dot{y}_t(\vartheta) &= \int_0^t \left[\dot{a}_\vartheta(\vartheta, s, y_s(\vartheta)) + a_y'(\vartheta, s, y_s(\vartheta))\,\dot{y}_s(\vartheta)\right] ds \\ &= \int_0^t \phi(\vartheta, s, t)\,\dot{a}_\vartheta(\vartheta, s, y_s(\vartheta))\, ds, \end{aligned}$$

where

$$\phi(\vartheta, s, t) = \exp\left(\int_s^t a_y'(\vartheta, v, y_v(\vartheta))\, dv\right).$$

Therefore, we obtain

$$\varepsilon^{-1}\left(y_t\left(\check{\vartheta}_{\tau,\varepsilon}\right) - Y_t\right) \Longrightarrow \dot{y}_t(\vartheta_0)\,\rho_\tau(\vartheta_0) - y_t^{(1)}(\vartheta_0).$$

In the second case, the estimator is the process and the calculation of the error is different. We have

$$y_t^\star - Y_t = y_t^\star - y_t(\vartheta_0) + y_t(\vartheta_0) - Y_t = y_t^\star - y_t(\vartheta_0) - \varepsilon y_t^{(1)}(\vartheta_0)(1 + o(1)).$$

The quantity $\delta_y(t) = y_t^\star - y_t(\vartheta_0)$ is estimated as follows:

$$\begin{aligned}
\delta_y(t) &= \delta_y(\tau) + \int_\tau^t \left[a\left(\vartheta^\star_{s,\varepsilon}, s, y^\star_s\right) - a\left(\vartheta_0, s, y_s(\vartheta_0)\right)\right] \mathrm{d}s \\
&= \delta_y(\tau) + \int_\tau^t \dot a_\vartheta\left(\tilde\vartheta_{s,\varepsilon}, s, y^\star_s\right)\left(\vartheta^\star_{s,\varepsilon} - \vartheta_0\right) \mathrm{d}s + \int_\tau^t a'_y\left(\vartheta^\star_{s,\varepsilon}, s, \tilde y_s\right) \delta_y(s)\, \mathrm{d}s \\
&= \varepsilon \dot y_t(\vartheta_0)\, \rho_\tau(\vartheta_0)\, \phi(\vartheta_0, 0, t) + \varepsilon \int_\tau^t \phi(\vartheta_0, s, t)\, \dot a_\vartheta(\vartheta_0, s, y_s(\vartheta_0)) \zeta^s_\tau(\vartheta_0)\, \mathrm{d}s \\
&\quad + o(\varepsilon),
\end{aligned}$$

where $\delta_y(\tau) = y_\tau\left(\check\vartheta_{\tau,\varepsilon}\right) - y_\tau(\vartheta_0)$.

Hence,

$$\begin{aligned}
\varepsilon^{-1}\left(y^\star_t - Y_t\right) \longrightarrow\ & \dot y_t(\vartheta_0)\, \rho_\tau(\vartheta_0)\, \phi(\vartheta_0, 0, t) - y^{(1)}_t(\vartheta_0) \\
&+ \int_\tau^t \phi(\vartheta_0, s, t)\, \dot a_\vartheta(\vartheta_0, s, y_s(\vartheta_0)) \zeta^s_\tau(\vartheta_0)\, \mathrm{d}s.
\end{aligned}$$

We can suppose that the approximation based on the extended Kalman filter equations (2.93) and (2.94) is better than these two and propose the following recurrent adaptive filter

$$\mathrm{d}m^\star_t = a\left(\vartheta^\star_{t,\varepsilon}, t, m^\star_t\right) \mathrm{d}t + \frac{f'_y\left(\vartheta^\star_{t,\varepsilon}, t, m^\star_t\right) \gamma^\star_t}{\sigma(t)^2}\left[\mathrm{d}X_t - f\left(\vartheta^\star_{t,\varepsilon}, t, m^\star_t\right) \mathrm{d}t\right], \tag{2.105}$$

$$\frac{\mathrm{d}\gamma^\star_t}{\mathrm{d}t} = 2a'_y\left(\vartheta^\star_{t,\varepsilon}, t, m^\star_t\right) \gamma^\star_t - \frac{f'_y\left(\vartheta^\star_{t,\varepsilon}, t, m^\star_t\right)^2 \gamma^{\star 2}_t}{\sigma(t)^2} + b(t)^2, \qquad \tau < t \le T, \tag{2.106}$$

with the initial values $m^\star_\tau = y_\tau\left(\check\vartheta_{\tau,\varepsilon}\right), \gamma^\star_\tau = \gamma_*\left(\check\vartheta_{\tau,\varepsilon}, \tau, y_\tau\left(\check\vartheta_{\tau,\varepsilon}\right)\right)$.

The One-step pseudo-MLE process $\vartheta^\star_{t,\varepsilon}, \tau < t \le T$ we write in the recurrent form too (see (2.51)). Denote $L(\vartheta, \vartheta, t) = \sigma(t)^{-2} \dot S_t(\vartheta, y)^2$, then (2.104) we can rewrite as follows

$$\begin{aligned}
\mathrm{d}\vartheta^\star_{t,\varepsilon} =\ & \mathrm{I}^t_\tau\left(\check\vartheta_{\tau,\varepsilon}\right)^{-1} L\left(\check\vartheta_{\tau,\varepsilon}, \check\vartheta_{\tau,\varepsilon}, t\right)\left[\check\vartheta_{\tau,\varepsilon} - \vartheta^\star_{t,\varepsilon}\right] \mathrm{d}t \\
&+ \mathrm{I}^t_\tau\left(\check\vartheta_{\tau,\varepsilon}\right)^{-1} \dot S_t\left(\check\vartheta_{\tau,\varepsilon}, y\right) \sigma(t)^{-2}\left[\mathrm{d}X_t - f\left(\check\vartheta_{\tau,\varepsilon}, s, m_s(\check\vartheta_{\tau,\varepsilon})\right) \mathrm{d}t\right].
\end{aligned}$$

To study the asymptotic of $\delta_m(t) = m^\star_t - m_t(\vartheta_0)$, we start with the asymptotic of $\delta_\gamma(t) = \gamma^\star_t - \gamma_*\left(\vartheta_0, t, m^t(\vartheta_0)\right)$. Remark that the Eq. (2.97) can not be used for estimation of the difference $\gamma^\star_t - \gamma_*\left(\vartheta_0, t, m^t(\vartheta_0)\right)$ because $\gamma^\star_t$ depends on the whole trajectory $\vartheta^\star_{s,\varepsilon}, \tau < s \le t$ and not on the last value $\vartheta^\star_{t,\varepsilon}$ only. Of course $\vartheta^\star_{s,\varepsilon} \to \vartheta_0, \tau < s \leq t$ and asymptotically it depends on ϑ_0. We will use notation $m^\star_t = m_t\left(\vartheta^\star_\varepsilon\right)$, which means that $m_t\left(\vartheta^\star_\varepsilon\right) = m_t\left(\vartheta^\star_{s,\varepsilon}, \tau < s \leq t\right)$, $\delta_{f'}(t) = f'_y\left(\vartheta^\star_{t,\varepsilon}, t, m^\star_t\right) - f'_y\left(\vartheta_0, t, m_t(\vartheta_0)\right)$ and $\delta_{a'}(t) = a'_y\left(\vartheta^\star_{t,\varepsilon}, t, m^\star_t\right) - a'_y\left(\vartheta_0, t, m_t(\vartheta_0)\right)$, $\gamma^*_t = \gamma_*\left(\vartheta_0, t, m^t(\vartheta_0)\right)$, $\gamma^\circ_t = \gamma_*\left(\vartheta_0, t, y^t(\vartheta_0)\right)$, $y^\circ_t = y_t(\vartheta_0)$. We have

$$\frac{d\delta_\gamma(t)}{dt} = 2a'_y\left(\vartheta^\star_{t,\varepsilon}, t, m^\star_t\right)\delta_\gamma(t) + 2\gamma^*_t\delta_{a'}(t)$$
$$- \frac{f'_y\left(\vartheta^\star_{t,\varepsilon}, t, m^\star_t\right)\gamma^\star_t + f'_y\left(\vartheta_0, t, m_t(\vartheta_0)\right)\gamma^*_t}{\sigma(t)^2}$$
$$\times\left[\gamma^\star_t\delta_{f'}(t) + f'_y\left(\vartheta_0, t, m_t(\vartheta_0)\right)\delta_\gamma(t)\right].$$

On the time interval $[0, \tau]$, we put $\gamma^\star_t = \gamma_*\left(\check\vartheta_{\tau,\varepsilon}, t, y_t\left(\check\vartheta_{\tau,\varepsilon}\right)\right)$ and therefore $\delta_\gamma(\tau) = \gamma^\star_\tau - \gamma_*\left(\vartheta_0, \tau, y_\tau(\vartheta_0)\right)$. This is a linear equation for $\delta_\gamma(\cdot)$, which can be solved. The solution has the following asymptotics:

$$\delta_\gamma(t) = \delta_\gamma(\tau)\,\Phi(\vartheta_0, 0, t)\,(1 + o(1))$$
$$+ 2\int_\tau^t \tilde\Phi(\vartheta_0, s, t)\left[\gamma^\circ_s\delta_{a'}(s) - \frac{f'_y\left(\vartheta_0, s, y^\circ_s\right)\gamma^{\circ 2}_s}{\sigma(s)^2}\delta_{f'}(s)\right]ds\,(1 + o(1))$$
$$= \left(\check\vartheta_{\tau,\varepsilon} - \vartheta_0\right)\dot\gamma_*\left(\vartheta_0, \tau, y_\tau(\vartheta_0)\right)\Phi(\vartheta_0, 0, t)\,(1 + o(1))$$
$$+ 2\varepsilon\int_\tau^t \tilde\Phi(\vartheta_0, s, t)\left[\gamma^\circ_s\dot a'_{y\vartheta}\left(\vartheta_0, s, y^\circ_s\right)\zeta^s_\tau(\vartheta_0) + \gamma^\circ_s a''_{yy}\left(\vartheta_0, s, y^\circ_s\right)v_s(\vartheta_0)\right.$$
$$\left. - \frac{f'_y\left(\vartheta_0, s, y^\circ_s\right)\gamma^{\circ 2}_s}{\sigma(s)^2}\left(\dot f'_{\vartheta y}\left(\vartheta_0, s, y^\circ_s\right)\zeta^s_\tau(\vartheta_0) + f''_{yy}\left(\vartheta_0, s, y^\circ_s\right)v_s(\vartheta_0)\right)\right]ds$$
$$\times(1 + o(1))$$
$$= \varepsilon\left[\tilde\rho_t(\vartheta_0) + \kappa^t_\tau(\vartheta_0)\right](1 + o(1)),$$

where

$$\tilde\Phi(\vartheta_0, s, t) = \exp\left(2\int_s^t\left[a'_y\left(\vartheta_0, v, y^\circ_v\right) - \frac{f'_y\left(\vartheta_0, v, y^\circ_v\right)^2\gamma^\circ_v}{\sigma(v)^2}\right]dv\right),$$
$$\tilde\rho(\vartheta_0) = \rho(\vartheta_0)\,\dot\gamma_*\left(\vartheta_0, \tau, y_\tau(\vartheta_0)\right)\Phi(\vartheta_0, 0, t).$$

The Gaussian process $\kappa^t_\tau(\vartheta_0)$, $\tau < t \leq T$ is defined by the last equality. Therefore,

$$\gamma^\star_t = \gamma^\circ_t + \varepsilon\left[\tilde\rho_t(\vartheta_0) + \kappa^t_\tau(\vartheta_0)\right](1 + o(1)).$$

The equation for the difference $\delta_m(t)$ is

$$\begin{aligned}
\mathrm{d}\delta_m(t) &= \left[a'_y\left(\vartheta^\star_{t,\varepsilon}, t, m^\star_t\right) - a'_y\left(\vartheta_0, t, m_t(\vartheta_0)\right)\right]\mathrm{d}t \\
&\quad + \varepsilon\frac{f'_y\left(\vartheta^\star_{t,\varepsilon}, t, m^\star_t\right)\gamma^*_t - f'_y\left(\vartheta_0, t, m_t(\vartheta_0)\right)\gamma^*_t}{\sigma(t)}\,\mathrm{d}W_t \\
&\quad - \frac{f'_y\left(\vartheta^\star_{t,\varepsilon}, t, m^\star_t\right)\gamma^\star_t - f'_y\left(\vartheta_0, t, m_t(\vartheta_0)\right)\gamma^*_t}{\sigma(t)^2} \\
&\quad \times\left[f\left(\vartheta_0, t, Y_t\right) - f\left(\vartheta^\star_{t,\varepsilon}, t, m^\star_t\right)\right]\,\mathrm{d}t \\
&\quad - \frac{f'_y\left(\vartheta_0, t, m_t(\vartheta_0)\right)\gamma^*_t}{\sigma(t)^2}\left[f\left(\vartheta^\star_{t,\varepsilon}, t, m^\star_t\right) - f\left(\vartheta_0, t, m_t(\vartheta_0)\right)\right]\,\mathrm{d}t.
\end{aligned}$$

Remark that it can be shown that

$$\varepsilon\int_0^T \frac{f'_y\left(\vartheta^\star_{t,\varepsilon}, t, m^\star_t\right)\gamma^\star_t - f'_y\left(\vartheta_0, t, m_t(\vartheta_0)\right)\gamma^\circ_t}{\sigma(t)}\,\mathrm{d}W_t = O\left(\varepsilon^2\right),$$

$$\begin{aligned}
&\frac{f'_y\left(\vartheta^\star_{t,\varepsilon}, t, m^\star_t\right)\gamma^\star_t - f'_y\left(\vartheta_0, t, m_t(\vartheta_0)\right)\gamma^\circ_t}{\sigma(t)^2}\left[f\left(\vartheta_0, t, Y_t\right) - f\left(\vartheta^\star_{t,\varepsilon}, t, m^\star_t\right)\right]\mathrm{d}t \\
&\quad = O\left(\varepsilon^2\right).
\end{aligned}$$

Therefore, omitting these two terms we obtain the equation

$$\begin{aligned}
\frac{\mathrm{d}\delta^*_m(t)}{\mathrm{d}t} &= a'_y\left(\vartheta^\star_{t,\varepsilon}, t, m^\star_t\right) - a'_y\left(\vartheta_0, t, m^*_t\right) \\
&\quad - \frac{f'_y\left(\vartheta_0, t, m^*_t\right)\gamma^*_t}{\sigma(t)^2}\left[f\left(\vartheta^\star_{t,\varepsilon}, t, m^\star_t\right) - f\left(\vartheta_0, t, m^*_t\right)\right] \\
&= \dot a'_{\vartheta y}(\tilde\vartheta_{t,\varepsilon}, t, m^\star_t)\left(\vartheta^\star_{t,\varepsilon} - \vartheta_0\right) + a''_{yy}(\tilde\vartheta_{t,\varepsilon}, t, \tilde m_t)\delta^*_m(t) \\
&\quad - \frac{f'_y\left(\vartheta_0, t, m_t(\vartheta_0)\right)\gamma^*_t}{\sigma(t)^2}\left[\dot f_\vartheta\left(\vartheta^\star_{t,\varepsilon}, t, m^\star_t\right)\left(\vartheta^\star_{t,\varepsilon} - \vartheta_0\right)\right. \\
&\quad \left. - f'_y\left(\vartheta_0, t, \bar m_t\right)\delta^*_m(t)\right] \\
&= \varepsilon\left[\dot a'_{\vartheta y}(\vartheta_0, t, y_t(\vartheta_0)) - \frac{f'_y\left(\vartheta_0, t, y_t(\vartheta_0)\right)\gamma^\circ_t\dot f_\vartheta\left(\vartheta_0, t, y_t(\vartheta_0)\right)}{\sigma(t)^2}\right]\zeta^t_\tau(\vartheta_0) \\
&\quad + \left[a''_{yy}(\vartheta_0, t, y_t(\vartheta_0)) - \frac{f'_y\left(\vartheta_0, t, y_t(\vartheta_0)\right)^2\gamma^\circ_t}{\sigma(t)^2}\right]\delta^*_m(t) + o(\varepsilon).
\end{aligned}$$

Let us denote

$$H_t(\vartheta_0) = \dot{a}'_{\vartheta y}(\vartheta_0, t, y_t(\vartheta_0)) - \frac{f'_y(\vartheta_0, t, y_t(\vartheta_0))\,\gamma_t^\circ\, f_\vartheta(\vartheta_0, t, y_t(\vartheta_0))}{\sigma(t)^2},$$

$$A_t(\vartheta_0) = a''_{yy}(\vartheta_0, t, y_t(\vartheta_0)) - \frac{f'_y(\vartheta_0, t, y_t(\vartheta_0))^2\,\gamma_t^\circ}{\sigma(t)^2}.$$

Therefore, we have the equation

$$\frac{\mathrm{d}\delta_m^*(t)}{\mathrm{d}t} = \varepsilon H_t(\vartheta_0)\,\zeta_\tau^t(\vartheta_0) + A_t(\vartheta_0)\,\delta_m^*(t) + o(\varepsilon),\quad \delta_m^*(\tau) = y_\tau\left(\check{\vartheta}_{\tau,\varepsilon}\right) - m_\tau(\vartheta_0).$$

The asymptotic of the initial value is

$$\begin{aligned} y_\tau\left(\check{\vartheta}_{\tau,\varepsilon}\right) - Y_\tau &= y_\tau\left(\check{\vartheta}_{\tau,\varepsilon}\right) - y_\tau(\vartheta_0) + y_\tau(\vartheta_0) - Y_\tau \\ &= \dot{y}_\tau(\tilde{\vartheta})\left(\check{\vartheta}_{\tau,\varepsilon} - \vartheta_0\right) + y_\tau(\vartheta_0) - Y_\tau \\ &= \varepsilon\Big[\dot{y}_\tau(\vartheta_0)\rho_\tau(\vartheta_0) - y_\tau^{(1)}(\vartheta_0)\Big](1 + o(1)) \end{aligned}$$

Let us denote

$$\hat{\Phi}(\vartheta_0, s, t) = \exp\left(\int_s^t A_v(\vartheta_0)\,\mathrm{d}v\right)$$

Then the error of approximation of $Y_t, \tau \le t \le T$ by $m_t^\star, \tau \le t \le T$ is asymptotically Gaussian process

$$\begin{aligned} \varepsilon^{-1}\left(m_t^\star - Y_t\right) &= y_t^{(1)}(\vartheta_0) + \Big[\dot{y}_\tau(\vartheta_0)\rho_\tau(\vartheta_0) - y_\tau^{(1)}(\vartheta_0)\Big]\hat{\Phi}(\vartheta_0, 0, t) \\ &\quad + \int_\tau^t \hat{\Phi}(\vartheta_0, s, t)\,H_s(\vartheta_0)\,\mathrm{d}s + o(1), \qquad \tau < t \le T. \end{aligned} \tag{2.107}$$

Therefore, we proved the following proposition.

Proposition 2.6 *Suppose that the conditions $\mathscr{E}_{1-5}$ and $\mathscr{E}_7$ are fulfilled then the error of approximation of Y_t by $m_{t,\varepsilon}^\star$ (EKF* (2.105), (2.106)*) is given by* (2.107).

2.4.2 *EKF-2*

Model of Observations

Let us consider once more the same nonlinear partially observed model

$$\mathrm{d}X_t = F(\vartheta, t, Y_t)\,\mathrm{d}t + \varepsilon\sigma(t)\,\mathrm{d}W_t, \qquad X_0 = 0, \quad 0 \le t \le T, \tag{2.108}$$

$$\mathrm{d}Y_t = A(\vartheta, t, Y_t)\,\mathrm{d}t + \varepsilon b(t)\,\mathrm{d}V_t, \qquad Y_0 = y_0. \tag{2.109}$$

We suppose that the functions $F(\vartheta, t, y)$, $A(\vartheta, t, y)$ have three continuous bounded derivatives on y and the unknown parameter $\vartheta \in \Theta = (\alpha, \beta)$. The functions $\sigma(\cdot)$ and $b(\cdot)$ are bounded on $[0, T]$ and $\sigma(t)^2 > c$, where $c > 0$.

Remark that the process $Y_t, 0 \le t \le T$ can be expanded by the powers of ε

$$Y_t = y_t(\vartheta_0) + \varepsilon y_t^{(1)} + O\left(\varepsilon^2\right).$$

Here

$$y_t^{(1)} = y_t^{(1)}(\vartheta_0) = \left.\frac{\mathrm{d}Y_t}{\mathrm{d}\varepsilon}\right|_{\varepsilon=0}$$

is a Gaussian process satisfying to the linear equation

$$\mathrm{d}y_t^{(1)} = A'_y(\vartheta_0, t, y_t(\vartheta_0))\, y_t^{(1)}\mathrm{d}t + b(t)\,\mathrm{d}V_t, \quad y_0^{(1)} = 0.$$

Therefore $Y_t - y_t(\vartheta_0) = \varepsilon y_t^{(1)} + O\left(\varepsilon^2\right)$.

Let us replace the initial model (2.108)–(2.109) by the linear w.r.t. Y_t model

$$\mathrm{d}X_t = F(\vartheta, t, y_t(\vartheta))\,\mathrm{d}t + F'_y(\vartheta, t, y_t(\vartheta))\,(Y_t - y_t(\vartheta))\,\mathrm{d}t + \varepsilon\sigma(t)\,\mathrm{d}W_t,$$

$$\mathrm{d}Y_t = A(\vartheta, t, y_t(\vartheta))\,\mathrm{d}t + A'_y(\vartheta, t, y_t(\vartheta))\,(Y_t - y_t(\vartheta))\,\mathrm{d}t + \varepsilon b(t)\,\mathrm{d}V_t, ,$$

where we omitted the terms of order ε^2 and which we write as follows:

$$\mathrm{d}X_t = g(\vartheta, t)\,\mathrm{d}t + f(\vartheta, t)\,Y_t\,\mathrm{d}t + \varepsilon\sigma(t)\,\mathrm{d}W_t, \qquad X_0 = 0, \tag{2.110}$$

$$\mathrm{d}Y_t = h(\vartheta, t)\,\mathrm{d}t + a(\vartheta, t)\,Y_t\,\mathrm{d}t + \varepsilon b(t)\,\mathrm{d}V_t, \qquad Y_0 = y_0, \tag{2.111}$$

where

$$g(\vartheta, t) = F(\vartheta, t, y_t(\vartheta)) - F'_y(\vartheta, t, y_t(\vartheta))\, y_t(\vartheta), \quad f(\vartheta, t) = F'_y(\vartheta, t, y_t(\vartheta)),$$

$$h(\vartheta, t) = A(\vartheta, t, y_t(\vartheta)) - A'_y(\vartheta, t, y_t(\vartheta))\, y_t(\vartheta), \quad a(\vartheta, t) = A'_y(\vartheta, t, y_t(\vartheta)).$$

Therefore, we are in the situation of misspecification and we use the wrong model. Remark that the difference between the solutions is of order ε^2. This can be verified with the help of Grönwall lemma.

We do not present below the strict statements and proofs but some sketchs of corresponding arguments, which can allow to prove the mentioned results.

Recall that

$$\frac{\mathrm{d}y_t(\vartheta)}{\mathrm{d}t} = A(\vartheta, t, y_t(\vartheta)), \qquad y_0(\vartheta) = y_0, \qquad 0 \le t \le T.$$

The EKF-2 studied in this section is the Kalman–Bucy filter of the model (2.110) and (2.111) given by the equations

$$\mathrm{d}m_t(\vartheta) = h(\vartheta, t)\,\mathrm{d}t + a(\vartheta, t)\,m_t(\vartheta)\,\mathrm{d}t$$
$$+ \frac{f(\vartheta, t)\,\gamma_*(\vartheta, t)}{\sigma(t)^2}\left[\mathrm{d}X_t - g(\vartheta, t)\,\mathrm{d}t - f(\vartheta, t)\,m_t(\vartheta)\,\mathrm{d}t\right],$$
$$\frac{\mathrm{d}\gamma_*(\vartheta, t)}{\mathrm{d}t} = 2a(\vartheta, t)\,\gamma_*(\vartheta, t) - \frac{f(\vartheta, t)^2\,\gamma_*(\vartheta, t)^2}{\sigma(t)^2} + b(t)^2,$$

with the initial values $m_0^* = y_0$, $\gamma_*(\vartheta, 0) = 0$.

Recall that the process $Y_t \rightarrow y_t(\vartheta_0)$ and $\varepsilon^{-1}(Y_t - y_t(\vartheta_0)) \rightarrow y_t^{(1)}(\vartheta_0)$ (see (2.99)). The random function $m_t(\vartheta)$ converges to the deterministic function $y_t(\vartheta, \vartheta_0)$, which is solution of the ODE

$$\frac{\mathrm{d}y_t(\vartheta, \vartheta_0)}{\mathrm{d}t} = h(\vartheta, t) + a(\vartheta, t)\,y_t(\vartheta, \vartheta_0)$$
$$+ \frac{f(\vartheta, t)\,\gamma_*(\vartheta, t)}{\sigma(t)^2}\left[F(\vartheta_0, t, y_t(\vartheta_0)) - g(\vartheta, t) - f(\vartheta, t)\,y_t(\vartheta, \vartheta_0)\right],$$

with the initial value $y_0(\vartheta, \vartheta_0) = y_0$. Let us verify that $y_t(\vartheta_0, \vartheta_0) = y_t(\vartheta_0)$. We have

$$\frac{\mathrm{d}y_t(\vartheta_0, \vartheta_0)}{\mathrm{d}t} = A(\vartheta_0, t, y_t(\vartheta_0)) + A'_y(\vartheta_0, t, y_t(\vartheta_0))\left[y_t(\vartheta_0, \vartheta_0) - y_t(\vartheta_0)\right]$$
$$- \frac{F'_y(\vartheta_0, t, y_t(\vartheta_0))^2\,\gamma_*(\vartheta_0, t)}{\sigma(t)^2}\left[y_t(\vartheta_0, \vartheta_0) - y_t(\vartheta_0)\right], \quad y_0(\vartheta_0, \vartheta_0) = y_0,$$
$$\frac{\mathrm{d}y_t(\vartheta_0)}{\mathrm{d}t} = A(\vartheta_0, t, y_t(\vartheta_0)), \qquad y_0(\vartheta_0) = y_0.$$

Hence, for $\delta_y(t) = y_t(\vartheta_0, \vartheta_0) - y_t(\vartheta_0)$ we obtain the linear equation

$$\frac{\mathrm{d}\delta_y(t)}{\mathrm{d}t} = \left[A'_y(\vartheta_0, t, y_t(\vartheta_0)) - \frac{F'_y(\vartheta_0, t, y_t(\vartheta_0))\,\gamma_*(\vartheta_0, t)}{\sigma(t)^2} F'_y(\vartheta, t, y_t(\vartheta))\right]\delta_y(t),$$

with the initial value $\delta_y(0) = 0$, whose solution is $\delta_y(t) \equiv 0$.

The equation for the difference $\delta_m(t) = m_t(\vartheta_0) - Y_t$ is

$$\begin{aligned}
\mathrm{d}\delta_m(t) &= [h(\vartheta_0, t) + a(\vartheta_0, t)\, m_t(\vartheta_0) - a(\vartheta_0, t, Y_t)]\,\mathrm{d}t \\
&\quad + \frac{f(\vartheta_0, t)\,\gamma_*(\vartheta_0, t)}{\sigma(t)^2}[F(\vartheta_0, t, Y_t) - g(\vartheta_0, t) - f(\vartheta_0, t)\, m_t(\vartheta_0)]\,\mathrm{d}t \\
&\quad + \varepsilon\frac{f(\vartheta_0, t)\,\gamma_*(\vartheta_0, t)}{\sigma(t)}\mathrm{d}W_t - \varepsilon b(t)\,\mathrm{d}V_t \\
&= \left[A'_y(\vartheta_0, t, y_t(\vartheta_0)) - \frac{F'(\vartheta_0, t, y_t(\vartheta_0))^2\,\gamma_*(\vartheta_0, t)}{\sigma(t)^2}\right]\delta_m(t)\,\mathrm{d}t \\
&\quad + \varepsilon\frac{F'(\vartheta_0, t, y_t(\vartheta_0))\,\gamma_*(\vartheta_0, t)}{\sigma(t)}\mathrm{d}W_t - \varepsilon b(t)\,\mathrm{d}V_t + o(\varepsilon), \\
m_0(\vartheta_0) - Y_0 &= 0.
\end{aligned}$$

Here the term $o(\varepsilon)$ is of order ε^2. The solution of it is

$$\begin{aligned}
\varepsilon^{-1}(m_t(\vartheta_0) - Y_t) &= \int_0^t \bar{\Phi}(\vartheta_0, s, t)\frac{F(\vartheta_0, s, y_s(\vartheta_0))\,\gamma_*(\vartheta_0, s)}{\sigma(s)}\mathrm{d}W_s \\
&\quad - \int_0^t \bar{\Phi}(\vartheta_0, s, t)\, b(s)\,\mathrm{d}V_s + o(\varepsilon) \equiv \varepsilon\pi_t^*(\vartheta_0) + o(\varepsilon)
\end{aligned}$$

with the corresponding function $\bar{\Phi}(\vartheta_0, t, s)$.

Therefore,

$$\varepsilon^{-1}\Big(m_t(\vartheta_0) - Y_t\Big) \longrightarrow \pi_t^*(\vartheta_0). \tag{2.112}$$

Estimators

The corresponding pseudo-LR $L\left(\vartheta, X^T\right)$, $\vartheta \in \Theta$ and pseudo-MLE $\hat{\vartheta}_\varepsilon$ are defined by the relations

$$\begin{aligned}
\ln L\left(\vartheta, X^T\right) &= \int_0^T \frac{g(\vartheta, t) + f(\vartheta, t)\, m_t(\vartheta)}{\varepsilon^2\sigma(t)^2}\mathrm{d}X_t \\
&\quad - \int_0^T \frac{[g(\vartheta, t) + f(\vartheta, t)\, m_t(\vartheta)]^2}{2\varepsilon^2\sigma(t)^2}\mathrm{d}t
\end{aligned}$$

and

$$L(\hat{\vartheta}_\varepsilon, X^T) = \sup_{\vartheta\in\Theta} L\left(\vartheta, X^T\right)$$

respectivly.

Notice that

$$\begin{aligned}\ln L\left(\vartheta, X^T\right) = &\int_0^T \frac{g\left(\vartheta, t, y_t\left(\vartheta\right)\right) + f\left(\vartheta, t\right) m_t\left(\vartheta\right)}{\varepsilon\sigma\left(t\right)} \mathrm{d}W_t \\ &- \int_0^T \frac{\left[g\left(\vartheta, t\right) + f\left(\vartheta, t\right) m_t\left(\vartheta\right) - F\left(\vartheta_0, t, Y_t\right)\right]^2}{2\varepsilon^2\sigma\left(t\right)^2} \mathrm{d}t \\ &+ \int_0^T \frac{F\left(\vartheta_0, t, Y_t\right)^2}{2\varepsilon^2\sigma\left(t\right)^2} \mathrm{d}t.\end{aligned}$$

To see if the p-MLE is consistent, we consider the following limit:

$$-2\varepsilon^2 \ln L\left(\vartheta, X^T\right) \longrightarrow H\left(\vartheta, \vartheta_0\right) - \int_0^T \frac{F\left(\vartheta_0, t, y_t\left(\vartheta_0\right)\right)^2}{\sigma\left(t\right)^2} \mathrm{d}t,$$

where the limit function $H\left(\vartheta, \vartheta_0\right)$ has the expression

$$H\left(\vartheta, \vartheta_0\right) = \int_0^T \frac{\left[g\left(\vartheta, t\right) + f\left(\vartheta, t\right) y_t\left(\vartheta, \vartheta_0\right) - F\left(\vartheta_0, t, y_t\left(\vartheta_0\right)\right)\right]^2}{\sigma\left(t\right)^2} \mathrm{d}t.$$

Remark, that if $\vartheta = \vartheta_0$, then we obtain the equality $H\left(\vartheta_0, \vartheta_0\right) = 0$. Therefore, in this misspecified problem, the condition of identifiability is:

For any $\nu > 0$ and any $\vartheta_0 \in \Theta$ the inequality

$$\inf_{|\vartheta-\vartheta_0|>\nu} H\left(\vartheta, \vartheta_0\right) > 0$$

holds.

Then as it was discussed above in EKF-1 case, it can be shown that the p-MLE is consistent and asymptotically normal. The normalized p-LR is (below $\vartheta_u = \vartheta_0 + \varepsilon u$)

$$Z_\varepsilon\left(\vartheta_0, u\right) = \frac{L\left(\vartheta_u, X^T\right)}{L\left(\vartheta_0, X^T\right)}, \qquad u \in \mathbb{U}_\varepsilon = \left(\frac{\alpha - \vartheta_0}{\varepsilon}, \frac{\beta - \vartheta_0}{\varepsilon}\right).$$

Introduce futher notations: $S_t\left(\vartheta_u, m\right) = g\left(\vartheta_u, t\right) + f\left(\vartheta_u, t\right) m_t\left(\vartheta_u\right), S_t\left(\vartheta_0, y\right) = g\left(\vartheta_0, t\right) + f\left(\vartheta_0, t\right) y_t\left(\vartheta_0\right)$. Then we obtain the representation

$$\ln Z_\varepsilon\left(\vartheta_0, u\right) = \int_0^T \frac{\left[S_t\left(\vartheta_u, m\right) - S_t\left(\vartheta_0, m\right)\right]}{\varepsilon\sigma\left(t\right)} \mathrm{d}W_t$$

$$-\int_0^T \frac{\left[S_t\left(\vartheta_u, m\right)-S_t\left(\vartheta_0, m\right)\right]^2}{2\varepsilon^2 \sigma(t)^2} \mathrm{d}t$$

$$-\int_0^T \frac{\left[S_t\left(\vartheta_u, m\right)-S_t\left(\vartheta_0, m\right)\right]\left[S_t\left(\vartheta_0, m\right)-f\left(\vartheta_0, t, Y_t\right)\right]}{\varepsilon^2 \sigma(t)^2} \mathrm{d}t .$$

Suppose that the functions below are sufficiently smooth w.r.t. ϑ to justify the expansions by the powers of ε

$$\begin{aligned} S_t\left(\vartheta_u, m\right)-S_t\left(\vartheta_0, m\right) &= u \varepsilon \dot{S}_t\left(\vartheta_0, y\right)(1+o(1)) \\ &= \varepsilon u\left[\dot{g}\left(\vartheta_0, t\right)+\dot{f}\left(\vartheta_0, t\right) y_t\left(\vartheta_0\right)+f\left(\vartheta_0, t\right) \dot{y}_t\left(\vartheta_0, \vartheta_0\right)\right](1+o(1)), \end{aligned} \tag{2.113}$$

$$\begin{aligned} S_t\left(\vartheta_0, m\right)-F\left(\vartheta_0, t, Y_t\right) &= F\left(\vartheta_0, t, y_t\left(\vartheta_0\right)\right)-F_y'\left(\vartheta_0, t, y_t\left(\vartheta_0\right)\right) y_t\left(\vartheta_0\right) \\ &\quad +F_y'\left(\vartheta_0, t, y_t\left(\vartheta_0\right)\right) m_t\left(\vartheta_0\right)-F\left(\vartheta_0, t, Y_t\right) \\ &= F_y'\left(\vartheta_0, t, y_t\left(\vartheta_0\right)\right)\left(m_t\left(\vartheta_0\right)-Y_t\right)(1+o(1)) \\ &= \varepsilon F_y'\left(\vartheta_0, t, y_t\left(\vartheta_0\right)\right) \pi_t\left(\vartheta_0\right)\ (1+o(1)) . \end{aligned} \tag{2.114}$$

Then the normalized p-LR has the representation

$$\begin{aligned} \ln Z_{\varepsilon}\left(\vartheta_0, u\right) &= u \int_0^T \frac{\dot{S}_s\left(\vartheta_0, y\right)}{\sigma(s)} \mathrm{d} W_s-\frac{u^2}{2} \int_0^T \frac{\dot{S}_s\left(\vartheta_0, y\right)^2}{\sigma(s)^2} \mathrm{d} s \\ &\quad +u \int_0^T \frac{\dot{S}_s\left(\vartheta_0, y\right) F_y'\left(\vartheta_0, s, y_s\left(\vartheta_0\right)\right) \pi_s\left(\vartheta_0\right)}{\sigma(s)^2} \mathrm{d} s+o(1) \\ &= u \Delta\left(\vartheta_0\right)-\frac{u^2}{2} \mathrm{I}\left(\vartheta_0\right)+u R\left(\vartheta_0\right)+o(1) \end{aligned}$$

with obvious notations.

The limit LR process is

$$Z\left(\vartheta_0, u\right)=\exp \left(u \Delta\left(\vartheta_0\right)-\frac{u^2}{2} \mathrm{I}\left(\vartheta_0\right)+u R\left(\vartheta_0\right)\right), \qquad u \in \mathcal{R}.$$

Suppose that the weak convergence

$$Z_{\varepsilon}(\cdot) \Longrightarrow Z(\cdot)$$

is already proved. Then the point $\hat{u}_{\varepsilon}$ of the maximum of $Z_{\varepsilon}(\cdot)$ converges to the Gaussian limit

$$\hat{u}_{\varepsilon} \Longrightarrow \hat{u}=\mathrm{I}\left(\vartheta_0\right)^{-1}\left[\Delta\left(\vartheta_0\right)+R\left(\vartheta_0\right)\right]=\zeta\left(\vartheta_0\right)$$

where

$$
\mathrm{I}\left(\vartheta_0\right)=\int_0^T \frac{\left[\dot{F}_\vartheta\left(\vartheta_0, t, y_t\left(\vartheta_0\right)\right)+F_y^{\prime}\left(\vartheta_0, t, y_t\left(\vartheta_0\right)\right) \dot{y}_t\left(\vartheta_0\right)\right]^2}{\sigma(t)^2} \mathrm{~d} t .
$$

We do not give the detailed proof of the asymptotic normality of the p-MLE

$$
\varepsilon^{-1}\left(\hat{\vartheta}_{\varepsilon}-\vartheta_0\right) \Longrightarrow \zeta\left(\vartheta_0\right) .
$$

This proof can be carried out following, for example, the proofs of the asymptotic normality in the problems of misspecifications considered in [118, Section 5.1] (see more details in Remark 2.14.)

One-Step pseudo-MLE

As preliminary estimator we take the same MDE

$$
\check{\vartheta}_{\tau, \varepsilon}=\arg \inf_{\vartheta \in \Theta} \int_0^\tau\left[X_t-x_t(\vartheta)\right]^2 \mathrm{~d} t, \qquad x_t(\vartheta)=\int_0^t F\left(\vartheta, s, y_s(\vartheta)\right) \mathrm{d} s
$$

constructed by the observations $X^\tau=\left(X_s, 0 \leq s \leq \tau\right)$.

Let us denote

$$
\mathrm{I}_\tau^t(\vartheta)=\int_\tau^t \frac{\dot{S}_s\left(\vartheta_0, y\right)^2}{\sigma(s)^2} \mathrm{~d} s
$$

The One-step pseudo-MLE process can be

$$
\vartheta_{t, \varepsilon}^{\star}=\check{\vartheta}_{\tau, \varepsilon}+\hat{\mathrm{I}}_\tau^t\left(\check{\vartheta}_{\tau, \varepsilon}\right)^{-1} \int_\tau^t \frac{\dot{S}_s\left(\check{\vartheta}_{\tau, \varepsilon}, y\right)}{\sigma(s)^2}\left[\mathrm{d} X_s-S_s\left(\check{\vartheta}_{\tau, \varepsilon}, y\right) \mathrm{d} s\right], \quad \tau<t \leq T. \tag{2.115}
$$

We have the equality

$$
\begin{aligned}
\frac{\vartheta_{t, \varepsilon}^{\star}-\vartheta_0}{\varepsilon}= & \frac{\check{\vartheta}_{\tau, \varepsilon}-\vartheta_0}{\varepsilon}+\mathrm{I}_\tau^t\left(\check{\vartheta}_{\tau, \varepsilon}\right)^{-1} \int_\tau^t \frac{\dot{S}_s\left(\check{\vartheta}_{\tau, \varepsilon}, y\right)}{\sigma(s)} \mathrm{d} W_s \\
& +\hat{\mathrm{I}}_\tau^t\left(\check{\vartheta}_{\tau, \varepsilon}\right)^{-1} \int_\tau^t \frac{\dot{S}_s\left(\check{\vartheta}_{\tau, \varepsilon}, y\right)}{\varepsilon \sigma(s)^2}\left[F\left(\vartheta_0, s, Y_s\right)-S_s\left(\check{\vartheta}_{\tau, \varepsilon}, m\right)\right] \mathrm{d} s .
\end{aligned}
$$

The MDE is consistent and therefore

$$\int_\tau^t \frac{\dot S_s\left(\check\vartheta_{\tau,\varepsilon},y\right)}{\sigma\left(s\right)}\mathrm{d}W_s\longrightarrow \int_\tau^t \frac{\dot S_s\left(\vartheta_0,y\right)}{\sigma\left(s\right)}\mathrm{d}W_s=\Delta_\tau^t\left(\vartheta_0\right).$$

Here $\Delta_\tau^t\left(\vartheta_0\right)\sim\mathcal{N}\left(0,\mathrm{I}_\tau^t\left(\vartheta_0\right)\right)$. Further by (2.113) and (2.114) we have the relations

$$\int_\tau^t \frac{\dot S_s\left(\check\vartheta_{\tau,\varepsilon},y\right)}{\varepsilon\sigma\left(s\right)^2}\left[F\left(\vartheta_0,s,Y_s\right)\right)-S_s\left(\vartheta_0,m\right)+S_s\left(\vartheta_0,m\right)-S_s\left(\check\vartheta_{\tau,\varepsilon},m\right)\right]\mathrm{d}s$$
$$=-\frac{\check\vartheta_{\tau,\varepsilon}-\vartheta_0}{\varepsilon}\mathrm{I}_\tau^t\left(\vartheta_0\right)\left(1+O\left(\varepsilon\right)\right)+R_\tau^t\left(\vartheta_0\right)\left(1+o\left(1\right)\right).$$

Therefore,

$$\frac{\vartheta_{t,\varepsilon}^\star-\vartheta_0}{\varepsilon}=\frac{\check\vartheta_{\tau,\varepsilon}-\vartheta_0}{\varepsilon}-\mathrm{I}_\tau^t\left(\check\vartheta_{\tau,\varepsilon}\right)^{-1}\frac{\check\vartheta_{\tau,\varepsilon}-\vartheta_0}{\varepsilon}\mathrm{I}_\tau^t\left(\vartheta_0\right)\left(1+O\left(\varepsilon\right)\right)$$
$$+\mathrm{I}_\tau^t\left(\check\vartheta_{\tau,\varepsilon}\right)^{-1}R_\tau^t\left(\vartheta_0\right)\left(1+o\left(1\right)\right)+\mathrm{I}_\tau^t\left(\check\vartheta_{\tau,\varepsilon}\right)^{-1}\int_\tau^t\frac{\dot S_s\left(\check\vartheta_{\tau,\varepsilon},y\right)}{\sigma\left(s\right)}\mathrm{d}W_s$$
$$\longrightarrow \mathrm{I}_\tau^t\left(\vartheta_0\right)^{-1}\left[\Delta_\tau^t\left(\vartheta_0\right)+R_\tau^t\left(\vartheta_0\right)\right]\equiv\zeta_\tau^t\left(\vartheta_0\right).$$

Adaptiv EKF-2

Once more we have the learning interval $[0,\tau]$, the MDE $\check\vartheta_{\tau,\varepsilon}$ and the One-step pseudo-MLE process (2.115). The adaptive filter is

$$\mathrm{d}m_t^\star=\left[h\left(\vartheta_\varepsilon^\star,t\right)+a\left(\vartheta_\varepsilon^\star,t\right)m_t^\star\right]\mathrm{d}t$$
$$+\frac{f\left(\vartheta_\varepsilon^\star,t\right)\gamma_t^\star}{\sigma\left(t\right)^2}\left[\mathrm{d}X_t-\left[g\left(\vartheta_\varepsilon^\star,t\right)+f\left(\vartheta_\varepsilon^\star,t\right)m_t^\star\right]\mathrm{d}t\right],$$
$$\frac{\mathrm{d}\gamma_t^\star}{\mathrm{d}t}=2a\left(\vartheta_\varepsilon^\star,t\right)\gamma_t^\star-\frac{f\left(\vartheta_\varepsilon^\star,t\right)^2\left(\gamma_t^\star\right)^2}{\sigma\left(t\right)^2}+b\left(t\right)^2,$$

with the initial values $m_\tau^\star\left(\vartheta\right)=y_\tau\left(\check\vartheta_{\tau,\varepsilon}\right),\gamma_\tau^\star=\gamma_*\left(\check\vartheta_{\tau,\varepsilon},\tau\right)$. We write $f\left(\vartheta_\varepsilon^\star,t\right)$ and not $f\left(\vartheta_{t,\varepsilon}^\star,t\right)$ because $f\left(\vartheta_\varepsilon^\star,t\right)$ and other functions depend on the trajectory $\vartheta_{s,\varepsilon}^\star,\tau<s\leq t$. Indeed, we have

$$f\left(\vartheta_\varepsilon^\star,t\right)=F_y'\left(\vartheta_{t,\varepsilon}^\star,t,y_t\left(\vartheta_\varepsilon^\star\right)\right),\quad y_t\left(\vartheta_\varepsilon^\star\right)=y_\tau\left(\check\vartheta_{\tau,\varepsilon}\right)+\int_\tau^t A\left(\vartheta_{s,\varepsilon}^\star,s,y_s\left(\vartheta_\varepsilon^\star\right)\right)\mathrm{d}s$$

and

$$
\begin{aligned}
y_t\left(\vartheta_\varepsilon^\star\right)-y_t\left(\vartheta_0\right)&=y_\tau\left(\check{\vartheta}_{\tau,\varepsilon}\right)-y_\tau\left(\vartheta_0\right)\\
&\qquad+\int_\tau^t\left[A\left(\vartheta_{s,\varepsilon}^\star,s,y_s\left(\vartheta_\varepsilon^\star\right)\right)-A\left(\vartheta_0,s,y_s\left(\vartheta_0\right)\right)\right]\mathrm{d}s\\
&=\varepsilon\dot{y}_t\left(\vartheta_0\right)\rho_\tau\left(\vartheta_0\right)+\int_\tau^t\dot{A}_\vartheta\left(\vartheta_0,s,y_s\left(\vartheta_0\right)\right)\left(\vartheta_{s,\varepsilon}^\star-\vartheta_0\right)\mathrm{d}s\\
&\qquad+\int_\tau^t A_y'\left(\vartheta_0,s,y_s\left(\vartheta_0\right)\right)\left(y_s\left(\vartheta_\varepsilon^\star\right)-y_s\left(\vartheta_0\right)\right)\mathrm{d}s+o\left(\varepsilon\right)\\
&=\varepsilon\dot{y}_t\left(\vartheta_0\right)\rho_\tau\left(\vartheta_0\right)\phi\left(\vartheta_0,0,t\right)\\
&\qquad+\varepsilon\int_\tau^t\phi\left(\vartheta_0,s,t\right)\dot{A}_\vartheta\left(\vartheta_0,s,y_s\left(\vartheta_0\right)\right)\zeta_\tau^s\left(\vartheta_0\right)\mathrm{d}s+o\left(\varepsilon\right)\\
&=\varepsilon\,\varpi_\tau^t\left(\vartheta_0\right)+o\left(\varepsilon\right).
\end{aligned}\tag{2.116}
$$

Consider the difference $\delta_m\left(t\right)=m_t^\star-m_t\left(\vartheta_0\right)$, where $m_t\left(\vartheta_0\right)=m_t^*$ is from the equation

$$
\begin{aligned}
\mathrm{d}m_t^*&=\left[h\left(\vartheta_0,t\right)+a\left(\vartheta_0,t\right)m_t^*\right]\mathrm{d}t+\varepsilon\frac{f\left(\vartheta_0,t\right)\gamma_*\left(\vartheta_0,t\right)}{\sigma\left(t\right)}\mathrm{d}W_t\\
&\quad+\frac{f\left(\vartheta_0,t\right)\gamma_*\left(\vartheta_0,t\right)}{\sigma\left(t\right)^2}\Big[F\left(\vartheta_0,t,Y_t\right)-\left[g\left(\vartheta_0,t\right)+f\left(\vartheta_0,t\right)m_t^*\right]\Big]\mathrm{d}t,\\
m_0^*&=y_0.
\end{aligned}
$$

We have

$$
\begin{aligned}
\mathrm{d}\delta_m\left(t\right)&=\Big[h\left(\vartheta_\varepsilon^\star,t\right)-h\left(\vartheta_0,t\right)+\left[a\left(\vartheta_\varepsilon^\star,t\right)-a\left(\vartheta_0,t\right)\right]m_t^\star+a\left(\vartheta_0,t\right)\delta_m\left(t\right)\Big]\mathrm{d}t\\
&\quad+\varepsilon\frac{f\left(\vartheta_\varepsilon^\star,t\right)\gamma_t^\star-f\left(\vartheta_0,t\right)\gamma_*\left(\vartheta_0,t\right)}{\sigma\left(t\right)}\mathrm{d}W_t\\
&\quad+\frac{f\left(\vartheta_\varepsilon^\star,t\right)\gamma_t^\star-f\left(\vartheta_0,t\right)\gamma_*\left(\vartheta_0,t\right)}{\sigma\left(t\right)^2}\\
&\quad\times\left[F\left(\vartheta_0,t,Y_t\right)-g\left(\vartheta_\varepsilon^\star,t\right)-f\left(\vartheta_\varepsilon^\star,t\right)m_t^\star\right]\mathrm{d}t\\
&\quad+\frac{f\left(\vartheta_0,t\right)\gamma_*\left(\vartheta_0,t\right)}{\sigma\left(t\right)^2}\\
&\quad\times\left[g\left(\vartheta_\varepsilon^\star,t\right)-f\left(\vartheta_\varepsilon^\star,t\right)m_t^\star-g\left(\vartheta_0,t\right)-f\left(\vartheta_0,t\right)m_t^*\right]\mathrm{d}t.
\end{aligned}
$$

In this equation, the second (stochastic integral) and third terms are of order ε^2. We omit these two terms and write

$$\frac{\mathrm{d}\delta_m^*(t)}{\mathrm{d}t}=h\left(\vartheta_\varepsilon^\star,t\right)-h\left(\vartheta_0,t\right)+\left[a\left(\vartheta_\varepsilon^\star,t\right)-a\left(\vartheta_0,t\right)\right]m_t^\star$$
$$+\frac{f\left(\vartheta_0,t\right)\gamma_*\left(\vartheta_0,t\right)}{\sigma\left(t\right)^2}\left[g\left(\vartheta_\varepsilon^\star,t\right)-g\left(\vartheta_0,t\right)+\left[f\left(\vartheta_\varepsilon^\star,t\right)-f\left(\vartheta_0,t\right)\right]m_t^\star\right]$$
$$+\left[a\left(\vartheta_0,t\right)-\frac{f\left(\vartheta_0,t\right)^2\gamma_*\left(\vartheta_0,t\right)}{\sigma\left(t\right)^2}\right]\delta_m^*\left(t\right),\quad \delta_m^*\left(\tau\right)=y_\tau\left(\check\vartheta_{\tau,\varepsilon}\right)-m_\tau\left(\vartheta_0\right).$$

Therefore,

$$\frac{y_t^\star-m_t\left(\vartheta_0\right)}{\varepsilon}=\left[\dot y_\tau\left(\vartheta_0\right)\rho_\tau\left(\vartheta_0\right)-\nu_\tau\left(\vartheta_0\right)-y_\tau^{(1)}\left(\vartheta_0\right)\right]\Phi\left(\vartheta_0,0,t\right)$$
$$+\int_\tau^t\Phi\left(\vartheta_0,s,t\right)\left[\dot h\left(\vartheta_0,s\right)+\dot a\left(\vartheta_0,s\right)y_s\left(\vartheta\right)\right]\mathrm{d}s+o\left(1\right)$$
$$+\int_\tau^t\Phi\left(\vartheta_0,s,t\right)\frac{f\left(\vartheta_0,s\right)\gamma_*\left(\vartheta_0,s\right)}{\sigma\left(s\right)^2}\left[\dot g\left(\vartheta_0,s\right)+\dot f\left(\vartheta_0,s\right)y_t\left(\vartheta_0\right)\right]\mathrm{d}s.$$

Here $\dot h\left(\vartheta_0,s\right),\dot a\left(\vartheta_0,s\right),\dot g\left(\vartheta_0,s\right),\dot f\left(\vartheta_0,s\right)$ are linear functionals of the Gaussian process $\zeta_\tau^s\left(\vartheta_0\right),\tau<s\leq t$. For example,

$$\begin{aligned}f\left(\vartheta_\varepsilon^\star,t\right)-f\left(\vartheta_0,t\right)&=F_y'\left(\vartheta_{t,\varepsilon}^\star,t,y_t\left(\vartheta_\varepsilon^\star\right)\right)-F_y'\left(\vartheta_0,t,y_t\left(\vartheta_0\right)\right)\\&=F_y'\left(\vartheta_{t,\varepsilon}^\star,t,y_t\left(\vartheta_\varepsilon^\star\right)\right)-F_y'\left(\vartheta_0,t,y_t\left(\vartheta_\varepsilon^\star\right)\right)+F_y'\left(\vartheta_0,t,y_t\left(\vartheta_\varepsilon^\star\right)\right)\\&\quad-F_y'\left(\vartheta_0,t,y_t\left(\vartheta_0\right)\right)\\&=\dot F_{\vartheta y}'\left(\bar\vartheta_{t,\varepsilon},t,y_t\left(\vartheta_\varepsilon^\star\right)\right)\left(\vartheta_{t,\varepsilon}^\star-\vartheta_0\right)+F_{yy}''\left(\vartheta_0,t,\bar y_t\right)\left(y_t\left(\vartheta_\varepsilon^\star\right)-y_t\left(\vartheta_0\right)\right)\\&=\varepsilon\left[\dot F_{\vartheta y}'\left(\vartheta_0,t,y_t\left(\vartheta_0\right)\right)\zeta_\tau^t\left(\vartheta_0\right)+F_{yy}''\left(\vartheta_0,t,y_t\left(\vartheta_0\right)\right)\varpi_\tau^t\left(\vartheta_0\right)\right]+o\left(\varepsilon\right),\end{aligned}$$

where we used (2.116). The similar relations we have for the other mentioned functions.

The asymptotic of the error $m_t^\star-Y_t$ follows from (2.112) and the asymptotic of $m_t^\star-m_t\left(\vartheta_0\right)$ described above. The limit Gaussian process can be written explicitly but we will not write it here to avoid cumbersom expressions.

Notice that the adaptive filter and related functions can be written in the recurrent form.

Remark 2.15 Interesting results were established in the work of Skorokhod and Khasminskii [177]. They studied there more general model we write in the particuar interesting for use case as follows:

$$\begin{aligned}\mathrm{d}X_t&=f\left(Y_t\right)\mathrm{d}t+\varepsilon\mathrm{d}W_t,\qquad X_0=0,\quad 0\leq t\leq 1,\\Y_t&=a\left(\vartheta_0,t\right)t+\varepsilon\xi_t,\end{aligned}$$

Here the functions $f(\cdot)$ and $a(\cdot)$ are supposed to be known and sufficently smooth. The Gaussian process $\xi_t, 0 \leq t \leq 1$ has mean zero and known correlation function $\mathbf{E}\xi_t\xi_s = r(s,t)$, $r(t,t) \leq c_1$. It is shown that the corresponding family of measures is LAN and then an asymptotically efficient One-step MLE of the unknown parameter is proposed. The construction of this estimator is in some sense close to considered in our work One-step MLE. First by observations on some small interval $[t_\delta, t_\delta + \delta] \in [0,1]$ a preliminary MDE $\vartheta_{\delta,\varepsilon}$ was constructed and then this estimator was used for writing the One-step MLE. Here $\varepsilon \to 0$ and $\delta \to 0$ but slowly providing the consistency of the MDE and the consistency and asymptotic normality of the One-step MLE.

The linear version of this model with $f(Y_t) = Y_t$ was studied by Khasminskii and Lazareva [94], where the asymptotic normality and asymptotic efficiency of the MLE were established. We return to this model of observations in the Sect. 7.2.

Remark 2.16 In the work [125] the following partially observed model was studied

$$\mathrm{d}X_t = F\left(\vartheta, t, Y^t\right)\mathrm{d}t + \mathrm{d}W_t, \qquad X_0 = 0, \quad 0 \leq t \leq T,$$

$$\mathrm{d}Y_t = A\left(\vartheta, t, Y^t\right)\mathrm{d}t + B\left(\vartheta, t, Y\right)\mathrm{d}V_t, \qquad Y_0 = y_0.$$

Here $Y^t = (Y_s, 0 \leq s \leq t)$, the functions $F(\vartheta, t, Y)$, $A(\vartheta, t, Y)$ and $B(\vartheta, t, Y)$ are supposed to be known and sufficiently smooth. The considered problem is: estimation of the parameter $\vartheta \in \Theta \subset \mathcal{R}^d$ by n independent observations $X^j = \left(X_t^j, 0 \leq t \leq T\right)$. The consistency and asymptotic normality of the MDE are established in the asymptotic $n \to \infty$.

2.5 Different Levels of Noises

2.5.1 Model of Observations

Consider the partially observed stochastic linear system

$$\mathrm{d}X_t = f(\vartheta, t)\, Y_t \mathrm{d}t + \varepsilon\sigma(t)\,\mathrm{d}W_t, \qquad X_0 = 0, \qquad 0 \leq t \leq T, \tag{2.117}$$

$$\mathrm{d}Y_t = a(\vartheta, t)\, Y_t \mathrm{d}t + \psi_\varepsilon b(\vartheta, t)\,\mathrm{d}V_t, \quad Y_0 = y_0, \tag{2.118}$$

where $W_t, 0 \leq t \leq T$ and $V_t, 0 \leq t \leq T$ are independent Wiener processes, $f(\cdot,\cdot)$, $a(\cdot,\cdot)$, $b(\cdot,\cdot)$ and $\sigma(\cdot)$ are known functions, $\varepsilon \in (0,1]$ and $\psi_\varepsilon \in (0,1]$ are noise intensities. The initial value y_0 is deterministic (known). The parameter $\vartheta \in \Theta \subset \mathcal{R}^d$ is unknown and the Gaussian process $Y^T = (Y_t, 0 \leq t \leq T)$ is unobservable (hidden). The set $\Theta \subset \mathcal{R}^d$ is open, convex and bounded. Suppose that the conditions $\mathscr{A}_*$ are fulfilled.

We have the same problem of the approximation of the unobserved component or more precizely, the estimation of the conditional expectation $m(\vartheta, t) = \mathbf{E}_\vartheta\left(Y_t|\mathfrak{F}_t^X\right)$ and propose the same solution of this problem with the help of the One-step MLE-process.

The asymptotic in this problem corresponds to $\varepsilon \to 0$ and $\psi_\varepsilon \to 0$ and the first question is of course the relation between ε and ψ_ε.

Remark, that if $y_0 = 0$, then the notations $\bar\varepsilon = \psi_\varepsilon^{-1}\varepsilon$, $\bar X_t = \psi_\varepsilon^{-1}X_t$, $\bar Y_t = \psi_\varepsilon^{-1}Y_t$ allow us to rewrite the system (2.117) ad (2.118) as

$$\mathrm{d}\bar X_t = f(\vartheta, t)\,\bar Y_t \mathrm{d}t + \bar\varepsilon\sigma(t)\,\mathrm{d}W_t, \qquad \bar X_0 = 0, \qquad 0 \le t \le T, \tag{2.119}$$

$$\mathrm{d}\bar Y_t = a(\vartheta, t)\,\bar Y_t \mathrm{d}t + b(\vartheta, t)\,\mathrm{d}V_t, \qquad \bar Y_0 = 0. \tag{2.120}$$

To obtain consistent estimators, we have to suppose that $\bar\varepsilon \to 0$, but in this case the system (2.119), (2.120) is a particular case (with $y_0 = 0$) of the system (3.1), (3.2) studied in the next chapter. To avoid this coincidence, we suppose that the initial value $y_0 \neq 0$. Then the limit ($\varepsilon = 0, \psi_\varepsilon = 0$) system (2.117), (2.118) is nondegenerate

$$\frac{\mathrm{d}x_t(\vartheta)}{\mathrm{d}t} = f(\vartheta, t)\,y_t(\vartheta), \; x_0 = 0, \qquad \frac{\mathrm{d}y_t(\vartheta)}{\mathrm{d}t} = a(\vartheta, t)\,y_t(\vartheta), \; y_0, \quad 0 \le t \le T.$$

Remark that in the case of the system (3.1), (3.2) the process $Y_t, 0 \le t \le T$ does not depend on ε, and therefore, the limit is the same Gaussian process.

The Kalman–Bucy filter for the model (2.117), (2.118) is

$$\begin{aligned}\mathrm{d}m(\vartheta, t) &= a(\vartheta, t)m(\vartheta, t)\,\mathrm{d}t\\ &\quad + \frac{f(\vartheta, t)\,\gamma(\vartheta, t)}{\varepsilon^2\sigma(t)^2}\left[\mathrm{d}X_t - f(\vartheta, t)\,m(\vartheta, t)\,\mathrm{d}t\right], \quad m(\vartheta, 0) = y_0,\end{aligned} \tag{2.121}$$

with Riccati equation

$$\frac{\partial\gamma(\vartheta, t)}{\partial t} = 2a(\vartheta, t)\gamma(\vartheta, t) - \frac{f(\vartheta, t)^2\,\gamma(\vartheta, t)^2}{\varepsilon^2\sigma(t)^2} + \psi_\varepsilon^2 b(\vartheta, t)^2, \qquad \gamma(\vartheta, 0) = 0. \tag{2.122}$$

The likelihood ratio function is

$$L\left(\vartheta, X^T\right) = \exp\left(\int_0^T \frac{M(\vartheta, t)}{\varepsilon^2\sigma(t)^2}\mathrm{d}X_t - \int_0^T \frac{M(\vartheta, t)^2}{2\varepsilon^2\sigma(t)^2}\mathrm{d}t\right), \qquad \vartheta \in \Theta. \tag{2.123}$$

Here $M(\vartheta,t)=f(\vartheta,t)\,m(\vartheta,t)$. Recall as well the representation of the observations (2.117)

$$
{\rm d}X_t=f(\vartheta_0,t)\,m(\vartheta_0,t)\,{\rm d}t+\varepsilon\sigma(t)\,{\rm d}\bar W_t,\qquad X_0=0\quad 0\leq t\leq T \tag{2.124}
$$

where $\bar W_t, 0\leq t\leq T$ is the *innovation* Wiener process (see Lemma 1.2).

2.5.2 Parameter Estimation

The properties of MLE and BE are defined by the asymptotic of the integral

$$
\int_0^T\frac{\left[M(\vartheta_0+\varphi_\varepsilon u,t)-M(\vartheta_0,t)\right]^2}{\varepsilon^2\,\sigma(t)^2}{\rm d}t\longrightarrow u^\top {\rm I}(\vartheta_0)\,u.
$$

Suppose that $\psi_\varepsilon=\varepsilon^\delta$, $\delta\in[0,1]$ and the normalizing function we write as follows: $\varphi_\varepsilon=\varepsilon^{h(\delta)}$. The rate of the MLE is given by the function φ_ε because in the regular statistical problems we have

$$
\varphi_\varepsilon^{-1}\left(\hat\vartheta_\varepsilon-\vartheta_0\right)\Longrightarrow\mathcal{N}\left(0,{\rm I}(\vartheta_0)^{-1}\right).
$$

The model (2.117), (2.118) corresponds to different statistical problems depending on the value of δ. Denote $a^*(\vartheta,t)=a(\vartheta,t)+\ell(\vartheta,t)$, where $\ell(\vartheta,t)=[\ln f(\vartheta,t)]'_t$.

Below, it will be shown (see (2.136)) that if $\varepsilon/\psi_\varepsilon\rightarrow 0$ then the following representation is valid

$$
\begin{aligned}
&\frac{M(\vartheta,t)-M(\vartheta_0,t)}{\varepsilon}\\
&\quad=\sqrt{\frac{\psi_\varepsilon}{\varepsilon}}\,\frac{\left[f(\vartheta,t)\,b(\vartheta,t)-f(\vartheta_0,t)\,b(\vartheta_0,t)\right]}{\sqrt{2b(\vartheta,t)\,f(\vartheta,t)}}\sqrt{\sigma(t)}\;\xi_{t,\varepsilon}\,(1+o(1))\\
&\qquad+\frac{1}{\psi_\varepsilon}\,\frac{\left[a^*(\vartheta,t)-a^*(\vartheta_0,t)\right]}{b(\vartheta,t)\,f(\vartheta,t)}f(\vartheta_0,t)\,\sigma(t)\,y_t(\vartheta_0)\,(1+o(1))\\
&\quad=\sqrt{\frac{\psi_\varepsilon}{\varepsilon}}L_1(\vartheta,\vartheta_0,t)\,\xi_{t,\varepsilon}\,(1+o(1))+\frac{1}{\psi_\varepsilon}L_2(\vartheta,\vartheta_0,t)\,(1+o(1))
\end{aligned}
$$

with obvious notation. Here $\xi_{t,\varepsilon}, t\in(\tau_\varepsilon<t\leq T)$ are Gaussian random variables, $\xi_{t,\varepsilon}\Longrightarrow\xi_t\sim\mathcal{N}(0,1)$. The random variables $\xi_t, 0<t\leq T$ are independent.

Note that we obtained five different problems depending on the relation between $1/\psi_\varepsilon$ and $\sqrt{\psi_\varepsilon/\varepsilon}$. Roughly speaking, this can be explained as follows. Suppose that $\delta\in(0,1)$. Then we can write

$$
\begin{aligned}
\int_0^T & \frac{\left[M\left(\vartheta_0+\varphi_{\varepsilon}\, u, t\right)-M\left(\vartheta_0, t\right)\right]^2}{\varepsilon^2\, \sigma(t)^2} \mathrm{d} t \\
& =u^2 \int_0^T\left[\sqrt{\varepsilon \psi_{\varepsilon}}\, \frac{\varphi_{\varepsilon}}{\varepsilon}\, \dot{L}_2\left(\vartheta_0, \vartheta_0, t\right)\ \xi_{t, \varepsilon}+\frac{\varepsilon}{\psi_{\varepsilon}} \frac{\varphi_{\varepsilon}}{\varepsilon} \dot{L}_1\left(\vartheta_0, \vartheta_0, t\right)\right]^2 \frac{\mathrm{d} t}{\sigma(t)^2} \\
& =u^2\, \frac{\psi_{\varepsilon} \varphi_{\varepsilon}^2}{\varepsilon} \int_0^T \dot{L}_1\left(\vartheta_0, \vartheta_0, t\right)^2 \frac{\mathrm{d} t}{\sigma(t)^2}+u^2\, \frac{\varphi_{\varepsilon}^2}{\psi_{\varepsilon}^2} \int_0^T \dot{L}_2\left(\vartheta_0, \vartheta_0, t\right)^2 \frac{\mathrm{d} t}{\sigma(t)^2} \\
& =u^2 \varphi_{\varepsilon}^2\, \frac{\psi_{\varepsilon}}{\varepsilon}\, \mathrm{I}_1\left(\vartheta_0\right)+u^2 \varphi_{\varepsilon}^2\, \frac{1}{\psi_{\varepsilon}^2}\, \mathrm{I}_2\left(\vartheta_0\right) \\
& =u^2\, \mathrm{I}_1\left(\vartheta_0\right)\, \mathbb{I}_{\{0<\delta \leq 1/3\}}+u^2\, \mathrm{I}_2\left(\vartheta_0\right)\, \mathbb{I}_{\{1/3 \leq \delta \leq 1\}} .
\end{aligned}
\qquad (2.125)
$$

Above we denoted $\dot{L}_i\left(\vartheta_0, \vartheta_0, t\right)=\partial L_i\left(\vartheta, \vartheta_0, t\right) / \partial \vartheta|_{\vartheta=\vartheta_0}$, $\mathrm{I}_i\left(\vartheta_0\right), i=1,2$ is Fisher information, set

$$
\varphi_{\varepsilon}=\varepsilon^{\delta}\, \mathbb{I}_{\{0<\delta \leq 1/3\}}+\varepsilon^{\frac{1-\delta}{2}}\, \mathbb{I}_{\{1/3<\delta \leq 1\}},
$$

used the relation

$$
\int_0^T \dot{L}_1\left(\vartheta_0, \vartheta_0, t\right)\ \dot{L}_2\left(\vartheta, \vartheta_0, t\right)\ \xi_{t, \varepsilon} \mathrm{d} t=o(1)
$$

and omitted once more the terms $o(1)$. Denote as well $\mathrm{I}_0\left(\vartheta_0\right)$ the Fisher information defined in (2.30) ($d=1$).

The correpondance between $\psi_{\varepsilon}, \varphi_{\varepsilon}$ and $\mathrm{I}_i\left(\vartheta_0\right), i=0,1,2$ is the follows:

1.	$\psi_{\varepsilon}=1,(\delta=0)$,	$\varphi_{\varepsilon}=\sqrt{\varepsilon}$,	$\mathrm{I}_1\left(\vartheta_0\right)$,
2.	$\psi_{\varepsilon}=\varepsilon^{\delta},(0<\delta<1/3)$,	$\varphi_{\varepsilon}=\varepsilon^{\frac{1-\delta}{2}}$,	$\mathrm{I}_1\left(\vartheta_0\right)$,
3.	$\psi_{\varepsilon}=\varepsilon^{1/3},(\delta=1/3)$,	$\varphi_{\varepsilon}=\varepsilon^{\frac{1}{3}}$,	$\mathrm{I}_3\left(\vartheta_0\right)=\mathrm{I}_1\left(\vartheta_0\right)+\mathrm{I}_2\left(\vartheta_0\right)$,
4.	$\psi_{\varepsilon}=\varepsilon^{\delta},(1/3<\delta<1)$,	$\varphi_{\varepsilon}=\varepsilon^{\delta}$,	$\mathrm{I}_2\left(\vartheta_0\right)$,
5.	$\psi_{\varepsilon}=\varepsilon,(\delta=1)$,	$\varphi_{\varepsilon}=\varepsilon$,	$\mathrm{I}_0\left(\vartheta_0\right)$.

The function $h(\delta), \delta \in[0,1]$ is given on Fig. 2.1. Thus, we have five different problems.

The case $\delta=1$ was studied in the Sect. 2.1. The case $\delta=0$ will be discussed in the Sect. 3.1 separately because this is the only system where the limit model ($\varepsilon=0$) contains a random process.

The value of φ_{ε} in the case $0<\delta<1/3$ seems to be a bit strange by the following reason. Equations (2.117), (2.118) can be called a *dynamic system perturbated by small noise.* It is natural to think that the smaller noise provides the smaller error of estimation. This is usualy true but not in this case! We see that the larger values of δ (smaller noise) give us the bigger error of estimation. The problem is not in mathematics but in the terminology. In this case, the useful information about ϑ is

Fig. 2.1 Function $h(\delta)$

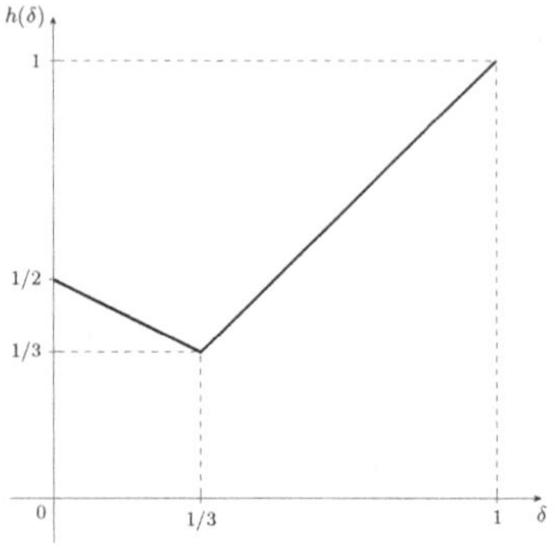

in the functions $b(\vartheta, \cdot)$ and $f(\vartheta, \cdot)$ and therefore ψ_ε in $\psi_\varepsilon b(\vartheta, \cdot)$ is no more a *noise* but can be called an *amplitude* of the signal $b(\vartheta, \cdot)$.

MDE

Below we describe the asymptotic properties of estimators in the case $\varepsilon/\psi_\varepsilon \to 0$ in two situations. First we consider the case with the fixed observation interval $[0, \tau]$ and then the case with the vanishing observation interval $[0, \tau_\varepsilon]$, where $\tau_\varepsilon \to 0$ as $\varepsilon \to 0$.

Fixed Observation Interval

The MDE $\check\vartheta_{\tau,\varepsilon}$ is defined by the equation

$$\int_0^\tau \left[X_t - x_t\left(\check\vartheta_\varepsilon\right)\right]^2 \mathrm{d}t = \inf_{\vartheta\in\Theta} \int_0^\tau \left[X_t - x_t(\vartheta)\right]^2 \mathrm{d}t. \tag{2.126}$$

Introduce the $d \times d$ matrix

$$\check{\mathbf{T}}(\vartheta) = \int_0^\tau \dot{\mathrm{x}}_t(\vartheta)\, \dot{\mathrm{x}}_t(\vartheta)^\top \mathrm{d}t, \qquad \vartheta \in \Theta,$$

Gaussian vector

$$\check\zeta(\vartheta) = \check{\mathbf{T}}(\vartheta)^{-1} \int_0^\tau \int_0^t b(\vartheta, r) \int_r^t f(\vartheta, s)\, \phi(\vartheta, r, s)\, \mathrm{d}s\, \mathrm{d}V_r\, \dot{\mathrm{x}}_t(\vartheta)\, \mathrm{d}t,$$

where

$$\phi(\vartheta, r, s) = \exp\left(\int_r^s a(\vartheta, v)\, \mathrm{d}v\right),$$

and the function

$$\breve{g}\left(\vartheta_0,\nu\right)\equiv\inf_{\left|\vartheta-\vartheta_0\right|>\nu}\int_0^\tau\left[x_t\left(\vartheta_0\right)-x_t\left(\vartheta\right)\right]^2\mathrm{d}t.$$

Recall that $\dot{\mathrm{x}}_t\left(\vartheta\right)=\partial x_t\left(\vartheta\right)/\partial\vartheta$.

Conditions $\breve{\mathscr{F}}$.

$\breve{\mathscr{F}}_1$. *The functions* $f\left(\vartheta,t\right),a\left(\vartheta,t\right),b\left(\vartheta,t\right);\vartheta\in\Theta,t\in\left[0,\tau\right]$ *have two continuous bounded derivatives w.r.t.* $\vartheta\in\bar{\Theta}$.

$\breve{\mathscr{F}}_2$. *The matrix* $\breve{\mathbf{T}}\left(\vartheta\right)$ *is uniformly nondegenerate*

$$\inf_{\vartheta\in\Theta}\inf_{\|\mathrm{e}\|=1,\mathrm{e}\in\mathcal{R}^d}\mathrm{e}\,\breve{\mathbf{T}}\left(\vartheta\right)\mathrm{e}^\top>0.$$

$\breve{\mathscr{F}}_2$. *For any* $\nu>0$

$$\inf_{\vartheta_0\in\Theta}\breve{g}\left(\vartheta_0,\nu\right)>0.$$

Theorem 2.14 *Let the conditions* $\breve{\mathscr{F}}$ *hold. Then the MDE is consistent, asymptotically normal*

$$\psi_\varepsilon^{-1}\left(\breve{\vartheta}_{\tau,\varepsilon}-\vartheta_0\right)\Longrightarrow\breve{\zeta}\left(\vartheta_0\right)$$

and the polynomial moments converge: for any $p>0$

$$\psi_\varepsilon^{-p}\mathbf{E}_{\vartheta_0}\left\|\breve{\vartheta}_{\tau,\varepsilon}-\vartheta_0\right\|^p\longrightarrow\mathbf{E}_{\vartheta_0}\left\|\breve{\zeta}\left(\vartheta_0\right)\right\|^p.$$

Proof This theorem is a particular case of Theorem 2.2. Note that

$$\begin{aligned}X_t&=\int_0^t y_0\,\phi\left(\vartheta_0,0,s\right)f\left(\vartheta_0,s\right)\mathrm{d}s+\varepsilon\int_0^t\sigma\left(s\right)\,\mathrm{d}W_s\\&\quad+\psi_\varepsilon\int_0^t f\left(\vartheta_0,s\right)\int_0^s\phi\left(\vartheta_0,r,s\right)b\left(\vartheta_0,r\right)\mathrm{d}V_r\,\mathrm{d}s\\&=x_t\left(\vartheta_0\right)+\psi_\varepsilon\int_0^t f\left(\vartheta_0,s\right)\int_0^s\phi\left(\vartheta_0,r,s\right)b\left(\vartheta_0,r\right)\mathrm{d}V_r\,\mathrm{d}s\left(1+O\left(\frac{\varepsilon}{\psi_\varepsilon}\right)\right).\end{aligned}$$

□

Vanishing Observation Interval

Let us describe the properties of MDE

$$\check\vartheta_{\tau_\varepsilon}=\arg\inf_{\vartheta\in\Theta}\left\|X_t-x_t\left(\vartheta\right)\right\|_\tau,\qquad \left\|X_t-x_t\left(\vartheta\right)\right\|_\tau^2=\int_0^{\tau_\varepsilon}\left[X_t-x_t\left(\vartheta\right)\right]^2{\rm d}t$$

constructed by the observations $X^{\tau_\varepsilon}=(X_t,0\leq t\leq\tau_\varepsilon)$ on the vanishing interval $[0,\tau_\varepsilon]$, where $\tau_\varepsilon\to 0$. We will need such estimator for the construction of One-step MLE process later. The parameter ϑ is one-dimensional $\vartheta\in\Theta=(\alpha,\beta)$, where α and β are finite.

Conditions $\mathscr{F}^\circ$.

$\mathscr{F}_1^\circ$. *The functions* $f\left(\vartheta,t\right),a\left(\vartheta,t\right),\vartheta\in\Theta,t\in[0,t_0]$ *have two continuous bounded derivatives w.r.t.* ϑ. *Here* $t_0>0$ *is some (small) real from the interval* $[0,T]$.

$\mathscr{F}_2^\circ$. *The functions* $f\left(\vartheta,0\right),\dot f\left(\vartheta,0\right),b\left(\vartheta,0\right)$ *and* $\sigma\left(0\right)$ *are separated from zero uniformly on* $\vartheta\in\Theta$.

$\mathscr{F}_3^\circ$. *The functions* $f\left(\vartheta,t\right),a\left(\vartheta,t\right),b\left(\vartheta,t\right),\vartheta\in\Theta,t\in[0,t_0]$ *have continuous bounded derivatives w.r.t.* t.

Let us denote $\kappa=\inf_{\vartheta\in\Theta}\left|\dot f\left(\vartheta,0\right)\right|$ and set

$$D\left(\vartheta\right)=y_0^{-2}\dot f\left(\vartheta,0\right)^{-2}\left(\frac{33}{140}f\left(\vartheta,0\right)^2b\left(\vartheta,0\right)^2+\frac{6}{5}\sigma\left(0\right)^2\right).$$

Proposition 2.7 *Let conditions* $\mathscr{F}^\circ$ *hold and* $\sqrt{\tau_\varepsilon}\varepsilon^{-1}\to\infty$. *Then the MDE* $\check\vartheta_{\tau_\varepsilon}$ *is consistent and asymptotically normal*

$$\sqrt{\frac{\psi_\varepsilon}{\varepsilon}}\left(\check\vartheta_{\tau_\varepsilon}-\vartheta_0\right)\Longrightarrow\mathcal{N}\left(0,D\left(\vartheta_0\right)\right).\qquad(2.127)$$

Proof To verify the consistency, we first obtain the relation as in (2.14)

$$\begin{aligned}\mathbf{P}_{\vartheta_0}\left(\left|\check\vartheta_{\tau_\varepsilon}-\vartheta_0\right|>\nu\right)&\leq\mathbf{P}_{\vartheta_0}\left(2\left\|X_t-x_t\left(\vartheta_0\right)\right\|_\tau>\check g_\tau\left(\vartheta_0,\nu\right)\right)\\&\leq\frac{4}{\check g_\tau\left(\vartheta_0,\nu\right)^2}\mathbf{E}_{\vartheta_0}\left\|X_t-x_t\left(\vartheta_0\right)\right\|_\tau^2,\end{aligned}\qquad(2.128)$$

and then consider the asymptotics of these integrals as $\tau_\varepsilon\to 0$ (see Lemma 2.2 and Remark 2.2 where the similar problems with $\psi_\varepsilon=\varepsilon$ were considered).

Let us compare the contributions of the terms containing Wiener processes $W_t,0\leq t\leq\tau_\varepsilon$ and $V_t,0\leq t\leq\tau_\varepsilon$.

The asymptotic ($t\to 0$) of one integral is

$$\begin{aligned}J_1\left(t\right)&=\int_0^t f\left(\vartheta_0,s\right)\int_0^s\phi\left(\vartheta_0,r,s\right)b\left(\vartheta_0,r\right){\rm d}V_r\,{\rm d}s\\&=\int_0^t\int_r^t f\left(\vartheta_0,s\right)\phi\left(\vartheta_0,r,s\right){\rm d}s\;b\left(\vartheta_0,r\right)\,{\rm d}V_r\\&=f\left(\vartheta_0,0\right)b\left(\vartheta_0,0\right)\int_0^t\left(t-r\right)\,{\rm d}V_r\left(1+o\left(1\right)\right)\end{aligned}$$

and

$$\mathbf{E}_{\vartheta_0}\int_0^{\tau_\varepsilon} J_1\left(t\right)^2 \mathrm{d}t = f\left(\vartheta_0,0\right)^2 b\left(\vartheta_0,0\right)^2 \int_0^{\tau_\varepsilon}\int_0^t \left(t-r\right)^2\,\mathrm{d}r\,\mathrm{d}t\left(1+o\left(1\right)\right)$$
$$=\frac{f\left(\vartheta_0,0\right)^2 b\left(\vartheta_0,0\right)^2}{12}\,\tau_\varepsilon^4\,\left(1+o\left(1\right)\right).$$

For the second integral, we have

$$\mathbf{E}_{\vartheta_0}\int_0^{\tau_\varepsilon}\left(\int_0^t \sigma\left(s\right)\mathrm{d}W_s\right)^2\mathrm{d}t=\frac{\sigma\left(0\right)^2}{2}\,\tau_\varepsilon^2\,\left(1+o\left(1\right)\right).$$

The function $\breve{g}_\tau\left(\vartheta_0,\nu\right)$ is estimated as follows. We have

$$x_t\left(\vartheta\right)-x_t\left(\vartheta_0\right)=y_0\int_0^t\left[f\left(\vartheta,s\right)\phi\left(\vartheta,0,s\right)-f\left(\vartheta_0,s\right)\phi\left(\vartheta_0,0,s\right)\right]\mathrm{d}s$$
$$=y_0\,t\,\left[f\left(\vartheta,0\right)-f\left(\vartheta_0,0\right)\right]\left(1+o\left(1\right)\right).$$

Thus, for the small τ_ε

$$\left\|x_t\left(\vartheta\right)-x_t\left(\vartheta_0\right)\right\|_\tau^2=\frac{y_0^2\tau_\varepsilon^3}{3}\left[f\left(\vartheta,0\right)-f\left(\vartheta_0,0\right)\right]^2\left(1+o\left(1\right)\right)$$
$$=\frac{y_0^2\tau_\varepsilon^3}{3}\,\dot f\left(\bar\vartheta,0\right)^2\left[\vartheta-\vartheta_0\right]^2\left(1+o\left(1\right)\right).$$

Therefore, for some $\varepsilon_0>0$ and $\varepsilon<\varepsilon_0$ we obtain the estimate

$$\breve{g}_\tau\left(\vartheta_0,\nu\right)\geq\frac{y_0\,\kappa}{2}\,\tau_\varepsilon^{3/2}\,\nu.$$

The RHS of the inequality (2.128) is

$$\breve{g}_\tau\left(\vartheta_0,\nu\right)^{-2}\mathbf{E}_{\vartheta_0}\left\|X_t\left(\vartheta\right)-x_t\left(\vartheta_0\right)\right\|_\tau^2\leq C\psi_\varepsilon^2\tau_\varepsilon+C\frac{\varepsilon^2}{\tau_\varepsilon}. \tag{2.129}$$

Any choice of the rate $\tau_\varepsilon\rightarrow 0$ such that $\varepsilon^2/\tau_\varepsilon\rightarrow 0$ makes the MDE consistent.

The asymptotic normality of this estimator will follow from the asymptotic normality of the ratio of two integrals

$$\breve{\vartheta}_{\tau_\varepsilon}-\vartheta_0=\frac{\int_0^{\tau_\varepsilon}\left[X_t-x_t\left(\vartheta_0\right)\right]\dot x_t\left(\vartheta_0\right)\mathrm{d}t}{\int_0^{\tau_\varepsilon}\dot x_t\left(\vartheta_0\right)^2\mathrm{d}t}\left(1+o\left(1\right)\right)$$

$$
\begin{aligned}
&= \frac{3\psi_\varepsilon\, f(\vartheta_0, 0)\, b(\vartheta_0, 0)}{\tau_\varepsilon^3\, y_0 \dot f(\vartheta_0, 0)} \int_0^{\tau_\varepsilon} t \int_0^t (t-r)\, \mathrm{d}V_r\, \mathrm{d}t\ \left(1 + o(1)\right) \\
&\quad + \frac{3\,\varepsilon\,\sigma(0)}{\tau_\varepsilon^3\, y_0\, \dot f(\vartheta_0, 0)} \int_0^{\tau_\varepsilon} t\, W_t\, \mathrm{d}t\ \left(1 + o(1)\right).
\end{aligned}
$$

Here we used the asymptotics

$$
\int_0^{\tau_\varepsilon} \dot x_t(\vartheta_0)^2\, \mathrm{d}t = \frac{y_0^2}{3}\, \dot f(\vartheta_0, 0)^2\, \tau_\varepsilon^3\ \left(1 + o(1)\right).
$$

Remark that

$$
\int_0^{\tau_\varepsilon} t \int_0^{\tau_\varepsilon} (t-r)\, \mathrm{d}V_r\, \mathrm{d}t = \int_0^{\tau_\varepsilon} \int_r^{\tau_\varepsilon} t\,(t-r)\, \mathrm{d}t\, \mathrm{d}V_r \sim \mathcal{N}\left(0, \frac{11\tau_\varepsilon^7}{420}\right) = \sqrt{\frac{11\tau_\varepsilon^7}{420}}\, \zeta_1,
$$

$$
\int_0^{\tau_\varepsilon} t\, W_t\, \mathrm{d}t \sim \mathcal{N}\left(0, \frac{2\tau_\varepsilon^5}{15}\right) = \sqrt{\frac{2\tau_\varepsilon^5}{15}}\, \zeta_2\left(1 + o(1)\right).
$$

Here $\zeta_1 \sim \mathcal{N}(0, 1)$ and $\zeta_2 \sim \mathcal{N}(0, 1)$ are independent r.v.'s.

Thus,

$$
\begin{aligned}
\check\vartheta_{\tau_\varepsilon} - \vartheta_0 &= \frac{f(\vartheta_0, 0)\, b(\vartheta_0, 0)}{y_0 \dot f(\vartheta_0, 0)} \sqrt{\frac{33}{140}}\, \zeta_1\, \psi_\varepsilon \sqrt{\tau_\varepsilon}\left(1 + o(1)\right) \\
&\quad + \frac{\sigma(0)}{y_0\, \dot f(\vartheta_0, 0)} \sqrt{\frac{6}{5}}\, \zeta_2\, \frac{\varepsilon}{\sqrt{\tau_\varepsilon}}\ \left(1 + o(1)\right).
\end{aligned}
$$

The equation of balance

$$
\psi_\varepsilon \sqrt{\tau_\varepsilon} = \frac{\varepsilon}{\sqrt{\tau_\varepsilon}}
$$

gives us the rate $\tau_\varepsilon = \varepsilon / \psi_\varepsilon$. Hence

$$
\sqrt{\frac{\psi_\varepsilon}{\varepsilon}}\left(\check\vartheta_{\tau_\varepsilon} - \vartheta_0\right) = \frac{f(\vartheta_0, 0)\, b(\vartheta_0, 0)}{y_0 \dot f(\vartheta_0, 0)} \sqrt{\frac{33}{140}}\, \zeta_1 + \frac{\sigma(0)}{y_0\, \dot f(\vartheta_0, 0)} \sqrt{\frac{6}{5}}\, \zeta_2 + o(1).
$$

Here the equality is understood in distribution. □

Remark 2.17 The Proposition 2.7 describes the properties of MDE in the case $|\dot f(\vartheta, 0)| > 0$. If $\dot f(\vartheta, 0) = 0$ and $|\dot a(\vartheta, 0)| > 0$, then the MDE $\check\vartheta_{\tau_\varepsilon}$ can be consistent but with the worse rate.

Note that in the case $f(\vartheta,t) = f(t)$, $a(\vartheta,t) = a(t)$ and $|\dot b(\vartheta,0)| > 0$, the consistent estimation is impossible for this model of observations.

Remark 2.18 Remark that if $\sigma(t) \equiv 0$ and the observed system is

$$\begin{aligned} \mathrm{d}X_t &= f(\vartheta_0,t)\,Y_t\,\mathrm{d}t, \qquad X_0 = 0, \qquad 0 \le t \le T,\\ \mathrm{d}Y_t &= a(\vartheta_0,t)\,Y_t\,\mathrm{d}t + \psi_\varepsilon b(\vartheta_0,t)\,\mathrm{d}V_t, \qquad Y_0 = y_0 \neq 0, \end{aligned}$$

then the problem of estimation ϑ_0 is singular. This means that even if $\psi_\varepsilon \equiv 1$ the value ϑ_0 can be "estimated" without error. This follows directly from the relations (2.128) and (2.129).

Remark 2.19 If $\delta > 1/3$, then the rate $\varphi_\varepsilon^{(f)} = \psi_\varepsilon = \varepsilon^\delta$ of convergence of MDE constructed by observations on the fixed interval is better than the rate $\varphi_\varepsilon^{(v)} = \sqrt{\varepsilon/\psi_\varepsilon} = \varepsilon^{(1-\delta)/2}$ of the same estimator constructed by the observations on the vanishing interval. Indeed, the comparison of the rates gives us

$$\frac{\varphi_\varepsilon^{(f)}}{\varphi_\varepsilon^{(v)}} = \psi_\varepsilon\sqrt{\frac{\psi_\varepsilon}{\varepsilon}} = \varepsilon^{\frac{3\delta-1}{2}} \longrightarrow 0. \qquad (2.130)$$

If $\delta < 1/3$, the situation changes, i.e., one can think that *less observation is better for the estimation of the unknown parameter*! We have to note that this is not a general property of the model but the property of the particular estimator. Mention that in the next sections the properties of the MLE and BE are studied under condition $\delta > 1/3$, $\delta = 1/3$ and $\delta < 1/3$. As it was mentioned above, this model of observations with $\delta < 1/3$ is in some sense more close to the model studied in Sect. 3.1.

Remark 2.20 Note that if $y_0 = 0$, then the MDE's proposed above with fixed and vanishing intervals cannot be consistent. This follows from the equality $y_t(\vartheta) \equiv 0$, which gives $x_t(\vartheta) \equiv 0$.

MLE and BE

Case $\delta \in (1/3, 1)$.

The MLE $\hat\vartheta_\varepsilon$ and BE $\tilde\vartheta_\varepsilon$ are defined by the usual relations

$$L\left(\hat\vartheta_\varepsilon, X^T\right) = \sup_{\vartheta\in\Theta} L\left(\vartheta, X^T\right),$$

$$\tilde\vartheta_\varepsilon = \int_\Theta \vartheta p\left(\vartheta|X^T\right)\mathrm{d}\vartheta, \qquad p\left(\vartheta|X^T\right) = \frac{p(\vartheta)\,L\left(\vartheta,X^T\right)}{\int_\Theta p(\theta)\,L\left(\theta,X^T\right)\mathrm{d}\theta}.$$

If this equation for the MLE has more than one solution, then any one of them can be taken as MLE. Here $p(\vartheta), \vartheta \in \Theta$ is some known continuous positive density function.

We study the properties of these estimators in the case $\psi_\varepsilon = \varepsilon^\delta \to 0$, where $\frac{1}{3} < \delta < 1$.

Denote $\ell(\vartheta, t) = \partial \ln f(\vartheta, t)/\partial t = f_t'(\vartheta, t)/f(\vartheta, t)$. Then the Fisher information in this problem can be written as follows

$$\mathrm{I}_2(\vartheta) = \int_0^T \frac{\left[\dot a(\vartheta, t) + \dot\ell(\vartheta, t)\right]^2 y_t(\vartheta)^2}{b(\vartheta, t)^2}\,\mathrm{d}t, \qquad \vartheta \in \Theta.$$

Define as well the function

$$G(\vartheta, \vartheta_0) = \int_0^T \frac{\left[a(\vartheta, t) - a(\vartheta_0, t) + \ell(\vartheta, t) - \ell(\vartheta_0, t)\right]^2 y_t(\vartheta_0)^2}{b(\vartheta, t)^2}\mathrm{d}t.$$

The properties of estimators will be derived under the following regularity conditions.

Conditions $\hat{\mathscr{F}}$.

$\hat{\mathscr{F}}_1$. *The functions* $f(\vartheta, t), a(\vartheta, t), b(\vartheta, t), t \in [0, T], \vartheta \in \Theta$ *and* $\sigma(t), t \in [0, T]$ *have continuous derivatives w.r.t. t.*

$\hat{\mathscr{F}}_2$. *The functions* $f(\vartheta, t), b(\vartheta, t), t \in [0, T], \vartheta \in \Theta$ *and* $\sigma(t), 0 \le t \le T$ *are bounded and separated from zero.*

$\hat{\mathscr{F}}_3$. *The functions* $f(\vartheta, t), a(\vartheta, t), t \in [0, T], \vartheta \in \Theta$ *are two times continuously differentiable w.r.t.* ϑ.

$\hat{\mathscr{F}}_4$. *The Fisher information is uniformly positive*

$$\inf_{\vartheta \in \Theta} \mathrm{I}_2(\vartheta) > 0.$$

$\hat{\mathscr{F}}_5$. *For any* $\nu > 0$

$$g(\nu) \equiv \inf_{\vartheta_0 \in \Theta} \inf_{|\vartheta - \vartheta_0| > \nu} G(\vartheta, \vartheta_0) > 0.$$

For the sake of simplicity and without loss of generality, we assume that the functions $f(\cdot), b(\cdot)$ and $\sigma(\cdot)$ are positive.

The family of measures $\left\{\mathbf{P}_\vartheta^{(\varepsilon)}, \vartheta \in \Theta\right\}$ induced by the observation process $X^T = (X_t, 0 \le t \le T)$ on the space $(\mathcal{C}[0, T], \mathfrak{B}_T)$ of continuous functions on $[0, T]$ is locally asymptotically normal (LAN) (see Theorem 1.13 and Lemma 2.8 below). Therefore, the mean-squared estimation error satisfies the Hajek's minimax lower bound: for any $\vartheta_0 \in \Theta$ and any estimator $\bar\vartheta_\varepsilon$

$$\lim_{\nu \to 0} \varliminf_{\varepsilon \to 0} \sup_{|\vartheta - \vartheta_0| \le \nu} \psi_\varepsilon^{-2} \mathbf{E}_\vartheta \left|\bar\vartheta_\varepsilon - \vartheta\right|^2 \ge \mathrm{I}_2(\vartheta_0)^{-1}.$$

The estimator ϑ_ε^* is called asymptotically efficient if for any $\vartheta_0 \in \Theta$

$$\lim_{\nu\to 0}\lim_{\varepsilon\to 0}\sup_{|\vartheta-\vartheta_0|\leq\nu}\psi_\varepsilon^{-2}\mathbf{E}_\vartheta\left|\vartheta_\varepsilon^*-\vartheta\right|^2=\mathrm{I}_2\left(\vartheta_0\right)^{-1}.$$

Theorem 2.15 *Let conditions $\hat{\mathscr{F}}$ hold. Then the MLE $\hat\vartheta_\varepsilon$ and BE are uniformly consistent, asymptotically normal*

$$\psi_\varepsilon^{-1}\left(\hat\vartheta_\varepsilon-\vartheta_0\right)\Longrightarrow\zeta\left(\vartheta_0\right)\sim\mathcal{N}\left(0,\mathrm{I}_2\left(\vartheta_0\right)^{-1}\right),\qquad\psi_\varepsilon^{-1}\left(\tilde\vartheta_\varepsilon-\vartheta_0\right)\Longrightarrow\zeta\left(\vartheta_0\right),$$

the convergence of moments holds,

$$\psi_\varepsilon^{-p}\mathbf{E}_{\vartheta_0}|\hat\vartheta_\varepsilon-\vartheta_0|^p\rightarrow\mathbf{E}_{\vartheta_0}|\zeta\left(\vartheta_0\right)|^p,\qquad\psi_\varepsilon^{-p}\mathbf{E}_{\vartheta_0}|\tilde\vartheta_\varepsilon-\vartheta_0|^p\rightarrow\mathbf{E}_{\vartheta_0}\left|\zeta\left(\vartheta_0\right)\right|^p,$$

for any $p>0$, and both estimators are asymptotically efficient.

Proof We verify the conditions of Theorem 1.10 and obtain the asymptotics of the MLE. Then by Theorem 1.12, we obtain the properties of BE too.

In the case $\vartheta=\vartheta_0$ we have

$$\mathrm{d}m\left(\vartheta_0,t\right)=a(\vartheta_0,t)m\left(\vartheta_0,t\right)\mathrm{d}t+\psi_\varepsilon\frac{\gamma_*\left(\vartheta_0,t\right)f\left(\vartheta_0,t\right)}{\sigma\left(t\right)}\,\mathrm{d}\bar W_t,\qquad m\left(\vartheta_0,0\right)=y_0.$$

Here $\bar W_t, 0\leq t\leq T$ is the innovation Wiener process (see (2.124)).

Define the normalized likelihood ratio process

$$Z_\varepsilon\left(\vartheta_0,u\right)=\frac{L\left(\vartheta_0+\psi_\varepsilon u,X^T\right)}{L\left(\vartheta_0,X^T\right)},\qquad u\in\mathbb{U}_\varepsilon=\left(\frac{\alpha-\vartheta_0}{\psi_\varepsilon},\frac{\beta-\vartheta_0}{\psi_\varepsilon}\right).$$

The substitution of the observations (2.124) in $Z_\varepsilon\left(\vartheta_0,u\right)$ gives us the representation (see (2.123))

$$\ln Z_\varepsilon\left(\vartheta_0,u\right)=\int_0^T\frac{M\left(\vartheta_u,t\right)-M\left(\vartheta_0,t\right)}{\varepsilon\sigma\left(t\right)}\,\mathrm{d}\bar W_t-\int_0^T\frac{\left[M\left(\vartheta_u,t\right)-M\left(\vartheta_0,t\right)\right]^2}{2\varepsilon^2\sigma\left(t\right)^2}\,\mathrm{d}t,$$

where $\vartheta_u=\vartheta_0+\psi_\varepsilon u$. Taylor formula yields

$$M\left(\vartheta_0+\psi_\varepsilon u,t\right)-M\left(\vartheta_0,t\right)=\psi_\varepsilon u\dot M\left(\vartheta_0,t\right)\left(1+O\left(\psi_\varepsilon\right)\right).$$

Recall that $\dot M\left(\vartheta,t\right)$ is the derivative w.r.t. ϑ of the random process $M\left(\vartheta,t\right)$,

$$\dot M\left(\vartheta,t\right)=\dot f\left(\vartheta,t\right)m\left(\vartheta,t\right)+f\left(\vartheta,t\right)\dot m\left(\vartheta,t\right).$$

Thus, we need to find the asymptotics of the derivative $\dot m(\vartheta,t)$ of $m(\vartheta,t)$ and the related derivative $\dot\gamma_*(\vartheta,t)$ of $\gamma_*(\vartheta,t)$, since the Eq. (2.121) for $m(\vartheta,t)$ depends on $\gamma_*(\vartheta,t)$.

We start with the asymptotics of Riccati equation. It will be convenient to introduce the function $\gamma_*(\vartheta,t)=\gamma(\vartheta,t)/(\varepsilon\psi_\varepsilon)$. Equations (2.121) and (2.122) can be rewritten as follows

$$\mathrm{d}m(\vartheta,t)=\left[a(\vartheta,t)-\frac{\psi_\varepsilon}{\varepsilon}\frac{\gamma_*(\vartheta,t)\,f(\vartheta,t)^2}{\sigma(t)^2}\right]m(\vartheta,t)\,\mathrm{d}t$$
$$+\frac{\psi_\varepsilon}{\varepsilon}\frac{\gamma_*(\vartheta,t)\,f(\vartheta,t)}{\sigma(t)^2}\,\mathrm{d}X_t,$$

$$\frac{\varepsilon}{\psi_\varepsilon}\frac{\partial\gamma_*(\vartheta,t)}{\partial t}=\frac{2\varepsilon}{\psi_\varepsilon}a(\vartheta,t)\gamma_*(\vartheta,t)-\frac{\gamma_*(\vartheta,t)^2\,f(\vartheta,t)^2}{\sigma(t)^2}+b(\vartheta,t)^2,$$

subject to initial conditions $m(\vartheta,0)=y_0$, $\gamma_*(\vartheta,0)=0$.

Let us denote

$$\gamma_0(\vartheta,t)=\frac{b(\vartheta,t)\,\sigma(t)}{f(\vartheta,t)}.$$

□

Lemma 2.6 *Let the conditions* $\hat{\mathscr{F}}_{1-2}$ *hold and* $\varepsilon\psi_\varepsilon{}^{-1}\to 0$. *Then, for any* $t_0\in(0,T]$,

$$\sup_{t_0\le t\le T}\left|\gamma_*(\vartheta,t)-\gamma_0(\vartheta,t)\right|=O\left(\frac{\varepsilon}{\psi_\varepsilon}\right).$$

Proof Introduce another Riccati equation

$$\frac{\partial\hat\gamma(t)}{\partial t}=2a_M\hat\gamma(t)-\frac{\psi_\varepsilon}{\varepsilon}\frac{\hat\gamma(t)^2\,f_m^2}{S^2}+\frac{\psi_\varepsilon}{\varepsilon}b_M^2,\qquad\hat\gamma(0)=0,\tag{2.131}$$

where we denoted

$$a_M=\sup_{0\le t\le T}\sup_{\vartheta\in\Theta}|a(\vartheta,t)|,\qquad f_m=\inf_{0\le t\le T}\inf_{\vartheta\in\Theta}f(\vartheta,t),$$
$$b_M=\sup_{0\le t\le T}\sup_{\vartheta\in\Theta}|b(\vartheta,t)|,\qquad \sigma_M=\sup_{0\le t\le T}\sigma(t).$$

Note that by condition $\hat{\mathscr{F}}_2$ we have $f_m>0$. Due to the comparison theorem for ordinary differential equations, the estimate $\gamma_*(\vartheta,t)\le\hat\gamma(t)$ holds for all $t\in[0,T]$. This bound is intuitive too, i.e., as replacing the "noise coefficient" $\sigma(t)$ by its maximal value should increase the estimation error. Further, as we take smaller coefficient f_m in the drift $f(\vartheta, t)\, Y_t$ the error increases as well. The solution of Riccati equation (2.131) with constant coefficients can be written explicitly (see (1.26))

$$\hat{\gamma}(t) = e^{-2R_\varepsilon t}\left[\frac{\psi_\varepsilon f_m^2\left(1-e^{-2R_\varepsilon t}\right)}{2\varepsilon R_\varepsilon \sigma_M^2} - \hat{\gamma}^{-1}\right]^{-1} + \hat{\gamma}$$

Here

$$R_\varepsilon = \left(a_m^2 + \frac{\psi_\varepsilon^2}{\varepsilon^2}\frac{b_M^2 f_m^2}{\sigma_M^2}\right)^{1/2} = \frac{\psi_\varepsilon}{\varepsilon}\frac{f_m b_M}{\sigma_M}\left(1 + O\left(\frac{\varepsilon^2}{\psi_\varepsilon^2}\right)\right)$$

and

$$\hat{\gamma} = \frac{\varepsilon}{\psi_\varepsilon}\frac{a_m\sigma_M^2}{f_m^2}\left[\left(1+\frac{\psi_\varepsilon^2}{\varepsilon^2}\frac{b_M^2 f_m^2}{a_m^2\sigma_M^2}\right)^{1/2} - 1\right] = \frac{b_M\sigma_M}{f_m}\left(1 + O\left(\frac{\varepsilon}{\psi_\varepsilon}\right)\right).$$

Therefore, for any $t_0 \in (0, T]$

$$\sup_{t_0\le t\le T}\left|\hat{\gamma}(t) - \frac{b_M\sigma_M}{f_m}\right| = O\left(\frac{\varepsilon}{\psi_\varepsilon}\right). \tag{2.132}$$

Let us write the Riccati equation in integral form

$$\frac{\varepsilon}{\psi_\varepsilon}\left[\gamma_*(\vartheta, t) - \gamma_*(\vartheta, t_0) + 2\int_{t_0}^{t} a(\vartheta, s)\,\gamma_*(\vartheta, s)\,\mathrm{d}s\right]$$
$$= -\int_{t_0}^{t}\frac{\gamma_*(\vartheta, s)^2 f(\vartheta, s)^2}{\sigma(s)^2}\mathrm{d}s + \int_{t_0}^{t} b(\vartheta, s)^2\,\mathrm{d}s.$$

As $\gamma_*(\vartheta, t) \le \hat{\gamma}(t)$ and $\hat{\gamma}(t)$ according to (2.132) is bounded, the left-hand side of the integral equation tends to 0. Hence,

$$\int_{t_0}^{t}\frac{\gamma_*(\vartheta, s)^2 f(\vartheta, s)^2}{\sigma(s)^2}\mathrm{d}s = \int_{t_0}^{t} b(\vartheta, s)^2\,\mathrm{d}s + O\left(\frac{\varepsilon}{\psi_\varepsilon}\right)$$

where $O\left(\frac{\varepsilon}{\psi_\varepsilon}\right)$ has continuous derivative w.r.t. t. All functions in these integrals are continuously differentiable on s, and therefore, it is possible to verify that

$$\gamma_*(\vartheta, t)^2 = \frac{b(\vartheta, t)^2\sigma(t)^2}{f(\vartheta, t)^2} + O\left(\frac{\varepsilon}{\psi_\varepsilon}\right) = \gamma_0(\vartheta, t)^2 + O\left(\frac{\varepsilon}{\psi_\varepsilon}\right).$$

□

It can be checked that all terms of order $O\left(\frac{\varepsilon}{\psi_\varepsilon}\right)$ above satisfy

$$\left|O\left(\frac{\varepsilon}{\psi_\varepsilon}\right)\right|\leq C\frac{\varepsilon}{\psi_\varepsilon},$$

where the constant $C>0$ can be chosen independent on ϑ.

Below $O_p\left(\frac{\varepsilon}{\psi_\varepsilon}\right)$ and $o_p\left(\frac{\varepsilon}{\psi_\varepsilon}\right)$ means that for any $p>1$

$$\mathbf{E}_{\vartheta_0}\left|O_p\left(\frac{\varepsilon}{\psi_\varepsilon}\right)\right|^p\leq C\left(\frac{\varepsilon}{\psi_\varepsilon}\right)^p,\qquad \mathbf{E}_{\vartheta_0}\left|o_p\left(\frac{\varepsilon}{\psi_\varepsilon}\right)\right|^p\leq c_\varepsilon\left(\frac{\varepsilon}{\psi_\varepsilon}\right)^p,\qquad c_\varepsilon\to 0,$$

where the constants $C>0$ and $c_\varepsilon>0$ do not depend on ϑ.

Lemma 2.7 *Let conditions $\hat{\mathscr{F}}_{1-3}$ hold. Then for any $u\in\mathbb{U}_\varepsilon$ and any $t\in(0,T]$ the relation*

$$M\left(\vartheta_u,t\right)-M\left(\vartheta_0,t\right)=\varepsilon\, u\,\frac{[\dot a\left(\vartheta_0,t\right)+\dot\ell\left(\vartheta_0,t\right)]y_t\left(\vartheta_0\right)\sigma\left(t\right)}{b\left(\vartheta_0,t\right)}\left(1+o\left(1\right)\right), \tag{2.133}$$

holds.

Proof By Lemma 2.6, we can replace the function $\gamma_*\left(\vartheta,t\right)$ by the function $\gamma_0\left(\vartheta,t\right)=b\left(\vartheta,t\right)\sigma\left(t\right)f\left(\vartheta,t\right)^{-1}$. Therefore, omitting the negligible terms, we obtain the equation

$$\begin{aligned}
\mathrm{d}m\left(\vartheta,t\right)&=\left[a(\vartheta,t)-\frac{\psi_\varepsilon}{\varepsilon}\frac{b\left(\vartheta,t\right)f\left(\vartheta,t\right)}{\sigma(t)}\right]m\left(\vartheta,t\right)\mathrm{d}t+\frac{\psi_\varepsilon}{\varepsilon}\frac{b\left(\vartheta,t\right)}{\sigma\left(t\right)}\mathrm{d}X_t\\
&=\left[a(\vartheta,t)-\frac{\psi_\varepsilon}{\varepsilon}B\left(\vartheta,t\right)\right]m\left(\vartheta,t\right)\mathrm{d}t+\frac{\psi_\varepsilon}{\varepsilon}\frac{b\left(\vartheta,t\right)}{\sigma\left(t\right)}f\left(\vartheta_0,t\right)m\left(\vartheta_0,t\right)\mathrm{d}t\\
&\quad+\psi_\varepsilon b\left(\vartheta,t\right)\mathrm{d}\bar W_t,\qquad m\left(\vartheta,t_0\right),\qquad t_0\leq t\leq T,
\end{aligned}$$

where we denoted $B\left(\vartheta,t\right)=b\left(\vartheta,t\right)f\left(\vartheta,t\right)\sigma\left(t\right)^{-1}$.

To write the equation for $M\left(\vartheta,t\right)$ we remark that

$$\mathrm{d}M\left(\vartheta,t\right)=f\left(\vartheta,t\right)\mathrm{d}m\left(\vartheta,t\right)+f_t'\left(\vartheta,t\right)m\left(\vartheta,t\right)\mathrm{d}t.$$

Thus, the equations for $M\left(\vartheta,t\right),t_0\leq t\leq T$ and $M\left(\vartheta_0,t\right),t_0\leq t\leq T$ are

$$\begin{aligned}
\mathrm{d}M\left(\vartheta,t\right)&=[a\left(\vartheta,t\right)+\ell\left(\vartheta,t\right)]M\left(\vartheta,t\right)\mathrm{d}t-\frac{\psi_\varepsilon}{\varepsilon}B\left(\vartheta,t\right)M\left(\vartheta,t\right)\mathrm{d}t\\
&\quad+\frac{\psi_\varepsilon}{\varepsilon}B\left(\vartheta,t\right)M\left(\vartheta_0,t\right)\mathrm{d}t+\psi_\varepsilon B\left(\vartheta,t\right)\sigma\left(t\right)\mathrm{d}\bar W_t,\qquad M\left(\vartheta,t_0\right),
\end{aligned}$$

$$\mathrm{d}M(\vartheta_0,t)=[a(\vartheta_0,t)+\ell(\vartheta_0,t)]M(\vartheta_0,t)\,\mathrm{d}t+\psi_\varepsilon B(\vartheta_0,t)\,\sigma(t)\,\mathrm{d}\bar{W}_t,\ M(\vartheta_0,t_0).$$

The particular values of $M(\vartheta,t_0)$ and $M(\vartheta_0,t_0)$ do not important because their contributions are exponentially negligible.

Let us denote $a^*(\vartheta,t)=a(\vartheta,t)+\ell(\vartheta,t)$, $\delta_M(t)=M(\vartheta,t)-M(\vartheta_0,t)$, $\delta_{a^*}(t)=a^*(\vartheta,t)-a^*(\vartheta_0,t)$ and $\delta_B(t)=B(\vartheta,t)-B(\vartheta_0,t)$. Then the equations for $M(\vartheta_0,t)$ and $\delta_M(t)$ are

$$\mathrm{d}M(\vartheta_0,t)=a^*(\vartheta_0,t)\,M(\vartheta_0,t)\,\mathrm{d}t+\psi_\varepsilon B(\vartheta_0,t)\,\sigma(t)\,\mathrm{d}\bar{W}_t,\qquad M(\vartheta_0,t_0),\tag{2.134}$$

$$\begin{aligned}\mathrm{d}\delta_M(t)&=a^*(\vartheta,t)\,\delta_M(t)\,\mathrm{d}t-\frac{\psi_\varepsilon}{\varepsilon}B(\vartheta,t)\,\delta_M(t)\,\mathrm{d}t+\delta_{a^*}(t)\,M(\vartheta_0,t)\,\mathrm{d}t\\&\quad+\psi_\varepsilon\delta_B(t)\,\sigma(t)\,\mathrm{d}\bar{W}_t,\qquad \delta_M(t_0)=M(\vartheta,t_0)-M(\vartheta_0,t_0).\end{aligned}$$

The solution of this equation is

$$\begin{aligned}\delta_M(t)&=\delta_M(0)\,\phi_\varepsilon(\vartheta,0,t)+\int_0^t\phi_\varepsilon(\vartheta,s,t)\,\delta_{a^*}(s)\,M(\vartheta_0,s)\,\mathrm{d}s\\&\quad+\int_0^t\phi_\varepsilon(\vartheta,s,t)\,a^*(\vartheta,s)\,\delta_M(s)\,\mathrm{d}s+\psi_\varepsilon\int_0^t\phi_\varepsilon(\vartheta,s,t)\,\delta_B(s)\,\sigma(s)\,\mathrm{d}\bar{W}_s.\end{aligned}$$

Here

$$\phi_\varepsilon(\vartheta,s,t)=\exp\left(-\frac{\psi_\varepsilon}{\varepsilon}\int_s^t B(\vartheta,v)\,\mathrm{d}v\right).$$

By Lemma 1.7, these integrals have the following asymptotics:

$$\begin{aligned}&\int_{t_0}^t\phi_\varepsilon(\vartheta,s,t)\,\delta_{a^*}(s)\,M(\vartheta_0,s)\,\mathrm{d}s\\&\qquad=f(\vartheta_0,t)\,y_t(\vartheta_0)\int_{t_0}^t\phi_\varepsilon(\vartheta,s,t)\,\delta_{a^*}(s)\,\mathrm{d}s\\&\qquad\qquad+[M(\vartheta_0,t)-f(\vartheta_0,t)\,y_t(\vartheta_0)]\int_{t_0}^t\phi_\varepsilon(\vartheta,s,t)\,\delta_{a^*}(s)\,\mathrm{d}s\\&\qquad\qquad+\int_{t_0}^t\phi_\varepsilon(\vartheta,s,t)\,\delta_{a^*}(s)\,[M(\vartheta_0,s)-M(\vartheta_0,t)]\,\mathrm{d}s\\&\qquad=\frac{\varepsilon}{\psi_\varepsilon}\frac{a^*(\vartheta,t)-a^*(\vartheta_0,t)}{B(\vartheta,t)}f(\vartheta_0,t)\,y_t(\vartheta_0)\left(1+O\left(\frac{\varepsilon}{\psi_\varepsilon}\right)\right),\end{aligned}\tag{2.135}$$

$$\int_{t_0}^{t}\phi_\varepsilon\left(\vartheta,s,t\right)a^*\left(\vartheta,s\right)\delta_M\left(s\right){\rm d}s=\frac{\varepsilon}{\psi_\varepsilon}\frac{a^*\left(\vartheta,t\right)}{B\left(\vartheta,t\right)}\delta_M\left(t\right)\left(1+O\left(\frac{\varepsilon}{\psi_\varepsilon}\right)\right),$$

$$\psi_\varepsilon\int_{t_0}^{t}\phi_\varepsilon\left(\vartheta,s,t\right)\delta_B\left(s\right)\sigma\left(s\right){\rm d}\bar W_s$$

$$=\sqrt{\varepsilon\psi_\varepsilon}\frac{\left[f\left(\vartheta,t\right)b\left(\vartheta,t\right)-f\left(\vartheta_0,t\right)b\left(\vartheta_0,t\right)\right]}{\sqrt{2B\left(\vartheta,t\right)}}\xi_{t,\varepsilon}\left(1+O\left(\sqrt{\frac{\varepsilon}{\psi_\varepsilon}}\right)\right).$$

Here

$$\xi_{t,\varepsilon}=\sqrt{2B\left(\vartheta,t\right)}\sqrt{\frac{\varepsilon}{\psi_\varepsilon}}\int_{t_0}^{t}\exp\left(-\frac{\psi_\varepsilon}{\varepsilon}\left(t-s\right)B\left(\vartheta,t\right)\right){\rm d}\bar W_s\Longrightarrow\xi_t,$$

where $\xi_t, t_0\leq t\leq T$ are independent r.v.'s, $\xi_t\sim\mathcal{N}\left(0,1\right)$ (see the proof of Lemma 1.7 for details).

Let us clarify the order $\varepsilon/\psi_\varepsilon$ of the residual in (2.135). When we replace $M\left(\vartheta_0,s\right)$ by $M\left(\vartheta_0,t\right)$ the difference is

$$\begin{aligned}M\left(\vartheta_0,t\right)-M\left(\vartheta_0,s\right)&=\int_s^t f\left(\vartheta_0,r\right)y_r\left(\vartheta_0\right){\rm d}r+\psi_\varepsilon\int_s^t B\left(\vartheta_0,r\right)\sigma\left(r\right){\rm d}\bar W_r\\&=f\left(\vartheta_0,t\right)y_t\left(\vartheta_0\right)\left(t-s\right)\left(1+O\left(t-s\right)\right)\\&\quad+\psi_\varepsilon B\left(\vartheta_0,t\right)\sigma\left(t\right)\left[\bar W_t-\bar W_s\right]\left(1+o\left(1\right)\right).\end{aligned}$$

Therefore,

$$\int_{t_0}^{t}\phi_\varepsilon\left(\vartheta,s,t\right)\delta_{a^*}\left(s\right)\left[M\left(\vartheta_0,s\right)-M\left(\vartheta_0,t\right)\right]{\rm d}s$$

$$=\psi_\varepsilon\delta_{a^*}\left(t\right)B\left(\vartheta_0,t\right)\sigma\left(t\right)\int_{t_0}^{t}\phi_\varepsilon\left(\vartheta,s,t\right)\left[\bar W_s-\bar W_t\right]{\rm d}s\left(1+o\left(1\right)\right)$$

$$=\psi_\varepsilon\left(\frac{\varepsilon}{\psi_\varepsilon}\right)^{3/2}O\left(1\right)=\left(\frac{\varepsilon^3}{\psi_\varepsilon}\right)^{1/2}O\left(1\right).$$

Here we used Lemma 1.7. We have as well the relations

$$\left[M\left(\vartheta_0,t\right)-f\left(\vartheta_0,t\right)y_t\left(\vartheta_0\right)\right]\int_{t_0}^{t}\phi_\varepsilon\left(\vartheta,s,t\right)\delta_{a^*}\left(s\right){\rm d}s=\varepsilon\,O\left(1\right)$$

and

$$f\left(\vartheta_0,t\right)y_t\left(\vartheta_0\right)\int_{t_0}^{t}\phi_\varepsilon\left(\vartheta,s,t\right)\delta_{a^*}\left(s\right){\rm d}s$$

$$
= \frac{\varepsilon}{\psi_\varepsilon} \frac{\left[a(\vartheta,t) - a(\vartheta_0,t) + \ell(\vartheta,t) - \ell(\vartheta_0,t)\right] f(\vartheta_0,t)\, y_t(\vartheta_0)}{B(\vartheta,t)}
\times \left(1 + O\left(\frac{\varepsilon}{\psi_\varepsilon}\right)\right)
$$

Finally, we obtain for any (small) real $t_0 > 0$ the equality

$$
\begin{aligned}
\delta_M(t)\left(1 - \frac{\varepsilon}{\psi_\varepsilon}\frac{a^*(\vartheta,t)}{B(\vartheta,t)}\right) &= \frac{\varepsilon}{\psi_\varepsilon}\frac{a^*(\vartheta,t) - a^*(\vartheta_0,t)}{B(\vartheta,t)} f(\vartheta_0,t)\, y_t(\vartheta_0) \\
&\quad \times \left(1 + O\left(\varepsilon^{1-\delta}\right)\right) \\
&\quad + \sqrt{\varepsilon\psi_\varepsilon}\frac{\left[f(\vartheta,t)\, b(\vartheta,t) - f(\vartheta_0,t)\, b(\vartheta_0,t)\right]}{\sqrt{2B(\vartheta,t)}}\xi_{t,\varepsilon}\left(1 + O\left(\varepsilon^{\frac{1-\delta}{2}}\right)\right), \\
&\quad t_0 \le t \le T,
\end{aligned}
$$

and

$$
\begin{aligned}
\delta_M(t) &= \frac{\varepsilon}{\psi_\varepsilon}\frac{a^*(\vartheta,t) - a^*(\vartheta_0,t)}{B(\vartheta,t)} f(\vartheta_0,t)\, y_t(\vartheta_0)\left(1 + O\left(\varepsilon^{1-\delta}\right)\right) \\
&\quad + \sqrt{\varepsilon\psi_\varepsilon}\frac{\left[f(\vartheta,t)\, b(\vartheta,t) - f(\vartheta_0,t)\, b(\vartheta_0,t)\right]}{\sqrt{2B(\vartheta,t)}}\xi_{t,\varepsilon}\left(1 + O\left(\varepsilon^{\frac{1-\delta}{2}}\right)\right),
\end{aligned}
\tag{2.136}
$$

where we omitted exponentially small term related with $\delta_M(t_0)$.

Thus, if we put $\vartheta = \vartheta_u = \vartheta_0 + \psi_\varepsilon u$ and use Taylor expansion, then we obtain

$$
\begin{aligned}
M(\vartheta_u,t) - M(\vartheta_0,t) &= \frac{\varepsilon}{\psi_\varepsilon}\frac{a^*(\vartheta_u,t) - a^*(\vartheta_0,t)}{B(\vartheta_u,t)} f(\vartheta_0,t)\, y_t(\vartheta_0)\left(1 + O\left(\varepsilon^{1-\delta}\right)\right) \\
&\quad + \sqrt{\varepsilon\psi_\varepsilon}\frac{\left[f(\vartheta_u,t)\, b(\vartheta_u,t) - f(\vartheta_0,t)\, b(\vartheta_0,t)\right]}{\sqrt{2B(\vartheta_u,t)}}\xi_{t,\varepsilon}\left(1 + O\left(\varepsilon^{\frac{1-\delta}{2}}\right)\right) \\
&= \varepsilon\, u\, \frac{\left[\dot a(\vartheta_0,t) + \dot\ell(\vartheta_0,t)\right] f(\vartheta_0,t)\, y_t(\vartheta_0)\, \sigma(t)}{f(\vartheta_u,t)\, b(\vartheta_u,t)}\left(1 + O\left(\varepsilon^{1-\delta}\right)\right) \\
&= \varepsilon\, u\, \frac{\left[\dot a(\vartheta_0,t) + \dot\ell(\vartheta_0,t)\right] y_t(\vartheta_0)\, \sigma(t)}{b(\vartheta_0,t)}\left(1 + O\left(\varepsilon^{1-\delta}\right)\right).
\end{aligned}
$$

□

Remark that these expansions allow us to write

$$
\varepsilon\psi_\varepsilon \int_{t_0}^{T} \frac{\left[f(\vartheta_u,t)\, b(\vartheta_u,t) - f(\vartheta_0,t)\, b(\vartheta_0,t)\right]^2}{2B(\vartheta_u,t)\, \varepsilon^2\, \sigma(t)^2}\, \xi_{t,\varepsilon}^2\, \mathrm{d}t
$$

$$= u^2 \frac{\psi_\varepsilon^2 \varepsilon \psi_\varepsilon}{\varepsilon^2} \int_{t_0}^{T} \frac{\left[\dot f(\vartheta_0,t)\, b(\vartheta_0,t) + f(\vartheta_0,t)\, \dot b(\vartheta_0,t)\right]^2}{2B(\vartheta_0,t)\,\sigma(t)^2}$$

$$\times\, \xi_{t,\varepsilon}^2\, \mathrm{d}t \left(1+\psi_\varepsilon u\, O_p(1)\right)$$

$$= u^2 \frac{\psi_\varepsilon^3}{\varepsilon} \left(1+\psi_\varepsilon u\, O_p(1)\right) O(1) \longrightarrow 0 \tag{2.137}$$

and

$$\frac{\varepsilon^2}{\psi_\varepsilon^2} \int_{t_0}^{T} \frac{\left[a^*(\vartheta_u,t) - a^*(\vartheta_0,t)\right]^2}{B(\vartheta_u,t)^2\, \varepsilon^2 \sigma(t)^2} f(\vartheta_0,t)^2\, y_t(\vartheta_0)^2\, \mathrm{d}t$$

$$= u^2 \int_{t_0}^{T} \frac{\dot a^*(\vartheta_0,t)^2\; y_t(\vartheta_0)^2}{b(\vartheta_0,t)^2}\, \mathrm{d}t\, (1+\psi_\varepsilon u\, O(1))$$

$$= u^2 \int_{t_0}^{T} \frac{\left[\dot a(\vartheta_0,t) + \dot \ell(\vartheta_0,t)\right]^2 y_t(\vartheta_0)^2}{b(\vartheta_0,t)^2}\, \mathrm{d}t\, (1+\psi_\varepsilon u\, O(1))$$

$$= u^2\, \mathrm{I}_2(\vartheta_0)\,(1+\psi_\varepsilon u\, O(1))\,(1+O(t_0))\,. \tag{2.138}$$

Here $O_p(1)$ means that all polynomial moments of $O_p(1)$ are bounded by a constant, which does not depend on ε.

Therefore,

$$\int_{t_0}^{T} \frac{\left[M(\vartheta_u,t) - M(\vartheta_0,t)\right]^2}{\varepsilon^2 \sigma(t)^2} \mathrm{d}t = u^2\, \mathrm{I}(\vartheta_0) \left(1+\psi_\varepsilon u\, O_p(1)\right)\left(1+O_p(t_0)\right).$$

To simplify the exposition, we will write $\mathrm{I}_2(\vartheta_0)$ instead of $\mathrm{I}_2(\vartheta_0)\,(1+O(t_0))$ because the difference between them

$$\int_{0}^{t_0} \frac{\left[\dot a(\vartheta_0,t) + \dot \ell(\vartheta_0,t)\right]^2 y_t(\vartheta_0)^2}{b(\vartheta_0,t)^2}\, \mathrm{d}t \le C\, t_0$$

can be made less than any ϵ, $0<\epsilon \ll 1$ by the choice of t_0.

Lemma 2.8 *Let the conditions $\hat{\mathscr{F}}_{1-4}$ hold. Then the family of measures $\left\{\mathbf{P}_\vartheta^{(\varepsilon)},\, \vartheta \in \Theta\right\}$ is uniformly on compacts $\mathbb{K} \in \Theta$ locally asymptotically normal in Θ, i.e., the normalized likelihood ratio $Z_\varepsilon(\vartheta_0,u)$ for all $\vartheta_0 \in \Theta$ admits the representation*

$$Z_\varepsilon(\vartheta_0,u) = \exp\left(u\Delta_\varepsilon\left(\vartheta_0, X^T\right) - \frac{u^2}{2}\mathrm{I}_2(\vartheta_0) + r_\varepsilon(\vartheta_0,u,X^T)\right), \qquad u \in \mathbb{U}_\varepsilon,$$

where

$$\Delta_\varepsilon\left(\vartheta_0, X^T\right) = \int_{t_0}^T \frac{[\dot a(\vartheta_0,t) + \dot\ell(\vartheta_0,t)]y_t(\vartheta_0)}{b(\vartheta_0,t)}\mathrm{d}\bar W_t = \Delta(\vartheta_0) \sim \mathcal{N}(0, \mathrm{I}_2(\vartheta_0)).$$

Here the last equality is in distribution and $r_\varepsilon(\vartheta_0, u, X^T) \to 0$ uniformly on $\vartheta_0 \in \mathbb{K}$.

Proof The process $Z_\varepsilon(\vartheta_0, u)$ by Lemma 2.7 and (2.137), (2.138) admits of the representation

$$\begin{aligned}\ln Z_\varepsilon(\vartheta_0, u) &= \frac{u\psi_\varepsilon}{\varepsilon}\int_0^T \dot M(\vartheta_0,t)\,\mathrm{d}\bar W_t - \frac{u^2\psi_\varepsilon^2}{2\varepsilon^2}\int_0^T \dot M(\vartheta_0,t)^2\,\mathrm{d}t + o_p(1)\\ &= u\int_0^T \frac{[\dot a(\vartheta_0,t) + \dot\ell(\vartheta_0,t)]y_t(\vartheta_0)}{b(\vartheta_0,t)}\mathrm{d}\bar W_t\\ &\quad - \frac{u^2}{2}\int_0^T \frac{[\dot a(\vartheta_0,t) + \dot\ell(\vartheta_0,t)]^2 y_t(\vartheta_0)^2}{b(\vartheta_0,t)^2}\mathrm{d}t + o_p(1)\\ &= u\Delta(\vartheta_0) - \frac{u^2}{2}\mathrm{I}_2(\vartheta_0) + o(1).\end{aligned}$$

□

Lemma 2.9 *Let conditions $\hat{\mathscr{F}}_{1-3}$ hold. Then there exists a constant $C > 0$ such that*

$$\sup_{\vartheta_0\in\mathbb{K}} \mathbf{E}_{\vartheta_0}\left[Z_\varepsilon(\vartheta_0, u_1)^{1/2} - Z_\varepsilon(\vartheta_0, u_2)^{1/2}\right]^2 \le C\,|u_2 - u_1|^2. \tag{2.139}$$

Proof By 1.11, we have

$$\begin{aligned}\mathbf{E}_{\vartheta_0}\left[Z_\varepsilon(\vartheta_0, u_1)^{1/2} - Z_\varepsilon(\vartheta_0, u_2)^{1/2}\right]^2 &\le \frac{1}{2\varepsilon^2}\int_0^T \frac{\mathbf{E}_{\vartheta_{u_2}}\delta_M(u,t)^2}{\sigma(t)^2}\mathrm{d}t\\ &\quad + \frac{1}{2\varepsilon^2}\int_0^T \frac{\mathbf{E}_{\vartheta_{u_1}}\delta_M(u,t)^2}{\sigma(t)^2}\mathrm{d}t,\end{aligned}\tag{2.140}$$

where $\delta_M(u,t) = M(\vartheta_0 + \psi_\varepsilon u_2, t) - M(\vartheta_0 + \psi_\varepsilon u_1, t)$.

Consider the term $\mathbf{E}_{\vartheta_{u_1}}\delta_M(u,t)^2$. The function $\delta_M(u,\cdot)$ with $\mathbf{P}_{\vartheta_{u_1}}$ probability one is solution of the equation

$$
\begin{aligned}
\mathrm{d}\delta_M(u,t) &= a^*\left(\vartheta_{u_2},t\right)\delta_M(u,t)\,\mathrm{d}t - \frac{\psi_\varepsilon}{\varepsilon}B\left(\vartheta_{u_2},t\right)\delta_M(u,t)\,\mathrm{d}t\\
&\quad + \delta_{a^*}(u,t)\,M\left(\vartheta_{u_1},t\right)\mathrm{d}t\\
&\quad + \psi_\varepsilon\delta_B(u,t)\,\sigma(t)\,\mathrm{d}\bar W_t,\ \ \delta_M(u,0) = \left[f\left(\vartheta_{u_2},0\right) - f\left(\vartheta_{u_1},0\right)\right]y_0.
\end{aligned}
$$

Here $\delta_{a^*}(u,t) = a^*\left(\vartheta_{u_2},t\right) - a^*\left(\vartheta_{u_1},t\right) = \psi_\varepsilon(u_2-u_1)\,\dot a^*\left(\vartheta_{\tilde u_1},t\right)$ and $\delta_B(u,t) = B\left(\vartheta_{u_2},t\right) - B\left(\vartheta_{u_1},t\right) = \psi_\varepsilon(u_2-u_1)\,\dot B\left(\vartheta_{\tilde u_1},t\right)$, $B(\vartheta,t) = \gamma_*(\vartheta,t)\,f(\vartheta,t)\,\sigma(t)^{-2}$.

We have the representation

$$
\begin{aligned}
\delta_M(u,t) &= \delta_M(u,0)\,\Phi\left(\vartheta_{u_2},0,t\right) + \int_0^t \Phi\left(\vartheta_{u_2},s,t\right)\delta_{a^*}(u,s)\,M\left(\vartheta_{u_1},s\right)\mathrm{d}s\\
&\quad + \psi_\varepsilon\int_0^t \Phi\left(\vartheta_{u_2},s,t\right)\delta_B(u,s)\,\sigma(s)\,\mathrm{d}\bar W_s
\end{aligned}
$$

where $\bar W_s, 0\leq s\leq T$ is the innovation Winer process (under the measure $\mathbf{P}_{\vartheta_{u_1}}$) and

$$
\Phi\left(\vartheta_{u_2},s,t\right) = \exp\left(-\frac{\psi_\varepsilon}{\varepsilon}\int_s^t B\left(\vartheta_{u_2},v\right)\mathrm{d}v + \int_s^t a^*\left(\vartheta_{u_2},v\right)\mathrm{d}v\right).
$$

There exist constants $c_*>0$ and $C_*>0$ such that for $t\geq t_0$ we have $B\left(\vartheta_{u_2},v\right) > c_*$ and

$$
\int_0^T \left|a^*(\vartheta,v)\right|\mathrm{d}v < C_*.
$$

Thus,

$$
\Phi\left(\vartheta_{u_2},s,t\right) \leq \exp\left(-\frac{\psi_\varepsilon}{\varepsilon}c_*(t-s) + C_*\right),
$$

where $c_*>0$ is some constant.

Remark that for $t\in(0,T]$

$$
\begin{aligned}
\varepsilon^{-1}\left|\delta_M(u,0)\right|\Phi\left(\vartheta_{u_2},0,t\right) &\leq C\,|u_2-u_1|\,y_0\left|\dot f(\tilde\vartheta,0)\right|\frac{\psi_\varepsilon}{\varepsilon}\,e^{-\psi_\varepsilon/\varepsilon\, c_* t}\\
&\leq C\,r_\varepsilon\,|u_2-u_1|, \qquad (2.141)
\end{aligned}
$$

where $r_\varepsilon\to 0$.

For any $m=1,2,\ldots$ and $t>0$, we have

$$
\begin{aligned}
\mathbf{E}_{\vartheta_{u_1}}\delta_M(u,t)^{2m} \leq C\, r_\varepsilon^{2m}\varepsilon^{2m}\left|u_2-u_1\right|^{2m}& \\
+C\mathbf{E}_{\vartheta_{u_1}}&\left(\int_0^t e^{-\frac{\psi_\varepsilon}{\varepsilon}c_*(t-s)}\delta_{a^*}(s)\,M\left(\vartheta_{u_1},s\right)\mathrm{d}s\right)^{2m} \\
+C\psi_\varepsilon^{2m}&\left(\int_0^t e^{-\frac{2\psi_\varepsilon}{\varepsilon}c_*(t-s)}\delta_B(s)^2\,\sigma\left(\vartheta_{u_1},s\right)^2\mathrm{d}s\right)^m .
\end{aligned}
$$

Here

$$
\begin{aligned}
&\mathbf{E}_{\vartheta_{u_1}}\left(\int_0^t e^{-\frac{\psi_\varepsilon}{\varepsilon}c_*(t-s)}\delta_{a^*}(s)\,M\left(\vartheta_{u_1},s\right)\mathrm{d}s\right)^{2m} \\
&\quad\leq\left(\int_0^t e^{-\frac{\psi_\varepsilon}{\varepsilon}c_*(t-s)}\mathrm{d}s\right)^{2m-1}\int_0^t e^{-\frac{\psi_\varepsilon}{\varepsilon}c_*(t-s)}\delta_{a^*}(s)^{2m}\,\mathbf{E}_{\vartheta_{u_1}}M\left(\vartheta_{u_1},s\right)^{2m}\mathrm{d}s \\
&\quad\leq C\psi_\varepsilon^{2m}\left(\frac{\varepsilon}{\psi_\varepsilon}\right)^{2m}\left|u_2-u_1\right|^{2m}=C\,\varepsilon^{2m}\left|u_2-u_1\right|^{2m}.
\end{aligned}
$$

The Gaussian random process $M\left(\vartheta_{u_1},\cdot\right)$ satisfies (2.134) where ϑ_0 is replaced by ϑ_{u_1} and $\bar{W}_t, 0\leq t\leq T$ is the innovation Wiener process under probability $\mathbf{P}_{\vartheta_{u_1}}$. Therefore, $M\left(\vartheta_{u_1},\cdot\right)$ has bounded polynomial moments.

For the second integral, we have

$$
\begin{aligned}
\psi_\varepsilon^{2m}&\left(\int_{t_0}^t e^{-\frac{2\psi_\varepsilon}{\varepsilon}c_*(t-s)}\delta_B(s)^2\,\sigma\left(\vartheta_{u_1},s\right)^2\mathrm{d}s\right)^m \\
&\leq C\psi_\varepsilon^{4m}\left|u_2-u_1\right|^{2m}\left(\int_{t_0}^t e^{-\frac{2\psi_\varepsilon}{\varepsilon}c_*(t-s)}\mathrm{d}s\right)^m \\
&\leq C\psi_\varepsilon^{4m}\left|u_2-u_1\right|^{2m}\left(\frac{\varepsilon}{\psi_\varepsilon}\right)^m \\
&=C\left(\varepsilon\psi_\varepsilon^3\right)^m\left|u_2-u_1\right|^{2m}.
\end{aligned}
$$

The similar estimates we have for the term $\mathbf{E}_{\vartheta_{u_2}}\delta_M(u,t)^2$ too. The substuitution of these estimates with $m=2$ in (2.140) proves (2.139). □

Lemma 2.10 *Assume that conditions $\hat{\mathscr{F}}$ hold, then for some constant $\kappa>0$ and any $N>0$, there exists a constant $C_N>0$ such that*

$$
\sup_{\vartheta_0\in\mathbb{K}}\mathbf{P}_{\vartheta_0}\left(Z_\varepsilon\left(\vartheta_0,u\right)\geq e^{-\kappa u^2}\right)\leq\frac{C_N}{|u|^N}.
$$

Proof Denote $\delta_M(t)=M\left(\vartheta_u,t\right)-M\left(\vartheta_0,t\right)$ and write

$$
\begin{aligned}
&\mathbf{P}_{\vartheta_0}\left(\ln Z_\varepsilon\left(\vartheta_0,u\right)\geq -\kappa u^2\right)\\
&\quad=\mathbf{P}_{\vartheta_0}\left(\int_0^T\frac{\delta_M\left(t\right)^2}{2\varepsilon\sigma\left(t\right)}\,\mathrm{d}\bar W_t-\int_0^T\frac{\delta_M\left(t\right)^2}{4\varepsilon^2\sigma\left(t\right)^2}\,\mathrm{d}t\geq -\frac{\kappa}{2}u^2\right)\\
&\quad\leq \mathbf{P}_{\vartheta_0}\left(\int_0^T\frac{\delta_M\left(t\right)}{2\varepsilon\sigma\left(t\right)}\,\mathrm{d}\bar W_t-\int_0^T\frac{\delta_M\left(t\right)^2}{8\varepsilon^2\sigma\left(t\right)^2}\,\mathrm{d}t\geq \frac{\kappa}{2}u^2\right)\\
&\qquad\qquad+\mathbf{P}_{\vartheta_0}\left(-\int_0^T\frac{\delta_M\left(t\right)^2}{8\varepsilon^2\sigma\left(t\right)^2}\,\mathrm{d}t\geq -\kappa u^2\right)\\
&\quad\leq e^{-\frac{\kappa}{2}u^2}+\mathbf{P}_{\vartheta_0}\left(\int_0^T\frac{\delta_M\left(t\right)^2}{\varepsilon^2\sigma\left(t\right)^2}\,\mathrm{d}t\leq 8\kappa u^2\right). \qquad (2.142)
\end{aligned}
$$

Here we used the equality

$$
\mathbf{E}_{\vartheta_0}\exp\left(\int_0^T\frac{\delta_M\left(t\right)}{2\varepsilon\sigma\left(t\right)}\mathrm{d}\bar W_t-\int_0^T\frac{\delta_M(t)^2}{8\varepsilon^2\sigma\left(t\right)^2}\mathrm{d}t\right)=1.
$$

As it follows from (2.136)

$$
\delta_M\left(t\right)=\delta_M^\circ\left(t\right)+O\left(\sqrt{\varepsilon\psi_\varepsilon}\right)
$$

where

$$
\delta_M^\circ\left(t\right)=\frac{\varepsilon}{\psi_\varepsilon}\frac{\left[a\left(\vartheta,t\right)-a\left(\vartheta_0,t\right)+\ell\left(\vartheta,t\right)-\ell\left(\vartheta_0,t\right)\right]f\left(\vartheta_0,t\right)y_t\left(\vartheta_0\right)\sigma\left(t\right)}{b\left(\vartheta,t\right)f\left(\vartheta,t\right)}.
$$

Set $\mu_s\left(\vartheta_0\right)=f\left(\vartheta_0,s\right)y_s\left(\vartheta_0\right)$ and remind the obvious equality (see (2.134))

$$
\begin{aligned}
\mathbf{E}_{\vartheta_0}M\left(\vartheta_0,s\right)&=y_0\,f\left(\vartheta_0,0\right)\exp\left(\int_0^s\left[a\left(\vartheta_0,v\right)+f\left(\vartheta_0,v\right)^{-1}f_v'\left(\vartheta_0,v\right)\right]\mathrm{d}v\right)\\
&=y_0\,f\left(\vartheta_0,s\right)\exp\left(\int_0^s a\left(\vartheta_0,r\right)\mathrm{d}v\right)=f\left(\vartheta_0,s\right)y_s\left(\vartheta_0\right)=\mu_s\left(\vartheta_0\right).
\end{aligned}
$$

We have the estimate (2.141) and therefore for any $t\in(0,T]$

$$
\left|\delta_M\left(0\right)\right|\Phi\left(\vartheta_0,0,t\right)\leq C\,e^{-\frac{\psi_\varepsilon}{\varepsilon}c_*t}\rightarrow 0.
$$

That is why we have omitted this term below.

Let us return to the last probability in (2.142)

$$
\begin{aligned}
&\mathbf{P}_{\vartheta_0}\left(-\int_0^T \frac{\delta_M(t)^2}{\varepsilon^2\sigma(t)^2}\mathrm{d}t \geq -8\kappa u^2\right)\\
&\qquad = \mathbf{P}_{\vartheta_0}\left(\frac{G(\vartheta_u,\vartheta_0)}{\psi_\varepsilon^2} - \int_0^T \frac{\delta_M(t)^2}{\varepsilon^2\sigma(t)^2}\mathrm{d}t \geq \frac{G(\vartheta_u,\vartheta_0)}{\psi_\varepsilon^2} - 8\kappa u^2\right)
\end{aligned}
\tag{2.143}
$$

Consider two sets $\mathbb{A} = (u : |\vartheta_u - \vartheta_0| \leq \nu)$, $\nu > 0$ and $\mathbb{A}^c = \mathbb{U}_\varepsilon \setminus \mathbb{A}$. For $u \in \mathbb{A}$ and sufficiently small ν, we can write

$$
\frac{G(\vartheta_u,\vartheta_0)}{\psi_\varepsilon^2} = u^2 \mathrm{I}_2(\vartheta_0)(1+o(1)) \geq \frac{u^2}{2}\mathrm{I}_2(\vartheta_0) \geq \frac{u^2}{2}\inf_{\vartheta\in\Theta}\mathrm{I}_2(\vartheta) = \frac{u^2}{2}\hat{\kappa}.
$$

Therefore,

$$
\frac{G(\vartheta_u,\vartheta_0)}{\psi_\varepsilon^2} - 8\kappa u^2 \geq \left[\frac{\hat{\kappa}}{2} - 8\kappa\right] u^2 = \kappa_1 u^2,
$$

where κ is such that $\kappa_1 > 0$.

Further, for any $m = 1, 2, \ldots$

$$
\begin{aligned}
&\mathbf{P}_{\vartheta_0}\left(\frac{G(\vartheta_u,\vartheta_0)}{\psi_\varepsilon^2} - \int_0^T \frac{\delta_M(t)^2}{\varepsilon^2\sigma(t)^2}\mathrm{d}t \geq \kappa_1 u^2\right)\\
&\quad \leq \mathbf{P}_{\vartheta_0}\left(\int_0^T \frac{\left|\delta_M(t)^2 - \delta_M^\circ(t)^2\right|}{\varepsilon^2\sigma(t)^2}\mathrm{d}t \geq \kappa_1 u^2\right)\\
&\quad \leq \frac{T^{2m-1}}{\kappa_1^{2m}u^{4m}}\int_0^T \frac{\mathbf{E}_{\vartheta_0}\left|\delta_M(t)^2 - \delta_M^\circ(t)^2\right|^{2m}}{\varepsilon^{4m}\sigma(t)^{4m}}\mathrm{d}t\\
&\quad \leq \frac{T^{2m-1}}{\kappa_1^{2m}u^{4m}}\int_0^T \frac{\left(\mathbf{E}_{\vartheta_0}\left|\delta_M(t) + \delta_M^\circ(t)\right|^{4m}\mathbf{E}_{\vartheta_0}\left|\delta_M(t) - \delta_M^\circ(t)\right|^{4m}\right)^{1/2}}{\varepsilon^{4m}\sigma(t)^{4m}}\mathrm{d}t.
\end{aligned}
\tag{2.144}
$$

We have the estimates

$$
\frac{\left|\delta_M^\circ(t)\right|^{4m}}{\varepsilon^{4m}} \leq \frac{C}{\psi_\varepsilon^{4m}}\psi_\varepsilon^{4m}\,|u|^{4m} = C\,|u|^{4m},
$$

$$
\frac{\mathbf{E}_{\vartheta_0}\left|\delta_M(t)\right|^{4m}}{\varepsilon^{4m}} \leq C\,|u|^{4m} + C\left(\frac{\psi_\varepsilon^3}{\varepsilon}\right)^{2m}|u|^{4m} \leq C\,|u|^{4m},
$$

$$\frac{\mathbf{E}_{\vartheta_0}\left|\delta_M(t)-\delta^{\circ}_M(t)\right|^{4m}}{\varepsilon^{4m}}\leq C\,\frac{\psi_\varepsilon^{4m}}{\varepsilon^{4m}}\left(\frac{\varepsilon}{\psi_\varepsilon}\right)^{6m}\;|u|^{4m}=C\,\left(\frac{\varepsilon}{\psi_\varepsilon}\right)^{2m}\;|u|^{4m}.$$

Remind that $\psi_\varepsilon^3/\varepsilon\to 0$.

Finally, we obtain for $u\in\mathbb{A}$

$$\mathbf{P}_{\vartheta_0}\left(-\int_0^T\frac{\delta_M(t)^2}{\varepsilon^2\sigma(t)^2}\mathrm{d}t\geq -8\kappa u^2\right)\leq C\frac{T^{2m}}{\kappa_1^{2m}}\;\left(\frac{\varepsilon}{\psi_\varepsilon}\right)^m\;\leq C\;|u|^{-\frac{(1-\delta)m}{\delta}},$$

where we used the relations

$$\vartheta_0+\psi_\varepsilon u\in(\alpha,\beta),\qquad |u|\leq\frac{\beta-\alpha}{\psi_\varepsilon}=\varepsilon^{-\delta}\left(\beta-\alpha\right),\qquad \varepsilon<\frac{(\beta-\alpha)^{\frac{1}{\delta}}}{|u|^{\frac{1}{\delta}}}. \tag{2.145}$$

Hence $\varepsilon^{m(1-\delta)}\leq C\,|u|^{-\frac{(1-\delta)m}{\delta}}$. Therefore if we set $m=N\delta/(1-\delta)$, then we obtain (2.140) for $u\in\mathbb{A}$.

Consider now the case $u\in\mathbb{A}^c$. We have

$$\frac{G(\vartheta_u,\vartheta_0)}{\psi_\varepsilon^2}\geq\frac{g(\nu)}{\psi_\varepsilon^2}=\frac{g(\nu)}{(u\psi_\varepsilon)^2}\,u^2\geq\frac{g(\nu)}{(\beta-\alpha)^2}\,u^2$$

and

$$\frac{G(\vartheta_u,\vartheta_0)}{\psi_\varepsilon^2}-8\kappa u^2\geq\left(\frac{g(\nu)}{(\beta-\alpha)^2}-8\kappa\right)u^2=\kappa_2\,u^2$$

where κ is such that $\kappa_2>0$ too.

First, we repeat the calculations as in (2.144) and obtain

$$\mathbf{P}_{\vartheta_0}\left(-\int_0^T\frac{\delta_M(t)^2}{\varepsilon^2\sigma(t)^2}\mathrm{d}t\geq -8\kappa u^2\right)$$

$$\leq\frac{T^{2m-1}}{\kappa_2^{2m}u^{4m}}\int_0^T\frac{\left(\mathbf{E}_{\vartheta_0}\left|\delta_M(t)+\delta^{\circ}_M(t)\right|^{4m}\mathbf{E}_{\vartheta_0}\left|\delta_M(t)-\delta^{\circ}_M(t)\right|^{4m}\right)^{1/2}}{\varepsilon^{4m}\sigma(t)^{4m}}\mathrm{d}t.$$

Then the relations (2.143) and (2.144) allow us to write the corresponding asymptotics

$$\frac{\left|\delta^{\circ}_M(t)\right|^{4m}}{\varepsilon^{4m}}\leq\frac{C}{\psi_\varepsilon^{4m}},$$

$$\frac{\mathbf{E}_{\vartheta_0}\left|\delta_M\left(t\right)\right|^{4m}}{\varepsilon^{4m}} \leq \frac{C}{\psi_\varepsilon^{4m}},$$

$$\frac{\mathbf{E}_{\vartheta_0}\left|\delta_M\left(t\right)-\delta_M^\circ\left(t\right)\right|^{4m}}{\varepsilon^{4m}} \leq C\,\frac{1}{\varepsilon^{4m}}\left(\frac{\varepsilon}{\psi_\varepsilon}\right)^{6m} = C\left(\frac{\varepsilon}{\psi_\varepsilon^3}\right)^{2m}.$$

Hence,

$$\mathbf{P}_{\vartheta_0}\left(-\int_0^T \frac{\delta_M\left(t\right)^2}{\varepsilon^2\sigma\left(t\right)^2}\mathrm{d}t \geq -8\kappa u^2\right) \leq C\frac{T^{2m}}{\kappa_2^{2m}u^{4m}}\,\frac{\varepsilon^m}{\psi_\varepsilon^{5m}} \leq C\,\frac{T^{2m}}{\kappa_2^{2m}\,\nu^{4m}}\left(\frac{\varepsilon}{\psi_\varepsilon}\right)^m$$
$$\leq C\,\varepsilon^{m(1-\delta)} \leq \frac{C}{\left|u\right|^N}.$$

□

The properties of the normalized likelihood ratio $Z_\varepsilon\left(\cdot\right)$ established in Lemmas 2.8–2.10 verify the sufficient conditions of the Theorem 1.10 (and 1.12), which, in turn, imply the properties of the MLE and BE claimed in Theorem 2.15.

Since the convergence of moments is uniform on compacts of Θ, we can prove the asymptotic efficiency of estimators as follows. The uniform convergence of moments for the MLE implies

$$\lim_{\varepsilon\to 0}\sup_{\left|\vartheta-\vartheta_0\right|\leq\nu}\psi_\varepsilon^{-1}\mathbf{E}_\vartheta\left[\hat\vartheta_\varepsilon-\vartheta\right]^2 = \sup_{\left|\vartheta-\vartheta_0\right|\leq\nu}\mathrm{I}_2\left(\vartheta\right)^{-1} \xrightarrow{\nu\to 0} \mathrm{I}_2\left(\vartheta_0\right)^{-1}.$$

The same convergence holds for BE $\tilde\vartheta_\varepsilon$.

Remark 2.21 The condition of identifiability requires that $G\left(\vartheta,\vartheta_0\right)\neq 0$ for any $\vartheta\neq\vartheta_0$ and any ϑ_0. This condition is not fulfilled if, for example, the functions $a\left(\vartheta,t\right)$ and $f\left(\vartheta,t\right)$ satisfy the equation

$$\frac{f_t'\left(\vartheta,t\right)}{f\left(\vartheta,t\right)} = -a\left(\vartheta,t\right)+h\left(t\right), \qquad 0\leq t\leq T,$$

where $h\left(t\right)$ is some continuous function, which does not depend on ϑ. This means that

$$f\left(\vartheta,t\right) = H\left(t\right)\exp\left(-\int_0^t a\left(\vartheta,s\right)\mathrm{d}s\right), \qquad H\left(t\right) = \exp\int_0^t h\left(s\right)\mathrm{d}s. \tag{2.146}$$

Therefore, for such functions $f\left(\vartheta,\cdot\right)$ and $a\left(\vartheta,\cdot\right)$ we have $G\left(\vartheta,\vartheta_0\right)=0$ for all $\vartheta\in\Theta$.

If the functions $f(\vartheta,\cdot)$ and $a(\vartheta,\cdot)$ satisfy the relation (2.146), then under the corresponding conditions we can obtain the representation

$$\begin{aligned}&\frac{M(\vartheta_u,t)-M(\vartheta_0,t)}{\varepsilon\sigma(t)}\\&\qquad=\sqrt{\frac{\psi_\varepsilon}{\varepsilon}}\frac{[f(\vartheta_u,t)\,b(\vartheta_u,t)-f(\vartheta_0,t)\,b(\vartheta_0,t)]}{\sqrt{2B(\vartheta_u,t)}\,\sigma(t)}\xi_{t,\varepsilon}(1+o(1)).\end{aligned}$$

This case is in some sense close to the considered below case with $\delta\in(0,1/3)$ and probably needs a special study.

Case $\delta\in(0,1/3)$ Let us return to the key expression (2.136). If $\delta\in(0,1/3)$, then

$$\begin{aligned}&\frac{M(\vartheta_u,t)-M(\vartheta_0,t)}{\varepsilon\sigma(t)}\\&\quad=\frac{1}{\psi_\varepsilon}\frac{a(\vartheta_u,t)-a(\vartheta_0,t)+\ell(\vartheta_u,t)-\ell(\vartheta_0,t)}{B(\vartheta_u,t)\,\sigma(t)}f(\vartheta_0,t)\,y_t(\vartheta_0)(1+o(1))\\&\qquad\qquad+\sqrt{\frac{\psi_\varepsilon}{\varepsilon}}\frac{[f(\vartheta_u,t)\,b(\vartheta_u,t)-f(\vartheta_0,t)\,b(\vartheta_0,t)]}{\sqrt{2B(\vartheta_u,t)}\,\sigma(t)}\xi_{t,\varepsilon}(1+o(1))\\&\quad=\sqrt{\frac{\psi_\varepsilon}{\varepsilon}}\frac{[f(\vartheta_u,t)\,b(\vartheta_u,t)-f(\vartheta_0,t)\,b(\vartheta_0,t)]}{\sqrt{2B(\vartheta_u,t)}\,\sigma(t)}\xi_{t,\varepsilon}(1+o(1))\\&\quad=u\,\varphi_\varepsilon\sqrt{\frac{\psi_\varepsilon}{\varepsilon}}\frac{\left[\dot f(\vartheta_0,t)\,b(\vartheta_0,t)+f(\vartheta_0,t)\,\dot b(\vartheta_0,t)\right]}{\sqrt{2B(\vartheta_0,t)}\,\sigma(t)}\xi_{t,\varepsilon}(1+o(1)).\end{aligned}$$

Therefore, we have to take $\varphi_\varepsilon=\sqrt{\varepsilon/\psi_\varepsilon}=\varepsilon^{(1-\delta)/2}$. The Fisher information in this case is

$$\mathrm{I}_1(\vartheta_0)=\int_0^T\frac{[\dot f(\vartheta_0,t)\,b(\vartheta_0,t)+f(\vartheta_0,t)\,\dot b(\vartheta_0,t)]^2}{2b(\vartheta_0,t)\,f(\vartheta_0,t)\,\sigma(t)}\,\mathrm{d}t.$$

Therefore, we obtained the same Fisher information as in Sect. 3.1, where it is supposed that $\psi_\varepsilon\equiv1$. Of course, the normalization of the MLE is different.

Define the function

$$G_1(\vartheta,\vartheta_0)=\int_0^T\frac{[f(\vartheta,t)\,b(\vartheta,t)-f(\vartheta_0,t)\,b(\vartheta_0,t)]^2}{2f(\vartheta,t)\,b(\vartheta,t)\,\sigma(t)}\mathrm{d}t$$

and introduce *Conditions* $\hat{\mathscr{F}}^*$

$\hat{\mathscr{F}}_1^*$. *The functions* $f(\vartheta,t),b(\vartheta,t),\vartheta\in\Theta,t\in[0,T]$ *and* $\sigma(t),t\in[0,T]$ *have continuous derivatives w.r.t. t.*

$\hat{\mathscr{F}}_2^*$. *The functions* $f(\vartheta, t), b(\vartheta, t), t \in [0,T], \vartheta \in \Theta$ *and* $\sigma(t), 0 \leq t \leq T$ *are bounded and separated from zero.*

$\hat{\mathscr{F}}_3^*$. *The functions* $f(\vartheta, t), b(\vartheta, t), t \in [0,T], \vartheta \in \Theta$ *are two times continuously differentiable w.r.t.* ϑ.

$\hat{\mathscr{F}}_4^*$. *The Fisher information is uniformly positive*

$$\inf_{\vartheta\in\Theta} \mathrm{I}_1(\vartheta) > 0.$$

$\hat{\mathscr{F}}_5^*$. *For any* $\nu > 0$

$$g(\nu) \equiv \inf_{\vartheta_0\in\Theta} \inf_{|\vartheta-\vartheta_0|>\nu} G_1(\vartheta, \vartheta_0) > 0.$$

Theorem 2.16 *Let the conditions* $\hat{\mathscr{F}}^*$ *hold and* $\delta \in (0, 1/3)$. *Then the MLE* $\hat\vartheta_\varepsilon$ *and BE are uniformly consistent, asymptotically normal*

$$\sqrt{\frac{\psi_\varepsilon}{\varepsilon}}\left(\hat\vartheta_\varepsilon - \vartheta_0\right) \Longrightarrow \zeta(\vartheta_0) \sim \mathcal{N}\left(0, \mathrm{I}_1(\vartheta_0)^{-1}\right), \quad \sqrt{\frac{\psi_\varepsilon}{\varepsilon}}\left(\tilde\vartheta_\varepsilon - \vartheta_0\right) \Longrightarrow \zeta(\vartheta_0),$$

the moments converge and the both estimators are asymptotically efficient.

Proof We give a sketch of the proof because it is quite close to the proof of Theorem 3.1. Denote $S(\vartheta,t) = f(\vartheta,t)\,b(\vartheta,t)$, $\varphi_\varepsilon = (\varepsilon/\psi_\varepsilon)^{1/2}$ and $\vartheta_u = \vartheta_0 + \varphi_\varepsilon u$. The normalized likelihood ratio $Z_\varepsilon(\vartheta_0, u) = L\left(\vartheta_0 + \varphi_\varepsilon u, X^T\right)/L\left(\vartheta_0, X^T\right), u \in \mathbb{U}_\varepsilon$ according to (2.125) admits the representation

$$\begin{aligned}
\ln Z_\varepsilon(\vartheta_0, u) &= \int_0^T \frac{M(\vartheta_u, t) - M(\vartheta_0, t)}{\varepsilon\sigma(t)} \mathrm{d}\bar W_t - \int_0^T \frac{[M(\vartheta_u, t) - M(\vartheta_0, t)]^2}{2\varepsilon^2\sigma(t)^2}\mathrm{d}t \\
&= \sqrt{\frac{\psi_\varepsilon}{\varepsilon}} \int_0^T \frac{S(\vartheta_u, t) - S(\vartheta_0, t)}{\sqrt{2S(\vartheta_u, t)\sigma(t)}}\, \xi_{t,\varepsilon}\, \mathrm{d}\bar W_t\,(1 + o(1)) \\
&\quad - \frac{\psi_\varepsilon}{\varepsilon} \int_0^T \frac{[S(\vartheta_u, t) - S(\vartheta_0, t)]^2}{4S(\vartheta_u, t)\sigma(t)}\, \xi_{t,\varepsilon}^2\, \mathrm{d}t\,(1 + o(1)) \\
&= u \int_0^T \frac{\dot S(\vartheta_0, t)}{\sqrt{2S(\vartheta_0, t)\sigma(t)}}\, \xi_{t,\varepsilon}\, \mathrm{d}\bar W_t\,(1 + o(1)) \\
&\quad - \frac{u^2}{2} \int_0^T \frac{\dot S(\vartheta_0, t)^2}{2S(\vartheta_0, t)\sigma(t)}\, \xi_{t,\varepsilon}^2\, \mathrm{d}t\,(1 + o(1)).
\end{aligned}$$

The Gaussian random variables

$$\xi_{t,\varepsilon} = \sqrt{2B(\vartheta_0, t)\frac{\psi_\varepsilon}{\varepsilon}} \int_0^t \exp\left(-\frac{\psi_\varepsilon}{\varepsilon}(t-s)\,B(\vartheta_0, t)\right)\mathrm{d}\bar W_s, \qquad 0 < t \leq T$$

are asymptotically independent and for any continuous function $R(t), 0 \leq t \leq T$ the relations

$$\int_0^T R(t)\, \xi_{t,\varepsilon}\, dt \longrightarrow 0, \qquad \int_0^T R(t)\, \xi_{t,\varepsilon}^2\, dt \longrightarrow \int_0^T R(t)\, dt$$

hold (see Remark 3.5).

Therefore, we have the limits:

$$\int_0^T \frac{\dot S(\vartheta_0,t)^2}{2S(\vartheta_0,t)\,\sigma(t)}\, \xi_{t,\varepsilon}^2\, dt \longrightarrow \int_0^T \frac{\dot S(\vartheta_0,t)^2}{2S(\vartheta_0,t)\,\sigma(t)}\, dt = \mathrm{I}_1(\vartheta_0)$$

and by the CLT

$$\begin{aligned}\Delta_\varepsilon\left(\vartheta_0, X^T\right) &\equiv \int_0^T \frac{\dot S(\vartheta_0,t)}{\sqrt{2S(\vartheta_0,t)\,\sigma(t)}}\, \xi_{t,\varepsilon}\, d\bar W_t \Longrightarrow \int_0^T \frac{\dot S(\vartheta_0,t)}{\sqrt{2S(\vartheta_0,t)\,\sigma(t)}}\, dw(t)\\ &= \Delta(\vartheta_0) \sim \mathcal{N}(0, \mathrm{I}_1(\vartheta_0)).\end{aligned}$$

Here $w(t), 0 \leq t \leq T$ is some Wiener process. Moreover, these convergences are uniform on compacts $\mathbb{K} \in \Theta$. Therefore we verified that the LR $Z_\varepsilon(\vartheta_0, u), u \in \mathbb{U}_\varepsilon$ have the representation

$$Z_\varepsilon(\vartheta_0, u) = \exp\left(u\Delta_\varepsilon\left(\vartheta_0, X^T\right) - \frac{u^2}{2}\mathrm{I}_1(\vartheta_0) + r_\varepsilon\right)$$

where $r_\varepsilon \to 0$.

The proof of the estimates

$$\sup_{\vartheta_0 \in \mathbb{K}} \mathbf{E}_{\vartheta_0}\left[Z_\varepsilon(\vartheta_0, u_2)^{1/2} - Z_\varepsilon(\vartheta_0, u_1)^{1/2}\right]^2 \leq C\, |u_2 - u_1|^2 \tag{2.147}$$

and

$$\sup_{\vartheta_0 \in \mathbb{K}} \mathbf{P}_{\theta_0}\left(Z_\varepsilon(\vartheta_0, u) > e^{-\kappa u^2}\right) \leq \frac{C_N}{|u|^N} \tag{2.148}$$

follows the same lines as the proofs of Lemmas 2.9 and 2.10. □

Case $\delta = 1/3$

In this case, we take $\varphi_\varepsilon = \varepsilon^{1/3}$ and the representation (2.125) is

$$
\begin{aligned}
&\frac{M(\vartheta_u,t)-M(\vartheta_0,t)}{\varepsilon\sigma(t)}\\
&=\frac{1}{\varepsilon^{1/3}}\frac{a(\vartheta_u,t)-a(\vartheta_0,t)+\ell(\vartheta_u,t)-\ell(\vartheta_0,t)}{B(\vartheta_u,t)\sigma(t)}f(\vartheta_0,t)\,y_t(\vartheta_0)\,(1+o(1))\\
&\quad+\frac{1}{\varepsilon^{1/3}}\frac{[f(\vartheta_u,t)b(\vartheta_u,t)-f(\vartheta_0,t)b(\vartheta_0,t)]}{\sqrt{2B(\vartheta_u,t)}\,\sigma(t)}\xi_{t,\varepsilon}(1+o(1))\\
&=u\left[\frac{\dot a^*(\vartheta_0,t)\,y_t(\vartheta_0)}{b(\vartheta_0,t)}+\frac{\dot S(\vartheta_0,t)}{\sqrt{2S(\vartheta_0,t)\sigma(t)}}\xi_{t,\varepsilon}\right](1+o(1)).
\end{aligned}
$$

Therefore, if we suppose that the corresponding version of the conditions $\hat{\mathscr{F}}$ and $\hat{\mathscr{F}}^*$ holds, then

$$
\begin{aligned}
\ln Z_\varepsilon(\vartheta_0,u)&=\int_0^T\frac{M(\vartheta_u,t)-M(\vartheta_0,t)}{\varepsilon\sigma(t)}\mathrm{d}\bar W_t-\int_0^T\frac{[M(\vartheta_u,t)-M(\vartheta_0,t)]^2}{\varepsilon^2\sigma(t)^2}\mathrm{d}t\\
&=u\,\Delta_\varepsilon\left(\vartheta_0,X^T\right)-\frac{u^2}{2}\,\mathrm{I}_3(\vartheta_0)+r_\varepsilon
\end{aligned}\tag{2.149}
$$

where $r_\varepsilon\rightarrow 0$,

$$
\begin{aligned}
\Delta_\varepsilon\left(\vartheta_0,X^T\right)&=\int_0^T\left[\frac{\dot a^*(\vartheta_0,t)\,y_t(\vartheta_0)}{b(\vartheta_0,t)}+\frac{\dot S(\vartheta_0,t)}{\sqrt{2S(\vartheta_0,t)\sigma(t)}}\xi_{t,\varepsilon}\right]\mathrm{d}\bar W_t\\
&\Longrightarrow\Delta(\vartheta_0)\sim\mathcal{N}(0,\mathrm{I}_3(\vartheta_0))\\
\mathrm{I}_3(\vartheta_0)&=\int_0^T\left[\frac{\dot a^*(\vartheta_0,t)^2\,y_t(\vartheta_0)^2}{b(\vartheta_0,t)^2}+\frac{\dot S(\vartheta_0,t)^2}{2S(\vartheta_0,t)\sigma(t)}\right]\mathrm{d}t.
\end{aligned}
$$

Remind that

$$
\int_0^T\frac{\dot a^*(\vartheta_0,t)\,\dot S(\vartheta_0,t)\,y_t(\vartheta_0)}{b(\vartheta_0,t)\sqrt{2S(\vartheta_0,t)\sigma(t)}}\xi_{t,\varepsilon}\,\mathrm{d}t\longrightarrow 0.
$$

The proofs of the estimates like (2.147), (2.148) once more follow the proofs of Lemmas 2.9,2.10. Just note that the function $G_3(\vartheta,\vartheta_0)$ in this case has the following form:

$$
\begin{aligned}
G_3(\vartheta,\vartheta_0)&=\int_0^T\frac{[a(\vartheta,t)-a(\vartheta_0,t)+\ell(\vartheta,t)-\ell(\vartheta_0,t)]^2}{f(\vartheta,t)^2\,b(\vartheta,t)^2}f(\vartheta_0,t)^2\,y_t(\vartheta_0)^2\,\mathrm{d}t\\
&\quad+\int_0^T\frac{[f(\vartheta,t)b(\vartheta,t)-f(\vartheta_0,t)b(\vartheta_0,t)]^2}{2f(\vartheta,t)b(\vartheta,t)\;\sigma(t)}\mathrm{d}t
\end{aligned}
$$

Having (2.149) and these estimates we obtain the consistency, asymptotic normality

$$\varepsilon^{-1/3}\left(\hat{\vartheta}_\varepsilon-\vartheta_0\right)\Longrightarrow\zeta\left(\vartheta_0\right)\sim\mathcal{N}\left(0,\mathrm{I}_3\left(\vartheta_0\right)^{-1}\right),\quad \varepsilon^{-1/3}\left(\tilde{\vartheta}_\varepsilon-\vartheta_0\right)\Longrightarrow\zeta\left(\vartheta_0\right),$$

convergence of moments and asymptotic efficiency of the MLE and BE.

2.5.3 One-Step MLE Processes

Case $\delta\in(1/3,1)$

The partially observed system is (2.117) and (2.118) with $\psi_\varepsilon=\varepsilon^\delta,\ \delta\in(1/3,1)$. The learning interval is $[0,\tau]$ and the preliminary MDE is

$$\check{\vartheta}_{\tau,\varepsilon}=\arg\inf_{\vartheta\in\Theta}\int_0^\tau\left[X_t-x_t\left(\vartheta\right)\right]^2\mathrm{d}t.$$

Note that if the conditions $\check{\mathscr{F}}$ hold, then by Theorem 2.14 the MDE is consistent and asymptotically normal

$$\psi_\varepsilon^{-1}\left(\check{\vartheta}_{\tau,\varepsilon}-\vartheta_0\right)\Longrightarrow\check{\zeta}\left(\vartheta_0\right)\sim\mathcal{N}\left(0,\check{D}\left(\vartheta_0\right)\right).$$

Introduce the random process $M\left(\check{\vartheta}_{\tau,\varepsilon},t\right),\tau\leq t\leq T$ by the equation

$$\mathrm{d}M\left(\check{\vartheta}_{\tau,\varepsilon},t\right)=A^*\left(\check{\vartheta}_{\tau,\varepsilon},t\right)M\left(\check{\vartheta}_{\tau,\varepsilon},t\right)\mathrm{d}t+\frac{\psi_\varepsilon}{\varepsilon}\tilde{B}\left(\check{\vartheta}_{\tau,\varepsilon},t\right)\,\mathrm{d}X_t \tag{2.150}$$

subject to initial value $M\left(\check{\vartheta}_{\tau,\varepsilon},\tau\right)$. Here

$$A^*\left(\check{\vartheta}_{\tau,\varepsilon},t\right)=a^*\left(\check{\vartheta}_{\tau,\varepsilon},t\right)-\frac{\psi_\varepsilon}{\varepsilon}\tilde{B}^*\left(\check{\vartheta}_{\tau,\varepsilon},t\right),$$

$$\tilde{B}\left(\check{\vartheta}_{\tau,\varepsilon},t\right)=\frac{\gamma_*\left(\check{\vartheta}_{\tau,\varepsilon},t\right)f\left(\check{\vartheta}_{\tau,\varepsilon},t\right)^2}{\sigma\left(t\right)^2}.$$

The initial value can be calculated as follows. The function $M\left(\vartheta,t\right),0\leq t\leq\tau$ by Lemma 1.3 has the representation

$$\begin{aligned}M\left(\vartheta,\tau\right)=&\,M\left(\vartheta,0\right)\Phi\left(\vartheta,0,\tau\right)+\frac{\psi_\varepsilon}{\varepsilon}\tilde{B}\left(\vartheta,\tau\right)\,X_\tau\\&-\frac{\psi_\varepsilon}{\varepsilon}\int_0^\tau\Phi\left(\vartheta,s,\tau\right)X_s\left[\tilde{B}_s'\left(\vartheta,s\right)-A^*\left(\vartheta,s\right)\right]\mathrm{d}s,\end{aligned}$$

where

$$\Phi(\vartheta, s, \tau) = \exp\left(\int_s^t A^*(\vartheta, v)\,\mathrm{d}v\right).$$

Therefore we can write

$$M\left(\check\vartheta_{\tau,\varepsilon}, \tau\right) = M\left(\check\vartheta_{\tau,\varepsilon}, 0\right)\Phi\left(\check\vartheta_{\tau,\varepsilon}, 0, \tau_\varepsilon\right) + \frac{\psi_\varepsilon}{\varepsilon}\tilde B\left(\check\vartheta_{\tau,\varepsilon}, \tau\right) X_{\tau,\varepsilon}$$
$$- \frac{\psi_\varepsilon}{\varepsilon}\int_0^{\tau_\varepsilon}\Phi\left(\check\vartheta_{\tau,\varepsilon}, s, \tau\right) X_s\left[\tilde B_s'\left(\check\vartheta_{\tau,\varepsilon}, s\right) - A^*\left(\check\vartheta_{\tau,\varepsilon}, s\right)\right]\mathrm{d}s. \tag{2.151}$$

Denote

$$F(\vartheta, t) = \frac{[\dot a(\vartheta, t) + \dot\ell(\vartheta, t)]y_t(\vartheta)}{b(\vartheta, t)\,\sigma(t)}$$

and Fisher information function function

$$\mathrm{I}_0^t(\vartheta) = \int_0^t \frac{[\dot a(\vartheta, t) + \dot\ell(\vartheta, t)]^2 y_s(\vartheta)^2}{b(\vartheta, s)^2}\mathrm{d}s, \qquad 0 \le t \le T.$$

We need *Conditions* $\mathscr{F}^\star$.

$\mathscr{F}_1^\star$. *The functions* $\dot a(\vartheta, t), \dot\ell(\vartheta, t), b(\vartheta, t), \sigma(t), \vartheta \in \Theta, t \in [0, T]$ *have continuous bounded derivatives w.r.t. t.*

$\mathscr{F}_2^\star$. *The Fisher information function* $\mathrm{I}_0^t(\vartheta), 0 < \tau < t \le T$ *is uniformly on* $\vartheta \in \Theta$ *positive for all* $t \in (\tau, T]$.

Note that $\mathrm{I}_0^t(\vartheta), \tau < t \le T$ is nondecreasing function of t and if $\inf_{\vartheta\in\Theta}\mathrm{I}_0^\tau(\vartheta) > 0$ then the condition $\mathscr{F}_2^\star$ is fulfilled.

Introduce the One-step MLE process $\vartheta_{t,\varepsilon}^\star, \tau \le t \le T$

$$\vartheta_{t,\varepsilon}^\star = \check\vartheta_{\tau,\varepsilon} + \mathrm{I}_0^t\left(\check\vartheta_{\tau,\varepsilon}\right)^{-1}\frac{\psi_\varepsilon}{\varepsilon}\Big\{F\left(\check\vartheta_{\tau,\varepsilon}, \tau\right)X_\tau - \int_0^\tau X_s\,F_s'\left(\check\vartheta_{\tau,\varepsilon}, s\right)\mathrm{d}s$$
$$- \int_0^\tau F\left(\check\vartheta_{\tau,\varepsilon}, s\right)M\left(\check\vartheta_{\tau,\varepsilon}, s\right)\mathrm{d}s + \int_\tau^t F\left(\check\vartheta_{\tau,\varepsilon}, s\right)\left[\mathrm{d}X_s - M\left(\check\vartheta_{\tau,\varepsilon}, s\right)\mathrm{d}s\right]\Big\}. \tag{2.152}$$

Theorem 2.17 *Let the conditions* $\check{\mathscr{F}}$, $\hat{\mathscr{F}}_{1-3}$ *and* $\mathscr{F}^\star$ *hold and* $\delta \in (1/3, 1)$. *Then the estimator process* $\vartheta_{t,\varepsilon}^\star, \tau < t \le T$ *is uniformly consistent, asymptotically normal*

$$\psi_\varepsilon^{-1}\left(\vartheta_{t,\varepsilon}^\star - \vartheta_0\right) \Longrightarrow \mathcal{N}\left(0, \mathrm{I}_0^t(\vartheta_0)^{-1}\right),$$

the moments converge and this estimator is asymptotically efficient.

Proof In the definition of the One-step estimator process, we used the relation between the stochastic and ordinary integrals

$$\int_0^\tau F(\vartheta, s)\,\mathrm{d}X_s = F(\vartheta, \tau)\; X_\tau - \int_0^\tau X_s\, F_s'(\vartheta, s)\,\mathrm{d}s.$$

Applying Taylor expansion at the point ϑ_0 we have

$$\begin{aligned}
&F\left(\check\vartheta_{\tau,\varepsilon}, \tau\right) X_\tau - \int_0^\tau X_s\, F_s'\left(\check\vartheta_{\tau,\varepsilon}, s\right)\mathrm{d}s - \int_0^\tau F\left(\check\vartheta_{\tau,\varepsilon}, s\right) M\left(\check\vartheta_{\tau,\varepsilon}, s\right)\mathrm{d}s\\
&\quad = F(\vartheta_0, \tau)\, X_\tau - \int_0^\tau X_s\, F_s'(\vartheta_0, s)\,\mathrm{d}s - \int_0^\tau F(\vartheta_0, s)\, M\left(\check\vartheta_{\tau,\varepsilon}, s\right)\mathrm{d}s\\
&\qquad + \left(\check\vartheta_{\tau,\varepsilon} - \vartheta_0\right)\Big[\dot F(\vartheta_0, \tau)\, X_\tau - \int_0^\tau X_s\, \dot F_s'(\vartheta_0, s)\,\mathrm{d}s\\
&\qquad - \int_0^\tau \dot F(\vartheta_0, s)\, M\left(\check\vartheta_{\tau,\varepsilon}, s\right)\mathrm{d}s\Big]\\
&\qquad + \left(\check\vartheta_{\tau,\varepsilon} - \vartheta_0\right)^2 O(1)\\
&\quad = \int_0^\tau F(\vartheta_0, s)\left[\mathrm{d}X_s - M\left(\check\vartheta_{\tau,\varepsilon}, s\right)\mathrm{d}s\right]\\
&\qquad + \left(\check\vartheta_{\tau,\varepsilon} - \vartheta_0\right)\int_0^\tau \dot F(\vartheta_0, s)\left[\mathrm{d}X_s - M\left(\check\vartheta_{\tau,\varepsilon}, s\right)\mathrm{d}s\right] + \left(\check\vartheta_{\tau,\varepsilon} - \vartheta_0\right)^2 O(1)\\
&\quad = \varepsilon\int_0^\tau F(\vartheta_0, s)\,\sigma(s)\,\mathrm{d}\bar W_s + \left(\vartheta_0 - \check\vartheta_{\tau,\varepsilon}\right)\int_0^\tau F(\vartheta_0, s)\,\dot M\left(\bar\vartheta_\varepsilon, s\right)\mathrm{d}s\\
&\qquad + \varepsilon\left(\check\vartheta_{\tau,\varepsilon} - \vartheta_0\right)\int_0^\tau \dot F(\vartheta_0, s)\,\sigma(s)\,\mathrm{d}\bar W_s + \frac{\varepsilon}{\psi_\varepsilon}\left(\check\vartheta_{\tau,\varepsilon} - \vartheta_0\right)^2 O(1)\\
&\quad = \varepsilon\int_0^\tau F(\vartheta_0, s)\,\sigma(s)\,\mathrm{d}\bar W_s + \frac{\varepsilon}{\psi_\varepsilon}\left(\vartheta_0 - \check\vartheta_{\tau,\varepsilon}\right)\mathrm{I}_0^\tau(\vartheta_0) + \left(\check\vartheta_{\tau,\varepsilon} - \vartheta_0\right)^2 O(1).
\end{aligned}$$

Here we used (2.133). For the second stochastic integral, we obtain the similar expansion directly

$$\begin{aligned}
&\int_\tau^t F\left(\check\vartheta_{\tau,\varepsilon}, s\right)\left[\mathrm{d}X_s - M\left(\check\vartheta_{\tau,\varepsilon}, s\right)\mathrm{d}s\right] = \int_\tau^t F(\vartheta_0, s)\left[\mathrm{d}X_s - M\left(\check\vartheta_{\tau,\varepsilon}, s\right)\mathrm{d}s\right]\\
&\qquad + \left(\check\vartheta_{\tau,\varepsilon} - \vartheta_0\right)\int_\tau^t \dot F(\vartheta_0, s)\left[\mathrm{d}X_s - M\left(\check\vartheta_{\tau,\varepsilon}, s\right)\mathrm{d}s\right] + \left(\check\vartheta_{\tau,\varepsilon} - \vartheta_0\right)^2 O(1)\\
&\quad = \varepsilon\int_\tau^t F(\vartheta_0, s)\,\sigma(s)\,\mathrm{d}\bar W_s + \int_\tau^t F(\vartheta_0, s)\left[M(\vartheta_0, s) - M\left(\check\vartheta_{\tau,\varepsilon}, s\right)\right]\mathrm{d}s
\end{aligned}$$

$$
\begin{aligned}
&+\left(\check\vartheta_{\tau,\varepsilon}-\vartheta_0\right)^2 O\left(1\right)\\
&=\varepsilon\int_\tau^t F\left(\vartheta_0,s\right)\sigma\left(s\right){\rm d}\bar W_s+\frac{\varepsilon}{\psi_\varepsilon}\left(\vartheta_0-\check\vartheta_{\tau,\varepsilon}\right)\int_\tau^t F\left(\vartheta_0,s\right)^2{\rm d}s\\
&+\frac{\varepsilon}{\psi_\varepsilon}\left(\check\vartheta_{\tau,\varepsilon}-\vartheta_0\right)^2 O\left(1\right).
\end{aligned}
$$

This implies

$$
\begin{aligned}
\frac{\vartheta_{t,\varepsilon}^\star-\vartheta_0}{\psi_\varepsilon}&=\frac{\check\vartheta_{\tau,\varepsilon}-\vartheta_0}{\psi_\varepsilon}+{\rm I}_\tau^t\left(\check\vartheta_{\tau,\varepsilon}\right)^{-1}\int_0^t\frac{\dot a^*\left(\vartheta_0,s\right)y_s\left(\vartheta_0\right)}{b\left(\vartheta_0,s\right)}{\rm d}\bar W_s\\
&\quad+\frac{1}{\psi_\varepsilon}\left(\vartheta_0-\check\vartheta_{\tau,\varepsilon}\right){\rm I}_0^t\left(\check\vartheta_{\tau,\varepsilon}\right)^{-1}{\rm I}_0^t\left(\vartheta_0\right)+\frac{\left(\check\vartheta_{\tau,\varepsilon}-\vartheta_0\right)^2}{\psi_\varepsilon}O\left(1\right)\\
&={\rm I}_\tau^t\left(\vartheta_0\right)^{-1}\int_0^t\frac{\dot a^*\left(\vartheta_0,s\right)y_s\left(\vartheta_0\right)}{b\left(\vartheta_0,s\right)}{\rm d}\bar W_s\left(1+o\left(1\right)\right)+\frac{\left(\check\vartheta_{\tau,\varepsilon}-\vartheta_0\right)^2}{\psi_\varepsilon}O\left(1\right)\\
&\Longrightarrow\eta_t^\star\left(\vartheta_0\right)={\rm I}_\tau^t\left(\vartheta_0\right)^{-1}\int_0^t\frac{\dot a^*\left(\vartheta_0,s\right)y_s\left(\vartheta_0\right)}{b\left(\vartheta_0,s\right)}{\rm d}w\left(s\right).
\end{aligned}
$$

The convergence of moments follows from the convergence of moments of $\check\vartheta_{\tau,\varepsilon}$ and conditions of Theorem 2.17. The uniform convergence of the second moments and the continuity of the Fisher information allow to verify the asymptotic efficiency too. □

Case $\delta\in(0,1/3)$

Remind (see (2.136)) that in this case $\varphi_\varepsilon=\varepsilon^{\frac{1-\delta}{2}}$

$$
\frac{M\left(\vartheta_0+\varphi_\varepsilon u,t\right)-M\left(\vartheta_0,t\right)}{\varepsilon\sigma\left(t\right)}=u\;\frac{\dot S\left(\vartheta_0,t\right)}{\sqrt{2B\left(\vartheta_0,t\right)}\,\sigma\left(t\right)}\;\xi_{t,\varepsilon}\left(1+O\left(\varphi_\varepsilon\right)\right)
$$

where $S\left(\vartheta,t\right)=f\left(\vartheta,t\right)b\left(\vartheta,t\right)$ and $B\left(\vartheta,t\right)=f\left(\vartheta,t\right)b\left(\vartheta,t\right)\sigma\left(t\right)^{-1}$. The Fisher information function is

$$
{\rm I}_\tau^t\left(\vartheta\right)=\int_\tau^t\frac{\dot S\left(\vartheta,s\right)^2}{2S\left(\vartheta,s\right)\sigma\left(s\right)}{\rm d}s,\qquad\qquad \tau<t\leq T. \tag{2.153}
$$

We have two preliminary MDEs. The first one $\check\vartheta_{\tau,\varepsilon}$ was constructed by the observations on the fixed learning interval $[0,\tau]$, and the second $\check\vartheta_{\tau_\varepsilon}$ by the observations on the vanishing learning interval $[0, \tau_\varepsilon]$, $\tau_\varepsilon = \varepsilon^{1-\delta} \to 0$. The rates of convergence are ε^δ for the first and $\varepsilon^{\frac{1-\delta}{2}}$ for the second estimator, respectively. The comparison of the rates (2.130) shows that the MDE $\check\vartheta_{\tau_\varepsilon}$ has better rate than $\check\vartheta_{\tau_\varepsilon}$. This leads to the following consequence: the One-step MLE process constructed with the help of $\check\vartheta_{\tau,\varepsilon}$ is consistent for the values $\delta \in (1/5, 1/3)$ and if we use $\check\vartheta_{\tau,\varepsilon}$, then it is consistent for all $\delta \in (0, 1/3)$. This is a bit unusual situation (see Remark 2.19), and below, we show why it happens.

Fixed Learning Interval

$\delta \in (1/5, 1/3)$. Let us fix the learning interval $[0, \tau]$, take $\delta \in (1/5, 1/3)$ and define the MDE $\check\vartheta_{\tau,\varepsilon}$ by the relation (2.126). Suppose as well that the conditions $\check{\mathscr{F}}$ hold. Then by Theorem 2.14 we have

$$\psi_\varepsilon^{-1}\left(\check\vartheta_{\tau,\varepsilon} - \vartheta_0\right) \Longrightarrow \mathcal{N}\left(0, D\left(\vartheta_0\right)\right).$$

For One-step MLE process, we need two stochastic processes $M\left(\check\vartheta_{\tau,\varepsilon}, t\right), \tau \le t \le T$ and $\dot M\left(\check\vartheta_{\tau,\varepsilon}, t\right), \tau \le t \le T$. These processes are defined by the equations

$$\mathrm{d}M\left(\check\vartheta_{\tau,\varepsilon}, t\right) = \left[a^*\left(\check\vartheta_{\tau,\varepsilon}, t\right) - \frac{\psi_\varepsilon}{\varepsilon} B\left(\check\vartheta_{\tau,\varepsilon}, t\right)\right] M\left(\check\vartheta_{\tau,\varepsilon}, t\right) \mathrm{d}t + \frac{\psi_\varepsilon}{\varepsilon} B\left(\check\vartheta_{\tau,\varepsilon}, t\right) \mathrm{d}X_t, \tag{2.154}$$

$$\begin{aligned}\mathrm{d}\dot M\left(\check\vartheta_{\tau,\varepsilon}, t\right) &= \left[a^*\left(\check\vartheta_{\tau,\varepsilon}, t\right) - \frac{\psi_\varepsilon}{\varepsilon} B\left(\check\vartheta_{\tau,\varepsilon}, t\right)\right] \dot M\left(\check\vartheta_{\tau,\varepsilon}, t\right) \mathrm{d}t + \frac{\psi_\varepsilon}{\varepsilon} \dot B\left(\check\vartheta_{\tau,\varepsilon}, t\right) \mathrm{d}X_t \\ &\quad + \left[\dot a^*\left(\check\vartheta_{\tau,\varepsilon}, t\right) - \frac{\psi_\varepsilon}{\varepsilon} \dot B\left(\check\vartheta_{\tau,\varepsilon}, t\right)\right] M\left(\check\vartheta_{\tau,\varepsilon}, t\right) \mathrm{d}t.\end{aligned} \tag{2.155}$$

The initial conditions are $M\left(\check\vartheta_{\tau,\varepsilon}, \tau\right) = 0$, $\dot M\left(\check\vartheta_{\tau,\varepsilon}, \tau\right) = 0$.

The One-step MLE process is

$$\vartheta_{t,\varepsilon}^\star = \check\vartheta_{\tau,\varepsilon} + \mathrm{I}_\tau^t\left(\check\vartheta_{\tau,\varepsilon}\right)^{-1} \int_\tau^t \frac{\dot M\left(\check\vartheta_{\tau,\varepsilon}, s\right)}{\varepsilon\, \psi_\varepsilon\, \sigma(s)^2} \left[\mathrm{d}X_s - M\left(\check\vartheta_{\tau,\varepsilon}, s\right) \mathrm{d}s\right].$$

We need *Conditions* $\mathscr{F}^\star$.

$\mathscr{F}_3^\star$. *The functions* $\dot b(\vartheta, t), \dot f(\vartheta, t), \vartheta \in \Theta, t \in [0, T]$ *have continuous bounded derivatives w.r.t. t.*

$\mathscr{F}_4^\star$. *The Fisher information function* $\mathrm{I}_\tau^t(\vartheta), 0 < \tau < t \le T$ *is uniformly on* $\vartheta \in \Theta$ *positive for all* $t \in (\tau, T]$.

Theorem 2.18 *Let the conditions* $\check{\mathscr{F}}$, $\hat{\mathscr{F}}_{1-3}$ *and* $\mathscr{F}_{3-4}^\star$ *hold and* $\delta \in (1/5, 1/3)$. *Then the estimator process* $\vartheta_{t,\varepsilon}^\star, \tau < t \le T$ *is uniformly consistent, asymptotically normal*

$$\varphi_\varepsilon^{-1}\left(\vartheta^\star_{t,\varepsilon}-\vartheta_0\right)\Longrightarrow\mathcal{N}\left(0,\mathrm{I}_0^t\left(\vartheta_0\right)^{-1}\right),$$

the moments converge and this estimator is asymptotically efficient.

Proof Consider the asymptotic behavior of this estimator. We have

$$\frac{\vartheta^\star_{t,\varepsilon}-\vartheta_0}{\varphi_\varepsilon}=\frac{\check\vartheta_{\tau,\varepsilon}-\vartheta_0}{\varphi_\varepsilon}+\mathrm{I}_\tau^t\left(\check\vartheta_{\tau,\varepsilon}\right)^{-1}\int_\tau^t\frac{\dot M\left(\check\vartheta_{\tau,\varepsilon},s\right)}{\varphi_\varepsilon\,\psi_\varepsilon\,\sigma\left(s\right)}\,\mathrm{d}\bar W_s$$

$$+\frac{\mathrm{I}_\tau^t\left(\check\vartheta_{\tau,\varepsilon}\right)^{-1}}{\varepsilon\varphi_\varepsilon\,\psi_\varepsilon}\int_\tau^t\frac{\dot M\left(\check\vartheta_{\tau,\varepsilon},s\right)}{\sigma\left(s\right)^2}\left[M\left(\vartheta_0,s\right)-M\left(\check\vartheta_{\tau,\varepsilon},s\right)\right]\,\mathrm{d}s.$$

The stochastic integral is asymptotically normal

$$\int_\tau^t\frac{\dot M\left(\check\vartheta_{\tau,\varepsilon},s\right)}{\varphi_\varepsilon\,\psi_\varepsilon\,\sigma\left(s\right)}\,\mathrm{d}\bar W_s=\int_\tau^t\frac{\dot S\left(\vartheta_0,s\right)\,\xi_{s,\varepsilon}}{\sqrt{2S\left(\vartheta_0,s\right)}\sigma\left(s\right)}\,\mathrm{d}\bar W_s\left(1+o\left(1\right)\right)$$

$$\Longrightarrow\int_\tau^t\frac{\dot S\left(\vartheta_0,s\right)}{\sqrt{2S\left(\vartheta_0,s\right)}\sigma\left(s\right)}\,\mathrm{d}w\left(s\right)\sim\mathcal{N}\left(0,\mathrm{I}_\tau^t\left(\vartheta_0\right)\right).$$

For the ordinary integral, we have

$$\frac{1}{\varepsilon\varphi_\varepsilon\,\psi_\varepsilon}\int_\tau^t\frac{\dot M\left(\check\vartheta_{\tau,\varepsilon},s\right)}{\sigma\left(s\right)^2}\left[M\left(\vartheta_0,s\right)-M\left(\check\vartheta_{\tau,\varepsilon},s\right)\right]\,\mathrm{d}s$$

$$=\frac{\vartheta_0-\check\vartheta_{\tau,\varepsilon}}{\varphi_\varepsilon}\frac{1}{\varepsilon\,\psi_\varepsilon}\int_\tau^t\frac{\dot M\left(\check\vartheta_{\tau,\varepsilon},s\right)^2}{\sigma\left(s\right)^2}\,\mathrm{d}s$$

$$+\frac{\left(\vartheta_0-\check\vartheta_{\tau,\varepsilon}\right)^2}{\varphi_\varepsilon}\frac{1}{\varepsilon\,\psi_\varepsilon}\int_\tau^t\frac{\dot M\left(\check\vartheta_{\tau,\varepsilon},s\right)\ddot M\left(\bar\vartheta_{\tau,\varepsilon},s\right)}{\sigma\left(s\right)^2}\,\mathrm{d}s$$

$$=\frac{\vartheta_0-\check\vartheta_{\tau,\varepsilon}}{\varphi_\varepsilon}\int_\tau^t\frac{\dot S\left(\vartheta_{\tau,\varepsilon},s\right)^2\xi_{s,\varepsilon}^2}{2S\left(\vartheta_0,s\right)\sigma\left(s\right)}\,\mathrm{d}s+\left(\vartheta_0-\check\vartheta_{\tau,\varepsilon}\right)O\left(1\right)+\frac{\left(\vartheta_0-\check\vartheta_{\tau,\varepsilon}\right)^2}{\varphi_\varepsilon}O\left(1\right)$$

$$=\frac{\vartheta_0-\check\vartheta_{\tau,\varepsilon}}{\varphi_\varepsilon}\left[\mathrm{I}_\tau^t\left(\vartheta_0\right)+\left(\vartheta_0-\check\vartheta_{\tau,\varepsilon}\right)O\left(1\right)\right]+\left(\vartheta_0-\check\vartheta_{\tau,\varepsilon}\right)O\left(1\right)$$

$$+\frac{\left(\vartheta_0-\check\vartheta_{\tau,\varepsilon}\right)^2}{\varphi_\varepsilon}O\left(1\right).$$

For the last term, we have the estimate

$$\frac{\left(\vartheta_0-\check\vartheta_{\tau,\varepsilon}\right)^2}{\varphi_\varepsilon}=\left[\frac{\left(\vartheta_0-\check\vartheta_{\tau,\varepsilon}\right)^2}{\psi_\varepsilon^2}\right]\frac{\psi_\varepsilon^2}{\varphi_\varepsilon}=\left[\frac{\left(\vartheta_0-\check\vartheta_{\tau,\varepsilon}\right)}{\psi_\varepsilon}\right]^2\varepsilon^{\frac{5\delta-1}{2}}\longrightarrow 0$$

because $\delta > 1/5$. Therefore,

$$\begin{aligned}\frac{\vartheta^\star_{t,\varepsilon}-\vartheta_0}{\varphi_\varepsilon}&=\frac{\check\vartheta_{\tau,\varepsilon}-\vartheta_0}{\varphi_\varepsilon}+\frac{\vartheta_0-\check\vartheta_{\tau,\varepsilon}}{\varphi_\varepsilon}+o(1)\\&\quad+\mathrm{I}_\tau^t\left(\vartheta_0\right)^{-1}\int_\tau^t\frac{\dot S\left(\vartheta_0,s\right)\,\xi_{s,\varepsilon}}{\sqrt{2S\left(\vartheta_0,s\right)\sigma\left(s\right)}}\,\mathrm{d}\bar W_s\left(1+o(1)\right)\\&\Longrightarrow\eta_t^\star\left(\vartheta_0\right)=\mathrm{I}_\tau^t\left(\vartheta_0\right)^{-1}\int_\tau^t\frac{\dot S\left(\vartheta_0,s\right)}{\sqrt{2S\left(\vartheta_0,s\right)\sigma\left(s\right)}}\,\mathrm{d}w\left(s\right)\end{aligned}$$

□

Remark 2.22 Note that the One-step MLE-process

$$\vartheta^\star_{t,\varepsilon}=\check\vartheta_{\tau,\varepsilon}+\frac{\mathrm{I}_\tau^t\left(\check\vartheta_{\tau,\varepsilon}\right)^{-1}}{\sqrt{\varepsilon\,\psi\varepsilon}}\int_\tau^t\frac{\dot S\left(\check\vartheta_{\tau,\varepsilon},s\right)}{\sqrt{2S(\check\vartheta_{\tau,\varepsilon},s)\sigma\left(s\right)^{3/2}}}\left[\mathrm{d}X_s-M\left(\check\vartheta_{\tau,\varepsilon},s\right)\mathrm{d}s\right]$$

is not consistent because in the corresponding ordinary integral we obtain the term $\xi_{s,\varepsilon}$ and not $\xi^2_{s,\varepsilon}$. Thus, this integral tends to zero.

Remark 2.23 It is possible to extend this construction for some values $\delta < 1/5$ if we introduce Two-step MLE-process (see Sect. 1.2), but in this case there can be difficulties with the numerical realization of the filtering equations.

Vanishing Learning Interval, $\delta\in(0,1/3)$
Suppose now that the MDE $\check\vartheta_{\tau_\varepsilon}$ is constructed by observations on the vanishing interval $[0,\tau_\varepsilon]$, where $\tau_\varepsilon=\varepsilon^{1-\delta}$. Then under conditions $\mathscr{F}^\circ$

$$\varepsilon^{-\frac{1-\delta}{2}}\left(\check\vartheta_{\tau_\varepsilon}-\vartheta_0\right)\Longrightarrow\mathcal{N}\left(0,D\left(\vartheta_0\right)\right).$$

The One-step MLE process is

$$\vartheta^\star_{t,\varepsilon}=\check\vartheta_{\tau_\varepsilon}+\frac{\mathrm{I}_{\tau_\varepsilon}^t\left(\check\vartheta_{\tau_\varepsilon}\right)^{-1}}{\varepsilon\,\psi_\varepsilon}\int_{\tau_\varepsilon}^t\frac{\dot M\left(\check\vartheta_{\tau_\varepsilon},s\right)}{\sigma\left(s\right)^2}\left[\mathrm{d}X_s-M\left(\check\vartheta_{\tau_\varepsilon},s\right)\mathrm{d}s\right],\qquad\tau<t\leq T.\tag{2.156}$$

The random processes $M\left(\check\vartheta_{\tau_\varepsilon},s\right)$ and $\dot M\left(\check\vartheta_{\tau_\varepsilon},s\right)$ are defined by the same Eqs. (2.154), (2.155).

The repetition of the same calculations gives us the convergence ($\varphi_\varepsilon = \varepsilon^{\frac{1-\delta}{2}}$)

$$\frac{\vartheta^\star_{t,\varepsilon} - \vartheta_0}{\varphi_\varepsilon} = \frac{\check\vartheta_{\tau_\varepsilon} - \vartheta_0}{\varphi_\varepsilon} + \frac{\vartheta_0 - \check\vartheta_{\tau_\varepsilon}}{\varphi_\varepsilon} + o(1) + \frac{\left(\vartheta_0 - \check\vartheta_{\tau_\varepsilon}\right)^2}{\varphi_\varepsilon} O(1)$$

$$+ \mathrm{I}^t_{\tau_\varepsilon}(\vartheta_0)^{-1} \int_{\tau_\varepsilon}^t \frac{\dot S(\vartheta_0, s)\, \xi_{s,\varepsilon}}{\sqrt{2S(\vartheta_0, s)\,\sigma(s)}}\, \mathrm{d}\bar W_s\,(1 + o(1))$$

$$\Longrightarrow \eta^\star_t(\vartheta_0) = \mathrm{I}^t_0(\vartheta_0)^{-1} \int_0^t \frac{\dot S(\vartheta_0, s)}{\sqrt{2S(\vartheta_0, s)\,\sigma(s)}}\, \mathrm{d}w(s)$$

because

$$\frac{\left(\vartheta_0 - \check\vartheta_{\tau_\varepsilon}\right)^2}{\varphi_\varepsilon} = \left[\sqrt{\frac{\psi_\varepsilon}{\varepsilon}}\left(\vartheta_0 - \check\vartheta_{\tau_\varepsilon}\right)\right]^2 \frac{\varepsilon}{\varphi_\varepsilon\psi_\varepsilon} = \left[\sqrt{\frac{\psi_\varepsilon}{\varepsilon}}\left(\vartheta_0 - \check\vartheta_{\tau_\varepsilon}\right)\right]^2 \varepsilon^{\frac{1-\delta}{2}} \longrightarrow 0. \tag{2.157}$$

One-Step MLE-Process, $\delta = 1/3$

This case can be treated as above. The learning interval is $[0, \tau_\varepsilon]$, $\tau_\varepsilon = \varepsilon^{2/3}$ and

$$\varepsilon^{-1/3}\left(\check\vartheta_{\tau_\varepsilon} - \vartheta_0\right) \Longrightarrow \mathcal{N}(0, D(\vartheta_0))$$

with the corresponding limit variance $D(\vartheta_0)$. The Fisher information function is

$$\mathrm{I}^t_\tau(\vartheta) = \int_\tau^t \frac{[\dot f(\vartheta, s)\, b(\vartheta, s) + f(\vartheta, s)\, \dot b(\vartheta, s)]^2}{2f(\vartheta, s)\, b(\vartheta, s)\, \sigma(s)}\, \mathrm{d}s$$

$$+ \int_\tau^t \frac{[\dot a(\vartheta, s) + \dot\ell(\vartheta, s)]^2 y_s(\vartheta)^2}{b(\vartheta, s)^2}\, \mathrm{d}s, \qquad \tau \le t \le T.$$

The One-step MLE process is defined by (2.156)

$$\vartheta^\star_{t,\varepsilon} = \check\vartheta_{\tau_\varepsilon} + \frac{\mathrm{I}^t_{\tau_\varepsilon}\left(\check\vartheta_{\tau_\varepsilon}\right)^{-1}}{\varepsilon^{4/3}} \int_{\tau_\varepsilon}^t \frac{\dot M\left(\check\vartheta_{\tau_\varepsilon}, s\right)}{\sigma(s)^2}\left[\mathrm{d}X_s - M\left(\check\vartheta_{\tau_\varepsilon}, s\right)\mathrm{d}s\right] \tag{2.158}$$

where the random processes $M\left(\check\vartheta_{\tau_\varepsilon}, s\right)$, $\tau_\varepsilon \le s \le T$ and $\dot M\left(\check\vartheta_{\tau_\varepsilon}, s\right)$, $\tau_\varepsilon \le s \le T$ are solutions of the equations like (2.154) and (2.155), where $\check\vartheta_{\tau,\varepsilon}$ is replaced by $\check\vartheta_{\tau_\varepsilon}$.

The estimator is asymptotically normal

$$\varepsilon^{-1/3}\left(\vartheta^\star_{t,\varepsilon}-\vartheta_0\right)\Longrightarrow \eta^\star_t\left(\vartheta_0\right)\sim\mathcal{N}\left(0,\mathrm{I}^t_0\left(\vartheta_0\right)^{-1}\right) \tag{2.159}$$

but there is a question: how to write the limit variable $\eta^\star_t\left(\vartheta_0\right)$? Recall the representation before the limit

$$\begin{aligned}\varepsilon^{-1/3}\left(\vartheta^\star_{t,\varepsilon}-\vartheta_0\right)=\mathrm{I}^t_{\tau_\varepsilon}\left(\vartheta_0\right)^{-1}&\left\{\int_{\tau_\varepsilon}^t\frac{[\dot f\left(\vartheta_0,s\right)b\left(\vartheta_0,s\right)+f\left(\vartheta_0,s\right)\dot b\left(\vartheta_0,s\right)]}{\sqrt{2f\left(\vartheta_0,s\right)b\left(\vartheta_0,s\right)\sigma\left(s\right)}}\,\mathrm{d}\bar W_s\right.\\&\left.+\int_{\tau_\varepsilon}^t\frac{[\dot a\left(\vartheta_0,s\right)+\dot\ell\left(\vartheta_0,s\right)]y_s\left(\vartheta_0\right)}{b\left(\vartheta_0,s\right)}\xi_{s,\varepsilon}\,\mathrm{d}\bar W_s\right\}+o\left(1\right),\end{aligned}$$

where the limit of $\xi_{s,\varepsilon},0<s\leq T$ is not a stochastic process, and in the both integrals there, we have the same innovation Wiener process. The limit random variable $\eta^\star_t\left(\vartheta_0\right)$ can be written with the help of two independent Wiener processes

$$\begin{aligned}\eta^\star_t\left(\vartheta_0\right)=\mathrm{I}^t_0\left(\vartheta_0\right)^{-1}&\left\{\int_0^t\frac{[\dot f\left(\vartheta_0,s\right)b\left(\vartheta_0,s\right)+f\left(\vartheta_0,s\right)\dot b\left(\vartheta_0,s\right)]}{\sqrt{2f\left(\vartheta_0,s\right)b\left(\vartheta_0,s\right)\sigma\left(s\right)}}\,\mathrm{d}w_1\left(s\right)\right.\\&\left.+\int_0^t\frac{[\dot a\left(\vartheta_0,s\right)+\dot\ell\left(\vartheta_0,s\right)]y_s\left(\vartheta_0\right)}{b\left(\vartheta_0,s\right)}\,\mathrm{d}w_2\left(s\right)\right\}\end{aligned} \tag{2.160}$$

because if we write the same Wiener process, say, $w\left(s\right),0\leq s\leq t$ in both integrals, then the limit variance of $\eta^\star_{t,\varepsilon}$ will be different.

Remark 2.24 Nonlinear partially observed systems with different levels of noises were studied by many authors because such systems are extensively used in different applied problems. The exact calculation of the conditional expectations for such models is in general impossible, and for approximation of the filters, the perturbation theory was widely used. The quality of the obtained filters was verified with the help of numerical simulations. Analytic approach was developped by Bensoussan [9, 10]. The linear systems presented in [10] as illustration of the properties of filters before the study of nonlinear systems are written as follows:

$$\mathrm{d}X_t=f\,Y_t\,\mathrm{d}t+\epsilon^{1-\gamma}\,\mathrm{d}W_t,\qquad X_0=0,\quad 0\leq t\leq T, \tag{2.161}$$

$$\mathrm{d}Y_t=a\,Y_t\,\mathrm{d}t+\epsilon^{\gamma}\,b\,\mathrm{d}V_t,\qquad Y_0, \tag{2.162}$$

where f,a,b are known parameters, $W_t,V_t;0\leq t\leq T$ are independent Wiener processes and the initial value Y_0 is a Gaussian random variable. The parameter $\gamma\in[0,1]$ and the properties of the filters are studied in two situations: $0\leq\gamma<1/2$ and $1/2\leq\gamma\leq 1$. Exhaustive exposition of the analytic results for nonlinear and these linear models are given in Chapter 5, [10].

To compare the system (2.161), (2.162) with studied here (2.117), (2.118) we denote $\varepsilon = \epsilon^{1-\gamma}$, then $\epsilon^{\gamma} = \varepsilon^{\frac{\gamma}{1-\gamma}}$ and this system becomes

$$dX_t = f\, Y_t\, dt + \varepsilon\, dW_t, \qquad X_0 = 0, \quad 0 \le t \le T, \tag{2.163}$$

$$dY_t = a\, Y_t\, dt + \varepsilon^{\delta}\, b\, dV_t, \qquad Y_0, \tag{2.164}$$

with $\delta = \frac{\gamma}{1-\gamma}$. Of course, this is particular case of the system (2.117), (2.118) for the values $\gamma \in (0, 1/2)$. Indeed, the case $0 < \gamma < 1/2$ corresponds to $0 < \delta < 1$ studied here, but in our model the initial value Y_0 is usually deterministic. The case $\gamma = 1/2\,(\delta = 1)$ was studied in the Sect. 2.1 and the case $\gamma = 0\,(\delta = 0)$ will be studied in the Sect. 3.1. The asymptotic of $\gamma_*(\vartheta, t) = \gamma(\vartheta, t) / (\varepsilon\psi_\varepsilon)$ described in Lemma 2.6 coincides with that of $P_\epsilon(t)/\epsilon$ shown in the Section 5.1.2 in [10]. Note that with such notation, we have $\varepsilon\psi_\varepsilon = \epsilon$.

The properties of estimators in the case of partially observed system (2.161), (2.162) where $1/2 < \gamma \le 1$ or $\delta > 1$ in (2.163), (2.164) are not yet discussed in this work. Remark that the asymptotic of the linear filters studied in the Section 5.3 [10] requires the initial value $Y_0 \sim \mathcal{N}(y_0, \epsilon P_0)$, $P_0 > 0$. This means that for the system (2.163), (2.164) and $\delta > 1$, the initial value has to be $Y_0 \sim \mathcal{N}\left(y_0, \varepsilon^{1+\delta} P_0\right)$, $P_0 > 0$.

Let us see what can be behavior of estimators if $\delta > 1$. The partially observed system is

$$dX_t = f(\vartheta, t)\, Y_t\, dt + \varepsilon\, \sigma(t)\, dW_t, \qquad X_0 = 0, \quad 0 \le t \le T,$$

$$dY_t = a(\vartheta, t)\, Y_t\, dt + \varepsilon^{\delta}\, b(\vartheta, t)\, dV_t, \quad Y_0 \sim \mathcal{N}\left(y_0, \varepsilon^{\frac{1+\delta}{2}} P_0\right), \; y_0 \ne 0,\, P_0 > 0,$$

where $f(\vartheta, t), \sigma(t), a(\vartheta, t), b(\vartheta, t); \vartheta \in \Theta, t \in [0, T]$ are known smooth w.r.t. ϑ functions, condition $\mathscr{A}_*$ is fulfilled and for simplicity of exposition $\Theta = (\alpha, \beta)$.

The Kalman–Bucy filter is given by the equations

$$dm(\vartheta, t) = a(\vartheta, t)\, m(\vartheta, t)\, dt + \frac{f(\vartheta, t)\, \gamma(\vartheta, t)}{\varepsilon^2 \sigma(t)^2}\left[dX_t - f(\vartheta, t)\, m(\vartheta, t)\, dt\right],$$

$$\frac{d\gamma(\vartheta, t)}{dt} = 2a(\vartheta, t)\, \gamma(\vartheta, t) - \frac{\gamma(\vartheta, t)^2 f(\vartheta, t)^2}{\varepsilon^2 \sigma(t)^2} + \varepsilon^{2\delta} b(\vartheta, t)^2,$$

$$\gamma(\vartheta, 0) = \varepsilon^{1+\delta} P_0,$$

where the initial value $m(\vartheta, 0) = y_0$. Following [10] denote $\gamma_*(\vartheta, t) = \varepsilon^{-(1+\delta)} \gamma(\vartheta, t)$, then the Riccati equation will be

$$\frac{d\gamma_*(\vartheta, t)}{dt} = 2a(\vartheta, t)\, \gamma_*(\vartheta, t) - \varepsilon^{\delta-1}\left[\frac{\gamma_*(\vartheta, t)^2 f(\vartheta, t)^2}{\sigma(t)^2} - b(\vartheta, t)^2\right],$$

$$\gamma_*(\vartheta, 0) = P_0$$

and it's solution $\gamma_*(\vartheta,t)=\gamma_0(\vartheta,t)\left(1+O\left(\varepsilon^{\delta-1}\right)\right)$, where

$$\gamma_0(\vartheta,t)=P_0\exp\left(2\int_0^t a(\vartheta,s)\,\mathrm{d}s\right).$$

The equation for $m(\vartheta,t)$ is

$$\begin{aligned}\mathrm{d}m(\vartheta,t)&=a(\vartheta,t)\,m(\vartheta,t)\,\mathrm{d}t\\&\quad+\varepsilon^{\delta-1}\frac{f(\vartheta,t)\,\gamma_0(\vartheta,t)}{\sigma(t)^2}\left[\mathrm{d}X_t-f(\vartheta,t)\,m(\vartheta,t)\,\mathrm{d}t\right],\qquad m(\vartheta,0)=y_0,\end{aligned}$$

where we omitted the term of order $O\left(\varepsilon^{2\delta-2}\right)$. Hence

$$\mathrm{d}m(\vartheta_0,t)=a(\vartheta_0,t)\,m(\vartheta_0,t)\,\mathrm{d}t+\varepsilon^{\delta}\frac{f(\vartheta_0,t)\,\gamma_0(\vartheta_0,t)}{\sigma(t)}\mathrm{d}\bar W_t,\qquad m(\vartheta_0,0)=y_0.$$

Then for the difference $\delta_m(t)=m(\vartheta_0,t)-Y_t$, we obtain the equation

$$\mathrm{d}\delta_m(t)=a(\vartheta_0,t)\,\delta_m(t)\,\mathrm{d}t+\varepsilon^{\delta}\frac{f(\vartheta_0,s)\,\gamma_0(\vartheta_0,t)}{\sigma(t)}\mathrm{d}\bar W_t-\varepsilon^{\delta}b(\vartheta_0,s)\,\mathrm{d}V_t$$

with the initial value $\delta_m(0)=y_0-Y_0\sim\mathcal{N}\left(0,\varepsilon^{1+\delta}P_0\right)$. Therefore

$$\mathbf{E}_{\vartheta_0}\left(m(\vartheta_0,t)-Y_t\right)^2\leq C\,\varepsilon^{1+\delta}.$$

Note that if we introduce the function

$$y_t(\vartheta)=m(\vartheta,t)|_{\varepsilon=0}=y_0\exp\left(\int_0^t a(\vartheta,s)\,\mathrm{d}s\right)\equiv y_0\exp\left(N(\vartheta,t)\right),$$

then we have $\mathbf{E}_{\vartheta_0}\left(y_t(\vartheta_0)-Y_t\right)^2\leq C\,\varepsilon^{1+\delta}$ too.

The Fisher information in this problem is

$$\mathrm{I}(\vartheta)=\int_0^T\frac{[\dot f(\vartheta,t)+f(\vartheta,t)\,\dot N(\vartheta,t)]^2y_t(\vartheta)^2}{\sigma(t)^2}\mathrm{d}t$$

and the identifiability condition is: for any $\nu>0$

$$\inf_{\vartheta_0\in\Theta}\inf_{|\vartheta-\vartheta_0|>\nu}\int_0^T\frac{[f(\vartheta,t)\,y_t(\vartheta)-f(\vartheta_0,t)\,y_t(\vartheta_0)]^2}{\sigma(t)^2}\mathrm{d}t>0.$$

Folowing the proof of Theorem 2.4, it can be shown that the MLE and BE are consistent, asymptotically normal

$$\varepsilon^{-1}(\hat{\vartheta}_\varepsilon - \vartheta_0) \Longrightarrow \mathcal{N}\left(0, \mathrm{I}(\vartheta_0)^{-1}\right), \qquad \varepsilon^{-1}(\tilde{\vartheta}_\varepsilon - \vartheta_0) \Longrightarrow \mathcal{N}\left(0, \mathrm{I}(\vartheta_0)^{-1}\right)$$

and asymptotically efficient.

The MDE

$$\check{\vartheta}_{\tau,\varepsilon} = \arg\inf_{\vartheta\in\Theta} \int_0^\tau \left[X_t - f(\vartheta,t)\, y_t(\vartheta)\right]^2 \mathrm{d}t, \qquad x_t(\vartheta) = \int_0^t f(\vartheta,s)\, y_s(\vartheta)\, \mathrm{d}s,$$

is as well consistent, asymptotically normal

$$\varepsilon^{-1}\left(\check{\vartheta}_\varepsilon - \vartheta_0\right) \Longrightarrow \mathcal{N}\left(0, D(\vartheta_0)\right)$$

and the moments converge. For the conditions and proofs, see Theorem 2.2.

The One-step MLE process and adaptive filter can be constructed similar to that of Sect. 2.1.

Note that the initial condition $Y_0 \sim \mathcal{N}\left(y_0, \varepsilon^{1+\delta} P_0\right)$ can be replaced by $Y_0 = y_0$ with the same asymptotic properties of the estimators.

2.5.4 Adaptive Filters

We present the detailed proof in the case $\delta \in (1/3, 1)$, and in the other cases, we just describe the results. The proofs follow the same arguments and are omitted.

Case $\delta \in (1/3, 1)$

Suppose that the learning interval is $[0, \tau]$, $0 < \tau < T$ and we have the MDE $\check{\vartheta}_{\tau,\varepsilon}$ constructed by the observations $X^\tau = (X_t, 0 \le t \le \tau)$. This estimator by Theorem 2.14 is asymptotically normal with the rate ψ_ε. This allow us to admit the values $\delta \in (1/3, 1)$ for the construction of the One-step MLE process $\vartheta^\star_{t,\varepsilon}$, $\tau < t \le T$ (see (2.152)). Recall that this estimator is asymptotically normal

$$\eta^\star_{t,\varepsilon} = \frac{\vartheta^\star_{t,\varepsilon} - \vartheta_0}{\psi_\varepsilon} \Longrightarrow \eta^\star_t(\vartheta_0) = \mathrm{I}_0^t(\vartheta_0)^{-1} \int_0^t \frac{[\dot{a}(\vartheta_0,s) + \dot{\ell}(\vartheta_0,s)]y_s(\vartheta_0)}{b(\vartheta_0,s)} \mathrm{d}w(s).$$

The adaptive filter is defined by the equation

$$\mathrm{d}m^\star_{t,\varepsilon} = \left[a(\vartheta^\star_{t,\varepsilon},t) - \frac{\psi_\varepsilon}{\varepsilon} \frac{b\left(\vartheta^\star_{t,\varepsilon},t\right) f\left(\vartheta^\star_{t,\varepsilon},t\right)}{\sigma(t)}\right] m^\star_{t,\varepsilon}\mathrm{d}t + \frac{\psi_\varepsilon}{\varepsilon} \frac{b\left(\vartheta^\star_{t,\varepsilon},t\right)}{\sigma(t)}\, \mathrm{d}X_t \tag{2.165}$$

for the values $\tau_* \le t \le T$, where $\tau_* \in (\tau, T)$. The initial value $m^\star_{\tau_*,\varepsilon}$ can be calculated similar to (2.151).

Proposition 2.8 *Let the conditions* $\check{\mathscr{F}}$, $\hat{\mathscr{F}}_{1-3}$ *and* $\mathscr{F}^\star$ *hold. Then*

$$\frac{m^\star_{t,\varepsilon} - m(\vartheta_0, t)}{\psi_\varepsilon} \Longrightarrow -\frac{\dot f(\vartheta_0, t)\ y_t(\vartheta_0)}{f(\vartheta_0, t)}\,\eta^\star_t(\vartheta_0).$$

Proof Note that from the equations for $m(\vartheta_0, t), 0 \le t \le T$ and $y_t(\vartheta_0), 0 \le t \le T$ we obtain the relation

$$m(\vartheta_0, t) - y_t(\vartheta_0) = \psi_\varepsilon \int_0^t \exp\left(\int_s^t a(\vartheta_0, v)\,\mathrm{d}v\right) b(\vartheta_0, s)\mathrm{d}\bar W_s = \psi_\varepsilon\, O(1). \tag{2.166}$$

To study of the error $\delta_{m^*}(t) = m^\star_{t,\varepsilon} - m(\vartheta_0, t)$ we remind the equation for $m(\vartheta_0, t)$

$$\mathrm{d}m(\vartheta_0, t) = \left[a(\vartheta_0, t) - \frac{\psi_\varepsilon}{\varepsilon}\frac{b(\vartheta_0, t)\, f(\vartheta_0, t)}{\sigma(t)}\right] m(\vartheta_0, t)\,\mathrm{d}t + \frac{\psi_\varepsilon}{\varepsilon}\frac{b(\vartheta_0, t)}{\sigma(t)}\,\mathrm{d}X_t$$

and consider the equation for difference $\delta_{m^*}(t)$

$$\begin{aligned}
\mathrm{d}\delta_{m^*}(t) &= A(\vartheta_0, t)\,\delta_{m^*}(t)\,\mathrm{d}t + \left[A\left(\vartheta^\star_{t,\varepsilon}, t\right) - A(\vartheta_0, t)\right] m^\star_{t,\varepsilon}\mathrm{d}t\\
&\quad + \frac{\psi_\varepsilon}{\varepsilon}\left[b\left(\vartheta^\star_{t,\varepsilon}, t\right) - b(\vartheta_0, t)\right]\frac{f(\vartheta_0, t)}{\sigma(t)} m(\vartheta_0, t)\,\mathrm{d}t\\
&\quad + \psi_\varepsilon\left[b\left(\vartheta^\star_{t,\varepsilon}, t\right) - b(\vartheta_0, t)\right]\mathrm{d}\bar W_t\\
&= A(\vartheta_0, t)\,\delta_{m^*}(t)\,\mathrm{d}t + \delta_A(t)\, m^\star_{t,\varepsilon}\mathrm{d}t + \frac{\psi_\varepsilon}{\varepsilon}\delta_b(t)\,\frac{f(\vartheta_0, t)}{\sigma(t)} m(\vartheta_0, t)\,\mathrm{d}t\\
&\quad + \psi_\varepsilon \delta_b(t)\,\mathrm{d}\bar W_t.
\end{aligned}$$

Here $\delta_A(t) = A\left(\vartheta^\star_{t,\varepsilon}, t\right) - A(\vartheta_0, t)$, $\delta_b(t) = b\left(\vartheta^\star_{t,\varepsilon}, t\right) - b(\vartheta_0, t)$ and

$$A(\vartheta, t) = a(\vartheta, t) - \frac{\psi_\varepsilon}{\varepsilon}\frac{b(\vartheta, t)\, f(\vartheta, t)}{\sigma(t)}$$

Denote as well

$$\Phi(\vartheta, s, t) = \exp\left(\int_s^t A(\vartheta, v)\,\mathrm{d}v\right), \qquad B(\vartheta, t) = \frac{b(\vartheta, t)\, f(\vartheta, t)}{\sigma(t)}.$$

Then

$$
\begin{aligned}
\delta_{m^*}(t) &= \delta_{m^*}(\tau)\,\Phi(\vartheta_0,0,t) + \int_\tau^t \Phi(\vartheta_0,s,t)\,\delta_A(s)\,m^\star_{s,\varepsilon}\mathrm{d}s \\
&\quad + \frac{\psi_\varepsilon}{\varepsilon}\int_\tau^t \Phi(\vartheta_0,s,t)\,\delta_b(t)\,\frac{f(\vartheta_0,s)}{\sigma(s)}m(\vartheta_0,s)\,\mathrm{d}s \\
&\quad + \psi_\varepsilon\int_\tau^t \Phi(\vartheta_0,s,t)\,\delta_b(t)\;\mathrm{d}\bar W_s \\
&= \delta_m(\tau)\,\Phi(\vartheta_0,0,t) \\
&\quad - \frac{\psi_\varepsilon}{\varepsilon}\int_\tau^t \Phi(\vartheta_0,s,t)\,\dot B(\vartheta_0,s)\left[\vartheta^\star_{s,\varepsilon}-\vartheta_0\right]m^\star_{s,\varepsilon}\mathrm{d}s\,(1+o(1)) \\
&\quad + \frac{\psi_\varepsilon}{\varepsilon}\int_\tau^t \Phi(\vartheta_0,s,t)\,\dot b(\vartheta_0,s)\left[\vartheta^\star_{s,\varepsilon}-\vartheta_0\right]\frac{f(\vartheta_0,s)}{\sigma(s)}m(\vartheta_0,s)\,\mathrm{d}s\,(1+o(1)) \\
&\quad + \psi_\varepsilon\int_\tau^t \Phi(\vartheta_0,s,t)\,\dot b(\vartheta_0,s)\left[\vartheta^\star_{s,\varepsilon}-\vartheta_0\right]\mathrm{d}\bar W_s\,(1+o(1)) \\
&= \delta_{m^*}(\tau)\,\Phi(\vartheta_0,0,t) - \frac{\psi_\varepsilon^2}{\varepsilon}\int_\tau^t \Phi(\vartheta_0,s,t)\,\dot B(\vartheta_0,s)\,\eta^\star_{s,\varepsilon}m^\star_{s,\varepsilon}\mathrm{d}s\,(1+o(1)) \\
&\quad + \frac{\psi_\varepsilon^2}{\varepsilon}\int_\tau^t \Phi(\vartheta_0,s,t)\,\dot b(\vartheta_0,s)\,\eta^\star_{s,\varepsilon}\frac{f(\vartheta_0,s)}{\sigma(s)}m(\vartheta_0,s)\,\mathrm{d}s\,(1+o(1)) \\
&\quad + \psi_\varepsilon^2\int_\tau^t \Phi(\vartheta_0,s,t)\,\dot b(\vartheta_0,s)\,\eta^\star_{s,\varepsilon}\;\mathrm{d}\bar W_s\,(1+o(1)).
\end{aligned}
\tag{2.167}
$$

The term $\delta_{m^*}(\tau)\,\Phi(\vartheta_0,0,t)$ can be omitted. The Lemma 1.7 and (1.58) allow us to write

$$
\delta_{m^*}(t) = -\psi_\varepsilon\,\frac{\dot f(\vartheta_0,t)}{f(\vartheta_0,t)}\;\eta^\star_{t,\varepsilon}\;y_t(\vartheta_0)\,(1+o(1)).
$$

Here we used the convergences $m^\star_{t,\varepsilon}\to y_t(\vartheta_0)$ and $m(\vartheta_0,t)\longrightarrow y_t(\vartheta_0)$ (2.166). To verify $m^\star_{t,\varepsilon}\to y_t(\vartheta_0)$ it is sufficient to prove that $\mathbf{E}_{\vartheta_0}\delta_{m^\star}(t)^2\to 0$. We have

$$
\begin{aligned}
\mathrm{d}\delta_{m^*}(t)^2 &= 2A(\vartheta_0,t)\,\delta_{m^*}(t)^2\,\mathrm{d}t + 2\frac{\psi_\varepsilon}{\varepsilon}\delta_b(t)\,\frac{f(\vartheta_0,t)}{\sigma(t)}m(\vartheta_0,t)\,\delta_{m^*}(t)\,\mathrm{d}t \\
&\quad + 2\delta_A(t)\,m^\star_{t,\varepsilon}\delta_{m^*}(t)\,\mathrm{d}t + \psi_\varepsilon^2\delta_b(t)^2\,\mathrm{d}t + 2\psi_\varepsilon\delta_b(t)\,\delta_{m^*}(t)\,\mathrm{d}\bar W_t.
\end{aligned}
$$

Then we replace $A(\vartheta_0,t)$ by $-\frac{\psi_\varepsilon}{\varepsilon}\kappa$ with $\kappa>0$, write the equation-inequality for $\mathbf{E}_{\vartheta_0}\delta_{m^\star}(t)^2$ and apply Grönwall's lemma . The terms like $\delta_b(t)\,m(\vartheta_0,t)\,\delta_{m^*}(t)$ we replace as follows

$$
2\left|\mathbf{E}_{\vartheta_0}\delta_b(t)\,m(\vartheta_0,t)\,\delta_{m^*}(t)\right| \le C\mathbf{E}_{\vartheta_0}\delta_b(t)^2\,m(\vartheta_0,t)^2 + \frac{1}{C}\mathbf{E}_{\vartheta_0}\delta_{m^*}(t)^2
$$

with sufficiently large constant $C > 0$ such that $\kappa > 1/C$.

Finally, we obtain

$$\psi_\varepsilon^{-1}\left(m^\star_{t,\varepsilon} - m\left(\vartheta_0, t\right)\right) \Longrightarrow -\frac{\dot f\left(\vartheta_0, t\right)\ y_t\left(\vartheta_0\right)}{f\left(\vartheta_0, t\right)}\ \eta^\star_t\left(\vartheta_0\right),$$

where

$$\eta^\star_t\left(\vartheta_0\right) = \mathrm{I}_0^t\left(\vartheta_0\right)^{-1}\int_0^t \frac{[\dot a\left(\vartheta_0, s\right) + \dot \ell\left(\vartheta_0, s\right)]y_s\left(\vartheta_0\right)}{b\left(\vartheta_0, s\right)}\,dw\left(s\right).$$

□

Remark 2.25 It is possible to prove the asymptotic efficiency of the adaptive filter with the lower bound

$$\lim_{\nu\to 0}\varliminf_{\varepsilon\to 0}\sup_{|\vartheta-\vartheta_0|<\nu} \psi_\varepsilon^{-2}\mathbf{E}_\vartheta\left(\bar m_{t,\varepsilon} - m\left(\vartheta, t\right)\right)^2 \geq S^\star_t\left(\vartheta_0\right)^2$$

where

$$S^\star_t\left(\vartheta_0\right)^2 = \frac{\dot f\left(\vartheta_0, t\right)^2\ y_t\left(\vartheta_0\right)^2}{f\left(\vartheta_0, t\right)^2}\ \mathrm{I}_0^t\left(\vartheta_0\right)^{-1}.$$

Case $\delta \in (1/5, 1/3)$

Consider the fixed learning interval $[0, \tau]$ and the MDE $\check\vartheta_{\tau,\varepsilon}$. Recall that the asymptotic of the derivative $\dot M\left(\vartheta, t\right)$ contains the random processes $\xi_{t,\varepsilon}$, and we can not use this approximation for the construction of One-step MLE process. The process $\dot M\left(\vartheta, t\right), \tau \leq t \leq T$ satisfies the equation

$$\mathrm{d}\dot M\left(\vartheta, t\right) = A^*\left(\vartheta, t\right)\dot M\left(\vartheta, t\right)\mathrm{d}t + \dot A^*\left(\vartheta, t\right) M\left(\vartheta, t\right)\mathrm{d}t + \frac{\psi_\varepsilon}{\varepsilon}\dot{\tilde B}\left(\vartheta, t\right)\ \mathrm{d}X_t.$$

The One-step MLE process is

$$\vartheta^\star_{t,\varepsilon} = \check\vartheta_{\tau,\varepsilon} + \mathrm{I}_\tau^t\left(\check\vartheta_{\tau,\varepsilon}\right)^{-1}\int_\tau^t \frac{\dot M\left(\check\vartheta_{\tau,\varepsilon}, s\right)}{\varepsilon^{1+\delta}\ \sigma\left(s\right)^2}\left[\mathrm{d}X_s - M\left(\check\vartheta_{\tau,\varepsilon}, s\right)\mathrm{d}s\right], \qquad \tau < t \leq T$$

where $M\left(\check\vartheta_{\tau,\varepsilon}, s\right), \tau \leq s \leq T$ is solution of the Eq. (2.150), the Fisher information function is given in (2.153). We suppose that conditions of Theorem 2.18 hold. The adaptive filter is the same as (2.165)

$$\mathrm{d}m^\star_{t,\varepsilon}=\left[a\left(\vartheta^\star_{t,\varepsilon},t\right)-\varepsilon^{-1+\delta}B\left(\vartheta^\star_{t,\varepsilon},t\right)\right]m^\star_{t,\varepsilon}\mathrm{d}t+\varepsilon^{-1+\delta}\frac{b\left(\vartheta^\star_{t,\varepsilon},t\right)}{\sigma\left(t\right)}\mathrm{d}X_t,\quad \tau<t\leq T,$$

but $\delta_{m^\star}(t)=m^\star_{t,\varepsilon}-m\left(\vartheta_0,t\right)$ has a different asymptotic. The representation (2.167) allows to write

$$\begin{aligned}\delta_{m^\star}(t)&=-\frac{\varphi_\varepsilon\psi_\epsilon}{\varepsilon}\int_\tau^t\Phi\left(\vartheta_0,s,t\right)\dot B\left(\vartheta_0,s\right)\ \eta^\star_{s,\varepsilon}m^\star_{s,\varepsilon}\mathrm{d}s\left(1+o\left(1\right)\right)\\&\quad+\frac{\varphi_\varepsilon\psi_\epsilon}{\varepsilon}\int_\tau^t\Phi\left(\vartheta_0,s,t\right)\dot b\left(\vartheta_0,s\right)\ \eta^\star_{s,\varepsilon}\frac{f\left(\vartheta_0,s\right)}{\sigma\left(s\right)}m\left(\vartheta_0,s\right)\mathrm{d}s\left(1+o\left(1\right)\right)\\&\quad+\varphi_\varepsilon\psi_\epsilon\int_\tau^t\Phi\left(\vartheta_0,s,t\right)\dot b\left(\vartheta_0,s\right)\ \eta^\star_{s,\varepsilon}\ \mathrm{d}\bar W_s\left(1+o\left(1\right)\right)\\&=-\varphi_\varepsilon\frac{\dot f\left(\vartheta_0,t\right)}{f\left(\vartheta_0,t\right)}\eta^\star_{t,\varepsilon}\ m\left(\vartheta_0,t\right)\left(1+o\left(1\right)\right).\end{aligned}$$

This implies the convergence

$$\frac{m^\star_{t,\varepsilon}-m\left(\vartheta_0,t\right)}{\varepsilon^{(1-\delta)/2}}\Longrightarrow-\frac{\dot f\left(\vartheta_0,t\right)}{f\left(\vartheta_0,t\right)}\eta^\star_t\left(\vartheta_0\right)\ y_t\left(\vartheta_0\right).$$

where

$$\eta^\star_t\left(\vartheta_0\right)=\mathrm{I}^t_\tau\left(\vartheta_0\right)^{-1}\int_\tau^t\frac{\dot f\left(\vartheta_0,s\right)b\left(\vartheta_0,s\right)+f\left(\vartheta_0,s\right)\dot b\left(\vartheta_0,s\right)}{\sqrt{2f\left(\vartheta_0,s\right)b\left(\vartheta_0,s\right)\sigma\left(s\right)}}\mathrm{d}w\left(s\right).$$

Recall that the random variables ξ_t and $\eta^\star_t\left(\vartheta_0\right)$ are independent.

Remark that it is possible to write the adaptive filter with the One-step MLE process constructed by the vanishing learning interval (see (2.156) and (2.157)).

Case $\delta=1/3$

The preliminary MDE $\check\vartheta_{\tau_\varepsilon}$ constructed by the observations on the vanishing learning interval $[0,\tau_\varepsilon]$, $\tau_\varepsilon=\varepsilon^{2/3}$ is asymptotically normal (see Proposition 2.7)

$$\varepsilon^{-1/3}\left(\check\vartheta_{\tau,\varepsilon}-\vartheta_0\right)\Longrightarrow\mathcal N\left(0,D\left(\vartheta_0\right)\right).$$

The corresponding One-step MLE process $\vartheta^\star_{t,\varepsilon}$, $\tau_\varepsilon<t\leq T$ defined in (2.158) is asymptotically normal (2.159).

The adaptive filter is

$$\mathrm{d}m^\star_{t,\varepsilon}=\left[a^*\left(\vartheta^\star_{t,\varepsilon},t\right)-\varepsilon^{-2/3}B\left(\vartheta^\star_{t,\varepsilon},t\right)\right]m^\star_{t,\varepsilon}\mathrm{d}t+\varepsilon^{-2/3}B\left(\vartheta^\star_{t,\varepsilon},t\right)\mathrm{d}X_t,\quad \tau<t\leq T,$$

where $B(\vartheta,t)=f(\vartheta,t)\,b(\vartheta,t)\,\sigma(t)^{-1}$ and $\tau\in(0,T)$.

For the error of approximation $\delta_{m^\star}(t)$ we have the representation

$$
\begin{aligned}
\delta_{m^\star}(t)&=-\varepsilon^{-1/3}\int_\tau^t \Phi(\vartheta_0,s,t)\,\dot B(\vartheta_0,s)\;\eta^\star_{s,\varepsilon}\,m^\star_{s,\varepsilon}{\rm d}s\,(1+o(1))\\
&\quad+\varepsilon^{-1/3}\int_\tau^t \Phi(\vartheta_0,s,t)\,\frac{\dot b(\vartheta_0,s)}{\sigma(s)}\;\eta^\star_{s,\varepsilon}\,f(\vartheta_0,s)m(\vartheta_0,s)\,{\rm d}s\,(1+o(1))\\
&\quad+\varepsilon^{2/3}\int_\tau^t \Phi(\vartheta_0,s,t)\,\dot b(\vartheta_0,s)\;\eta^\star_{s,\varepsilon}\,{\rm d}\bar W_s\,(1+o(1))\\
&=-\varepsilon^{1/3}\,\frac{\dot B(\vartheta_0,t)}{B(\vartheta_0,t)}\;\eta^\star_{t,\varepsilon}\,m^\star_{t,\varepsilon}\,(1+o(1))\\
&\quad+\varepsilon^{1/3}\frac{\dot b(\vartheta_0,t)}{B(\vartheta_0,t)\,\sigma(s)}\;\eta^\star_{t,\varepsilon}\,f(\vartheta_0,t)m(\vartheta_0,t)\,(1+o(1))\\
&\quad+\varepsilon^{1/3}\frac{\dot b(\vartheta_0,t)\sqrt{\sigma(t)}}{\sqrt{2b(\vartheta_0,t)\,f(\vartheta_0,t)}}\;\eta^\star_{t,\varepsilon}\,\xi_{t,\varepsilon}\;(1+o(1))\\
&=-\varepsilon^{1/3}\,\frac{\dot f(\vartheta_0,t)}{f(\vartheta_0,t)}\;\eta^\star_{t,\varepsilon}\,y_t(\vartheta_0)+\varepsilon^{1/3}\frac{\dot b(\vartheta_0,t)\sqrt{\sigma(t)}}{\sqrt{2b(\vartheta_0,t)\,f(\vartheta_0,t)}}\;\eta^\star_{t,\varepsilon}\,\xi_{t,\varepsilon}\\
&\quad\times(1+o(1))\,.
\end{aligned}
$$

Finally,

$$
\begin{aligned}
\frac{m^\star_{t,\varepsilon}-m(\vartheta_0,t)}{\varepsilon^{1/3}}&=\left[\frac{\dot b(\vartheta_0,t)\sqrt{\sigma(t)}}{\sqrt{2b(\vartheta_0,t)\,f(\vartheta_0,t)}}\;\xi_{t,\varepsilon}\;-\frac{\dot f(\vartheta_0,t)\;\,y_t(\vartheta_0)}{f(\vartheta_0,t)}\right]\eta^\star_{t,\varepsilon}\,(1+o(1))\\
&\Longrightarrow\left[\frac{\dot b(\vartheta_0,t)\sqrt{\sigma(t)}}{\sqrt{2b(\vartheta_0,t)\,f(\vartheta_0,t)}}\;\xi_t\;-\;\frac{\dot f(\vartheta_0,t)\;\,y_t(\vartheta_0)}{f(\vartheta_0,t)}\right]\eta^\star_t(\vartheta_0)\,,
\end{aligned}
$$

where $\eta_t(\vartheta_0)$ is defined in (2.160).

Chapter 3
Low Noise Observations

In this chapter, we consider partially observed diffusion processes with small diffusion coefficients in the observation equations only. The models depend on some unknown finite-dimensional parameters, and the construction of adaptive filters is realized in three steps: first, we propose preliminary estimators of the unknown parameters based on observations over a learning interval; then, these estimators are used for the construction of the One-step MLE process; and finally, this estimator process is substituted into the equations of filtration. The preliminary estimators are based on the calculation of the asymptotics of the quadratic variation of the derivative of the observed process. First, we study the linear Gaussian model; then, we discuss the generalization of the obtained results to conditionally Gaussian models. Finally, we propose an adaptive filter for one nonlinear system.

3.1 Linear Model

3.1.1 *Model of Observations*

Consider a non-homogeneous partially observed linear system described by the equations

$$\mathrm{d}X_t = f\left(\vartheta, t\right) Y_t \,\mathrm{d}t + \varepsilon\sigma\left(t\right) \mathrm{d}W_t, \qquad X_0 = 0, \tag{3.1}$$

$$\mathrm{d}Y_t = a\left(\vartheta, t\right) Y_t \,\mathrm{d}t + b\left(\vartheta, t\right) \mathrm{d}V_t, \qquad Y_0 = y_0, \tag{3.2}$$

where $f(\cdot), \sigma(\cdot), a(\cdot)$ and $b(\cdot)$ are known, smooth functions, while $W_t, 0 \leq t \leq T$ and $V_t, 0 \leq t \leq T$ are two independent Wiener processes. The process $X^T = (X_t, 0 \leq t \leq T)$ is observed, and the process $Y^T = (Y_t, 0 \leq t \leq T)$ is *hidden*. We are interested by the same problems: estimation of the finite-dimensional parameter

Y. A. Kutoyants, *Hidden Markov Processes and Adaptive Filtering*, Springer Series in Statistics, https://doi.org/10.1007/978-3-032-00052-1_3

$\vartheta \in \Theta \subset \mathcal{R}^d$ and the construction of adaptive filter. Below, we study the MME, MLE, BE, and One-step MLE-process and propose an adaptive filter.

As before, we suppose that the condition $\mathscr{A}_*$ is fulfilled. Recall it here.

Condition $\mathscr{A}_*$. *The functions* $f(\vartheta,t), a(\vartheta,t), b(\vartheta,t), \sigma(t); t, \vartheta \in [0,T]\times\Theta$ *are bounded and* $\sigma(t)^2 \geq c > 0$.

The main difference between this and considered before in Chap. 2 models is in the volatility coefficient of the hidden process. There it was $\varepsilon^\delta b(\vartheta,t)$, $\delta \in (0,1]$, and now it is $b(\vartheta,t)$. This essentially changes the construction of the preliminary estimators. The limit model ($\varepsilon = 0$) corresponds to the observations of the integral of the hidden process and the possibility of the consistent estimation of unknown parameter is equivalent to the possibility to estimate without error this parameter by the observations of Gaussian process.

Limit Model ($\varepsilon = 0$)

Let us show how an estimator ϑ_ε^* of the true value ϑ_0 could be obtained without error by means of the observations $X^T\big|_{\varepsilon=0} = x^T = (x_t, 0 \leq t \leq T)$. Consider the system of the equations

$$\frac{\mathrm{d}x_t}{\mathrm{d}t} = f(\vartheta_0,t)\,Y_t, \qquad x_0 = 0, \qquad 0 \leq t \leq T,$$

$$\mathrm{d}Y_t = a(\vartheta_0,t)\,Y_t\mathrm{d}t + b(\vartheta_0,t)\,\mathrm{d}V_t, \qquad Y_0 = y_0.$$

Suppose that the function $f(\vartheta,t)$ has a continuous derivative with respect to t (denoted $f_t'(\vartheta,t)$). Then the process x_t', $0 \leq t \leq T$ is described by the following stochastic differential equation

$$\begin{aligned}\mathrm{d}x_t' &= f_t'(\vartheta_0,t)\,Y_t\mathrm{d}t + f(\vartheta_0,t)\,\mathrm{d}Y_t\\ &= \left[f_t'(\vartheta_0,t) + f(\vartheta_0,t)\,a(\vartheta_0,t)\right]Y_t\mathrm{d}t + f(\vartheta_0,t)\,b(\vartheta_0,t)\,\mathrm{d}V_t, \qquad x_0' = 0.\end{aligned}$$

Since the volatility function depends on the unknown parameter ϑ_0 through the product $S(\vartheta_0,t) = f(\vartheta_0,t)\,b(\vartheta_0,t)$, it is clear that consistent estimation is feasible. It will be shown that the identifiability condition and the Fisher information have to be based mainly on the function $S(\vartheta_0,t)$. For example if $f(\vartheta,t) = \vartheta$ and $b(\vartheta,t) = 1/\vartheta$, then the consistent estimation is impossible, and the (limit) Fisher information is equal to 0.

Recall that by Itô's formula, we have the following relation for the quadratic variation:

$$x_t'^2 = 2\int_0^t x_s'\mathrm{d}x_s + \int_0^t b(\vartheta_0,s)^2 f(\vartheta_0,s)^2\,\mathrm{d}s$$

and the function

$$\Psi_t = x_t'^2 - 2\int_0^t x_s' \mathrm{d}x_s' = \int_0^t b\left(\vartheta_0, s\right)^2 f\left(\vartheta_0, s\right)^2 \mathrm{d}s \equiv K\left(\vartheta_0, t\right)$$

is deterministic. Under mild identifiability conditions, the observed function Ψ_t defines ϑ_0 for any $t \in (0, T]$ without error. For example, the estimator ϑ^* defined by the equation $K\left(\vartheta^*, t\right) = \Psi_t$ is without error, i.e., $\vartheta^* = \vartheta_0$.

If $f\left(\vartheta, t\right) = f\left(t\right)$, $b\left(\vartheta, t\right)^2 = h\left(t\right) + \vartheta g\left(t\right)$, where all functions and ϑ are positive, then the "estimator"

$$\vartheta_t^* = \left(\int_0^t g\left(s\right) f\left(s\right)^2 \mathrm{d}s\right)^{-1} \left[\Psi_t - \int_0^t h\left(s\right) f\left(s\right)^2 \mathrm{d}s\right] = \vartheta_0.$$

Therefore, the consistent estimation by observations (3.1) is not excluded.

Note that if we consider the limit model, then the measures that correspond to observations with different ϑ are singular. Moreover, Ψ_t does not depend on $a\left(\vartheta_0, t\right)$.

Consistent Estimator

One way to estimate ϑ using observations from (3.1) is by estimating first the function Ψ_t and then using the estimator, $\Psi_{t,\varepsilon}$, to estimate ϑ, i.e., to solve for some τ the equation

$$\Psi_{\tau,\varepsilon} = \int_0^\tau b(\breve\vartheta_{\tau,\varepsilon}, s)^2 f(\breve\vartheta_{\tau,\varepsilon}, s)^2 \mathrm{d}s,$$

where $\breve\vartheta_{\tau,\varepsilon}$ is the MME. It can be called as well the *method of substitution estimator*. Of course, to verify the consistency of $\breve\vartheta_{\tau,\varepsilon}$, we need the condition of identifiability. If the unknown parameter is multidimensional, then it is possible to define the estimator as

$$\breve\vartheta_{\tau,\varepsilon} = \arg\inf_{\vartheta\in\Theta} \int_0^\tau \left[\Psi_{s,\varepsilon} - K\left(\vartheta, s\right)\right]^2 \mathrm{d}s.$$

Therefore, to estimate ϑ, we first estimate the function $\Psi_t, 0 \leq t \leq \tau$ and then using this estimator and One-step procedure, we estimate the parameter ϑ. Below, we realize this program.

3.1.2 Semiparametric Estimation

Suppose that we have the partially observed system

$$\mathrm{d}X_t = f(t)\,Y_t\mathrm{d}t + \varepsilon\sigma(t)\,\mathrm{d}W_t, \qquad X_0 = 0, \qquad 0 \le t \le T, \tag{3.3}$$

$$\mathrm{d}Y_t = a(t)\,Y_t\mathrm{d}t + b(t)\,\mathrm{d}V_t, \qquad Y_0 = y_0. \tag{3.4}$$

Introduce the condition

$\bar{\mathscr{G}}$. *The functions* $a(\cdot), b(\cdot), f(\cdot)$ *and* $\sigma(\cdot)$ *satisfy the conditions* $\mathscr{A}_*$ *and are continuously differentiable.*

Suppose that the functions $a(\cdot), b(\cdot), f(\cdot)$ and $\sigma(\cdot)$ are unknown and consider the problem of estimation of the function

$$\Psi_\tau = \int_0^\tau f(t)^2\, b(t)^2\,\mathrm{d}t, \qquad 0 < \tau \le T$$

by observations X^τ. Remind that this is quadratic variation of the derivative of the process X_τ at the point $\varepsilon = 0$.

Introduce the statistic

$$\Psi_{\tau,\varepsilon} = \sum_{i=0}^{N_{\tau,\varepsilon}-1} \left(\frac{X_{t_{i+1}+\delta_\varepsilon} - X_{t_{i+1}}}{\delta_\varepsilon} - \frac{X_{t_i+\delta_\varepsilon} - X_{t_i}}{\delta_\varepsilon} \right)^2, \qquad 0 < \tau \le T.$$

Here $t_i = i\varphi_\varepsilon$, $N_{\tau,\varepsilon} = \left[\frac{\tau}{\varphi_\varepsilon}\right]$, the rates $\varphi_\varepsilon \to 0$, $\delta_\varepsilon \to 0$ will be defind later. Just note that as the first step is derivation, we wait that the rate $\delta_\varepsilon \to 0$ has to be faster than the step of discretization $\varphi_\varepsilon \to 0$.

Let us explain why this statistic can be a consistent estimator of Ψ_τ. We have

$$\frac{X_{t_{i+1}+\delta_\varepsilon} - X_{t_{i+1}}}{\delta_\varepsilon} = \frac{1}{\delta_\varepsilon} \int_{t_{i+1}}^{t_{i+1}+\delta_\varepsilon} f(t)\,Y_t\,\mathrm{d}t + \frac{\varepsilon}{\delta_\varepsilon} \int_{t_{i+1}}^{t_{i+1}+\delta_\varepsilon} \sigma(t)\,\mathrm{d}W_t.$$

Hence, if we take $\varepsilon\delta_\varepsilon^{-1/2} \to 0$, then

$$\frac{X_{t_{i+1}+\delta_\varepsilon} - X_{t_{i+1}}}{\delta_\varepsilon} = f(t_{i+1})\,Y_{t_{i+1}} + o(1), \qquad \frac{X_{t_i+\delta_\varepsilon} - X_{t_i}}{\delta_\varepsilon} = f(t_i)\,Y_{t_i} + o(1).$$

Further, using the smoothness of the functions, we can write formally (!)

$$\begin{aligned}\Psi_{\tau,\varepsilon} &= \sum_{i=0}^{N_{\tau,\varepsilon}-1} \left(f(t_{i+1})\,Y_{t_{i+1}} - f(t_i)\,Y_{t_i} \right)^2 + o(1) \\ &= \sum_{i=0}^{N_{\tau,\varepsilon}-1} f(t_i)^2 \left(Y_{t_{i+1}} - Y_{t_i} \right)^2 + o(1)\end{aligned}$$

$$
\begin{aligned}
&= \sum_{i=0}^{N_{t,\varepsilon}-1} f(t_i)^2 \left(\int_{t_i}^{t_{i+1}} a(s)\, Y_s \mathrm{d}s + \int_{t_i}^{t_{i+1}} b(s)\, \mathrm{d}V_s\right)^2 + o(1) \\
&= \sum_{i=0}^{N_{\tau,\varepsilon}-1} f(t_i)^2 \left(\int_{t_i}^{t_{i+1}} b(s)\, \mathrm{d}V_s\right)^2 + o(1) \\
&= \sum_{i=0}^{N_{\tau,\varepsilon}-1} f(t_i)^2\, b(t_i)^2 \left(V_{t_{i+1}} - V_{t_i}\right)^2 + o(1) \\
&= \sum_{i=0}^{N_{\tau,\varepsilon}-1} f(t_i)^2\, b(t_i)^2\ (t_{i+1} - t_i) + o(1) \\
&\longrightarrow \int_0^\tau f(s)^2\, b(s)^2\, \mathrm{d}s = \Psi_\tau .
\end{aligned}
$$

Therefore, this statistic can be a consistent estimator of the integral Ψ_τ.

Proposition 3.1 *Let the conditions $\bar{\mathscr{G}}$ hold and chose $\delta_\varepsilon = \varepsilon, \varphi_\varepsilon = \varepsilon^{2/3}$. Then for any $p > 0$, there exists a constant $C > 0$ such that*

$$
\mathbf{E}\left|\Psi_{\tau,\varepsilon} - \Psi_\tau\right|^p \le C\tau^{p/2}\varepsilon^{p/3}. \tag{3.5}
$$

Proof Let us consider first the simplest case just to see how this rate of convergence can be obtained. Put $f(t) \equiv 1, a(t) \equiv 0, \sigma(t) \equiv 1, b(t) \equiv 1, \tau = 1$. Then $\Psi_\tau = 1$ and the observed process is

$$
\mathrm{d}X_t = V_t \mathrm{d}t + \varepsilon \mathrm{d}W_t, \qquad X_0 = 0, \qquad 0 \le t \le 1.
$$

Denote

$$
\xi_{t_{i+1},\varepsilon} = \frac{1}{\sqrt{\delta_\varepsilon}}\left[W_{t_{i+1}+\delta_\varepsilon} - W_{t_{i+1}}\right] \sim \mathcal{N}(0,1), \qquad \Delta\xi_{t_i,\varepsilon} = \xi_{t_{i+1},\varepsilon} - \xi_{t_i,\varepsilon},
$$

$$
\eta_{t_{i+1},\varepsilon} = \frac{1}{\delta_\varepsilon^{3/2}} \int_{t_{i+1}}^{t_{i+1}+\delta_\varepsilon} \left[V_s - V_{t_{i+1}}\right] \mathrm{d}s \sim \mathcal{N}(0, 1/3), \qquad \Delta\eta_{t_i,\varepsilon} = \eta_{t_{i+1},\varepsilon} - \eta_{t_i,\varepsilon}.
$$

Then we can write

$$
\begin{aligned}
\bar\Psi_{\tau,\varepsilon} &= \sum_{i=0}^{N_{\tau,\varepsilon}} \left(\frac{1}{\delta_\varepsilon}\int_{t_{i+1}}^{t_{i+1}+\delta_\varepsilon} V_s \mathrm{d}s + \frac{\varepsilon}{\sqrt{\delta_\varepsilon}}\xi_{t_{i+1},\varepsilon} - \frac{1}{\delta_\varepsilon}\int_{t_i}^{t_i+\delta_\varepsilon} V_s \mathrm{d}s - \frac{\varepsilon}{\sqrt{\delta_\varepsilon}}\xi_{t_i,\varepsilon}\right)^2 \\
&= \sum_{i=0}^{N_{\tau,\varepsilon}} \left(V_{t_{i+1}} - V_{t_i} + \sqrt{\delta_\varepsilon}\Delta\eta_{t_i,\varepsilon} + \frac{\varepsilon}{\sqrt{\delta_\varepsilon}}\Delta\xi_{t_i,\varepsilon}\right)^2
\end{aligned}
$$

$$= \sum_{i=0}^{N_{\tau,\varepsilon}} \left[V_{t_{i+1}} - V_{t_i}\right]^2 + \delta_\varepsilon \sum_{i=0}^{N_{\tau,\varepsilon}} \Delta\eta_{t_{i,\varepsilon}}^2 + \frac{\varepsilon^2}{\delta_\varepsilon} \sum_{i=0}^{N_{\tau,\varepsilon}} \Delta\xi_{t_{i,\varepsilon}}^2$$

$$+ 2\sqrt{\delta_\varepsilon} \sum_{i=0}^{N_{\tau,\varepsilon}} \left[V_{t_{i+1}} - V_{t_i}\right] \Delta\eta_{t_{i,\varepsilon}}$$

$$+ \frac{2\varepsilon}{\sqrt{\delta_\varepsilon}} \sum_{i=0}^{N_{\tau,\varepsilon}} \left[V_{t_{i+1}} - V_{t_i}\right] \Delta\xi_{t_{i,\varepsilon}} + 2\varepsilon \sum_{i=0}^{N_{\tau,\varepsilon}} \Delta\eta_{t_{i,\varepsilon}} \Delta\xi_{t_{i,\varepsilon}}. \tag{3.6}$$

By Itô formula

$$\sum_{i=0}^{N_{\tau,\varepsilon}} \left[V_{t_{i+1}} - V_{t_i}\right]^2 = \sum_{i=0}^{N_{\tau,\varepsilon}} (t_{i+1} - t_i) + 2 \sum_{i=0}^{N_{\tau,\varepsilon}} \int_{t_i}^{t_{i+1}} \left(V_s - V_{t_i}\right) \mathrm{d}V_s = 1 + \sqrt{\varphi_\varepsilon}\; O(1)$$

because

$$\mathbf{E} \left(\sum_{i=0}^{N_{\tau,\varepsilon}} \int_{t_i}^{t_{i+1}} \left(V_s - V_{t_i}\right) \mathrm{d}V_s \right)^2 = \sum_{i=0}^{N_{\tau,\varepsilon}} \int_{t_i}^{t_{i+1}} (s - t_i)\, \mathrm{d}s = \frac{\varphi_\varepsilon}{2}.$$

For the second term, we have

$$\varphi_\varepsilon \sum_{i=0}^{N_{\tau,\varepsilon}} \Delta\eta_{t_{i,\varepsilon}}^2 \longrightarrow \frac{2}{3}, \qquad \delta_\varepsilon \sum_{i=0}^{N_{\tau,\varepsilon}} \Delta\eta_{t_{i,\varepsilon}}^2 = \frac{\delta_\varepsilon}{\varphi_\varepsilon}\, \varphi_\varepsilon \sum_{i=0}^{N_{\tau,\varepsilon}} \Delta\eta_{t_{i,\varepsilon}}^2 = \frac{\delta_\varepsilon}{\varphi_\varepsilon} \left(\frac{2}{3} + O(1)\right).$$

We have as well

$$\varphi_\varepsilon \sum_{i=0}^{N_{\tau,\varepsilon}} \Delta\xi_{t_{i,\varepsilon}}^2 \longrightarrow 2, \qquad \frac{\varepsilon^2}{\delta_\varepsilon \varphi_\varepsilon}\, \varphi_\varepsilon \sum_{i=0}^{N_{\tau,\varepsilon}} \Delta\xi_{t_{i,\varepsilon}}^2 = \frac{\varepsilon^2}{\delta_\varepsilon \varphi_\varepsilon} \left(2 + O(1)\right)$$

For the other terms in (3.6), we obtain the relations

$$4\varepsilon^2 \mathbf{E} \left(\sum_{i=0}^{N_{\tau,\varepsilon}} \Delta\xi_{t_{i,\varepsilon}} \Delta\eta_{t_{i,\varepsilon}} \right)^2 = \frac{16\varepsilon^2}{3\varphi_\varepsilon}, \qquad \varepsilon \sum_{i=0}^{N_{\tau,\varepsilon}} \Delta\xi_{t_{i,\varepsilon}} \Delta\eta_{t_{i,\varepsilon}} = \frac{\varepsilon}{\sqrt{\varphi_\varepsilon}}\; O(1),$$

$$\frac{4\varepsilon^2}{\delta_\varepsilon} \mathbf{E} \left(\sum_{i=0}^{N_{\tau,\varepsilon}} \left[V_{t_{i+1}} - V_{t_i}\right] \Delta\xi_{t_{i,\varepsilon}} \right)^2 = \frac{8\varepsilon^2}{\delta_\varepsilon},$$

$$\frac{\varepsilon}{\sqrt{\delta_\varepsilon}} \sum_{i=0}^{N_{\tau,\varepsilon}} \left[V_{t_{i+1}} - V_{t_i}\right] \Delta\xi_{t_i,\varepsilon} = \frac{\varepsilon}{\sqrt{\delta_\varepsilon}}\; O\left(1\right).$$

Remark that

$$\begin{aligned}
&\mathbf{E}\left[V_{t_{i+1}} - V_{t_i}\right] \Delta\eta_{t_i,\varepsilon} \\
&= \mathbf{E}\left[V_{t_i} - V_{t_{i+1}}\right] \eta_{t_i,\varepsilon} = \frac{1}{\delta_\varepsilon^{3/2}} \mathbf{E}\left[V_{t_i} - V_{t_{i+1}}\right] \int_{t_i}^{t_i+\delta_\varepsilon} \left[V_s - V_{t_i}\right] \mathrm{d}s \\
&= -\frac{1}{\delta_\varepsilon^{3/2}} \int_{t_i}^{t_i+\delta_\varepsilon} \mathbf{E}\left[V_s - V_{t_i}\right]^2 \mathrm{d}s = -\frac{\sqrt{\delta_\varepsilon}}{2}
\end{aligned}$$

and therefore,

$$\sqrt{\delta_\varepsilon} \sum_{i=0}^{N_{\tau,\varepsilon}} \left[V_{t_{i+1}} - V_{t_i}\right] \Delta\eta_{t_i,\varepsilon} = -\frac{\delta_\varepsilon}{\varphi_\varepsilon} \left(\frac{1}{2} + o\left(1\right)\right).$$

Therefore, the main terms are

$$\Psi_{\tau,\varepsilon} = 1 + \sqrt{\varphi_\varepsilon}\; O\left(1\right) + \frac{\delta_\varepsilon}{\varphi_\varepsilon}\; O\left(1\right) + \frac{\varepsilon^2}{\delta_\varepsilon \varphi_\varepsilon}\; O\left(1\right).$$

If we write two equations of balance for $\delta_\varepsilon = \varepsilon^q$ and $\varphi_\varepsilon = \varepsilon^l$

$$\sqrt{\varphi_\varepsilon} = \frac{\delta_\varepsilon}{\varphi_\varepsilon}, \qquad \frac{\delta_\varepsilon}{\varphi_\varepsilon} = \frac{\varepsilon^2}{\delta_\varepsilon \varphi_\varepsilon},$$

then we obtain the values $q = 1$ and $l = 2/3$, i.e., $\delta_\varepsilon = \varepsilon$ and $\varphi_\varepsilon = \varepsilon^{2/3}$.

Now we repeat the given above expansions with description of the corresponding errors. As usual throughout the proof, C denote "generic" positive constants, which can vary from formula to formula (and even in the same formula). We have

$$\begin{aligned}
\frac{X_{t_{i+1}+\delta_\varepsilon} - X_{t_{i+1}}}{\delta_\varepsilon} &= f\left(t_{i+1}\right) Y_{t_{i+1}} + \frac{\varepsilon}{\delta_\varepsilon} \int_{t_{i+1}}^{t_{i+1}+\delta_\varepsilon} \sigma\left(s\right)\, \mathrm{d}W_s \\
&\quad + \frac{1}{\delta_\varepsilon} \int_{t_{i+1}}^{t_{i+1}+\delta_\varepsilon} \left[f\left(s\right) Y_s - f\left(t_{i+1}\right) Y_{t_{i+1}}\right]\, \mathrm{d}s \\
&= f\left(t_i\right) Y_{t_{i+1}} + \left[f\left(t_{i+1}\right) - f\left(\vartheta, t_i\right)\right] Y_{t_{i+1}} + \frac{\varepsilon}{\sqrt{\delta_\varepsilon}} R_{1,\varepsilon} \\
&\quad + \frac{f\left(t_{i+1}\right)}{\delta_\varepsilon} \int_{t_{i+1}}^{t_{i+1}+\delta_\varepsilon} \left[Y_s - Y_{t_{i+1}}\right]\, \mathrm{d}s
\end{aligned}$$

$$+\frac{1}{\delta_\varepsilon}\int_{t_{i+1}}^{t_{i+1}+\delta_\varepsilon}\left[f\left(s\right)-f\left(t_{i+1}\right)\right]Y_s\,\mathrm{d}s. \tag{3.7}$$

The following estimates are obtained using elementary calculations

$$\mathbf{E}R_{1,\varepsilon}^2=\mathbf{E}\left(\frac{1}{\sqrt{\delta_\varepsilon}}\int_{t_{i+1}}^{t_{i+1}+\delta_\varepsilon}\sigma\left(s\right)\,\mathrm{d}W_s\right)^2\leq C,$$

$$\mathbf{E}\left(\left[f\left(t_{i+1}\right)-f\left(t_i\right)\right]Y_{t_{i+1}}\right)^2\leq C\varphi_\varepsilon^2,$$

$$\mathbf{E}\left(\frac{f\left(t_{i+1}\right)}{\delta_\varepsilon}\int_{t_{i+1}}^{t_{i+1}+\delta_\varepsilon}\left[Y_s-Y_{t_{i+1}}\right]\,\mathrm{d}s\right)^2\leq C\delta_\varepsilon,$$

$$\mathbf{E}\left(\frac{1}{\delta_\varepsilon}\int_{t_{i+1}}^{t_{i+1}+\delta_\varepsilon}\left[f\left(s\right)-f\left(t_{i+1}\right)\right]Y_s\,\mathrm{d}s\right)^2$$
$$\leq\frac{1}{\delta_\varepsilon}\int_{t_{i+1}}^{t_{i+1}+\delta_\varepsilon}\left(s-t_{i+1}\right)^2f'\left(\tilde{s}\right)^2\mathbf{E}Y_s^2\,\mathrm{d}s\leq C\delta_\varepsilon^2.$$

Therefore,

$$\mathbf{E}\left(\frac{X_{t_{i+1}+\delta_\varepsilon}-X_{t_{i+1}}}{\delta_\varepsilon}-\frac{X_{t_i+\delta_\varepsilon}-X_{t_i}}{\delta_\varepsilon}-f\left(t_i\right)\left[Y_{t_{i+1}}-Y_{t_i}\right]\right)^2$$
$$\leq C\frac{\varepsilon^2}{\delta_\varepsilon}+C\delta_\varepsilon+C\varphi_\varepsilon^2. \tag{3.8}$$

These are estimates for the asymptotic derivatives of X_t at the points t_i and t_{i+1} and of their difference.

The quadratic variation is estimated as follows:

$$Q_{\tau,\varepsilon}=\sum_{i=0}^{N_{\tau,\varepsilon}-1}f\left(t_i\right)^2\left(Y_{t_{i+1}}-Y_{t_i}\right)^2$$
$$=\sum_{i=0}^{N_{\tau,\varepsilon}-1}f\left(t_i\right)^2\left(\int_{t_i}^{t_{i+1}}a\left(s\right)\mathrm{d}s+\int_{t_i}^{t_{i+1}}b\left(s\right)\mathrm{d}V_s\right)^2$$
$$=\sum_{i=0}^{N_{\tau,\varepsilon}-1}f\left(t_i\right)^2\left(\int_{t_i}^{t_{i+1}}a\left(s\right)\mathrm{d}s\right)^2+\sum_{i=0}^{N_{\tau,\varepsilon}-1}f\left(t_i\right)^2\left(\int_{t_i}^{t_{i+1}}b\left(s\right)\mathrm{d}V_s\right)^2$$
$$+2\sum_{i=0}^{N_{\tau,\varepsilon}-1}f\left(t_i\right)^2\int_{t_i}^{t_{i+1}}a\left(s\right)\mathrm{d}s\int_{t_i}^{t_{i+1}}b\left(s\right)\mathrm{d}V_s,$$

where

$$\sum_{i=0}^{N_{\tau,\varepsilon}-1} f(t_i)^2 \left(\int_{t_i}^{t_{i+1}} a(s)\,\mathrm{d}s\right)^2 \leq \varphi_\varepsilon \sum_{i=0}^{N_{\tau,\varepsilon}-1} f(t_i)^2 \int_{t_i}^{t_{i+1}} a(s)^2\,\mathrm{d}s \leq C\tau\varphi_\varepsilon,$$

and

$$\mathbf{E}\left(2\sum_{i=0}^{N_{\tau,\varepsilon}-1} f(t_i)^2 \int_{t_i}^{t_{i+1}} a(s)\,\mathrm{d}s \int_{t_i}^{t_{i+1}} b(s)\,\mathrm{d}V_s\right)^2 \leq C\tau\varphi_\varepsilon^2.$$

For the stochastic integral, we can write

$$\begin{aligned}
&\sum_{i=0}^{N_{\tau,\varepsilon}-1} f(t_i)^2 \left(\int_{t_i}^{t_{i+1}} b(s)\,\mathrm{d}V_s\right)^2 \\
&\quad= \sum_{i=0}^{N_{\tau,\varepsilon}-1} f(t_i)^2 \left(b(t_i)\left[V_{t_{t_{i+1}}} - V_{t_{t_i}}\right] + \int_{t_i}^{t_{i+1}} [b(s) - b(t_i)]\,\mathrm{d}V_s\right)^2.
\end{aligned}$$

Here

$$\begin{aligned}
\mathbf{E}\sum_{i=0}^{N_{\tau,\varepsilon}-1} f(t_i)^2 \left(\int_{t_i}^{t_{i+1}} [b(s) - b(t_i)]\,\mathrm{d}V_s\right)^2
&= \sum_{i=0}^{N_{\tau,\varepsilon}-1} f(t_i)^2 \int_{t_i}^{t_{i+1}} [b(s) - b(t_i)]^2\,\mathrm{d}s \\
&= \sum_{i=0}^{N_{\tau,\varepsilon}-1} f(t_i)^2 \int_{t_i}^{t_{i+1}} (s - t_i)^2\, b'(\tilde{s})^2\,\mathrm{d}s \leq C\tau\varphi_\varepsilon^2.
\end{aligned}$$

Hence,

$$\mathbf{E}\left|Q_{\tau,\varepsilon} - \sum_{i=0}^{N_{\tau,\varepsilon}-1} f(t_i)^2\, b(t_i)^2 \left[V_{t_{i+1}} - V_{t_i}\right]^2\right| \leq C\tau\varphi_\varepsilon. \tag{3.9}$$

Finally,

$$K_{\tau,\varepsilon} = \sum_{i=0}^{N_{\tau,\varepsilon}-1} f(t_i)^2\, b(t_i)^2 \frac{\left[V_{t_{t_{i+1}}} - V_{t_{t_i}}\right]^2}{t_{i+1} - t_i}(t_{i+1} - t_i)$$

$$= \sum_{i=0}^{N_{\tau,\varepsilon}-1} f(t_i)^2\, b(t_i)^2\, \xi_i^2\, (t_{i+1}-t_i)$$

$$= \sum_{i=0}^{N_{\tau,\varepsilon}-1} f(t_i)^2\, b(t_i)^2\, (t_{i+1}-t_i) + \sum_{i=0}^{N_{\tau,\varepsilon}-1} f(t_i)^2\, b(t_i)^2 \left(\xi_i^2 - 1\right)(t_{i+1}-t_i)\,.$$

Here ξ_i, $i = 0, \ldots, N_{\tau,\varepsilon}$ are independent standard Gaussian random variables, $\xi_i \sim \mathcal{N}(0, 1)$. Therefore,

$$\mathbf{E}\left(K_{\tau,\varepsilon} - \sum_{i=0}^{N_{\tau,\varepsilon}-1} f(t_i)^2\, b(t_i)^2\, (t_{i+1}-t_i)\right)^2$$

$$= 2\varphi_\varepsilon \sum_{i=0}^{N_{\tau,\varepsilon}-1} f(t_i)^4\, b(t_i)^4\, (t_{i+1}-t_i) \le C\tau\varphi_\varepsilon$$

and

$$\left|\Psi_\tau - \sum_{i=0}^{N_{\tau,\varepsilon}-1} f(t_i)^2\, b(t_i)^2\, (t_{i+1}-t_i)\right| \le C\tau\varphi_\varepsilon. \tag{3.10}$$

The obtained estimates allow us to write

$$\mathbf{E}\left|\Psi_{\tau,\varepsilon} - \sum_{i=0}^{N_{\tau,\varepsilon}-1} f(t_i)^2 \left[Y_{t_{i+1}} - Y_{t_i}\right]^2\right| \le C\sqrt{\tau}\left[\frac{\delta_\varepsilon}{\varphi_\varepsilon} + \sqrt{\varphi_\varepsilon} + \frac{\varepsilon^2}{\delta_\varepsilon\varphi_\varepsilon}\right].$$

If we put $\delta_\varepsilon = \varepsilon^q$, $\varphi_\varepsilon = \varepsilon^l$, then obtain the same equations of balance

$$\sqrt{\varphi_\varepsilon} = \frac{\delta_\varepsilon}{\varphi_\varepsilon} = \varphi_\varepsilon = \frac{\varepsilon^2}{\delta_\varepsilon\varphi_\varepsilon}$$

gives us the values $q = 1$ and $l = 2/3$, i.e., $\delta_\varepsilon = \varepsilon$ and $\varphi_\varepsilon = \varepsilon^{2/3}$. Hence,

$$\mathbf{E}\left|\Psi_{\tau,\varepsilon} - \sum_{i=0}^{N_{\tau,\varepsilon}-1} f(t_i)^2 \left[Y_{t_{i+1}} - Y_{t_i}\right]^2\right| \le C\,\tau^{1/2}\,\varepsilon^{1/3}.$$

This estimate together with (3.9) and (3.10) allows us to write

$$\mathbf{E}_\vartheta\left|\Psi_{\tau,\varepsilon} - \Psi_\tau\right| \le C\,\tau^{1/2}\,\varepsilon^{1/3}.$$

The similar calculations provide us the estimate of the polynomial moments too. □

Remark that it is possible to prove the asymptotic normality

$$\tau^{-1/2}\varepsilon^{-1/3}\left(\Psi_{\tau,\varepsilon}-\Psi_\tau\right)\Longrightarrow\mathcal{N}\left(0,D^2\right)$$

with some limit variance D^2. The estimate (3.5) is an intermediate result, which will be used in the next section for the construction of the preliminary estimator for One-step MLE process.

Remark 3.1 The obtained rate $\varepsilon^{1/3}$ is not optimal but is sufficient for the construction of the preliminary estimator below. An interesting result concerning estimation of the function Ψ_τ by observations (3.3) and (3.4) was obtained by Reiss [169]. Supposing that $a(t)\equiv 0$, $\sigma(t)\equiv\sigma$, $f(t)\equiv 1$ it was shown that this model of observations is asymptotically equivalent (in Le Cam's sense) to the model of observations

$$\mathrm{d}X_t=\sqrt{2b(t)}\,\mathrm{d}t+\sigma^{1/2}\varepsilon^{1/2}\mathrm{d}W_t.$$

This equivalence shows why the rate $\varepsilon^{1/2}$ is "more natural" for the model (3.3) and (3.4). Just note that for the further application of the estimator $\Psi_{\tau,\varepsilon}$ in the construction of One-step MLE process, the rate $\varepsilon^{1/3}$ is sufficient.

The considered problem is in some sense close to that well studied in financial econometrics problem of integrated volatility

$$\Psi_\tau=\int_0^\tau b_s^2\,\mathrm{d}s$$

estimation in the case of the "hidden" process

$$Y_t=Y_0+\int_0^t a_s\,\mathrm{d}s+\int_0^t b_s\,\mathrm{d}V_s,\qquad 0\le t\le T$$

and observations

$$X_{t_i}=Y_{t_i}+\epsilon_{t_i}$$

Here $a_s, 0\le s\le T$ and $b_s, 0\le s\le T$ are some adapted processes.

There is a large variety of the statements of the problems (parameters estimation, functions estimation, different limits, etc). We can mention here the works of Gloter and Jacod [60, 61], Jacod et al. [74], Nakakita et al. [156], Podolskij and Vetter [167] and the references therein.

Remark 3.2 The proposed estimator $\Psi_{\tau,\varepsilon}$ is consistent in the more general model of observations. Suppose, for example, that the partially observed system is

$$\mathrm{d}X_t = F(t, X_t, Y_t)\,\mathrm{d}t + \varepsilon\sigma(t, X_t)\,\mathrm{d}W_t, \qquad X_0 = 0, \qquad 0 \le t \le T,$$
$$\mathrm{d}Y_t = P(t, X_t, Y_t)\,\mathrm{d}t + Q(t, X_t, Y_t)\,\mathrm{d}V_s, \qquad Y_0.$$

Then the limit system is

$$x_t' = \frac{\partial x_t}{\partial t} = F(t, x_t, Y_t), \qquad x_0 = 0,$$
$$\mathrm{d}Y_t = P(t, x_t, Y_t)\,\mathrm{d}t + Q(t, x_t, Y_t)\,\mathrm{d}V_s, \qquad Y_0$$

and

$$\mathrm{d}x_t' = \Big[F_t'(t, x_t, Y_t) + F_x'(t, x_t, Y_t)\,x_t' + F_y'(t, x_t, Y_t)\,P(t, x_t, Y_t)\Big]\,\mathrm{d}t$$
$$+ \frac{1}{2}F_{yy}''(t, x_t, Y_t)\,Q(t, x_t, Y_t)^2\,\mathrm{d}t + F_y'(t, x_t, Y_t)\,Q(t, x_t, Y_t)\,\mathrm{d}V_t, \; x_0' = 0.$$

The statistic $\Psi_{\tau,\varepsilon}$ converges to the quadratic variation of the process, and the similar proof can show that

$$\Psi_{\tau,\varepsilon} \longrightarrow \int_0^\tau F_y'(t, x_t, Y_t)^2\,Q(t, x_t, Y_t)^2\,\mathrm{d}t.$$

We return to this model of observations later in this chapter.

3.1.3 Parameter Estimation

MME

Let us return to the parametric partially observed linear system

$$\mathrm{d}X_t = f(\vartheta, t)\,Y_t\,\mathrm{d}t + \varepsilon\sigma(t)\,\mathrm{d}W_t, \qquad X_0 = 0, \qquad 0 \le t \le T, \tag{3.11}$$
$$\mathrm{d}Y_t = a(\vartheta, t)\,Y_t\,\mathrm{d}t + b(\vartheta, t)\,\mathrm{d}V_t, \qquad Y_0 = y_0 \tag{3.12}$$

The process $X^T = (X_t, 0 \le t \le T)$ is observed, the process $Y^T = (Y_t, 0 \le t \le T)$ is *hidden* and the Wiener processes $V^T = (V_t, 0 \le t \le T)$ and $W^T = (W_t, 0 \le t \le T)$ as before are independent. The functions $b(\cdot)$ and $f(\cdot)$ are supposed to be known. The functions $a(\cdot)$ and $\sigma(\cdot)$ do not used in the construction of the estimator. The parameter $\vartheta \in \Theta \subset \mathcal{R}^d$ is unknown and has to be estimated by observations X^T. The set Θ is open, convex, and bounded.

Fix some $\tau \in (0, T]$ and define the function

$$\Psi_t(\vartheta) = \int_0^t f(\vartheta, s)^2 b(\vartheta, s)^2 \, ds, \qquad 0 < t \leq \tau, \qquad \vartheta \in \Theta \subset \mathcal{R}^d.$$

As estimator of this function, we take the nonparametric estimator

$$\Psi_{t,\varepsilon} = \sum_{i=0}^{N_{t,\varepsilon}-1} \left(\frac{X_{t_{i+1}+\varepsilon} - X_{t_{i+1}}}{\varepsilon} - \frac{X_{t_i+\varepsilon} - X_{t_i}}{\varepsilon} \right)^2, \qquad 0 < t \leq \tau.$$

Here $t_{i+1} - t_i = \varepsilon^{1/3}$, $t_0 = 0$ and $N_{t,\varepsilon} = \left[t\varepsilon^{-1/3}\right]$.

The MME $\vartheta^*_{\tau,\varepsilon}$ can be defined as follows. Let us fix d points $0 < \tau_1 < \tau_2 < \ldots < \tau_d = \tau$ and denote the vectors $\Psi^*_{\tau,\varepsilon} = \left(\Psi_{\tau_1,\varepsilon}, \Psi_{\tau_2,\varepsilon}, \ldots, \Psi_{\tau_d,\varepsilon}\right)^\top$ and $\Psi^*_\tau(\vartheta) = \left(\Psi_{\tau_1}(\vartheta), \Psi_{\tau_2}(\vartheta), \ldots, \Psi_{\tau_d}(\vartheta)\right)^\top$. The estimator is defined as solution of the system of equations

$$\Psi^*_{\tau,\varepsilon} = \Psi^*_\tau(\vartheta^*_{\tau,\varepsilon}).$$

If this equation has no solution, then we put $\vartheta^*_{\tau,\varepsilon} = \vartheta^\circ$, where ϑ° is some value $\notin \bar{\Theta}$.

Note that having the estimator $\Psi_{t,\varepsilon}, 0 \leq t \leq \tau$, we can define the MDE $\check{\vartheta}_{\tau,\varepsilon}$ too by the equation

$$\int_0^\tau \left[\Psi_{t,\varepsilon} - \Psi_t(\check{\vartheta}_{\tau,\varepsilon})\right]^2 dt = \inf_{\vartheta \in \Theta} \int_0^\tau \left[\Psi_{t,\varepsilon} - \Psi_t(\vartheta)\right]^2 dt.$$

If these equations have more than one solution, then any one of them can be taken as the estimator of ϑ.

Introduce the notation

$$g(\vartheta_0, \nu) = \inf_{\|\vartheta - \vartheta_0\| \geq \nu} \|\Psi_\tau(\vartheta) - \Psi_\tau(\vartheta_0)\|, \qquad \mathbf{T}(\vartheta_0) = \Psi^*_\tau(\vartheta_0)\dot{\Psi}^*_\tau(\vartheta_0)^\top$$

and *Conditions* $\mathscr{G}^*$:

$\mathscr{G}^*_1$. *The functions* $a(\vartheta, t), b(\vartheta, t), f(\vartheta, t), \sigma(t), (\vartheta, t) \in \bar{\Theta} \times [0, T]$ *have continuous derivatives w.r.t.* t *and two continuous derivatives w.r.t.* ϑ.

$\mathscr{G}^*_2$. *The matrix* $\mathbf{T}(\vartheta_0)$ *is uniformly nondegenerate*

$$\inf_{\vartheta_0 \in \Theta} \inf_{\|\mathrm{e}\|=1, \mathrm{e} \in \mathcal{R}^d} \mathrm{e}^\top \mathbf{T}(\vartheta_0)\, \mathrm{e} > 0.$$

$\mathscr{G}^*_3$. *For any* $\nu > 0$

$$\inf_{\vartheta_0 \in \Theta} g(\vartheta_0, \nu) > 0.$$

Proposition 3.2 *Let the conditions $\mathscr{G}^*$ hold. Then the MME $\vartheta^*_{\tau,\varepsilon}$ is uniformly consistent and for any $p \geq 2$, there exists a constant $C = C(p) > 0$ such that*

$$\sup_{\vartheta_0\in\Theta} \varepsilon^{-p/3}\,\tau^{p/2}\mathbf{E}_{\vartheta_0}\left\|\vartheta^*_{\tau,\varepsilon}-\vartheta_0\right\|^p \leq C. \tag{3.13}$$

Proof Remark that if $\vartheta^*_{\tau,\varepsilon} \in \Theta$, then it satisfies the equation

$$\inf_{\vartheta\in\Theta}\left\|\Psi^*_{\tau,\varepsilon}-\Psi^*_\tau(\vartheta)\right\| = \|\Psi^*_{\tau,\varepsilon}-\Psi^*_\tau(\vartheta^*_{\tau,\varepsilon})\| = 0.$$

Below, we use elementary inequalities and the estimate (3.5)

$$\begin{aligned}
&\sup_{\vartheta_0\in\Theta}\mathbf{P}_{\vartheta_0}\left(\|\vartheta^*_{\tau,\varepsilon}-\vartheta_0\| > \nu\right)\\
&\quad= \sup_{\vartheta_0\in\Theta}\mathbf{P}_{\vartheta_0}\left(\inf_{\|\vartheta-\vartheta_0\|<\nu}\left\|\Psi^*_{\tau,\varepsilon}-\Psi^*_\tau(\vartheta)\right\| > \inf_{\|\vartheta-\vartheta_0\|\geq\nu}\left\|\Psi^*_{\tau,\varepsilon}-\Psi^*_\tau(\vartheta)\right\|\right)\\
&\quad\leq \sup_{\vartheta_0\in\Theta}\mathbf{P}_{\vartheta_0}\left(\left\|\Psi^*_{\tau,\varepsilon}-\Psi^*_\tau(\vartheta_0)\right\| + \inf_{\|\vartheta-\vartheta_0\|<\nu}\left\|\Psi^*_\tau(\vartheta)-\Psi^*_\tau(\vartheta_0)\right\|\right.\\
&\qquad\quad\left. > \inf_{\|\vartheta-\vartheta_0\|\geq\nu}\left\|\Psi^*_\tau(\vartheta)-\Psi^*_\tau(\vartheta_0)\right\| - \left\|\Psi^*_{\tau,\varepsilon}-\Psi^*_\tau(\vartheta_0)\right\|\right)\\
&\quad= \sup_{\vartheta_0\in\Theta}\mathbf{P}_{\vartheta_0}\left(2\left\|\Psi^*_{\tau,\varepsilon}-\Psi^*_\tau(\vartheta_0)\right\| > g(\vartheta_0,\nu)\right)\\
&\quad\leq \frac{2^{2N}}{\inf_{\vartheta_0\in\Theta} g(\vartheta_0,\nu)^{2N}}\sup_{\vartheta_0\in\Theta}\mathbf{E}_{\vartheta_0}\left\|\Psi^*_{\tau,\varepsilon}-\Psi^*_\tau(\vartheta_0)\right\|^{2N}\\
&\quad\leq \frac{C\tau^N\varepsilon^{2N/3}}{\inf_{\vartheta_0\in\Theta} g(\vartheta_0,\nu)^{2N}} \longrightarrow 0.
\end{aligned}$$

Therefore, the estimator $\vartheta^*_{\tau,\varepsilon}$ is uniformly consistent.

Recall that the MME ϑ is one of the solutions of the MMEq

$$\left(\Psi^*_{\tau,\varepsilon}-\Psi^*_\tau(\check\vartheta_{\tau,\varepsilon})\right)^\top\dot\Psi^*_\tau(\check\vartheta_{\tau,\varepsilon}) = 0.$$

Using the consistency of this estimator, we can write

$$\begin{aligned}
0 &= \left(\Psi^*_{\tau,\varepsilon}-\Psi^*_\tau(\vartheta_0)+\Psi^*_\tau(\vartheta_0)-\Psi^*_\tau(\check\vartheta_{\tau,\varepsilon})\right)^\top\dot\Psi^*_\tau(\check\vartheta_{\tau,\varepsilon})\\
&= \left(\Psi^*_{\tau,\varepsilon}-\Psi^*_\tau(\vartheta_0)-\dot\Psi^*_\tau(\check\vartheta_0)^\top(\check\vartheta_{\tau,\varepsilon}-\vartheta_0)+o\left(\varepsilon^{2/3}\right)\right)^\top\left(\dot\Psi^*_\tau(\vartheta_0)+o\left(\varepsilon^{1/3}\right)\right)\\
&= \left(\Psi^*_{\tau,\varepsilon}-\Psi^*_\tau(\vartheta_0)\right)^\top\dot\Psi^*_\tau(\vartheta_0)-(\check\vartheta_{\tau,\varepsilon}-\vartheta_0)\dot\Psi^*_\tau(\check\vartheta_0)^\top\dot\Psi^*_\tau(\vartheta_0)+o\left(\varepsilon^{2/3}\right).
\end{aligned}$$

Therefore,

$$\begin{aligned}
&\varepsilon^{-1/3}(\check{\vartheta}_{\tau,\varepsilon} - \vartheta_0) \\
&= \left(\dot{\Psi}_\tau^*(\check{\vartheta}_0)^\top \dot{\Psi}_\tau^*(\vartheta_0)\right)^{-1} \varepsilon^{-1/3} \left(\Psi_{\tau,\varepsilon}^* - \Psi_\tau^*(\vartheta_0)\right)^\top \dot{\Psi}_\tau^*(\vartheta_0) + O\left(\varepsilon^{1/3}\right) \\
&= \mathbf{T}(\vartheta_0)^{-1}\, \varepsilon^{-1/3} \left(\Psi_{\tau,\varepsilon}^* - \Psi_\tau^*(\vartheta_0)\right)^\top \dot{\Psi}_\tau^*(\vartheta_0) + O\left(\varepsilon^{1/3}\right).
\end{aligned}$$

As the matrix $\mathbf{T}(\vartheta_0)$ is uniformly nondegenerate and we have (3.5), these allow us to write the estimate: for any $p \geq 2$

$$\varepsilon^{-p/3} \sup_{\vartheta_0 \in \Theta} \mathbf{E}_{\vartheta_0} \left\| \vartheta_{\tau,\varepsilon}^* - \vartheta_0 \right\|^p \leq C\, \varepsilon^{-p/3} \sup_{\vartheta_0 \in \Theta} \mathbf{E}_{\vartheta_0} \left\| \Psi_{\tau,\varepsilon}^* - \Psi_\tau^*(\vartheta_0) \right\|^p \leq C \tau^{p/2}.$$

□

Example 3.1 Suppose that we have the model of observations (3.11) and (3.12), where $f(\vartheta, t) = \vartheta f(t)$, $\vartheta \in (\alpha, \beta)$, $\alpha > 0$, $b(\vartheta, t) = b(t)$ and all corresponding conditions are fulfilled. Then

$$\Psi_\tau(\vartheta) = \vartheta^2 \int_0^\tau f(t)^2\, b(t)^2\, \mathrm{d}t$$

and the MME is

$$\vartheta_{\tau,\varepsilon}^* = \sqrt{\Psi_{\tau,\varepsilon}} \left(\int_0^\tau f(t)^2\, b(t)^2\, \mathrm{d}t\right)^{-1/2}.$$

This estimator is consistent and has rate of convergence $\varepsilon^{1/3}$ (Proposition 3.2).

Example 3.2 Consider the model (3.11) and (3.12), where $f(\vartheta, t) = f(t)$ and

$$b(\vartheta, t) = \left(h(t) + \sum_{k=1}^d \vartheta_k g_k(t)\right)^{1/2}.$$

Here the functions $h(\cdot)$ and $g_k(\cdot)$, $k = 1, \ldots, d$ are such that

$$\inf_{t \in [0,\tau]} \inf_{\vartheta \in \Theta} \left(h(t) + \sum_{k=1}^d \vartheta_k g_k(t)\right) > 0.$$

Then

$$\Psi_{\tau_k}(\vartheta) = \int_0^{\tau_k} f(s)^2\, h(s)\, \mathrm{d}s + \sum_{j=1}^d \vartheta_j \int_0^{\tau_k} f(s)^2\, g_j(s)\, \mathrm{d}s$$

$$= H_k + \sum_{j=1}^{d} \vartheta_j G_{jk}, \qquad\qquad k = 1, \ldots, d.$$

Suppose that the matrix $\mathbf{G} = \left(G_{jk}\right)_{d\times d}$ is nondegenerate. Then we have the relation

$$\vartheta = \mathbf{G}^{-1}\big(\Psi_\tau(\vartheta) - \mathrm{H}\big), \qquad\qquad \mathrm{H} = (H_1, \ldots, H_d)^\top$$

with the obvious notations.

Hence, the MME is

$$\vartheta^*_{\tau,\varepsilon} = \mathbf{G}^{-1}\big(\Psi_{\tau,\varepsilon} - \mathrm{H}\big)$$

and this estimator has the properties described in the Proposition 3.2.

MLE and BE

Recall the main definitions. We suppose that the conditions $\mathscr{A}_*$ are fulfilled. Therefore the measures $\left\{\mathbf{P}^\varepsilon_\vartheta, \vartheta \in \Theta\right\}$ induced by the observations (3.1) in the measurable space $(\mathcal{C}[0,T], \mathfrak{B})$ of continuous functions on $[0,T]$ are equivalent. The likelihood ratio function is given by

$$L(\vartheta, X^T) = \exp\left\{\int_0^T \frac{f(\vartheta,t)\, m(\vartheta,t)}{\varepsilon^2\sigma(t)^2}\,\mathrm{d}X_t - \int_0^T \frac{f(\vartheta,t)^2\, m(\vartheta,t)^2}{2\varepsilon^2\sigma(t)^2}\,\mathrm{d}t\right\}, \quad \vartheta \in \Theta.$$

The conditional expectation $m(\vartheta,t) = \mathbf{E}_\vartheta\left(Y_t | X_s, 0 \le s \le t\right)$ satisfies the Kalman–Bucy equations

$$\mathrm{d}m(\vartheta,t) = \left[a(\vartheta,t) - \frac{\gamma(\vartheta,t)\, f(\vartheta,t)^2}{\varepsilon^2\sigma(t)^2}\right] m(\vartheta,t)\,\mathrm{d}t + \frac{\gamma(\vartheta,t)\, f(\vartheta,t)}{\varepsilon^2\sigma(t)^2}\mathrm{d}X_t \tag{3.14}$$

where $m(\vartheta,0) = y_0$ (see (1.17)). The function $\gamma(\vartheta,t) = \mathbf{E}_\vartheta\left(m(\vartheta,t) - Y_t\right)^2$ is the solution of Riccati equation

$$\frac{\partial\gamma(\vartheta,t)}{\partial t} = 2a(\vartheta,t)\,\gamma(\vartheta,t) - \frac{\gamma(\vartheta,t)^2\, f(\vartheta,t)^2}{\varepsilon^2\sigma(t)^2} + b(\vartheta,t)^2, \tag{3.15}$$

with initial value $\gamma(\vartheta,0) = 0$ (see (1.18)).

Then the MLE $\hat\vartheta_\varepsilon$ and BE $\tilde\vartheta_\varepsilon$ are defined by the usual relations

$$L(\hat{\vartheta}_\varepsilon, X^T) = \sup_{\vartheta\in\Theta} L(\vartheta, X^T), \qquad \tilde{\vartheta}_\varepsilon = \mathbf{E}(\vartheta | X^T) = \frac{\int_\Theta \vartheta\, p(\vartheta)L(\vartheta, X^T)\,\mathrm{d}\vartheta}{\int_\Theta p(\vartheta)L(\vartheta, X^T)\,\mathrm{d}\vartheta}. \tag{3.16}$$

In the case of BE, we suppose that ϑ is a random vector with the density $p(\vartheta)$, $\vartheta \in \Theta$. This density is supposed to be positive continuous function on Θ.

The model under study is of special interest. Suppose, for instance, that $d = 1$. Note that, for all $t \in (0, T]$,

$$\lim_{\varepsilon\to 0}\Big[f(\vartheta, t)\,\dot{m}(\vartheta, t) + \dot{f}(\vartheta, t)\,m(\vartheta, t)\Big] = 0,$$

where the dot represents the derivative with respect to ϑ (see Lemma 3.5 below). Therefore, as $\varepsilon \to 0$, the quantity, which was asymptotically Fisher information in Sect. 2.1

$$\int_0^T \sigma(t)^{-2}\left[\frac{\partial}{\partial\vartheta}\Big(f(\vartheta, t)\,m_t(\vartheta)\Big)\right]^2 \mathrm{d}t \longrightarrow 0.$$

In addition, we have the convergence in distribution

$$\varepsilon^{-1/2}\Big[f(\vartheta, t)\,\dot{m}(\vartheta, t) + \dot{f}(\vartheta, t)\,m(\vartheta, t)\Big] \Longrightarrow n(\vartheta, t)\,\xi_t,$$

where $n(\vartheta, t)$ is a deterministic function and $\{\xi_t, t \in (0, T]\}$ is a family of i.i.d. Gaussian random variables, where $\xi_t \sim \mathcal{N}(0, 1)$. At first blush, one would expect that, as $\varepsilon \to 0$,

$$\varepsilon^{-1}\int_0^T \sigma(t)^{-2}\left[\frac{\partial}{\partial\vartheta}\Big(f(\vartheta, t)\,m_t(\vartheta)\Big)\right]^2 \mathrm{d}t \longrightarrow \int_0^T n(\vartheta, t)^2\,\xi_t^2\,\mathrm{d}t.$$

As it happens, however, this last integral does not exist, and we have instead, as $\varepsilon \to 0$,

$$\varepsilon^{-1}\int_0^T \sigma(t)^{-2}\left[\frac{\partial}{\partial\vartheta}\Big(f(\vartheta, t)\,m_t(\vartheta)\Big)\right]^2 \mathrm{d}t \longrightarrow \int_0^T n(\vartheta, t)^2\,\mathrm{d}t \equiv \mathrm{I}(\vartheta).$$

The quantity $\mathrm{I}(\vartheta)$ is the Fisher information for this model of observations.

The next particular issue concerns the existence of consistent estimators, especially since, in the limit, an integral of the Gaussian process $(f(\vartheta, t)\,Y_t, 0 \le t \le T)$ is observed. We have already discussed this situation at the beginning of this section and a consistent estimator was constructed (Proposition 3.2).

At this point, we recall that our objective is to estimate ϑ for the model under study and describe the properties of the MLE $\hat{\vartheta}_\varepsilon$ and BE $\tilde{\vartheta}_\varepsilon$ as $\varepsilon \to 0$. In what follows, the true value of the parameter is denoted by ϑ_0.

Recall the notation $S(\vartheta, t) = f(\vartheta, t)\, b(\vartheta, t)$ and set

$$G(\vartheta, \vartheta_0) = \int_0^T \frac{[f(\vartheta, t)\, b(\vartheta, t) - f(\vartheta_0, t)\, b(\vartheta_0, t)]^2}{2b(\vartheta, t)\, f(\vartheta, t)\, \sigma(t)} \mathrm{d}t$$
$$= \int_0^T \frac{[S(\vartheta, t) - S(\vartheta_0, t)]^2}{2S(\vartheta, t)\, \sigma(t)} \mathrm{d}t,$$

$$\mathbf{I}(\vartheta) = \int_0^T \frac{f(\vartheta, t)\, b(\vartheta, t)}{2\sigma(t)} \left(\frac{\dot{\mathrm{f}}(\vartheta, t)}{f(\vartheta, t)} + \frac{\dot{\mathrm{b}}(\vartheta, t)}{b(\vartheta, t)}\right) \left(\frac{\dot{\mathrm{f}}(\vartheta, t)}{f(\vartheta, t)} + \frac{\dot{\mathrm{b}}(\vartheta, t)}{b(\vartheta, t)}\right)^\top \mathrm{d}t$$
$$= \int_0^T \frac{S(\vartheta, t)}{2\sigma(t)} \left[\frac{\partial}{\partial\vartheta} \ln S(\vartheta, t)\right] \left[\frac{\partial}{\partial\vartheta} \ln S(\vartheta, t)\right]^\top \mathrm{d}t.$$

The regularity conditions, which are introduced in Theorem 3.1, enable us to verify that the family of measures $\left\{\mathbf{P}_\vartheta^{(\varepsilon)}, \vartheta \in \Theta\right\}$ is locally asymptotically normal (LAN) on Θ (see Lemma 3.6 below). This leads directly to the Hájek's lower bound for the mean square risk of all estimators $\bar{\vartheta}_\varepsilon$, viz.

$$\lim_{\delta\to 0} \varliminf_{\varepsilon\to 0} \sup_{\|\vartheta-\vartheta_0\|<\delta} \varepsilon^{-1} \mathbf{E}_\vartheta \|\bar{\vartheta}_\varepsilon - \vartheta\|^2 \geq \mathbf{E}_{\vartheta_0} \|\zeta(\vartheta_0)\|^2, \ \ \zeta(\vartheta_0) \sim \mathcal{N}\left(0, \mathbf{I}(\vartheta_0)^{-1}\right). \tag{3.17}$$

Therefore, we can give the definition of asymptotically efficient estimator as follows. The estimator $\bar{\vartheta}_\varepsilon$ is asymptotically efficient if, for any $\vartheta_0 \in \Theta$, it satisfies

$$\lim_{\delta\to 0} \lim_{\varepsilon\to 0} \sup_{\|\vartheta-\vartheta_0\|<\delta} \varepsilon^{-1} \mathbf{E}_\vartheta \|\bar{\vartheta}_\varepsilon - \vartheta\|^2 = \mathbf{E}_{\vartheta_0} \|\zeta(\vartheta_0)\|^2.$$

Introduce the *Conditions* $\hat{\mathscr{G}}$:

$\hat{\mathscr{G}}_0$. *The set* $\Theta \subset \mathcal{R}^d$ *is open, convex, and bounded.*

$\hat{\mathscr{G}}_1$. *The functions* $f(\vartheta, t), a(\vartheta, t), b(\vartheta, t); (t, \vartheta) \in [0, T] \times \bar{\Theta}$ *are bounded and have two continuous bounded derivatives with respect to* ϑ.

$\hat{\mathscr{G}}_2$. *The functions* $f(\vartheta, t), \sigma(t), b(\vartheta, t), (t, \vartheta) \in [0, T] \times \Theta$ *have continuous derivatives w.r.t.* t *and are separated from zero:*

$$\kappa = \inf_{\vartheta\in\Theta} \inf_{0\le t\le T} f(\vartheta, t) > 0, \qquad \inf_{0\le t\le T} \sigma(t) > 0, \qquad \inf_{\vartheta\in\Theta} \inf_{0\le t\le T} b(\vartheta, t) > 0.$$

$\hat{\mathscr{G}}_3$. *For any* $\nu > 0$,

$$\inf_{\vartheta_0\in\Theta}\ \inf_{\|\vartheta-\vartheta_0\|>\nu} G(\vartheta,\vartheta_0)>0.$$

$\hat{\mathscr{G}}_4$. *The Fisher information matrix is uniformly nondegenerate:*

$$\inf_{\vartheta\in\Theta}\ \inf_{\|\mathrm{e}\|=1,\,\mathrm{e}\in\mathcal{R}^d} \mathrm{e}^\top \mathbf{I}(\vartheta)\,\mathrm{e}>0.$$

Theorem 3.1 *Suppose that the conditions $\hat{\mathscr{G}}$ hold. Then the MLE $\hat\vartheta_\varepsilon$ and BE $\tilde\vartheta_\varepsilon$ are uniformly on compacts $\mathbb{K}\subset\Theta$ consistent, asymptotically normal, i.e.,*

$$\frac{(\hat\vartheta_\varepsilon-\vartheta_0)}{\sqrt{\varepsilon}}\Longrightarrow\zeta(\vartheta_0)\sim\mathcal{N}\left(0,\mathbf{I}(\vartheta_0)^{-1}\right),\qquad \frac{(\tilde\vartheta_\varepsilon-\vartheta_0)}{\sqrt{\varepsilon}}\Longrightarrow\zeta(\vartheta_0),$$

and the moments converge: for any $p\geq 0$ uniformly on $\mathbb{K}\subset\Theta$

$$\lim_{\varepsilon\to 0}\varepsilon^{p/2}\mathbf{E}_{\vartheta_0}\|(\hat\vartheta_\varepsilon-\vartheta_0)\|^p=\mathbf{E}_{\vartheta_0}\|\zeta(\vartheta_0)\|^p,$$

$$\lim_{\varepsilon\to 0}\varepsilon^{p/2}\mathbf{E}_{\vartheta_0}\|(\tilde\vartheta_\varepsilon-\vartheta_0)\|^p=\mathbf{E}_{\vartheta_0}\|\zeta(\vartheta_0)\|^p$$

and both estimators are asymptotically efficient.

Proof The proof of this theorem consists of the verification of the conditions $\mathscr{I}\mathscr{K}_d$ of Theorems 1.10 and 1.12. The normalized LR is

$$Z_\varepsilon(\vartheta_0,u)=\frac{L(\vartheta_0+\sqrt{\varepsilon}u,X^T)}{L\left(\vartheta_0,X^T\right)},\qquad u\in\mathbb{U}_\varepsilon=\left\{u\ :\ \vartheta_0+\sqrt{\varepsilon}u\in\Theta\right\}.$$

Let us now denote $M(\vartheta,t)=f(\vartheta,t)\,m(\vartheta,t)$ and $\vartheta_u=\vartheta_0+\sqrt{\varepsilon}u$, where $m(\vartheta,t)$ is the solution of Eq. (3.14). Then the log-likelihood ratio is

$$\ln Z_\varepsilon(\vartheta_0,u)=\int_0^T\frac{M(\vartheta_u,t)-M(\vartheta_0,t)}{\varepsilon\sigma(t)}\,\mathrm{d}\bar W_t-\int_0^T\frac{[M(\vartheta_u,t)-M(\vartheta_0,t)]^2}{2\varepsilon^2\sigma(t)^2}\,\mathrm{d}t.$$

Here we used the innovation representation (1.21)

$$\mathrm{d}X_t=f(\vartheta,t)\,m(\vartheta,t)\,\mathrm{d}t+\varepsilon\sigma(t)\,\mathrm{d}\bar W_t.$$

We need several technical results to show that

$$\int_0^T\frac{[M(\vartheta,t)-M(\vartheta_0,t)]}{\sigma(t)}\,\mathrm{d}\bar W_t\longrightarrow 0,\tag{3.18}$$

$$\int_0^T\frac{[M(\vartheta,t)-M(\vartheta_0,t)]^2}{\varepsilon\sigma(t)^2}\,\mathrm{d}t\longrightarrow G(\vartheta,\vartheta_0),\tag{3.19}$$

$$\int_0^T \frac{[M(\vartheta_u,t)-M(\vartheta_0,t)]}{\varepsilon\sigma(t)}\,\mathrm{d}\bar W_t \Longrightarrow \mathcal{N}\left(0, u^\top \mathbf{I}(\vartheta_0)\,u\right),$$

$$\int_0^T \frac{[M(\vartheta_u,t)-M(\vartheta_0,t)]^2}{\varepsilon^2\sigma(t)^2}\,\mathrm{d}t \longrightarrow u^\top \mathbf{I}(\vartheta_0)\,u. \tag{3.20}$$

□

Lemma 3.1 *For any* $t_0 \in (0, T]$, *we have*

$$\sup_{t_0\le t\le T}\left|\frac{\gamma(\vartheta,t)}{\varepsilon}-\frac{b(\vartheta,t)\,\sigma(t)}{f(\vartheta,t)}\right| = O(\varepsilon). \tag{3.21}$$

Proof The proof of this lemma is close to the proof of Lemma 2.6. The only difference is in the normalization, but as it is short we can remind it here. Recall that the function $\gamma_*(\vartheta,t)=\varepsilon^{-1}\mathbf{E}_\vartheta\,[m(\vartheta,t)-Y_t]^2$ satisfies the Riccati equation (3.15), i.e.,

$$\varepsilon\,\frac{\partial\gamma_*(\vartheta,t)}{\partial t} = 2\varepsilon a(\vartheta,t)\,\gamma_*(\vartheta,t) - \frac{\gamma_*(\vartheta,t)^2 f(\vartheta,t)^2}{\sigma(t)^2} + b(\vartheta,t)^2, \quad \gamma_*(\vartheta,0)=0.$$

To verify (3.21), we introduce the equation

$$\frac{\partial\gamma^*(t)}{\partial t} = 2a\gamma^*(t) - \frac{\gamma^*(t)^2\kappa^2}{\varepsilon^2 S^2} + B^2, \quad \gamma^*(0)=0,$$

where $a(\vartheta,t)\le a$, $\sigma(t)\le S$ and $b(\vartheta,t)\le B$ for all $t\in[0,T]$, $\vartheta\in\Theta$. By means of the comparison theorem for ordinary differential equations, we find that, for all $t\in[0,T]$, $\gamma(\vartheta,t)\le\gamma^*(t)$.

Furthermore, the solution $\gamma^*(t)$ can be written explicitly; see (1.26). Indeed,

$$\gamma^*(t) = e^{-2rt}\left\{\frac{1}{\gamma^*(0)-\hat\gamma}+\frac{\kappa_*^2}{2r\varepsilon^2}(1-e^{-2rt})\right\}^{-1}+\hat\gamma,$$

where

$$\kappa_* = \kappa S^{-1}, \qquad r = (a^2 + B^2\kappa_*^2\varepsilon^{-2})^{1/2},$$

$$\hat\gamma = (a\varepsilon^2/\kappa_*^2)\left\{\sqrt{1+B^2\kappa_*^2/(a^2\varepsilon^2)}-1\right\}.$$

Then it is easy to show that for any $t_0\in(0,T]$, the difference $\gamma^*(t)-\hat\gamma$ is exponentially small and the following representations hold true:

$$r = \frac{B\kappa_*}{\varepsilon}(1+O(\varepsilon)), \quad \hat\gamma = \frac{B\varepsilon}{\kappa_*}(1+O(\varepsilon)), \quad \sup_{t_0\le t\le T}\left|\gamma^*(t)-\frac{B\varepsilon}{\kappa_*}\right| = O(\varepsilon^2).$$

Hence, for $t > t_0$ and $\varepsilon \in (0, \varepsilon_0]$, we have $0 \le \gamma_* (\vartheta, t) = \gamma (\vartheta, t)/\varepsilon \le 2B/\kappa_*$. In a similar way, we obtain $\gamma_* (\vartheta, t) \ge b/(2K)$ and

$$\gamma (\vartheta, t) - \gamma (\vartheta, t_0) - \int_{t_0}^{t} a (\vartheta, s) \gamma (\vartheta, s) \, ds$$
$$= - \int_{t_0}^{t} \frac{\gamma_* (\vartheta, s)^2 f (\vartheta, s)^2}{\sigma (s)^2} ds + \int_{t_0}^{t} b (\vartheta, s)^2 \, ds,$$

where the left-hand part tends to zero uniformly on $t \in [t_0, T]$. Then for $t > t_0$, the following relations hold true:

$$\int_{t_0}^{t} \left[\frac{\gamma_* (\vartheta, s)^2 f (\vartheta, s)^2}{\sigma (s)^2} - b (\vartheta, s)^2 \right] ds = \varepsilon P_\varepsilon (t) .$$

All functions here are continuously differentiable and therefore,

$$\gamma_* (\vartheta, s) = \frac{b (\vartheta, s) \sigma (s)}{f (\vartheta, s)} + O (\varepsilon) .$$

□

Note that for $s > 0$ we have

$$\int_{s}^{t} \left[a (\vartheta, v) - \frac{\gamma (\vartheta, v) f (\vartheta, v)^2}{\varepsilon^2 \sigma (v)^2} \right] dv$$
$$= \int_{s}^{t} \left[a (\vartheta, v) - \frac{b (\vartheta, v) f (\vartheta, v)}{\varepsilon \sigma (v)} \right] dv \, (1 + O (\varepsilon))$$
$$= -\frac{1}{\varepsilon} \int_{s}^{t} \frac{b (\vartheta, v) f (\vartheta, v)}{\sigma (v)} dv \, (1 + O (\varepsilon)) .$$

Introduce the random variables

$$\bar\xi_{t,\varepsilon} = \sqrt{\frac{2\varepsilon}{q_t}} \int_0^t e^{-\frac{t-s}{\varepsilon} q_t} d\bar W_s = \sqrt{2} \int_0^{\frac{tq_t}{\varepsilon}} e^{-r} d\bar w_{t,\varepsilon} (r) , \qquad t \in (0, T]$$

where $q_t = b (\vartheta, t) f (\vartheta, t) \sigma (t)^{-1}$ and remark that

$$\bar\xi_{t,\varepsilon} \Longrightarrow \bar\xi_t = \sqrt{2} \int_0^{\infty} e^{-r} d\bar w_t (r) \sim \mathcal{N} (0, 1) .$$

where we changed the variable $s = t - r\varepsilon/q_t$ and $\bar w_{t,\varepsilon} (r) , 0 \le r \le tq_t/\varepsilon$ is the corresponding Winer process. Of course, $\bar w_t (r) , r \ge 0$ is also Wiener process.

Lemma 3.2 *With the notation* $S(\vartheta,t)=f(\vartheta,t)\,b(\vartheta,t)$, *we have the representation*

$$\varepsilon\ln\frac{L(\vartheta,X^T)}{L\left(\vartheta_0,X^T\right)}=-\int_0^T\frac{\left[S(\vartheta,t)-S(\vartheta_0,t)\right]^2}{4S(\vartheta,t)\,\sigma(t)}\,\bar\xi_{t,\varepsilon}^2\,\mathrm{d}t+O_p(\sqrt{\varepsilon}).$$

Proof First, we have

$$\begin{aligned}&f(\vartheta,t)\,m(\vartheta,t)-f(\vartheta_0,t)\,m(\vartheta_0,t)\\&\quad=f(\vartheta,t)\,\delta_m(t)+\left[f(\vartheta,t)-f(\vartheta_0,t)\right]m(\vartheta_0,t),\end{aligned}$$

where $\delta_m(t)=m(\vartheta,t)-m(\vartheta_0,t)$. The equation for $\delta_m(t)$ is obtained as follows. We have

$$\begin{aligned}\mathrm{d}m(\vartheta,t)&=a(\vartheta,t)\,m(\vartheta,t)\,\mathrm{d}t+\frac{\gamma_*(\vartheta,t)\,f(\vartheta,t)}{\sigma(t)}\mathrm{d}\bar W_t\\&\quad+\frac{\gamma_*(\vartheta,t)\,f(\vartheta,t)}{\varepsilon\sigma(t)^2}\left[f(\vartheta_0,t)\,m(\vartheta_0,t)-f(\vartheta,t)\,m(\vartheta,t)\right]\mathrm{d}t,\\\mathrm{d}m(\vartheta_0,t)&=a(\vartheta_0,t)\,m(\vartheta_0,t)\,\mathrm{d}t+\frac{\gamma_*(\vartheta_0,t)\,f(\vartheta_0,t)}{\sigma(t)}\,\mathrm{d}\bar W_t.\end{aligned}\tag{3.22}$$

Second, recall that by (3.21), $\gamma_*(\vartheta,t)\,f(\vartheta,t)/\sigma(t)=b(\vartheta,t)\,(1+O(\varepsilon))$.

To simplify the notation, we omit $O(\varepsilon)$ and set $m_t=m(\vartheta,t)$,$m_{0,t}=m(\vartheta_0,t)$ $a_t=a(\vartheta,t)$, $f_t=f(\vartheta,t)$, $a_{0,t}=a(\vartheta_0,t)$, $f_{0,t}=f(\vartheta_0,t)$, $b_t=b(\vartheta,t)$, $b_{0,t}=b(\vartheta_0,t)$ and $\sigma_t=\sigma(t)$. It is easy to verify that, as the underlying Gaussian processes have bounded variances, the reminders $O_p(\varepsilon)$ and $O_p(\sqrt{\varepsilon})$ satisfy

$$\mathbf{E}_{\vartheta_0}|O_p(\varepsilon)|^{2p}\leq C_1\,\varepsilon^{2p},\qquad\mathbf{E}_{\vartheta_0}|O_p(\sqrt{\varepsilon})|^{2p}\leq C_2\,\varepsilon^{p},$$

for any $p>1$, where the constants $C_1>0$, $C_2>0$ do not depend necessarily on t and ϑ. Of course, in the case of deterministic $O(\varepsilon)$, the previous inequalities hold true. Another simplification: the approximation of $\gamma(\vartheta,t)/\varepsilon$ by $\gamma_*(\vartheta,t)$ is valid on any interval $[t_0,T]$ with $t_0>0$, but in the integrals, we will use it on the intervals $(0,T]$ because the integrals on $[0,t_0]$ can be made as small as we want by chosing small $t_0>0$. Hence,

$$\begin{aligned}\mathrm{d}m_t&=a_tm_t\mathrm{d}t+b_t\mathrm{d}\bar W_t+\frac{b_t}{\varepsilon\sigma_t}\left(f_{0,t}m_{0,t}-f_tm_t\right)\mathrm{d}t,\\\mathrm{d}m_{0,t}&=a_{0,t}m_{0,t}\mathrm{d}t+b_{0,t}\mathrm{d}\bar W_t\end{aligned}$$

and

$$\mathrm{d}\delta_m(t) = \left(a_t - \frac{b_t f_t}{\varepsilon\sigma_t}\right)\delta_m(t)\,\mathrm{d}t + \left(b_t - b_{0,t}\right)\mathrm{d}\bar{W}_t$$
$$+\left\{\frac{b_t\left(f_{0,t}-f_t\right)}{\varepsilon\sigma_t} + a_t - a_{0,t}\right\} m_{0,t}\mathrm{d}t,$$

where $\delta_m(0) = 0$. The solution of this equation is (below $q_t = b_t f_t \sigma_t^{-1}$, and we omit the terms of order $O(\varepsilon)$)

$$\delta_m(t) = \int_0^t \exp\left\{-\frac{1}{\varepsilon}\int_s^t q_v\mathrm{d}v\right\}\left\{\frac{b_s\left(f_{0,s}-f_s\right)}{\varepsilon\sigma_s} + a_s - a_{0,s}\right\} m_{0,s}\,\mathrm{d}s$$
$$+\int_0^t \exp\left\{-\frac{1}{\varepsilon}\int_s^t q_v\mathrm{d}v\right\}\left(b_s - b_{0,s}\right)\mathrm{d}\bar{W}_s$$
$$= \frac{b_t\left(f_{0,t}-f_t\right)}{\sigma_t}\frac{\sigma_t}{b_t f_t} m_{0,t} + \sqrt{\varepsilon} Q_\varepsilon(t) = \frac{f_{0,t}-f_t}{f_t} m_{0,t} + \sqrt{\varepsilon}\, Q_\varepsilon(t),$$

where

$$Q_\varepsilon(t) = \frac{b_t\left(f_{0,t}-f_t\right)}{\sigma_t\varepsilon^{3/2}}\int_0^t \exp\left\{-\frac{1}{\varepsilon}\int_s^t q_v\mathrm{d}v\right\}\left(m_{0,s}-m_{0,t}\right)\,\mathrm{d}s$$
$$+\frac{\left(b_t - b_{0,t}\right)}{\sqrt{\varepsilon}}\int_0^t \exp\left\{-\frac{1}{\varepsilon}\int_s^t q_v\mathrm{d}v\right\}\mathrm{d}\bar{W}_s$$
$$= \frac{b_t\left(f_{0,t}-f_t\right)}{\sigma_t} I_{1,\varepsilon}(t) + \left(b_t - b_{0,t}\right) I_{2,\varepsilon}(t).$$

Therefore, we obtain the relations

$$m(\vartheta, t) = \frac{f(\vartheta_0, t)}{f(\vartheta, t)}\, m(\vartheta_0, t) + O_p(\sqrt{\varepsilon}),$$

which proves (3.18) and

$$M(\vartheta, t) - M(\vartheta_0, t) = f_t\delta_m(t) + \left(f_t - f_{0,t}\right) m_{0,t}$$
$$= f_t\frac{f_{0,t}-f_t}{f_t} m_{0,t} + \left(f_t - f_{0,t}\right) m_{0,t} + \sqrt{\varepsilon} f_t Q_\varepsilon(t) = \sqrt{\varepsilon} f_t Q_\varepsilon(t).$$

We focus now on the study of the integrals

$$I_{1,\varepsilon}(t) = \frac{1}{\varepsilon^{3/2}}\int_0^t \exp\left\{-\frac{(t-s)\,q_t}{\varepsilon}\right\}\left(m_{0,s}-m_{0,t}\right)\mathrm{d}s,$$

$$I_{2,\varepsilon}(t)=\frac{1}{\varepsilon^{1/2}}\int_0^t \exp\left\{-\frac{(t-s)\,q_t}{\varepsilon}\right\}\mathrm{d}\bar W_s.$$

For the values $s=t-u\varepsilon$, we have

$$\begin{aligned}m_{0,s}-m_{0,t}&=\int_s^t a_{0,v}m_{0,v}\mathrm{d}v-\int_s^t b_{0,v}\mathrm{d}\bar W_v+O_p(\varepsilon)\\&=-\sqrt{\varepsilon}\,b_{0,t}\frac{(\bar W_t-\bar W_{t-u\varepsilon})}{\sqrt{\varepsilon}}+O_p(\varepsilon)=\sqrt{\varepsilon}\,b_{0,t}\tilde W_{t,\varepsilon}(u)+O_p(\varepsilon),\end{aligned}$$

where $\tilde W_{t,\varepsilon}(u)=\varepsilon^{-1/2}\left(\bar W_{t-u\varepsilon}-\bar W_t\right)$. Therefore,

$$\begin{aligned}I_{1,\varepsilon}(t)=b_{0,t}\int_0^t e^{-uq_t}\tilde W_{t,\varepsilon}(u)\,\mathrm{d}u&=\frac{b_{0,t}}{q_t^{3/2}}\int_0^{\frac{tq_t}{\varepsilon}} e^{-r}\mathrm{d}\bar w_{t,\varepsilon}(r)+O_p(\varepsilon)\\&=\frac{b_{0,t}}{q_t^{3/2}2^{1/2}}\,\bar\xi_{t,\varepsilon}+O_p(\varepsilon),\end{aligned}$$

where $\bar w_{t,\varepsilon}(r),0\leq r\leq tq_t\varepsilon^{-1}$ is a Wiener process and remind that

$$\bar\xi_{t,\varepsilon}=\sqrt{2}\int_0^{\frac{tq_t}{\varepsilon}} e^{-r}\mathrm{d}\bar w_{t,\varepsilon}(r)\Longrightarrow\bar\xi_t=\sqrt{2}\int_0^\infty e^{-r}\mathrm{d}\bar w_t(r)\sim\mathcal{N}(0,1).$$

Here $\{\bar w_t(r),r\geq 0\},t\in(0,T]$ are independent Wiener processes.

Similarly, we obtain $I_{2,\varepsilon}(t)=\bar\xi_{t,\varepsilon}/\left(q_t^{1/2}\sqrt{2}\right)+O_p(\varepsilon)$ and finally, by omitting $O_p(\varepsilon)$, we have

$$\begin{aligned}Q_\varepsilon(t)&=\left\{\frac{b_t\left(f_t-f_{0,t}\right)}{\sigma_t}\frac{b_{0,t}}{q_t^{3/2}}+\frac{\left(b_t-b_{0,t}\right)}{q_t^{1/2}}\right\}\frac{\bar\xi_{t,\varepsilon}}{\sqrt{2}}\\&=\frac{b_{0,t}}{q_t^{1/2}}\left\{\frac{\left(f_t-f_{0,t}\right)}{f_t}+\frac{\left(b_t-b_{0,t}\right)}{b_{0,t}}\right\}\frac{\bar\xi_{t,\varepsilon}}{\sqrt{2}}\\&=\frac{b_{0,t}\sqrt{\sigma_t}}{\sqrt{2b_tf_t}}\left(\frac{b_t}{b_{0,t}}-\frac{f_{0,t}}{f_t}\right)\bar\xi_{t,\varepsilon}=\frac{\sqrt{\sigma_t}\left(b_tf_t-f_{0,t}b_{0,t}\right)}{f_t^{3/2}\sqrt{2b_t}}\,\bar\xi_{t,\varepsilon}.\end{aligned}$$

This allows us to write

$$\begin{aligned}G_\varepsilon(\vartheta,\vartheta_0)&=\int_0^T\frac{\left[M(\vartheta,t)-M(\vartheta_0,t)\right]^2}{\varepsilon\sigma(t)^2}\mathrm{d}t=\int_0^T\frac{f_t^2\,Q_\varepsilon(t)^2}{2\sigma(t)^2}\,\mathrm{d}t+O_p(\varepsilon)\\&=\int_0^T\frac{\left[f(\vartheta,t)\,b(\vartheta,t)-f(\vartheta_0,t)\,b(\vartheta_0,t)\right]^2}{2b(\vartheta,t)\,f(\vartheta,t)\,\sigma(t)}\,\bar\xi_{t,\varepsilon}^2\,\mathrm{d}t+O_p(\varepsilon).\end{aligned}$$

□

The next step is to verify the convergence (3.19), i.e., $G_\varepsilon(\vartheta, \vartheta_0) \to G(\vartheta, \vartheta_0)$.

Remark 3.3 Recall that $\bar\xi_{t,\varepsilon} \Longrightarrow \bar\xi_t$, where the variables $\bar\xi_t$ are mutually independent for all $t \in (0, T]$. Therefore, the integral

$$\int_0^T R(t)\,\bar\xi_t^2\, \mathrm{d}t$$

with continuous function $R(t) \geq 0$ does not exist, and there is no limit such that

$$S_\varepsilon = \int_0^T R(t)\,\bar\xi_{t,\varepsilon}^2\, \mathrm{d}t \longrightarrow \int_0^T R(t)\,\bar\xi_t^2\, \mathrm{d}t.$$

The limit of the function $G_\varepsilon(\vartheta, \vartheta_0)$ is given in following lemma.

Lemma 3.3 *For a given continuous function $R(t) \geq 0$, we have the mean square convergences*

$$\lim_{\varepsilon\to 0}\int_0^T R(t)\,\bar\xi_{t,\varepsilon}\, \mathrm{d}t = 0, \tag{3.23}$$

$$\lim_{\varepsilon\to 0}\int_0^T R(t)\,\bar\xi_{t,\varepsilon}^2\, \mathrm{d}t = \int_0^T R(t)\, \mathrm{d}t. \tag{3.24}$$

Proof First, let us explain why the limit like (3.24) has such a form. The Gaussian processes $(\bar\xi_{t,\varepsilon}, t \in [0, T])$ with $\varepsilon > 0$ are continuous and the integral S_ε could be well approximated by the sum

$$S_{n,\varepsilon} = \frac{T}{n}\sum_{j=1}^{n} R(t_j)\bar\xi_{t_j,\varepsilon}^2,$$

where $t_j = jT/n$. Then we have, as $\varepsilon \to 0$,

$$S_{n,\varepsilon} \Longrightarrow S_n = \frac{T}{n}\sum_{j=1}^{n} R(t_j)\bar\xi_{t_j}^2$$

because the vector $(\bar\xi_{t_1,\varepsilon}, \ldots, \bar\xi_{t_n,\varepsilon})$ converges in distribution to the Gaussian vector $(\bar\xi_{t_1}, \ldots, \bar\xi_{t_n})$ with independent components. By the LLN, the limit as $n \to \infty$ is,

$$S_n \longrightarrow S_0 = \int_0^T R(t)\, \mathrm{d}t.$$

Indeed we have

$$\mathbf{E}_{\vartheta_0} S_n = \frac{T}{n}\sum_{j=1}^{n} R(t_j) \to \int_0^T R(t)\,\mathrm{d}t,$$

$$\mathbf{E}_{\vartheta_0}\left(S_n - \mathbf{E}_{\vartheta_0} S_n\right)^2 = \frac{T^2}{n^2}\sum_{j=1}^{n}\sum_{i=1}^{n} R(t_j)R(t_i)\mathbf{E}_{\vartheta_0}(\bar\xi_{t_j}^2 - 1)(\bar\xi_{t_i}^2 - 1)$$

$$= \frac{CT^2}{n^2}\sum_{j=1}^{n} R(t_j)^2 \le \frac{CT}{n} \longrightarrow 0.$$

Obviously, the two-step proof is not satisfactory and is given just to explain the form of the limit. The convergence could be verified directly by computing the expectation

$$\mathbf{E}_{\vartheta_0}\left\{\int_0^T R(t)\,(\bar\xi_{t,\varepsilon}^2 - 1)\,\mathrm{d}t\right\}^2 = o(1),$$

since the correlation between $\bar\xi_{t_2,\varepsilon}$ and $\bar\xi_{t_3,\varepsilon}$ for any $t_2 \neq t_3$ is exponentially small. For our purpose, it is sufficient to study the integrals

$$J_\varepsilon(t) = \sqrt{\frac{2q_t}{\varepsilon}}\int_0^t \exp\left(-\frac{(t-s)\,q_t}{\varepsilon}\right)\mathrm{d}W_s, \qquad \mathbf{E}_{\vartheta_0} J_\varepsilon(t) = 0,$$

$$\mathbf{E}_{\vartheta_0} J_\varepsilon(t)^2 = \left(1 - \exp\left(-\frac{2tq_t}{\varepsilon}\right)\right),$$

since

$$\bar\xi_{t,\varepsilon} = \sqrt{2}\int_0^{tq_t/\varepsilon} e^{-r}\ \mathrm{d}w_{t,\varepsilon}(r) = J_\varepsilon(t).$$

For the Gaussian process $J_\varepsilon(t)$ and as $\varepsilon \to 0$, we have

$$\mathbf{E}_{\vartheta_0} J_\varepsilon(t_2)\,J_\varepsilon(t_3) = \frac{2\sqrt{q_{t_2}q_{t_3}}}{\varepsilon}\exp\left(-\frac{t_2q_{t_2}+t_3q_{t_3}}{\varepsilon}\right)\int_0^{t_2\wedge t_3}\exp\left\{\frac{s\,(q_{t_2}+q_{t_3})}{\varepsilon}\right\}\mathrm{d}s$$

$$= \frac{2\sqrt{q_{t_2}q_{t_2}}}{q_{t_2}+q_{t_3}}\exp\left(-\frac{t_2q_{t_2}+t_3q_{t_3}}{\varepsilon}\right)$$

$$\times\left[\exp\left\{\frac{(t_2\wedge t_3)\,(q_{t_2}+q_{t_3})}{\varepsilon}\right\} - 1\right]$$

$$= \frac{2\sqrt{q_{t_2}q_{t_3}}}{q_{t_2}+q_{t_3}}\left[\exp\left\{\frac{(t_1\wedge t_2)\,(q_{t_2}+q_{t_3}) - t_2q_{t_2} - t_3q_{t_3}}{\varepsilon}\right\}\right.$$

$$
\left. - \exp\left(-\frac{t_2 q_{t_2} + t_3 q_{t_3}}{\varepsilon}\right)\right]
$$

$$
\leq \frac{2\sqrt{q_{t_2} q_{t_3}}}{q_{t_2} + q_{t_3}} \exp\left(-\frac{|t_2 - t_3|\, q_{t_2 \vee t_3}}{\varepsilon}\right)
$$

$$
\leq \frac{2\sqrt{q_{t_2} q_{t_3}}}{q_{t_2} + q_{t_3}} \exp\left(-\frac{|t_2 - t_3|\, q}{\varepsilon}\right) \longrightarrow 0,
$$

where $q = \inf_{t_0 \leq t \leq T} q_t$.

Recall that if we consider two Gaussian random variables $\bar{\xi}, \eta \sim \mathcal{N}(0, 1)$ and $\mathbf{E}\bar{\xi}\eta = \rho$, then $\mathbf{E}(\bar{\xi}^2 - 1)(\eta^2 - 1) = 2\rho^2$. Hence

$$
\mathbf{E}_{\vartheta_0} S_\varepsilon^2 = \int_0^T \int_0^T R(t)\, R(s)\, \mathbf{E}_{\vartheta_0} \left\{(\bar{\xi}_{t,\varepsilon}^2 - 1)(\bar{\xi}_{s,\varepsilon}^2 - 1)\right\} \mathrm{d}t\mathrm{d}s
$$

$$
\leq C \int_0^T \int_0^T \exp\left(-2\,|t - s|\, q_{t\vee s}/\varepsilon\right) \mathrm{d}t\mathrm{d}s \leq C\varepsilon.
$$

This proves (3.19).

Moreover, for any integer $p > 1$, we have

$$
\mathbf{E}_{\vartheta_0} \left| \int_0^T R(t)\, (\bar{\xi}_{t,\varepsilon}^2 - 1)\mathrm{d}t \right|^{2p} \leq C\varepsilon^p. \tag{3.25}
$$

The last estimate follows from Lemma 1.6. □

Therefore, we have the estimate

$$
\mathbf{E}_{\vartheta_0} \left| G_\varepsilon(\vartheta, \vartheta_0) - G(\vartheta, \vartheta_0) \right|^{2p} \leq C\varepsilon^p.
$$

Lemma 3.4 *There exists a constant $\hat{\kappa} > 0$ such that, for any $N > 1$ and some constant $C_N > 0$ the estimate*

$$
\mathbf{P}_{\vartheta_0}\left(Z_\varepsilon(\vartheta_0, u) > e^{-\hat{\kappa}\|u\|^2}\right) \leq \frac{C_N}{\|u\|^{2N}}
$$

holds.

Proof The function $G(\vartheta_0 + h, \vartheta_0)$, $\|h\| \leq \nu$ for sufficiently small $\nu > 0$ has the representation

$$
G(\vartheta_0 + h, \vartheta_0) = h^\top \mathbf{I}(\vartheta_0)\, h\, (1 + o(1))
$$

$$
\geq \|h\|^2 \inf_{\|e\|=1} e^\top \mathbf{I}(\vartheta_0)\, e\, (1 + o(1)) \geq c_* \|h\|^2
$$

with some constant $c_* > 0$.

However, for $\|h\| > \nu$, it follows from the identifiability condition that

$$G\left(\vartheta_0 + h, \vartheta_0\right) \geq g_\nu \geq g_\nu \frac{\|h\|^2}{\sup_{\vartheta,\vartheta'\in\Theta} \|\vartheta - \vartheta'\|^2} \geq c_{**} \|h\|^2,$$

where $g_\nu = \inf_{\vartheta_0\in\Theta} \inf_{\|\vartheta-\vartheta_0\|>\nu} G\left(\vartheta, \vartheta_0\right) > 0$. Therefore, if we put $\kappa^\circ = c_* \wedge c_{**}$, then

$$G\left(\vartheta, \vartheta_0\right) \geq \kappa^\circ \|\vartheta - \vartheta_0\|^2, \tag{3.26}$$

for all $\vartheta, \vartheta_0 \in \Theta$. Hence $G(\vartheta_0 + \sqrt{\varepsilon}u, \vartheta_0) \geq \varepsilon\kappa^\circ \|u\|^2$.

Let us now denote $\delta_M\left(u, t\right) = M\left(\vartheta_u, t\right) - M\left(\vartheta_0, t\right)$ and $G_u = G\left(\vartheta_u, \vartheta_0\right)$. Then we obtain

$$\mathbf{P}_{\vartheta_0}\left\{\frac{G\left(\vartheta_u, \vartheta_0\right)}{\varepsilon} - \int_0^T \frac{\delta_M\left(u, t\right)^2}{\varepsilon^2\sigma\left(t\right)^2} \mathrm{d}t \geq \frac{G_u}{\varepsilon} - \hat{\kappa}\|u\|^2\right\}$$

$$\leq \mathbf{P}_{\vartheta_0}\left[\int_0^T \left[\frac{\left[S\left(\vartheta_u, t\right) - S\left(\vartheta_0, t\right)\right]^2}{2\varepsilon^2\sigma\left(t\right) S\left(\vartheta_u, t\right)} - \frac{\delta_M\left(u, t\right)^2}{\varepsilon\sigma\left(t\right)^2}\right] \mathrm{d}t \geq \left(\kappa^\circ - \hat{\kappa}\right)\|u\|^2\right]$$

$$\leq \frac{1}{\left|\kappa^\circ - \hat{\kappa}\right|^{2N} \|u\|^{4N}} \mathbf{E}_{\vartheta_0}\left|\int_0^T \left[\frac{\left[S\left(\vartheta_u, t\right) - S\left(\vartheta_0, t\right)\right]^2}{4\varepsilon\sigma\left(t\right) S\left(\vartheta_u, t\right)} - \frac{\delta_M\left(u, t\right)^2}{2\varepsilon\sigma\left(t\right)^2}\right] \mathrm{d}t\right|^{2N},$$

where the constant $\hat{\kappa}$ is such that $\kappa^\circ - \hat{\kappa} > 0$, and we used Chebyshev's inequality.

We have the estimates

$$\left|\int_0^T \left[\frac{\left[S\left(\vartheta_u, t\right) - S\left(\vartheta_0, t\right)\right]^2}{4\sigma\left(t\right) S\left(\vartheta_u, t\right)} - \frac{\delta_M\left(u, t\right)^2}{2\varepsilon\sigma\left(t\right)^2}\right] \mathrm{d}t\right|$$

$$\leq \left|\int_0^T \frac{\left[S\left(\vartheta_u, t\right) - S\left(\vartheta_0, t\right)\right]^2}{4\sigma\left(t\right) S\left(\vartheta_u, t\right)} \left(1 - \bar{\xi}_{t,\varepsilon}^2\right) \mathrm{d}t\right| + O_p\left(\varepsilon\right)$$

$$\leq \left|\int_0^T \int_0^1 \frac{\left[u^\top \dot{S}\left(\vartheta_{u(r)}, t\right) u\right]^2}{4\sigma\left(t\right) S\left(\vartheta_u, t\right)} \mathrm{d}r \left(1 - \bar{\xi}_{t,\varepsilon}^2\right) \mathrm{d}t\right| \varepsilon + O_p\left(\varepsilon\right).$$

Therefore,

$$\mathbf{E}_{\vartheta_0}\left|\int_0^T \left[\frac{\left[S\left(\vartheta_u, t\right) - S\left(\vartheta_0, t\right)\right]^2}{4\varepsilon\sigma\left(t\right) S\left(\vartheta_u, t\right)} - \frac{\delta_M\left(u, t\right)^2}{2\varepsilon^2\sigma\left(t\right)^2}\right] \mathrm{d}t\right|^{2N} \leq C\|u\|^{4N}\varepsilon^N \leq C\|u\|^{2N} \tag{3.27}$$

because $\left\|\sqrt{\varepsilon}u\right\| \leq \sup_{\vartheta,\vartheta'\in\Theta}\left\|\vartheta - \vartheta'\right\|$.

We further use the estimates (3.26)–(3.27) and obtain

$$
\begin{aligned}
&\mathbf{P}_{\vartheta_0}\{Z_\varepsilon(\vartheta_0,u) > e^{-\hat\kappa\|u\|^2}\}\\
&= \mathbf{P}_{\vartheta_0}\left\{\int_0^T \frac{\delta_M(u,t)}{\varepsilon\sigma(t)}\mathrm{d}\bar W_t - \int_0^T \frac{\delta_M(u,t)^2}{2\varepsilon^2\sigma(t)^2}\mathrm{d}t > -\hat\kappa\|u\|^2\right\}\\
&= \mathbf{P}_{\vartheta_0}\left\{\int_0^T \frac{c_1\delta_M(u,t)}{\varepsilon\sigma(t)}\mathrm{d}\bar W_t - \int_0^T \frac{c_1^2\delta_M(u,t)^2}{2\varepsilon^2\sigma(t)^2}\mathrm{d}t > \frac{(c_1-c_1^2)G_u}{2\varepsilon} - \frac{c_1\hat\kappa}{2}\|u\|^2\right\}\\
&+ \mathbf{P}_{\vartheta_0}\left\{\frac{(c_1-c_1^2)G_u}{2\varepsilon} - \frac{c_1-c_1^2}{2}\int_0^T \frac{\delta_M(u,t)^2}{\varepsilon^2\sigma(t)^2}\mathrm{d}t > \frac{(c_1-c_1^2)G_u}{2\varepsilon} - \frac{c_1\hat\kappa}{2}\|u\|^2\right\}\\
&\le \mathbf{P}_{\vartheta_0}\left\{c_1\int_0^T \frac{\delta_M(u,t)}{\varepsilon\sigma(t)}\mathrm{d}\bar W_t - \frac{c_1^2}{2}\int_0^T \frac{\delta_M(u,t)^2}{\varepsilon^2\sigma(t)^2}\mathrm{d}t > \kappa^*\|u\|^2\right\}\\
&+ \mathbf{P}_{\vartheta_0}\left\{\frac{G_u}{\varepsilon} - \int_0^T \frac{\delta_M(u,t)^2}{\varepsilon^2\sigma(t)^2}\mathrm{d}t > \mu\|u\|^2\right\}\\
&\le e^{-\kappa^*\|u\|^2} + \frac{1}{\mu^{2N}\|u\|^{4N}}\mathbf{E}_{\vartheta_0}\left|\frac{G_u}{\varepsilon} - \int_0^T \frac{\delta_M(u,t)^2}{\varepsilon\sigma(t)^2}\mathrm{d}t\right|^{2N} \le \frac{C_N}{\|u\|^{2N}},
\end{aligned}
$$

where $\kappa^* = (c_1 - c_1^2)\hat\kappa/2 - c_1\hat\kappa/2$ and $\mu = \kappa_* - c_1\hat\kappa/(c_1 - c_1^2)$, and the constant $c_1 < 1$ is such that $\kappa^* > 0$ and $\mu > 0$. □

The Eq. (3.14) we write as follows:

$$
\mathrm{d}m(\vartheta,t) = A(\vartheta,t)\,m(\vartheta,t)\,\mathrm{d}t + B(\vartheta,t)\,\mathrm{d}X_t, \qquad m(\vartheta,0) = y_0,
$$

where we denoted

$$
A(\vartheta,t) = a(\vartheta,t) - \frac{f(\vartheta,t)^2\gamma_*(\vartheta,t)}{\varepsilon\sigma(t)^2}, \qquad B(\vartheta,t) = \frac{f(\vartheta,t)\gamma_*(\vartheta,t)}{\varepsilon\sigma(t)^2}.
$$

Recall that by Lemma 3.1

$$
A(\vartheta,t) = -\frac{1}{\varepsilon}\frac{f(\vartheta,t)\,b(\vartheta,t)}{\sigma(t)}(1+O(\varepsilon)), \qquad B(\vartheta,t) = \frac{1}{\varepsilon}\frac{b(\vartheta,t)}{\sigma(t)}(1+O(\varepsilon))
$$

We replace this equation for $m(\vartheta,t)$ by

$$
\mathrm{d}m_*(\vartheta,t) = A_*(\vartheta,t)\,m_*(\vartheta,t)\,\mathrm{d}t + B_*(\vartheta,t)\,\mathrm{d}X_t, \qquad m_*(\vartheta,0) = y_0,
$$

where

$$A_*(\vartheta,t)=-\frac{1}{\varepsilon}\frac{f(\vartheta,t)\,b(\vartheta,t)}{\sigma(t)}=-\frac{q(\vartheta,t)}{\varepsilon}=-\frac{q_t}{\varepsilon},\qquad B_*(\vartheta,t)=\frac{1}{\varepsilon}\frac{b(\vartheta,t)}{\sigma(t)}.$$

Following the proof of Lemma 1.3, we obtain the system of equations for the vector of derivatives

$$\mathrm{d}\dot{\mathrm{m}}_*(\vartheta,t)=A_*(\vartheta,t)\,\dot{\mathrm{m}}_*(\vartheta,t)\,\mathrm{d}t+\dot{\mathrm{A}}_*(\vartheta,t)\,m_*(\vartheta,t)\,\mathrm{d}t+\dot{\mathrm{B}}_*(\vartheta,t)\,\mathrm{d}X_t.$$

Its solution is

$$\dot{\mathrm{m}}_*(\vartheta,t)=\int_0^t\Phi_*(\vartheta,s,t)\,\dot{\mathrm{A}}_*(\vartheta,s)\,m(\vartheta,s)\,\mathrm{d}s+\int_0^t\Phi_*(\vartheta,s,t)\,\dot{\mathrm{B}}_*(\vartheta,s)\,\mathrm{d}X_s.$$

Here

$$\Phi_*(\vartheta,s,t)=\exp\left(-\frac{1}{\varepsilon}\int_s^t\frac{f(\vartheta,v)\,b(\vartheta,v)}{\sigma(v)}\mathrm{d}v\right).$$

If $\vartheta=\vartheta_0$, then

$$\dot{\mathrm{m}}_*(\vartheta_0,t)=-\frac{1}{\varepsilon}\int_0^t\Phi_*(\vartheta_0,s,t)\,\frac{\dot{\mathrm{f}}(\vartheta_0,s)\,b(\vartheta_0,s)}{\sigma(s)}m(\vartheta_0,s)\,\mathrm{d}s$$
$$+\int_0^t\Phi_*(\vartheta_0,s,t)\,\dot{\mathrm{b}}(\vartheta_0,s)\,\mathrm{d}\bar W_s. \tag{3.28}$$

Note that difference between $m(\vartheta,t)$ and $m_*(\vartheta,t)$ is of order $O(\varepsilon)$. The same is true for the difference between $\dot{\mathrm{m}}(\vartheta,t)$ and $\dot{\mathrm{m}}_*(\vartheta,t)$. It follows from the linear structure of the filtration equations.

The asymptotics of $\dot{\mathrm{m}}_*(\vartheta_0,t)$ is described in the next lemma.

Lemma 3.5 *For any $t\in(t_0,T]$ we have the representation*

$$\dot{\mathrm{m}}_*(\vartheta_0,t)+\frac{\dot{\mathrm{f}}(\vartheta_0,t)}{f(\vartheta_0,t)}m_*(\vartheta_0,t)$$
$$=\sqrt{\varepsilon}\sqrt{\frac{b(\vartheta_0,t)\,\sigma(t)}{2f(\vartheta_0,t)}}\left\{\frac{\dot{\mathrm{f}}(\vartheta_0,t)}{f(\vartheta_0,t)}+\frac{\dot{\mathrm{b}}(\vartheta_0,t)}{b(\vartheta_0,t)}\right\}\xi_{t,\varepsilon}(t)+o_p(\sqrt{\varepsilon}),$$

as $\varepsilon\to0$ and, therefore, $\dot{\mathrm{M}}(\vartheta_0,t)\to0$.

Proof The solution (3.28) can be written as follows:

$$\dot{\mathrm{m}}_*(\vartheta_0,t)=\int_0^t\exp\left(-\frac{1}{\varepsilon}\int_s^t q_v\mathrm{d}v\right)\dot{\mathrm{b}}(\vartheta_0,s)\,\mathrm{d}\bar W_s$$

$$- \int_0^t \exp\left(-\frac{1}{\varepsilon}\int_s^t q_v \mathrm{d}v\right) \frac{b(\vartheta_0, s)\,\dot{\mathrm{f}}(\vartheta_0, s)}{\varepsilon\sigma(s)} m_*(\vartheta_0, s)\,\mathrm{d}s + O_p(\varepsilon)$$

$$= -\frac{\dot{\mathrm{f}}(\vartheta_0, t)}{f(\vartheta_0, t)} m_*(\vartheta_0, t) + \sqrt{\varepsilon} R_\varepsilon(t) + O_p(\varepsilon),$$

where we used (1.56) to write

$$\int_0^t \exp\left(-\frac{1}{\varepsilon}\int_s^t q_v \mathrm{d}v\right) \frac{b(\vartheta_0, s)\,\dot{\mathrm{f}}(\vartheta_0, s)}{\varepsilon\sigma(s)} m_*(\vartheta_0, t)\,\mathrm{d}s$$

$$= \frac{b(\vartheta_0, t)\,\dot{\mathrm{f}}(\vartheta_0, t)}{\sigma(t)\,q(\vartheta_0, t)} m_*(\vartheta_0, t) + O_p(\varepsilon) = \frac{\dot{\mathrm{f}}(\vartheta_0, t)}{f(\vartheta_0, t)} m_*(\vartheta_0, t) + O_p(\varepsilon),$$

and denoted

$$R_\varepsilon(t) = \frac{1}{\sqrt{\varepsilon}} \int_0^t \exp\left(-\frac{1}{\varepsilon}\int_s^t q_v \mathrm{d}v\right) \dot{\mathrm{b}}(\vartheta_0, s)\,\mathrm{d}\bar{W}_s$$

$$+ \int_0^t \exp\left(-\frac{1}{\varepsilon}\int_s^t q_v \mathrm{d}v\right) \frac{b(\vartheta_0, s)\,\dot{\mathrm{f}}(\vartheta_0, s)}{\varepsilon^{3/2}\sigma(s)} \{m_*(\vartheta_0, t) - m_*(\vartheta_0, s)\}\,\mathrm{d}s.$$

In the last equation, we use the relation.

Then the limit of the function $R_\varepsilon(t)$ is computed as follows. By means of (1.57), the first integral has the following asymptotic behavior:

$$\frac{1}{\sqrt{\varepsilon}} \int_0^t \exp\left(-\frac{1}{\varepsilon}\int_s^t q_v \mathrm{d}v\right) \dot{\mathrm{b}}(\vartheta_0, s)\,\mathrm{d}\bar{W}_s$$

$$= \frac{\dot{\mathrm{b}}(\vartheta_0, t)}{\sqrt{\varepsilon}} \int_0^t \exp\left(-\frac{t-s}{\varepsilon} q_t\right) \mathrm{d}\bar{W}_s + O_p(\sqrt{\varepsilon})$$

$$= \frac{\dot{\mathrm{b}}(\vartheta_0, t)\sqrt{\sigma(t)}}{\sqrt{2b(\vartheta_0, t)\,f(\vartheta_0, t)}}\, \xi_{t,\varepsilon} + O_p(\sqrt{\varepsilon}).$$

For the second integral, we first note that for the values $s = t - r\varepsilon/q_t$ we have

$$m(\vartheta_0, t) - m(\vartheta_0, s) = \sqrt{\varepsilon/q_t}\, b(\vartheta_0, t)\, w_{t,\varepsilon}(r) + O_p(\varepsilon).$$

Second, we write

$$\int_0^t \exp\left(-\frac{1}{\varepsilon}\int_s^t q_v \mathrm{d}v\right) \frac{b(\vartheta_0, s)\,\dot{\mathrm{f}}(\vartheta_0, s)}{\varepsilon^{3/2}\sigma(s)} [m(\vartheta_0, t) - m(\vartheta_0, s)]\,\mathrm{d}s$$

$$= \frac{b(\vartheta_0, t)\,\dot{\mathrm{f}}(\vartheta_0, t)}{\varepsilon^{3/2}\sigma(t)} \int_0^t \exp\left(-\frac{t-s}{\varepsilon} q_t\right) [m(\vartheta_0, t) - m(\vartheta_0, s)]\,\mathrm{d}s + O_p(\sqrt{\varepsilon})$$

$$
= \frac{b(\vartheta_0,t)^2\,\dot{f}(\vartheta_0,t)}{q_t^{3/2}\sigma(t)}\int_0^{\frac{tq_t}{\varepsilon}} e^{-r}\mathrm{d}w_{t,\varepsilon}(r) + O_p(\sqrt{\varepsilon})
$$

$$
= \sqrt{\frac{b(\vartheta_0,t)\,\sigma(t)}{2f(\vartheta_0,t)}}\;\frac{\dot{f}(\vartheta_0,t)}{f(\vartheta_0,t)}\,\bar{\xi}_{t,\varepsilon}(t) + O_p(\sqrt{\varepsilon}).
$$

Therefore, we obtain the expression

$$
R_\varepsilon(t) = \frac{\dot{b}(\vartheta_0,t)\sqrt{\sigma(t)}}{\sqrt{2b(\vartheta_0,t)\,f(\vartheta_0,t)}}\,\bar{\xi}_{t,\varepsilon} + \sqrt{\frac{b(\vartheta_0,t)\,\sigma(t)}{2f(\vartheta_0,t)}}\;\frac{\dot{f}(\vartheta_0,t)}{f(\vartheta_0,t)}\,\bar{\xi}_{t,\varepsilon}(t) + O(\sqrt{\varepsilon})
$$

$$
= \sqrt{\frac{b(\vartheta_0,t)\,\sigma(t)}{2f(\vartheta_0,t)}}\left\{\frac{\dot{f}(\vartheta_0,t)}{f(\vartheta_0,t)} + \frac{\dot{b}(\vartheta_0,t)}{b(\vartheta_0,t)}\right\}\bar{\xi}_{t,\varepsilon}(t) + O(\sqrt{\varepsilon}).
$$

□

Corollary 3.1 *The empirical Fisher information satisfies the relations*

$$
\mathbf{I}_\varepsilon(\vartheta_0) = \int_0^T \frac{\dot{M}(\vartheta_0,t)\,\dot{M}(\vartheta_0,t)^\top}{\varepsilon\sigma(t)^2}\,\mathrm{d}t
$$

$$
= \int_0^T \frac{f(\vartheta_0,t)\,b(\vartheta_0,t)}{2\sigma(t)}\left\{\frac{\dot{f}(\vartheta_0,t)}{f(\vartheta_0,t)} + \frac{\dot{b}(\vartheta_0,t)}{b(\vartheta_0,t)}\right\}\left\{\frac{\dot{f}(\vartheta_0,t)}{f(\vartheta_0,t)} + \frac{\dot{b}(\vartheta_0,t)}{b(\vartheta_0,t)}\right\}^\top \bar{\xi}_{t,\varepsilon}^2\,\mathrm{d}t
$$

$$
+ o(1)
$$

$$
\longrightarrow \int_0^T \frac{f(\vartheta_0,t)\,b(\vartheta_0,t)}{2\sigma(t)}\left\{\frac{\dot{f}(\vartheta_0,t)}{f(\vartheta_0,t)} + \frac{\dot{b}(\vartheta_0,t)}{b(\vartheta_0,t)}\right\}\left\{\frac{\dot{f}(\vartheta_0,t)}{f(\vartheta_0,t)} + \frac{\dot{b}(\vartheta_0,t)}{b(\vartheta_0,t)}\right\}^\top \mathrm{d}t
$$

$$
= \int_0^T \frac{S(\vartheta_0,t)}{2\sigma(t)}\left\{\frac{\partial \ln S(\vartheta_0,t)}{\partial\vartheta_0}\right\}\left\{\frac{\partial \ln S(\vartheta_0,t)}{\partial\vartheta_0}\right\}^\top \mathrm{d}t = \mathbf{I}(\vartheta_0).
$$

Let us denote

$$
\Delta_\varepsilon\left(\vartheta_0, X^T\right) = \frac{1}{\sqrt{\varepsilon}}\int_0^T \frac{\left[\dot{f}(\vartheta_0,t)\,m(\vartheta_0,t) + f(\vartheta_0,t)\,\dot{m}(\vartheta_0,t)\right]}{\sigma(t)}\,\mathrm{d}\bar{W}_t,
$$

$$
Z_\varepsilon(\vartheta_0,u) = \frac{L\left(\vartheta_0+\sqrt{\varepsilon}\,u, X^T\right)}{L\left(\vartheta_0, X^T\right)}, \qquad u \in \mathbb{U}_\varepsilon = \left(u\ :\ \vartheta_0+\sqrt{\varepsilon}\,u \in \Theta\right).
$$

The following lemma shows that this statistical model is regular.

Lemma 3.6 *The family of measures* $\{\mathbf{P}_\vartheta^\varepsilon, \vartheta \in \Theta\}$ *is LAN, i.e., for any* $\vartheta_0 \in \Theta$, *we have the representation*

$$Z_\varepsilon\left(\vartheta_0,u\right)=\exp\left(u^\top\Delta_\varepsilon(\vartheta_0,X^T)-\frac{1}{2}u^\top\mathbf{I}\left(\vartheta_0\right)u+\rho_\varepsilon\right),$$

where $\rho_\varepsilon\to 0$ *and* $\Delta_\varepsilon(\vartheta_0,X^T)\Longrightarrow\mathcal{N}\left(0,\mathbf{I}_0\left(\vartheta_0\right)\right)$.

Proof We have ($\vartheta_u=\vartheta_0+\sqrt{\varepsilon}\,u$)

$$\ln Z_\varepsilon\left(\vartheta_0,u\right)=\int_0^T\frac{M(\vartheta_u,t)-M\left(\vartheta_0,t\right)}{\varepsilon\sigma\left(t\right)}\,\mathrm{d}\bar W_t-\int_0^T\frac{\left[M(\vartheta_u,t)-M\left(\vartheta_0,t\right)\right]^2}{2\varepsilon^2\sigma\left(t\right)^2}\,\mathrm{d}t,$$

where

$$M(\vartheta_0+\sqrt{\varepsilon}u,t)-M\left(\vartheta_0,t\right)=\sqrt{\varepsilon}\,u^\top\dot{\mathrm{M}}\left(\vartheta_0,t\right)\left(1+o_p(1)\right).$$

Moreover,

$$\begin{aligned}\int_0^T\frac{\left[M(\vartheta_0+\sqrt{\varepsilon}u,t)-M(\vartheta_0,t)\right]^2}{\varepsilon^2\sigma\left(t\right)^2}\,\mathrm{d}t&=\frac{1}{\varepsilon}\int_0^T\frac{\left(u^\top\dot{\mathrm{M}}\left(\vartheta_0,t\right)\right)^2}{\sigma\left(t\right)^2}\,\mathrm{d}t\,\left(1+o_p\left(1\right)\right)\\&=u^\top\mathbf{I}_\varepsilon\left(\vartheta_0\right)u\,\left(1+o_p\left(1\right)\right)\longrightarrow u^\top\mathrm{I}\left(\vartheta_0\right)u,\end{aligned}$$

which leads directly to (3.20). The asymptotic normality of $\Delta_\varepsilon(\vartheta,X^T)$ follows from the CLT for stochastic integrals 1.7. □

Moreover, it can be shown that these two convergences are uniform on compacts $\mathbb{K}\subset\Theta$.

Due to the LAN property, we obtain the lower bound (3.17).

Lemma 3.7 *For any* $m>d/2$, *there exists a constant* $C>0$ *such that for any* $R>0$ *and* $u_1,u_2\in\mathbb{U}_\varepsilon$ *satisfying* $\|u_1\|\leq R$, $\|u_2\|\leq R$, *we have the estimate*

$$\sup_{\vartheta_0\in\mathbb{K}}\mathbf{E}_{\vartheta_0}\left|Z_\varepsilon\left(\vartheta_0,u_2\right)^{\frac{1}{2m}}-Z_\varepsilon\left(\vartheta_0,u_1\right)^{\frac{1}{2m}}\right|^{2m}\leq C\left(1+R^{6m}\right)\|u_2-u_1\|^{2m}.$$

Proof Let us denote $\vartheta_{u_i}=\vartheta_0+\sqrt{\varepsilon}u_i$ and set $\delta_M\left(u_1,u_2,t\right)=M\left(\vartheta_{u_2},t\right)-M\left(\vartheta_{u_1},t\right)$. By Lemma 1.1

$$\begin{aligned}\mathbf{E}_{\vartheta_0}\left|Z_\varepsilon\left(u_2,\vartheta_0\right)^{\frac{1}{2m}}-Z_\varepsilon\left(u_1,\vartheta_0\right)^{\frac{1}{2m}}\right|^{2m}&\leq C_1T^{2m-1}\mathbf{E}_{\vartheta_{u_1}}\int_0^T\frac{\delta_M\left(u_1,u_2,t\right)^{4m}}{\varepsilon^{4m}\sigma\left(t\right)^{4m}}\mathrm{d}t\\&\quad+C_2T^{m-\frac{1}{2}}\mathbf{E}_{\vartheta_{u_1}}\int_0^T\frac{\delta_M\left(u_1,u_2,t\right)^{2m}}{\varepsilon^{2m}\sigma\left(t\right)^{2m}}\mathrm{d}t.\end{aligned}$$

Recall that the random processes $M\left(\vartheta,t\right),\dot{\mathrm{M}}\left(\vartheta,t\right),0\leq t\leq T$ have the following stochastic differentials (2.43) and (2.46)

$$\begin{aligned} \mathrm{d}M(\vartheta,t) &= \bar{A}(\vartheta,t)\, M(\vartheta,t)\,\mathrm{d}t + \bar{B}(\vartheta,t)\,\mathrm{d}X_t, \\ M(\vartheta,0) &= y_0 f(\vartheta,0), \quad 0\le t\le T, \\ \mathrm{d}\dot{\mathrm{M}}(\vartheta,t) &= \bar{A}(\vartheta,t)\,\dot{\mathrm{M}}(\vartheta,t)\,\mathrm{d}t + \dot{\bar{\mathrm{A}}}(\vartheta,t)\, M(\vartheta,t)\,\mathrm{d}t + \dot{\bar{\mathrm{B}}}(\vartheta,t)\,\mathrm{d}X_t, \end{aligned}$$

where $\dot{\mathrm{M}}(\vartheta,0) = y_0\dot{\mathrm{f}}(\vartheta,0)$, and

$$\bar{A}(\vartheta,t) = \frac{f'(\vartheta,t)}{f(\vartheta,t)} + a(\vartheta,t) - \frac{f(\vartheta,t)^2\gamma(\vartheta,t)}{\varepsilon^2\sigma(t)^2}, \qquad \bar{B}(\vartheta,t) = \frac{f(\vartheta,t)^2\gamma(\vartheta,t)}{\varepsilon^2\sigma(t)^2}.$$

By Lemma 3.1, we have the following asymptotics

$$\begin{aligned} \bar{A}(\vartheta,t) &= -\frac{1}{\varepsilon}\,\frac{f(\vartheta,t)\,b(\vartheta,t)}{\sigma(t)}\,(1+O(\varepsilon)), \\ \bar{B}(\vartheta,t) &= \frac{1}{\varepsilon}\,\frac{f(\vartheta,t)\,b(\vartheta,t)}{\sigma(t)}\,(1+O(\varepsilon)). \end{aligned}$$

Let us put

$$\begin{aligned} \bar{A}_*(\vartheta,t) &= -\frac{1}{\varepsilon}\,\frac{f(\vartheta,t)\,b(\vartheta,t)}{\sigma(t)}, \qquad \bar{B}_*(\vartheta,t) = \frac{1}{\varepsilon}\,\frac{f(\vartheta,t)\,b(\vartheta,t)}{\sigma(t)}, \\ q_t = q(\vartheta,t) &= \frac{f(\vartheta,t)\,b(\vartheta,t)}{\sigma(t)}. \end{aligned}$$

Introduce the random processes

$$\begin{aligned} \mathrm{d}M_*(\vartheta,t) &= -\frac{q(\vartheta,t)}{\varepsilon}\,M_*(\vartheta,t)\,\mathrm{d}t + \frac{q(\vartheta,t)}{\varepsilon}\,\mathrm{d}X_t, \qquad M_*(\vartheta,0) = y_0 f(\vartheta,0), \\ \mathrm{d}\dot{\mathrm{M}}_*(\vartheta,t) &= -\frac{q(\vartheta,t)}{\varepsilon}\dot{\mathrm{M}}_*(\vartheta,t)\,\mathrm{d}t - \frac{\dot{\mathrm{q}}(\vartheta,t)}{\varepsilon}M_*(\vartheta,t)\,\mathrm{d}t + \frac{\dot{\mathrm{q}}(\vartheta,t)}{\varepsilon}\mathrm{d}X_t \end{aligned}$$

with $\dot{\mathrm{M}}_*(\vartheta,0) = y_0\dot{\mathrm{f}}(\vartheta,0)$. Remark that as the both equations are linear, we have the relations $M(\vartheta,t) - M_*(\vartheta,t) = O(\varepsilon)$ and $\dot{\mathrm{M}}(\vartheta,t) - \dot{\mathrm{M}}_*(\vartheta,t) = O(\varepsilon)$.

By Lemma 1.7 with $\mathbf{P}_{\vartheta_{u_1}}$ probability 1

$$\begin{aligned} \dot{\mathrm{M}}_*(\vartheta_u,t) &= -\frac{\dot{\mathrm{q}}(\vartheta_u,t)}{q(\vartheta_u,t)}\left[M_*(\vartheta_u,t) - M\left(\vartheta_{u_1},t\right)\right] + \sqrt{\varepsilon}\frac{\dot{\mathrm{q}}(\vartheta_u,t)}{\sqrt{2q(\vartheta_u,t)}}\,\bar{\xi}_{t,\varepsilon} + O\left(\sqrt{\varepsilon}\right) \\ &= \|u-u_1\|\sqrt{\varepsilon}O(1) + O\left(\sqrt{\varepsilon}\right) \end{aligned}$$

Denote $\vartheta(v) = \vartheta_{u_1} + \sqrt{\varepsilon}v\,(u_2-u_1)$, then we can write

$$
\begin{aligned}
\delta_M(u_1,u_2,t) &= \sqrt{\varepsilon}\,(u_2-u_1)^\top \int_0^1 \dot{\mathrm{M}}(\vartheta(v),t)\,\mathrm{d}v\\
&= \sqrt{\varepsilon}\,(u_2-u_1)^\top \dot{\mathrm{M}}_*\left(\vartheta_{u_1},t\right) + O(\varepsilon)\\
&\quad + \sqrt{\varepsilon}\,(u_2-u_1)^\top \int_0^1 \left(\dot{\mathrm{M}}_*(\vartheta(v),t) - \dot{\mathrm{M}}_*\left(\vartheta_{u_1},t\right)\right)\mathrm{d}v.
\end{aligned}
$$

Hence,

$$
\mathbf{E}_{\vartheta_{u_1}} \int_0^T \frac{\delta_M(u_1,u_2,t)^{4m}}{\varepsilon^{4m}\sigma(t)^{4m}}\mathrm{d}t \le C\,\|u_2-u_1\|^{4m} + C\,\|u_2-u_1\|^{8m}
$$

and

$$
\begin{aligned}
\mathbf{E}_{\vartheta_0}\left|Z_\varepsilon(u_2,\vartheta_0)^{\frac{1}{2m}} - Z_\varepsilon(u_1,\vartheta_0)^{\frac{1}{2m}}\right|^{2m} &\le C\,\|u_2-u_1\|^{2m} + C\,\|u_2-u_1\|^{8m}\\
&\le C\left(1+R^{6m}\right)\|u_2-u_1\|^{2m}.
\end{aligned}
$$

The properties of the normalized likelihood ratio established in Lemmas 3.4–3.6 correspond well to the conditions $\mathscr{I}\mathscr{K}_d$ of Theorems 1.10 and 1.12, and this means that the asymptotic properties of the MLE and BE (Theorem 3.1) are obtained. Note that all convergences are uniform on compacts $\mathbb{K} \subset \Theta$. □

Remark 3.4 Note that if the initial value Y_0 is a Gaussian random variable, say, $\mathcal{N}(y_0, p_0)$ and $m(\vartheta_0, t)$ is solution of the Eq. (3.22) with an arbitrary deterministic initial value m_0, then there exist the constants $C > 0$ and $\kappa > 0$ such that

$$
\mathbf{E}_{\vartheta_0}\left|Y_t - m(\vartheta_0,t)\right|^2 \le C\left(\varepsilon + e^{-\kappa t/\varepsilon}\left|y_0 - m_0\right|^2\right).
$$

For the proof, see Lemma 5.1.2 in [10]. The convergence (3.21) as well discussed in Section 5.1.2 there. See more about comparison of linear models considered in Sections 5.1 and 5.2 in [10] and in Sects. 2.1, 2.5, and 3.1 of this work in Remark 2.24.

One-Step MLE Process

We have the same partially observed linear system

$$
\begin{aligned}
\mathrm{d}X_t &= f(\vartheta,t)\,Y_t\mathrm{d}t + \varepsilon\sigma(t)\,\mathrm{d}W_t, &\qquad X_0 &= 0, \qquad 0 \le t \le T,\\
\mathrm{d}Y_t &= a(\vartheta,t)\,Y_t\mathrm{d}t + b(\vartheta,t)\,\mathrm{d}V_t, & Y_0 &= 0.
\end{aligned}
\tag{3.29}
$$

As before, the process $X^T = (X_t, 0 \le t \le T)$ is observable and Y^T is hidden. All functions $a(\cdot), b(\cdot), f(\cdot)$ and $\sigma(\cdot)$ are supposed to be known. The parameter $\vartheta \in \Theta \subset \mathcal{R}^d$ is unknown and has to be estimated by observations X^T.

Remind that the MLE $\hat\vartheta_\varepsilon$ and BE $\tilde\vartheta_\varepsilon$ are defined by the relations (3.16) and their calculation requires solutions $m(\vartheta, \cdot), \vartheta \in \Theta, \gamma(\vartheta, \cdot), \vartheta \in \Theta$ of the Eqs. (3.14) and (3.15). Let us remind these equations here for the convinience of references.

$$\mathrm{d}m(\vartheta,t) = a(\vartheta,t)\, m(\vartheta,t)\,\mathrm{d}t + \frac{\gamma(\vartheta,t)\, f(\vartheta,t)}{\varepsilon^2\sigma(t)^2}\left[\mathrm{d}X_t - f(\vartheta,t)\, m(\vartheta,t)\,\mathrm{d}t\right] \tag{3.30}$$

with the initial value $m(\vartheta, 0) = y_0$. The function $\gamma(\vartheta,t) = \mathbf{E}_\vartheta\left(Y_t - m(\vartheta,t)\right)^2$ is solution of Riccati equation

$$\frac{\partial\gamma(\vartheta,t)}{\partial t} = 2a(\vartheta,t)\,\gamma(\vartheta,t) - \frac{\gamma(\vartheta,t)^2 f(\vartheta,t)^2}{\varepsilon^2\sigma(t)^2} + b(\vartheta,t)^2, \tag{3.31}$$

with the initial value $\gamma(\vartheta, 0) = 0$.

As in Sect. 2.2, our goal is to construct an approximation $m_t^\star$ of the process $m(\vartheta_0, t)$ (adaptive filter) by replacing ϑ in the Eqs. (3.30) and (3.31) by some estimator-process $\bar\vartheta_{t,\varepsilon}, 0 < t \le T$. Therefore, we need a sequence of estimators $\bar\vartheta_{t,\varepsilon}, 0 < t \le T$ which are measurable w.r.t. observations up to the time t for each $t \in (0, T]$. To avoid the evident numerical difficulties related with the calculations of MLE $\hat\vartheta_{t,\varepsilon}, 0 < t \le T$ we consider the same approach as in the Sect. 2.2 based on the Fisher-score device and MME $\vartheta^*_{\tau,\varepsilon}$ as preliminary estimator. Therefore, in some sense we repeat the procedure realized in the Sect. 2.5.4, but, of course, there are some technical differences due to the difference between these two models. The proposed One-step MLE process allows us to obtain an estimator process that is easy to calculate, has good asymptotic properties, and can be used for the construction of adaptive filter.

Let us fix a learning interval $[0, \tau]$ and remind the definition of MME $\vartheta^*_{\tau,\varepsilon}$. Introduce two functions

$$\Psi_t(\vartheta) = \int_0^t f(\vartheta,s)^2\, b(\vartheta,s)^2\,\mathrm{d}s, \qquad \vartheta\in\Theta, \qquad 0\le t\le\tau,$$

$$\Psi_{t,\varepsilon} = \sum_{i=0}^{N_{t,\varepsilon}-1}\left(\frac{X_{t_{i+1}+\varepsilon} - X_{t_{i+1}}}{\varepsilon} - \frac{X_{t_i+\varepsilon} - X_{t_i}}{\varepsilon}\right)^2, \qquad 0 < t \le \tau.$$

Here $t_{i+1} - t_i = \varepsilon^{1/3}, t_0 = 0$ and $N_{t,\varepsilon} = \left[t\varepsilon^{-1/3}\right]$. Take d points $0 < \tau_1 < \tau_2 < \ldots < \tau_d = \tau$ and denote the vectors $\mathbf{\Psi}^*_{\tau,\varepsilon} = \left(\Psi_{\tau_1,\varepsilon}, \Psi_{\tau_2,\varepsilon}, \ldots, \Psi_{\tau_d,\varepsilon}\right)^\top$ and $\mathbf{\Psi}^*_\tau(\vartheta) = \left(\Psi_{\tau_1}(\vartheta), \Psi_{\tau_2}(\vartheta), \ldots, \Psi_{\tau_d}(\vartheta)\right)^\top$. The MME $\vartheta^*_{\tau,\varepsilon}$ is defined as solution of the system of equations

$$\Psi^*_{\tau,\varepsilon} = \Psi^*_\tau(\vartheta^*_{\tau,\varepsilon}).$$

We suppose below that the conditions $\mathscr{G}^*$ hold and therefore, by Proposition 3.2, the MME is consistent and the polynomial moments are finite (3.13).

The Fisher-score function for this model is

$$\frac{\partial}{\partial\vartheta}\ln L\left(\vartheta, X^T\right) = \int_0^T \frac{\dot{M}(\vartheta,t)}{\varepsilon^2\sigma(t)^2}\left[\mathrm{d}X_t - M(\vartheta,t)\,\mathrm{d}t\right]$$

Consider the construction of One-step MLE process in the case of observations (3.29). Suppose that for some fixed small value $\tau > 0$ we already have MME $\vartheta^*_{\tau,\varepsilon}$ constructed by the observations $X^\tau = (X_t, 0 \le t \le \tau)$. The solution of the Eq. (3.30) on the time interval $[0,\tau]$ can be written as follows:

$$\begin{aligned} m(\vartheta,t) &= y_0\Phi(\vartheta,0,t) + \Phi(\vartheta,0,t)\int_0^t \Phi(\vartheta,0,s)^{-1}B(\vartheta,s)\,\mathrm{d}X_s \\ &= y_0\Phi(\vartheta,0,t) + \Phi(\vartheta,0,t)\int_0^t H(\vartheta,s)\,\mathrm{d}X_s, \end{aligned}$$

where

$$\Phi(\vartheta,0,t) = \exp\left(\int_0^t A(\vartheta,v)\,\mathrm{d}v\right), \qquad A(\vartheta,t) = a(\vartheta,t) - \frac{\gamma(\vartheta,t)f(\vartheta,t)^2}{\varepsilon^2\sigma(t)^2},$$

$$B(\vartheta,s) = \frac{\gamma(\vartheta,s)f(\vartheta,s)}{\varepsilon^2\sigma(s)^2}, \qquad H(\vartheta,s) = \Phi(\vartheta,0,s)^{-1}\frac{\gamma(\vartheta,s)f(\vartheta,s)}{\varepsilon^2\sigma(s)^2}.$$

We cannot put $\vartheta^*_{\tau,\varepsilon}$ in $m(\vartheta,t)$ because the corresponding stochastic integral Itô is not defined. Note that the function $H(\vartheta,s)$ is continuously differentiable w.r.t. s, hence the following relation

$$\int_0^\tau H(\vartheta,s)\,\mathrm{d}X_s = X_\tau H(\vartheta,\tau) - \int_0^\tau X_s H'(\vartheta,s)\,\mathrm{d}s$$

holds (see details in Lemma 1.3). Now we can put

$$\begin{aligned} m\left(\vartheta^*_{\tau,\varepsilon},\tau\right) &= y_0\Phi\left(\vartheta^*_{\tau,\varepsilon},0,\tau\right) + \Phi\left(\vartheta^*_{\tau,\varepsilon},0,\tau\right)X_\tau H\left(\vartheta^*_{\tau,\varepsilon},\tau\right) \\ &\quad - \Phi\left(\vartheta^*_{\tau,\varepsilon},0,\tau\right)\int_0^\tau X_s H'\left(\vartheta^*_{\tau,\varepsilon},s\right)\mathrm{d}s. \end{aligned} \tag{3.32}$$

Let us define the random process $\hat{m}(t)$, $\tau \le t \le T$ by the equation

$$\mathrm{d}\hat{m}(t) = a\left(\vartheta^*_{\tau,\varepsilon},t\right)\hat{m}(t)\,\mathrm{d}t + \frac{\gamma\left(\vartheta^*_{\tau,\varepsilon},t\right)f\left(\vartheta^*_{\tau,\varepsilon},t\right)}{\varepsilon^2\sigma(t)^2}\left[\mathrm{d}X_t - f\left(\vartheta^*_{\tau,\varepsilon},t\right)\hat{m}(t)\,\mathrm{d}t\right]$$

with the initial value $\hat m(\tau)=m\left(\vartheta^*_{\tau,\varepsilon},\tau\right)$. The solution of this equation is

$$\hat m(t)=m\left(\vartheta^*_{\tau,\varepsilon},\tau\right)\Phi\left(\vartheta^*_{\tau,\varepsilon},\tau,t\right)+\int_\tau^t\Phi\left(\vartheta^*_{\tau,\varepsilon},s,t\right)\frac{\gamma\left(\vartheta^*_{\tau,\varepsilon},s\right)f\left(\vartheta^*_{\tau,\varepsilon},s\right)}{\varepsilon^2\sigma(s)^2}{\rm d}X_s.$$

We need as well the equation for the derivative $\dot{\rm m}(\vartheta,t)$ on the interval $[\tau,T]$ at the point $\vartheta=\vartheta^*_{\tau,\varepsilon}$ (denoted $\dot{\rm m}\left(\vartheta^*_{\tau,\varepsilon},t\right)=\hat{\dot{\rm m}}(t)$), which we obtain by formal differentiation of the Eq. (3.30)

$${\rm d}\hat{\dot{\rm m}}(t)=A\left(\vartheta^*_{\tau,\varepsilon},t\right)\hat{\dot{\rm m}}(t)\,{\rm d}t+\dot{\rm A}\left(\vartheta^*_{\tau,\varepsilon},t\right)\hat m(t)\,{\rm d}t+\dot{\rm B}\left(\vartheta^*_{\tau,\varepsilon},t\right){\rm d}X_t,\qquad \tau\leq t\leq T$$

with the initial value $\hat{\dot{\rm m}}(\tau)=\dot{\rm m}\left(\vartheta^*_{\tau,\varepsilon},\tau\right)$, which can be calculated similar to the given above representation of the corresponding stochastic integral like (3.32).

Introduce notations:

$$M\left(\vartheta^*_{\tau,\varepsilon},t\right)=f\left(\vartheta^*_{\tau,\varepsilon},t\right)\hat m(t),\quad \dot{\rm M}\left(\vartheta^*_{\tau,\varepsilon},t\right)=\dot{\rm f}\left(\vartheta^*_{\tau,\varepsilon},t\right)\hat m(t)+f\left(\vartheta^*_{\tau,\varepsilon},t\right)\hat{\dot{\rm m}}(t),$$

$$\mathbf{I}^t_\tau(\vartheta)=\int_\tau^t\frac{\dot{\rm S}(\vartheta,s)\,\dot{\rm S}(\vartheta,s)^\top}{2S(\vartheta,s)\,\sigma(s)}{\rm d}s,\qquad \eta_{t,\varepsilon}=\frac{\vartheta^\star_{t,\varepsilon}-\vartheta_0}{\sqrt{\varepsilon}},$$

$$\eta_t(\vartheta_0)=\mathbf{I}^t_\tau(\vartheta_0)^{-1}\int_\tau^t\frac{\dot{\rm S}(\vartheta_0,s)}{\sqrt{2S(\vartheta_0,s)\,\sigma(s)}}{\rm d}w(s),\qquad \tau<t\leq T,$$

where $w(s),\tau\leq s\leq T$ is some Wiener process.

Introduce the One-step MLE process

$$\vartheta^\star_{t,\varepsilon}=\vartheta^*_{\tau,\varepsilon}+\mathbf{I}^t_\tau(\vartheta^*_{\tau,\varepsilon})^{-1}\int_\tau^t\frac{\dot{\rm M}(\vartheta^*_{\tau,\varepsilon},s)}{\varepsilon\sigma(s)^2}\left[{\rm d}X_s-M(\vartheta^*_{\tau,\varepsilon},s){\rm d}s\right],\qquad \tau<t\leq T,$$

and *Conditions* $\mathscr{G}^\star$:

$\mathscr{G}^\star_1$. *The functions* $f(\vartheta,t),b(\vartheta,t),\sigma(t);(\vartheta,t)\in\bar\Theta\times[0,T]$ *are separated from zero.*

$\mathscr{G}^\star_2$. *The conditions* $\mathscr{G}^*$ *of Proposition 3.2 hold.*

$\mathscr{G}^\star_3$. *For any* $t_0\in(\tau,T]$

$$\inf_{\vartheta_0\in\Theta}\ \inf_{\|{\rm e}\|=1,{\rm e}\in\mathcal{R}^d}{\rm e}^\top\mathbf{I}^{t_0}_\tau(\vartheta_0)\,{\rm e}>0.$$

Theorem 3.2 *Let the assumptions* $\mathscr{G}^\star$ *hold. Then the One-step MLE-process* $\vartheta^\star_{t,\varepsilon}$, $\tau<t\leq T$ *has the following properties:*

1. *It is uniformly consistent: for any* $t_0\in(\tau,T]$ *and any* $\nu>0$

$$\mathbf{P}_{\vartheta_0}\left(\sup_{t_0\le t\le T}\left\|\vartheta^\star_{t,\varepsilon}-\vartheta_0\right\|\ge\nu\right)\longrightarrow 0,$$

2. *The random process* $\eta_{t,\varepsilon}=\varepsilon^{-1/2}\left(\vartheta^\star_{t,\varepsilon}-\vartheta_0\right), t_0\le t\le T$ *converges in distribution in the measurable space* $(\mathcal{C}[t_0,T],\mathfrak{B}])$ *to the Gaussian process* $\eta_t\left(\vartheta_0\right), t_0\le t\le T$

$$\eta_{\cdot,\varepsilon}\Longrightarrow\eta_\cdot\left(\vartheta_0\right),\qquad\qquad \eta_t\left(\vartheta_0\right)\sim\mathcal{N}\left(0,\mathbf{I}^t_\tau\left(\vartheta_0\right)^{-1}\right),$$

3. *All polynomial moments converge: for any* $p>0$

$$\varepsilon^{-p/2}\mathbf{E}_{\vartheta_0}\|\vartheta^\star_{t,\varepsilon}-\vartheta_0\|^p\longrightarrow\mathbf{E}_{\vartheta_0}\|\eta_t\left(\vartheta_0\right)\|^p.$$

Proof The uniform consistency follows from the representation like (2.49)

$$\begin{aligned}\sup_{t_0\le t\le T}\left\|\vartheta^\star_{t,\varepsilon}-\vartheta_0\right\|&\le\left\|\vartheta^*_{\tau,\varepsilon}-\vartheta_0\right\|+C\varepsilon\sup_{t_0\le t\le T}\left\|\int_{\tau_\varepsilon}^t\frac{\dot{\mathbf{M}}\left(\vartheta^*_{\tau,\varepsilon},s\right)}{\sigma\left(s\right)}\mathrm{d}W_s\right\|\\&\quad+C\int_{\tau_\varepsilon}^T\left\|\dot{\mathbf{M}}\left(\vartheta^*_{\tau,\varepsilon},s\right)\right\|\left|M\left(\vartheta_0,s\right)-M\left(\vartheta^*_{\tau,\varepsilon},s\right)\right|\mathrm{d}s\end{aligned}$$

as follows:

$$\begin{aligned}\mathbf{P}_{\vartheta_0}\left(\sup_{t_0\le t\le T}\left\|\vartheta^\star_{t,\varepsilon}-\vartheta_0\right\|\ge\nu\right)&\le\mathbf{P}_{\vartheta_0}\left(\left\|\vartheta^*_{\tau,\varepsilon}-\vartheta_0\right\|\ge\frac{\nu}{3}\right)\\&\quad+\mathbf{P}_{\vartheta_0}\left(\sup_{t_0\le t\le T}\left\|\int_\tau^t\frac{\dot{\mathbf{M}}(\vartheta^*_{\tau,\varepsilon},s)}{\sigma\left(s\right)}\mathrm{d}\bar W_s\right\|\ge\frac{\nu}{C\varepsilon}\right)\\&\quad+\mathbf{P}_{\vartheta_0}\left(\left\|\vartheta^*_{\tau,\varepsilon}-\vartheta_0\right\|\int_\tau^T\left\|\frac{\dot{\mathbf{M}}(\vartheta^*_{\tau,\varepsilon},s)\dot{\mathbf{M}}(\tilde\vartheta_{\tau,\varepsilon},s)^\top}{\varepsilon\sigma\left(s\right)^2}\right\|\mathrm{d}s\ge\frac{\nu}{C}\right).\end{aligned}$$

Here we used the condition $\mathscr{G}^\star_3$. The first probability of the RHS tends to zero thanks to the consistency of $\vartheta^*_{\tau,\varepsilon}$. The second and third probabilities tend to zero because the corresponding integrals have finite moments. Therefore, the estimator $\vartheta^\star_{t,\varepsilon}$ is uniformly on $t\in[t_0,T]$ consistent.

To prove the convergence in distribution, we have to study the following expression:

$$\frac{\vartheta^\star_{t,\varepsilon}-\vartheta_0}{\sqrt{\varepsilon}}=\frac{\left(\vartheta^*_{\tau,\varepsilon}-\vartheta_0\right)}{\sqrt{\varepsilon}}+\mathbf{I}^t_\tau(\vartheta^*_{\tau,\varepsilon})^{-1}\int_\tau^t\frac{\dot{\mathbf{M}}(\vartheta^*_{\tau,\varepsilon},s)}{\sqrt{\varepsilon}\sigma\left(s\right)}\mathrm{d}\bar W_s$$

$$+\mathbf{I}_\tau^t\left(\vartheta_{\tau,\varepsilon}^*\right)^{-1}\int_\tau^t \frac{\dot{\mathbf{M}}(\vartheta_{\tau,\varepsilon}^*,s)}{\varepsilon^{3/2}\sigma\left(s\right)^2}\left[f\left(\vartheta_0,s\right)m\left(\vartheta_0,s\right)-M(\vartheta_{\tau,\varepsilon}^*,s)\right]\mathrm{d}s.$$

We have the relations

$$\mathbf{I}_\tau^t\left(\vartheta_{\tau,\varepsilon}^*\right)^{-1}=\mathbf{I}_\tau^t\left(\vartheta_0\right)^{-1}+O\left(\left\|\vartheta_{\tau,\varepsilon}^*-\vartheta_0\right\|\right),$$

$$f\left(\vartheta_0,s\right)m\left(\vartheta_0,s\right)-M(\vartheta_{\tau,\varepsilon}^*,s)=\left(\vartheta_0-\vartheta_{\tau,\varepsilon}^*\right)^\top\dot{\mathbf{M}}\left(\vartheta_0,t\right)+O\left(\left\|\vartheta_{\tau,\varepsilon}^*-\vartheta_0\right\|\right),$$

$$\int_\tau^t \frac{\dot{\mathbf{M}}(\vartheta_{\tau,\varepsilon}^*,s)}{\sqrt{\varepsilon}\sigma\left(s\right)}\mathrm{d}\bar{W}_s=\int_\tau^t \frac{\dot{\mathbf{M}}(\vartheta_0,s)}{\sqrt{\varepsilon}\sigma\left(s\right)}\mathrm{d}\bar{W}_s+O\left(\left\|\vartheta_{\tau,\varepsilon}^*-\vartheta_0\right\|\right).$$

Therefore,

$$\begin{aligned}&\mathbf{I}_\tau^t\left(\vartheta_{\tau,\varepsilon}^*\right)^{-1}\int_\tau^t \frac{\dot{\mathbf{M}}(\vartheta_{\tau,\varepsilon}^*,s)}{\varepsilon^{3/2}\sigma\left(s\right)^2}\left[f\left(\vartheta_0,s\right)m\left(\vartheta_0,s\right)-M(\vartheta_{\tau,\varepsilon}^*,s)\right]\mathrm{d}s\\&\qquad=\mathbf{I}_\tau^t\left(\vartheta_0\right)^{-1}\int_\tau^t \frac{\dot{\mathbf{M}}(\vartheta_0,s)\dot{\mathbf{M}}(\vartheta_0,s)^\top}{\varepsilon^{3/2}\sigma\left(s\right)^2}\mathrm{d}s\left(\vartheta_0-\vartheta_{\tau,\varepsilon}^*\right)+O\left(\left\|\vartheta_{\tau,\varepsilon}^*-\vartheta_0\right\|\right).\end{aligned}$$

The random process $M\left(\vartheta,t\right)$ is solution of the stochastic differential equation

$$\begin{aligned}\mathrm{d}M\left(\vartheta,t\right)&=f'\left(\vartheta,t\right)m\left(\vartheta,t\right)\mathrm{d}t+a\left(\vartheta,t\right)M\left(\vartheta,t\right)\mathrm{d}t-\frac{\gamma\left(\vartheta,t\right)f\left(\vartheta,t\right)^2}{\varepsilon^2\sigma\left(t\right)^2}M\left(\vartheta,t\right)\mathrm{d}t\\&\qquad+\frac{\gamma\left(\vartheta,t\right)f\left(\vartheta,t\right)^2}{\varepsilon^2\sigma\left(t\right)^2}\mathrm{d}X_t\\&=-q_\varepsilon\left(\vartheta,t\right)M\left(\vartheta,t\right)\mathrm{d}t+\frac{\gamma\left(\vartheta,t\right)f\left(\vartheta,t\right)^2}{\varepsilon^2\sigma\left(t\right)^2}\mathrm{d}X_t,\end{aligned}$$

where we denoted

$$q_\varepsilon\left(\vartheta,t\right)=\frac{\gamma\left(\vartheta,t\right)f\left(\vartheta,t\right)^2}{\varepsilon^2\sigma\left(t\right)^2}-a\left(\vartheta,t\right)-\frac{f'\left(\vartheta,t\right)}{f\left(\vartheta,t\right)}.$$

The initial value is $M\left(\vartheta,\tau\right)=f\left(\vartheta,\tau\right)m\left(\vartheta,\tau\right)$.

The asymptotic behavior of $M(\vartheta,t)$ and $\dot{\mathbf{M}}(\vartheta,t)$ was already described above in the proof of Theorem 3.1. Let us remind some of them here. The function $\gamma_*\left(\vartheta,t\right)=\varepsilon^{-1}\gamma\left(\vartheta,t\right)$ is bounded and for any $t_0>0$ uniformly on $t\in\left[t_0,T\right]$ converges to the function $\gamma_0\left(\vartheta,t\right)=b\left(\vartheta,t\right)\sigma\left(t\right)f\left(\vartheta,t\right)^{-1}$ (Lemma 3.1). It was shown that we can replace the function $q_\varepsilon\left(\vartheta,t\right)$ by the function $q_*\left(\vartheta,t\right)/\varepsilon$, where

$$q_*\left(\vartheta,t\right)=\frac{\gamma_0\left(\vartheta,t\right)f\left(\vartheta,t\right)^2}{\sigma\left(t\right)^2}=\frac{b\left(\vartheta,t\right)f\left(\vartheta,t\right)}{\sigma\left(t\right)}.$$

We rewrite the equation for $M(\vartheta, t)$ without changing notation

$$\mathrm{d}M(\vartheta, t) = -\frac{1}{\varepsilon} q_*(\vartheta, t)\left[M(\vartheta, t) - M(\vartheta_0, t)\right]\mathrm{d}t + q_*(\vartheta, t)\,\sigma(t)\,\mathrm{d}\bar{W}_t$$

and the equation for the derivative

$$\begin{aligned}\mathrm{d}\dot{\mathrm{M}}(\vartheta, t) &= -\frac{1}{\varepsilon} q_*(\vartheta, t)\,\dot{\mathrm{M}}(\vartheta, t)\,\mathrm{d}t \\ &\quad + \frac{1}{\varepsilon}\dot{\mathrm{q}}_*(\vartheta, t)\left[M(\vartheta_0, t) - M(\vartheta, t)\right]\mathrm{d}t + \dot{\mathrm{q}}_*(\vartheta, t)\,\sigma(t)\,\mathrm{d}\bar{W}_t.\end{aligned}$$

As the contribution of the initial value in $\dot{\mathrm{M}}(\vartheta, t)$ is exponentially small to simplify the exposition, we will omit it below. We have

$$\begin{aligned}\dot{\mathrm{M}}(\vartheta, t) &= \frac{1}{\varepsilon}\int_\tau^t \phi(\vartheta, s, t)\,\dot{\mathrm{q}}_*(\vartheta, s)\left[M(\vartheta_0, s) - M(\vartheta, s)\right]\mathrm{d}s \\ &\quad + \int_\tau^t \phi(\vartheta, s, t)\,\dot{\mathrm{q}}_*(\vartheta, s)\,\sigma(s)\,\mathrm{d}\bar{W}_s, \end{aligned} \tag{3.33}$$

where

$$\phi_*(\vartheta, s, t) = \exp\left(-\varepsilon^{-1}\int_s^t q_*(\vartheta, v)\,\mathrm{d}v\right).$$

The asymptotics of $\dot{\mathrm{M}}(\vartheta, t)$ is described with the help of the Lemma 1.7. Note that

$$\dot{\mathrm{M}}(\vartheta_0, t) = \int_\tau^t \phi_*(\vartheta_0, s, t)\dot{\mathrm{q}}_*(\vartheta_0, s)\sigma(s)\,\mathrm{d}\bar{W}_s,$$

and the asymptotic of stochastic integral by (1.57) is

$$\begin{aligned}&\int_\tau^t \phi_*(\vartheta, s, t)\dot{\mathrm{q}}_*(\vartheta, s)\sigma(s)\,\mathrm{d}\bar{W}_s \\ &\quad = \sqrt{\varepsilon}\sqrt{b(\vartheta, t)\sigma(t) f(\vartheta, t)}\left(\frac{\dot{\mathrm{b}}(\vartheta, t)}{b(\vartheta, t)} + \frac{\dot{\mathrm{f}}(\vartheta, t)}{f(\vartheta, t)}\right)\frac{\bar{\xi}_{t,\varepsilon}}{\sqrt{2}}\left(1 + O\left(\sqrt{\varepsilon}\right)\right).\end{aligned}$$

Hence,

$$\begin{aligned}&\varepsilon^{-1/2}\dot{\mathrm{M}}(\vartheta_0, t) \\ &\quad = \sqrt{b(\vartheta_0, t)\sigma(t) f(\vartheta_0, t)}\left(\frac{\dot{\mathrm{b}}(\vartheta_0, t)}{b(\vartheta_0, t)} + \frac{\dot{\mathrm{f}}(\vartheta_0, t)}{f(\vartheta_0, t)}\right)\frac{\bar{\xi}_{t,\varepsilon}}{\sqrt{2}}\left(1 + O\left(\sqrt{\varepsilon}\right)\right)\end{aligned}$$

$$\Longrightarrow \sqrt{b(\vartheta_0,t)\sigma\left(t\right)f(\vartheta_0,t)}\,\left(\frac{\dot{\mathrm{b}}(\vartheta_0,t)}{b(\vartheta_0,t)}+\frac{\dot{\mathrm{f}}(\vartheta_0,t)}{f(\vartheta_0,t)}\right)\frac{\bar{\xi}_t}{\sqrt{2}}.$$

The first integral in (3.33), we write as follows:

$$\begin{aligned}
&\frac{1}{\varepsilon}\int_\tau^t \phi_*\left(\vartheta,s,t\right)\left[M\left(\vartheta_0,s\right)-M\left(\vartheta,s\right)\right]\dot{\mathrm{q}}_*\left(\vartheta,s\right)\mathrm{d}s\\
&\qquad=\frac{\left(\vartheta_0-\vartheta\right)^\top}{\varepsilon}\int_\tau^t \phi_*\left(\vartheta,s,t\right)\dot{\mathrm{M}}(\tilde{\vartheta},s)\dot{\mathrm{q}}_*\left(\vartheta,s\right)\mathrm{d}s\\
&\qquad=\frac{\left(\vartheta_0-\vartheta\right)^\top}{\varepsilon}\int_\tau^t \phi_*\left(\vartheta,s,t\right)\dot{\mathrm{M}}(\vartheta_0,s)\dot{\mathrm{q}}_*\left(\vartheta,s\right)\mathrm{d}s+\|\vartheta_0-\vartheta\|^2\,O\left(1\right)\\
&\qquad=\sqrt{\varepsilon}\,\|\vartheta_0-\vartheta\|\,O\left(1\right)+\left(\vartheta_0-\vartheta\right)^2 O\left(1\right).
\end{aligned}$$

where $\left\|\tilde{\vartheta}-\vartheta_0\right\|\leq\|\vartheta-\vartheta_0\|$.

Further,

$$\begin{aligned}
&\mathrm{I}_\tau^t\left(\vartheta_{\tau,\varepsilon}^*\right)^{-1}\int_\tau^t\frac{\dot{\mathrm{M}}(\vartheta_{\tau,\varepsilon}^*,s)}{\varepsilon^{3/2}\sigma\left(s\right)^2}\left[M\left(\vartheta_0,s\right)-M(\vartheta_{\tau,\varepsilon}^*,s)\right]\mathrm{d}s\\
&=\mathrm{I}_\tau^t\left(\vartheta_{\tau,\varepsilon}^*\right)^{-1}\int_\tau^t\frac{\dot{\mathrm{M}}(\vartheta_{\tau,\varepsilon}^*,s)\dot{\mathrm{M}}(\vartheta_{\tau,\varepsilon}^*,s)^\top}{\varepsilon^{3/2}\sigma\left(s\right)^2}\mathrm{d}s\left(\vartheta_0-\vartheta_{\tau,\varepsilon}^*\right)\left(1+O\left(\varepsilon^{1/3}\right)\right)\\
&=-\sqrt{\varepsilon}\,\mathrm{I}_\tau^t\left(\vartheta_0\right)^{-1}\int_\tau^t\frac{S(\vartheta_0,s)}{2\sigma\left(s\right)}\left[\frac{\dot{\mathrm{f}}(\vartheta_0,s)}{f\left(\vartheta_0,s\right)}+\frac{\dot{\mathrm{b}}(\vartheta_0,s)}{b(\vartheta_0,s)}\right]\\
&\qquad\times\left[\frac{\dot{\mathrm{f}}(\vartheta_0,s)}{f\left(\vartheta_0,s\right)}+\frac{\dot{\mathrm{b}}(\vartheta_0,s)}{b(\vartheta_0,s)}\right]^\top\bar{\xi}_{s,\varepsilon}^2\mathrm{d}s\left(\vartheta_{\tau,\varepsilon}^*-\vartheta_0\right)\left(1+O\left(\varepsilon^{1/3}\right)\right).
\end{aligned}$$

Recall that by Lemma 3.3

$$\begin{aligned}
&\int_\tau^t\frac{S(\vartheta_0,s)}{2\sigma\left(s\right)}\left[\frac{\dot{\mathrm{f}}(\vartheta_0,s)}{f\left(\vartheta_0,s\right)}+\frac{\dot{\mathrm{b}}(\vartheta_0,s)}{b(\vartheta_0,s)}\right]\left[\frac{\dot{\mathrm{f}}(\vartheta_0,s)}{f\left(\vartheta_0,s\right)}+\frac{\dot{\mathrm{b}}(\vartheta_0,s)}{b(\vartheta_0,s)}\right]^\top\bar{\xi}_{s,\varepsilon}^2\mathrm{d}s\\
&\quad\longrightarrow\int_\tau^t\frac{S(\vartheta_0,s)}{2\sigma\left(s\right)}\left[\frac{\dot{\mathrm{f}}(\vartheta_0,s)}{f\left(\vartheta_0,s\right)}+\frac{\dot{\mathrm{b}}(\vartheta_0,s)}{b(\vartheta_0,s)}\right]\left[\frac{\dot{\mathrm{f}}(\vartheta_0,s)}{f\left(\vartheta_0,s\right)}+\frac{\dot{\mathrm{b}}(\vartheta_0,s)}{b(\vartheta_0,s)}\right]^\top\mathrm{d}s=\mathrm{I}_\tau^t\left(\vartheta_0\right).
\end{aligned}\tag{3.34}$$

Therefore,

$$
\begin{aligned}
&\frac{(\vartheta_{\tau,\varepsilon}^{*}-\vartheta_0)}{\sqrt{\varepsilon}}+\mathrm{I}_\tau^t(\vartheta_{\tau,\varepsilon}^{*})^{-1}\int_\tau^t \frac{\dot{\mathrm{M}}(\vartheta_{\tau,\varepsilon}^{*},s)}{\varepsilon^{3/2}\sigma(s)^2}\left[M(\vartheta_0,s)-M(\vartheta_{\tau,\varepsilon}^{*},s)\right]\mathrm{d}s\\
&\quad=\left[1-\mathrm{I}_\tau^t(\vartheta_0)^{-1}\int_\tau^t\frac{S(\vartheta_0,s)}{\sigma(s)}\left[\frac{\dot{\mathrm{f}}(\vartheta_0,s)}{f(\vartheta_0,s)}+\frac{\dot{\mathrm{b}}(\vartheta_0,s)}{b(\vartheta_0,s)}\right]\right.\\
&\qquad\left.\times\left[\frac{\dot{\mathrm{f}}(\vartheta_0,s)}{f(\vartheta_0,s)}+\frac{\dot{\mathrm{b}}(\vartheta_0,s)}{b(\vartheta_0,s)}\right]^\top\frac{\bar{\xi}_{s,\varepsilon}^2}{2}\mathrm{d}s\frac{(\vartheta_{\tau,\varepsilon}^{*}-\vartheta_0)}{\sqrt{\varepsilon}}\left(1+O\left(\varepsilon^{1/3}\right)\right)\right]\\
&\quad=\frac{(\vartheta_{\tau,\varepsilon}^{*}-\vartheta_0)}{\sqrt{\varepsilon}}O\left(\varepsilon^{1/3}\right)=\frac{O\left(\varepsilon^{2/3}\right)}{\sqrt{\varepsilon}}\longrightarrow 0. \qquad (3.35)
\end{aligned}
$$

The convergence (3.34) is sufficient for the asymptotic normality of the stochastic integral

$$
\int_\tau^t\frac{\dot{\mathrm{M}}(\vartheta_{\tau,\varepsilon}^{*},s)}{\sqrt{\varepsilon}\sigma(s)}\mathrm{d}\bar{W}_s=\int_\tau^t\frac{\dot{\mathrm{M}}(\vartheta_0,s)}{\sqrt{\varepsilon}\sigma(s)}\mathrm{d}\bar{W}_s\left(1+O\left(\varepsilon^{1/3}\right)\right)\Longrightarrow\mathcal{N}\left(0,\mathrm{I}_\tau^t(\vartheta_0)\right).
$$

Finally, we obtained the asymptotic normality of One-step MLE process: for any $t>\tau$

$$
\frac{\vartheta_{t,\varepsilon}^{\star}-\vartheta_0}{\sqrt{\varepsilon}}\Longrightarrow\eta_t(\vartheta_0)\sim\mathcal{N}\left(0,\mathrm{I}_\tau^t(\vartheta_0)^{-1}\right).
$$

To prove the convergence of moments, we slightly modify the calculations in (3.35):

$$
\begin{aligned}
&\frac{(\vartheta_{\tau,\varepsilon}^{*}-\vartheta_0)}{\sqrt{\varepsilon}}+\mathrm{I}_\tau^t(\vartheta_{\tau,\varepsilon}^{*})^{-1}\int_\tau^t \frac{\dot{\mathrm{M}}(\vartheta_{\tau,\varepsilon}^{*},s)}{\varepsilon^{3/2}\sigma(s)^2}\left[M(\vartheta_0,s)-M(\vartheta_{\tau,\varepsilon}^{*},s)\right]\mathrm{d}s\\
&\quad=\frac{(\vartheta_{\tau,\varepsilon}^{*}-\vartheta_0)}{\sqrt{\varepsilon}}\left[1-\mathrm{I}_\tau^t(\vartheta_{\tau,\varepsilon}^{*})^{-1}\int_\tau^t\frac{S(\vartheta_{\tau,\varepsilon}^{*},s)}{2\sigma(s)}\left[\frac{\dot{\mathrm{f}}(\vartheta_{\tau,\varepsilon}^{*},s)}{f(\vartheta_{\tau,\varepsilon}^{*},s)}+\frac{\dot{\mathrm{b}}(\vartheta_{\tau,\varepsilon}^{*},s)}{b(\vartheta_{\tau,\varepsilon}^{*},s)}\right]\right.\\
&\qquad\left.\times\left[\frac{\dot{\mathrm{f}}(\vartheta_{\tau,\varepsilon}^{*},s)}{f(\vartheta_{\tau,\varepsilon}^{*},s)}+\frac{\dot{\mathrm{b}}(\vartheta_{\tau,\varepsilon}^{*},s)}{b(\vartheta_{\tau,\varepsilon}^{*},s)}\right]^\top\mathrm{d}s\left(1+O(\varepsilon^{\frac{1}{3}})\right)\right]\\
&\qquad+\frac{(\vartheta_{\tau,\varepsilon}^{*}-\vartheta_0)}{\sqrt{\varepsilon}}\mathrm{I}_\tau^t(\vartheta_{\tau,\varepsilon}^{*})^{-1}\int_\tau^t\frac{S(\vartheta_{\tau,\varepsilon}^{*},s)}{2\sigma(s)}\left[\frac{\dot{\mathrm{f}}(\vartheta_{\tau,\varepsilon}^{*},s)}{f(\vartheta_{\tau,\varepsilon}^{*},s)}+\frac{\dot{\mathrm{b}}(\vartheta_{\tau,\varepsilon}^{*},s)}{b(\vartheta_{\tau,\varepsilon}^{*},s)}\right]\\
&\qquad\times\left[\frac{\dot{\mathrm{f}}(\vartheta_{\tau,\varepsilon}^{*},s)}{f(\vartheta_{\tau,\varepsilon}^{*},s)}+\frac{\dot{\mathrm{b}}(\vartheta_{\tau,\varepsilon}^{*},s)}{b(\vartheta_{\tau,\varepsilon}^{*},s)}\right]^\top\left\{\bar{\xi}_{s,\varepsilon}^2-1\right\}\mathrm{d}s\left(1+O(\varepsilon^{\frac{1}{3}})\right)
\end{aligned}
$$

$$= \frac{\left\| \vartheta^*_{\tau,\varepsilon} - \vartheta_0 \right\|^2}{\sqrt{\varepsilon}} O(1) + \left(\vartheta^*_{\tau,\varepsilon} - \vartheta_0\right) O(1).$$

The terms $\varepsilon^{-1/2}\left(\vartheta^*_{\tau,\varepsilon} - \vartheta_0\right)^2$, $O(1)$ and the integral containing $\xi^2_{s,\varepsilon} - 1$ have bounded all polynomial moments. Therefore, for any $p > 1$ there exists a constant $C = C_p > 0$ such that

$$\mathbf{E}_{\vartheta_0} \left\| \frac{\vartheta^\star_{t,\varepsilon} - \vartheta_0}{\sqrt{\varepsilon}} \right\|^p \leq C.$$

The presented proof here can be used for verification of the convergence of the vectors: for any $K = 1, 2, \ldots$ and $t_0 < t_1 < \ldots < t_K \leq T$

$$\left(\eta_{t_1,\varepsilon}, \ldots, \eta_{t_K,\varepsilon}\right) \Longrightarrow \left(\eta_{t_1}(\vartheta_0), \ldots, \eta_{t_K}(\vartheta_0)\right).$$

It is sufficient to study the convergence of the sum: for any $\lambda_1, \ldots, \lambda_K$

$$\sum_{k=1}^K \lambda_k \eta_{t_k,\varepsilon} \Longrightarrow \sum_{k=1}^K \lambda_k \eta_{t_k}(\vartheta_0).$$

The estimate

$$\mathbf{E}_{\vartheta_0} \left|\eta_{t_1,\varepsilon} - \eta_{t_2,\varepsilon}\right|^4 \leq C_1 \left|t_1 - t_2\right|^4 + C_2 \left|t_1 - t_2\right|^2$$

with some constants $C_1 > 0$ and $C_2 > 0$ can be verified by direct calculations.

Having these properties of $\eta_{t,\varepsilon}$, we obtain the weak convergence of the random process $\eta_{t,\varepsilon}, t_0 \leq t \leq T$ to the random process $\eta_t(\vartheta_0), t_0 \leq t \leq T$ [15]. □

Remark 3.5 Note that if $d = 1$ and $\inf_{\vartheta\in\Theta} |\dot S(\vartheta, \tau)| > 0$, then the condition $\mathscr{G}^\star_3$ is fulfilled for any $t_0 > \tau$.

Remark 3.6 The proposed above construction of the One-step MLE $\vartheta^\star_{T,\varepsilon}$ and the One-step MLE process $\vartheta^\star_{t,\varepsilon}, \tau < t \leq T$ are not asymptotically efficient because the limit variance of the MLE $\hat\vartheta_\varepsilon$ is $\mathrm{I}_0^T(\vartheta_0)^{-1}$ and the limit variance of $\vartheta^\star_{T,\varepsilon}$ is $\mathrm{I}_\tau^T(\vartheta_0)^{-1}$. Choosing sufficiently small τ the ratio

$$\frac{\mathrm{I}_0^T(\vartheta_0)}{\mathrm{I}_\tau^T(\vartheta_0)} \geq 1$$

can be made close to 1.

Another possibility is to consider the sequence of problems with $\tau = \tau_\varepsilon \to 0$ but slowly. The main condition on the rate of convergence of a preliminary estimator $\bar\vartheta_{\tau_\varepsilon,\varepsilon}$ is

$$\frac{\left(\bar\vartheta_{\tau_\varepsilon,\varepsilon}-\vartheta_0\right)^2}{\sqrt{\varepsilon}}\longrightarrow 0.$$

The MME $\vartheta^*_{\tau_\varepsilon,\varepsilon}$ satisfies to this condition

$$\frac{\left(\vartheta^*_{\tau,\varepsilon}-\vartheta_0\right)^2}{\sqrt{\varepsilon}}\approx\frac{\varepsilon^{2/3}}{\tau\varepsilon^{1/2}}=\frac{\varepsilon^{1/6}}{\tau}\longrightarrow 0.$$

Therefore, if we take $\tau=\tau_\varepsilon=\varepsilon^{1/12}$, then

$$\frac{\left(\vartheta^*_{\tau_\varepsilon,\varepsilon}-\vartheta_0\right)^2}{\sqrt{\varepsilon}}\approx\varepsilon^{1/12}\longrightarrow 0$$

and we have the asymptotic normality of the corresponding One-step estimator: for any $t>0$

$$\frac{\vartheta^\star_{t,\varepsilon}-\vartheta_0}{\sqrt{\varepsilon}}\Longrightarrow\eta_t\left(\vartheta_0\right)=\frac{1}{\mathrm{I}_0^t(\vartheta_0)}\int_0^t\frac{\dot S\left(\vartheta_0,s\right)}{\sqrt{2S\left(\vartheta_0,s\right)\sigma\left(s\right)}}\mathrm{d}w\left(s\right)\sim\mathcal{N}\left(0,\mathrm{I}_0^t(\vartheta_0)^{-1}\right).$$

3.1.4 Adaptive Filter

Consider the same partially observed two-dimensional linear stochastic system

$$\begin{aligned}\mathrm{d}X_t&=f\left(\vartheta,t\right)Y_t\mathrm{d}t+\varepsilon\sigma\left(t\right)\mathrm{d}W_t, &\quad X_0&=0, &\quad 0\le t\le T,\\ \mathrm{d}Y_t&=a\left(\vartheta,t\right)Y_t\mathrm{d}t+b\left(\vartheta,t\right)\mathrm{d}V_t, &\quad Y_0&=y_0,\end{aligned}\tag{3.36}$$

where $\vartheta\in\Theta$ is unknown parameter. For simplicity, we suppose that the parameter ϑ is one-dimensional, $\Theta=(\alpha,\beta)$. The main problem as usual is the estimation of the hidden component Y_t by observations $X^t=(X_s,0\le s\le t)$.

Let us recall the notations:

$$\eta_t\left(\vartheta_0\right)=\mathrm{I}_0^t\left(\vartheta_0\right)^{-1}\int_0^t\frac{\dot S\left(\vartheta_0,s\right)}{\sqrt{2S\left(\vartheta_0,s\right)\sigma\left(s\right)}}\mathrm{d}w\left(s\right),\quad \bar\xi_{t,\varepsilon}=\sqrt{\frac{2q_t}{\varepsilon}}\int_0^{\frac{tq_t}{\varepsilon}}e^{-v}\mathrm{d}\bar W_{t-\frac{v\varepsilon}{q_t}}$$

$$\eta_{t,\varepsilon}\left(\vartheta_0\right)=\frac{\vartheta^\star_{t,\varepsilon}-\vartheta_0}{\sqrt{\varepsilon}}=\frac{1}{\mathrm{I}_0^t\left(\vartheta_0\right)}\int_0^t\frac{\dot S\left(\vartheta_0,s\right)}{\sqrt{2S\left(\vartheta_0,s\right)\sigma\left(s\right)}}\bar\xi_{s,\varepsilon}\,\mathrm{d}\bar W_s\left(1+O\left(\sqrt{\varepsilon}\right)\right),$$

$$\mathrm{d}M\left(\vartheta,t\right)=-\frac{b\left(\vartheta,t\right)f\left(\vartheta,t\right)}{\varepsilon\sigma\left(t\right)}M\left(\vartheta,t\right)\mathrm{d}t+\frac{b\left(\vartheta,t\right)f\left(\vartheta,t\right)}{\varepsilon\sigma\left(t\right)}\mathrm{d}X_t,$$

$$
\begin{aligned}
\mathrm{d}\dot{M}(\vartheta,t) &= -\frac{b(\vartheta,t)\,f(\vartheta,t)}{\varepsilon\sigma(t)}\dot{M}(\vartheta,t)\,\mathrm{d}t \\
&\quad + \frac{\dot{b}(\vartheta,t)\,f(\vartheta,t)+\dot{f}(\vartheta,t)\,b(\vartheta,t)}{\varepsilon\sigma(t)}\left[\mathrm{d}X_t - M(\vartheta,t)\,\mathrm{d}t\right].
\end{aligned}
$$

The adaptive filter is constructed as follows.

1. *Fix the learning interval* $[0,\tau_\varepsilon]$, *where* $\tau_\varepsilon=\varepsilon^{1/12}$ *with the observations* $X^{\tau_\varepsilon}=(X_t, 0\leq t\leq\tau_\varepsilon)$.
2. *Put* $t_0=0, t_{i+1}=t_i+\varepsilon^{1/3}$, $N_{\tau_\varepsilon,\varepsilon}=\left[\varepsilon^{-1/4}\right]$ *and calculate the statistic*

$$
\hat{\Psi}_{\tau_\varepsilon,\varepsilon}=\sum_{i=0}^{N_{\tau_\varepsilon,\varepsilon}-1}\left(\frac{X_{t_{i+1}+\varepsilon}-X_{t_{i+1}}-X_{t_i+\varepsilon}+X_{t_i}}{\varepsilon}\right)^2.
$$

3. *Solve the equation*

$$
\int_0^{\tau_\varepsilon} f\left(\vartheta^*_{\tau_\varepsilon},t\right)^2 b\left(\vartheta^*_{\tau_\varepsilon},t\right)^2\mathrm{d}t=\hat{\Psi}_{\tau_\varepsilon,\varepsilon}
$$

whose solution $\vartheta^*_{\tau_\varepsilon}$ *is the MME of the parameter* ϑ.

4. *Introduce the One-step MLE process*

$$
\vartheta^\star_{t,\varepsilon}=\vartheta^*_{\tau_\varepsilon}+\frac{1}{\mathrm{I}^t_{\tau_\varepsilon}\left(\vartheta^*_{\tau_\varepsilon}\right)}\int_{\tau_\varepsilon}^t\frac{\dot{M}\left(\vartheta^*_{\tau_\varepsilon},s\right)}{\varepsilon\sigma(s)^2}\left[\mathrm{d}X_s-M\left(\vartheta^*_{\tau_\varepsilon},s\right)\mathrm{d}s\right],\qquad \tau_\varepsilon<t\leq T,
$$

where $M\left(\vartheta^*_{\tau_\varepsilon},t\right)=f\left(\vartheta^*_{\tau_\varepsilon},t\right)m\left(\vartheta^*_{\tau_\varepsilon},t\right), \tau_\varepsilon<t\leq T$ *and* $\dot{M}\left(\vartheta^*_{\tau_\varepsilon},t\right), \tau_\varepsilon<t\leq T$ *are solutions of the equations*

$$
\begin{aligned}
\mathrm{d}M\left(\vartheta^*_{\tau_\varepsilon},t\right) &= -\frac{b\left(\vartheta^*_{\tau_\varepsilon},t\right)f\left(\vartheta^*_{\tau_\varepsilon},t\right)}{\varepsilon\sigma(t)}M\left(\vartheta^*_{\tau_\varepsilon},t\right)\mathrm{d}t+\frac{b\left(\vartheta^*_{\tau_\varepsilon},t\right)f\left(\vartheta^*_{\tau_\varepsilon},t\right)}{\varepsilon\sigma(t)}\mathrm{d}X_t,\\
\mathrm{d}\dot{M}\left(\vartheta^*_{\tau_\varepsilon},t\right) &= -\frac{b\left(\vartheta^*_{\tau_\varepsilon},t\right)f\left(\vartheta^*_{\tau_\varepsilon},t\right)}{\varepsilon\sigma(t)}\dot{M}\left(\vartheta^*_{\tau_\varepsilon},t\right)\mathrm{d}t\\
&\quad+\frac{\dot{b}\left(\vartheta^*_{\tau_\varepsilon},t\right)f\left(\vartheta^*_{\tau_\varepsilon},t\right)+\dot{f}\left(\vartheta^*_{\tau_\varepsilon},t\right)b(\vartheta,t)}{\varepsilon\sigma(t)}\left[\mathrm{d}X_t-M\left(\vartheta^*_{\tau_\varepsilon},t\right)\mathrm{d}t\right].
\end{aligned}
$$

5. *The adaptive filter is given by the equation*

$$
\mathrm{d}m^\star_{t,\varepsilon}=\left[a\left(\vartheta^\star_{t,\varepsilon},t\right)-\frac{b\left(\vartheta^\star_{t,\varepsilon},t\right)f\left(\vartheta^\star_{t,\varepsilon},t\right)}{\varepsilon\sigma(t)}\right]m^\star_{t,\varepsilon}\mathrm{d}t+\frac{b\left(\vartheta^\star_{t,\varepsilon},t\right)}{\varepsilon\sigma(t)}\mathrm{d}X_t,
$$

$$
t_0\leq t\leq T \tag{3.37}
$$

subject to the initial value $m^\star_{t_0,\varepsilon}=m\left(\check{\vartheta}_{\tau_\varepsilon},t_0\right)$.

Note that the calculation of $m^\star_{t,\varepsilon}$ does not require the solution of Riccati equation.

Let us discuss the errors of approximations: $Y_t - m(\vartheta_0, t)$ and $m^\star_{t,\varepsilon} - m(\vartheta_0, t)$. Recall that Y_t is solution of the equation

$$\mathrm{d}Y_t = a(\vartheta_0, t)\, Y_t \mathrm{d}t + b(\vartheta_0, t)\, \mathrm{d}V_t, \qquad Y_0 = y_0, \quad 0 \le t \le T.$$

Introduce the independent random variables

$$\xi^{(1)}_{t,\varepsilon} = \sqrt{\frac{2q_t}{\varepsilon}} \int_0^t e^{-\frac{t-s}{\varepsilon} q_t} \mathrm{d}V_s, \qquad \xi^{(2)}_{t,\varepsilon} = \sqrt{\frac{2q_t}{\varepsilon}} \int_0^t e^{-\frac{t-s}{\varepsilon} q_t} \mathrm{d}W_s, \qquad t \in (0, T]$$

where $q_t = b(\vartheta_0, t)\, f(\vartheta_0, t)\, \sigma(t)^{-1}$. Recall that they are asymptotically normal

$$\xi^{(1)}_{t,\varepsilon} \Longrightarrow \xi^{(1)}_t \sim \mathcal{N}(0, 1), \qquad \xi^{(2)}_{t,\varepsilon} \Longrightarrow \xi^{(2)}_t \sim \mathcal{N}(0, 1),$$

where $\xi^{(1)}_t, \xi^{(2)}_t, 0 < t \le T$ are independent random variables. These random variables are independent of the Gaussian process $\eta_t(\vartheta_0)$, $0 < t \le T$ too.

Introduce the condition

$\mathscr{G}^\star_4$. *The functions $f(\vartheta, t)$ and $b(\vartheta, t)$ are such that*

$$\inf_{\vartheta \in \Theta} \left| \dot f(\vartheta, 0)\, b(\vartheta, 0) + f(\vartheta, 0)\, \dot b(\vartheta, 0) \right| > 0.$$

Note that under this condition, we have $\dot\Psi_\tau(\vartheta) > 0$ and $\mathrm{I}^\tau_0(\vartheta) > 0$ for any $\tau > 0$.

Theorem 3.3 *Let the conditions $\mathscr{A}_*$, $\mathscr{G}^\star$ and $\mathscr{G}^\star_4$ hold. Then, for any (small) $t_0 > 0$ and $t \in (t_0, T]$, we have the following relations.*

1. If ϑ_0 is known then

$$\frac{Y_t - m(\vartheta_0, t)}{\sqrt{\varepsilon}} = \sqrt{\frac{b(\vartheta_0, t)\, \sigma(t)}{2f(\vartheta_0, t)}} \left(\xi^{(1)}_{t,\varepsilon} - \xi^{(2)}_{t,\varepsilon}\right) \left(1 + o\left(\varepsilon^{1/2}\right)\right). \tag{3.38}$$

2. If ϑ_0 is unknown, then we have the representation

$$\frac{m^\star_t - m(\vartheta_0, t)}{\sqrt{\varepsilon}} = -\frac{\dot f(\vartheta_0, t)}{f(\vartheta_0, t)} \eta_{t,\varepsilon}\, Y_t \Longrightarrow -\frac{\dot f(\vartheta_0, t)}{f(\vartheta_0, t)} \eta_t(\vartheta_0)\, Y_t. \tag{3.39}$$

and

$$\mathbf{E}_{\vartheta_0} \left(\eta_{t,\varepsilon}(\vartheta_0)\, Y_t\right)^2 \longrightarrow \mathbf{E}_{\vartheta_0} \eta_t(\vartheta_0)^2\, \mathbf{E}_{\vartheta_0} Y_t^2 = \mathrm{I}^t_0(\vartheta_0)^{-1}\, \mathbf{E}_{\vartheta_0} Y_t^2. \tag{3.40}$$

Proof Suppose that ϑ_0 is known, and we are interested by the error $Y_t - m(\vartheta_0, t)$. Of course, we have immediately: for any $t \in (0, T]$

$$\mathbf{E}_{\vartheta_0}\left(Y_t - m\left(\vartheta_0, t\right)\right)^2 = \gamma\left(\vartheta_0, t\right) = \frac{b\left(\vartheta_0, t\right)\sigma\left(t\right)}{f\left(\vartheta_0, t\right)}\varepsilon\left(1 + o\left(\varepsilon\right)\right).$$

Further, using Eqs. (3.36) and (3.30) and asymptotics of $\gamma\left(\vartheta, t\right)$, we can write the equation for $\delta_Y\left(t\right) = Y_t - m\left(\vartheta_0, t\right)$

$$\begin{aligned}
\mathrm{d}\delta_Y\left(t\right) &= a\left(\vartheta_0, t\right)\delta_Y\left(t\right)\mathrm{d}t - \frac{\gamma\left(\vartheta_0, t\right) f\left(\vartheta_0, t\right)^2}{\varepsilon^2\sigma\left(t\right)^2}\delta_Y\left(t\right)\mathrm{d}t \\
&\qquad + b\left(\vartheta_0, t\right)\mathrm{d}V_t - \frac{\gamma\left(\vartheta_0, t\right) f\left(\vartheta_0, t\right)}{\varepsilon\sigma\left(t\right)}\mathrm{d}W_t \\
&= -q_\varepsilon\left(\vartheta_0, t\right)\delta_Y\left(t\right)\mathrm{d}t + b\left(\vartheta_0, t\right)\mathrm{d}V_t - \frac{\gamma\left(\vartheta_0, t\right) f\left(\vartheta_0, t\right)}{\varepsilon\sigma\left(t\right)}\mathrm{d}W_t \\
&= -\frac{1}{\varepsilon}q_*\left(\vartheta_0, t\right)\delta_Y\left(t\right)\mathrm{d}t + b\left(\vartheta_0, t\right)\mathrm{d}V_t - b\left(\vartheta_0, t\right)\mathrm{d}W_t,
\end{aligned}$$

where we omitted the term $O\left(\varepsilon\right)$.

Hence, for any $t > 0$ we have the relation

$$Y_t - m\left(\vartheta_0, t\right) = \sqrt{\frac{b\left(\vartheta_0, t\right)\sigma\left(t\right)}{2f\left(\vartheta_0, t\right)}}\left(\xi_{t,\varepsilon}^{(1)} - \xi_{t,\varepsilon}^{(2)}\right)\sqrt{\varepsilon}\left(1 + O\left(\sqrt{\varepsilon}\right)\right).$$

The equation for $m_{t,\varepsilon}^\star$ and $m_{0,t} = m\left(\vartheta_0, t\right)$ are

$$\begin{aligned}
\mathrm{d}m_{t,\varepsilon}^\star &= \left[a\left(\vartheta_{t,\varepsilon}^\star, t\right) - \frac{b\left(\vartheta_{t,\varepsilon}^\star, t\right) f\left(\vartheta_{t,\varepsilon}^\star, t\right)}{\varepsilon\sigma\left(t\right)}\right] m_{t,\varepsilon}^\star\mathrm{d}t + \frac{b\left(\vartheta_{t,\varepsilon}^\star, t\right) f\left(\vartheta_0, t\right)}{\varepsilon\sigma\left(t\right)} m_{0,t}\mathrm{d}t \\
&\qquad + b\left(\vartheta_{t,\varepsilon}^\star, t\right)\mathrm{d}\bar{W}_t, \qquad m_{t_0,\varepsilon}^\star, \qquad t_0 \le t \le T
\end{aligned}$$

and

$$\begin{aligned}
\mathrm{d}m_{0,t} &= \left[a\left(\vartheta_0, t\right) - \frac{b\left(\vartheta_0, t\right) f\left(\vartheta_0, t\right)}{\varepsilon\sigma\left(t\right)}\right] m_{0,t}\mathrm{d}t + \frac{b\left(\vartheta_0, t\right) f\left(\vartheta_0, t\right)}{\varepsilon\sigma\left(t\right)} m_{0,t}\mathrm{d}t \\
&\qquad + b\left(\vartheta_0, t\right)\mathrm{d}\bar{W}_t, \qquad m_{0,t_0}, \qquad t_0 \le t \le T.
\end{aligned}$$

Recall the notation

$$A_*\left(\vartheta, t\right) = -\frac{b\left(\vartheta, t\right) f\left(\vartheta, t\right)}{\varepsilon\sigma\left(t\right)}, \qquad S\left(\vartheta, t\right) = b\left(\vartheta, t\right) f\left(\vartheta, t\right), \quad q_t = \frac{S\left(\vartheta, t\right)}{\sigma\left(t\right)},$$

$$\Phi_*\left(\vartheta, s, t\right) = \exp\left(\int_s^t A_*\left(\vartheta, v\right)\mathrm{d}v\right), \qquad \eta_{t,\varepsilon} = \varepsilon^{-1/2}\left(\vartheta_{t,\varepsilon}^\star - \vartheta_0\right).$$

The equation for $\delta_m\left(t\right) = m_{t,\varepsilon}^\star - m_{0,t}$ is

$$
\begin{aligned}
\mathrm{d}\delta_m(t) &= a(\vartheta_0,t)\,\delta_m(t)\,\mathrm{d}t + \left[a\left(\vartheta^\star_{t,\varepsilon},t\right) - a(\vartheta_0,t)\right] m^\star_{t,\varepsilon}\mathrm{d}t + A_*(\vartheta_0,t)\,\delta_m(t)\,\mathrm{d}t\\
&\quad + \left[A_*\left(\vartheta^\star_{t,\varepsilon},t\right) - A_*(\vartheta_0,t)\right] m^\star_{t,\varepsilon}\mathrm{d}t\\
&\quad + \frac{\left[b\left(\vartheta^\star_{t,\varepsilon},t\right) - b(\vartheta_0,t)\right] f(\vartheta_0,t)}{\varepsilon\sigma(t)} m_{0,t}\mathrm{d}t\\
&\quad + \left[b\left(\vartheta^\star_{t,\varepsilon},t\right) - b(\vartheta_0,t)\right]\mathrm{d}\bar W_t, \qquad \delta_m(t_0), \qquad t_0 \le t \le T.
\end{aligned}
$$

The solution of this equation can be written as follows:

$$
\begin{aligned}
\delta_m(t) &= \delta_m(t_0)\,\Phi_*(\vartheta_0,t_0,t) + \int_{t_0}^t \Phi_*(\vartheta_0,s,t)\,a(\vartheta_0,s)\,\delta_m(s)\,\mathrm{d}s\\
&\quad + \int_{t_0}^t \Phi_*(\vartheta_0,s,t)\left[a\left(\vartheta^\star_{s,\varepsilon},s\right) - a(\vartheta_0,s)\right] m^\star_{s,\varepsilon}\mathrm{d}s\\
&\quad + \int_{t_0}^t \Phi_*(\vartheta_0,s,t)\left[A_*\left(\vartheta^\star_{s,\varepsilon},s\right) - A_*(\vartheta_0,s)\right] m^\star_{s,\varepsilon}\mathrm{d}s\\
&\quad + \int_{t_0}^t \Phi_*(\vartheta_0,s,t)\,\frac{\left[b\left(\vartheta^\star_{s,\varepsilon},s\right) - b(\vartheta_0,s)\right] f(\vartheta_0,s)}{\varepsilon\sigma(s)} m_{0,s}\mathrm{d}s\\
&\quad + \int_{t_0}^t \Phi_*(\vartheta_0,s,t)\left[b\left(\vartheta^\star_{s,\varepsilon},s\right) - b(\vartheta_0,s)\right]\mathrm{d}\bar W_s.
\end{aligned}
$$

The first term in the RHS of this equation is exponentially small for any $t > t_0$ and will be omitted. It is possible to show that $\mathbf{E}_{\vartheta_0}\delta_m(t)^2 \le C\varepsilon$. For the verification, it is sufficient to write the equation for $\delta_m(t)^2$, then to use Grönwall's lemma (Lemma 1.5) for $\mathbf{E}_{\vartheta_0}\delta_m(t)^2$. Therefore, the second term by Lemma 1.7 is of order $O\left(\varepsilon^{3/2}\right)$. The same order has the third term too because $a(\vartheta^\star_{s,\varepsilon},s) - a(\vartheta_0,s) = \dot a(\bar\vartheta_{s,\varepsilon},s)\eta_{t,\varepsilon}\sqrt{\varepsilon}$ and we have

$$
\begin{aligned}
&\int_{t_0}^t \Phi_*(\vartheta_0,s,t)\left[a\left(\vartheta^\star_{s,\varepsilon},s\right) - a(\vartheta_0,s)\right] m^\star_{s,\varepsilon}\mathrm{d}s\\
&\qquad = \sqrt{\varepsilon}\int_{t_0}^t \Phi_*(\vartheta_0,s,t)\,\dot a\left(\bar\vartheta_{s,\varepsilon},s\right)\,\eta_{s,\varepsilon}\, m^\star_{s,\varepsilon}\,\mathrm{d}s\\
&\qquad = \sqrt{\varepsilon}\,\eta_{t,\varepsilon}\, m^\star_{t,\varepsilon}\int_{t_0}^t \Phi_*(\vartheta_0,s,t)\,\dot a\left(\bar\vartheta_{s,\varepsilon},s\right)\,\mathrm{d}s\left(1 + O\left(\sqrt{\varepsilon}\right)\right)\\
&\qquad = \varepsilon^{3/2}\frac{\dot a(\vartheta_0,t)\,\sigma(t)}{b(\vartheta_0,t)\,f(\vartheta_0,t)}\,\eta_{t,\varepsilon}\, m^\star_{t,\varepsilon}\,\left(1 + O\left(\sqrt{\varepsilon}\right)\right)
\end{aligned}
$$

For the next term, we have

$$
\begin{aligned}
\int_{t_0}^{t} & \Phi_*(\vartheta, s, t)\left[A_*\left(\vartheta^\star_{s,\varepsilon}, s\right) - A_*(\vartheta_0, s)\right] m^\star_{s,\varepsilon} \mathrm{d}s \\
&= -\frac{1}{\varepsilon}\int_{t_0}^{t} e^{-\frac{t-s}{\varepsilon}q_t}\frac{\left[S\left(\vartheta^\star_{s,\varepsilon}, s\right) - S(\vartheta_0, s)\right]}{\sigma(s)} m_{0,s}\,\mathrm{d}s\left(1 + O\left(\sqrt{\varepsilon}\right)\right) \\
&= -\frac{\dot S(\vartheta_0, t)}{\sqrt{\varepsilon}\,\sigma(t)}\int_{t_0}^{t} e^{-\frac{t-s}{\varepsilon}q_t}\,\eta_{s,\varepsilon} Y_s\,\mathrm{d}s\left(1 + O\left(\sqrt{\varepsilon}\right)\right) \\
&= -\sqrt{\varepsilon}\,\frac{\left[\dot b(\vartheta_0, t) f(\vartheta_0, t) + b(\vartheta_0, t)\dot f(\vartheta_0, t)\right]}{b(\vartheta_0, t) f(\vartheta_0, t)}\,\eta_{t,\varepsilon} Y_t\,\left(1 + O\left(\sqrt{\varepsilon}\right)\right) \\
&= -\sqrt{\varepsilon}\left[\frac{\dot b(\vartheta_0, t)}{b(\vartheta_0, t)} + \frac{\dot f(\vartheta_0, t)}{f(\vartheta_0, t)}\right]\eta_{t,\varepsilon} Y_t\,\left(1 + O\left(\sqrt{\varepsilon}\right)\right).
\end{aligned}
$$

Here we used the relations $\delta_m(s) = O\left(\sqrt{\varepsilon}\right)$ and $\delta_Y(s) = O\left(\sqrt{\varepsilon}\right)$. We have as well

$$
\begin{aligned}
\frac{1}{\varepsilon}\int_{t_0}^{t} & \Phi_*(\vartheta, s, t)\,\frac{\left[b\left(\vartheta^\star_{s,\varepsilon}, s\right) - b(\vartheta_0, s)\right] f(\vartheta_0, s)}{\sigma(s)} m_{0,s}\mathrm{d}s \\
&= \frac{\dot b(\vartheta_0, t) f(\vartheta_0, t)}{\sqrt{\varepsilon}\sigma(t)}\int_{t_0}^{t} e^{-\frac{t-s}{\varepsilon}q_t}\,\eta_{s,\varepsilon} Y_s\,\mathrm{d}s\left(1 + O\left(\sqrt{\varepsilon}\right)\right) \\
&= \sqrt{\varepsilon}\,\frac{\dot b(\vartheta_0, t)}{b(\vartheta_0, t)}\,\eta_{t,\varepsilon} Y_t\left(1 + O\left(\sqrt{\varepsilon}\right)\right).
\end{aligned}
$$

The order of the last integral is obtained as follows.

$$
\begin{aligned}
\mathbf{E}_{\vartheta_0} & \left(\int_{t_0}^{t} \Phi_*(\vartheta_0, s, t)\left[b\left(\vartheta^\star_{s,\varepsilon}, s\right) - b(\vartheta_0, s)\right]\mathrm{d}\bar W_s\right)^2 \\
&= \int_{t_0}^{t} \Phi_*(\vartheta_0, s, t)^2\,\mathbf{E}_{\vartheta_0}\left[b\left(\vartheta^\star_{s,\varepsilon}, s\right) - b(\vartheta_0, s)\right]^2\mathrm{d}s \\
&\leq C\int_{t_0}^{t} \Phi_*(\vartheta_0, s, t)^2\,\mathbf{E}_{\vartheta_0}\left|\vartheta^\star_{s,\varepsilon} - \vartheta_0\right|^2\mathrm{d}s \\
&\leq C\,\varepsilon\int_{t_0}^{t} \Phi_*(\vartheta_0, s, t)^2\,\mathrm{d}s \leq C\,\varepsilon^2.
\end{aligned}
$$

Hence,

$$
\int_{t_0}^{t} \Phi_*(\vartheta_0, s, t)\left[b\left(\vartheta^\star_{s,\varepsilon}, s\right) - b(\vartheta_0, s)\right]\mathrm{d}\bar W_s = O(\varepsilon).
$$

Finally, we obtain

$$\varepsilon^{-1/2}\delta_m(t) = -\frac{\dot f(\vartheta_0,t)}{f(\vartheta_0,t)}\,\eta_{t,\varepsilon}\,Y_t\left(1+O\left(\sqrt{\varepsilon}\right)\right) \Longrightarrow -\frac{\dot f(\vartheta_0,t)}{f(\vartheta_0,t)}\,\eta_t(\vartheta_0)\,Y_t.$$

Consider now $\mathbf{E}_{\vartheta_0}\delta_m(t)^2$. The vector $\left(\eta_{t,\varepsilon},\,Y_t\right)\Rightarrow\left(\eta_t(\vartheta_0),\,Y_t\right)$ and we have to estimate the correlation $\mathbf{E}_{\vartheta_0}\left[\eta_{t,\varepsilon}\,Y_t\right]$.

Note that

$$\mathbf{E}_{\vartheta_0}\left[\eta_{t,\varepsilon}\,Y_t\right] = \mathbf{E}_{\vartheta_0}\left[\eta_{t,\varepsilon}\,\mathbf{E}_{\vartheta_0}\left(Y_t|\mathfrak{F}_t^X\right)\right] = \mathbf{E}_{\vartheta_0}\left[\eta_{t,\varepsilon}\,m(\vartheta_0,t)\right]$$

and

$$m(\vartheta_0,t) = y_0\phi(\vartheta_0,0,t) + \int_0^t \phi(\vartheta_0,s,t)\,b(\vartheta_0,s)\,\mathrm{d}\bar W_s,$$

where

$$\phi(\vartheta_0,s,t) = \exp\left(\int_s^t a(\vartheta_0,v)\,\mathrm{d}v\right).$$

Therefore,

$$\begin{aligned}
&\mathbf{E}_{\vartheta_0}\left[\eta_{t,\varepsilon}\,Y_t\right]\\
&= \mathrm{I}_0^t(\vartheta_0)^{-1}\,\mathbf{E}_{\vartheta_0}\int_0^t \frac{\dot S(\vartheta_0,s)}{\sqrt{2S(\vartheta_0,s)\,\sigma(s)}}\,\bar\xi_{s,\varepsilon}\,\mathrm{d}\bar W_s \int_0^t \phi(\vartheta_0,s,t)\,b(\vartheta_0,s)\,\mathrm{d}\bar W_s\\
&= \mathrm{I}_0^t(\vartheta_0)^{-1}\int_0^t \frac{\dot S(\vartheta_0,s)}{\sqrt{2S(\vartheta_0,s)\,\sigma(s)}}\phi(\vartheta_0,s,t)\,b(\vartheta_0,s)\;\mathbf{E}_{\vartheta_0}\bar\xi_{s,\varepsilon}\,\mathrm{d}s = 0,
\end{aligned}$$

where we omitted terms of order $O(\varepsilon)$. Therefore,

$$\lim_{\varepsilon\to 0}\mathbf{E}_{\vartheta_0}\left[\eta_{t,\varepsilon}\,Y_t\right] = 0$$

and we obtain the convergence (3.40) too. □

Remark 3.7 As it follows from Theorem 3.3 we have the weak convergences

$$\frac{Y_t - m_{t,\varepsilon}^\star}{\sqrt{\varepsilon}} \Longrightarrow \frac{\dot f(\vartheta_0,t)}{f(\vartheta_0,t)}\,\eta_t(\vartheta_0)\,Y_t + \sqrt{\frac{b(\vartheta_0,t)\,\sigma(t)}{2f(\vartheta_0,t)}}\,\left(\xi_t^{(2)} - \xi_t^{(1)}\right).$$

Recall that the Gaussian process $\eta_t(\vartheta_0), 0 < t \leq T$ and Gaussian r.v.'s $\xi_t^{(1)}, \xi_t^{(2)}, 0 < t \leq T$ are independent.

The variance of the limit law is

$$\Gamma_*(\vartheta_0,t)=\frac{b(\vartheta_0,t)\,\sigma(t)}{f(\vartheta_0,t)}+\frac{\dot f(\vartheta_0,t)^2\,\mathbf{E}_{\vartheta_0}Y_t^2}{f(\vartheta_0,t)^2\,\mathrm{I}_0^t(\vartheta_0)}=\gamma_*(\vartheta_0,t)+\frac{\dot f(\vartheta_0,t)^2\,\mathbf{E}_{\vartheta_0}Y_t^2}{f(\vartheta_0,t)^2\,\mathrm{I}_0^t(\vartheta_0)}$$

in the second case.

For the integral linear and quadratic functionals, we have the following limits : for any $t_0>0$

$$\int_{t_0}^{T}\frac{Y_t-m(\vartheta_0,t)}{\sqrt{\varepsilon}}\,\mathrm{d}t\longrightarrow 0,$$

$$\int_{t_0}^{T}\left(\frac{Y_t-m(\vartheta_0,t)}{\sqrt{\varepsilon}}\right)^2\mathrm{d}t\longrightarrow\int_{t_0}^{T}\frac{b(\vartheta_0,t)\,\sigma(t)}{f(\vartheta_0,t)}\mathrm{d}t,$$

and

$$\int_{t_0}^{T}\frac{Y_t-m_{t,\varepsilon}^{\star}}{\sqrt{\varepsilon}}\,\mathrm{d}t\Longrightarrow\int_{t_0}^{T}\frac{\dot f(\vartheta_0,t)}{f(\vartheta_0,t)}\,\eta_t(\vartheta_0)\,Y_t\mathrm{d}t,$$

$$\int_{t_0}^{T}\left(\frac{Y_t-m_{t,\varepsilon}^{\star}}{\sqrt{\varepsilon}}\right)^2\mathrm{d}t\Longrightarrow\int_{t_0}^{T}\left[\frac{\dot f(\vartheta_0,t)^2}{f(\vartheta_0,t)^2}\,\eta_t(\vartheta_0)^2\,Y_t^2+\frac{b(\vartheta_0,t)\,\sigma(t)}{f(\vartheta_0,t)}\right]\mathrm{d}t.$$

The proofs follow from Lemma 3.3.

Remark 3.8 It is interesting to note that if $f(\vartheta,t)=f(t)$ and the function $b(\vartheta,t)$ depends on unknown parameter ϑ, then the limit representation and the limit variance in both cases coincide. This means that *the unobserved component Y_t is approximated by $m_{t,\varepsilon}^{\star}$ with the same asymptotic precision as if the value ϑ is known.*

3.1.5 Asymptotic Efficiency

Consider the question of asymptotic optimality of the estimation of the conditional expectation $m(\vartheta,t)$. Remind that $m(\vartheta,t)$ itself is mean-squared optimal estimator of Y_t.

Let us denote

$$\mathcal{S}_t^{\star}(\vartheta)^2=\frac{\dot f(\vartheta,t)^2}{f(\vartheta,t)^2}\mathbf{E}_{\vartheta}\left[\eta_t(\vartheta)\,Y_t\right]^2=\frac{\dot f(\vartheta,t)^2\,\mathbf{E}_{\vartheta}Y_t^2}{f(\vartheta,t)^2\,\mathrm{I}_0^t(\vartheta)}.$$

Here

$$\mathbf{E}_{\vartheta}Y_t^2=y_0^2\phi(\vartheta,0,t)^2+\int_0^t\phi(\vartheta,s,t)^2\,b(\vartheta,s)^2\,\mathrm{d}s.$$

We have the lower minimax bound: for any $t \in (0, T]$ and any $\vartheta_0 \in \Theta$

$$\lim_{\nu \to} \varliminf_{\varepsilon \to 0} \sup_{|\vartheta - \vartheta_0| \leq \nu} \varepsilon^{-1} \mathbf{E}_\vartheta \left| \bar m(t) - m(\vartheta, t) \right|^2 \geq \mathcal{S}_t^\star(\vartheta_0)^2, \tag{3.41}$$

where $\bar m(t)$ is an arbitrary estimator of $m(\vartheta, t)$. Let us remind the construction of the lower bound. We have elementary estimates

$$\begin{aligned}\sup_{|\vartheta - \vartheta_0| \leq \nu} \mathbf{E}_\vartheta \left| \bar m(t) - m(\vartheta, t) \right|^2 &\geq \int_{\vartheta_0 - \nu}^{\vartheta_0 + \nu} \mathbf{E}_\vartheta \left| \bar m(t) - m(\vartheta, t) \right|^2 p(\vartheta)\, \mathrm{d}\vartheta \\ &\geq \int_{\vartheta_0 - \nu}^{\vartheta_0 + \nu} \mathbf{E}_\vartheta \left| \tilde m(t) - m(\vartheta, t) \right|^2 p(\vartheta)\, \mathrm{d}\vartheta.\end{aligned}$$

Here we introduced a continuous positive density $p(\vartheta), \vartheta \in (\vartheta_0 - \delta, \vartheta_0 + \delta)$ and $\tilde m(t)$ is Bayesian estimator, which corresponds to this density $p(\cdot)$ and quadratic loss function, i.e.,

$$\tilde m(t) = \int_{\vartheta_0 - \nu}^{\vartheta_0 + \nu} m(\vartheta, t)\, p\left(\vartheta | X^t\right) \mathrm{d}\vartheta = \frac{\displaystyle\int_{\vartheta_0 - \nu}^{\vartheta_0 + \nu} m(\vartheta, t)\, p(\vartheta)\, L\left(\vartheta, X^t\right) \mathrm{d}\vartheta}{\displaystyle\int_{\vartheta_0 - \nu}^{\vartheta_0 + \nu} p(\vartheta)\, L\left(\vartheta, X^t\right) \mathrm{d}\vartheta}.$$

Below, we change the variables $\vartheta = \vartheta_u = \vartheta_0 + \sqrt{\varepsilon} u$ and denote

$$Z_{t,\varepsilon}(\vartheta_0, u) = \frac{L\left(\vartheta_0 + \sqrt{\varepsilon} u, X^t\right)}{L\left(\vartheta_0, X^t\right)}, \qquad \mathbb{U}_\varepsilon = \left(-\nu \varepsilon^{-1/2}, \nu \varepsilon^{-1/2}\right).$$

We can write

$$\begin{aligned}\frac{\displaystyle\int_{\vartheta_0 - \nu}^{\vartheta_0 + \nu} m(\vartheta, t)\, p(\vartheta)\, L\left(\vartheta, X^t\right) \mathrm{d}\vartheta}{\displaystyle\int_{\vartheta_0 - \nu}^{\vartheta_0 + \nu} p(\vartheta)\, L\left(\vartheta, X^t\right) \mathrm{d}\vartheta} &= \frac{\displaystyle\int_{\vartheta_0 - \nu}^{\vartheta_0 + \nu} m(\vartheta, t)\, p(\vartheta) \frac{L\left(\vartheta, X^t\right)}{L\left(\vartheta_0, X^t\right)} \mathrm{d}\vartheta}{\displaystyle\int_{\vartheta_0 - \nu}^{\vartheta_0 + \nu} p(\vartheta) \frac{L\left(\vartheta, X^t\right)}{L\left(\vartheta_0, X^t\right)} \mathrm{d}\vartheta} \\ &= \int_{\mathbb{U}_\varepsilon} m(\vartheta_u, t)\, p(\vartheta_u)\, Z_{t,\varepsilon}(\vartheta_0, u)\, \mathrm{d}u \left(\int_{\mathbb{U}_\varepsilon} p(\vartheta_u)\, Z_{t,\varepsilon}(\vartheta_0, u)\, \mathrm{d}u\right)^{-1} \\ &= m(\vartheta_0, t) + \sqrt{\varepsilon}\, \dot m(\vartheta_0, t) \frac{\displaystyle\int_{\mathbb{U}_\varepsilon} u p(\vartheta_0)\, Z_{t,\varepsilon}(\vartheta_0, u)\, \mathrm{d}u}{\displaystyle\int_{\mathbb{U}_\varepsilon} p(\vartheta_0)\, Z_{t,\varepsilon}(\vartheta_0, u)\, \mathrm{d}u} (1 + o(1))\end{aligned}$$

$$
= m(\vartheta_0, t) - \sqrt{\varepsilon}\, \frac{\dot f(\vartheta_0, t)}{f(\vartheta_0, t)}\, m(\vartheta_0, t)\, \frac{\displaystyle\int_{\mathbb{U}_\varepsilon} u\, Z_{t,\varepsilon}(\vartheta_0, u)\, \mathrm{d}u}{\displaystyle\int_{\mathbb{U}_\varepsilon} Z_{t,\varepsilon}(\vartheta_0, u)\, \mathrm{d}u}\, (1 + o(1)),
$$

where we used the expansion

$$
\begin{aligned}
m\left(\vartheta_0 + \sqrt{\varepsilon}u, t\right) &= m(\vartheta_0, t) + u\dot m(\vartheta_0, t)\, \sqrt{\varepsilon}\,(1 + o(1)) \\
&= m(\vartheta_0, t) - u\, \frac{\dot f(\vartheta_0, t)}{f(\vartheta_0, t)} m(\vartheta_0, t)\, \sqrt{\varepsilon}\,(1 + o(1)).
\end{aligned}
$$

Therefore,

$$
\begin{aligned}
\frac{\tilde m(t) - m(\vartheta_0, t)}{\sqrt{\varepsilon}} &= -\frac{\dot f(\vartheta_0, t)}{f(\vartheta_0, t)}\, m(\vartheta_0, t)\, \frac{\displaystyle\int_{\mathbb{U}_\varepsilon} u\, Z_{t,\varepsilon}(\vartheta_0, u)\, \mathrm{d}u}{\displaystyle\int_{\mathbb{U}_\varepsilon} Z_{t,\varepsilon}(\vartheta_0, u)\, \mathrm{d}u}\, (1 + o(1)) \\
&= -\frac{\dot f(\vartheta_0, t)}{f(\vartheta_0, t)}\, Y_t\, \frac{\displaystyle\int_{\mathbb{U}_\varepsilon} u\, Z_{t,\varepsilon}(\vartheta_0, u)\, \mathrm{d}u}{\displaystyle\int_{\mathbb{U}_\varepsilon} Z_{t,\varepsilon}(\vartheta_0, u)\, \mathrm{d}u}\, (1 + o(1)).
\end{aligned}
$$

According to Theorem 3.1, the Bayes estimators constructed by observations X^T are asymptotically normal

$$
\frac{\tilde\vartheta_\varepsilon - \vartheta_0}{\sqrt{\varepsilon}} = \frac{\displaystyle\int_{\mathbb{U}_\varepsilon} u\, Z_{T,\varepsilon}(\vartheta_0, u)\, \mathrm{d}u}{\displaystyle\int_{\mathbb{U}_\varepsilon} Z_{T,\varepsilon}(\vartheta_0, u)\, \mathrm{d}u}\, (1 + o(1)) \Longrightarrow \eta_T(\vartheta_0)
$$

and all polynomial moments converge.

Hence,

$$
\frac{\tilde m(t) - m(\vartheta_0, t)}{\sqrt{\varepsilon}} \Longrightarrow -\frac{\dot f(\vartheta_0, t)}{f(\vartheta_0, t)}\, \eta_t(\vartheta_0)\, Y_t
$$

and we have the convergence ($\varepsilon \to 0$)

$$
\begin{aligned}
&\int_{\vartheta_0-\nu}^{\vartheta_0+\nu} \mathbf{E}_\vartheta \frac{|\tilde m(t) - m(\vartheta, t)|^2}{\varepsilon}\, p(\vartheta)\, \mathrm{d}\vartheta \\
&\qquad \longrightarrow \int_{\vartheta_0-\nu}^{\vartheta_0+\nu} \frac{\dot f(\vartheta, t)^2}{f(\vartheta, t)^2} \mathbf{E}_\vartheta \Big(\eta_t(\vartheta)\, Y_t\Big)^2 p(\vartheta)\, \mathrm{d}\vartheta
\end{aligned}
$$

$$=\int_{\vartheta_0-\nu}^{\vartheta_0+\nu} \mathcal{S}_t^\star\left(\vartheta\right)^2 p\left(\vartheta\right) \mathrm{d}\vartheta.$$

Therefore, we obtained the relation

$$\underline{\lim}_{\varepsilon\to 0} \sup_{|\vartheta-\vartheta_0|\leq\nu} \varepsilon^{-1}\mathbf{E}_\vartheta \left|\bar m\left(t\right)-m\left(\vartheta,t\right)\right|^2 \geq \int_{\vartheta_0-\nu}^{\vartheta_0+\nu} \mathcal{S}_t^\star\left(\vartheta\right)^2 p\left(\vartheta\right) \mathrm{d}\vartheta.$$

The function $\mathcal{S}_t^\star\left(\vartheta\right)^2$ is continuous on ϑ and we have the limit

$$\lim_{\nu\to 0}\int_{\vartheta_0-\nu}^{\vartheta_0+\nu} \mathcal{S}_t^\star\left(\vartheta\right)^2 p\left(\vartheta\right) \mathrm{d}\vartheta = \mathcal{S}_t^\star\left(\vartheta_0\right)^2 .$$

The bound (3.41) is proved.

From relations (3.38) and (3.39), we obtain

$$\frac{m_t^\star - m\left(\vartheta_0,t\right)}{\sqrt{\varepsilon}} \Longrightarrow -\frac{\dot f\left(\vartheta_0,t\right)}{f\left(\vartheta_0,t\right)}\,\eta_t\left(\vartheta_0\right)\,Y_t. \tag{3.42}$$

Therefore, the variance of the limit law is $\mathcal{S}_t^\star\left(\vartheta_0\right)^2$. It can be shown that the convergence (3.42) is uniform on compacts $\mathbb{K}\subset\Theta$, and we have

$$\lim_{\nu\to 0}\lim_{\varepsilon\to 0} \sup_{|\vartheta-\vartheta_0|\leq\nu} \varepsilon^{-1}\mathbf{E}_\vartheta \left|m_t^\star - m\left(\vartheta,t\right)\right|^2 = \mathcal{S}_t^\star\left(\vartheta_0\right)^2 ,$$

i.e., $m_t^\star$ is asymptotically efficient (minimax) estimator of $m\left(\vartheta,t\right)$.

Example 3.3 Consider the model of observations of Example 3.3

$$\mathrm{d}X_t = f\left(t\right)Y_t\mathrm{d}t + \varepsilon\sigma\left(t\right)\mathrm{d}W_t, \qquad X_0=0, \qquad 0\leq t\leq T,$$
$$\mathrm{d}Y_t = a\left(t\right)Y_t\mathrm{d}t + \sqrt{h\left(t\right)+\vartheta g\left(t\right)}\mathrm{d}V_t, \quad Y_0=y_0$$

where the functions $f\left(\cdot\right),\sigma\left(\cdot\right),a\left(\cdot\right),h\left(\cdot\right),g\left(\cdot\right)$ have continuous derivatives on t, are strictly positive and $\vartheta\in\left(\alpha,\beta\right)$, where $\alpha>0$. It is easy to see that the conditions $\mathscr{A}_*, \mathscr{G}^\star, \mathscr{G}_4^\star$ in this case are fulfilled.

The preliminary estimator is:

$$\vartheta_{\tau_\varepsilon,\varepsilon}^* = \left(\Psi_{\tau_\varepsilon,\varepsilon} - \int_0^{\tau_\varepsilon} f\left(t\right)^2 h\left(t\right)\mathrm{d}t\right)\left(\int_0^{\tau_\varepsilon} f\left(t\right)^2 g\left(t\right)\mathrm{d}t\right)^{-1}$$

where $\tau_\varepsilon=\varepsilon^{1/12}$, Fisher information

$$
\mathrm{I}_{\tau_\varepsilon}^{t}\left(\vartheta\right)=\int_{\tau_\varepsilon}^{t}\frac{f\left(s\right)g\left(s\right)^2}{8\left[h\left(s\right)+\vartheta g\left(s\right)\right]^{3/2}\sigma\left(s\right)}\mathrm{d}s.
$$

The random process $M(\vartheta^*_{\tau_\varepsilon,\varepsilon},t)$ is solution of the equation

$$
\mathrm{d}M(\vartheta^*_{\tau_\varepsilon,\varepsilon},t)=\frac{f\left(t\right)\left[h\left(t\right)+\vartheta^*_{\tau_\varepsilon,\varepsilon}g\left(t\right)\right]^{1/2}}{\varepsilon\sigma\left(t\right)}\left[\mathrm{d}X_t-M(\vartheta^*_{\tau_\varepsilon,\varepsilon},t)\mathrm{d}t\right]
$$

on the interval $[\tau_\varepsilon,T]$.

For the derivative, we have the following equation:

$$
\begin{aligned}
\mathrm{d}\dot M(\vartheta^*_{\tau_\varepsilon,\varepsilon},t)=&-\frac{f\left(t\right)\left[h\left(t\right)+\vartheta^*_{\tau_\varepsilon,\varepsilon}g\left(t\right)\right]^{1/2}}{\varepsilon\sigma\left(t\right)}\dot M(\vartheta^*_{\tau_\varepsilon,\varepsilon},t)\mathrm{d}t\\
&+\frac{f\left(t\right)g\left(t\right)}{2\varepsilon\sigma\left(t\right)\left[h\left(t\right)+\vartheta^*_{\tau_\varepsilon,\varepsilon}g\left(t\right)\right]^{1/2}}\left[\mathrm{d}X_t-M(\vartheta^*_{\tau_\varepsilon,\varepsilon},t)\mathrm{d}t\right].
\end{aligned}
$$

The initial values are defined by the relation like (3.32). The One-step MLE process has the same form

$$
\vartheta^\star_{t,\varepsilon}=\vartheta^*_{\tau_\varepsilon,\varepsilon}+\frac{1}{\mathrm{I}_{\tau_\varepsilon}^{t}\left(\vartheta^*_{\tau_\varepsilon,\varepsilon}\right)}\int_{\tau_\varepsilon}^{t}\frac{\dot M(\vartheta^*_{\tau_\varepsilon,\varepsilon},s)}{\varepsilon\sigma\left(s\right)^2}\left[\mathrm{d}X_s-M(\vartheta^*_{\tau_\varepsilon,\varepsilon},s)\mathrm{d}s\right],\qquad \tau_\varepsilon<t\leq T.
$$

The adaptive filter (3.37) is

$$
\mathrm{d}m^\star_{t,\varepsilon}=\frac{\left[h\left(t\right)+\vartheta^\star_{t,\varepsilon}g\left(t\right)\right]^{1/2}}{\varepsilon\sigma\left(t\right)}\left[\mathrm{d}X_t-f\left(t\right)m^\star_{t,\varepsilon}\mathrm{d}t\right],\qquad \tau_\varepsilon<t\leq T.
$$

For the initial value $m^\star_{\tau_\varepsilon,\varepsilon}=\hat m_{\tau_\varepsilon}$, we write

$$
\hat m_{\tau_\varepsilon}=y_0\phi\left(\vartheta^*_{\tau_\varepsilon,\varepsilon},\tau_\varepsilon\right)+\phi\left(\vartheta^*_{\tau_\varepsilon,\varepsilon},\tau_\varepsilon\right)\left(X_{\tau_\varepsilon}H\left(\vartheta^*_{\tau_\varepsilon,\varepsilon},\tau_\varepsilon\right)-\int_0^{\tau_\varepsilon}X_sH'\left(\vartheta^*_{\tau_\varepsilon,\varepsilon},s\right)\mathrm{d}s\right).
$$

Here

$$
\phi\left(\vartheta,s\right)=\exp\left(-\int_0^s q_\varepsilon\left(\vartheta,v\right)\mathrm{d}v\right),\qquad q_\varepsilon\left(\vartheta,t\right)=-a\left(\vartheta,t\right)+\frac{\gamma\left(\vartheta,t\right)f\left(\vartheta,t\right)^2}{\varepsilon^2\sigma\left(t\right)^2},
$$

$$
H\left(\vartheta,s\right)=\phi\left(\vartheta,s\right)^{-1}\frac{\gamma\left(\vartheta,s\right)f\left(\vartheta,s\right)}{\varepsilon^2\sigma\left(s\right)^2}.
$$

Recall that $m^\star_{t,\varepsilon}$ is asymptotically efficient estimator of the process $m\left(\vartheta_0,t\right)$.

3.2 Conditionally Gaussian System

3.2.1 *Model of Observations*

Consider the conditionally Gaussian processes (2.91) and (2.92) but with the noise in observation equation only.

$$\mathrm{d}X_t = f(\vartheta, t, X_t)\, Y_t\, \mathrm{d}t + \varepsilon\sigma(t, X_t)\,\mathrm{d}W_t, \qquad X_0 = 0, \quad 0 \le t \le T,$$
$$\mathrm{d}Y_t = a(\vartheta, t, X_t)\, Y_t\, \mathrm{d}t + b(\vartheta, t)\,\mathrm{d}V_t, \qquad Y_0 = y_0.$$

The functions $f(\cdot), \sigma(\cdot), a(\cdot), b(\cdot)$ are known and ϑ is an unknown parameter. Recall the conditions from Sect. 2.3, which we slightly modify because now $b(\vartheta, t, x) = b(\vartheta, t)$.

Conditions $\mathscr{H}$.

$\mathscr{H}_0$. *The unknown parameter* $\vartheta \in \Theta \subset \mathcal{R}$, *where* $\Theta = (\alpha, \beta)$.

$\mathscr{H}_1$. *The functions* $f(\vartheta, t, x), \sigma(t, x), a(\vartheta, t, x); \vartheta \in \Theta, t \in [0, T], x \in \mathcal{R}$ *have continuous bounded derivatives w.r.t.* x.

$\mathscr{H}_2$. *There exists a constant* $c > 0$ *which does not depend on* t *and* x *such that*

$$\sigma(t, x)^2 > c.$$

$\mathscr{H}_3$. *There exists a constant* $C > 0$ *which does not depend on* ϑ, t *and* x *such that*

$$|f(\vartheta, t, x)| + |a(\vartheta, t, x)| < C.$$

The limit ($\varepsilon = 0$) model is

$$x_t'(\vartheta) = f(\vartheta, t, x_t)\, y_t, \qquad x_0 = 0, \qquad 0 \le t \le T,$$
$$\mathrm{d}y_t(\vartheta) = a(\vartheta, t, x_t)\, y_t(\vartheta)\,\mathrm{d}t + b(\vartheta, t)\,\mathrm{d}V_t, \qquad y_0.$$

Here $\left(x^T, y^T\right) = (x_t(\vartheta), y_t(\vartheta), 0 \le t \le T)$ are random processes and $x_t'(\vartheta) = \frac{\partial x_t}{\partial t}$. The process

$$y_t(\vartheta) = y_0\phi_t + \phi_t \int_0^t \phi_s^{-1} b(\vartheta, s)\,\mathrm{d}V_s, \qquad \phi_t = \exp\left(\int_0^t a(\vartheta, s, x_s)\,\mathrm{d}s\right).$$

Recall that by Theorem 2.11 the equations of optimal filter are

$$dm(\vartheta, t) = A(\vartheta, t, X)\, m(\vartheta, t)\,\mathrm{d}t + B(\vartheta, t, X)\,\mathrm{d}X_t, \qquad m(\vartheta, 0) = y_0,$$
$$\frac{\partial\gamma(\vartheta, t, X)}{\partial t} = 2a(\vartheta, t, X_t)\,\gamma(\vartheta, t, X) - \frac{f(\vartheta, t, X_t)^2\,\gamma(\vartheta, t, X)^2}{\varepsilon^2\sigma(t, X_t)^2} + b(\vartheta, t)^2,$$

where $\gamma(\vartheta, 0, X) = 0$ and we used the notation

$$A(\vartheta, t, X) = a(\vartheta, t, X_t) - \frac{f(\vartheta, t, X_t)^2 \gamma(\vartheta, t, X)}{\varepsilon^2 \sigma(t, X_t)^2},$$

$$B(\vartheta, t, X) = \frac{f(\vartheta, t, X_t) \gamma(\vartheta, t, X)}{\varepsilon^2 \sigma(t, X_t)^2}.$$

Let us verify the convergences $X_t \to x_t$ and $Y_t \to y_t$.

Proposition 3.3 *Let the conditions $\mathscr{H}_1$, $\mathscr{H}_3$ hold. Then*

$$\sup_{0\leq t\leq T} \mathbf{E}_{\vartheta_0} |X_t - x_t|^2 \longrightarrow 0, \qquad \sup_{0\leq t\leq T} \mathbf{E}_{\vartheta_0} |Y_t - y_t|^2 \longrightarrow 0. \tag{3.43}$$

Proof We have the following equations for $\delta_x(t) = X_t - x_t$, $\delta_y(t) = Y_t - y_t$, $\delta_f(t) = f(\vartheta_0, s, X_s) - f(\vartheta_0, s, x_s)$ and $\delta_a(t) = a(\vartheta_0, s, X_s) - a(\vartheta_0, s, x_s)$

$$\mathrm{d}\delta_x(t) = f(\vartheta, t, X_t)\,\delta_y(t)\,\mathrm{d}t + \delta_f(t)\,y_t \mathrm{d}t + \varepsilon\sigma(t, X_t)\,\mathrm{d}W_t, \qquad \delta_x(t) = 0,$$

$$\frac{\mathrm{d}\delta_y(t)}{\mathrm{d}t} = a(\vartheta, t, X_t)\,\delta_y(t) + \delta_a(t)\,y_t, \qquad \delta_y(0) = 0.$$

The solution of the second equation is the random function

$$\delta_y(t) = \phi_t \int_0^t \phi_s^{-1} \delta_a(s)\, y_s \mathrm{d}s.$$

Substitution of this function in the first equation gives us the expression

$$\delta_x(t) = \int_0^t f(\vartheta, s, X_s)\,\phi_s \int_0^s \phi_r^{-1} \delta_a(r)\, y_r \mathrm{d}r\, \mathrm{d}s + \int_0^t \delta_f(s)\, y_s \mathrm{d}s + \varepsilon \int_0^t \sigma(s, X_s)\,\mathrm{d}W_s.$$

By conditions $\mathscr{H}_1$, $\mathscr{H}_3$, we can write

$$\begin{aligned}
&|\delta_x(t)| \\
&\leq C \int_0^t \int_0^s |\delta_a(r)|\, |y_r|\, \mathrm{d}r\, \mathrm{d}s + \int_0^t \left|\delta_f(s)\right| |y_s|\, \mathrm{d}s + \varepsilon \left| \int_0^t \sigma(s, X_s)\,\mathrm{d}W_s \right| \\
&\leq C \int_0^t \int_0^s |\delta_x(r)|\, |y_r|\, \mathrm{d}r\, \mathrm{d}s + C \int_0^t |\delta_x(s)|\, |y_s|\, \mathrm{d}s + \varepsilon \left| \int_0^t \sigma(s, X_s)\,\mathrm{d}W_s \right| \\
&\leq C \int_0^t |\delta_x(s)|\, |y_s|\, \mathrm{d}s + \varepsilon \left| \int_0^t \sigma(s, X_s)\,\mathrm{d}W_s \right|.
\end{aligned}$$

Thus, by Grönwall's lemma , we obtain

$$\left|\delta_x\left(t\right)\right| \leq \varepsilon \left|\int_0^t \sigma\left(s, X_s\right) \mathrm{d}W_s\right| + \varepsilon \int_0^t \hat{\phi}_s^t \left|\int_0^s \sigma\left(r, X_r\right) \mathrm{d}W_r\right| \mathrm{d}s \tag{3.44}$$

where (below, we use the estimate $|a\left(\vartheta, t, x\right)| < C_a$)

$$\hat{\phi}_s^t = \exp\left(C_a \int_s^t |y_r|\,\mathrm{d}r\right) \leq \exp\left(C_a \int_0^T |y_r|\,\mathrm{d}r\right).$$

Now the convergence $X_t \to x_t$ follows directly from the estimate (3.44).
Further,

$$\begin{aligned}\mathbf{E}_\vartheta \delta_x\left(t\right)^2 &\leq C\,\varepsilon^2 + C\varepsilon^2 \int_0^t \mathbf{E}_\vartheta \left(\hat{\phi}_s^t\right)^2 \left(\int_0^s \sigma\left(r, X_r\right) \mathrm{d}W_r\right)^2 \mathrm{d}s\\ &\leq C\varepsilon^2 \int_0^t \left[\mathbf{E}_\vartheta \left(\hat{\phi}_0^T\right)^4 \mathbf{E}_\vartheta \left(\int_0^s \sigma\left(r, X_r\right) \mathrm{d}W_r\right)^4\right]^{1/2} \mathrm{d}s.\end{aligned}$$

We have

$$\mathbf{E}_\vartheta \left(\hat{\phi}_0^T\right)^4 = \mathbf{E}_\vartheta \exp\left(4TC_a\, \frac{1}{T} \int_0^T |y_r|\,\mathrm{d}r\right) \leq \frac{1}{T} \int_0^T \mathbf{E}_\vartheta \exp\left(4TC_a\, |y_r|\right)\mathrm{d}r < C$$

because $y_s, 0 \leq s \leq T$ is a Gaussian process with finite variance.
Therefore,

$$\mathbf{E}_\vartheta \delta_x\left(t\right)^2 \leq C\,\varepsilon^2.$$

Moreover, applying the similar estimate, it can be shown that for any $p = 2, 3, \ldots$ there exists a constant $C = C\left(p\right)$ such that

$$\mathbf{E}_\vartheta \delta_x\left(t\right)^{2p} \leq C\,\varepsilon^{2p}. \tag{3.45}$$

For the process $\delta_y\left(t\right)$, the equation

$$\delta_y\left(t\right) = \int_0^t a\left(\vartheta, s, X_s\right) \delta_y\left(s\right) \mathrm{d}s + \int_0^t \delta_a\left(s\right) y_s \mathrm{d}s$$

allows to write

$$\mathbf{E}_\vartheta \delta_y\left(t\right)^2 \leq C \int_0^t \mathbf{E}_\vartheta \delta_y\left(s\right)^2 \mathrm{d}s + C \int_0^t \left[\mathbf{E}_\vartheta \delta_x\left(s\right)^4 \mathbf{E}_\vartheta y_s^4\right]^{1/2} \mathrm{d}s.$$

Now (3.43) follows from (3.45) and Grönwall lemma. □

3.2.2 Parameter Estimator

Quadratic Variation

As before, we have to estimate ϑ and then using this estimator to write the adaptive Kalman filter. The true value is denoted ϑ_0. The preliminary estimator is based on the quadratic variation of the trend coefficient of the observed process. This quadratic variation on the interval $[0, \tau]$ is

$$\Psi_{\tau,\varepsilon}^{*} = \lim_{\max_i (t_{i+1}-t_i)\to 0} \sum_{0\leq t_i<t_{i+1}\leq\tau} \left[f\left(\vartheta_0, t_{i+1}, X_{t_{i+1}}\right) Y_{t_{i+1}} - f\left(\vartheta_0, t_i, X_{t_i}\right) Y_{t_i}\right]^2$$

$$= \varepsilon^2 \int_0^\tau f_x'\left(\vartheta_0, t, X_t\right)^2 Y_t^2 \sigma\left(t\right)^2 \mathrm{d}t + \int_0^\tau f\left(\vartheta_0, t, X_t\right)^2 b\left(\vartheta_0, t\right)^2 \mathrm{d}t.$$

The limit of the quadratic variation is

$$\lim_{\varepsilon\to 0} \Psi_{\tau,\varepsilon}^{*} = \int_0^\tau f\left(\vartheta_0, t, x_t\right)^2 b\left(\vartheta_0, t\right)^2 \mathrm{d}t \equiv \Psi_\tau\left(\vartheta_0\right),$$

where $\Psi_\tau\left(\vartheta_0\right)$ is a random variable.

Introduce the statistic

$$\Psi_{\tau,\varepsilon} = \sum_{i=0}^{N_{\tau,\varepsilon}-1} \left(\frac{X_{t_{i+1}+\delta_\varepsilon} - X_{t_{i+1}}}{\delta_\varepsilon} - \frac{X_{t_i+\delta_\varepsilon} - X_{t_i}}{\delta_\varepsilon}\right)^2,$$

where $t_{i+1} = t_i + \varphi_\varepsilon$, $\delta_\varepsilon = \varepsilon$, $\varphi_\varepsilon = \varepsilon^{1/3}$ and $N_{\tau,\varepsilon} = \left[\tau\varepsilon^{-1/3}\right]$ ($[A]$ means integer part of A).

Using calculations similar to that of Sect. 3.1.2 (see (3.7)–(3.8)), we obtain the same estimate

$$\mathbf{E}_{\vartheta_0}\left(\frac{X_{t_{i+1}+\delta_\varepsilon} - X_{t_{i+1}}}{\delta_\varepsilon} - \frac{X_{t_i+\delta_\varepsilon} - X_{t_i}}{\delta_\varepsilon} - f\left(\vartheta_0, t_i, X_{t_i}\right)\left[Y_{t_{i+1}} - Y_{t_i}\right]\right)^2$$

$$\leq C\frac{\varepsilon^2}{\delta_\varepsilon} + C\delta_\varepsilon + C\varphi_\varepsilon^2.$$

In the calculation of the quadratic variation

$$Q_{\tau,\varepsilon} = \sum_{i=0}^{N_{\tau,\varepsilon}-1} f\left(\vartheta_0, t_i, X_{t_i}\right)^2 \left[Y_{t_{i+1}} - Y_{t_i}\right]^2$$

we as well follow the same lines.

Denote

$$Q^*_{\tau,\varepsilon} = \sum_{i=0}^{N_{\tau,\varepsilon}-1} f\left(\vartheta_0, t_i, x_{t_i}\right)^2 \left[y_{t_{i+1}} - y_{t_i}\right]^2$$

Then using (3.44), it can be shown that

$$Q_{\tau,\varepsilon} - Q^*_{\tau,\varepsilon} \longrightarrow 0$$

and

$$Q^*_{\tau,\varepsilon} \longrightarrow \int_0^\tau f\left(\vartheta_0, t, x_t\right)^2 b\left(\vartheta_0, t\right)^2 \mathrm{d}t.$$

Finally, we obtain the convergence

$$\Psi_{\tau,\varepsilon} \longrightarrow \Psi_\tau\left(\vartheta_0\right) = \int_0^\tau f\left(\vartheta_0, t, x_t\right)^2 b\left(\vartheta_0, t\right)^2 \mathrm{d}t$$

and the estimate

$$\mathbf{E}_{\vartheta_0} \left|\Psi_{\tau,\varepsilon} - \Psi_\tau\left(\vartheta_0\right)\right|^p \leq C\varepsilon^{p/3}.$$

MME

Of course, we cannot use $\Psi_\tau\left(\vartheta\right), \vartheta \in \Theta$ for construction of MME because its calculation requires the knowledge of $x_t, 0 \leq t \leq \tau$. If $d = 1$, then the MME $\vartheta^*_{\tau,\varepsilon}$ can be defined by the equation

$$\Psi_{\tau,\varepsilon} = \Psi_\tau\left(\vartheta^*_{\tau,\varepsilon}, X^\tau\right),$$

where

$$\Psi_\tau\left(\vartheta, X^\tau\right) = \int_0^\tau f\left(\vartheta, t, X_t\right)^2 b\left(\vartheta, t\right)^2 \mathrm{d}t.$$

Example 3.4 If the volatility coefficient of the hidden equation is $b\left(\vartheta, t\right)^2 = h\left(t\right) + \vartheta g\left(t\right)$ and $f\left(\vartheta, t, x\right) = f\left(t, x\right)$ (all functions and the parameter are positive), then the MME

$$\vartheta^*_{\tau,\varepsilon} = \left(\int_0^\tau f\left(t, X_t\right)^2 g\left(t, X_t\right) \mathrm{d}t\right)^{-1} \left(\Psi_{\tau,\varepsilon} - \int_0^\tau f\left(t, X_t\right)^2 h\left(t, X_t\right) \mathrm{d}t\right) \longrightarrow \vartheta_0,$$

i.e., is consistent.

In the case $d > 1$, we can take d different values $0 < \tau_1 < \ldots < \tau_d < T$ and to define the MME $\vartheta^*_{\tau,\varepsilon}$ as solution of the system of equations

$$\Psi_{\tau_k,\varepsilon} = \Psi_{\tau_k}\left(\vartheta^*_{\tau,\varepsilon}, X^{\tau_k}\right), \qquad k = 1, \ldots, d.$$

Of course, to have the consistency of this estimator, we need to suppose that the limit system

$$\Psi_{\tau_k}(\vartheta_0) = \Psi_{\tau_k}(\vartheta), \qquad k = 1, \ldots, d$$

has a unique solution $\vartheta = \vartheta_0$ for all $\vartheta_0 \in \Theta$.

Consider the one-dimensional case $d = 1$. The limit equation is

$$\int_0^\tau f(\vartheta, t, x_t)^2 b(\vartheta, t)^2 \,\mathrm{d}t = \int_0^\tau f(\vartheta_0, t, x_t)^2 b(\vartheta_0, t)^2 \,\mathrm{d}t, \qquad \vartheta \in \Theta.$$

Here $x_t = x_t(\vartheta_0)$. Therefore, it is sufficient to suppose that the function $\psi(\vartheta, t, x) \equiv f(\vartheta_0, t, x)\, b(\vartheta_0, t)$ is strictly monotone in ϑ. For example, let $|\dot\psi(\vartheta, t, x)| > \kappa$ where $\kappa > 0$ does not depend of ϑ, t, x. The proof of the consistency under this condition is elementary. We have

$$\Psi_{\tau,\varepsilon} = \Psi_\tau\left(\vartheta^*_{\tau,\varepsilon}, X^\tau\right) = \Psi_\tau\left(\vartheta_0, X^\tau\right) + \left(\vartheta^*_{\tau,\varepsilon} - \vartheta_0\right)\dot\Psi_\tau\left(\bar\vartheta, X^\tau\right).$$

Hence,

$$\vartheta^*_{\tau,\varepsilon} - \vartheta_0 = \frac{\Psi_{\tau,\varepsilon} - \Psi_\tau\left(\vartheta_0, X^\tau\right)}{\dot\Psi_\tau\left(\bar\vartheta, X^\tau\right)}$$

and we obtain

$$\mathbf{E}_{\vartheta_0}|\vartheta^*_{\tau,\varepsilon} - \vartheta_0|^p \le C\mathbf{E}_{\vartheta_0}\left|\Psi_{\tau,\varepsilon} - \Psi_\tau\left(\vartheta_0, X^\tau\right)\right|^p \le C\varepsilon^{p/3}.$$

MLE and BE

To study the MLE $\hat\vartheta_\varepsilon$ and BE $\tilde\vartheta_\varepsilon$, we recall the definitions of these estimators (3.16) where the LR formula is

$$L\left(\vartheta, X^T\right) = \exp\left(\int_0^T \frac{M\left(\vartheta, t, X^t\right)}{\varepsilon^2\sigma(t, X_t)^2}\,\mathrm{d}X_t - \int_0^T \frac{M\left(\vartheta, t, X^t\right)^2}{2\varepsilon^2\sigma(t, X_t)^2}\,\mathrm{d}t\right), \qquad \theta \in \Theta.$$

Here $M\left(\vartheta, t, X^t\right) = f(\vartheta, t, X_t)\, m(\vartheta, t)$ and for simplicity we take $d = 1$.

Recall the notation $x_t = x_t(\vartheta_0)$, $S(\vartheta, t, x) = f(\vartheta, t, x)\, b(\vartheta, t)$ and set

$$G\left(\vartheta, \vartheta_0, x^T\right) = \int_0^T \frac{[f(\vartheta, t, x_t)\, b(\vartheta, t) - f(\vartheta_0, t, x_t)\, b(\vartheta_0, t)]^2}{2b(\vartheta, t)\, f(\vartheta, t, x_t)\, \sigma(t, x_t)} \mathrm{d}t$$

$$= \int_0^T \frac{[S(\vartheta, t, x_t) - S(\vartheta_0, t, x_t)]^2}{2S(\vartheta, t, x_t)\, \sigma(t, x_t)} \mathrm{d}t,$$

$$\mathrm{I}\left(\vartheta, x^T\right) = \int_0^T \frac{f(\vartheta, t, x_t)\, b(\vartheta, t)}{2\sigma(t, x_t)} \left(\frac{\dot f(\vartheta, t, x_t)}{f(\vartheta, t, x_t)} + \frac{\dot b(\vartheta, t)}{b(\vartheta, t)}\right)^2 \mathrm{d}t$$

$$= \int_0^T \frac{S(\vartheta, t, x_t)}{2\sigma(t, x_t)} \left[\frac{\partial}{\partial\vartheta} \ln S(\vartheta, t, x_t)\right]^2 \mathrm{d}t.$$

Remark that $S(\vartheta, t, x_t)$ depends on ϑ_0 too.

Regularity conditions $\mathscr{H}$:

$\hat{\mathscr{H}}_1$. *The functions $f(\vartheta, t, x), \sigma(t, x)$ have bounded derivatives in x and two bounded drivatives in ϑ.*

$\hat{\mathscr{H}}_2$. *The functions $f(\vartheta, t, x), \sigma(t, x), b(\vartheta, t)$ are bounded and separated from zero, i.e.,*

$$f(\vartheta, t, x) > c_1, \qquad \sigma(t, x) > c_2, \qquad b(\vartheta, t) > c_3,$$

where the constants $c_1 > 0, c_2 > 0, c_3 > 0$ do not depend on ϑ, t, x.

$\hat{\mathscr{H}}_3$. *There exists a constant $c_4 > 0$ such that for all $\vartheta \in \Theta, t \in [0, T], x \in \mathcal{R}$ the derivative*

$$|\dot S(\vartheta, t, x)| > c_4.$$

Let us denote $\mathfrak{F}_T^x$ the σ-algebra generated by the limit process $x_t(\vartheta_0), 0 \leq t \leq T$ and introduce conditionally Gaussian random variable $\zeta_*(\vartheta_0)$:

$$\mathcal{L}\left(\zeta_*(\vartheta_0) \middle| \mathfrak{F}_T^x\right) = \mathcal{N}\left(0, \mathrm{I}\left(\vartheta_0, x^T\right)^{-1}\right).$$

Below, it is shown (see (3.47)) that the normalized likelihood ratio

$$Z_\varepsilon(\vartheta_0, u) = \frac{L\left(\vartheta_0 + \sqrt{\varepsilon}\, u, X^T\right)}{L\left(\vartheta_0, X^T\right)}, \qquad u \in \mathbb{U}_\varepsilon = \left(\frac{\alpha - \vartheta_0}{\sqrt{\varepsilon}}, \frac{\beta - \vartheta_0}{\sqrt{\varepsilon}}\right)$$

admits the representation

$$Z_\varepsilon(\vartheta_0, u) = \exp\left(u\Delta_\varepsilon(\vartheta_0, X) - \frac{u^2}{2}\mathrm{I}\left(\vartheta_0, x^T\right) + r_\varepsilon\right)$$

where $r_\varepsilon \to 0$ and

$$\Delta_\varepsilon\left(\vartheta_0, X^T\right) = \int_0^T \frac{\dot S\left(\vartheta_0, t, x_t\right)\xi_{t,\varepsilon}}{\sqrt{2\sigma\left(t, x_t\right)S\left(\vartheta_0, t, x_t\right)}}{\rm d}\bar W_t$$

$$\Longrightarrow \Delta_0\left(\vartheta_0, x^T\right) \equiv \int_0^T \frac{\dot S\left(\vartheta_0, t, x_t\right)}{\sqrt{2\sigma\left(t, x_t\right)S\left(\vartheta_0, t, x_t\right)}}{\rm d}w_t .$$

The family of measures $\left(\mathbf{P}_\vartheta^{(\varepsilon)}, \vartheta \in \Theta\right)$ admitting such representation is called *local asymptotic mixing normal* (LAMN) . The lower bound on the risks of all estimators established by Davis [40] and Jeganathan [78]: for any estimator $\bar\vartheta_\varepsilon$ and any $\vartheta_0 \in \Theta$

$$\lim_{\nu\to 0}\varliminf_{\varepsilon\to 0}\sup_{|\vartheta-\vartheta_0|<\nu} \varepsilon^{-1}\mathbf{E}_\vartheta\left[\bar\vartheta_\varepsilon - \vartheta_0\right]^2 \geq \mathbf{E}_{\vartheta_0}\zeta_*\left(\vartheta_0\right)^2 .$$

The estimator ϑ_ε is called asymptotically efficient if the equality

$$\lim_{\nu\to 0}\varliminf_{\varepsilon\to 0}\sup_{|\vartheta-\vartheta_0|<\nu} \varepsilon^{-1}\mathbf{E}_\vartheta\left[\vartheta_\varepsilon - \vartheta_0\right]^2 = \mathbf{E}_{\vartheta_0}\zeta_*\left(\vartheta_0\right)^2$$

holds for all $\vartheta_0 \in \Theta$.

Theorem 3.4 *Let the regularity conditions $\hat{\mathscr{H}}$ hold. Then the MLE $\hat\vartheta_\varepsilon$ and BE $\tilde\vartheta_\varepsilon$ are unifromly consistent, uniformly on compacts $\mathbb{K} \subset \Theta$ asymptotically mixing normal*

$$\frac{\hat\vartheta_\varepsilon - \vartheta_0}{\sqrt{\varepsilon}} \Longrightarrow \zeta_*\left(\vartheta_0\right), \qquad \frac{\tilde\vartheta_\varepsilon - \vartheta_0}{\sqrt{\varepsilon}} \Longrightarrow \zeta_*\left(\vartheta_0\right),$$

the polynomial moments converge: for any $p > 0$

$$\lim_{\varepsilon\to 0}\varepsilon^{-p/2}\mathbf{E}_{\vartheta_0}\left|\hat\vartheta_\varepsilon - \vartheta_0\right|^p = \mathbf{E}_{\vartheta_0}\left|\zeta_*\left(\vartheta_0\right)\right|^p ,$$

and the both estimators are asymptotically efficient.

Proof The proof follows the same steps as the proof of the Theorem 3.1. Therefore, we have to study the properties of the normalized LR function is ($\vartheta_u = \vartheta_0 + \sqrt{\varepsilon}\, u$)

$$Z_\varepsilon\left(\vartheta_0, u\right) = \exp\left(\int_0^T \frac{M\left(\vartheta_u, t, X^t\right) - M\left(\vartheta_0, t, X^t\right)}{\varepsilon\sigma\left(t, X_t\right)}{\rm d}\bar W_t\right.$$

$$\left. - \int_0^T \frac{\left[M\left(\vartheta_u, t, X^t\right) - M\left(\vartheta_0, t, X^t\right)\right]^2}{2\varepsilon^2\sigma\left(t, X_t\right)^2}{\rm d}t\right), \qquad u \in \mathbb{U}_\varepsilon .$$

Note that it is possible to follow the proof of the Lemma 3.1 and to show that under conditions $\hat{\mathscr{H}}_{1-2}$ the relations

$$\sup_{t_0\leq t\leq T}\left|\frac{\gamma\left(\vartheta,t,X\right)}{\varepsilon}-\frac{b\left(\vartheta,t\right)\sigma\left(t,X_t\right)}{f\left(\vartheta,t,X_t\right)}\right|=O\left(\varepsilon\right),$$

$$\sup_{t_0\leq t\leq T}\left|\frac{b\left(\vartheta,t\right)\sigma\left(t,X_t\right)}{f\left(\vartheta,t,X_t\right)}-\frac{b\left(\vartheta,t\right)\sigma\left(t,x_t\right)}{f\left(\vartheta,t,x_t\right)}\right|=O\left(\varepsilon\right)$$

hold.

Let us replace $\gamma\left(\vartheta,t,X_t\right)$ by $\varepsilon b\left(\vartheta,t\right)\sigma\left(t,X_t\right)f\left(\vartheta,t,X_t\right)^{-1}$. This will simplify the calculations, but we can do it for the values $t\in\left[t_0,T\right]$ only. That is why we consider the LR function with the integrals on the time intervals $\left[t_0,T\right]$. The value t_0 can be chosen as $t_{0,\varepsilon}\rightarrow 0$ but sufficiently slowely. Then the limit integrals will be on the interval $\left[0,T\right]$. Moreover, we will write integrals on $\left[0,T\right]$.

Therefore, omitting the terms of order $O\left(\varepsilon\right)$ we obtain the equations for $m_u\left(t\right)=m\left(\vartheta_u,t\right)$ and $m_0\left(t\right)$

$$\begin{aligned}\mathrm{d}m_u\left(t\right)&=\frac{b\left(\vartheta_u,t\right)}{\varepsilon\sigma\left(t,X_t\right)}\left[f\left(\vartheta_0,t,X_t\right)m_0\left(t\right)-f\left(\vartheta_u,t,X_t\right)m_u\left(t\right)\right]\mathrm{d}t\\&\quad+a\left(\vartheta_u,t,X_t\right)m_u\left(t\right)\mathrm{d}t+b\left(\vartheta_u,t\right)\mathrm{d}\bar W_t,\qquad m_u\left(t_0\right),\\\mathrm{d}m_0\left(t\right)&=a\left(\vartheta_0,t,X_t\right)m_0\left(t\right)\mathrm{d}t+b\left(\vartheta_0,t\right)\mathrm{d}\bar W_t,\qquad m_0\left(t_0\right).\end{aligned}$$

Remark that to study the difference $\delta_m\left(t\right)=m_u\left(t\right)-m_0\left(t\right)$, we cannot use here Lemma 3.5 because the corresponding function

$$F\left(v\right)=q_v=\frac{b\left(\vartheta,v\right)f\left(\vartheta,v,X_v\right)}{\sigma\left(v,X_v\right)}$$

in the stochastic integral

$$\int_0^t\exp\left(-\frac{1}{\varepsilon}\int_s^t q_v\,\mathrm{d}v\right)\mathrm{d}\bar W_s$$

"depends on the future" of the Wiener process $\bar W_s$. Therefore, we need another proof.

The equation for $\delta_m\left(t\right)$ is

$$\begin{aligned}\mathrm{d}\delta_m\left(t\right)=&-\frac{1}{\varepsilon}\left[\frac{b\left(\vartheta_u,t\right)f\left(\vartheta_u,t,X_t\right)}{\sigma\left(t,X_t\right)}-\varepsilon a\left(\vartheta_u,t,X_t\right)\right]\delta_m\left(t\right)\mathrm{d}t+\delta_b\left(t\right)\mathrm{d}\bar W_t\\&-\frac{1}{\varepsilon}\left[\frac{b\left(\vartheta_u,t\right)\delta_f\left(t\right)}{\sigma\left(t,X_t\right)}-\varepsilon\delta_a\left(t\right)\right]m_0\left(t\right)\mathrm{d}t,\qquad\delta_m\left(t_0\right).\end{aligned}$$

Here $\delta_f(t) = f(\vartheta_u, t, X_t) - f(\vartheta_0, t, X_t), \delta_a(t) = a(\vartheta_u, t, X_t) - a(\vartheta_0, t, X_t), \delta_b(t) = b(\vartheta_u, t) - b(\vartheta_0, t)$. Omitting once more the terms of order $O(\varepsilon)$, we obtain the equation

$$\mathrm{d}\delta_m(t) = -\frac{1}{\varepsilon} A(\vartheta_u, t, X_t)\,\delta_m(t)\,\mathrm{d}t - \frac{1}{\varepsilon}\frac{b(\vartheta_u, t)\,\delta_f(t)}{\sigma(t, X_t)} m_0(t)\,\mathrm{d}t + \delta_b(t)\,\mathrm{d}\bar{W}_t,$$

where $A(\vartheta_u, t, X_t) = f(\vartheta_u, t, X_t)\,b(\vartheta_u, t)\sigma(t, X_t)^{-1}$.

Introduce the random process

$$z_t = \delta_m(t) - \int_{t_0}^{t} \delta_b(s)\,\mathrm{d}\bar{W}_s \equiv \delta_m(t) - H_t.$$

Then the equation for z_t is

$$\frac{\mathrm{d}z_t}{\mathrm{d}t} = -\frac{1}{\varepsilon} A(\vartheta_u, t, X_t)\,z_t - \frac{1}{\varepsilon} A(\vartheta_u, t, X_t)\,H_t - \frac{1}{\varepsilon}\frac{b(\vartheta_u, t)\,\delta_f(t)}{\sigma(t, X_t)} m_0(t), \quad z_{t_0}.$$

The solution of this linear equation is the random function

$$\begin{aligned} z_t = z_{t_0}\Phi(\vartheta_u, t_0, t) &- \frac{1}{\varepsilon}\int_{t_0}^{t} \Phi(\vartheta_u, s, t)\,A(\vartheta_u, s, X_s)\,H_s\mathrm{d}s \\ &- \frac{1}{\varepsilon}\int_{t_0}^{t} \Phi(\vartheta_u, s, t)\,\frac{b(\vartheta_u, s)\,\delta_f(s)}{\sigma(s, X_s)} m_0(s)\,\mathrm{d}s, \end{aligned}$$

where

$$\Phi(\vartheta_u, s, t) = \exp\left(-\frac{1}{\varepsilon}\int_{s}^{t} A(\vartheta_u, r, X_r)\,\mathrm{d}r\right).$$

We have the estimate

$$A(\vartheta_u, s, X_s) \geq q_* = \frac{c_1 c_3}{C_\sigma}, \qquad \sigma(t, x) \leq C_\sigma.$$

Therefore,

$$|z_{t_0}|\,\Phi(\vartheta_u, t_0, t) = e^{-\frac{q_*}{\varepsilon}(t-t_0)}\,O(1)$$

and this term we omit below.

Further,

$$\begin{aligned} A(\vartheta_u, s, X_s) - A(\vartheta_u, t, X_t) = A_t'(\vartheta_u, t, X_t)(s-t) \\ + A_x'(\vartheta_u, t, X_t)(X_s - X_t) + O(\varepsilon). \end{aligned}$$

Notice that for the small values of $t-s$ we have

$$\begin{aligned}X_t - X_s &= \int_s^t f(\vartheta_0, r, X_r)\, Y_r \mathrm{d}r + \varepsilon \int_s^t \sigma(r, X_r)\, \mathrm{d}W_r \\ &= (t-s)\, O(1) + \varepsilon\, |t-s|^{1/2}\, O(1).\end{aligned}$$

We have

$$\begin{aligned}&\frac{1}{\varepsilon}\int_{t_0}^t \Phi(\vartheta_u, s, t)\, A(\vartheta_u, s, X_s)\, H_s \mathrm{d}s \\ &\qquad = H_t - \frac{1}{\varepsilon}\int_{t_0}^t \Phi(\vartheta_u, s, t)\, A(\vartheta_u, s, X_s)\, [H_s - H_t]\, \mathrm{d}s.\end{aligned}$$

Now we can apply the elements of the proof of Lemma 1.7 and obtain the relations

$$\begin{aligned}&\frac{1}{\varepsilon}\int_{t_0}^t \Phi(\vartheta_u, s, t)\, A(\vartheta_u, s, X_s) \int_s^t \delta_b(r)\, \mathrm{d}\bar W_r\, \mathrm{d}s \\ &\quad = \frac{1}{\varepsilon}\int_{t_0}^t \Phi(\vartheta_u, s, t)\, A(\vartheta_u, s, X_s)\, \delta_b(s) \left[\bar W_t - \bar W_s\right] \mathrm{d}s \left(1 + O\left(\sqrt{\varepsilon}\right)\right) \\ &\quad = \frac{1}{\varepsilon} A(\vartheta_u, t, X_t)\, \delta_b(t) \int_{t_0}^t \Phi(\vartheta_u, s, t) \left[\bar W_t - \bar W_s\right] \mathrm{d}s \left(1 + O\left(\sqrt{\varepsilon}\right)\right) \\ &\quad = \frac{1}{\varepsilon} A(\vartheta_u, t, X_t)\, \delta_b(t) \int_{t_0}^t e^{-\frac{(t-s)}{\varepsilon} A(\vartheta_u, t, X_t)} \left[\bar W_t - \bar W_s\right] \mathrm{d}s \left(1 + O\left(\sqrt{\varepsilon}\right)\right) \\ &\quad = \sqrt{\varepsilon} A(\vartheta_u, t, X_t)\, \delta_b(t)\, \varepsilon^{-3/2} \int_{t_0}^t e^{-\frac{(t-s)}{\varepsilon} A(\vartheta_u, t, X_t)} \left[\bar W_t - \bar W_s\right] \mathrm{d}s \left(1 + O\left(\sqrt{\varepsilon}\right)\right) \\ &\quad = \varepsilon\, u\, \frac{\dot b(\vartheta_0, t)}{\sqrt{2A(\vartheta_u, t, X_t)}}\, \xi_{t,\varepsilon}(\vartheta_u) \left(1 + O\left(\sqrt{\varepsilon}\right)\right),\end{aligned}$$

where we denoted

$$\begin{aligned}\xi_{t,\varepsilon}(\vartheta_u) &= \varepsilon^{-3/2} A(\vartheta_u, t, X_t)^{3/2} \sqrt{2} \int_{t_0}^t e^{-\frac{(t-s)}{\varepsilon} A(\vartheta_u, s, X_s)} \left[\bar W_t - \bar W_s\right] \mathrm{d}s \\ &= A(\vartheta_u, t, X_t)^{3/2} \sqrt{2} \int_0^{\frac{t-t_0}{\varepsilon}} e^{-vA(\vartheta_u, t - v\varepsilon, X_{t-v\varepsilon})} \frac{\left[\bar W_t - \bar W_{t-\varepsilon v}\right]}{\sqrt{\varepsilon}}\, \mathrm{d}v \\ &= A(\vartheta_u, t, X_t)^{3/2} \sqrt{2} \int_0^{\frac{t-t_0}{\varepsilon}} e^{-vA(\vartheta_u, t - v\varepsilon, X_{t-v\varepsilon})} w_{t,\varepsilon}(v)\, \mathrm{d}v.\end{aligned}$$

Note that $A(\vartheta_u, t - v\varepsilon, X_{t-v\varepsilon})$ and the Wiener process $w_{t,\varepsilon}(v)$ for all $t \in (t_0, T]$ and $\varepsilon > 0$ are independent.

Consider the integrals

$$J_\varepsilon = 2\int_0^{\frac{t-t_0}{\varepsilon}}\int_0^{\frac{t-t_0}{\varepsilon}} e^{-vA(\vartheta_u,t-v\varepsilon,X_{t-v\varepsilon})}e^{-rA(\vartheta_u,t-r\varepsilon,X_{t-r\varepsilon})}w_{t,\varepsilon}(v)\,w_{t,\varepsilon}(r)\,\mathrm{d}r\,\mathrm{d}v.$$

Let $r < v$ and denote $A_{t-\varepsilon v} = A(\vartheta_u, t - v\varepsilon, X_{t-v\varepsilon})$ then

$$\begin{aligned}
&\mathbf{E}_{\vartheta_0} e^{-vA_{t-\varepsilon v}}e^{-rA_{t-\varepsilon r}}w_{t,\varepsilon}(v)\,w_{t,\varepsilon}(r)\\
&\quad= \mathbf{E}_{\vartheta_0}\left[e^{-vA_{t-\varepsilon v}}e^{-rA_{t-\varepsilon r}}\mathbf{E}_{\vartheta_0}\left(w_{t,\varepsilon}(v)\,w_{t,\varepsilon}(r)\,|\mathfrak{F}^X_{t-\varepsilon r}\right)\right]\\
&\quad= r\mathbf{E}_{\vartheta_0}\left[e^{-vA_{t-\varepsilon v}}e^{-rA_{t-\varepsilon r}}\right].
\end{aligned}$$

Hence,

$$\begin{aligned}
\mathbf{E}_{\vartheta_0} J_\varepsilon &= 2\mathbf{E}_{\vartheta_0}\int_0^{\frac{t-t_0}{\varepsilon}}\int_0^{\frac{t-t_0}{\varepsilon}} (r\wedge v)\,e^{-vA_{t-\varepsilon v}}e^{-rA_{t-\varepsilon r}}\mathrm{d}r\,\mathrm{d}v\\
&= 2\mathbf{E}_{\vartheta_0}\int_0^{\frac{t-t_0}{\varepsilon}}\int_0^{\frac{t-t_0}{\varepsilon}} (r\wedge v)\,e^{-vA_t}e^{-rA_t}\mathrm{d}r\,\mathrm{d}v\,(1+O(\varepsilon))\\
&= 2\mathbf{E}_{\vartheta_0}A_t^{-3}\int_0^{\frac{t-t_0}{\varepsilon}A_t}\int_0^{\frac{t-t_0}{\varepsilon}A_t} (x\wedge y)\,e^{-x}e^{-y}\mathrm{d}x\,\mathrm{d}y\,(1+O(\varepsilon))\\
&= \mathbf{E}_{\vartheta_0}A_t^{-3}\,(1+O(\varepsilon)).
\end{aligned}$$

Remark that if A_t is deterministic, then $\xi_{t,\varepsilon}(\vartheta_u) \Rightarrow \xi_t \sim \mathcal{N}(0,1)$. In our case we have

$$\begin{aligned}
\xi_{t,\varepsilon}(\vartheta_u) &\Longrightarrow A(\vartheta_0, t, x_t)^{3/2}\sqrt{2}\int_0^\infty e^{-vA(\vartheta_0,t,x_t)}w_t(v)\,\mathrm{d}v\\
&= \sqrt{2}\int_0^\infty e^{-x}W_t(x)\,\mathrm{d}x \equiv \xi_t.
\end{aligned}$$

The random variables $\xi_t, t_0 \le t \le T$ are i.i.d. $\mathcal{N}(0,1)$. Applying the similar arguments, we write

$$\begin{aligned}
&\frac{1}{\varepsilon}\int_{t_0}^t \Phi(\vartheta_u, s, t)\,\frac{b(\vartheta_u, s)}{\sigma(s, X_s)}\left[f(\vartheta_u, s, X_s) - f(\vartheta_0, s, X_s)\right]m_0(s)\,\mathrm{d}s\\
&\quad= \frac{b(\vartheta_u, t)}{\sigma(t, X_t)}\left[f(\vartheta_u, t, X_t) - f(\vartheta_0, t, X_t)\right]m_0(t)\,\frac{\sigma(t, X_t)}{f(\vartheta_u, t, X_t)\,b(\vartheta_u, t)}
\end{aligned}$$

$$
\begin{aligned}
&+\frac{b\left(\vartheta_u,t\right)}{\sigma\left(t,X_t\right)}\left[f\left(\vartheta_u,t,X_t\right)-f\left(\vartheta_0,t,X_t\right)\right]\\
&\times\frac{1}{\varepsilon}\int_{t_0}^{t}\Phi\left(\vartheta_u,s,t\right)\left[m_0\left(s\right)-m_0\left(t\right)\right]\mathrm{d}s\\
=&\frac{f\left(\vartheta_u,t,X_t\right)-f\left(\vartheta_0,t,X_t\right)}{f\left(\vartheta_u,t,X_t\right)}m_0\left(t\right)\\
&-\frac{b\left(\vartheta_u,t\right)}{\sigma\left(t,X_t\right)}\left[f\left(\vartheta_u,t,X_t\right)-f\left(\vartheta_0,t,X_t\right)\right]\\
&\times\frac{1}{\varepsilon}\int_{t_0}^{t}\Phi\left(\vartheta_u,s,t\right)\int_{s}^{t}b\left(\vartheta_0,s\right)\mathrm{d}\bar W_s\,\mathrm{d}s\\
=&\frac{f\left(\vartheta_u,t,X_t\right)-f\left(\vartheta_0,t,X_t\right)}{f\left(\vartheta_u,t,X_t\right)}m_0\left(t\right)\\
&-\frac{b\left(\vartheta_u,t\right)b\left(\vartheta_0,t\right)}{\sigma\left(t,X_t\right)}\left[f\left(\vartheta_u,t,X_t\right)-f\left(\vartheta_0,t,X_t\right)\right]\\
&\times\frac{1}{\varepsilon}\int_{t_0}^{t}e^{-\frac{t-s}{\varepsilon}A_t}\left[\bar W_t-\bar W_s\right]\mathrm{d}s\\
=&\frac{f\left(\vartheta_u,t,X_t\right)-f\left(\vartheta_0,t,X_t\right)}{f\left(\vartheta_u,t,X_t\right)}m_0\left(t\right)\\
&-\sqrt{\varepsilon}\frac{\left[f\left(\vartheta_u,t,X_t\right)-f\left(\vartheta_0,t,X_t\right)\right]}{f\left(\vartheta_u,t,X_t\right)}b\left(\vartheta_0,t\right)\\
&\times\sqrt{\frac{\sigma\left(t,X_t\right)}{2f\left(\vartheta_u,t,X_t\right)b\left(\vartheta_u,t\right)}}\,\xi_{t,\varepsilon}\left(\vartheta_u\right)\\
=&\frac{f\left(\vartheta_u,t,X_t\right)-f\left(\vartheta_0,t,X_t\right)}{f\left(\vartheta_u,t,X_t\right)}m_0\left(t\right)\\
&-\varepsilon\,u\frac{\dot f\left(\vartheta_u,t,x_t\right)\;b\left(\vartheta_0,t\right)}{f\left(\vartheta_u,t,x_t\right)\sqrt{2A\left(\vartheta_u,t,x_t\right)}}\,\xi_{t,\varepsilon}\left(\vartheta_u\right).
\end{aligned}
$$

Therefore, we obtain the relations

$$
\begin{aligned}
z_t=&-H_t-\frac{f\left(\vartheta_u,t,X_t\right)-f\left(\vartheta_0,t,X_t\right)}{f\left(\vartheta_u,t,X_t\right)}m_0\left(t\right)\\
&+\sqrt{\varepsilon}\;\frac{\delta_f\left(t\right)\;b\left(\vartheta_0,t\right)+f\left(\vartheta_u,t,X_t\right)\;\delta_b\left(t\right)}{f\left(\vartheta_u,t,X_t\right)\sqrt{2A\left(\vartheta_u,t,X_t\right)}}\;\xi_{t,\varepsilon}\left(\vartheta_u\right)\left(1+o\left(1\right)\right)\\
=&-H_t-\frac{f\left(\vartheta_u,t,X_t\right)-f\left(\vartheta_0,t,X_t\right)}{f\left(\vartheta_u,t,X_t\right)}m_0\left(t\right)
\end{aligned}
$$

$$
+\,\varepsilon u\,\frac{\dot f\left(\vartheta_0,t,x_t\right)\,b\left(\vartheta_0,t\right)+f\left(\vartheta_0,t,x_t\right)\,\dot b\left(\vartheta_0,t\right)}{f\left(\vartheta_0,t,x_t\right)\sqrt{2A\left(\vartheta_0,t,x_t\right)}}\,\xi_{t,\varepsilon}\left(\vartheta_u\right)\left(1+o\left(1\right)\right),
$$

and

$$
\begin{aligned}
\delta_M\left(t\right)&=M\left(\vartheta_u,t,X^t\right)-M\left(\vartheta_0,t,X^t\right)\\
&=\sqrt{\varepsilon}\;\frac{\delta_f\left(t\right)\,b\left(\vartheta_0,t\right)+f\left(\vartheta_0,t,X_t\right)\,\delta_b\left(t\right)}{\sqrt{2A\left(\vartheta_0,t,X_t\right)}}\;\xi_{t,\varepsilon}\left(\vartheta_u\right)\left(1+o\left(1\right)\right)\\
&=\varepsilon u\frac{\dot S(\vartheta_0,t,X_t)}{\sqrt{2A\left(\vartheta_0,t,X_t\right)}}\;\xi_{t,\varepsilon}\left(\vartheta_0\right)\left(1+O\left(1\right)\right).
\end{aligned}
$$

We have the asymptotic approximations

$$
\begin{aligned}
&\frac{M\left(\vartheta_u,t,X^t\right)-M\left(\vartheta_0,t,X^t\right)}{\varepsilon\sigma\left(t,X_t\right)}\\
&\qquad=\frac{1}{\sqrt{\varepsilon}}\,\frac{\delta_f\left(t\right)\,b\left(\vartheta_0,t\right)+f\left(\vartheta_0,t,X_t\right)\,\delta_b\left(t\right)}{\sqrt{2S\left(\vartheta_0,t,X_t\right)\sigma\left(t,x_t\right)}}\;\xi_{t,\varepsilon}\left(\vartheta_u\right)\left(1+o\left(1\right)\right)\\
&\qquad=u\,\frac{\dot S\left(\vartheta_0,t,X_t\right)}{\sqrt{2S\left(\vartheta_0,t,X_t\right)\sigma\left(t,X_t\right)}}\;\xi_{t,\varepsilon}\left(1+o\left(1\right)\right)\\
&\qquad=u\,\frac{\dot S\left(\vartheta_0,t,x_t\right)}{\sqrt{2S\left(\vartheta_0,t,x_t\right)\sigma\left(t,x_t\right)}}\;\xi_{t,\varepsilon}\left(1+o\left(1\right)\right).
\end{aligned}
\tag{3.46}
$$

These relations allow us to write (Lemma 3.15)

$$
\begin{aligned}
&\int_0^T\frac{\left[M\left(\vartheta_u,t,X^t\right)-M\left(\vartheta_0,t,X^t\right)\right]^2}{2\varepsilon^2\sigma\left(t,X_t\right)^2}\mathrm{d}t\\
&\qquad=\frac{u^2}{2}\int_0^T\frac{\dot S\left(\vartheta_0,t,x_t\right)^2\xi_{t,\varepsilon}^2}{2\sigma\left(t,x_t\right)S\left(\vartheta_0,t,x_t\right)}\mathrm{d}t\left(1+o\left(1\right)\right)\\
&\qquad\longrightarrow\frac{u^2}{2}\mathrm{I}\left(\vartheta_0,x^T\right),\\
&\int_0^T\frac{M\left(\vartheta_u,t,X^t\right)-M\left(\vartheta_0,t,X^t\right)}{\varepsilon\sigma\left(t,X_t\right)}\,\mathrm{d}\bar W_t\\
&\qquad=u\int_0^T\frac{\dot S\left(\vartheta_0,t,x_t\right)\xi_{t,\varepsilon}}{\sqrt{2\sigma\left(t,x_t\right)S\left(\vartheta_0,t,x_t\right)}}\mathrm{d}\bar W_t\left(1+o\left(1\right)\right)\\
&\qquad=u\Delta_\varepsilon\left(\vartheta_0,X\right)\Longrightarrow u\;\zeta_*\left(\vartheta_0\right)\,\mathrm{I}\left(\vartheta_0,x^T\right).
\end{aligned}
$$

Therefore, $Z_\varepsilon\left(\vartheta_0,u\right)$ admits the representation

$$Z_\varepsilon(\vartheta_0, u) = \exp\left(u\Delta_\varepsilon(\vartheta_0, X) - \frac{u^2}{2}\mathrm{I}\left(\vartheta_0, x^T\right) + r_\varepsilon\right) \tag{3.47}$$

where $r_\varepsilon \to 0$.

We have

$$\begin{aligned}\mathbf{P}_{\vartheta_0}\left(Z_\varepsilon(\vartheta_0, u) > e^{-cu^2}\right) &\leq \mathbf{P}_{\vartheta_0}\left(\frac{1}{2}\int_0^T N_t \mathrm{d}\bar{W}_t - \frac{1}{8}\int_0^T N_t^2 \mathrm{d}t > \bar{c}u^2\right)\\ &\quad + \mathbf{P}_{\vartheta_0}\left(-\frac{1}{8}\int_0^T N_t^2 \mathrm{d}t > -(c+\bar{c})u^2\right)\\ &\leq e^{-\bar{c}u^2} + \mathbf{P}_{\vartheta_0}\left(\int_0^T N_t^2 \mathrm{d}t < 8(c+\bar{c})u^2\right).\end{aligned}$$

The constants $c > 0, \bar{c} > 0$ will be chosen later and here we denoted (see (3.46))

$$N_t = \frac{M\left(\vartheta_u, t, X^t\right) - M\left(\vartheta_0, t, X^t\right)}{\varepsilon\sigma(t, X_t)} = u\,\frac{\dot{S}(\vartheta_0, t, X_t)}{\sqrt{2S(\vartheta_0, t, X_t)\,\sigma(t, X_t)}}\,\xi_{t,\varepsilon}(1 + o(1)).$$

Below, we use the conditions $\hat{\mathscr{H}}$

$$\begin{aligned}\int_0^T N_t^2 \mathrm{d}t &= u^2\int_0^T \frac{\dot{S}(\vartheta, t, X_t)^2}{2f(\vartheta, t, X_t)\,b(\vartheta, t)\,\sigma(t, X_t)}\,\xi_{t,\varepsilon}^2\,\mathrm{d}t\,(1 + o(1))\\ &\geq \frac{u^2\,c_4^2}{4\,C_f\,C_b\,C_\sigma}\int_0^T \xi_{t,\varepsilon}^2\,\mathrm{d}t = \frac{u^2\,c_4^2 T}{4\,C_f\,C_b\,C_\sigma} + \frac{u^2\,c_4^2}{4\,C_f\,C_b\,C_\sigma}\int_0^T\left[\xi_{t,\varepsilon}^2 - 1\right]\mathrm{d}t\\ &= c_* T\,u^2 + c_*\,u^2\int_0^T\left[\xi_{t,\varepsilon}^2 - 1\right]\mathrm{d}t.\end{aligned}$$

Hence,

$$\begin{aligned}\mathbf{P}_{\vartheta_0}\left(\int_0^T N_t^2 \mathrm{d}t < 8(c+\bar{c})u^2\right) &\leq \mathbf{P}_{\vartheta_0}\left(\int_0^T\left[\xi_{t,\varepsilon}^2 - 1\right]\mathrm{d}t < -\left(T - 8\frac{(c+\bar{c})}{c_*}\right)\right)\\ &\leq \mathbf{P}_{\vartheta_0}\left(\left|\int_0^T\left[\xi_{t,\varepsilon}^2 - 1\right]\mathrm{d}t\right| > c^*\right).\end{aligned}$$

Let take such c and $\bar{c}$ that $c^* = T - 8\frac{(c+\bar{c})}{c_*} > 0$. Then applying (3.25), we obtain for any $p > 1$

$$\mathbf{P}_{\vartheta_0}\left(\int_0^T N_t^2 \mathrm{d}t < 8(c+\bar{c})u^2\right) \leq C\,\varepsilon^p.$$

Therefore, for any compact $\mathbb{K}\subset\Theta$

$$\sup_{\vartheta_0\in\mathbb{K}}\mathbf{P}_{\vartheta_0}\left(Z_\varepsilon\left(\vartheta_0,u\right)>e^{-cu^2}\right)\leq e^{-\bar c u^2}+C\,\varepsilon^p. \tag{3.48}$$

Using the equalities $|u|=|\vartheta_u-\vartheta_0|\,\varepsilon^{-1/2}$ and $\varepsilon=(\vartheta_u-\vartheta_0)^2\,u^{-2}$ we obtain

$$\mathbf{P}_{\vartheta_0}\left(Z_\varepsilon\left(\vartheta_0,u\right)>e^{-cu^2}\right)\leq e^{-\bar c u^2}+\frac{C}{|u|^{2p}}.$$

The inequality

$$\sup_{\vartheta_0\in\mathbb{K}}\mathbf{E}_{\vartheta_0}\left|Z_\varepsilon\left(\vartheta_0,u_2\right)^{1/2}-Z_\varepsilon\left(\vartheta_0,u_1\right)^{1/2}\right|^2\leq C\left|u_2-u_1\right|^2 \tag{3.49}$$

can be proved as follows. First applying (1.11) and then the representation (3.46), we obtain

$$\begin{aligned}
&\mathbf{E}_{\vartheta_0}\left|Z_\varepsilon\left(\vartheta_0,u_2\right)^{1/2}-Z_\varepsilon\left(\vartheta_0,u_1\right)^{1/2}\right|^2\\
&\quad\leq\frac{1}{2}\int_0^T\mathbf{E}_{\vartheta_1}\frac{\left[M\left(\vartheta_{u_2},t,X_t\right)-M\left(\vartheta_{u_1},t,X_t\right)\right]^2}{\varepsilon^2\sigma\left(t,X_t\right)^2}\mathrm{d}t\\
&\quad\quad+\frac{1}{2}\int_0^T\mathbf{E}_{\vartheta_2}\frac{\left[M\left(\vartheta_{u_2},t,X_t\right)-M\left(\vartheta_{u_1},t,X_t\right)\right]^2}{\varepsilon^2\sigma\left(t,X_t\right)^2}\mathrm{d}t\leq C\left|u_2-u_1\right|^2.
\end{aligned}$$

Now the properties of the estimators $\hat\vartheta_\varepsilon$ and $\tilde\vartheta_\varepsilon$ follow from the Theorems 1.10 and 1.12 because the process $Z_\varepsilon\left(\vartheta_0,\cdot\right)$ satisfies (3.47), (3.48), and (3.49). □

3.2.3 One-Step MLE

The construction and study of the properties of the One-step MLE-process are similar to that of described in the Sect. 3.1.3. Suppose that we already have the preliminary MME $\vartheta^*_{\tau,\varepsilon}$ satisying (3.13). Define the limit Fisher information function

$$\mathrm{I}_0^t\left(\vartheta_0,x^t\right)=\int_0^t\frac{S\left(\vartheta_0,s,x_s\right)}{2\sigma\left(s,x_s\right)}\left[\frac{\partial}{\partial\vartheta_0}\ln S\left(\vartheta_0,s,x_s\right)\right]^2\mathrm{d}s,\qquad 0<t\leq T$$

and empirical Fisher information function

$$\mathrm{I}_\tau^t\left(\vartheta, X^t\right)=\int_\tau^t \frac{S\left(\vartheta, s, X_s\right)}{2 \sigma\left(s, X_s\right)}\left[\frac{\partial}{\partial \vartheta} \ln S\left(\vartheta, s, X_s\right)\right]^2 \mathrm{d} s, \qquad \tau<t \leq T .$$

The One-step MLE process is

$$\vartheta_{t, \varepsilon}^{\star}=\vartheta_{\tau, \varepsilon}^*+\frac{1}{\mathrm{I}_\tau^t\left(\vartheta_{\tau, \varepsilon}^*, X^t\right)} \int_\tau^t \frac{\dot{M}\left(\vartheta_{\tau, \varepsilon}^*, s, X^s\right)}{\varepsilon \sigma\left(s, X_s\right)^2}\left[\mathrm{d} X_s-M\left(\vartheta_{\tau, \varepsilon}^*, s, X^s\right) \mathrm{d} s\right],$$

$$\tau<t \leq T$$

Here the random processes $M(\vartheta, s, X^s)=f\left(\vartheta, s, X_s\right) m(\vartheta, s), \tau \leq t \leq T$ and $\dot{M}(\vartheta, s, X^s)=\dot{f}\left(\vartheta, s, X_s\right) m(\vartheta, s)+f\left(\vartheta, s, X_s\right) \dot{m}(\vartheta, s), \tau \leq t \leq T$ are obtained above (see Theorem 3.2).

The estimates obtained above in the proof of Theorem 3.4 allow us to follow the proof of the Theorem 3.2 and under the abovementioned regulatity conditions to prove the asymptotic mixing normality of the One-step MLE, i.e.,

$$\frac{\vartheta_{t, \varepsilon}^{\star}-\vartheta_0}{\sqrt{\varepsilon}} \Longrightarrow \zeta_t\left(\vartheta_0\right), \qquad \mathcal{L}\left(\zeta_t\left(\vartheta_0\right) \mid \mathfrak{F}_t^x\right) \sim \mathcal{N}\left(0, \mathrm{I}_\tau^t\left(\vartheta_0, x^t\right)^{-1}\right).$$

3.2.4 *Adaptive Filter*

The adaptive EKF-1 $m_t^{\star}, \tau \leq t \leq T$ we propose by the following equation

$$\mathrm{d} m_t^{\star}=-\frac{b\left(\vartheta_{t, \varepsilon}^{\star}, t\right) f\left(\vartheta_{t, \varepsilon}^{\star}, t, X_t\right)}{\varepsilon \sigma\left(t, X_t\right)} m_t^{\star} \mathrm{d} t+\frac{b\left(\vartheta_{t, \varepsilon}^{\star}, t\right)}{\varepsilon \sigma\left(t, X_t\right)} \mathrm{d} X_t, \quad m(\vartheta, \tau)=y_0,$$

$$\tau \leq t \leq T .$$

Recall that the contribution of the initial value is negligible. We suppose that it is possible to prove the theorem similar to Theorem 3.3, but we will not do it here because the proof will be too close to the given proof of that theorem.

3.3 Extended Kalman Filter

3.3.1 *Model of Observations*

Let us consider the nonlinear partially observed system

$$\mathrm{d} X_t=f\left(\vartheta, t, Y_t\right) \mathrm{d} t+\varepsilon \sigma(t) \mathrm{d} W_t, \qquad X_0=0, \qquad 0 \leq t \leq T, \tag{3.50}$$

$$\mathrm{d}Y_t = a(\vartheta, t, Y_t)\,\mathrm{d}t + b(\vartheta, t, Y_t)\,\mathrm{d}V_t, \qquad Y_0 = y_0, \tag{3.51}$$

where, as before, the Wiener processes $W_t, V_t, 0 \leq t \leq T$ are independent, the functions $f(\cdot), \sigma(\cdot), a(\cdot), b(\cdot)$ and the initial value y_0 are known. The unknown parameter $\vartheta \in \Theta = (\alpha, \beta)$.

We have the partially observed system (3.50)–(3.51) with the small noise ($\varepsilon \in (0,1), \varepsilon \to 0$) in observation equation only. The equations of EKF similar to (2.93)–(2.94) are

$$\mathrm{d}m_t(\vartheta) = \left[a(\vartheta, t, m_t(\vartheta)) - \frac{f'_y(\vartheta, t, m_t(\vartheta))\, f(\vartheta, t, m_t(\vartheta))\,\gamma_\varepsilon(\vartheta, t)}{\varepsilon^2 \sigma(t)^2}\right]\mathrm{d}t$$
$$+ \frac{f'_y(\vartheta, t, m_t(\vartheta))\,\gamma_\varepsilon(\vartheta, t)}{\varepsilon^2 \sigma(t)^2}\,\mathrm{d}X_t, \qquad m_0(\vartheta) = y_0, \tag{3.52}$$

$$\frac{\mathrm{d}\gamma_\varepsilon(\vartheta, t)}{\mathrm{d}t} = 2a'_y(\vartheta, t, m_t(\vartheta))\,\gamma_\varepsilon(\vartheta, t) - \frac{f'_y(\vartheta, t, m_t(\vartheta))^2\,\gamma_\varepsilon(\vartheta, t)^2}{\varepsilon^2 \sigma(t)^2}$$
$$+ b(\vartheta, t, m_t(\vartheta))^2, \qquad \gamma_\varepsilon(\vartheta, 0) = 0. \tag{3.53}$$

Note that if $f(\vartheta, t, y) = f(\vartheta, t)\,y$ and $a(\vartheta, t, y) = a(\vartheta, t)\,y$, then the Eqs. (3.52)–(3.53) coincide with the equations of Kalman–Bucy (3.14)–(3.15).

Introduce some notations

$$\dot g_\vartheta(\vartheta, t, y) = \frac{\partial g(\vartheta, t, y)}{\partial \vartheta}, \qquad g'_t(\vartheta, t, y) = \frac{\partial g(\vartheta, t, y)}{\partial t},$$
$$g'_y(\vartheta, t, y) = \frac{\partial g(\vartheta, t, y)}{\partial y},$$
$$\dot g'_{\vartheta y}(\vartheta, t, y) = \frac{\partial^2 g(\vartheta, t, y)}{\partial \vartheta\, \partial t}, \qquad \ldots.$$

$g(\vartheta, t, y) \in \mathcal{C}^{2,b}_\vartheta$ means that the function $g(\cdot)$ has two continuous bounded derivatives with respect to ϑ etc. If the function $g(\cdot)$ is bounded, then we denote C_g such constant that $|g(\vartheta, t, y)| \leq C_g$.

Conditions $\mathscr{I}$.

$\mathscr{I}_1$. *The functions $f(\vartheta, t, y), a(\vartheta, t, y), b(\vartheta, t, y), \vartheta \in \Theta, t \in [0, T], y \in \mathcal{R}$ have continuous bounded derivatives on t, $f(\vartheta, t, y)$ has two continuous bounded derivatives in ϑ and $a(\vartheta, t, y)$ has has continuous bounded derivative on ϑ.*

$\mathscr{I}_2$. *The functions $f'_y(\vartheta, t, y)$, $\sigma(t)$, $b(\vartheta, t, y)$ are separated from 0, i.e., there exist constants $c_y > 0, c_\sigma > 0, c_b > 0$ such that*

$$f_y'(\vartheta, t, y) > c_y, \qquad \sigma(t) > c_\sigma, \qquad b(\vartheta, t, y) > c_b.$$

$\mathscr{I}_3$. *The functions* $f(\vartheta, t, y), b(\vartheta, t, y), f_y'(\vartheta, t, y) \in C_\vartheta^{2,b}$. *There exists a constants* $c_4 > 0, c_5 > 0$ *such that for all* $\vartheta \in \Theta$ *the derivatives*

$$\dot{f}_{y\vartheta}'(\vartheta, t, y) > c_4, \qquad \dot{b}(\vartheta, t, y) > c_5.$$

It can be shown that the solution $\gamma_\varepsilon(\vartheta, t)$ of Riccati equation (3.53) for the values $t \in [t_0, T]$ and any $t_0 > 0$ has the following limit

$$\lim_{\varepsilon \to 0} \left| \frac{f_y'(\vartheta, t, m_t(\vartheta))\, \gamma_\varepsilon(\vartheta, t)}{\varepsilon} - b(\vartheta, t, m_t(\vartheta))\, \sigma(t) \right| = 0.$$

To explain how such limit can be verified, we denote $\gamma_\varepsilon^*(\vartheta, t) = \varepsilon^{-1}\gamma_\varepsilon(\vartheta, t)$. Then we obtain the equation

$$\begin{aligned}\varepsilon\gamma_\varepsilon^*(\vartheta, t) &= \varepsilon\gamma_\varepsilon^*(\vartheta, t_0) + 2\varepsilon \int_{t_0}^{t} a'(\vartheta, s, m_s(\vartheta))\, \gamma_\varepsilon^*(\vartheta, s)\, \mathrm{d}s \\ &\quad - \int_{t_0}^{t} \left[\frac{f_y'(\vartheta, s, m_s(\vartheta))^2\, \gamma_\varepsilon^*(\vartheta, s)^2}{\sigma(s)^2} - b(\vartheta, s, m_s(\vartheta))^2 \right] \mathrm{d}s.\end{aligned}$$

As $\varepsilon \to 0$ we obtain the following convergence for all $t \in [t_0, T]$

$$\int_{t_0}^{t} \left[\frac{f_y'(\vartheta, s, m_s(\vartheta))^2\, \gamma_\varepsilon^*(\vartheta, s)^2}{\sigma(s)^2} - b(\vartheta, s, m_s(\vartheta))^2 \right] \mathrm{d}s \longrightarrow 0.$$

Introduce the notation:

$$\gamma_*(\vartheta, t) = \frac{b(\vartheta, t, m_t(\vartheta))\, \sigma(t)}{f_y'(\vartheta, t, m_t(\vartheta))}, \qquad \gamma_*^\circ(\vartheta, t) = \frac{b\left(\vartheta, t, m_t^\circ(\vartheta)\right) \sigma(t)}{f_y'\left(\vartheta, t, m_t^\circ(\vartheta)\right)}.$$

Here $m_t^\circ(\vartheta)$ is the limit of $m_t(\vartheta)$ as $\varepsilon \to 0$. It will be descibed below. It follows as well from the modified proof of Lemmas 3.1 and 3.2. The conditions $\mathscr{I}_1, \mathscr{I}_2$ are sufficient for the proof. For these values of t we replace $\varepsilon^{-1}\gamma_\varepsilon(\vartheta, t)$ by $\gamma_*(\vartheta, t)$ in all further study of this model. This will essentially simplify calculations related with the adaptive EKF.

The equation for estimate $m_t(\vartheta)$ of Y_t becames

$$\begin{aligned}\mathrm{d}m_t(\vartheta) &= a(\vartheta, t, m_t(\vartheta))\, \mathrm{d}t + \frac{f_y'(\vartheta, t, m_t(\vartheta))\, \gamma_*(\vartheta, t)}{\varepsilon\sigma(t)^2} \\ &\quad \times \left[\mathrm{d}X_t - f(\vartheta, t, m_t(\vartheta))\, \mathrm{d}t\right]\end{aligned}$$

$$= a(\vartheta, t, m_t(\vartheta))\,dt + \frac{b(\vartheta, t, m_t(\vartheta))}{\varepsilon\sigma(t)}\left[dX_t - f(\vartheta, t, m_t(\vartheta))\,dt\right],\ t_0 \le t \le T. \tag{3.54}$$

Consider the limits in probability of the random functions $m_t(\vartheta), \dot m_t(\vartheta) = \frac{\partial}{\partial\vartheta} m_t(\vartheta), t_0 \le t \le T$, which we denote as $m_t^\circ(\vartheta), \dot m_t^\circ(\vartheta), t_0 \le t \le T$.

Lemma 3.8 *Let conditions $\mathscr{I}$ hold. Then for any $t_0 \in (0, T]$, the following relations hold*

$$f\left(\vartheta, t, m_t^\circ(\vartheta)\right) = f(\vartheta_0, t, Y_t), \qquad m_t^\circ(\vartheta_0) = Y_t, \tag{3.55}$$

$$\dot m_t^\circ(\vartheta) = -\frac{\dot f_\vartheta\left(\vartheta, t, m_t^\circ(\vartheta)\right)}{f_y'\left(\vartheta, t, m_t^\circ(\vartheta)\right)}, \qquad \dot m_t^\circ(\vartheta_0) = -\frac{\dot f_\vartheta(\vartheta_0, t, Y_t)}{f_y'(\vartheta_0, t, Y_t)}. \tag{3.56}$$

Proof The limit $m_t^\circ(\vartheta), t_0 < t \le T$ can be described as follows. We have

$$dm_t(\vartheta) = a(\vartheta, t, m_t(\vartheta))\,dt + \frac{b(\vartheta, t, m_t(\vartheta))}{\varepsilon\sigma(t)}\left[f(\vartheta_0, t, Y_t) - f(\vartheta, t, m_t(\vartheta))\right]dt + b(\vartheta, t, m_t(\vartheta))\,dW_t. \tag{3.57}$$

Hence,

$$\int_{t_0}^t \frac{b(\vartheta, s, m_s(\vartheta))}{\sigma(s)}\left[f(\vartheta_0, s, Y_s) - f(\vartheta, s, m_s(\vartheta))\right]ds$$
$$= \varepsilon\left[m_t(\vartheta) - m_{t_0}(\vartheta) - \int_{t_0}^t a(\vartheta, s, m_s(\vartheta))\,ds - \int_{t_0}^t b(\vartheta, s, m_s(\vartheta))\,dW_s\right]$$

and the limit as $\varepsilon \to 0$ is: for all $t \in [t_0, T]$

$$\int_{t_0}^t \frac{b\left(\vartheta, s, m_s^\circ(\vartheta)\right)}{\sigma(s)}\left[f(\vartheta_0, s, Y_s) - f\left(\vartheta, s, m_s^\circ(\vartheta)\right)\right]ds = 0. \tag{3.58}$$

The functions $b(\vartheta, s) > c_b, \sigma(t) < C_\sigma$, and this equality is possible iff

$$f(\vartheta_0, s, Y_s) = f\left(\vartheta, s, m_s^\circ(\vartheta)\right), \qquad t_0 \le s \le T. \tag{3.59}$$

By condition $\mathscr{I}_2$, the function $f(\vartheta, t, y)$ is strictly monotone (increasing) in y and therefore, the equation $f(\vartheta, t, y) = z$ has a unique solution $y = \bar h(\vartheta, t, z)$. Hence, the following representation

$$m_t^\circ(\vartheta) = \bar h(\vartheta, t, f(\vartheta_0, t, Y_t)) = h(\vartheta, \vartheta_0, t, Y_t), \qquad t_0 \le t \le T$$

holds. The function $h(\cdot)$ is defined by the last equality. Note that $h(\vartheta_0, \vartheta_0, t, Y_t) = Y_t$. Therefore $m_s^\circ(\vartheta_0) = Y_s$.

Recall that the relation similar to (3.59) we had in the linear case (3.1)–(3.2)

$$f(\vartheta_0, s)\, m_s(\vartheta_0) = f(\vartheta, s)\, m_s^\circ(\vartheta), \qquad t_0 \le s \le T.$$

We seek the stochastic differential of the process $m_t^\circ(\vartheta)$

$$\mathrm{d}m_t^\circ(\vartheta) = A_t \mathrm{d}t + D_t \mathrm{d}V_t, \qquad m_{t_0}^\circ(\vartheta),$$

i.e., we have to define the random processes $A_t, D_t, t_0 \le t \le T$. This differential we write with the help of (3.55) as follows

$$\begin{aligned}
f_t'\left(\vartheta, t, m_t^\circ(\vartheta)\right)\mathrm{d}t &+ f_y'\left(\vartheta, t, m_t^\circ(\vartheta)\right)\mathrm{d}m_t^\circ(\vartheta) + \frac{1}{2} f_{yy}''\left(\vartheta, t, m_t^\circ(\vartheta)\right) D_t^2 \mathrm{d}t\\
&= f_t'(\vartheta_0, t, Y_t)\,\mathrm{d}t + f_y'(\vartheta_0, t, Y_t)\, a(\vartheta_0, t, Y_t)\,\mathrm{d}t\\
&\quad + \frac{1}{2} f_{yy}''(\vartheta_0, t, Y_t)\, b(\vartheta_0, t, Y_t)^2\, \mathrm{d}t\\
&\quad + f_y'(\vartheta_0, t, Y_t)\, b(\vartheta_0, t, Y_t)\, \mathrm{d}V_t.
\end{aligned}$$

Hence, A_t and D_t are defined by the equation on the time interval $[t_0, T]$

$$\mathrm{d}m_t^\circ(\vartheta) = A\left(\vartheta, t, m_t^\circ(\vartheta)\right)\mathrm{d}t + \frac{f_y'(\vartheta_0, t, Y_t)\, b(\vartheta_0, t, Y_t)}{f_y'\left(\vartheta, t, m_t^\circ(\vartheta)\right)}\mathrm{d}V_t, \qquad m_{t_0}^\circ(\vartheta),$$

i.e.,

$$\begin{aligned}
A_t = A(\vartheta, t, y) &= \frac{f_t'(\vartheta_0, t, Y_t) - f_t'(\vartheta, t, y) + f_y'(\vartheta_0, t, Y_t)\, a(\vartheta_0, t, Y_t)}{f_y'(\vartheta, t, y)}\\
&\quad + \frac{f_{yy}''(\vartheta_0, t, Y_t)\, b(\vartheta_0, t, Y_t)^2 - f_{yy}''(\vartheta_0, t, y)\, D_t^2}{2 f_y'(\vartheta, t, y)},
\end{aligned}$$

$$D_t = \frac{f_y'(\vartheta_0, t, Y_t)\, b(\vartheta_0, t, Y_t)}{f_y'\left(\vartheta, t, m_t^\circ(\vartheta)\right)}.$$

The equation for $\dot m_t(\vartheta)$ we obtain by the formal differentiaton of the Eq. (3.57)

$$\begin{aligned}
\mathrm{d}\dot m_t(\vartheta) &= \left[\dot a_\vartheta(\vartheta, t, m_t(\vartheta)) + a_y'(\vartheta, t, m_t(\vartheta))\, \dot m_t(\vartheta)\right]\mathrm{d}t + \dot B_\vartheta(\vartheta, t)\,\mathrm{d}W_t\\
&\quad + \frac{\dot B_\vartheta(\vartheta, t)}{\varepsilon\sigma(t)}\left[f(\vartheta_0, t, Y_t) - f(\vartheta, t, m_t(\vartheta))\right]\mathrm{d}t
\end{aligned}$$

$$
-\frac{B(\vartheta,t)}{\varepsilon\sigma(t)}\left[\dot f_\vartheta(\vartheta,t,m_t(\vartheta))+f_y'(\vartheta,t,m_t(\vartheta))\,\dot m_t(\vartheta)\right]\mathrm{d}t,
$$

where

$$
\dot B_\vartheta(\vartheta,t)=\dot b_\vartheta(\vartheta,t,m_t(\vartheta))+b_y'(\vartheta,t,m_t(\vartheta))\,\dot m_t(\vartheta).
$$

Therefore,

$$
\begin{aligned}
\int_{t_0}^{t}\Bigg[\frac{\dot B_\vartheta(\vartheta,s)}{\sigma(s)}&\left[f(\vartheta_0,s,Y_s)-f(\vartheta,s,m_s(\vartheta))\right]\Bigg]\mathrm{d}s\\
&-\int_{t_0}^{t}\frac{b(\vartheta,s,m_s(\vartheta))}{\sigma(s)}\left[\dot f_\vartheta(\vartheta,s,m_s(\vartheta))+f_y'(\vartheta,t,m_t(\vartheta))\,\dot m_t(\vartheta)\right]\mathrm{d}s\\
=\varepsilon\Bigg[\dot m_t(\vartheta)-\dot m_{t_0}(\vartheta)&-\int_{t_0}^{t}\left[\dot a_\vartheta(\vartheta,s,m_t(\vartheta))+a_y'(\vartheta,t,m_t(\vartheta))\,\dot m_t(\vartheta)\right]\mathrm{d}s\Bigg].
\end{aligned}
$$

The first integral converges to zero (see (3.58)), and we obtain the limit

$$
\int_{t_0}^{t}\frac{b\left(\vartheta,s,m_s^\circ(\vartheta)\right)}{\sigma(s)}\left[\dot f_\vartheta\left(\vartheta,s,m_s^\circ(\vartheta)\right)+f_y'\left(\vartheta,s,m_s^\circ(\vartheta)\right)\dot m_s^\circ(\vartheta)\right]\mathrm{d}s=0,
$$

$$
t_0<t\le T.
$$

Hence,

$$
\dot f_\vartheta\left(\vartheta,t,m_t^\circ(\vartheta)\right)+f_y'\left(\vartheta,t,m_t^\circ(\vartheta)\right)\dot m_t^\circ(\vartheta)=0,\quad \dot m_t^\circ(\vartheta)=-\frac{\dot f_\vartheta\left(\vartheta,t,m_t^\circ(\vartheta)\right)}{f_y'\left(\vartheta,t,m_t^\circ(\vartheta)\right)}. \tag{3.60}
$$

The last equality can be obtained as well by the differentiation in ϑ of the Eq. (3.59). If $\vartheta=\vartheta_0$, then

$$
\dot m_t^\circ(\vartheta_0)=-\frac{\dot f_\vartheta\left(\vartheta_0,t,m_t^\circ(\vartheta_0)\right)}{f_y'\left(\vartheta_0,t,m_t^\circ(\vartheta_0)\right)}=-\frac{\dot f_\vartheta(\vartheta_0,t,Y_t)}{f_y'(\vartheta_0,t,Y_t)}.
$$

□

Let us study the differences $m_t(\vartheta)-m_t^\circ(\vartheta)$ and $\dot m_t(\vartheta)-\dot m_t^\circ(\vartheta)$. Introduce notation

$$
\Gamma^\circ(\vartheta,t)=\sqrt{\frac{b\left(\vartheta,t,m_t^\circ(\vartheta)\right)\sigma(t)}{2f_y'\left(\vartheta,t,m_t^\circ(\vartheta)\right)}},\qquad \Gamma^\circ(\vartheta_0,t)=\sqrt{\frac{b(\vartheta_0,t,Y_t)\,\sigma(t)}{2f_y'(\vartheta_0,t,Y_t)}},,
$$

$$\Delta^{\circ}(\vartheta,t)=\frac{b(\vartheta_0,t,Y_t)\,f_y'(\vartheta_0,t,Y_t)}{b\left(\vartheta,t,m_t^{\circ}(\vartheta)\right)f_y'\left(\vartheta,t,m_t^{\circ}(\vartheta)\right)},\qquad \Delta^{\circ}(\vartheta_0,t)=1,$$

$$\dot B_{\vartheta}(\vartheta,t)=\dot b(\vartheta,t,m_t(\vartheta))+b_y'(\vartheta,t,m_t(\vartheta))\,\dot m_t(\vartheta),$$

$$\dot B_{\vartheta}^{\circ}(\vartheta,t)=\dot b\left(\vartheta,t,m_t^{\circ}(\vartheta)\right)-\frac{b_y'\left(\vartheta,t,m_t^{\circ}(\vartheta)\right)\dot f_{\vartheta}\left(\vartheta,t,m_t^{\circ}(\vartheta)\right)}{f_y'\left(\vartheta,t,m_t^{\circ}(\vartheta)\right)},$$

$$R^{\circ}(\vartheta,t)=\frac{f_{yy}''\left(\vartheta,t,m_t^{\circ}(\vartheta)\right)\dot f_{\vartheta}\left(\vartheta_0,t,m_t^{\circ}(\vartheta)\right)}{f_y'\left(\vartheta_0,t,m_t^{\circ}(\vartheta)\right)}-\dot f_{\vartheta y}'\left(\vartheta,t,m_t^{\circ}(\vartheta)\right)$$
$$-\frac{\dot B_{\vartheta}^{\circ}(\vartheta,t)}{b(\vartheta,t)}f_y'\left(\vartheta,t,m_t^{\circ}(\vartheta)\right),$$

$$\Pi^{\circ}(\vartheta,t)=\frac{R^{\circ}(\vartheta,t)}{f_y'\left(\vartheta,t,m_t^{\circ}(\vartheta)\right)}+\frac{\dot B_{\vartheta}^{\circ}(\vartheta,t)^2}{b\left(\vartheta,t,m_t^{\circ}(\vartheta)\right)^2},$$

$$\Pi^{\circ}(\vartheta_0,t)=\frac{R^{\circ}(\vartheta_0,t)}{f_y'(\vartheta_0,t,Y_t)}+\frac{\dot B_{\vartheta}^{\circ}(\vartheta_0,t)^2}{b(\vartheta_0,t,Y_t)^2},$$

$$\Lambda^{\circ}(\vartheta,t)=\frac{b(\vartheta_0,t,Y_t)\,f_y'(\vartheta_0,t,Y_t)}{b\left(\vartheta,t,m_t^{\circ}(\vartheta)\right)f_y'\left(\vartheta,t,m_t^{\circ}(\vartheta)\right)}\left[\frac{\dot B_{\vartheta}^{\circ}(\vartheta,t)\,\dot f_{y\vartheta}'\left(\vartheta,t,m_t^{\circ}(\vartheta)\right)}{b\left(\vartheta,t,m_t^{\circ}(\vartheta)\right)f_y'\left(\vartheta,t,m_t^{\circ}(\vartheta)\right)}\right.$$
$$\left.-\frac{R^{\circ}(\vartheta,t)}{f_y'\left(\vartheta,t,m_t^{\circ}(\vartheta)\right)}-\frac{\dot B_{\vartheta}^{\circ}(\vartheta,t)\,f_{yy}''\left(\vartheta,t,m_t^{\circ}(\vartheta)\right)\dot f_{\vartheta}\left(\vartheta,t,m_t^{\circ}(\vartheta)\right)}{b\left(\vartheta,t,m_t^{\circ}(\vartheta)\right)f_y'\left(\vartheta,t,m_t^{\circ}(\vartheta)\right)^2}\right],$$

$$\Lambda^{\circ}(\vartheta_0,t)=\frac{\dot B_{\vartheta}^{\circ}(\vartheta_0,t)\,\dot f_{y\vartheta}'(\vartheta_0,t,Y_t)}{b(\vartheta_0,t,Y_t)\,f_y'(\vartheta_0,t,Y_t)}-\frac{\dot B_{\vartheta}^{\circ}(\vartheta_0,t)\,f_{yy}''(\vartheta_0,t,Y_t)\,\dot f_{\vartheta}(\vartheta_0,t,Y_t)}{b(\vartheta_0,t,Y_t)\,f_y'(\vartheta_0,t,Y_t)^2}$$
$$-\frac{R^{\circ}(\vartheta_0,t)}{f_y'(\vartheta_0,t,Y_t)}$$

$$\xi_{t,\varepsilon}=\sqrt{2}\int_0^{\varepsilon^{\nu-1}q(t)}e^{-u}\mathrm{d}w_{t,\varepsilon}(u),\qquad \eta_{t,\varepsilon}=\sqrt{2}\int_0^{\varepsilon^{\nu-1}q(t)}e^{-u}\mathrm{d}v_{t,\varepsilon}(u),$$

where $\nu\in(1/2,1)$, $w_{t,\varepsilon}(u),u\le 0$ and $v_{t,\varepsilon}(u),u\le 0$ are independent Wiener processes.

Proposition 3.4 *Let conditions* $\mathscr{I}$ *hold. Then for any* $t_0\in(0,T]$, *the following relations hold:*

$$\frac{m_t(\vartheta)-m_t^{\circ}(\vartheta)}{\Gamma^{\circ}(\vartheta,t)\sqrt{\varepsilon}}=\xi_{t,\varepsilon}-\Delta^{\circ}(\vartheta,t)\,\eta_{t,\varepsilon}+O\left(\varepsilon^{\frac{\nu}{2}}\right),\tag{3.61}$$

$$\frac{m_t(\vartheta_0)-Y_t}{\Gamma^{\circ}(\vartheta_0,t)\sqrt{\varepsilon}}=\xi_{t,\varepsilon}-\eta_{t,\varepsilon}+O\left(\varepsilon^{\frac{\nu}{2}}\right),$$

$$\frac{\dot m_t(\vartheta)-\dot m_t^\circ(\vartheta)}{\Gamma^\circ(\vartheta,t)\sqrt{\varepsilon}}\Longrightarrow \Pi^\circ(\vartheta,t)\,\xi_t+\Lambda^\circ(\vartheta,t)\,\eta_t. \tag{3.62}$$

Proof The Eq. (3.54) can be written as follows:

$$\begin{aligned}
\mathrm{d}m_t(\vartheta)&\\
&=a(\vartheta,t,m_t(\vartheta))\,\mathrm{d}t+\frac{b(\vartheta,t,m_t(\vartheta))}{\varepsilon\sigma(t)}\left[f(\vartheta_0,t,Y_t)-f\left(\vartheta,t,m_t^\circ(\vartheta)\right)\right]\mathrm{d}t\\
&\quad+\frac{b(\vartheta,t,m_t(\vartheta))}{\varepsilon\sigma(t)}\left[f\left(\vartheta,t,m_t^\circ(\vartheta)\right)-f(\vartheta,t,m_t(\vartheta))\right]+b(\vartheta,t,m_t(\vartheta))\,\mathrm{d}W_t\\
&=a(\vartheta,t,m_t(\vartheta))\,\mathrm{d}t+\frac{b(\vartheta,t,m_t(\vartheta))}{\varepsilon\sigma(t)}\left[f\left(\vartheta,t,m_t^\circ(\vartheta)\right)-f(\vartheta,t,m_t(\vartheta))\right]\mathrm{d}t\\
&\qquad+b(\vartheta,t,m_t(\vartheta))\,\mathrm{d}W_t\\
&=a(\vartheta,t,m_t(\vartheta))\,\mathrm{d}t-\frac{b(\vartheta,t,m_t(\vartheta))}{\varepsilon\sigma(t)}f_y'(\vartheta,t,\tilde m_t(\vartheta))\left[m_t(\vartheta)-m_t^\circ(\vartheta)\right]\mathrm{d}t\\
&\qquad+b(\vartheta,t,m_t(\vartheta))\,\mathrm{d}W_t,
\end{aligned}$$

where $\left|\tilde m_t(\vartheta)-m_t^\circ(\vartheta)\right|\leq\left|m_t(\vartheta)-m_t^\circ(\vartheta)\right|$. Therefore,

$$\begin{aligned}
\mathrm{d}\left[m_t(\vartheta)-m_t^\circ(\vartheta)\right]&=\left[a(\vartheta,t,m_t(\vartheta))-A\left(\vartheta,t,m_t^\circ(\vartheta)\right)\right]\mathrm{d}t\\
&\quad-\frac{f_y'(\vartheta_0,t,Y_t)\,b(\vartheta_0,t,Y_t)}{f_y'(\vartheta,t,m_t(\vartheta))}\mathrm{d}V_t+b(\vartheta,t,m_t(\vartheta))\,\mathrm{d}W_t\\
&\quad-\frac{b(\vartheta,t,m_t(\vartheta))}{\varepsilon\sigma(t)}f_y'(\vartheta,t,\tilde m_t(\vartheta))\left[m_t(\vartheta)-m_t^\circ(\vartheta)\right]\mathrm{d}t.
\end{aligned} \tag{3.63}$$

Introduce the random process $Z_t=m_t(\vartheta)-m_t^\circ(\vartheta)-C_t+F_t$, where

$$C_t=\int_{t_0}^t b(\vartheta,s,m_s(\vartheta))\,\mathrm{d}W_s,\qquad F_t=\int_{t_0}^t\frac{f_y'(\vartheta_0,s,Y_s)\,b(\vartheta_0,s,Y_s)}{f_y'(\vartheta,s,m_s(\vartheta))}\mathrm{d}V_s.$$

The equation for Z_t is

$$\begin{aligned}
\mathrm{d}Z_t&=\left[a(\vartheta,t,m_t(\vartheta))-A\left(\vartheta,t,m_t^\circ(\vartheta)\right)\right]\,\mathrm{d}t\\
&\quad-\frac{b(\vartheta,t,m_t(\vartheta))}{\varepsilon\sigma(t)}f_y'(\vartheta,t,\tilde m_t(\vartheta))\,Z_t\,\mathrm{d}t\\
&\quad+\frac{b(\vartheta,t,m_t(\vartheta))}{\varepsilon\sigma(t)}f_y'(\vartheta,t,\tilde m_t(\vartheta))\left[F_t-C_t\right]\,\mathrm{d}t.
\end{aligned}$$

Therefore, the solution of this equation is

$$Z_t = Z_{t_0}\Psi(t,t_0) + \frac{1}{\varepsilon}\int_{t_0}^{t}\Psi(t,s)\,\frac{b(\vartheta,s,m_s(\vartheta))}{\sigma(s)}\,f_y'(\vartheta,s,\tilde m_s(\vartheta))\,[F_s - C_s]\,\mathrm{d}s$$

$$+ \int_{t_0}^{t}\Psi(t,s)\left[a(\vartheta,s,m_s(\vartheta)) - A\left(\vartheta,s,m_s^\circ(\vartheta)\right)\right]\,\mathrm{d}s$$

$$= F_t - C_t + \frac{1}{\varepsilon}\int_{t_0}^{t}\Psi(t,s)\,\frac{b(\vartheta,s,m_s(\vartheta))}{\sigma(s)}\,f_y'(\vartheta,s,\tilde m_s(\vartheta))\,[F_s - F_t]\,\mathrm{d}s$$

$$+ \int_{t_0}^{t}\Psi(t,s)\left[a(\vartheta,s,m_s(\vartheta)) - A\left(\vartheta,s,m_s^\circ(\vartheta)\right)\right]\,\mathrm{d}s$$

$$+ \frac{1}{\varepsilon}\int_{t_0}^{t}\Psi(t,s)\,\frac{b(\vartheta,s,m_s(\vartheta))}{\sigma(s)}\,f_y'(\vartheta,s,\tilde m_s(\vartheta))\,[C_t - C_s]\,\mathrm{d}s + Z_{t_0}\Psi(t,t_0),$$

where we used the abbreviations

$$\Psi(t,s) = \exp\left(-\frac{1}{\varepsilon}\int_s^t q(v)\,\mathrm{d}v\right),\qquad q(t) = \frac{b(\vartheta,t,m_t(\vartheta))}{\sigma(t)}\,f_y'(\vartheta,t,\tilde m_t(\vartheta))$$

and the following property of the integral ($t > t_0$).

$$\frac{1}{\varepsilon}\int_{t_0}^{t}\Psi(t,s)\,\frac{b(\vartheta,s,m_s(\vartheta))}{\sigma(s)}\,f_y'(\vartheta,s,\tilde m_s(\vartheta))\,\mathrm{d}s$$

$$= \Psi(t,0)\,\frac{1}{\varepsilon}\int_{t_0}^{t}\Psi(s,0)^{-1}\,\frac{b(\vartheta,s,m_s(\vartheta))}{\sigma(s)}\,f_y'(\vartheta,s,\tilde m_s(\vartheta))\,\mathrm{d}s$$

$$= \Psi(t,0)\int_{t_0}^{t}\mathrm{d}\Psi(s,0)^{-1} = 1 - \Psi(t,t_0)$$

$$= 1 + O\left(e^{-\frac{c}{\varepsilon}(t-t_0)}\right),$$

with some $c > 0$.

Denote as well

$$q_*(t) = \frac{b(\vartheta,t,m_t(\vartheta))}{\sigma(t)}\,f_y'(\vartheta,t,m_t(\vartheta)),\qquad \Psi_*(t,s) = \exp\left(-\frac{1}{\varepsilon}(t-s)\,q_*(t)\right).$$

We have the estimates ($t > t_0$)

$$Z_{t_0}\Psi(t,t_0) = O(\varepsilon),$$

$$\int_{t_0}^{t}\Psi(t,s)\left[a(\vartheta,s,m_s(\vartheta)) - A\left(\vartheta,s,m_s^\circ(\vartheta),Y_s\right)\right]\,\mathrm{d}s = O(\varepsilon).$$

To obtain these estimates and to study the other integrals below, we use the following well-known device, for example, in the proof of Lemma 1.7. Let $G(s)$, $t_0 \leq s \leq T$ be a bounded function and $t_\varepsilon = t - \varepsilon^\nu$, where $\nu \in (1/2, 1)$. We have the estimate

$$q(t) \geq \frac{c_b c_y}{C_\sigma} = \kappa_* > 0,$$

where the constants c_b, c_y are from the condition $\mathscr{I}_2$ and $\sigma(t) \leq C_\sigma$. Hence,

$$\Psi(t,s) \leq e^{-\frac{\kappa_*}{\varepsilon}(t-s)}$$

and

$$\int_{t_0}^{t_\varepsilon} \Psi(t,s)\, G(s)\, \mathrm{d}s \leq \int_{t_0}^{t_\varepsilon} e^{-\frac{\kappa_*}{\varepsilon}(t-s)} G(s)\, \mathrm{d}s$$
$$\leq e^{-\frac{\kappa_*}{\varepsilon^{1-\nu}}} \int_{t_0}^{t_\varepsilon} G(s)\, \mathrm{d}s = O\left(e^{-\frac{\kappa_*}{\varepsilon^{1-\nu}}}\right).$$

Then

$$\int_{t_0}^{t} \Psi(t,s) \frac{b(\vartheta, s, m_s(\vartheta))}{\sigma(s)} f_y'(\vartheta, s, \tilde m_s(\vartheta))\, G(s)\, \mathrm{d}s$$
$$= \int_{t_0}^{t_\varepsilon} \Psi(t,s) \frac{b(\vartheta, s, m_s(\vartheta))}{\sigma(s)} f_y'(\vartheta, s, \tilde m_s(\vartheta))\, G(s)\, \mathrm{d}s$$
$$+ \int_{t_\varepsilon}^{t} \Psi(t,s) \frac{b(\vartheta, s, m_s(\vartheta))}{\sigma(s)} f_y'(\vartheta, s, \tilde m_s(\vartheta))\, G(s)\, \mathrm{d}s$$
$$= \int_{t_\varepsilon}^{t} \Psi(t,s) \frac{b(\vartheta, s, m_s(\vartheta))}{\sigma(s)} f_y'(\vartheta, s, \tilde m_s(\vartheta))\, G(s)\, \mathrm{d}s + O\left(e^{-\frac{\kappa_*}{\varepsilon^{1-\nu}}}\right).$$

As the terms like $O\left(e^{-\frac{\kappa_*}{\varepsilon^{1-\nu}}}\right)$ will be added to more large errors, we will write $O(\varepsilon)$ to simplify the exposition.

If the function $G(\cdot)$ is Hölder of order $\mu \in (0,1]$, then

$$\left|\frac{1}{\varepsilon}\int_{t_\varepsilon}^{t} \Psi(t,s) \frac{b(\vartheta, s, m_s(\vartheta))}{\sigma(s)} f_y'(\vartheta, s, \tilde m_s(\vartheta))\, G(s)\, \mathrm{d}s - G(t)\right|$$
$$\leq \frac{1}{\varepsilon}\int_{t_\varepsilon}^{t} \Psi(t,s) \frac{b(\vartheta, s, m_s(\vartheta))}{\sigma(s)} f_y'(\vartheta, s, \tilde m_s(\vartheta))\, \left|G(s) - G(t)\right|\, \mathrm{d}s$$
$$\leq \frac{C}{\varepsilon}\int_{t_\varepsilon}^{t} e^{-\frac{\kappa_*}{\varepsilon}(t-s)} \left|t-s\right|^\mu \mathrm{d}s \leq C\int_0^{\frac{\kappa_*(t-t_\varepsilon)}{\varepsilon}} e^{-u} u^\mu \mathrm{d}u\, \varepsilon^\mu \leq C\,\varepsilon^\mu.$$

These estimates allow us to write for such functions $G(\cdot)$ the representation

$$\frac{1}{\varepsilon}\int_{t_0}^{t}\Psi(t,s)\,\frac{b(\vartheta,s,m_s(\vartheta))}{\sigma(s)}\,f'_y(\vartheta,s,\tilde m_s(\vartheta))\,G(s)\,\mathrm{d}s = G(t) + O\left(\varepsilon^{\mu}\right).$$

We have on the interval $[t_\varepsilon, t]$ the estimates

$$\begin{aligned}
&|q_*(t) - q(s)| \\
&\quad = \left|\frac{b(\vartheta,t,m_t(\vartheta))}{\sigma(t)}\,f'_y(\vartheta,t,m_t(\vartheta)) - \frac{b(\vartheta,s,m_s(\vartheta))}{\sigma(s)}\,f'_y(\vartheta,s,\tilde m_s(\vartheta))\right| \\
&\quad \le C\,|t-s| + C\left|f'_y(\vartheta,t,m_t(\vartheta)) - f'_y(\vartheta,t,\tilde m_s(\vartheta))\right| \\
&\quad \le C\,\varepsilon^{\nu} + C\,|m_t(\vartheta) - m_s(\vartheta)| + C\left|m_s(\vartheta) - m_s^{\circ}(\vartheta)\right|.
\end{aligned}$$

Hence,

$$\begin{aligned}
&m_t(\vartheta) - m_t^{\circ}(\vartheta) \\
&\quad = \frac{1}{\varepsilon}\int_{t_\varepsilon}^{t}\Psi(t,s)\,\frac{b(\vartheta,s,m_s(\vartheta))}{\sigma(s)}\,f'_y(\vartheta,s,\tilde m_s(\vartheta))\,[F_s - F_t]\,\mathrm{d}s \\
&\qquad + \frac{1}{\varepsilon}\int_{t_\varepsilon}^{t}\Psi(t,s)\,\frac{b(\vartheta,s,m_s(\vartheta))}{\sigma(s)}\,f'_y(\vartheta,s,\tilde m_s(\vartheta))\,[C_t - C_s]\,\mathrm{d}s + O(\varepsilon).
\end{aligned}$$

We estimate the order of $m_t(\vartheta) - m_t^{\circ}(\vartheta)$ first. We have

$$\begin{aligned}
\mathbf{E}_{\vartheta_0}\left|m_t(\vartheta) - m_t^{\circ}(\vartheta)\right| &\le \frac{C}{\varepsilon}\int_{t_\varepsilon}^{t}\Psi_*(t,s)\,\mathbf{E}_{\vartheta_0}\left(|F_s - F_t| + |C_s - C_t|\right)\mathrm{d}s + O(\varepsilon) \\
&\le \frac{C}{\varepsilon}\int_{t_\varepsilon}^{t} e^{-\frac{\kappa_*}{\varepsilon}(t-s)}\,|t-s|^{1/2}\,\mathrm{d}s + O(\varepsilon) = O\left(\sqrt{\varepsilon}\right).
\end{aligned}$$

Therefore, $m_t(\vartheta) - m_t^{\circ}(\vartheta) = O\left(\sqrt{\varepsilon}\right)$ and

$$\begin{aligned}
&\frac{1}{t-t_\varepsilon}\int_{t_\varepsilon}^{t}\frac{b(\vartheta,v,m_v(\vartheta))}{\sigma(v)}\,f'_y(\vartheta,v,\tilde m_v(\vartheta))\,\mathrm{d}v \\
&\qquad\qquad \longrightarrow \frac{b\left(\vartheta,t,m_t^{\circ}(\vartheta)\right)}{\sigma(t)}\,f'_y\left(\vartheta,t,m_t^{\circ}(\vartheta)\right) = q^{\circ}(t).
\end{aligned}$$

We have as well

$$\frac{1}{\varepsilon}\int_{s}^{t}\frac{b(\vartheta,r,m_r(\vartheta))}{\sigma(r)}\,f'_y(\vartheta,r,\tilde m_r(\vartheta))\,\mathrm{d}r - \frac{(t-s)}{\varepsilon}\,q(t)$$

$$= \frac{(t-s)}{\varepsilon}\left[O\left(\varepsilon^{\nu}\right)+O\left(\varepsilon^{\nu/2}\right)+O\left(\sqrt{\varepsilon}\right)\right]=\frac{(t-s)}{\varepsilon^{1-\frac{\nu}{2}}}O\left(1\right)=\frac{(t-s)}{\varepsilon}h_{\varepsilon},$$

$$h_{\varepsilon}=O\left(\varepsilon^{\frac{\nu}{2}}\right).$$

To study these integrals, we use the elements of the proof of Lemma 1.7, which allow us to write

$$\frac{1}{\varepsilon}\int_{t_{\varepsilon}}^{t}\Psi\left(t,s\right)\frac{b\left(\vartheta,s,m_s\left(\vartheta\right)\right)}{\sigma\left(s\right)}f_y'\left(\vartheta,s,\tilde m_s\left(\vartheta\right)\right)\left[C_t-C_s\right]\mathrm{d}s$$

$$=\frac{1}{\varepsilon}\int_{t_{\varepsilon}}^{t}\Psi\left(t,s\right)\frac{b\left(\vartheta,s,m_s\left(\vartheta\right)\right)}{\sigma\left(s\right)}f_y'\left(\vartheta,s,\tilde m_s\left(\vartheta\right)\right)\int_s^t b\left(\vartheta,r,m_r\left(\vartheta\right)\right)\mathrm{d}W_r\;\mathrm{d}s$$

$$=\frac{1}{\varepsilon}\int_{t_{\varepsilon}}^{t}e^{-\frac{1}{\varepsilon}(t-s)[q(t)+h_{\varepsilon}]}\frac{b\left(\vartheta,s,m_s\left(\vartheta\right)\right)}{\sigma\left(s\right)}f_y'\left(\vartheta,s,\tilde m_s\left(\vartheta\right)\right)\int_s^t b\left(\vartheta,r,m_r\left(\vartheta\right)\right)\mathrm{d}W_r\;\mathrm{d}s$$

$$=\frac{b\left(\vartheta,t,m_t\left(\vartheta\right)\right)}{\varepsilon\sigma\left(t\right)}f_y'\left(\vartheta,t,\tilde m_t\left(\vartheta\right)\right)\int_{t_{\varepsilon}}^{t}e^{-\frac{(t-s)}{\varepsilon}q(t)}$$

$$\int_s^t b\left(\vartheta,r,m_r\left(\vartheta\right)\right)\mathrm{d}W_r\;\mathrm{d}s\left(1+O\left(\varepsilon^{\frac{\nu}{2}}\right)\right)$$

$$=\frac{b\left(\vartheta,t,m_t\left(\vartheta\right)\right)^2}{\varepsilon\sigma\left(t\right)}f_y'\left(\vartheta,t,\tilde m_t\left(\vartheta\right)\right)\int_{t_{\varepsilon}}^{t}e^{-\frac{(t-s)}{\varepsilon}q(t)}\left[W_t-W_s\right]\;\mathrm{d}s\left(1+O\left(\varepsilon^{\frac{\nu}{2}}\right)\right)$$

$$=\sqrt{\varepsilon}\sqrt{\frac{b\left(\vartheta,t,m_t\left(\vartheta\right)\right)\sigma\left(t\right)}{f_y'\left(\vartheta,t,\tilde m_t\left(\vartheta\right)\right)}}\int_0^{\frac{(t-t_{\varepsilon})q(t)}{\varepsilon}}e^{-u}\,w_{t,\varepsilon}\left(u\right)\,\mathrm{d}u\left(1+O\left(\varepsilon^{\frac{\nu}{2}}\right)\right)$$

$$=\sqrt{\varepsilon}\sqrt{\frac{b\left(\vartheta,t,m_t\left(\vartheta\right)\right)\sigma\left(t\right)}{2f_y'\left(\vartheta,t,\tilde m_t\left(\vartheta\right)\right)}}\;\xi_{t,\varepsilon}\left(1+O\left(\varepsilon^{\frac{\nu}{2}}\right)\right),$$

where we changed the variable $s=t-\frac{\varepsilon}{q(t)}u$ and denoted Wiener processes $w_{t,\varepsilon}\left(u\right)=\sqrt{\frac{q(t)}{\varepsilon}}\left[W_t-W_{t-\frac{\varepsilon u}{q(t)}}\right]$.

Note that

$$\xi_{t,\varepsilon}=\sqrt{2}\int_0^{\frac{(t-t_{\varepsilon})q(t)}{\varepsilon}}e^{-u}\,w_{t,\varepsilon}\left(u\right)\,\mathrm{d}u=\sqrt{2}\int_0^{\frac{(t-t_{\varepsilon})q(t)}{\varepsilon}}e^{-u}\,\mathrm{d}w_{t,\varepsilon}\left(u\right)+O\left(e^{-\frac{c}{\varepsilon}}\right)$$

$$\Longrightarrow\xi_t=\sqrt{2}\int_0^{\infty}e^{-u}\,\mathrm{d}w_t\left(u\right)\sim\mathcal{N}\left(0,1\right).$$

Here $w_t\left(\cdot\right)$ is some Wiener process and the Wiener processes $w_{t_1}\left(\cdot\right)$ and $w_{t_2}\left(\cdot\right)$ for any $t_1\neq t_2$ are independent. Therefore, the random variables ξ_t, $t_0<t\leq T$ are independent.

For the second integral, we have the similar relations

$$
\begin{aligned}
&\frac{1}{\varepsilon}\int_{t_\varepsilon}^{t}\Psi(t,s)\frac{b(\vartheta,s,m_s(\vartheta))}{\sigma(s)}f_y'(\vartheta,s,\tilde m_s(\vartheta))\int_s^t\frac{f_y'(\vartheta_0,r,Y_r)\,b(\vartheta_0,r,Y_r)}{f_y'(\vartheta,r,m_r(\vartheta))}\mathrm{d}V_r\,\mathrm{d}s\\
&\quad=\frac{b(\vartheta,t,\bar m_t(\vartheta))\,f_y'(\vartheta,t,\tilde m_t(\vartheta))\,f_y'(\vartheta_0,t,Y_t)\,b(\vartheta_0,t,Y_t)}{\varepsilon\sigma(t)\,f_y'(\vartheta,t,\bar m_t(\vartheta))}\\
&\qquad\times\int_{t_\varepsilon}^{t}e^{-\frac{1}{\varepsilon}(t-s)q(t)}\left[V_t-V_s\right]\mathrm{d}s\left(1+O\left(\varepsilon^{\frac{\nu}{2}}\right)\right)\\
&\quad=\sqrt{\varepsilon}\,\frac{b(\vartheta_0,t,Y_t)\,f_y'(\vartheta_0,t,Y_t)}{b(\vartheta,t,\bar m_t(\vartheta))\,f_y'(\vartheta,t,m_t(\vartheta))}\sqrt{\frac{b(\vartheta,t,\bar m_t(\vartheta))\,\sigma(t)}{2f_y'(\vartheta,t,\tilde m_t(\vartheta))}}\,\bar\eta_{t,\varepsilon}\\
&\qquad\times\left(1+O\left(\varepsilon^{\frac{\nu}{2}}\right)\right)\\
&\quad=\sqrt{\varepsilon}\,\Delta(\vartheta,t)\sqrt{\frac{b(\vartheta,t,\bar m_t(\vartheta))\,\sigma(t)}{2f_y'(\vartheta,t,\tilde m_t(\vartheta))}}\,\eta_{t,\varepsilon}\left(1+O\left(\varepsilon^{\frac{\nu}{2}}\right)\right).
\end{aligned}
$$

Here

$$
\bar\eta_{t,\varepsilon}=\sqrt{2}\int_0^{\frac{(t-t_\varepsilon)q(t)}{\varepsilon}}e^{-u}\,v_{t,\varepsilon}(u)\,\mathrm{d}u,\qquad v_{t,\varepsilon}(u)=\sqrt{\frac{q(t)}{\varepsilon}}\left[V_t-V_{t-\frac{\varepsilon u}{q(t)}}\right]
$$

and

$$
\begin{aligned}
\bar\eta_{t,\varepsilon}&=\sqrt{2}\int_0^{\frac{(t-t_\varepsilon)q(t)}{\varepsilon}}e^{-u}\,\mathrm{d}v_{t,\varepsilon}(u)+O\left(e^{-\frac{c}{\varepsilon}}\right)=\eta_{t,\varepsilon}+O\left(e^{-\frac{c}{\varepsilon}}\right)\\
&\Longrightarrow\eta_t=\sqrt{2}\int_0^\infty e^{-u}\,\mathrm{d}v_t(u)\sim\mathcal{N}(0,1).
\end{aligned}
$$

The Wiener processes $w_t(\cdot)$ and $v_t(\cdot)$ are independent and the random variables $\eta_{t_1},\eta_{t_2},\ t_0\le t\le T$ for any $t_1\neq t_2$ are independent too.

We obtained the representation

$$
\begin{aligned}
&\frac{m_t(\vartheta)-m_t^\circ(\vartheta)}{\sqrt{\varepsilon}}\\
&=\sqrt{\frac{\sigma(t)\,b(\vartheta,t,\bar m_t(\vartheta))}{2f_y'(\vartheta,t,\tilde m_t(\vartheta))}}\left[\xi_{t,\varepsilon}-\frac{b(\vartheta_0,t,Y_t)\,f_y'(\vartheta_0,t,Y_t)}{b(\vartheta,t,\bar m_t(\vartheta))f_y'(\vartheta,t,m_t(\vartheta))}\eta_{t,\varepsilon}\right]+O\left(\varepsilon^{\frac{\nu}{2}}\right)\\
&=\sqrt{\frac{\sigma(t)\,b\left(\vartheta,t,m_t^\circ(\vartheta)\right)}{f_y'\left(\vartheta,t,m_t^\circ(\vartheta)\right)}}\left[\xi_{t,\varepsilon}-\frac{b(\vartheta_0,t,Y_t)\,f_y'(\vartheta_0,t,Y_t)}{b\left(\vartheta,t,m_t^\circ(\vartheta)\right)f_y'\left(\vartheta,t,m_t^\circ(\vartheta)\right)}\eta_{t,\varepsilon}\right]+O\left(\varepsilon^{\frac{\nu}{2}}\right)\\
&=\Gamma^\circ(\vartheta,t)\left[\xi_{t,\varepsilon}-\Delta^\circ(\vartheta,t)\,\eta_{t,\varepsilon}\right]+O\left(\varepsilon^{\frac{\nu}{2}}\right).
\end{aligned}
$$

Note that as $m_t(\vartheta) - m_t^\circ(\vartheta) \approx \sqrt{\varepsilon}$, we replaced in this expression $m_t(\vartheta)$ and $\tilde m_t(\vartheta)$ by $m_t^\circ(\vartheta)$.

If $\vartheta = \vartheta_0$, then $m_t^\circ(\vartheta_0) = Y_t$ (see (3.55)) and

$$\begin{aligned} m_t(\vartheta_0) - Y_t &= \Gamma^\circ(\vartheta_0, t)\left[\xi_{t,\varepsilon} - \Delta^\circ(\vartheta_0, t)\,\eta_{t,\varepsilon}\right]\sqrt{\varepsilon}\left(1 + O\left(\varepsilon^{\frac{\nu}{2}}\right)\right)\\ &= \Gamma^\circ(\vartheta_0, t)\left[\xi_{t,\varepsilon} - \eta_{t,\varepsilon}\right]\sqrt{\varepsilon}\left(1 + O\left(\varepsilon^{\frac{\nu}{2}}\right)\right). \end{aligned}$$

Therefore, we can formally evaluate the error

$$\begin{aligned} \mathbf{E}_{\vartheta_0}\left(m_t(\vartheta_0) - Y_t\right)^2 &= \sigma(t)\mathbf{E}_{\vartheta_0}\left[\frac{b(\vartheta_0, t, Y_t)}{f_y'(\vartheta_0, t, Y_t)}\right]\varepsilon + O\left(\varepsilon^{1+\frac{\nu}{2}}\right)\\ &= \mathbf{E}_{\vartheta_0}\gamma_*(\vartheta_0, t)\;\varepsilon + O\left(\varepsilon^{1+\frac{\nu}{2}}\right). \end{aligned}$$

Hence, the function $\mathbf{E}_{\vartheta_0}\gamma_*(\vartheta_0, t)$ asymptoticaly plays the same role as in the linear case, i.e., describes the mean-squared error of estimation Y_t by $m_t(\vartheta_0)$.

Let us denote

$$P(\vartheta, t, y, z) = a(\vartheta, t, y) - A(\vartheta, t, z)$$

and consider now the difference $\dot m_t(\vartheta) - \dot m_t^\circ(\vartheta)$. The derivative in ϑ of (3.63) yields the equation

$$\begin{aligned} \mathrm{d}\left[\dot m_t(\vartheta) - \dot m_t^\circ(\vartheta)\right] &= \dot P_\vartheta\left(\vartheta, t, m_t(\vartheta), m_t^\circ(\vartheta)\right)\mathrm{d}t\\ &\quad + P_y'\left(\vartheta, t, m_t(\vartheta), m_t^\circ(\vartheta)\right)\dot m_t(\vartheta)\,\mathrm{d}t\\ &\quad + P_z'\left(\vartheta, t, m_t(\vartheta), m_t^\circ(\vartheta)\right)\dot m_t^\circ(\vartheta)\,\mathrm{d}t\\ &\quad + \frac{f_y'(\vartheta_0, t, Y_t)\,b(\vartheta_0, t, Y_t)}{f_y'(\vartheta, t, m_t(\vartheta))^2}\left[\dot f_{y\vartheta}'(\vartheta, t, m_t(\vartheta)) + f_{yy}''(\vartheta, t, m_t(\vartheta))\,\dot m_t(\vartheta)\right]\mathrm{d}V_t\\ &\quad - \frac{\left[\dot B_\vartheta(\vartheta, t)\,f_y'(\vartheta, t, \tilde m_t(\vartheta)) + b(\vartheta, t, m_t(\vartheta))\,\dot f_{\vartheta y}'(\vartheta, t, \tilde m_t(\vartheta))\right]}{\varepsilon\sigma(t)}\\ &\quad \times\left[m_t(\vartheta) - m_t^\circ(\vartheta)\right]\mathrm{d}t\\ &\quad - \frac{b(\vartheta, t, m_t(\vartheta))}{\varepsilon\sigma(t)}\left[f_y'(\vartheta, t, m_t(\vartheta))\,\dot m_t(\vartheta) - f_y'\left(\vartheta, t, m_t^\circ(\vartheta)\right)\dot m_t^\circ(\vartheta)\right]\mathrm{d}t\\ &\quad + \dot B(\vartheta, t)\,\mathrm{d}W_t\\ &= Q(\vartheta, t)\,\mathrm{d}t + \dot B_\vartheta(\vartheta, t)\,\mathrm{d}W_t - \frac{b(\vartheta, t, m_t(\vartheta))}{\varepsilon\sigma(t)}f_{yy}''(\vartheta, t, \bar m_t)\,\dot m_t(\vartheta) \end{aligned}$$

$$
\begin{aligned}
&\times\left[m_t(\vartheta)-m_t^\circ(\vartheta)\right]\mathrm{d}t\\
&-\frac{\left[\dot B_\vartheta(\vartheta,t)\,f_y'(\vartheta,t,\tilde m_t(\vartheta))+b(\vartheta,t,m_t(\vartheta))\,\dot f_{\vartheta y}'(\vartheta,t,\tilde m_t(\vartheta))\right]}{\varepsilon\sigma(t)}\\
&\times\left[m_t(\vartheta)-m_t^\circ(\vartheta)\right]\mathrm{d}t\\
&+\frac{f_y'(\vartheta_0,t,Y_t)\,b(\vartheta_0,t,Y_t)}{f_y'(\vartheta,t,m_t(\vartheta))^2}\left[\dot f_{y\vartheta}'(\vartheta,t,m_t(\vartheta))+f_{yy}''(\vartheta,t,m_t(\vartheta))\,\dot m_t(\vartheta)\right]\mathrm{d}V_t\\
&-\frac{b(\vartheta,t,m_t(\vartheta))}{\varepsilon\sigma(t)}f_y'\left(\vartheta,t,m_t^\circ(\vartheta)\right)\left[\dot m_t(\vartheta)-\dot m_t^\circ(\vartheta)\right]\mathrm{d}t\left(1+O\left(\sqrt{\varepsilon}\right)\right)
\end{aligned}
$$

with obvious notation for $Q(\vartheta,t)$. Here we used the relations

$$
\begin{aligned}
f(\vartheta_0,t,Y_t)-f(\vartheta,t,m_t(\vartheta))&=f\left(\vartheta,t,m_t^\circ(\vartheta)\right)-f(\vartheta,t,m_t(\vartheta))\\
&=f_y'(\vartheta,t,\tilde m_t)\left(m_t^\circ(\vartheta)-m_t(\vartheta)\right)\\
&=f_y'\left(\vartheta,t,m_t^\circ\right)\left(m_t^\circ(\vartheta)-m_t(\vartheta)\right)\left(1+O\left(\sqrt{\varepsilon}\right)\right).
\end{aligned}
$$

Introduce the function

$$
\begin{aligned}
R(\vartheta,t)=&-f_{yy}''(\vartheta,t,\bar m_t)\,\dot m_t(\vartheta)-\frac{\dot B_\vartheta(\vartheta,t)}{b(\vartheta,t,m_t(\vartheta))}f_y'(\vartheta,t,\tilde m_t(\vartheta))\\
&-\dot f_{\vartheta y}'(\vartheta,t,\tilde m_t(\vartheta)).
\end{aligned}
$$

Then we can write

$$
\begin{aligned}
\mathrm{d}\left[\dot m_t(\vartheta)-\dot m_t^\circ(\vartheta)\right]&=Q(\vartheta,t)\,\mathrm{d}t+\frac{b(\vartheta,t,m_t(\vartheta))\,R(\vartheta,t)}{\varepsilon\sigma(t)}\left[m_t(\vartheta)-m_t^\circ(\vartheta)\right]\mathrm{d}t\\
&+\frac{f_y'(\vartheta_0,t,Y_t)\,b(\vartheta_0,t,Y_t)}{f_y'(\vartheta,t,m_t(\vartheta))^2}\left[\dot f_{y\vartheta}'(\vartheta,t,m_t(\vartheta))+f_{yy}''(\vartheta,t,m_t(\vartheta))\,\dot m_t(\vartheta)\right]\mathrm{d}V_t\\
&-\frac{b(\vartheta,t,m_t(\vartheta))}{\varepsilon\sigma(t)}f_y'\left(\vartheta,t,m_t^\circ(\vartheta)\right)\left[\dot m_t(\vartheta)-\dot m_t^\circ(\vartheta)\right]\mathrm{d}t\\
&+\dot B_\vartheta(\vartheta,t)\,\mathrm{d}W_t\left(1+O\left(\sqrt{\varepsilon}\right)\right).
\end{aligned}
$$

Introduce the notations

$$
\begin{aligned}
\dot C_t&=\int_{t_0}^{t}\dot B_\vartheta(\vartheta,s)\,\mathrm{d}W_s,\\
\dot F_t&=\int_{t_0}^{t}\frac{f_y'(\vartheta_0,s,Y_s)\,b(\vartheta_0,s,Y_s)}{f_y'(\vartheta,s,m_s(\vartheta))^2}\\
&\quad\times\left[\dot f_{y\vartheta}'(\vartheta,s,m_s(\vartheta))+f_{yy}''(\vartheta,s,m_s(\vartheta))\,\dot m_s(\vartheta)\right]\mathrm{d}V_s.
\end{aligned}
$$

Then for the process $\dot{Z}_t = \dot{m}_t(\vartheta) - \dot{m}_t^\circ(\vartheta) - \dot{C}_t - \dot{F}_t$ we obtain the equation

$$\begin{aligned} \mathrm{d}\dot{Z}_t &= \frac{b(\vartheta, t, m_t(\vartheta))\, R(\vartheta, t)}{\varepsilon\sigma(t)} \left[m_t(\vartheta) - m_t^\circ(\vartheta)\right] \mathrm{d}t \\ &\quad - \frac{b(\vartheta, t, m_t(\vartheta))}{\varepsilon\sigma(t)} f_y'\left(\vartheta, t, m_t^\circ(\vartheta)\right) \dot{Z}_t \mathrm{d}t + Q(\vartheta, t)\,\mathrm{d}t \\ &\quad - \frac{b(\vartheta, t, m_t(\vartheta))}{\varepsilon\sigma(t)} f_y'\left(\vartheta, t, m_t^\circ(\vartheta)\right) \left[\dot{C}_t + \dot{F}_t\right] \mathrm{d}t. \end{aligned}$$

Therefore, $(t > t_0)$

$$\begin{aligned} \dot{Z}_t &= \int_{t_0}^{t} \Psi(t, s)\, Q(\vartheta, s)\,\mathrm{d}s \\ &\quad + \frac{1}{\varepsilon}\int_{t_0}^{t} \Psi(t, s) \frac{b(\vartheta, s, m_s(\vartheta))\, R(\vartheta, s)}{\sigma(s)} \left[m_s(\vartheta) - m_s^\circ(\vartheta)\right] \mathrm{d}s \\ &\quad + \dot{Z}_{t_0}\Psi(t, 0) - \frac{1}{\varepsilon}\int_{t_0}^{t} \Psi(t, s) \frac{\dot{B}_\vartheta(\vartheta, s)}{\sigma(s)} f_y'\left(\vartheta, s, m_s^\circ(\vartheta)\right) \left[\dot{C}_s + \dot{F}_s\right] \mathrm{d}s \\ &= \frac{R(\vartheta, t)}{f_y'\left(\vartheta, t, m_t^\circ(\vartheta)\right)} \left[m_t(\vartheta) - m_t^\circ(\vartheta)\right] - \dot{C}_t - \dot{F}_t \\ &\quad + \frac{1}{\varepsilon}\int_{t_\varepsilon}^{t} \Psi(t, s) \frac{\dot{B}_\vartheta(\vartheta, s)}{\sigma(s)} f_y'\left(\vartheta, s, m_s^\circ(\vartheta)\right) \left[\dot{C}_t - \dot{C}_s\right] \mathrm{d}s \\ &\quad + \frac{1}{\varepsilon}\int_{t_\varepsilon}^{t} \Psi(t, s) \frac{\dot{B}_\vartheta(\vartheta, s)}{\sigma(s)} f_y'\left(\vartheta, s, m_s^\circ(\vartheta)\right) \left[\dot{F}_t - \dot{F}_s\right] \mathrm{d}s + O(\varepsilon). \end{aligned}$$

Here we used the slightly different notation

$$\Psi(t, s) = \exp\left(-\frac{1}{\varepsilon}\int_s^t q(r)\,\mathrm{d}r\right), \qquad q(t) = \frac{b\left(\vartheta, t, m_t^\circ(\vartheta)\right)}{\sigma(t)} f_y'(\vartheta, t, m_t^\circ(\vartheta))$$

and the estimates

$$\dot{Z}_{t_0}\Psi(t, 0) = O\left(e^{-\frac{c}{\varepsilon}}\right), \qquad \int_{t_\varepsilon}^{t} \Psi(t, s)\, Q(\vartheta, s)\,\mathrm{d}s = O(\varepsilon).$$

Further,

$$\frac{1}{\varepsilon}\int_{t_0}^{t}\Psi(t,s)\,\frac{\dot{B}_\vartheta(\vartheta,s)}{\sigma(s)}\,f_y'\left(\vartheta,s,m_s^\circ(\vartheta)\right)\int_s^t\dot{B}(\vartheta,r)\,\mathrm{d}W_r\,\mathrm{d}s$$
$$=\frac{\dot{B}_\vartheta(\vartheta,t)^2}{\varepsilon\sigma(t)}\,f_y'\left(\vartheta,t,m_t^\circ(\vartheta)\right)\int_{t_\varepsilon}^{t}\Psi(t,s)\left[W_t-W_s\right]\mathrm{d}s\left(1+O\left(\varepsilon^{\frac{\nu}{2}}\right)\right)$$
$$=\sqrt{\varepsilon}\,\frac{\dot{B}_\vartheta(\vartheta,t)^2}{\sigma(t)}\,f_y'\left(\vartheta,t,m_t^\circ(\vartheta)\right)q(t)^{-3/2}\,\xi_{t,\varepsilon}\left(1+O\left(\varepsilon^{\frac{\nu}{2}}\right)\right)$$
$$=\sqrt{\varepsilon}\,\frac{\dot{B}_\vartheta(\vartheta,t)^2}{b\left(\vartheta,t,m_t^\circ(\vartheta)\right)^2}\sqrt{\frac{b\left(\vartheta,t,m_t^\circ(\vartheta)\right)\sigma(t)}{2f_y'\left(\vartheta,t,m_t^\circ(\vartheta)\right)}}\,\xi_{t,\varepsilon}\left(1+O\left(\varepsilon^{\frac{\nu}{2}}\right)\right),$$

and

$$\frac{1}{\varepsilon}\int_{t_0}^{t}\Psi(t,s)\,\frac{\dot{B}_\vartheta(\vartheta,s)}{\sigma(s)}\,f_y'\left(\vartheta,s,m_s^\circ(\vartheta)\right)\left[\dot{F}_t-\dot{F}_s\right]\mathrm{d}s$$
$$=\sqrt{\varepsilon}\,\frac{\dot{B}_\vartheta(\vartheta,t)\,f_y'(\vartheta_0,t,Y_t)\,b(\vartheta_0,t,Y_t)}{\sqrt{2}\sigma(t)\,f_y'(\vartheta,t,m_t(\vartheta))\,q(t)^{3/2}}\left[\dot{f}_{y\vartheta}'(\vartheta,t,m_t(\vartheta))\right.$$
$$\left.+f_{yy}''(\vartheta,t,m_t(\vartheta))\,\dot{m}_t(\vartheta)\right]\eta_{t,\varepsilon}\left(1+O\left(\varepsilon^{\frac{\nu}{2}}\right)\right)$$
$$=\sqrt{\varepsilon}\,\frac{\dot{B}_\vartheta(\vartheta,t)\,f_y'(\vartheta_0,t,Y_t)\,b(\vartheta_0,t,Y_t)\sqrt{\sigma(t)}}{\sqrt{2}b\left(\vartheta,t,m_t^\circ(\vartheta)\right)^{3/2}f_y'\left(\vartheta,t,m_t^\circ(\vartheta)\right)^{7/2}}\left[\dot{f}_{y\vartheta}'\left(\vartheta,t,m_t^\circ(\vartheta)\right)\right.$$
$$\left.f_y'\left(\vartheta,t,m_t^\circ(\vartheta)\right)-f_{yy}''\left(\vartheta,t,m_t^\circ(\vartheta)\right)\dot{f}_\vartheta\left(\vartheta,t,m_t^\circ(\vartheta)\right)\right]\eta_{t,\varepsilon}\left(1+O\left(\varepsilon^{\frac{\nu}{2}}\right)\right).$$

These relations allow us to write

$$\dot{m}_t(\vartheta)-\dot{m}_t^\circ(\vartheta)=\frac{R(\vartheta,t)\,\Gamma^\circ(\vartheta,t)}{f_y'\left(\vartheta,t,m_t^\circ(\vartheta)\right)}$$
$$\left[\xi_{t,\varepsilon}-\frac{b(\vartheta_0,t,Y_t)\,f_y'(\vartheta_0,t,Y_t)}{b\left(\vartheta,t,m_t^\circ(\vartheta)\right)f_y'\left(\vartheta,t,m_t^\circ(\vartheta)\right)}\,\eta_{t,\varepsilon}\right]\sqrt{\varepsilon}$$
$$+\frac{\dot{B}_\vartheta(\vartheta,t)^2}{b\left(\vartheta,t,m_t^\circ(\vartheta)\right)^2}\sqrt{\frac{b\left(\vartheta,t,m_t^\circ(\vartheta)\right)\sigma(t)}{2f_y'\left(\vartheta,t,m_t^\circ(\vartheta)\right)}}\,\xi_{t,\varepsilon}\sqrt{\varepsilon}$$
$$+\frac{\dot{B}_\vartheta(\vartheta,t)\,f_y'(\vartheta_0,t,Y_t)\,b(\vartheta_0,t,Y_t)\sqrt{\sigma(t)}}{\sqrt{2}b\left(\vartheta,t,m_t^\circ(\vartheta)\right)^{3/2}f_y'\left(\vartheta,t,m_t^\circ(\vartheta)\right)^{7/2}}$$
$$\left[\dot{f}_{y\vartheta}'\left(\vartheta,t,m_t^\circ(\vartheta)\right)f_y'\left(\vartheta,t,m_t^\circ(\vartheta)\right)\right.$$
$$\left.-f_{yy}''\left(\vartheta,t,m_t^\circ(\vartheta)\right)\dot{f}_\vartheta\left(\vartheta,t,m_t^\circ(\vartheta)\right)\right]\eta_{t,\varepsilon}\sqrt{\varepsilon}+O\left(\varepsilon^{\frac{1+\nu}{2}}\right).$$

Hence,

$$
\begin{aligned}
\frac{\dot m_t(\vartheta)-\dot m_t^\circ(\vartheta)}{\Gamma^\circ(\vartheta,t)\sqrt{\varepsilon}} &= \frac{R^\circ(\vartheta,t)}{f_y'\left(\vartheta,t,m_t^\circ(\vartheta)\right)}\left[\xi_{t,\varepsilon}-\frac{b(\vartheta_0,t,Y_t)\,f_y'(\vartheta_0,t,Y_t)}{b\left(\vartheta,t,m_t^\circ(\vartheta)\right)f_y'\left(\vartheta,t,m_t^\circ(\vartheta)\right)}\eta_{t,\varepsilon}\right]\\
&\quad+\frac{\dot b_\vartheta\left(\vartheta,t,m_t^\circ(\vartheta)\right)f_y'(\vartheta_0,t,Y_t)\,b\left(\vartheta_0,t,m_t^\circ(\vartheta)\right)}{b(\vartheta,t)^2\,f_y'\left(\vartheta,t,m_t^\circ(\vartheta)\right)}\left[\frac{\dot f_{y\vartheta}'\left(\vartheta,t,m_t^\circ(\vartheta)\right)}{f_y'\left(\vartheta,t,m_t^\circ(\vartheta)\right)}\right.\\
&\quad\left.-\frac{f_{yy}''\left(\vartheta,t,m_t^\circ(\vartheta)\right)\dot f_\vartheta\left(\vartheta,t,m_t^\circ(\vartheta)\right)}{f_y'\left(\vartheta,t,m_t^\circ(\vartheta)\right)^2}\right]\eta_{t,\varepsilon}+\frac{\dot B_\vartheta(\vartheta,t)^2}{b\left(\vartheta,t,m_t^\circ(\vartheta)\right)^2}\xi_{t,\varepsilon}+O\left(\varepsilon^{\frac{\nu}{2}}\right)\\
&=\left[\frac{R^\circ(\vartheta,t)}{f_y'\left(\vartheta,t,m_t^\circ(\vartheta)\right)}+\frac{\dot B_\vartheta(\vartheta,t)^2}{b\left(\vartheta,t,m_t^\circ(\vartheta)\right)^2}\right]\xi_{t,\varepsilon}\\
&\quad+\frac{b(\vartheta_0,t,Y_t)\,f_y'(\vartheta_0,t,Y_t)}{b\left(\vartheta,t,m_t^\circ(\vartheta)\right)f_y'\left(\vartheta,t,m_t^\circ(\vartheta)\right)}\left[\frac{\dot B_\vartheta(\vartheta,t)\,\dot f_{y\vartheta}'\left(\vartheta,t,m_t^\circ(\vartheta)\right)}{b\left(\vartheta,t,m_t^\circ(\vartheta)\right)f_y'\left(\vartheta,t,m_t^\circ(\vartheta)\right)}\right.\\
&\quad\left.-\frac{R^\circ(\vartheta,t)}{f_y'\left(\vartheta,t,m_t^\circ(\vartheta)\right)}-\frac{\dot B_\vartheta(\vartheta,t)\,f_{yy}''\left(\vartheta,t,m_t^\circ(\vartheta)\right)\dot f_{y\vartheta}'\left(\vartheta,t,m_t^\circ(\vartheta)\right)}{b\left(\vartheta,t,m_t^\circ(\vartheta)\right)f_y'\left(\vartheta,t,m_t^\circ(\vartheta)\right)^2}\right]\eta_{t,\varepsilon}\\
&\quad+O\left(\varepsilon^{\frac{\nu}{2}}\right)\\
&=\Pi^\circ(\vartheta,t)\;\xi_{t,\varepsilon}+\Lambda^\circ(\vartheta,t)\;\eta_{t,\varepsilon}+O\left(\varepsilon^{\frac{\nu}{2}}\right).
\end{aligned}
$$

Therefore, if $\vartheta=\vartheta_0$, then this expression is simplified and we have

$$
\begin{aligned}
&\frac{\dot m_t(\vartheta_0)-\dot m_t^\circ(\vartheta_0)}{\Gamma^\circ(\vartheta_0,t)\sqrt{\varepsilon}}\\
&\quad\Longrightarrow\frac{R^\circ(\vartheta_0,t)}{f_y'(\vartheta_0,t,Y_t)}\left[\xi_t-\eta_t\right]+\frac{\dot B_\vartheta^\circ(\vartheta_0,t)^2}{b(\vartheta_0,t,Y_t)^2}\xi_t\\
&\qquad+\frac{\dot B_\vartheta^\circ(\vartheta_0,t)}{b(\vartheta_0,t,Y_t)}\left[\frac{\dot f_{y\vartheta}'(\vartheta_0,t,Y_t)}{f_y'(\vartheta_0,t,Y_t)}-\frac{f_{yy}''(\vartheta_0,t,Y_t)\,\dot f_\vartheta(\vartheta_0,t,Y_t)}{f_y'(\vartheta_0,t,Y_t)^2}\right]\eta_t\\
&\quad=\left[\frac{R^\circ(\vartheta_0,t)}{f_y'(\vartheta_0,t,Y_t)}+\frac{\dot B_\vartheta^\circ(\vartheta_0,t)^2}{b(\vartheta_0,t,Y_t)^2}\right]\xi_t+\left[\frac{\dot B_\vartheta^\circ(\vartheta_0,t)\,\dot f_{y\vartheta}'(\vartheta_0,t,Y_t)}{b(\vartheta_0,t,Y_t)\,f_y'(\vartheta_0,t,Y_t)}\right.\\
&\qquad\left.-\frac{\dot B_\vartheta^\circ(\vartheta_0,t)\,f_{yy}''(\vartheta_0,t,Y_t)\,\dot f_\vartheta(\vartheta_0,t,Y_t)}{b(\vartheta_0,t,Y_t)\,f_y'(\vartheta_0,t,Y_t)^2}-\frac{R^\circ(\vartheta_0,t)}{f_y'(\vartheta_0,t,Y_t)}\right]\eta_t\\
&\quad=\Pi^\circ(\vartheta_0,t)\,\xi_t+\Lambda^\circ(\vartheta_0,t)\,\eta_t.
\end{aligned}
$$

□

Corollary 3.2 *The representations* (3.61) *and* (3.62) *allow us to write*

$$\begin{aligned}
\frac{1}{\Gamma^\circ(\vartheta,t)\sqrt{\varepsilon}}\frac{\mathrm{d}f(\vartheta,t,m_t(\vartheta))}{\mathrm{d}\vartheta} &= K\left(\vartheta,t,m_t^\circ(\vartheta)\right)\left[\xi_{t,\varepsilon}-\Delta^\circ(\vartheta,t)\ \eta_{t,\varepsilon}\right]\\
&\quad + f_y'(\vartheta,t,m_t(\vartheta))\left[\Pi^\circ(\vartheta,t)\ \xi_{t,\varepsilon}+\Lambda^\circ(\vartheta,t)\ \eta_{t,\varepsilon}\right]+O\left(\varepsilon^{\frac{\nu}{2}}\right)\\
&=\left[K\left(\vartheta,t,m_t^\circ(\vartheta)\right)+f_y'(\vartheta,t,m_t(\vartheta))\,\Pi^\circ(\vartheta,t)\right]\xi_{t,\varepsilon}+O\left(\varepsilon^{\frac{\nu}{2}}\right)\\
&\quad+\left[f_y'(\vartheta,t,m_t(\vartheta))\,\Lambda^\circ(\vartheta,t)-K\left(\vartheta,t,m_t^\circ(\vartheta)\right)\Delta^\circ(\vartheta,t)\right]\eta_{t,\varepsilon}.
\end{aligned} \tag{3.64}$$

Here

$$K(\vartheta,t,y)=\dot f_{\vartheta y}'(\vartheta,t,y)-\frac{f_{yy}''(\vartheta,t,y)}{f_y'(\vartheta,t,y)}.$$

Indeed we have

$$\begin{aligned}
\frac{\mathrm{d}}{\mathrm{d}\vartheta}f(\vartheta,t,m_t(\vartheta)) &= \frac{\mathrm{d}}{\mathrm{d}\vartheta}\left[f(\vartheta,t,m_t(\vartheta))-f\left(\vartheta_0,t,m_t^\circ(\vartheta_0)\right)\right]\\
&=\frac{\mathrm{d}}{\mathrm{d}\vartheta}\left[f(\vartheta,t,m_t(\vartheta))-f\left(\vartheta,t,m_t^\circ(\vartheta)\right)\right]\\
&=\dot f_\vartheta(\vartheta,t,m_t(\vartheta))-\dot f_\vartheta\left(\vartheta,t,m_t^\circ(\vartheta)\right)\\
&\qquad+f_y'(\vartheta,t,m_t(\vartheta))\,\dot m_t(\vartheta)-f_y'\left(\vartheta,t,m_t^\circ(\vartheta)\right)\dot m_t^\circ(\vartheta)\\
&=\dot f_{\vartheta y}'(\vartheta,t,\tilde m_t)\left[m_t(\vartheta)-m_t^\circ(\vartheta)\right]+f_y'(\vartheta,t,m_t(\vartheta))\left[\dot m_t(\vartheta)-\dot m_t^\circ(\vartheta)\right]\\
&\quad+\left[f_y'(\vartheta,t,m_t(\vartheta))-f_y'\left(\vartheta,t,m_t^\circ(\vartheta)\right)\right]\dot m_t(\vartheta)\\
&=\left[\dot f_{\vartheta y}'(\vartheta,t,\tilde m_t)+f_{yy}''(\vartheta,t,\tilde m_t)\,\dot m_t(\vartheta)\right]\left[m_t(\vartheta)-m_t^\circ(\vartheta)\right]\\
&\qquad+f_y'(\vartheta,t,m_t(\vartheta))\left[\dot m_t(\vartheta)-\dot m_t^\circ(\vartheta)\right]\\
&=\left[\dot f_{\vartheta y}'\left(\vartheta,t,m_t^\circ(\vartheta)\right)+f_{yy}''\left(\vartheta,t,m_t^\circ(\vartheta)\right)\dot m_t(\vartheta)\right]\left[m_t(\vartheta)-m_t^\circ(\vartheta)\right]\\
&\qquad+f_y'\left(\vartheta,t,m_t^\circ(\vartheta)\right)\left[\dot m_t(\vartheta)-\dot m_t^\circ(\vartheta)\right]\left(1+O\left(\sqrt{\varepsilon}\right)\right)\\
&=\left[\dot f_{\vartheta y}'\left(\vartheta,t,m_t^\circ(\vartheta)\right)-\frac{f_{yy}''\left(\vartheta,t,m_t^\circ(\vartheta)\right)}{f_y'\left(\vartheta,t,m_t^\circ(\vartheta)\right)}\right]\left[m_t(\vartheta)-m_t^\circ(\vartheta)\right]\left(1+O\left(\sqrt{\varepsilon}\right)\right)\\
&\qquad+f_y'\left(\vartheta,t,m_t^\circ(\vartheta)\right)\left[\dot m_t(\vartheta)-\dot m_t^\circ(\vartheta)\right]\left(1+O\left(\sqrt{\varepsilon}\right)\right)
\end{aligned}$$

$$= K\left(\vartheta, t, m_t^{\circ}(\vartheta)\right)\left[m_t(\vartheta) - m_t^{\circ}(\vartheta)\right]\left(1 + O\left(\sqrt{\varepsilon}\right)\right)$$
$$+ f_y'\left(\vartheta, t, m_t^{\circ}(\vartheta)\right)\left[\dot m_t(\vartheta) - \dot m_t^{\circ}(\vartheta)\right]\left(1 + O\left(\sqrt{\varepsilon}\right)\right).$$

The substitution of (3.61) *and* (3.62) *here gives* (3.64).

3.3.2 Empirical Quadratic Variation

Introduce the same statistic as in the linear case ($\delta_\varepsilon = \varepsilon, t_{i+1} = t_i + \varphi_\varepsilon, \varphi_\varepsilon = \varepsilon^{2/3}, N_{\tau,\varepsilon} = \left[\frac{\tau}{\varphi_\varepsilon}\right]$)

$$\Psi_{\tau,\varepsilon} = \sum_{i=0}^{N_{\tau,\varepsilon}-1}\left(\frac{X_{t_{i+1}+\delta_\varepsilon} - X_{t_{i+1}}}{\delta_\varepsilon} - \frac{X_{t_i+\delta_\varepsilon} - X_{t_i}}{\delta_\varepsilon}\right)^2, \qquad 0 < \tau \le T.$$

Remind that

$$\frac{X_{t_{i+1}+\delta_\varepsilon} - X_{t_{i+1}}}{\delta_\varepsilon} \qquad \text{and} \qquad \frac{X_{t_i+\delta_\varepsilon} - X_{t_i}}{\delta_\varepsilon}$$

are *asymptotic derivatives* ($\delta_\varepsilon \to 0$) of X_t at the points t_{i+1} and t_i respectively and the sum $\Psi_{\tau,\varepsilon}$ is the quadratic variation of these "derivatives."

Note that $\Psi_{\tau,\varepsilon}$ converges to the limit of the quadratic variation of the trend coefficient $f(\vartheta_0, t, Y_t)$. To obtain the expression for this limit, we first write the formal expansions and later in the proof of Proposition 3.5 we give the detailed proof. The quadratic variation of $f(\vartheta_0, t, Y_t)$ is the limit of the following sum

$$\sum_{i=0}^{N_{\tau,\varepsilon}-1}\left[f\left(\vartheta_0, t_{i+1}, Y_{t_{i+1}}\right) - f\left(\vartheta_0, t_i, Y_{t_i}\right)\right]^2$$
$$= \sum_{i=0}^{N_{\tau,\varepsilon}-1}\left[f\left(\vartheta_0, t_i, Y_{t_{i+1}}\right) - f\left(\vartheta_0, t_i, Y_{t_i}\right)\right]^2 + o(1)$$
$$= \sum_{i=0}^{N_{\tau,\varepsilon}-1} f_y'(\vartheta_0, t_i, \tilde Y_{t_i})^2\left(Y_{t_{i+1}} - Y_{t_i}\right)^2 + o(1)$$
$$= \sum_{i=0}^{N_{\tau,\varepsilon}-1} f_y'(\vartheta_0, t_i, Y_{t_i})^2\, b\left(\vartheta_0, t_i, Y_{t_i}\right)^2\left(V_{t_{i+1}} - V_{t_i}\right)^2 + o(1)$$
$$\longrightarrow \int_0^\tau f_y'(\vartheta_0, t, Y_t)^2\, b(\vartheta_0, t, Y_t)^2\,\mathrm{d}t = \Psi_\tau(\vartheta_0).$$

Here the function $\Psi_\tau(\vartheta), \vartheta \in \Theta$ is defined by the last equality.

Let us denote

$$\pi_1 = 2\int_0^\tau \sigma(t)^2\,dt, \qquad \pi_2(\vartheta_0) = \frac{2}{3}\int_0^\tau f_y'(\vartheta_0,t,Y_t)^2\, b(\vartheta_0,t,Y_t)^2\,dt,$$

$$\pi_3(\vartheta_0) = \int_0^\tau f_y'(\vartheta_0,t,Y_t)^2\, b(\vartheta_0,t,Y_t)^2\,dv_t,\ \ \pi(\vartheta_0) = \pi_1 + \pi_2(\vartheta_0) + \pi_3(\vartheta_0),$$

where $v_t, 0 \le t \le \tau$ is some Wiener process.

Details of this convergence are given in the following lemma.

Proposition 3.5 *Let the condition $\mathscr{I}_1$, holds, then*

$$\Psi_{\tau,\varepsilon} = \Psi_\tau(\vartheta_0) + \pi_\varepsilon\,\varepsilon^{1/3} + O\left(\varepsilon^{2/3}\right), \qquad \varepsilon^{-1/3}\left(\Psi_{\tau,\varepsilon} - \Psi_\tau(\vartheta_0)\right) \Longrightarrow \pi(\vartheta_0)$$

and for any $p > 0$, there exists a constant $C = C_p > 0$ such that

$$\sup_{\vartheta_0\in\Theta} \mathbf{E}_{\vartheta_0}\left|\Psi_{\tau,\varepsilon} - \Psi_\tau(\vartheta_0)\right|^p \le C\,\varepsilon^{p/3}.$$

The random variables π_ε are defined below in (3.65)–(3.66).

Proof The proof follows the main steps of the proof of the Proposition 3.1. To simplify the exposition we omit for instant ϑ_0.

Using elementary expansions, we write

$$\begin{aligned}
&\frac{X_{t_i+\delta_\varepsilon} - X_{t_i}}{\delta_\varepsilon}\\
&\quad= \frac{1}{\delta_\varepsilon}\int_{t_i}^{t_i+\delta_\varepsilon} f(t,Y_t)\,dt + \frac{\varepsilon}{\delta_\varepsilon}\int_{t_i}^{t_i+\delta_\varepsilon}\sigma(t)\,dW_t\\
&\quad= f\left(t_i,Y_{t_i}\right) + \sigma(t_i)\frac{W_{t_i+\delta_\varepsilon} - W_{t_i}}{\sqrt{\delta_\varepsilon}}\frac{\varepsilon}{\sqrt{\delta_\varepsilon}} + \int_{t_i}^{t_i+\delta_\varepsilon}\left[\sigma(t) - \sigma(t_i)\right]dW_t\frac{\varepsilon}{\delta_\varepsilon}\\
&\qquad+ \frac{1}{\delta_\varepsilon}\int_{t_i}^{t_i+\delta_\varepsilon}\left[f(t,Y_t) - f\left(t_i,Y_{t_i}\right)\right]dt\\
&\quad= f\left(t_i,Y_{t_i}\right) + \sigma(t_i)\,\xi_{t_i,\varepsilon}^*\sqrt{\varepsilon} + \frac{1}{\delta_\varepsilon}\int_{t_i}^{t_i+\delta_\varepsilon} f_y'\left(t_i,Y_{t_i}\right)\left[Y_t - Y_{t_i}\right]dt,\\
&\qquad+ \frac{1}{\delta_\varepsilon}\int_{t_i}^{t_i+\delta_\varepsilon}\left[f(t,Y_t) - f\left(t_i,Y_{t_i}\right)\right]dt + O\left(\varepsilon^{3/2}\right)
\end{aligned}$$

$$= f\left(t_i, Y_{t_i}\right) + \sigma\left(t_i\right) \xi^*_{t_i,\varepsilon} \sqrt{\varepsilon} + \frac{1}{\delta_\varepsilon} \int_{t_i}^{t_i+\delta_\varepsilon} f'_y\left(t_i, Y_{t_i}\right) \int_{t_i}^{t} a\left(s, Y_s\right) \,\mathrm{d}s \,\mathrm{d}t$$

$$+ \frac{1}{\delta_\varepsilon} \int_{t_i}^{t_i+\delta_\varepsilon} f'_y\left(t_i, Y_{t_i}\right) \int_{t_i}^{t} b\left(s, Y_s\right) \,\mathrm{d}V_s \,\mathrm{d}t + O\left(\varepsilon^{3/2}\right),$$

$$= f\left(t_i, Y_{t_i}\right) + \sigma\left(t_i\right) \xi^*_{t_i,\varepsilon} \sqrt{\varepsilon} + f'_y\left(t_i, Y_{t_i}\right) b\left(t_i, Y_{t_i}\right) \eta^*_{t_i,\varepsilon} \sqrt{\varepsilon} + O\left(\varepsilon\right).$$

Therefore we have as well the representations

$$\frac{X_{t_{i+1}+\delta_\varepsilon} - X_{t_{i+1}}}{\delta_\varepsilon} = f\left(t_{i+1}, Y_{t_{i+1}}\right) + \sigma\left(t_{i+1}\right) \xi^*_{t_{i+1},\varepsilon} \sqrt{\varepsilon}$$

$$+ f'_y\left(t_{i+1}, Y_{t_{i+1}}\right) b\left(t_{i+1}, Y_{t_{i+1}}\right) \eta^*_{t_{i+1},\varepsilon} \sqrt{\varepsilon} + O\left(\varepsilon\right)$$

and

$$\frac{X_{t_{i+1}+\delta_\varepsilon} - X_{t_{i+1}}}{\delta_\varepsilon} - \frac{X_{t_i+\delta_\varepsilon} - X_{t_i}}{\delta_\varepsilon} = f\left(t_{i+1}, Y_{t_{i+1}}\right) - f\left(t_i, Y_{t_i}\right) + \sigma\left(t_{i+1}\right) \xi^*_{t_{i+1},\varepsilon} \sqrt{\varepsilon}$$

$$- \sigma\left(t_i\right) \xi^*_{t_i,\varepsilon} \sqrt{\varepsilon} + f'_y\left(t_{i+1}, Y_{t_{i+1}}\right) b\left(t_{i+1}, Y_{t_{i+1}}\right) \eta^*_{t_{i+1},\varepsilon} \sqrt{\varepsilon}$$

$$- f'_y\left(t_i, Y_{t_i}\right) b\left(t_i, Y_{t_i}\right) \eta^*_{t_i,\varepsilon} \sqrt{\varepsilon} + O\left(\varepsilon\right)$$

$$= f\left(t_i, Y_{t_{i+1}}\right) - f\left(t_i, Y_{t_i}\right) + f'_t\left(t_i, Y_{t_i}\right) \varepsilon^{2/3} + \sigma\left(t_i\right) \Delta\xi^*_{t_i,\varepsilon} \varepsilon^{1/2}$$

$$+ f'_y\left(t_i, Y_{t_i}\right) b\left(t_i, Y_{t_i}\right) \Delta\eta^*_{t_i,\varepsilon} \varepsilon^{1/2} + O\left(\varepsilon^{5/6}\right)$$

$$= f'_y\left(t_i, Y_{t_i}\right) \left[Y_{t_{i+1}} - Y_{t_i}\right] + \sigma\left(t_i\right) \Delta\xi^*_{t_i,\varepsilon} \varepsilon^{1/2}$$

$$+ f'_y\left(t_i, Y_{t_i}\right) b\left(t_i, Y_{t_i}\right) \Delta\eta^*_{t_i,\varepsilon} \varepsilon^{1/2} + O\left(\varepsilon^{2/3}\right)$$

$$= f'_y\left(t_i, Y_{t_i}\right) b\left(t_i, Y_{t_i}\right) \left[V_{t_{i+1}} - V_{t_i}\right] + \sigma\left(t_i\right) \Delta\xi^*_{t_i,\varepsilon} \varepsilon^{1/2}$$

$$+ f'_y\left(t_i, Y_{t_i}\right) b\left(t_i, Y_{t_i}\right) \Delta\eta^*_{t_i,\varepsilon} \varepsilon^{1/2} + O\left(\varepsilon^{2/3}\right).$$

Hence

$$\Psi_{\tau,\varepsilon} = \sum_{i=0}^{N_{\tau,\varepsilon}-1} f'_y\left(t_i, Y_{t_i}\right)^2 b\left(t_i, Y_{t_i}\right)^2 \left[V_{t_{i+1}} - V_{t_i}\right]^2 + \sum_{i=0}^{N_{\tau,\varepsilon}-1} \sigma\left(t_i\right)^2 \left(\Delta\xi^*_{t_i,\varepsilon}\right)^2 \varepsilon$$

$$+ \sum_{i=0}^{N_{\tau,\varepsilon}-1} f'_y\left(t_i, Y_{t_i}\right)^2 b\left(t_i, Y_{t_i}\right)^2 \left(\Delta\eta^{*2}_{t_i,\varepsilon}\right) \varepsilon$$

$$+ 2 \sum_{i=0}^{N_{\tau,\varepsilon}-1} f'_y\left(t_i, Y_{t_i}\right) b\left(t_i, Y_{t_i}\right) \sigma\left(t_i\right) \Delta\eta^{*2}_{t_i,\varepsilon} \Delta\xi^*_{t_i,\varepsilon} \varepsilon$$

$$+2\sum_{i=0}^{N_{\tau,\varepsilon}-1} f_y'\left(t_i,Y_{t_i}\right)^2 b\left(t_i,Y_{t_i}\right)^2\left[V_{t_{i+1}}-V_{t_i}\right]\Delta\eta_{t_i,\varepsilon}\,\varepsilon^{1/2}$$

$$+2\sum_{i=0}^{N_{\tau,\varepsilon}-1} f_y'\left(t_i,Y_{t_i}\right) b\left(t_i,Y_{t_i}\right)\sigma\left(t_i\right)\left[V_{t_{i+1}}-V_{t_i}\right]\Delta\xi^*_{t_i,\varepsilon}\,\varepsilon^{1/2}.$$

Note that

$$\pi_{1,\varepsilon}\equiv\varphi_\varepsilon\sum_{i=0}^{N_{\tau,\varepsilon}-1}\sigma\left(t_i\right)^2\left(\Delta\xi^*_{t_i,\varepsilon}\right)^2\longrightarrow 2\int_0^\tau\sigma\left(t\right)^2\mathrm{d}t=\pi_1\left(\vartheta_0\right), \tag{3.65}$$

$$\varepsilon\sum_{i=0}^{N_{\tau,\varepsilon}-1}\sigma\left(t_i\right)^2\left(\Delta\xi^*_{t_i,\varepsilon}\right)^2=2\int_0^\tau\sigma\left(t\right)^2\mathrm{d}t\left(1+o\left(1\right)\right)\,\varepsilon^{1/3},$$

$$\pi_{2,\varepsilon}\equiv\varphi_\varepsilon\sum_{i=0}^{N_{\tau,\varepsilon}-1} f_y'\left(t_i,Y_{t_i}\right)^2 b\left(t_i,Y_{t_i}\right)^2\left(\Delta\eta^{*2}_{t_i,\varepsilon}\right)$$

$$\longrightarrow\frac{2}{3}\int_0^\tau f_y'\left(t,Y_t\right)^2 b\left(t,Y_t\right)^2\mathrm{d}t=\pi_2\left(\vartheta_0\right),$$

$$\varepsilon\sum_{i=0}^{N_{\tau,\varepsilon}-1} f_y'\left(t_i,Y_{t_i}\right)^2 b\left(t_i\right)^2\left(\Delta\eta^{*2}_{t_i,\varepsilon}\right)=\frac{2}{3}\int_0^\tau f_y'\left(t,Y_t\right)^2 b\left(t,Y_t\right)^2\mathrm{d}t\left(1+o\left(1\right)\right)\,\varepsilon^{1/3}$$

and

$$\sum_{i=0}^{N_{\tau,\varepsilon}-1} f_y'\left(t_i,Y_{t_i}\right) b\left(t_i,Y_{t_i}\right)\sigma\left(t_i\right)\Delta\eta^*_{t_i,\varepsilon}\Delta\xi^*_{t_i,\varepsilon}\,\varepsilon=O\left(\varepsilon^{2/3}\right).$$

It is easy to verify that

$$\mathbf{E}\left(\sum_{i=0}^{N_{\tau,\varepsilon}-1} f_y'\left(t_i,Y_{t_i}\right) b\left(t_i,Y_{t_i}\right)\sigma\left(t_i\right)\Delta\eta^*_{t_i,\varepsilon}\Delta\xi^{*2}_{t_i,\varepsilon}\right)$$

$$=\sum_{i=0}^{N_{\tau,\varepsilon}-1}\mathbf{E}f_y'\left(t_i,Y_{t_i}\right)^2 b\left(t_i,Y_{t_i}\right)^2\sigma\left(t_i\right)^2\Delta\eta^{*2}_{t_i,\varepsilon}\Delta\xi^{*2}_{t_i,\varepsilon}$$

$$\leq C\sum_{i=0}^{N_{\tau,\varepsilon}-1}\mathbf{E}\Delta\eta^{*2}_{t_i,\varepsilon}\,\mathbf{E}\Delta\xi^{*2}_{t_i,\varepsilon}\leq C\,\varepsilon^{-2/3}.$$

Further, we have

$$\sum_{i=0}^{N_{\tau,\varepsilon}-1} f_y'\left(t_i, Y_{t_i}\right) b\left(t_i, Y_{t_i}\right) \sigma\left(t_i\right) \Delta\xi^*_{t_i,\varepsilon}\left[V_{t_{i+1}} - V_{t_i}\right] = \int_0^\tau H_\varepsilon\left(t\right) \mathrm{d}V_t,$$

where the step function $H_\varepsilon\left(t\right)$ is

$$H_\varepsilon\left(t\right) = f_y'\left(t_i, Y_{t_i}\right) b\left(t_i, Y_{t_i}\right) \sigma\left(t_i\right) \Delta\xi^*_{t_i,\varepsilon}\, \mathbb{I}_{\left\{t_i \le t < t_{i+1}\right\}}.$$

Recall that $W_t, 0 \le t \le T$ and $V_t, 0 \le t \le T$ are independent Wiener processes. Therefore,

$$\begin{aligned}\int_0^\tau H_\varepsilon\left(t\right)^2 \mathrm{d}t &= \sum_{i=0}^{N_{\tau,\varepsilon}-1} f_y'\left(t_i, Y_{t_i}\right)^2 b\left(t_i, Y_{t_i}\right)^2 \sigma\left(t_i\right)^2 \Delta\xi^{*2}_{t_i,\varepsilon}\left(t_{i+1} - t_i\right)\\ &\longrightarrow 2\int_0^\tau f_y'\left(t, Y_t\right)^2 b\left(t, Y_t\right)^2 \sigma\left(t\right)^2 \mathrm{d}t\end{aligned}$$

because

$$\begin{aligned}&\mathbf{E}\left(\sum_{i=0}^{N_{\tau,\varepsilon}-1} f_y'\left(t_i, Y_{t_i}\right)^2 b\left(t_i, Y_{t_i}\right)^2 \sigma\left(t_i\right)^2 \left(\Delta\xi^{*2}_{t_i,\varepsilon} - 2\right)\left(t_{i+1} - t_i\right)\right)^2\\ &\quad= \sum_{i=0}^{N_{\tau,\varepsilon}-1} \mathbf{E}\left(\left(f_y'\left(t_i, Y_{t_i}\right)^4 b\left(t_i, Y_{t_i}\right)^4 \sigma\left(t_i\right)^4 \left(\Delta\xi^{*2}_{t_i,\varepsilon} - 2\right)\right)^2\right)\left(t_{i+1} - t_i\right)^2\\ &\quad\le C\,\varepsilon^{2/3}.\end{aligned}$$

Therefore,

$$\int_0^\tau H_\varepsilon\left(t\right) \mathrm{d}V_t \Longrightarrow \tilde\pi\left(\vartheta_0\right), \qquad \mathbf{E}\tilde\pi\left(\vartheta_0\right)^2 = 2\int_0^\tau f_y'\left(t, Y_t\right)^2 b\left(t, Y_t\right)^2 \sigma\left(t\right))^2 \mathrm{d}t$$

and we obtained

$$\sum_{i=0}^{N_{\tau,\varepsilon}-1} f_y'\left(t_i, Y_{t_i}\right) b\left(t_i, Y_{t_i}\right) \sigma\left(t_i\right) \Delta\xi^*_{t_i,\varepsilon}\left[V_{t_{i+1}} - V_{t_i}\right]\sqrt{\varepsilon} = O\left(\sqrt{\varepsilon}\right).$$

For estimation of the next term, we denote $G\left(t\right) = f_y'\left(t, Y_t\right)^2 b\left(t, Y_t\right)^2$ and write

$$\varepsilon^{3/2} \sum_{i=0}^{N_{\tau,\varepsilon}-1} f_y'\left(t_i, Y_{t_i}\right)^2 b\left(t_i, Y_{t_i}\right)^2 \left[V_{t_{i+1}} - V_{t_i}\right] \Delta\eta^*_{t_i,\varepsilon}$$

$$= \sum_{i=0}^{N_{\tau,\varepsilon}-1} G(t_i)\left[V_{t_{i+1}} - V_{t_i}\right]\left[\int_{t_{i+1}}^{t_{i+1}+\varepsilon} \left[V_s - V_{t_{i+1}}\right] \mathrm{d}s - \int_{t_i}^{t_i+\varepsilon} \left[V_s - V_{t_i}\right] \mathrm{d}s\right]$$

$$= \sum_{i=0}^{N_{\tau,\varepsilon}-1} G(t_i)\left[V_{t_{i+1}} - V_{t_i}\right]\int_{t_{i+1}}^{t_{i+1}+\varepsilon} \left[V_s - V_{t_{i+1}}\right] \mathrm{d}s$$

$$- \sum_{i=0}^{N_{\tau,\varepsilon}-1} G(t_i)\left[V_{t_{i+1}} - V_{t_i}\right]\int_{t_i}^{t_i+\varepsilon} \left[V_s - V_{t_i}\right] \mathrm{d}s \equiv S_{1,\varepsilon} - S_{2,\varepsilon}$$

with obvious notations. We have

$$\mathbf{E}S_{1,\varepsilon}^2 = \mathbf{E}\left(\sum_{i=0}^{N_{\tau,\varepsilon}-1} G(t_i)\left[V_{t_{i+1}} - V_{t_i}\right]\int_{t_{i+1}}^{t_{i+1}+\varepsilon} \left[V_s - V_{t_{i+1}}\right] \mathrm{d}s\right)^2$$

$$= \sum_{i=0}^{N_{\tau,\varepsilon}-1} \mathbf{E}G(t_i)^2\left[V_{t_{i+1}} - V_{t_i}\right]^2 \mathbf{E}\left(\int_{t_{i+1}}^{t_{i+1}+\varepsilon} \left[V_s - V_{t_{i+1}}\right] \mathrm{d}s\right)^2 \leq C\,\varepsilon^3.$$

For the second sum $S_{2,\varepsilon}$, we first note that $V_{t_{i+1}} - V_{t_i+\varepsilon}$ and $\int_{t_i}^{t_i+\varepsilon}\left[V_s - V_{t_i}\right]\mathrm{d}s$ are independent and can be estimated as the sum $S_{1,\varepsilon}$. Therefore, it is sufficient to study

$$S_{3,\varepsilon} = \sum_{i=0}^{N_{\tau,\varepsilon}-1} G(t_i)\left[V_{t_i+\varepsilon} - V_{t_i}\right]\int_{t_i}^{t_i+\varepsilon} \left[V_s - V_{t_i}\right] \mathrm{d}s.$$

We have

$$\mathbf{E}S_{3,\varepsilon}^2 \leq C \sum_{i=0}^{N_{\tau,\varepsilon}-1} \mathbf{E}\left(\int_{t_i}^{t_i+\varepsilon} \left[V_{t_i+\varepsilon} - V_{t_i}\right]\left[V_s - V_{t_i}\right] \mathrm{d}s\right)^2$$

$$\leq C\varepsilon \sum_{i=0}^{N_{\tau,\varepsilon}-1} \int_{t_i}^{t_i+\varepsilon} \mathbf{E}\left[V_{t_i+\varepsilon} - V_{t_i}\right]^2\left[V_s - V_{t_i}\right]^2 \mathrm{d}s \leq C\,\varepsilon^{10/3}.$$

Finally, we obtain

$$\varepsilon^{1/2}\sum_{i=0}^{N_{\tau,\varepsilon}-1} f_y'\left(t_i, Y_{t_i}\right)^2 b\left(t_i, Y_{t_i}\right)^2\left[V_{t_{i+1}} - V_{t_i}\right]\Delta\eta_{t_i,\varepsilon}^* = O\left(\varepsilon^{2/3}\right).$$

Consider now the term

$$
\begin{aligned}
&\sum_{i=0}^{N_{\tau,\varepsilon}-1} G\left(t_i\right)\left[V_{t_{i+1}} - V_{t_i}\right]^2 \\
&\quad = \sum_{i=0}^{N_{\tau,\varepsilon}-1} G\left(t_i\right)\left[t_{i+1} - t_i\right] + 2\sum_{i=0}^{N_{\tau,\varepsilon}-1} G\left(t_i\right)\int_{t_i}^{t_{i+1}}\left[V_s - V_{t_i}\right]\mathrm{d}V_s \\
&\quad = \sum_{i=0}^{N_{\tau,\varepsilon}-1} G\left(t_i\right)\left[t_{i+1} - t_i\right] + \varepsilon^{1/3}\int_0^\tau R_\varepsilon\left(s\right)\mathrm{d}V_s,
\end{aligned}
$$

where we denoted

$$
R_\varepsilon\left(s\right) = \varepsilon^{-1/3} G\left(t_i\right)\left[V_s - V_{t_i}\right]\,\mathbb{1}_{\left\{t_i \le s < t_{i+1}\right\}}.
$$

We have

$$
\begin{aligned}
\int_0^\tau R_\varepsilon\left(s\right)^2\mathrm{d}s &= \varepsilon^{-2/3}\sum_{i=0}^{N_{\tau,\varepsilon}-1} G\left(t_i\right)^2\int_{t_i}^{t_{i+1}}\left[V_s - V_{t_i}\right]^2\mathrm{d}s \\
&= \sum_{i=0}^{N_{\tau,\varepsilon}-1} G\left(t_i\right)^2 \mu_{i,\varepsilon}\left(t_{i+1} - t_i\right)\left(1 + o\left(1\right)\right) \longrightarrow \frac{1}{2}\int_0^\tau G\left(s\right)^2\mathrm{d}s.
\end{aligned}
$$

Here

$$
\mu_{i,\varepsilon} = \varepsilon^{-4/3}\int_{t_i}^{t_{i+1}}\left[V_s - V_{t_i}\right]^2\mathrm{d}s = \int_0^1 v_{i,\varepsilon}\left(r\right)^2\mathrm{d}r, \qquad i = 0, 1, \ldots
$$

are i.i.d. random variables with mean $\frac{1}{2}$ and $v_{i,\varepsilon}\left(r\right), 0 \le r \le 1$ are independent Wiener processes.

Therefore,

$$
\begin{aligned}
\pi_{3,\varepsilon} &\equiv \varepsilon^{-1/3}\, 2\sum_{i=0}^{N_{\tau,\varepsilon}-1} G\left(t_i\right)\int_{t_i}^{t_{i+1}}\left[V_s - V_{t_i}\right]\mathrm{d}V_s \Longrightarrow \pi_3\left(\vartheta_0\right), \\
\mathbf{E}\pi_3\left(\vartheta_0\right)^2 &= \mathbf{E}\int_0^\tau f_y'\left(t, Y_t\right)^4 b\left(t, Y_t\right)^4\mathrm{d}t.
\end{aligned}
$$

The random variable $\pi_3\left(\vartheta_0\right)$ can be written in the integral form

$$
\pi_3\left(\vartheta_0\right) = \int_0^\tau f_y'\left(t, Y_t\right)^2 b\left(t, Y_t\right)^2\mathrm{d}v_t,
$$

where $v_t, 0 \le t \le \tau$ is some Wiener process.

For the last term, we write

$$\begin{aligned}
&\sum_{i=0}^{N_{\tau,\varepsilon}-1} f_y'(t_i,Y_{t_i})^2 b\left(t_i,Y_{t_i}\right)^2(t_{i+1}-t_i)-\int_0^{\tau} f_y'(t,Y_t)^2 b\left(t,Y_t\right)^2 \mathrm{d}t\\
&=\sum_{i=0}^{N_{\tau,\varepsilon}-1}\int_{t_i}^{t_{i+1}}\left[f_y'(t,Y_t)^2 b\left(t,Y_t\right)^2-f_y'(t_i,Y_{t_i})^2 b\left(t_i,Y_{t_i}\right)^2\right]\mathrm{d}t\\
&=2\sum_{i=0}^{N_{\tau,\varepsilon}-1}\int_{t_i}^{t_{i+1}} f_y'(t,Y_t)b\left(t,Y_t\right)\left[f_{yy}''(t,\tilde Y_t)b\left(t,Y_t\right)\right.\\
&\qquad\left.+f_y'(t,Y_t)b_y'(t,\tilde Y_t)\right]\int_{t_i}^{t} b\left(s,Y_s\right)\mathrm{d}V_s\,\mathrm{d}t+O\left(\varepsilon^{2/3}\right)\\
&=2\sum_{i=0}^{N_{\tau,\varepsilon}-1}\int_{t_i}^{t_{i+1}} f_y'(t,Y_t)b\left(t,Y_t\right)^2\left[f_{yy}''(t,Y_t)b\left(t,Y_t\right)\right.\\
&\qquad\left.+f_y'(t,Y_t)b_y'\left(t,Y_t\right)\right]\left[V_t-V_{t_i}\right]\,\mathrm{d}t+O\left(\varepsilon^{2/3}\right)\\
&=2\varepsilon\sum_{i=0}^{N_{\tau,\varepsilon}-1} f_y'(t_i,Y_{t_i})b\left(t_i,Y_{t_i}\right)^2\left[f_{yy}''(t_i,Y_{t_i})b\left(t_i,Y_{t_i}\right)\right.\\
&\qquad\left.+f_y'(t_i,Y_{t_i})b_y'\left(t_i,Y_{t_i}\right)\right]\tilde\eta_{i,\varepsilon}+O\left(\varepsilon^{2/3}\right)\\
&=\varepsilon^{1/3}\sum_{i=0}^{N_{\tau,\varepsilon}-1} L_\varepsilon\left(t_i\right)\tilde\eta_{i,\varepsilon}\left(t_{i+1}-t_i\right)+O\left(\varepsilon^{2/3}\right),
\end{aligned}$$

where the two last lines define $L_\varepsilon\left(t_i\right)$ and we denoted

$$\tilde\eta_{i,\varepsilon}=\varepsilon^{-1}\int_{t_i}^{t_{i+1}}\left[V_t-V_{t_i}\right]\,\mathrm{d}t\quad\sim\quad\mathcal{N}\left(0,1/3\right).$$

We have

$$\mathbf{E}\left(\sum_{i=0}^{N_{\tau,\varepsilon}-1} L_\varepsilon\left(t_i\right)\tilde\eta_{i,\varepsilon}\left(t_{i+1}-t_i\right)\right)^2=\frac{1}{3}\sum_{i=0}^{N_{\tau,\varepsilon}-1}\mathbf{E}L_\varepsilon\left(t_i\right)^2\left(t_{i+1}-t_i\right)^2=O\left(\varepsilon^{2/3}\right).$$

Hence,

$$\varepsilon^{1/3}\sum_{i=0}^{N_{\tau,\varepsilon}-1} L_\varepsilon\left(t_i\right)\tilde\eta_{i,\varepsilon}\left(t_{i+1}-t_i\right)+O\left(\varepsilon^{2/3}\right)=O\left(\varepsilon^{2/3}\right).$$

Finally,

$$\Psi_{\tau,\varepsilon}=\Psi_\tau\left(\vartheta_0\right)+\pi_\varepsilon\,\varepsilon^{1/3}+O\left(1\right)\,\varepsilon^{2/3},\qquad \pi_\varepsilon=\pi_{1,\varepsilon}+\pi_{2,\varepsilon}+\pi_{3,\varepsilon},\tag{3.66}$$

$$\varepsilon^{-1/3}\left(\Psi_{\tau,\varepsilon}-\Psi_\tau\left(\vartheta_0\right)\right)\Longrightarrow\pi\left(\vartheta_0\right).\tag{3.67}$$

The term $O\left(1\right)$ here has finite all polynomial moments. □

3.3.3 MME

The method of moments estimator (MME) ϑ_ε^* can be introduced by the equation

$$\Psi_{\tau,\varepsilon}=\Psi_\tau\left(\vartheta_{\tau,\varepsilon}^*\right),\tag{3.68}$$

where

$$\Psi_{\tau,\varepsilon}=\sum_{i=i_0}^{N_{\tau,\varepsilon}-1}\left(\frac{X_{t_{i+1}+\varepsilon}-X_{t_{i+1}}}{\varepsilon}-\frac{X_{t_i+\varepsilon}-X_{t_i}}{\varepsilon}\right)^2,\tag{3.69}$$

Here $0<t_0<\tau<T$, $t_{i+1}-t_i=\varepsilon^{2/3}$, $i_0=\left[\frac{t_0}{\varepsilon^{2/3}}\right]$, $N_{\tau,\varepsilon}=\left[\frac{\tau}{\varepsilon^{2/3}}\right]$, $[A]$ is integer part of A.

The random function

$$\Psi_\tau\left(\vartheta\right)=\int_{t_0}^{\tau}f_y'(\vartheta,t,m_t\left(\vartheta\right))^2\,b\left(\vartheta,t,m_t\left(\vartheta\right)\right)^2{\rm d}t,\qquad \vartheta\in\Theta$$

is continuous with probability 1. The random process $m_t\left(\vartheta\right),t_0\leq t\leq T$ is solution of the equation

$$
{\rm d}m_t\left(\vartheta\right)=a\left(\vartheta,t,m_t\left(\vartheta\right)\right){\rm d}t+\frac{b\left(\vartheta,t,m_t\left(\vartheta\right)\right)}{\varepsilon\sigma\left(t\right)}\left[{\rm d}X_t-f\left(\vartheta,t,m_t\left(\vartheta\right)\right){\rm d}t\right].\tag{3.70}
$$

Here the initial value $m_{t_0}\left(\vartheta\right)$ is solution of the system (3.52)–(3.53).

We have to make several remarks concerning the definition (3.68)–(3.70) of the MME $\vartheta_{\tau,\varepsilon}^*$. For a given ε and observed trajectory $X^\tau=\left(X_t,0\leq t\leq\tau\right)$ we obtain a random interval $[\psi_m,\psi_M]$, where

$$\psi_m=\inf_{\vartheta\in\Theta}\Psi_\tau\left(\vartheta\right),\qquad \psi_M=\sup_{\vartheta\in\Theta}\Psi_\tau\left(\vartheta\right).$$

If $\Psi_{\tau,\varepsilon}\not\in[\psi_m,\psi_M]$, then (3.68) has no solution and has to be replaced by the definition

$$\vartheta^*_{\tau,\varepsilon} = \alpha\,\mathrm{1\!I}_{\{\Psi_{\tau,\varepsilon}<\psi_m\}} + \bar\vartheta_{\tau,\varepsilon}\,\mathrm{1\!I}_{\{\psi_m<\Psi_{\tau,\varepsilon}<\psi_M\}} + \beta\,\mathrm{1\!I}_{\{\Psi_{\tau,\varepsilon}>\psi_M\}},\qquad \Psi_{\tau,\varepsilon} = \Psi_\tau\left(\bar\vartheta_{\tau,\varepsilon}\right).$$

Note that in the case of the consistent estimation, the probabilities of the events $\Psi_{\tau,\varepsilon} < \psi_m$ and $\Psi_{\tau,\varepsilon} > \psi_M$ are asymptotically negligible.

The Eq. (3.70) can be simplified as follows. The initial value $m_{t_0}(\vartheta)$ can be replaced by any other value, say, y_0 and the trend coefficient $a(\vartheta, t, m_t(\vartheta))$ can be omitted. Let us explain why (3.70) can be replaced by

$$\mathrm{d}m^*_t(\vartheta) = \frac{b\left(\vartheta, t, m^*_t(\vartheta)\right)}{\varepsilon\sigma(t)}\left[\mathrm{d}X_t - f\left(\vartheta, t, m^*_t(\vartheta)\right)\mathrm{d}t\right],\ m^*_{t_0}(\vartheta) = y_0,\ t_0 \le t \le \tau. \tag{3.71}$$

Consider the difference $\delta_m(t) = m^*_t(\vartheta) - m_t(\vartheta)$ and the equation for $\delta_m(t)^4$

$$\begin{aligned}\frac{\mathrm{d}\delta_m(t)^4}{\mathrm{d}t} &= -\frac{4b\left(\vartheta, t, m^*_t(\vartheta)\right)}{\varepsilon\sigma(t)}\delta_m(t)^3\left[f\left(\vartheta, t, m^*_t(\vartheta)\right) - f\left(\vartheta, t, m_t(\vartheta)\right)\right]\\ &\quad - 4\delta_m(t)^3\, a\left(\vartheta, t, m_t(\vartheta)\right).\end{aligned}$$

We have

$$\begin{aligned}\mathbf{E}_{\vartheta_0}\delta_m(t)^4 &= \mathbf{E}_{\vartheta_0}\left(m^*_{t_0}(\vartheta) - y_0\right)^4 - 4\int_{t_0}^t \mathbf{E}_{\vartheta_0}\delta_m(s)^3 a\left(\vartheta, s, m_s(\vartheta)\right)\mathrm{d}s\\ &\quad - \int_{t_0}^t \frac{4}{\varepsilon\sigma(s)}\mathbf{E}_{\vartheta_0}\delta_m(s)^3 b\left(\vartheta, s, m^*_s(\vartheta)\right)\left[f\left(\vartheta, s, m^*_s(\vartheta)\right) - f\left(\vartheta, s, m_s(\vartheta)\right)\right]\mathrm{d}s\\ &= \mathbf{E}_{\vartheta_0}\left(m^*_{t_0}(\vartheta) - y_0\right)^4 - \int_{t_0}^t \frac{4}{\varepsilon\sigma(s)}\mathbf{E}_{\vartheta_0}\delta_m(s)^4 b\left(\vartheta, s, m_s(\vartheta)\right) f'_y\left(\vartheta, s, \bar m_s\right)\mathrm{d}s -\\ &\quad - 4\int_{t_0}^t \mathbf{E}_{\vartheta_0}\delta_m(s)^3 a\left(\vartheta, s, m_s(\vartheta)\right)\mathrm{d}s\\ &\le C - \frac{4c_b c_y}{\varepsilon C_\sigma}\int_{t_0}^t \mathbf{E}_{\vartheta_0}\delta_m(s)^4\,\mathrm{d}s + 4\int_{t_0}^t \mathbf{E}_{\vartheta_0}\delta_m(s)^4\,\mathrm{d}s\\ &\quad + 4\int_{t_0}^t \mathbf{E}_{\vartheta_0} a\left(\vartheta, s, m_s(\vartheta)\right)^4\mathrm{d}s\\ &\le C - \frac{2c_b c_y}{\varepsilon C_\sigma}\int_{t_0}^t \mathbf{E}_{\vartheta_0}\delta_m(s)^4\,\mathrm{d}s\end{aligned}$$

for sufficiently small ε. Here following [163], Lemma 2.1 we used the estimate $fg \le f^k + g^m$, where $f > 0$, $g > 0$ and $k^{-1} + m^{-1} = 1$, which allows us to write $\mathbf{E}_{\vartheta_0}\left|\delta_m(s)^3 a\left(\vartheta, s, m_s(\vartheta)\right)\right| \le \mathbf{E}_{\vartheta_0}\delta_m(s)^4 + \mathbf{E}_{\vartheta_0} a\left(\vartheta, s, m_s(\vartheta)\right)^4$, i.e., $k = 4/3$ and $m = 4$.

Therefore,

$$\mathbf{E}_{\vartheta_0}\left(m_t^*(\vartheta)-m_t(\vartheta)\right)^4\leq Ce^{-\frac{2c_bc_y}{\varepsilon C_\sigma}(t-t_0)}$$

and this proves that the difference between the solutions of (3.70) and (3.71) is asymptotically negligible. This is typical result for such short memory stochastic systems (see Corollary 2.3 and remarks just after in [163]).

Therefore, the statistic $\Psi_\tau(\vartheta)$ can be replaced by

$$\Psi_\tau^*(\vartheta)=\int_{t_0}^{\tau}f_y'(\vartheta,t,m_t^*(\vartheta))^2\,b\left(\vartheta,t,m_t^*(\vartheta)\right)^2{\rm d}t,\qquad \vartheta\in\Theta \tag{3.72}$$

with the corresponding definition of the MME $\vartheta_{\tau,\varepsilon}^*$.

We will need the estimator $\vartheta_{\tau,\varepsilon}^*$ in the next section for the construction of the adaptive EKF and the direct substitution of it in the EKF leads to the equation with the stochastic integral

$$\int_{t_0}^{t}\frac{b\left(\vartheta_{\tau,\varepsilon}^*,s,m_s^*\left(\vartheta_{\tau,\varepsilon}^*\right)\right)}{\sigma(s)}{\rm d}X_s.$$

As the estimator $\vartheta_{\tau,\varepsilon}^*$ depends on the whole trajectory X^τ this Itô integral is not well defined. To avoid this problem, we consider construction of this estimator for the values $0<t_0<\tau<T$, where t_0 and τ are small and the adaptive filter $m_t^\star$ will be given for the values $t\in[\tau,T]$.

Let us denote

$$\Psi_\tau^\circ(\vartheta)=\int_{t_0}^{\tau}f_y'(\vartheta,t,m_t^\circ(\vartheta))^2\,b\left(\vartheta,t,m_t^\circ(\vartheta)\right)^2{\rm d}t,\qquad \vartheta\in\Theta,$$

$$\begin{aligned}H(\vartheta,t,y,z)&=2\left[\dot f_{\vartheta y}'(\vartheta,t,y)+f_{yy}''(\vartheta,t,y)\,z\right]f_y'(\vartheta,t,y)b(\vartheta,t,y)^2\\&\quad+2\left[\dot b_\vartheta(\vartheta,t,y)+b_y'(\vartheta,t,y)\,z\right]f_y'(\vartheta,t,y)^2b(\vartheta,t,y),\\H^\circ(\vartheta,t,y)&=2\left[\dot f_{\vartheta y}'(\vartheta,t,y)f_y'(\vartheta,t,y)-f_{yy}''(\vartheta,t,y)\,\dot f_\vartheta(\vartheta,t,y)\right]b(\vartheta,t,y)^2\\&\quad+2\left[\dot b_\vartheta(\vartheta,t,y)\,f_y'(\vartheta,t,y)-b_y'(\vartheta,t,y)\,\dot f_\vartheta(\vartheta,t,y)\right]\\&\quad\times f_y'(\vartheta,t,y)b(\vartheta,t,y).\end{aligned}$$

The MME is defined by the Eqs. (3.68), (3.69), (3.71), and (3.72).

Introduce the conditions:

$\mathscr{I}_4$. *There exists a constant* $c_H>0$, *which does not depend on* $\vartheta\in\Theta,t\in[t_0,\tau],y\in\mathcal{R}$ *such that*

$$H^\circ(\vartheta,t,y)\geq c_H.$$

$\mathscr{I}_5$. *The derivatives* $\dot f''_{\vartheta yy}(\vartheta,t,y)$ *and* $f^{(3)}_{yyy}(\vartheta,t,y)$ *are bounded.*

Proposition 3.6 *Let conditions* $\mathscr{I}_1-\mathscr{I}_5$ *hold. Then the MME* ϑ^*_ε *is consistent and*

$$\varepsilon^{-1/3}\left(\vartheta^*_{\tau,\varepsilon}-\vartheta_0\right)\Longrightarrow\frac{\pi\left(\vartheta_0\right)}{\int_{t_0}^{\tau}H^\circ\left(\vartheta_0,t,Y_t\right)\mathrm{d}t}.$$

Proof To prove the consistency, we write as usual in such problems the following chaine of relations: for any $\nu>0$

$$\begin{aligned}
&\mathbf{P}_{\vartheta_0}\left(\inf_{|\vartheta-\vartheta_0|<\nu}\left|\vartheta^*_{\tau,\varepsilon}-\vartheta_0\right|>\inf_{|\vartheta-\vartheta_0|\geq\nu}\left|\vartheta^*_{\tau,\varepsilon}-\vartheta_0\right|\right)\\
&\quad=\mathbf{P}_{\vartheta_0}\left(\inf_{|\vartheta-\vartheta_0|<\nu}\left|\Psi_{\tau,\varepsilon}-\Psi^*_\tau\left(\vartheta\right)\right|>\inf_{|\vartheta-\vartheta_0|\geq\nu}\left|\Psi_{\tau,\varepsilon}-\Psi^*_\tau\left(\vartheta\right)\right|\right)\\
&\quad\leq\mathbf{P}_{\vartheta_0}\left(\inf_{|\vartheta-\vartheta_0|<\nu}\left(\left|\Psi_{\tau,\varepsilon}-\Psi^*_\tau\left(\vartheta_0\right)\right|+\left|\Psi^*_\tau\left(\vartheta\right)-\Psi^*_\tau\left(\vartheta_0\right)\right|\right)\right.\\
&\qquad\qquad\left.>\inf_{|\vartheta-\vartheta_0|\geq\nu}\left(\left|\Psi^*_\tau\left(\vartheta\right)-\Psi^*_\tau\left(\vartheta_0\right)\right|-\left|\Psi_{\tau,\varepsilon}-\Psi^*_\tau\left(\vartheta_0\right)\right|\right)\right)\\
&\quad=\mathbf{P}_{\vartheta_0}\left(2\left|\Psi_{\tau,\varepsilon}-\Psi^*_\tau\left(\vartheta_0\right)\right|\geq\inf_{|\vartheta-\vartheta_0|\geq\nu}\left|\Psi^*_\tau\left(\vartheta\right)-\Psi^*_\tau\left(\vartheta_0\right)\right|\right).
\end{aligned}$$

Recall that $m^*_t\left(\vartheta\right)=m^\circ_t\left(\vartheta\right)+O\left(\sqrt{\varepsilon}\right)$, $\dot m^*_t\left(\vartheta\right)=\dot m^\circ_t\left(\vartheta\right)+O\left(\sqrt{\varepsilon}\right)$ (see (3.61), (3.62)) and $\dot m^\circ_t\left(\vartheta\right)=-\dot f_\vartheta\left(\vartheta,t,m^\circ_t\left(\vartheta\right)\right)f'_y\left(\vartheta,t,m^\circ_t\left(\vartheta\right)\right)^{-1}$ (see (3.60)). Note as well that the derivatives of $\Psi^*_\tau\left(\vartheta\right)$ and $\Psi^\circ_\tau\left(\vartheta\right)$ are

$$\dot\Psi_\tau\left(\vartheta\right)=\int_{t_0}^{\tau}H\left(\vartheta,t,m^*_t\left(\vartheta\right),\dot m^*_t\left(\vartheta\right)\right)\mathrm{d}t,\quad\dot\Psi^\circ_\tau\left(\vartheta\right)=\int_{t_0}^{\tau}H^\circ\left(\vartheta,t,m^\circ_t\left(\vartheta\right)\right)\mathrm{d}t,$$

$$\dot\Psi^\circ_\tau\left(\vartheta_0\right)=\int_{t_0}^{\tau}H^\circ\left(\vartheta_0,t,m^\circ_t\left(\vartheta_0\right)\right)\mathrm{d}t=\int_{t_0}^{\tau}H^\circ\left(\vartheta_0,t,Y_t\right)\mathrm{d}t.$$

Therefore,

$$\begin{aligned}
\Psi^*_\tau\left(\vartheta\right)-\Psi^*_\tau\left(\vartheta_0\right)&=\Psi^\circ_\tau\left(\vartheta\right)-\Psi^\circ_\tau\left(\vartheta_0\right)+O\left(\sqrt{\varepsilon}\right)=\left(\vartheta-\vartheta_0\right)\dot\Psi^\circ_\tau(\tilde\vartheta)\\
&=\left(\vartheta-\vartheta_0\right)\int_{t_0}^{\tau}H^\circ(\tilde\vartheta,t,m^\circ_t(\tilde\vartheta))\mathrm{d}t+O\left(1\right)\sqrt{\varepsilon}
\end{aligned}$$

and

$$\mathbf{P}_{\vartheta_0}\left(2\left|\Psi_{\tau,\varepsilon}-\Psi^*_\tau\left(\vartheta_0\right)\right|\geq\inf_{|\vartheta-\vartheta_0|\geq\nu}\left|\Psi^*_\tau\left(\vartheta\right)-\Psi^*_\tau\left(\vartheta_0\right)\right|\right)$$

$$
\begin{aligned}
&= \mathbf{P}_{\vartheta_0}\left(2\left|\Psi_{\tau,\varepsilon}-\Psi_\tau^*(\vartheta_0)\right| \geq \inf_{|\vartheta-\vartheta_0|\geq\nu} |\vartheta-\vartheta_0| \left|\int_{t_0}^{\tau} H^\circ(\tilde\vartheta,t,m_t^\circ(\tilde\vartheta))\mathrm{d}t\right|\right)\\
&= \mathbf{P}_{\vartheta_0}\left(\left|\Psi_{\tau,\varepsilon}-\Psi_\tau^*(\vartheta_0)\right| \geq \frac{\nu\, c_H}{4}(\tau-t_0)\right)\\
&= \mathbf{P}_{\vartheta_0}\left(O\left(\varepsilon^{1/3}\right) \geq \frac{\nu\, c_H}{4}(\tau-t_0)\right) \longrightarrow 0.
\end{aligned}
$$

The left and right sides of the equation $\Psi_{\tau,\varepsilon} = \Psi_\tau^*\left(\vartheta_{\tau,\varepsilon}^*\right)$ can be written as follows:

$$
\Psi_{\tau,\varepsilon} = \Psi_\tau(\vartheta_0) + \pi\varepsilon^{1/3} + O\left(\varepsilon^{2/3}\right) = \Psi_\tau^\circ(\vartheta_0) + \pi\varepsilon^{1/3} + O\left(\varepsilon^{1/2}\right)
$$

and $\Psi_\tau^*\left(\vartheta_{\tau,\varepsilon}^*\right) = \Psi_\tau^\circ\left(\vartheta_{\tau,\varepsilon}^*\right) + O\left(\sqrt{\varepsilon}\right) = \Psi_\tau^\circ(\vartheta_0) + \left(\vartheta_{\tau,\varepsilon}^* - \vartheta_0\right)\dot\Psi_\tau^\circ(\tilde\vartheta) + O\left(\sqrt{\varepsilon}\right)$. Thus

$$
\varepsilon^{-1/3}\left(\vartheta_{\tau,\varepsilon}^*-\vartheta_0\right) = \frac{\varepsilon^{-1/3}\left(\Psi_{\tau,\varepsilon}-\Psi_\tau^\circ(\vartheta_0)\right)}{\dot\Psi_\tau^\circ(\vartheta_0)+O\left(\sqrt{\varepsilon}\right)} = \frac{\pi(\vartheta_0)}{\int_{t_0}^{\tau} H^\circ(\vartheta_0,t,Y_t)\,\mathrm{d}t} + O\left(\varepsilon^{1/6}\right).
$$

□

Remark 3.9 Let us see what can be in this case the limit of pseudo Fisher information

$$
\int_\tau^t \frac{1}{\varepsilon\sigma(s)^2}\left(\frac{\mathrm{d}}{\mathrm{d}\vartheta} f(\vartheta,s,m_s(\vartheta))\right)^2 \mathrm{d}s.
$$

Denote

$$
\begin{aligned}
L(\vartheta_0,s,y) &= f_y'(\vartheta_0,s,y)^{-1} K(\vartheta_0,s,y) + \Pi^\circ(\vartheta_0,s,y),\\
N(\vartheta_0,s,y) &= f_y'(\vartheta_0,s,y)^{-1} K(\vartheta_0,s,y) + \Lambda^\circ(\vartheta_0,s,y).
\end{aligned}
$$

Applying (3.61)–(3.62), and (3.64) we obtain

$$
\begin{aligned}
&\mathrm{I}_\tau^t(\vartheta_0,\varepsilon)\\
&\equiv \int_\tau^t \frac{b(\vartheta_0,s,Y_s)\, f_y'(\vartheta_0,s,Y_s)}{2\sigma(s)}\left[L(\vartheta_0,s,Y_s)\,\xi_{s,\varepsilon} - N(\vartheta_0,s,Y_s)\,\eta_{s,\varepsilon}\right]^2 \mathrm{d}s\\
&\longrightarrow \int_\tau^t \frac{b(\vartheta_0,s,Y_s)\, f_y'(\vartheta_0,s,Y_s)}{2\sigma(s)}\left[L(\vartheta_0,s,Y_s)^2+N(\vartheta_0,s,Y_s)^2\right]\mathrm{d}s \equiv \mathrm{I}_\tau^t(\vartheta_0).
\end{aligned}
$$

3.3.4 Adaptive EKF

Recall that we have the partially observed nonlinear model

$$\begin{aligned} \mathrm{d}X_t &= f(\vartheta, t, Y_t)\,\mathrm{d}t + \varepsilon\sigma(t)\,\mathrm{d}W_t, \qquad X_0 = 0, \qquad 0 \le t \le T, \\ \mathrm{d}Y_t &= a(\vartheta, t, Y_t)\,\mathrm{d}t + b(\vartheta, t, Y_t)\,\mathrm{d}V_t, \qquad Y_0 = y_0, \end{aligned}$$

where the observed process is $X^T = (X_t, 0 \le t \le T)$ and the diffusion process Y^T is hidden. The parameter $\vartheta \in \Theta = (\alpha, \beta)$ is unknown, and we have to construct an approximation $m^\star_{t,\varepsilon}$ of the unobserved component Y_t. We wait that this approximation has to be consistent ($m^\star_{t,\varepsilon} - Y_t \to 0$) and its calculation has to be relatively simple.

The observation interval $[0, T]$ is subdivided on three intervals: $[0, t_0], [t_0, \tau]$ and $[\tau, T]$. The observations on $[0, t_0]$ are neglected because our problem involves a boundary layer at $t = 0$. During this time interval, the solution $\varepsilon^{-1}\gamma_\varepsilon(\vartheta, t)$ of the Riccati equation (3.53) exponentially fast converges to the value $\gamma_*(\vartheta, t)$. Therefore, starting since t_0 we can replace $\varepsilon^{-1}\gamma_\varepsilon(\vartheta, t)$ in the Eqs. (3.52) and (3.53) by $\gamma_*(\vartheta, t)$. This simplifies the futher calculations. By the observations $X_t, t_0 \le t \le \tau$, we construct the MME $\vartheta^*_{\tau,\varepsilon}$ (see (3.68), (3.69), (3.71) and (3.72)).

The adaptive EKF $m^\star_t, \tau \le t \le T$ can be

$$\mathrm{d}m^\star_t = a\left(\vartheta^*_{\tau,\varepsilon}, t, m^\star_t\right)\mathrm{d}t + \frac{b\left(\vartheta^*_{\tau,\varepsilon}, t, m^\star_t\right)}{\varepsilon\sigma(t)}\left[\mathrm{d}X_t - f\left(\vartheta^*_{\tau,\varepsilon}, t, m^\star_t\right)\mathrm{d}t\right], \qquad m^\star_{t_0} = y_0.$$

Theorem 3.5 *Let conditions $\mathscr{I}_{1-5}$ hold. Then for $t \in (\tau, T)$ the following limit holds*

$$\varepsilon^{-1/3}\left(m^\star_t - Y_t\right) \Longrightarrow -\frac{\dot f_\vartheta(\vartheta_0, t, Y_t)}{f'_y(\vartheta_0, t, Y_t)}\frac{\pi(\vartheta_0)}{\int_{t_0}^{\tau} H^\circ(\vartheta_0, t, Y_t)\,\mathrm{d}t}. \tag{3.73}$$

Proof Note that $m^\star_t = m_t\left(\vartheta^*_{\tau,\varepsilon}\right)$ and consider the difference $m^\star_t - Y_t$. We have

$$\begin{aligned} m^\star_t - Y_t &= m_t\left(\vartheta^*_{\tau,\varepsilon}\right) - m_t(\vartheta_0) + m_t(\vartheta_0) - Y_t \\ &= \dot m_t\left(\bar\vartheta\right)\left(\vartheta^*_{\tau,\varepsilon} - \vartheta_0\right) + m_t(\vartheta_0) - m^\circ_t(\vartheta_0) \\ &= \dot m_t(\vartheta_0)\left(\vartheta^*_{\tau,\varepsilon} - \vartheta_0\right)\left(1 + O\left(\varepsilon^{1/3}\right)\right) \\ &\quad + \Gamma^\circ(\vartheta_0, t)\left[\xi_{t,\varepsilon} - \eta_{t,\varepsilon}\right]\sqrt{\varepsilon}\,(1 + o(1)) \\ &= \dot m^\circ_t(\vartheta_0)\left(\vartheta^*_{\tau,\varepsilon} - \vartheta_0\right)\left(1 + O\left(\varepsilon^{1/3}\right)\right) \\ &\quad + \Gamma^\circ(\vartheta_0, t)\left[\xi_{t,\varepsilon} - \eta_{t,\varepsilon}\right]\sqrt{\varepsilon}\,(1 + o(1)) \end{aligned}$$

$$= -\frac{\dot{f}_\vartheta\left(\vartheta_0,t,Y_t\right)}{f'_y\left(\vartheta_0,t,Y_t\right)}\left(\vartheta^*_{\tau,\varepsilon}-\vartheta_0\right)\left(1+O\left(\varepsilon^{1/3}\right)\right)$$

$$+\Gamma^\circ\left(\vartheta_0,t\right)\left[\xi_{t,\varepsilon}-\eta_{t,\varepsilon}\right]\sqrt{\varepsilon}\left(1+o\left(1\right)\right).$$

Therefore,

$$\varepsilon^{-1/3}\left(m^\star_t-Y_t\right)\Longrightarrow-\frac{\dot{f}_\vartheta\left(\vartheta_0,t,Y_t\right)}{f'_y\left(\vartheta_0,t,Y_t\right)}\frac{\pi\left(\vartheta_0\right)}{\int_{t_0}^{\tau}H^\circ\left(\vartheta_0,t,Y_t\right)\mathrm{d}t}.$$

This proves (3.73). □

The construction of the asymptotically optimal filter is an open problem.

Chapter 4
Hidden Ergodic O-U Process

We consider the problem of adaptive filtering for ergodic Ornstein-Uhlenbeck process observed in the presence of white Gaussian noise. The asymptotic normality and convergence of moments of the EMM, MLE, and BE are established. Preliminary estimators of the unknown parameters are constructed by the observations on the learning interval with the help of the method of moments. These estimators are then used to define the One-step MLE-process, and substituting of this estimator process into the equations of Kalman-Bucy filtering yields the recursive adaptive filter. The properties of all estimators and the approximation error of the conditional expectation of the Ornstein-Uhlenbeck process are described in the asymptotics of large samples. The asymptotic efficiency of the proposed adaptive filter is discussed too.

4.1 Model of Observations

Consider the partially observed homogeneous system

$$\mathrm{d}X_t = f(\vartheta)\, Y_t\, \mathrm{d}t + \sigma\, \mathrm{d}W_t, \qquad X_0, \qquad 0 \le t \le T, \tag{4.1}$$

$$\mathrm{d}Y_t = -a(\vartheta)\, Y_t\, \mathrm{d}t + b(\vartheta)\, \mathrm{d}V_t, \qquad Y_0, \tag{4.2}$$

where $W_t, t \ge 0$ and $V_t, t \ge 0$ are independent Wiener processes. As usual $X^T = (X_t, 0 \le t \le T)$ are observations and $Y^T = (Y_t, 0 \le t \le T)$ is a *hidden* Ornstein-Uhlenbeck process. The function $a(\cdot)$ in this chapter is always positive, and therefore the Ornstein-Uhlenbeck process has ergodic properties with the invariant Gaussian density with zero mean and the variance $b(\vartheta)^2/2a(\vartheta)$.

The Gaussian random variables X_0, Y_0 are independent of the Wiener processes. The functions $f(\cdot), a(\cdot), b(\cdot)$ and the constant $\sigma > 0$ are supposed to be known. The unknown parameter $\vartheta \in \Theta$ can be one-, two-, or three-dimensional. For

Y. A. Kutoyants, *Hidden Markov Processes and Adaptive Filtering*, Springer Series in Statistics, https://doi.org/10.1007/978-3-032-00052-1_4

example, as particular cases we consider the models: **F** $f(\vartheta) = \vartheta, a(\vartheta) = a, b(\vartheta) = b$, **AB** $f(\vartheta) = f, a(\vartheta) = \theta_1, b(\vartheta) = \theta_2$ and therefore $\vartheta = (\theta_1, \theta_2)$, etc.

We are interested in the asymptotic ($T \to \infty$) properties of estimators and study the corresponding adaptive filter.

The Kalman-Bucy filter for $m(\vartheta, t) = \mathbf{E}_\vartheta \left(Y_t | X^t\right)$ and for the error of estimation $\gamma(\vartheta, t) = \mathbf{E}_\vartheta \left((Y_t - m(\vartheta, t))^2 | X^t\right)$ is

$$\mathrm{d}m(\vartheta, t) = -\left[a(\vartheta) + \frac{\gamma(\vartheta, t) f(\vartheta)^2}{\sigma^2}\right] m(\vartheta, t)\,\mathrm{d}t + \frac{\gamma(\vartheta, t) f(\vartheta)}{\sigma^2} \mathrm{d}X_t, \quad m(\vartheta, 0), \tag{4.3}$$

$$\frac{\partial \gamma(\vartheta, t)}{\partial t} = -2a(\vartheta)\gamma(\vartheta, t) - \frac{\gamma(\vartheta, t)^2 f(\vartheta)^2}{\sigma^2} + b(\vartheta)^2, \qquad \gamma(\vartheta, 0). \tag{4.4}$$

Recall that the initial conditions for these equations are $m(\vartheta, 0) = \mathbf{E}_\vartheta (Y_0 | X_0)$ and $\gamma(\vartheta, 0) = \mathbf{E}_\vartheta \left((Y_0 - m(\vartheta, 0))^2 | X_0\right)$.

Our goal is to construct an adaptive filter $m^\star_{t,T}, 0 < t \le T$, which provides a good approximation of the conditional expectation $m(\vartheta, t), 0 < t \le T$. Thus the first step is to find a good estimator of ϑ, which can be easily calculated for each t and can be used for construction of $m^\star_{t,T}, 0 < t \le T$. Natural candidates are the MLE $\hat\vartheta_t, 0 < t \le T$ or the Bayesian estimator (BE) $\tilde\vartheta_t, 0 < t \le T$, defined by the equations

$$L(\hat\vartheta_t, X^t) = \sup_{\vartheta \in \Theta} L\left(\vartheta, X^t\right), \qquad \tilde\vartheta_t = \frac{\int_\Theta \vartheta p(\vartheta) L\left(\vartheta, X^t\right) \mathrm{d}\vartheta}{\int_\Theta p(\vartheta) L(\vartheta, X^t)\,\mathrm{d}\vartheta},$$

where $p(\vartheta), \vartheta \in \Theta$ is a prior density and

$$L\left(\vartheta, X^t\right) = \exp\left\{\int_0^t \frac{f(\vartheta) m(\vartheta, s)}{\sigma^2} \mathrm{d}X_s - \int_0^t \frac{f(\vartheta)^2 m(\vartheta, s)^2}{2\sigma^2} \mathrm{d}s\right\}$$

is the likelihood function. However, neither is suitable for our purposes since their computation or even numerical approximation would require solving Eqs. (4.3)–(4.4) for numerous values of the parameter ϑ and the time variable $t \in (0, T]$. Such a direct approach is clearly computationally inefficient.

The estimation of $m(\vartheta, t), t \in (0, T]$ is realized following the program, which was applied several times in the preceding chapters. Remind it here.

1. *Construct a preliminary estimator* $\bar\vartheta_\tau$ *by observations on a relatively small interval* $[0, \tau]$.
2. *Use* $\bar\vartheta_\tau$ *to define the One-step MLE-process* $\vartheta^\star_t, \tau < t \le T$.

3. *Approximate* $m(\vartheta, t)$ *by* $m_t^\star$ *obtained by solving the K-B equations with* ϑ *being replaced by* $\vartheta_t^\star, \tau < t \leq T$.
4. *Estimate the error* $m_t^\star - m(\vartheta, t), \tau < t \leq T$.

This program defines an approximation procedure which can be carried out on-line except the time interval $[0, \tau]$ of asymptotically negligible length because we take $\tau/T \to 0$.

Stationary Solution of Riccati Equation

Consider the asymptotic behavior of the solution of (4.4). Below we repeat the calculations of Example 1.8. Let us denote

$$r(\vartheta) = \left(a(\vartheta)^2 + \frac{f(\vartheta)^2 b(\vartheta)^2}{\sigma^2}\right)^{1/2}, \qquad \gamma_*(\vartheta) = \frac{\sigma^2 [r(\vartheta) - a(\vartheta)]}{f(\vartheta)^2}.$$

Recall that in the case of constant coefficients in (4.1)–(4.2) the Riccati equation (4.4) has an explicit solution (1.26)

$$\gamma(\vartheta, t) = e^{-2r(\vartheta)t} \left[\frac{1}{\gamma(\vartheta, 0) - \gamma_*(\vartheta)} + \frac{f(\vartheta)^2}{2r(\vartheta)\sigma^2}\left(1 - e^{-2r(\vartheta)t}\right)\right]^{-1} + \gamma_*(\vartheta), \tag{4.5}$$

where $\gamma(\vartheta, 0) \neq \gamma_*(\vartheta)$ is assumed. In view of (4.5) there exists a constant $C > 0$ such that

$$|\gamma(\vartheta, t) - \gamma_*(\vartheta)| \leq C\, e^{-2r(\vartheta)t} \longrightarrow 0, \qquad \text{as} \quad t \longrightarrow \infty.$$

In some problems, this exponential convergence allows us to consider the stationary regime corresponding to $\gamma(\vartheta, 0) = \gamma_*(\vartheta)$, without loss of generality. This simplifies the calculations in an essential way.

If $m_0 \sim \mathcal{N}(0, \gamma_*(\vartheta))$, i.e., the initial value $\gamma(\vartheta, 0) = \gamma_*(\vartheta)$, then

$$\frac{\partial \gamma(\vartheta, t)}{\partial t} \equiv 0, \qquad \gamma(\vartheta, t) \equiv \gamma_*(\vartheta) \qquad \text{for all} \quad t > 0.$$

The value $\gamma_*(\theta)$ is a positive solution z of the equation

$$-2a(\vartheta)z - \frac{z^2 f(\vartheta)^2}{\sigma^2} + b(\vartheta)^2 = 0,$$

which is obtained from the Riccati equation (4.4):

$$-2a(\vartheta)\gamma_*(\vartheta) - \frac{f(\vartheta)^2 \gamma_*(\vartheta)^2}{\sigma^2} + b(\vartheta)^2 = 0.$$

Below we will use an additional notation $\Gamma(\vartheta) = \gamma_*(\vartheta)\sigma^{-2}f(\vartheta)^2 = r(\vartheta) - a(\vartheta)$. In this case the system (4.3)–(4.4) is reduced to a single equation

$$\mathrm{d}m(\vartheta,t) = -\left[a(\vartheta) + \Gamma(\vartheta)\right] m(\vartheta,t)\,\mathrm{d}t + \Gamma(\vartheta)\, f(\vartheta)^{-1}\,\mathrm{d}X_t, \qquad m(\vartheta,0) = m_0. \tag{4.6}$$

Stationary Representations

To simplify the calculations of mathematical expectations we will use the representations (here ϑ_0 is the true value)

$$m(\vartheta_0,t) = m_0 e^{-a(\vartheta_0)t} + \frac{\gamma_*(\vartheta_0)\, f(\vartheta_0)}{\sigma}\int_0^t e^{-a(\vartheta_0)(t-s)}\mathrm{d}\bar W_s,$$

$$m_0 = \frac{\gamma_*(\vartheta_0)\, f(\vartheta_0)}{\sigma}\int_{-\infty}^0 e^{a(\vartheta_0)s}\mathrm{d}\bar W_s,$$

where $\bar W_s, s \in \mathcal{R}$ is a two-sided Wiener process. This means that we added an independent Wiener process $\tilde W_s, s \geq 0$ and set $\bar W_t = \tilde W_{-t}, t \leq 0$. Recall that the *innovation* stochastic process $\left\{\bar W_t,\ t \geq 0\right\}$, defined by the equation

$$\sigma\,\bar W_t = X_t - X_0 - \int_0^t f(\vartheta_0)\, m(\vartheta_0,s)\,\mathrm{d}s,$$

is a standard Wiener process (see Lemma 1.2).

If $\vartheta \neq \vartheta_0$, then the process $m(\vartheta,t), t \geq 0$ has the stochastic differential

$$\begin{aligned}\mathrm{d}m(\vartheta,t) &= -r(\vartheta)\, m(\vartheta,t)\,\mathrm{d}t + \Gamma(\vartheta)\, f(\vartheta)^{-1} f(\vartheta_0)\, m(\vartheta_0,t)\,\mathrm{d}t\\ &\quad + \Gamma(\vartheta)\, f^{-1}\sigma\mathrm{d}\bar W_t,\end{aligned}$$

subject to the initial condition $m(\vartheta,0)$. It will be more convenient to work with the stationary stochastic process $M(\vartheta,t) = f(\vartheta)m(\vartheta,t)$ satisfying the equation

$$\mathrm{d}M(\vartheta,t) = -r(\vartheta)\, M(\vartheta,t)\,\mathrm{d}t + \Gamma(\vartheta)\, M(\vartheta_0,t)\,\mathrm{d}t + \Gamma(\vartheta)\,\sigma\mathrm{d}\bar W_t,$$

subject to the initial condition $M(\vartheta,0)$.

To derive its representation in the integral form we consider two cases separately. First, suppose that $r(\vartheta) \neq a(\vartheta_0)$, then

$$\begin{aligned}M(\vartheta,t) &= \Gamma(\vartheta)\int_{-\infty}^t e^{-r(\vartheta)(t-s)} M(\vartheta_0,s)\,\mathrm{d}s + \Gamma(\vartheta)\,\sigma\int_{-\infty}^t e^{-r(\vartheta)(t-s)}\mathrm{d}\bar W_s\\ &= \Gamma(\vartheta)\,\Gamma(\vartheta_0)\,\sigma\int_{-\infty}^t e^{-r(\vartheta)(t-s)}\int_{-\infty}^s e^{-a(\vartheta_0)(s-q)}\mathrm{d}\bar W_q\,\mathrm{d}s\end{aligned}$$

$$
\begin{aligned}
&\quad + \Gamma(\vartheta)\,\sigma \int_{-\infty}^{t} e^{-r(\vartheta)(t-s)} \mathrm{d}\bar{W}_s \\
&= \frac{\Gamma(\vartheta)\,\Gamma(\vartheta_0)\,\sigma}{r(\vartheta) - a(\vartheta_0)} \int_{-\infty}^{t} e^{-a(\vartheta_0)(t-s)} \mathrm{d}\bar{W}_s \\
&\quad + \frac{\Gamma(\vartheta)\,\sigma\,(r(\vartheta) - r(\vartheta_0))}{r(\vartheta) - a(\vartheta_0)} \int_{-\infty}^{t} e^{-r(\vartheta)(t-s)} \mathrm{d}\bar{W}_s. \qquad (4.7)
\end{aligned}
$$

Here we used the Fubini theorem

$$
\begin{aligned}
&\int_{-\infty}^{t} e^{-r(\vartheta)(t-s)} \int_{-\infty}^{s} e^{-a(\vartheta_0)(s-q)} \mathrm{d}\bar{W}_q \,\mathrm{d}s \\
&\qquad = e^{-r(\vartheta)t} \int_{-\infty}^{t} e^{a(\vartheta_0)q} \int_{q}^{t} e^{(r(\vartheta)-a(\vartheta_0))s} \mathrm{d}s\, \mathrm{d}\bar{W}_q \\
&\qquad = \frac{1}{r(\vartheta) - a(\vartheta_0)} \int_{-\infty}^{t} \left[e^{-a(\vartheta_0)(t-q)} - e^{-r(\vartheta)(t-q)} \right] \mathrm{d}\bar{W}_q.
\end{aligned}
$$

This representation corresponds to the initial value

$$
\begin{aligned}
M(\vartheta, 0) = &\frac{\Gamma(\vartheta)\,\Gamma(\vartheta_0)\,\sigma}{r(\vartheta) - a(\vartheta_0)} \int_{-\infty}^{0} e^{a(\vartheta_0)s} \mathrm{d}\bar{W}_s \\
&+ \frac{\Gamma(\vartheta)\,\sigma\,(r(\vartheta) - r(\vartheta_0))}{r(\vartheta) - a(\vartheta_0)} \int_{-\infty}^{0} e^{r(\vartheta)s} \mathrm{d}\bar{W}_s.
\end{aligned}
$$

If $r(\vartheta) = a(\vartheta_0)$, then

$$
\begin{aligned}
\int_{-\infty}^{t} e^{-a(\vartheta_0)(t-s)} \int_{-\infty}^{s} e^{-a(\vartheta_0)(s-q)} \mathrm{d}\bar{W}_q \,\mathrm{d}s &= e^{-a(\vartheta_0)t} \int_{-\infty}^{t} \int_{-\infty}^{s} e^{a(\vartheta_0)q} \mathrm{d}\bar{W}_q \,\mathrm{d}s \\
&= \int_{-\infty}^{t} e^{-a(\vartheta_0)(t-s)} \,(t-s)\, \mathrm{d}\bar{W}_s
\end{aligned}
$$

and

$$
M(\vartheta, t) = \Gamma(\vartheta)\,\sigma \int_{-\infty}^{t} e^{-a(\vartheta_0)(t-s)} \left[\gamma(\vartheta_0)(t-s) + 1\right] \mathrm{d}\bar{W}_s.
$$

These expressions will allow us calculate moments involving $M(\vartheta, t)$ and $\dot{M}(\vartheta, t)$. The dot denotes the derivation w.r.t. ϑ.

4.2 Parameter Estimation

4.2.1 MME

Consider the partially observed system

$$\mathrm{d}X_t = f\, Y_t\, \mathrm{d}t + \sigma\, \mathrm{d}W_t, \quad 0 \le t \le T, \tag{4.8}$$

$$\mathrm{d}Y_t = -a\, Y_t\, \mathrm{d}t + b\, \mathrm{d}V_t, \quad t \ge 0, \tag{4.9}$$

subject to the initial conditions X_0 and Y_0. The process $X_t, t \ge 0$ is not ergodic and thus the method of moments estimators cannot be based on the integrals such as

$$\frac{1}{T}\int_0^T g\left(X_t\right) \mathrm{d}t.$$

Let us first assume that $\vartheta = (f, a, b) \in \Theta$. The set Θ is a bounded, convex, closed subset of $\mathcal{R}^3$ and $a > 0$. For simplicity, we assume that T is integer.

We will use the following notations:

$$R_{1,T} = \frac{1}{T}\sum_{k=1}^{T}\left[X_k - X_{k-1}\right]^2, \qquad \Phi_1\left(\vartheta\right) = \frac{f^2b^2}{a^3}\left[e^{-a} - 1 + a\right] + \sigma^2,$$

$$R_{2,T} = \frac{1}{T}\sum_{k=2}^{T}\left[X_k - X_{k-1}\right]\left[X_{k-1} - X_{k-2}\right], \quad \Phi_2\left(\vartheta\right) = \frac{f^2b^2}{2a^3}\left[1 - e^{-a}\right]^2,$$

$$K_{1,1}\left(\vartheta\right) = \frac{2f^4b^4}{a^6}\left[e^{-a} - 1 + a\right]^2 + \frac{f^4b^4e^{4a}\left[1 - e^{-a}\right]^3}{a^6\left[1 + e^{-a}\right]} + \frac{4f^2\sigma^2b^2}{a^3}\left[e^{-a} - 1 + a\right] + 2\sigma^4,$$

$$R_T = \left(R_{1,T}, R_{2,T}\right)^\top, \qquad \Phi\left(\vartheta\right) = \left(\Phi_1\left(\vartheta\right), \Phi_2\left(\vartheta\right)\right)^\top, \quad \xi_* = \left(\xi_1, \xi_2\right)^\top,$$

$$\mathbf{K}\left(\vartheta\right) = \begin{pmatrix} K_{1,1}\left(\vartheta\right) & K_{1,2}\left(\vartheta\right) \\ K_{2,1}\left(\vartheta\right) & K_{2,2}\left(\vartheta\right) \end{pmatrix}, \quad R_T = \left(R_{1,T}, R_{2,T}\right)^\top,$$

$$\Phi\left(\vartheta\right) = \left(\Phi_1\left(\vartheta\right), \Phi_2\left(\vartheta\right)\right)^\top.$$

Here $\xi_* \sim \mathcal{N}\left(0, \mathbf{K}\left(\vartheta_0\right)\right)$ and the undefined entries of the matrix $\mathbf{K}\left(\vartheta\right)$ will be made precise later.

The MMEs are

$$f_T^* = \left(\frac{a^3}{b^2\left[e^{-a}-1+a\right]T}\sum_{k=1}^{T}\left([X_k - X_{k-1}]^2 - \sigma^2\right)\right)^{1/2},$$

$$b_T^* = \left(\frac{a^3}{f^2\left[e^{-a}-1+a\right]T}\sum_{k=1}^{T}\left([X_k - X_{k-1}]^2 - \sigma^2\right)\right)^{1/2},$$

$$a_T^* = H\left(\frac{1}{f^2b^2T}\sum_{k=1}^{T}\left([X_k - X_{k-1}]^2 - \sigma^2\right)\right),$$

where $H(y)$ is the inverse function of $y = h(x) = x^{-3}\left(x - 1 + e^{-x}\right)$. Below we show that $h(x), x > 0$ is strictly decreasing.

Proposition 4.1 *The statistic R_T has the following properties.*

1. *Convergence to $\Phi(\vartheta)$, uniformly on Θ: for any $\nu > 0$*

$$\lim_{T\to\infty}\sup_{\vartheta\in\Theta}\mathbf{P}_\vartheta\left(\left\|R_T - \Phi(\vartheta)\right\| \geq \nu\right) = 0.$$

2. *Asymptotic normality, uniform on compacts $\mathbb{K} \subset \Theta$:*

$$\sqrt{T}\left(R_T - \Phi(\vartheta)\right) \Longrightarrow \xi_*, \tag{4.10}$$

3. *Convergence of moments: for any $p > 0$ uniformly on $\mathbb{K}$*

$$\lim_{T\to\infty} T^{p/2}\mathbf{E}_\vartheta\left\|R_T - \Phi(\vartheta)\right\|^p = \mathbf{E}_\vartheta\left|\xi_*\right|^p,$$

and there exists a constant $C > 0$ such that

$$\sup_{\vartheta\in\mathbb{K}} T^{p/2}\mathbf{E}_\vartheta\left\|R_T - \Phi(\vartheta)\right\|^p \leq C. \tag{4.11}$$

Proof Using the equality (4.8) we write

$$R_{1,T} = \frac{1}{T}\sum_{k=1}^{T}[X_k - X_{k-1}]^2 = \frac{1}{T}\sum_{k=1}^{T}\left[\int_{k-1}^{k}\mathrm{d}X_s\right]^2$$

$$= \frac{f^2}{T}\sum_{k=1}^{T}\eta_k^2 + \frac{2f\sigma}{T}\sum_{k=1}^{T}\eta_k\left[W_k - W_{k-1}\right] + \frac{\sigma^2}{T}\sum_{k=1}^{T}[W_k - W_{k-1}]^2,$$

where

$$\eta_k = \int_{k-1}^{k} Y_t \, \mathrm{d}t.$$

We have

$$\mathbf{E}_{\vartheta_0} R_{1,T} = \frac{f^2}{T} \sum_{k=1}^{T} \mathbf{E}_\vartheta \eta_k^2 + \sigma^2$$

since Y^T and W^T are independent. The process Y^T satisfies

$$Y_t = Y_0 e^{-at} + b \int_0^t e^{-a(t-s)} \mathrm{d}V_s.$$

Hence

$$\begin{aligned}
\mathbf{E}_\vartheta Y_t Y_s &= \mathbf{E}_{\vartheta_0} Y_0^2 e^{-a(t+s)} + b^2 e^{-a(t+s)} \int_0^{t \wedge s} e^{2ar} \mathrm{d}r \\
&= \left[d^2 - \frac{b^2}{2a} \right] e^{-a(t+s)} + \frac{b^2}{2a} e^{-a|t-s|}
\end{aligned}$$

and

$$\begin{aligned}
\mathbf{E}_\vartheta \eta_k^2 &= \int_{k-1}^{k} \int_{k-1}^{k} \mathbf{E}_\vartheta Y_t Y_s \, \mathrm{d}s \mathrm{d}t \\
&= \left[d^2 - \frac{b^2}{2a} \right] \left(\int_{k-1}^{k} e^{-at} \mathrm{d}t \right)^2 + \frac{b^2}{2a} \int_{k-1}^{k} \int_{k-1}^{k} e^{-a|t-s|} \mathrm{d}s \mathrm{d}t \\
&= \left[\frac{d^2}{a^2} - \frac{b^2}{2a^3} \right] \left[e^a - 1 \right]^2 e^{-2ak} + \frac{b^2}{a^3} \left[e^{-a} - 1 + a \right].
\end{aligned}$$

Therefore

$$\begin{aligned}
\mathbf{E}_\vartheta R_{1,T} &= \left[e^a - 1 \right]^2 \frac{f^2 \left[2ad^2 - b^2 \right]}{2a^3 T} \sum_{k=1}^{T} e^{-2ak} + \frac{f^2 b^2}{a^3} \left[e^{-a} - 1 + a \right] + \sigma^2 \\
&= \frac{f^2 b^2}{a^3} \left[e^{-a} - 1 + a \right] + \sigma^2 + r_T, \qquad \left| r_T \right| \leq \frac{C}{T}.
\end{aligned}$$

For simplicity of calculations we assume that the process $Y_t, t \geq 0$ is stationary. Then

$$\mathbf{E}_\vartheta R_{1,T} = \frac{f^2 b^2}{a^3} \left[e^{-a} - 1 + a \right] + \sigma^2.$$

The covariance is $(j < k)$

$$\mathbf{E}_\vartheta \eta_k\eta_j = b^2\int_{k-1}^{k}\int_{j-1}^{j}\mathbf{E}_\vartheta Y_t Y_s \mathrm{d}s\mathrm{d}t = b^2\int_{k-1}^{k}\int_{j-1}^{j} e^{-a(t+s)}\int_{-\infty}^{t\wedge s} e^{2au}\mathrm{d}u\mathrm{d}s\mathrm{d}t$$

$$= \frac{b^2}{2a}\int_{k-1}^{k}\int_{j-1}^{j} e^{-a|t-s|}\mathrm{d}s\mathrm{d}t = \frac{b^2e^a}{a^3}\left(1-e^{-a}\right)^2 e^{-a|k-j|}.$$

Hence

$$\sup_{\vartheta\in\Theta}\mathbf{E}_\vartheta \eta_k\eta_j \leq C\, e^{-a|k-j|} \tag{4.12}$$

and this Gaussian sequence $\eta_1, \eta_2, \ldots$ is exponential mixing. By the law of large numbers

$$R_{1,T} \longrightarrow \Phi_1(\vartheta) = \frac{f^2b^2}{a^3}\left[e^{-a}-1+a\right]+\sigma^2,$$

as $T\rightarrow\infty$.

Then we can write

$$\mathbf{E}_\vartheta\left(R_{1,T}-\mathbf{E}_\vartheta R_{1,T}\right)^2$$

$$= \mathbf{E}_\vartheta\left(\frac{f^2}{T}\sum_{k=1}^{T}\left[\eta_k^2-\mathbf{E}_\vartheta\eta_k^2\right]+\frac{2f\sigma}{T}\sum_{k=1}^{T}\eta_k\zeta_k+\frac{\sigma^2}{T}\sum_{k=1}^{T}\left(\zeta_k^2-1\right)\right)^2$$

$$= \frac{f^4}{T^2}\sum_{k=1}^{T}\sum_{j=1}^{T}\mathbf{E}_\vartheta\left(\left[\eta_k^2-\mathbf{E}_\vartheta\eta_k^2\right]\left[\eta_j^2-\mathbf{E}_\vartheta\eta_j^2\right]\right)$$

$$+\frac{4f^2\sigma^2}{T^2}\sum_{k=1}^{T}\sum_{j=1}^{T}\mathbf{E}_\vartheta\left(\eta_k\eta_j\zeta_k\zeta_j\right)$$

$$+\frac{\sigma^4}{T^2}\sum_{k=1}^{T}\sum_{j=1}^{T}\mathbf{E}_\vartheta\left(\zeta_k^2-1\right)\left(\zeta_j^2-1\right)$$

$$= \frac{f^4}{T^2}\sum_{k=1}^{T}\sum_{j=1}^{T}\mathbf{E}_\vartheta\left(\left[\eta_k^2-\mathbf{E}_\vartheta\eta_k^2\right]\left[\eta_j^2-\mathbf{E}_\vartheta\eta_j^2\right]\right)+\frac{4f^2\sigma^2}{T^2}\sum_{k=1}^{T}\mathbf{E}_\vartheta\eta_k^2$$

$$+\frac{\sigma^4}{T^2}\sum_{k=1}^{T}\mathbf{E}_\vartheta\left(\zeta_k^2-1\right)^2$$

$$\leq \frac{f^4}{T^2}\sum_{k=1}^{T}\sum_{j=1}^{T}\mathbf{E}_\vartheta\left(\left[\eta_k^2-\mathbf{E}_\vartheta\eta_k^2\right]\left[\eta_j^2-\mathbf{E}_\vartheta\eta_j^2\right]\right)+\frac{4f^2\sigma^2b^2}{a^3T}\left[a-1+e^{-a}\right]$$

$$+\frac{2\sigma^4}{T}.$$

By using (4.9), the last sum can be shown to satisfy an exponential bound similar to (4.12). Indeed, note that

$$\mathbf{E}_\vartheta\left(\left[\eta_k^2-\mathbf{E}_\vartheta\eta_k^2\right]\left[\eta_j^2-\mathbf{E}_\vartheta\eta_j^2\right]\right)=\mathbf{E}_\vartheta\left[\eta_k^2\eta_j^2\right]-\left(\mathbf{E}_\vartheta\left(\eta_1^2\right)\right)^2,$$

where $\mathbf{E}_\vartheta\eta_1^2=b^2a^{-3}\left[e^{-a}-1+a\right]$. We have

$$\mathbf{E}_\vartheta\left[\eta_k^2\eta_j^2\right]=\int_{k-1}^{k}\int_{k-1}^{k}\int_{j-1}^{j}\int_{j-1}^{j}\mathbf{E}_\vartheta Y_tY_sY_rY_q\ \mathrm{d}t\mathrm{d}s\mathrm{d}r\mathrm{d}q$$
$$=b^4\int_{k-1}^{k}\int_{k-1}^{k}\int_{j-1}^{j}\int_{j-1}^{j}e^{-a(t+s+r+q)}\mathbf{E}_\vartheta I\left(t\right)I\left(s\right)I\left(r\right)I\left(q\right)\ \mathrm{d}t\mathrm{d}s\mathrm{d}r\mathrm{d}q.$$

Suppose that $j\leq k-1$. Then using the elementary properties of the stochastic integral we can write

$$\begin{aligned}
&\mathbf{E}_\vartheta I\left(t\right)I\left(s\right)I\left(r\right)I\left(q\right)\\
&=\mathbf{E}_\vartheta\left(I\left(r\right)I\left(q\right)\mathbf{E}_\vartheta\left(I\left(t\right)I\left(s\right)|\mathfrak{F}_{r\vee q}^Y\right)\right)\\
&=\mathbf{E}_\vartheta\left(I\left(r\right)I\left(q\right)\mathbf{E}_\vartheta\left(I\left(s\wedge t\right)^2|\mathfrak{F}_{r\vee q}^Y\right)\right)\\
&=\mathbf{E}_\vartheta\left(I\left(r\right)I\left(q\right)\mathbf{E}_\vartheta\left(\left[I\left(s\wedge t\right)-I\left(r\vee q\right)+I\left(r\vee q\right)\right]^2|\mathfrak{F}_{r\vee q}^Y\right)\right)\\
&=\mathbf{E}_\vartheta\left(I\left(r\right)I\left(q\right)\left(\int_{r\vee q}^{s\wedge t}e^{2au}\mathrm{d}u+I\left(r\vee q\right)^2\right)\right)\\
&=\frac{1}{4a^2}\left[e^{2a(s\wedge t)}-e^{2a(r\vee q)}\right]e^{2a(r\wedge q)}+\mathbf{E}_\vartheta\left(I\left(r\wedge q\right)I\left(r\vee q\right)^3\right)\\
&=\frac{1}{4a^2}\left[e^{2a[(s\wedge t)+(r\wedge q)]}-e^{2a(r+q)}\right]+\frac{3}{4a^2}e^{2a(r+q)}\\
&=\frac{1}{4a^2}\left[e^{2a[(s\wedge t)+(r\wedge q)]}+2e^{2a(r+q)}\right].
\end{aligned}$$

Here we defined the σ-algebra $\mathfrak{F}_t^Y=\sigma\left(Y_s,s\leq t\right)$ and used the equality

$$\mathbf{E}_\vartheta\left(I\left(r\wedge q\right)I\left(r\vee q\right)^3\right)=\mathbf{E}_\vartheta\left(I\left(r\wedge q\right)\left[I\left(r\vee q\right)-I\left(r\wedge q\right)+I\left(r\wedge q\right)\right]^3\right)$$

$$= 3\mathbf{E}_{\vartheta}\left(I\left(r\wedge q\right)\right)^{2}\mathbf{E}_{\vartheta}\left(\left[I\left(r\vee q\right)-I\left(r\wedge q\right)\right]^{2}\right)+\mathbf{E}_{\vartheta}\left(I\left(r\wedge q\right)\right)^{4}$$
$$= \frac{3}{4a^{2}}e^{2a(r\wedge q)}\left[e^{2a(r\vee q)}-e^{2a(r\wedge q)}\right]+\frac{3}{4a^{2}}e^{4a(r\wedge q)}=\frac{3}{4a^{2}}e^{2a(r+q)}.$$

Therefore

$$e^{-a(t+s+r+q)}\mathbf{E}_{\vartheta}I\left(t\right)I\left(s\right)I\left(r\right)I\left(q\right)=\frac{1}{4a^{2}}e^{-a|t-s|-a|r-q|}+\frac{1}{2a^{2}}e^{-a(t+s)}e^{a(r+q)}$$

and

$$\int_{k-1}^{k}\int_{k-1}^{k}\int_{j-1}^{j}\int_{j-1}^{j}e^{-a(t+s+r+q)}\mathbf{E}_{\vartheta}I\left(t\right)I\left(s\right)I\left(r\right)I\left(q\right)\,\mathrm{d}t\mathrm{d}s\mathrm{d}r\mathrm{d}q$$
$$=\frac{1}{4a^{2}}\int_{k-1}^{k}\int_{k-1}^{k}\int_{j-1}^{j}\int_{j-1}^{j}e^{-a|t-s|-a|r-q|}\,\mathrm{d}t\mathrm{d}s\mathrm{d}r\mathrm{d}q$$
$$+\frac{1}{2a^{2}}\int_{k-1}^{k}\int_{k-1}^{k}\int_{j-1}^{j}\int_{j-1}^{j}e^{-a(t+s-r-q)}\,\mathrm{d}t\mathrm{d}s\mathrm{d}r\mathrm{d}q$$
$$=\frac{1}{a^{6}}\left[e^{-a}-1+a\right]^{2}+\frac{e^{2a}\left[1-e^{-a}\right]^{4}}{2a^{6}}\,e^{-2a|k-j|}.$$

Hence if $j\leq k-1$, then

$$\mathbf{E}_{\vartheta}\left(\left[\eta_{k}^{2}-\mathbf{E}_{\vartheta}\eta_{k}^{2}\right]\left[\eta_{j}^{2}-\mathbf{E}_{\vartheta}\eta_{j}^{2}\right]\right)=\frac{b^{4}e^{2a}\left[1-e^{-a}\right]^{4}}{2a^{6}}\,e^{-2a|k-j|}\leq C\,e^{-2a|k-j|}.$$

If $k=j$, then

$$\mathbf{E}_{\vartheta}\eta_{k}^{4}=3\left(\mathbf{E}_{\vartheta}\eta_{k}^{2}\right)^{2}=\frac{3b^{4}}{a^{6}}\left[e^{-a}-1+a\right]^{2}$$

and

$$\mathbf{E}_{\vartheta}\left[\eta_{k}^{2}-\mathbf{E}_{\vartheta}\eta_{k}^{2}\right]^{2}=\frac{2b^{4}}{a^{6}}\left[e^{-a}-1+a\right]^{2}.$$

Therefore

$$\frac{1}{T}\sum_{k=1}^{T}\sum_{j=1}^{T}\mathbf{E}_{\vartheta}\left(\left[\eta_{k}^{2}-\mathbf{E}_{\vartheta}\eta_{k}^{2}\right]\left[\eta_{j}^{2}-\mathbf{E}_{\vartheta}\eta_{j}^{2}\right]\right)$$

$$= \frac{2b^4}{a^6}\left[e^{-a}-1+a\right]^2+\frac{b^4e^{2a}\left[1-e^{-a}\right]^4}{2a^6T}\sum_{k=1}^{T}\sum_{j\neq k}^{T}e^{-2a|k-j|}$$

$$\longrightarrow \frac{2b^4}{a^6}\left[e^{-a}-1+a\right]^2+\frac{b^4e^{4a}\left[1-e^{-a}\right]^3}{a^6\left[1+e^{-a}\right]}.$$

Here we used the identities

$$\sum_{j=1,j\neq k}^{T} e^{-2a|k-j|}=\frac{2e^{2a}}{1-e^{-2a}}-\frac{e^{-2ak}+e^{-2a(T-k+1)}}{1-e^{-2a}}$$

and

$$\frac{\left[1-e^{-a}\right]^4}{\left(1-e^{-a}\right)\left(1+e^{-a}\right)}=\frac{\left[1-e^{-a}\right]^3}{\left(1+e^{-a}\right)}.$$

Thus we obtained the bound

$$\sup_{\vartheta\in\mathbb{K}}\mathbf{E}_\vartheta\left(R_{1,T}-\mathbf{E}_\vartheta R_{1,T}\right)^2\leq\frac{C}{T}$$

and the convergence

$$T\mathbf{E}_\vartheta\left(R_{1,T}-\mathbf{E}_\vartheta R_{1,T}\right)^2\longrightarrow\frac{2f^4b^4}{a^6}\left[e^{-a}-1+a\right]^2+\frac{f^4b^4e^{4a}\left[1-e^{-a}\right]^3}{a^6\left[1+e^{-a}\right]}$$
$$+\frac{4f^2\sigma^2b^2}{a^3}\left[e^{-a}-1+a\right]+2\sigma^4=K_{1,1}(\vartheta).$$

The estimate (4.11) is obtained directly, by applying Rosenthal's inequality for time series. As the random variables $\left|\sqrt{T}\left(R_{1,T}-\Phi_1(\vartheta)\right)\right|^p$ are uniformly integrable, the convergence of polynomial moments holds as well.

By the CLT for exponentially mixing time series we obtain the convergence

$$\sqrt{T}\left(R_{1,T}-\Phi_1(\vartheta)\right)\Longrightarrow\xi_1\sim\mathcal{N}\left(0,K_{1,1}(\vartheta)\right).$$

Consider now the second component

$$R_{2,T}=\frac{1}{T}\sum_{k=2}^{T}\left[f\eta_k+\sigma\left(W_k-W_{k-1}\right)\right]\left[f\eta_{k-1}+\sigma\left(W_{k-1}-W_{k-2}\right)\right]$$

$$= \frac{f^2}{T}\sum_{k=2}^{T}\eta_k\eta_{k-1} + \frac{f\sigma}{T}\sum_{k=2}^{T}\left[\eta_k\zeta_{k-1}+\eta_{k-1}\zeta_k\right] + \frac{\sigma^2}{T}\sum_{k=2}^{T}\zeta_k\zeta_{k-1}.$$

Recall that

$$\mathbf{E}_\vartheta\,\eta_k\zeta_{k-1}=0,\qquad \mathbf{E}_\vartheta\,\eta_{k-1}\zeta_k=0,\qquad \mathbf{E}_\vartheta\,\zeta_k\zeta_{k-1}=0$$

and, by the LLN,

$$\frac{1}{T}\sum_{k=2}^{T}\eta_k\zeta_{k-1}\longrightarrow 0,\qquad \frac{1}{T}\sum_{k=2}^{T}\eta_{k-1}\zeta_k\to 0,\qquad \frac{1}{T}\sum_{k=2}^{T}\zeta_k\zeta_{k-1}\longrightarrow 0.$$

Further,

$$\mathbf{E}_\vartheta R_{2,T} = \frac{f^2}{T}\sum_{k=2}^{T}\mathbf{E}_\vartheta\,\eta_k\eta_{k-1} = \frac{f^2b^2}{2aT}\sum_{k=2}^{T}\int_{k-1}^{k}\left(\int_{k-2}^{k-1}e^{-a|t-s|}\mathrm{d}s\right)\mathrm{d}t$$
$$\longrightarrow \frac{f^2b^2}{2a^3}\left[1-e^{-a}\right]^2 \equiv \Phi_2(\vartheta).$$

For the variance of $R_{2,T}$ we have the following expression:

$$\mathbf{E}_\vartheta\left(R_{2,T}-\mathbf{E}_\vartheta R_{2,T}\right)^2$$
$$= \frac{f^4}{T^2}\sum_{k=2}^{T}\sum_{j=2}^{T}\mathbf{E}_\vartheta\left[\eta_k\eta_{k-1}-\mathbf{E}_\vartheta\,\eta_k\eta_{k-1}\right]\left[\eta_j\eta_{j-1}-\mathbf{E}_\vartheta\,\eta_j\eta_{j-1}\right]$$
$$+\frac{f^2\sigma^2}{T^2}\sum_{k=2}^{T}\sum_{j=2}^{T}\mathbf{E}_\vartheta\left[\eta_k\zeta_{k-1}+\eta_{k-1}\zeta_k\right]\left[\eta_j\zeta_{j-1}+\eta_{j-1}\zeta_j\right]$$
$$+\frac{\sigma^4}{T^2}\sum_{k=2}^{T}\sum_{j=2}^{T}\mathbf{E}_\vartheta\,\zeta_k\zeta_{k-1}\zeta_j\zeta_{j-1}$$
$$= \frac{f^4}{T^2}\sum_{k=2}^{T}\sum_{j=2}^{T}\left(\mathbf{E}_\vartheta\left[\eta_k\eta_{k-1}\eta_j\eta_{j-1}\right]-\left[\mathbf{E}_\vartheta\,\eta_k\eta_{k-1}\right]^2\right)+\frac{\sigma^4\,(T-2)}{T^2}$$
$$+\frac{f^2\sigma^2}{T^2}\sum_{k=2}^{T}\mathbf{E}_\vartheta\left[\eta_k^2+2\eta_k\eta_{k-2}+\eta_{k-1}^2\right].$$

The values of these expectations are already calculated above

$$f^2\mathbf{E}_\vartheta \eta_k^2 = \Phi_1(\vartheta), \quad f^2\mathbf{E}_\vartheta \eta_k\eta_{k-1} = \Phi_2(\vartheta), \quad f^2\mathbf{E}_\vartheta \eta_k\eta_{k-2} = e^{-a}\Phi_2(\vartheta).$$

The calculation of expectation $\mathbf{E}_\vartheta\left[\eta_k\eta_{k-1}\eta_j\eta_{j-1}\right]$ can be carried out following the same lines as it was done above and thus

$$\sup_{\vartheta\in\mathbb{K}} \mathbf{E}_\vartheta \left(R_{2,T} - \mathbf{E}_\vartheta R_{2,T}\right)^2 \leq \frac{C}{T}.$$

By the LLN

$$R_{2,T} \longrightarrow \Phi_2(\vartheta)$$

and by the CLT

$$\sqrt{T}\left(R_{2,T} - \Phi_2(\vartheta)\right) \Longrightarrow \xi_2.$$

Moreover, a direct calculation shows that for any λ and μ

$$\sqrt{T}\Big(\lambda\left(R_{1,T} - \Phi_1(\vartheta)\right) + \mu\left(R_{2,T} - \Phi_2(\vartheta)\right)\Big) \Longrightarrow \lambda\xi_1 + \mu\xi_2,$$

which proves the asymptotic normality (4.10). These calculations are somewhat cumbersome but essentially similar to those detailed above for $K_{1,1}(\vartheta)$ and therefore will be omitted. For the same reason, we will also omit the calculations related to the entries $K_{1,2}(\vartheta) = K_{2,1}(\vartheta)$ and $K_{2,2}(\vartheta)$ of the matrix $\mathbf{K}(\vartheta)$.

Note that for any $\nu > 0$

$$\sup_{\vartheta\in\mathbb{K}} \mathbf{P}_\vartheta\left(\|R_T - \Phi(\vartheta)\| \geq \nu\right) \leq \nu^{-2}\sup_{\vartheta\in\mathbb{K}} \mathbf{E}_\vartheta \|R_T - \Phi(\vartheta)\|^2 \leq \frac{C}{\nu^2 T}.$$

□

Estimation of the Parameters *f*, *b*, and *a*

The asymptotic normality of the estimators f_T^*, b_T^*, a_T^* can be deduced from Proposition 4.1 as follows. The estimators f_T^* and b_T^* satisfy

$$\sqrt{T}\left(f_T^* - f_0\right) \Rightarrow \xi_f \sim \mathcal{N}\left(0, D_f(f_0)^2\right),$$

$$D_f(f)^2 = \frac{a^3\, K_{1,1}(f)}{4b^2\left(e^{-a} - 1 + a\right)\left(\Phi_1(f) - \sigma^2\right)},$$

$$\sqrt{T}\left(b_T^* - b_0\right) \Rightarrow \xi_b \sim \mathcal{N}\left(0, D_b(b_0)^2\right),$$

$$D_b(b)^2 = \frac{a^3 K_{1,1}(b)}{4f^2\left(e^{-a}-1+a\right)\left(\Phi_1(b)-\sigma^2\right)}.$$

The asymptotic normality of a_T^* is obtained as follows. Suppose that $\vartheta = a \in \Theta = (\alpha, \beta), \alpha > 0$ and the values of $f \neq 0$ and $b \neq 0$ are known.

First we verify that the function $\Phi_1(a), a \in \Theta$ is strictly decreasing. Define the function $h(x) = x^{-3}\left(x-1+e^{-x}\right), x > 0$ with the derivative $h'(x) = x^{-4}[3-2x - (3+x)e^{-x}]$. Note that the function $g(x) = 3-2x-(3+x)e^{-x}$ vanishes at zero and has derivative $g'(x) = -2+(2+x)e^{-x}$ with $g'(0) = 0$. Finally, $g''(x) = -(1+x)e^{-x} < 0$. Hence $g'(x) < 0$, $g(x) < 0$ and $h'(x) < 0$ for $x > 0$. As the function $\Phi_1(a), a \in \Theta$ is strictly decreasing, the equation

$$R_{1,T} = \Phi_1(a), \qquad a \in \Theta$$

has a unique solution $a = a_T^*$ for the values $\Phi_1(\beta) < R_{1,T} < \Phi_1(\alpha)$.

$$\begin{aligned} a_T^* = H\left(\frac{R_{1,T}-\sigma^2}{f^2b^2}\right) &= H\left(\frac{\Phi_1(a_0)-\sigma^2}{f^2b^2}\right) \\ &\quad + H'\left(\frac{\Phi_1(a_0)-\sigma^2}{f^2b^2}\right)\frac{\left(R_{1,T}-\Phi_1(a_0)\right)}{f^2b^2} \\ &= a_0 + \frac{R_{1,T}-\Phi_1(a_0)}{h'(a_0)\,f^2b^2} + o\left(T^{-1}\right). \end{aligned}$$

Hence

$$\sqrt{T}\left(a_T^*-a_0\right) \Longrightarrow \xi_a \sim \mathcal{N}\left(0, D_a(a_0)^2\right), \qquad D_a(a_0)^2 = \frac{K_{1,1}(a_0)}{h'(a_0)^2 f^4b^4}.$$

As the derivative $h'(x), x > \alpha$ is separated from zero, for any $p \geq 2$ there exists a constant $C > 0$ such that

$$\sup_{a_0\in\mathbb{K}} T^{p/2}\mathbf{E}_{a_0}\left|a_T^*-a_0\right|^p \leq C$$

and the convergence of moments

$$\lim_{T\to\infty} T^{p/2}\mathbf{E}_{a_0}\left|a_T^*-a_0\right|^p = \mathbf{E}_{a_0}\left|\xi_a\right|^p$$

holds.

More General Model

Estimation of one-dimensional parameters in all the settings considered above are particular cases of a slightly more general model

$$\begin{aligned} \mathrm{d}X_t &= f(\vartheta)\, Y_t \mathrm{d}t + \sigma \mathrm{d}W_t, && 0 \leq t \leq T, \\ \mathrm{d}Y_t &= -a(\vartheta)\, Y_t \mathrm{d}t + b(\vartheta)\, \mathrm{d}V_t, && t \geq 0 \end{aligned}$$

subject to the initial conditions X_0 and Y_0, where $f(\vartheta), a(\vartheta), b(\vartheta), \vartheta \in \Theta = (\alpha, \beta)$ are known smooth functions. In order to estimate the parameter ϑ from the sample $X^T = (X_t, 0 \leq t \leq T)$ we can use the MME

$$R_{1,T} = \frac{1}{T}\sum_{k=1}^{T} [X_k - X_{k-1}]^2, \qquad \Psi(\vartheta) = \frac{f(\vartheta)^2 b(\vartheta)^2}{a(\vartheta)^2}\left[e^{-a(\vartheta)} - 1 + a(\vartheta)\right],$$

$$\vartheta_T^* = \arg\inf_{\vartheta\in\Theta} \left|R_{1,T} - \Psi(\vartheta)\right|,$$

Proposition 4.2 *Let the functions $f(\vartheta), a(\vartheta), b(\vartheta), \vartheta \in [\alpha, \beta]$ have two continuous derivatives, and*

$$\inf_{\vartheta\in\Theta} \left|\dot\Psi(\vartheta)\right| > 0.$$

Then the MME $\check\vartheta_T$ is uniformly consistent, and for any $p \geq 2$ there exists a constant $C = C_p$ such that

$$\sup_{\vartheta_0\in\mathbb{K}} T^{p/2}\mathbf{E}_{\vartheta_0}\left|\check\vartheta_T - \vartheta_0\right|^p \leq C.$$

Proof Note that for any $\nu > 0$

$$\begin{aligned} g(\vartheta_0, \nu) &= \inf_{|\vartheta-\vartheta_0|>\nu} |\Psi(\vartheta) - \Psi(\vartheta_0)| \geq \inf_{|\vartheta-\vartheta_0|>\nu} |\vartheta - \vartheta_0| \left|\dot\Psi(\bar\vartheta)\right| \\ &\geq \nu \inf_{\vartheta\in\Theta} \left|\dot\Psi(\vartheta)\right| \geq \check\kappa\, \nu, \end{aligned}$$

where $\check\kappa > 0$. Thus the consistency follows by the usual calculation:

$$\begin{aligned} \mathbf{P}_{\theta_0}\left(\left|\check\vartheta_T - \vartheta_0\right| \geq \nu\right) &= \mathbf{P}_{\theta_0}\left(\inf_{|\vartheta-\vartheta_0|\leq\nu} \left|R_{1,T} - \Psi(\vartheta)\right| \geq \inf_{|\vartheta-\vartheta_0|>\nu} \left|R_{1,T} - \Psi(\vartheta)\right|\right) \\ &\leq \mathbf{P}_{\theta_0}\left(2\left|R_{1,T} - \Psi(\vartheta_0)\right| \geq \inf_{|\vartheta-\vartheta_0|>\nu} |\Psi(\vartheta) - \Psi(\vartheta_0)|\right) \end{aligned}$$

$$\leq \frac{4}{g\left(\vartheta_0,\nu\right)^2}\mathbf{E}_{\vartheta_0}\left|\check\vartheta_T-\vartheta_0\right|^2\leq \frac{C}{Tg\left(\vartheta_0,\nu\right)^2}\to 0.$$

With probability tending to 1, the estimator $\check\vartheta_T$ is the solution of the equation

$$0=R_{1,T}-\Psi\left(\check\vartheta_T\right)=R_{1,T}-\Psi\left(\vartheta_0\right)-\left(\check\vartheta_T-\vartheta_0\right)\dot\Psi\left(\bar\vartheta_T\right)$$

where $\left|\bar\vartheta_T-\vartheta_0\right|\leq\left|\check\vartheta_T-\vartheta_0\right|$ and therefore

$$\sqrt{T}\left(\check\vartheta_T-\vartheta_0\right)=\frac{\sqrt{T}\left(R_{1,T}-\Psi\left(\vartheta_0\right)\right)}{\dot\Psi\left(\bar\vartheta_T\right)}.$$

Hence (see (4.11))

$$T^{p/2}\mathbf{E}_{\vartheta_0}\left|\check\vartheta_T-\vartheta_0\right|^p\leq\frac{T^{p/2}\mathbf{E}_{\vartheta_0}\left|R_{1,T}-\Psi\left(\vartheta_0\right)\right|^p}{\inf_{\vartheta\in\Theta}\left|\dot\Psi\left(\vartheta\right)\right|}\leq C.$$

□

Two-Dimensional Parameter $\vartheta=(a,f)$

Suppose that

$$\vartheta=\left(\theta_1,\theta_2\right)^\top=\left(a,f\right)^\top\in\Theta=\left(\alpha_a,\beta_a\right)\times\left(\alpha_f,\beta_f\right),\qquad \alpha_a>0\qquad \alpha_f>0.$$

The MME $\vartheta_T^*=\left(\theta_{1,T}^*,\theta_{2,T}^*\right)^\top=\left(a_T^*,f_T^*\right)^\top$ is defined as a solution $\vartheta=\vartheta_T^*$ to the system of equations

$$\Phi_1\left(\vartheta\right)=R_{1,T},\qquad \Phi_2\left(\vartheta\right)=R_{2,T},\qquad \vartheta\in\Theta$$

under the usual assumption that these equations have a solution $\vartheta_T^*\in\Theta$.

Recall that

$$\Phi_1\left(\vartheta\right)=\frac{f^2b^2}{a^3}\left[e^{-a}-1+a\right]+\sigma^2,\qquad \Phi_2\left(\vartheta\right)=\frac{f^2b^2}{2a^3}\left[1-e^{-a}\right]^2.$$

Therefore this system can be solved in two steps as follows: first, the equation

$$\frac{e^{-a_T^*}-1+a_T^*}{\left(1-e^{-a_T^*}\right)^2}=\frac{R_{1,T}-\sigma^2}{2R_{2,T}},\tag{4.13}$$

is solved and its solution a_T^* is used to define the second estimator

$$f_T^* = \frac{2(a_T^*)^{3/2} R_{2,T}^{1/2}}{b^2 \left(1 - e^{-a_T^*}\right)^2}. \tag{4.14}$$

Let us show that the function

$$h\left(x\right) = \frac{e^{-x} - 1 + x}{\left(1 - e^{-x}\right)^2}, \qquad x > 0$$

is strictly increasing. Its derivative is $h'\left(x\right) = \left(1 - e^{-x}\right)^{-3} g\left(x\right)$ where the function $g\left(x\right) = 1 - 2xe^{-x} - e^{-2x}$ vanishes at zero and has the derivative $g'\left(x\right) = 2e^{-x}\left(x - 1 + e^{-x}\right) > 0$ for $x > 0$. Therefore the first equation has a unique solution a_T^* and the second equation gives an explicit expression for the estimate f_T^*.

Note that

$$h\left(a_0\right) = \frac{\Phi_1\left(\vartheta_0\right) - \sigma^2}{2\Phi_2\left(\vartheta_0\right)}, \qquad \frac{R_{1,T} - \sigma^2}{2R_{2,T}} \longrightarrow h\left(a_0\right), \qquad a_T^* \longrightarrow a_0,$$

$$f_0 = \frac{2a_0^{3/2}\Phi_2\left(\vartheta_0\right)^{1/2}}{b^2\left(1 - e^{-a_0}\right)}, \qquad \frac{2(a_T^*)^{3/2} R_{2,T}^{1/2}}{b^2\left(1 - e^{-a_T^*}\right)^2} \longrightarrow f_0.$$

Therefore

$$h\left(a_T^*\right) = h\left(a_0\right) + h'\left(a_0\right)\left(a_T^* - a_0\right) + o\left(\left(a_T^* - a_0\right)^2\right) = \frac{R_{1,T} - \sigma^2}{2R_{2,T}}$$

and using the convergence (4.10) we obtain

$$\begin{aligned}
&\sqrt{T}\left(a_T^* - a_0\right)\\
&= h'\left(a_0\right)^{-1}\sqrt{T}\left(\frac{R_{1,T} - \sigma^2}{2R_{2,T}} - \frac{\Phi_1\left(\vartheta_0\right) - \sigma^2}{2\Phi_2\left(\vartheta_0\right)}\right) + o\left(\sqrt{T}\left(a_T^* - a_0\right)^2\right)\\
&= \frac{\sqrt{T}\left(R_{1,T} - \Phi_1\left(\vartheta_0\right)\right)}{2h'\left(a_0\right)\Phi_2\left(\vartheta_0\right)} - \frac{\left(\Phi_1\left(\vartheta_0\right) - \sigma^2\right)}{2h'\left(a_0\right)\Phi_2\left(\vartheta_0\right)}\sqrt{T}\left(R_{2,T} - \Phi_2\left(\vartheta_0\right)\right) + o\left(1\right)\\
&\Longrightarrow \frac{\xi_1}{2h'\left(a_0\right)\Phi_2\left(\vartheta_0\right)} - \frac{\left(\Phi_1\left(\vartheta_0\right) - \sigma^2\right)\xi_2}{2h'\left(a_0\right)\Phi_2\left(\vartheta_0\right)} \equiv \xi_3.
\end{aligned}$$

For the second estimator we have

$$\sqrt{T}\left(f_T^* - f_0\right) = \sqrt{T}\left(\frac{2\left(a_T^*\right)^{3/2} R_{2,T}^{1/2}}{b^2\left(1 - e^{-a_T^*}\right)^2} - \frac{2a_0^{3/2}\Phi_2\left(\vartheta_0\right)^{1/2}}{b^2\left(1 - e^{-a_0}\right)}\right)$$

$$= n'\left(a_0\right)\Phi_2\left(\vartheta_0\right)^{1/2}\sqrt{T}\left(a_T^* - a_0\right) + \frac{n\left(a_0\right)\sqrt{T}\left(R_{2,T} - \Phi_2\left(\vartheta_0\right)\right)}{2\Phi_2\left(\vartheta_0\right)^{1/2}} + o\left(1\right)$$

$$\Longrightarrow n'\left(a_0\right)\Phi_2\left(\vartheta_0\right)^{1/2}\xi_3 + \frac{n\left(a_0\right)}{2\Phi_2\left(\vartheta_0\right)^{1/2}}\xi_2 \equiv \xi_4.$$

Let $\xi^* = \left(\xi_3, \xi_4\right)^\top$ denote the Gaussian vector with the covariance matrix $\mathbf{Q}\left(\vartheta_0\right)$. We proved the following proposition.

Proposition 4.3 *The MME ϑ_T^* is uniformly consistent, asymptotically normal, uniformly on compacts $\mathbb{K} \subset \Theta$:*

$$\sqrt{T}\left(\vartheta_T^* - \vartheta_0\right) \Longrightarrow \xi^* \sim \mathcal{N}\left(0, \mathbf{Q}\left(\vartheta_0\right)\right),$$

the polynomial moments converge and the bound

$$\sup_{\vartheta_0 \in \mathbb{K}} T^{p/2}\mathbf{E}_{\vartheta_0}\left\|\vartheta_T^* - \vartheta_0\right\|^p \leq C \tag{4.15}$$

holds.

The proof of estimate (4.15) can be carried out using standard arguments.

Two-Dimensional Parameter $\boldsymbol{\vartheta = (a, b)}$

There is no real difference between estimation of f or b because the model depends on the product fb. Therefore for the definition of estimator a_T^* see (4.13) and the estimator b_T^* is almost the same as in (4.14), where we replace b with f

$$b_T^* = \frac{2(a_T^*)^{3/2} R_{2,T}^{1/2}}{f^2\left(1 - e^{-a_T^*}\right)^2}.$$

4.2.2 MLE and BE

Remind that we are given a two-dimensional diffusion process $\{X_t, Y_t,\ t \geq 0\}$

$$\mathrm{d}Y_t = -a\left(\vartheta\right)Y_t\,\mathrm{d}t + b\left(\vartheta\right)\,\mathrm{d}V_t, \qquad Y_0, \quad 0 \leq t \leq T,$$

$$\mathrm{d}X_t = \quad f(\vartheta)\, Y_t\, \mathrm{d}t + \sigma\, \mathrm{d}W_t, \qquad X_0 = 0,$$

where $\{V_t, W_t, t \geq 0\}$ are two independent Wiener processes, the functions $a(\cdot)$, $b(\cdot)$, $f(\cdot)$ and the constant $\sigma \neq 0$ are known and ϑ is a finite-dimensional (unknown) parameter. The random variable Y_0 is $\mathfrak{F}_0$-measurable and Gaussian.

This model of observations allows us to study estimators in the five different situations:

F. $f(\vartheta) = \vartheta, \quad a(\vartheta) = a, \quad b(\vartheta) = b,$
A. $f(\vartheta) = f, \quad a(\vartheta) = \vartheta, \quad b(\vartheta) = b,$
B. $f(\vartheta) = f, \quad a(\vartheta) = a, \quad b(\vartheta) = \vartheta,$
AB. $f(\vartheta) = f, \quad a(\vartheta) = \theta_1, \quad b(\vartheta) = \theta_2$, i.e., $\vartheta = (\theta_1, \theta_2)^\top$.
AF. $f(\vartheta) = \theta_2, \quad a(\vartheta) = \theta_1, \quad b(\vartheta) = b$, i.e., $\vartheta = (\theta_1, \theta_2)^\top$.

Let us denote $\mathbf{P}_\vartheta^{(T)}$ the measure induced in the measurable space of continuous on $[0, T]$ functions $\left(\mathcal{C}_T, \mathfrak{B}_T\right)$ by the process X^T and $\mathbf{P}^{(T)}$ the measure, which corresponds to the process X^T with $f(\vartheta) \equiv 0$. The likelihood ratio function

$$L\left(\vartheta, X^T\right) = \frac{\mathrm{d}\mathbf{P}_\vartheta^{(T)}}{\mathrm{d}\mathbf{P}^{(T)}}\left(X^T\right), \qquad \vartheta \in \Theta$$

has the following form:

$$L\left(\vartheta, X^T\right) = \exp\left\{\int_0^T \frac{M(\vartheta, t)}{\sigma^2}\, \mathrm{d}X_t - \int_0^T \frac{M(\vartheta, t)^2}{2\sigma^2} \mathrm{d}t\right\}.$$

The MLE $\hat{\vartheta}_T$ and BE $\tilde{\vartheta}_T$ are defined by the relations

$$L(\hat{\vartheta}_T, X^T) = \sup_{\vartheta \in \Theta} L\left(\vartheta, X^T\right), \qquad \tilde{\vartheta}_T = \frac{\int_\Theta \vartheta p(\vartheta)\, L\left(\vartheta, X^T\right) \mathrm{d}\vartheta}{\int_\Theta p(\vartheta)\, L\left(\vartheta, X^T\right) \mathrm{d}\vartheta}. \tag{4.16}$$

Here $M(\vartheta, t) = f(\vartheta)\, m(\vartheta, t)$ and $p(\cdot)$ is continuous positive density function on Θ.

To construct the MLE of the parameter ϑ we have to calculate the function $\{M(\vartheta, t),\ 0 \leq t \leq T\},\ \vartheta \in \Theta$ defined by the equation

$$\mathrm{d}M(\vartheta, t) = -r(\vartheta)\, M(\vartheta, t)\, \mathrm{d}t + \Gamma(\vartheta)\, \mathrm{d}X_t, \qquad M(\vartheta, 0). \tag{4.17}$$

It will be convenient to consider the cases **F**, **A**, **B** and **AB**, **AF** separately. Therefore we study first the estimation of one-dimensional parameter and then we consider the cases **AB**, **AF** (two-dimensional parameter).

One-Dimensional Parameter

Suppose that the parameter ϑ is one-dimensional and $\Theta = (\alpha, \beta)$. Introduce the quantity

$$\mathrm{I}(\vartheta) = \frac{\dot a(\vartheta)^2}{2a(\vartheta)} - \frac{2\dot a(\vartheta)\,\dot r(\vartheta)}{r(\vartheta) + a(\vartheta)} + \frac{\dot r(\vartheta)^2}{2r(\vartheta)}, \tag{4.18}$$

which will play the role of Fisher information. As before the dot means the derivation w.r.t. ϑ.

Conditions $\mathscr{J}$.

$\mathscr{J}_1$. *The functions $a(\vartheta)$, $b(\vartheta)$, $f(\vartheta)$, $\vartheta \in [\alpha, \beta]$ are twice continuously differentiable in $\vartheta \in \bar\Theta$.*

$\mathscr{J}_2$. *These functions satisfy the conditions: $b(\vartheta) \neq 0$, $f(\vartheta) \neq 0$ and $a(\vartheta) > 0$ for all $\vartheta \in [\alpha, \beta]$.*

$\mathscr{J}_3$. *For all $\nu > 0$ we have*

$$\inf_{\vartheta_0 \in \Theta} \inf_{|\vartheta - \vartheta_0| > \nu} \Big(|a(\vartheta) - a(\vartheta_0)| + |r(\vartheta) - r(\vartheta_0)| \Big) > 0. \tag{4.19}$$

$\mathscr{J}_4$. *The following condition holds*

$$\inf_{\vartheta \in \Theta} \Big(|\dot a(\vartheta)| + |\dot r(\vartheta)| \Big) > 0. \tag{4.20}$$

Under these regularity conditions we have the lower bound on the mean square risks of all estimators (Hájek's bound)

$$\lim_{\delta \to 0} \varliminf_{T \to \infty} \sup_{|\vartheta - \vartheta_0| \le \delta} T\mathbf{E}_\vartheta \left|\bar\vartheta_T - \vartheta\right|^2 \ge \mathrm{I}(\vartheta_0)^{-1}.$$

We call the estimator $\bar\vartheta_T$ asymptotically efficient if for all $\vartheta_0 \in \Theta$

$$\lim_{\delta \to 0} \lim_{T \to \infty} \sup_{|\vartheta - \vartheta_0| \le \delta} T\mathbf{E}_\vartheta \left|\bar\vartheta_T - \vartheta\right|^2 = \mathrm{I}(\vartheta_0)^{-1}.$$

Theorem 4.1 *Let the conditions $\mathscr{J}$ hold. Then the MLE $\hat\vartheta_T$ and BE $\tilde\vartheta_T$ are uniformly consistent, uniformly on compacts $\mathbb{K} \subset \Theta$ asymptotically normal*

$$\sqrt{T}\left(\hat\vartheta_T - \vartheta_0\right) \Longrightarrow \zeta \sim \mathcal{N}\left(0, \mathrm{I}(\vartheta_0)^{-1}\right), \quad \sqrt{T}\left(\tilde\vartheta_T - \vartheta_0\right) \Longrightarrow \zeta,$$

polynomial moments converge: for any $p \ge 2$

$$\lim_{T \to \infty} T^{p/2}\mathbf{E}_{\vartheta_0} \left|\hat\vartheta_T - \vartheta_0\right|^p = \mathbf{E}_{\vartheta_0} |\zeta|^p, \quad \lim_{T \to \infty} T^{p/2}\mathbf{E}_{\vartheta_0} \left|\tilde\vartheta_T - \vartheta_0\right|^p = \mathbf{E}_{\vartheta_0} |\zeta|^p,$$

and both estimators are asymptotically efficient.

Proof The proof of this theorem is based on two general results Theorems 1.10 and 1.12. Therefore we verify the conditions $\mathscr{K}_1$, i.e., we establish the corresponding properties of the normalized likelihood ratio process (below $\varphi_T = T^{-1/2}$)

$$Z_T(\vartheta_0, u) = \frac{L\left(\vartheta_0 + \varphi_T u, X^T\right)}{L\left(\vartheta_0, X^T\right)}, \qquad u \in \mathbb{U}_T = \left(\sqrt{T}(\alpha - \vartheta_0), \sqrt{T}(\beta - \vartheta_0)\right).$$

Note that under condition (4.20) $\inf_{\vartheta \in \Theta} \mathrm{I}(\vartheta) > 0$ because we have the following representation of the Fisher information

$$\mathrm{I}(\vartheta) = \left(\frac{\dot a(\vartheta)}{\sqrt{2a(\vartheta)}} - \frac{\dot r(\vartheta)}{\sqrt{2r(\vartheta)}}\right)^2 + \frac{2\dot a(\vartheta)\dot r(\vartheta)\left[\sqrt{a(\vartheta)} - \sqrt{r(\vartheta)}\right]^2}{\sqrt{a(\vartheta) r(\vartheta)}\left[a(\vartheta) + r(\vartheta)\right]},$$

where $\sqrt{r(\vartheta)} > \sqrt{a(\vartheta)}$ and if $\dot a(\vartheta)\dot r(\vartheta) < 0$, then from the condition (4.20) of this theorem and (4.18) follows that $\mathrm{I}(\vartheta) > 0$. □

Lemma 4.1 *The random process $Z_T(\vartheta_0, u)$, $u \in \mathbb{U}_T$ admits the representation*

$$Z_T(\vartheta_0, u) = \exp\left(u\Delta_T\left(\vartheta_0, X^T\right) - \frac{u^2}{2}\mathrm{I}(\vartheta_0) + \varepsilon_T\left(\vartheta_0, u, X^T\right)\right), \qquad u \in \mathbb{U}_T, \tag{4.21}$$

where $\varepsilon_T\left(\vartheta_0, u, X^T\right) \to 0$ and

$$\Delta_T\left(\vartheta_0, X^T\right) = \frac{\varphi_T}{\sigma^2}\int_0^T \dot M(\vartheta_0, t)\left[\mathrm{d}X_t - M(\vartheta_0, t)\,\mathrm{d}t\right] \Longrightarrow \mathcal{N}(0, \mathrm{I}(\vartheta_0)).$$

Both convergences are uniform on compacts $\mathbb{K}$, i.e., the family of measures is ULAN.

Proof We have ($\vartheta_u = \vartheta + \varphi_T u$)

$$\ln Z_T(\vartheta_0, u) = \int_0^T \frac{M(\vartheta_u, t) - M(\vartheta_0, t)}{\sigma}\mathrm{d}\bar W_t - \int_0^T \frac{\left(M(\vartheta_u, t) - M(\vartheta_0, t)\right)^2}{2\sigma^2}\mathrm{d}t.$$

To study the difference $M(\vartheta_u, t) - M(\vartheta_0, t)$ we write three equations

$$\mathrm{d}M(\vartheta_0, t) = -r(\vartheta_0)M(\vartheta_0, t)\,\mathrm{d}t + \Gamma(\vartheta_0)M(\vartheta_0, t)\,\mathrm{d}t + \Gamma(\vartheta_0)\sigma\,\mathrm{d}\bar W_t,$$

$$\mathrm{d}M(\vartheta_u, t) = -r(\vartheta_u)M(\vartheta_u, t)\,\mathrm{d}t + \Gamma(\vartheta_u)M(\vartheta_0, t)\,\mathrm{d}t + \Gamma(\vartheta_u)\sigma\,\mathrm{d}\bar W_t,$$

$$\mathrm{d}\dot M(\vartheta_0, t) = -r(\vartheta_0)\dot M(\vartheta_0, t)\,\mathrm{d}t - \dot a(\vartheta_0)M(\vartheta_0, t)\,\mathrm{d}t + \dot\Gamma(\vartheta_0)\sigma\,\mathrm{d}\bar W_t.$$

Then we denote $\delta(F_t, u) = F(\vartheta_u, t) - F(\vartheta_0, t) - u\varphi_T\dot F(\vartheta_0, t)$, where $F(\vartheta, t), 0 \le t \le T$ is some function and write the equation for $\delta(M_t, u)$

$$\mathrm{d}\delta\left(M_t,u\right)=-r\left(\vartheta_0\right)\delta\left(M_t,u\right)\mathrm{d}t+\delta\left(\Gamma,u\right)M\left(\vartheta_0,t\right)\mathrm{d}t-\delta\left(r,u\right)M\left(\vartheta_u,t\right)\mathrm{d}t$$
$$-u\varphi_T\dot r\left(\vartheta_0\right)\left[M\left(\vartheta_0,t\right)-M\left(\vartheta_u,t\right)\right]\mathrm{d}t+\delta\left(\Gamma,u\right)\sigma\mathrm{d}\bar W_t,$$
$$\delta_0\left(M,u\right)=0.$$

The solution of this equation is

$$\begin{aligned}\delta\left(M_t,u\right)&=\delta\left(\Gamma,u\right)\int_0^t e^{-r(\vartheta_0)(t-s)}M\left(\vartheta_0,s\right)\mathrm{d}s\\&\quad-\delta\left(r,u\right)\int_0^t e^{-r(\vartheta_0)(t-s)}M\left(\vartheta_u,s\right)\mathrm{d}s\\&\quad-u\varphi_T\dot r\left(\vartheta_0\right)\int_0^t e^{-r(\vartheta_0)(t-s)}\left[M\left(\vartheta_0,s\right)-M\left(\vartheta_u,s\right)\right]\mathrm{d}s\\&\quad+\delta\left(\Gamma,u\right)\int_0^t e^{-r(\vartheta_0)(t-s)}\sigma\mathrm{d}\bar W_s.\end{aligned}$$

Using the estimates $\left|\delta\left(\Gamma,u\right)\right|\leq Cu^2\varphi_T^2$, $\left|\delta\left(r,u\right)\right|\leq Cu^2\varphi_T^2$ and

$$\sup_{\vartheta_0\in\mathbb{K}}\mathbf{E}_{\vartheta_0}\left|M\left(\vartheta_0,s\right)-M\left(\vartheta_u,s\right)\right|^2\leq Cu^2\varphi_T^2 \tag{4.22}$$

by direct calculations we obtain the following inequality

$$\sup_{\vartheta_0\in\mathbb{K}}\mathbf{E}_{\vartheta_0}\delta\left(M_t,u\right)^2\leq Cu^4\varphi_T^4, \tag{4.23}$$

where the constant $C>0$ does not depend on t. Therefore

$$\int_0^T\left[M\left(\vartheta_u,t\right)-M\left(\vartheta_0,t\right)-u\varphi_T\dot M\left(\vartheta_0,t\right)\right]^2\mathrm{d}t\longrightarrow 0$$

and we have

$$\int_0^T\left[M\left(\vartheta_u,t\right)-M\left(\vartheta_0,t\right)\right]^2\mathrm{d}t=\frac{u^2}{T}\int_0^T\dot M\left(\vartheta_0,t\right)^2\mathrm{d}t\left(1+O\left(T^{-1/2}\right)\right).$$

The Gaussian process $\dot M\left(\vartheta_0,\cdot\right)$ has the representation

$$\begin{aligned}\dot M\left(\vartheta_0,t\right)&=\dot M\left(\vartheta,0\right)e^{-r(\vartheta_0)t}+\dot r\left(\vartheta_0\right)\sigma\int_0^t e^{-r(\vartheta_0)(t-s)}\mathrm{d}\bar W_s\\&\quad-\dot a\left(\vartheta_0\right)\sigma\int_0^t e^{-a(\vartheta_0)(t-s)}\mathrm{d}\bar W_s.\end{aligned}$$

The *stationary version* of this representation is

$$\dot{M}^*\left(\vartheta_0,t\right)=\dot{r}\left(\vartheta_0\right)\sigma\int_{-\infty}^{t}e^{-r(\vartheta_0)(t-s)}\mathrm{d}\bar{W}_s-\dot{a}\left(\vartheta_0\right)\sigma\int_{-\infty}^{t}e^{-a(\vartheta_0)(t-s)}\mathrm{d}\bar{W}_s.$$

Note that direct calculation gives us the equality

$$\sigma^{-2}\mathbf{E}_{\vartheta_0}\dot{M}^*\left(\vartheta_0,t\right)^2=\mathrm{I}\left(\vartheta_0\right).$$

The covariance function of $\dot{M}^*\left(\vartheta_0,t\right)^2$ satisfies the inequality

$$\mathbf{E}_{\vartheta_0}\left(\dot{M}^*\left(\vartheta_0,t\right)^2-\sigma^2\mathrm{I}\left(\vartheta_0\right)\right)\left(\dot{M}^*\left(\vartheta_0,s\right)^2-\sigma^2\mathrm{I}\left(\vartheta_0\right)\right)\leq Ce^{-c|t-s|}$$

with some positive constants c, C. Therefore $\dot{M}\left(\vartheta_0,\cdot\right)$ has ergodic properties and the estimate: for any $p\geq 2$ there exists a constant $C_p>0$ such that

$$\sup_{\vartheta_0\in\mathbb{K}}\mathbf{E}_{\vartheta_0}\left|\frac{1}{T}\int_0^T\left[\sigma^{-2}\dot{M}\left(\vartheta_0,t\right)^2-\mathrm{I}\left(\vartheta_0\right)\right]\mathrm{d}t\right|^p\leq\frac{C_p}{T^{p/2}}\tag{4.24}$$

holds.

The proof of this estimate follows from Lemma 1.6 (see also Proposition 18 in [109]).

Below we suppose that $M^*\left(\vartheta_0,t\right)=M\left(\vartheta_0,t\right)$ and $\dot{M}^*\left(\vartheta_0,t\right)=\dot{M}\left(\vartheta_0,t\right)$.

Let us denote $y_t=\sigma^{-2}\dot{M}\left(\vartheta_0,t\right)^2-\mathrm{I}\left(\vartheta_0\right)$. Then (4.24) follows from Lemma 1.6.

The uniform convergence (4.24) of the integral

$$\frac{1}{T\sigma^2}\int_0^T\dot{M}\left(\vartheta_0,t\right)^2\mathrm{d}t\longrightarrow\mathrm{I}\left(\vartheta_0\right)$$

is a sufficient condition for the uniform asymptotic normality of the stochastic integral

$$\Delta_T\left(\vartheta_0,X^T\right)=\frac{1}{\sigma\sqrt{T}}\int_0^T\dot{M}\left(\vartheta_0,t\right)\mathrm{d}\bar{W}_t\Longrightarrow\mathcal{N}\left(0,\mathrm{I}\left(\vartheta_0\right)\right).$$

Therefore we obtained the representation (4.21). □

Lemma 4.2 *There exists a constant* $C>0$ *such that*

$$\sup_{\vartheta_0\in\mathbb{K}}\mathbf{E}_{\vartheta_0}\left|Z_T\left(\vartheta_0,u_2\right)^{1/2}-Z_T\left(\vartheta_0,u_1\right)^{1/2}\right|^2\leq C\left|u_2-u_1\right|^2\tag{4.25}$$

Proof This inequality follows from the Lemma 1.1 and the estimate (4.22)

$$
\begin{aligned}
&\mathbf{E}_{\vartheta_0}\left|Z_T\left(\vartheta_0,u_2\right)^{1/2}-Z_T\left(\vartheta_0,u_1\right)^{1/2}\right|^2\\
&\quad\leq\frac{1}{16\sigma^2T}\int_0^T\mathbf{E}_{\vartheta_{u_2}}\left[M\left(\vartheta_{u_2}\right)-M\left(\vartheta_{u_1}\right)\right]^2\mathrm{d}t\\
&\quad+\frac{1}{16\sigma^2T}\int_0^T\mathbf{E}_{\vartheta_{u_1}}\left[M\left(\vartheta_{u_2}\right)-M\left(\vartheta_{u_1}\right)\right]^2\mathrm{d}t\\
&\quad\leq C\left|u_2-u_1\right|^2.
\end{aligned}
$$

□

Lemma 4.3 *For any compact $\mathbb{K}\subset\Theta$ there exists a constant $\kappa=\kappa\left(\mathbb{K}\right)>0$ such that for any $N>0$ we have*

$$
\sup_{\vartheta_0\in\mathbb{K}}\mathbf{P}_{\vartheta}\left(Z_T\left(\vartheta_0,u\right)>e^{-\kappa u^2}\right)\leq\frac{C}{|u|^N},\tag{4.26}
$$

where the constant $C=C\left(N\right)>0$.

Proof We can write (below $\delta_t\left(u\right)=M\left(\vartheta_u,t\right)-M\left(\vartheta_0,t\right)$ and $q>0$)

$$
\begin{aligned}
\mathbf{P}_{\vartheta_0}\left(Z_T\left(\vartheta_0,u\right)>e^{-\kappa u^2}\right)&=\mathbf{P}_{\vartheta_0}\left(\ln Z_T\left(\vartheta_0,u\right)>-\kappa u^2\right)\\
&=\mathbf{P}_{\vartheta_0}\left(\int_0^T\frac{q\delta_t\left(u\right)}{\sigma}\mathrm{d}\bar W_t-\int_0^T\frac{q\delta_t\left(u\right)^2}{2\sigma^2}\mathrm{d}t>-\kappa qu^2\right).
\end{aligned}\tag{4.27}
$$

Using the stationary representations for $M\left(\vartheta_u,t\right)$ and $M\left(\vartheta_0,t\right)$ we obtain the following representation for the difference of stationary versions of these processes: if $r\left(\vartheta_u\right)\neq a\left(\vartheta_0\right)$

$$
\begin{aligned}
M\left(\vartheta_u,t\right)-M\left(\vartheta_0,t\right)&=\frac{\Gamma\left(\vartheta_0\right)\sigma\left[a\left(\vartheta_0\right)-a\left(\vartheta_u\right)\right]}{r\left(\vartheta_u\right)-a_0}\int_{-\infty}^t e^{-a(\vartheta_0)(t-s)}\mathrm{d}\bar W_s\\
&\quad+\frac{\gamma\left(\vartheta_u\right)\sigma\left[r\left(\vartheta_u\right)-r\left(\vartheta_0\right)\right]}{r\left(\vartheta_u\right)-a\left(\vartheta_0\right)}\int_{-\infty}^t e^{-r(\vartheta_u)(t-s)}\mathrm{d}\bar W_s\\
&=A\left(\vartheta_u,\vartheta_0\right)\sigma\int_{-\infty}^t e^{-a(\vartheta_0)(t-s)}\mathrm{d}\bar W_s+B\left(\vartheta_u,\vartheta_0\right)\sigma\int_{-\infty}^t e^{-r(\vartheta_u)(t-s)}\mathrm{d}\bar W_s\\
&=\int_{-\infty}^t H\left(\vartheta_u,\vartheta_0,s,t\right)\mathrm{d}\bar W_s
\end{aligned}
$$

and if $r\left(\vartheta_u\right)=a\left(\vartheta_0\right)$ we have

$$M(\vartheta_u,t)-M(\vartheta_0,t)=\sigma\Gamma(\vartheta_u)\Gamma(\vartheta_0)\int_{-\infty}^{t}e^{-a(\vartheta_0)(t-s)}(t-s)\,\mathrm{d}\bar W_s$$
$$+\sigma\left[\Gamma(\vartheta_u)-\Gamma(\vartheta_0)\right]\int_{-\infty}^{t}e^{-a(\vartheta_0)(t-s)}\mathrm{d}\bar W_s.$$

Here

$$A(\vartheta,\vartheta_0)=\frac{\Gamma(\vartheta_0)\left[a(\vartheta_0)-a(\vartheta)\right]}{r(\vartheta)-a_0},\qquad B(\vartheta,\vartheta_0)=\frac{\Gamma(\vartheta)\left[r(\vartheta)-r(\vartheta_0)\right]}{r(\vartheta)-a(\vartheta_0)},$$
$$H(\vartheta,\vartheta_0,s,t)=A(\vartheta,\vartheta_0)\,e^{-a(\vartheta_0)(t-s)}+B(\vartheta,\vartheta_0)\,e^{-r(\vartheta_u)(t-s)}$$

Let $r(\vartheta_u)\neq a(\vartheta_0)$ then

$$Q_1(\vartheta_u,\vartheta_0)=\sigma^{-2}\mathbf{E}_{\vartheta_0}\left(M(\vartheta_u,t)-M(\vartheta_0,t)\right)^2=\int_0^t H(\vartheta,\vartheta_0,s,t)^2\,\mathrm{d}s=0$$

if $A(\vartheta_u,\vartheta_0)=0$ and $B(\vartheta_u,\vartheta_0)=0$. These equalities imply $a(\vartheta_u)=a(\vartheta_0)$ and $r(\vartheta_u)=r(\vartheta_0)$, but by condition $\mathscr{I}_3$ this is impossible.

If $r(\vartheta_u)=a(\vartheta_0)$, then

$$\sigma^{-2}\mathbf{E}_{\vartheta}\left(M(\vartheta_u,t)-M(\vartheta_0,t)\right)^2$$
$$=\frac{\Gamma(\vartheta_u)^2\Gamma(\vartheta_0)^2}{4a(\vartheta_0)^3}+\frac{\left[\Gamma(\vartheta_u)-\Gamma(\vartheta_0)\right]^2}{2a(\vartheta_0)}+\frac{\Gamma(\vartheta_u)\Gamma(\vartheta_0)\left[\Gamma(\vartheta_u)-\Gamma(\vartheta_0)\right]}{a(\vartheta_0)^2}$$
$$=\frac{\Gamma(\vartheta_u)^2\Gamma(\vartheta_0)^2}{2a(\vartheta_0)}\left(z^2+\frac{1}{a(\vartheta_0)}z+\frac{1}{2a(\vartheta_0)^2}\right)\equiv Q_2(\vartheta_u,\vartheta_0),$$

where we denoted

$$z=\frac{1}{\Gamma(\vartheta_0)}-\frac{1}{\Gamma(\vartheta_u)}.$$

Remark that the equation

$$z^2+\frac{1}{a(\vartheta_0)}z+\frac{1}{2a(\vartheta_0)^2}=0$$

has no real solutions. Hence the function $Q_2(\vartheta_u,\vartheta_0)>0$.

We see that the function $Q(\vartheta,\vartheta_0)=\sigma^{-2}\mathbf{E}_{\vartheta}\left(M(\vartheta,t)-M(\vartheta_0,t)\right)^2,\ \vartheta\in\Theta$ is continuous in ϑ and strictly positive for $\vartheta\neq\vartheta_0$. Hence for the values ϑ from the set

$$\mathbb{A}=\left(\vartheta_*\ :\ a\left(\vartheta_*\right)^2+\frac{f\left(\vartheta_*\right)^2 b\left(\vartheta_*\right)^2}{\sigma^2}=a\left(\vartheta_0\right)^2\right)$$

we have $Q_1\left(\vartheta_n,\vartheta_0\right)\longrightarrow Q_2\left(\vartheta_*,\vartheta_0\right)$ as $\vartheta_n\longrightarrow\vartheta_*$ and we can denote $Q\left(\vartheta,\vartheta_0\right)=Q_1\left(\vartheta,\vartheta_0\right)\mathbb{I}_{\{\vartheta\in\mathbb{A}^c\}}+Q_2\left(\vartheta,\vartheta_0\right)\mathbb{I}_{\{\vartheta\in\mathbb{A}\}}$. Moreover, we have

$$g\left(\vartheta_0,\nu\right)=\inf_{|\vartheta-\vartheta_0|\geq\nu}\sigma^{-2}\mathbf{E}_{\vartheta_0}\left(M\left(\vartheta,t\right)-M\left(\vartheta_0,t\right)\right)^2>0$$

for any (small) $\nu>0$. If $g\left(\vartheta_0,\nu\right)=0$, then there exists $\vartheta'\neq\vartheta_0$ such that $Q_1\left(\vartheta',\vartheta_0\right)=0$ or $Q_2\left(\vartheta',\vartheta_0\right)=0$. Let us denote $c_1=\inf_{\vartheta_0\in\Theta}g\left(\vartheta_0,\nu\right)>0$. Then for $\left|\vartheta_u-\vartheta_0\right|\geq\nu$

$$\frac{1}{\sigma^2}\int_0^T\mathbf{E}_{\vartheta_0}\left[M\left(\vartheta_u,t\right)-M\left(\vartheta_0,t\right)\right]^2\mathrm{d}t\geq c_1T\geq c_1\frac{u^2}{(\beta-\alpha)^2}=\kappa_1u^2.$$

Here we used the estimate $|u|\,T^{-1/2}\leq\beta-\alpha$.

Further, for $\left|\vartheta_u-\vartheta_0\right|\leq\nu$ and sufficiently small $\nu>0$

$$\frac{1}{T\sigma^2}\int_0^T\mathbf{E}_{\vartheta_0}\left[M\left(\vartheta_u,t\right)-M\left(\vartheta_0,t\right)\right]^2\mathrm{d}t\geq\frac{1}{2}\mathrm{I}\left(\vartheta_0\right)\,u^2\geq\kappa_2u^2,$$

where $\kappa_2>0$ can be chosen not depending on ϑ_0.

Therefore if we put $\kappa=\kappa_1\wedge\kappa_2$, then we obtain the estimate

$$\frac{1}{T\sigma^2}\int_0^T\mathbf{E}_{\vartheta_0}\left[M\left(\vartheta_u,t\right)-M\left(\vartheta_0,t\right)\right]^2\mathrm{d}t\geq\kappa u^2,\tag{4.28}$$

which holds for all $u\in\mathbb{U}_T$.

We have by (4.23) (law of large numbers)

$$\sup_{\vartheta_0\in\mathbb{K}}\mathbf{E}_{\vartheta_0}\left(\frac{1}{T}\int_0^T\left[\left(M\left(\vartheta,t\right)-M\left(\vartheta_0,t\right)\right)^2-Q\left(\vartheta,\vartheta_0\right)\sigma^2\right]\mathrm{d}t\right)^2\longrightarrow 0.$$

Hence

$$\frac{1}{T\sigma^2}\int_0^T\left(M\left(\vartheta,t\right)-M\left(\vartheta_0,t\right)\right)^2\mathrm{d}t=Q\left(\vartheta,\vartheta_0\right)+R_T,\quad R_T\rightarrow 0.$$

Moreover, the higher moments converge too.

Hence, for any $N\geq 1$ by (4.24)

$$\mathbf{E}_{\vartheta_0}\left|\frac{1}{\sqrt{T}}\int_0^T\left(\left[M\left(\vartheta,t\right)-M\left(\vartheta_0,t\right)\right]^2-Q\left(\vartheta,\vartheta_0\right)\right)\mathrm{d}t\right|^{2N}\leq C.$$

In this case we have as well

$$\sup_{\vartheta_0\in\mathbb{K}} \mathbf{E}_{\vartheta_0}\left|\frac{1}{\sqrt{T}}\int_0^T\left(\left[M\left(\vartheta,t\right)-M\left(\vartheta_0,t\right)\right]^2-Q\left(\vartheta,\vartheta_0\right)\right)\mathrm{d}t\right|^{2N}\leq C.$$

Let us denote $\bar\delta_t\left(u\right)^2=\mathbf{E}_\vartheta\left(m_t\left(\vartheta+T^{-1/2}u\right)-m_t\left(\vartheta\right)\right)^2$ and return to (4.27). We have

$$\begin{aligned}
&\int_0^T\frac{q\delta_t\left(u\right)}{\sigma}\mathrm{d}\bar W_t-\int_0^T\frac{q\delta_t\left(u\right)^2}{2\sigma^2}\mathrm{d}t+\int_0^T\frac{(q-q^2)\bar\delta_t\left(u\right)^2}{2\sigma^2}\mathrm{d}t\\
&\quad=\int_0^T\frac{q\delta_t\left(u\right)}{\sigma}\mathrm{d}\bar W_t-\int_0^T\frac{q^2\delta_t\left(u\right)^2}{2\sigma^2}\mathrm{d}t-\int_0^T\frac{\left(q-q^2\right)\left[\delta_t\left(u\right)^2-\bar\delta_t\left(u\right)^2\right]}{2\sigma^2}\mathrm{d}t.
\end{aligned}$$

The last integral we denote as $D_T\left(u\right)$.

Then

$$\begin{aligned}
&\mathbf{P}_\vartheta\left(\int_0^T\frac{q\delta_t\left(u\right)}{\sigma}\mathrm{d}\bar W_t-\int_0^T\frac{q\delta_t\left(u\right)^2}{2\sigma^2}\mathrm{d}t\geq -qcu^2\right)\\
&\quad=\mathbf{P}_\vartheta\left(\int_0^T\frac{q\delta_t\left(u\right)}{\sigma}\mathrm{d}\bar W_t-\int_0^T\frac{q^2\delta_t\left(u\right)^2}{2\sigma^2}\mathrm{d}t-D_T\left(u\right)\right.\\
&\qquad\qquad\left.\geq\int_0^T\frac{(q-q^2)\bar\delta_t\left(u\right)^2}{2\sigma^2}\mathrm{d}t-qcu^2\right)\\
&\quad\leq\mathbf{P}_\vartheta\left(\int_0^T\frac{q\delta_t\left(u\right)}{\sigma}\mathrm{d}\bar W_t-\int_0^T\frac{q^2\delta_t\left(u\right)^2}{2\sigma^2}\mathrm{d}t-D_T\left(u\right)\right.\\
&\qquad\qquad\left.\geq\left[\frac{(q-q^2)\kappa}{\sigma^2}-qc\right]u^2\right).
\end{aligned}$$

Let us take such c and q that $2\kappa_*=(q-q^2)\kappa\sigma^{-2}-qc>0$. Then

$$\begin{aligned}
&\mathbf{P}_\vartheta\left(\int_0^T\frac{q\delta_t\left(u\right)}{\sigma}\mathrm{d}\bar W_t-\int_0^T\frac{q^2\delta_t\left(u\right)^2}{2\sigma^2}\mathrm{d}t-D_T\left(u\right)\geq 2\kappa_*u^2\right)\\
&\quad\leq\mathbf{P}_\vartheta\left(\int_0^T\frac{q\delta_t\left(u\right)}{\sigma}\mathrm{d}\bar W_t-\int_0^T\frac{q^2\delta_t\left(u\right)^2}{2\sigma^2}\mathrm{d}t\geq\kappa_*u^2\right)\\
&\qquad+\mathbf{P}_\vartheta\left(\left|D_T\left(u\right)\right|\geq\kappa_*u^2\right)
\end{aligned}$$

$$\leq e^{-\kappa_* u^2} \mathbf{E} \exp\left(\int_0^T \frac{q\delta_t(u)}{\sigma} d\bar{W}_t - \int_0^T \frac{q^2\delta_t(u)^2}{2\sigma^2} dt\right) + \frac{\mathbf{E}\left|D_T(u)\right|^N}{\left(\kappa_* u^2\right)^N}$$

$$\leq e^{-\kappa_* u^2} + \frac{C}{|u|^{2N}} \leq \frac{C}{|u|^{2N}}.$$

□

Having uniform on compacts LAN and two estimates (4.25) and (4.26) we can apply the same general theorems and obtain the desired properties of the MLE and BE described in Theorem 4.1.

Note as well that LAN gives us the lower bound on the risks of all estimators, and the convergence of moments together with the continuity of the Fisher information provide the asymptotic efficiency of the MLE and BE.

Case F, *i.e.*, $\vartheta = f$

Consider the particular case $\vartheta = f$. Then $r(\vartheta) = \left(a^2 + b^2\vartheta^2\sigma^{-2}\right)^{1/2}$ and the Fisher information is

$$\mathrm{I}(\vartheta) = \frac{b^4\vartheta^2}{2\sigma^4 r(\vartheta)^3}.$$

Hence by Theorem 4.1 the MLE and BE are consistent, asymptotically normal

$$\sqrt{T}\left(\hat{\vartheta}_T - \vartheta_0\right) \Longrightarrow \mathcal{N}\left(0, \frac{2\sigma^4 r(\vartheta_0)^3}{b^4\vartheta_0^2}\right)$$

and asymptotically efficient.

Case A, *i.e.*, $\vartheta = a$

We have $r(\vartheta) = \left(\vartheta^2 + b^2 f^2\sigma^{-2}\right)^{1/2}$ and the Fisher information is

$$\mathrm{I}(\vartheta) = \frac{\left[r(\vartheta)^2 - \vartheta^2\right]^2 + r(\vartheta)\,\vartheta\left[r(\vartheta) - \vartheta\right]^2}{2\vartheta r(\vartheta)^3\left[r(\vartheta) + \vartheta\right]} = \frac{\Gamma(\vartheta)^2\left[\left(r(\vartheta) + \vartheta\right)^2 + r(\vartheta)\,\vartheta\right]}{2\vartheta r(\vartheta)^3\left[r(\vartheta) + \vartheta\right]}, \tag{4.29}$$

both estimators are asymptotically normal and asymptotically efficient. Say,

$$\sqrt{T}\left(\hat{\vartheta}_T - \vartheta_0\right) \Longrightarrow \mathcal{N}\left(0, \mathrm{I}(\vartheta_0)^{-1}\right).$$

Case B, *i.e.*, $\vartheta = b$

The Fisher information is

$$\mathrm{I}(\vartheta) = \frac{\vartheta^2 f^4}{2r(\vartheta)^3 \sigma^4} \tag{4.30}$$

and the MLE and BE have the corresponding asymptotic properties.

Case AB, i.e., $\vartheta = (\theta_1, \theta_2,)^\top$, $\vartheta_0 = (\theta_{1,0}, \theta_{2,0},)^\top$, $\theta_{1,0} = a_0$, $\theta_{2,0} = b_0$

As before, we suppose that $\vartheta \in \Theta = (\alpha_a, \beta_a) \times (\alpha_b, \beta_b)$, where $\alpha_a > 0, \alpha_b > 0$. The system is

$$\begin{aligned} \mathrm{d}X_t &= f Y_t \mathrm{d}t + \sigma \mathrm{d}W_t, \qquad X_0 = 0, \qquad t \geq 0 \\ \mathrm{d}Y_t &= -\theta_1 Y_t \mathrm{d}t + \theta_2 \mathrm{d}V_t, \qquad Y_0, \end{aligned}$$

and the observations are $X^T = (X_t, 0 \leq t \leq T)$.

To calculate Fisher information matrix $\mathbf{I}(\vartheta)$ we first note that the values $\mathrm{I}_{1,1}(\vartheta)$ and $\mathrm{I}_{2,2}(\vartheta)$ were already calculated above (see (4.29) and (4.30)) and for the calculation of the terms $\mathrm{I}_{1,2}(\vartheta) = \mathrm{I}_{2,1}(\vartheta)$ we use the stationary representation for the derivatives

$$\begin{aligned} \dot{M}_a^*(\vartheta_0, t) &= \frac{\theta_{1,0}\sigma}{r(\vartheta_0)} \int_{-\infty}^t e^{-r(\vartheta_0)(t-s)} \mathrm{d}\bar{W}_s - \sigma \int_{-\infty}^t e^{-a(\vartheta_0)(t-s)} \mathrm{d}\bar{W}_s, \\ \dot{M}_b^*(\vartheta_0, t) &= \frac{f^2 \theta_{2,0}}{r(\vartheta_0)\sigma} \int_{-\infty}^t e^{-r(\vartheta_0)(t-s)} \mathrm{d}\bar{W}_s. \end{aligned}$$

It gives us the value

$$\mathrm{I}_{1,2}(\vartheta_0) = \sigma^{-2} \mathbf{E}_{\vartheta_0} \dot{M}_a^*(\vartheta_0, t)\, \dot{M}_b^*(\vartheta_0, t) = \frac{f^2 \theta_{2,0} \left(\theta_{1,0}^2 + r(\vartheta_0)\theta_{1,0} - 2r(\vartheta_0)^2\right)}{2r(\vartheta_0)^3 \left(r(\vartheta_0) + \theta_{1,0}\right)\sigma^2}.$$

Therefore Fisher information matrix $\mathbf{I}(\vartheta_0)$ is

$$\begin{pmatrix} \dfrac{\Gamma(\vartheta_0)^2 \left[r(\vartheta_0)\theta_{1,0} + \left(r(\vartheta_0) + \theta_{1,0}\right)^2\right]}{2\theta_{1,0} r(\vartheta_0)^3 \left[r(\vartheta_0) + \theta_{1,0}\right]}, & \dfrac{f^2 \theta_{2,0} \left(\theta_{1,0}^2 + r(\vartheta_0)\theta_{1,0} - 2r(\vartheta_0)\right)}{2r(\vartheta_0)^3 \left(r(\vartheta_0) + \theta_{1,0}\right)\sigma^2} \\ \dfrac{f^2 \theta_{2,0} \left(\theta_{1,0}^2 + r(\vartheta_0)\theta_{1,0} - 2r(\vartheta_0)\right)}{2r(\vartheta_0)^3 \left(r(\vartheta_0) + \theta_{1,0}\right)\sigma^2}, & \dfrac{\theta_{2,0}^2 f^4}{2r(\vartheta_0)^3 \sigma^4} \end{pmatrix}. \tag{4.31}$$

The determinant of the Fisher information matrix with a slightly simplified but obvious notation can be calculated as follows:

$$
\begin{aligned}
&\mathrm{Det}\left(\mathbf{I}\left(\vartheta_0\right)\right)\\
&=\frac{b^2 f^4}{4ar^6\left(r+a\right)^2\sigma^4}\left[\left(r-a\right)^2\left(ar+\left(r+a\right)^2\right)\left(r+a\right)-a\left(a^2+ar-2r^2\right)^2\right]\\
&=\frac{b^2 f^4\left(r-a\right)^2}{4ar^6\left(r+a\right)^2\sigma^4}\left[\left(ar+\left(r+a\right)^2\right)\left(r+a\right)-a\left(a+r+r\right)^2\right]\\
&=\frac{b^2 f^4\left(r-a\right)^2}{4ar^6\left(r+a\right)^2\sigma^4}\left[\left(r+a\right)\left\{ar+\left(r+a\right)^2-a\left(r+a\right)-2ar\right\}-ar^2\right]\\
&=\frac{b^2 f^4\left(r-a\right)^2}{4ar^3\left(r+a\right)^2\sigma^4}=\frac{\theta_{2,0}^2 f^4\Gamma\left(\vartheta_0\right)^2}{4\theta_{1,0}r\left(\vartheta_0\right)^3\left(r\left(\vartheta_0\right)+\theta_{1,0}\right)^2\sigma^4}.
\end{aligned}
\tag{4.32}
$$

Therefore

$$
\inf_{\vartheta_0\in\Theta}\mathrm{Det}\big(\mathbf{I}\left(\vartheta_0\right)\big)>0
$$

and the Fisher information matrix is uniformly nondegenerate

$$
\inf_{\vartheta_0\in\Theta}\ \inf_{\mathbf{e}\in\mathcal{R}^2,\|\mathbf{e}\|=1}\mathbf{e}^\top\mathbf{I}\left(\vartheta_0\right)\mathbf{e}>0.
$$

The Hajek's bound is

$$
\lim_{\nu\to0}\varliminf_{T\to\infty}\sup_{\|\vartheta-\vartheta_0\|\leq\nu}T\mathbf{E}_\vartheta\left\|\bar\vartheta_T-\vartheta\right\|^2\geq\mathbf{E}_{\vartheta_0}\left\|\zeta\right\|^2,\qquad\zeta\sim\mathcal{N}\left(0,\mathbf{I}\left(\vartheta_0\right)^{-1}\right)
\tag{4.33}
$$

and asymptotically efficient estimators are defined as estimators for which we have equality here for all $\vartheta_0\in\Theta$.

Theorem 4.2 *The MLE $\hat\vartheta_T$ and BE $\tilde\vartheta_T$ are uniformly consistent, uniformly on compacts $\mathbb{K}\subset\Theta$ asymptotically normal*

$$
\sqrt{T}\left(\hat\vartheta_T-\vartheta_0\right)\Longrightarrow\zeta,\quad\sqrt{T}\left(\tilde\vartheta_T-\vartheta_0\right)\Longrightarrow\zeta,
$$

polynomial moments converge: for any $p\geq2$

$$
\lim_{T\to\infty}T^{p/2}\mathbf{E}_{\vartheta_0}\left\|\hat\vartheta_T-\vartheta_0\right\|^p=\mathbf{E}_{\vartheta_0}\left\|\zeta\right\|^p,\ \lim_{T\to\infty}T^{p/2}\mathbf{E}_{\vartheta_0}\left\|\tilde\vartheta_T-\vartheta_0\right\|^p=\mathbf{E}_{\vartheta_0}\left\|\zeta\right\|^p,
$$

and both estimators are asymptotically efficient, i.e., these estimators satisfy the equality in (4.33) *for all* $\vartheta_0 \in \Theta$.

Proof The proof of this theorem consists in the verification of the following two-dimensional versions of Lemmas 4.1, 4.2, and 4.3:

1. For any $\mathbf{u} = (u_1, u_2)^\top \in \mathbb{U}_T$

$$Z_T(\vartheta_0, \mathbf{u}) = \frac{L\left(\theta_{1,0} + u_1\varphi_T, \theta_{2,0} + u_2\varphi_T, X^T\right)}{L\left(\theta_{1,0}, \theta_{2,0}, X^T\right)}$$
$$= \exp\left(\mathbf{u}^\top \Delta_T\left(\vartheta_0, X^T\right) - \frac{1}{2}\mathbf{u}^\top \mathbf{I}(\vartheta_0)\,\mathbf{u} + \varepsilon_T\left(\vartheta_0, \mathbf{u}, X^T\right)\right), \tag{4.34}$$

where $\varepsilon_T\left(\vartheta_0, \mathbf{u}, X^T\right) \to 0$ uniformly on compacts $\mathbb{K} \in \Theta$ and the vector

$$\Delta_T\left(\vartheta_0, X^T\right) = \frac{1}{\sigma\sqrt{T}}\int_0^T \dot{\mathbb{M}}(\vartheta_0, t)\,\mathrm{d}\bar{W}_t \Longrightarrow \mathcal{N}(0, \mathbf{I}(\vartheta_0)). \tag{4.35}$$

2. There exist a constant $C > 0$ such that for any $R > 0$ and $\|\mathbf{u}_i\| \leq R, i = 1, 2$ the estimate

$$\sup_{\vartheta_0 \in \mathbb{K}} \mathbf{E}_{\vartheta_0}\left|Z_T(\vartheta_0, \mathbf{u}_2)^{1/4} - Z_T(\vartheta_0, \mathbf{u}_1)^{1/4}\right|^4 \leq C\left(1 + R^4\right)\|\mathbf{u}_2 - \mathbf{u}_1\|^4$$

holds.

3. There exist a constant $\kappa > 0$ such that for any $N > 0$

$$\sup_{\vartheta_0 \in \mathbb{K}} \mathbf{P}_{\vartheta_0}\left(\vartheta_0, Z_T(\mathbf{u}) > e^{-\kappa\|\mathbf{u}\|^2}\right) \leq \frac{C}{\|\mathbf{u}\|^N} \tag{4.36}$$

where the constant $C = C(N) >$.

To prove (4.34) we repeat the same calculations as in the one-dimensional case and for $\delta(M_t, \mathbf{u}) = M(\vartheta_{\mathbf{u}}, t) - M(\vartheta_0, t) - \varphi_T \mathbf{u}^\top \dot{M}(\vartheta_0, t)$ obtain the estimate like (4.22)

$$\sup_{\vartheta_0 \in \mathbb{K}} \mathbf{E}_{\vartheta_0}\delta(M_t, \mathbf{u})^2 \leq \frac{C}{T^2}\|\mathbf{u}\|^4.$$

By the law of large numbers, which can be verified by calculations like (4.24), we obtain the limit in probability

$$\frac{1}{T}\int_0^T \dot{M}(\vartheta_0, t)\,\dot{M}(\vartheta_0, t)^\top \mathrm{d}t \longrightarrow \mathbf{I}(\vartheta_0).$$

This limit is a sufficient condition for the asymptotic normality (4.35). These relations provide the representation (4.34).

By Lemma 1.1, estimate like (4.23), and condition $\|\mathbf{u}_i\| \leq R$ we have the estimate

$$\begin{aligned}
&\mathbf{E}_{\vartheta_0} \left| Z_T\left(\vartheta_0, \mathbf{u}_2\right)^{1/4} - Z_T\left(\vartheta_0, \mathbf{u}_1\right)^{1/4}\right|^4 \\
&\quad \leq CT^3 \mathbf{E}_{\vartheta_{\mathbf{u}_2}} \int_0^T \left[M\left(\vartheta_{\mathbf{u}_2}, t\right) - M\left(\vartheta_{\mathbf{u}_1}, t\right)\right]^8 \mathrm{d}t \\
&\quad + CT^3 \mathbf{E}_{\vartheta_{\mathbf{u}_1}} \int_0^T \left[M\left(\vartheta_{\mathbf{u}_1}, t\right) - M\left(\vartheta_{\mathbf{u}_1}, t\right)\right]^8 \mathrm{d}t \\
&\quad + CT \mathbf{E}_{\vartheta_{\mathbf{u}_2}} \int_0^T \left[M\left(\vartheta_{\mathbf{u}_1}, t\right) - M\left(\vartheta_{\mathbf{u}_1}, t\right)\right]^4 \mathrm{d}t \\
&\quad + CT \mathbf{E}_{\vartheta_{\mathbf{u}_1}} \int_0^T \left[M\left(\vartheta_{\mathbf{u}_1}, t\right) - M\left(\vartheta_{\mathbf{u}_1}, t\right)\right]^4 \mathrm{d}t \\
&\quad \leq CT^4 \varphi_T^8 \left\|\mathbf{u}_2 - \mathbf{u}_1\right\|^8 + CT^2 \varphi_T^4 \left\|\mathbf{u}_2 - \mathbf{u}_1\right\|^4 \\
&\quad \leq C\left(1 + R^4\right) \left\|\mathbf{u}_2 - \mathbf{u}_1\right\|^4 .
\end{aligned}$$

To verify the estimate (4.36) we once more repeat the corresponding proof of (4.26).

Locally for $\|\mathbf{u}\| \leq \nu\sqrt{T}$ we use the nondegeneracy of the Fisher information and write

$$\begin{aligned}
\sigma^{-2} \mathbf{E}_{\vartheta_0} \int_0^T \left[M\left(\vartheta_{\mathbf{u}}, t\right) - M\left(\vartheta_0, t\right)\right]^2 \mathrm{d}t &= \mathbf{u}^\top \mathbf{I}\left(\vartheta_0\right) \mathbf{u}\left(1 + o\left(1\right)\right) \\
&\geq \frac{1}{2} \mathbf{u}^\top \mathbf{I}\left(\vartheta_0\right) \mathbf{u} = \frac{1}{2} \|\mathbf{u}\|^2 \inf_{\|\mathbf{e}\|=1, \mathbf{e}\in\mathcal{R}^2} \mathbf{I}\left(\vartheta_0\right) \geq \kappa_1 \|\mathbf{u}\|^2 .
\end{aligned}$$

To obtain the second estimate we remark that the condition (4.19) is always fulfilled, because, if $|a - a_0| + |r(\vartheta) - r(\vartheta_0)| = 0$, then this implies the equalities $a = a_0$ and $b = b_0$. Therefore we have the same estimate

$$2\kappa_2 = \inf_{|\vartheta_{\mathbf{u}} - \vartheta_0| > \nu} Q_1\left(\vartheta_u, \vartheta_0\right) + \inf_{|\vartheta_{\mathbf{u}} - \vartheta_0| > \nu} Q_2\left(\vartheta_u, \vartheta_0\right) > 0$$

and the corresponding lower estimate

$$\mathbf{E}_{\vartheta_0} \int_0^T \left[M\left(\vartheta_{\mathbf{u}}, t\right) - M\left(\vartheta_0, t\right)\right]^2 \mathrm{d}t \geq \kappa \|\mathbf{u}\|^2$$

see (4.28).

The rest of the proof coincides with the proof of Theorem 4.1. □

Case AF, i.e., $\vartheta = (\theta_1, \theta_2)$, $\theta_1 = a, \theta_2 = f$

The partially observed system is

$$\begin{aligned} \mathrm{d}X_t &= \theta_2 Y_t \mathrm{d}t + \sigma\, \mathrm{d}W_t, \qquad X_0 = 0, \qquad 0 \leq t \leq T, \\ \mathrm{d}Y_t &= -\theta_1 Y_t \mathrm{d}t + b\, \mathrm{d}V_t, \qquad Y_0. \end{aligned}$$

The derivatives of $M^*(\vartheta, t)$ are almost the same as above

$$\begin{aligned} \dot M_a^*(\vartheta_0, t) &= \frac{\theta_{1,0}\sigma}{r(\vartheta_0)} \int_{-\infty}^t e^{-r(\vartheta_0)(t-s)} \mathrm{d}\bar W_s - \sigma \int_{-\infty}^t e^{-a(\vartheta_0)(t-s)} \mathrm{d}\bar W_s, \\ \dot M_f^*(\vartheta_0, t) &= \frac{b^2\theta_{2,0}}{r(\vartheta_0)\sigma} \int_{-\infty}^t e^{-r(\vartheta_0)(t-s)} \mathrm{d}\bar W_s. \end{aligned}$$

Therefore Fisher information matrix $\mathbf{I}(\vartheta_0)$ is almost the same too

$$\begin{pmatrix} \dfrac{\Gamma(\vartheta_0)^2\left[r(\vartheta_0)\theta_{1,0} + \left(r(\vartheta_0) + \theta_{1,0}\right)^2\right]}{2\theta_{1,0} r(\vartheta_0)^3\left[r(\vartheta_0) + \theta_{1,0}\right]}, & \dfrac{b^2\theta_{2,0}\left(\theta_{1,0}^2 + r(\vartheta_0)\theta_{1,0} - 2r(\vartheta_0)\right)}{2r(\vartheta_0)^3\left(r(\vartheta_0) + \theta_{1,0}\right)\sigma^2} \\ \dfrac{b^2\theta_{2,0}\left(\theta_{1,0}^2 + r(\vartheta_0)\theta_{1,0} - 2r(\vartheta_0)\right)}{2r(\vartheta_0)^3\left(r(\vartheta_0) + \theta_{1,0}\right)\sigma^2}, & \dfrac{\theta_{2,0}^2 b^4}{2r(\vartheta_0)^3\sigma^4} \end{pmatrix}.$$

Due to the similitude of these two models of observations there is no need to present a proof of the following theorem.

Theorem 4.3 *The MLE $\hat\vartheta_T$ and BE $\tilde\vartheta_T$ are uniformly consistent, uniformly on compacts $\mathbb{K} \subset \Theta$ asymptotically normal*

$$\sqrt{T}\left(\hat\vartheta_T - \vartheta_0\right) \Longrightarrow \zeta, \quad \sqrt{T}\left(\tilde\vartheta_T - \vartheta_0\right) \Longrightarrow \zeta,$$

polynomial moments converge: for any $p \geq 2$

$$\begin{aligned} \lim_{T\to\infty} T^{p/2}\mathbf{E}_{\vartheta_0}\left\|\hat\vartheta_T - \vartheta_0\right\|^p &= \mathbf{E}_{\vartheta_0}\|\zeta\|^p, \\ \lim_{T\to\infty} T^{p/2}\mathbf{E}_{\vartheta_0}\left\|\tilde\vartheta_T - \vartheta_0\right\|^p &= \mathbf{E}_{\vartheta_0}\|\zeta\|^p, \end{aligned}$$

and both estimators are asymptotically efficient, i.e., these estimators satisfy the equality in the corresponding modification of (4.33) *for all $\vartheta_0 \in \Theta$.*

4.2.3 One-Step MLE-Process

Let us return to the partially observed system

$$
\begin{aligned}
\mathrm{d}X_t &= f(\vartheta)\, Y_t\, \mathrm{d}t + \sigma\, \mathrm{d}W_t, \qquad X_0, \qquad 0 \le t \le T,\\
\mathrm{d}Y_t &= -a(\vartheta)\, Y_t\, \mathrm{d}t + b(\vartheta)\, \mathrm{d}V_t, \qquad Y_0
\end{aligned}
$$

and remind the problem. The initial values X_0, Y_0 are Gaussian and do not depend of the independent Wiener processes W^T and V^T. At the beginning the unknown parameter ϑ is one-dimensional, $\vartheta \in \Theta = (\alpha, \beta)$ and its estimation is to be based on the sample $X^T = (X_t, 0 \le t \le T)$. The MLE and BE described in Theorem 4.1 have convenient asymptotic properties and, moreover, are asymptotically efficient. However, as explained above, they are computationally inefficient. For example, the construction of the BE (4.16) requires knowledge of $L\left(\vartheta, X^T\right)$ for all $\vartheta \in \Theta$. Even if we assume that the system is in the stationary regime with the initial value $m(\vartheta, 0) \sim \mathcal{N}(0, \gamma_*(\vartheta))$, so that solving the Riccati equation can be avoided, we will still need the trajectories $m(\vartheta, t), 0 \le t \le T, \vartheta \in \Theta$.

The goal of this section is to study One-step MLE-process which enjoys many good properties. We will slightly simplify the calculations by assuming that the system is already in the stationary regime, i.e., the initial value $\gamma(\vartheta, 0)$ of the Riccati equation is $\gamma(\vartheta, 0) = \gamma_*(\vartheta)$ and therefore $\gamma(\vartheta, t) \equiv \gamma_*(\vartheta)$ for all $t \ge 0$. Moreover, we define $M(\vartheta, t) = f(\vartheta) m(\vartheta, t)$. Recall that $\Gamma(\vartheta) = \gamma_*(\vartheta) f(\vartheta)^2 \sigma^{-2} = r(\vartheta) - a(\vartheta), r(\vartheta) = \left(a(\vartheta)^2 + b(\vartheta)^2 f(\vartheta)^2 \sigma^{-2}\right)^{-1/2}$. Then the equations for $M(\vartheta, t), t \ge 0$ and $\dot{M}(\vartheta, t), t \ge 0$ read

$$
\begin{aligned}
\mathrm{d}M(\vartheta, t) &= -r(\vartheta)\, M(\vartheta, t)\, \mathrm{d}t + \Gamma(\vartheta)\, \mathrm{d}X_t,\\
\mathrm{d}\dot{M}(\vartheta, t) &= -r(\vartheta)\, \dot{M}(\vartheta, t)\, \mathrm{d}t - \dot{r}(\vartheta)\, M(\vartheta, t)\, \mathrm{d}t + \dot{\Gamma}(\vartheta)\, \mathrm{d}X_t,
\end{aligned}
$$

subject to the initial values $M(\vartheta, 0)$ and $\dot{M}(\vartheta, 0)$. The integral stationary presentation of $M(\vartheta, t), t \ge 0$ is given in (4.7). The integral presentation for the derivative $\dot{M}(\vartheta, t), t \ge 0$ can be obtained by direct differentiation of Eq. (4.7). The initial values $M(\vartheta, 0), \dot{M}(\vartheta, 0)$ can also be derived from these representations.

The One-step MLE-process is constructed as usual by means of the following two-stage procedure. First, we construct an MME $\vartheta^*_{\tau_T}$ using the observations X^{τ_T} on a relatively small *learning* interval $\left[0, \tau_T\right], \tau_T = \left[T^\delta\right], \delta \in \left(\frac{1}{2}, 1\right)$. The scoring method is then applied to this preliminary estimator to define the One-step MLE-process. Here $[a]$ stands for the integer part of $a \in \mathcal{R}$.

Define

$$
R_{1,\tau_T} = \frac{1}{\tau_T} \sum_{k=1}^{\tau_T} \left[X_k - X_{k-1}\right]^2, \quad \Psi(\vartheta) = \frac{f(\vartheta)^2 b(\vartheta)^2}{a(\vartheta)^2} \left(e^{a(\vartheta)} - 1 + a(\vartheta)\right) + \sigma^2,
$$

$$\vartheta^*_{\tau_T} = \arg\inf_{\vartheta\in\Theta} \left|R_{1,\tau_T} - \Psi(\vartheta)\right|, \qquad \mathrm{I}(\vartheta) = \frac{\dot a(\vartheta)^2}{2a(\vartheta)} - \frac{2\dot a(\vartheta)\,\dot r(\vartheta)}{a(\vartheta)+r(\vartheta)} + \frac{\dot r(\vartheta)^2}{2r(\vartheta)}$$

and the estimator-process

$$\vartheta^\star_{t,T} = \vartheta^*_{\tau_T} + \frac{1}{\mathrm{I}(\vartheta^*_{\tau_T})\left(t-\tau_T\right)} \int_{\tau_T}^{t} \frac{\dot M(\vartheta^*_{\tau_T}, s)}{\sigma^2}\left[\mathrm{d}X_s - M(\vartheta^*_{\tau_T}, s)\mathrm{d}s\right], \quad \tau_T < t \le T. \tag{4.37}$$

The random processes $M(\vartheta^*_{\tau_T}, t)$, $\dot M(\vartheta^*_{\tau_T}, t)$, $\tau_T \le t \le T$ satisfy the equations

$$\mathrm{d}M(\vartheta^*_{\tau_T}, t) = -r(\vartheta^*_{\tau_T})M(\vartheta^*_{\tau_T}, t)\mathrm{d}t + \Gamma(\vartheta^*_{\tau_T})\mathrm{d}X_t,$$

$$\mathrm{d}\dot M(\vartheta^*_{\tau_T}, t) = -r(\vartheta^*_{\tau_T})\dot M(\vartheta^*_{\tau_T}, t)\mathrm{d}t - \dot r(\vartheta^*_{\tau_T})M(\vartheta^*_{\tau_T}, t)\mathrm{d}t + \dot\Gamma(\vartheta^*_{\tau_T})\mathrm{d}X_t.$$

The initial values $M(\vartheta^*_{\tau_T}, \tau_T)$, $\dot M(\vartheta^*_{\tau_T}, \tau_T)$ are defined as follows. We have

$$\begin{aligned} M(\vartheta, \tau_T) &= M(\vartheta, 0)\, e^{-r(\vartheta)\tau_T} + \Gamma(\vartheta)e^{-r(\vartheta)\tau_T}\int_0^{\tau_T} e^{r(\vartheta)s}\mathrm{d}X_s \\ &= M(\vartheta, 0)\, e^{-r(\vartheta)\tau_T} + \Gamma(\vartheta)\left[X_{\tau_T} - r(\vartheta)\int_0^{\tau_T} e^{-r(\vartheta)(\tau_T-s)}X_s\mathrm{d}s\right], \end{aligned}$$

and the initial value is

$$\begin{aligned} M(\vartheta^*_{\tau_T}, \tau_T) &= M(\vartheta^*_{\tau_T}, 0)e^{-r(\vartheta^*_{\tau_T})\tau_T} \\ &\quad + \Gamma(\vartheta^*_{\tau_T})\left[X_{\tau_T} - r(\vartheta^*_{\tau_T})\int_0^{\tau_T} e^{-r(\vartheta^*_{\tau_T})(\tau_T-s)}X_s\mathrm{d}s\right]. \end{aligned} \tag{4.38}$$

For $\dot M(\vartheta^*_{\tau_T}, \tau_T)$ we have a similar representation

$$\begin{aligned} \dot M(\vartheta^*_{\tau_T}, \tau_T) &= \left[\dot M(\vartheta^*_{\tau_T}, 0) - M(\vartheta^*_{\tau_T}, 0)\dot r(\vartheta^*_{\tau_T})\tau_T\right]e^{-r(\vartheta^*_{\tau_T})\tau_T} + \dot\Gamma(\vartheta^*_{\tau_T})X_{\tau_T} \\ &\quad - \left[\dot\Gamma(\vartheta^*_{\tau_T})r(\vartheta^*_{\tau_T}) + \Gamma(\vartheta^*_{\tau_T})\dot r(\vartheta^*_{\tau_T})\right]\int_0^{\tau_T} e^{-r(\vartheta^*_{\tau_T}(\tau_T-s))}X_s\mathrm{d}s \\ &\quad - \Gamma(\vartheta^*_{\tau_T})\dot r(\vartheta^*_{\tau_T})r(\vartheta^*_{\tau_T})\int_0^{\tau_T} e^{-r(\vartheta^*_{\tau_T})(\tau_T-s)}\left(\tau_T - s\right)X_s\mathrm{d}s. \end{aligned} \tag{4.39}$$

Conditions $\mathscr{K}^\star$.

$\mathscr{K}_1^\star$. *The functions* $f(\vartheta), a(\vartheta), b(\vartheta), \vartheta \in [\alpha,\beta]$ *are positive and twice continuously differentiable.*
$\mathscr{K}_2^\star$. $\inf_{\vartheta\in\Theta}|\dot\Psi(\vartheta)| > 0$.
$\mathscr{K}_3^\star$. $\inf_{\vartheta\in\Theta}(|\dot a(\vartheta)| + |\dot r(\vartheta)|) > 0$.

Let us change the variables to $t = vT, v \in [\varepsilon_T, 1], \varepsilon_T = \tau_T T^{-1} = T^{-1+\delta} \to 0$ and denote $\vartheta_T^\star(v) = \vartheta_{vT,T}^\star, \varepsilon_T < v \le 1$.

Theorem 4.4 *Let the conditions* $\mathscr{K}^\star$ *hold. Then the One-step MLE-process* $\vartheta_T^\star(v), \varepsilon_T < v \le 1$ *with* $\delta \in (1/2, 1)$ *is*

1. *Consistent, uniformly on compacts* $\mathbb{K} \subset \Theta$*: for any* $\nu > 0$ *and any* $\varepsilon_0 \in (0, 1]$

$$\lim_{T\to\infty}\sup_{\vartheta_0\in\mathbb{K}} \mathbf{P}_{\vartheta_0}\left(\sup_{\varepsilon_0\le v\le 1}\left|\vartheta_T^\star(v) - \vartheta_0\right| > \nu\right) = 0.$$

2. *It is uniformly on compacts asymptotically normal*

$$\sqrt{T}\left(\vartheta_T^\star(v) - \vartheta_0\right) \Longrightarrow \zeta_v \sim \mathcal{N}\left(0, v^{-1}\mathrm{I}(\vartheta_0)^{-1}\right).$$

3. *The moments converge: for any* $p \ge 1$

$$T^{p/2}\mathbf{E}_{\vartheta_0}\left|\vartheta_T^\star(v) - \vartheta_0\right|^p \longrightarrow \mathbf{E}_{\vartheta_0}|\zeta_v|^p.$$

The random process $\eta_T(v) = v\sqrt{T\mathrm{I}(\vartheta_0)}\left(\vartheta_T^\star(v) - \vartheta_0\right), \varepsilon_0 \le v \le 1$ *converges in distribution in the measurable space* $(\mathcal{C}[\varepsilon_0, 1], \mathfrak{B})$ *to the Wiener process* $W(v), \varepsilon_0 \le v \le 1$.

Proof Let us denote $\mathrm{I}_m = \inf_{\vartheta\in\Theta}\mathrm{I}(\vartheta) > 0$, suppose that $\varepsilon_T < \varepsilon_0$ and consider the probability

$$\begin{aligned}\mathbf{P}_{\vartheta_0}&\left(\sup_{\varepsilon_0\le v\le 1}\left|\vartheta_T^\star(v) - \vartheta_0\right| > \nu\right) \le \mathbf{P}_{\vartheta_0}\left(|\vartheta_{\tau_T}^* - \vartheta_0| > \frac{\nu}{3}\right)\\ &+\mathbf{P}_{\vartheta_0}\left(\frac{1}{\mathrm{I}_m T(\varepsilon_0 - \varepsilon_T)}\sup_{\varepsilon_0\le v\le 1}\left|\int_{\varepsilon_T}^{v}\frac{\dot M(\vartheta_{\tau_T}^*, uT)}{\sigma}\mathrm{d}\bar W_{uT}\right| \ge \frac{\nu}{3}\right)\\ &+\mathbf{P}_{\vartheta_0}\left(\frac{1}{\mathrm{I}_m(\varepsilon_0 - \varepsilon_T)}\int_{\varepsilon_T}^{1}\frac{\left|\dot M(\vartheta_{\tau_T}^*, uT)\left[M(\vartheta_0, uT) - M(\vartheta_{\tau_T}^*, uT)\right]\right|}{\sigma^2}\mathrm{d}u \ge \frac{\nu}{3}\right).\end{aligned}$$

For the first probability due to (4.11) we have

$$\mathbf{P}_{\vartheta_0}\left(\left|\vartheta^*_{\tau_T}-\vartheta_0\right|>\frac{\nu}{3}\right)\leq\frac{9}{\nu^2}\mathbf{E}_{\vartheta_0}|\vartheta^*_{\tau_T}-\vartheta_0|^2\leq\frac{C}{\nu^2\tau_T}\to 0.$$

The second probability satisfies

$$\begin{aligned}\mathbf{P}_{\vartheta_0}&\left(\sup_{\varepsilon_0\leq v\leq 1}\left|\int_{\varepsilon_T}^{v}\frac{\dot M(\vartheta^*_{\tau_T},uT)}{\sigma}\mathrm{d}\bar W_{uT}\right|\geq\frac{\nu \mathrm{I}_m T\left(\varepsilon_0-\varepsilon_T\right)}{3}\right)\\&\qquad\leq\frac{C}{\nu^2\mathrm{I}_m^2T\left(\varepsilon_0-\varepsilon_T\right)^2}\int_{\varepsilon_T}^{1}\frac{\mathbf{E}_{\vartheta_0}\dot M(\vartheta^*_{\tau_T},uT)^2}{\sigma^2}\mathrm{d}u\\&\qquad\leq\frac{C}{\nu^2\mathrm{I}_m^2T\left(\varepsilon_0-\varepsilon_T\right)^2}\longrightarrow 0.\end{aligned}$$

Further

$$\begin{aligned}&\mathbf{E}_{\vartheta_0}\left|\dot M(\vartheta^*_{\tau_T},uT)\left[M\left(\vartheta_0,uT\right)-M(\vartheta^*_{\tau_T},uT)\right]\right|\\&\quad=\mathbf{E}_{\vartheta_0}\left|\dot M(\vartheta^*_{\tau_T},uT)\dot M\left(\bar\vartheta_{\tau_T},uT\right)\left[\vartheta^*_{\tau_T}-\vartheta_0\right]\right|\\&\quad\leq\left(\mathbf{E}_{\vartheta_0}\left|\dot M(\vartheta^*_{\tau_T},uT)\dot M\left(\bar\vartheta_{\tau_T},uT\right)\right|^2\mathbf{E}_{\vartheta_0}[\vartheta^*_{\tau_T}-\vartheta_0]^2\right)^{1/2}\\&\quad\leq C\left(\mathbf{E}_{\vartheta_0}[\vartheta^*_{\tau_T}-\vartheta_0]^2\right)^{1/2}\leq CT^{-\delta/2}\to 0.\end{aligned}$$

Therefore the last probability tends to zero as well. All estimates are uniform on compacts. The estimator $\vartheta^\star_T(v)$ has the representation

$$\begin{aligned}\sqrt{T}\left(\vartheta^\star_T(v)-\vartheta_0\right)&=\sqrt{T}\left(\vartheta^\star_{\tau_T}(v)-\vartheta_0\right)+\frac{\sqrt{T}}{\mathrm{I}(\vartheta^*_{\tau_T})\left(t-\tau_T\right)}\int_{\tau_T}^{t}\frac{\dot M(\vartheta^*_{\tau_T},s)}{\sigma}\mathrm{d}\bar W_s\\&\quad+\frac{\sqrt{T}}{\mathrm{I}(\vartheta^*_{\tau_T})\left(t-\tau_T\right)}\int_{\tau_T}^{t}\frac{\dot M(\vartheta^*_{\tau_T},s)}{\sigma^2}\left[M(\vartheta_0,s)-M(\vartheta^*_{\tau_T},s)\right]\mathrm{d}s\\&=\frac{1}{\mathrm{I}(\vartheta^*_{\tau_T})\left(v-\varepsilon_T\right)}\frac{1}{\sqrt{T}}\int_{\tau_T}^{t}\frac{\dot M(\vartheta^*_{\tau_T},s)}{\sigma}\mathrm{d}\bar W_s\\&\quad+\frac{\sqrt{T}\left(\vartheta^*_{\tau_T}-\vartheta_0\right)}{\mathrm{I}(\vartheta^*_{\tau_T})}\left(\mathrm{I}(\vartheta^*_{\tau_T})-\frac{1}{(t-\tau_T)}\int_{\tau_T}^{t}\frac{\dot M(\vartheta^*_{\tau_T},s)\dot M(\bar\vartheta_{\tau_T},s)}{\sigma^2}\mathrm{d}s\right).\end{aligned}$$

The stochastic integral satisfies

$$\mathbf{E}_{\vartheta_0}\left(\frac{1}{\sqrt{T}}\int_{\tau_T}^{t}\frac{\dot M(\vartheta^*_{\tau_T},s)}{\sigma}{\rm d}\bar W_s-\frac{1}{\sqrt{T}}\int_{\tau_T}^{t}\frac{\dot M(\vartheta_0,s)}{\sigma}{\rm d}\bar W_s\right)^2$$

$$=\frac{1}{T}\int_{\tau_T}^{t}\frac{\mathbf{E}_{\vartheta_0}\left[\dot M(\vartheta^*_{\tau_T},s)-\dot M(\vartheta_0,s)\right]^2}{\sigma^2}{\rm d}s\leq C\frac{t}{T}\mathbf{E}_{\vartheta_0}\left[\vartheta^*_{\tau_T}-\vartheta_0\right]^2$$

$$\leq C\frac{t}{T}T^{-\delta}\to 0$$

and for a fixed v

$$\frac{1}{\sqrt{T}}\int_{\tau_T}^{vT}\frac{\dot M(\vartheta_0,s)}{\sigma}{\rm d}\bar W_s\Longrightarrow\mathcal{N}\left(0,v{\rm I}\left(\vartheta_0\right)\right)$$

since, by the LLN,

$$\frac{1}{T}\int_{\tau_T}^{vT}\frac{\dot M(\vartheta_0,s)^2}{\sigma^2}{\rm d}s\longrightarrow v{\rm I}\left(\vartheta_0\right).$$

Therefore

$$\frac{1}{{\rm I}(\vartheta^*_{\tau_T})\left(v-\varepsilon_T\right)\sqrt{T}}\int_{\tau_T}^{t}\frac{\dot M(\vartheta^*_{\tau_T},s)}{\sigma}{\rm d}\bar W_s\Longrightarrow\mathcal{N}\left(0,v^{-1}{\rm I}\left(\vartheta_0\right)^{-1}\right).\tag{4.40}$$

For the last term we can write

$$\int_{\tau_T}^{t}\frac{\dot M(\vartheta^*_{\tau_T},s)\dot M(\bar\vartheta_{\tau_T},s)}{\sigma^2}{\rm d}s$$

$$=\int_{\tau_T}^{t}\frac{\dot M(\vartheta^*_{\tau_T},s)^2}{\sigma^2}{\rm d}s+\left(\bar\vartheta_{\tau_T}-\vartheta^*_{\tau_T}\right)\int_{\tau_T}^{t}\frac{\dot M(\vartheta^*_{\tau_T},s)\ddot M(\bar{\bar\vartheta}_{\tau_T},s)}{\sigma^2}{\rm d}s,$$

where $\left|\bar\vartheta_{\tau_T}-\vartheta^*_{\tau_T}\right|\leq|\vartheta^*_{\tau_T}-\vartheta_0|$. Hence

$$\frac{\sqrt{T}\left(\vartheta^*_{\tau_T}-\vartheta_0\right)}{{\rm I}\left(\vartheta^*_{\tau_T}\right)}\left({\rm I}\left(\vartheta^*_{\tau_T}\right)-\frac{1}{\left(t-\tau_T\right)}\int_{\tau_T}^{t}\frac{\dot M(\vartheta^*_{\tau_T},s)\dot M(\bar\vartheta_{\tau_T},s)}{\sigma^2}{\rm d}s\right)$$

$$=\frac{\sqrt{T}\left(\vartheta^*_{\tau_T}-\vartheta_0\right)}{{\rm I}\left(\vartheta^*_{\tau_T}\right)}\left({\rm I}\left(\vartheta^*_{\tau_T}\right)-\frac{1}{\left(t-\tau_T\right)}\int_{\tau_T}^{t}\frac{\dot M(\vartheta^*_{\tau_T},s)^2}{\sigma^2}{\rm d}s\right)$$

$$+\frac{\sqrt{T}\left(\vartheta^*_{\tau_T}-\vartheta_0\right)\left(\bar\vartheta_{\tau_T}-\vartheta_0\right)}{\mathrm{I}(\vartheta^*_{\tau_T})\left(t-\tau_T\right)}\int_{\tau_T}^{t}\frac{\dot M(\vartheta^*_{\tau_T},s)\ddot M(\bar{\bar\vartheta}_{\tau_T},s)}{\sigma^2}\mathrm{d}s.$$

Note that

$$\sqrt{T}\left|\left(\vartheta^*_{\tau_T}-\vartheta_0\right)\left(\bar\vartheta_{\tau_T}-\vartheta_0\right)\right|\leq\sqrt{T}\left|\vartheta^*_{\tau_T}-\vartheta_0\right|^2=O\left(T^{1/2-\delta}\right)\longrightarrow 0.$$

Similar estimates imply

$$\mathrm{I}(\vartheta^*_{\tau_T})-\frac{1}{\left(t-\tau_T\right)}\int_{\tau_T}^{t}\frac{\dot M(\vartheta^*_{\tau_T},s)^2}{\sigma^2}\mathrm{d}s=\mathrm{I}(\vartheta_0)-\frac{1}{\left(t-\tau_T\right)}\int_{\tau_T}^{t}\frac{\dot M(\vartheta_0,s)^2}{\sigma^2}\mathrm{d}s$$
$$+O\left(\vartheta^*_{\tau_T}-\vartheta_0\right).$$

Further

$$\mathbf{E}_{\vartheta_0}\left(\frac{1}{\left(t-\tau_T\right)}\int_{\tau_T}^{t}\left[\frac{\dot M(\vartheta_0,s)^2}{\sigma^2}-\mathrm{I}(\vartheta_0)\right]\mathrm{d}s\right)^2\leq\frac{C}{t}$$

and thus we obtain

$$\frac{\sqrt{T}\left(\vartheta^*_{\tau_T}-\vartheta_0\right)}{\mathrm{I}(\vartheta^*_{\tau_T})}\left(\mathrm{I}(\vartheta^*_{\tau_T})-\frac{1}{\left(t-\tau_T\right)}\int_{\tau_T}^{t}\frac{\dot M(\vartheta^*_{\tau_T},s)\dot M(\bar\vartheta_{\tau_T},s)}{\sigma^2}\mathrm{d}s\right)\longrightarrow 0.$$

This convergence and (4.40) yield the asymptotic normality.

The convergence of moments follows from (4.11) and boundedness of all moments of the derivatives of $M(\vartheta,s)$. Thus we proved that the process $\eta_T(v)\,,\varepsilon_0,\leq v\leq 1$ admits the representation

$$\eta_T(v)=\frac{1}{\sigma\sqrt{\mathrm{I}(\vartheta_0)\,T}}\int_0^{vT}\dot M(\vartheta_0,s)\,\mathrm{d}\bar W_s+h_T(v)\,, \tag{4.41}$$

where the residuals vanish asymptotically: $h_T(v)=h_T(v,\vartheta_0,\vartheta^*_{\tau_T})\to 0$ uniformly on $\vartheta_0\in\mathbb{K}$

$$\sup_{\vartheta_0\in\mathbb{K}}\mathbf{E}_{\vartheta_0}h_T(v,\vartheta_0,\vartheta^*_{\tau_T})^2\to 0$$

and such that

$$\mathbf{E}_{\vartheta_0}\left|\eta_T(v_2)-\eta_T(v_1)\right|^4\leq C\left|v_2-v_1\right|^2+C\left|v_2-v_1\right|^4\leq C\left|v_2-v_1\right|^2. \tag{4.42}$$

To verify the convergence of two-dimensional distributions $\eta_T(v_1), \eta_T(v_2)$ we consider the linear combination $\langle \lambda, \eta_T \rangle = \lambda_1 \eta_T(v_1) + \lambda_2 \eta_T(v_2)$ and note that

$$\langle \lambda, \eta_T \rangle = \frac{1}{\sigma \sqrt{\mathrm{I}(\vartheta_0)\, T}} \int_0^T \left[\lambda_1 \mathbb{I}_{\{s<v_1 T\}} + \lambda_2 \mathbb{I}_{\{s<v_2 T\}}\right] \dot{M}(\vartheta_0, s)\, \mathrm{d}\bar{W}_s + \langle \lambda, h_T \rangle$$

where $\langle \lambda, h_T \rangle \to 0$ and

$$\begin{aligned}
&\frac{1}{\sigma^2 \mathrm{I}(\vartheta_0)\, T} \int_0^T \left[\lambda_1 \mathbb{I}_{\{s<v_1 T\}} + \lambda_2 \mathbb{I}_{\{s<v_2 T\}}\right]^2 \dot{M}(\vartheta_0, s)^2\, \mathrm{d}s \\
&\quad = \frac{\lambda_1^2}{\sigma^2 \mathrm{I}(\vartheta_0)\, T} \int_0^{v_1 T} \dot{M}(\vartheta_0, s)^2\, \mathrm{d}s + \frac{\lambda_2^2}{\sigma^2 \mathrm{I}(\vartheta_0)\, T} \int_0^{v_2 T} \dot{M}(\vartheta_0, s)^2\, \mathrm{d}s \\
&\qquad + \frac{2\lambda_1 \lambda_2}{\sigma^2 \mathrm{I}(\vartheta_0)\, T} \int_0^{(v_1 \wedge v_2)T} \dot{M}(\vartheta_0, s)^2\, \mathrm{d}s \\
&\quad \longrightarrow \lambda_1^2 v_1 + 2\lambda_1 \lambda_2 (v_1 \wedge v_2) + \lambda_2^2 v_2
\end{aligned}$$

since

$$\frac{v}{\sigma^2 \mathrm{I}(\vartheta_0)\, T v} \int_0^{vT} \dot{M}(\vartheta_0, s)^2\, \mathrm{d}s \longrightarrow v.$$

Therefore by the CLT

$$\langle \lambda, \eta_T \rangle \Longrightarrow \langle \lambda, W \rangle = \lambda_1 W(v_1) + \lambda_2 W(v_2).$$

The convergence of finite-dimensional distributions

$$\left(\eta_T(v_1), \ldots, \eta_T(v_k)\right) \Longrightarrow \left(W(v_1), \ldots, W(v_k)\right)$$

can be verified in a similar way. This convergence and (4.42) provide the weak convergence $\eta_T(\cdot) \Longrightarrow W(\cdot)$. □

Remark 4.1 The representation (4.37) of the One-step MLE-process can be given in the recursive form as follows. For the values $t \in (\tau_T, T]$

$$\left(t - \tau_T\right) \vartheta^\star_{t,T} = \left(t - \tau_T\right) \vartheta^*_{\tau_T} + \frac{1}{\mathrm{I}\left(\vartheta^*_{\tau_T}\right)} \int_{\tau_T}^t \frac{\dot{M}(\vartheta^*_{\tau_T}, s)}{\sigma^2} \left[\mathrm{d}X_s - M(\vartheta^*_{\tau_T}, s)\mathrm{d}s\right].$$

Hence

$$\mathrm{d}\vartheta^\star_{t,T} = \frac{\vartheta^*_{\tau_T} - \vartheta^\star_{t,T}}{t - \tau_T} \mathrm{d}t + \frac{\dot{M}(\vartheta^*_{\tau_T}, t)}{\mathrm{I}\left(\vartheta^*_{\tau_T}\right)\sigma^2 \left(t - \tau_T\right)} \left[\mathrm{d}X_t - M(\vartheta^*_{\tau_T}, t)\mathrm{d}t\right]. \tag{4.43}$$

To avoid the singularities in this formula the value $t - \tau_T$ can be replaced with $t - \tau_T + \varepsilon^*$, where $\varepsilon^* > 0$ is some small number. This modification, of course, will not change the asymptotic behavior of $\vartheta^\star_{t,T}$.

Estimation of the Parameter f

If $f(\vartheta) = \vartheta \in \Theta = (\alpha_f, \beta_f), \alpha_f > 0$, $a(\vartheta) = a > 0$ and $b(\vartheta) = b > 0$, then the preliminary estimator is

$$\vartheta^*_{\tau_T} = \left(\frac{a^3}{b^2\left[e^{-a} - 1 + a\right]\tau_T}\sum_{k=1}^{\tau_T}\left([X_k - X_{k-1}]^2 - \sigma^2\right)\right)^{1/2}.$$

The processes $M(\vartheta, t), \dot M(\vartheta, t), t \geq 0$ are solutions to the equations

$$\mathrm{d}M(\vartheta, t) = -r(\vartheta) M(\vartheta, t)\,\mathrm{d}t + \Gamma(\vartheta)\,\mathrm{d}X_t, \qquad t \geq 0,$$

$$\mathrm{d}\dot M(\vartheta, t) = -r(\vartheta)\dot M(\vartheta, t)\,\mathrm{d}t + \frac{b^2\vartheta}{r(\vartheta)\sigma^2}\,\mathrm{d}X_t,$$

where $r(\vartheta) = \left(a^2 + b^2\vartheta^2\sigma^{-2}\right)^{1/2}$ and $\Gamma(\vartheta) = r(\vartheta) - a$.

The One-step MLE-process is

$$\vartheta^\star_{t,T} = \vartheta^*_{\tau_T} + \frac{2\sigma^3 r(\vartheta^*_{\tau_T})^3}{b^4\vartheta^{*2}_{\tau_T}(t - \tau_T)}\int_{\tau_T}^{t}\dot M(\vartheta^*_{\tau_T}, s)\left[\mathrm{d}X_s - M(\vartheta^*_{\tau_T}, s)\mathrm{d}s\right], \quad \tau_T < t \leq T.$$

According to Theorem 4.4 this estimator satisfies all of its assertions and, in particular, the asymptotic normality

$$\sqrt{T}\left(\vartheta^\star_T(v) - \vartheta_0\right) \Longrightarrow \mathcal{N}\left(0, \frac{2\sigma^4 r(\vartheta_0)^3}{v\, b^4\vartheta_0^2}\right).$$

Estimation of the Parameter b

If $b(\vartheta) = \vartheta \in \Theta = (\alpha_b, \beta_b), \alpha_b > 0$, $f(\vartheta) = f > 0$ and $a(\vartheta) = a > 0$, then the preliminary estimator and One-step MLE-process are

$$\vartheta^*_{\tau_T} = \left(\frac{a^3}{f^2\left[e^{-a} - 1 + a\right]\tau_T}\sum_{k=1}^{\tau_T}\left([X_k - X_{k-1}]^2 - \sigma^2\right)\right)^{1/2},$$

$$\vartheta^\star_{t,T} = \vartheta^*_{\tau_T} + \frac{2\sigma^3 r(\vartheta^*_{\tau_T})^3}{f^4\,\vartheta^{*2}_{\tau_T}(t - \tau_T)}\int_{\tau_T}^{t}\dot M(\vartheta^*_{\tau_T}, s)\left[\mathrm{d}X_s - M(\vartheta^*_{\tau_T}, s)\mathrm{d}s\right], \quad \tau_T < t \leq T$$

with the corresponding processes $M(\vartheta, t), t \geq 0$ and $\dot M(\vartheta, t), t \geq 0$. Then

$$\sqrt{T}\left(\vartheta_T^\star(v)-\vartheta_0\right)\Longrightarrow\mathcal{N}\left(0,\frac{2\sigma^4 r(\vartheta_0)^3}{v f^4\vartheta_0^2}\right).$$

Estimation of the Parameter a
If $a(\vartheta)=\vartheta\in\Theta=(\alpha_a,\beta_a),\alpha_a>0,\ f(\vartheta)=f>0$ and $a(\vartheta)=a>0$, then the preliminary estimator is

$$\vartheta_{\tau_T}^*=\arg\inf_{\vartheta\in\Theta}\left|\frac{1}{\tau_T}\sum_{k=1}^{\tau_T}\left[X_k-X_{k-1}\right]^2-\Psi(\vartheta)\right|,$$

$$\Psi(\vartheta)=\frac{f^2b^2}{\vartheta^3}\left[e^{-\vartheta}-1+\vartheta\right]+\sigma^2.$$

The Fisher information $\mathrm{I}(\vartheta)$ defined in (4.29) reads

$$\mathrm{I}(\vartheta_0)=\frac{\Gamma(\vartheta_0)\left[r(\vartheta_0)+\vartheta_0+r(\vartheta_0)\vartheta_0\Gamma(\vartheta_0)\right]}{2\vartheta_0 r(\vartheta_0)^3\left[r(\vartheta_0)+\vartheta_0\right]},$$

and the One-step MLE-process is given by (4.37). It is asymptotically normal

$$\sqrt{T}\left(\vartheta_T^\star(v)-\vartheta_0\right)\Longrightarrow\mathcal{N}\left(0,v^{-1}\mathrm{I}(\vartheta_0)^{-1}\right).$$

Estimation of the Two-Dimensional Parameter (a,f)
The partially observed system is

$$\mathrm{d}X_t=\theta_2 Y_t\mathrm{d}t+\sigma\mathrm{d}W_t,\qquad 0\le t\le T,$$
$$\mathrm{d}Y_t=-\theta_1 Y_t\mathrm{d}t+b\mathrm{d}V_t,\qquad t\ge 0,$$

subject to the initial conditions X_0 and Y_0, where $b\neq 0$ is known and $\vartheta=(a,f)=(\theta_1,\theta_2)\in\Theta=(\alpha_a,\beta_a)\times\left(\alpha_f,\beta_f\right),\alpha_a>0,\alpha_f>0$ is to be estimated.

Recall that the preliminary MME $\vartheta_{\tau_T}^*=(\theta_{1,\tau_T}^*,\theta_{2,\tau_T}^*)=(a_{\tau_T}^*,f_{\tau_T}^*)$ can be computed as follows:

$$\frac{e^{-a_{\tau_T}^*}-1+a_{\tau_T}^*}{(1-e^{-a_{\tau_T}^*})^2}=\frac{R_{1,\tau_T}-\sigma^2}{2R_{2,\tau_T}},\qquad R_{1,\tau_T}=\frac{1}{\tau_T}\sum_{k=1}^{\tau_T}\left[X_k-X_{k-1}\right]^2,$$

$$f_{\tau_T}^*=\frac{2(a_{\tau_T}^*)^{3/2}R_{2,\tau_T}^{1/2}}{b^2(1-e^{-a_{\tau_T}^*})^2},\qquad R_{2,\tau_T}=\frac{1}{\tau_T}\sum_{k=2}^{\tau_T}\left[X_k-X_{k-1}\right]\left[X_{k-1}-X_{k-2}\right].$$

The Fisher information matrix $\mathbf{I}(\vartheta_0)$ is

$$\begin{pmatrix} \frac{\Gamma(\vartheta_0)^2\left[r(\vartheta_0)\theta_{1,0}+(r(\vartheta_0)+\theta_{1,0})^2\right]}{2\theta_{1,0}r(\vartheta_0)^3\left[r(\vartheta_0)+\theta_{1,0}\right]}, & \frac{b^2\theta_{2,0}\left(\theta_{1,0}^2+r(\vartheta_0)\theta_{1,0}-2r(\vartheta_0)\right)}{2r(\vartheta_0)^3\left(r(\vartheta_0)+\theta_{1,0}\right)\sigma^2} \\ \frac{b^2\theta_{2,0}\left(\theta_{1,0}^2+r(\vartheta_0)\theta_{1,0}-2r(\vartheta_0)\right)}{2r(\vartheta_0)^3\left(r(\vartheta_0)+\theta_{1,0}\right)\sigma^2}, & \frac{\theta_{2,0}^2b^4}{2r(\vartheta_0)^3\sigma^4} \end{pmatrix}$$

and it is nondegenerate. For its computation we used the stationary representations for the derivatives $\dot{M}_f(\vartheta_0, t)$ and $\dot{M}_a(\vartheta_0, t)$. One-step MLE-process is the vector-process $\vartheta^\star_{t,T}$, $\tau_T < t \leq T$ defined by the expression

$$\vartheta^\star_{t,T} = \vartheta^*_{\tau_T} + \frac{\mathbf{I}(\vartheta^*_{\tau_T})^{-1}}{\sigma\left(t-\tau_T\right)}\int_{\tau_T}^{t}\dot{M}(\vartheta^*_{\tau_T}, s)\left[\mathrm{d}X_s - M(\vartheta^*_{\tau_T}, s)\mathrm{d}s\right].$$

Here $M(\vartheta^*_{\tau_T}, s)$, $\tau_T \leq t \leq T$ and vector $\dot{M}(\vartheta^*_{\tau_T}, s) = (\dot{M}_a(\vartheta^*_{\tau_T}, t), \dot{M}_f(\vartheta^*_{\tau_T}, t))^\top$, $\tau_T \leq t \leq T$ are solutions to the equations

$$\begin{aligned} \mathrm{d}M(\vartheta^*_{\tau_T}, t) &= -r(\vartheta^*_{\tau_T})M(\vartheta^*_{\tau_T}, t)\mathrm{d}t + \Gamma(\vartheta^*_{\tau_T})\mathrm{d}X_t, \\ \mathrm{d}\dot{M}_a(\vartheta^*_{\tau_T}, t) &= -r(\vartheta^*_{\tau_T})\dot{M}_a(\vartheta^*_{\tau_T}, t)\mathrm{d}t - \dot{r}_a(\vartheta^*_{\tau_T})M(\vartheta^*_{\tau_T}, t)\mathrm{d}t + \dot{\Gamma}_a(\vartheta^*_{\tau_T})\mathrm{d}X_t, \\ \mathrm{d}\dot{M}_f(\vartheta^*_{\tau_T}, t) &= -r(\vartheta^*_{\tau_T})\dot{M}_f(\vartheta^*_{\tau_T}, t)\mathrm{d}t - \dot{r}_f(\vartheta^*_{\tau_T})M(\vartheta^*_{\tau_T}, t)\mathrm{d}t + \dot{\Gamma}_f(\vartheta^*_{\tau_T})\mathrm{d}X_t, \end{aligned}$$

The initial values $M(\vartheta^*_{\tau_T}, \tau_T)$, $\dot{M}_a(\vartheta^*_{\tau_T}, \tau_T)$, and $\dot{M}_f(\vartheta^*_{\tau_T}, \tau_T)$ are calculated with the help of the representations (4.38) and (4.39) given above.

Of course this expression for One-step MLE-process can replaced by the recursive version similar to (4.43):

$$\mathrm{d}\vartheta^\star_{t,T} = \frac{\vartheta^*_{\tau_T} - \vartheta^\star_{t,T}}{t-\tau_T}\mathrm{d}t + \frac{\mathbf{I}\left(\vartheta^*_{\tau_T}\right)^{-1}\dot{M}(\vartheta^*_{\tau_T}, t)}{\sigma^2\left(t-\tau_T\right)}\left[\mathrm{d}X_t - M(\vartheta^*_{\tau_T}, t)\mathrm{d}t\right].$$

Proposition 4.4 *One-step MLE-process $\vartheta^\star_{t,T}$, $\tau_T < t \leq T$ is uniformly consistent, asymptotically ($t = vT$, $T \to \infty$) normal*

$$\sqrt{t}\left(\vartheta^\star_{t,T} - \vartheta_0\right) \Longrightarrow \zeta(\vartheta_0) \sim \mathcal{N}\left(0, \mathbf{I}(\vartheta_0)^{-1}\right)$$

and the polynomial moments converge.

The proof is similar to the proof of Theorem 4.4.

4.3 Adaptive Filter

4.3.1 Filter

Consider the same partially observed linear system

$$\mathrm{d}X_t = f(\vartheta)\, Y_t \mathrm{d}t + \sigma \mathrm{d}W_t, \qquad 0 \le t \le T,$$
$$\mathrm{d}Y_t = -a(\vartheta)\, Y_t \mathrm{d}t + b(\vartheta)\, \mathrm{d}V_t, \qquad t \ge 0,$$

subject to the initial conditions X_0 and Y_0, where $W_t, t \ge 0$ and $V_t, t \ge 0$ are independent Wiener processes. Here the functions $f(\vartheta), a(\vartheta), b(\vartheta), \vartheta \in \Theta$ are known, positive, and smooth. The value of the parameter $\vartheta \in \Theta = (\alpha, \beta)$ is unknown.

The conditional expectation $m(\vartheta, t) = \mathbf{E}_\vartheta(Y_t | X_s, 0 \le s \le t)$ satisfies the Kalman-Bucy equations

$$\mathrm{d}m(\vartheta, t) = -\left[a(\vartheta) + \frac{\gamma(\vartheta, t) f(\vartheta)^2}{\sigma^2}\right] m(\vartheta, t)\, \mathrm{d}t + \frac{\gamma(\vartheta, t) f(\vartheta)}{\sigma^2} \mathrm{d}X_t, \; m(\vartheta, 0),$$

$$\frac{\partial \gamma(\vartheta, t)}{\partial t} = -2a(\vartheta)\gamma(\vartheta, t) - \frac{\gamma(\vartheta, t)^2 f(\vartheta)^2}{\sigma^2} + b(\vartheta)^2, \qquad \gamma(\vartheta, 0).$$

Solving these equations requires knowledge of the parameter value ϑ. Below we consider the problem of approximation $m(\vartheta, t), 0 \le t \le T$ based on observations $X^T = (X_t, 0 \le t \le T)$. The main idea is to use the recursive One-step MLE-process on the time interval $t \in (\tau_T, T]$

$$\mathrm{d}\vartheta^\star_{t,T} = \frac{\vartheta^*_{\tau_T} - \vartheta^\star_{t,T}}{t - \tau_T} \mathrm{d}t + \frac{\mathrm{I}\left(\vartheta^*_{\tau_T}\right)^{-1} \dot{M}(\vartheta^*_{\tau_T}, t)}{\sigma^2\left(t - \tau_T\right)} \left[\mathrm{d}X_t - M(\vartheta^*_{\tau_T}, t)\mathrm{d}t\right],$$

where $\vartheta^*_{\tau_T}$ is an MME constructed using the observations $X^{\tau_T} = \left(X_s, 0 \le s \le \tau_T\right)$, $\tau_T = T^\delta, \delta \in (1/2, 1)$, $M(\vartheta, s) = f(\vartheta)\, m(\vartheta, s)$ and to replace ϑ with $\vartheta^\star_{t,T}$ in the filtering equations.

There are several possibilities for approximation of the random process $m(\vartheta_0, t), 0 \le t \le T$. Consider the random process $\hat{m}_{t,T}, \tau_T < t \le T$ obtained by solving the equation

$$\mathrm{d}\hat{m}_{t,T} = -a\left(\vartheta^\star_{t,T}\right) \hat{m}_{t,T} \mathrm{d}t + \frac{\hat{\gamma}_{t,T} f\left(\vartheta^\star_{t,T}\right)}{\sigma^2} \left[\mathrm{d}X_t - f\left(\vartheta^\star_{t,T}\right) \hat{m}_{t,T} \mathrm{d}t\right], \quad \hat{m}_{\tau_T, T},$$

where

$$\frac{\partial \hat{\gamma}_{t,T}}{\partial t}=-2a\left(\vartheta^\star_{t,T}\right)\hat{\gamma}_{t,T}-\frac{\hat{\gamma}^2_{t,T}f\left(\vartheta^\star_{t,T}\right)^2}{\sigma^2}+b\left(\vartheta^\star_{t,T}\right)^2,\qquad \hat{\gamma}_{\tau_T,T}=\gamma(\vartheta^*_{\tau_T},\tau_T).$$

Of course, a special attention has to be paid to the initial value $\hat{m}_{\tau_T,T}$ (see (4.38)).

Another possibility is to use the explicit expression for the solution of the Eq. (4.6)

$$\gamma\left(\vartheta,t\right)=e^{-2r(\vartheta)t}\left[\frac{1}{\gamma\left(\vartheta,0\right)-\gamma_*\left(\vartheta\right)}+\frac{f\left(\vartheta\right)^2}{2r\left(\vartheta\right)\sigma^2}\left(1-e^{-2r(\vartheta)t}\right)\right]^{-1}+\gamma_*\left(\vartheta\right),$$

where

$$r\left(\vartheta\right)=\left(a\left(\vartheta\right)^2+\frac{f\left(\vartheta\right)^2b\left(\vartheta\right)^2}{\sigma^2}\right)^{1/2},\qquad \gamma_*\left(\vartheta\right)=\frac{\sigma^2\left[r\left(\vartheta\right)-a\left(\vartheta\right)\right]}{f\left(\vartheta\right)^2}.$$

The approximation of $m\left(\vartheta,t\right)$ can be realized by the random process $\hat{m}_T\left(t\right),\ \tau_T<t\leq T$

$$\mathrm{d}\hat{m}_T\left(t\right)=-a\left(\vartheta^\star_{t,T}\right)\hat{m}_T\left(t\right)\mathrm{d}t+\frac{\gamma\left(\vartheta^\star_{t,T},t\right)f\left(\vartheta^\star_{t,T}\right)}{\sigma^2}\left[\mathrm{d}X_t-f\left(\vartheta^\star_{t,T}\right)\hat{m}_T\left(t\right)\mathrm{d}t\right],$$

with the initial value $m_T\left(\tau_T\right)$.

The third possibility to write the adaptive filter is to use the exponential convergence of the solution $\gamma\left(\vartheta,t\right)$ of the Riccati equation to the steady-state value $\gamma_*\left(\vartheta\right)$ and the approximation of $m\left(\vartheta,t\right)$ can be done with the help of the solution $m^\star_{t,T},\ \tau_T<t\leq T$ of the equation

$$\mathrm{d}m^\star_{t,T}=-r\left(\vartheta^\star_{t,T}\right)m^\star_{t,T}\mathrm{d}t+B\left(\vartheta^\star_{t,T}\right)\,\mathrm{d}X_t,\tag{4.44}$$

subject to the initial condition $m^\star_{\tau_T,T}$, where we denoted $B\left(\vartheta\right)=\left[r\left(\vartheta\right)-a\left(\vartheta\right)\right]/f\left(\vartheta\right)$. To define $m^\star_{\tau_T,T}$ we can use the equality similar to (4.38), i.e., we let

$$m^\star_{\tau_T,T}=m(\vartheta^*_{\tau_T},0)e^{-r(\vartheta^*_{\tau_T})\tau_T}+B(\vartheta^*_{\tau_T})\left[X_{\tau_T}-r(\vartheta^*_{\tau_T})\int_0^{\tau_T}e^{-r(\vartheta^*_{\tau_T})(\tau_T-s)}X_s\,\mathrm{d}s\right].$$

The asymptotic results will be the same if we take, say, $m^\star_{\tau_T,T}=y_0$ because the contribution of the initial value is asymptotically negligible.

Let us study the adaptive filter (4.44).

Define the functions

$$K_1\left(\vartheta_0\right)=-\frac{\left[\dot{a}\left(\vartheta_0\right)+\dot{f}\left(\vartheta_0\right)B\left(\vartheta_0\right)\right]\sigma}{f\left(\vartheta_0\right)\sqrt{2a(\vartheta_0)\mathrm{I}\left(\vartheta_0\right)}},$$

$$K_2(\vartheta_0)=\frac{\dot r(\vartheta_0)\,\sigma}{f(\vartheta_0)\sqrt{2r(\vartheta_0)\mathrm{I}(\vartheta_0)}},\quad R_{1,2}(\vartheta_0)=\frac{2K_1(\vartheta_0)\,K_2(\vartheta_0)\sqrt{a(\vartheta_0)\,r(\vartheta_0)}}{a(\vartheta_0)+r(\vartheta_0)},$$

and the random processes

$$\xi_T^{(1)}(v)=\sqrt{2a(\vartheta_0)\,T}\int_{v-\nu_T}^{v}e^{-a(\vartheta_0)(v-u)T}\mathrm{d}W_T(u)\Longrightarrow\xi^{(1)}(v)\sim\mathcal{N}(0,1),$$
$$v\in[\varepsilon_0,1],$$
$$\xi_T^{(2)}(v)=\sqrt{2r(\vartheta_0)\,T}\int_{v-\nu_T}^{v}e^{-r(\vartheta_0)(v-u)T}\mathrm{d}W_T(u)\Longrightarrow\xi^{(2)}(v)\sim\mathcal{N}(0,1),$$
$$v\in[\varepsilon_0,1].$$

Here $\varepsilon_0\in(0,1)$, $\nu_T=T^{-\kappa}$, $\kappa\in(0,1)$ and

$$\mathbf{E}_{\vartheta_0}\xi^{(1)}(v)\,\xi^{(2)}(v)=\frac{2\sqrt{a(\vartheta_0)\,r(\vartheta_0)}}{a(\vartheta_0)+r(\vartheta_0)},\qquad \mathbf{E}_{\vartheta_0}\xi^{(1)}(v_1)\,\xi^{(2)}(v_2)=0$$

if $v_1\neq v_2$ since

$$\mathbf{E}_{\vartheta_0}\xi_T^{(1)}(v_1)\,\xi_T^{(2)}(v_2)$$
$$=\sqrt{r(\vartheta_0)\,a(\vartheta_0)}T\int_{(v_1-\nu_T)\vee(v_2-\nu_T)}^{v_1\wedge v_2}e^{-[r(\vartheta_0)+a(\vartheta_0)](v-u)T}\,\mathbb{I}_{\mathbb{A}_1\cap\mathbb{A}_2}\mathrm{d}u=0$$

for $|v_2-v_1|>\nu_T\to0$ and the sets $\mathbb{A}_i=\left\{u:v_i-\nu_T\leq u\leq v_i\right\}$.

Denote the 2×2 matrix $\rho(\vartheta_0)$ with the components $\rho_{i,j}(\vartheta_0)=\mathbf{E}_{\vartheta_0}\xi^{(j)}(v)\,\xi^{(j)}(v)$ and the Gaussian vector $\xi(v)=\left(\xi^{(1)}(v),\xi^{(2)}(v)\right)\sim\mathcal{N}(0,\rho(\vartheta_0))$.

Recall that the random process $\left(\eta_T(v),v\in[\varepsilon_0,1]\right)$ converges in distribution to the Wiener process $(W(v),v\in[\varepsilon_0,1])$ for any $\varepsilon_0\in(0,1)$ (Theorem 4.4). The direct calculations reveal that

$$\mathbf{E}_{\vartheta_0}\xi_T^{(i)}(v)\,\eta_T(v)\longrightarrow\mathbf{E}_{\vartheta_0}\xi^{(i)}(v)\,W(v)=0.$$

Therefore the vectors $\xi(v)$ and $W(v)$ are independent and the vectors $\xi^{(1)}(v_1)$, $\xi^{(2)}(v_2)$ for $v_1\neq v_2$ are independent as well.

Theorem 4.5 *Let the assumptions of Theorem 4.4 hold. Then for $v\in[\varepsilon_0,1]$, the convergence in distribution holds*

$$\sqrt{T}\left(m_{vT,T}^{\star}-m(\vartheta_0,vT)\right)\Longrightarrow\frac{W(v)}{v}\left[K_1(\vartheta_0)\,\xi^{(1)}(v)+K_2(\vartheta_0)\,\xi^{(2)}(v)\right]\tag{4.45}$$

as well as the convergence of moments

$$T\,\mathbf{E}_{\vartheta_0}\left(m^\star_{\upsilon T,T}-m\left(\vartheta_0,\upsilon T\right)\right)^2\longrightarrow\frac{1}{\upsilon}\Big[K_1\left(\vartheta_0\right)^2+K_2\left(\vartheta_0\right)^2+2R_{1,2}\left(\vartheta_0\right)\Big].\tag{4.46}$$

Proof Let us denote

$$r_m=\inf_{\vartheta\in\Theta}r\left(\vartheta\right),\qquad\qquad B_M=\sup_{\vartheta\in\Theta}B\left(\vartheta\right).$$

We will need the following two estimates. The first one is based on the formula

$$\begin{aligned}m^\star_{t,T}&=m^\star_{\tau_T,T}e^{-\int_{\tau_T}^t r(\vartheta^\star_{s,T})\mathrm{d}s}+f\left(\vartheta_0\right)\int_{\tau_T}^t e^{-\int_{\tau_T}^t r(\vartheta^\star_{s,T})\mathrm{d}s}B\left(\vartheta^\star_{s,T}\right)m\left(\vartheta_0,s\right)\mathrm{d}s\\&\quad+\sigma\int_{\tau_T}^t e^{-\int_{\tau_T}^t r(\vartheta^\star_{s,T})\mathrm{d}s}B\left(\vartheta^\star_{s,T}\right)\mathrm{d}\bar W_s\end{aligned}$$

which implies

$$\begin{aligned}\mathbf{E}_{\vartheta_0}\left|m^\star_{t,T}\right|^2&\leq 3e^{-r_m t}\mathbf{E}_{\vartheta_0}\left|m^\star_{\tau_T,T}\right|^2+3\sigma^2B_M^2\int_{\tau_T}^t e^{-r_m(t-s)}\mathrm{d}s\\&\quad+3f\left(\vartheta_0\right)^2B_M^2\int_{\tau_T}^t e^{-r_m(t-s)}\int_{\tau_T}^t e^{-r_m(t-q)}\mathbf{E}_{\vartheta_0}m\left(\vartheta_0,s\right)m\left(\vartheta_0,q\right)\\&\quad\times\mathrm{d}s\mathrm{d}q\leq C,\end{aligned}$$

where the constant $C>0$ does not depend on t. Similarly, for any $p\geq 2$ there exists a constant $C=C\left(p\right)>0$ such that

$$\sup_{\vartheta_0\in\mathbb{K}}\sup_{\tau_T\leq t\leq T}\mathbf{E}_{\vartheta_0}\left|m^\star_{t,T}\right|^p\leq C.$$

To obtain the second estimate we can use the representations for the differences $\Delta_{t,T}=m^\star_{t,T}-m\left(\vartheta_0,t\right)$ and $\Delta_{\upsilon T,T}$

$$\begin{aligned}\Delta_{t,T}&=\int_{\tau_T}^t e^{-r(\vartheta_0)(t-s)}\left[r\left(\vartheta_0\right)-r\left(\vartheta^\star_{s,T}\right)\right]m^\star_{s,T}\mathrm{d}s\\&\quad+\int_{\tau_T}^t e^{-r(\vartheta_0)(t-s)}\left[B\left(\vartheta^\star_{s,T}\right)-B\left(\vartheta_0\right)\right]f\left(\vartheta_0\right)m\left(\vartheta_0,s\right)\mathrm{d}s\\&\quad+\sigma\int_{\tau_T}^t e^{-r(\vartheta_0)(t-s)}\left[B\left(\vartheta^\star_{s,T}\right)-B\left(\vartheta_0\right)\right]\mathrm{d}\bar W_s\end{aligned}$$

$$
\begin{aligned}
&= T\int_{\varepsilon_T}^{v} e^{-r(\vartheta_0)(v-u)T}\left[r\left(\vartheta_0\right)-r\left(\vartheta_{uT,T}^{\star}\right)\right]m_{uT,T}^{\star}\mathrm{d}u\\
&\quad + T\int_{\varepsilon_T}^{v} e^{-r(\vartheta_0)(v-u)T}\left[B\left(\vartheta_{uT,T}^{\star}\right)-B\left(\vartheta_0\right)\right]f\left(\vartheta_0\right)m\left(\vartheta_0,uT\right)\mathrm{d}u\\
&\quad + \sqrt{T}\sigma\int_{\varepsilon_T}^{v} e^{-r(\vartheta_0)(v-u)T}\left[B\left(\vartheta_{uT,T}^{\star}\right)-B\left(\vartheta_0\right)\right]\mathrm{d}W_T\left(u\right)\equiv\Delta_{vT,T},
\end{aligned}
$$

where $\varepsilon_T = T^{\delta-1}$ and $W_T(u) = T^{-1/2}\bar{W}_{uT}$ is a Wiener process.

Let us define $\nu_T = T^{-\kappa}$, $\kappa\in(0,1)$. Then

$$
\begin{aligned}
T^{3/2}&\int_{\varepsilon_T}^{v-\nu_T} e^{-r(\vartheta_0)(v-u)T}\left|\left[r\left(\vartheta_0\right)-r\left(\vartheta_{uT,T}^{\star}\right)\right]m_{uT,T}^{\star}\right|\mathrm{d}u\\
&\qquad\leq CT^{3/2}e^{-r(\vartheta_0)T^{1-\kappa}}\int_{\varepsilon_T}^{v-\nu_T}\left|m_{uT,T}^{\star}\right|\mathrm{d}u\to 0.
\end{aligned}
$$

We consider $\Delta_{vT,T}$ for $v\in(\varepsilon_0,1)$, where $\varepsilon_0\in(0,1)$. Therefore it suffices to consider the integrals

$$
\begin{aligned}
\sqrt{T}\Delta_{vT,T} &= T^{3/2}\int_{v-\nu_T}^{v} e^{-r(\vartheta_0)(v-u)T}\left[r\left(\vartheta_0\right)-r\left(\vartheta_{uT,T}^{\star}\right)\right]m_{uT,T}^{\star}\mathrm{d}u\\
&\quad + T^{3/2}\int_{v-\nu_T}^{v} e^{-r(\vartheta_0)(v-u)T}\left[B\left(\vartheta_{uT,T}^{\star}\right)-B\left(\vartheta_0\right)\right]f\left(\vartheta_0\right)m\left(\vartheta_0,uT\right)\mathrm{d}u\\
&\quad + \sigma T\int_{v-\nu_T}^{v} e^{-r(\vartheta_0)(v-u)T}\left[B\left(\vartheta_{uT,T}^{\star}\right)-B\left(\vartheta_0\right)\right]\mathrm{d}\bar{W}_T\left(u\right)+o\left(1\right).
\end{aligned}
$$

We have

$$
\begin{aligned}
&\mathbf{E}_{\vartheta_0}\left(\Delta_{t,T}\right)^2\\
&\quad\leq 3\mathbf{E}_{\vartheta_0}\left(\int_{\tau_T}^{t} e^{-r(\vartheta_0)(t-s)}\left[r\left(\vartheta_0\right)-r\left(\vartheta_{s,T}^{\star}\right)\right]m_{s,T}^{\star}\mathrm{d}s\right)^2\\
&\qquad + 3\mathbf{E}_{\vartheta_0}\left(\int_{\tau_T}^{t} e^{-r(\vartheta_0)(t-s)}\left[B\left(\vartheta_{s,T}^{\star}\right)-B\left(\vartheta_0\right)\right]f\left(\vartheta_0\right)m\left(\vartheta_0,s\right)\mathrm{d}s\right)^2\\
&\qquad + 3\sigma^2\int_{\tau_T}^{t} e^{-2r(\vartheta_0)(t-s)}\mathbf{E}_{\vartheta_0}\left[B\left(\vartheta_{s,T}^{\star}\right)-B\left(\vartheta_0\right)\right]^2\mathrm{d}s.
\end{aligned}
$$

Recall that the moments of One-step MLE-process and the derivatives $\dot{r}(\vartheta)$ and $\dot{B}(\vartheta)$ are bounded. Hence

$$\mathbf{E}_{\vartheta_0}\left|r\left(\vartheta_0\right)-r\left(\vartheta^\star_{uT,T}\right)\right|^2\leq C\mathbf{E}_{\vartheta_0}\left|\vartheta_0-\vartheta^\star_{uT,T}\right|^2\leq CT^{-1},$$
$$\mathbf{E}_{\vartheta_0}\left|B\left(\vartheta^\star_{uT,T}\right)-B\left(\vartheta_0\right)\right|^2\leq C\mathbf{E}_{\vartheta_0}\left|\vartheta^\star_{uT,T}-\vartheta_0\right|^2\leq CT^{-1}$$

and using elementary estimates we obtain

$$\mathbf{E}_{\vartheta_0}\left(\Delta_{t,T}\right)^2=o\left(1\right)\qquad\text{and}\qquad m^\star_{t,T}=m\left(\vartheta_0,t\right)+O\left(\delta_T\right),$$

where we denoted $\delta_T=\vartheta^\star_{uT,T}-\vartheta_0$.

We are interested in obtaining the asymptotic ($T\to\infty$) of the estimation errors $\Delta_{t,T}$ and $T\mathbf{E}_{\vartheta_0}\Delta^2_{t,T}$. To this end, consider the re-parametrization $t=vT$ and $\Delta_{vT,T}$. Note that

$$\begin{aligned}
&f\left(\vartheta_0\right)\left[B(\vartheta^\star_{uT,T})-B\left(\vartheta_0\right)\right]\\
&\quad=\frac{f\left(\vartheta_0\right)\Gamma(\vartheta^\star_{uT,T})}{f(\vartheta^\star_{uT,T})}-\Gamma(\vartheta_0)\\
&\quad=\Gamma(\vartheta^\star_{uT,T})-\Gamma\left(\vartheta_0\right)+\frac{\left(f\left(\vartheta_0\right)-f(\vartheta^\star_{uT,T})\right)}{f(\vartheta^\star_{uT,T})}\Gamma(\vartheta^\star_{uT,T})
\end{aligned}$$

and

$$\begin{aligned}
&\left[r\left(\vartheta_0\right)-r\left(\vartheta^\star_{uT,T}\right)\right]m^\star_{uT,T}+f\left(\vartheta_0\right)\left[B(\vartheta^\star_{uT,T})-B\left(\vartheta_0\right)\right]m\left(\vartheta_0,uT\right)\\
&\quad=\left[r\left(\vartheta_0\right)-r\left(\vartheta^\star_{uT,T}\right)\right]m\left(\vartheta_0,uT\right)\left(1+O\left(\delta_T\right)\right)\\
&\qquad+\left[\Gamma(\vartheta^\star_{uT,T})-\Gamma\left(\vartheta_0\right)\right]m\left(\vartheta_0,uT\right)\\
&\qquad+\frac{\left(f\left(\vartheta_0\right)-f(\vartheta^\star_{uT,T})\right)}{f(\vartheta^\star_{uT,T})}\Gamma(\vartheta^\star_{uT,T})m\left(\vartheta_0,uT\right)\\
&\quad=-\left[\dot a\left(\vartheta_0\right)+\frac{\dot f\left(\vartheta_0\right)}{f\left(\vartheta_0\right)}\Gamma\left(\vartheta_0\right)\right]m\left(\vartheta_0,uT\right)\delta_T+O\left(\delta_T^2\right).
\end{aligned}$$

These equations imply

$$\begin{aligned}
\sqrt{T}\Delta_{vT,T}&=T^{3/2}\int_{v-\nu_T}^{v}e^{-r(\vartheta_0)(v-u)T}\left[r\left(\vartheta_0\right)-r\left(\vartheta^\star_{uT,T}\right)\right]m\left(\vartheta_0,uT\right)\\
&\quad\times{\rm d}u\left(1+o\left(\delta_T\right)\right)\\
&\quad+T^{3/2}\int_{v-\nu_T}^{v}e^{-r(\vartheta_0)(v-u)T}\left[\frac{f\left(\vartheta_0\right)\Gamma\left(\vartheta^\star_{uT,T}\right)}{f(\vartheta^\star_{uT,T})}-\Gamma\left(\vartheta_0\right)\right]m\left(\vartheta_0,uT\right)\\
&\quad\times{\rm d}u\left(1+o\left(\delta_T\right)\right)
\end{aligned}$$

$$
\begin{aligned}
&+\sigma T\int_{v-\nu_T}^{v} e^{-r(\vartheta_0)(v-u)T}\left[B\left(\vartheta^\star_{uT,T}\right)-B\left(\vartheta_0\right)\right]\mathrm{d}\bar W_T\left(u\right)\\
&=-\left[\dot a\left(\vartheta_0\right)+\frac{\dot f\left(\vartheta_0\right)}{f\left(\vartheta_0\right)}\Gamma\left(\vartheta_0\right)\right]T^{3/2}\\
&\quad\times\int_{v-\nu_T}^{v} e^{-r(\vartheta_0)(v-u)T}\left(\vartheta^\star_{uT,T}-\vartheta_0\right)m\left(\vartheta_0,uT\right)\mathrm{d}u\\
&\quad+\dot B\left(\vartheta_0\right)\sigma T\int_{v-\nu_T}^{v} e^{-r(\vartheta_0)(v-u)T}\left(\vartheta^\star_{uT,T}-\vartheta_0\right)\mathrm{d}\bar W_T\left(u\right)+o\left(1\right)\\
&=C\left(\vartheta_0\right)T\int_{v-\nu_T}^{v} e^{-r(\vartheta_0)(v-u)T}u^{-1}\eta_T\left(u\right)m\left(\vartheta_0,uT\right)\mathrm{d}u\\
&\quad+D\left(\vartheta_0\right)\sqrt{T}\int_{v-\nu_T}^{v} e^{-r(\vartheta_0)(v-u)T}u^{-1}\eta_T\left(u\right)\mathrm{d}\bar W_T\left(u\right)+o\left(1\right),
\end{aligned}
$$

where

$$
C\left(\vartheta_0\right)=-\left[\dot a\left(\vartheta_0\right)+\frac{\dot f\left(\vartheta_0\right)}{f\left(\vartheta_0\right)}\Gamma\left(\vartheta_0\right)\right]\mathrm{I}\left(\vartheta_0\right)^{-1/2},\qquad D\left(\vartheta_0\right)=\dot B\left(\vartheta_0\right)\mathrm{I}\left(\vartheta_0\right)^{-1/2}\sigma.
$$

Thus

$$
\begin{aligned}
&\mathbf{E}_{\vartheta_0}T\int_{v-\nu_T}^{v} e^{-r(\vartheta_0)(v-u)T}\left|u^{-1}\eta_T\left(u\right)-v^{-1}\eta_T\left(v\right)\right|\left|m\left(\vartheta_0,uT\right)\right|\mathrm{d}u\\
&\leq T\int_{v-\nu_T}^{v} e^{-r(\vartheta_0)(v-u)T}\left(\mathbf{E}_{\vartheta_0}\left|u^{-1}\eta_T\left(u\right)-v^{-1}\eta_T\left(v\right)\right|^2\mathbf{E}_{\vartheta_0}\left|m\left(\vartheta_0,uT\right)\right|^2\right)^{1/2}\mathrm{d}u\\
&\leq CT\int_{v-\nu_T}^{v} e^{-r(\vartheta_0)(v-u)T}\left|v-u\right|^{1/2}\mathrm{d}u\leq C\nu_T^{1/2}\longrightarrow 0,
\end{aligned}
$$

where we used estimate (4.42).

Therefore

$$
\begin{aligned}
&T\int_{v-\nu_T}^{v} e^{-r(\vartheta_0)(v-u)T}u^{-1}\eta_T\left(u\right)m\left(\vartheta_0,uT\right)\mathrm{d}u\\
&\qquad=v^{-1}\eta_T\left(v\right)T\int_{v-\nu_T}^{v} e^{-r(\vartheta_0)(v-u)T}m\left(\vartheta_0,uT\right)\mathrm{d}u+O(\nu_T^{1/2}).
\end{aligned}
$$

The substitution of

$$
m\left(\vartheta_0,uT\right)=\frac{\Gamma\left(\vartheta_0\right)\sigma}{f\left(\vartheta_0\right)}\sqrt{T}\int_{-\infty}^{u} e^{-a(\vartheta_0)(u-z)T}\mathrm{d}\bar W_T\left(z\right)
$$

into the last integral gives us the following expression:

$$
\begin{aligned}
T\int_{v-\nu_T}^{v} & e^{-r(\vartheta_0)(v-u)T} m\left(\vartheta_0, uT\right) \mathrm{d}u \\
&= \frac{\Gamma\left(\vartheta_0\right)\sigma}{f\left(\vartheta_0\right)} T^{3/2}\int_{v-\nu_T}^{v} e^{-r(\vartheta_0)(v-u)T}\int_{-\infty}^{u} e^{-a(\vartheta_0)(u-z)T}\mathrm{d}\bar W_T\left(z\right)\,\mathrm{d}u \\
&= \frac{\Gamma\left(\vartheta_0\right)\sigma}{f\left(\vartheta_0\right)} T^{3/2} e^{-r(\vartheta_0)vT}\int_{-\infty}^{v} e^{a(\vartheta_0)zT}\int_{(v-\nu_T)\vee z}^{v} e^{\Gamma(\vartheta_0)uT}\mathrm{d}u\,\mathrm{d}\bar W_T\left(z\right) \\
&= \frac{\sigma\sqrt{T}}{f\left(\vartheta_0\right)} e^{-r(\vartheta_0)vT}\int_{-\infty}^{v} e^{a(\vartheta_0)zT}\left[e^{\Gamma(\vartheta_0)vT} - e^{\Gamma(\vartheta_0)[(v-\nu_T)\vee z]T}\right]\mathrm{d}\bar W_T\left(z\right) \\
&= \frac{\sigma\sqrt{T}}{f\left(\vartheta_0\right)}\int_{v-\nu_T}^{v}\left[e^{-a(\vartheta_0)(v-z)T} - e^{-r(\vartheta_0)(v-z)T}\right]\mathrm{d}\bar W_T\left(z\right)\left(1+o\left(1\right)\right).
\end{aligned}
$$

Note that

$$
\begin{aligned}
\mathbf{E}_{\vartheta_0} & \left(\sqrt{T}\int_{-\infty}^{v-\nu_T} e^{-r(\vartheta_0)(v-z)T}\,\mathrm{d}\bar W_T\left(z\right)\right)^2 \\
&= T\int_{-\infty}^{v-\nu_T} e^{-2r(\vartheta_0)(v-z)T}\,\mathrm{d}z = \frac{1}{2r\left(\vartheta_0\right)} e^{-2r(\vartheta_0)\nu_T T} \longrightarrow 0.
\end{aligned}
$$

Thus we obtained the representation

$$
\begin{aligned}
T\int_{v-\nu_T}^{v} & e^{-r(\vartheta_0)(v-u)T} m\left(\vartheta_0, uT\right) \mathrm{d}u \\
&= \frac{\sigma\sqrt{T}}{f\left(\vartheta_0\right)}\int_{v-\nu_T}^{v}\left[e^{-a(\vartheta_0)(v-z)T} - e^{-r(\vartheta_0)(v-z)T}\right]\mathrm{d}\bar W_T\left(z\right)\left(1+o\left(1\right)\right).
\end{aligned}
$$

Let us denote

$$
Q_T\left(v\right) = \sqrt{T}\int_{v-\nu_T}^{v} e^{-r(\vartheta_0)(v-u)T} u^{-1}\eta_T\left(u\right)\mathrm{d}\bar W_T\left(u\right),
$$

$$
U_T\left(v\right) = \eta_T\left(v\right)\sqrt{T}\int_{v-\nu_T}^{v} e^{-r(\vartheta_0)(v-u)T}\mathrm{d}\bar W_T\left(u\right),
$$

$$
g\left(\vartheta_0, zT\right) = \frac{\dot M\left(\vartheta_0, zT\right)}{\sigma\sqrt{\mathrm{I}\left(\vartheta_0\right)}},
$$

and calculate $\mathbf{E}_{\vartheta_0}\left[Q_T\left(v\right) - v^{-1}U_T\left(v\right)\right]^2$. Recall that $\mathbf{E}_{\vartheta_0} g\left(\vartheta_0, zT\right)^2 = 1$ and

$$\eta_T(v) = \frac{1}{\sigma\sqrt{\mathrm{I}(\vartheta_0)}} \int_{\varepsilon_T}^{v} \dot{M}(\vartheta_0, zT)\, d\bar{W}_T(z) = \int_{\varepsilon_T}^{v} g(\vartheta_0, zT)\, d\bar{W}_T(z).$$

We have

$$\begin{aligned}
\mathbf{E}_{\vartheta_0} Q_T(v)^2 &= T \int_{v-\nu_T}^{v} e^{-2r(\vartheta_0)(v-u)T} u^{-2} \mathbf{E}_{\vartheta_0} \eta_T(u)^2\, du \\
&= T \int_{v-\nu_T}^{v} e^{-2r(\vartheta_0)(v-u)T} u^{-1} du \left(1 + O(\varepsilon_T)\right) \\
&= \frac{1}{2r(\vartheta_0)\, v} + T \int_{v-\nu_T}^{v} e^{-2r(\vartheta_0)(v-u)T} \frac{(v-u)}{vu} du + O(\varepsilon_T) \\
&= \frac{1}{2r(\vartheta_0)\, v} + O(\nu_T) + O(\varepsilon_T).
\end{aligned}$$

Note that the integrals

$$I_{\varepsilon_T}^{v-\nu_T} = \int_{\varepsilon_T}^{v-\nu_T} g(\vartheta_0, zT) d\bar{W}_T(z) \text{ and } J_{v-\nu_T}^{v} = \sqrt{T} \int_{v-\nu_T}^{v} e^{-r(\vartheta_0)(v-u)T} d\bar{W}_T(u)$$

are independent random variables. This allows us to write

$$\begin{aligned}
\mathbf{E}_{\vartheta_0} U_T(v)^2 &= \mathbf{E}_{\vartheta_0} \left(\int_{\varepsilon_T}^{v} g(\vartheta_0, zT) d\bar{W}_T(z)\right)^2 \left(\sqrt{T} \int_{v-\nu_T}^{v} e^{-r(\vartheta_0)(v-u)T} d\bar{W}_T(u)\right)^2 \\
&= \mathbf{E}_{\vartheta_0} \left(I_{\varepsilon_T}^{v-\nu_T} + I_{v-\nu_T}^{v}\right)^2 \left(J_{v-\nu_T}^{v}\right)^2 \\
&= \mathbf{E}_{\vartheta_0} \left(I_{\varepsilon_T}^{v-\nu_T}\right)^2 \mathbf{E}_{\vartheta_0} \left(J_{v-\nu_T}^{v}\right)^2 + 2\mathbf{E}_{\vartheta_0} \left(I_{\varepsilon_T}^{v-\nu_T} I_{v-\nu_T}^{v}\right) \left(J_{v-\nu_T}^{v}\right)^2 \\
&\quad + \mathbf{E}_{\vartheta_0} \left(I_{v-\nu_T}^{v}\right)^2 \left(J_{v-\nu_T}^{v}\right)^2.
\end{aligned}$$

We have the following expressions for the first term

$$\mathbf{E}_{\vartheta_0} \left(I_{\varepsilon_T}^{v-\nu_T}\right)^2 \mathbf{E}_{\vartheta_0} \left(J_{v-\nu_T}^{v}\right)^2 = \left(v - \nu_T - \varepsilon_T\right) \frac{1}{2r(\vartheta_0)} (1 + o(1)),$$

for the second term

$$\begin{aligned}
\mathbf{E}_{\vartheta_0} \left(I_{\varepsilon_T}^{v-\nu_T} I_{v-\nu_T}^{v}\right) \left(J_{v-\nu_T}^{v}\right)^2 &\leq \left[\mathbf{E}_{\vartheta_0} \left(I_{v-\nu_T}^{v}\right)^2 \mathbf{E}_{\vartheta_0} \left(I_{\varepsilon_T}^{v-\nu_T}\right)^2 \left(J_{v-\nu_T}^{v}\right)^4\right]^{1/2} \\
&\leq C \nu_T^{1/2},
\end{aligned}$$

and for the third term

$$\mathbf{E}_{\vartheta_0}\left(I^{v}_{v-\nu_T}\right)^2\left(J^{v}_{v-\nu_T}\right)^2 \leq \left[\mathbf{E}_{\vartheta_0}\left(I^{v}_{v-\nu_T}\right)^4\mathbf{E}_{\vartheta_0}\left(J^{v}_{v-\nu_T}\right)^4\right]^{1/2}$$

$$\leq C\left[\mathbf{E}_{\vartheta_0}\left(\int_{v-\nu_T}^{v} g(\vartheta_0, zT)^2 \mathrm{d}z\right)^2\right]^{1/2} \leq C\nu_T.$$

Hence

$$\mathbf{E}_{\vartheta_0}\left[Q_T\left(v\right)-v^{-1}U_T\left(v\right)\right]^2 = O\left(\varepsilon_T\right)+O(\nu_T^{1/2}) \longrightarrow 0.$$

Thus we obtained the following representation for the derivative

$$\begin{aligned}\dot m\left(\vartheta_0, vT\right) &= -\frac{\sigma\left[\dot a\left(\vartheta_0\right)+B\left(\vartheta_0\right)\dot f\left(\vartheta_0\right)\right]}{f\left(\vartheta_0\right)\sqrt{2a\left(\vartheta_0\right)\mathrm{I}\left(\vartheta_0\right)}}\,\xi_T^{(1)}\left(v\right)\\ &\quad+\frac{\sigma\dot r\left(\vartheta_0\right)}{f\left(\vartheta_0\right)\sqrt{2r\left(\vartheta_0\right)\mathrm{I}\left(\vartheta_0\right)}}\,\xi_T^{(2)}\left(v\right)+o\left(1\right)\\ &= K_1\left(\vartheta_0\right)\xi_T^{(1)}\left(v\right)+K_2\left(\vartheta_0\right)\xi_T^{(2)}\left(v\right)+o\left(1\right).\end{aligned}$$

Now the limits (4.45) and (4.46) follow from the representation

$$\sqrt{T}\left(m^{\star}_{vT,T}-m\left(\vartheta_0, t\right)\right) = \frac{\eta_T\left(v\right)}{v}\sum_{i=1}^{2}K_i\left(\vartheta_0\right)\xi_T^{(i)}\left(v\right)+o\left(1\right).$$

and the properties of $\eta_T\left(v\right), \xi_T^{(i)}\left(v\right), i = 1, 2$ □

Remark 4.2 Suppose that the unknown parameter is $\vartheta = b$ and write the adaptive filter for the system

$$\begin{aligned}\mathrm{d}X_t &= f\,Y_t\,\mathrm{d}t+\sigma\,\mathrm{d}W_t, && X_0, && 0\leq t\leq T,\\ \mathrm{d}Y_t &= -a\,Y_t\,\mathrm{d}t+\vartheta\,\mathrm{d}V_t, && Y_0, && t\geq 0,\end{aligned}$$

in the recurrent form.

Recall that $\tau_T = T^{\delta}, \delta\in(\frac{1}{2}, 1)$, the preliminary MME and recursive One-step MLE process are

$$\vartheta^*_{\tau_T} = \left(\frac{a^3}{f^2\left[e^{-a}-1+a\right]\tau_T}\sum_{k=1}^{\tau_T}\left(\left[X_k-X_{k-1}\right]^2-\sigma^2\right)\right)^{1/2},$$

$$\mathrm{d}\vartheta^{\star}_{t,T} = \frac{\vartheta^{*}_{\tau_T} - \vartheta^{\star}_{t,T}}{t-\tau_T}\,\mathrm{d}t + \frac{2\sigma^3\, r(\vartheta^{*}_{\tau_T})^3\, \dot M(\vartheta^{*}_{\tau_T},t)}{f^4\,\vartheta^{*2}_{\tau_T}\left(t-\tau_T\right)}\left[\mathrm{d}X_t - M(\vartheta^{*}_{\tau_T},t)\mathrm{d}t\right],$$

$$\tau_T \quad < t \le T.$$

The equations for $M(\vartheta^{*}_{\tau_T},t)$, $\tau_T < t \le T$ and $\dot M(\vartheta^{*}_{\tau_T},t)$, $\tau_T < t \le T$ are

$$\mathrm{d}M(\vartheta^{*}_{\tau_T},t) = -r(\vartheta^{*}_{\tau_T})M(\vartheta^{*}_{\tau_T},t)\mathrm{d}t + \Gamma(\vartheta^{*}_{\tau_T})\mathrm{d}X_t, \qquad M(\vartheta^{*}_{\tau_T},0),$$

$$\mathrm{d}\dot M(\vartheta^{*}_{\tau_T},t) = -r(\vartheta^{*}_{\tau_T})\dot M(\vartheta^{*}_{\tau_T},t)\mathrm{d}t - \dot r(\vartheta^{*}_{\tau_T})M(\vartheta^{*}_{\tau_T},t)\mathrm{d}t + \dot\Gamma(\vartheta^{*}_{\tau_T})\mathrm{d}X_t,\ \dot M(\vartheta^{*}_{\tau_T},0),$$

where the initial values $M(\vartheta^{*}_{\tau_T},0)$ and $\dot M(\vartheta^{*}_{\tau_T},0)$ are calculated by formulas (4.38) and (4.39), respectively.

Remark 4.3 The approximation of $m\,(\vartheta,t)$ in the case of two-dimensional parameter, say, in the case **AB**: $\vartheta = (a,b)^\top = (\theta_1,\theta_2)$ with the Fisher information matrix can be written too, but the expressions will be more cumbersome, while the proof is close to the above proof in the one-dimensional case.

4.3.2 *Asymptotic Efficiency*

We have the same partially observed system

$$\mathrm{d}X_t = f\,(\vartheta)\; Y_t\,\mathrm{d}t + \sigma\,\mathrm{d}W_t, \qquad 0 \le t \le T,$$

$$\mathrm{d}Y_t = -a\,(\vartheta)\; Y_t\,\mathrm{d}t + b\,(\vartheta)\,\mathrm{d}V_t, \qquad t \ge 0,$$

subject to the initial conditions X_0 and Y_0, where the functions $f\,(\cdot)\,, a\,(\cdot)\,, b\,(\cdot)$ satisfy the conditions of Theorem 4.1. The adaptive filter is given by (4.44), and we would like to show that the error of approximation $\mathbf{E}_\vartheta\left|m^{\star}_{t,T} - m\,(\vartheta,t)\right|^2$ is asymptotically minimal. As before we propose a lower minimax bound on the risks of all estimators $\bar m_t$ assuming that their calculation is based on the observations up to time t, i.e., these estimators depend on $X^t = (X_s, 0 \le s \le t)$.

By Theorem 4.1 the MLE and BE are consistent, asymptotically normal, and the polynomial moments converge uniformly on compacts $\mathbb{K} \subset \Theta$. Moreover, the BE admits the representation similar to (4.41)

$$\sqrt{t}\left(\tilde\vartheta_t - \vartheta_0\right) = \frac{1}{\sigma \mathrm{I}\,(\vartheta_0)\,\sqrt{t}}\int_0^t \dot M\,(\vartheta_0,s)\,\mathrm{d}\bar W_s + o\,(1)\,.$$

Denote

$$\mathcal{S}^{\star}(\vartheta)^2 = K_1(\vartheta)^2 + K_2(\vartheta)^2 + R_{1,2}(\vartheta).$$

The following theorem proposes the lower bound for adaptive filter.

Theorem 4.6 *Let the conditions $\mathscr{K}$ hold. Then for any estimator $\bar m_{t,T}$ of $m(\vartheta,t)$ we have the following lower bound (below $t = \upsilon T$)*

$$\lim_{\nu\to 0}\varliminf_{T\to\infty}\sup_{|\vartheta-\vartheta_0|\le\nu} t\,\mathbf{E}_\vartheta\left|\bar m_{t,T} - m(\vartheta,t)\right|^2 \ge \mathcal{S}^{\star}(\vartheta_0)^2.$$

Proof This is a version of Theorem 1.14 already discussed in the Remark 1.7. Let us remind here the details in this large samples case. We have the elementary estimate

$$\sup_{|\vartheta-\vartheta_0|\le\nu}\mathbf{E}_\vartheta\left|\bar m_{t,T} - m(\vartheta,t)\right|^2 \ge \int_{\vartheta_0-\nu}^{\vartheta_0+\nu}\mathbf{E}_\vartheta\left|\bar m_{t,T} - m(\vartheta,t)\right|^2 p_\nu(\vartheta)\,\mathrm{d}\vartheta.$$

Here the function $p_\nu(\vartheta)$, $\vartheta_0-\nu<\vartheta<\vartheta_0+\nu$ is a positive continuous density. If we denote by $\tilde m_t$ the Bayesian estimator of $m(\vartheta,t)$, which corresponds to this density $p_\nu(\cdot)$, then

$$\tilde m_t = \int_{\vartheta_0-\nu}^{\vartheta_0+\nu} m(\theta,t)\,p_\nu\left(\theta|X^t\right)\mathrm{d}\theta,\qquad p_\nu\left(\theta|X^t\right) = \frac{p_\nu(\theta)\,L\left(\theta,X^t\right)}{\int_{\vartheta_0-\nu}^{\vartheta_0+\nu} p_\nu(\theta)\,L\left(\theta,X^t\right)\mathrm{d}\theta}$$

and

$$\int_{\vartheta_0-\nu}^{\vartheta_0+\nu}\mathbf{E}_\vartheta\left|\bar m_{t,T} - m(\vartheta,t)\right|^2 p_\nu(\vartheta)\,\mathrm{d}\vartheta \ge \int_{\vartheta_0-\nu}^{\vartheta_0+\nu}\mathbf{E}_\vartheta\left|\tilde m_t - m(\vartheta,t)\right|^2 p_\nu(\vartheta)\,\mathrm{d}\vartheta.$$

The asymptotic behavior of BE $\tilde m_t$ can be described as follows (below $\theta_u = \vartheta + \varphi_t u$, $\varphi_t = t^{-1/2}$, $\mathbb{U}_\nu = \left(\sqrt{t}\,(\vartheta_0-\nu-\vartheta),\sqrt{t}\,(\vartheta_0+\nu-\vartheta)\right)$):

$$\tilde m_t = \frac{\int_{\vartheta_0-\nu}^{\vartheta_0+\nu} m(\theta,t)\,p_\nu(\theta)\,L\left(\theta,X^t\right)\mathrm{d}\theta}{\int_{\vartheta_0-\nu}^{\vartheta_0+\nu} p_\nu(\theta)\,L\left(\theta,X^t\right)\mathrm{d}\theta} = \frac{\int_{\mathbb{U}_\nu} m(\theta_u,t)\,p_\nu(\theta_u)\,L\left(\theta_u,X^t\right)\mathrm{d}u}{\int_{\mathbb{U}_\nu} p_\nu(\theta_u)\,L\left(\theta_u,X^t\right)\mathrm{d}u}$$

$$= m(\vartheta,t) + \varphi_t\dot m(\vartheta,t)\frac{\int_{\mathbb{U}_\nu} u\,p_\nu(\theta_u)\,\dfrac{L\left(\theta_u,X^t\right)}{L\left(\vartheta,X^t\right)}\mathrm{d}u}{\int_{\mathbb{U}_\nu} p_\nu(\theta_u)\,\dfrac{L\left(\theta_u,X^t\right)}{L\left(\vartheta,X^t\right)}\mathrm{d}u}\left(1+o(1)\right)$$

$$= m(\vartheta,t) + \varphi_t \dot m(\vartheta,t)\frac{\displaystyle\int_{\mathbb{U}_\nu} u p_\nu(\vartheta) Z_t(\vartheta,u)\,du}{\displaystyle\int_{\mathbb{U}_\nu} p_\nu(\vartheta) Z_t(\vartheta,u)\,du}(1+o(1)).$$

Hence

$$\sqrt{t}\,(\tilde m_t - m(\vartheta,t)) = \dot m(\vartheta,t)\frac{\displaystyle\int_{\mathbb{U}_\nu} u Z_t(\vartheta,u)\,du}{\displaystyle\int_{\mathbb{U}_\nu} Z_t(\vartheta,u)\,du}(1+o(1))$$

$$= \dot m(\vartheta,t)\frac{\Delta_t\left(\vartheta,X^t\right)}{\mathrm{I}(\vartheta)}(1+o(1)).$$

Here

$$\Delta_t\left(\vartheta,X^t\right) = \frac{1}{\sigma\sqrt{t}}\int_0^t \dot M(\vartheta,s)\,\mathrm{d}\bar W_s \Longrightarrow \Delta \sim \mathcal{N}(0,\mathrm{I}(\vartheta))$$

Recall that

$$Z_t(\vartheta,u) \Longrightarrow Z(\vartheta,u) = \exp\left(u\Delta(\vartheta) - \frac{u^2}{2}\mathrm{I}(\vartheta)\right), \qquad \frac{\displaystyle\int_{\mathcal{R}} u Z(\vartheta,u)\,du}{\displaystyle\int_{\mathcal{R}} Z(\vartheta,u)\,du} = \frac{\Delta(\vartheta)}{\mathrm{I}(\vartheta)}.$$

We have to calculate the limit

$$\lim_{t\to\infty} \mathbf{E}_\vartheta\left[\dot m(\vartheta,t)\,\Delta_t\left(\vartheta,X^t\right)\right]^2.$$

The equation for the derivative is

$$\mathrm{d}\dot m(\vartheta,t) = -\left[\dot a(\vartheta) + B(\vartheta)\dot f(\vartheta)\right] m(\vartheta,t)\,\mathrm{d}t - r(\vartheta)\,\dot m(\vartheta,t)\,\mathrm{d}t + \dot B(\vartheta)\,\sigma\mathrm{d}\bar W_t.$$

Its solution

$$\dot m(\vartheta,t) = -\left[\dot a(\vartheta) + B(\vartheta)\dot f(\vartheta)\right]\int_0^t e^{-r(\vartheta)(t-s)} m(\vartheta,s)\,\mathrm{d}s$$

$$+ \dot B(\vartheta)\,\sigma\int_0^t e^{-r(\vartheta)(t-s)}\mathrm{d}\bar W_s$$

$$= -\left[\dot a(\vartheta) + B(\vartheta)\dot f(\vartheta)\right] B(\vartheta)\,\sigma\int_0^t e^{-r(\vartheta)(t-s)}\int_0^s e^{-a(\vartheta)(s-v)}\mathrm{d}\bar W_v\mathrm{d}s$$

$$+\dot{B}(\vartheta)\sigma\int_0^t e^{-r(\vartheta)(t-s)}\mathrm{d}\bar{W}_s.$$

We have

$$\int_0^t e^{-r(\vartheta)(t-s)}\int_0^s e^{-a(\vartheta)(s-v)}\mathrm{d}\bar{W}_v\,\mathrm{d}s$$
$$=\frac{1}{r(\vartheta)-a(\vartheta)}\int_0^t\left[e^{-a(\vartheta)(t-s)}-e^{-r(\vartheta)(t-s)}\right]\mathrm{d}\bar{W}_s.$$

Hence

$$\dot{m}(\vartheta,t)=-\frac{\left[\dot{a}(\vartheta)+B(\vartheta)\dot{f}(\vartheta)\right]\sigma}{f(\vartheta)}\int_0^t e^{-a(\vartheta)(t-s)}\mathrm{d}\bar{W}_s$$
$$+\frac{\dot{r}(\vartheta)\sigma}{f(\vartheta)}\int_0^t e^{-r(\vartheta)(t-s)}\mathrm{d}\bar{W}_s$$

and

$$\mathbf{E}_\vartheta\dot{m}(\vartheta,t)^2=\frac{\left[\dot{a}(\vartheta)+B(\vartheta)\dot{f}(\vartheta)\right]^2\sigma^2}{2a(\vartheta)f(\vartheta)^2}+\frac{\dot{r}(\vartheta)^2\sigma^2}{2r(\vartheta)f(\vartheta)^2}$$
$$-\frac{2\left[\dot{a}(\vartheta)+B(\vartheta)\dot{f}(\vartheta)\right]\dot{r}(\vartheta)\sigma^2}{f(\vartheta)^2\left[a(\vartheta)+r(\vartheta)\right]}$$

Remark that in the proof of Theorem 4.5 it was shown that asymptotically $\dot{m}(\vartheta,t)$ and $\Delta_t\left(\vartheta,X^t\right)$ are independent and

$$\mathbf{E}_\vartheta\left[\dot{m}(\vartheta,t)\Delta_t\left(\vartheta,X^t\right)\right]^2=\mathbf{E}_\vartheta\dot{m}(\vartheta,t)^2\;\mathbf{E}_\vartheta\Delta_t\left(\vartheta,X^t\right)^2(1+o(1))$$
$$\longrightarrow\frac{\left[\dot{a}(\vartheta)+B(\vartheta)\dot{f}(\vartheta)\right]^2\sigma^2}{2a(\vartheta)f(\vartheta)^2\mathrm{I}(\vartheta)}+\frac{\dot{r}(\vartheta)^2\sigma^2}{2r(\vartheta)f(\vartheta)^2\mathrm{I}(\vartheta)}$$
$$-\frac{2\left[\dot{a}(\vartheta)+B(\vartheta)\dot{f}(\vartheta)\right]\dot{r}(\vartheta)\sigma^2}{f(\vartheta)^2\left[a(\vartheta)+r(\vartheta)\right]\mathrm{I}(\vartheta)}$$
$$=\mathcal{S}^\star(\vartheta)^2.$$

Therefore we obtained the convergence

$$t\mathbf{E}_\vartheta\left(\tilde{m}_t-m(\vartheta,t)\right)^2\rightarrow\lim_{t\to\infty}t\mathbf{E}_\vartheta\left[\dot{m}(\vartheta,t)^2\left(\tilde{\vartheta}_t-\vartheta\right)^2\right]=\mathcal{S}^\star(\vartheta)^2.$$

It can be shown that this convergence is uniform on compacts $\mathbb{K}\subset(\vartheta_0-\nu,\vartheta_0+\nu)$ as well.

Thus

$$t\int_{\vartheta_0-\nu}^{\vartheta_0+\nu}\mathbf{E}_\vartheta\left|\tilde m_t-m\left(\vartheta,t\right)\right|^2 p_\nu\left(\vartheta\right)\mathrm{d}\vartheta\longrightarrow\int_{\vartheta_0-\nu}^{\vartheta_0+\nu}\mathcal{S}^\star\left(\vartheta\right)^2 p_\nu\left(\vartheta\right)\mathrm{d}\vartheta$$

and, as $\nu\rightarrow 0$,

$$\int_{\vartheta_0-\nu}^{\vartheta_0+\nu}\mathcal{S}^\star\left(\vartheta\right)^2 p_\nu\left(\vartheta\right)\mathrm{d}\vartheta\longrightarrow\mathcal{S}^\star\left(\vartheta_0\right)^2.$$

□

The estimator $m^\circ_{t,T},\ \tau_T<t\leq T$ is called *asymptotically efficient* if for all $\vartheta_0\in\Theta$, $t=vT$ and any $v\in[\varepsilon_0,1]$

$$\lim_{\nu\rightarrow 0}\lim_{T\rightarrow\infty}\sup_{|\vartheta-\vartheta_0|\leq\nu} t\mathbf{E}_\vartheta\left|m^\circ_{t,T}-m\left(\vartheta,t\right)\right|^2=\mathcal{S}^\star\left(\vartheta_0\right)^2.$$

Here $\varepsilon_0\in(0,1)$.

Theorem 4.7 *Let the conditions $\mathscr{K}^\star$ hold. Then the estimator $m^\star_t$, $\tau_T<t\leq T$ is asymptotically efficient.*

Proof The comparison of the limit in (4.46) and $\mathcal{S}^\star\left(\vartheta_0\right)^2$ shows that

$$\mathcal{S}^\star\left(\vartheta_0\right)^2=K_1\left(\vartheta_0\right)^2+K_1\left(\vartheta_0\right)^2+2R_{12}\left(\vartheta_0\right).$$

□

Chapter 5
Hidden Telegraph Process

The problem of adaptive filtering for the two-state telegraph process observed in white Gaussian noise is considered. We assume that the infinitesimal matrix depends on two unknown parameters. The construction of the adaptive Wonham filter begins with the estimation of these parameters based on observations over a relatively small learning interval using the method of moments. These estimators are then used to define the One-step MLE-process. Substituting this estimator process into the filtering equation provides the adaptive filter for the conditional expectation of the unobserved component. We then consider a two-state process based on the adaptive filter as an estimator for the telegraph process and describe the asymptotic behavior of the estimation error. The asymptotic properties of the MME, MLE, and BE estimators are also described.

5.1 Model of Observations

The observed process $X^T = (X_t, 0 \leq t \leq T)$ satisfies the SDE

$$\mathrm{d}X_t = Y_t\,\mathrm{d}t + \mathrm{d}W_t, \qquad X_0 = 0, \qquad 0 \leq t \leq T, \tag{5.1}$$

where $Y_t, 0 \leq t \leq T$ is *telegraph process* and $W_t, 0 \leq t \leq T$ is independent of $Y_t, 0 \leq t \leq T$ Wiener process. Recall that the telegraph process $Y_t, t \geq 0$ is a memoryless continuous-time stochastic Markov process that shows two distinct values $y_1 = a$ and $y_2 = b$. It models burst noise (also called *popcorn noise* or *random telegraph signal*). This process can be described by the transition probabilities $P_{y_i y_j}(t) = \mathbf{P}\left(Y_t = y_j | Y_0 = y_i\right)$:

$$P_{aa}(t) = \frac{\mu}{\lambda+\mu} + \frac{\lambda}{\lambda+\mu} e^{-(\lambda+\mu)t}, \quad P_{ba}(t) = \frac{\lambda}{\lambda+\mu} - \frac{\lambda}{\lambda+\mu} e^{-(\lambda+\mu)t},$$

Y. A. Kutoyants, *Hidden Markov Processes and Adaptive Filtering*, Springer Series in Statistics, https://doi.org/10.1007/978-3-032-00052-1_5

$$P_{ab}(t)=\frac{\mu}{\lambda+\mu}-\frac{\mu}{\lambda+\mu}e^{-(\lambda+\mu)t},\quad P_{bb}(t)=\frac{\lambda}{\lambda+\mu}+\frac{\mu}{\lambda+\mu}e^{-(\lambda+\mu)t}. \tag{5.2}$$

Here the parameters $\lambda\geq 0$ and $\mu\geq 0$ are transition rates, i.e., λ is the transition rate for going from state a to state b and μ is the transition rate for going from state b to state a.

It follows from (5.2) that

$$\begin{aligned}K(s)=\mathbf{E}[Y_tY_{t+s}]&=\left(\frac{a\mu+b\lambda}{\lambda+\mu}\right)^2+(b-a)^2\frac{\lambda\mu}{(\lambda+\mu)^2}e^{-(\lambda+\mu)s}\\&=\left(\bar{Y}\right)^2+De^{-(\lambda+\mu)s},\end{aligned} \tag{5.3}$$

where

$$\bar{Y}=\mathbf{E}Y_t=\frac{a\mu+b\lambda}{\lambda+\mu},\qquad D=(b-a)^2\frac{\lambda\mu}{(\lambda+\mu)^2}. \tag{5.4}$$

The stationary distribution of the process $Y_t, t\geq 0$ is given by

$$\mathbf{P}(Y_t=a)=\frac{\mu}{\lambda+\mu},\qquad \mathbf{P}(Y_t=b)=\frac{\lambda}{\lambda+\mu}.$$

Let $\mathfrak{F}_t^Y\subset\mathfrak{F}$ be the family of σ-algebras of events $\{Y_s=y_i, 0\leq s\leq t, i=1,2\}$. For some constant $K>0$, any $A<T$ and for all $s>A, t>0$ the inequality

$$\left|\mathbf{E}\left(Y_{s+T}Y_{t+T}|\mathfrak{F}_A^Y\right)-\mathbf{E}\left(Y_sY_t\right)\right|<Ke^{-(\lambda+\mu)(T-A)}$$

holds. More about telegraph processes see in [98].

The parameters a, b, λ, μ can be considered as unknown, but we will mainly study this model of observations with unknown λ and μ. Therefore we estimate $\vartheta=(\lambda,\mu)=(\theta_1,\theta_2)$. The true value of ϑ is denoted ϑ_0. Now the probabilities and mathematical expectations we denote depending on $\vartheta\in\Theta$, where $\Theta=(\alpha,\beta)\times(\alpha,\beta), \alpha>0, \beta<\infty$.

By the *Innovation lemma* (see Lemma 1.2), the stochastic process (5.1) admits the representation

$$\mathrm{d}X_t=m(\vartheta_0,t)\,\mathrm{d}t+\mathrm{d}\bar{W}_t,\qquad X_0,\qquad 0\leq t\leq T, \tag{5.5}$$

where $m(\vartheta,t)$ is the conditional expectation

$$m(\vartheta,t)=\mathbf{E}_\vartheta\left(Y_t|\,\mathfrak{F}_t^X\right)=a\mathbf{P}_\vartheta\left(Y_t=a|\,\mathfrak{F}_t^X\right)+b\mathbf{P}_\vartheta\left(Y_t=b|\,\mathfrak{F}_t^X\right).$$

Here $\mathfrak{F}_t^X$ is the σ-algebra of events generated by the observations up to time t, i.e., $\mathfrak{F}_t^X := \sigma\left(X_s, 0 \leq s \leq t\right)$ and $\bar{W}_t, 0 \leq t \leq T$ is the *innovation* Wiener process. Let us denote

$$\pi\left(\vartheta, t\right) = \mathbf{P}_\vartheta\left(Y_t = a \middle| \mathfrak{F}_t^X\right).$$

Then $\mathbf{P}_\vartheta\left(Y_t = b \middle| \mathfrak{F}_t^X\right) = 1 - \pi\left(\vartheta, t\right)$ and

$$m\left(\vartheta, t\right) = b + \left(a - b\right)\pi\left(\vartheta, t\right).$$

The random process $\pi\left(\vartheta, t\right), 0 \leq t \leq T$ satisfies the following Wonham equation [197] (see as well [141], Theorem 9.1 and equation (9.23) therein)

$$\begin{aligned} \mathrm{d}\pi\left(\vartheta, t\right) = & \left[\mu - \left(\lambda + \mu\right)\pi\left(\vartheta, t\right)\right. \\ & \left. + \pi\left(\vartheta, t\right)\left(1 - \pi\left(\vartheta, t\right)\right)\left(b - a\right)\left(b + \left(a - b\right)\pi\left(\vartheta, t\right)\right)\right]\mathrm{d}t \\ & + \pi\left(\vartheta, t\right)\left(1 - \pi\left(\vartheta, t\right)\right)\left(a - b\right)\,\mathrm{d}X_t, \qquad \pi\left(\vartheta, 0\right) = \pi_0, \end{aligned} \tag{5.6}$$

where $\pi_0 \in (0, 1)$.

Denote by $\left\{\mathbf{P}_\vartheta^{(T)}, \vartheta \in \Theta\right\}$ the family of measures, induced by the process $X^T = \left(X_t, 0 \leq t \leq T\right)$ from (5.1) on the measurable space of realizations $\left(\mathcal{C}\left[0, T\right], \mathfrak{B}_T\right)$ (continuous functions on $[0, T]$ with Borelian σ-algebra) for different values of the parameter $\vartheta \in \Theta$. These measures are equivalent and the likelihood ratio function

$$L\left(\vartheta, X^T\right) = \frac{\mathrm{d}\mathbf{P}_\vartheta^{(T)}}{\mathrm{d}\mathbf{P}_0^{(T)}}\left(X^T\right), \qquad \vartheta \in \Theta,$$

can be written as follows:

$$L\left(\vartheta, X^T\right) = \exp\left(\int_0^T m\left(\vartheta, t\right)\mathrm{d}X_t - \frac{1}{2}\int_0^T m\left(\vartheta, t\right)^2\,\mathrm{d}t\right). \tag{5.7}$$

Here $\mathbf{P}_0^{(T)}$ is the measure corresponding to observations X^T with $Y_t \equiv 0$, i.e., $X_t = W_t, 0 \leq t \leq T$.

Remark that as $\lambda_0 > 0$ and $\mu_0 > 0$, the process $\pi\left(\vartheta_0, t\right) = \pi^\circ \in (0, 1)$ is ergodic with two reflecting boundaries at 0 and 1 and has the invariant density

$$f\left(\vartheta_0, x\right) = \frac{D\left(\vartheta_0\right)}{x^2\left(1 - x\right)^2}\left(\frac{x}{1 - x}\right)^{\gamma\left(\mu_0 - \lambda_0\right)}\exp\left\{-\frac{\gamma\mu_0}{x} - \frac{\gamma\lambda_0}{1 - x}\right\}, \qquad x \in (0, 1), \tag{5.8}$$

where we denote $\gamma = 2\left(a - b\right)^{-2}$ and the normalizing constant $D\left(\vartheta_0\right)$ is

$$D\left(\vartheta_0\right)^{-1}=\int_0^1 \frac{1}{x^2\left(1-x\right)^2}\left(\frac{x}{1-x}\right)^{\gamma\left(\mu_0-\lambda_0\right)} \exp\left\{-\frac{\gamma\mu_0}{x}-\frac{\gamma\lambda_0}{1-x}\right\}\,\mathrm{d}x. \tag{5.9}$$

Let us denote $g=\theta_1+\theta_2$ and $h=a-b$. Then the Wonham equation (5.6) becomes

$$\begin{aligned}\mathrm{d}\pi\left(\vartheta,t\right)&=\left[\theta_2-g\pi\left(\vartheta,t\right)-h\pi\left(\vartheta,t\right)\left(1-\pi\left(\vartheta,t\right)\right)\left(b+h\pi\left(\vartheta,t\right)\right)\right]\mathrm{d}t\\&\quad+h\pi\left(\vartheta,t\right)\left(1-\pi\left(\vartheta,t\right)\right)\,\mathrm{d}X_t,\qquad \pi\left(\vartheta,0\right).\end{aligned} \tag{5.10}$$

We will need as well the limit theorems for $\pi\left(\vartheta,t\right)$ and for its derivatives

$$\dot\pi_\lambda\left(\vartheta,t\right)=\frac{\partial\pi\left(t,\vartheta\right)}{\partial\theta_1},\qquad \dot\pi_\mu\left(\vartheta,t\right)=\frac{\partial\pi\left(t,\vartheta\right)}{\partial\theta_2}.$$

Using the standard arguments it can be shown that the stochastic process $\pi\left(\vartheta,t\right)$ has $\mathcal{L}_2$ continuous derivatives w.r.t. θ_1 and θ_2 (see the proof of Lemma 1.3 and the estimate (1.40)). The equations for the processes $\dot\pi_\lambda\left(\vartheta,t\right)$ and $\dot\pi_\mu\left(\vartheta,t\right)$ we obtain by the formal differentiation of the Wonham equation (5.10)

$$\begin{aligned}\mathrm{d}\dot\pi_\lambda\left(\vartheta,t\right)&=-\,\pi\left(\vartheta,t\right)\,\mathrm{d}t-\dot\pi_\lambda\left(\vartheta,t\right)\Big[g+h\left(1-2\pi\left(\vartheta,t\right)\right)\left[b+h\pi\left(\vartheta,t\right)\right]\\&\quad+h^2\pi\left(\vartheta,t\right)\left(1-\pi\left(\vartheta,t\right)\right)\Big]\mathrm{d}t+h\dot\pi_\lambda\left(\vartheta,t\right)\left(1-2\pi\left(\vartheta,t\right)\right)\mathrm{d}X_t,\\ \mathrm{d}\dot\pi_\mu\left(\vartheta,t\right)&=\left[1-\pi\left(\vartheta,t\right)\right]\mathrm{d}t-\dot\pi_\mu\left(\vartheta,t\right)\Big[g+h\left(1-2\pi\left(\vartheta,t\right)\right)\left[b+h\pi\left(\vartheta,t\right)\right]\\&\quad+h^2\pi\left(\vartheta,t\right)\left(1-\pi\left(\vartheta,t\right)\right)\Big]\mathrm{d}t+h\dot\pi_\mu\left(\vartheta,t\right)\left(1-2\pi\left(\vartheta,t\right)\right)\,\mathrm{d}X_t.\end{aligned}$$

Let us study the moments of the processes $\dot\pi_\lambda\left(\vartheta,t\right)$ and $\dot\pi_\mu\left(\vartheta,t\right)$.

Lemma 5.1 *For any $p=1,2,\ldots$ if*

$$\frac{\left(a-b\right)^2}{\alpha}<\frac{4}{p+1},$$

then there exists a constant $C_{\pi,p}$ which does not depend on t,ϑ_0 such that

$$\sup_{\vartheta\in\Theta}\mathbf{E}_{\vartheta_0}\left|\dot\pi_\lambda\left(\vartheta,t\right)\right|^p<C_{\pi,p},\qquad \sup_{\vartheta\in\Theta}\mathbf{E}_{\vartheta_0}\left|\dot\pi_\mu\left(\vartheta,t\right)\right|^p<C_{\pi,p}. \tag{5.11}$$

Particularly

$$C_{\pi,1}=\frac{1}{2\alpha-(a-b)^2},\qquad C_{\pi,2}=\frac{2}{\left(2\alpha-(a-b)^2\right)\left(4\alpha-3\,(a-b)^2\right)}. \tag{5.12}$$

Proof The substitution of the observations (5.5) in the equations for $\dot{\pi}_\lambda(\vartheta,t)$ and $\dot{\pi}_\mu(\vartheta,t)$ gives us the system

$$\begin{aligned}
\mathrm{d}\dot{\pi}_\lambda(\vartheta,t) &= -\pi(\vartheta,t)\,\mathrm{d}t + h\dot{\pi}_\lambda(\vartheta,t)\,(1-2\pi(\vartheta,t))\,\mathrm{d}\bar{W}_t - g\dot{\pi}_\lambda(\vartheta,t)\,\mathrm{d}t\\
&\quad - h^2\dot{\pi}_\lambda(\vartheta,t)\Big[(1-2\pi(\vartheta,t))\,[\pi(\vartheta,t)-\pi(\vartheta_0,t)]+\pi(\vartheta,t)\,(1-\pi(\vartheta,t))\Big]\mathrm{d}t,
\end{aligned} \tag{5.13}$$

$$\begin{aligned}
\mathrm{d}\dot{\pi}_\mu(\vartheta,t) &= [1-\pi(\vartheta,t)]\,\mathrm{d}t + h\dot{\pi}_\mu(\vartheta,t)\,(1-2\pi(\vartheta,t))\,\mathrm{d}\bar{W}_t - g\dot{\pi}_\mu(\vartheta,t)\,\mathrm{d}t\\
&\quad - h^2\dot{\pi}_\mu(\vartheta,t)\Big[(1-2\pi(\vartheta,t))\,[\pi(\vartheta,t)-\pi(\vartheta_0,t)]+\pi(\vartheta,t)\,(1-\pi(\vartheta,t))\Big]\mathrm{d}t
\end{aligned}$$

If we set $\vartheta=\vartheta_0$, $g_0=\theta_{0,1}+\theta_{0,2}$, then the system becomes

$$\mathrm{d}\pi(\vartheta_0,t) = [\mu-g_0\pi(\vartheta_0,t)]\,\mathrm{d}t + h\pi(\vartheta_0,t)\,(1-\pi(\vartheta_0,t))\;\mathrm{d}\bar{W}_t,$$

$$\begin{aligned}
\mathrm{d}\dot{\pi}_\lambda(\vartheta_0,t) &= -\pi(\vartheta_0,t)\;\mathrm{d}t - \dot{\pi}_\lambda(\vartheta_0,t)\Big[g_0+h^2\pi(\vartheta_0,t)\,(1-\pi(\vartheta_0,t))\Big]\mathrm{d}t\\
&\quad + h\dot{\pi}_\lambda(\vartheta_0,t)\,(1-2\pi(\vartheta_0,t))\;\mathrm{d}\bar{W}_t,
\end{aligned} \tag{5.14}$$

$$\begin{aligned}
\mathrm{d}\dot{\pi}_\mu(\vartheta_0,t) &= [1-\pi(\vartheta_0,t)]\,\mathrm{d}t - \dot{\pi}_\mu(\vartheta_0,t)\Big[g_0+h^2\pi(\vartheta_0,t)\,(1-\pi(\vartheta_0,t))\Big]\mathrm{d}t\\
&\quad + h\dot{\pi}_\mu(\vartheta_0,t)\,(1-2\pi(\vartheta_0,t))\;\mathrm{d}\bar{W}_t.
\end{aligned} \tag{5.15}$$

Denote

$$\psi_t(\pi,\pi^\circ) = (1-2\pi(\vartheta,t))\,[\pi(\vartheta,t)-\pi(\vartheta_0,t)]+\pi(\vartheta,t)\,(1-\pi(\vartheta,t)).$$

The minimum of this function is

$$\min_{0\le\pi,\pi^\circ\le 1}\psi_t(\pi,\pi^\circ) = -1. \tag{5.16}$$

Hence for the solution of (5.13) we can write the estimate

$$\begin{aligned}
&\mathbf{E}_{\vartheta_0}\left|\dot{\pi}_\lambda(\vartheta,s)\right|\\
&\quad=\mathbf{E}_{\vartheta_0}\int_0^t \pi(\vartheta,t)\exp\left\{-\int_s^t\left[g+h^2\psi_v(\pi,\pi^\circ)\right]\mathrm{d}v\right\}\\
&\qquad\times\exp\left\{\int_s^t h\,(1-2\pi(\vartheta,v))\,\mathrm{d}\bar{W}_v-\frac{1}{2}\int_s^t h^2\,(1-2\pi(\vartheta,v))^2\,\mathrm{d}v\right\}\,\mathrm{d}s
\end{aligned}$$

$$\leq \int_0^t \exp\left\{-\int_s^t \left[g-h^2\right]\mathrm{d}v\right\}\,\mathrm{d}s$$

$$= \frac{1}{g-h^2}\left[1-e^{-(g-h^2)t}\right].$$

Here we used the estimates $\pi\left(\vartheta,t\right)\leq 1$, (5.16) and the equality

$$\mathbf{E}_{\vartheta_0}\exp\left\{\int_s^t h\left(1-2\pi\left(\vartheta,v\right)\right)\mathrm{d}\bar{W}_v-\frac{1}{2}\int_s^t h^2\left(1-2\pi\left(\vartheta,v\right)\right)^2\mathrm{d}v\right\}=1. \tag{5.17}$$

Therefore the first estimate of the derivative is

$$\sup_{\vartheta\in\Theta}\mathbf{E}_{\vartheta_0}\left|\dot{\pi}_\lambda\left(\vartheta,t\right)\right| < \sup_{\vartheta\in\Theta}\frac{1}{\theta_1+\theta_2-h^2}\leq\frac{1}{2\alpha-(a-b)^2}\equiv C_{\pi,1}.$$

The same estimate we have for the derivative $\dot{\pi}_\mu\left(\vartheta,t\right)$. The sufficient condition for the finiteness of these expectations is

$$\frac{(a-b)^2}{\alpha}<2.$$

For the second moment of the derivative, similar estimates are the following. We have

$$\begin{aligned}\mathrm{d}\dot{\pi}_\lambda\left(\vartheta,t\right)^2 &= 2\dot{\pi}_\lambda\left(\vartheta,t\right)\mathrm{d}\dot{\pi}_\lambda\left(\vartheta,t\right)+\dot{\pi}_\lambda\left(\vartheta,t\right)^2h^2\left(1-2\pi\left(\vartheta,t\right)\right)^2\mathrm{d}t\\ &= -2\pi\left(\vartheta,t\right)\dot{\pi}_\lambda\left(\vartheta,t\right)\mathrm{d}t+2h\dot{\pi}_\lambda\left(\vartheta,t\right)^2\left(1-2\pi\left(\vartheta,t\right)\right)\mathrm{d}\bar{W}_t\\ &\quad -\pi_\lambda\left(\vartheta,t\right)^2\left[2g+h^2\phi_t\left(\pi,\pi^\circ\right)\right]\mathrm{d}t,\end{aligned}$$

where

$$\begin{aligned}\phi_t\left(\pi,\pi^\circ\right) &= 2\left(1-2\pi\left(\vartheta,t\right)\right)\left(\pi\left(\vartheta,t\right)-\pi\left(\vartheta_0,t\right)\right)+2\pi\left(\vartheta,t\right)\left(1-\pi\left(\vartheta,t\right)\right)\\ &\quad -\left(1-2\pi\left(\vartheta,t\right)\right)^2.\end{aligned}$$

Therefore

$$\begin{aligned}\dot{\pi}_\lambda\left(\vartheta,t\right)^2 &= -2\int_0^t \pi\left(\vartheta,s\right)\dot{\pi}_\lambda\left(\vartheta,s\right)\exp\left\{-\int_s^t\left[2g+h^2\phi_v\left(\pi,\pi^\circ\right)\right]\mathrm{d}v\right\}\\ &\quad\times\exp\left\{\int_s^t 2h\left(1-2\pi\left(\vartheta,v\right)\right)\mathrm{d}\bar{W}_v-\int_s^t 2h^2\left(1-2\pi\left(\vartheta,v\right)\right)^2\mathrm{d}v\right\}\mathrm{d}s.\end{aligned}$$

Remark that

$$\min_{0\leq\pi,\pi^{\circ}\leq1}\ \phi_t\left(\pi,\pi^{\circ}\right)=-3.$$

Hence

$$\begin{aligned}\mathbf{E}_{\vartheta_0}\dot{\pi}_\lambda\left(\vartheta,t\right)^2&\leq 2\int_0^t \mathbf{E}_{\vartheta_0}\left|\dot{\pi}_\lambda\left(\vartheta,s\right)\right|\exp\Bigl\{-\left[2g-3h^2\right](t-s)\Bigr\}\mathrm{d}s\\&\leq\frac{2}{g-h^2}\int_0^t \exp\Bigl\{-\left[2g-3h^2\right](t-s)\Bigr\}\mathrm{d}s.\end{aligned}$$

Here we used the equality (5.17) and the relations

$$\begin{aligned}&\mathbf{E}_{\vartheta_0}\left[\left|\dot{\pi}_\lambda\left(\vartheta,s\right)\right|\mathbf{E}_{\vartheta_0}\left(\exp\left\{\int_s^t 2H\left(\vartheta,v\right))\mathrm{d}\bar{W}_v-\int_s^t 2H\left(\vartheta,v\right)^2\mathrm{d}v\right\}\middle|\mathfrak{F}_s^X\right)\right]\\&\qquad=\mathbf{E}_{\vartheta_0}\left|\dot{\pi}_\lambda\left(\vartheta,s\right)\right|\leq\frac{1}{g-h^2},\end{aligned}$$

where $H\left(\vartheta,v\right)=h\left(1-2\pi\left(\vartheta,v\right)\right)$. Finally

$$\sup_{\vartheta\in\Theta}\mathbf{E}_{\vartheta_0}\dot{\pi}_\lambda\left(\vartheta,t\right)^2\leq\sup_{\vartheta\in\Theta}\frac{2}{\left(g-h^2\right)\left(2g-3h^2\right)}\leq\frac{2}{\left(2\alpha-h^2\right)\left(4\alpha-3h^2\right)}\equiv C_{\pi,2}.$$

The same estimate we have for $\mathbf{E}_{\vartheta_0}\dot{\pi}_\mu\left(\vartheta,t\right)^2$.

Suppose that $(p+1)\,h^2<4\alpha$ and we already proved (5.11) for the $p-1$-th moments. Consider the p-th moment. The equation for $\pi_\lambda\left(\vartheta,t\right)^p$ is

$$\begin{aligned}\mathrm{d}\dot{\pi}_\lambda\left(\vartheta,t\right)^p&=-p\pi_\lambda\left(\vartheta,t\right)\dot{\pi}_\lambda\left(\vartheta,t\right)^{p-1}\mathrm{d}t+ph\dot{\pi}_\lambda\left(\vartheta,t\right)^p\left(1-2\pi_\lambda\left(\vartheta,t\right)\right)\mathrm{d}\bar{W}_t\\&\quad-p\dot{\pi}_\lambda\left(\vartheta,t\right)^p\left[g+h^2\psi_t\left(\pi,\pi^{\circ},p\right)\right]\mathrm{d}t,\end{aligned}$$

where

$$\begin{aligned}\psi_t\left(\pi,\pi^{\circ},p\right)&=\left(1-2\pi\left(\vartheta,t\right)\right)\left(\pi\left(\vartheta,t\right)-\pi\left(\vartheta_0,t\right)\right)+\pi\left(\vartheta,t\right)\left(1-\pi\left(\vartheta,t\right)\right)\\&\quad-\frac{(p-1)}{2}\left(1-2\pi\left(\vartheta,t\right)\right)^2.\end{aligned}$$

We have

$$\begin{aligned}&\dot{\pi}_\lambda\left(\vartheta,t\right)^p\\&\quad=-p\int_0^t\pi\left(\vartheta,s\right)\dot{\pi}_\lambda\left(\vartheta,s\right)^{p-1}\exp\left\{-p\int_s^t\left[g+h^2\phi_v\left(\pi,\pi^{\circ},p\right)\right]\mathrm{d}v\right\}\\&\qquad\times\exp\left\{\int_s^t 2hp\left(1-2\pi\left(\vartheta,v\right)\right)\mathrm{d}\bar{W}_v-\frac{1}{2}\int_s^t p^2h^2\left(1-2\pi\left(\vartheta,v\right)\right)^2\mathrm{d}v\right\}\mathrm{d}s\end{aligned}$$

and

$$\inf_{0\leq \pi,\pi^\circ\leq 1} \phi_v\left(\pi,\pi^\circ,p\right)=-\frac{p+1}{2}.$$

Thus

$$\begin{aligned}\mathbf{E}_{\vartheta_0}\left|\dot{\pi}_\lambda\left(\vartheta,t\right)\right|^p&\leq pC_{\pi,p-1}\int_0^t\exp\left\{-\left[pg-\frac{p\left(p+1\right)}{2}h^2\right]\left(t-s\right)\right\}\mathrm{d}s\\&\leq\frac{2\,C_{\pi,p-1}}{2\,g-\left(p+1\right)h^2}\equiv C_{\pi,p}.\end{aligned}$$

□

Introduce the matrix of the second derivatives

$$\ddot{\pi}\left(\vartheta,t\right)=\begin{pmatrix}\ddot{\pi}_{\lambda\lambda}\left(\vartheta,t\right), & \ddot{\pi}_{\lambda\mu}\left(\vartheta,t\right)\\ \ddot{\pi}_{\lambda\mu}\left(\vartheta,t\right), & \ddot{\pi}_{\mu\mu}\left(\vartheta,t\right)\end{pmatrix}.$$

and its Euclidean norm

$$\left\|\ddot{\pi}\left(\vartheta,t\right)\right\|_E=\left(\pi_{\lambda\lambda}\left(\vartheta,t\right)^2+2\pi_{\lambda\mu}\left(\vartheta,t\right)^2+\pi_{\mu\mu}\left(\vartheta,t\right)^2\right)^{1/2}$$

Lemma 5.2 *Let the condition* $5\left(a-b\right)^2<4\alpha$ *holds. Then*

$$\sup_{\vartheta\in\Theta}\mathbf{E}_{\vartheta_0}\left\|\ddot{\pi}\left(\vartheta,t\right)\right\|_E^2\leq C, \tag{5.18}$$

where the constant $C>0$ *does not depend on* ϑ_0 *and* t.

Proof The equations for $\ddot{\pi}_{\lambda\lambda}\left(\vartheta,t\right)$ and $\ddot{\pi}_{\lambda\lambda}\left(\vartheta,t\right)^2$ are

$$\begin{aligned}\mathrm{d}\ddot{\pi}_{\lambda\lambda}\left(\vartheta,t\right)&=-2\dot{\pi}_\lambda\left(\vartheta,t\right)\,\mathrm{d}t+h\left[\ddot{\pi}_{\lambda\lambda}\left(\vartheta,t\right)\left(1-2\pi\left(\vartheta,t\right)\right)-2\dot{\pi}_\lambda\left(\vartheta,t\right)^2\right]\mathrm{d}\bar{W}_t\\&\quad-h^2\ddot{\pi}_\lambda\left(\vartheta,t\right)\left[\left(1-2\pi\left(\vartheta,t\right)\right)\left[\pi\left(\vartheta,t\right)-\pi\left(\vartheta_0,t\right)\right]+\pi\left(\vartheta,t\right)\left(1-\pi\left(\vartheta,t\right)\right)\right]\mathrm{d}t\\&\quad-g\ddot{\pi}_\lambda\left(\vartheta,t\right)\mathrm{d}t-2h^2\dot{\pi}_\lambda\left(\vartheta,t\right)^2\left[1-3\pi\left(\vartheta,t\right)-\pi\left(\vartheta_0,t\right)\right]\mathrm{d}t,\\ \ddot{\pi}_{\lambda\lambda}\left(\vartheta,t\right)^2&=-4\int_0^t\ddot{\pi}_{\lambda\lambda}\left(\vartheta,s\right)\dot{\pi}_\lambda\left(\vartheta,s\right)\mathrm{d}s-4h\int_0^t\ddot{\pi}_{\lambda\lambda}\left(\vartheta,s\right)\dot{\pi}_\lambda\left(\vartheta,s\right)\mathrm{d}\bar{W}_s\\&\quad-4\int_0^t\ddot{\pi}_{\lambda\lambda}\left(\vartheta,s\right)\dot{\pi}_\lambda\left(\vartheta,s\right)^2\left[1-3\pi\left(\vartheta,s\right)+\pi\left(\vartheta_0,s\right)\right]\mathrm{d}s\\&\quad+2h\int_0^t\ddot{\pi}_{\lambda\lambda}\left(\vartheta,s\right)^2\left(1-2\pi\left(\vartheta,s\right)\right)\mathrm{d}\bar{W}_s-2g\int_0^t\ddot{\pi}_{\lambda\lambda}\left(\vartheta,s\right)^2\mathrm{d}s\end{aligned}$$

$$
\begin{aligned}
&- 2h^2 \int_0^t \ddot{\pi}_{\lambda\lambda}(\vartheta, s)^2 \left[(1 - 2\pi(\vartheta, t)) \left[\pi(\vartheta, s) - \pi(\vartheta_0, s)\right]\right. \\
&\left.+\pi(\vartheta, s)(1 - \pi(\vartheta, s))\right] \mathrm{d}s.
\end{aligned}
$$

Thus

$$
\begin{aligned}
\mathbf{E}_{\vartheta_0}\ddot{\pi}_{\lambda\lambda}(\vartheta, t)^2 &\le 4\int_0^t \mathbf{E}_{\vartheta_0}\left|\ddot{\pi}_{\lambda\lambda}(\vartheta, s)\right|\left[\left|\dot{\pi}_{\lambda}(\vartheta, s)\right| + \dot{\pi}_{\lambda}(\vartheta, s)^2\right]\mathrm{d}s \\
&\quad - 2\left(g - h^2\right)\int_0^t \mathbf{E}_{\vartheta_0}\ddot{\pi}_{\lambda\lambda}(\vartheta, s)^2\,\mathrm{d}s \\
&\le 4\sqrt{2}\int_0^t \left(\mathbf{E}_{\vartheta_0}\ddot{\pi}_{\lambda\lambda}(\vartheta, s)^2\right)^{1/2}\left(\mathbf{E}_{\vartheta_0}\dot{\pi}_{\lambda}(\vartheta, s)^2 + \mathbf{E}_{\vartheta_0}\dot{\pi}_{\lambda}(\vartheta, s)^4\right)^{1/2}\mathrm{d}s \\
&\quad - 2\left(g - h^2\right)\int_0^t \mathbf{E}_{\vartheta_0}\ddot{\pi}_{\lambda\lambda}(\vartheta, s)^2\,\mathrm{d}s \\
&\le C\int_0^t \left(\mathbf{E}_{\vartheta_0}\ddot{\pi}_{\lambda\lambda}(\vartheta, s)^2\right)^{1/2}\mathrm{d}s - 2\left(g - h^2\right)\int_0^t \mathbf{E}_{\vartheta_0}\ddot{\pi}_{\lambda\lambda}(\vartheta, s)^2\,\mathrm{d}s.
\end{aligned}
$$

Here we used (5.16) and the estimates (5.11) with $p = 2$ and $p = 4$. Similar estimates can be obtained for $\ddot{\pi}_{\lambda\mu}(\vartheta, s)$ and $\ddot{\pi}_{\mu\mu}(\vartheta, s)$.

Now (5.18) follows from the elementary calculations. □

Remark 5.1 By this Lemma there exists a constant $C > 0$ such that

$$
\mathbf{E}_{\vartheta_0}\left\|\dot{\pi}(\vartheta, t) - \dot{\pi}(\vartheta_0, t)\right\|^2 \le C\left\|\vartheta - \vartheta_0\right\|^2. \tag{5.19}
$$

5.2 Parameter Estimation

5.2.1 *MME*

Let us first address the problem of constructing a $\sqrt{T}$-consistent estimator of the unknown parameter ϑ by the method of moments. Recall that we observe the stochastic process (5.1) in continuous time, where $W_t, 0 \le t \le T$ is a standard Wiener process, X_0 is an independent random initial condition, and $Y_t, t \ge 0$ is the stationary Markov process with two states a and b and infinitesimal rate matrix

$$
\begin{pmatrix} -\lambda & \lambda \\ \mu & -\mu \end{pmatrix}.
$$

The processes $Y(t), t \ge 0$ and $W_t, t \ge 0$ are independent.

Estimation of λ and μ

Suppose that the values of a and b are known and we have to estimate $\vartheta = (\lambda, \mu) = (\theta_1, \theta_2)$. For simplicity we assume that T is an integer and impose the condition

$$\theta_1, \theta_2 \in \Theta = (\alpha, \beta) \times (\alpha, \beta), \tag{5.20}$$

where α and β are some positive constants. The true value is $\vartheta_0 = \left(\theta_{1,0}, \theta_{2,0}\right)$.

To introduce the MME $\vartheta_T^* = \left(\theta_{1,T}^*, \theta_{2,T}^*\right)$ we need the following objects:

- The function

$$\Phi\left(x\right) = \frac{1}{x} - \frac{1}{x^2}\left(1 - e^{-x}\right), \qquad x > 0. \tag{5.21}$$

- The statistics

$$R_{1,T} = \frac{1}{T}\sum_{k=0}^{T-1}\left[X_{k+1} - X_k\right]^2 - 1, \tag{5.22}$$

$$R_{2,T} = \left(\frac{X_T}{T} - a\right)\left(b - \frac{X_T}{T}\right). \tag{5.23}$$

- The random variable s_T defined as a solution of the equation

$$R_{1,T} = \left(\frac{X_T}{T}\right)^2 + 2R_{2,T}\,\Phi\left(s_T\right), \tag{5.24}$$

- The event $\mathbb{A}_T$ =“*the equation* (5.24) *has a solution* $s_T \in [2\alpha, 2\beta]$”.
- The random variable

$$q_T = s_T\ \mathbb{I}_{\{\mathbb{A}_T\}} + (\alpha + \beta)\ \mathbb{I}_{\{\mathbb{A}_T^c\}}. \tag{5.25}$$

Define the estimator $\vartheta_T^* = \left(\theta_{1,T}^*, \theta_{2,T}^*\right)$ where

$$\theta_{1,T}^* = \frac{\frac{X_T}{T} - a}{b - a}\,q_T; \qquad \theta_{2,T}^* = \frac{b - \frac{X)_T}{T}}{b - a}\,q_T. \tag{5.26}$$

Properties of these estimators are given in the following theorem.

Theorem 5.1 *The estimators in* (5.26) *satisfy*

$$\mathbf{E}_{\vartheta_0}\left[\sqrt{T}\left(\theta_{1,T}^* - \theta_{1,0}\right)\right]^2 < C, \qquad \mathbf{E}_{\vartheta_0}\left[\sqrt{T}\left(\theta_{2,T}^* - \theta_{2,0}\right)\right]^2 < C \tag{5.27}$$

with a constant $C > 0$, independent of T and ϑ_0.

Proof The proof is split below into several steps.

The next lemma gives $\sqrt{T}$-consistent estimator for $\bar{Y}$ (see (5.4)).

Lemma 5.3 *The estimator X_T/T is uniformly consistent for $\bar{Y}$:*

$$\mathbf{E}_{\vartheta_0}\left(\frac{X_T}{T}-\bar{Y}\right)^2\leq\frac{C}{T},\qquad T>0 \tag{5.28}$$

with a constant $C > 0$ independent of ϑ and T. □

Proof Using (5.22) we obtain the following relations:

$$\begin{aligned}\mathbf{E}_{\vartheta_0}\left(\frac{X_T}{T}-\bar{Y}\right)^2&=\mathbf{E}_{\vartheta_0}\left|\frac{1}{T}\int_0^T\left[Y_t-\bar{Y}\right]\mathrm{d}t+\frac{W_T}{T}\right|^2\\&=\frac{1}{T}+\frac{1}{T^2}\mathbf{E}_{\vartheta_0}\left|\int_0^T\left[Y_t-\bar{Y}\right]\mathrm{d}t\right|^2\\&\leq\frac{1}{T}\left(1+\frac{2\theta_{1,0}\theta_{2,0}}{\left(\theta_{1,0}+\theta_{2,0}\right)^3}(b-a)^2\right)\leq\frac{1}{T}\left(1+\frac{\beta^2}{4\alpha^3}(b-a)^2\right).\end{aligned}$$

□

Corollary 5.1 *Existence of consistent estimators for $\frac{\theta_{1,0}}{\theta_{1,0}+\theta_{2,0}}$ and $\frac{\theta_{2,0}}{\theta_{1,0}+\theta_{2,0}}$ follows from* (5.23) *and Lemma 5.3.*

Indeed, from the equality

$$\bar{Y}=\frac{\theta_{1,0}}{\theta_{1,0}+\theta_{2,0}}b+\frac{\theta_{2,0}}{\theta_{1,0}+\theta_{2,0}}a$$

and Lemma 5.3 we obtain

$$\begin{aligned}&\mathbf{E}_{\vartheta_0}\left[\sqrt{T}\left(\frac{T^{-1}X_T-a}{b-a}-\frac{\theta_{1,0}}{\theta_{1,0}+\theta_{2,0}}\right)\right]^2<C,\\&\mathbf{E}_{\vartheta_0}\left[\sqrt{T}\left(\frac{b-T^{-1}X_T}{b-a}-\frac{\theta_{2,0}}{\theta_{1,0}+\theta_{2,0}}\right)\right]^2<C.\end{aligned} \tag{5.29}$$

The statistic

$$\frac{X_T}{T}=\frac{1}{T}\int_0^T Y_t\mathrm{d}t+\frac{W_T}{T}$$

is a sum of a bounded random variable and an independent normal random variable $N\left(0, T^{-1}\right)$. Hence $R_{2,T}$ defined in (5.23) satisfies

$$\mathbf{E}_{\vartheta_0}\left[\sqrt{T}\left(R_{2,T}-D\right)\right]^2<C, \tag{5.30}$$

where the constant $C>0$ does not depend on T and ϑ. The constant D is defined in (5.4).

Note that from the condition (5.20) we have

$$\frac{\theta_{1,0}\theta_{2,0}}{\left(\theta_{1,0}+\theta_{2,0}\right)^2}>\frac{\alpha^2}{4\beta^2}$$

and we easily obtain the bound (5.30) for the estimator

$$\tilde{R}_{2,T}=\max\left\{R_{2,T},(b-a)^2\frac{\alpha^2}{8\beta^2}\right\}. \tag{5.31}$$

Lemma 5.4 *The following equality holds*

$$\mathbf{E}_{\vartheta_0}R_{1,T}=\bar{Y}^2+2D\Phi\left(\theta_{1,0}+\theta_{2,0}\right)$$

and under condition (5.28) *we also have*

$$\mathbf{E}_{\vartheta}\left[\sqrt{T}\left(R_{1,T}-\mathbf{E}_{\vartheta}R_{1,T}\right)\right]^2<C.$$

Proof By (5.3) and stationarity of the process Y_t we obtain

$$\begin{aligned}\mathbf{E}_{\vartheta_0}R_{1,T}&=\mathbf{E}_{\vartheta_0}\left[X_1-X_0\right]^2-1=\mathbf{E}_{\vartheta}\int_0^1\int_0^1 Y_sY_t\mathrm{d}s\mathrm{d}t+1-1\\&=\bar{Y}^2+2D\Phi\left(\theta_{1,0}+\theta_{2,0}\right).\end{aligned}$$

Denote

$$\eta_k=\int_k^{k+1}Y_t\,\mathrm{d}t,\qquad\qquad \zeta_k=W_{k+1}-W_k.$$

Further, the equality

$$R_{1,T}-\mathbf{E}_{\vartheta_0}R_{1,T}=\frac{1}{T}\sum_{k=0}^{T-1}\left(\eta_k^2-\mathbf{E}_{\vartheta_0}\eta_k^2\right)+\frac{2}{T}\sum_{k=0}^{T-1}\eta_k\zeta_k+\frac{1}{T}\sum_{k=0}^{T-1}\left(\zeta_k^2-1\right)$$

implies

$$
\mathbf{E}_{\vartheta_0}\left(R_{1,T}-\mathbf{E}_{\vartheta_0}R_{1,T}\right)^2\leq\frac{3}{T^2}\mathbf{E}_{\vartheta_0}\left(\sum_{k=0}^{T-1}\left(\eta_k^2-\mathbf{E}_{\vartheta_0}\eta_k^2\right)\right)^2+\frac{12}{T^2}\mathbf{E}_{\vartheta_0}\left(\sum_{k=0}^{T-1}\eta_k\zeta_k\right)^2
$$
$$
+\frac{3}{T^2}\mathbf{E}_{\vartheta_0}\left(\sum_{k=0}^{T-1}\left(\zeta_k^2-1\right)\right)^2:=3J_1+12J_2+3J_3. \tag{5.32}
$$

By stationarity of Y_t

$$
\begin{aligned}
J_1&=\frac{1}{T^2}\mathbf{E}_{\vartheta_0}\left(\sum_{k=0}^{T-1}\left(\eta_k^2-\mathbf{E}_{\vartheta_0}\eta_k^2\right)\right)^2\\
&=\frac{1}{T^2}\sum_{k=0}^{T-1}\sum_{j=0}^{T-1}\mathbf{E}_{\vartheta_0}\left(\eta_k^2-\mathbf{E}_{\vartheta_0}\eta_k^2\right)\left(\eta_j^2-\mathbf{E}_{\vartheta_0}\eta_j^2\right)\\
&=\frac{1}{T^2}\sum_{k,j=0}^{T-1}\left\{\int_0^1\int_0^1\int_0^1\int_0^1\mathbf{E}_{\vartheta_0}\left(Y_sY_t\right.\right.\\
&\qquad\left.\left.\mathbf{E}_{\vartheta_0}\left[Y_{|i-j|+s_1}Y_{|i-j|+t_1}|\mathfrak{F}_1^Y\right]\right)\mathrm{d}s\,\mathrm{d}t\,\mathrm{d}s_1\,\mathrm{d}t_1-\left(\mathbf{E}_{\vartheta}\eta_0^2\right)^2\right\}.
\end{aligned} \tag{5.33}
$$

The bound (5.25) gives

$$
\left|\mathbf{E}_{\vartheta_0}\left[Y_{|i-j|+s_1}Y_{|i-j|+t_1}|\mathfrak{F}_1^Y\right]-\mathbf{E}_{\vartheta_0}Y_{s_1}Y_{t_1}\right|\leq K\,e^{-(\theta_{1,0}+\theta_{2,0})|j-i|},
$$

and, using (5.33) and (5.20), we obtain

$$
J_1\leq\frac{K}{T^2}\sum_{k,j=0}^{T-1}e^{-(\theta_{1,0}+\theta_{2,0})|j-i|}\leq\frac{K_1}{T}.
$$

The obvious bounds

$$
J_2=\frac{1}{T^2}\mathbf{E}_{\vartheta_0}\left(\sum_{k=0}^{T-1}\eta_k\zeta_k\right)^2=\frac{1}{T^2}\sum_{k=0}^{T-1}\mathbf{E}_{\vartheta}\eta_k^2=\frac{\mathbf{E}_{\vartheta_0}\eta_0^2}{T}\leq\frac{K}{T},
$$
$$
J_3=\frac{1}{T^2}\mathbf{E}_{\vartheta_0}\left(\sum_{k=0}^{T-1}\left[\zeta_k^2-1\right]\right)^2=\frac{1}{T^2}\sum_{k=0}^{T-1}\mathbf{E}_{\vartheta_0}\left[\zeta_k^2-1\right]^2\leq\frac{K}{T}
$$

and (5.32) imply the second assertion of the lemma. □

Lemma 5.5 *The function $\Phi(x)$ (see (5.21)) has the following properties:*

$$\lim_{x\to 0+} \Phi(x) = \frac{1}{2}, \tag{5.34}$$

$$\lim_{x\to\infty} \Phi(x) = 0, \tag{5.35}$$

$$\Phi'(x) < 0, \quad \text{for} \quad x > 0. \tag{5.36}$$

Proof Expanding (5.21) into power series, we get

$$\Phi(x) = \frac{1}{2} - \frac{x}{3} + \frac{x^2}{4} - \frac{x^3}{5} + \dots$$

$$\Phi'(x) = -\left(\frac{1}{3} - \frac{2x}{4}\right) - \left(\frac{3x^2}{5!} - \frac{4x^3}{6!}\right) - \dots$$

which give the limits (5.34) and (5.35) and the bound (5.36) for $x < 2$. For $x \geq 2$ it follows from the explicit expression for this derivative

$$\Phi'(x) = \frac{1}{x^2}\left(\frac{2}{x} - 1\right) - \left(\frac{2}{x^3} + \frac{1}{x^2}\right)e^{-x}.$$

□

Let us consider the Eq. (5.24) for s_T, where $R_{1,T}$ and $R_{2,T}$ are defined in (5.22) and (5.23), respectively. By Lemma 5.5 this equation has at most one solution. Recall that on the event $\mathbb{A}_T$, Eq. (5.24) does have a solution and consider the statistic q_T defined in (5.25) (here α, β are the constants from the condition (5.20)).

Lemma 5.6 *The estimate q_T is $\sqrt{T}$-consistent for $\theta_{1,0} + \theta_{2,0}$:*

$$\mathbf{E}_{\vartheta_0}\left[\sqrt{T}\left(q_T - (\theta_{1,0} + \theta_{2,0})\right)\right]^2 < C,$$

with a constant $C > 0$, independent of ϑ and T.

Proof It follows from Lemmas 5.3 and 5.4 that

$$R_{1,T} = \bar{Y}^2 + 2D\Phi\left(\theta_{1,0} + \theta_{2,0}\right) + \varepsilon_1(T). \tag{5.37}$$

Here and below $\varepsilon_i(T), i = 1, 2\dots$ satisfy

$$\mathbf{E}_{\vartheta_0}\left(\sqrt{T}\varepsilon_i(T)\right)^2 < C.$$

By Lemma 5.3, estimates (5.30) and (5.31) and boundedness of $\Phi(x)$ we also obtain

$$R_{1,T} = \bar{Y}^2 + 2\tilde{R}_{2,T}\Phi\left(\theta_{1,0} + \theta_{2,0}\right) + \varepsilon_2(T).$$

It follows from (5.27) and (5.37) that on the event $\mathbb{A}_T$

$$2\tilde{R}_{2,T}\Phi(s_T) = 2\tilde{R}_{2,T}\Phi\left(\theta_{1,0} + \theta_{2,0}\right) + \varepsilon_3(T).$$

Since $\tilde{R}_{2,T}$ is bounded away from zero by a positive constant (see Corollary 5.1), the latter equality and Lemma 5.5 imply that on $\mathbb{A}_T$

$$\Phi(s_T) - \Phi\left(\theta_{1,0} + \theta_{2,0}\right) = \varepsilon_4(T).$$

Therefore by Lemma 5.5 we obtain

$$\mathbf{E}_{\vartheta_0}\left\{\mathbb{I}_{\{\mathbb{A}_T\}}\sqrt{T}\left(q_T - \left(\theta_{1,0} + \theta_{2,0}\right)\right)\right\}^2 < C. \tag{5.38}$$

If $\omega \in \mathbb{A}_T^c$, then the equation

$$\bar{Y}^2 + 2D\Phi\left(\theta_{1,0} + \theta_{2,0}\right) + \varepsilon_3(T) = \left(\frac{X(T)}{T}\right)^2 + 2\tilde{R}_{2,T}\Phi(x) \tag{5.39}$$

has no solution $x \in [2\alpha, 2\beta]$.

It follows from (5.39), Lemma 5.3 and Corollary 5.1 that the equation

$$\Phi(x) = \Phi\left(\theta_{1,0} + \theta_{2,0}\right) + \varepsilon_4(T)$$

has no solution for $x \in [2\alpha, 2\beta]$. This means that

$$\left[\bar{Y}^2 + 2D\Phi\left(\theta_{1,0} + \theta_{2,0}\right) + \varepsilon_3 - \left(\bar{Y} + \varepsilon_0\right)^2\right](2D + 2\varepsilon_5)^{-1} \notin [\Phi(2\alpha), \Phi(2\beta)],$$

where we used the notations $\varepsilon_0 = T^{-1}X_T - \bar{Y}$ and $\varepsilon_5 = \tilde{R}_{2,T} - D$.

Hence we can write

$$\mathbb{A}_T^c \subset \{|\varepsilon_0(T)| > d_1\} \cup \{|\varepsilon_3(T)| > d_2\} \cup \{|\varepsilon_5(T)| > d_3\}$$

for some positive constants d_1, d_2, and d_3, which do not depend on T. This inclusion implies

$$\mathbf{P}_{\vartheta_0}\left(\mathbb{A}_T^c\right) \le \mathbf{P}_{\vartheta_0}\{|\varepsilon_0(T)| > d_1\} + \mathbf{P}_{\vartheta_0}\{|\varepsilon_3(T)| > d_2\} + \mathbf{P}_{\vartheta_0}\{|\varepsilon_5(T)| > d_3\} < \frac{C}{T}$$

and the claim of Lemma 5.6 holds by (5.38). □

Proof of Theorem 5.1 Let us now apply obtained results to prove $\sqrt{T}$-consistency of the estimators defined in (5.26). To this end, the obvious equality

$$\theta_{1,T}^{*}-\theta_{1,0}=q_T\frac{\frac{X(T)}{T}-a}{b-a}-q_T\frac{\theta_{1,0}}{\theta_{1,0}+\theta_{2,0}}+\frac{\theta_{1,0}}{\theta_{1,0}+\theta_{2,0}}\left(q_T-\left(\theta_{1,0}+\theta_{2,0}\right)\right),$$

the bounds (5.29) and Lemma 5.6 give

$$\begin{aligned}&\mathbf{E}_{\vartheta_0}\left(\sqrt{T}\left(\theta_{1,T}^{*}-\theta_{1,0}\right)\right)^2\\&\quad\leq 2\mathbf{E}_{\vartheta_0}\left[\sqrt{T}q_T\left(\frac{\frac{X(T)}{T}-a}{b-a}-\frac{\theta_{1,0}}{\theta_{1,0}+\theta_{2,0}}\right)\right]^2\\&\quad+2\left(\frac{\theta_{1,0}}{\theta_{1,0}+\theta_{2,0}}\right)^2\mathbf{E}_{\vartheta_0}\left[\sqrt{T}\left(q_T-\left(\theta_{1,0}+\theta_{2,0}\right)\right)\right]^2<C.\end{aligned}$$

The second inequality in (5.27) is shown similarly, and thus the MME $\vartheta_T^{*}=\left(\theta_{1,T}^{*},\theta_{2,T}^{*}\right)$ is $\sqrt{T}$-consistent. □

Estimation of *a* and *b*

Suppose that the values of $\lambda>0$ and $\mu>0$ are known and we have to estimate $\vartheta=(a,b)$ with distinct a and b by observations (5.1). We suppose that $\vartheta\in\Theta=(\alpha_a,\beta_a)\times(\alpha_b,\beta_b)$. As $(\alpha_a,\beta_a)\cap(\alpha_b,\beta_b)=\emptyset$ we assume for simplicity that $\alpha_a>\beta_b$. As usual, the true value is denoted $\vartheta_0=(a_0,b_0)$.

It was shown above that

$$\bar{Y}_T=\frac{X_T}{T}\longrightarrow\bar{Y}=\frac{a_0\mu+b_0\lambda}{\lambda+\mu}$$

and

$$R_{1,T}=\frac{1}{T}\sum_{k=0}^{T-1}\left[X_{k+1}-X_k\right]^2-1\longrightarrow\bar{Y}^2+2\frac{\lambda\mu\Phi\left(\lambda+\mu\right)}{(\lambda+\mu)^2}\left(b_0-a_0\right)^2.$$

Introduce two statistics

$$S_{1,T}=(1+\gamma)\frac{X_T}{T},\qquad\gamma=\frac{\lambda}{\mu},$$

and

$$S_{2,T}=\frac{\left[R_{1,T}-\bar{Y}_T^2\right](\lambda+\mu)^2}{2\lambda\,\mu\,\Phi\left(\lambda+\mu\right)}.$$

Now we have the convergence

$$S_{1,T} \longrightarrow a_0 + \gamma b_0, \qquad\qquad S_{2,T} \longrightarrow (a_0 - b_0)^2 .$$

The MME ϑ_T^* is defined as solution of the equations

$$a_T^* + \gamma b_T^* = S_{1,T},$$
$$a_T^* - b_T^* = \sqrt{\tilde S_{2,T}},$$

where

$$\tilde S_{2,T} = (\alpha_a - \beta_b)^2 \vee S_{2,T}.$$

Recall that $a > b$ and $a - b > \alpha_a - \beta_b$. Therefore

$$a_T^* = \frac{S_{1,T} + \gamma\sqrt{\tilde S_{2,T}}}{1+\gamma}, \qquad\qquad b_T^* = \frac{S_{1,T} - \sqrt{\tilde S_{2,T}}}{1+\gamma}. \tag{5.40}$$

For simplicity of exposition we suppose that $\vartheta_T^* \in \Theta$.

Theorem 5.2 *The estimators a_T^* and b_T^* in (5.40) satisfy*

$$\mathbf{E}_{\vartheta_0}\left[\sqrt{T}\left(a_T^* - a_0\right)\right]^2 < C, \qquad\qquad \mathbf{E}_{\vartheta_0}\left[\sqrt{T}\left(b_T^* - b_0\right)\right]^2 < C.$$

with a constant $C > 0$, independent of T.

Proof The proof follows directly from Lemmas 5.3 and 5.4 and standard arguments. □

5.2.2 MLE and BE

The model of observation is the same

$$\mathrm{d}X_t = Y_t \mathrm{d}t + \mathrm{d}W_t, \qquad X_0, \qquad 0 \le t \le T,$$

where $Y_t, t \ge 0$ is telegraph process with two states a and b and the infinitesimal transition rate matrix

$$\begin{pmatrix} -\lambda & \lambda \\ \mu & -\mu \end{pmatrix}.$$

Suppose that the values $\lambda = \theta_1$ and $\mu = \theta_2$ are unknown and we have to estimate $\vartheta = (\theta_1, \theta_2) \in \Theta$ by the observations $X^T = (X_t, 0 \le t \le T)$. The set $\Theta = (\alpha, \beta) \times$

$(\alpha, \beta), \alpha > 0$. Recall that the likelihood ratio function is given by the formula (5.7) and the MLE $\hat{\vartheta}_T$ and BE $\tilde{\vartheta}_T$ as usual are defined by the relations

$$L(\hat{\vartheta}_T, X^T) = \sup_{\vartheta \in \Theta} L\left(\vartheta, X^T\right), \qquad \tilde{\vartheta}_T = \frac{\int_\Theta \vartheta p\,(\vartheta)\, L\left(\vartheta, X^T\right) \mathrm{d}\vartheta}{\int_\Theta p\,(\vartheta)\, L\left(\vartheta, X^T\right) \mathrm{d}\vartheta}. \tag{5.41}$$

In the case of BE we suppose that ϑ is a random vector with the density $p\,(\vartheta)\,, \vartheta \in \Theta$, which is a positive, continuous function on Θ.

To study the MLE, BE, and One-step MLE-process we will need the following limits:

$$\lim_{T\to\infty} \frac{1}{T} \int_0^T \left[\pi\,(\vartheta, t) - \pi\,(\vartheta_0, t)\right]^2 \mathrm{d}t = \Pi\,(\vartheta, \vartheta_0)\,, \tag{5.42}$$

$$\lim_{T\to\infty} \frac{1}{T} \int_0^T \dot{\pi}_\lambda\,(\vartheta_0, t)^2\, \mathrm{d}t = \Pi_\lambda\,(\vartheta_0)\,, \tag{5.43}$$

$$\lim_{T\to\infty} \frac{1}{T} \int_0^T \dot{\pi}_\mu\,(\vartheta_0, t)^2\, \mathrm{d}t = \Pi_\mu\,(\vartheta_0)\,, \tag{5.44}$$

$$\lim_{T\to\infty} \frac{1}{T} \int_0^T \dot{\pi}_\lambda\,(\vartheta_0, t)\, \dot{\pi}_\mu\,(\vartheta_0, t)\, \mathrm{d}t = \Pi_{\lambda,\mu}\,(\vartheta_0)\,. \tag{5.45}$$

Remark that $\Pi\,(\vartheta, \vartheta_0) < 1$ and by Lemma 5.2

$$\sup_{\vartheta_0\in\Theta} \Pi_\lambda\,(\vartheta_0) < C_{\pi,2}, \qquad \sup_{\vartheta_0\in\Theta} \Pi_\mu\,(\vartheta_0) < C_{\pi,2}, \qquad \sup_{\vartheta_0\in\Theta} \left|\Pi_{\lambda,\mu}\,(\vartheta_0)\right| < C_{\pi,2}.$$

The constant $C_{\pi,2}$ was defined before in (5.11) and (5.12). The first limits (5.42)–(5.44) can be justified with the help of the results obtained in the works [37] and [36]. In particular, the proofs of the limits like (5.43) and (5.44) in [36] are based on one modification of Lemma 1.6. Recall that this lemma was published in [90]. The same arguments can be used for the proof of (5.45).

Another approach to verify these limits is the following. Note that the process $(\pi\,(\vartheta_0, t)\,, \dot{\pi}_\lambda\,(\vartheta_0, t)\,, \dot{\pi}_\mu\,(\vartheta_0, t)\,, t \geq 0)$ is a degenerate diffusion with one Wiener process. The existence of a unique invariant measure can be proved if one verifies the Hörmander conditions for this system [70, 186]. We leave this rather technical verification for the future, and for now we simply assume that the system parameters are such that this convergence takes place.

Conditions $\mathscr{L}$.

$\mathscr{L}_0$. *The unknown parameter* $\vartheta = (\lambda, \mu) \in \Theta = (\alpha, \beta) \times (\alpha, \beta), \alpha > 0$.
$\mathscr{L}_1$. *The parameters* a, b, α *satisfy the condition* $9\,(a - b)^2 < 4\alpha$.

$\mathscr{L}_2$. *The diffusion process* $\left(\pi\left(\vartheta_0,t\right),\dot\pi_\lambda\left(\vartheta_0,t\right),\dot\pi_\mu\left(\vartheta_0,t\right)\right), t\geq 0$ *has a unique invariant measure* $\mathbf{Q}\left(\vartheta_0,\mathrm{d}x,\mathrm{d}y,\mathrm{d}z\right)$ *and the convergences in probability* (5.43)–(5.45) *are uniform on compacts* $\mathbb{K}\subset\Theta$.

The Fisher information matrix in this problem is

$$\mathbf{I}(\vartheta)=h^2\begin{pmatrix}\Pi_\lambda\left(\vartheta\right), & \Pi_{\lambda,\mu}\left(\vartheta\right)\\ \Pi_{\lambda,\mu}\left(\vartheta\right), & \Pi_\mu\left(\vartheta\right)\end{pmatrix}=\begin{pmatrix}\mathrm{I}_{11}\left(\vartheta\right), \mathrm{I}_{12}\left(\vartheta\right)\\ \mathrm{I}_{12}\left(\vartheta\right), \mathrm{I}_{22}\left(\vartheta\right)\end{pmatrix}.$$

Here

$$\Pi_\lambda\left(\vartheta\right)=\int_{\mathcal{R}^3}y^2\mathbf{Q}\left(\vartheta,\mathrm{d}x,\mathrm{d}y,\mathrm{d}z\right),\qquad \Pi_{\lambda,\mu}\left(\vartheta\right)=\int_{\mathcal{R}^3}yz\mathbf{Q}\left(\vartheta,\mathrm{d}x,\mathrm{d}y,\mathrm{d}z\right).$$

It will be shown below that under conditions $\mathscr{L}$ this matrix is nondegenerate and that the corresponding family of measures is LAN. Therefore we have Hájek's lower bound on the risks of all estimators

$$\lim_{\nu\rightarrow 0}\varliminf_{T\rightarrow\infty}\sup_{\left\|\vartheta-\vartheta_0\right\|\leq\nu}T\mathbf{E}_\vartheta\left\|\bar\vartheta_T-\vartheta\right\|^2\geq\mathbf{E}_{\vartheta_0}\left\|\zeta\left(\vartheta_0\right)\right\|^2,\ \zeta\left(\vartheta_0\right)\sim\mathcal{N}\left(0,\mathbf{I}(\vartheta_0)^{-1}\right)$$

for all $\vartheta_0\in\Theta$.

We call an estimator ϑ_T^* asymptotically efficient if

$$\lim_{\nu\rightarrow 0}\lim_{T\rightarrow\infty}\sup_{\left\|\vartheta-\vartheta_0\right\|\leq\nu}T\mathbf{E}_\vartheta\left\|\vartheta_T^*-\vartheta\right\|^2=\mathbf{E}_{\vartheta_0}\left\|\zeta\left(\vartheta_0\right)\right\|^2$$

for all $\vartheta_0\in\Theta$.

Theorem 5.3 *Let the conditions* $\mathscr{L}$ *hold. Then the MLE* $\hat\vartheta_T$ *and BE* $\tilde\vartheta_T$ *are uniformly on compacts consistent, asymptotically normal*

$$\sqrt{T}\left(\hat\vartheta_T-\vartheta_0\right)\Longrightarrow\zeta\left(\vartheta_0\right),\qquad \sqrt{T}\left(\tilde\vartheta_T-\vartheta_0\right)\Longrightarrow\zeta\left(\vartheta_0\right),$$

the polynomial moments converge, and estimators are asymptotically efficient.

Remark 5.2 This theorem was proved by P. Chigansky (Theorem 1 in [36]) in the case where $Y_t, t\geq 0$ is Markov chain with finite number $(d>1)$ of states and transition rates $\Lambda=\left(\lambda_{ij}\right), i,j\in\{1,\ldots,d\}$ depending on a one-dimensional parameter ϑ. The case $a=1, b=0$ $(d=2)$ was considered as example. It is mentioned that "*The result stated in Theorem 1 is extendable to the vector parameter space* Θ*, since the key properties ... do not depend on the dimension of* ϑ*.*" Moreover, in this example the author does not use the conditions $\mathscr{L}_1$ and $\mathscr{L}_2$. We present here partially the proof of Theorem 5.3. It contains simple proofs of the first two lemmas and the last lemma is taken from [36] without proof. The detailed proof of Theorem 1 in [36] is too technical, is essentially based on the other work [37], and is out of the scope of this work.

Proof Introduce the normalized likelihood ratio

$$Z_T(\vartheta_0, u) = \exp\left(\int_0^T [m(\vartheta_u, t) - m(\vartheta_0, t)]\, d\bar{W}_t - \frac{1}{2}\int_0^T [m(\vartheta_u, t) - m(\vartheta_0, t)]^2\, dt\right).$$

Here $\vartheta_u = \vartheta_0 + T^{-1/2}u$ and

$$u \in \mathbb{U}_T = \left(\left(\alpha - \theta_{1,0}\right)\sqrt{T}, \left(\beta - \theta_{1,0}\right)\sqrt{T}\right) \times \left(\left(\alpha - \theta_{2,0}\right)\sqrt{T}, \left(\beta - \theta_{2,0}\right)\sqrt{T}\right).$$

To verify that this process satisfies the conditions $\mathscr{I}\mathscr{H}_2$ of Theorems 1.10 and 1.12 we need some auxiliary results.

Denote by $\dot{\mathrm{m}}(\vartheta, t) = \left(\dot{m}_\lambda(\vartheta, t), \dot{m}_\mu(\vartheta, t)\right)^\top$ and $\varphi_T = T^{-1/2}$. □

Lemma 5.7 *Let the conditions $\mathscr{L}$ hold. Then the family $\mathbf{P}_\vartheta, \vartheta \in \Theta$ is uniformly LAN, i.e., the likelihood ratio $Z_T(\vartheta_0, u), u \in \mathbb{U}_T$ admits the representation*

$$Z_T(\vartheta_0, u) = \exp\left\{u^\top \Delta_T\left(\vartheta_0, X^T\right) - \frac{1}{2}u^\top \mathbf{I}(\vartheta_0)\, u + r_T\left(\vartheta_0, u, X^T\right)\right\},$$

where $r_T\left(\vartheta_0, u, X^T\right) \longrightarrow 0$ in probability and the vector

$$\Delta_T\left(\vartheta_0, X^T\right) = \varphi_T \int_0^T \dot{\mathrm{m}}(\vartheta_0, t)\left[dX_t - m(\vartheta_0, t)\, dt\right] \Longrightarrow \mathcal{N}\left(0, \mathbf{I}(\vartheta_0)\right). \tag{5.46}$$

Here the convergences of $r_T(\cdot)$ and $\Delta_T(\cdot)$ are uniform on compacts $\mathbb{K} \subset \Theta$.

Proof We have

$$\int_0^T \left[m\left(\vartheta_0 + \varphi_T u, t\right) - m(\vartheta_0, t) - \varphi_T u^\top \dot{\mathrm{m}}(\vartheta_0, t)\right] d\bar{W}_t + u\Delta_T\left(\vartheta_0, X^T\right).$$

By condition $\mathscr{L}_2$ the following matrix converges in probability

$$\frac{1}{T}\int_0^T \dot{\mathrm{m}}(\vartheta_0, t)\,\dot{\mathrm{m}}(\vartheta_0, t)^\top dt \longrightarrow \mathbf{I}(\vartheta_0)$$

uniformly on compacts, and therefore by the CLT 1.7 the convergence (5.46) takes place. Further, the Newton-Leibniz formula

$$m\left(\vartheta_0 + \varphi_T u, t\right) - m(\vartheta_0, t) = \varphi_T \int_0^1 u^\top \dot{\mathrm{m}}\left(\vartheta_0 + \varphi_T s u, t\right) ds$$

allows us to write

$$\begin{aligned}
&\mathbf{E}_{\vartheta_0}\left(\int_0^T\left[m\left(\vartheta_0+\varphi_T u,t\right)-m\left(\vartheta_0,t\right)-\varphi_T u^\top\dot{\mathrm{m}}\left(\vartheta_0,t\right)\right]\mathrm{d}\bar{W}_t\right)^2\\
&\qquad=\int_0^T\mathbf{E}_{\vartheta_0}\left[m\left(\vartheta_0+\varphi_T u,t\right)-m\left(\vartheta_0,t\right)-\varphi_T u^\top\dot{\mathrm{m}}\left(\vartheta_0,t\right)\right]^2\mathrm{d}t\\
&\qquad=\varphi_T^2\int_0^T\mathbf{E}_{\vartheta_0}\left[\int_0^1 u^\top\left[\dot{\mathrm{m}}\left(\vartheta_0+\varphi_T su,t\right)-\dot{\mathrm{m}}\left(\vartheta_0,t\right)\mathrm{d}s\right]\right]^2\mathrm{d}t\\
&\qquad\leq\varphi_T^2\left\|u\right\|^2\int_0^T\int_0^1\mathbf{E}_{\vartheta_0}\left\|\dot{\mathrm{m}}\left(\vartheta_0+\varphi_T su,t\right)-\dot{\mathrm{m}}\left(\vartheta_0,t\right)\right\|^2\mathrm{d}s\mathrm{d}t\\
&\qquad\leq C\left\|u\right\|^4\varphi_T^2\longrightarrow 0.
\end{aligned}$$

Here we used the estimate (5.19).

Similar calculations provide the uniform compacts convergence

$$\int_0^T\mathbf{E}_{\vartheta_0}\left[m\left(\vartheta_0+\varphi_T u,t\right)-m\left(\vartheta_0,t\right)\right]^2\mathrm{d}t\longrightarrow u^\top\mathbf{I}\left(\vartheta_0\right)u.$$

□

Denote $\vartheta_u=\vartheta_0+\varphi_T u$.

Lemma 5.8 *Let the condition $\mathscr{L}_2$ hold. Then there exists a constant $C_*>0$ such that for any $R>0$ and for the values u_1,u_2 satisfying $|u_1|+|u_2|\leq R$ the estimate*

$$\sup_{\vartheta_0\in\mathbb{K}}\mathbf{E}_{\vartheta_0}\left|Z_T^{1/4}\left(\vartheta_0,u_2\right)-Z_T^{1/4}\left(\vartheta_0,u_1\right)\right|^4\leq C\left(1+R^4\right)\left\|u_2-u_1\right\|^4$$

holds.

Proof By Lemma 1.1 we have the estimate

$$\begin{aligned}
&\mathbf{E}_{\vartheta_0}\left|Z_T^{1/4}\left(\vartheta_0,u_2\right)-Z_T^{1/4}\left(\vartheta_0,u_1\right)\right|^4\\
&\qquad\leq C_{1,2}T^3\int_0^T\mathbf{E}_{\vartheta_{u_2}}\left|m\left(\vartheta_{u_2},t\right)-m\left(\vartheta_{u_1},t\right)\right|^8\mathrm{d}t\\
&\qquad\quad+C_{2,2}T^{3/2}\int_0^T\mathbf{E}_{\vartheta_{u_2}}\left|m\left(\vartheta_{u_2},t\right)-m\left(\vartheta_{u_1},t\right)\right|^4\mathrm{d}t\\
&\qquad=C_{1,2}T^{2p-1}h^{4p}\int_0^T\mathbf{E}_{\vartheta_{u_2}}\left|\pi\left(\vartheta_{u_2},t\right)-\pi\left(\vartheta_{u_1},t\right)\right|^8\mathrm{d}t
\end{aligned}$$

$$
\begin{aligned}
&+ C_{2,2}T^{p-1/2}h^{2p}\int_0^T \mathbf{E}_{\vartheta_{u_2}}\left|\pi\left(\vartheta_{u_2},t\right)-\pi\left(\vartheta_{u_1},t\right)\right|^4 \mathrm{d}t\\
&\leq C\;\|u_2-u_1\|^8 + C\;\|u_2-u_1\|^4\\
&\leq C\left(1+R^4\right)\|u_2-u_1\|^4 .
\end{aligned}
$$

Here we used the estimate (5.19). □

Lemma 5.9 *Let the conditions $\mathscr{L}$ hold. Then for any integer $p \geq 1$ there exist constants $\kappa > 0$, $C > 0$ such that*

$$
\sup_{\vartheta_0\in\mathbb{K}} \mathbf{P}_{\vartheta_0}\left(Z_T\left(\vartheta_0,u\right)\geq e^{-\kappa\|u\|}\right)\leq \frac{C}{\|u\|^{2p}}.
$$

For the proof see the slight modification of the proof of Lemma 6 in [36]. Particularly for this modification the estimate

$$
\kappa \equiv \inf_{\vartheta\in\mathbb{K}}\ \inf_{\|\mathrm{e}\|=1,\ \mathrm{e}\in\mathcal{R}^2} \mathrm{e}^\top \mathbf{I}\left(\vartheta\right)\mathrm{e} > 0 \tag{5.47}
$$

is needed.

Lemma 5.10 *Let the conditions $\mathscr{L}$ hold. Then the Fisher information matrix $\mathbf{I}\left(\vartheta\right)$, $\vartheta\in\Theta$ is uniformly nondegenerate* (5.47).

Proof This proof follows the same lines as the proof of positivity of the Fisher information in [36].

It suffices to show invertibility of the matrix

$$
\begin{pmatrix} \mathbf{E}_{\vartheta_0} y_t^2, & \mathbf{E}_{\vartheta_0}\, y_t z_t\\ \mathbf{E}_{\vartheta_0} y_t z_t, & \mathbf{E}_{\vartheta_0} z_t^2 \end{pmatrix},
$$

where $y_t = \dot\pi_\lambda\left(\vartheta_0,t\right)$ and $z_t = \dot\pi_\mu\left(\vartheta_0,t\right)$ are stationary solutions of (5.14) and (5.15), respectively. If this matrix is degenerate, then

$$
\mathbf{E}_{\vartheta_0} y_t^2\ \mathbf{E}_{\vartheta_0} z_t^2 = \left(\mathbf{E}_{\vartheta_0} y_t z_t\right)^2 .
$$

By the Cauchy-Schwartz inequality

$$
\left(\mathbf{E}_{\vartheta_0} y_t z_t\right)^2 \leq \mathbf{E}_{\vartheta_0} y_t^2\ \mathbf{E}_{\vartheta_0} z_t^2
$$

with equality if and only if $z_t = c y_t$ with some constant $c \neq 0$. Therefore in the case of equality we have $\mathbf{E}_{\vartheta_0}\left(z_t - c y_t\right)^2 = 0$.

Introduce a new process $\delta_{c,\dot\pi}\left(t\right) = c y_t - z_t$ as a solution of the equation

$$d\delta_{c,\dot{\pi}}(t) = [x_t(1-c)-1]\,dt - \left[g_0 + h^2 x_t(1-x_t)\right]\delta_{c,\dot{\pi}}(t)\,dt$$
$$+h(1-2x_t)\,\delta_{c,\dot{\pi}}(t)\,d\bar{W}_t,$$

where $\delta_{c,\dot{\pi}}(t)$ and $x_t = \pi(\vartheta_0, t)$ are stationary. Denote $\Pi(\vartheta_0) = \mathbf{E}_{\vartheta_0}\pi(\vartheta_0, t)^2$.

Further, following Section 4 of [36], we can write

$$\delta_{c,\dot{\pi}}(t) = \delta_{c,\dot{\pi}}(0)\,e^{-g_0 t} + \int_0^t e^{-g_0(t-s)}\,[x_s(1-c)-1]\,ds$$
$$-h^2\int_0^t e^{-g_0(t-s)} x_s(1-x_s)\,\delta_{c,\dot{\pi}}(s)\,ds$$
$$+h\int_0^t e^{-g_0(t-s)}(1-2x_s)\,\delta_{c,\dot{\pi}}(s)\,d\bar{W}_s.$$

Hence

$$\mathbf{E}_{\vartheta_0}\left(\int_0^t e^{-g_0(t-s)}\,[x_s(1-c)-1]\,ds\right)^2 \leq 4\left(1+e^{-2g_0 t}\right)\mathbf{E}_{\vartheta_0}\left(\delta_{c,\dot{\pi}}(t)^2\right)$$
$$+\frac{4h^4}{g_0}\int_0^t e^{-g_0(t-s)}\frac{1}{16}\mathbf{E}_{\vartheta_0}\left(\delta_{c,\dot{\pi}}(s)^2\right)ds$$
$$+4h^2\int_0^t e^{-2g_0(t-s)}\mathbf{E}_{\vartheta_0}\left(\delta_{c,\dot{\pi}}(s)^2\right)ds$$
$$\leq C\mathbf{E}_{\vartheta_0}\left(\delta_{c,\dot{\pi}}(s)^2\right)$$

with some constant $C > 0$ which does not depend on t. By stationarity $\mathbf{E}_{\vartheta_0}\left(\delta_{c,\dot{\pi}}(s)^2\right)$ does not depend on t and hence nondegeneracy of the matrix follows, if we can show that

$$\lim_{t\to\infty}\mathbf{E}_{\vartheta_0}\left(\int_0^t e^{-g_0(t-s)}\,[x_s(1-c)-1]\,ds\right)^2 > 0,$$

for all c. For the stochastic process

$$\zeta_t = \int_0^t e^{-g_0(t-s)}\,[x_s(1-c)-1]\,ds$$

elementary calculations show that for all ϑ_0 and c

$$\lim_{t\to\infty}\mathbf{E}_{\vartheta_0}\zeta_t^2 = \frac{\mathbf{E}_{\vartheta_0}[\pi(\vartheta_0,t)(1-c)-1]^2}{4g_0^2} > 0,$$

where $\pi(\vartheta_0, t)$ is a random variable with the density of the invariant distribution (5.8). Thus the Fisher information matrix $\mathrm{I}(\vartheta_0)$ is nondegenerate for all $\vartheta_0 \in \Theta$ and we have the estimate (5.47).

The properties of $Z_T(\vartheta_0, u)$, $u \in \mathbb{U}_T$ established in Lemmas 5.7–5.9 correspond well to required conditions $\mathscr{I}\mathscr{K}_2$ of Theorems 1.10 and 1.12, and therefore, Theorem 5.3 is proved. □

5.2.3 One-Step MLE-Process

The construction of adaptive filter requires an estimator-process $\bar\vartheta_t, 0 < t \leq T$, where $\bar\vartheta_t$ depends on the observations up to time t. Recall once more that calculation of the MLE-process $\hat\vartheta_t, 0 < t \leq T$ and BE-process $\tilde\vartheta_t, 0 < t \leq T$ according to the definitions (5.41) is a computationally hard problem, because (5.41) requires solving the family of equations (5.10) for many values of $\vartheta \in \Theta$ and for all $t \in (0, T]$. Computational complexity can be reduced as follows.

First we chose the *learning interval* $[0, \tau_T]$ with $\tau_T = [T^\delta], \frac{1}{2} < \delta < 1$. Here $[A]$ is integer part of A. Then by observations $X^{\tau_T} = (X_t, 0 \leq t \leq \tau_T)$ we construct the MME $\vartheta^*_{\tau_T} = (\theta^*_{1,\tau_T}, \theta^*_{2,\tau_T})$ by relations (5.26)

$$\theta^*_{1,\tau_T} = (b-a)^{-1}\left(\frac{X_{\tau_T}}{\tau_T} - a\right) q_{\tau_T}; \qquad \theta^*_{2,\tau_T} = (b-a)^{-1}\left(b - \frac{X_{\tau_T}}{\tau_T}\right) q_{\tau_T},$$

where q_{τ_T}, roughly speaking, is solution of the equation (see (5.24))

$$\Phi(q_{\tau_T}) = \left(2R_{2,\tau_T}\right)^{-1}\left[R_{1,\tau_T} - \left(\frac{X_{\tau_T}}{\tau_T}\right)^2\right], \qquad \Phi(x) = \frac{e^{-x} - 1 + x}{x^2}.$$

Here the statistics R_{1,τ_T}, R_{2,τ_T} are (see (5.22) and (5.23))

$$R_{1,\tau_T} = \frac{1}{\tau_T}\sum_{k=0}^{\tau_T - 1}[X_{k+1} - X_k]^2 - 1, \qquad R_{2,\tau_T} = \left(\frac{X_{\tau_T}}{\tau_T} - a\right)\left(b - \frac{X_{\tau_T}}{\tau_T}\right).$$

Recall that by Theorem 5.1

$$\sup_{\vartheta\in\Theta} \mathbf{E}_\vartheta \left\| \sqrt{\tau_T}(\vartheta^*_{\tau_T} - \vartheta)\right\|^2 \leq C,$$

where the constant $C > 0$ does not depend on τ_T.

Introduce the conditions $\mathscr{L}^\star$

$\mathscr{L}^\star_0$. *The unknown parameter* $\vartheta = (\lambda, \mu) \in \Theta = (\alpha, \beta) \times (\alpha, \beta), \alpha > 0.$

$\mathscr{L}_1^\star$. *The parameters a, b, α satisfy the condition $5\,(a-b)^2 < 4\alpha$.*
$\mathscr{L}_2^\star$. *The convergences in probability* (5.43)–(5.45) *are uniform on compacts* $\mathbb{K} \subset \Theta$.

The One-step MLE-process is

$$\vartheta^\star_{t,T} = \vartheta^*_{\tau_T} + t^{-1/2}\bar{\mathbf{I}}^t_{\tau_T}(\vartheta^*_{\tau_T})_+^{-1}\Delta_t(\vartheta^*_{\tau_T}, X^t), \qquad \tau_T \le t \le T.$$

The score-function $\Delta_t(\vartheta, X^t)$ and empirical Fisher information matrix $\bar{\mathbf{I}}^t_{\tau_T}(\vartheta)$ are

$$\Delta_t\left(\vartheta, X^t\right) = \frac{1}{\sqrt{t}}\int_{\tau_T}^t \dot{\mathrm{m}}\left(\vartheta, s\right)\,\left[\mathrm{d}X_s - m\left(\vartheta, s\right)\mathrm{d}s\right],$$

$$\bar{\mathbf{I}}^t_{\tau_T}(\vartheta) = \frac{1}{t}\int_{\tau_T}^t \dot{\mathrm{m}}(\vartheta, s)\dot{\mathrm{m}}(\vartheta, s)^\top \mathrm{d}s = \frac{(a-b)^2}{t}\int_{\tau_T}^t \dot{\pi}\left(\vartheta, s\right)\dot{\pi}\left(\vartheta, s\right)^\top \mathrm{d}s.$$

The inverse matrix $\bar{\mathbf{I}}^t_{\tau_T}(\vartheta^*_{\tau_T})_+^{-1}$ is understood as follows: if

$$\inf_{\mathrm{e}\in\mathcal{R}^2, \|\mathrm{e}\|=1} \mathrm{e}^\top\mathbf{I}^t_{\tau_T}(\vartheta^*_{\tau_T})\mathrm{e} \ge \frac{\kappa}{2},$$

then we put $\bar{\mathbf{I}}^t_{\tau_T}(\vartheta^*_{\tau_T})_+^{-1} = \bar{\mathbf{I}}^t_{\tau_T}(\vartheta^*_{\tau_T})^{-1}$ otherwise $\bar{\mathbf{I}}^t_{\tau_T}(\vartheta^*_{\tau_T})_+^{-1} = 0$. The value κ is from (5.47).

Change the variable $t = vT \to \infty$, $\tau_T = o\left(t\right)$, $v \in (0, 1]$. Then by condition $\mathscr{L}_2^\star$ the uniform convergences

$$\Delta_t\left(\vartheta, X^t\right) \Longrightarrow \mathcal{N}\left(0, \mathbf{I}\left(\vartheta\right)\right), \qquad \bar{\mathbf{I}}^t_{\tau_T}(\vartheta) \longrightarrow \mathbf{I}\left(\vartheta\right),$$

hold.

Let us change the variable $v = tT^{-1} \in [0, 1]$ and introduce the stochastic process $\vartheta^\star_T\left(v\right)$, $v_\delta \le v \le 1$, where $\vartheta^\star_T\left(v\right) = \vartheta^\star_{vT,T}$ and $v_\delta = T^{\delta-1} \to 0$. Below ϑ_0 denotes the true value of the parameter and set

$$\eta^\star_T\left(v\right) = v\sqrt{T}\mathbf{I}(\vartheta_0)^{1/2}\left(\vartheta^\star_T\left(v\right) - \vartheta_0\right), \qquad \epsilon_0 \le v \le 1.$$

Theorem 5.4 *Let the conditions $\mathscr{L}^\star$ hold. Then the One-step MLE-process is consistent: for any $\nu > 0$ and any $v \in (0, 1]$*

$$\sup_{\vartheta_0\in\mathbb{K}} \mathbf{P}_{\vartheta_0}\left\{\left\|\vartheta^\star_T\left(v\right) - \vartheta_0\right\| > \nu\right\} \longrightarrow 0$$

is asymptotically normal

$$\sqrt{vT}\left(\vartheta^\star_T\left(v\right) - \vartheta_0\right) \Longrightarrow \mathcal{N}\left(0, \mathbf{I}(\vartheta_0)^{-1}\right) \tag{5.48}$$

and the finite-dimensional distributions of the process $\eta_T^\star(v)$, $\epsilon_0 \leq v \leq 1$ *converge to the finite-dimensional distributions of the two-dimensional standard Wiener process* $W(v)$, $\epsilon_0 \leq v \leq 1$, *where* $\epsilon_0 \in (0, 1]$.

Proof We have

$$\mathbf{P}_{\vartheta_0}\left\{\left\|\vartheta_T^\star(v)-\vartheta_0\right\|>\nu\right\}\leq \mathbf{P}_{\vartheta_0}\left\{\|\vartheta_{\tau_T}^*-\vartheta_0\|>\nu/2\right\}$$
$$+\mathbf{P}_{\vartheta_0}\left\{\|t^{-1/2}\bar{\mathbf{I}}_{\tau_T}^t(\vartheta_{\tau_T}^*)_+^{-1}\Delta_t(\vartheta_{\tau_T}^*,X^t)\|>\nu/2\right\}.$$

The first probability tends to zero since the MME $\vartheta_{\tau_T}^*$ is uniformly consistent. For the second probability we have

$$\mathbf{P}_{\vartheta_0}\left\{\|t^{-1/2}\bar{\mathbf{I}}_{\tau_T}^t(\vartheta_{\tau_T}^*)_+^{-1}\Delta_t(\vartheta_{\tau_T}^*,X^t)\|>\nu/2\right\}$$
$$\leq \mathbf{P}_{\vartheta_0}\left\{\frac{\kappa}{t}\left\|\int_{\tau_T}^t \dot{\mathrm{m}}(\vartheta_{\tau_T}^*,s)\,\mathrm{d}\bar W_s\right\|>\nu\right\}$$
$$\leq \frac{\kappa^2}{\nu^2t^2}\int_{\tau_T}^t \mathbf{E}_{\vartheta_0}\left\|\dot{\mathrm{m}}(\vartheta_{\tau_T}^*,s)\right\|^2\,\mathrm{d}s\leq \frac{C}{t}\longrightarrow 0.$$

To prove (5.48) let us write

$$\sqrt{vT}\left(\vartheta_T^\star(\tau)-\vartheta_0\right)=\sqrt{vT}\left(\vartheta_{\tau_T}^*-\vartheta_0\right)+\frac{\mathbf{I}_{\tau_T}^{vT}(\vartheta_{\tau_T}^*)_+^{-1}}{\sqrt{vT}}\int_{\tau_T}^{vT}\dot{\mathrm{m}}(\vartheta_{\tau_T}^*,s)\,\mathrm{d}\bar W_s$$
$$+\frac{\mathbf{I}_{\tau_T}^{vT}(\vartheta_{\tau_T}^*)_+^{-1}}{\sqrt{vT}}\int_{\tau_T}^{vT}\dot{\mathrm{m}}(\vartheta_{\tau_T}^*,s)\left[m(\vartheta_0,s)-m(\vartheta_{\tau_T}^*,s)\right]\mathrm{d}s,$$

and note that

$$\mathbf{E}_{\vartheta_0}\left\|\frac{1}{\sqrt{vT}}\int_{\tau_T}^{vT}\left[\dot{\mathrm{m}}(\vartheta_{\tau_T}^*,s)-\dot{\mathrm{m}}(\vartheta_0,s)\right]\mathrm{d}\bar W_s\right\|^2$$
$$\leq \frac{1}{vT}\int_{\tau_T}^{vT}\mathbf{E}_{\vartheta_0}\left\|\dot{\mathrm{m}}(\vartheta_{\tau_T}^*,s)-\dot{\mathrm{m}}(\vartheta_0,s)\right\|^2\mathrm{d}s\longrightarrow 0$$

as $T\rightarrow\infty$. By the CLT (see, Theorem 1.7) the convergence in distribution

$$\frac{1}{\sqrt{vT}}\int_{\tau_T}^{vT}\dot{\mathrm{m}}(\vartheta_0,s)\,\mathrm{d}\bar W_s\Longrightarrow\mathcal{N}\left(0,\mathbf{I}(\vartheta_0)\right)$$

holds, since by the LLN

$$\frac{1}{\upsilon T}\int_{\tau_T}^{\upsilon T} \dot{\mathrm{m}}(\vartheta_0,s)\,\dot{\mathrm{m}}(\vartheta_0,s)^\top \mathrm{d}t \longrightarrow \mathbf{I}(\vartheta_0).$$

Further, let us denote $V_{\tau_T}^* = \sqrt{\upsilon T}\left(\vartheta_{\tau_T}^* - \vartheta_0\right)$, then we can write

$$\begin{aligned} V_{\tau_T}^* &+ \frac{\mathbf{I}_{\tau_T}^{\upsilon T}(\vartheta_{\tau_T}^*)_+^{-1}}{\sqrt{\upsilon T}}\int_{\tau_T}^{\upsilon T} \dot{\mathrm{m}}(\vartheta_{\tau_T}^*,s)\left[m(\vartheta_0,s)-m(\vartheta_{\tau_T}^*,s)\right]\mathrm{d}s \\ &= \mathbf{I}_{\tau_T}^{\upsilon T}(\vartheta_{\tau_T}^*)_+^{-1}\left(\mathbf{I}_{\tau_T}^{\upsilon T}(\vartheta_{\tau_T}^*) - \frac{1}{\upsilon T}\int_0^1\int_{\tau_T}^{\upsilon T} \dot{\mathrm{m}}(\vartheta_{\tau_T}^*,s)\dot{\mathrm{m}}(\vartheta_r^*,s)^\top \mathrm{d}r\,\mathrm{d}s\right) V_{\tau_T}^*, \end{aligned}$$

where $\vartheta_r = \vartheta_{\tau_T} + r\left(\vartheta_{\tau_T} - \vartheta_0\right)$. The estimate

$$\left\|\dot{\mathrm{m}}(\vartheta_r,s) - \dot{\mathrm{m}}(\vartheta_{\tau_T},s)\right\| \le C\left\|\vartheta_{\tau_T} - \vartheta_0\right\|$$

(see (5.18)) and the equality

$$\mathrm{I}_{\tau_T}^{\upsilon T}(\vartheta_{\tau_T}^*) = \frac{1}{\upsilon T}\int_{\tau_T}^{\upsilon T} \dot{\mathrm{m}}(\vartheta_{\tau_T}^*,s)\dot{\mathrm{m}}(\vartheta_{\tau_T}^*,s)^\top \mathrm{d}s$$

give

$$\begin{aligned} V_{\tau_T}^* &+ \frac{\mathrm{I}_{\upsilon_\delta}^{\upsilon T}(\vartheta_{\tau_T}^*)^{-1}}{\sqrt{\upsilon T}}\int_{\tau_T}^{\upsilon T} \dot{\mathrm{m}}(\vartheta_{\tau_T}^*,s)\left[m(\vartheta_0,s)-m(\vartheta_{\tau_T}^*,s)\right]\mathrm{d}s \\ &= \sqrt{\upsilon T}\|\vartheta_{\tau_T}^* - \vartheta_0\|^2 O(1) = T^{\frac{1}{2}-\delta}O(1) \longrightarrow 0, \end{aligned}$$

as $T \to \infty$.

Introduce the vectors $\lambda_1,\ldots,\lambda_k$, where $\lambda_j \in \mathcal{R}^2$. To prove the convergence

$$\left(\lambda_1^\top \eta_T^\star(\upsilon_1),\ldots,\lambda_k^\top \eta_T^\star(\upsilon_k)\right) \Longrightarrow \left(\lambda_k^\top W(\upsilon_1),\ldots,\lambda_1^\top W(\upsilon_k)\right)$$

it is sufficient to verify the convergence of the sum

$$\sum_{j=1}^k a_j\lambda_j^\top \eta_T^\star(\upsilon_j) \Longrightarrow \sum_{j=1}^k a_j\lambda_j^\top W(\upsilon_j). \tag{5.49}$$

Here $a_1,\ldots,a_k$ are real numbers. The proof of the convergence (5.49) follows the same steps as it was done to prove the convergence (5.48). □

Note that, by the Theorem 5.2, the limit covariance matrix of the One-step MLE-process coincides with the covariance of the asymptotically efficient MLE.

5.3 Adaptive Filter

Recall that the observed process is (5.1), where the telegraph process $Y_t, t \geq 0$ depends on the unknown parameter $\vartheta = (\lambda, \mu) = (\theta_1, \theta_2)$. To estimate the unobserved $Y_t, t \geq 0$ we first estimate the conditional expectation $m(\vartheta, t) = \mathbf{E}_\vartheta\left(Y_t|\mathfrak{F}_t^X\right)$. Having the estimator $m_t^\star$ of $m(\vartheta, t)$ we define the estimator of the current state Y_t (following [93]) the two-state process

$$Y_t^\star = \begin{cases} a \ \text{if} \ \pi_t^\star > \frac{1}{2} \\ b \ \text{if} \ \pi_t^\star \leq \frac{1}{2} \end{cases}$$

The true value is $\vartheta_0 = \left(\theta_{1,0}, \theta_{2,0}\right)$. The construction of the adaptive filter follows the same lines as it was done in the preceding chapters.

1. *The learning interval is* $\left[0, \tau_T\right]$, *where* $\tau_T = \left[T^\delta\right], \delta \in (1/2, 1)$. *Here* $[A]$ *means the integer part of* A.
2. *The statistics* R_{1,τ_T}, R_{2,τ_T} *and equation for* $q_{\tau_T} = s_{\tau_T}$ *defined by the relations (5.22), (5.24), and (5.23) respectively. Therefore* q_{τ_T} *is solution of the following equation:*

$$2\Psi\left(q_{\tau_T}\right)\left(\frac{X_{\tau_T}}{\tau_T} - a\right)\left(b - \frac{X_{\tau_T}}{\tau_T}\right) = \frac{1}{\tau_T}\sum_{k=0}^{\tau_T - 1}\left[X_{k+1} - X_k\right]^2 - 1 - \left(\frac{X_{\tau_T}}{\tau_T}\right)^2.$$

3. *The MME* $\vartheta_{\tau_T}^* = \left(\theta_{1,\tau_T}^*, \theta_{2,\tau_T}^*\right)$ *is*

$$\theta_{1,\tau_T}^* = \frac{\tau_T^{-1} X_{\tau_T} - a}{b - a} q_{\tau_T}, \qquad \theta_{2,\tau_T}^* = \frac{b - \tau_T^{-1} X_{\tau_T}}{b - a} q_{\tau_T},$$

where

$$\Psi(x) = \frac{x - 1 + e^{-x}}{x^2}, \qquad x > 0.$$

4. *Equations for* $\pi_t^* = \pi(\vartheta_{\tau_T}^*, t)$, $\dot\pi_{1,t}^* = \dot\pi_\lambda(\vartheta_{\tau_T}^*, t)$, $\dot\pi_{2,t}^* = \dot\pi_\mu(\vartheta_{\tau_T}^*, t)$, $\tau_T \leq t \leq T$ *are*

$$\mathrm{d}\pi_t^* = \left[\theta_{2,\tau_T}^* - (\theta_{1,\tau_T}^* + \theta_{2,\tau_T}^*)\pi_t^* + \pi_t^* \left(1 - \pi_t^*\right)(b-a)\left(b + (a-b)\,\pi_t^*\right)\right]\mathrm{d}t$$
$$+ \pi_t^* \left(1 - \pi_t^*\right)(a-b)\ \mathrm{d}X_t, \qquad \pi_{\tau_T}^* = \pi_0, \tag{5.50}$$

$$\mathrm{d}\dot\pi_{1,t}^* = -\pi_t^*\,\mathrm{d}t - \dot\pi_{1,t}^*\left[(\theta_{1,\tau_T}^* + \theta_{2,\tau_T}^*) + h\,(1-2\pi)\left[b + h\pi_t^*\right] + h^2\pi_t^*\left(1 - \pi_t^*\right)\right]\mathrm{d}t$$
$$+ h\dot\pi_{1,t}^*\left(1 - 2\pi_t^*\right)\mathrm{d}X_t, \qquad \dot\pi_{1,\tau_T}^* = \pi_1,$$

$$\mathrm{d}\dot\pi_{2,t}^* = \left[1 - \pi_t^*\right]\mathrm{d}t - \dot\pi_{2,t}^*\left[(\theta_{1,\tau_T}^* + \theta_{2,\tau_T}^*) + h\left(1 - 2\pi_t^*\right)[b + h\pi]\right.$$
$$+ h\dot\pi_{2,t}^*\left(1 - 2\pi_t^*\right)\mathrm{d}X_t, \qquad \dot\pi_{2,\tau_T}^* = \pi_2 \tag{5.51}$$

5. *Introduce random processes* $m(\vartheta_{\tau_T}^*, t)$, $\dot{\mathrm{m}}(\vartheta_{\tau_T}^*, t)$, *where the vector-process* $\dot{\mathrm{m}}(\vartheta_{\tau_T}^*, t) = (\dot m_1(\vartheta_{\tau_T}^*, t), \dot m_2(\vartheta_{\tau_T}^*, t))^\top$ *and*

$$m(\vartheta_{\tau_T}^*, t) = b + (a-b)\,\pi_t^*, \quad \dot m_1(\vartheta_{\tau_T}^*, t) = (a-b)\,\dot\pi_{1,t}^*,$$
$$\dot m_2(\vartheta_{\tau_T}^*, t) = (a-b)\,\dot\pi_{2,t}^*, \qquad \tau_T \le t \le T.$$

6. *Define the empirical Fisher information matrix*

$$\mathrm{I}_{\tau_T}^t(\vartheta_{\tau_T}^*) = \frac{1}{t}\int_{\tau_T}^t \dot{\mathrm{m}}(\vartheta_{\tau_T}^*, s)\dot{\mathrm{m}}(\vartheta_{\tau_T}^*, s)^\top \mathrm{d}s.$$

7. *The One-step MLE-process* $\vartheta_{t,T}^\star = (\theta_{1,t}^\star, \theta_{2,t}^\star)^\top$, $\tau_T < t \le T$ *is given by the equality*

$$\vartheta_{t,T}^\star = \vartheta_{\tau_T}^* + t^{-1/2}\mathrm{I}_{\tau_T}^t(\vartheta_{\tau_T}^*)^{-1}\int_{\tau_T}^t \dot{\mathrm{m}}(\vartheta_{\tau_T}^*, s)\left[\mathrm{d}X_s - m(\vartheta_{\tau_T}^*, s)\mathrm{d}s\right].$$

8. *The adaptive filter for the conditional expectation* $m(\vartheta_0, t)$ *is defined as follows:* $m_{t,T}^\star = b + (a-b)\,\pi_{t,T}^\star$, $\tau_T < t \le T$, *where*

$$\mathrm{d}\pi_{t,T}^\star = \left[\theta_{2,t}^\star - (\theta_{1,t}^\star + \theta_{2,t}^\star)\pi_{t,T}^\star + \pi_{t,T}^\star(1 - \pi_{t,T}^\star)\,(b-a)\,(b + (a-b)\,\pi_{t,T}^\star)\right]\mathrm{d}t$$
$$+ (a-b)\,\pi_t^\star(1 - \pi_{t,T}^\star)\ \mathrm{d}X_t, \qquad \pi_{\tau_T}^\star. \tag{5.52}$$

9. *The current state of* Y_t *is estimated by the given above relation*

$$Y_{t,T}^\star = \begin{cases} a \ \text{ if } \ \pi_{t,T}^\star > \dfrac{1}{2} \\ b \ \text{ if } \ \pi_{t,T}^\star \le \dfrac{1}{2} \end{cases}.$$

10. *The last step is the study of the asymptotics of errors* $\mathbf{E}_{\vartheta_0}\left(m^\star_{t,T} - m\left(\vartheta_0, t\right)\right)^2$ *and* $\mathbf{E}_{\vartheta_0}\left(Y^\star_{t,T} - Y_t\right)^2$ *as* $T \to \infty$.

Remark 5.3 There is the problem of the definition of the initial values $\pi^*_{\tau_T}, \dot\pi^*_{1,\tau_T}, \dot\pi^*_{2,\tau_T}$ because we cannot use the equations like (5.50)–(5.51) on the time interval $\left[0, \tau_T\right]$. The stochastic Itô integral

$$\int_0^{\tau_T} \pi(\vartheta^*_{\tau_T}, t)(1 - \pi(\vartheta^*_{\tau_T}, t))\, \mathrm{d}X_t$$

is not defined. But we remark that the processes $\pi^*_t, \dot\pi^*_{1,t}, \dot\pi^*_{2,t}$ have ergodic properties and the limits of the corresponding integrals like

$$\bar{\mathrm{I}}^t_{\tau_T}(\vartheta^*_{\tau_T}) \longrightarrow \mathrm{I}\left(\vartheta_0\right)$$

in the proofs do not depend on the initial values. Therefore, we can take any values $\pi^*_{\tau_T}, \dot\pi^*_{1,\tau_T}, \dot\pi^*_{2,\tau_T}$ as initial values and the limits will be the same.

Consider the asymptotic ($t = vT \to \infty$) behavior of the error of approximation

$$\delta_m\left(t\right) = m^\star_t - m\left(\vartheta_0, t\right) = \left(a - b\right)\left(\pi^\star_t - \pi\left(\vartheta_0, t\right)\right) = \left(a - b\right)\delta_{\pi^\star}\left(t\right)$$

on the interval $[\epsilon_0 T, T]$ where (small) $\epsilon_0 \in (0, 1)$ and $v \in [\epsilon_0, 1]$.

Recall the notation $h = a - b$, $g = \theta_1 + \theta_2$ and substitute the observations

$$\mathrm{d}X_t = (b + h\pi_t)\,\mathrm{d}t + \mathrm{d}\bar W_t$$

in (5.52). Here $\pi_t = \pi\left(\vartheta_0, t\right)$. Then we obtain the representation ($\pi^\star_t = \pi^\star_{t,T}$)

$$\mathrm{d}\pi^\star_t = \left[\theta^\star_{2,t} - g^\star_t\pi^\star_t + h^2\pi^\star_t(1 - \pi^\star_t)(\pi_t - \pi^\star_t)\right]\mathrm{d}t + h\pi^\star_t(1 - \pi^\star_t)\,\mathrm{d}\bar W_t.$$

The equation for π_t is

$$\mathrm{d}\pi_t = \left[\theta_{2,0} - g_0\pi_t\right]\mathrm{d}t + h\pi_t(1 - \pi_t)\,\mathrm{d}\bar W_t.$$

Therefore $\delta_{\pi^\star}\left(t\right)$ satisfies the equation

$$\begin{aligned}
\mathrm{d}\delta_{\pi^\star}\left(t\right) &= \left[\theta^\star_{2,t} - \theta_{2,0} + \left(g^\star_t - g_0\right)\pi^\star_t\right]\mathrm{d}t \\
&\quad - \left[g_0 + h^2\pi^\star_t(1 - \pi^\star_t)\right]\delta_{\pi^\star}\left(t\right)\mathrm{d}t + h\left[1 - \left(\pi^\star_t + \pi_t\right)\right]\delta_{\pi^\star}\left(t\right)\mathrm{d}\bar W_t \\
&= \left[\theta^\star_{2,t} - \theta_{2,0} + \left(g^\star_t - g_0\right)\pi^\star_t\right]\mathrm{d}t - \psi^\star\left(\vartheta_0, \pi^\star_t\right)\delta_{\pi^\star}\left(t\right)\mathrm{d}t \\
&\quad + \sigma\left(\pi^\star_t, \pi_{\pi^\star}\right)\delta_{\pi^\star}\left(t\right)\mathrm{d}\bar W_t
\end{aligned}$$

with obvious notations. The solution of this equation is

$$\delta_{\pi^\star}(t)=\delta_{\pi^\star}\left(\tau_T\right)\Phi\left(\tau_T,t\right)+\int_{\tau_T}^{t}\Phi\left(s,t\right)\left[\theta_{2,s}^\star-\theta_{2,0}+\left(g_s^\star-g_0\right)\pi_s^\star\right]\mathrm{d}s$$

where

$$\Phi\left(s,t\right)=\exp\left(-\int_s^t\left[\psi^\star\left(\vartheta_0,\pi_r^\star\right)+\frac{1}{2}\sigma\left(\pi_r^\star,\pi_r\right)^2\right]\mathrm{d}r+\int_s^t\sigma\left(\pi_r^\star,\pi_r\right)\mathrm{d}\bar{W}_r\right).$$

Note that

$$\psi^\star\left(\vartheta_0,\pi_t^\star\right)=g_0+h^2\pi_t^\star(1-\pi_t^\star)>2\alpha>0$$

and

$$\begin{aligned}\mathbf{E}_{\vartheta_0}\Phi\left(s,t\right)&\leq e^{-2\alpha(t-s)}\mathbf{E}_{\vartheta_0}\exp\left(-\frac{1}{2}\int_s^t\sigma\left(\pi_r^\star,\pi_r\right)^2\mathrm{d}r+\int_s^t\sigma\left(\pi_r^\star,\pi_r\right)\mathrm{d}\bar{W}_r\right)\\&=e^{-2\alpha(t-s)}.\end{aligned}$$

Hence

$$\mathbf{E}_{\vartheta_0}\left|\delta_{\pi^\star}\left(\tau_T\right)\right|\Phi\left(\tau_T,t\right)\leq e^{-2\alpha(t-\tau_T)}$$

and we can omit this term. Further

$$\begin{aligned}\mathbf{E}_{\vartheta_0}\left|\delta_{\pi^\star}(t)\right|&\leq\int_{\tau_T}^t\mathbf{E}_{\vartheta_0}\Phi\left(s,t\right)\left|\theta_{2,s}^\star-\theta_{2,0}+\left(g_s^\star-g_0\right)\pi_s^\star\right|\mathrm{d}s\\&=\int_{\tau_T}^t\mathbf{E}_{\vartheta_0}\left(\left|\theta_{2,s}^\star-\theta_{2,0}+\left(g_s^\star-g_0\right)\pi_s^\star\right|\mathbf{E}_{\vartheta_0}\left(\Phi\left(s,t\right)|\mathfrak{F}_s^X\right)\right)\mathrm{d}s\\&\leq\int_{\tau_T}^t e^{-2\alpha(t-s)}\mathbf{E}_{\vartheta_0}\left|\theta_{2,s}^\star-\theta_{2,0}+\left(g_s^\star-g_0\right)\pi_s^\star\right|\mathrm{d}s\\&\leq\int_{\tau_T}^t e^{-2\alpha(t-s)}\mathbf{E}_{\vartheta_0}\left|\theta_{2,s}^\star-\theta_{2,0}\right|+\mathbf{E}_{\vartheta_0}\left|g_s^\star-g_0\right|\mathrm{d}s\\&\leq C\int_{\tau_T}^t e^{-2\alpha(t-s)}s^{-1/2}\mathrm{d}s\leq\frac{C}{\alpha\sqrt{t}}\longrightarrow 0.\end{aligned}\tag{5.53}$$

Therefore the estimator $\pi_t^\star\rightarrow\pi\left(\vartheta_0,t\right)$ as $t=vT\longrightarrow\infty$ for all $v\in[\epsilon_0,1]$.

Remark 5.4 Similar problem was considered by Khasminskii and Lazareva [93] for the model of observations

$$\mathrm{d}X_t = Y_t\,\mathrm{d}t + \sigma\,\mathrm{d}W_t, \qquad X_0 = 0, \qquad 0 \le t \le T,$$

but with $a = 1, b = 0$ and known $\mu > 0$, $\lambda > 0$. In this case Wonham equation for $\pi(t) = \mathbf{P}\left(Y_t = 0 | \mathfrak{F}_t^X\right)$ is

$$\mathrm{d}\pi(t) = \left[\mu - (\lambda + \mu)\,\pi(t)\right]\mathrm{d}t + \sigma^{-1}\pi(t)\,(1 - \pi(t))\left[\mathrm{d}X_t - (1 - \pi(t))\,\mathrm{d}t\right]$$

and the estimator of Y_t is

$$\hat{Y}_t = \begin{cases} 1 & \text{if } \pi(t) > \frac{1}{2} \\ 0 & \text{if } \pi(t) \le \frac{1}{2}. \end{cases}$$

It was shown that

$$\lim_{t\to\infty} \mathbf{E}_{\vartheta_0}\left(\hat{Y}_t - Y_t\right)^2 = R_0(\lambda, \mu, \sigma) \tag{5.54}$$

and the function $R_0(\lambda, \mu, \sigma) = R_0\left(\lambda\sigma^2, \mu\sigma^2, 1\right)$. Moreover, for the function $R_0(x, y, 1)$ the explicit expression is given (see (9)–(11) in [93]). Just note that we have a slightly different notation and to apply $R_0(\cdot, \cdot, \cdot)$ in our case ($a = 0, b = 1$) the function $R_0(\lambda, \mu, \sigma)$ has to be written as $R_0\left(\mu\sigma^2, \lambda\sigma^2, 1\right)$.

To study the error $\mathbf{E}_{\vartheta_0}\left(Y_{t,T}^\star - Y_t\right)^2$ we note that

$$\begin{aligned}\mathbf{E}_{\vartheta_0}\left(Y_{t,T}^\star - Y_t\right)^2 &= \mathbf{E}_{\vartheta_0}\left(Y_{t,T}^\star - \hat{Y}_t\right)^2 + \mathbf{E}_{\vartheta_0}\left(\hat{Y}_t - Y_t\right)^2 \\ &\quad + 2\mathbf{E}_{\vartheta_0}\left(Y_{t,T}^\star - \hat{Y}_t\right)\left(\hat{Y}_t - Y_t\right).\end{aligned}$$

Here $\hat{Y}_t = a\,1_{\{\pi_t^\circ > 1/2\}} + b\,1_{\{\pi_t^\circ \le 1/2\}}$, where $\pi_t^\circ = \pi(\vartheta_0, t)$.

We have

$$\begin{aligned}&\mathbf{E}_{\vartheta_0}\left(Y_{t,T}^\star - \hat{Y}_t\right)^2 \\ &= \mathbf{E}_{\vartheta_0}\left(a\left[1_{\{\pi_t^\star > 1/2\}} - 1_{\{\pi_t^\circ > 1/2\}}\right] + b\left[1_{\{\pi_t^\star \le 1/2\}} - 1_{\{\pi_t^\circ \le 1/2\}}\right]\right)^2 \\ &\le 2a^2\mathbf{E}_{\vartheta_0}\left[1_{\{\pi_t^\star > 1/2\}} - 1_{\{\pi_t^\circ > 1/2\}}\right]^2 + 2b^2\mathbf{E}_{\vartheta_0}\left[1_{\{\pi_t^\star \le 1/2\}} - 1_{\{\pi_t^\circ \le 1/2\}}\right]^2.\end{aligned}$$

Further

$$\begin{aligned}\mathbf{E}_{\vartheta_0}\left[1_{\{\pi_t^\star>1/2\}}-1_{\{\pi_t^\circ>1/2\}}\right]^2&=\mathbf{E}_{\vartheta_0}\left|1_{\{\pi_t^\star>1/2\}}-1_{\{\pi_t^\circ>1/2\}}\right|\\&=\mathbf{E}_{\vartheta_0}\left|1_{\{\pi_t^\circ+\delta_{\pi^\star}(t)>1/2\}}-1_{\{\pi_t^\circ>1/2\}}\right|\\&\leq\mathbf{E}_{\vartheta_0}\,1_{\{|\pi_t^\circ-1/2|\leq|\delta_{\pi^\star}(t)|\}}\\&=\mathbf{P}_{\vartheta_0}\left(\left|\pi_t^\circ-1/2\right|\leq|\delta_{\pi^\star}(t)|\right)\\&\leq\mathbf{P}_{\vartheta_0}\left(\left|\pi_t^\circ-1/2\right|\leq\varepsilon\right)+\mathbf{P}_{\vartheta_0}\left(|\delta_{\pi^\star}(t)|\geq\varepsilon\right)\end{aligned}$$

for any $\varepsilon > 0$. The process $\pi_t^\circ, t \geq 0$ is stationary with the invariant density $f(\vartheta_0, x)$ (see (5.9) and (5.10)). Thus

$$\mathbf{P}_{\vartheta_0}\left(\left|\pi_t^\circ-1/2\right|\leq\varepsilon\right)=\int_{1/2-\varepsilon}^{1/2+\varepsilon}f(\vartheta_0,x)\ \mathrm{d}x\leq C\varepsilon$$

where the constant $C > 0$ can be chosen not depending on ϑ_0. According to (5.53) we have as well

$$\mathbf{P}_{\vartheta_0}\left(|\delta_{\pi^\star}(t)|\geq\varepsilon\right)\leq\varepsilon^{-1}\mathbf{E}_{\vartheta_0}\,|\delta_{\pi^\star}(t)|\leq\frac{C}{\varepsilon\,\alpha}\,\frac{1}{\sqrt{t}}\,(1+o(1))\,.$$

For the term $\mathbf{E}_{\vartheta_0}\left[1_{\{\pi_t^\star\leq1/2\}}-1_{\{\pi_t^\circ\leq1/2\}}\right]^2$ we have a similar estimate.

Hence

$$\mathbf{E}_{\vartheta_0}\left(Y_{t,T}^\star-\hat{Y}_t\right)^2\leq C\varepsilon+\frac{C}{\varepsilon\,\alpha}\,\frac{1}{\sqrt{t}}\,(1+o(1))$$

and

$$\mathbf{E}_{\vartheta_0}\left|\left(Y_{t,T}^\star-\hat{Y}_t\right)(\hat{Y}_t-Y_t)\right|\leq\left(\mathbf{E}_{\vartheta_0}(\hat{Y}_t-Y_t)\right)^2\left(\mathbf{E}_{\vartheta_0}(Y_{t,T}^\star-\hat{Y}_t)^2\right)^{1/2}.$$

Therefore if we put $\varepsilon = t^{-1/4}$, then

$$\mathbf{E}_{\vartheta_0}\left(Y_{t,T}^\star-\hat{Y}_t\right)^2=\mathbf{E}_{\vartheta_0}\left(Y_t-\hat{Y}_t\right)^2+O\left(t^{-1/8}\right).$$

Further,

$$\begin{aligned}\mathbf{E}_{\vartheta_0}\left(Y_t-\hat{Y}_t\right)^2&=\mathbf{E}_{\vartheta_0}Y_t^2+\mathbf{E}_{\vartheta_0}\hat{Y}_t^2-2\mathbf{E}_{\vartheta_0}[Y_t\hat{Y}_t]\\&=\mathbf{E}_{\vartheta_0}Y_t^2+\mathbf{E}_{\vartheta_0}\hat{Y}_t^2-2\mathbf{E}_{\vartheta_0}\hat{Y}_t\left[a\pi_t^\circ+b\left(1-\pi_t^\circ\right)\right]\end{aligned}$$

because $\mathbf{E}_{\vartheta_0}[Y_t\hat Y_t]=\mathbf{E}_{\vartheta_0}\left[\hat Y_t\mathbf{E}_{\vartheta_0}\left(Y_t|\mathfrak{F}_t^X\right)\right]$. According to (5.3) we can write

$$\mathbf{E}_{\vartheta_0}Y_t^2=\frac{(a\mu_0+b\lambda_0)^2+\lambda_0\mu_0\,(b-a)^2}{(\lambda_0+\mu_0)^2}\equiv A\,(\vartheta_0)\,.$$

We have as well

$$\begin{aligned}\mathbf{E}_{\vartheta_0}\hat Y_t^2&=a^2\mathbf{P}_{\vartheta_0}\left(\pi_t^\circ>1/2\right)+b^2\mathbf{P}_{\vartheta_0}\left(\pi_t^\circ\leq 1/2\right)\\&=a^2\int_{1/2}^1 f\,(\vartheta_0,x)\,\mathrm{d}x+b^2\int_0^{1/2}f\,(\vartheta_0,x)\,\mathrm{d}x\end{aligned}$$

and

$$\begin{aligned}\mathbf{E}_{\vartheta_0}\hat Y_t\left[a\pi_t^\circ+b\left(1-\pi_t^\circ\right)\right]&=a^2\int_{1/2}^1 xf\,(\vartheta_0,x)\,\mathrm{d}x+ab\int_{1/2}^1(1-x)\,f\,(\vartheta_0,x)\,\mathrm{d}x\\&\quad+ab\int_0^{1/2}xf\,(\vartheta_0,x)\,\mathrm{d}x+b^2\int_0^{1/2}(1-x)\,f\,(\vartheta_0,x)\,\mathrm{d}x.\end{aligned}$$

Thus

$$\begin{aligned}\mathbf{E}_{\vartheta_0}\hat Y_t^2-2\mathbf{E}_{\vartheta_0}\hat Y_t\left[a\pi_t^\circ+b\left(1-\pi_t^\circ\right)\right]&=\left(a^2-2ab\right)\int_{1/2}^1 f\,(\vartheta_0,x)\,\mathrm{d}x\\&\quad-b^2\int_0^{1/2}f\,(\vartheta_0,x)\,\mathrm{d}x+2\left(ab-a^2\right)\int_{1/2}^1 xf\,(\vartheta_0,x)\,\mathrm{d}x\\&\quad+2\left(b^2-ab\right)\int_0^{1/2}xf\,(\vartheta_0,x)\,\mathrm{d}x\equiv B\,(\vartheta_0)\,.\end{aligned}$$

Therefore the following result is valid.

Theorem 5.5 *Let the conditions* $\mathscr{L}^\star$ *hold. Then*

$$\lim_{T\to\infty}\mathbf{E}_{\vartheta_0}\left(Y_{t,T}^\star-Y_t\right)^2=A\,(\vartheta_0)+B\,(\vartheta_0)\,. \tag{5.55}$$

Recall that the adaptive filter is constructed in the "scheme of series," i.e., we have a family of problems indexed by T and $T\to\infty$. This means that for each T we take a learning interval $\left[0,T^\delta\right]$, $\delta\in(1/2,1)$ and propose an adaptive filter on the time interval $[\epsilon_0 T,T]$ with small positive ϵ_0. The asymptotic $t\to\infty$ in (5.55) means that $t=vT$, where $v\in[\epsilon_0,1]$.

Of course, the limit (5.55) coincides with the limit (5.54) if we put $a=0, b=1, \lambda=\lambda_0$, $\mu=\mu_0$, and $\sigma=1$.

Remark 5.5 Another statement of the problem and lower bound for the probability $\inf_{\bar{X}_t} \mathbf{P}\left(\bar{X}_t \neq X_t\right)$ in the asymptotic $t \rightarrow \infty$ was proposed in the work of Khasminskii and Zeitouni [95]. Here $\bar{X}_t$ is an arbitrary estimator of X_t, and there is no unknown parameter.

Chapter 6
Hidden AR Process

An autoregressive (AR) Gaussian time series is observed in the presence of additive white Gaussian noise. It is assumed that some of the parameters of the system are unknown. The goal is to construct an adaptive Kalman filter and describe the errors of approximation in the asymptotic framework of large samples. This filter is realized in several steps. First, the model of observations and the corresponding Kalman filter are introduced. Then, the asymptotic properties of the estimators obtained by the method of moments are described. In the next step, these estimators are used to construct the One-step MLE-process. The final step involves substituting the MLE-process into the Kalman filtering equations. The question of the asymptotic efficiency of this approximation is also discussed. The asymptotic properties of the MME, MLE, and BE estimators are described as well.

6.1 Model of Observations

We are given a partially observed linear system

$$X_t = f\, Y_{t-1} + \sigma\, w_t, \qquad X_0, \qquad t = 1, 2, \ldots, \tag{6.1}$$

$$Y_t = a\, Y_{t-1} + b\, v_t, \qquad Y_0, \tag{6.2}$$

where $X^T = (X_0, X_1, \ldots, X_T)$ are the observations, and $Y_t, t \geq 0$ is a hidden autoregressive (AR) process. Here, $w_t, t \geq 1$ and $v_t, t \geq 1$ are standard Gaussian independent random variables, i.e., $w_t \sim \mathcal{N}(0, 1)$, $v_t \sim \mathcal{N}(0, 1)$. The initial values are $X_0 \sim \mathcal{N}\left(0, d_x^2\right)$ and $Y_0 \sim \mathcal{N}\left(0, d_y^2\right)$, and their correlation coefficient is $\rho_{xy} = \mathbf{E} X_0 Y_0$. The system is defined by the parameters $a, b, f, \sigma^2, d_x^2, d_y^2$. It will be convenient, for instance, to denote $\vartheta = \left(a, b, f, \sigma^2\right)$.

Y. A. Kutoyants, *Hidden Markov Processes and Adaptive Filtering*, Springer Series in Statistics, https://doi.org/10.1007/978-3-032-00052-1_6

Remark that this model can be considered as discrete-time approximation of the continuous time model studied in Chap. 4:

$$\begin{aligned} \mathrm{d}X(s) &= \bar{f}Y(s)\,\mathrm{d}s + \sigma\,\mathrm{d}W(s), \qquad && X(0), \quad 0 \le s \le S, \\ \mathrm{d}Y(s) &= -\bar{a}Y(s)\,\mathrm{d}s + \bar{b}\,\mathrm{d}V(s), \qquad && Y(0). \end{aligned}$$

Consider the corresponding discrete-time model

$$\begin{aligned} X(s_k) - X(s_{k-1}) &= \bar{f}\delta Y(s_{k-1}) + \sigma\,[W(s_k) - W(s_{k-1})],\ X(0),\ k = 1, \ldots, K, \\ Y(s_k) &= (1 - \bar{a}\delta)\,Y(s_{k-1}) + \bar{b}\,[V(s_k) - V(s_{k-1})] \end{aligned}$$

where $s_k = s_{k-1} + \delta$, $K = \left[\frac{S}{\delta}\right]$ and $\delta > 0$. Introduce notations

$$\begin{aligned} &X_k = \delta^{-1/2}[X(s_k) - X(s_{k-1})], \quad w_k = \delta^{-1/2}[W(s_k) - W(s_{k-1})] \sim \mathcal{N}(0,1), \\ &Y_k = Y(s_k), \qquad v_k = \delta^{-1/2}[V(s_k) - V(s_{k-1})] \sim \mathcal{N}(0,1), \\ &f = \delta^{1/2}\bar{f}, \qquad a = 1 - \delta\bar{a}, \qquad b = \delta^{1/2}\bar{b}. \end{aligned}$$

Now the partially observed discrete-time system is

$$\begin{aligned} X_k &= f\,X_{k-1} + \sigma\,w_k, \qquad && X_0, \quad k = 1, \ldots, K, \\ Y_k &= a\,Y_{k-1} + b\,v_k, \qquad && Y_0 \end{aligned}$$

which coincides with (6.1) and (6.2) if we put $k = t$, $K = T$.

The σ-algebra generated by the first $t+1$ observations $X_0, X_1, \ldots, X_t$ is denoted by $\mathfrak{F}_t^X$. The conditional expectation $m(\vartheta, t) = \mathbf{E}_\vartheta\left(Y_t | \mathfrak{F}_t^X\right)$ according to the equations of Kalman filter (see [83] or Theorem 13.4 in [142]) satisfies the following equation:

$$m(\vartheta, t) = a\,m(\vartheta, t-1) + \frac{af\gamma(\vartheta, t-1)}{\sigma^2 + f^2\gamma(\vartheta, t-1)}\,[X_t - fm(\vartheta, t-1)], \quad t \ge 1. \tag{6.3}$$

The initial value is $m(\vartheta, 0) = \mathbf{E}_\vartheta(Y_0 | X_0)$.

The mean squared error $\gamma(\vartheta, t) = \mathbf{E}_\vartheta(Y_t - m(\vartheta, t))^2$ is described by the equation

$$\gamma(\vartheta, t) = a^2\gamma(\vartheta, t-1) + b^2 - \frac{a^2 f^2 \gamma(\vartheta, t-1)^2}{\sigma^2 + f^2\gamma(\vartheta, t-1)}, \qquad t \ge 1 \tag{6.4}$$

with the initial value $\gamma(\vartheta, 0) = \mathbf{E}_\vartheta(Y_0 - m(\vartheta, 0))^2$.

We suppose that the observations X^T are given and that some of the parameters are unknown, but their values always satisfy the condition:

$$\mathscr{M}_0: \qquad a^2 \in [0,1), \qquad b^2 > 0, \qquad f^2 > 0, \qquad \sigma^2 > 0. \tag{6.5}$$

The abovementioned condition is uniform in the sense that, in case of the unknown parameter being like $f \in \Theta = (\alpha_f, \beta_f)$, then $\alpha_f > 0$ or $\beta_f < 0$. If, for example, the unknown parameter is $a \in \Theta = (\alpha_a, \beta_a)$, then by condition $\mathscr{M}_0$ we have $\alpha_a > -1$ and $\beta_a < 1$. Our goal in situations like this is to propose an approximation of $m(\vartheta, t), t \geq 1$, as well as describe the error of it in the asymptotic of *large samples*, i.e., as $T \to \infty$.

The consistent estimation of the parameters $\left(d_x^2, d_y^2\right)$ is impossible, and we suppose that these parameters are known. If these values are unknown, then the system (6.3)–(6.4) will be solved with some wrong initial values, but due to the robustness of the solutions, the difference between solutions with true and wrong initial values under condition $\mathscr{M}_0$ is asymptotically negligible. Note as well that the consistent estimation of the parameters $\vartheta = (f, b)$ or $\vartheta = (f, a, b)$ is impossible because the model (6.1)–(6.2) depends on the product fb.

If some of the mentioned parameters are unknown, then, of course, we cannot use (6.3)–(6.4) for calculation of $m(\vartheta, t), t \geq 1$.

We suppose that the observations $X^T = (X_0, X_1, \ldots, X_T)$ are given and that some of the parameters are unknown, but their values always satisfy the condition $\mathscr{M}_0$. Our goal is to propose an approximation of $m(\vartheta, t), t \geq 1$ in such situations and to describe the error of approximation in the asymptotic of *large samples*, i.e., as $T \to \infty$.

As before, the proposed program consists of several steps. First on some learning interval $\left[0, \tau_T\right]$ of negligible length ($\tau_T/T \to 0$) we construct a consistent preliminary estimator $\vartheta^*_{\tau_T}$. Then this estimator is used for defining the One-step MLE-process $\vartheta^\star_T = \left(\vartheta^\star_{t,T}, t = \tau_T + 1, \ldots, T\right)$ and finally the approximation $m^\star_T = (m^\star_{t,T}, t = \tau_T + 1, \ldots, T)$ is obtained by substituting $\vartheta^\star_T$ in the slightly modified Eqs. (6.3)–(6.4). The last step is to evaluate the error $m^\star_{t,T} - m(\vartheta_0, t)$.

Remark that the function $\gamma(\vartheta, t)$ converges to the value

$$\gamma_*(\vartheta) = \frac{f^2b^2 - \sigma^2\left(1-a^2\right)}{2f^2} + \frac{1}{2}\left[\left(\frac{\sigma^2\left(1-a^2\right)}{f^2} - b^2\right)^2 + \frac{4b^2\sigma^2}{f^2}\right]^{1/2} \tag{6.6}$$

as $t \to \infty$ (see Example 3 in Section 14.4, [142]). The value $\gamma_*(\vartheta)$ is obtained as a positive solution of the Eq. (6.4), where we put $\gamma(\vartheta, t) = \gamma(\vartheta, t-1) = \gamma_*(\vartheta)$, which becomes

$$\gamma_*(\vartheta)^2+\left[\frac{\sigma^2\left(1-a^2\right)}{f^2}-b^2\right]\gamma_*(\vartheta)-\frac{b^2\sigma^2}{f^2}=0.$$

Below we study the asymptotic ($T \rightarrow \infty$) properties of the estimators. To simplify the exposition, we suppose that the initial value $\gamma(\vartheta,0)=\gamma_*(\vartheta)$. Then for any $t\geq 1$ we have $\gamma(\vartheta,t)=\gamma_*(\vartheta)$. Of course, this is the condition on the correlation between X_0 and Y_0 and the values d_x^2, d_y^2.

Therefore the Eq. (6.3) is replaced by the Kalman filter described by the equation

$$m_t(\vartheta)=a\,m_{t-1}(\vartheta)+\frac{af\gamma_*(\vartheta)}{\sigma^2+f^2\gamma_*(\vartheta)}\left[X_t-fm_{t-1}(\vartheta)\right],\quad t\geq 1 \tag{6.7}$$

with the corresponding initial value, providing $\gamma(\vartheta,0)=\gamma_*(\vartheta)$. Recall that Eq. (6.3) is stable w.r.t. the initial value, i.e., for the wrong initial condition the difference $m(\vartheta,t)-m_t(\vartheta)\rightarrow 0$.

Note as well that if we denote ϑ_0 the true value, then $\zeta_0(\vartheta_0)\sim\mathcal{N}(0,1)$ and

$$\zeta_t(\vartheta_0)=\frac{X_t-f_0m_{t-1}(\vartheta_0)}{\sqrt{\sigma_0^2+f_0^2\gamma_*(\vartheta_0)}}\sim\mathcal{N}(0,1),\qquad t\geq 1$$

are i.i.d. Gaussian random variables (see Theorem 13.5 in [142]). This means that the equation of observations (6.1) can be written as follows:

$$X_t=f_0m_{t-1}(\vartheta_0)+\sqrt{\sigma_0^2+f_0^2\gamma_*(\vartheta_0)}\,\zeta_t(\vartheta_0),\qquad t\geq 1.$$

Using this representation, we can rewrite the Eq. (6.7) too

$$\begin{aligned}m_t(\vartheta)&=a\,m_{t-1}(\vartheta)+\frac{af\gamma_*(\vartheta)}{\sigma^2+f^2\gamma_*(\vartheta)}\left[f_0m_{t-1}(\vartheta_0)-fm_{t-1}(\vartheta)\right]\\&\quad+\frac{af\gamma_*(\vartheta)\sqrt{\sigma_0^2+f_0^2\gamma_*(\vartheta_0)}}{\sigma^2+f^2\gamma_*(\vartheta)}\zeta_t(\vartheta_0),\quad t\geq 1.\end{aligned} \tag{6.8}$$

We will often use the following notation:

$$M(\vartheta,t)=fm(\vartheta,t),\qquad \Gamma(\vartheta)=\frac{f^2\gamma_*(\vartheta)}{\sigma^2+f^2\gamma_*(\vartheta)},\qquad P(\vartheta)=\sigma^2+\Gamma(\vartheta),$$

$$B(\vartheta,\vartheta_0)=\frac{a\Gamma(\vartheta)\sqrt{P(\vartheta_0)}}{P(\vartheta)},\qquad A(\vartheta)=\frac{a\sigma^2}{P(\vartheta)}.$$

The Eq. (6.8) can be re-written as follows:

$$M_t(\vartheta) = A(\vartheta)\ M_{t-1}(\vartheta) + (a - A(\vartheta))\,M_{t-1}(\vartheta_0) + B(\vartheta,\vartheta_0)\,\zeta_t(\vartheta_0), \quad t \ge 1. \tag{6.9}$$

If $\vartheta = \vartheta_0$, then this equation becomes

$$M_t(\vartheta_0) = a_0 M_{t-1}(\vartheta_0) + \frac{a_0 \Gamma(\vartheta_0)}{\sqrt{P(\vartheta_0)}}\,\zeta_t(\vartheta_0), \quad t \ge 1.$$

As $\zeta_t(\vartheta_0), t \ge 1$ are independent Gaussian random variables, the likelihood function is

$$\begin{aligned} L\left(\vartheta, X^T\right) &= \left(2\pi\left(\sigma^2 + f^2\gamma_*(\vartheta)\right)\right)^{-T/2} \exp\left(-\frac{1}{2}\sum_{t=1}^{T}\frac{(X_t - f m_{t-1}(\vartheta))^2}{\sigma^2 + f^2\gamma_*(\vartheta)}\right)\\ &= \left(\frac{1}{2\pi P(\vartheta)}\right)^{T/2} \exp\left(-\frac{1}{2}\sum_{t=1}^{T}\frac{(X_t - M_{t-1}(\vartheta))^2}{P(\vartheta)}\right), \qquad \vartheta \in \Theta. \end{aligned} \tag{6.10}$$

Here Θ is an open, bounded, convex set of the possible values of the parameter ϑ.

Our goal is to obtain a good recurrent approximation $m_t^\star$, $0 < t \le T$, of the process $m(\vartheta, t)$, $0 < t \le T$, in the homogeneous partially observed system (6.1)–(6.2), and to examine the question of asymptotic efficiency of adaptive filters. This approximation is realized by the following program:

1. *Calculate a preliminary estimator* $\bar\vartheta_\tau$ *on relatively small interval* $[0, \tau]$.
2. *Construct the One-step MLE-process* $\vartheta_{t,T}^\star$, $\tau < t \le T$ *by using* $\bar\vartheta_\tau$.
3. *As an approximation of* $m(\vartheta_0, t)$, $\tau < t \le T$, *we propose* $m_t^\star$, $\tau < t \le T$ *which has been obtained with the help of Kalman equation* (6.9), *where* ϑ *is replaced by* $\vartheta_{t,T}^\star$, *where* $\tau < t \le T$.
4. *Estimate the error* $m_t^\star - m(\vartheta_0, t)$, $\tau < t \le T$.
5. *Discuss the asymptotic efficiency of the adaptive filter.*

This means that we have no online approximation on the time interval $[0, \tau]$, but $\tau/T \to 0$. Recall that the One-step MLE-process used here is the well-known Le Cam's One-step MLE, wherein we treat the upper limit of the sum (i.e., time t) as a variable.

We propose the adaptive filter for the model (6.1)–(6.2) (steps 1–5). The exposition is in some sense similar to the exposition in the continuous time case of Chap. 4.

6.2 Parameter Estimation

6.2.1 *MME*

The method of moments for this model of observations is well known. Usually it is based on the empirical moments of the observed process, see, e.g., [176]. We follow the procedure proposed in Chap. 5, where a similar problem of adaptive filtration for continuous time system was considered, and the MMEs are constructed on the basis of the increments of the observed process. There is no essential difference between these two solutions of the problem.

Three statistics are introduced:

$$S_{1,T}\left(X^T\right)=\frac{1}{T}\sum_{t=2}^{T}\left(X_t-X_{t-1}\right)^2,$$

$$S_{2,T}\left(X^T\right)=\frac{1}{T}\sum_{t=3}^{T}\left(X_t-X_{t-1}\right)\left(X_{t-1}-X_{t-2}\right),$$

$$S_{3,T}\left(X^T\right)=\frac{1}{T}\sum_{t=4}^{T}\left(X_t-X_{t-1}\right)\left(X_{t-2}-X_{t-3}\right)$$

and their asymptotic ($T\rightarrow\infty$) behaviors are studied. We always suppose that the condition $\mathscr{M}_0$ is fulfilled, and the true value of the parameter $\vartheta=\left(a,b,f,\sigma^2\right)$ is denoted as ϑ_0. This means that $\Theta=\left(\alpha_a,\beta_a\right)\times\left(\alpha_b,\beta_b\right)\times\left(\alpha_f,\beta_f\right)\times\left(\alpha_\sigma,\beta_\sigma\right)$, where $\alpha_a>-1,\beta_a<1,\alpha_b>0,\alpha_f>0,\alpha_\sigma>0$. We do not estimate this four-dimensional parameter, and this is a temporary notation.

Denote

$$\Phi_1\left(\vartheta\right)=\frac{2f^2b^2}{1+a}+2\sigma^2,\qquad \Phi_2\left(\vartheta\right)=\frac{f^2b^2a\left(1-a\right)}{\left(1+a\right)}-\sigma^2,$$

$$\Phi_3\left(\vartheta\right)=\frac{f^2b^2a\left(a-1\right)}{\left(1+a\right)}.$$

Lemma 6.1 *We have the limits*

$$S_{1,T}\left(X^T\right)\longrightarrow\Phi_1\left(\vartheta_0\right),\tag{6.11}$$

$$S_{2,T}\left(X^T\right)\longrightarrow\Phi_2\left(\vartheta_0\right),\tag{6.12}$$

$$S_{3,T}\left(X^T\right)\longrightarrow\Phi_3\left(\vartheta_0\right),\tag{6.13}$$

and there exist constants $C_1>0$, $C_2>0$, $C_3>0$ *such that*

$$\mathbf{E}_{\vartheta_0}\left|S_{1,T}\left(X^T\right)-\Phi_1\left(\vartheta_0\right)\right|^2\leq\frac{C_1}{T},\quad \mathbf{E}_{\vartheta_0}\left|S_{2,T}\left(X^T\right)-\Phi_2\left(\vartheta_0\right)\right|^2\leq\frac{C_2}{T},\tag{6.14}$$

$$\mathbf{E}_{\vartheta_0}\left|S_{3,T}\left(X^T\right)-\Phi_3\left(\vartheta_0\right)\right|^2\leq\frac{C_3}{T}.\tag{6.15}$$

Proof According to (6.1)

$$\frac{1}{T}\sum_{t=2}^T\left(X_t-X_{t-1}\right)^2=\frac{f_0^2}{T}\sum_{t=2}^T\left(Y_{t-1}-Y_{t-2}\right)^2+\frac{\sigma_0^2}{T}\sum_{t=2}^T\left(w_t-w_{t-1}\right)^2$$
$$+\frac{2f_0\sigma_0}{T}\sum_{t=2}^T\left(Y_{t-1}-Y_{t-2}\right)\left(w_t-w_{t-1}\right).$$

The Gaussian time series $Y_t, t\geq 1$ is exponentially mixing with the stationary (invariant) Gaussian distribution $\mathcal{N}\left(0,\frac{b_0^2}{1-a_0^2}\right)$ and it is independent of $w_t, t\geq 1$. Therefore by the law of large numbers we have the convergences

$$\frac{f_0^2}{T}\sum_{t=2}^T\left(Y_{t-1}-Y_{t-2}\right)^2=\frac{f_0^2}{T}\sum_{t=2}^T\left(\left(a_0-1\right)Y_{t-2}+b_0v_{t-1}\right)^2$$
$$=\frac{f_0^2\left(1-a_0\right)^2}{T}\sum_{t=2}^T Y_{t-2}^2-\frac{2b_0\left(1-a_0\right)f_0^2}{T}\sum_{t=2}^T Y_{t-2}v_{t-1}+\frac{f_0^2b_0^2}{T}\sum_{t=2}^T v_{t-1}^2$$
$$\longrightarrow\frac{f_0^2b_0^2\left(1-a_0\right)^2}{1-a_0^2}+f_0^2b_0^2=\Phi_1\left(\vartheta_0\right)-2\sigma_0^2,$$

$$\frac{2f_0\sigma_0}{T}\sum_{t=2}^T\left(Y_{t-1}-Y_{t-2}\right)\left(w_t-w_{t-1}\right)\longrightarrow 0,$$

$$\frac{\sigma_0^2}{T}\sum_{t=2}^T\left(w_t-w_{t-1}\right)^2\longrightarrow 2\sigma_0^2,$$

which proves (6.11).

To prove (6.12) we write

$$\left(X_t-X_{t-1}\right)\left(X_{t-1}-X_{t-2}\right)$$
$$=\left[f_0\left(Y_{t-1}-Y_{t-2}\right)+\sigma_0\left(w_t-w_{t-1}\right)\right]\left[f_0\left(Y_{t-2}-Y_{t-3}\right)+\sigma_0\left(w_{t-1}-w_{t-2}\right)\right]$$
$$=\left[f_0\left(a_0-1\right)Y_{t-2}+f_0b_0v_{t-1}+\sigma_0\left(w_t-w_{t-1}\right)\right]$$

$$\times \frac{1}{a_0}\left[f_0\left(a_0-1\right)Y_{t-2}+f_0b_0a_0v_{t-2}+\sigma_0a_0\left(w_{t-1}-w_{t-2}\right)\right].$$

Therefore

$$\frac{1}{T}\sum_{t=3}^{T}\left(X_t-X_{t-1}\right)\left(X_{t-1}-X_{t-2}\right)=\frac{f_0^2\left(a_0-1\right)^2}{a_0T}\sum_{t=3}^{T}Y_{t-2}^2$$

$$+\frac{f_0^2b_0\left(a_0-1\right)}{a_0T}\sum_{t=3}^{T}Y_{t-2}v_{t-2}+\frac{\sigma_0^2}{T}\sum_{t=3}^{T}\left(w_t-w_{t-1}\right)\left(w_{t-1}-w_{t-2}\right)+o\left(1\right).$$

Here $o\left(1\right)$ means $L_2(\mathbf{P}_{\vartheta_0})$ convergence. We have

$$\frac{1}{T}\sum_{t=3}^{T}Y_{t-2}v_{t-2}=\frac{a_0}{T}\sum_{t=3}^{T}Y_{t-3}v_{t-2}+\frac{b_0}{T}\sum_{t=3}^{T}v_{t-2}^2\longrightarrow b_0,$$

$$\frac{1}{T}\sum_{t=3}^{T}\left(w_t-w_{t-1}\right)\left(w_{t-1}-w_{t-2}\right)\longrightarrow -1.$$

Finally

$$\frac{1}{T}\sum_{t=4}^{T}\left(X_t-X_{t-1}\right)\left(X_{t-1}-X_{t-2}\right)$$

$$\longrightarrow \frac{f_0^2b_0^2\left(a_0-1\right)^2}{a_0\left(1-a_0^2\right)}+\frac{f_0^2b_0^2\left(a_0-1\right)}{a_0}-\sigma^2=\Phi_2\left(\vartheta_0\right).$$

The last convergence (6.13) is obtained by using similar arguments as follows:

$$\begin{aligned}&\left(X_t-X_{t-1}\right)\left(X_{t-2}-X_{t-3}\right)\\&\quad=\left[f_0\left(Y_{t-1}-Y_{t-2}\right)+\sigma_0\left(w_t-w_{t-1}\right)\right]\left[f_0\left(Y_{t-3}-Y_{t-4}\right)+\sigma_0\left(w_{t-2}-w_{t-3}\right)\right]\\&\quad=\left[f_0\left(a_0Y_{t-2}-Y_{t-2}\right)+f_0b_0v_{t-1}+\sigma_0\left(w_t-w_{t-1}\right)\right]\\&\qquad\times\frac{1}{a_0}\left[f_0\left(a_0Y_{t-3}-Y_{t-3}\right)+f_0b_0v_{t-3}+a_0\sigma_0\left(w_{t-2}-w_{t-3}\right)\right].\end{aligned}$$

Here we used Eq. (6.2) and the equality $a_0Y_{t-4}=Y_{t-3}-b_0v_{t-3}$. Further, as $w_t, t\geq 1$ are independent of $Y_t, t\geq 1$ and $v_t, t\geq 1$ we can write

$$S_{3,T}\left(X^T\right)=\frac{1}{a_0T}\sum_{t=4}^{T}\left[f_0\left(a_0-1\right)Y_{t-2}+f_0b_0v_{t-1}\right]$$

$$\times \left[f_0\left(a_0-1\right)Y_{t-3}+f_0b_0v_{t-3}\right]+o\left(1\right)$$

$$=\frac{1}{a_0T}\sum_{t=4}^{T}\left[f_0\left(a_0-1\right)a_0Y_{t-3}+f\left(a_0-1\right)v_{t-2}+f_0b_0v_{t-1}\right]$$

$$\times\left[f_0\left(a_0-1\right)Y_{t-3}+f_0b_0v_{t-3}\right]+o\left(1\right)$$

$$=\frac{f_0^2\left(a_0-1\right)^2}{T}\sum_{t=4}^{T}Y_{t-3}^2+\frac{f_0^2\left(a_0-1\right)b_0}{T}\sum_{t=3}^{T}Y_{t-3}v_{t-3}+o\left(1\right)$$

$$=\frac{f_0^2\left(a_0-1\right)^2}{T}\sum_{t=4}^{T}Y_{t-3}^2+\frac{f_0^2b_0^2\left(a_0-1\right)}{T}\sum_{t=4}^{T}v_{t-3}^2+o\left(1\right)$$

$$\longrightarrow\frac{f_0^2b_0^2\left(1-a_0\right)}{1+a_0}+f_0^2b_0^2\left(a_0-1\right)=\Phi_3\left(\vartheta_0\right).$$

As the Gaussian AR processes $Y_t, t\geq 1$ and $Y_t^2, t\geq 1$ have exponentially decreasing correlation functions, the convergences (6.14) and (6.15) follow from the standard arguments. For the higher moments, see Rosenthal-type inequalities [41]. □

Remark 6.1 According to the proofs, it follows that the estimates (6.14) and (6.15) are valid uniformly on Θ, i.e.,

$$\sup_{\vartheta_0\in\Theta}\mathbf{E}_{\vartheta_0}\left|S_{1,T}\left(X^T\right)-\Phi_1\left(\vartheta_0\right)\right|^2\leq\frac{C}{T},\qquad\sup_{\vartheta_0\in\Theta}\mathbf{E}_{\vartheta_0}\left|S_{2,T}\left(X^T\right)-\Phi_2\left(\vartheta_0\right)\right|^2\leq\frac{C}{T}\tag{6.16}$$

$$\sup_{\vartheta_0\in\Theta}\mathbf{E}_{\vartheta_0}\left|S_{3,T}\left(X^T\right)-\Phi_3\left(\vartheta_0\right)\right|^2\leq\frac{C}{T}.\tag{6.17}$$

Remark 6.2 More detailed analysis allows for verifying the asymptotic normality

$$\sqrt{T}\left(S_{1,T}\left(X^T\right)-\Phi_1\left(\vartheta_0\right)\right)\Longrightarrow\mathcal{N}\left(0,D_1\left(\vartheta_0\right)^2\right),$$

$$\sqrt{T}\left(S_{2,T}\left(X^T\right)-\Phi_2\left(\vartheta_0\right)\right)\Longrightarrow\mathcal{N}\left(0,D_2\left(\vartheta_0\right)^2\right),$$

$$\sqrt{T}\left(S_{3,T}\left(X^T\right)-\Phi_3\left(\vartheta_0\right)\right)\Longrightarrow\mathcal{N}\left(0,D_3\left(\vartheta_0\right)^2\right)$$

but we do not prove these convergences because we need these MMEs just for the construction of One-step MLE-processes, and the estimates (6.16) and (6.17) are sufficient for these problems.

Remark 6.3 All parameters of the model (6.1)–(6.2) can be estimated with the help of the introduced statistics $S_{1,T}\left(X^T\right), S_{2,T}\left(X^T\right), S_{3,T}\left(X^T\right)$. Below, the method of moments estimators (MME) of the parameters f, a, b, σ^2 are proposed, and their asymptotic behavior is described. The only forbidden couple of parameters is $\vartheta = (f, b)$. In this case, consistent estimation of ϑ is impossible. Indeed, the stationary AR process $Y_t, t = \dots, -1, 0, 1, \dots$ admits the representation $Y_t = b\sum_{k=0}^{\infty} a^k v_{t-k}$, and therefore, the observed process is

$$X_t = fb\sum_{k=0}^{\infty} a^k v_{t-k} + \sigma w_t, \qquad t = ..., -1, 0, 1, \dots.$$

Here we introduced the sequence $v_t, t = ..., -1, 0, 1, \dots$ of i.i.d. standard Gaussian r.v.'s. The model depends on the product fb, and therefore, these parameters cannot be estimated separately.

Estimation of the Parameter f

Suppose that the parameters a, b, σ^2 are known and we have to estimate $\vartheta = f \in (\alpha_f, \beta_f), \alpha_f > 0$. Then the MME can be defined as follows:

$$f_T^* = \alpha_f \mathbb{I}_{\{\mathbb{B}_{1,T}\}} + \bar{f}_T \mathbb{I}_{\{\mathbb{B}_{2,T}\}} + \beta_f \mathbb{I}_{\{\mathbb{B}_{3,T}\}}. \tag{6.18}$$

Here

$$\bar{f}_T = \left(\frac{\left(S_{1,T}\left(X^T\right) - 2\sigma^2\right)(1+a)}{2b^2}\right)^{1/2},$$

$$\mathbb{B}_{1,T} = \left\{\text{The event}: \quad S_{1,T}\left(X^T\right) \le \frac{2\alpha_f^2 b^2}{1+a} + 2\sigma^2\right\},$$

$$\mathbb{B}_{2,T} = \left\{\text{The event}: \quad \frac{2\alpha_f^2 b^2}{1+a} + 2\sigma^2 < S_{1,T}\left(X^T\right) \le \frac{2\beta_f^2 b^2}{1+a} + 2\sigma^2\right\},$$

$$\mathbb{B}_{3,T} = \left\{\text{The event}: \quad S_{1,T}\left(X^T\right) \ge \frac{2\beta_f^2 b^2}{1+a} + 2\sigma^2\right\}.$$

Therefore $f_T^* \in \left[\alpha_f, \beta_f\right]$. As

$$S_{1,T}\left(X^T\right) \longrightarrow \Phi_1(\vartheta_0) = \frac{2f_0^2 b^2}{1+a} + 2\sigma^2$$

the probabilities

$$\mathbf{P}_{f_0}\left(\mathbb{B}_{1,T}\right)\longrightarrow 0,\qquad \mathbf{P}_{f_0}\left(\mathbb{B}_{2,T}\right)\longrightarrow 1,\quad \mathbf{P}_{f_0}\left(\mathbb{B}_{3,T}\right)\longrightarrow 0,$$

and below, we omit the representations like (6.18) for the other MMEs.

It is easy to see that by Lemma 6.1 and Remark 6.1 the MME f_T^* is consistent, i.e., $f_T^*\to f_0$.

Let us verify the upper bound

$$\sup_{f_0\in\mathbb{K}}\mathbf{E}_{f_0}\left|f_T^*-f_0\right|^2\leq\frac{C}{T}.$$

Put $\eta_T=\sqrt{T}\left(S_{1,T}\left(X^T\right)-\Phi_1\left(\vartheta_0\right)\right)$. Note that thanks to the definition (6.18) of f_T^* it is sufficient to study the statistic $S_{1,T}\left(X_T\right)$ on the set $\mathbb{B}_2$ only, and therefore we have the estimates

$$\frac{2\alpha_f^2b^2}{1+a}+2\sigma^2\leq S_{1,T}\left(X_T\right)\leq\frac{2\beta_f^2b^2}{1+a}+2\sigma^2. \tag{6.19}$$

Then

$$\begin{aligned}
f_T^*&=\sqrt{\frac{1+a}{2b^2}}\left[\Phi_1\left(\vartheta_0\right)-2\sigma^2+T^{-1/2}\eta_T\right]^{1/2}\\
&=\sqrt{\frac{1+a}{2b^2}}\left[\Phi_1\left(\vartheta_0\right)-2\sigma^2\right]^{1/2}\\
&\quad+\sqrt{\frac{1+a}{8b^2}}\left[\Phi_1\left(\vartheta_0\right)-2\sigma^2+sT^{-1/2}\eta_T\right]^{-1/2}T^{-1/2}\eta_T\\
&=f_0+\sqrt{\frac{1+a}{8b^2}}\left[\Phi_1\left(\vartheta_0\right)-2\sigma^2+sT^{-1/2}\eta_T\right]^{-1/2}T^{-1/2}\eta_T,
\end{aligned}$$

where $s\in(0,1)$ and therefore

$$\left|f_T^*-f_0\right|\leq\frac{(1+a)}{2b^2\alpha_f}\,T^{-1/2}\eta_T.$$

Here we used (6.19). Therefore, by Lemma 6.1

$$\sup_{f_0\in\mathbb{K}}\mathbf{E}_{f_0}\left|f_T^*-f_0\right|^2\leq C\,T^{-1}\sup_{f_0\in\mathbb{K}}\mathbf{E}_{f_0}\left|\eta_T\right|^2\leq C\,T^{-1}.$$

Remark 6.4 Of course, the statistics $S_{2,T}\left(X^T\right)$ and $S_{3,T}\left(X^T\right)$ as well can be used for the construction of the MME of f.

Estimation of the Parameters b, a, and σ^2

The estimation of b is almost the same as the estimation of f because these parameters in $\Phi_1(\vartheta)$, $\Phi_2(\vartheta)$, and $\Phi_3(\vartheta)$ are in the products bf only. Therefore

$$b_T^* = \left(\frac{\left(S_{1,T}\left(X^T\right) - 2\sigma^2\right)(1+a)}{2f^2}\right)^{1/2} \longrightarrow b_0, \qquad \sup_{b_0\in\mathbb{K}} \mathbf{E}_{b_0}\left|b_T^* - b_0\right|^2 \leq \frac{C}{T}. \tag{6.20}$$

The corresponding estimators of a and σ^2 are

$$a_T^* = \frac{2f^2b^2}{S_{1,T}\left(X^T\right) - 2\sigma^2} - 1, \qquad \sigma_T^{2*} = \frac{1}{2}S_{1,T}\left(X^T\right) - \frac{f^2b^2}{1+a}.$$

Estimation of the Parameter $\vartheta = (a, f)$

The MME $\vartheta_T^* = \left(a_T^*, f_T^*\right)$ is solution of the system of equations

$$S_{1,T}\left(X^T\right) = \Phi_1\left(\vartheta_T^*\right), \qquad S_{2,T}\left(X^T\right) = \Phi_2\left(\vartheta_T^*\right)$$

and has the following form and properties

$$a_T^* = \frac{S_{1,T}\left(X^T\right) + S_{2,T}\left(X^T\right) - \sigma^2}{S_{1,T}\left(X^T\right) - 2\sigma^2}, \qquad \sup_{(a_0,f_0)\in\mathbb{K}} \mathbf{E}_{a_0,f_0}\left|a_T^* - a_0\right|^2 \leq \frac{C}{T},$$

$$f_T^* = \left(\frac{S_{1,T}\left(X^T\right)\left(1+a_T^*\right) - 2\sigma^2}{2b^2}\right)^{1/2}, \qquad \sup_{(a_0,f_0)\in\mathbb{K}} \mathbf{E}_{a_0,f_0}\left|f_T^* - f_0\right|^2 \leq \frac{C}{T}.$$

Therefore by Lemma 6.1, we have

$$\vartheta_T^* = \left(a_T^*, f_T^*\right) \longrightarrow (a_0, f_0), \qquad \sup_{\vartheta_0\in\mathbb{K}} \mathbf{E}_{\vartheta_0}\left\|\vartheta_T^* - \vartheta_0\right\|^2 \leq \frac{C}{T}. \tag{6.21}$$

Estimation of the Parameter $\vartheta = \left(a, f, \sigma^2\right)$

In this case, we have to solve three equations

$$S_{1,T}\left(X^T\right) = \frac{2f^2b^2}{1+a} + 2\sigma^2, \qquad S_{2,T}\left(X^T\right) = \frac{f^2b^2(a-1)}{1+a} - \sigma^2,$$

$$S_{3,T}\left(X^T\right)=\frac{f^2b^2a\,(a-1)}{(1+a)}.$$

The MME $\vartheta_T^*=\left(a_T^*,f_T^*,\sigma_T^{2*}\right)$ is the following solution of this system:

$$a_T^*=\frac{2S_{3,T}\left(X^T\right)}{S_{1,T}\left(X^T\right)+2S_{2,T}\left(X^T\right)}+1,\qquad f_T^*=\frac{S_{3,T}\left(X^T\right)\left(1+a_T^*\right)}{b^2a_T^*\left(a_T^*-1\right)},$$

$$\sigma_T^{2*}=\frac{1}{2}S_{1,T}\left(X^T\right)-\frac{\left(f_T^*\right)^2b^2}{1+a_T^*}.$$

Once more, by Lemma 6.1 we have the consistency of ϑ_T^* and the bound like (6.21)

$$\sup_{\vartheta_0\in\mathbb{K}}\mathbf{E}_{\vartheta_0}\left\|\vartheta_T^*-\vartheta_0\right\|^2\leq\frac{C}{T}.$$

We have almost similar results in the case of estimation $\vartheta=\left(a,b,\sigma^2\right)$.

6.2.2 *Fisher Information*

The MMEs studied above are consistent but not asymptotically efficient. That is why we propose below the construction of a one-step MLE-process, which allow us to solve two problems: first, we obtain asymptotically efficient estimators of these parameters, and second, we describe the approximation of the conditional expectation $m\left(\vartheta,t\right),t\geq1$.

The construction of One-step MLE-process requires the knowledge of Fisher information. That is why we calculate below the Fisher information related to different parameters. To do this, we recall first some known properties of $\gamma\left(\vartheta,t\right)$ and describe the likelihood ratio function for this model.

Recall the studied model (6.1) and (6.2) of partially observed time series where $X^T=\left(X_0,X_1,\ldots,X_T\right)$ are observations and autoregressive process (AR) $Y_t,t\geq0$ is a hidden process. Here $w_t,t\geq1$ and $v_t,t\geq1$ are independent standard Gaussian random variables, i.e., $w_t\sim\mathcal{N}\left(0,1\right)$, $v_t\sim\mathcal{N}\left(0,1\right)$. The initial values are $X_0\sim\mathcal{N}\left(0,d_x^2\right)$ and $Y_0\sim\mathcal{N}\left(0,d_y^2\right)$. The system is defined by the parameters $a,b,f,\sigma^2,d_x^2,d_y^2$. We suppose that some of these parameters are unknown and have to be estimated by observations X^T.

Unknown Parameter b

We start with one-dimensional case, say, $\vartheta = b \in \Theta = (\alpha_b, \beta_b), \alpha_b > 0$. As above, we suppose that the system is (6.1), (6.2), and $f \neq 0, a^2 \in [0, 1)$.

The notation:

$$\Gamma(\vartheta) = f^2 \gamma_*(\vartheta), \qquad P(\vartheta) = \sigma^2 + \Gamma(\vartheta), \qquad A(\vartheta) = \frac{a\sigma^2}{P(\vartheta)},$$

$$B(\vartheta, \vartheta_0) = \frac{a\Gamma(\vartheta)\sqrt{P(\vartheta_0)}}{P(\vartheta)}, \qquad \dot{B}_b(\vartheta_0, \vartheta_0) = \left.\frac{\partial B(\vartheta, \vartheta_0)}{\partial \vartheta}\right|_{\vartheta=\vartheta_0} = \frac{a\sigma^2 \dot{\Gamma}_b(\vartheta_0)}{P(\vartheta_0)^{3/2}}.$$

Note that $\inf_{\vartheta \in \Theta} \dot{\Gamma}_b(\vartheta_0) > 0$ (see (6.39) below). Another inequality that we will use is

$$\sup_{\vartheta \in \Theta} |A(\vartheta)| < 1. \tag{6.22}$$

Let us remind the well-known fact that it is always fulfilled if $a^2 \in [0, 1)$ because it is a product of $|a| < 1$ and $\sigma^2 / \left(\sigma^2 + \Gamma(\vartheta)\right) < 1$.

Proposition 6.1 *The Fisher information is*

$$\mathrm{I}_b(\vartheta_0) = \frac{\dot{P}_b(\vartheta_0)^2 \left[P(\vartheta_0)^2 + a^2\sigma^4\right]}{2P(\vartheta_0)^2 \left[P(\vartheta_0)^2 - a^2\sigma^4\right]}. \tag{6.23}$$

Proof The Fisher score is (see (6.10))

$$\begin{aligned}
\frac{\partial \ln L\left(\vartheta, X^T\right)}{\partial \vartheta} &= -\frac{\partial}{\partial \vartheta} \sum_{t=1}^{T} \left[\frac{(X_t - M_{t-1}(\vartheta))^2}{2P(\vartheta)} + \frac{1}{2} \ln P(\vartheta)\right] \\
&= \sum_{t=1}^{T} \left[\frac{[X_t - M_{t-1}(\vartheta)]}{P(\vartheta)} \dot{M}_{t-1}(\vartheta) + \frac{[X_t - M_{t-1}(\vartheta)]^2 \dot{P}_b(\vartheta)}{2P(\vartheta)^2} - \frac{\dot{P}_b(\vartheta)}{2P(\vartheta)}\right] \\
&= \sum_{t=1}^{T} \left[\frac{[X_t - M_{t-1}(\vartheta)]}{\sqrt{P(\vartheta)}} \frac{\dot{M}_{t-1}(\vartheta)}{\sqrt{P(\vartheta)}} + \frac{[X_t - M_{t-1}(\vartheta)]^2}{P(\vartheta)} \frac{\dot{P}_b(\vartheta)}{2P(\vartheta)} - \frac{\dot{P}_b(\vartheta)}{2P(\vartheta)}\right] \\
&= \frac{1}{\sqrt{P(\vartheta)}} \sum_{t=1}^{T} \left[\zeta_t(\vartheta) \dot{M}_{t-1}(\vartheta) + \left[\zeta_t(\vartheta)^2 - 1\right] \frac{\dot{P}_b(\vartheta)}{2\sqrt{P(\vartheta)}}\right].
\end{aligned}$$

Here $\dot{M}_{t-1}(\vartheta) = \partial M_{t-1}(\vartheta) / \partial \vartheta$, $\dot{P}_b(\vartheta) = \partial P(\vartheta) / \partial \vartheta$ and

$$\zeta_t(\vartheta_0) = \frac{X_t - M_{t-1}(\vartheta_0)}{\sqrt{\sigma^2 + \Gamma(\vartheta_0)}} = \frac{X_t - M_{t-1}(\vartheta_0)}{\sqrt{P(\vartheta_0)}}, \qquad t \geq 1$$

are independent standard Gaussian random variables ($\zeta_t(\vartheta_0) \sim \mathcal{N}(0,1)$).

The equation for derivative $\dot M_t(\vartheta), t \geq 1$ we obtain by formal differentiation of the Eq. (6.9)

$$\dot M_t(\vartheta) = A(\vartheta)\,\dot M_{t-1}(\vartheta) + \dot A_b(\vartheta)\left[M_{t-1}(\vartheta) - M_{t-1}(\vartheta_0)\right] + \dot B_b(\vartheta,\vartheta_0)\,\zeta_t(\vartheta_0)$$

with the initial value $\dot M_0(\vartheta)$. If $\vartheta = \vartheta_0$, then

$$\dot M_t(\vartheta_0) = A(\vartheta_0)\,\dot M_{t-1}(\vartheta_0) + \dot B_b(\vartheta_0,\vartheta_0)\;\zeta_t(\vartheta_0), \qquad \dot M_0(\vartheta_0), \qquad t \geq 1. \tag{6.24}$$

The stationary version of the process $\dot m_t(\vartheta_0), t \geq 1$ can be written as a sum of i.i.d. variables

$$\dot M_t(\vartheta_0) = \dot B_b(\vartheta_0,\vartheta_0)\sum_{k=0}^{\infty} A(\vartheta_0)^k\,\zeta_{t-k}(\vartheta_0),$$

$$\dot M_0(\vartheta_0) = \dot B_b(\vartheta_0,\vartheta_0)\sum_{k=0}^{\infty} A(\vartheta_0)^k\,\zeta_{-k}(\vartheta_0),$$

where we introduced i.i.d. r.v.'s $\zeta_k(\vartheta_0) \sim \mathcal{N}(0,1), k = 0,-1,-2,\ldots$. The real process $\dot M_t(\vartheta_0),\ 0 \leq t \leq T$ has similar representation but with the finite sum. As we are interested in asymptotic ($T \to \infty$) properties of estimators, we replace the finite sum by the infinite sum and remark that the difference between these two representations is asymptotically negligible. We have

$$\begin{aligned}\mathbf{E}_{\vartheta_0}\dot M_t(\vartheta_0)^2 &= \dot B_b(\vartheta_0,\vartheta_0)^2\sum_{k=0}^{\infty} A(\vartheta_0)^{2k} = \frac{\dot B_b(\vartheta_0,\vartheta_0)^2}{1 - A(\vartheta_0)^2}\\ &= \frac{a^2\sigma^4\dot P(\vartheta_0)^2}{P(\vartheta_0)^3\left(1 - A(\vartheta_0)^2\right)}.\end{aligned}$$

This moment can be calculated directly from (6.24) using the stationarity of $\dot M_t(\vartheta_0)$. We have

$$\mathbf{E}_{\vartheta_0}\dot M_t(\vartheta_0)^2 = A(\vartheta_0)^2\,\mathbf{E}_{\vartheta_0}\dot M_{t-1}(\vartheta_0)^2 + \dot B_b(\vartheta_0,\vartheta_0)^2\,\mathbf{E}_{\vartheta_0}\zeta_t(\vartheta_0)^2.$$

As $\mathbf{E}_{\vartheta_0}\dot M_t(\vartheta_0)^2 = \mathbf{E}_{\vartheta_0}\dot M_{t-1}(\vartheta_0)^2$ and $\mathbf{E}_{\vartheta_0}\zeta_t(\vartheta_0)^2 = 1$ we obtain

$$\mathbf{E}_{\vartheta_0}\dot M_t(\vartheta_0)^2 = \frac{\dot B_b(\vartheta_0,\vartheta_0)^2}{1 - A(\vartheta_0)^2}.$$

For the second moment of the score-function, we have the following expression:

$$
\begin{aligned}
&\mathbf{E}_{\vartheta_0}\left[\frac{\partial \ln L\left(\vartheta, X^T\right)}{\partial \vartheta}\right]^2_{\vartheta=\vartheta_0}\\
&=\frac{1}{P\left(\vartheta_0\right)}\mathbf{E}_{\vartheta_0}\left(\sum_{t=1}^{T}\left[\zeta_t\left(\vartheta_0\right)\dot M_{t-1}\left(\vartheta_0\right)+\frac{1}{2}\left[\zeta_t\left(\vartheta_0\right)^2-1\right]\dot P_b\left(\vartheta_0\right)P\left(\vartheta_0\right)^{-1/2}\right]\right)^2\\
&=\frac{1}{P\left(\vartheta_0\right)}\mathbf{E}_{\vartheta_0}\sum_{t=1}^{T}\left(\zeta_t\left(\vartheta_0\right)\dot M_{t-1}\left(\vartheta_0\right)+\frac{1}{2}\left[\zeta_t\left(\vartheta_0\right)^2-1\right]\dot P_b\left(\vartheta_0\right)P\left(\vartheta_0\right)^{-1/2}\right)^2
\end{aligned}
$$

because (below $t > s$ and for simplicity we omit $\frac{1}{2}P\left(\vartheta_0\right)^{-1/2}$)

$$
\begin{aligned}
&\mathbf{E}_{\vartheta_0}\left(\zeta_t\left(\vartheta_0\right)\dot M_{t-1}\left(\vartheta_0\right)+\left[\zeta_t\left(\vartheta_0\right)^2-1\right]\dot P_b\left(\vartheta_0\right)\right)\\
&\qquad\qquad\times\left(\zeta_s\left(\vartheta_0\right)\dot M_{s-1}\left(\vartheta_0\right)+\left[\zeta_s\left(\vartheta_0\right)^2-1\right]\dot P_b\left(\vartheta_0\right)\right)\\
&\quad=\mathbf{E}_{\vartheta_0}\zeta_t\left(\vartheta_0\right)\dot M_{t-1}\left(\vartheta_0\right)\zeta_{s-1}\left(\vartheta_0\right)\dot M_{s-1}\left(\vartheta_0\right)\\
&\qquad+\dot P_b\left(\vartheta_0\right)\mathbf{E}_{\vartheta_0}\zeta_t\left(\vartheta_0\right)\dot M_{t-1}\left(\vartheta_0\right)\left[\zeta_s\left(\vartheta_0\right)^2-1\right]\\
&\qquad+\dot P_b\left(\vartheta_0\right)\mathbf{E}_{\vartheta_0}\zeta_s\left(\vartheta_0\right)\dot M_{s-1}\left(\vartheta_0\right)\left[\zeta_t\left(\vartheta_0\right)^2-1\right]\\
&\qquad+\dot P_b\left(\vartheta_0\right)^2\mathbf{E}_{\vartheta_0}\left[\zeta_t\left(\vartheta_0\right)^2-1\right]\left[\zeta_s\left(\vartheta_0\right)^2-1\right]=0.
\end{aligned}
$$

Here we used equalities, such as

$$
\begin{aligned}
&\mathbf{E}_{\vartheta_0}\left[\zeta_t\left(\vartheta_0\right)\dot M_{t-1}\left(\vartheta_0\right)\zeta_s\left(\vartheta_0\right)\dot M_{s-1}\left(\vartheta_0\right)\right]\\
&\qquad=\mathbf{E}_{\vartheta_0}\left[\dot M_{t-1}\left(\vartheta_0\right)\zeta_s\left(\vartheta_0\right)\dot M_{s-1}\left(\vartheta_0\right)\mathbf{E}_{\vartheta_0}\left(\zeta_t\left(\vartheta_0\right)|\mathfrak{F}^X_{t-1}\right)\right]=0.
\end{aligned}
$$

Further

$$
\begin{aligned}
&\mathbf{E}_{\vartheta_0}\left(\zeta_t\left(\vartheta_0\right)\dot M_{t-1}\left(\vartheta_0\right)+\left[\zeta_t\left(\vartheta_0\right)^2-1\right]\frac{\dot P_b\left(\vartheta_0\right)}{2\sqrt{P\left(\vartheta_0\right)}}\right)^2\\
&=\mathbf{E}_{\vartheta_0}\zeta_t\left(\vartheta_0\right)^2\dot M_{t-1}\left(\vartheta_0\right)^2+\frac{\dot P_b\left(\vartheta_0\right)^2}{4P\left(\vartheta_0\right)}\mathbf{E}_{\vartheta_0}\left[\zeta_t\left(\vartheta_0\right)^2-1\right]^2\\
&=\mathbf{E}_{\vartheta_0}\dot M_{t-1}\left(\vartheta_0\right)^2+\frac{\dot P_b\left(\vartheta_0\right)^2}{2P\left(\vartheta_0\right)}=\frac{\dot B_b\left(\vartheta_0,\vartheta_0\right)^2}{1-A\left(\vartheta_0\right)^2}+\frac{\dot P_b\left(\vartheta_0\right)^2}{2P\left(\vartheta_0\right)}\\
&=\frac{a^2\sigma^4\dot P_b\left(\vartheta_0\right)^2}{P\left(\vartheta_0\right)^3\left(1-A\left(\vartheta_0\right)^2\right)}+\frac{\dot P_b\left(\vartheta_0\right)^2}{2P\left(\vartheta_0\right)}=\frac{\dot P_b\left(\vartheta_0\right)^2}{2P\left(\vartheta_0\right)}\left[\frac{2a^2\sigma^4}{P\left(\vartheta_0\right)^2-a^2\sigma^4}+1\right].
\end{aligned}
$$

Therefore

$$\frac{1}{T}\mathbf{E}_{\vartheta_0}\left[\frac{\partial \ln L\left(\vartheta, X^T\right)}{\partial \vartheta}\right]^2_{\vartheta=\vartheta_0} = \mathrm{I}_b\left(\vartheta_0\right).$$

□

Remark 6.5 If the unknown parameter is $\vartheta = f$ and all other parameters are known, then the filtration equations are almost the same

$$M_t\left(\vartheta\right) = a\,M_{t-1}\left(\vartheta\right) + \frac{a\Gamma\left(\vartheta\right)}{\sigma^2+\Gamma\left(\vartheta\right)}\left[X_t - M_{t-1}\left(\vartheta\right)\right],\ M_t\left(\vartheta\right) = \vartheta m_t\left(\vartheta\right),\ t\geq 1,$$

$$\Gamma\left(\vartheta\right) = \frac{1}{2}\left[\vartheta^2 b^2 - \sigma^2\left(1-a^2\right)\right] + \frac{1}{2}\left[\left(\sigma^2\left(1-a^2\right) - \vartheta^2 b^2\right)^2 + 4\vartheta^2 b^2\sigma^2\right]^{1/2}.$$

It is easy to see that the score-function has the same form, but of course the derivation now is w.r.t. $\vartheta = f$

$$\frac{\partial L(\vartheta, X^T)}{\partial\vartheta} = \frac{1}{\sqrt{P\left(\vartheta\right)}}\sum_{t=1}^T\left[\zeta_t\left(\vartheta\right)\dot M_{t-1}\left(\vartheta\right) + \left[\zeta_t\left(\vartheta\right)^2 - 1\right]\frac{\dot P_f\left(\vartheta\right)}{2\sqrt{P\left(\vartheta\right)}}\right].$$

Here $P\left(\vartheta\right) = \sigma^2 + \Gamma\left(\vartheta\right)$.

To calculate the Fisher information, we have to use the following equation for the derivative $\dot M_t\left(\vartheta\right)$:

$$\dot M_t\left(\vartheta_0\right) = A\left(\vartheta_0\right)\dot M_{t-1}\left(\vartheta_0\right) + \dot B_f\left(\vartheta_0, \vartheta_0\right)\zeta_t\left(\vartheta_0\right), \qquad t\geq 1$$

with the functions $A\left(\vartheta\right) = a\sigma^2/P\left(\vartheta\right)$, and $B_f\left(\vartheta, \vartheta_0\right) = a\Gamma\left(\vartheta\right)\sqrt{P\left(\vartheta_0\right)}P\left(\vartheta\right)^{-1}$.

This gives us the Fisher information

$$\mathrm{I}_f\left(\vartheta_0\right) = \frac{\dot P_f\left(\vartheta_0\right)^2\left[P\left(\vartheta_0\right)^2 + a^2\sigma^4\right]}{2P\left(\vartheta_0\right)^2\left[P\left(\vartheta_0\right)^2 - a^2\sigma^4\right]}. \tag{6.25}$$

Unknown Parameter *a*

Suppose that the partially observed system is (6.1) and (6.2), the values of $f > 0, \sigma^2 > 0, b > 0$ are known and parameter $\vartheta = a \in \Theta = \left(\alpha_a, \beta_a\right), -1 < \alpha_a < \beta_a < 1$ is unknown. The Kalman filter for $M_t\left(\vartheta\right) = \vartheta m_t\left(\vartheta\right)$ is given by the relations

$$M_t\left(\vartheta\right) = \left[\vartheta - \frac{\vartheta\Gamma\left(\vartheta\right)}{\sigma^2+\Gamma\left(\vartheta\right)}\right]M_{t-1}\left(\vartheta\right) + \frac{\vartheta\Gamma\left(\vartheta\right)}{\sigma^2+\Gamma\left(\vartheta\right)}X_t,$$

$$= \frac{\vartheta\sigma^2}{\sigma^2+\Gamma(\vartheta)} M_{t-1}(\vartheta) + \frac{\vartheta\Gamma(\vartheta)}{\sigma^2+\Gamma(\vartheta)} X_t,$$

$$= A(\vartheta)\, M_{t-1}(\vartheta) + \frac{\vartheta\Gamma(\vartheta)}{P(\vartheta)} M_{t-1}(\vartheta_0) + B(\vartheta,\vartheta_0)\,\zeta_t(\vartheta_0), \qquad t \geq 1,$$

$$\Gamma(\vartheta) = \frac{1}{2}\left[f^2b^2 - \sigma^2\left(1-\vartheta^2\right)\right] + \frac{1}{2}\left[\left(\sigma^2\left(1-\vartheta^2\right) - f^2b^2\right)^2 + 4b^2\sigma^2\right]^{1/2}.$$

Here $A(\vartheta) = \vartheta\sigma^2/P(\vartheta)$, $B(\vartheta,\vartheta_0) = \vartheta\Gamma(\vartheta)\sqrt{P(\vartheta_0)}P(\vartheta)^{-1}$, and $\zeta_t(\vartheta_0), t \geq 1$ are independent standard Gaussian random variables.

We present the corresponding score-function and Fisher information without detailed proofs.

We have

$$\frac{\partial \ln L\left(\vartheta, X^T\right)}{\partial\vartheta} = \frac{1}{\sqrt{P(\vartheta)}} \sum_{t=1}^{T} \left[\zeta_t(\vartheta)\, \dot M_{t-1}(\vartheta) + \left[\zeta_t(\vartheta)^2 - 1\right] \frac{\dot P_a(\vartheta)}{2\sqrt{P(\vartheta)}}\right].$$

The equation for $\dot M_t(\vartheta_0) = f\,\partial m_t(\vartheta_0)/\partial a$ is

$$\dot M_t(\vartheta_0) = A(\vartheta_0)\, \dot M_{t-1}(\vartheta_0) + M_{t-1}(\vartheta_0) + \dot B_a(\vartheta_0,\vartheta_0)\,\zeta_t(\vartheta_0), \qquad t \geq 1,$$

where

$$\dot B_a(\vartheta_0,\vartheta_0) = \frac{\Gamma(\vartheta_0)\,P(\vartheta_0) + \vartheta_0\sigma^2\dot P_a(\vartheta_0)}{P(\vartheta_0)^{3/2}}.$$

Therefore, using the stationarity of all processes, we write

$$\begin{aligned}\mathbf{E}_{\vartheta_0}\dot M_t(\vartheta_0)^2 &= A(\vartheta_0)^2\,\mathbf{E}_{\vartheta_0}\dot M_{t-1}(\vartheta_0)^2 + \mathbf{E}_{\vartheta_0}M_{t-1}(\vartheta_0)^2 + \dot B_a(\vartheta_0,\vartheta_0)^2\\ &\quad + 2A(\vartheta_0)\,\mathbf{E}_{\vartheta_0}\dot M_{t-1}(\vartheta_0)\,M_{t-1}(\vartheta_0)\\ &= \left(1 - A(\vartheta_0)^2\right)^{-1}\Big[\mathbf{E}_{\vartheta_0}M_{t-1}(\vartheta_0)^2 + \dot B_a(\vartheta_0,\vartheta_0)^2\\ &\quad + 2A(\vartheta_0)\,\mathbf{E}_{\vartheta_0}\dot M_{t-1}(\vartheta_0)\,M_{t-1}(\vartheta_0)\Big].\end{aligned}$$

The equations for $M_t(\vartheta_0)$ and $\dot M_t(\vartheta_0)$ allow us to calculate the following moments

$$\mathbf{E}_{\vartheta_0}M_{t-1}(\vartheta_0)^2 = \frac{\vartheta_0^2\Gamma(\vartheta_0)^2}{P(\vartheta_0)\left(1-\vartheta_0^2\right)},$$

$$\mathbf{E}_{\vartheta_0}\dot M_{t-1}(\vartheta_0)\,M_{t-1}(\vartheta_0)$$

$$= \frac{\vartheta_0 \Gamma(\vartheta_0)^2}{P(\vartheta_0)^2 (1 - \vartheta_0 A(\vartheta_0))} \left[\frac{\vartheta_0^2 P(\vartheta_0)}{1 - \vartheta_0^2} + P(\vartheta_0) + \frac{\vartheta_0 \sigma^2 \dot{\Gamma}_a(\vartheta_0)}{\Gamma(\vartheta_0)} \right].$$

Hence

$$\begin{aligned}
&\mathbf{E}_{\vartheta_0} \dot{M}_t(\vartheta_0)^2 \\
&= \left(1 - A(\vartheta_0)^2\right)^{-1} \left[\frac{\vartheta_0^2 \Gamma(\vartheta_0)^2}{P(\vartheta_0)\left(1 - \vartheta_0^2\right)} + \frac{\left[\Gamma(\vartheta_0) P(\vartheta_0) + \sigma^2 \dot{\Gamma}_a(\vartheta_0)\right]^2}{P(\vartheta_0)^3} \right. \\
&\left. + \frac{2A(\vartheta_0)\vartheta_0 \Gamma(\vartheta_0)^2}{P(\vartheta_0)^2 (1 - \vartheta_0 A(\vartheta_0))} \left(\frac{\vartheta_0^2 P(\vartheta_0)}{1 - \vartheta_0^2} + P(\vartheta_0) + \frac{\vartheta_0 \sigma^2 \dot{\Gamma}_a(\vartheta_0)}{\Gamma(\vartheta_0)} \right) \right] \equiv Q(\vartheta_0).
\end{aligned}$$

Recall that the Fisher information is

$$\begin{aligned}
\mathrm{I}_a(\vartheta_0) &= \frac{1}{T} \mathbf{E}_{\vartheta_0} \left(\left. \frac{\partial \ln L\left(\vartheta, X^T\right)}{\partial \vartheta} \right|_{\vartheta = \vartheta_0} \right)^2 \\
&= \frac{1}{T P(\vartheta_0)} \sum_{t=1}^{T} \mathbf{E}_{\vartheta_0} \left[\zeta_t(\vartheta_0) \dot{M}_{t-1}(\vartheta_0) + \left[\zeta_t(\vartheta_0)^2 - 1\right] \frac{\dot{P}(\vartheta_0)}{2\sqrt{P(\vartheta_0)}} \right]^2 \\
&= \frac{1}{P(\vartheta_0)} \left[\mathbf{E}_{\vartheta_0} \dot{M}_{t-1}(\vartheta_0)^2 + \frac{\dot{P}_a(\vartheta_0)^2}{2P(\vartheta_0)} \right] = \frac{2Q(\vartheta_0) + \dot{\Gamma}_a(\vartheta_0)^2}{2P(\vartheta_0)^2}.
\end{aligned} \tag{6.26}$$

Unknown Parameter σ^2

The corresponding notations are

$$A(\vartheta) = \frac{a\vartheta}{P(\vartheta)}, \qquad B(\vartheta, \vartheta_0) = a\sqrt{P(\vartheta_0)} - \frac{a\vartheta\sqrt{P(\vartheta_0)}}{P(\vartheta)},$$

$$\dot{B}_{\sigma^2}(\vartheta_0, \vartheta_0) = \frac{a\left[\vartheta_0 \dot{\Gamma}_{\sigma^2}(\vartheta_0) - \Gamma(\vartheta_0)\right]}{P(\vartheta_0)^{3/2}}, \qquad P(\vartheta) = \vartheta + \Gamma(\vartheta).$$

The Kalman filter, derivative of it on ϑ, and Fisher score-function are (see (6.9))

$$M_t(\vartheta) = A(\vartheta)\, M_{t-1}(\vartheta) + (a - A(\vartheta))\, M_{t-1}(\vartheta_0) + B(\vartheta, \vartheta_0)\, \zeta_t(\vartheta_0),$$

$$\dot{M}_t(\vartheta_0) = A(\vartheta_0)\, \dot{M}_{t-1}(\vartheta_0) + \dot{B}_{\sigma^2}(\vartheta_0, \vartheta_0)\, \zeta_t(\vartheta_0), \quad t \geq 1$$

and

$$\frac{\partial L(\vartheta, X^T)}{\partial \vartheta} = \frac{1}{\sqrt{P(\vartheta)}} \sum_{t=1}^{T} \left[\zeta_t(\vartheta) \dot{M}_{t-1}(\vartheta) + \left[\zeta_t(\vartheta)^2 - 1\right] \frac{\dot{P}_{\sigma^2}(\vartheta)}{2\sqrt{P(\vartheta)}} \right],$$

respectively. Therefore we can write

$$\begin{aligned}\frac{1}{T}\mathbf{E}_{\vartheta_0}\left[\frac{\partial L(\vartheta, X^T)}{\partial \vartheta}\right]^2\Bigg|_{\vartheta=\vartheta_0} &= \frac{1}{P(\vartheta_0)}\left[\mathbf{E}_{\vartheta_0}\dot{M}_{t-1}(\vartheta)^2 + \frac{\dot{P}_{\sigma^2}(\vartheta_0)^2}{2P(\vartheta_0)}\right] \\ &= \frac{a^2\left[\vartheta_0\dot{\Gamma}_{\sigma^2}(\vartheta_0) - \Gamma(\vartheta_0)\right]^2}{P(\vartheta_0)^4\left(1 - A(\vartheta_0)^2\right)} + \frac{\dot{P}_{\sigma^2}(\vartheta_0)^2}{2P(\vartheta_0)^2}.\end{aligned}$$

This gives us the Fisher information

$$\mathrm{I}_{\sigma^2}(\vartheta_0) = \frac{a^2\left[\vartheta_0\dot{\Gamma}_{\sigma^2}(\vartheta_0) - \Gamma(\vartheta_0)\right]^2}{P(\vartheta_0)^4\left(1 - A(\vartheta_0)^2\right)} + \frac{\dot{P}_{\sigma^2}(\vartheta_0)^2}{2P(\vartheta_0)^2}. \tag{6.27}$$

Unknown Parameter $\vartheta = \left(f, \sigma^2\right)$

The unknown parameter $\vartheta = \left(f, \sigma^2\right) = (\theta_1, \theta_2)$ and the partially observed system are

$$\begin{aligned} X_t &= \theta_1 Y_{t-1} + \sqrt{\theta_2}\, w_t, \qquad X_0, \qquad t = 1, 2, \ldots, \\ Y_t &= a\, Y_{t-1} + b\, v_t, \qquad Y_0. \end{aligned}$$

For the Fisher information matrix we have the representation

$$\mathbf{I}(\vartheta_0) = \begin{pmatrix} I_{11}(\vartheta_0), & I_{12}(\vartheta_0) \\ I_{21}(\vartheta_0), & I_{22}(\vartheta_0) \end{pmatrix},$$

where

$$\begin{aligned} I_{11}(\vartheta_0) &= \frac{\dot{B}_f(\vartheta_0, \vartheta_0)^2}{P(\vartheta_0)\left(1 - A(\vartheta_0)^2\right)} + \frac{\dot{P}_f^2}{2P(\vartheta_0)^2}, \\ I_{12}(\vartheta_0) &= \frac{\dot{B}_f(\vartheta_0, \vartheta_0)\, \dot{B}_{\sigma^2}(\vartheta_0, \vartheta_0)}{P(\vartheta_0)\left(1 - A(\vartheta_0)^2\right)} + \frac{\dot{P}_f \dot{P}_{\sigma^2}}{2P(\vartheta_0)^2}, \\ I_{22}(\vartheta_0) &= \frac{\dot{B}_{\sigma^2}(\vartheta_0, \vartheta_0)^2}{P(\vartheta_0)\left(1 - A(\vartheta_0)^2\right)} + \frac{\dot{P}_{\sigma^2}^2}{2P(\vartheta_0)^2}. \end{aligned}$$

To verify if this matrix is nondegenerate, i.e., that

$$\operatorname{Det}\left(\mathbf{I}\left(\vartheta_0\right)\right)=I_{11}\left(\vartheta_0\right) I_{22}\left(\vartheta_0\right)-I_{12}\left(\vartheta_0\right)^2 \neq 0.$$

we write it as follows:

$$\operatorname{Det}\left(\mathbf{I}\left(\vartheta_0\right)\right)=\left\|\mathrm{V}_1\right\|^2\left\|\mathrm{V}_2\right\|^2-\left\langle\mathrm{V}_1, \mathrm{V}_2\right\rangle^2$$

where we introduced the vectors

$$\mathrm{V}_1=\frac{1}{\sqrt{P\left(\vartheta_0\right)}}\left(\frac{\dot{B}_f\left(\vartheta_0, \vartheta_0\right)}{\sqrt{1-A\left(\vartheta_0\right)^2}}, \frac{\dot{P}_f\left(\vartheta_0\right)}{\sqrt{2}}\right),$$

$$\mathrm{V}_2=\frac{1}{\sqrt{P\left(\vartheta_0\right)}}\left(\frac{\dot{B}_{\sigma^2}\left(\vartheta_0, \vartheta_0\right)}{\sqrt{1-A\left(\vartheta_0\right)^2}}, \frac{\dot{P}_{\sigma^2}\left(\vartheta_0\right)}{\sqrt{2}}\right).$$

The equality $\operatorname{Det}\left(\mathbf{I}\left(\vartheta_0\right)\right)=0$ is possible iff $\mathrm{V}_1=c \mathrm{V}_2$ with some constant c. The last equality is equivalent to two equalities

$$\dot{B}_f\left(\vartheta_0, \vartheta_0\right)=c \dot{B}_{\sigma^2}\left(\vartheta_0, \vartheta_0\right), \qquad \dot{P}_f\left(\vartheta_0\right)=c \dot{P}_{\sigma^2}\left(\vartheta_0\right).$$

Hence

$$\theta_{0,2} \dot{\Gamma}_f\left(\vartheta_0\right)=c\left[\theta_{0,2} \dot{\Gamma}_{\sigma^2}\left(\vartheta_0\right)-\Gamma\left(\vartheta_0\right)\right], \qquad \dot{\Gamma}_f\left(\vartheta_0\right)=c\left[1+\dot{\Gamma}_{\sigma^2}\left(\vartheta_0\right)\right]$$

and $\Gamma\left(\vartheta_0\right)=-\theta_{0,2}$. Of course, the equality $f^2 \gamma_*(\vartheta)=-\sigma^2$ is impossible, and this leads to the conclusion that $\operatorname{Det}\left(\mathbf{I}\left(\vartheta_0\right)\right) \neq 0$.

Therefore, we verified that this Fisher information matrix is always nondegenerate.

Unknown Parameter $\vartheta=\left(f, a, \sigma^2\right)$

Consider the system

$$\begin{aligned}
X_t & =\theta_1 Y_{t-1}+\sqrt{\theta_3}\, w_t, \qquad X_0, \qquad t=1,2, \ldots, \\
Y_t & =\theta_2 Y_{t-1}+b\, v_t, \qquad Y_0 .
\end{aligned}$$

Here the unknown parameter is $\vartheta=\left(\theta_1, \theta_2, \theta_3\right) \in \Theta=\left(\alpha_f, \beta_f\right) \times\left(\alpha_a, \beta_a\right) \times \left(\alpha_{\sigma^2}, \beta_{\sigma^2}\right)$. Let us denote $M(\vartheta, t)=\theta_1 m_t(\vartheta), \Gamma(\vartheta)=\theta_1^2 \gamma_*(\vartheta)$. We write the Kalman filter as follows:

$$\begin{aligned}
M(\vartheta, t) & =\frac{\theta_2 \theta_3}{P(\vartheta)} M(\vartheta, t-1)+\frac{\theta_2 \Gamma(\vartheta)}{P(\vartheta)} X_t \\
& =A(\vartheta) M(\vartheta, t-1)+E(\vartheta) M\left(\vartheta_0, t-1\right)+B\left(\vartheta, \vartheta_0\right) \zeta_t\left(\vartheta_0\right), \quad t \geq 1,
\end{aligned}$$

$$\Gamma(\vartheta)=\frac{1}{2}\left[\theta_1^2b^2-\theta_3\left(1-\theta_2^2\right)\right]+\frac{1}{2}\left[\left(\theta_3\left(1-\theta_2^2\right)-\theta_1^2b^2\right)^2+4b^2\theta_3\right]^{1/2},$$

where $P(\vartheta)=\theta_3+\Gamma(\vartheta)$,

$$A(\vartheta)=\frac{\theta_2\theta_3}{P(\vartheta)},\qquad E(\vartheta)=\frac{\theta_2\Gamma(\vartheta)}{P(\vartheta)},\qquad B(\vartheta,\vartheta_0)=\frac{\theta_2\Gamma(\vartheta)\sqrt{P(\vartheta_0)}}{P(\vartheta)}.$$

The equations for derivatives $\dot M_f(\vartheta_0,t)$, $\dot M_a(\vartheta_0,t)$, $\dot M_{\sigma^2}(\vartheta_0,t)$ are

$$\begin{aligned}\dot M_f(\vartheta_0,t)&=A(\vartheta_0)\,\dot M_f(\vartheta_0,t-1)+\dot B_f(\vartheta_0,\vartheta_0)\,\zeta_t(\vartheta_0),\\ \dot M_a(\vartheta_0,t)&=A(\vartheta_0)\,\dot M_a(\vartheta_0,t-1)+M(\vartheta_0,t-1)+\dot B_a(\vartheta_0,\vartheta_0)\,\zeta_t(\vartheta_0),\\ \dot M_{\sigma^2}(\vartheta_0,t)&=A(\vartheta_0)\,\dot M_{\sigma^2}(\vartheta_0,t-1)+\dot B_{\sigma^2}(\vartheta_0,\vartheta_0)\,\zeta_t(\vartheta_0),\qquad t\geq 1\end{aligned}\tag{6.28}$$

where recall that

$$M(\vartheta_0,t-1)=\theta_{0,2}M(\vartheta_0,t-2)+B(\vartheta_0,\vartheta_0)\,\zeta_{t-1}(\vartheta_0),$$

$$\dot B_f(\vartheta_0,\vartheta_0)=\frac{\theta_{0,2}\theta_{0,3}\dot P_f(\vartheta_0)}{P(\vartheta_0)^{3/2}},\qquad \dot B_a(\vartheta_0,\vartheta_0)=\frac{\Gamma(\vartheta_0)\,P(\vartheta_0)+\theta_{0,2}\dot P_a(\vartheta_0)}{P(\vartheta_0)^{3/2}},$$

$$\dot B_{\sigma^2}(\vartheta_0,\vartheta_0)=\frac{\theta_{0,2}\left[\theta_{0,3}\dot\Gamma_{\sigma^2}(\vartheta_0)-\Gamma(\vartheta_0)\right]}{P(\vartheta_0)^{3/2}}.$$

Here the true value is $\vartheta_0=\left(\theta_{0,1},\theta_{0,2},\theta_{0,3}\right)$ and we used the relation $A(\vartheta)+E(\vartheta)=\theta_2$.

The vector of score-function is

$$\begin{aligned}&\left.\frac{\partial L(\vartheta,X^T)}{\partial\theta_1}\right|_{\vartheta=\vartheta_0}\\ &\quad=\frac{1}{\sqrt{P(\vartheta_0)}}\sum_{t=1}^{T}\left[\zeta_t(\vartheta_0)\,\dot M_f(\vartheta_0,t-1)+\left[\zeta_t(\vartheta_0)^2-1\right]\frac{\dot P_f(\vartheta_0)}{2\sqrt{P(\vartheta_0)}}\right],\\ &\left.\frac{\partial L(\vartheta,X^T)}{\partial\theta_2}\right|_{\vartheta=\vartheta_0}\\ &\quad=\frac{1}{\sqrt{P(\vartheta_0)}}\sum_{t=1}^{T}\left[\zeta_t(\vartheta_0)\,\dot M_a(\vartheta_0,t-1)+\left[\zeta_t(\vartheta_0)^2-1\right]\frac{\dot P_a(\vartheta_0)}{2\sqrt{P(\vartheta_0)}}\right],\\ &\left.\frac{\partial L(\vartheta,X^T)}{\partial\theta_3}\right|_{\vartheta=\vartheta_0}\end{aligned}$$

$$= \frac{1}{\sqrt{P(\vartheta_0)}} \sum_{t=1}^{T} \left[\zeta_t(\vartheta_0) \dot{M}_{\sigma^2}(\vartheta_0, t-1) + \left[\zeta_t(\vartheta_0)^2 - 1\right] \frac{\dot{P}_{\sigma^2}(\vartheta_0)}{2\sqrt{P(\vartheta_0)}} \right].$$

Using the stationarity of the underlying processes, we write

$$\mathbf{E}_{\vartheta_0} \left(\left. \frac{\partial L(\vartheta, X^T)}{\partial \theta_1} \right|_{\vartheta=\vartheta_0} \left. \frac{\partial L(\vartheta, X^T)}{\partial \theta_2} \right|_{\vartheta=\vartheta_0} \right)$$

$$= \frac{T}{P(\vartheta_0)} \left[\mathbf{E}_{\vartheta_0} \dot{M}_f(\vartheta_0, t-1) \dot{M}_a(\vartheta_0, t-1) + \frac{\dot{P}_f(\vartheta_0) \dot{P}_a(\vartheta_0)}{2P(\vartheta_0)} \right].$$

Further

$$\mathbf{E}_{\vartheta_0} \left[\dot{M}_a(\vartheta_0, t-1) \dot{M}_f(\vartheta_0, t-1) \right] = A(\vartheta_0)^2 \mathbf{E}_{\vartheta_0} \left[\dot{M}_a(\vartheta_0, t-2) \dot{M}_f(\vartheta_0, t-2) \right]$$
$$+ A(\vartheta_0) \mathbf{E}_{\vartheta_0} \left[M(\vartheta_0, t-2) \dot{M}_f(\vartheta_0, t-2) \right] + \dot{B}_a(\vartheta_0, \vartheta_0) \dot{B}_f(\vartheta_0, \vartheta_0)$$
$$= \frac{1}{(1 - A(\vartheta_0)^2)} \left[\frac{A(\vartheta_0) \dot{B}_f(\vartheta_0, \vartheta_0) B(\vartheta_0, \vartheta_0)}{(1 - \theta_{0,2} A(\vartheta_0))} + \dot{B}_a(\vartheta_0, \vartheta_0) \dot{B}_f(\vartheta_0, \vartheta_0) \right]$$
$$= \frac{A(\vartheta_0) B(\vartheta_0, \vartheta_0) \dot{B}_f(\vartheta_0, \vartheta_0) + \left(1 - \theta_{0,2} A(\vartheta_0)\right) \dot{B}_a(\vartheta_0, \vartheta_0) \dot{B}_f(\vartheta_0, \vartheta_0)}{\left(1 - \theta_{0,2} A(\vartheta_0)\right) \left(1 - A(\vartheta_0)^2\right)}$$
$$\equiv K_f(\vartheta_0)$$

because

$$\mathbf{E}_{\vartheta_0} \left[M(\vartheta_0, t-2) \dot{M}_f(\vartheta_0, t-2) \right] = \frac{B(\vartheta_0, \vartheta_0) \dot{B}_f(\vartheta_0, \vartheta_0)}{\left(1 - \theta_{0,2} A(\vartheta_0)\right)}.$$

Hence

$$I_{12}(\vartheta_0) = \frac{2P(\vartheta_0) K(\vartheta_0) + \dot{P}_a(\vartheta_0) \dot{P}_f(\vartheta_0)}{2P(\vartheta_0)^2}.$$

The next component is

$$\frac{1}{T} \mathbf{E}_{\vartheta_0} \left(\left. \frac{\partial L(\vartheta, X^T)}{\partial \theta_1} \right|_{\vartheta=\vartheta_0} \left. \frac{\partial L(\vartheta, X^T)}{\partial \theta_3} \right|_{\vartheta=\vartheta_0} \right)$$

$$= \frac{\mathbf{E}_{\vartheta_0} \dot{M}_f(\vartheta_0, t-1) \dot{M}_{\sigma^2}(\vartheta_0, t-1)}{P(\vartheta_0)} + \frac{\dot{P}_f(\vartheta_0) \dot{P}_{\sigma^2}(\vartheta_0)}{2P(\vartheta_0)^2}$$

$$= \frac{\dot{B}_f(\vartheta_0,\vartheta_0)\,\dot{B}_{\sigma^2}(\vartheta_0,\vartheta_0)}{P(\vartheta_0)\left(1-A(\vartheta_0)^2\right)} + \frac{\dot{P}_f(\vartheta_0)\,\dot{P}_{\sigma^2}(\vartheta_0)}{2P(\vartheta_0)^2} = I_{13}(\vartheta_0).$$

And the last component is similar to $I_{12}(\vartheta_0)$

$$I_{23}(\vartheta_0) = \frac{2P(\vartheta_0)\,K_{\sigma^2}(\vartheta_0) + \dot{P}_a(\vartheta_0)\,\dot{P}_{\sigma^2}(\vartheta_0)}{2P(\vartheta_0)^2},$$

where

$$K_{\sigma^2}(\vartheta_0)$$
$$= \frac{A(\vartheta_0)\,B(\vartheta_0,\vartheta_0)\,\dot{B}_{\sigma^2}(\vartheta_0,\vartheta_0) + \left(1-\theta_{0,2}A(\vartheta_0)\right)\dot{B}_a(\vartheta_0,\vartheta_0)\,\dot{B}_{\sigma^2}(\vartheta_0,\vartheta_0)}{\left(1-\theta_{0,2}A(\vartheta_0)\right)\left(1-A(\vartheta_0)^2\right)}.$$

The Fisher information matrix is

$$\mathbf{I}(\vartheta_0) = \begin{pmatrix} I_{11}(\vartheta_0), & I_{12}(\vartheta_0), & I_{13}(\vartheta_0) \\ I_{21}(\vartheta_0), & I_{22}(\vartheta_0), & I_{23}(\vartheta_0) \\ I_{31}(\vartheta_0), & I_{32}(\vartheta_0), & I_{33}(\vartheta_0) \end{pmatrix}, \tag{6.29}$$

where the values $I_{11}(\vartheta_0)$, $I_{22}(\vartheta_0)$, and $I_{33}(\vartheta_0)$ are given in (6.25), (6.26), and (6.27), respectively.

Remark 6.6 Let us mention one model which can be called *hidden threshold autoregressive* (Hidden TAR) process. It was discussed in [29]. Consider a two-dimensional time series

$$X_{t+1} = \rho_1\,Y_t\,\mathbb{1}_{\{X_t<\vartheta\}} + \rho_2\,Y_t\,\mathbb{1}_{\{X_t\geq\vartheta\}} + w_{t+1}, \qquad t=0,1,\ldots,T-1,$$
$$Y_{t+1} = a\,Y_t + v_{t+1},$$

where $w_t, t\geq 1$ and $v_t, t\geq 1$ are two independent sequences of random variables such that $w_t \sim \mathcal{N}\left(0,\sigma^2\right)$ and $v_t \sim \mathcal{N}\left(0,b^2\right)$ and $X_0 = Y_0 = 0$. Suppose that $\mathbf{P}\left(Y_0 \middle| X_0\right) \sim \mathcal{N}\left(0,d^2\right)$, and only the values $X_0, X_1, \ldots, X_T$ are observed.

The values $\rho_1 \neq \rho_2$, a, $\sigma^2>0$, $b^2>0$ are known, $|a|<1$ and the parameter $\vartheta\in\Theta=(\alpha,\beta)$ has to be estimated.

This model is a particular case of conditionally Gaussian partially observed systems discussed in Chapter 13 of [141]. Introduce the notation

$$A(x,\vartheta) = \rho_1\,\mathbb{1}_{\{x<\vartheta\}} + \rho_2\,\mathbb{1}_{\{x\geq\vartheta\}}, \qquad m_t(\vartheta) = \mathbf{E}_\vartheta\left(Y_t \middle| \mathfrak{F}_t^X\right),$$
$$\gamma_t(\vartheta) = \mathbf{E}_\vartheta\left(\left[Y_t - m_t(\vartheta)\right]^2 \middle| \mathfrak{F}_t^X\right).$$

Here the σ-algebra $\mathfrak{F}_t^X = \sigma(X_0, X_1, \ldots, X_t)$.

Then we can write the equations of Kalman filter for conditionally Gaussian time series (see Section 13.4 in [141])

$$m_{t+1}(\vartheta) = a\, m_t(\vartheta) + \frac{aA(X_t,\vartheta)\,\gamma_t(\vartheta)}{\sigma^2 + A(X_t,\vartheta)^2\,\gamma_t(\vartheta)}\left[X_{t+1} - A(X_t,\vartheta)\, m_t(\vartheta)\right], \qquad m_0,$$

$$\gamma_{t+1}(\vartheta) = a^2\gamma_t(\vartheta) + b^2 - \frac{\left[a\,A(X_t,\vartheta)\,\gamma_t(\vartheta)\right]^2}{\sigma^2 + A(X_t,\vartheta)^2\,\gamma_t(\vartheta)}, \qquad \gamma_0 = \mathbf{E}\,(Y_0 - m_0)^2,$$

where $m_0 = \mathbf{E}\left(Y_0 \middle| X_0\right)$. Moreover, the random variables X_t admit the representation (innovation)

$$X_{t+1} = A(X_t,\vartheta)\, m_t(\vartheta) + \left[\sigma^2 + A(X_t,\vartheta)^2\,\gamma_t(\vartheta)\right]^{1/2} \bar{w}_{t+1},$$

where $\bar{w}_1, \bar{w}_2, \ldots, \bar{w}_n$ are i.i.d. $\mathcal{N}(0,1)$ (see Theorem 13.5 in [141]).

The likelihood function is

$$L\left(\vartheta, X^T\right) = (2\pi)^{-T/2} \exp\left\{-\frac{1}{2}\sum_{t=0}^{T-1} \frac{\left[X_{t+1} - A(X_t,\vartheta)\, m_t(\vartheta)\right]^2}{\sigma^2 + A(X_t,\vartheta)\,\gamma_t(\vartheta)}\right\}, \qquad \vartheta \in \Theta$$

and the MLE $\hat{\vartheta}_T$ and BE $\tilde{\vartheta}_T$ are defined by the well-known equations

$$L(\hat{\vartheta}_T, X^T) = \sup_{\vartheta\in\Theta} L\left(\vartheta, X^T\right), \qquad \tilde{\vartheta}_T = \frac{\int_\Theta \vartheta\, p(\vartheta)\, L\left(\vartheta, X^T\right) \mathrm{d}\vartheta}{\int_\Theta p(\vartheta)\, L\left(\vartheta, X^T\right) \mathrm{d}\vartheta}.$$

Here $p(\vartheta)$, $\vartheta \in \Theta$ is continuous positive density.

The study of the asymptotic ($T \to \infty$) of the MLE and BE for this model is more complicated than in the case of hidden AR process considered above. Let us give some hints.

Note that the functions $m_t(\vartheta)$ and $\gamma_t(\vartheta)$ are continuous with respect to ϑ and $A(X_t,\vartheta)$ has jumps. In this case, the likelihood function is no longer piece-wise constant as in the case of TAR process, but continuously varying between jumps. Now consider the normalized likelihood ratio

$$Z_T(u) = \frac{L\left(\vartheta_0 + \frac{u}{T}, X^T\right)}{L\left(\vartheta_0, X^T\right)}, \qquad u \in \mathbb{U}_n = \left(T(\alpha - \vartheta_0), T(\beta - \vartheta_0)\right).$$

For $u > 0$, one has

$$
\begin{aligned}
&-\ln Z_T(\vartheta_0, u)\\
&=\sum_{t=0}^{T-1}\left[\frac{[X_{t+1}-A(X_t,\vartheta_u)\,m_t(\vartheta_u)]^2}{B_t(\vartheta_u)^2}-\frac{[X_{t+1}-A(X_t,\vartheta_0)\,m_t(\vartheta_0)]^2}{B_t(\vartheta_0)^2}\right]\\
&=\sum_{t=0}^{T-1}\left[\frac{[A(X_t,\vartheta_0)\,m_t(\vartheta_0)-A(X_t,\vartheta_u)\,m_t(\vartheta_u)+B_t(\vartheta_0)\;\bar{w}_{t+1}]^2}{B_t(\vartheta_u)^2}-\bar{w}_{t+1}^2\right]\\
&=\sum_{t=0}^{T-1}\left(\left[\varrho\;\mathbb{I}_{\{\vartheta_0\le X_t<\vartheta_0+\frac{u}{n}\}}\frac{m_t(\vartheta_0)}{B_t(\vartheta_u)}+\frac{B_t(\vartheta_0)}{B_t(\vartheta_u)}\bar{w}_{t+1}\right]^2-\bar{w}_{t+1}^2\right)+o(1)\\
&=\sum_{t=0}^{T-1}\left(\varrho^2\frac{m_t(\vartheta_0)^2}{B_t(\vartheta_0)^2}+2\varrho\frac{m_t(\vartheta_0)}{B_t(\vartheta_0)}\;\bar{w}_{t+1}\right)\mathbb{I}_{\{\vartheta_0\le X_t<\vartheta_0+\frac{u}{n}\}}+o(1)\\
&=\varrho^2\,S_T^{(1)}+2\varrho\,S_T^{(2)}+o(1)\,,
\end{aligned}
$$

where

$$
\vartheta_u=\vartheta_0+\frac{u}{T},\qquad \varrho=\rho_2-\rho_1,\qquad B_t(\vartheta)^2=\sigma^2+A(X_t,\vartheta)\,\gamma_t(\vartheta)\,.
$$

The following relations were used.

$$
\begin{aligned}
&A(X_t,\vartheta_0)-A(X_t,\vartheta_u)=(\rho_2-\rho_1)\;\mathbb{I}_{\{\vartheta_0\le X_t<\vartheta_0+\frac{u}{T}\}},\\
&m_t(\vartheta_u)=m_t(\vartheta_0)+O\left(\frac{1}{T}\right).
\end{aligned}
$$

Now study the sums

$$
S_T^{(1)}=\sum_{t=0}^{T-1}\mu_t^2\;\mathbb{I}_{\{\vartheta_0\le X_t<\vartheta_0+\frac{u}{T}\}},\qquad S_T^{(2)}=\sum_{t=0}^{T-1}\mu_t\;\bar{w}_{t+1}\;\mathbb{I}_{\{\vartheta_0\le X_t<\vartheta_0+\frac{u}{T}\}}
$$

with $\mu_t=\frac{m_t(\vartheta_0)}{B_t(\vartheta_0)}$. Note that

$$
\mathbf{E}_{\vartheta_0}\left(\mu_t\;\bar{w}_{t+1}\;\mathbb{I}_{\{\vartheta_0\le X_t<\vartheta_0+\frac{u}{n}\}}|\mathfrak{F}_t^X\right)=\mu_t\;\mathbb{I}_{\{\vartheta_0\le X_t<\vartheta_0+\frac{u}{T}\}}\mathbf{E}_{\vartheta_0}\left(\bar{w}_{t+1}|\mathfrak{F}_t^X\right)=0.
$$

As the value u is fixed, the window $\mathbb{V}_T=\left[\vartheta_0,\vartheta_0+\frac{u}{T}\right]$ becomes more and more narrow, and asymptotically there are only a finite (random) number of terms remaining. The time series $m_t(\vartheta_0)$, due to the condition $|a|<1$, is strong mixing. Therefore the limit values $\xi_l=\xi_l(\vartheta_0)$ of $\mu_t(\vartheta_0)\,,t\ge 1$ which rest in the limit in the sums become independent random variables, which are independent of $\tilde{w}_{l+1}$ because the distance in time between any two values ξ_l and ξ_k is "equal to infinity."

To conclude, one needs to establish limit theorems for the following sums:

$$S_T^{(1)} \Longrightarrow \sum_{l=1}^{N_+(u)} \xi_l^2, \qquad S_T^{(2)} \Longrightarrow \sum_{l=1}^{N_+(u)} \xi_l \, \tilde{w}_{l+1},$$

where $N_+ (u), u \geq 0$ is a Poisson process, ξ_l and $\tilde{w}_{l+1}$ are independent random variables. This yields the convergence ($u > 0$)

$$Z_T (\vartheta_0, u) \Longrightarrow Z (\vartheta_0, u) = \exp \left\{ -\frac{1}{2} \sum_{l=1}^{N_+(u)} \left(\varrho^2 \xi_l^2 + 2\varrho \, \xi_l \, \tilde{w}_{l+1} \right) \right\}.$$

Similar convergence can be proved for the values $u < 0$ too.

This convergence together with the other two usual estimates

$$\mathbf{E}_{\vartheta_0} \left| Z_T (\vartheta_0, u_2)^{1/2} - Z_T (\vartheta_0, u_1)^{1/2} \right| \leq C \left| u_2 - u_1 \right|,$$

$$\mathbf{P}_{\vartheta_0} \left(Z_T (\vartheta_0, u) > e^{-\kappa |u|} \right) \leq \frac{C}{|u|^N}$$

provide the corresponding properties of the BE (Theorem 1.11).

The MLE according to the Ibragimov-Khasminskii program for discontinuous likelihood ratios requires a special study (see Chapter 5, [73]). One has to prove the weak convergence of $Z_n (\vartheta_0, \cdot)$ to $Z (\vartheta_0, \cdot)$ in the space $(\mathcal{D}_0 (\mathcal{R}), \mathfrak{B})$ of discontinuous, vanishing at the infinity functions.

Of course what is written here is not a proof, and study of hidden TAR process by this approach or any other is an open problem. Similar results but for TAR processes can be found in [27–30, 38].

6.2.3 One-Step MLE-Process

We consider the construction of a one-step MLE-process in the case of unknown parameter $\vartheta = b \in \Theta = (\alpha_b, \beta_b), \alpha_b > 0$. Let us fix a learning interval of observations $X^{\tau_T} = \left(X_0, X_1, \ldots, X_{\tau_T} \right)$, where $\tau_T = \left[T^\delta \right], \delta \in \left(\frac{1}{2}, 1 \right)$ and $[A]$ here is the integer part of A. As preliminary estimator we take the MME $\vartheta^*_{\tau_T} = b^*_{\tau_T}$ defined in (6.20). One-step MLE-process is

$$\vartheta^\star_{t,T} = \vartheta^*_{\tau_T} + \frac{1}{\mathrm{I}_b(\vartheta^*_{\tau_T}) \left(t - \tau_T \right)} \sum_{s=\tau_T+1}^{t} \left[\frac{\left[X_s - f m_{s-1}(\vartheta^*_{\tau_T}) \right]}{P(\vartheta^*_{\tau_T})} f \dot{m}_{s-1}(\vartheta^*_{\tau_T}) \right.$$

$$+\left(\left[X_s - fm_{s-1}(\vartheta^*_{\tau_T})\right]^2 - P(\vartheta^*_{\tau_T})\right)\frac{\dot P(\vartheta^*_{\tau_T})}{2P(\vartheta^*_{\tau_T})^2}\Bigg], \qquad t\in\left[\tau_T+2, T\right]. \tag{6.30}$$

Theorem 6.1 *If* $t = [vT], v \in (0,1]$ *and* $T \to \infty$, *then the following convergences*

$$\sqrt{t}\left(\vartheta^\star_{t,T}-\vartheta_0\right) \Longrightarrow \mathcal{N}\left(0, \mathrm{I}_b(\vartheta_0)^{-1}\right), \qquad t\mathbf{E}_{\vartheta_0}\left(\vartheta^\star_{t,T}-\vartheta_0\right)^2 \longrightarrow \mathrm{I}_b(\vartheta_0)^{-1}$$

hold uniformly on compacts $\mathbb{K}\subset\Theta$.

Proof To prove this theorem, we need the bounds on the first and second derivatives of $m_t(\vartheta)$ presented in the next lemma.

Lemma 6.2 *For any* $p>1$ *there exist constants* $C_1>0, C_2>0$ *not depending on* $\vartheta_0\in\Theta$ *and* $t\geq 1$ *such that for*

$$\sup_{\vartheta\in\Theta}\mathbf{E}_{\vartheta_0}\left|\dot m_t(\vartheta)\right|^p < C_1, \qquad \sup_{\vartheta\in\Theta}\mathbf{E}_{\vartheta_0}\left|\ddot m_t(\vartheta)\right|^p < C_2.$$

□

Proof The equation for the second derivative $\ddot m(\vartheta, t)$ is

$$\begin{aligned}\ddot m_t(\vartheta) &= A(\vartheta)\,\ddot m_{t-1}(\vartheta) + 2\dot A(\vartheta)\,\dot m_{t-1}(\vartheta) + \ddot A(\vartheta)\,m_{t-1}(\vartheta)\\ &\quad - \ddot A(\vartheta)\,m_{t-1}(\vartheta_0) + \ddot B^*(\vartheta,\vartheta_0)\,\zeta_t(\vartheta_0).\end{aligned}$$

Note that here $B^*(\vartheta,\vartheta_0) = af\gamma_*(\vartheta)\sqrt{P(\vartheta_0)}P(\vartheta)^{-1}$.

Recall that the stationary versions of the random functions $m(\vartheta_0, t-1)$, $m(\vartheta, t-1)$, $\dot m(\vartheta, t-1)$, and $\ddot m(\vartheta, t-1)$ are

$$m_{t-1}(\vartheta_0) = B^*(\vartheta_0,\vartheta_0)\sum_{k=0}^\infty a^k\zeta_{t-k-1},$$

$$\begin{aligned}m_{t-1}(\vartheta) &= A(\vartheta)\,m_{t-2}(\vartheta) + [a-A(\vartheta)]\,B^*(\vartheta_0,\vartheta_0)\sum_{k=0}^\infty a^k\zeta_{t-k-2}\\ &\quad + B^*(\vartheta,\vartheta_0)\,\zeta_{t-k-1}\\ &= [a-A(\vartheta)]\,B^*(\vartheta_0,\vartheta_0)\sum_{j=0}^\infty A(\vartheta)^j\sum_{k=0}^\infty a^k\zeta_{t-j-k-2}\\ &\quad + B^*(\vartheta,\vartheta_0)\sum_{j=0}^\infty A(\vartheta)^j\,\zeta_{t-j-1},\end{aligned}$$

$$\dot{m}_{t-1}(\vartheta) = B^*(\vartheta_0, \vartheta_0)\,\dot{A}(\vartheta)\,A(\vartheta)^{-1}\sum_{j=0}^{\infty}\left[aj - A(\vartheta)(j+1)\right]A(\vartheta)^j$$

$$\times\sum_{k=0}^{\infty} a^k \zeta_{t-j-k-2}$$

$$+\sum_{j=0}^{\infty}\left[\dot{B}^*(\vartheta,\vartheta_0) + jB^*(\vartheta,\vartheta_0)\,\dot{A}(\vartheta)\,A(\vartheta)^{-1}\right]A(\vartheta)^j\,\zeta_{t-j-1},$$

$$\ddot{m}_t(\vartheta) = \sum_{j=0}^{\infty}\left[F_0(\vartheta,\vartheta_0) + jF_1(\vartheta,\vartheta_0) + j^2F_2(\vartheta,\vartheta_0)\right]A(\vartheta)^j\sum_{k=0}^{\infty}a^k\zeta_{t-j-k-2}$$

$$+\sum_{j=0}^{\infty}\left[H_0(\vartheta,\vartheta_0) + jH_1(\vartheta,\vartheta_0) + j^2H_2(\vartheta,\vartheta_0)\right]A(\vartheta)^j\,\zeta_{t-j-1}.$$

The bounded functions $F_i(\vartheta,\vartheta_0)$, $H_i(\vartheta,\vartheta_0)$, $i = 0, 1, 2$, can be easily calculated by the formal differentiation of $\dot{m}_{t-1}(\vartheta)$.

From these representations, (6.22), condition $a^2 \in (0, 1)$ and boundedness of the functions $F_i(\cdot,\cdot)$, $H_i(\cdot,\cdot)$ it follows that the Gaussian processes $\dot{m}_{t-1}(\vartheta)$ and $\ddot{m}_{t-1}(\vartheta)$ have bounded variances and fast-decreasing covariance functions.

For example,

$$\mathbf{E}_{\vartheta_0}\left(\sum_{j=0}^{\infty} j^2 A(\vartheta)^j \sum_{k=0}^{\infty} a^k \zeta_{t-j-k-2}\right)^2$$

$$= \mathbf{E}_{\vartheta_0}\left(\sum_{j=0}^{\infty} j^2 A(\vartheta)^j \sum_{k=0}^{\infty} a^k \zeta_{t-j-k-2}\sum_{l=0}^{\infty} l^2 A(\vartheta)^l \sum_{m=0}^{\infty} a^m \zeta_{t-m-l-2}\right)$$

$$= \sum_{j=0}^{\infty} j^2 A(\vartheta)^j \sum_{l=0}^{\infty} l^2 A(\vartheta)^l \sum_{k=0}^{\infty}\sum_{m=0}^{\infty} a^k a^m \mathbf{E}_{\vartheta_0}\left(\zeta_{t-j-k-2}\zeta_{t-m-l-2}\right)$$

$$= \sum_{j=0}^{\infty} j^2 A(\vartheta)^j \sum_{l=0}^{\infty} l^2 A(\vartheta)^l a^{j-l} \sum_{k\geq l-j\geq 0}^{\infty} a^{2k}$$

$$= \sum_{j=0}^{\infty} \frac{j^2 A(\vartheta)^j}{1-a^2} \sum_{l\geq j}^{\infty} l^2 A(\vartheta)^l a^{j-l} a^{2(l-j)}$$

$$= \frac{1}{1-a^2}\sum_{j=0}^{\infty} j^2 A(\vartheta)^j a^{-j} \sum_{l\geq j}^{\infty} l^2\left[aA(\vartheta)\right]^l.$$

Recall that

$$[\ln aA(\vartheta)]^2 \sum_{l\geq j}^{\infty} l^2 [aA(\vartheta)]^l = \frac{\partial^2}{\partial x^2} \sum_{l\geq j}^{\infty} [aA(\vartheta)]^{xl}\Big|_{x=1}$$

$$= \frac{\partial^2}{\partial x^2}\left(\frac{[aA(\vartheta)]^{xj}}{1-[aA(\vartheta)]^x}\right)\Bigg|_{x=1}.$$

Therefore the moments $\mathbf{E}_{\vartheta_0}\dot{m}_t(\vartheta)^2$ and $\mathbf{E}_{\vartheta_0}\ddot{m}_t(\vartheta)^2$ can be calculated exactly and the constants C_1, C_2 in (6.29) can be chosen not depending on $\vartheta_0 \in \Theta$ and $t \geq 0$. □

Consider the normalized difference

$$\sqrt{t}\left(\vartheta^{\star}_{t,T} - \vartheta_0\right) = \sqrt{t}\left(\vartheta^*_{\tau_T} - \vartheta_0\right)$$

$$+ \frac{\sqrt{t}}{\mathrm{I}_b(\vartheta^*_{\tau_T})\left(t-\tau_T\right)} \sum_{s=\tau_T+1}^{t} \left[\frac{\left[X_s - fm_{s-1}(\vartheta^*_{\tau_T})\right]}{P(\vartheta^*_{\tau_T})} f\dot{m}_{s-1}(\vartheta^*_{\tau_T})\right.$$

$$\left. + \left(\left[X_s - fm_{s-1}(\vartheta^*_{\tau_T})\right]^2 - P(\vartheta^*_{\tau_T})\right)\frac{\dot{P}(\vartheta^*_{\tau_T})}{2P(\vartheta^*_{\tau_T})^2}\right], \qquad t \in \left[\tau_T + 2, T\right].$$

We can write

$$X_s - fm_{s-1}(\vartheta^*_{\tau_T}) = X_s - fm_{s-1}(\vartheta_0) + f\left[m_{s-1}(\vartheta_0) - m_{s-1}(\vartheta^*_{\tau_T})\right]$$

$$= \sqrt{P(\vartheta_0)}\,\zeta_s(\vartheta_0) - (\vartheta^*_{\tau_T} - \vartheta_0) f\dot{m}_{s-1}(\vartheta_0) - \frac{f}{2}(\vartheta^*_{\tau_T} - \vartheta_0)^2 \ddot{m}_{s-1}(\tilde{\vartheta}),$$

$$P(\vartheta^*_{\tau_T}) = P(\vartheta_0) + (\vartheta^*_{\tau_T} - \vartheta_0)\dot{P}(\vartheta_0) + \frac{1}{2}(\vartheta^*_{\tau_T} - \vartheta_0)^2 \ddot{P}(\tilde{\vartheta}),$$

$$\mathrm{I}_b(\vartheta^*_{\tau_T}) = \mathrm{I}_b(\vartheta_0) + O\left(\vartheta^*_{\tau_T} - \vartheta_0\right),$$

$$\left[X_s - fm_{s-1}(\vartheta^*_{\tau_T})\right]^2 - P(\vartheta^*_{\tau_T}) = P(\vartheta_0)\left[\zeta_s(\vartheta_0)^2 - 1\right] - (\vartheta^*_{\tau_T} - \vartheta_0)\dot{P}(\vartheta_0)$$

$$- 2(\vartheta^*_{\tau_T} - \vartheta_0)\sqrt{P(\vartheta_0)}\zeta_s(\vartheta_0)\,\dot{m}_{s-1}(\vartheta_0) + O\left((\vartheta^*_{\tau_T} - \vartheta_0)^2\right).$$

We used here generic notation $\tilde{\vartheta}$ for corresponding intermediate points in Taylor expansions. The substitution of these relations in (6.30) yields

$$\vartheta^{\star}_{t,T} = \vartheta^*_{\tau_T}$$

$$+ \sum_{s=\tau_T+1}^{t} \frac{\left[2\zeta_s(\vartheta_0) f \dot{m}_{s-1}(\vartheta_0)\sqrt{P(\vartheta_0)} + \left[\zeta_s(\vartheta_0)^2 - 1\right] \dot{P}(\vartheta_0)\right]}{2\left(t-\tau_T\right) \mathrm{I}_b(\vartheta_0) P(\vartheta_0)}$$

$$\times \left(1 + o\left(\vartheta^*_{\tau_T} - \vartheta_0\right)\right)$$

$$- \frac{(\vartheta^*_{\tau_T} - \vartheta_0)}{\left(t-\tau_T\right) \mathrm{I}_b(\vartheta_0)} \sum_{s=\tau_T+1}^{t} \left(\frac{f^2 \dot{m}_{s-1}(\vartheta_0)^2}{P(\vartheta_0)} + \frac{\dot{P}(\vartheta_0)^2}{2P(\vartheta_0)^2}\right) + o\left((\vartheta^*_{\tau_T} - \vartheta_0)^2\right).$$

Note that if $t \to \infty$, then by LLN and by the CLT uniformly on compacts $\mathbb{K} \subset \Theta$ we have

$$R_{1,T} = \frac{1}{t} \sum_{s=\tau_T+1}^{t} \left(\frac{f^2 \dot{m}_{s-1}(\vartheta_0)^2}{P(\vartheta_0)} + \frac{\dot{P}(\vartheta_0)^2}{2P(\vartheta_0)^2}\right) \longrightarrow \mathrm{I}_b(\vartheta_0),$$

$$R_{2,T} = \frac{1}{\sqrt{t}} \sum_{s=\tau_T+1}^{t} \left(\mathrm{I}_b(\vartheta_0) - \frac{f^2 \dot{m}_{s-1}(\vartheta_0)^2}{P(\vartheta_0)} - \frac{\dot{P}(\vartheta_0)^2}{2P(\vartheta_0)^2}\right) \Longrightarrow \mathcal{N}\left(0, \mathrm{D}(\vartheta_0)^2\right),$$

$$R_{3,T} = \frac{1}{t} \mathbf{E}_{\vartheta_0} \left[\sum_{s=\tau_T+1}^{t} \left(\frac{f^2 \dot{m}_{s-1}(\vartheta_0)^2}{P(\vartheta_0)} + \frac{\dot{P}(\vartheta_0)^2}{2P(\vartheta_0)^2} - \mathrm{I}_b(\vartheta_0)\right)\right]^2$$

$$= \frac{1}{tP(\vartheta_0)^2} \mathbf{E}_{\vartheta_0} \left[\sum_{s=\tau_T+1}^{t} \left(f^2 \left[\dot{m}_{s-1}(\vartheta_0)^2 - \mathbf{E}_{\vartheta_0} \dot{m}_{s-1}(\vartheta_0, t)^2\right]\right)\right]^2 \le C,$$

where the constant $C > 0$ does not depend on t and $\vartheta_0 \in \mathbb{K}$. The particular form of the limit variance $\mathrm{D}(\vartheta_0)^2$ is not interesting for us because we just use its boundedness $\sup_{\vartheta_0 \in \Theta} \mathrm{D}(\vartheta_0)^2 < \infty$. We have as well the uniform convergence

$$\frac{1}{\sqrt{t}} \sum_{s=\tau_T+1}^{t} \frac{\left[2\zeta_s(\vartheta_0) f \dot{m}_{s-1}(\vartheta_0)\sqrt{P(\vartheta_0)} + \left[\zeta_s(\vartheta_0)^2 - 1\right] \dot{P}(\vartheta_0)\right]}{2\mathrm{I}_b(\vartheta_0) P(\vartheta_0)}$$

$$\Longrightarrow \mathcal{N}\left(0, \mathrm{I}_b(\vartheta_0)^{-1}\right).$$

Therefore, if we put $t = vT$, $v \in (T^{\delta-1}, 1]$, then

$$
\begin{aligned}
\sqrt{t}\left(\vartheta_{t,T}^{\star}-\vartheta_0\right) &= \sqrt{t}\left(\vartheta_{\tau_T}^{*}-\vartheta_0\right)\\
&+\sum_{s=\tau_T+1}^{t}\frac{\left[2\zeta_s(\vartheta_0)f\dot m_{s-1}(\vartheta_0)\sqrt{P(\vartheta_0)}+\left[\zeta_s(\vartheta_0)^2-1\right]\dot P(\vartheta_0)\right]}{2\left(\sqrt{t}-\tau_Tt^{-1/2}\right)\mathrm{I}_b(\vartheta_0)P(\vartheta_0)}\\
&\times\left(1+O\left(\vartheta_{\tau_T}^{*}-\vartheta_0\right)\right)\\
&-\frac{\sqrt{t}(\vartheta_{\tau_T}^{*}-\vartheta_0)}{(t-\tau_T)\,\mathrm{I}_b(\vartheta_0)}\sum_{s=\tau_T+1}^{t}\left(\frac{f^2\dot m_{s-1}(\vartheta_0)^2}{P(\vartheta_0)}+\frac{\dot P(\vartheta_0)^2}{2P(\vartheta_0)^2}\right)\\
&+O\left(\sqrt{t}(\vartheta_{\tau_T}^{*}-\vartheta_0)^2\right).
\end{aligned}
$$

Remind that $\delta\in\left(\frac{1}{2},1\right)$ and the MME $\vartheta_{\tau_T}^{*}$ satisfies (6.20). Hence, for any fixed $v\in(0,1]$

$$
\sqrt{t}\mathbf{E}_{\vartheta_0}\left|\vartheta_{\tau_T}^{*}-\vartheta_0\right|^2\leq\frac{C\sqrt{t}}{\tau_T^2}\leq\frac{C\sqrt{v}T^{1/2}}{T^{\delta}}\longrightarrow 0.
$$

Further,

$$
\begin{aligned}
&\frac{\sqrt{t}\left(\vartheta_{\tau_T}^{*}-\vartheta_0\right)}{\mathrm{I}_b(\vartheta_0)}\left[\mathrm{I}_b(\vartheta_0)-\frac{1}{t}\sum_{s=\tau_T+1}^{t}\left(\frac{f^2\dot m_{s-1}(\vartheta_0)^2}{P(\vartheta_0)}+\frac{\dot P(\vartheta_0)^2}{2P(\vartheta_0)^2}\right)\right]\\
&\quad=\frac{\sqrt{t}\left(\vartheta_{\tau_T}^{*}-\vartheta_0\right)}{\mathrm{I}_b(\vartheta_0)}\frac{1}{t}\sum_{s=\tau_T+1}^{t}\left[\frac{f^2\dot B^{*}(\vartheta_0,\vartheta_0)^2}{\left[1-A(\vartheta_0)^2\right]P(\vartheta_0)}-\frac{f^2\dot m_{s-1}(\vartheta_0)^2}{P(\vartheta_0)}\right]\\
&\quad=\frac{\sqrt{t}\left(\vartheta_{\tau_T}^{*}-\vartheta_0\right)f^2}{\mathrm{I}_b(\vartheta_0)P(\vartheta_0)}\frac{1}{t}\sum_{s=\tau_T+1}^{t}\left[\frac{\dot B^{*}(\vartheta_0,\vartheta_0)^2}{\left[1-A(\vartheta_0)^2\right]}-\dot m_{s-1}(\vartheta_0)^2\right]\\
&\quad\longrightarrow 0.
\end{aligned}
$$

Therefore for any fixed $v\in(0,1)$ and $t=vT$ uniformly on $\vartheta_0\in\mathbb{K}$

$$
\sqrt{t}\left(\vartheta_{t,T}^{\star}-\vartheta_0\right)\Longrightarrow\mathcal{N}\left(0,\mathrm{I}_b(\vartheta_0)^{-1}\right)\quad\text{and}\quad t\mathbf{E}_{\vartheta_0}\left(\vartheta_{t,T}^{\star}-\vartheta_0\right)^2\longrightarrow\mathrm{I}_b(\vartheta_0)^{-1}.
$$

□

Remark 6.7 The One-step MLE-processes in the cases $\vartheta=f$, $\vartheta=a$, $\vartheta=\sigma^2$ can be constructed following the same lines. We do not present the corresponding calculations because it will be mainly repetition of the given above proof.

If the unknown parameter is three-dimensional, say, $\vartheta = \left(f, a, \sigma^2\right)^\top = (\theta_1, \theta_2, \theta_3)^\top$, then the One-step MLE-process $\vartheta^\star_{t,T} = \left(\theta^\star_{1,t,T}, \theta^\star_{2,t,T}, \theta^\star_{3,t,T}\right)^\top$ can be constructed as follows. Introduce MMEs $\vartheta^*_{\tau_T} = (\theta^*_{1,\tau_T}, \theta^*_{2,\tau_T}, \theta^*_{3,\tau_T})^\top$, $\theta^*_{1,\tau_T} = f^*_{\tau_T}$ and $\theta^*_{2,\tau_T} = a^*_{\tau_T}$, $\theta^*_{3,\tau_T} = \sigma^{2*}_{\tau_T}$, where $\tau_T = \left[T^\delta\right]$, $\delta \in (1/2, 1]$ and

$$\theta^*_{2,\tau_T} = \frac{2S_{3,T}\left(X^T\right)}{S_{1,T}\left(X^T\right) + 2S_{2,T}\left(X^T\right)} + 1, \qquad \theta^*_{1,\tau_T} = \frac{S_{3,T}\left(X^T\right)(1+\theta^*_{2,\tau_T})}{b^2\theta^*_{2,\tau_T}(\theta^*_{2,\tau_T}-1)}, \tag{6.31}$$

$$\theta^*_{3,\tau_T} = \frac{1}{2}S_{1,T}\left(X^T\right) - \frac{(\theta^*_{1,\tau_T})^2 b^2}{1+\theta^*_{2,\tau_T}}. \tag{6.32}$$

Denote

$$M(\vartheta, t) = \theta_1 m_t(\vartheta),\ \dot{\mathrm{M}}(\vartheta, t) = \left(\dot M_f(\vartheta, t), \dot M_a(\vartheta, t), \dot M_{\sigma^2}(\vartheta, t)\right)^\top, \tag{6.33}$$

$$\dot{\mathrm{P}}(\vartheta) = \left(\dot P_f(\vartheta), \dot P_a(\vartheta), \dot P_{\sigma^2}(\vartheta)\right)^\top.$$

The equations for $\dot M_f(\vartheta, t)$, $\dot M_a(\vartheta, t)$, and $\dot M_{\sigma^2}(\vartheta, t)$ are given above (see (6.28)). Recall that the Fisher information matrix $\mathbf{I}(\vartheta)$ has an explicit form (6.29). Suppose that this matrix is uniformly on $\vartheta \in \Theta$ nondegenerate. Then the One-step MLE-process is given by the equality

$$\vartheta^\star_{t,T} = \vartheta^*_{\tau_T} + \frac{\mathbf{I}(\vartheta^*_{\tau_T})^{-1}}{t-\tau_T}\sum_{s=\tau_T+1}^{t}\left[\frac{\left[X_s - M(\vartheta^*_{\tau_T}, s-1)\right]}{P(\vartheta^*_{\tau_T})}\dot{\mathrm{M}}(\vartheta^*_{\tau_T}, s-1)\right.$$
$$\left. + \left(\left[X_s - M(\vartheta^*_{\tau_T}, s-1)\right]^2 - P(\vartheta^*_{\tau_T})\right)\frac{\dot{\mathrm{P}}(\vartheta^*_{\tau_T})}{2P(\vartheta^*_{\tau_T})^2}\right], \qquad t \in \left[\tau_T + 2, T\right]. \tag{6.34}$$

It can be shown following the same proof as in the one-dimensional case above, that for any $v \in (0, T]$, if we put $t = vT$, then the normalized difference is asymptotically normal:

$$\sqrt{t}\left(\vartheta^\star_{t,T} - \vartheta_0\right) \Longrightarrow \xi(\vartheta_0) \sim \mathcal{N}\left(0, \mathbf{I}(\theta_0)^{-1}\right)$$

and

$$t\mathbf{E}_{\vartheta_0}\left\|\vartheta^\star_{t,T} - \vartheta_0\right\|^2 \longrightarrow \mathbf{E}_{\vartheta_0}\left\|\xi(\vartheta_0)\right\|^2.$$

Remark 6.8 The question of nondegeneracy of the Fisher information matrix is open. It is possible to suppose that under the made assumptions on the range of parameter ϑ it is always nondegenerate, but the proof of it can be a nontrivial problem.

6.2.4 MLE and BE

Consider once more the model of observations (6.1)–(6.2), where the unknown parameter is $\vartheta = b \in \Theta = (\alpha_b, \beta_b), \alpha_b > 0$. Below we study the MLE $\hat\vartheta_T$ and BE $\tilde\vartheta_T$ defined by the usual relations

$$L(\hat\vartheta_T, X^T) = \sup_{\vartheta\in\Theta} L(\hat\vartheta, X^T), \qquad \tilde\vartheta_T = \frac{\int_\Theta \vartheta p(\vartheta) L(\hat\vartheta, X^T)\mathrm{d}\vartheta}{\int_\Theta p(\vartheta) L(\hat\vartheta, X^T)\mathrm{d}\vartheta}.$$

Here $p(\vartheta), \vartheta \in \Theta$ is continuous, positive on Θ density a priori.

Recall the notation $P(\vartheta) = \sigma^2 + f^2\gamma_*(\vartheta)$, $A(\vartheta) = a\sigma^2/P(\vartheta)^2$,

$$B^*(\vartheta, \vartheta_0) = \frac{af\gamma_*(\vartheta)\sqrt{P(\vartheta_0)}}{P(\vartheta)},$$

$$\dot B^*(\vartheta_0, \vartheta_0) = \left.\frac{\partial B^*(\vartheta, \vartheta_0)}{\partial\vartheta}\right|_{\vartheta=\vartheta_0} = \frac{af\sigma^2\dot\gamma_*(\vartheta_0)}{P(\vartheta_0)^{3/2}}.$$

$$\mathrm{I}_b(\vartheta_0) = \frac{f^2\dot B^*(\vartheta_0, \vartheta_0)^2}{\left[1 - A(\vartheta_0)^2\right]P(\vartheta_0)} + \frac{\dot P(\vartheta_0)^2}{2P(\vartheta_0)^2} = \frac{f^4\dot\gamma_*(\vartheta)^2}{2P(\vartheta_0)}\left[\frac{P(\vartheta_0)^2 + a^2\sigma^4}{P(\vartheta_0)^2 - a^2\sigma^4}\right].$$

Below it is shown (see (6.35)–(6.36)) that the family of measures is LAN. Therefore, we have Hájek's lower bound: for any estimator $\bar\vartheta_T$ and any $\vartheta_0 \in \Theta$

$$\lim_{\nu\to0}\varliminf_{T\to\infty}\sup_{|\vartheta-\vartheta_0|\le\nu} T\mathbf{E}_\vartheta\left(\bar\vartheta_T - \vartheta\right)^2 \ge \mathrm{I}_b(\vartheta_0)^{-1}.$$

This bound allows us to give the following definition.

Define the asymptotically efficient estimator ϑ_T° as estimator for which the equality

$$\lim_{\nu\to0}\lim_{T\to\infty}\sup_{|\vartheta-\vartheta_0|\le\nu} T\mathbf{E}_\vartheta\left(\vartheta_T^\circ - \vartheta\right)^2 = \mathrm{I}_b(\vartheta_0)^{-1}.$$

holds for all $\vartheta_0 \in \Theta$.

Theorem 6.2 *The MLE and BE are uniformly consistent, asymptotically normal,*

$$\sqrt{T}\left(\hat{\vartheta}_T-\vartheta_0\right)\Longrightarrow \mathcal{N}\left(0,\mathrm{I}_b\left(\vartheta_0\right)^{-1}\right),\quad \sqrt{T}\left(\tilde{\vartheta}_T-\vartheta_0\right)\Longrightarrow \mathcal{N}\left(0,\mathrm{I}_b\left(\vartheta_0\right)^{-1}\right),$$

the polynomial moments converge uniformly on compacts, and both estimators are asymptotically efficient.

Proof We will prove the results with the help of the general Theorems 1.10 and 1.12, i.e., we verify the conditions $\mathscr{I}\mathscr{K}_1$ of these theorems for this model of observations. Introduce the normalized ($\varphi_T=T^{-1/2}$) likelihood ratio function

$$Z_T\left(u\right)=\frac{L(\vartheta_0+\varphi_T u,X^T)}{L(\vartheta_0,X^T)},\qquad u\in\mathbb{U}_T=\left(\sqrt{T}\left(\alpha_T-\vartheta_0\right),\sqrt{T}\left(\beta_T-\vartheta_0\right)\right).$$

Therefore we have to prove the following properties of the process $Z_T\left(u\right), u\in\mathbb{U}_T$.

1. *The process $Z_T\left(\cdot\right)$ admits the representation ULAN*

$$Z_T\left(u\right)=\exp\left(u\Delta_T(\vartheta_0,X^T)-\frac{u^2}{2}\mathrm{I}_b\left(\vartheta_0\right)+r_T(\vartheta_0,u,X^T)\right),\qquad u\in\mathbb{U}_T,\tag{6.35}$$

where uniformly on compacts $\mathbb{K}\subset\Theta$ the convergence $r_T(\vartheta_0,u,X^T)\to 0$ in probability holds,

$$\Delta_T(\vartheta_0,X^T)=\frac{1}{\sqrt{TP\left(\vartheta_0\right)}}\sum_{t=1}^{T}\left[\zeta_t\left(\vartheta_0\right)f\dot{m}_{t-1}\left(\vartheta_0\right)+\left[\zeta_t\left(\vartheta_0\right)^2-1\right]\frac{\dot{P}\left(\vartheta_0\right)}{2\sqrt{P\left(\vartheta_0\right)}}\right]$$

and uniformly on $\mathbb{K}$

$$\Delta_T(\vartheta_0,X^T)\Longrightarrow\mathcal{N}\left(0,\mathrm{I}_b\left(\vartheta_0\right)\right).\tag{6.36}$$

2. *There exists constant $C>0$ such that*

$$\sup_{\vartheta_0\in\mathbb{K}}\mathbf{E}_{\vartheta_0}\left|Z_T\left(u_2\right)^{1/2}-Z_T\left(u_1\right)^{1/2}\right|^2\leq C\left|u_1-u_2\right|^2.\tag{6.37}$$

3. *There exists a constant $\kappa>0$ and for any $N>0$ there is a constant $C=C\left(N\right)>0$ such that*

$$\sup_{\vartheta_0\in\mathbb{K}}\mathbf{P}_{\vartheta_0}\left(Z_T\left(u\right)>e^{-\kappa u^2}\right)\leq\frac{C}{\left|u\right|^N}.\tag{6.38}$$

Let us verify (6.35)–(6.38). Using Taylor expansions, we can write (below $\vartheta_u = \vartheta_0 + \varphi_T u$)

$$\ln Z_T(u) = -\frac{1}{2}\sum_{t=1}^{T}\left[\frac{(X_t - f m_{t-1}(\vartheta_u))^2}{P(\vartheta_u)} + \ln\frac{P(\vartheta_u)}{P(\vartheta_0)} - \frac{(X_t - f m_{t-1}(\vartheta_0))^2}{P(\vartheta_0)}\right]$$

$$= \frac{u}{\sqrt{T}}\sum_{t=1}^{T}\left[\frac{\zeta_t(\vartheta_0)\, f\dot{m}_{t-1}(\vartheta_0)}{\sqrt{P(\vartheta_0)}} + \frac{\left(\zeta_t(\vartheta_0)^2 - 1\right)\dot{P}(\vartheta_0)}{2P(\vartheta_0)}\right]$$

$$- \frac{u^2}{2T}\left[\frac{f^2\dot{m}_{t-1}(\vartheta_0)^2}{P(\vartheta_0)} + \frac{\zeta_t(\vartheta_0)^2\,\dot{P}(\vartheta_0)^2}{P(\vartheta_0)^2} - \frac{\dot{P}(\vartheta_0)^2}{2P(\vartheta_0)^2}\right] + o\left(T^{-1/2}\right).$$

Now the representation (6.35)–(6.36) follows from the CLT and the LLN:

$$\frac{1}{\sqrt{T}}\sum_{t=1}^{T}\left[\frac{\zeta_t(\vartheta_0)\, f\dot{m}_{t-1}(\vartheta_0)}{\sqrt{P(\vartheta_0)}} + \frac{\left(\zeta_t(\vartheta_0)^2 - 1\right)\dot{P}(\vartheta_0)}{2P(\vartheta_0)}\right] \Longrightarrow \mathcal{N}\left(0, \mathrm{I}_b(\vartheta_0)\right),$$

$$\frac{1}{T}\sum_{t=1}^{T}\left[\frac{f^2\dot{m}_{t-1}(\vartheta_0)^2}{P(\vartheta_0)} + \frac{\zeta_t(\vartheta_0)^2\,\dot{P}(\vartheta_0)^2}{P(\vartheta_0)^2} - \frac{\dot{P}(\vartheta_0)^2}{2P(\vartheta_0)^2}\right] \longrightarrow \mathrm{I}_b(\vartheta_0).$$

Let us denote

$$\Pi(u) = -\frac{1}{2}\sum_{t=1}^{T}\left[\frac{(X_t - f m_{t-1}(\vartheta_u))^2}{P(\vartheta_u)} + \ln\frac{P(\vartheta_u)}{P(\vartheta_0)} - \frac{(X_t - f m_{t-1}(\vartheta_0))^2}{P(\vartheta_0)}\right]$$

and calculate its derivative

$$\dot{\Pi}(u) = \frac{\partial}{\partial u}\Pi(u) = \frac{\varphi_T}{2}\sum_{t=1}^{T}\left[\frac{2\left(X_t - f m_{t-1}(\vartheta_u)\right) f\dot{m}_{t-1}(\vartheta_u)}{P(\vartheta_u)} - \frac{\dot{P}(\vartheta_u)}{P(\vartheta_u)}\right.$$
$$\left. + \frac{(X_t - f m_{t-1}(\vartheta_u))^2\,\dot{P}(\vartheta_u)}{P(\vartheta_u)^2}\right].$$

Note that $X_t - f m_{t-1}(\vartheta_u) = \zeta_t(\vartheta_u)\sqrt{P(\vartheta_u)}, t \geq 1$, where $\zeta_t(\vartheta_u), t \geq 1$ under measure $\mathbf{P}_{\vartheta_u}$ are i.i.d. r.v.'s $\zeta_t(\vartheta_u) \sim \mathcal{N}(0, 1)$. Therefore, with $\mathbf{P}_{\vartheta_u}$ probability 1 we have

$$\dot{\Pi}(u) = \varphi_T\sum_{t=1}^{T}\left[\frac{\zeta_t(\vartheta_u)\, f\dot{m}_{t-1}(\vartheta_u)}{\sqrt{P(\vartheta_u)}} + \frac{\left[\zeta_t(\vartheta_u)^2 - 1\right]\dot{P}(\vartheta_u)}{2P(\vartheta_u)}\right]$$

and

$$
\begin{aligned}
&\mathbf{E}_{\vartheta_u}\dot{\Pi}(u)^2\\
&=\varphi_T^2\sum_{t=1}^{T}\mathbf{E}_{\vartheta_u}\left[\frac{\zeta_t(\vartheta_u)\,f\dot{m}_{t-1}(\vartheta_u)}{\sqrt{P(\vartheta_u)}}+\frac{\left[\zeta_t(\vartheta_u)^2-1\right]\dot{P}(\vartheta_u)}{2P(\vartheta_u)}\right]^2\\
&=\varphi_T^2\sum_{t=1}^{T}\left[\frac{f^2\mathbf{E}_{\vartheta_u}\left[\zeta_t(\vartheta_u)^2\,\dot{m}_{t-1}(\vartheta_u)^2\right]}{P(\vartheta_u)}+\frac{\dot{P}(\vartheta_u)^2\,\mathbf{E}_{\vartheta_u}\left[\zeta_t(\vartheta_u)^2-1\right]^2}{4P(\vartheta_u)^2}\right]\\
&=\frac{1}{T}\sum_{t=1}^{T}\left[\frac{f^2\mathbf{E}_{\vartheta_u}\left[\dot{m}_{t-1}(\vartheta_u)^2\right]}{P(\vartheta_u)}+\frac{\dot{P}(\vartheta_u)^2}{2P(\vartheta_u)^2}\right]\\
&=\frac{f^2\dot{B}^*(\vartheta_u,\vartheta_u)^2}{P(\vartheta_u)\left(1-A(\vartheta_u)\right)^2}+\frac{\dot{P}(\vartheta_u)^2}{2P(\vartheta_u)^2}\leq C.
\end{aligned}
$$

Hence we can write

$$
\begin{aligned}
\mathbf{E}_{\vartheta_0}\left|Z_T(u_2)^{1/2}-Z_T(u_1)^{1/2}\right|^2&=\mathbf{E}_{\vartheta_0}\left|\int_{u_1}^{u_2}\frac{\partial}{\partial u}Z_T(u)^{1/2}\,\mathrm{d}u\right|^2\\
&\leq\frac{(u_2-u_1)}{4}\int_{u_1}^{u_2}\mathbf{E}_{\vartheta_0}Z_T(u)\,\dot{\Pi}(u)^2\,\mathrm{d}u\\
&=\frac{(u_2-u_1)}{4}\int_{u_1}^{u_2}\mathbf{E}_{\vartheta_u}\dot{\Pi}(u)^2\,\mathrm{d}u\leq C\,(u_2-u_1)^2,
\end{aligned}
$$

where the constant $C>0$ can be chosen not depending on ϑ_0. The estimate (6.37) is proved too.

To verify the last estimate (6.38) we write

$$
\begin{aligned}
\mathbf{P}_{\vartheta_0}\left(Z_T(u)>e^{-\kappa u^2}\right)&=\mathbf{P}_{\vartheta_0}\left(\Pi(u)>-\kappa u^2\right)\\
&=\mathbf{P}_{\vartheta_0}\Big(\Pi(u)-\mathbf{E}_{\vartheta_0}\Pi(u)>-\kappa u^2-\mathbf{E}_{\vartheta_0}\Pi(u)\Big).
\end{aligned}
$$

Note that

$$
\begin{aligned}
-2\mathbf{E}_{\vartheta_0}\Pi(u)=\sum_{t=1}^{T}&\frac{\mathbf{E}_{\vartheta_0}\left[\zeta_t(\vartheta_0)\sqrt{P(\vartheta_0)}-f\left[m_{t-1}(\vartheta_u)-m_{t-1}(\vartheta_0)\right]\right]^2}{P(\vartheta_u)}\\
&+T\ln\frac{P(\vartheta_u)}{P(\vartheta_0)}-\sum_{t=1}^{T}\mathbf{E}_{\vartheta_0}\zeta_t(\vartheta_0)^2
\end{aligned}
$$

$$
= \frac{f^2}{P(\vartheta_u)} \sum_{t=1}^{T} \mathbf{E}_{\vartheta_0} \left[m_{t-1}(\vartheta_u) - m_{t-1}(\vartheta_0)\right]^2
$$

$$
+ T\left(\frac{P(\vartheta_0)}{P(\vartheta_u)} - 1 - \ln \frac{P(\vartheta_0)}{P(\vartheta_u)}\right).
$$

Consider two regions of u. First suppose that $\left|u\varphi_T\right| \leq \nu$ with some small $\nu > 0$. Then using expansions at the vicinity of ϑ_0 we can write

$$
-\mathbf{E}_{\vartheta_0}\Pi(u) = \left(\frac{u^2 f^2 \dot{B}^*(\vartheta_0,\vartheta_0)^2}{2P(\vartheta_0)\left(1 - A(\vartheta_0)^2\right)} + \frac{u^2 \dot{P}(\vartheta_0)^2}{2P(\vartheta_0)^2}\right)(1 + o(\nu))
$$

$$
= \frac{u^2}{2}\mathrm{I}_b(\vartheta_0)(1 + o(\nu)) \geq \frac{u^2}{4}\mathrm{I}_b(\vartheta_0) \geq \kappa_1 u^2
$$

for sufficiently small ν. Remark that

$$
\inf_{\vartheta\in\Theta} \mathrm{I}_b(\vartheta) > 0
$$

and the constant κ_1 can be chosen not depending on ϑ_0.

Let $|u\varphi_T| \geq \nu$. Consider the difference of the two equations

$$
m_{t-1}(\vartheta_0) = a m_{t-2}(\vartheta_0) + B^*(\vartheta_0,\vartheta_0)\zeta_{t-1}(\vartheta_0),
$$

$$
m_{t-1}(\vartheta_u) = A(\vartheta_u) m_{t-2}(\vartheta_u) + [a - A(\vartheta_u)] m_{t-2}(\vartheta_0) + B^*(\vartheta_u,\vartheta_0)\zeta_{t-1}(\vartheta_0),
$$

and write

$$
m_{t-1}(\vartheta_u) - m_{t-1}(\vartheta_0) = A(\vartheta_u)\left[m_{t-2}(\vartheta_u) - m_{t-2}(\vartheta_0)\right]
$$

$$
+ \left[B^*(\vartheta_u,\vartheta_0) - B^*(\vartheta_0,\vartheta_0)\right]\zeta_{t-1}(\vartheta_0).
$$

Therefore

$$
\mathbf{E}_{\vartheta_0}\left[m_{t-1}(\vartheta_u) - m_{t-1}(\vartheta_0)\right]^2 = A(\vartheta_u)^2 \mathbf{E}_{\vartheta_0}\left[m_{t-2}(\vartheta_u) - m_{t-2}(\vartheta_0)\right]^2
$$

$$
+ \left[B^*(\vartheta_u,\vartheta_0) - B^*(\vartheta_0,\vartheta_0)\right]^2.
$$

We suppose that $m_t(\vartheta_u)$ and $m_t(\vartheta_0)$ are stationary processes. Hence

$$
\mathbf{E}_{\vartheta_0}\left[m_{t-1}(\vartheta_u) - m_{t-1}(\vartheta_0)\right]^2 = \frac{\left[B^*(\vartheta_u,\vartheta_0) - B^*(\vartheta_0,\vartheta_0)\right]^2}{1 - A(\vartheta_u)^2}
$$

and

$$-2\mathbf{E}_{\vartheta_0}\Pi(u)=\frac{Tf^2\left[B^*(\vartheta_u,\vartheta_0)-B^*(\vartheta_0,\vartheta_0)\right]^2}{\left(1-A(\vartheta_u)^2\right)P(\vartheta_u)}$$
$$+T\left(\frac{P(\vartheta_0)}{P(\vartheta_u)}-1-\ln\frac{P(\vartheta_0)}{P(\vartheta_u)}\right).$$

Denote

$$G(\vartheta_u,\vartheta_0)=\frac{f^2\left[B^*(\vartheta_u,\vartheta_0)-B^*(\vartheta_0,\vartheta_0)\right]^2}{2\left(1-A(\vartheta_u)^2\right)P(\vartheta_u)}+\frac{1}{2}\left(\frac{P(\vartheta_0)}{P(\vartheta_u)}-1-\ln\frac{P(\vartheta_0)}{P(\vartheta_u)}\right)$$

and show that

$$g(\vartheta_0,\nu)=\inf_{|\vartheta-\vartheta_0|\geq\nu}G(\vartheta,\vartheta_0)>0.$$

To do this, it is sufficient to verify that

$$\inf_{|\vartheta-\vartheta_0|\geq\nu}|P(\vartheta)-P(\vartheta_0)|>0.$$

As $|P(\vartheta)-P(\vartheta_0)|=f^2|\gamma_*(\vartheta)-\gamma_*(\vartheta_0)|$ we check the condition $\dot\gamma_*(\vartheta)>0$. We have (see (6.6))

$$\dot\gamma_*(\vartheta)=\vartheta+\frac{\left(4\left[f^2\vartheta^2-\sigma^2\left(1-a^2\right)\right]f^2\vartheta+8\vartheta\sigma^2f^2\right)}{4f^2\left[\left(\sigma^2\left(1-a^2\right)-f^2\vartheta^2\right)^2+4\vartheta^2\sigma^2f^2\right]^{1/2}}$$
$$=\vartheta+\frac{\vartheta\left[f^2\vartheta^2+\sigma^2\left(1+a^2\right)\right]}{\left[\left(\sigma^2\left(1-a^2\right)-f^2\vartheta^2\right)^2+4\vartheta^2\sigma^2f^2\right]^{1/2}}>0 \tag{6.39}$$

and

$$\inf_{\vartheta_0\in\Theta}g(\vartheta_0,\nu)>0.$$

Therefore, using the relation $T\geq u^2/\left(\beta_b-\alpha_b\right)^2$ we can write

$$\inf_{|u|\geq\nu\sqrt{T}}\left(-\mathbf{E}_{\vartheta_0}\Pi(u)\right)\geq g(\vartheta_0,\nu)\,T\geq\frac{g(\vartheta_0,\nu)\,u^2}{\left(\beta_b-\alpha_b\right)^2}\geq\kappa_2u^2.$$

We have

$$\Pi(u)-\mathbf{E}_{\vartheta_0}\Pi(u)=\sum_{t=1}^{T}\left[\zeta_t(\vartheta_0)^2-1\right]\left(\frac{P(\vartheta_u)-P(\vartheta_0)}{2P(\vartheta_u)}\right)$$

$$
\begin{aligned}
&+\sum_{t=1}^{T}\frac{f\sqrt{P(\vartheta_0)}}{P(\vartheta_u)}\zeta_t(\vartheta_0)\left[m_{t-1}(\vartheta_u)-m_{t-1}(\vartheta_0)\right]\\
&-\sum_{t=1}^{T}\frac{f^2\left(\left[m_{t-1}(\vartheta_u)-m_{t-1}(\vartheta_0)\right]^2-\mathbf{E}_{\vartheta_0}\left[m_{t-1}(\vartheta_u)-m_{t-1}(\vartheta_0)\right]^2\right)}{2P(\vartheta_u)}\\
&=\Pi_1(u)+\Pi_2(u)-\Pi_3(u),
\end{aligned}
$$

where for any integer $N\geq 1$

$$
\begin{aligned}
\mathbf{E}_{\vartheta_0}\left|\Pi_1(u)\right|^{2N}&=\mathbf{E}_{\vartheta_0}\left|\sum_{t=1}^{T}\left[\zeta_t(\vartheta_0)^2-1\right]\left(\frac{P(\vartheta_u)-P(\vartheta_0)}{2P(\vartheta_u)}\right)\right|^{2N}\\
&=\left|u\varphi_T\dot P(\tilde\vartheta_u)\right|^{2N}\mathbf{E}_{\vartheta_0}\left|\sum_{t=1}^{T}\left[\zeta_t(\vartheta_0)^2-1\right]\right|^{2N}\leq C\left|u\right|^{2N},
\end{aligned}
\tag{6.40}
$$

$$
\begin{aligned}
\mathbf{E}_{\vartheta_0}\left|\Pi_2(u)\right|^{2N}&=\frac{f^{2N}P(\vartheta_0)^N}{P(\vartheta_u)^{2N}}\mathbf{E}_{\vartheta_0}\left|\sum_{t=1}^{T}\zeta_t(\vartheta_0)\left[m_{t-1}(\vartheta_u)-m_{t-1}(\vartheta_0)\right]\right|^{2N}\\
&\leq C\left|u\right|^{2N}T^{-N}\mathbf{E}_{\vartheta_0}\left|\sum_{t=1}^{T}\zeta_t(\vartheta_0)\,\dot m_{t-1}(\tilde\vartheta_u)\right|^{2N}\leq C\left|u\right|^{2N},
\end{aligned}
$$

and

$$
\begin{aligned}
\mathbf{E}_{\vartheta_0}\left|\Pi_3(u)\right|^{2N}&=\frac{f^{4N}\left|u\right|^{4N}\varphi_T^{4N}}{2^{2N}P(\vartheta_u)^{2N}}\mathbf{E}_{\vartheta_0}\left|\sum_{t=1}^{T}\left[\dot m_{t-1}(\tilde\vartheta_u)^2-\mathbf{E}_{\vartheta_0}\dot m_{t-1}(\tilde\vartheta_u)^2\right]\right|^{2N}\\
&\leq C\left|u\right|^{4N}\varphi_T^{4N}T^N=C\left|u\right|^{2N}\frac{\left|u\right|^{2N}}{T}\leq C\left|u\right|^{2N}.
\end{aligned}
\tag{6.41}
$$

Here we twice used the estimate

$$
\mathbf{E}_{\vartheta_0}\left|\sum_{t=1}^{T}K_t(\vartheta)\right|^{2N}\leq CT^N,
$$

which is valid for centered Gaussian time series $K_t(\vartheta), t\geq 1$ with bounded variance and exponential mixing. We used as well the relation $|u\varphi_T|<\beta_b-\alpha_b$. Remark that all constants $C>0$ in (6.40)–(6.41) can be chosen not depending of ϑ_0.

Denote $\kappa=2(\kappa_1\wedge\kappa_2)$. Now by Tchebyshev inequality and (6.40)–(6.41)

$$\mathbf{P}_{\vartheta_0}\left(Z_T(u) > e^{-\kappa u^2}\right) \leq \mathbf{P}_{\vartheta_0}\left(\left|\Pi(u) - \mathbf{E}_{\vartheta_0}\Pi(u)\right| \geq \frac{\kappa u^2}{2}\right)$$
$$\leq \mathbf{P}_{\vartheta_0}\left(|\Pi_1(u)| \geq \frac{\kappa u^2}{6}\right) + \mathbf{P}_{\vartheta_0}\left(|\Pi_2(u)| \geq \frac{\kappa u^2}{6}\right) + \mathbf{P}_{\vartheta_0}\left(|\Pi_3(u)| \geq \frac{\kappa u^2}{6}\right)$$
$$\leq \frac{6^{2N}}{\kappa u^{2N}}\mathbf{E}_{\vartheta_0}|\Pi_1(u)|^{2N} + \frac{6^{2N}}{\kappa u^{2N}}\mathbf{E}_{\vartheta_0}|\Pi_2(u)|^{2N}\frac{6^{2N}}{\kappa u^{2N}}\mathbf{E}_{\vartheta_0}|\Pi_3(u)|^{2N} \leq \frac{C}{u^{2N}}.$$

Therefore the conditions (6.35)–(6.38) are verified and the estimators $\hat{\vartheta}_T$ and $\tilde{\vartheta}_T$ according to Theorems 1.10 and 1.12 have all the properties mentioned in Theorem 6.2. □

6.3 Adaptive Filter

We are given the partially observed system (6.1) and (6.2). Recall that if all parameters of the model $\vartheta = \left(f, \sigma^2, a, b\right)$ are known, then the stationary version $m_t(\vartheta)$ of the conditional expectation $m(\vartheta, t) = \mathbf{E}_\vartheta(Y_t|X_s, s \leq t)$ satisfies the equation (see (6.7) and (6.6))

$$m_t(\vartheta) = am_{t-1}(\vartheta) + \frac{af\gamma_*(\vartheta)}{\sigma^2 + f^2\gamma_*(\vartheta)}\left[X_t - fm_{t-1}(\vartheta)\right], \quad m_0(\vartheta), \quad t \geq 1, \tag{6.42}$$

$$\gamma_*(\vartheta) = \frac{1}{2f^2}\left[f^2b^2 - \sigma^2\left(1 - a^2\right)\right]$$
$$+ \frac{1}{2f^2}\left[\left(\sigma^2\left(1 - a^2\right) - b^2f^2\right)^2 + 4f^2b^2\sigma^2\right]^{1/2}.$$

Consider the problem of approximation of the random function $m_t(\vartheta), t \geq 1$ when one of the parameters is unknown.

6.3.1 Recursive Filter

Suppose that the values $f > 0$, $a^2 \in [0, 1)$ and $\sigma^2 > 0$ are known and the only unknown parameter is $\vartheta = b \in \Theta = (\alpha_b, \beta_b)$, $\alpha_b > 0$. The proposed below construction is a direct analog of similar solutions discussed in preceding chapters for continuous time models. First by observations $X_0, X_1, \ldots, X_{\tau_T}$ we calculate the MME $\vartheta^*_{\tau_T} = b^*_{\tau_T}$, $\tau_T = \left[T^\delta\right]$, $\delta \in (\frac{1}{2}, 1)$ (see (6.20))

$$\vartheta_{\tau_T}^* = 2^{-1/2} f^{-1}\left[\left(\frac{1}{\tau_T}\sum_{s=1}^{\tau_T}[X_s - X_{s-1}]^2 - 2\sigma^2\right)(1+a)\right]^{1/2}.$$

Then we define the One-step MLE-process (see (6.30))

$$\begin{aligned}\vartheta_{t,T}^\star &= \vartheta_{\tau_T}^* + \frac{1}{\mathrm{I}_b(\vartheta_{\tau_T}^*)\left(t-\tau_T\right)}\sum_{s=\tau_T+1}^{t}\left[\frac{\left[X_s - f m_{s-1}(\vartheta_{\tau_T}^*)\right]}{P(\vartheta_{\tau_T}^*)} f\dot m_{s-1}(\vartheta_{\tau_T}^*)\right.\\ &\quad\left.+\left(\left[X_s - f m_{s-1}(\vartheta_{\tau_T}^*)\right]^2 - P(\vartheta_{\tau_T}^*)\right)\frac{\dot P(\vartheta_{\tau_T}^*)}{2P(\vartheta_{\tau_T}^*)^2}\right],\qquad t\in\left[\tau_T+1, T\right].\end{aligned} \tag{6.43}$$

Here $m_{s-1}(\vartheta_{\tau_T}^*, s-1), s \geq \tau_T + 1$ satisfies the eq. (6.42), where b is replaced by $\vartheta_{\tau_T}^*$ and $\dot m_{s-1}(\vartheta_{\tau_T}^*, s-1), s \geq \tau_T + 1$ is obtained by differentiation of (6.42) on b and once more replacing b by $\vartheta_{\tau_T}^*$. Therefore for $s > \tau_T + 1$ we can write

$$m_{s-1}(\vartheta_{\tau_T}^*) = P(\vartheta_{\tau_T}^*)^{-1}\left[a\sigma^2 m_{s-2}(\vartheta_{\tau_T}^*) + af\gamma_*(\vartheta_{\tau_T}^*)X_{s-1}\right], \tag{6.44}$$

$$\begin{aligned}\dot m_{s-1}(\vartheta_{\tau_T}^*) &= P(\vartheta_{\tau_T}^*)^{-2} a\sigma^2\left[P(\vartheta_{\tau_T}^*)\dot m_{s-2}(\vartheta_{\tau_T}^*, s-2) - f^2\dot\gamma_*(\vartheta_{\tau_T}^*)\, m_{s-2}(\vartheta_{\tau_T}^*)\right]\\ &\quad + \frac{af\sigma^2\dot\gamma_*(\vartheta_{\tau_T}^*)}{P(\vartheta_{\tau_T}^*)^2} X_{s-1}.\end{aligned}$$

Here $P(\vartheta) = \sigma^2 + f^2\gamma_*(\vartheta)$ and the Fisher information is

$$\mathrm{I}_b(\vartheta) = \frac{\dot P_b(\vartheta_0)^2\left[P(\vartheta_0)^2 + a^2\sigma^4\right]}{2P(\vartheta_0)^2\left[P(\vartheta_0)^2 - a^2\sigma^4\right]},$$

where $A(\vartheta) = P(\vartheta)^{-1} a\sigma^2$ (see (6.23)).

The adaptive Kalman filter we introduce with the help of the process $m_{t,T}^\star$ defined as follows:

$$m_{t,T}^\star = P(\vartheta_{t-1,T}^\star)^{-1}\left[a\sigma^2 m_{t-1,T}^\star + af\gamma_*(\vartheta_{t-1,T}^\star)X_t\right],\qquad t\in\left[\tau_T+1, T\right]. \tag{6.45}$$

The initial value $m_{\tau_T,T}^\star = m_{\tau_T,T}(\vartheta_{\tau_T}^*)$, where $m_{\tau_T,T}(\vartheta_{\tau_T}^*)$ is solution of the eq. (6.44). We have to compare $m_{t,T}^\star$ with $m(\vartheta_0, t)$ for large values of T. Below we denote $\eta_{t,T} = \sqrt{t}\left[\vartheta_{t,T}^\star - \vartheta_0\right]$ and $\dot B^*(\vartheta_0, \vartheta_0) = -f^{-1}\sqrt{P(\vartheta_0)}\dot A(\vartheta_0) = P(\vartheta_0)^{-3/2} af\sigma^2\dot\gamma_*(\vartheta_0)$.

Theorem 6.3 *Let $t = [\upsilon T]$, $\upsilon \in (0, 1]$, $k = t - \tau_T + 2$ and $T \to \infty$. Then the following relations hold:*

$$\sqrt{t}\left[m^{\star}_{t,T} - m_t(\vartheta_0)\right] = \dot{B}^*(\vartheta_0, \vartheta_0) \sum_{m=0}^{k} A(\vartheta_0)^m \, \eta_{t-m-1,T} \, \zeta_{t-m}(\vartheta_0) + o(1),$$

$$t\mathbf{E}_{\vartheta_0}\left[m^{\star}_{t,T} - m_t(\vartheta_0)\right]^2 \longrightarrow S^{\star}_b(\vartheta_0)^2 \equiv \frac{\dot{B}^*(\vartheta_0, \vartheta_0)^2}{\mathrm{I}_b(\vartheta_0)\left(1 - A(\vartheta_0)^2\right)}. \tag{6.46}$$

Here $o(1)$ and (6.46) converge uniformly on compacts $\mathbb{K} \subset \Theta$.

Proof Recall that

$$m_t(\vartheta_0) = A(\vartheta_0) m_{t-1}(\vartheta_0) + f^{-1}\left[a - A(\vartheta_0)\right] X_t,$$
$$m^{\star}_{t,T} = A(\vartheta^{\star}_{t-1,T}) m^{\star}_{t-1,T} + f^{-1}\left[a - A(\vartheta^{\star}_{t-1,T})\right] X_t.$$

Therefore for $\delta_{t,T} = \sqrt{t}\left[m^{\star}_{t,T} - m_t(\vartheta_0)\right]$ we have the equation

$$\begin{aligned}
\delta_{t,T} &= A(\vartheta^{\star}_{t-1,T})\delta_{t-1,T} + \sqrt{t}\left[A(\vartheta^{\star}_{t-1,T}) - A(\vartheta_0)\right] m_{t-1}(\vartheta_0) \\
&\quad + f^{-1}\sqrt{t}\left[A(\vartheta_0) - A(\vartheta^{\star}_{t-1,T})\right] X_t \\
&= A(\vartheta^{\star}_{t-1,T})\delta_{t-1,T} + f^{-1}\sqrt{P(\vartheta_0)}\sqrt{t}\left[A(\vartheta_0) - A(\vartheta^{\star}_{t-1,T})\right] \zeta_t(\vartheta_0) \\
&= A(\vartheta^{\star}_{t-1,T})\delta_{t-1,T} - f^{-1}\sqrt{P(\vartheta_0)}\dot{A}(\tilde{\vartheta}_{t-1})\sqrt{t}\left[\vartheta^{\star}_{t-1,T} - \vartheta_0\right] \zeta_t(\vartheta_0) \\
&= A(\vartheta_0)\, \delta_{t-1,T} + \dot{B}^*(\vartheta_0, \vartheta_0)\, \eta_{t-1,T}\, \zeta_t(\vartheta_0) + \varepsilon_t, \qquad t \in \left[\tau_T + 1, T\right],
\end{aligned}$$

where $\varepsilon_t = O\left(\vartheta^{\star}_{t-1,T} - \vartheta_0\right) = O\left(T^{-1/2}\right)$.

Let us denote $A \equiv A(\vartheta_0)$, $\dot{B}^* \equiv \dot{B}^*(\vartheta_0, \vartheta_0)$, $\zeta_t = \zeta_t(\vartheta_0)$ and make $k = t - \tau_T + 2$ iterations

$$\begin{aligned}
\delta_{t,T} &= A\, \delta_{t-1,T} + \dot{B}^*\, \eta_{t-1,T}\, \zeta_t + \varepsilon_t \\
&= A^2\, \delta_{t-2,T} + A\dot{B}^*\, \eta_{t-2,T}\, \zeta_{t-1} + \dot{B}^*\, \eta_{t-1,T}\, \zeta_t + A\varepsilon_{t-1} + \varepsilon_t \\
&= \ldots\ldots\ldots\ldots\ldots\ldots\ldots\ldots\ldots\ldots\ldots\ldots\ldots\ldots \\
&= A^{k+1}\delta_{t-k,T} + \dot{B}^* \sum_{m=0}^{k} A^m \eta_{t-m-1,T}\, \zeta_{t-m} + \sum_{m=0}^{k} A^m \varepsilon_{t-m}.
\end{aligned}$$

Therefore for the large values of t and $k = t - \tau_T + 2$ we have the representation,

$$\delta_{t,T}=\dot B^*\sum_{m=0}^{k}A^m\eta_{t-m-1,T}\,\zeta_{t-m}+o\left(1\right).$$

Note that $\eta_{t-m-1,T}$ and ζ_{t-m} are independent. Hence, if $t=[vT]$, $T\to\infty$, then

$$\begin{aligned}\mathbf{E}_{\vartheta_0}\delta_{t,T}^2&=\dot B^{*2}\sum_{m=0}^{k}\sum_{l=0}^{k}A^{m+l}\mathbf{E}_{\vartheta_0}\left[\eta_{t-m-1,T}\,\zeta_{t-m}\eta_{t-l-1,T}\,\zeta_{t-l}\right]+o\left(1\right)\\&=\dot B^{*2}\sum_{m=0}^{k}A^{2m}\mathbf{E}_{\vartheta_0}\left[\eta_{t-m-1,T}^2\zeta_{t-m}^2\right]+o\left(1\right)\\&=\dot B^{*2}\sum_{m=0}^{k}A^{2m}\mathbf{E}_{\vartheta_0}\left[\eta_{t-m-1,T}^2\right]+o\left(1\right)\\&=\frac{\dot B^{*2}}{\mathrm{I}_b\left(\vartheta_0\right)}\sum_{m=0}^{k}A^{2m}\frac{t}{t-m-1}+o\left(1\right)\\&\longrightarrow\frac{\dot B^*\left(\vartheta_0,\vartheta_0\right)^2}{\mathrm{I}\left(\vartheta_0\right)}\sum_{m=0}^{\infty}A^{2m}=\frac{\dot B^*\left(\vartheta_0,\vartheta_0\right)^2}{\mathrm{I}_b\left(\vartheta_0\right)\left(1-A\left(\vartheta_0\right)^2\right)}.\end{aligned}$$

□

Remark 6.9 The adaptive Kalman filter is given by the relations (6.43)–(6.44), where the only One-step MLE-process (6.43) is in nonrecurrent form. Let us write this estimator in recursive form too. Denote

$$\begin{aligned}S_{t,T}(\vartheta_{\tau_T}^*)=\frac{1}{\mathrm{I}_b(\vartheta_{\tau_T}^*)\left(t-\tau_T\right)}\sum_{s=\tau_T+1}^{t}&\left[\frac{\left[X_s-fm_{s-1}(\vartheta_{\tau_T}^*)\right]}{P(\vartheta_{\tau_T}^*)}f\dot m_{s-1}(\vartheta_{\tau_T}^*)\right.\\&\left.+\left(\left[X_s-fm_{s-1}(\vartheta_{\tau_T}^*)\right]^2-P(\vartheta_{\tau_T}^*)\right)\frac{\dot P(\vartheta_{\tau_T}^*)}{2P(\vartheta_{\tau_T}^*)^2}\right].\end{aligned}$$

Then we can write recursive One-step MLE-process as follows:

$$\begin{aligned}\vartheta_{t,T}^{\star}=\vartheta_{\tau_T}^*+S_{t,T}(\vartheta_{\tau_T}^*)=\frac{\vartheta_{\tau_T}^*}{t-\tau_T}+\frac{\left(t-1-\tau_T\right)}{\left(t-\tau_T\right)}\left[\vartheta_{\tau_T}^*+S_{t-1,T}(\vartheta_{\tau_T}^*)\right]\\+\frac{1}{\mathrm{I}_b(\vartheta_{\tau_T}^*)\left(t-\tau_T\right)}\left[\frac{\left[X_t-fm_{t-1}(\vartheta_{\tau_T}^*)\right]}{P(\vartheta_{\tau_T}^*)}f\dot m_{t-1}(\vartheta_{\tau_T}^*)\right.\end{aligned}$$

$$+\left(\left[X_t - fm_{t-1}(\vartheta^*_{\tau_T})\right]^2 - P(\vartheta^*_{\tau_T})\right)\frac{\dot P(\vartheta^*_{\tau_T})}{2P(\vartheta^*_{\tau_T})^2}\Bigg]$$

$$= \frac{\vartheta^*_{\tau_T}}{t-\tau_T} + \left(1 - \frac{1}{t-\tau_T}\right)\vartheta^\star_{t-1,T} + \frac{\left[X_t - fm_{t-1}(\vartheta^*_{\tau_T})\right] f\dot m_{t-1}(\vartheta^*_{\tau_T})}{\mathrm{I}_b(\vartheta^*_{\tau_T})\left(t-\tau_T\right)P(\vartheta^*_{\tau_T})}$$

$$+ \frac{\left(\left[X_t - fm_{t-1}(\vartheta^*_{\tau_T})\right]^2 - P(\vartheta^*_{\tau_T})\right)\dot P(\vartheta^*_{\tau_T})}{2\mathrm{I}_b(\vartheta^*_{\tau_T})\left(t-\tau_T\right)P(\vartheta^*_{\tau_T})^2}, \qquad t \in \left[\tau_T + 1, T\right].$$

Therefore the recursive equation for the One-step MLE-process is

$$\vartheta^\star_{t,T} = \frac{\vartheta^*_{\tau_T}}{t-\tau_T} + \left(1 - \frac{1}{t-\tau_T}\right)\vartheta^\star_{t-1,T} + \frac{\left[X_t - fm_{t-1}(\vartheta^*_{\tau_T})\right] f\dot m_{t-1}(\vartheta^*_{\tau_T})}{\mathrm{I}_b(\vartheta^*_{\tau_T})\left(t-\tau_T\right)P(\vartheta^*_{\tau_T})}$$

$$+ \frac{\left(\left[X_t - fm_{t-1}(\vartheta^*_{\tau_T})\right]^2 - P(\vartheta^*_{\tau_T})\right)\dot P(\vartheta^*_{\tau_T})}{2\mathrm{I}_b(\vartheta^*_{\tau_T})\left(t-\tau_T\right)P(\vartheta^*_{\tau_T})^2}, \qquad t \in \left[\tau_T + 1, T\right]. \tag{6.47}$$

Now the adaptive Kalman filter (6.44)–(6.45), and (6.47) is in recurrent form.

Remark 6.10 The cases $\vartheta = f$, $\vartheta = a$, and $\vartheta = \sigma^2$ can be studied similarly. If we consider the three-dimensional cases, say, $\vartheta = \left(f, a, \sigma^2\right)$, then the construction of adaptive Kalman filter is close to the given above. For One-step MLE-process we need the preliminary MMEs (6.31)–(6.32) and the vector of derivatives (6.33). The Fisher information matrix is given in (6.29) and the adaptive Kalman filter has the same form as in (6.45), where f, a, σ^2 are replaced by the corresponding components of One-step MLE-process (6.34), i.e.,

$$m^\star_{t,T} = \frac{\theta^\star_{2,t-1,T}\left[\theta^\star_{3,t-1,T} m^\star_{t-1,T} + \theta^\star_{1,t-1,T}\gamma_*(\vartheta^\star_{t-1,T})X_t\right]}{\theta^\star_{3,t-1,T} + \left(\theta^\star_{1,t-1,T}\right)^2\gamma_*(\vartheta^\star_{t-1,T})}, \qquad t \in \left[\tau_T + 1, T\right].$$

Of course, the asymptotics of the difference $m^\star_{t,T} - m\left(\vartheta_0, t\right)$ can be described too using Taylor expansions of the function $\gamma_*\left(\vartheta\right)$ and the limits of $\theta^\star_{i,t-1,T}$, $i = 1, 2, 3$, but the arising expressions are cumbersome, and we will not do it here.

Remark 6.11 It is possible as well to verify the asymptotic normality

$$\sqrt{t}\left(m^\star_{t,T} - m\left(\vartheta_0, t\right)\right) \Longrightarrow \mathcal{N}\left(0, S^\star_b\left(\vartheta_0\right)^2\right).$$

6.3.2 Asymptotic Efficiency

We have the same system (6.1) and (6.2), where the parameters f, a, b, σ^2 satisfy the condition $\mathscr{M}_0$.

The adaptive filter is given by (6.44), and we would like to know if the error of approximation $\mathbf{E}_\vartheta \left|m^\star_{t,T} - m(\vartheta, t)\right|^2$ is asymptotically minimal? As usual in such situations we propose a lower minimax bound on the risks of all estimators $\bar m_t$ supposing that $\bar m_t$ is based on the observations up to time t, i.e., these estimators depend on $X^t = (X_s, s = 0, 1, \ldots, t)$.

Recall some notations

$$A(\vartheta_0) = \frac{a\sigma^2}{\sigma^2 + f^2\gamma_*(\vartheta_0)}, \qquad B^*(\vartheta_0, \vartheta_0) = \frac{a\sigma^2 f\dot\gamma_*(\vartheta_0)}{\left[\sigma^2 + f^2\gamma_*(\vartheta_0)\right]^{3/2}}$$

and the equations for $\dot m(\vartheta_0, \cdot)$

$$\dot m(\vartheta, t) = A(\vartheta_0)\,\dot m(\vartheta_0, t-1) + B^*(\vartheta_0, \vartheta_0)\,\zeta_t(\vartheta_0).$$

Recall as well the asymptotic representations for $\eta^\star_{t,T}(\vartheta_0) = \sqrt{vT}\left(\vartheta^\star_{t,T} - \vartheta_0\right)$, $t = vT, v \in (0, 1]$

$$\eta^\star_{t,T}(\vartheta_0) = \frac{\mathrm{I}(\vartheta_0)^{-1}}{\sqrt{tP(\vartheta_0)}} \sum_{s=1}^{t} \Big[\zeta_s(\vartheta_0)\, f\dot m(\vartheta_0, s-1) + \left[\zeta_s(\vartheta_0)^2 - 1\right] \frac{\dot P(\vartheta_0)}{2\sqrt{P(\vartheta_0)}}\Big] + o(1). \tag{6.48}$$

Similar representations for the MLE $\hat\eta_t(\vartheta_0) = \sqrt{t}\left(\hat\vartheta_t - \vartheta_0\right)$, $v \in (0, 1]$ and BE $\tilde\eta_t(\vartheta_0) = \sqrt{t}\left(\bar\vartheta_t - \vartheta_0\right)$, $t = vT, v \in (0, 1]$ are

$$\hat\eta_t(\vartheta_0) = \frac{\mathrm{I}(\vartheta_0)^{-1}}{\sqrt{tP(\vartheta_0)}} \sum_{s=1}^{t} \Big[\zeta_s(\vartheta_0)\, f\dot m(\vartheta_0, s-1) + \left[\zeta_s(\vartheta_0)^2 - 1\right] \frac{\dot P(\vartheta_0)}{2\sqrt{P(\vartheta_0)}}\Big] + o(1),$$

$$\tilde\eta_t(\vartheta_0) = \frac{\mathrm{I}(\vartheta_0)^{-1}}{\sqrt{tP(\vartheta_0)}} \sum_{s=1}^{t} \Big[\zeta_s(\vartheta_0)\, f\dot m(\vartheta_0, s-1) + \left[\zeta_s(\vartheta_0)^2 - 1\right] \frac{\dot P(\vartheta_0)}{2\sqrt{P(\vartheta_0)}}\Big] + o(1). \tag{6.49}$$

The properties (6.35)–(6.38) of the normalized LR $Z_T(\cdot)$ established in the Theorem 6.2 correspond to the conditions of Theorems 3.1.2 and 3.2.2 in [73], where such representations were proved in the general case.

Introduce the limit (6.46)

$$S_b^\star(\vartheta)^2 = \lim_{T\to\infty} \mathbf{E}_\vartheta\left[\dot m(\vartheta,t)^2\,\tilde\eta_t(\vartheta)^2\right] = \frac{\dot B^*(\vartheta,\vartheta)^2}{\mathrm{I}(\vartheta)\left(1-A(\vartheta)^2\right)}. \tag{6.50}$$

As the asymptotic representations of $\tilde\eta_{t,T}(\vartheta)$ and $\eta_{t,T}^\star(\vartheta)$ are similar (see (6.48) and (6.49)) hence the limits of (6.48) and (6.50) coincide too.

Theorem 6.4 *Let the conditions of Theorem 6.2 be fulfilled. Then we have the following lower minimax bound : for any estimator $\bar m_{t,T}$ of $m(\vartheta,t)$ (below $t=vT$)*

$$\lim_{\nu\to 0}\varliminf_{T\to\infty}\sup_{|\vartheta-\vartheta_0|\le\nu} t\mathbf{E}_\vartheta\left|\bar m_{t,T}-m(\vartheta,t)\right|^2 \ge S_b^\star(\vartheta_0)^2.$$

Proof The proof was already discussed before in the Introduction. It is quite short, and we repeat it here for convenience. We have the elementary estimate

$$\sup_{|\vartheta-\vartheta_0|\le\nu}\mathbf{E}_\vartheta\left|\bar m_{t,T}-m(\vartheta,t)\right|^2 \ge \int_{\vartheta_0-\nu}^{\vartheta_0+\nu}\mathbf{E}_\vartheta\left|\bar m_{t,T}-m(\vartheta,t)\right|^2 p_\nu(\vartheta)\,\mathrm{d}\vartheta.$$

Here the function $p_\nu(\vartheta)$, $\vartheta_0-\nu<\vartheta<\vartheta_0+\nu$ is a positive continuous density on the interval $[\vartheta_0-\nu,\vartheta_0+\nu]$. If we denote $\tilde m_t$ Bayesian estimator of $m(\vartheta,t)$, which corresponds to this density $p_\nu(\cdot)$, then

$$\tilde m_t = \int_{\vartheta_0-\nu}^{\vartheta_0+\nu} m(\theta,t)\,p_\nu\left(\theta|X^t\right)\mathrm{d}\theta,\qquad p_\nu\left(\theta|X^t\right) = \frac{p_\nu(\theta)\,L\left(\theta,X^t\right)}{\int_{\vartheta_0-\nu}^{\vartheta_0+\nu}p_\nu(\theta)\,L\left(\theta,X^t\right)\mathrm{d}\theta}$$

and

$$\int_{\vartheta_0-\nu}^{\vartheta_0+\nu}\mathbf{E}_\vartheta\left|\bar m_{t,T}-m(\vartheta,t)\right|^2 p_\nu(\vartheta)\,\mathrm{d}\vartheta \ge \int_{\vartheta_0-\nu}^{\vartheta_0+\nu}\mathbf{E}_\vartheta\left|\tilde m_t-m(\vartheta,t)\right|^2 p_\nu(\vartheta)\,\mathrm{d}\vartheta.$$

The asymptotic behavior of BE $\tilde m_t$ can be described as follows (below $\theta_u=\vartheta+\varphi_t u$, $\varphi_t=t^{-1/2}$, $\mathbb{U}_\nu=\left(\sqrt{t}\,(\vartheta_0-\nu-\vartheta),\sqrt{t}\,(\vartheta_0+\nu-\vartheta)\right)$):

$$\tilde m_t = \frac{\int_{\vartheta_0-\nu}^{\vartheta_0+\nu}m(\theta,t)\,p_\nu(\theta)\,L\left(\theta,X^t\right)\mathrm{d}\theta}{\int_{\vartheta_0-\nu}^{\vartheta_0+\nu}p_\nu(\theta)\,L\left(\theta,X^t\right)\mathrm{d}\theta} = \frac{\int_{\mathbb{U}_\nu}m(\theta_u,t)\,p_\nu(\theta_u)\,L\left(\theta_u,X^t\right)\mathrm{d}u}{\int_{\mathbb{U}_\nu}p_\nu(\theta_u)\,L\left(\theta_u,X^t\right)\mathrm{d}u}$$

$$= m(\vartheta,t)+\varphi_t\dot m(\vartheta,t)\frac{\int_{\mathbb{U}_\nu}u\,p_\nu(\theta_u)\,\frac{L(\theta_u,X^t)}{L(\vartheta,X^t)}\mathrm{d}u}{\int_{\mathbb{U}_\nu}p_\nu(\theta_u)\,\frac{L(\theta_u,X^t)}{L(\vartheta,X^t)}\mathrm{d}u}\left(1+o(1)\right)$$

$$= m(\vartheta, t) + \varphi_t \dot{m}(\vartheta, t) \frac{\int_{\mathbb{U}_\nu} u p_\nu(\vartheta) Z_t(u)\,\mathrm{d}u}{\int_{\mathbb{U}_\nu} p_\nu(\vartheta) Z_t(u)\,\mathrm{d}u} (1 + o(1)).$$

Hence

$$\sqrt{t}\,(\tilde{m}_t - m(\vartheta, t)) = \dot{m}(\vartheta, t) \frac{\int_{\mathbb{U}_\nu} u Z_t(u)\,\mathrm{d}u}{\int_{\mathbb{U}_\nu} Z_t(u)\,\mathrm{d}u} (1 + o(1))$$

$$= \dot{m}(\vartheta, t) \frac{\Delta_t\left(\vartheta, X^t\right)}{\mathrm{I}(\vartheta)} (1 + o(1)),$$

where

$$\Delta_t\left(\vartheta, X^t\right) = \frac{1}{\sqrt{tP(\vartheta_0)}} \sum_{s=1}^{t} \left[\zeta_s(\vartheta_0) f \dot{m}_{s-1}(\vartheta_0) + \left[\zeta_s(\vartheta_0)^2 - 1\right] \frac{\dot{P}(\vartheta_0)}{2\sqrt{P(\vartheta_0)}} \right].$$

Recall that $Z_t(u) \Rightarrow Z(u) = \exp\left(u\Delta_t(\vartheta) - \frac{u^2}{2}\mathrm{I}(\vartheta)\right)$ and

$$\frac{\int_{\mathcal{R}} u Z(u)\,\mathrm{d}u}{\int_{\mathcal{R}} Z(u)\,\mathrm{d}u} = \frac{\Delta_t(\vartheta)}{\mathrm{I}(\vartheta)}.$$

Moreover the uniform on compacts $\mathbb{K} \subset (\vartheta_0 - \nu, \vartheta_0 + \nu)$ convergence of moments of BE allows us to write

$$t\mathbf{E}_\vartheta\,(\tilde{m}_t - m(\vartheta, t))^2 \to \lim_{t\to\infty} t\mathbf{E}_\vartheta \left[\dot{m}(\vartheta, t)^2 (\tilde{\vartheta}_t - \vartheta)^2\right] = S_b^\star(\vartheta)^2$$

holds too.

The detailed proof of the above written relations can be found in the proofs of Theorems 3.2.1 and 3.2.2 in [73].

Therefore

$$t \int_{\vartheta_0-\nu}^{\vartheta_0+\nu} \mathbf{E}_\vartheta \left|\tilde{m}_t - m(\vartheta, t)\right|^2 p_\nu(\vartheta)\,\mathrm{d}\vartheta \longrightarrow \int_{\vartheta_0-\nu}^{\vartheta_0+\nu} S_b^\star(\vartheta)^2\, p_\nu(\vartheta)\,\mathrm{d}\vartheta$$

and as $\nu \to 0$

$$\int_{\vartheta_0-\nu}^{\vartheta_0+\nu} S_b^\star(\vartheta)^2\, p_\nu(\vartheta)\,\mathrm{d}\vartheta \longrightarrow S_b^\star(\vartheta_0)^2.$$

□

We call the estimator $m_{t,T}^\circ$, $\tau_T < t \leq T$ *asymptotically efficient* if for all $\vartheta_0 \in \Theta$, $t = vT$, any $v \in [\varepsilon_0, 1]$

$$\lim_{\nu\to 0}\lim_{T\to\infty}\sup_{|\vartheta-\vartheta_0|\leq\nu} t\mathbf{E}_\vartheta \left|m^\circ_{t,T} - m\left(\vartheta, t\right)\right|^2 = S^\star_b\left(\vartheta_0\right)^2.$$

Here $\varepsilon_0 \in (0, 1)$.

Theorem 6.5 *The estimator* $m^\star_t$, $\tau_T < t \leq T$ *is asymptotically efficient.*

The proof follows from the uniform convergence (6.46) of Theorem 6.3.

Remark 6.12 It is possible to realize similar construction (adaptive filter + lower bound + asymptotic efficiency) for all other parameters, including the multidimensional parameter cases.

Chapter 7
Miscellaneous

We consider several problems where the approach developed in this work can be useful for their solutions. The first one is related to localizing a source on a plane using observations from K detectors. It is assumed that the detectors observe Gaussian signals from the source in the presence of white Gaussian noise. The unknown parameters are the coordinates of the source. Several variations of the problem are studied. First, we assume that the source is fixed and starts emitting at some known moment ($t = 0$). The position of the source is determined by the times required for the signals to reach the detectors. The properties of the estimators of the position depend significantly on the form of the signal front (smooth function, function with cusp-type singularity, or discontinuous function), and we describe the properties of the MLE and BE estimators for such models in the asymptotic of small noise in both equations. Then, we consider a similar problem with a moving source and discuss other possible generalizations. In the next model the deterministic signal depending on some unknown finite dimensional parameter is present in the hidden equation, and we study the problem of its estimation. The partially observed system related to the security prices process depending on unknown parameters is considered in the Sect. 7.3. The equations of adaptive filtering and adaptive extrapolation are proposed. Further the change point problems for partially observed linear system are discussed in three situations: change point in the observation equations, in the state equation, and in both equations. The properties of different estimators and the possibilities of adaptive filtering are discussed. In the last section the problem concerns the approximation of the solution of backward stochastic differential equation (BSDE) in the situation where the solution of the forward equation with unknown volatility is observed in WGN. The proposed approximation algorithm is based on the One-step MLE-process. The question of asymptotic optimality of such approximation is discussed.

Y. A. Kutoyants, *Hidden Markov Processes and Adaptive Filtering*, Springer Series in Statistics, https://doi.org/10.1007/978-3-032-00052-1_7

7.1 Source Localization

7.1.1 Fixed Source

Model of Observations

Consider the problem of estimation of the position $\vartheta_0 = (x_0, y_0)^\top$ of the source $\mathcal{S}_0$ by the observations of the signals from this source received by K detectors $\mathcal{D}_1, \ldots, \mathcal{D}_K$ on the plane (see Fig. 7.1).

If we denote $\mathrm{D}_k = (x_k, y_k)^\top \in \mathcal{R}^2$ the coordinates of $\mathcal{D}_k$ and suppose that the source starts emission at the moment $t = 0$, then the signal arrives at this detector at the moment $\tau_k(\vartheta_0) = \nu^{-1} \|\mathrm{D}_k - \vartheta_0\|$. Here $\nu > 0$ is the rate of propagation of the signals and $\|\cdot\|$ is Euclidean norm in $\mathcal{R}^2$. The set $\Theta \subset \mathcal{R}^2$ is supposed to be open, convex, and bounded.

The k-th detector receives the signal $Y_k = (Y_k(t), 0 \le t \le T)$ from the source $\mathcal{S}_0$ and additive Gaussian noise according to the equation

$$\mathrm{d}X_k(t) = f_k(t)\,\bar{\psi}(t - \tau_k(\vartheta_0))\, Y_k(t)\,\mathrm{d}t + \varepsilon\sigma_k(t)\,\mathrm{d}W_k(t), \qquad X_k(0) = 0. \tag{7.1}$$

Here $f_k(\cdot), \sigma_k(\cdot)$, and $\bar{\psi}(\cdot)$ are known functions and $W_k(\cdot), k = 1, \ldots, K$ are independent Wiener processes.

The signals $Y_k(\cdot), k = 1, \ldots, K$ satisfy the linear stochastic differential equations

$$\mathrm{d}Y_k(t) = a_k(t)\, Y_k(t)\,\mathrm{d}t + \varepsilon b_k(t)\,\mathrm{d}V_k(t), \qquad Y_k(0) = y_{k,0} \neq 0. \tag{7.2}$$

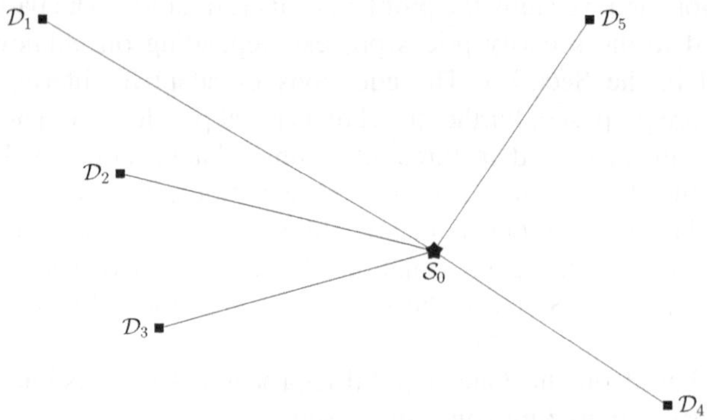

Fig. 7.1 Model of observations. $\mathcal{S}_0$ is position of the source and $\mathcal{D}_k, k = 1, \ldots, 5$ are positions of the sensors

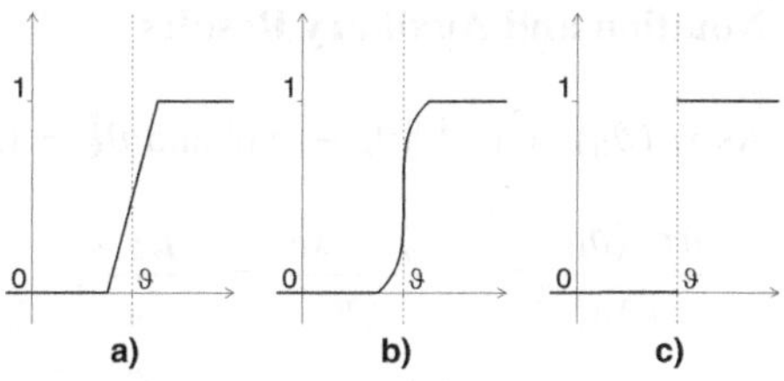

Fig. 7.2 Examples: (**a**) $\psi_\delta(\cdot)$, (**b**) $\psi_{\delta,\kappa}(\cdot)$ and (**c**) $\psi(\cdot)$

The functions $a_k(\cdot)$ and $b_k(\cdot)$ are known, and the Wiener processes $V_k(\cdot), k = 1, \ldots, K$ are independent. The Wiener processes $W_k(\cdot), k = 1, \ldots, K$ and $V_k(\cdot), k = 1, \ldots, K$ are supposed to be independent too.

The parameter $\varepsilon > 0$ controls the level of noises in both equations. We study the properties of estimators of ϑ_0 in the asymptotic of *small noise*, i.e., as $\varepsilon \to 0$. This is equivalent to the situation with *large signal*, which is rather reasonable in many real situations. This problem is quite close to the inverse problem, where we have K sources $\mathcal{D}_1, \ldots, \mathcal{D}_K$ of signals with known positions and known moments of emission and one detector $\mathcal{S}_0$. The detector receives K signals X^K and has to estimate its own position. This is a typical situation in the global positioning system (GPS/INS). The algorithms calculating the positions of different objects (cars, jets, ships, etc.) used in GPS/ISN are based essentially on the adaptive Kalman filtering theory, see, e.g., [2, 63, 72, 146, 191] and references therein.

The function $\bar\psi(t) \in [0, 1]$, $\bar\psi(t) = 0$ for $t < 0$ and describes the front of the signal just after it arrives; more precisely, it describes the transition of the value of the received signal from 0 ($\bar\psi(0) = 0$) at the moment of arriving to its real value ($\bar\psi(\delta) = 1$). Here $\delta > 0$ is the width of the front. We consider three different cases: *smooth* $\psi_\delta(t) = t\delta^{-1}\mathbb{I}_{\{0\le t\le\delta\}} + \mathbb{I}_{\{t>\delta\}}$, *cusp* type $\psi_{\delta,\kappa}(t) = \frac{1}{2}\left(1 + \text{sgn}(2t-\delta)\left|\frac{2t}{\delta} - 1\right|^{\kappa}\right)\mathbb{I}_{\{0\le t\le\delta\}} + \mathbb{I}_{\{t>\delta\}}$, and *change-point* type $\psi(t) = \mathbb{I}_{\{t>0\}}$ ($\delta = 0$) respectively. The examples of such functions are given in Fig. 7.2.

The parameter $\delta > 0$ is known and small. In the cusp case $\kappa \in (0, \frac{1}{2})$.

In this work we consider the problem of estimation ϑ_0 by the observations $X^T = (X_1, \ldots, X_K)$, where $X_k = (X_k(t), 0 \le t \le T)$. The processes $Y_k(\cdot)$ are nonobservable.

We suppose that the following conditions are always satisfied in this work.

Conditions $\mathscr{N}$.

$\mathscr{N}_1$. *The functions $f_k(\cdot), \sigma_k(\cdot), k = 1, \ldots, K$ are bounded and separated from 0 and the functions , $a_k(\cdot), b_k(\cdot), k = 1, \ldots, K$ are bounded.*

$\mathscr{N}_2$. *The set $\Theta \subset \mathcal{R}^2$ is open, convex, bounded, and such that*

$$0 < \min_{1\le k\le K} \inf_{\vartheta\in\Theta} \tau_k(\vartheta) < \max_{1\le k\le K} \sup_{\vartheta\in\Theta} \tau_k(\vartheta) < T.$$

$\mathscr{N}_3$. *There are at least three detectors which are not on the same line.*

By the condition 2 all $\tau_k(\vartheta) \in (\tau_\alpha, \tau_\beta)$, where $0 < \tau_\alpha < \tau_\beta < T$.

Notation and Auxiliary Results

As $\tau_k(\vartheta_0) = \nu^{-1}\|\vartheta_k - \vartheta_0\|$ and $\vartheta_k = (x_k, y_k)^\top$, $\vartheta_0 = (x_0, y_0)^\top$ we can write,

$$\frac{\partial \tau_k(\vartheta_0)}{\partial x_0} = -\frac{x_k - x_0}{\nu\rho_k} = -\frac{\mu_{k,x}}{\nu}, \qquad \frac{\partial \tau_k(\vartheta_0)}{\partial y_0} = -\frac{y_k - y_0}{\nu\rho_k} = -\frac{\mu_{k,y}}{\nu},$$

$$\rho_k = \|\vartheta_k - \vartheta_0\|, \qquad \mu_k = \left(\mu_{k,x}, \mu_{k,y}\right)^\top, \qquad \|\mu_k\| = 1,$$

$$h_k(t) = \frac{f_k(t)}{\sigma_k(t)}, \quad \mathbb{B}_k^+ = (w: \quad \langle\mu_k, w\rangle > 0), \ \mathbb{B}_k^- = (w: \quad \langle\mu_k, w\rangle < 0). \tag{7.3}$$

Here $a^\top$ means transposition of the vector a and $\langle a, b\rangle$ is scalar product. We have

$$\tau_k(\vartheta_0 + \nu\varphi_\varepsilon w) = \tau_k(\vartheta_0) - \varphi_\varepsilon\langle\mu_k, w\rangle + O\left(\varphi_\varepsilon^2\right), \qquad w \in \mathcal{R}^2, \quad \varphi_\varepsilon \to 0.$$

The random processes $m_k(\tau_k, \cdot)$, where $\tau_k = \tau_k(\vartheta)$, satisfy Kalman-Bucy filtration equations (see (1.31) and (1.32))

$$m_k(\tau_k, t) = y_k(t), \qquad 0 \le t \le \tau_k(\vartheta),$$

$$\mathrm{d}m_k(\tau_k, t) = \left[a_k(t) - \gamma_k(\tau_k, t)\, h_k(t)^2\, \bar\psi(t - \tau_k)^2\right] m_k(\tau_k, t)\,\mathrm{d}t$$

$$+ \frac{\gamma_k(\tau_k, t)\, h_k(t)\, \bar\psi(t - \tau_k)}{\sigma_k(t)}\,\mathrm{d}X_k(t), \quad \tau_k(\vartheta) \le t \le T. \tag{7.4}$$

Here we denoted $y_k(t) = \mathbf{E}_\vartheta\left(Y_k(t)\,|X_k^t\right)$ solution of equation (7.2) as $\varepsilon = 0$

$$y_k(t) = y_{k,0}\exp\left\{\int_0^t a_k(s)\,\mathrm{d}s\right\}, \quad 0 \le t \le T.$$

We have the representation

$$\mathrm{d}X_t = f_k(t)\,\bar\psi(t - \tau_k(\vartheta_0))\, m_k(\vartheta_0, t)\,\mathrm{d}t + \varepsilon\sigma_k(t)\,\mathrm{d}\bar W_k(t), \quad X_0 = 0,$$

where ϑ_0 is the true value and $\bar W_k(\cdot)$ is *innovation* Wiener process defined by Eq. (7.5). The function

$$\gamma_k(\tau_k, t) = \varepsilon^{-2}\mathbf{E}_\vartheta\left(m_k(\tau_k, t) - Y_k(t)\right)^2$$

is solution of the Riccati equation

$$\gamma_k(\tau_k, t) = \int_0^t \exp\left\{2\int_s^t a_k(r)\,\mathrm{d}r\right\} b_k(s)^2\,\mathrm{d}s, \quad 0 \le t \le \tau_k(\vartheta),$$

$$\frac{\partial \gamma_k(\tau_k,t)}{\partial t} = 2a_k(t)\,\gamma_k(\tau_k,t) + b_k(t)^2 - \gamma_k(\tau_k,t)^2\, f_k(t)^2\, \bar{\psi}(t-\tau_k)^2, \quad \tau_k(\vartheta) \le t \le T.$$

Note that the functions $\gamma_k(\tau_k,\cdot), k = 1,\ldots,K$ are bounded and do not depend on ε.

We denote $x_k(\cdot,\vartheta_0)$, the limit ($\varepsilon = 0$) deterministic function of the random process (7.1). Then we obtain the following expression

$$x_k(t,\vartheta_0) = \mathbb{I}_{\{t>\tau_k(\vartheta_0)\}} \int_{\tau_k(\vartheta_0)}^{t} f_k(s)\, y_k(s)\, \bar{\psi}(s-\tau_k(\vartheta_0))\,\mathrm{d}s.$$

Let us study the random process $m_k(\tau_k,t)$ as function of τ_k. Consider the difference $\Delta m_k(\tau,t) = m_k(\tau,t) - m_k(\tau_0,t)$, $\Delta\gamma_k(\tau,t) = \gamma_k(\tau,t) - \gamma_k(\tau_0,t)$ where $\tau = \tau_k(\vartheta), \tau_0 = \tau_k(\vartheta_0)$. We take the *worst* (discontinuous) function $\bar{\psi}(t-\tau) = \psi(t-\tau) = \mathbb{I}_{\{t\ge\tau\}}$. We omit for instance the index k in other functions too.

Recall that for $t \in [0,T]$ we have the equality

$$\mathrm{d}m(\tau_0,t) = a(t)\,m(\tau_0,t)\,\mathrm{d}t + \varepsilon\gamma(\tau_0,t)\,h(t)\,\mathbb{I}_{\{t\ge\tau_0\}}\mathrm{d}\bar{W}(t), \tag{7.5}$$

with initial value $y_0 \neq 0$.

Lemma 7.1 *Let the conditions $\mathscr{N}$ hold and $\tau < \tau_0$, then*

$$\mathbf{E}_{\vartheta_0}|\Delta m(\tau,t)|^2 \le C\,|\tau_0-\tau|\,\mathbb{I}_{\{\tau\le t\le\tau_0\}} + \varepsilon^2 C\,|\tau-\tau_0|^2\,\mathbb{I}_{\{\tau_0\le t\le T\}}, \tag{7.6}$$

$$|\gamma(\tau,t) - \gamma(\tau_0)| \le C\,|\tau_0-\tau|\,\mathbb{I}_{\{t\ge\tau\}}, \tag{7.7}$$

where the constants $C > 0$ do not depend on ε.

Proof Suppose that $\tau < \tau_0$. The case $\tau > \tau_0$ can be considered by a similar way. From (7.4) and (7.5) we obtain

$$\Delta m(\tau,t) = [m(\tau,t) - y(t)]\,\mathbb{I}_{\{\tau\le t\le\tau_0\}}, \qquad 0 \le t \le \tau_0,$$

$$\mathrm{d}\Delta m(\tau,t) = S(\tau,t)\,\Delta m(\tau,t)\,\mathrm{d}t + \varepsilon h(t)\,\Delta\gamma(\tau,t)\;\mathrm{d}\bar{W}(t), \quad \tau_0 < t \le T,$$

where we put $S(\tau,t) = a(t) - \gamma(\tau,t)\,h(t)^2$.

Then we obtain the representation

$$\Delta m(\tau,t) = [m(\tau,t) - y(t)]\,\mathbb{I}_{\{\tau\le t\le\tau_0\}} + \mathbb{I}_{\{t>\tau_0\}}\,[m(\tau,\tau_0) - y(\tau_0)]\,e^{\int_{\tau_0}^{t} S(\tau,v)\mathrm{d}v}$$

$$+\,\varepsilon\,\mathbb{I}_{\{t>\tau_0\}}\int_{\tau_0}^{t} e^{\int_s^t S(\tau,v)\mathrm{d}v}h\left(s\right)\Delta\gamma\left(\tau,s\right)\,\mathrm{d}\bar{W}\left(s\right), \tag{7.8}$$

where $\Delta\gamma\left(\tau,t\right)=\gamma\left(\tau,t\right)-\gamma\left(\tau_0,t\right)$. For the first term and $t\in\left[\tau,\tau_0\right]$ we have

$$\begin{aligned}m\left(\tau,t\right)-y\left(t\right)&=y\left(\tau\right)e^{\int_\tau^t S(\tau,v)\mathrm{d}v}-y\left(t\right)\\&\quad+\varepsilon\int_\tau^t e^{\int_\tau^s S(\tau,v)\mathrm{d}v}\gamma\left(\tau,s\right)h\left(s\right)\mathrm{d}\bar{W}\left(s\right)\\&=\varepsilon\int_\tau^t e^{\int_\tau^s S(\tau,v)\mathrm{d}v}\gamma\left(\tau,s\right)h\left(s\right)\mathrm{d}\bar{W}\left(s\right).\end{aligned}$$

Let us write the corresponding equation for $\Delta\gamma\left(\tau,t\right)$. Denote

$$g\left(t\right)=\int_0^t e^{2\int_s^t a(v)\mathrm{d}v}b\left(s\right)^2\mathrm{d}s.$$

Then we have (see (7.6)) $\gamma\left(\tau_0,t\right)=g\left(t\right),0\leq t\leq\tau_0$,

$$\frac{\partial\gamma\left(\tau_0,t\right)}{\partial t}=2a\left(t\right)\gamma\left(\tau_0,t\right)+b\left(t\right)^2-h\left(t\right)^2\gamma\left(\tau_0,t\right)^2,\qquad\tau_0\leq t\leq T$$

and $\gamma\left(\tau,t\right)=g\left(t\right),0\leq t\leq\tau$,

$$\frac{\partial\gamma\left(\tau,t\right)}{\partial t}=2a\left(t\right)\gamma\left(\tau,t\right)+b\left(t\right)^2-h\left(t\right)^2\gamma\left(\tau,t\right)^2,\qquad\tau\leq t\leq T.$$

As $\tau_0>\tau$ we obtain: $\Delta\gamma\left(\tau,t\right)=0,0\leq t\leq\tau$, $\Delta\gamma\left(\tau,t\right)=\gamma\left(\tau,t\right)-g\left(t\right),\tau\leq t\leq\tau_0$,

$$\frac{\partial\Delta\left(\tau,t\right)}{\partial t}=2a\left(t\right)\Delta\left(\tau,t\right)-h\left(t\right)^2\left[\gamma\left(\tau,t\right)+\gamma\left(\tau_0,t\right)\right]\Delta\gamma\left(\tau,t\right),\ \ \tau_0\leq t\leq T$$

with initial value $\Delta\gamma\left(\tau,\tau_0\right)=\gamma\left(\tau,\tau_0\right)-g\left(\tau_0\right)$. Hence for $t\in\left[\tau_0,T\right]$ we have

$$\Delta\gamma\left(\tau,t\right)=\Delta\gamma\left(\tau,\tau_0\right)\exp\left\{\int_{\tau_0}^t\left[2a\left(s\right)-h\left(s\right)^2\left[\gamma\left(\tau,t\right)+\gamma\left(\tau_0,t\right)\right]\right]\mathrm{d}s\right\}$$

and $\left|\Delta\gamma\left(\tau,t\right)\right|\leq\left|\gamma\left(\tau,\tau_0\right)-g\left(\tau_0\right)\right|$.

Recall that the function $\gamma\left(\tau,t\right)$ is bounded. On the interval $\left[\tau,\tau_0\right]$ we have

$$\frac{\partial\Delta\gamma\left(\tau,t\right)}{\partial t}=2a\left(t\right)\Delta\gamma\left(\tau,t\right)-\gamma\left(\tau,t\right)^2h\left(t\right)^2,\qquad\Delta\gamma\left(\tau,\tau\right)=0$$

and

$$\Delta\gamma\left(\tau,\tau_0\right)=-\int_{\tau}^{\tau_0}e^{2\int_s^t a(v)\mathrm{d}v}\gamma\left(\tau,s\right)^2h\left(s\right)^2\mathrm{d}s.$$

Therefore $\left|\Delta\gamma\left(\tau,\tau_0\right)\right|\leq C\left|\tau-\tau_0\right|, \qquad \left|\Delta\gamma\left(\tau,t\right)\right|\leq C\left|\tau-\tau_0\right|.$ □

This estimate and (7.8) allow us to write (7.6).

MLE and BE

Under the made assumptions the measures corresponding to the observations X^T for different values of $\vartheta\in\Theta$ are equivalent and the likelihood ratio function $L\left(\vartheta,X^K\right), \vartheta\in\Theta$ is given by the formula

$$\ln L\left(\vartheta,X^K\right)=\sum_{k=1}^{K}\int_0^T\frac{G_k\left(\tau_k\left(\vartheta\right),t\right)}{\varepsilon^2\sigma_k\left(t\right)}\,\mathrm{d}X_k\left(t\right)-\sum_{k=1}^{K}\int_0^T\frac{G_k\left(\tau_k\left(\vartheta\right),t\right)^2}{2\varepsilon^2}\,\mathrm{d}t. \tag{7.9}$$

We use notations: $G_k\left(\tau_k,t\right)=f_k\left(t\right)M_k\left(\tau_k\left(\vartheta\right),t\right)\sigma_k\left(t\right)^{-1}=h_k\left(t\right)M_k\left(\tau_k\left(\vartheta\right),t\right)$, $\tau_k=\tau_k\left(\vartheta\right)$, $M_k\left(\tau_k,t\right)=\bar\psi\left(t-\tau_k\right)m_k\left(\tau_k,t\right)$, where $m_k\left(\tau_k,\cdot\right)$ are conditional expectations

$$m_k\left(\tau_k,t\right)=\mathbf{E}_\vartheta\left(Y_k\left(t\right)|X_k\left(s\right),0\leq s\leq t\right).$$

The MLE $\hat\vartheta_\varepsilon$ and BE $\tilde\vartheta_\varepsilon$ for quadratic loss function are defined by the relations

$$L(\hat\vartheta_\varepsilon,X^K)=\sup_{\vartheta\in\Theta}L\left(\vartheta,X^K\right),\qquad \tilde\vartheta_\varepsilon=\frac{\displaystyle\int_\Theta\vartheta p\left(\vartheta\right)L\left(\vartheta,X^K\right)\mathrm{d}\vartheta}{\displaystyle\int_\Theta p\left(\vartheta\right)L\left(\vartheta,X^K\right)\mathrm{d}\vartheta}. \tag{7.10}$$

Here $p\left(\vartheta\right), \vartheta\in\Theta$ is prior density which is supposed to be known, continuous, and positive function on Θ.

Below we examine the asymptotic ($\varepsilon\rightarrow 0$) behavior of the MLE $\hat\vartheta_\varepsilon$ and BE $\tilde\vartheta_\varepsilon$. It is shown that the mean squared error $\mathbf{E}_{\vartheta_0}\|\hat\vartheta_\varepsilon-\vartheta_0\|^2\sim C\varepsilon^\gamma$ where $\gamma>0$ depends on the regularity of the front of the signals. We show that the rates are: *in smooth case* $\mathbf{E}_{\vartheta_0}\|\hat\vartheta_\varepsilon-\vartheta_0\|^2\sim C\varepsilon^2$, *in cusp-type case* $\mathbf{E}_{\vartheta_0}\|\hat\vartheta_\varepsilon-\vartheta_0\|^2\sim C\varepsilon^{\frac{4}{2\kappa+1}}$, and *in change-point case* $\mathbf{E}_{\vartheta_0}\|\hat\vartheta_\varepsilon-\vartheta_0\|^2\sim C\varepsilon^4$.

The proofs in all three cases are based on two general Theorems 1.9 and 1.11. The proofs of our results consist in the verification of the properties of the corresponding likelihood ratio random fields.

Smooth Case

We have the observations $X^K = (X_1, \ldots, X_K)$, where $X_k = (X_k(t), 0 \leq t \leq T)$ and $Y_k(\cdot), k = 1, \ldots, K$ satisfy the Eqs. (7.1) and (7.2) respectively. The function $\bar{\psi}(\cdot)$ is

$$\bar{\psi}(t - \tau_k) = \psi_\delta(t - \tau_k) = \frac{(t - \tau_k)}{\delta} \mathbb{I}_{\{0 \leq t - \tau_k \leq \delta\}} + \mathbb{I}_{\{t \geq \delta + \tau_k\}}.$$

We have to estimate $\vartheta_0 = (x_0, y_0)^\top$ by observations $X^K = (X_1, \ldots, X_K)$. We write $\tau_k = \tau_k(\vartheta)$, $\theta = (x, y)^\top$ and for the derivative $\dot{m}_k(\tau_k, t) = \partial m_k(\tau_k, t) / \partial \tau_k$ we obtain the equations:

$$\begin{aligned}
\dot{m}_k(\tau_k, t) &= 0, \qquad\qquad 0 \leq t \leq \tau_k(\vartheta), \\
\mathrm{d}\dot{m}_k(\tau_k, t) &= \Big[a_k(t) - \Gamma_k(\tau_k, t) h_k(t)^2 \psi_\delta(t - \tau_k)\Big] \dot{m}_k(\tau_k, t)\,\mathrm{d}t \\
&\quad - \Big[\dot{\Gamma}_k(\tau_k, t)\psi_\delta(t - \tau_k) - \Gamma_k(\tau_k, t)\delta^{-1}\Big] h_k(t)^2 m_k(\tau_k, t)\,\mathrm{d}t \\
&\quad + \frac{\dot{\Gamma}_k(\tau_k, t) h_k(t)}{\sigma_k(t)} \mathrm{d}X_t, \qquad \tau_k(\vartheta) < t \leq \tau_k(\vartheta) + \delta, \\
\mathrm{d}\dot{m}_k(\tau_k, t) &= \Big[a_k(t) - \gamma_k(\tau_k, t) h_k(t)^2\Big] \dot{m}_k(\tau_k, t)\,\mathrm{d}t + \frac{\dot{\gamma}_k(\tau_k, t) h_k(t)}{\sigma_k(t)} \mathrm{d}X_t, \\
&\quad - \dot{\gamma}_k(\tau_k, t) h_k(t)^2 m_k(\tau_k, t)\,\mathrm{d}t, \qquad \tau_k(\vartheta) + \delta < t \leq T,
\end{aligned} \tag{7.11}$$

where we denoted $\Gamma_k(\tau_k, t) = \gamma_k(\tau_k, t)\psi_\delta(t - \tau_k)$ and used the equality $\partial \psi_\delta(t - \tau_k)/\partial \tau_k = -\delta^{-1} \mathbb{I}_{\{0 \leq t - \tau_k \leq \delta\}}$.

The solution of equation (7.11) at point $\varepsilon = 0$, $\vartheta = \vartheta_0$ we denote $z_k(t, \vartheta_0), 0 \leq t \leq T$. The function $z_k(t, \vartheta_0)$ satisfies the equations

$$\begin{aligned}
\frac{\partial z_k(t, \vartheta_0)}{\partial t} &= 0, \qquad\qquad 0 \leq t \leq \tau_k(\vartheta_0), \\
\frac{\partial z_k(t, \vartheta_0)}{\partial t} &= \left[a_k(t) - \Gamma_k(\tau_k(\vartheta_0), t) h_k(t)^2 \frac{(t - \tau_k(\vartheta_0))}{\delta}\right] z_k(t, \vartheta_0) \\
&\quad + \Gamma_k(\tau_k(\vartheta_0), t) h_k(t)^2 \delta^{-1} y_k(t), \qquad \tau_k(\vartheta_0) < t \leq \tau_k(\vartheta_0) + \delta, \\
\frac{\partial z_k(t, \vartheta_0)}{\partial t} &= \Big[a_k(t) - \gamma_k(\tau_k(\vartheta_0), t) h_k(t)^2\Big] z_k(t, \vartheta_0), \quad \tau_k(\vartheta_0) + \delta < t \leq T.
\end{aligned}$$

Of course the solutions of this linear equations can be written explicitly.

Introduce two deterministic functions

$$\dot M_{k,x}^{o}\left(\tau_k\left(\vartheta_0\right),t\right)=\nu\left.\frac{\partial\left[m_k\left(\tau_k\left(\vartheta\right),t\right)\psi_\delta\left(t-\tau_k\left(\vartheta\right)\right)\right]}{\partial\tau_k}\right|_{\vartheta=\vartheta_0,\varepsilon=0}\frac{\partial\tau_k\left(\vartheta_0\right)}{\partial x_0}$$

$$=\left[\delta^{-1}y_k\left(t\right)1\!\!\mathrm{I}_{\{\tau_k(\vartheta_0)\le t\le\tau_k(\vartheta_0)+\delta\}}-z_k\left(t,\vartheta_0\right)\psi_\delta\left(t-\tau_k\left(\vartheta_0\right)\right)\right]\mu_{k,x},$$

$$\dot M_{k,y}^{o}\left(\tau_k\left(\vartheta_0\right),t\right)=\nu\left.\frac{\partial\left[m_k\left(\tau_k\left(\vartheta\right),t\right)\psi_\delta\left(t-\tau_k\left(\vartheta\right)\right)\right]}{\partial\tau_k}\right|_{\vartheta=\vartheta_0,\varepsilon=0}\frac{\partial\tau_k\left(\vartheta_0\right)}{\partial y_0}$$

$$=\left[\delta^{-1}y_k\left(t\right)1\!\!\mathrm{I}_{\{\tau_k(\vartheta_0)\le t\le\tau_k(\vartheta_0)+\delta\}}-z_k\left(t,\vartheta_0\right)\psi_\delta\left(t-\tau_k\left(\vartheta_0\right)\right)\right]\mu_{k,y}.$$

and 2×2 Fisher information matrix $\mathbf{I}\left(\vartheta_0\right)=\left(I\left(\vartheta_0\right)_{l,m}\right)$

$$I_{1,1}\left(\vartheta_0\right)=\sum_{k=1}^{K}\int_{\tau_k(\vartheta_0)}^{T}h_k\left(t\right)^2\dot M_{k,x}^{o}\left(\tau_k\left(\vartheta_0\right),t\right)^2\mathrm{d}t,$$

$$I_{2,2}\left(\vartheta_0\right)=\sum_{k=1}^{K}\int_{\tau_k(\vartheta_0)}^{T}h_k\left(t\right)^2\dot M_{k,y}^{o}\left(\tau_k\left(\vartheta_0\right),t\right)^2\mathrm{d}t,$$

$$I_{1,2}\left(\vartheta_0\right)=\mathrm{I}_{2,1}\left(\vartheta_0\right)=\sum_{k=1}^{K}\int_{\tau_k(\vartheta_0)}^{T}h_k\left(t\right)^2\dot M_{k,x}^{o}\left(\tau_k\left(\vartheta_0\right),t\right)\dot M_{k,y}^{o}\left(\tau_k\left(\vartheta_0\right),t\right)\mathrm{d}t.$$

The family of measures which corresponds to the solutions of equation (7.1) with different $\vartheta\in\Theta$ is locally asymptotically normal (LAN) (see Lemma 7.2 below). Therefore, we have the Hájek's lower minimax bound on the risks of all estimators $\bar\vartheta_\varepsilon$:

$$\lim_{\delta\to 0}\varliminf_{\varepsilon\to 0}\sup_{\|\vartheta-\vartheta_0\|\le\delta}\varepsilon^{-2}\mathbf{E}_\vartheta\left\|\bar\vartheta_\varepsilon-\vartheta\right\|^2\ge\mathbf{E}_{\vartheta_0}\left\|\zeta\left(\vartheta_0\right)\right\|^2,\ \zeta\left(\vartheta_0\right)\sim\mathcal{N}\left(0,\mathbf{I}\left(\vartheta_0\right)^{-1}\right).\tag{7.12}$$

As usual, we call the estimator ϑ_ε^* asymptotically efficient if for this estimator and all $\vartheta_0\in\Theta$ we have equality in (7.12).

The properties of the estimators are given in the following theorem.

Theorem 7.1 *Let the conditions $\mathscr{N}$ be fulfilled and the Fisher information matrix be uniformly nondegenerate*

$$\hat\kappa=\inf_{\vartheta_0\in\Theta}\inf_{\mathrm{e}\in\mathcal{R}^2,\|\mathrm{e}\|=1}\mathrm{e}^\top\mathbf{I}\left(\vartheta_0\right)\mathrm{e}>0.$$

Then the MLE $\hat\vartheta_\varepsilon$ and BE $\tilde\vartheta_\varepsilon$ are uniformly on compacts $\mathbb{K}\in\Theta$ consistent, asymptotically normal

$$\varepsilon^{-1}\left(\hat{\vartheta}_\varepsilon-\vartheta_0\right)\Longrightarrow\mathcal{N}\left(0,\mathbf{I}\left(\vartheta_0\right)^{-1}\right),\qquad\varepsilon^{-1}\left(\tilde{\vartheta}_\varepsilon-\vartheta_0\right)\Longrightarrow\mathcal{N}\left(0,\mathbf{I}\left(\vartheta_0\right)^{-1}\right),$$

the moments converge: for any $p>0$

$$\varepsilon^{-p}\mathbf{E}_{\vartheta_0}\|\hat{\vartheta}_\varepsilon-\vartheta_0\|^p\longrightarrow\mathbf{E}_{\vartheta_0}\|\zeta\left(\vartheta_0\right)\|^p,\qquad\varepsilon^{-p}\mathbf{E}_{\vartheta_0}\|\tilde{\vartheta}_\varepsilon-\vartheta_0\|^p\longrightarrow\mathbf{E}_{\vartheta_0}\|\zeta\left(\vartheta_0\right)\|^p,$$

where $\zeta\left(\vartheta_0\right)\sim\mathcal{N}\left(0,\mathbf{I}\left(\vartheta_0\right)^{-1}\right)$. *Both estimators are asymptotically efficient.*

Proof The proof of this theorem is based on Theorems 1.10 and 1.12. Below we verify the properties $\mathscr{I}\mathscr{K}_2$ of the normalized likelihood ratio random field

$$Z_\varepsilon\left(\vartheta_0,w\right)=\frac{L\left(\vartheta_0+\varepsilon w,X^K\right)}{L\left(\vartheta_0,X^K\right)},\qquad w=\left(u,v\right)\in\mathbb{W}_\varepsilon=\left\{w:\ \vartheta_0+\varepsilon w\in\Theta\right\},$$

which are the conditions of the mentioned theorems. Introduce the random field

$$Z\left(\vartheta_0,w\right)=\exp\left\{w^\top\Delta\left(\vartheta_0\right)-\frac{1}{2}w^\top\mathbf{I}\left(\vartheta_0\right)w\right\},\quad\Delta\left(\vartheta_0\right)\sim\mathcal{N}\left(0,\mathbf{I}\left(\vartheta_0\right)\right),\quad w\in\mathcal{R}^2.$$

We prove three lemmas below. □

Lemma 7.2 *The finite-dimensional distributions of the random fields* $Z_\varepsilon\left(\vartheta_0,\cdot\right)$ *converge to the finite-dimensional distributions of* $Z\left(\vartheta_0,\cdot\right)$, *and this convergence is uniform on compacts* $\mathbb{K}\in\Theta$.

Proof Denote: $D_{k,\varepsilon}\left(\vartheta_0,w,t\right)=\varepsilon^{-1}\left[M_k\left(\tau_k\left(\vartheta_w\right),t\right)-M_k\left(\tau_k\left(\vartheta_0\right),t\right)\right]$, $\vartheta_w=\vartheta_0+\varepsilon w$.

Then the log-likelihood ratio admits the representation (see (7.7))

$$\begin{aligned}\ln Z_\varepsilon\left(\vartheta_0,w\right)=&\sum_{k=1}^K\int_{\tau_k(\vartheta_w)\wedge\tau_k(\vartheta_0)}^T h_k\left(t\right)D_{k,\varepsilon}\left(\vartheta_0,w,t\right)\mathrm{d}\bar{W}_k\left(t\right)\\&-\frac{1}{2}\sum_{k=1}^K\int_{\tau_k(\vartheta_w)\wedge\tau_k(\vartheta_0)}^T h_k\left(t\right)^2D_{k,\varepsilon}\left(\vartheta_0,w,t\right)^2\mathrm{d}t.\end{aligned}$$

The random processes $m_k\left(\tau_k,t\right)$ are differentiable with probability 1 w.r.t. τ_k and the derivative $\dot{m}_k\left(\vartheta_0,t\right)$ satisfies the estimate

$$\mathbf{E}_{\vartheta_0}\left|\dot{m}_k\left(\tau_k\left(\vartheta_0\right),t\right)-z_k\left(\vartheta_0,t\right)\right|^2\leq C\varepsilon^2.$$

We have to write the difference of the equations for $\dot{m}_k\left(\tau_k\left(\vartheta_0\right),t\right)$ and $z_k\left(\vartheta_0,t\right)$ and then after simple transformations use the Grönwall's lemma .

By Taylor formula we obtain the relation

$$D_{k,\varepsilon}\left(\vartheta_0,w,t\right)=\langle\dot{M}_k^o\left(\tau_k\left(\vartheta_0\right),t\right),w\rangle+o\left(1\right)$$

where $\dot{M}_k^o(\tau_k(\vartheta_0), t) = \left(\dot{M}_{k,x}^o(\tau_k(\vartheta_0), t), \dot{M}_{k,y}^o(\tau_k(\vartheta_0), t)\right)^\top$. Introduce the random vector

$$\Delta_\varepsilon(\vartheta_0) = \sum_{k=1}^{K} \int_{\tau_k(\vartheta_0)}^{T} h_k(t)\, \dot{M}_k^o(\tau_k(\vartheta_0), t)\, \mathrm{d}\bar{W}_k(t).$$

Recall that $\bar{W}_k(t)$ depends on ε. Then we have

$$\begin{aligned}\int_{\tau_k(\vartheta_w)\wedge\tau_k(\vartheta_0)}^{T} & h_k(t)\, D_{k,\varepsilon}(\vartheta_0, w, t)\ \mathrm{d}\bar{W}_k(t) \\ &= \int_{\tau_k(\vartheta_0)}^{T} h_k(t)\, \langle \dot{M}_k^o(\tau_k(\vartheta_0), t), w\rangle\, \mathrm{d}\bar{W}_k(t) + o(1)\end{aligned}$$

because $\tau_k(\vartheta_w) = \tau_k(\vartheta_0) + O(\varepsilon)$ and

$$\begin{aligned}\int_{\tau_k(\vartheta_w)\wedge\tau_k(\vartheta_0)}^{T} & h_k(t)^2\, D_{k,\varepsilon}(\vartheta_0, w, t)^2\, \mathrm{d}t \\ &= \int_{\tau_k(\vartheta_0)}^{T} h_k(t)^2\, \langle \dot{M}_k^o(\tau_k(\vartheta_0), t), w\rangle^2 \mathrm{d}t + o(1).\end{aligned}$$

Note that

$$\sum_{k=1}^{K} \int_{\tau_k(\vartheta_0)}^{T} h_k(t)^2\, \langle \dot{M}_k^o(\tau_k(\vartheta_0), t), w\rangle^2 \mathrm{d}t = w^\top \mathbf{I}(\vartheta_0)\, w.$$

Hence for the normalized likelihood ratio function we obtain the representation called LAN (local asymptotic normality)

$$Z_\varepsilon(\vartheta_0, w) = \exp\left\{\langle w, \Delta_\varepsilon(\vartheta_0)\rangle - \frac{1}{2} w^\top \mathbf{I}(\vartheta_0)\, w + o(1)\right\},$$

where

$$\Delta_\varepsilon(\vartheta_0) \sim \mathcal{N}(0, \mathbf{I}(\vartheta_0)).$$

This representation of $Z_\varepsilon(\vartheta_0, \cdot)$ provides the convergence of finite-dimensional distributions. Moreover, it can be shown that all convergences are uniform on compacts $\mathbb{K} \subset \Theta$. □

Lemma 7.3 *There exists constant $C > 0$ such that for any $R > 0$ and $\|w_1\| + \|w_2\| < R$ we have*

$$\sup_{\vartheta_0\in\mathbb{K}} \mathbf{E}_{\vartheta_0}\left|Z_\varepsilon^{1/4}\left(\vartheta_0,w_2\right)-Z_\varepsilon^{1/4}\left(\vartheta_0,w_1\right)\right|^4\leq C\left(1+R^4\right)\left\|w_2-w_1\right\|^4. \tag{7.13}$$

Proof The estimate (1.10) allows us to write

$$\begin{aligned}
&\mathbf{E}_{\vartheta_0}\left|Z_\varepsilon^{1/4}\left(\vartheta_0,w_2\right)-Z_\varepsilon^{1/4}\left(\vartheta_0,w_1\right)\right|^4\\
&\qquad\leq CT^3K^3\sum_{k=1}^K\int_{\tau_k(\vartheta_w)\wedge\tau_k(\vartheta_0)}^T \mathbf{E}_{\vartheta_{w_2}}\left[D_{k,\varepsilon}\left(\vartheta_0,w_1,w_2,t\right)\right]^8 \mathrm{d}t\\
&\qquad\quad+CT^{\frac32}K^{\frac32}\sum_{k=1}^K\int_{\tau_k(\vartheta_w)\wedge\tau_k(\vartheta_0)}^T \mathbf{E}_{\vartheta_{w_2}}\left[D_{k,\varepsilon}\left(\vartheta_0,w_1,w_2,t\right)\right]^4 \mathrm{d}t,
\end{aligned}$$

where $D_{k,\varepsilon}\left(\vartheta_0,w_1,w_2,t\right)=D_{k,\varepsilon}\left(\vartheta_0,w_2,t\right)-D_{k,\varepsilon}\left(\vartheta_0,w_1,t\right)$.

Further, recall that $\vartheta_w=\vartheta_0+\varepsilon w$ and denote $\vartheta\left(s\right)=\vartheta_0+\varepsilon w_1+\varepsilon s\left(w_2-w_1\right)$, then we can write

$$\begin{aligned}
&D_{k,\varepsilon}\left(\vartheta_0,w_1,w_2,t\right)\\
&\quad=\varepsilon^{-1}\left[\psi_\delta\left(t-\tau_k\left(\vartheta_{w_2}\right)\right)m_k\left(\tau_k\left(\vartheta_{w_2}\right),t\right)-\psi_\delta\left(t-\tau_k\left(\vartheta_{w_1}\right)\right)m_k\left(\tau_k\left(\vartheta_{w_1}\right),t\right)\right]\\
&\quad=\nu^{-1}\int_0^1\left[\dot\psi_\delta\left(t-\tau_k\left(\vartheta\left(s\right)\right)\right)m_k\left(\tau_k\left(\vartheta\left(s\right),t\right)\right)\right.\\
&\qquad\qquad\left.+\psi_\delta\left(t-\tau_k\left(\vartheta\left(s\right)\right)\right)\dot m_k\left(\tau_k\left(\vartheta\left(s\right)\right),t\right)\right]\langle\mu_k,w_2-w_1\rangle \mathrm{d}s.
\end{aligned}$$

Therefore

$$\sum_{k=1}^K\int_0^T \mathbf{E}_{\vartheta_0+\varepsilon w_2}h_k\left(t\right)^8 D_{k,\varepsilon}\left(\vartheta_0,w_1,w_2,t\right)^8 \mathrm{d}t\leq C\left\|w_2-w_1\right\|^8,$$

$$\sum_{k=1}^K\int_0^T \mathbf{E}_{\vartheta_0+\varepsilon w_2}h_k\left(t\right)^4 D_{k,\varepsilon}\left(\vartheta_0,w_1,w_2,t\right)^4 \mathrm{d}t\leq C\left\|w_2-w_1\right\|^4$$

and we obtain (7.13). □

Lemma 7.4 *For any compact $\mathbb{K}\subset\Theta$ and any $N>0$ there exist constant $C_*>0$ such that*

$$\sup_{\vartheta_0\in\mathbb{K}} \mathbf{E}_{\vartheta_0}Z_\varepsilon\left(\vartheta_0,w\right)^{1/2}\leq\frac{C_*}{\left\|w\right\|^N}. \tag{7.14}$$

Proof Consider the integral

$$\int_0^T h_k(t)^2 D_{k,\varepsilon}(\vartheta_0, w, t)^2 \,\mathrm{d}t = \int_0^T \frac{h_k(t)^2}{\varepsilon^2} \left[M(\tau_k(\vartheta_w), t) - M(\tau_k(\vartheta_0), t)\right]^2 \mathrm{d}t.$$

Denote

$$\begin{aligned} &m_{k,w}(t) = m_k(\tau_k(\vartheta_w), t), \quad \Gamma_{k,w}(t) = \Gamma_k(\tau_k(\vartheta_w), t), \\ &\psi_{\delta,k,w}(t) = \psi_\delta(t - \tau_k(\vartheta_w)). \end{aligned}$$

Random functions $m_{k,0}(t)$ and $m_{k,w}(t)$ satisfy the equations

$$\begin{aligned} m_{k,0}(t) &= y_k(t), & 0 \le t \le \tau_k(\vartheta_0), \\ \mathrm{d}m_{k,0}(t) &= a_k(t)\, m_{k,0}(t)\,\mathrm{d}t + \varepsilon\Gamma_{k,0}(t)\, h_k(t)\,\mathrm{d}\bar{W}_t, & \tau_k(\vartheta_0) < t \le T, \\ m_{k,w}(t) &= y_k(t), & 0 \le t \le \tau_k(\vartheta_w), \\ \mathrm{d}m_{k,w}(t) &= \left[a_k(t) - \Gamma_{k,w}(t)\, h_k(t)^2\, \psi_{\delta,k,w}(t)\right] m_{k,w}(t)\,\mathrm{d}t \\ &\quad + \Gamma_{k,w}(t)\, h_k(t)^2\, \psi_{\delta,k,0}(t)\, m_{k,0}(t)\,\mathrm{d}t \\ &\quad + \varepsilon\Gamma_{k,w}(t)\, h_k(t)\,\mathrm{d}\bar{W}_t, & \tau_k(\vartheta_w) < t \le T. \end{aligned}$$

The solution of the first equation for $\varepsilon = 0$ is denoted as $y_k(t) = y_k(\vartheta_0, t)$ and given in (7.6). The solution $m_{k,w}(t)$ for $\varepsilon = 0$ and $t \in [\tau_k(\vartheta_w), T]$ (we denote it as $y_k(\vartheta_w, t) = y_k(\vartheta_w, \vartheta_0, t)$) satisfies the equation

$$\begin{aligned} \frac{\partial y_k(\vartheta_w, t)}{\partial t} &= \left[a_k(t) - \Gamma_{k,w}(t)\, h_k(t)^2\, \psi_{\delta,k,w}(t)\right] y_k(\vartheta_w, t) \\ &\quad + \Gamma_{k,w}(t)\, h_k(t)^2\, \psi_{\delta,k,0}(t)\, y_k(t), \quad y_k(\vartheta_w, \tau_k(\vartheta_w)) = y_k(\tau_k(\vartheta_w)). \end{aligned}$$

If we put $q_k(\vartheta_0, t) = m_k(\tau_k(\vartheta_0), t) - y_k(t)$ and $q_k(\vartheta_w, t) = m_k(\tau_k(\vartheta_w), t) - y_k(\vartheta_w, t)$, then for these differences it is possible to verify the relations

$$\begin{aligned} q_k(\vartheta_0, t) &= \varepsilon \int_0^t e^{\int_s^t a_k(r)\mathrm{d}r}\, \Gamma_{k,0}(s)\, h_k(s)\,\mathrm{d}\bar{W}_s, \\ q_k(\vartheta_w, t) &= \varepsilon \int_0^t e^{\int_s^t Q_k(\vartheta_w, r)\mathrm{d}r}\, \Gamma_{k,w}(s)\, h_k(s)\,\mathrm{d}\bar{W}_s \\ &\quad + \int_0^t e^{\int_s^t Q_k(\vartheta_w, r)\mathrm{d}r}\, \Gamma_{k,w}(\vartheta_w, s)\, h_k(s)^2\, \psi_{\delta,k}(0, s)\, q_k(s)\,\mathrm{d}s, \end{aligned}$$

where $Q_k(\vartheta_w, t) = a_k(t) - \Gamma_k(\vartheta_w, t)\, h_k(t)^2\, \psi_\delta(t - \tau_k(\vartheta_w))$.

Thus we have

$$m_k(\vartheta_0, t) - y_k(t) = \varepsilon\xi_1(t), \qquad m_k(\vartheta_w, t) - y_k(\vartheta_w, t) = \varepsilon\xi_2(t),$$

where $\xi_1(t)$ and $\xi_2(t)$ are Gaussian processes with bounded variances.

Therefore, as it follows from the proof of Lemma (2.5), it is sufficient to consider the deterministic integral

$$F(\vartheta_w, \vartheta_0) = \sum_{k=1}^{K} \int_0^T \frac{h_k(t)^2}{\varepsilon^2} \left[\psi_{\delta,k,w}(t)\, y_k(\vartheta_w, t) - \psi_{\delta,k,0}(t)\, y_k(t)\right]^2 \mathrm{d}t$$

and to show that there exists a constant $\kappa_* > 0$ such that

$$F(\vartheta_w, \vartheta_0) \geq \kappa_* \left\|w\right\|^2. \tag{7.15}$$

Remind that for $\left\|w\right\| < R$

$$F(\vartheta_w, \vartheta_0) = w^\top \mathbf{I}(\vartheta_0)\, w\, (1 + o(1)).$$

Hence for some $\varepsilon_1 > 0$ and $\varepsilon_1 \left\|w\right\| \leq \varepsilon_1 R = d$ and all $\varepsilon \leq \varepsilon_1$ we can write

$$F(\vartheta_w, \vartheta_0) \geq \frac{1}{2} w^\top \mathbf{I}(\vartheta_0)\, w \geq \frac{\left\|w\right\|^2}{2} \inf_{\|\mathrm{e}\|=1} \mathrm{e}^\top \mathrm{I}(\vartheta_0)\, \mathrm{e} \geq \frac{\hat\kappa}{4} \left\|w\right\|^2.$$

Here we denoted $\hat\kappa > 0$ the constant from the condition of the Theorem 7.1.

Introduce the function

$$G(\vartheta_*, \vartheta_0) = \sum_{k=1}^{K} \int_0^T h_k(t)^2 \left[\psi_\delta(t - \tau_k(\vartheta_*))\, y_k(\vartheta_*, t)\right.$$
$$\left. - \psi_\delta(t - \tau_k(\vartheta_0))\, y_k(\vartheta_0, t)\right]^2 \mathrm{d}t,$$

where $\vartheta_* = (x_*, y_*) \in \Theta$ and $\left\|\vartheta_* - \vartheta_0\right\| \geq d$. We have to show that

$$\tilde\kappa = \inf_{\vartheta_*: \|\vartheta_* - \vartheta_0\| \geq d} G(\vartheta_*, \vartheta_0) > 0.$$

Suppose that there exists ϑ_* such that $G(\vartheta_*, \vartheta_0) = 0$ and $\left\|\vartheta_* - \vartheta_0\right\| \geq d$. Then we have $\tau_k(\vartheta_*) = \tau_k(\vartheta_0)$ and $y_k(\vartheta_*, t) = y_k(\vartheta_0, t)$ for all k. This means that for two different positions of the source ϑ_* and ϑ_0 we have k equalities

$$\left\|\vartheta_k - \vartheta_*\right\| = \left\|\vartheta_k - \vartheta_0\right\|, \qquad k = 1, \ldots, K$$

but as we have at least three detectors not on the same line, such equalities for all k are impossible to have. This is the condition of identifiability.

Therefore for $\varepsilon \left\|w\right\| > d$ there exists a constant $c_* > 0$ such that

$$F(\vartheta_w, \vartheta_0) \geq \varepsilon^{-2} G(\vartheta_0 + \varepsilon w, \vartheta_0) \geq \frac{\tilde\kappa}{\varepsilon^2} \geq c_*^2 \tilde\kappa \left\|w\right\|^2$$

because

$$\|w\| \leq \varepsilon^{-1} \sup_{\vartheta,\vartheta'\in\Theta} \left\|\vartheta-\vartheta'\right\| \qquad \text{and} \qquad \varepsilon^{-1} \geq \frac{\|w\|}{\sup_{\vartheta,\vartheta'\in\Theta}\|\vartheta-\vartheta'\|} = c_*\,\|w\|\,.$$

Therefore we obtained (7.15). Now the proof of the estimate (7.14) follows from the proof of Lemma (2.5). □

The properties of the normalized likelihood ratio function $Z_\varepsilon(\vartheta_0, w)\,, w \in \mathbb{W}_n$ established in Lemmas 7.2–7.4 correspond to conditions $\mathscr{I}\mathscr{K}_2$ of Theorems 1.10 and 1.12.

Cusp-Type Case

The cusp-type models are proposed as an alternative to change-point models with the following motivation. The real signals in technical devices cannot have discontinuous characteristics. Say, electrical current cannot have pure jump at the moment of signal arrival. Usually we have continuous curves but with strong increasing at the moment of signal arrival. We presented different types of increasing functions: smooth, cusp-type, and change-point type. The cusp-type model with mean square rate of MLE $\varepsilon^{4/(2\kappa+1)}, 0 < \kappa < 1/2$ is intermediate between smooth model case with the rate of mean square error ε^2 ($\kappa = 1/2$) and change-point type models with the rate of mean square error ε^4 ($\kappa = 0$). We suppose that the cusp-type model with κ close to 0 is better approximation of real “change-point” situation, because it proposes a continuous curve close in $L_2[0, T]$ to discontinuous curve.

We return to the considered above model of K detectors with observations

$$\mathrm{d}X_k(t) = f_k(t)\,\psi_{\delta,\kappa}(t-\tau_k(\vartheta))\,Y_k(t)\,\mathrm{d}t + \varepsilon\sigma_k(t)\,\mathrm{d}W_t, \quad 0 \leq t \leq T$$

where $X_k(0) = 0, k = 1, \ldots, K$ and the function

$$\psi_{\delta,\kappa}(t) = \frac{1}{2}\left(1+\mathrm{sgn}\,(2t-\delta)\left|\frac{2t}{\delta}-1\right|^{\kappa}\right)\mathbb{I}_{\{0\leq t\leq\delta\}} + \mathbb{I}_{\{\delta\leq t\leq T\}},$$

see Fig. 7.1 and the case (**b**) on Fig. 7.2. The *hidden* Gaussian processes $Y_k(\cdot)\,, k = 1, \ldots, K$, as before, satisfy the same linear equations

$$\mathrm{d}Y_k(t) = a_k(t)\,Y_k(t)\,\mathrm{d}t + \varepsilon b_k(t)\,\mathrm{d}V_k(t)\,, \quad Y_k(0) = y_{k,0} > 0, \quad 0 \leq t \leq T.$$

We suppose that the conditions $\mathscr{N}$ are fulfilled and as before we study the properties of estimators as $\varepsilon \to 0$, i.e., *small noise* asymptotic.

The delay $\tau_k(\vartheta) = \rho_k/\nu = \nu^{-1}\|\vartheta_k - \vartheta_0\|$, and we have to estimate the position $\vartheta_0 = (x_0, y_0)$ by observations $X^K = (X_1, \ldots, X_K)$. Here the process $X_k = (X_k(t)\,, 0 \leq t \leq T)$.

The likelihood ratio function and estimators $\hat{\vartheta}_\varepsilon$, $\tilde{\vartheta}_\varepsilon$ are defined by the same relations (7.9)–(7.10).

To describe the asymptotic behavior of these estimators we need notations (7.3) and the following ones:

$$\pi_k(\vartheta_0) = h_k(\tau_k(\vartheta_0 + \delta/2))\, y_k(\tau_k(\vartheta_0) + \delta/2),$$

$$\pi_k = \pi_k(\vartheta_0), \quad Q(\kappa)^2 = \int_{\mathcal{R}} \left[\operatorname{sgn}(s-1)|s-1|^\kappa - \operatorname{sgn}(s)|s|^\kappa\right]^2 ds,$$

$$\varphi_\varepsilon = \varepsilon^{\frac{1}{H}} Q(\kappa)^{-\frac{1}{H}} 2^{\frac{1-\kappa}{H}} \delta^{\frac{\kappa}{H}}, \qquad H = \kappa + \frac{1}{2}.$$

Introduce the limit likelihood ratio random field

$$Z(\vartheta_0, w) = \exp\left\{\sum_{k=1}^K \left[\pi_k W_k^{H,+}(\langle\mu_k, w\rangle) - \frac{\pi_k^2}{2}|\langle\mu_k, w\rangle|^{2H}\right] \mathbb{I}_{\{\mathbb{B}_k^+\}}\right.$$
$$\left. + \sum_{k=1}^K \left[\pi_k W_k^{H,-}(-\langle\mu_k, w\rangle) - \frac{\pi_k^2}{2}|\langle\mu_k, w\rangle|^{2H}\right] \mathbb{I}_{\{\mathbb{B}_k^-\}}\right\}, \qquad w \in \mathcal{R}^2.$$

Here $W_k^{H,+}(s), s \geq 0, k = 1, \ldots, K$ and $W_k^{H,-}(s), s \geq 0, k = 1, \ldots, K$ are independent fractional Brownian motions with Hurst parameter H, i.e., independent Gaussian processes with properties: $\mathbf{E}W_k^{H,+}(s) = 0$ and

$$\mathbf{E}W_k^{H,+}(s_1)\, W_k^{H,+}(s_2) = \frac{1}{2}\left(|s_1|^{2H} + |s_2|^{2H} - |s_1 - s_2|^{2H}\right).$$

The limit distributions of the estimators are given by the random vectors $\hat{\mathrm{w}}$ and $\tilde{\mathrm{w}}$ defined by the following relations

$$Z\left(\vartheta_0, \hat{\mathrm{w}}(\vartheta_0)\right) = \sup_{w \in \mathcal{R}^2} Z(\vartheta_0, w), \qquad \tilde{\mathrm{w}}(\vartheta_0) = \frac{\int_{\mathcal{R}^2} w Z(\vartheta_0, w)\, dw}{\int_{\mathcal{R}^2} Z(\vartheta_0, w)\, dw}.$$

The random variable $\hat{w}(\vartheta_0)$ is a unique solution of this equation (see [166]).

We have the following minimax lower bound for mean square errors of any estimator

$$\lim_{\delta_* \to 0} \varliminf_{\varepsilon \to 0} \sup_{\|\vartheta - \vartheta_0\| \leq \delta_*} \varphi_\varepsilon^{-2} \mathbf{E}_\vartheta \left\|\bar{\vartheta}_\varepsilon - \vartheta\right\|^2 \geq \mathbf{E}_{\vartheta_0} \|\tilde{\mathrm{w}}\|^2. \tag{7.16}$$

This is a particular case of Theorem I.9.1 in [73]. To prove it we first suppose that it is already established that the BE is consistent and the polynomial moments converge uniformly on compacts $\mathbb{K} \subset \Theta$ (see Theorem 7.2 below). Then the bound (7.16) is obtained as follows. We have

$$\sup_{\|\vartheta-\vartheta_0\|\leq\nu} \mathbf{E}_\vartheta \left\|\bar\vartheta_\varepsilon - \vartheta\right\|^2 \geq \int_{\mathbb{S}(\vartheta_0,\nu)} \mathbf{E}_\vartheta \left\|\bar\vartheta_\varepsilon - \vartheta\right\|^2 p_\nu\left(\vartheta\right) \mathrm{d}\vartheta$$

$$\geq \int_{\mathbb{S}(\vartheta_0,\nu)} \mathbf{E}_\vartheta \|\tilde\vartheta_\varepsilon - \vartheta\|^2 \, p_\nu\left(\vartheta\right) \mathrm{d}\vartheta. \tag{7.17}$$

Here the set $\mathbb{S}\left(\vartheta_0, \nu\right) = \left(\vartheta \,:\, \|\vartheta - \vartheta_0\| \leq \nu\right)$, the function $p_\nu\left(\vartheta\right), \vartheta \in \mathbb{S}\left(\vartheta_0, \nu\right)$ is a continuous positive probability density on the set $\mathbb{S}\left(\vartheta_0, \nu\right)$ and $\tilde\vartheta_\varepsilon$ is the BE which corresponds to this density. From the uniform convergence of moments we have

$$\varphi_\varepsilon^{-2} \int_{\mathbb{S}(\vartheta_0,\nu)} \mathbf{E}_\vartheta \|\tilde\vartheta_\varepsilon - \vartheta\|^2 p_\nu\left(\vartheta\right) \mathrm{d}\vartheta \longrightarrow \int_{\mathbb{S}(\vartheta_0,\nu)} \mathbf{E}_\vartheta \|\tilde{\mathrm{w}}\left(\vartheta\right)\|^2 p_\nu\left(\vartheta\right) \mathrm{d}\vartheta,$$

and from the continuity of the function $\mathbf{E}_\vartheta \|\tilde u\left(\vartheta\right)\|^2$ it follows that

$$\lim_{\nu\to 0} \int_{\mathbb{S}(\vartheta_0,\nu)} \mathbf{E}_\vartheta \|\tilde{\mathrm{w}}\left(\vartheta\right)\|^2 p_\nu\left(\vartheta\right) \mathrm{d}\vartheta = \mathbf{E}_{\vartheta_0} \|\tilde{\mathrm{w}}\left(\vartheta_0\right)\|^2. \tag{7.18}$$

We call the estimator $\bar\vartheta_\varepsilon$ asymptotically efficient if for this estimator there is equality in (7.16) for all $\vartheta_0 \in \Theta$.

Theorem 7.2 *Suppose that the conditions $\mathscr{N}$ are fulfilled, then the MLE $\hat\vartheta_\varepsilon$ and BE $\tilde\vartheta_\varepsilon$ are uniformly on compacts consistent and have different limit distributions*

$$\varphi_\varepsilon^{-1}(\hat\vartheta_\varepsilon - \vartheta_0) \Longrightarrow \hat{\mathrm{w}}\left(\vartheta_0\right), \qquad \varphi_\varepsilon^{-1}(\tilde\vartheta_\varepsilon - \vartheta_0) \Longrightarrow \tilde{\mathrm{w}}\left(\vartheta_0\right),$$

we have uniform on compacts convergence of moments: for any $p > 0$

$$\varphi_\varepsilon^{-p}\mathbf{E}_{\vartheta_0}\|\hat\vartheta_\varepsilon - \vartheta_0\|^p \longrightarrow \mathbf{E}_{\vartheta_0} \left\|\hat{\mathrm{w}}\left(\vartheta_0\right)\right\|^p, \; \varphi_\varepsilon^{-p}\mathbf{E}_{\vartheta_0}\|\tilde\vartheta_\varepsilon - \vartheta_0\|^p \longrightarrow \mathbf{E}_{\vartheta_0} \|\tilde{\mathrm{w}}\left(\vartheta_0\right)\|^p$$

and BE are asymptotically efficient.

Proof We verify the conditions $\mathscr{IK}_2$ of the general Theorems 1.9 and 1.11. These conditions are given in terms of the properties of the normalized likelihood ratio random field

$$Z_\varepsilon\left(\vartheta_0, w\right) = \frac{L\left(\vartheta_0 + \nu\varphi_\varepsilon w, X^K\right)}{L\left(\vartheta_0, X^K\right)}, \qquad w \in \mathbb{W}_\varepsilon = \left(w :\, \vartheta_0 + \nu\varphi_\varepsilon w \in \Theta\right).$$

□

We do it with the help of three lemmas.

Lemma 7.5 *The finite-dimensional distributions of the random fields $Z_\varepsilon(\vartheta_0,\cdot)$ converge to the finite-dimensional distributions of $Z(\vartheta_0,\cdot)$, and this convergence is uniform on compacts $\mathbb{K}\in\Theta$.*

Proof The normalized log-likelihood ratio random field is

$$\ln Z_\varepsilon(\vartheta_0,w)=\sum_{k=1}^{K}\int_0^T\frac{h_k(t)}{\varepsilon}\left[M_k(\tau_k(\vartheta_w),t)-M_k(\tau_k(\vartheta_0),t)\right]\mathrm{d}\bar W_k(t)$$
$$-\sum_{k=1}^{K}\int_0^T\frac{h_k(t)^2}{2\varepsilon^2}\left[M_k(\tau_k(\vartheta_w),t)-M_k(\tau_k(\vartheta_0),t)\right]^2\mathrm{d}t,$$

where we used notation $\vartheta_w=\vartheta_0+\nu\varphi_\varepsilon w$,

$$\psi_{\delta,\kappa,k}(w,t)=\psi_{\delta,\kappa}(t-\tau_k(\vartheta_w)),\qquad m_{k,w}(t)=m(\tau_k(\vartheta_w),t),$$
$$M_k(\tau_k(\vartheta_w),t)=\psi_{\delta,\kappa,k}(w,t)\,m_{k,w}(t)-\psi_{\delta,\kappa,k}(0,t)\,m_{k,0}(t).$$

We have

$$M_k(\tau_k(\vartheta_w),t)-M_k(\tau_k(\vartheta_0),t)=\left[\psi_{\delta,\kappa,k}(w,t)-\psi_{\delta,\kappa,k}(0,t)\right]m_{k,w}(t)$$
$$+\psi_{\delta,\kappa,k}(0,t)\left[m_{k,w}(t)-m_{k,0}(t)\right].$$

It can be shown that

$$\varepsilon^{-2}\mathbf{E}_{\vartheta_0}\left|m_{k,w}(t)-m_{k,0}(t)\right|^2\leq C\,\varepsilon^{-2}\varphi_\varepsilon^2=C\,\varepsilon^{\frac{1-2\kappa}{\kappa+\frac12}}\to 0,$$
$$\mathbf{E}_{\vartheta_0}\left|m_{k,0}(t)-y_k(t)\right|^2\leq C\,\varepsilon^2.$$

For example, to verify the first estimate we have to write the equations for $m_{k,w}(t)$ and $m_{k,0}(t)$, take the difference $\left|m_{k,w}(t)-m_{k,0}(t)\right|$, and then use Grönwall's lemma and Cauchy-Schwartz inequality. Hence to study

$$\varepsilon^{-2}\int_0^T h_k(t)^2\left[M_k(\tau_k(\vartheta_w),t)-M_k(\tau_k(\vartheta_0),t)\right]^2\mathrm{d}t$$

it is sufficient to find the limit of the integral

$$H_{k,\varepsilon}(w)=\varepsilon^{-2}\int_0^T h_k(t)^2\left[\psi_{\delta,\kappa,k}(w,t)-\psi_{\delta,\kappa,k}(0,t)\right]^2 m_{k,w}(t)^2\,\mathrm{d}t$$
$$=\varepsilon^{-2}\int_0^T h_k(t)^2\left[\psi_{\delta,\kappa,k}(w,t)-\psi_{\delta,\kappa,k}(0,t)\right]^2 m_{k,0}(t)^2\,\mathrm{d}t\,(1+o(1))$$

$$= \varepsilon^{-2} \int_0^T h_k(t)^2 \left[\psi_{\delta,\kappa,k}(w,t) - \psi_{\delta,\kappa,k}(0,t)\right]^2 y_k(t)^2 \,\mathrm{d}t \,(1 + o(1)).$$

Suppose that $w \in \mathbb{B}_k^-$, i.e., for small ε we have

$$\tau_k(\vartheta_0 + \nu\varphi_\varepsilon w) = \tau_k(\vartheta_0) - \varphi_\varepsilon \langle \mu_k, w\rangle + O\left(\varphi_\varepsilon^2\right) > \tau_k(\vartheta_0).$$

Therefore $H_{k,\varepsilon}(w) = \hat{H}_{k,\varepsilon}(w)(1 + o(1))$, where

$$\hat{H}_{k,\varepsilon}(w) = \varepsilon^{-2} \int_{\tau_k(\vartheta_0)-\varphi_\varepsilon\langle\mu_k,w\rangle}^{\tau_k(\vartheta_0)+\delta} h_k(t)^2 \left[\psi_{\delta,\kappa,k}(w,t) - \psi_{\delta,\kappa,k}(0,t)\right]^2 y_k(t)^2 \,\mathrm{d}t.$$

In this integral we change the variables several times:

$$t = s + \tau_k(\vartheta_0), \qquad s = \frac{\delta}{2}(r+1), \qquad r = -\frac{2\varphi_\varepsilon \langle \mu_k, w\rangle}{\delta} q,$$

obtain $h_k(t) = h_k(\tau_k(\vartheta_0 + \delta/2)) + o(1)$, and $y_k(t) = y_k(\tau_k(\vartheta_0) + \delta/2) + o(1)$ and

$$\begin{aligned}
\hat{H}_{k,\varepsilon}(w) &= \varepsilon^{-2} \int_{\tau_k(\vartheta_0)-\varphi_\varepsilon\langle\mu_k,w\rangle}^{\tau_k(\vartheta_0)+\delta} h_k(t)^2 \left[\psi_{\delta,\kappa,k}(w,t) - \psi_{\delta,\kappa,k}(0,t)\right]^2 y_k(t)^2 \,\mathrm{d}t \\
&= \frac{\pi_k^2}{Q(\kappa)^2} \left|\langle \mu_k, w\rangle\right|^{2\kappa+1} \int_{\mathcal{R}} \left[\operatorname{sgn}(q-1)\left|q-1\right|^\kappa - \operatorname{sgn}(q)\left|q\right|^\kappa\right]^2 \mathrm{d}q \,(1 + o(1)) \\
&= \pi_k^2 \left|\langle \mu_k, w\rangle\right|^{2\kappa+1} (1 + o(1)).
\end{aligned}$$

Recall here that

$$\begin{aligned}
\pi_k^2 &= h_k(\tau_k(\vartheta_0 + \delta/2))^2 \, y_k(\tau_k(\vartheta_0) + \delta/2)^2, \\
Q(\kappa)^2 &= \int_{\mathcal{R}} \left[\operatorname{sgn}(q-1)\left|q-1\right|^\kappa - \operatorname{sgn}(q)\left|q\right|^\kappa\right]^2 \mathrm{d}q.
\end{aligned}$$

Therefore we obtained the relation

$$\begin{aligned}
\varepsilon^{-2} \int_0^T & h_k(t)^2 \left[M_k(\tau_k(\vartheta_w), t) - M_k(\tau_k(\vartheta_0), t)\right]^2 \mathrm{d}t \\
&= \pi_k(\vartheta_0)^2 \, Q(\kappa, \vartheta_0)^2 \left|\langle \mu_k, w\rangle\right|^{2\kappa+1} (1 + o(1)).
\end{aligned}$$

For stochastic integral we have similar relations

$$
\begin{aligned}
\int_0^T & \frac{h_k(t)}{\varepsilon}\left[M_k(\tau_k(\vartheta_w),t)-M_k(\tau_k(\vartheta_0),t)\right]\mathrm{d}\bar{W}_k(t)\\
&=\frac{\pi_k}{Q(\kappa)}\int_{-\varphi_\varepsilon^{-1}}^{\varphi_\varepsilon^{-1}}\Big[\operatorname{sgn}(q-\langle\mu_k,w\rangle)\left|q-\langle\mu_k,w\rangle\right|^{\kappa}\\
&\qquad\qquad-\operatorname{sgn}(q)\left|q\right|^{\kappa}\Big]\mathrm{d}w_{k,\varepsilon}(q)\,(1+o(1))\\
&=\pi_k(\vartheta_0)\,W_{k,\varepsilon}(\langle\mu_k,w\rangle)\,(1+o(1))\,.
\end{aligned}
$$

Here $w_{k,\varepsilon}(\cdot)$ are independent two-sided Wiener processes obtained from the $\bar{W}_k(\cdot)$ after the mentioned change of variables and the corresponding normalization as follows:

$$
\begin{aligned}
\bar{W}_k(t) &\longrightarrow \bar{W}_k(s+\tau_k(\vartheta_0))-\bar{W}_k(\tau_k(\vartheta_0))\\
&\longrightarrow \tilde{W}_k(r)=\sqrt{\frac{2}{\delta}}\left[\bar{W}_k\left(\frac{\delta}{2}(r+1)\right)-\bar{W}_k\left(\frac{\delta}{2}\right)\right]\\
&\longrightarrow w_{k,\varepsilon}(\cdot)=\sqrt{\frac{\delta}{2\varphi_\varepsilon}}\,\tilde{W}_k\left(\frac{2\varphi_\varepsilon}{\delta}q\right).
\end{aligned}
$$

The Gaussian random function

$$
\begin{aligned}
&W_{k,\varepsilon}(\langle\mu_k,w\rangle)\\
&\quad=Q(\kappa)^{-1}\int_{-\varphi_\varepsilon^{-1}}^{\varphi_\varepsilon^{-1}}\left[\operatorname{sgn}(q-\langle\mu_k,w\rangle)\left|q-\langle\mu_k,w\rangle\right|^{\kappa}-\operatorname{sgn}(q)\left|q\right|^{\kappa}\right]\mathrm{d}w_{k,\varepsilon}(q)
\end{aligned}
$$

has the properties

$$
\begin{aligned}
&\mathbf{E}_{\vartheta_0}W_{k,\varepsilon}(\langle\mu_k,w\rangle)=0,\qquad \mathbf{E}_{\vartheta_0}W_{k,\varepsilon}(\langle\mu_k,w\rangle)^2=\left|\langle\mu_k,w\rangle\right|^{2H}(1+o(1)),\\
&\mathbf{E}_{\vartheta_0}W_{k,\varepsilon}(\langle\mu_k,w_1\rangle)\,W_{k,\varepsilon}(\langle\mu_k,w_2\rangle)\\
&\quad=\frac{1}{2}\left(\left|\langle\mu_k,w_1\rangle\right|^{2H}+\left|\langle\mu_k,w_2\rangle\right|^{2H}-\left|\langle\mu_k,w_2-w_1\rangle\right|^{2H}\right)(1+o(1))\,.
\end{aligned}
$$

Therefore we have the convergence

$$
W_{k,\varepsilon}(\left|\langle\mu_k,w\rangle\right|)\Longrightarrow W_k^{H,-}(\left|\langle\mu_k,w\rangle\right|),\qquad w\in\mathbb{B}_k^-.
$$

For $w\in\mathbb{B}_k^+$ we have similar limits. Therefore the one-dimensional distributions of $Z_\varepsilon(\vartheta_0,w)$ converge to the one-dimensional distributions of $Z(\vartheta_0,w)$. As it follows from the given proof this convergence is uniform on compacts $\mathbb{K}\subset\Theta$. To prove the convergence of finite-dimensional distributions

$$\left(Z_\varepsilon\left(\vartheta_0,w_1\right),\ldots,Z_\varepsilon\left(\vartheta_0,w_N\right)\right)\Longrightarrow\left(Z\left(\vartheta_0,w_1\right),\ldots,Z\left(\vartheta_0,w_N\right)\right)$$

we can use the same steps as in the proof given above. □

Lemma 7.6 *For any compact $\mathbb{K}\subset\Theta$ there exists constant $C>0$ such that for any $R>0$ and $\|w_1\|+\|w_2\|<R$ we have*

$$\sup_{\vartheta_0\in\mathbb{K}}\mathbf{E}_{\vartheta_0}\left|Z_\varepsilon^{1/4}\left(\vartheta_0,w_2\right)-Z_\varepsilon^{1/4}\left(\vartheta_0,w_1\right)\right|^4\leq C\left(1+R^4\right)\|w_2-w_1\|^{4\kappa+2}. \tag{7.19}$$

Proof Using (1.10) we write

$$\begin{aligned}
\mathbf{E}_{\vartheta_0}\left|Z_\varepsilon^{1/4}\left(\vartheta_0,w_2\right)-Z_\varepsilon^{1/4}\left(\vartheta_0,w_1\right)\right|^4&\leq C\sum_{k=1}^K\mathbf{E}_{\vartheta_0+\varphi_\varepsilon w_1}\left|V_T-1\right|^4\\
&\leq C\mathbf{E}_{\vartheta_0+\varphi_\varepsilon w_1}\left(\sum_{k=1}^K\int_0^T V_th_t^2D_{k,\varepsilon}\left(\vartheta_0,w_1,w_2,t\right)^2\mathrm{d}t\right)^4\\
&\qquad+C\mathbf{E}_{\vartheta_0+\varphi_\varepsilon w_1}\left(\sum_{k=1}^K\int_0^T V_th_tD_{k,\varepsilon}\left(\vartheta_0,w_1,w_2,t\right)\mathrm{d}\bar W_k\left(t\right)\right)^4\\
&\leq C\sum_{k=1}^K\mathbf{E}_{\vartheta_0+\varphi_\varepsilon w_1}\left(\int_0^T V_tD_{k,\varepsilon}\left(\vartheta_0,w_1,w_2,t\right)^2\mathrm{d}t\right)^4\\
&\qquad+C\sum_{k=1}^K\mathbf{E}_{\vartheta_0+\varphi_\varepsilon w_1}\left(\int_0^T V_t^2D_{k,\varepsilon}\left(\vartheta_0,w_1,w_2,t\right)^2\mathrm{d}t\right)^2.
\end{aligned}$$

Introduce notation $\vartheta_1=\vartheta_0+\nu\varphi_\varepsilon w_1$, $\vartheta_2=\vartheta_0+\nu\varphi_\varepsilon w_2$

$$\begin{aligned}
\Delta\psi_k\left(t\right)&=\psi_{\delta,\kappa}\left(t-\tau_k\left(\vartheta_1\right)\right)-\psi_{\delta,\kappa}\left(t-\tau_k\left(\vartheta_2\right)\right),\\
\Delta m_k\left(t\right)&=m\left(\tau_k\left(\vartheta_1\right),t\right)-m\left(\tau_k\left(\vartheta_2\right),t\right),
\end{aligned}$$

and write

$$\varepsilon D_{k,\varepsilon}\left(\vartheta_0,w_1,w_2,t\right)=m\left(\tau_k\left(\vartheta_1\right),t\right)\Delta\psi_k\left(t\right)+\psi_{\delta,\kappa}\left(t-\tau_k\left(\vartheta_2\right)\right)\Delta m\left(t\right).$$

We have

$$\mathbf{E}_{\vartheta_1}\left(\int_0^T V_t^2D_{k,\varepsilon}\left(\vartheta_0,w_1,w_2,t\right)^2\mathrm{d}t\right)^2$$

$$\leq \frac{C}{\varepsilon^2}\mathbf{E}_{\vartheta_1}\left(\int_0^T\left[\Delta\psi_k(t)^2\, V_t^2 m\left(\tau_k\left(\vartheta_1\right),t\right)^2+V_t^2\Delta m_k(t)^2\right]\mathrm{d}t\right)^2.$$

Recall that $m\left(\tau_k\left(\vartheta_1\right),t\right), 0\leq t\leq T$ is Gaussian process with bounded variance. Hence $(2ab\leq a^2+b^2)$

$$\begin{aligned}2\mathbf{E}_{\vartheta_1}V_s^2V_t^2 m\left(\tau_k\left(\vartheta_1\right),t\right)^2 m\left(\tau_k\left(\vartheta_1\right),s\right)^2&\\ \leq \mathbf{E}_{\vartheta_1}V_t^4 m\left(\tau_k\left(\vartheta_1\right),t\right)^4+\mathbf{E}_{\vartheta_1}V_s^4 m\left(\tau_k\left(\vartheta_1\right),s\right)^4&\\ =\mathbf{E}_{\vartheta_2} m\left(\tau_k\left(\vartheta_1\right),t\right)^4+\mathbf{E}_{\vartheta_2} m\left(\tau_k\left(\vartheta_1\right),s\right)^4\leq C,&\end{aligned}$$

and

$$\begin{aligned}2\mathbf{E}_{\vartheta_1}V_s^2V_t^2\Delta m_k(t)^2\,\Delta m_k(s)^2&\leq \mathbf{E}_{\vartheta_1}V_s^4\Delta m_k(t)^4+\mathbf{E}_{\vartheta_1}V_s^4\Delta m_k(s)^4\\ &\leq \mathbf{E}_{\vartheta_2}\Delta m_k(t)^4+\mathbf{E}_{\vartheta_2}\Delta m_k(s)^4.\end{aligned}$$

Note that the Gaussian process $m_k(\tau,t)$ is mean square differentiable w.r.t. τ and we have

$$\begin{aligned}\mathbf{E}_{\vartheta}\Delta m_k(s)^4&=\mathbf{E}_{\vartheta}\left|m\left(\tau_k\left(\vartheta_1\right),t\right)-m\left(\tau_k\left(\vartheta_2\right),t\right)\right|^4\\ &\leq C\left|\tau_k\left(\vartheta_1\right)-\tau_k\left(\vartheta_2\right)\right|^4\leq C\varphi_\varepsilon^4\left\|w_2-w_1\right\|^4.\end{aligned}$$

These estimates allow us to write

$$\begin{aligned}&\mathbf{E}_{\vartheta_1}\left(\int_0^T V_t^2 D_{k,\varepsilon}\left(\vartheta_0,w_1,w_2,t\right)^2\mathrm{d}t\right)^2\\ &\quad=\varepsilon^{-4}\mathbf{E}_{\vartheta_1}\int_0^T\int_0^T V_t^2V_s^2 D_{k,\varepsilon}\left(\vartheta_0,w_1,w_2\right)^2 D_{k,\varepsilon}\left(\vartheta_0,w_1,w_2,s\right)^2\mathrm{d}t\,\mathrm{d}s\\ &\quad\leq C\varepsilon^{-4}\left(\int_0^T\Delta\psi_k(t)^2\,\mathrm{d}t\right)^2+\varepsilon^{-4}\varphi_\varepsilon^4\left\|w_2-w_1\right\|^4\\ &\quad\leq C\frac{\varphi_\varepsilon^{4\kappa+2}}{\varepsilon^4}\left\|w_2-w_1\right\|^{4\kappa+2}\left(\int_{\mathcal{R}}\left[\operatorname{sgn}(s-1)\left|s-1\right|^\kappa-\operatorname{sgn}(s)\left|s\right|^\kappa\right]^2\mathrm{d}s\right)^2\\ &\qquad\qquad+C+\varepsilon^{-4}\varphi_\varepsilon^4\left\|w_2-w_1\right\|^4\\ &\quad\leq C\ \left\|w_2-w_1\right\|^{4\kappa+2}+C\,\varepsilon^{\frac{2-4\kappa}{\kappa+\frac12}}\left\|w_2-w_1\right\|^4\\ &\quad\leq C\left(1+R^{2-4\kappa}\right)\left\|w_2-w_1\right\|^{4\kappa+2}\end{aligned}$$

for $\left\|w_1\right\|+\left\|w_2\right\|<R$.

For the term

$$\mathbf{E}_{\vartheta_1}\left(\int_0^T V_t D_{k,\varepsilon}(\vartheta_0, w_1, w_2, t)^2\,\mathrm{d}t\right)^4$$

we use similar relations like

$$\begin{aligned}&\mathbf{E}_{\vartheta_1} V_{t_1} V_{t_2} V_{t_3} V_{t_4} m(\tau_k(\vartheta_1), t_1)\, m(\tau_k(\vartheta_1), t_2)\, m(\tau_k(\vartheta_1), t_3)\, m(\tau_k(\vartheta_1), t_4)\\ &\quad\leq C\sum_{i=1}^4 \mathbf{E}_{\vartheta_1} V_{t_i}^4 m(\tau_k(\vartheta_1), t_i)^4 = C\sum_{i=1}^4 \mathbf{E}_{\vartheta_i} m(\tau_k(\vartheta_1), t_i)^4 \leq C,\end{aligned}$$

and obtain the relations

$$\begin{aligned}&\sum_{k=1}^K \mathbf{E}_{\vartheta_1}\left(\int_0^T V_t D_{k,\varepsilon}(\vartheta_0, w_1, w_2, t)^2\,\mathrm{d}t\right)^4\\ &\qquad\leq C\frac{\varphi_\varepsilon^{8\kappa+4}}{\varepsilon^8}\left\|w_2-w_1\right\|^{8\kappa+4} + C\frac{\varphi_\varepsilon^8}{\varepsilon^8}\left\|w_2-w_1\right\|^8\\ &\qquad\leq C\left\|w_2-w_1\right\|^{8\kappa+4} + C\varepsilon^{\frac{4-8\kappa}{\kappa+\frac12}}\left\|w_2-w_1\right\|^8\\ &\qquad\leq C\left(1+R^4\right)\left\|w_2-w_1\right\|^{4\kappa+2}.\end{aligned}$$

Therefore we obtained (7.19). □

Lemma 7.7 *For any compact* $\mathbb{K}\subset\Theta$ *and any* $N>0$ *there exists constant* $C_*>0$ *such that*

$$\sup_{\vartheta_0\in\mathbb{K}} \mathbf{E}_{\vartheta_0} Z_\varepsilon(\vartheta_0, w)^{1/2} \leq \frac{C_*}{\|w\|^{2HN}}.$$

Proof We follow the same steps as in the proof of Lemma 7.4. The local relation is (below $\vartheta_w=\vartheta_0+\nu\varphi_\varepsilon w$)

$$\begin{aligned}F(\vartheta_w,\vartheta_0) &= \sum_{k=1}^K\int_0^T \frac{h_k(t)^2}{\varepsilon^2}\left[\psi_{\delta,\kappa}(t-\tau_k(\vartheta_w))\, y_k(\vartheta_w,t)\right.\\ &\qquad\left.-\psi_{\delta,\kappa}(t-\tau_k(\vartheta_0))\, y_k(\vartheta_0,t)\right]^2\mathrm{d}t\\ &= \sum_{k=1}^K\int_0^T \frac{h_k(t)^2}{\varepsilon^2}\left[\psi_{\delta,\kappa}(t-\tau_k(\vartheta_w))\right.\\ &\qquad\left.-\psi_{\delta,\kappa}(t-\tau_k(\vartheta_0))\right]^2 y_k(\vartheta_0,t)^2\,\mathrm{d}t\,(1+o(1))\end{aligned}$$

$$= \sum_{k=1}^{K} \pi_k^2 \left|\langle \mu_k, w\rangle\right|^{2H} (1 + o(1)) .$$

Therefore there exist $\varepsilon_1 > 0$ and $d > 0$ such that $\varepsilon_1 \|w\| < d$ and all $\varepsilon < \varepsilon_1$ we have

$$F(\vartheta_w, \vartheta_0) \geq \sum_{k=1}^{K} \pi_k^2 \left|\langle \mu_k, w\rangle\right|^{2H} (1 + o(1)) \geq \frac{1}{2} \sum_{k=1}^{K} \pi_k^2 \left|\langle \mu_k, w\rangle\right|^{2H} \geq \hat{c} \|w\|^{2H} .$$

To verify the last inequality with some $\hat{c} > 0$ we write

$$\sum_{k=1}^{K} \pi_k^2 \left|\langle \mu_k, w\rangle\right|^{2H} = \sum_{k=1}^{K} \pi_k^2 \left|\langle \mu_k, e\rangle\right|^{2H} \|w\|^{2H} ,$$

where we denoted $e = w/ \|w\|$. We have

$$\hat{c} = \inf_{e \in \mathcal{R}^2, \|e\|=1} \sum_{k=1}^{K} \pi_k^2 \left|\langle \mu_k, e\rangle\right|^{2H} > 0.$$

If $\hat{c} = 0$, then there exists a vector e_* such that all scalar products $\langle \mu_k, e_*\rangle = 0, k = 1, \ldots, K$. As $K \geq 3$ and there exist at least three detectors not on the same line, the vector e_*, which is orthogonal to all vectors $\mu_k, k = 1, \ldots, K$ does not exists.

For $\|\vartheta - \vartheta_0\| \geq d$ we have

$$\inf_{\vartheta_0 \in \Theta} \inf_{\|\vartheta - \vartheta_0\| \geq d} \varepsilon^2 F(\vartheta, \vartheta_0) > 0.$$

The proof of this inequality is the same as the proof of similar relation in the Lemma 7.4.

Hence there exists $\tilde{c} > 0$ such that

$$F(\vartheta, \vartheta_0) \geq \tilde{c} \|w\|^{2H} .$$

The end of the proof is the same as in Lemma 7.4. □

The properties of the normalized likelihood ratio random field $Z_\varepsilon(\vartheta_0, \cdot)$ established in Lemmas 7.5–7.7 are sufficient to cite Theorems 1.59 and 1.60, where the mentioned properties of estimators were established under such conditions.

Change-Point Case

Let us consider once more the problem of one source detection with K detectors (see Fig. 7.1). The observations $X^K = (X_1, \ldots, X_K)$, where $X_k = (X_k(t), 0 \le t \le T)$ and

$$dX_k(t) = f_k(t)\,\psi(t - \tau_k(\vartheta))\,Y_k(t)\,dt + \varepsilon\sigma_k(t)\,dW_t, \quad X_k(0) = 0, \quad 0 \le t \le T.$$

Here the function

$$\psi(t) = \mathbb{I}_{\{t \ge 0\}},$$

corresponds to the case (**c**) in Fig. 7.2. The *hidden* Gaussian processes $Y_k(\cdot), k = 1, \ldots, K$, satisfy the same equations

$$dY_k(t) = a_k(t)\,Y_k(t)\,dt + \varepsilon b_k(t)\,dV_k(t), \quad Y_k(0) = y_{k,0}, \quad 0 \le t \le T.$$

As before we suppose that the conditions $\mathscr{N}$ are fulfilled. The parameter $\varepsilon \in (0, 1)$ is small and the asymptotic is the same: $\varepsilon \to 0$.

The delay $\tau_k(\vartheta) = \rho_k/\nu = \nu^{-1}\|\vartheta_k - \vartheta_0\|$ and we have to estimate the position $\vartheta_0 = (x_0, y_0)^\top$ by observations X^K.

The likelihood ratio function is

$$L\left(\vartheta, X^K\right) = \exp\left\{\sum_{k=1}^{K}\int_{\tau_k(\vartheta)}^{T} \frac{f_k(t)\,m_k(\tau_k(\vartheta), t)}{\varepsilon^2\sigma_k(t)^2}\,dX_k(t) - \sum_{k=1}^{K}\int_{\tau_k(\vartheta)}^{T} \frac{f_k(t)^2\,m_k(\tau_k(\vartheta), t)^2}{2\varepsilon^2\sigma_k(t)^2}\,dt\right\}, \quad \vartheta \in \Theta.$$

The estimators $\hat{\vartheta}_\varepsilon, \tilde{\vartheta}_\varepsilon$ are defined by the relations (7.10).

We need the same notations (7.3) and

$$\pi_k(\vartheta_0) = h_k(\tau_k(\vartheta_0))\,y_k(\tau_k(\vartheta_0)), \qquad \pi_k = \pi_k(\vartheta_0).$$

The normalizing function is $\varphi_\varepsilon = \varepsilon^2\nu$, and the limit likelihood ratio random field for this model is

$$Z(\vartheta_0, w) = \exp\left\{\sum_{k=1}^{K}\left[\pi_k W_k^+(\langle\mu_k, w\rangle) - \frac{\pi_k^2}{2}\,|\langle\mu_k, w\rangle|\right]\mathbb{I}_{\{\mathbb{B}_k^+\}} + \sum_{k=1}^{K}\left[\pi_k W_k^-(-\langle\mu_k, w\rangle) - \frac{\pi_k^2}{2}\,|\langle\mu_k, w\rangle|\right]\mathbb{I}_{\{\mathbb{B}_k^-\}}\right\}, \quad w \in \mathcal{R}^2.$$

Here $W_k^+(s), s \geq 0, k = 1, \ldots, K$ and $W_k^-(s), s \geq 0, k = 1, \ldots, K$ are independent two-sided Brownian motions.

The limit random vectors $\hat{\mathrm{w}}(\vartheta_0)$ and $\tilde{\mathrm{w}}(\vartheta_0)$ are defined by the same relations

$$Z\left(\vartheta_0, \hat{\mathrm{w}}(\vartheta_0)\right) = \sup_{w \in \mathcal{R}^2} Z(\vartheta_0, w), \qquad \tilde{\mathrm{w}}(\vartheta_0) = \frac{\int_{\mathcal{R}^2} w Z(\vartheta_0, w)\, dw}{\int_{\mathcal{R}^2} Z(\vartheta_0, w)\, dw}$$

but the limit distributions of the estimators will be given by the random vectors $\hat{\mathrm{u}}(\vartheta_0) = \nu \hat{\mathrm{w}}(\vartheta_0)$ and $\tilde{\mathrm{u}}(\vartheta_0) = \nu \tilde{\mathrm{w}}(\vartheta_0)$, respectively.

The asymptotically efficient estimators in this problem are defined with the help of the following minimax lower bound

$$\lim_{h \to 0} \underline{\lim}_{\varepsilon \to 0} \sup_{\|\vartheta - \vartheta_0\| \leq h} \varepsilon^{-4} \mathbf{E}_\vartheta \|\bar{\vartheta}_\varepsilon - \vartheta\|^2 \geq \mathbf{E}_{\vartheta_0} \|\tilde{\mathrm{u}}(\vartheta_0)\|^2.$$

For the proof see (7.17)–(7.18).

We call the estimator ϑ_ε^* asymptotically efficient if for this estimator we have

$$\lim_{h \to 0} \lim_{\varepsilon \to 0} \sup_{\|\vartheta - \vartheta_0\| \leq h} \varepsilon^{-4} \mathbf{E}_\vartheta \|\vartheta_\varepsilon^* - \vartheta\|^2 = \mathbf{E}_{\vartheta_0} \|\tilde{\mathrm{u}}(\vartheta_0)\|^2$$

for all $\vartheta_0 \in \Theta$.

Theorem 7.3 *Suppose that the conditions are fulfilled, then the MLE $\hat{\vartheta}_\varepsilon$ and BE $\tilde{\vartheta}_\varepsilon$ are uniformly on compacts consistent and have different limit distributions*

$$\varepsilon^{-2}\left(\hat{\vartheta}_\varepsilon - \vartheta_0\right) \Longrightarrow \hat{\mathrm{u}}(\vartheta_0), \qquad \varepsilon^{-2}\left(\tilde{\vartheta}_\varepsilon - \vartheta_0\right) \Longrightarrow \tilde{\mathrm{u}}(\vartheta_0),$$

the moments converge: for any $p > 0$

$$\varepsilon^{-2p} \mathbf{E}_{\vartheta_0} \|\hat{\vartheta}_\varepsilon - \vartheta_0\|^p \longrightarrow \mathbf{E}_{\vartheta_0} \|\hat{\mathrm{u}}(\vartheta_0)\|^p, \quad \varepsilon^{-2p} \mathbf{E}_{\vartheta_0} \|\tilde{\vartheta}_\varepsilon - \vartheta_0\|^p \longrightarrow \mathbf{E}_{\vartheta_0} \|\tilde{\mathrm{u}}(\vartheta_0)\|^p$$

and BE are asymptotically efficient.

Proof Introduce the normalized likelihood ratio

$$Z_\varepsilon(\vartheta_0, w) = \frac{L\left(\vartheta_0 + \nu\varepsilon^2 w, X^K\right)}{L\left(\vartheta_0, X^K\right)}, \qquad w \in \mathbb{W}_\varepsilon = \left(w : \vartheta_0 + \nu\varepsilon^2 w \in \Theta\right).$$

□

Once more we have to prove three lemmas, which verify $\mathscr{I}\mathscr{K}_2$ and then to cite Theorems 1.10 and 1.12.

Lemma 7.8 *The finite-dimensional distributions of the random fields* $Z_\varepsilon(\vartheta_0,\cdot)$ *converge to the finite-dimensional distributions of* $Z(\vartheta_0,\cdot)$*, and this convergence is uniform on compacts* $\mathbb{K}\in\Theta$.

Proof Suppose that $w\in\mathbb{B}_k^-$. Then

$$\tau_k(\vartheta_0+\varphi_\varepsilon w)=\tau_k(\vartheta_0)-\varphi_\varepsilon\langle\mu_k,w\rangle+O\left(\varphi_\varepsilon^2\right)>\tau_k(\vartheta_0)$$

and

$$\begin{aligned}
\ln Z_\varepsilon&(\vartheta_0,w)\\
&=\sum_{k=1}^K\int_{\tau_k(\vartheta_0)}^T\frac{h_k(t)\left[\mathbb{I}_{\{t>\tau_k(\vartheta_w)\}}m_k(\tau_k(\vartheta_w),t)-m_k(\tau_k(\vartheta_w),t)\right]}{\varepsilon\sigma_k(t)}\mathrm{d}\bar W_k(t)\\
&-\sum_{k=1}^K\int_{\tau_k(\vartheta_0)}^T\frac{h_k(t)^2\left[\mathbb{I}_{\{t>\tau_k(\vartheta_w)\}}m_k(\tau_k(\vartheta_w),t)-m_k(\tau_k(\vartheta_w),t)\right]^2}{2\varepsilon^2\sigma_k(t)^2}\mathrm{d}t.
\end{aligned}$$

For ordinary integral we can write

$$\begin{aligned}
\int_{\tau_k(\vartheta_0)}^T&\frac{h_k(t)^2\left[\mathbb{I}_{\{t>\tau_k(\vartheta_w)\}}m_k(\tau_k(\vartheta_w),t)-m_k(\tau_k(\vartheta_w),t)\right]^2}{\varepsilon^2\sigma_k(t)^2}\mathrm{d}t\\
&=\int_{\tau_k(\vartheta_0)}^{\tau_k(\vartheta_0+\varphi_\varepsilon w)}\frac{h_k(t)^2\,m_k(\tau_k(\vartheta_w),t)^2}{\varepsilon^2\sigma_k(t)^2}\mathrm{d}t\,(1+o(1))\\
&=\frac{h_k(\tau_k(\vartheta_0))^2}{\varepsilon^2\sigma_k(\tau_k(\vartheta_0))^2}\int_{\tau_k(\vartheta_0)}^{\tau_k(\vartheta_0+\varphi_\varepsilon w)}m_k(\tau_k(\vartheta_w),t)^2\,\mathrm{d}t\,(1+o(1))\\
&=\frac{h_k(\tau_k(\vartheta_0))^2}{\varepsilon^2\sigma_k(\tau_k(\vartheta_0))^2}\int_{\tau_k(\vartheta_0)}^{\tau_k(\vartheta_0+\varphi_\varepsilon w)}m_k(\tau_k(\vartheta_0),t)^2\,\mathrm{d}t\,(1+o(1))\\
&=\frac{h_k(\tau_k(\vartheta_0))^2}{\varepsilon^2\sigma_k(\tau_k(\vartheta_0))^2}\int_{\tau_k(\vartheta_0)}^{\tau_k(\vartheta_0)-\varphi_\varepsilon\langle\mu_k,w\rangle}y_k(t)^2\,\mathrm{d}t\,(1+o(1))\\
&=\frac{h_k(\tau_k(\vartheta_0))^2\,y_k(\tau_k(\vartheta_0))^2}{\sigma_k(\tau_k(\vartheta_0))^2}\frac{\varphi_\varepsilon}{\varepsilon^2}\,\left|\langle\mu_k,w\rangle\right|(1+o(1))\\
&\qquad\longrightarrow\pi_k(\vartheta_0)^2\,\left|\langle\mu_k,w\rangle\right|.
\end{aligned}$$

Hence

$$\sum_{k=1}^K\int_{\tau_k(\vartheta_0)}^T\frac{h_k(t)^2\left[\mathbb{I}_{\{t>\tau_k(\vartheta_w)\}}m_k(\tau_k(\vartheta_w),t)-m_k(\tau_k(\vartheta_w),t)\right]^2}{2\varepsilon^2\sigma_k(t)^2}\mathrm{d}t$$

$$\longrightarrow \sum_{k=1}^{K} \pi_k \left(\vartheta_0\right)^2 \ \left|\langle \mu_k, w\rangle\right| .$$

Similar calculations for stochastic integral lead to the following relations

$$\begin{aligned}
&\int_{\tau_k(\vartheta_0)}^{T} \frac{h_k\left(t\right)\left[\mathbb{I}_{\{t>\tau_k(\vartheta_w)\}} m_k\left(\tau_k\left(\vartheta_w\right), t\right) - m_k\left(\tau_k\left(\vartheta_w\right), t\right)\right]}{\varepsilon \sigma_k\left(t\right)} \mathrm{d}\bar{W}_k\left(t\right) \\
&\qquad = \int_{\tau_k(\vartheta_0)}^{\tau_k(\vartheta_0+\varphi_\varepsilon w)} \frac{h_k\left(t\right) m_k\left(\tau_k\left(\vartheta_w\right), t\right)}{\varepsilon \sigma_k\left(t\right)} \mathrm{d}\bar{W}_k\left(t\right) t \left(1 + o\left(1\right)\right) \\
&\qquad = \frac{h_k\left(\tau_k\left(\vartheta_0\right)\right)}{\varepsilon \sigma_k\left(\tau_k\left(\vartheta_0\right)\right)} \int_{\tau_k(\vartheta_0)}^{\tau_k(\vartheta_0+\varphi_\varepsilon w)} m_k\left(\tau_k\left(\vartheta_w\right), t\right) \mathrm{d}\bar{W}_k\left(t\right) \left(1 + o\left(1\right)\right) \\
&\qquad = \frac{h_k\left(\tau_k\left(\vartheta_0\right)\right)}{\varepsilon \sigma_k\left(\tau_k\left(\vartheta_0\right)\right)} \int_{\tau_k(\vartheta_0)}^{\tau_k(\vartheta_0+\varphi_\varepsilon w)} m_k\left(\tau_k\left(\vartheta_0\right), t\right) \mathrm{d}\bar{W}_k\left(t\right) \left(1 + o\left(1\right)\right) \\
&\qquad = \frac{h_k\left(\tau_k\left(\vartheta_0\right)\right)}{\varepsilon \sigma_k\left(\tau_k\left(\vartheta_0\right)\right)} \int_{\tau_k(\vartheta_0)}^{\tau_k(\vartheta_0)-\varphi_\varepsilon\langle \mu_k, w\rangle} y_k\left(t\right) \mathrm{d}\bar{W}_k\left(t\right) \left(1 + o\left(1\right)\right) \\
&\qquad = \pi_k\left(\vartheta_0\right) \frac{\bar{W}_k\left(\tau_k\left(\vartheta_0\right) - \varphi_\varepsilon\langle \mu_k, w\rangle\right) - \bar{W}_k\left(\tau_k\left(\vartheta_0\right)\right)}{\varepsilon} \left(1 + o\left(1\right)\right) \\
&\qquad = \pi_k\left(\vartheta_0\right) \tilde{W}_{k,\varepsilon}\left(\left|\langle \mu_k, w\rangle\right|\right) \left(1 + o\left(1\right)\right) \Longrightarrow \pi_k\left(\vartheta_0\right) W_k^-\left(\left|\langle \mu_k, w\rangle\right|\right)
\end{aligned}$$

and

$$\begin{aligned}
&\sum_{k=1}^{K} \int_{\tau_k(\vartheta_0)}^{T} \frac{h_k\left(t\right)\left[\mathbb{I}_{\{t>\tau_k(\vartheta_w)\}} m_k\left(\tau_k\left(\vartheta_w\right), t\right) - m_k\left(\tau_k\left(\vartheta_w\right), t\right)\right]}{\varepsilon \sigma_k\left(t\right)} \mathrm{d}\bar{W}_k\left(t\right) \\
&\qquad \Longrightarrow \sum_{k=1}^{K} \pi_k\left(\vartheta_0\right) W_k^-\left(\left|\langle \mu_k, w\rangle\right|\right) .
\end{aligned}$$

Therefore we obtained convergence of one-dimensional distributions. The proof of the convergence of multidimensional distributions follows the same steps but is cumbersome. □

Lemma 7.9 *For any compact* $\mathbb{K}$ *there exists constant* $C > 0$ *such that for any* $R > 0$ *and* $\|w_1\| + \|w_2\| < R$ *we have*

$$\sup_{\vartheta_0 \in \mathbb{K}} \mathbf{E}_{\vartheta_0} \left| Z_\varepsilon^{1/8}\left(\vartheta_0, w_2\right) - Z_\varepsilon^{1/8}\left(\vartheta_0, w_1\right)\right|^8 \leq C \left(1 + R^8\right) \left\|w_2 - w_1\right\|^4 .$$

Proof Suppose that w_1 and w_2 are such that $\tau_k(\vartheta_{w_1}) < \tau_k(\vartheta_{w_2})$ and denote

$$\Delta M_k\left(w_1,w_2,t\right)=1\!\!1_{\{t>\tau_k(\vartheta_{w_2})\}}m_k\left(\tau_k(\vartheta_{w_2}),t\right)-1\!\!1_{\{t>\tau_k(\vartheta_{w_1})\}}m_k\left(\tau_k(\vartheta_{w_1}),t\right).$$

Using once more (7.14) we write

$$\begin{aligned}
\mathbf{E}_{\vartheta_0}&\left|Z_\varepsilon^{1/8}\left(\vartheta_0,w_2\right)-Z_\varepsilon^{1/8}\left(\vartheta_0,w_1\right)\right|^8=\mathbf{E}_{\vartheta_0+\varphi_\varepsilon w_1}\left|V_T-1\right|^8\\
&\leq C\varepsilon^{-16}\mathbf{E}_{\vartheta_{w_1}}\left(\sum_{k=1}^K\int_0^T V_t h_t^2\Delta M_k\left(w_1,w_2,t\right)^2\,\mathrm{d}t\right)^8\\
&\qquad+C\varepsilon^{-8}\mathbf{E}_{\vartheta_{w_1}}\left(\sum_{k=1}^K\int_0^T V_t h_t\Delta M_k\left(w_1,w_2,t\right)\mathrm{d}\bar W_k\left(t\right)\right)^8\\
&\leq C\varepsilon^{-16}\sum_{k=1}^K\mathbf{E}_{\vartheta_{w_1}}\left(\int_0^T V_t\Delta M_k\left(w_1,w_2,t\right)^2\,\mathrm{d}t\right)^8\\
&\qquad+C\varepsilon^{-8}\sum_{k=1}^K\mathbf{E}_{\vartheta_{w_1}}\left(\int_0^T V_t^2\Delta M_k\left(w_1,w_2,t\right)^2\,\mathrm{d}t\right)^4.
\end{aligned}$$

Here

$$\begin{aligned}
V_t=\exp&\left\{\sum_{k=1}^K\int_0^t\frac{h_k\left(s\right)}{8\varepsilon}\Delta M_k\left(w_1,w_2,s\right)\mathrm{d}\bar W_k\left(s\right)\right.\\
&\qquad\left.-\sum_{k=1}^K\int_0^t\frac{h_k\left(s\right)^2}{16\varepsilon^2}\Delta M_k\left(w_1,w_2,s\right)^2\,\mathrm{d}s\right\}.
\end{aligned}$$

We have

$$\begin{aligned}
\int_0^T V_t^2\Delta M_k\left(w_1,w_2,t\right)^2\,\mathrm{d}t&=\int_{\tau_k(\vartheta_{w_1})}^{\tau_k(\vartheta_{w_2})}V_t^2 m_k\left(\tau_k\left(\vartheta_{w_1}\right),t\right)^2\mathrm{d}t\\
&\quad+\int_{\tau_k(\vartheta_{w_2})}^T V_t^2\left[m_k\left(\tau_k\left(\vartheta_{w_2}\right),t\right)-m_k\left(\tau_k\left(\vartheta_{w_1}\right),t\right)\right]^2\mathrm{d}t.
\end{aligned}$$

Let us denote $m_k\left(t\right)=m_k\left(\tau_k\left(\vartheta_{w_1}\right),t\right)$, then

$$\begin{aligned}
\mathbf{E}_{\vartheta_{w_1}}&V_{t_1}^2 m_k\left(t_1\right)^2V_{t_2}^2 m_k\left(t_2\right)^2V_{t_3}^2 m_k\left(t_3\right)^2V_{t_4}^2 m_k\left(t_4\right)^2\\
&\leq C\sum_{i=1}^4\mathbf{E}_{\vartheta_{w_1}}V_{t_i}^8 m_k\left(t_i\right)^8=C\sum_{i=1}^4\mathbf{E}_{\vartheta_{w_2}}m_k\left(t_i\right)^8\leq C.
\end{aligned}$$

Further

$$\mathbf{E}_{\vartheta_{w_1}}\left(\int_{\tau_k(\vartheta_{w_1})}^{\tau_k(\vartheta_{w_2})} V_t^2 m_k\left(\tau_k\left(\vartheta_{w_1}\right),t\right)^2 \mathrm{d}t\right)^4$$

$$=\int_{\tau_k(\vartheta_{w_1})}^{\tau_k(\vartheta_{w_2})}\int_{\tau_k(\vartheta_{w_1})}^{\tau_k(\vartheta_{w_2})}\int_{\tau_k(\vartheta_{w_1})}^{\tau_k(\vartheta_{w_2})}\int_{\tau_k(\vartheta_{w_1})}^{\tau_k(\vartheta_{w_2})}\prod_{i=1}^{4}\mathbf{E}_{\vartheta_{w_1}} V_{t_i}^2 m_k\left(\tau_k\left(\vartheta_{w_1}\right),t_i\right)^2 \mathrm{d}t_i$$

$$\leq C\left|\tau_k(\vartheta_{w_2}-\tau_k(\vartheta_{w_1}\right|^4 \leq C\varepsilon^8\left\|w_2-w_1\right\|^4$$

and

$$\mathbf{E}_{\vartheta_{w_1}}\left(\int_{\tau_k(\vartheta_{w_2})}^{T} V_t^2\left[m_k\left(\tau_k\left(\vartheta_{w_2}\right),t\right)-m_k\left(\tau_k\left(\vartheta_{w_1}\right),t\right)\right]^2 \mathrm{d}t\right)^4$$

$$\leq\left(T-\tau_k(\vartheta_{w_2})\right)^3\int_{\tau_k(\vartheta_{w_2})}^{T}\mathbf{E}_{\vartheta_{w_1}} V_t^8\left[m_k\left(\tau_k\left(\vartheta_{w_2}\right),t\right)-m_k\left(\tau_k\left(\vartheta_{w_1}\right),t\right)\right]^8 \mathrm{d}t$$

$$=T^3\int_{\tau_k(\vartheta_{w_2})}^{T}\mathbf{E}_{\vartheta_{w_2}}\left[m_k\left(\tau_k\left(\vartheta_{w_2}\right),t\right)-m_k\left(\tau_k\left(\vartheta_{w_1}\right),t\right)\right]^8 \mathrm{d}t$$

$$\leq C\left\|\tau_k\left(\vartheta_{w_2}\right)-\tau_k\left(\vartheta_{w_1}\right)\right\|^8 \leq C\varepsilon^{16}\left\|w_2-w_1\right\|^8.$$

Therefore for $\|w_1\|<R$, $\|w_2\|<R$ we obtained

$$\varepsilon^{-8}\sum_{k=1}^{K}\mathbf{E}_{\vartheta_{w_1}}\left(\int_0^T V_t^2\Delta M_k\left(w_1,w_2,t\right)^2 \mathrm{d}t\right)^4$$

$$\leq C\left\|w_2-w_1\right\|^4+C\varepsilon^8\left\|w_2-w_1\right\|^8\leq C\left(1+R^4\right)\left\|w_2-w_1\right\|^4.$$

Using similar arguments it is possible to verify the estimate

$$\varepsilon^{-16}\sum_{k=1}^{K}\mathbf{E}_{\vartheta_{w_1}}\left(\int_0^T V_t\Delta M_k\left(w_1,w_2,t\right)^2 \mathrm{d}t\right)^8$$

$$\leq C\left\|w_2-w_1\right\|^8+C\varepsilon^{16}\left\|w_2-w_1\right\|^{16}\leq C\left(1+R^8\right)\left\|w_2-w_1\right\|^8.$$

□

Lemma 7.10 *For any compact* $\mathbb{K}$ *and any* $N>0$ *There exist constant* $C_*>0$ *such that*

$$\sup_{\vartheta_0\in\mathbb{K}} \mathbf{E}_{\vartheta_0} Z_\varepsilon\left(\vartheta_0,w\right)^{1/2} \leq \frac{C_*}{\|w\|^N}.$$

Proof Once more we follow the proof of Lemma 7.4. First we show that for some $d>0$ and $\|\vartheta_w-\vartheta_0\|\leq d$ we have the estimate

$$\sum_{k=1}^K\int_0^T \frac{h_t^2}{\varepsilon^2}\left[\mathbb{I}_{\{t\geq\tau_k(\vartheta_w)\}}y_k\left(\vartheta_w,t\right)-\mathbb{I}_{\{t\geq\tau_k(\vartheta_0)\}}y_k\left(t\right)\right]^2{\rm d}t\geq \bar c\left\|w\right\|$$

with some constant $\bar c>0$. Then for $\vartheta_0\in\mathbb{K}$

$$\inf_{\|\vartheta-\vartheta_0\|>d}\sum_{k=1}^K\int_0^T h_t^2\left[\mathbb{I}_{\{t\geq\tau_k(\vartheta)\}}y_k\left(\vartheta,t\right)-\mathbb{I}_{\{t\geq\tau_k(\vartheta_0)\}}y_k\left(t\right)\right]^2{\rm d}t>0. \tag{7.20}$$

Consider the values $\varepsilon\left\|w\right\|\leq d$ and $\tau_k\left(\vartheta_w\right)>\tau_k\left(\vartheta_0\right)$

$$\begin{aligned}
F_k\left(\vartheta_w,\vartheta_0\right)&=\varepsilon^{-2}\int_0^T h_t^2\left[\mathbb{I}_{\{t\geq\tau_k(\vartheta_w)\}}y_k\left(\vartheta_w,t\right)-\mathbb{I}_{\{t\geq\tau_k(\vartheta_0)\}}y_k\left(t\right)\right]^2{\rm d}t\\
&\geq C\varepsilon^{-2}\int_0^T\left[\mathbb{I}_{\{t\geq\tau_k(\vartheta_w)\}}y_k\left(\vartheta_w,t\right)-\mathbb{I}_{\{t\geq\tau_k(\vartheta_0)\}}y_k\left(t\right)\right]^2{\rm d}t\\
&\geq C\varepsilon^{-2}\int_{\tau_k(\vartheta_0)}^{\tau_k(\vartheta_w)} y_k\left(t\right)^2{\rm d}t+C\varepsilon^{-2}\int_{\tau_k(\vartheta_w)}^{T}\left[y_k\left(\vartheta_w,t\right)-y_k\left(t\right)\right]^2{\rm d}t\\
&\geq C\varepsilon^{-2}y_k\left(\tau_k\left(\vartheta_0\right)\right)^2\left[\tau_k\left(\vartheta_w\right)-\tau_k\left(\vartheta_0\right)\right]\left(1+o\left(1\right)\right)-C\varepsilon^{-2}\left\|\vartheta_w-\vartheta_0\right\|^2\\
&\geq Cy_k\left(\tau_k\left(\vartheta_0\right)\right)^2\left|\langle\mu_k,w\rangle\right|\left(1+o\left(1\right)\right)-C\varepsilon^2\left\|w\right\|^2.
\end{aligned}$$

Therefore

$$\begin{aligned}
\sum_{k=1}^K F_k\left(\vartheta_w,\vartheta_0\right)&\geq C\sum_{k=1}^K y_k\left(\tau_k\left(\vartheta_0\right)\right)^2\left|\langle\mu_k,w\rangle\right|\left(1+o\left(1\right)\right)-Cd\varepsilon\left\|w\right\|\\
&\geq C\sum_{k=1}^K\left|\langle\mu_k,w\rangle\right|\left(1+o\left(1\right)\right)-Cd\varepsilon\left\|w\right\|.
\end{aligned}$$

Once more we have the estimate

$$\sum_{k=1}^K\left|\langle\mu_k,w\rangle\right|=\left\|w\right\|\sum_{k=1}^K\left|\langle\mu_k,e\rangle\right|\geq\check c\left\|w\right\|.$$

Hence

$$\sum_{k=1}^{K} F_k\left(\vartheta_w,\vartheta_0\right) \geq \left(C\check{c} - Cd\varepsilon\right)\left\|w\right\| \geq c_1\left\|w\right\|$$

and we obtain (7.20) for sufficiently small ε.

Suppose that we have equality ($=0$) in (7.20). Then there exists ϑ_* such that $\left\|\vartheta_* - \vartheta_0\right\| > 0$ and

$$\sum_{k=1}^{K}\int_0^T h_t^2\left[\mathbb{I}_{\{t\geq\tau_k(\vartheta_*)\}}y_k\left(\vartheta_*,t\right) - \mathbb{I}_{\{t\geq\tau_k(\vartheta_0)\}}y_k\left(t\right)\right]^2 \mathrm{d}t = 0.$$

This is possible if and only if $\tau_k\left(\vartheta_*\right) = \tau_k\left(\vartheta_0\right)$ for all $k = 1,\ldots,K$. These equalities simultaneously are impossible if there are at least three detectors not on the same line. Therefore (7.20) is valid.

Now the properties of the MLE and BE follow from Theorems 1.9 and 1.11. □

LSE

Let us consider the different statements of the problem. Suppose that we have the same model of observations (7.1) and (7.2), but the estimators $\tau_{k,\varepsilon}^*$ of the time of arrive $\tau_k\left(\vartheta_0\right)$ of the signals are calculated for each detector. Suppose as well that these estimators are consistent and have known limit distributions, i.e., $\tau_{k,\varepsilon} = \tau_k\left(\vartheta_0\right) + \varphi_\varepsilon\xi_{k,\varepsilon}\left(\vartheta_0\right)$, where φ_ε is the rate of convergence and $\xi_{k,\varepsilon}\left(\vartheta_0\right) \Longrightarrow \xi_k\left(\vartheta_0\right)$. As it was shown above the rates can be $\varphi_\varepsilon = \varepsilon$ (smooth case), $\varphi_\varepsilon = \varepsilon^{\frac{2}{2\kappa+1}}$ (cusp case), and $\varphi_\varepsilon = \varepsilon^2$ (change-point case).

We have the convergence $\nu\tau_{k,\varepsilon} \longrightarrow \rho_k\left(\vartheta_0\right)$. Therefore it is possible to construct a *least squares estimator* (LSE) $\check{\vartheta}_\varepsilon$ with the help of the definition

$$Q\left(\check{\vartheta}_\varepsilon\right) = \inf_{\vartheta\in\Theta} Q\left(\vartheta\right)$$

where

$$Q\left(\vartheta\right) = \sum_{k=1}^{K}\left[\nu^2\tau_{k,\varepsilon}^2 - \rho_k\left(\vartheta\right)^2\right]^2.$$

Let us study this estimator, but first we simplified the expression for $Q\left(\vartheta\right)$. We have

$$Q\left(\vartheta\right) = \sum_{k=1}^{K}\left[\nu^2\tau_{k,\varepsilon}^2 - \left(x_k-\theta_1\right)^2 - \left(y_k-\theta_2\right)^2\right]^2$$

$$= \sum_{k=1}^{K} \left[\nu^2 \tau_{k,\varepsilon}^2 - \left(x_k^2 + y_k^2 \right) + 2\theta_1 x_k + 2\theta_2 y_k - \left(\theta_1^2 + \theta_2^2 \right) \right]^2$$

$$= \sum_{k=1}^{K} \left[\nu^2 \tau_{k,\varepsilon}^2 - r_k^2 + 2\theta_1 x_k + 2\theta_2 y_k - 2\vartheta_3 \right]^2$$

Here we denoted $r_k^2 = x_k^2 + y_k^2$ and introduced a new variable $\vartheta_3 = \frac{1}{2}\left(\theta_1^2 + \theta_2^2\right)$. Now $\vartheta = (\theta_1, \theta_2, \theta_3)$ and the system of three equations $\partial Q(\vartheta)/\partial\vartheta = 0$ allow us to write

$$\sum_{k=1}^{K} \left[\nu^2 \tau_{k,\varepsilon}^2 - r_k^2 + 2\theta_1 x_k + 2\theta_2 y_k - 2\vartheta_3 \right] x_k = 0,$$

$$\sum_{k=1}^{K} \left[\nu^2 \tau_{k,\varepsilon}^2 - r_k^2 + 2\theta_1 x_k + 2\theta_2 y_k - 2\vartheta_3 \right] y_k = 0,$$

$$\sum_{k=1}^{K} \left[\nu^2 \tau_{k,\varepsilon}^2 - r_k^2 + 2\theta_1 x_k + 2\theta_2 y_k - 2\vartheta_3 \right] = 0.$$

This system is equivalent to

$$\theta_1 \sum_{k=1}^{K} x_k^2 + \theta_2 \sum_{k=1}^{K} x_k y_k - \vartheta_3 \sum_{k=1}^{K} x_k = \frac{1}{2} \sum_{k=1}^{K} \left[r_k^2 - \nu^2 \tau_{k,\varepsilon}^2 \right] x_k,$$

$$\theta_1 \sum_{k=1}^{K} x_k y_k + \theta_2 \sum_{k=1}^{K} y_k^2 - \vartheta_3 \sum_{k=1}^{K} y_k = \frac{1}{2} \sum_{k=1}^{K} \left[r_k^2 - \nu^2 \tau_{k,\varepsilon}^2 \right] y_k,$$

$$\theta_1 \sum_{k=1}^{K} x_k + \theta_2 \sum_{k=1}^{K} y_k - \vartheta_3 K = \frac{1}{2} \sum_{k=1}^{K} \left[r_k^2 - \nu^2 \tau_{k,\varepsilon}^2 \right].$$

Let us introduce the notation $\bar{x} = \frac{1}{K}\sum_{k=1}^{K} x_k$, $\overline{xy} = \frac{1}{K}\sum_{k=1}^{K} x_k y_k$, and so on. Then if we solve the last equation w.r.t. ϑ_3 and substitute in the first two, then we obtain the following system for the LSE $\check{\vartheta}_\varepsilon = \left(\check{\theta}_{1,\varepsilon}, \check{\theta}_{2,\varepsilon}\right)^\top$

$$\check{\theta}_{1,\varepsilon} \left[\overline{x^2} - (\bar{x})^2 \right] + \check{\theta}_{2,\varepsilon} \left[\overline{xy} - \bar{x}\bar{y} \right] = \frac{1}{2} \left[\overline{\left(r^2 - \nu^2 \tau_\varepsilon^2\right) x} - \overline{\left(r^2 - \nu^2 \tau_\varepsilon^2\right)}\, \bar{x} \right],$$

$$\check{\theta}_{1,\varepsilon} \left[\overline{xy} - \bar{x}\bar{y} \right] + \check{\theta}_{2,\varepsilon} \left[\overline{y^2} - (\bar{y})^2 \right] = \frac{1}{2} \left[\overline{\left(r^2 - \nu^2 \tau_\varepsilon^2\right) y} - \overline{\left(r^2 - \nu^2 \tau_\varepsilon^2\right)}\, \bar{y} \right]. \tag{7.21}$$

This system in matrix form and its solution is

$$\mathbf{A}\check{\vartheta}_\varepsilon^* = z_\varepsilon, \qquad \check{\vartheta}_\varepsilon^* = \mathbf{A}^{-1} z_\varepsilon,$$

where the elements of the vector $z_\varepsilon = \left(z_{1,\varepsilon}, z_{2,\varepsilon}\right)^\top$ are the right-hand parts of the system (7.21).

The equality

$$\text{Det}\,(\mathbf{A}) = \overline{(x-\bar{x})^2}\;\overline{(y-\bar{y})^2} - \left(\overline{(x-\bar{x})(y-\bar{y})}\right)^2 = 0$$

is possible iff there exists a constant c such that $y - \bar{y} = c\ (x - \bar{x})$, and this equality means that all detectors are on the same line. By condition $\mathscr{N}_3$ of the problem such situations are excluded and therefore $\text{Det}\,(\mathbf{A}) \neq 0$ and the estimator $\check{\vartheta}_\varepsilon$ is well defined.

The direct calculations give us the convergences

$$z_{1,\varepsilon} \longrightarrow z_{1,0} = \theta_{0,1}\left[\bar{x^2} - (\bar{x})^2\right] + \theta_{0,2}\left[\overline{xy} - \bar{x}\,\bar{y}\right],$$

$$z_{2,\varepsilon} \longrightarrow z_{2,0} = \theta_{0,1}\left[\overline{xy} - \bar{x}\,\bar{y}\right] + \theta_{0,2}\left[\bar{y^2} - (\bar{y})^2\right].$$

and

$$\varphi_\varepsilon^{-1}\left(\check{\vartheta}_\varepsilon^* - \vartheta_0\right) \Longrightarrow \mathbf{A}^{-1}\eta\left(\vartheta_0\right),$$

where the vector $\eta\left(\vartheta_0\right) = \left(\eta_1\left(\vartheta_0\right), \eta_2\left(\vartheta_0\right)\right)^\top$ is

$$\eta_1\left(\vartheta_0\right) = -\overline{\nu^2\tau\left(\vartheta_0\right) x\,\xi\left(\vartheta_0\right)} = -\frac{\nu^2}{K}\sum_{k=1}^K \tau_k\left(\vartheta_0\right) x_k\,\xi_k\left(\vartheta_0\right),$$

$$\eta_2\left(\vartheta_0\right) = -\overline{\nu^2\tau\left(\vartheta_0\right) y\,\xi\left(\vartheta_0\right)} = -\frac{\nu^2}{K}\sum_{k=1}^K \tau_k\left(\vartheta_0\right) y_k\,\xi_k\left(\vartheta_0\right).$$

Therefore, if the estimators $\tau_{k,\varepsilon}^*, k = 1, \ldots, K$ are asymptotically normal with the rate ε, then the estimator $\check{\vartheta}_\varepsilon$ is asymptotically normal as well

$$\varepsilon^{-1}\left(\check{\vartheta}_\varepsilon^* - \vartheta_0\right) \Longrightarrow \mathcal{N}\left(0, \mathbf{K}\left(\vartheta_0\right)\right)$$

and the covariance function $\mathbf{K}\left(\vartheta_0\right)$ has an explicit expression.

7.1.2 Moving Source

Model of Observations

The model of observations is almost the same as in Sect. 7.1

$$\mathrm{d}X_k(t) = f_k(\vartheta, t)\, Y_k(t)\,\mathrm{d}t + \varepsilon\sigma_k(t)\,\mathrm{d}W_k(t), \quad X_k(0) = 0, \quad 0 \le t \le T, \tag{7.22}$$

$$\mathrm{d}Y_k(t) = a_k(t)\, Y_k(t)\,\mathrm{d}t + \varepsilon b_k(t)\,\mathrm{d}V_k(t), \qquad Y_k(0) = y_k(0) \neq 0. \tag{7.23}$$

Here $f_k(\cdot), \sigma_k(\cdot), a_k(\cdot), b_k(\cdot)$ are known bounded functions, the function $\sigma_k(\cdot)$ is separated from 0, and $W_k(\cdot)$ and $V_k(\cdot)$ are independent Wiener processes, $k = 1, \ldots, K$.

The observations are $X^K = (X_1, \ldots, X_K)$, where $X_k = (X_k(t), 0 \le t \le T)$. We suppose that the source is moving and its position at the moment t denoted $\vartheta(t) = (x_t(\vartheta), y_t(\vartheta))$ is given by the equations

$$x_t(\vartheta) = \theta_1 + \theta_2 t,$$
$$y_t(\vartheta) = \theta_3 + \theta_4 t.$$

The unknown parameter is $\vartheta = (\theta_1, \theta_2, \theta_3, \theta_4)$, and the distance between the source and the k-th detector $D_k = x_k, y_k$ is $\rho_{k,t}(\vartheta) = \|D_k - \vartheta(t)\|$. The functions

$$f_k(\vartheta, t) = \frac{f_k\, S_k(t)}{\rho_{k,t}(\vartheta)^2}, \qquad k = 1, \ldots, K.$$

Here $f_k > 0$ and the functions $S_k(t), k = 1, \ldots, K$ are separated from 0.

The Kalman-Bucy filter for the conditional expectation $m_k(\vartheta, t)$ is

$$\mathrm{d}m_k(\vartheta, t) = \left[a_k(t) - \frac{f_k(\vartheta, t)^2\, \gamma_k^*(\vartheta, t)}{\sigma_k(t)^2}\right] m_k(\vartheta, t)\,\mathrm{d}t + \frac{f_k(\vartheta, t)\, \gamma_k^*(\vartheta, t)}{\sigma_k(t)^2}\mathrm{d}X_k(t), \tag{7.24}$$

$$\frac{\partial \gamma_k^*(\vartheta, t)}{\partial t} = 2a_k(t)\, \gamma_k^*(\vartheta, t) - \frac{f_k(\vartheta, t)^2\, \gamma_k^*(\vartheta, t)^2}{\sigma_k(t)^2} + b_k(t)^2. \tag{7.25}$$

The initial values are $m_k(\vartheta, 0) = y_k(0) \neq 0$, $\gamma_k^*(\vartheta, 0) = 0$.

Preliminary Estimators

Consider the construction of the consistent preliminary estimators of ϑ. Introduce statistics

$$\Phi_{k,\varepsilon}(\tau)=\frac{1}{f_k}\int_0^\tau\frac{\mathrm{d}X_k(t)}{S_k(t)\,y_k(t)},\qquad k=1,\ldots,K.$$

Here

$$y_k(t)=\mathbf{E}_{\vartheta_0}Y_k(t)=y_k(0)\exp\left(\int_0^\tau a_k(s)\,\mathrm{d}s\right),\qquad k=1,\ldots,K.$$

We have the relations

$$\Phi_k(\vartheta_0,\tau)=\mathbf{E}_{\vartheta_0}\Phi_{k,\varepsilon}(\tau)=\int_0^\tau\frac{\mathrm{d}t}{\rho_{k,t}(\vartheta_0)^2},\qquad k=1,\ldots,K$$

$$\begin{aligned}\mathbf{E}_{\vartheta_0}\left(\Phi_{k,\varepsilon}(\tau)-\int_0^\tau\frac{\mathrm{d}t}{\rho_{k,t}(\vartheta_0)^2}\right)^2&=\varepsilon^2\int_0^\tau\frac{\sigma_k(t)^2}{f_k^2S_k(t)^2\,y_k(t)^2}\mathrm{d}t\\&\quad+\mathbf{E}_{\vartheta_0}\left(\int_0^\tau\frac{Y_k(t)-y_k(t)}{\rho_{k,t}(\vartheta_0)^2}\mathrm{d}t\right)^2\\&=\varepsilon^2 R_k(\vartheta_0,\tau),\end{aligned}$$

where $\sup_{\vartheta\in\Theta}R_k(\vartheta,\tau)<C$. Recall that

$$Y_k(t)-y_k(t)=\varepsilon\int_0^t\frac{\psi_k(t)\,b_k(s)}{\psi_k(s)}\mathrm{d}V_k(s),\quad \psi_k(t)=\exp\left(\int_0^t a_k(v)\,\mathrm{d}v\right).\tag{7.26}$$

Therefore

$$\Phi_{k,\varepsilon}(\tau)\longrightarrow\int_0^\tau\frac{\mathrm{d}t}{\rho_{k,t}(\vartheta_0)^2},$$

$$\eta_{k,\varepsilon}(\tau)\equiv\varepsilon^{-1}\big(\Phi_{k,\varepsilon}(\tau)-\Phi_k(\vartheta_0,\tau)\big)\longrightarrow\mathcal{N}(0,R_k(\vartheta_0,\tau)).$$

As the random processes $X_k, k=1\ldots,K$ are independent, the vector $\eta_\varepsilon(\tau)=\big(\eta_{1,\varepsilon}(\tau),\ldots,\eta_{K,\varepsilon}(\tau)\big)^\top$ is asymptotically normal

$$\eta_\varepsilon(\tau)\Longrightarrow\mathcal{N}(0,\mathbf{R}_k(\vartheta_0,\tau)),\tag{7.27}$$

where the matrix $\mathbf{R}_k(\vartheta_0,\tau)$ is diagonal.

Consider two possibilities of the construction of the preliminary estimators for One-step MLE-process. The first estimator has rate of convergence ε and the second the rate ε^{κ}, where $\kappa < 1$.

LSE-1

We calculate this integral $\Phi_k(\vartheta_0, \tau)$ following [33]. Introduce abbreviations:

$$\rho_{k,0}(\vartheta)^2 = (x_k - \theta_1)^2 + (y_k - \theta_3)^2, \qquad \theta_v^2 = \theta_2^2 + \theta_4^2,$$

$$q_k(\vartheta) = \theta_2(x_k - \theta_1) + \theta_4(y_k - \theta_3),$$

$$c_k(\vartheta) = \left(\rho_{k,0}(\vartheta)^2\,\theta_v^2 - q_k(\vartheta)^2\right)^{1/2} = \left|\theta_2(y_k - \theta_3) - \theta_4(x_k - \theta_1)\right|,$$

$$\rho_{k,t}(\vartheta)^2 = \theta_v^2 t^2 - 2q_k(\vartheta)\,t + \rho_{k,0}(\vartheta)^2 = \theta_v^2\left[t^2 - 2\frac{q_k(\vartheta)}{\theta_v^2}t + \frac{\rho_{k,0}(\vartheta)^2}{\theta_v^2}\right]$$

$$= \theta_v^2\left[\left(t - \frac{q_k(\vartheta)}{\theta_v^2}\right)^2 + \frac{c_k(\vartheta)^2}{\theta_v^4}\right] = \frac{c_k(\vartheta)^2}{\theta_v^2}\left[\frac{\theta_v^4}{c_k(\vartheta)^2}\left(t - \frac{q_k(\vartheta)}{\theta_v^2}\right)^2 + 1\right].$$

Therefore, if $c_k(\vartheta) \neq 0$, then

$$\int_0^{\tau} \frac{\mathrm{d}t}{\rho_{k,t}(\vartheta)^2} = \frac{\theta_v^2}{c_k(\vartheta)^2}\int_{-\frac{q_k(\vartheta)}{\theta_v^2}}^{\tau - \frac{q_k(\vartheta)}{\theta_v^2}} \frac{\mathrm{d}s}{\frac{\theta_v^4}{c_k(\vartheta)^2}s^2 + 1} = \frac{1}{c_k(\vartheta)}\int_{-\frac{q_k(\vartheta)}{c_k(\vartheta)}}^{\frac{\tau\theta_v^2 - q_k(\vartheta)}{c_k(\vartheta)}} \frac{\mathrm{d}z}{z^2 + 1}$$

$$= \frac{1}{c_k(\vartheta)}\left[\arctan\left(\frac{\tau\theta_v^2 - q_k(\vartheta)}{c_k(\vartheta)}\right) + \arctan\left(\frac{q_k(\vartheta)}{c_k(\vartheta)}\right)\right].$$

Note that if $c_k(\vartheta) = 0$, then we have

$$\int_0^{\tau} \frac{\mathrm{d}t}{\rho_{k,t}(\vartheta)^2} = \frac{\tau\theta_v^2}{q_k(\vartheta) - \tau\theta_v^2},$$

but in this problem $c_k(\vartheta) \neq 0$. Indeed, if $c_k(\vartheta) = \left|\theta_2(y_k - \theta_3) - \theta_4(x_k - \theta_1)\right| = 0$, then at the moment

$$t_k^* = \frac{x_k - \theta_1}{\theta_2} = \frac{y_k - \theta_3}{\theta_4},$$

the source will be at the point (x_k, y_k). Remind that this situation is in contradiction with the condition that the position of the source cannot coincide with the position of any detector.

We have

$$\Phi_k(\vartheta,\tau)=\frac{1}{c_k(\vartheta)}\left[\arctan\left(\frac{\tau\theta_v^2-q_k(\vartheta)}{c_k(\vartheta)}\right)+\arctan\left(\frac{q_k(\vartheta)}{c_k(\vartheta)}\right)\right],\qquad \vartheta\in\Theta.$$

Define the LSE $\check\vartheta_\varepsilon$ as solution of the equation

$$\sum_{k=1}^{K}\left[\Phi_{k,\varepsilon}(\tau)-\Phi_k\left(\check\vartheta_\varepsilon,\tau\right)\right]^2=\inf_{\vartheta\in\Theta}\sum_{k=1}^{K}\left[\Phi_{k,\varepsilon}(\tau)-\Phi_k(\vartheta,\tau)\right]^2.$$

For its calculation the system of four equations is used

$$\frac{\partial}{\partial\vartheta}\sum_{k=1}^{K}\left[\Phi_{k,\varepsilon}(\tau)-\Phi_k(\vartheta,\tau)\right]^2\Big|_{\vartheta=\vartheta_\varepsilon^*}=0 \tag{7.28}$$

with the corresponding identifiability condition: for any $\delta>0$

$$g(\delta)\equiv\inf_{\vartheta_0\in\Theta}\inf_{\|\vartheta-\vartheta_0\|_4>\delta}\|\mathbf{\Phi}(\vartheta_0,\tau)-\mathbf{\Phi}(\vartheta,\tau)\|_K>0. \tag{7.29}$$

Here we denoted $\mathbf{\Phi}(\vartheta,\tau)=\left(\Phi_1(\vartheta,\tau),\ldots,\Phi_K(\vartheta,\tau)\right)^\top$.

Introduce the matrices

$$\dot{\mathbf{\Phi}}(\vartheta,\tau)=\left(\frac{\partial\mathbf{\Phi}(\vartheta,\tau)}{\partial\vartheta}\right)_{4\times K},\qquad \mathbf{R}(\vartheta_0,\tau)=\left(R_k(\vartheta_0,\tau)\,\delta_{kk}\right)_{K\times K},$$

$$\mathbf{J}(\vartheta,\tau)_{4\times 4}=\dot{\mathbf{\Phi}}(\vartheta,\tau)\,\dot{\mathbf{\Phi}}(\vartheta,\tau)^\top,$$

$$\mathbf{D}(\vartheta,\tau)=\mathbf{J}(\vartheta,\tau)^{-1}\,\dot{\mathbf{\Phi}}(\vartheta,\tau)\,\mathbf{R}(\vartheta_0,\tau)\,\dot{\mathbf{\Phi}}(\vartheta,\tau)\,\mathbf{J}(\vartheta,\tau)^{-1}.$$

Proposition 7.1 *Assume that the conditions* (7.29) *and*

$$\inf_{\vartheta\in\Theta}\inf_{\mathrm{e}\in\mathcal{R}^4,\|\mathrm{e}\|_4=1}\mathrm{e}^\top\mathbf{J}(\vartheta,\tau)\,\mathrm{e}>0.$$

are fulfilled, then the LSE ϑ_ε^* is uniformly consistent and asymptotically normal

$$\varepsilon^{-1}\left(\vartheta_\varepsilon^*-\vartheta_0\right)\Longrightarrow\zeta^*(\vartheta_0)\sim\mathcal{N}\left(0,\mathbf{D}(\vartheta_0,\tau)\right). \tag{7.30}$$

Proof The condition (7.29) allows us to prove the consistency as follows:

$$\begin{aligned}&\mathbf{P}_{\vartheta_0}\left(\left\|\vartheta_\varepsilon^*-\vartheta_0\right\|_4\geq\delta\right)\\&\quad=\mathbf{P}_{\vartheta_0}\left(\inf_{\|\vartheta-\vartheta_0\|_4<\delta}\|\mathbf{\Phi}(\vartheta,\tau)-\mathbf{\Phi}_\varepsilon(\tau)\|_K\geq\inf_{\|\vartheta-\vartheta_0\|_4\geq\delta}\|\mathbf{\Phi}(\vartheta,\tau)-\mathbf{\Phi}_\varepsilon(\tau)\|_K\right)\\&\quad\leq\mathbf{P}_{\vartheta_0}\left(\|\mathbf{\Phi}(\vartheta_0,\tau)-\mathbf{\Phi}_\varepsilon(\tau)\|_K\geq g(\delta)-\|\mathbf{\Phi}(\vartheta_0,\tau)-\mathbf{\Phi}_\varepsilon(\tau)\|_K\right)\end{aligned}$$

$$= \mathbf{P}_{\vartheta_0}\left(2\left\|\boldsymbol{\Phi}\left(\vartheta_0,\tau\right)-\boldsymbol{\Phi}_\varepsilon\left(\tau\right)\right\|_K \geq g\left(\delta\right)\right)$$
$$\leq \frac{4}{g\left(\delta\right)^2}\sum_{k=1}^{K}\mathbf{E}_{\vartheta_0}\left[\Phi_k\left(\vartheta_0,\tau\right)-\Phi_{k,\varepsilon}\left(\tau\right)\right]^2 \leq C\,\varepsilon^2.$$

Equation (7.28)

$$\sum_{k=1}^{K}\frac{\partial\Phi_k\left(\vartheta,\tau\right)}{\partial\vartheta}\left[\Phi_{k,\varepsilon}\left(\tau\right)-\Phi_k\left(\vartheta,\tau\right)\right]=0$$

we expand at the vicinity of the true value

$$\sum_{k=1}^{K}\frac{\partial\Phi_k\left(\vartheta_\varepsilon^*,\tau\right)}{\partial\vartheta}\left[\Phi_{k,\varepsilon}\left(\tau\right)-\Phi_k\left(\vartheta_0,\tau\right)+\Phi_k\left(\vartheta_0,\tau\right)-\Phi_k\left(\vartheta_\varepsilon^*,\tau\right)\right]$$
$$=\sum_{k=1}^{K}\frac{\partial\Phi_k\left(\vartheta_\varepsilon^*,\tau\right)}{\partial\vartheta}\left[\Phi_{k,\varepsilon}\left(\tau\right)-\Phi_k\left(\vartheta_0,\tau\right)-\left(\vartheta_\varepsilon^*-\vartheta_0\right)^\top\frac{\partial\Phi_k\left(\vartheta_0,\tau\right)}{\partial\vartheta}\left(1+o\left(1\right)\right)\right]$$
$$=\sum_{k=1}^{K}\frac{\partial\Phi_k\left(\vartheta_\varepsilon^*,\tau\right)}{\partial\vartheta}\left[\Phi_{k,\varepsilon}\left(\tau\right)-\Phi_k\left(\vartheta_0,\tau\right)\right]$$
$$-\sum_{k=1}^{K}\frac{\partial\Phi_k\left(\vartheta_\varepsilon^*,\tau\right)}{\partial\vartheta}\left(\vartheta_\varepsilon^*-\vartheta_0\right)^\top\frac{\partial\Phi_k\left(\vartheta_0,\tau\right)}{\partial\vartheta}\left(1+o\left(1\right)\right)=0$$

Hence

$$\varepsilon^{-1}\left(\vartheta_\varepsilon^*-\vartheta_0\right)=\mathbf{J}\left(\vartheta_0,\tau\right)^{-1}\dot{\boldsymbol{\Phi}}\left(\vartheta_0,\tau\right)\eta_\varepsilon\left(\tau\right)\left(1+o\left(1\right)\right)$$

and from (7.27) we obtain the convergence (7.30). □

LSE-2

Introduce the statistic

$$\Lambda_{k,\varepsilon}\left(\tau\right)=\frac{1}{\delta_\varepsilon}\int_{\tau-\delta_\varepsilon}^{\tau}\frac{\mathrm{d}X_k\left(t\right)}{f_k S_k\left(t\right)y_k\left(t\right)}$$

where $\delta_\varepsilon=\varepsilon^{2/3}$ and $\tau-\delta_\varepsilon>0$. Denote

$$H_k\left(\vartheta,\tau\right)=\frac{\left(\theta_{0,1}-x_k\right)\theta_{0,2}+\left(\theta_{0,3}-y_k\right)\theta_{0,4}+\left[\theta_{0,2}^2+\theta_{0,4}^2\right]\tau}{\rho_{k,\tau}\left(\vartheta_0\right)^4},$$

$$\tilde{H}_k(\vartheta_0,\tau)=\left(x_k-\theta_{0,1}\right)\theta_{0,2}+\left(y_k-\theta_{0,3}\right)\theta_{0,4}-\left[\theta_{0,2}^2+\theta_{0,4}^2\right]\tau,$$

$$F_k(\vartheta,\tau)=\frac{\sigma_k(\tau)}{f_kS_k(\tau)\,y_k(\tau)},\qquad \tilde{F}_k(\vartheta,\tau)=\frac{\rho_{k,\tau}(\vartheta_0)^4\,\sigma_k(\tau)}{f_kS_k(\tau)\,y_k(\tau)}.$$

Proposition 7.2 *For any $k=1,\ldots,K$*

$$\mathbf{E}_{\vartheta_0}\left(\Lambda_{k,\varepsilon}(\tau)-\frac{1}{\rho_{k,\tau}(\vartheta_0)^2}\right)^2\leq C\,\varepsilon^{4/3},$$

$$\varepsilon^{-2/3}\left(\Lambda_{k,\varepsilon}(\tau)-\frac{1}{\rho_{k,\tau}(\vartheta_0)^2}\right)\Longrightarrow H_k(\vartheta_0,\tau)+F_k(\vartheta,\tau)\ \xi_k,$$

$$\varepsilon^{-2/3}\left(\Lambda_{k,\varepsilon}(\tau)^{-1}-\rho_{k,\tau}(\vartheta_0)^2\right)\Longrightarrow \tilde{H}_k(\vartheta_0,\tau)-\tilde{F}_k(\vartheta,\tau)\ \xi_k. \tag{7.31}$$

Here $\xi_k,k=1,\ldots,K$ are independent random variables, $\xi_k\sim\mathcal{N}(0,1)$.

Proof We have

$$\mathbf{E}_{\vartheta_0}\left(\Lambda_{k,\varepsilon}(\tau)-\frac{1}{\rho_{k,\tau}(\vartheta_0)^2}\right)^2=\mathbf{E}_{\vartheta_0}\left(\frac{1}{\delta_\varepsilon}\int_{\tau-\delta_\varepsilon}^{\tau}\frac{[Y_k(t)-y_k(t)]}{y_k(t)\,\rho_{k,t}(\vartheta_0)^2}\mathrm{d}t\right.$$
$$\left.+\frac{1}{\delta_\varepsilon}\int_{\tau-\delta_\varepsilon}^{\tau}\left[\frac{\rho_{k,\tau}(\vartheta_0)^2-\rho_{k,t}(\vartheta_0)^2}{\rho_{k,\tau}(\vartheta_0)^2\,\rho_{k,t}(\vartheta_0)^2}\right]\mathrm{d}t+\frac{\varepsilon}{\delta_\varepsilon}\int_{\tau-\delta_\varepsilon}^{\tau}\frac{\sigma_k(t)\,\mathrm{d}W_k(t)}{f_kS_k(t)\,y_k(t)}\right)^2$$
$$\leq C\,\varepsilon^2+C\,\delta_\varepsilon^2+C\,\frac{\varepsilon^2}{\delta_\varepsilon}.$$

Here we used the equality (7.26) and the relations

$$\frac{1}{\delta_\varepsilon}\int_{\tau-\delta_\varepsilon}^{\tau}\frac{[Y_k(t)-y_k(t)]}{y_k(t)\,\rho_{k,t}(\vartheta_0)^2}\mathrm{d}t=\frac{\varepsilon}{\delta_\varepsilon}\int_{\tau-\delta_\varepsilon}^{\tau}\frac{\psi_k(t)}{y_k(t)\,\rho_{k,t}(\vartheta_0)^2}\int_0^t\frac{b_k(s)}{\psi_k(s)}\mathrm{d}V_s\,\mathrm{d}t$$
$$=\frac{\varepsilon}{\delta_\varepsilon}\int_0^{\tau}\frac{b_k(s)}{\psi_k(s)}\int_{s\vee(\tau-\delta_\varepsilon)}^{\tau}\frac{\psi_k(t)}{y_k(t)\,\rho_{k,t}(\vartheta_0)^2}\mathrm{d}t\,\mathrm{d}V_s$$

and

$$\rho_{k,\tau}(\vartheta_0)^2-\rho_{k,t}(\vartheta_0)^2=2\left[\left(x_k-\vartheta_{0,1}\right)\vartheta_{0,2}+\left(y_k-\vartheta_{0,3}\right)\vartheta_{0,4}\right](t-\tau)$$
$$+\left[\vartheta_{0,2}^2+\vartheta_{0,4}^2\right]\left(\tau^2-t^2\right).$$

The rate $\delta_\varepsilon=\varepsilon^{2/3}$ we obtain from the equation of balance $\delta_\varepsilon^2=\varepsilon^2/\delta_\varepsilon$ Further, as $\delta_\varepsilon\to 0$ we can write

$$
\begin{aligned}
\Lambda_{k,\varepsilon}(\tau) - \frac{1}{\rho_{k,\tau}(\vartheta_0)^2} &= \frac{\varepsilon}{\delta_\varepsilon}\int_{\tau-\delta_\varepsilon}^{\tau} \frac{\psi_k(t)}{y_k(t)\,\rho_{k,t}(\vartheta)^2}\int_0^t \frac{b_k(s)}{\psi_k(s)}\mathrm{d}V_k(s)\;\mathrm{d}t \\
&+\frac{1}{\delta_\varepsilon}\int_{\tau-\delta_\varepsilon}^{\tau}\frac{\rho_{k,\tau}(\vartheta_0)^2-\rho_{k,t}(\vartheta_0)^2}{\rho_{k,\tau}(\vartheta_0)^2\,\rho_{k,t}(\vartheta)^2}\mathrm{d}t+\frac{\varepsilon}{\delta_\varepsilon}\int_{\tau-\delta_\varepsilon}^{\tau}\frac{\sigma_k(t)}{f_kS_k(t)\,y_k(t)}\mathrm{d}W_k(t) \\
&=\delta_\varepsilon\,H_k(\vartheta_0,\tau)\,(1+o(1))+\frac{\varepsilon}{\sqrt{\delta_\varepsilon}}F_k(\vartheta_0,\tau)\,\frac{W_k(\tau)-W_k(\tau-\delta_\varepsilon)}{\sqrt{\delta_\varepsilon}}\,(1+o(1)) \\
&+\varepsilon\,\frac{\psi_k(\tau)}{y_k(\tau)\,\rho_{k,\tau}(\vartheta)^2}\int_0^\tau\frac{b_k(s)}{\psi_k(s)}\mathrm{d}V_k(s)\,(1+o(1))\,.
\end{aligned}
$$

Hence

$$
\begin{aligned}
\varepsilon^{-2/3}\left(\Lambda_{k,\varepsilon}(\tau)-\frac{1}{\rho_{k,\tau}(\vartheta_0)^2}\right) &= H_k(\vartheta_0,\tau)+F_k(\vartheta_0,\tau)\,\xi_{k,\varepsilon}+o(1) \\
&\Longrightarrow H_k(\vartheta_0,\tau)+F_k(\vartheta_0,\tau)\,\xi_k.
\end{aligned}
$$

Therefore we have the convergences

$$
Z_{k,\varepsilon}(\tau)=\Lambda_{k,\varepsilon}(\tau)^{-1}\longrightarrow\rho_{k,\tau}(\vartheta_0)^2
$$

and

$$
\begin{aligned}
\varepsilon^{-2/3}\left(Z_{k,\varepsilon}(\tau)-\rho_{k,\tau}(\vartheta_0)^2\right) &= \varepsilon^{-2/3}\left(\Lambda_{k,\varepsilon}(\tau)^{-1}-\rho_{k,\tau}(\vartheta_0)^2\right) \\
&=\frac{\rho_{k,\tau}(\vartheta_0)^2}{\Lambda_{k,\varepsilon}(\tau)}\,\varepsilon^{-2/3}\left(\frac{1}{\rho_{k,\tau}(\vartheta_0)^2}-\Lambda_{k,\varepsilon}(\tau)\right) \\
&\Longrightarrow\rho_{k,\tau}(\vartheta_0)^4\left[-H_k(\vartheta_0,\tau)-F_k(\vartheta_0,\tau)\,\xi_k\right] \\
&=\tilde H_k(\vartheta_0,\tau)-\tilde F_k(\vartheta_0,\tau)\,\xi_k \\
&\equiv\tilde\eta_k(\vartheta_0,\tau)\sim\mathcal{N}\left(\tilde H_k(\vartheta_0,\tau),\,\tilde F_k(\vartheta_0,\tau)^2\right).
\end{aligned}
$$

□

Using once more the least squares approach we can define the LSE $\check\vartheta_\varepsilon$

$$
\begin{aligned}
&\sum_{k=1}^K\left[Z_{k,\varepsilon}(\tau)-\left(x_k-\check\theta_{1,\varepsilon}-\check\theta_{2,\varepsilon}\tau\right)^2-\left(y_k-\check\theta_{3,\varepsilon}-\check\theta_{4,\varepsilon}\tau\right)^2\right]^2 \\
&\qquad=\inf_{\vartheta\in\Theta}\sum_{k=1}^K\left[Z_{k,\varepsilon}(\tau)-(x_k-\theta_1-\theta_2\tau)^2-(y_k-\theta_3-\theta_4\tau)^2\right]^2
\end{aligned}
\tag{7.32}
$$

and calculate it as a solution of the system of four equations similar to (7.28)

$$\frac{\partial}{\partial \vartheta} \sum_{k=1}^{K} \left[Z_{k,\varepsilon}(\tau) - (x_k - \theta_1 - \theta_2 \tau)^2 - (y_k - \theta_3 - \theta_4 \tau)^2 \right]^2 \bigg|_{\vartheta = \check{\vartheta}_\varepsilon} = 0. \qquad (7.33)$$

The obtained system is too complicated, and we propose some simplifications of the model.

Introduce new variables $\lambda = (\lambda_1, \lambda_2)^\top$, $\lambda_1 = \theta_1 + \theta_2 \tau_1$, $\lambda_2 = \theta_3 + \theta_4 \tau_1$ and $\mu = (\mu_1, \mu_2)^\top$, $\mu_1 = \theta_1 + \theta_2 \tau_2$, $\mu_2 = \theta_3 + \theta_4 \tau_2$. Here $\tau_1 \neq \tau_2$. Then we write

$$\begin{aligned} Z_{k,\varepsilon}(\tau_1) &- (x_k - \theta_1 - \theta_2 \tau)^2 - (y_k - \theta_3 - \theta_4 \tau)^2 \\ &= Z_{k,\varepsilon}(\tau_1) - \left(x_k^2 + y_k^2 \right) + 2x_k \lambda_1 + 2y_k \lambda_2 - \left(\lambda_1^2 + \lambda_2^2 \right) \\ &= Z_{k,\varepsilon}(\tau_1) - r_k^2 + 2x_k \lambda_1 + 2y_k \lambda_2 - 2\lambda_3. \end{aligned}$$

Here we denoted $r_k^2 = x_k^2 + y_k^2$ and introduced a new independent variable $\lambda_3 = \frac{1}{2}\left(\lambda_1^2 + \lambda_2^2\right)$. Similar transformation we do with Eq. (7.33) where we take τ_2 and the unknown parameter denoted as $\mu = (\mu_1, \mu_2, \mu_3)$. Now we replace the system of four equations (7.33) by two systems of three equations. We return to the situation discussed in Sect. 7.1. For example, in the case of the parameter λ we have the equation

$$\frac{\partial}{\partial \lambda} \sum_{k=1}^{K} \left[Z_{k,\varepsilon}(\tau_1) - r_k^2 + 2x_k \lambda_1 + 2y_k \lambda_2 - 2\lambda_3 \right]^2 = 0,$$

which we write in the equivalent form

$$\lambda_1 \sum_{k=1}^{K} x_k^2 + \lambda_2 \sum_{k=1}^{K} x_k y_k - \lambda_3 \sum_{k=1}^{K} x_k = \frac{1}{2} \sum_{k=1}^{K} \left[r_k^2 - Z_{k,\varepsilon}(\tau_1) \right] x_k,$$

$$\lambda_1 \sum_{k=1}^{K} x_k y_k + \lambda_2 \sum_{k=1}^{K} y_k^2 - \lambda_3 \sum_{k=1}^{K} y_k = \frac{1}{2} \sum_{k=1}^{K} \left[r_k^2 - Z_{k,\varepsilon}(\tau_1) \right] y_k,$$

$$\lambda_1 \sum_{k=1}^{K} x_k + \lambda_2 \sum_{k=1}^{K} y_k - \lambda_3 K = \frac{1}{2} \sum_{k=1}^{K} \left[r_k^2 - Z_{k,\varepsilon}(\tau_1) \right].$$

Once more the last equation we solve w.r.t. λ_3 and substitute in the first two equations. Then we obtain two equations

$$\left[\overline{x^2}-(\bar{x})^2\right]\lambda_1+[\overline{xy}-\bar{x}\bar{y}]\,\lambda_2=\bar{z}_{1,\varepsilon},$$
$$[\overline{xy}-\bar{x}\bar{y}]\,\lambda_1+[\overline{y^2}-(\bar{y})^2]\lambda_2=\bar{z}_{2,\varepsilon} \tag{7.34}$$

with the corresponding random elements

$$\bar{z}_{1,\varepsilon}=\frac{1}{2}\left[\overline{\left(r^2-Z_\varepsilon(\tau_1)\right)x}-\overline{\left(r^2-Z_\varepsilon(\tau_1)\right)}\,\bar{x}\right],$$
$$\bar{z}_{2,\varepsilon}=\frac{1}{2}\left[\overline{\left(r^2-Z_\varepsilon(\tau_1)\right)y}-\overline{\left(r^2-Z_\varepsilon(\tau_1)\right)}\,\bar{y}\right].$$

We have the same matrix

$$\mathbf{A}=\begin{pmatrix}(\overline{x^2}-\bar{x}^2), & (\overline{xy}-\bar{x}\bar{y})\\ (\overline{xy}-\bar{x}\bar{y}), & (\overline{y^2}-\bar{y}^2)\end{pmatrix}$$

and the vector $\bar{z}_\varepsilon=\left(\bar{z}_{1,\varepsilon},\bar{z}_{2,\varepsilon}\right)^\top$. Then the system (7.34) and its solution can be written as follows:

$$\mathbf{A}\check{\lambda}_\varepsilon=\bar{z}_\varepsilon,\qquad\qquad \check{\lambda}_\varepsilon=\mathbf{A}^{-1}\bar{z}_\varepsilon. \tag{7.35}$$

Recall that the matrix $\mathbf{A}$ is nondegenerate because the equality $\mathrm{Det}\,(\mathbf{A})=0$ is possible iff there exists a constant c such that $x_k-\bar{x}=c\,(y_k-\bar{y})\,,k=1,\ldots,K$, and this equality means that all detectors are on the same line. By condition of the problem such situations are excluded, and therefore $\mathrm{Det}\,(\mathbf{A})\neq 0$ and the estimator $\check{\lambda}_\varepsilon$ is well defined by Eq. (7.35).

As $Z_{k,\varepsilon}(\tau_1)\longrightarrow\rho_{k,\tau_1}(\vartheta_0)^2$ after elementary calculations we obtain the convergences

$$\bar{z}_{1,\varepsilon}\longrightarrow\bar{z}_{1,0}=\frac{1}{2}\left[\overline{\left(r^2-\rho_{\tau_1}^2\right)x}-\overline{\left(r^2-\rho_{\tau_1}^2\right)}\bar{x}\right]$$
$$=\lambda_{0,1}\left[\overline{x^2}-(\bar{x})^2\right]+\lambda_{0,2}\left[\overline{yx}-(\bar{x}\bar{y})\right],$$
$$\bar{z}_{2,\varepsilon}\longrightarrow\bar{z}_{0,0}=\frac{1}{2}\overline{\left(r^2-\rho_{\tau_1}^2\right)y}-\overline{\left(r^2-\rho_{\tau_1}^2\right)}\bar{y}$$
$$=\lambda_{0,1}\left[\overline{yx}-(\bar{x}\bar{y})\right]+\lambda_{0,2}\left[\overline{y^2}-(\bar{y})^2\right]$$

and

$$\varepsilon^{-2/3}\left(\bar{z}_{1,\varepsilon}-\bar{z}_{1,0}\right)\Longrightarrow\overline{x\eta\left(\vartheta_0,\tau_1\right)},$$
$$\varepsilon^{-2/3}\left(\bar{z}_{2,\varepsilon}-\bar{z}_{2,0}\right)\Longrightarrow\overline{y\eta\left(\vartheta_0,\tau_1\right)}.$$

These provide us the consistency and asymptotic normality of the LSE $\check{\lambda}_\varepsilon$ because

$$\mathbf{A}^{-1}\tilde{z}_0 = \lambda_0, \qquad \varepsilon^{-2/3}\left(\check{\lambda}_{1,\varepsilon} - \lambda_{1,0}\right) \Longrightarrow \mathbf{A}^{-1}\left(\overline{\frac{x\,\tilde{\eta}\,(\vartheta_0, \tau_1)}{y\,\tilde{\eta}\,(\vartheta_0, \tau_1)}}\right).$$

If we consider the second equation

$$\frac{\partial}{\partial \mu}\sum_{k=1}^{K}\left[Z_{k,\varepsilon}\,(\tau_2) - r_k^2 + 2x_k\mu_1 + 2y_k\mu_2 - 2\mu_3\right]^2 = 0$$

and repeat the calculations, then the final equation will be

$$\check{\mu}_\varepsilon = \mathbf{A}^{-1}z'_\varepsilon. \tag{7.36}$$

with the same matrix $\mathbf{A}$ and the corresponding definition of the vector z'_ε.

The relations between ϑ and λ and μ allow us to write the LSE $\check{\vartheta}_\varepsilon = \check{\vartheta}_{\tau_1,\tau_2,\varepsilon}$

$$\begin{aligned} \check{\theta}_{1,\varepsilon} &= \frac{\check{\mu}_{1,\varepsilon}\tau_1 - \check{\lambda}_{1,\varepsilon}\tau_2}{\tau_1 - \tau_2}, & \check{\theta}_{2,\varepsilon} &= \frac{\check{\lambda}_{1,\varepsilon} - \check{\mu}_{1,\varepsilon}}{\tau_1 - \tau_2}, \\ \check{\theta}_{3,\varepsilon} &= \frac{\check{\mu}_{2,\varepsilon}\tau_1 - \check{\lambda}_{2,\varepsilon}\tau_2}{\tau_1 - \tau_2}, & \check{\theta}_{4,\varepsilon} &= \frac{\check{\lambda}_{2,\varepsilon} - \check{\mu}_{2,\varepsilon}}{\tau_1 - \tau_2}. \end{aligned} \tag{7.37}$$

Therefor we obtain the consistency and asymptotic normality of the LSE-2

$$\varepsilon^{-2/3}\left(\check{\vartheta}_\varepsilon - \vartheta_0\right) \Longrightarrow \mathcal{N}\left(0, \mathbf{K}\,(\vartheta_0)\right).$$

Here $\mathbf{K}\,(\vartheta_0)$ is 4×4 matrix with bounded elements.

One-Step MLE

As usual for One-step construction we fix a learning interval $[0, \tau]$, then by observations $X^\tau = \left(X_k^\tau, k = 1, \ldots, K\right)$ we define the LSE $\check{\vartheta}_\varepsilon = \check{\vartheta}_{\tau,\varepsilon}$. This can be LSE-1 (see (7.27)) or LSE-2 with $0 < \tau_2 < \tau_1 = \tau$ (see (7.37)). The next quantity which we need is the Fisher information matrix.

Recall that the observations (7.22) can be written as follows (see (1.21)):

$$\mathrm{d}X_k\,(t) = f_k\,(\vartheta, t)\,m_k\,(\vartheta, t)\,\mathrm{d}t + \varepsilon\sigma_k\,(t)\,\mathrm{d}\bar{W}_k\,(t)\,, \qquad X_k\,(0) = 0.$$

To calculate the Fisher information matrix we need $\dot{m}_k\,(\vartheta, t) = \partial m_k\,(\vartheta, t)\,/\partial\vartheta$. The K systems of eight equations we obtain by differentiating Eqs. (7.24) and (7.25)

$$\begin{aligned}
&\mathrm{d}\dot{m}_k(\vartheta,t)\\
&\quad=-\sigma_k(t)^{-2}\Big[2f_k(\vartheta,t)\,\dot{f}_k(\vartheta,t)\,\gamma_k^*(\vartheta,t)+f_k(\vartheta,t)^2\,\dot{\gamma}_k^*(\vartheta,t)\Big]m_k(\vartheta,t)\,\mathrm{d}t\\
&\qquad+\Big(a_k(t)-\sigma_k(t)^{-2}\Big[f_k(\vartheta,t)^2\,\gamma_k^*(\vartheta,t)\Big]\Big)\dot{m}_k(\vartheta,t)\,\mathrm{d}t\\
&\qquad+\sigma_k(t)^{-2}\left[\dot{f}_k(\vartheta,t)\,\gamma_k^*(\vartheta,t)+f_k(\vartheta,t)\,\dot{\gamma}_k^*(\vartheta,t)\right]\mathrm{d}X_k(t), \qquad (7.38)\\
&\frac{\partial\dot{\gamma}_k^*(\vartheta,t)}{\partial t}\\
&\quad=-2\sigma_k(t)^{-2}\Big[f_k(\vartheta,t)\,\dot{f}_k(\vartheta,t)\,\gamma_k^*(\vartheta,t)^2+f_k(\vartheta,t)^2\,\gamma_k^*(\vartheta,t)\,\dot{\gamma}_*(\vartheta,t)\Big]\\
&\qquad+2a_k(t)\,\dot{\gamma}_k^*(\vartheta,t), \qquad\qquad \dot{\gamma}_k^*(\vartheta,0)=0.
\end{aligned}$$

The limit ($\varepsilon=0$) equations for $m_k(\vartheta_0,t)|_{\varepsilon=0}=y_k(\vartheta_0,t)$ and $\dot{m}_k(\vartheta_0,t)|_{\varepsilon=0}=\dot{y}_k(\vartheta_0,t)$ are

$$\begin{aligned}
y_k(\vartheta_0,t)&=y_k(0)\exp\left(\int_0^t a_k(s)\,\mathrm{d}s\right),\\
\frac{\partial\dot{y}_k(\vartheta_0,t)}{\partial t}&=-\sigma_k(t)^{-2}\,f_k(\vartheta_0,t)\,\dot{f}_k(\vartheta_0,t)\,\gamma_k^*(\vartheta_0,t)\,y_k(\vartheta_0,t)\\
&\quad+\Big(a_k(t)-\sigma_k(t)^{-2}\,f_k(\vartheta_0,t)^2\,\gamma_k^*(\vartheta_0,t)\Big)\dot{y}_k(\vartheta_0,t), \quad \dot{y}_k(\vartheta_0,0)=0.
\end{aligned}$$

Denote $\dot{\mathrm{M}}_k^\circ(\vartheta_0,t)=\left[\dot{f}_k(\vartheta_0,t)\,y_k(\vartheta_0,t)+f_k(\vartheta_0,t)\,\dot{y}_k(\vartheta_0,t)\right]$ the limit ($\varepsilon=0$) solution of (7.38). Then introduce the Fisher information matrix

$$\mathbf{I}_\tau^t(\vartheta_0)=\int_\tau^t\frac{\dot{\mathrm{M}}_k^\circ(\vartheta_0,s)^\top\,\dot{\mathrm{M}}_k^\circ(\vartheta_0,s)}{\sigma_k(s)^2}\,\mathrm{d}s.$$

The One-step MLE-process $\vartheta_{t,\varepsilon}^\star,\ \tau<t\leq T$ is

$$\vartheta_{t,\varepsilon}^\star=\check{\vartheta}_{\tau,\varepsilon}+\mathbf{I}_\tau^t\left(\check{\vartheta}_{\tau,\varepsilon}^*\right)^{-1}\int_\tau^t\frac{\dot{\mathrm{M}}_k^\circ\left(\check{\vartheta}_{\tau,\varepsilon},s\right)}{\sigma_k(s)^2}\left[\mathrm{d}X_k(s)-f_k\left(\vartheta_\varepsilon^*,s\right)m_k\left(\check{\vartheta}_{\tau,\varepsilon},s\right)\mathrm{d}s\right]. \tag{7.39}$$

Theorem 7.4 *Suppose that the Fisher information for some $\tau\in(0,T]$ and any $t\in(\tau,T]$ is uniformly nondegenerate*

$$\inf_{\vartheta\in\Theta}\ \inf_{\mathrm{e}\in\mathcal{R}^4,\|\mathrm{e}\|_4=1}\mathrm{e}^\top\,\mathbf{I}_\tau^t(\vartheta)\,\mathrm{e}>0,$$

and the identifiability condition (7.28) *for any $\delta>0$ and any $\tau<T$ holds. Then the One-step MLE-process $\vartheta_{t,\varepsilon}^\star,\ \tau<t\leq T$ for any $t\in(\tau,T]$ is consistent and*

asymptotically normal

$$\varepsilon^{-1}\left(\vartheta^\star_{t,\varepsilon}-\vartheta_0\right)\Longrightarrow\mathcal{N}\left(0,\mathbf{I}^t_\tau\left(\vartheta_0\right)^{-1}\right). \tag{7.40}$$

The proof is similar to the proof of Theorem 2.5. Recall that

$$\begin{aligned}\frac{\vartheta^\star_{t,\varepsilon}-\vartheta_0}{\varepsilon}&=\frac{\check\vartheta_{\tau,\varepsilon}-\vartheta_0}{\varepsilon}+\mathbf{I}^t_\tau\left(\check\vartheta_{\tau,\varepsilon}\right)^{-1}\int_\tau^t\frac{\dot{\mathbf{M}}^\circ_k\left(\check\vartheta_{\tau,\varepsilon},s\right)}{\sigma_k\left(s\right)}\,\mathrm{d}\bar W_k\left(s\right)\\&\quad+\mathbf{I}^t_\tau\left(\check\vartheta_{\tau,\varepsilon}\right)^{-1}\int_\tau^t\frac{\dot{\mathbf{M}}^\circ_k\left(\check\vartheta_{\tau,\varepsilon},s\right)}{\varepsilon\sigma_k\left(s\right)^2}\left[f_k\left(\vartheta_0,s\right)m_k\left(\vartheta_0,s\right)\right.\\&\quad\left.-f_k\left(\check\vartheta_{\tau,\varepsilon},s\right)m_k\left(\check\vartheta_{\tau,\varepsilon},s\right)\right]\mathrm{d}s.\end{aligned}$$

To prove (7.40) we have to verify that

$$\mathbf{I}^t_\tau\left(\check\vartheta_{\tau,\varepsilon}\right)^{-1}\int_\tau^t\frac{\dot{\mathbf{M}}^\circ_k\left(\check\vartheta_{\tau,\varepsilon},s\right)}{\sigma_k\left(s\right)}\,\mathrm{d}\bar W_k\left(s\right)\Longrightarrow\mathcal{N}\left(0,\mathbf{I}^t_\tau\left(\vartheta_0,\tau\right)^{-1}\right)$$

and that

$$\begin{aligned}\mathbf{I}^t_\tau\left(\check\vartheta_{\tau,\varepsilon}\right)^{-1}\int_\tau^t\frac{\dot{\mathbf{M}}^\circ_k\left(\check\vartheta_{\tau,\varepsilon},s\right)}{\varepsilon\sigma_k\left(s\right)^2}&\left[f_k\left(\vartheta_0,s\right)m_k\left(\vartheta_0,s\right)-f_k\left(\check\vartheta_{\tau,\varepsilon},s\right)m_k\left(\check\vartheta_{\tau,\varepsilon},s\right)\right]\mathrm{d}s\\&=\frac{\vartheta_0-\check\vartheta_{\tau,\varepsilon}}{\varepsilon}+o\left(1\right).\end{aligned}$$

Note that the rate $\varepsilon^{2/3}$ of LSE-2 is sufficient for the proof of this theorem because what we need is the convergence

$$\varepsilon^{-1}\left(\check\vartheta_{\tau,\varepsilon}-\vartheta_0\right)^2\longrightarrow 0.$$

All processes and functions are sufficiently smooth, and the proof can be carried out using direct calculations.

Remark 7.1 It is possible to consider the same problem with $\tau_\varepsilon\rightarrow 0$ but slowly, i.e., such that $\varepsilon^{-1}\left(\check\vartheta_{\varepsilon,\tau_\varepsilon}-\vartheta_0\right)^2\rightarrow 0$. Then the One-step MLE-process $\vartheta^\star_{t,\varepsilon},\tau<t\leq T$ will be asymptotically efficient with the limit covariance matrix $\mathbf{I}^t_0\left(\vartheta_0,\tau\right)^{-1}$.

Source Tracking

Suppose that the conditions of Theorem 7.4 are satisfied and therefore the One-step MLE-process $\vartheta^\star_{t,\varepsilon}=\left(\theta^\star_{1,t,\varepsilon},\theta^\star_{2,t,\varepsilon},\theta^\star_{3,t,\varepsilon},\theta^\star_{4,t,\varepsilon}\right),\tau<t\leq T$ is asymptotically normal (7.40). The trajectory of the source can be estimated as follows.

Recall the construction in the case of LSE-2. Introduce two statistics

$$Z_{k,\varepsilon}\left(\tau_1\right)=\left(\frac{1}{\delta_\varepsilon}\int_{\tau_1-\delta_\varepsilon}^{\tau_1}\frac{{\rm d}X_k\left(t\right)}{f_kS_k\left(t\right)y_k\left(t\right)}\right)^{-1},$$

$$Z_{k,\varepsilon}\left(\tau_2\right)=\left(\frac{1}{\delta_\varepsilon}\int_{\tau_2-\delta_\varepsilon}^{\tau_2}\frac{{\rm d}X_k\left(t\right)}{f_kS_k\left(t\right)y_k\left(t\right)}\right)^{-1}.$$

Then calculate the LSE's $\check\lambda_\varepsilon,\check\mu_\varepsilon$

$$\check\lambda_\varepsilon=\begin{pmatrix}\left(\bar{x^2}-\bar x^2\right), & \left(\overline{xy}-\bar x\bar y\right)\\ \left(\overline{xy}-\bar x\bar y\right), & \left(\bar{y^2}-\bar y^2\right)\end{pmatrix}^{-1}\begin{pmatrix}\overline{r^2-Z_\varepsilon\left(\tau_1\right)x}-\overline{r^2-Z_\varepsilon\left(\tau_1\right)}\,\bar x\\ \overline{r^2-Z_\varepsilon\left(\tau_1\right)y}-\overline{r^2-Z_\varepsilon\left(\tau_1\right)}\,\bar y\end{pmatrix},$$

$$\check\mu_\varepsilon=\begin{pmatrix}\left(\bar{x^2}-\bar x^2\right), & \left(\overline{xy}-\bar x\bar y\right)\\ \left(\overline{xy}-\bar x\bar y\right), & \left(\bar{y^2}-\bar y^2\right)\end{pmatrix}^{-1}\begin{pmatrix}\overline{r^2-Z_\varepsilon\left(\tau_2\right)x}-\overline{r^2-Z_\varepsilon\left(\tau_2\right)}\,\bar x\\ \overline{r^2-Z_\varepsilon\left(\tau_2\right)y}-\overline{r^2-Z_\varepsilon\left(\tau_2\right)}\,\bar y\end{pmatrix}$$

The preliminary estimator $\check\vartheta_\varepsilon$ we obtain as the linear combination of $\check\lambda_\varepsilon$,$\check\mu_\varepsilon$ (see (7.37))

$$\check\theta_{1,\varepsilon}=\frac{\check\mu_{1,\varepsilon}\tau_1-\check\lambda_{1,\varepsilon}\tau_2}{\tau_1-\tau_2},\qquad \check\theta_{2,\varepsilon}=\frac{\check\lambda_{1,\varepsilon}-\check\mu_{1,\varepsilon}}{\tau_1-\tau_2},$$

$$\check\theta_{3,\varepsilon}=\frac{\check\mu_{2,\varepsilon}\tau_1-\check\lambda_{2,\varepsilon}\tau_2}{\tau_1-\tau_2},\qquad \check\theta_{4,\varepsilon}=\frac{\check\lambda_{2,\varepsilon}-\check\mu_{2,\varepsilon}}{\tau_1-\tau_2}.$$

Having $\check\vartheta_\varepsilon$ we introduce the One-step MLE-process (see (7.39))

$$\vartheta^\star_{t,\varepsilon}=\check\vartheta_{\tau,\varepsilon}+\mathbf{I}^t_\tau\left(\check\vartheta_{\tau,\varepsilon}\right)^{-1}\int_\tau^t\frac{\dot{\mathbf{M}}^\circ_k\left(\check\vartheta_{\tau,\varepsilon},s\right)}{\sigma_k\left(s\right)^2}\left[{\rm d}X_k\left(s\right)-f_k\left(\check\vartheta_\varepsilon,s\right)m_k\left(\check\vartheta_{\tau,\varepsilon},s\right){\rm d}s\right].$$

The last step is to construct the estimator of the trajectory of the source:

$$\vartheta^\star_\varepsilon\left(t\right)=\left(x^\star_t,y^\star_t\right),\qquad \tau<t\leq T,$$

where $x^\star_t=x_t\left(\theta^\star_{1,t,\varepsilon},\theta^\star_{2,t,\varepsilon}\right),\ y^\star_t=y_t\left(\theta^\star_{3,t,\varepsilon},\theta^\star_{4,t,\varepsilon}\right)$

$$x_t\left(\theta^\star_{1,t,\varepsilon},\theta^\star_{2,t,\varepsilon}\right)=\theta^\star_{1,t,\varepsilon}+\theta^\star_{2,t,\varepsilon}t,$$
$$y_t\left(\theta^\star_{3,t,\varepsilon},\theta^\star_{4,t,\varepsilon}\right)=\theta^\star_{3,t,\varepsilon}+\theta^\star_{4,t,\varepsilon}t.$$

The true value is $\vartheta_0=\left(\theta_{0,1},\theta_{0,2},\theta_{0,3},\theta_{0,4}\right)^\top$. Therefore the estimator of the line on the plane is not a line but a continuous sequence of lines with decreasing error of estimation of the coefficients.

We have

$$\begin{aligned}\varepsilon^{-1}\left(\vartheta^\star_\varepsilon(t)-\vartheta_0(t)\right)&=\Big(\varepsilon^{-1}\left(x_t\left(\theta^\star_{1,t,\varepsilon},\theta^\star_{2,t,\varepsilon}\right)-x_t\left(\theta_{0,1},\theta_{0,2}\right)\right),\\&\qquad\varepsilon^{-1}\left(y_t\left(\theta^\star_{3,t,\varepsilon},\theta^\star_{4,t,\varepsilon}\right)-y_t\left(\theta_{0,3},\theta_{0,4}\right)\right)\Big)^\top\\&\Longrightarrow\Big(\left(\zeta_{1,t}+\zeta_{2,t}t\right),\left(\zeta_{3,t}+\zeta_{4,t}t\right)\Big)^\top.\end{aligned}$$

The vector

$$\zeta_t=\left(\zeta_{1,t},\zeta_{2,t},\zeta_{3,t},\zeta_{4,t}\right)^\top\sim\mathcal{N}\left(0,\mathbf{I}^t_\tau\left(\vartheta_0\right)^{-1}\right),\qquad \tau<t\le T.$$

Note that $\mathbf{E}_{\vartheta_0}\zeta^2_{i,t}=I^t_\tau\left(\vartheta_0\right)^{-1}_{ii}$, $i=1,\ldots,4$.

MLE and BE

The likelihood ratio function in the case of observations (7.22)–(7.23) is

$$\ln L\left(\vartheta,X^K\right)=\sum_{k=1}^K\left(\int_0^T\frac{M_k\left(\vartheta,t\right)}{\varepsilon^2\sigma_k\left(t\right)^2}\,\mathrm{d}X_k\left(t\right)-\int_0^T\frac{M_k\left(\vartheta,t\right)^2}{2\varepsilon^2\sigma_k\left(t\right)^2}\,\mathrm{d}t\right),\qquad\vartheta\in\Theta.$$

Here $M_k\left(\vartheta,t\right)=f_k\left(\vartheta,t\right)m_k\left(\vartheta,t\right)$, $f_k\left(\vartheta,t\right)=\rho_{k,t}\left(\vartheta\right)^{-2}f_kS_k\left(t\right)$, $m_k\left(\vartheta,t\right)$ is solution of equation (7.24), $\rho_{k,t}\left(\vartheta\right)^2=\left(x_k-\theta_1-\theta_2t\right)^2+\left(y_k-\theta_3-\theta_4t\right)^2$. The set $\Theta\subset\mathcal{R}^4$ is open, convex, and bounded. We suppose that at least three detectors are not on the same line.

Recall that the MLE $\hat\vartheta_\varepsilon$ and BE $\tilde\vartheta_\varepsilon$ are defined by the relations

$$L\left(\hat\vartheta_\varepsilon,X^K\right)=\sup_{\vartheta\in\Theta}L\left(\vartheta,X^K\right),\qquad\tilde\vartheta_\varepsilon=\frac{\displaystyle\int_\Theta\vartheta p\left(\vartheta\right)L\left(\vartheta,X^K\right)\mathrm{d}\vartheta}{\displaystyle\int_\Theta p\left(\vartheta\right)L\left(\vartheta,X^K\right)\mathrm{d}\vartheta}.$$

This model of observations is a particular case of the model (2.1)–(2.2). Therefore it is sufficient to check the conditions $\mathscr{I}\mathscr{K}_4$ of Theorems 1.10 and 1.12.

The condition of identifiability is: for any $\delta>0$

$$g(\vartheta_0, \delta) = \inf_{\|\vartheta-\vartheta_0\|>\delta} \sum_{k=1}^{K} \int_0^T \left(\frac{f_k(\vartheta, t)\, y_k(\vartheta, \vartheta_0, t) - f_k(\vartheta_0, t)\, y_k(\vartheta_0, t)}{\sigma_k(t)} \right)^2 \mathrm{d}t > 0,$$

where $y_k(\vartheta, \vartheta_0, t) = m_k(\vartheta, t)|_{\varepsilon=0}$.

Suppose that for some $\delta > 0$ we have $g(\vartheta_0, \delta) = 0$, then there exists at least one value ϑ^* such that $\|\vartheta^* - \vartheta_0\| \geq \delta$ and

$$f_k\left(\vartheta^*, t\right) y_k\left(\vartheta^*, \vartheta_0, t\right) = f_k(\vartheta_0, t)\, y_k(\vartheta_0, t), \qquad 0 \leq t \leq T, \quad k = 1, \ldots, K. \tag{7.41}$$

The Eq. (7.24) as $\varepsilon = 0$ becomes

$$\begin{aligned}
\frac{\partial y_k(\vartheta^*, \vartheta_0, t)}{\partial t} \\
&= A_k\left(\vartheta^*, t\right) y_k\left(\vartheta^*, \vartheta_0, t\right) + \frac{f_k(\vartheta^*, t)\, \gamma_k^*(\vartheta^*, t)}{\sigma_k(t)^2} f_k(\vartheta_0, t)\, y_k(\vartheta_0, t) \\
&= A_k\left(\vartheta^*, t\right) y_k\left(\vartheta^*, \vartheta_0, t\right) + \frac{f_k(\vartheta^*, t)^2\, \gamma_k^*(\vartheta^*, t)}{\sigma_k(t)^2} y_k\left(\vartheta^*, \vartheta^*, t\right) \\
&= a_k(t)\, y_k\left(\vartheta^*, \vartheta_0, t\right), \qquad y_k\left(\vartheta^*, \vartheta_0, 0\right) = y_k(0).
\end{aligned}$$

Here we used the equality (7.41) and denoted

$$A_k\left(\vartheta^*, t\right) = a_k(t) - \frac{f_k(\vartheta^*, t)^2\, \gamma_k^*(\vartheta^*, t)}{\sigma_k(t)^2}.$$

Therefore $y_k(\vartheta^*, \vartheta_0, t) \equiv y_k(\vartheta_0, t)$, and from (7.41) it follows that

$$f_k\left(\vartheta^*, t\right) \equiv f_k(\vartheta_0, t), \qquad \rho_{k,t}\left(\vartheta^*\right) \equiv \rho_{k,t}(\vartheta_0), \quad t \in [0, T], \quad k = 1, \ldots, K.$$

These relations imply the equality $\vartheta^* = \vartheta_0$ which is in contradiction with the condition $\|\vartheta^* - \vartheta_0\| \geq \delta$.

The Fisher information matrix is

$$\mathbf{I}(\vartheta_0)_{4\times4} = \int_0^T \frac{\dot{\mathbf{M}}_k^\circ(\vartheta_0, s)^\top \dot{\mathbf{M}}_k^\circ(\vartheta_0, s)}{\sigma_k(s)^2}\, \mathrm{d}s.$$

Let us see more detailed expression of it. Introduce notations

$$\ell_k(x, \vartheta, t) = x_k - \theta_1 - \theta_2 t, \quad \ell_k(y, \vartheta, t) = y_k - \theta_3 - \theta_4 t,$$

$$p_k = p_k(\vartheta, t) = \rho_{k,t}(\vartheta)^2, \qquad N_k(p_k, t) = \frac{f_k(\vartheta, t)\, y_k(\vartheta, \vartheta_0, t)}{\sigma_k(t)},$$

$$\dot{N}_k(p_k,t)=\frac{\partial N_k(p_k,t)}{\partial p_k},\qquad \nu_k(\vartheta_0,t)=4\dot{N}_k(p_k(\vartheta_0),t)^2.$$

Define as well the inner product and norm

$$\langle q,s\rangle_*=\sum_{k=1}^{K}\int_0^T q_k(t)\,s_k(t)\,\nu_k(\vartheta_0,t)\,\mathrm{d}t,\quad \|q\|_*^2=\sum_{k=1}^{K}\int_0^T q_k(t)^2\,\nu_k(\vartheta_0,t)\,\mathrm{d}t.$$

We have four vectors

$$\ell_1(t)=\Big(\ell_k(x,\vartheta_0,t),\quad t\in[0,T],\quad k=1,\ldots,K\Big),$$

$$\ell_2(t)=\Big(\ell_k(x,\vartheta_0,t)\,t,\quad t\in[0,T],\quad k=1,\ldots,K\Big),$$

$$\ell_3(t)=\Big(\ell_k(y,\vartheta_0,t),\quad t\in[0,T],\quad k=1,\ldots,K\Big),$$

$$\ell_4(t)=\Big(\ell_k(y,\vartheta_0,t)\,t,\quad t\in[0,T],\quad k=1,\ldots,K\Big).$$

For example, the inner product of two such vectors is

$$\langle\ell_i(\cdot),\ell_j(\cdot)\rangle_*=\sum_{k=1}^{K}\int_0^T \ell_i(t)\,\ell_j(t)\;\nu_k(\vartheta_0,t)\,\mathrm{d}t.$$

Then the Fisher information is the Gram matrix with the following expression

$$\mathbf{I}(\vartheta_0)=\begin{pmatrix}\|\ell_1(\cdot)\|_*^2, & \langle\ell_1(\cdot)\,\ell_2(\cdot)\rangle_*, & \langle\ell_1(\cdot),\ell_3(\cdot)\rangle_*, & \langle\ell_1(\cdot),\ell_4(\cdot)\rangle_*\\ \langle\ell_1(\cdot)\,\ell_2(\cdot)\rangle_*, & \|\ell_2(\cdot)\|_*^2, & \langle\ell_2(\cdot)\,\ell_2(\cdot)\rangle_*, & \langle\ell_2(\cdot)\,\ell_4(\cdot)\rangle_*,\\ \langle\ell_1(\cdot),\ell_3(\cdot)\rangle_*, & \langle\ell_2(\cdot),\ell_3(\cdot)\rangle_*, & \|\ell_3(\cdot)\|_*^2, & \langle\ell_3(\cdot),\ell_4(\cdot)\rangle_*,\\ \langle\ell_1(\cdot),\ell_4(\cdot)\rangle_*, & \langle\ell_2(\cdot),\ell_4(\cdot)\rangle_*, & \langle\ell_3(\cdot),\ell_4(\cdot)\rangle_*, & \|\ell_4(\cdot)\|_*^2\end{pmatrix}.$$

This matrix is degenerate if the vectors $\ell_1(\cdot),\ldots,\ell_4(\cdot)$ are linearly dependent, i.e., there exist constants $c_1,\ldots,c_4$ $(c_i\neq 0)$ such that $\sum_{i=1}^4 c_i\,\ell_i(t)=0$. The last equality is possible iff all detectors are on the same line. Therefore $\mathrm{Det}(\mathbf{I}(\vartheta_0))\neq 0$ and the conditions $\mathscr{I}\mathscr{K}_4$ of Theorems 1.10 and 1.12 are fulfilled. Hence the MLE and BE are uniformly consistent, asymptotically normal

$$\varepsilon^{-1}(\hat{\vartheta}_\varepsilon-\vartheta_0)\Longrightarrow\zeta(\vartheta_0)\sim\mathcal{N}\Big(0,\mathbf{I}(\vartheta_0)^{-1}\Big),\qquad \varepsilon^{-1}(\tilde{\vartheta}_\varepsilon-\vartheta_0)\Longrightarrow\zeta(\vartheta_0),$$

for any $p>0$

$$\lim_{\varepsilon\to 0}\varepsilon^{-p}\mathbf{E}_{\vartheta_0}\|\hat{\vartheta}_\varepsilon-\vartheta_0\|^p=\mathbf{E}_{\vartheta_0}\|\zeta(\vartheta_0)\|^p,$$

$$\lim_{\varepsilon\to 0}\varepsilon^{-p}\mathbf{E}_{\vartheta_0}\|\tilde{\vartheta}_\varepsilon-\vartheta_0\|^p=\mathbf{E}_{\vartheta_0}\|\zeta(\vartheta_0)\|^p,$$

and both estimators are asymptotically efficient.

7.1.3 *Some Other Problems*

We consider below several models of source localization, which are close to already studied above. That is why we do not give the detailed proofs but clarify some basic technical features, which can be used for the description of the properties of estimators of source positions.

One Signal

We have the same problem with K detectors with coordinates $D_k = (x_k, y_k), k = 1, \ldots, K$ on the plane and one source at (unknown) point $\vartheta_0 = (x_0, y_0)$. In the two sections above we considered situations where each detectors received independent signals from the same source. Sometimes it is reasonable to suppose that all detectors receive the same signal process

$$\mathrm{d}Y_t = a(t) Y(t)\, \mathrm{d}t + \varepsilon b(t)\, \mathrm{d}V_t, \qquad Y_0.$$

Therefore the observed process $X_k = (X_k(t), 0 \le t \le T)$ of the k-th detector has the stochastic differential

$$\mathrm{d}X_k(t) = f_k(\vartheta_0, t) Y(t)\, \mathrm{d}t + \varepsilon\sigma_k(t)\, \mathrm{d}W_k(t), \qquad X_k(0) = 0, \qquad 0 \le t \le T.$$

The family of Kalman-Bucy filters for the conditional expectations $m(\vartheta, t) = \mathbf{E}_\vartheta\left(Y(t) | \mathfrak{F}_t^X\right), \vartheta \in \Theta$ is

$$\mathrm{d}m(\vartheta, t) = a(t) m(\vartheta, t)\,\mathrm{d}t + \gamma_*(\vartheta, t) \sum_{k=1}^{K} \frac{f_k(\vartheta, t)}{\sigma_k(t)^2} \left[\mathrm{d}X_k(t) - f_k(\vartheta, t) m(\vartheta, t)\,\mathrm{d}t\right]$$

$$= A(\vartheta, t) m(\vartheta, t)\,\mathrm{d}t + \gamma_*(\vartheta, t) \sum_{k=1}^{K} \frac{f_k(\vartheta, t)}{\sigma_k(t)^2}\, \mathrm{d}X_k(t), \qquad m(\vartheta, 0),$$

$$\frac{\partial \gamma_*(\vartheta, t)}{\partial \vartheta} = 2a(t)\gamma(\vartheta, t) - \gamma_*(\vartheta, t)^2 \sum_{k=1}^{K} \frac{f_k(\vartheta, t)^2}{\sigma_k(t)^2} + b(t)^2, \qquad \gamma_*(\vartheta, 0) = 0.$$

Here $\vartheta = (x, y) = (\theta_1, \theta_2)$, $\gamma_*(\vartheta, t) = \varepsilon^{-2}\mathbf{E}_\vartheta(Y(t) - m(\vartheta, t))^2$ and

$$A(\vartheta, t) = a(t) - \gamma_*(\vartheta, t) \sum_{k=1}^{K} \frac{f_k(\vartheta, t)^2}{\sigma_k(t)^2}.$$

Recall that $\mathfrak{F}_t^X$ is σ-algebra generated by the observations $X_1(s), \ldots, X_K(s), 0 \le s \le t$.

If the source started emission at the known moment, say, $t = 0$ and the k-th detector receives the signal with time delay $\tau_k(\vartheta_0) = \nu^{-1}\|D_k - \vartheta_0\|$, then the function

$$f_k(\vartheta, t) = a_k(t)\,\bar{\psi}\left(t - \tau_k(\vartheta)\right).$$

The function $a_k(\cdot)$ is positive and bounded, and the transition function $\bar{\psi}_k(\cdot) \in [0, 1]$ forms the front of the signal. We considered in the beginning of this section three cases: smooth function $\psi_\delta(t - \tau_k) = (t - \tau_k - \delta)\,\delta^{-1}\,\mathbb{I}_{\{0 \le t-\tau_k \le \delta\}} + \mathbb{I}_{\{t-\tau_k \ge \delta\}}$, the function with cusp-type singularity

$$\begin{aligned}&\psi_{\delta,k}(t - \tau_k)\\&\quad = \frac{1}{2}\left[1 + \operatorname{sgn}\left(2(t - \tau_k) - \delta\right)\left|\frac{2(t - \tau_k)}{\delta} - 1\right|^{\kappa}\right]\mathbb{I}_{\{0 \le t-\tau_k \le \delta\}} + \mathbb{I}_{\{t-\tau_k \ge \delta\}}\end{aligned}$$

and discontinuous function $\psi(t - \tau_k) = \mathbb{I}_{\{t \ge \tau_k\}}$.

It is possible to study the properties of the same estimators and the construction of adaptive filter as it was done in Sect. 7.1.

Notice that the limits of the $Y(t)$, $X_k(t)$ and $m(\vartheta, t)$ are

$$y_t = y_0 \exp\left(\int_0^t a(s)\,\mathrm{d}s\right), \qquad x_k(\vartheta_0, t) = \int_0^t f_k(\vartheta_0, s)\,y_s\,\mathrm{d}s,$$

$$\begin{aligned}y_t(\vartheta, \vartheta_0) &= y_0\Phi(\vartheta, 0, t)\\&\quad + \sum_{k=1}^{K}\int_{\tau_k(\vartheta)\vee\tau_k(\vartheta_0)}^{t}\Phi(\vartheta, s, t)\,\frac{\gamma_*(\vartheta, s)\,f_k(\vartheta, s)\,f_k(\vartheta_0, s)\,y_s}{\sigma_k(s)^2}\,\mathrm{d}s\end{aligned}$$

respectively. Here

$$\Phi(\vartheta, s, t) = \exp\left(\int_s^t A(\vartheta, v)\,\mathrm{d}v\right).$$

The likelihood ratio is

$$L\left(\vartheta, X^K\right) = \exp\left(\sum_{k=1}^{K}\int_{\tau_k(\vartheta)}^{T}\frac{M_k(\vartheta, t)}{\varepsilon^2\sigma_k(t)^2}\,\mathrm{d}X_k(t) - \sum_{k=1}^{K}\int_{\tau_k(\vartheta)}^{T}\frac{M_k(\vartheta, t)^2}{2\varepsilon^2\sigma_k(t)^2}\,\mathrm{d}t\right),$$
$$\vartheta \in \Theta,$$

where $M_k(\vartheta, t) = f_k(\vartheta, t)\,m(\vartheta, t)$.

Let us discuss some technical features of this model in the smooth case. To calculate the Fisher information matrix $\mathbf{I}(\vartheta_0)_{2\times 2}$ we note that

$$\begin{aligned}\left.\frac{\partial M_k(\vartheta,t)}{\partial\vartheta}\right|_{\varepsilon=0} &= \frac{\partial f_k(\vartheta,t)}{\partial\vartheta}\, y_t(\vartheta,\vartheta_0,t) + f_k(\vartheta,t)\,\frac{\partial y_t(\vartheta,\vartheta_0,t)}{\partial\vartheta}\\ &= \frac{\partial f_k(\vartheta,t)}{\partial\tau_k}\, y_t(\vartheta,\vartheta_0,t)\,\frac{\partial\tau_k(\vartheta)}{\partial\vartheta}\\ &\quad + f_k(\vartheta,t)\sum_{l=1}^{K}\frac{\partial y_t(\vartheta,\vartheta_0,t)}{\partial\tau_l}\frac{\partial\tau_l(\vartheta)}{\partial\vartheta}\\ &= -\frac{\partial f_k(\vartheta,t)}{\partial\tau_k}\, y_t(\vartheta,\vartheta_0,t)\,\frac{\mu_k}{\nu} - f_k(\vartheta,t)\sum_{l=1}^{K}\frac{\partial y_t(\vartheta,\vartheta_0,t)}{\partial\tau_l}\frac{\mu_l}{\nu}.\end{aligned}$$

Recall that $\partial\tau_l(\vartheta)/\partial\vartheta = -\nu^{-1}\mu_l$ where ν is the known rate of propagation of the signal $Y(\cdot)$, $\mu_l = \left(\mu_{l,x},\mu_{l,y}\right)^\top$ is a unite vector, $\partial\tau_l(\vartheta)/\partial\theta_1 = -\nu^{-1}\mu_{l,x}$, $\partial\tau_l(\vartheta)/\partial\theta_2 = -\nu^{-1}\mu_{l,y}$ where $\mu_{l,x} = (x_l-\theta_1)\,\|D_l-\vartheta\|^{-1}$ and $\mu_{l,y} = (y_l-\theta_2)\,\|D_l-\vartheta\|^{-1}$. Of course, $\mu_l=\mu_l(\vartheta)$.

Put

$$\dot{\mathbf{M}}_k^\circ(\vartheta_0,t) = \left.\frac{\partial M_k(\vartheta,t)}{\partial\vartheta}\right|_{\varepsilon=0,\vartheta=\vartheta_0}.$$

Then

$$\mathbf{I}(\vartheta_0) = \sum_{k=1}^{K}\int_{\tau_k(\vartheta_0)}^{T}\frac{\dot{\mathbf{M}}_k^\circ(\vartheta_0,t)\,\dot{\mathbf{M}}_k^\circ(\vartheta_0,t)^\top}{\sigma_k(t)^2}\,\mathrm{d}t.$$

For example,

$$\begin{aligned}&I(\vartheta_0)_{1,1}\\ &= \sum_{k=1}^{K}\frac{1}{\nu^2}\int_{\tau_k(\vartheta_0)}^{T}\left(\frac{\partial f_k(\vartheta_0,t)}{\partial\vartheta}\, y_t\,\mu_{k,x}(\vartheta_0) + f_k(\vartheta_0,t)\sum_{l=1}^{K}\frac{\partial y_t}{\partial\tau_l}\,\mu_{l,x}(\vartheta_0)\right)^2\mathrm{d}t.\end{aligned}$$

Here

$$\frac{\partial y_t}{\partial\tau_l} = \left.\frac{\partial y_t(\vartheta,\vartheta_0)}{\partial\tau_l(\vartheta)}\right|_{\vartheta=\vartheta_0}.$$

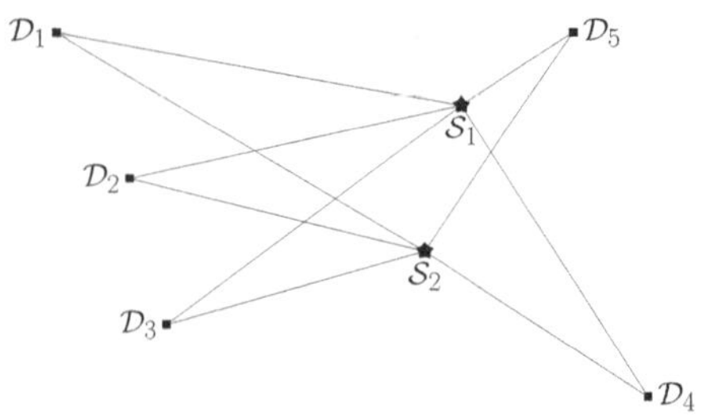

Fig. 7.3 Model of observations. $\mathcal{S}_1, \mathcal{S}_2$ are positions of the sources and $\mathcal{D}_k, k = 1, \ldots, 5$ are positions of the sensors

Introduce the function

$$g(\vartheta_0, d) = \inf_{\|\vartheta - \vartheta_0\| > d} \sum_{k=1}^{K} \int_0^T \frac{[f_k(\vartheta, t)\, y_t(\vartheta, \vartheta_0) - f_k(\vartheta_0, t)\, y_t]^2}{\sigma_k(t)^2}\, \mathrm{d}t.$$

The condition of identifiability: for any $d > 0$ the function $g(\vartheta_0, d) > 0$. Remark that in this problem this condition is always satisfied. The verification is the same as in the proof of Lemma 7.4.

Therefore if we suppose that the Fisher information is nondegenerate, then the slight modification of the proof of Theorem 7.1 will allow us to prove the consistency, asymptotic normality, convergence of moments, and asymptotic efficiency of the MLE and BE of the parameter ϑ.

Remark 7.2 Of course, it is possible to describe the properties of the BE in the cases with cusp-type and change-point type singularities and to present for BE theorems similar to Theorems 7.2 and 7.3, respectively. The properties of the LR established in Lemmas 7.2–7.10 are sufficient for the theorems on BE in these situations.

Two Sources

Let us study the problem of localization of two sources $\mathcal{S}_1, \mathcal{S}_2$ by $K \geq 4$ detectors (see Fig. 7.3).

The positions of the detectors we denote $(x_k, y_k), k = 1, \ldots, K$ and the positions of the sources $\vartheta_1 = (\theta_{1,1}, \theta_{1,2})$ and $\vartheta_2 = (\theta_{2,1}, \theta_{2,2})$. The unknown parameter we denote as $\vartheta = (\vartheta_1, \vartheta_2)$ and the true value is ϑ_0. The observations $X_k(t), 0 \leq t \leq T, k = 1, \ldots, T$ are given by the stochastic differentials

$$\mathrm{d}X_k(t) = f_k(\vartheta_1, t)\, Y_k^{(1)}(t)\, \mathrm{d}t + g_k(\vartheta_2, t)\, Y_k^{(2)}(t)\, \mathrm{d}t + \varepsilon \sigma_k(t)\, \mathrm{d}W_k(t),$$
$$X_k(0) = 0.$$

The signals $Y_k^{(1)}(\cdot), Y_k^{(2)}(\cdot)$ satisfy the equations

$$\mathrm{d}Y_k^{(1)}(t) = a_k(t)\,Y_k^{(1)}(t)\,\mathrm{d}t + \varepsilon b_k(t)\,\mathrm{d}V_k^{(1)}(t), \quad Y_k(0) = y_k^{(1)} \neq 0,$$
$$\mathrm{d}Y_k^{(2)}(t) = c_k(t)\,Y_k^{(2)}(t)\,\mathrm{d}t + \varepsilon d_k(t)\,\mathrm{d}V_k^{(2)}(t), \quad Z_k(0) = y_k^{(2)} \neq 0.$$

The Kalman-Bucy filter for the conditional expectations $m_k(t) \equiv m_k(\vartheta, t) = \mathbf{E}_\vartheta\left(Y_k^{(1)}(t)\,|\mathfrak{F}_t^X\right)$, $n_k(t) \equiv n_k(\vartheta, t) = \mathbf{E}_\vartheta\left(Y_k^{(2)}(t)\,|\mathfrak{F}_t^X\right)$ we write omitting ϑ:

$$\begin{aligned}\mathrm{d}m_k(t) &= a_k(t)\,m_k(t)\,\mathrm{d}t + \frac{\gamma_{k,1,1}(t)\,f_k(t) + \gamma_{k,1,2}(t)\,g_k(t)}{\sigma_k(t)^2}\left[\mathrm{d}X_k(t)\right.\\ &\qquad \left. - \left(f_k(t)\,m_k(t) + g_k(t)\,n_k(t)\right)\mathrm{d}t\right], \qquad m_k(0) = y_k^{(1)},\end{aligned} \tag{7.42}$$

$$\begin{aligned}\mathrm{d}n_k(t) &= c_k(t)\,n_k(t)\,\mathrm{d}t + \frac{\gamma_{k,2,1}(t)\,f_k(t) + \gamma_{k,2,2}(t)\,g_k(t)}{\sigma_k(t)^2}\left[\mathrm{d}X_k(t)\right.\\ &\qquad \left. - \left(f_k(t)\,m_k(t) + g_k(t)\,n_k(t)\right)\mathrm{d}t\right], \qquad n_k(0) = y_k^{(2)}.\end{aligned} \tag{7.43}$$

Here $\gamma_{k,i,j}(t) = \gamma_{k,i,j}(\vartheta, t)$ are elements of the 2×2 matrix $\gamma_k(\vartheta, t) = \gamma_k(t)$

$$\gamma_{k,1,1}(\vartheta, t) = \varepsilon^{-2}\mathbf{E}_\vartheta\left[Y_k^{(1)}(t) - m_k(\vartheta, t)\right]^2,$$

$$\gamma_{k,2,2}(\vartheta, t) = \varepsilon^{-2}\mathbf{E}_\vartheta\left[Y_k^{(2)}(t) - n_k(\vartheta, t)\right]^2,$$

$$\gamma_{k,1,2}(\vartheta, t) = \gamma_{k,2,1}(\vartheta, t) = \varepsilon^{-2}\mathbf{E}_\vartheta\left[\left(Y_k^{(1)}(t) - m_k(\vartheta, t)\right)\left(Y_k^{(2)}(t) - n_k(\vartheta, t)\right)\right].$$

This matrix is the solution of the Riccati equation

$$\begin{aligned}\frac{\partial \gamma_k(t)}{\partial t} &= P_k(t)\,\gamma_k(t) + \gamma_k(t)\,P_k(t)^\top + \gamma_k(t)\,P_k(t)^\top P_k(t)\,\gamma_k(t)\,\sigma_k(t)^{-2}\\ &\quad + Q_k(t).\end{aligned}$$

The initial value is $\gamma_k(0) = 0$, and we introduced the vectors $P_k(t) = (a_k(t), c_k(t))^\top$, $Q_k(t) = (b_k(t), d_k(t))^\top$.

Suppose that the intensity of the signals is decreasing function of the distance between the sources and the detectors, i.e.,

$$f_k(\vartheta_1, t) = \frac{f_k S_k(t)}{\rho_k(\vartheta_1)^2}, \qquad g_k(\vartheta_2, t) = \frac{g_k R_k(t)}{\rho_k(\vartheta_2)^2}.$$

Thus the limits are

$$y_k^{(1)}(t)=y_k^{(1)}\exp\left(\int_0^t a_k(s)\,\mathrm{d}s\right),\qquad y_k^{(2)}(t)=y_k^{(2)}\exp\left(\int_0^t c_k(s)\,\mathrm{d}s\right),$$

$$x_k(\vartheta,t)=\frac{f_k}{\rho_k(\vartheta_1)^2}\int_0^t S_k(s)\,y_k^{(1)}(s)\;\mathrm{d}s+\frac{g_k}{\rho_k(\vartheta_2)^2}\int_0^t R_k(s)\,y_k^{(2)}(s)\;\mathrm{d}s.$$

To define the preliminary estimators of the parameters ϑ_1 and ϑ_2 we fix some (small) $\tau\in(0,T)$ and take two functions $F_k(t)\,,G_k(t)\,,0\leq t\leq\tau$ such that

$$\int_0^\tau G_k(t)\int_0^t S_k(s)\,y_k^{(1)}(s)\,\mathrm{d}s\;\mathrm{d}t=0,\qquad \int_0^\tau F_k(t)\int_0^t R_k(s)\,y_k^{(2)}(s)\,\mathrm{d}s\;\mathrm{d}t=0,$$

$$K_k(\tau)=\int_0^\tau F_k(t)\int_0^t S_k(s)\,y_k^{(1)}(s)\,\mathrm{d}s\;\mathrm{d}t\neq 0,$$

$$L_k(\tau)=\int_0^\tau G_k(t)\int_0^t R_k(s)\,y_k^{(2)}(s)\,\mathrm{d}s\;\mathrm{d}t\neq 0.$$

This can be done if we suppose that $S_k(t)\,y_k^{(1)}(t)\neq R_k(t)\,y_k^{(2)}(t)\,,0\leq t\leq\tau$. Therefore

$$\int_0^\tau F_k(t)\,x_k(\vartheta,s)\,\mathrm{d}s=\frac{f_kK_k(\tau)}{\rho_k(\vartheta_1)^2},\qquad \int_0^\tau G_k(t)\,x_k(\vartheta,s)\,\mathrm{d}s=\frac{g_kL_k(\tau)}{\rho_k(\vartheta_2)^2}$$

and

$$\rho_k(\vartheta_1)^2=f_kK_k(\tau)\left(\int_0^\tau F_k(t)\,x_k(\vartheta,s)\,\mathrm{d}s\right)^{-1},$$

$$\rho_k(\vartheta_2)^2=f_kL_k(\tau)\left(\int_0^\tau G_k(t)\,x_k(\vartheta,s)\,\mathrm{d}s\right)^{-1}.$$

Introduce the statistics

$$Z_{k,\varepsilon}^{(1)}(\tau)=f_kK_k(\tau)\left(\int_0^\tau F_k(t)\,X_k(s)\,\mathrm{d}s\right)^{-1},$$

$$Z_{k,\varepsilon}^{(2)}(\tau)=g_kL_k(\tau)\left(\int_0^\tau G_k(t)\,X_k(s)\,\mathrm{d}s\right)^{-1}.$$

If one of this integrals takes value 0, we put $Z_{k,\varepsilon}^{(i)}(\tau)=0$, but as the limits are $Z_{k,\varepsilon}^{(1)}(\tau)\to\rho_k\left(\vartheta_{0,1}\right)^2,\ Z_{k,\varepsilon}^{(2)}(\tau)\to\rho_k\left(\vartheta_{0,2}\right)^2$, the probabilities of such events tend to zero.

It is easy to see that

$$\varepsilon^{-1}\left(Z_{k,\varepsilon}^{(1)}(\tau)-\rho_k\left(\vartheta_{0,1}\right)^2\right)=O\left(1\right),\qquad \varepsilon^{-1}\left(Z_{k,\varepsilon}^{(2)}(\tau)-\rho_k\left(\vartheta_{0,2}\right)^2\right)=O\left(1\right)$$

Moreover, these limits have Gaussian distributions which can be described too.

The LSEs $\vartheta_{1,\varepsilon}^*$ and $\vartheta_{2,\varepsilon}^*$ we define separately following the same steps as it was done above studying the LSE-2 (see (7.32)–(7.36)). Consider the construction of $\vartheta_{1,\varepsilon}^*$. Introduce the sum

$$\begin{aligned}N\left(\vartheta_1\right)&=\sum_{k=1}^{K}\left[Z_{k,\varepsilon}^{(1)}(\tau)-\left(x_k-\theta_{1,1}\right)^2-\left(y_k-\theta_{1,2}\right)^2\right]^2\\&=\sum_{k=1}^{K}\left[Z_{k,\varepsilon}^{(1)}(\tau)-\left(x_k^2+y_k^2\right)+2\theta_{1,1}x_k+2\theta_{1,2}y_k-\left(\theta_{1,1}^2+\theta_{1,2}^2\right)\right]^2\\&=\sum_{k=1}^{K}\left[Z_{k,\varepsilon}^{(1)}(\tau)-r_k^2+2\theta_{1,1}x_k+2\theta_{1,2}y_k-2\theta_{1,3}\right]^2\end{aligned}$$

where we denoted $r_k^2=x_k^2+y_k^2$ and introduced a new variable $\theta_{1,3}=\left(\theta_{1,1}^2+\theta_{1,2}^2\right)/2$. The minimization of this sum w.r.t. variables $\theta_{1,1},\theta_{1,2},\theta_{1,3}$ leads to the equations

$$\begin{aligned}\theta_{1,1}\sum_{k=1}^{K}x_k^2+\theta_{1,2}\sum_{k=1}^{K}x_ky_k-\theta_{1,3}\sum_{k=1}^{K}x_k&=\frac{1}{2}\sum_{k=1}^{K}\left[Z_{k,\varepsilon}^{(1)}(\tau)-r_k^2\right]x_k,\\\theta_{1,1}\sum_{k=1}^{K}x_ky_k+\theta_{1,2}\sum_{k=1}^{K}y_k^2-\theta_{1,3}\sum_{k=1}^{K}y_k&=\frac{1}{2}\sum_{k=1}^{K}\left[Z_{k,\varepsilon}^{(1)}(\tau)-r_k^2\right]y_k,\\\theta_{1,1}\sum_{k=1}^{K}x_k+\theta_{1,2}\sum_{k=1}^{K}y_k-\theta_{1,3}K&=\frac{1}{2}\sum_{k=1}^{K}\left[Z_{k,\varepsilon}^{(1)}(\tau)-r_k^2\right].\end{aligned}$$

From the last equation we obtain the expression for $\theta_{1,3}$ and substitute it in the first two equations. Thus the system is reduced to

$$\begin{aligned}\theta_{1,1}\left[\overline{x^2}-(\bar{x})^2\right]+\theta_{1,2}\left[\overline{xy}-\bar{x}\,\bar{y}\right]&=\frac{1}{2}\overline{\left[Z_{\varepsilon}^{(1)}(\tau)-r^2\right]x}-\frac{1}{2}\overline{\left[Z_{\varepsilon}^{(1)}(\tau)-r^2\right]}\,\bar{x},\\\theta_{1,1}\left[\overline{xy}-\bar{x}\,\bar{y}\right]+\theta_{1,2}\left[\overline{y^2}-(\bar{y})^2\right]&=\frac{1}{2}\overline{\left[Z_{\varepsilon}^{(1)}(\tau)-r^2\right]y}-\frac{1}{2}\overline{\left[Z_{\varepsilon}^{(1)}(\tau)-r^2\right]}\,\bar{y},\end{aligned}$$

where we used the notation $\overline{xy}=\frac{1}{K}\sum_{k=1}^{K}x_ky_k$ and so on. As the detectors are not on the same line the determinant of this system is nondegenerate and the LSE

$\vartheta^*_{1,\varepsilon} = \left(\vartheta^*_{1,1,\varepsilon}, \vartheta^*_{1,2,\varepsilon}\right)^\top$ is well defined, is consistent, and $\varepsilon^{-1}\left(\vartheta^*_{1,\varepsilon} - \vartheta_{0,1}\right)$ is asymptotically normal.

Similar study allows us to obtain the equations for $\vartheta^*_{2,\varepsilon}$ and to establish the same asymptotic properties. Hence the LSE $\vartheta^*_\varepsilon = \left(\vartheta^*_{1,\varepsilon}, \vartheta^*_{2,\varepsilon}\right)$ of the positions of two sources is consistent and asymptotically normal.

Let us denote $M_k(\vartheta, t) = f_k(\vartheta_1, t) m_k(\vartheta, t) + g_k(\vartheta_2, t) n_k(\vartheta, t)$ and $\dot{\mathrm{M}}_k(\vartheta, t) = \partial M_k(\vartheta, t)/\partial\vartheta$. The derivatives of $m_k(\vartheta, t)$ and $n_k(\vartheta, t)$ are obtained by the formal differentiation of Eqs. (7.42) and (7.43). Note that, for example, $m_k(\vartheta, t)$ depends on ϑ_2 too. Denote as well $\dot{\mathrm{M}}^\circ_k(\vartheta_0) = \dot{\mathrm{M}}_k(\vartheta, t)\big|_{\varepsilon=0,\vartheta=\vartheta_0}$. The Fisher information 4×4 matrix is

$$\mathbf{I}^t_\tau(\vartheta_0) = \sum_{k=1}^K \int_\tau^t \frac{\dot{\mathrm{M}}^\circ_k(\vartheta_0, s)\,\dot{\mathrm{M}}^\circ_k(\vartheta_0, s)^\top}{\sigma_k(s)^2}\,\mathrm{d}s.$$

We suppose that this matrix is uniformly on $\vartheta_0 \in \Theta$ nondegenerate. The One-step MLE-process has the *usual* expression

$$\vartheta^\star_{t,\varepsilon} = \vartheta^*_\varepsilon + \mathbf{I}^t_\tau\left(\vartheta^*_\varepsilon\right)^{-1} \sum_{k=1}^K \int_\tau^t \frac{\dot{\mathrm{M}}_k\left(\vartheta^*_\varepsilon, s\right)}{\sigma_k(s)^2}\left[\mathrm{d}X_k(s) - M_k\left(\vartheta^*_\varepsilon, s\right)\mathrm{d}s\right], \quad \tau < t \le T,$$

and it can be shown that this estimator is asymptotically normal

$$\varepsilon^{-1}\left(\vartheta^\star_{t,\varepsilon} - \vartheta_0\right) \Longrightarrow \mathcal{N}\left(0, \mathbf{I}^t_\tau(\vartheta_0)^{-1}\right).$$

If we take $\tau = \tau_\varepsilon \to 0$ but sufficiently slowly, then it is possible to prove the asymptotic efficiency of the One-step MLE

$$\varepsilon^{-1}\left(\vartheta^\star_{T,\varepsilon} - \vartheta_0\right) \Longrightarrow \mathcal{N}\left(0, \mathbf{I}^T_0(\vartheta_0)^{-1}\right).$$

The *Poissonian version* of such proof is given in [31], and in the Gaussian case the main steps of the proof are the same.

Large Samples

We have the same as in Sect. 7.1.1 structure on the plane with K detectors located at the points $D_k = (x_k, y_k)$, $k = 1, \ldots, K$ and one source S at the (unknown) point $\vartheta = (x_0, y_0) = (\theta_1, \theta_2)$. We suppose that at least three detectors are not on the same line. The observations are $X^T = \left(X^T_1, \ldots, X^T_K\right)$, where the stochastic process $X_k = (X_k(t), 0 \le t \le T)$ satisfies the system of equations

$$
\begin{aligned}
\mathrm{d}X_k(t) &= f_k(\vartheta)\, Y_k(t)\, \mathrm{d}t + \sigma_k\, \mathrm{d}W_k(t), \qquad X_k(0), \quad 0 \le t \le T, \\
\mathrm{d}Y_k(t) &= -a_k Y_k(t)\, \mathrm{d}t + b_k\, \mathrm{d}V_k(t), \qquad Y_k(0).
\end{aligned}
$$

Here $W_k(\cdot), V_k(\cdot), k = 1, \ldots, K$ are independent Wiener processes, the function

$$
f_k(\vartheta) = \frac{f_k}{\rho_k(\vartheta_0)^2},
$$

$f_k \neq 0$, $a_k > 0$, $b_k \neq 0$, $\sigma_k \neq 0$ are known constants. This model of observations is close to that of Chap. 4, and the proposed estimation here is motivated by the results obtained there.

The Kalman-Bucy filter is

$$
\mathrm{d}m_k(\vartheta, t) = -a_k m_k(\vartheta, t)\, \mathrm{d}t + \frac{f_k(\vartheta)\, \gamma_k(\vartheta, t)}{\sigma_k^2} \left[\mathrm{d}X_k(t) - f_k(\vartheta)\, m_k(\vartheta, t)\, \mathrm{d}t\right], \tag{7.44}
$$

$$
\frac{\partial \gamma_k(\vartheta, t)}{\partial t} = -2a_k \gamma_k(\vartheta, t) - \frac{f_k(\vartheta)^2\, \gamma_k(\vartheta, t)^2}{\sigma_k^2} + b_k^2.
$$

Recall that this Riccati equation has an explicit solution which exponentially converges to the steady-state value (see (4.5))

$$
\gamma_k^*(\vartheta) = \frac{a_k \sigma_k^2}{f_k(\vartheta)^2} \left[\sqrt{1 + \frac{b_k^2 f_k(\vartheta)^2}{a_k^2 \sigma_k^2}} - 1\right]. \tag{7.45}
$$

Therefore to simplify the exposition we put $\gamma_k(\vartheta, 0) = \gamma_k^*(\vartheta)$, and this provides us the equality $\gamma(\vartheta, t) \equiv \gamma_k^*(\vartheta)$ for all $t > 0$.

Thus Eq. (7.44) became

$$
\mathrm{d}m_k(\vartheta, t) = -a_k m_k(\vartheta, t)\, \mathrm{d}t + \frac{f_k(\vartheta)\, \gamma_k^*(\vartheta)}{\sigma_k^2} \left[\mathrm{d}X_k(t) - f_k(\vartheta)\, m_k(\vartheta, t)\, \mathrm{d}t\right].
$$

To construct the preliminary estimator we introduce the learning interval $\left[0, \tau_T\right]$, where $\tau_T = \left[T^\delta\right]$ (here $\delta \in (1/2, 1)$ and $[A]$ means entire part of A). Recall that the statistic

$$
\begin{aligned}
R_{k,\tau_T} &= \frac{1}{\tau_T} \sum_{j=1}^{\tau_T} \left[X_k(j) - X_k(j-1)\right]^2 \longrightarrow \Phi_k(\vartheta) \\
&= \frac{f_k^2 b_k^2}{\rho_k(\vartheta)^4\, a_k^3} \left[e^{-a_k} - 1 + a_k\right] + \sigma_k^2.
\end{aligned}
$$

Therefore we can introduce the statistics

$$Z_{k,\tau_T} = \left(\frac{f_k^2 b_k^2 \left[e^{-a_k} - 1 + a_k \right]}{\left(R_{k,\tau_T} - \sigma_k^2 \right) a_k^3} \right)^{1/2} \longrightarrow \rho_k \left(\vartheta_0 \right)^2 .$$

If $R_{k,\tau_T} < \sigma_k^2$, we put $Z_{k,\tau_T} = 0$. This statistic is an MME of the distance $\rho_k \left(\vartheta_0 \right)^2$. From Proposition 4.8 it follows that this MME is asymptotically normal

$$\sqrt{\tau_T} \left(Z_{k,\tau_T} - \rho_k \left(\vartheta_0 \right)^2 \right) \Longrightarrow \mathcal{N} \left(0, D \left(\vartheta_0 \right) \right),$$

where the limit variance $D \left(\vartheta_0 \right)$ can be calculated.

The LSE $\vartheta_{\tau_T}^*$ is defined by the minimization of the sum

$$\sum_{k=1}^{K} \left[Z_{k,\tau_T} - \left(x_k - \theta_1 \right)^2 - \left(y_k - \theta_2 \right)^2 \right]^2 .$$

This gives us the system

$$\theta_1 \left[\overline{x^2} - (\bar{x})^2 \right] + \theta_2 \left[\overline{xy} - \bar{x}\bar{y} \right] = \frac{1}{2} \overline{\left[Z_{\tau_T} - r^2 \right] x} - \frac{1}{2} \overline{\left[Z_{\tau_T} - r^2 \right]}\, \bar{x},$$

$$\theta_1 \left[\overline{xy} - \bar{x}\bar{y} \right] + \theta_2 \left[\overline{y^2} - (\bar{y})^2 \right] = \frac{1}{2} \overline{\left[Z_{\tau_T} - r^2 \right] y} - \frac{1}{2} \overline{\left[Z_{\tau_T} - r^2 \right]}\, \bar{y}.$$

The solution of it is the LSE $\vartheta_{\tau_T}^*$. It can be shown that this estimator is consistent and asymptotically normal

$$\sqrt{\tau_T} \left(\vartheta_{\tau_T}^* - \vartheta_0 \right) \Longrightarrow \mathcal{N} \left(0, D_* \left(\vartheta_0 \right) \right),$$

where the limit variance $D_* \left(\vartheta_0 \right)$ can be written explicitly.

Let us denote $M_k \left(\vartheta \right) = f_k \left(\vartheta \right) m_k \left(\vartheta, t \right)$. The vector of derivatives $\dot{M}_k \left(\vartheta \right) = \partial M_k \left(\vartheta \right) / \partial \vartheta = \dot{f}_k \left(\vartheta \right) m_k \left(\vartheta, t \right) + f_k \left(\vartheta \right) \dot{m}_k \left(\vartheta, t \right)$ can be calculated using Eqs. (7.44) and (7.45).

To write the Fisher information matrix we use the equality (4.18) and denote

$$P_k \left(\vartheta \right) = \frac{2 f_k^4 b_k^4}{\sigma_k^4 \left(a_k + f_k^2 b_k^2 \sigma_k^{-2} \rho_k \left(\vartheta \right)^{-4} \right)^{3/2}}.$$

Introduce the inner product and norm with weights $P_k \left(\vartheta \right), k = 1, \ldots, K$ in the space $\mathcal{R}^K$

$$\langle Q, H\rangle_\vartheta = \sum_{k=1}^{K} Q_k H_k P_k(\vartheta), \qquad \|Q\|_\vartheta^2 = \sum_{k=1}^{K} Q_k^2 P_k(\vartheta).$$

Then for the Fisher information matrix we obtain the following expression

$$\mathbf{I}(\vartheta_0) = \begin{pmatrix} \left\|\mathrm{x} - \theta_{0,1}\right\|_{\vartheta_0}^2 & \langle(\mathrm{x} - \theta_{0,1})(\mathrm{y} - \theta_{0,2})\rangle_{\vartheta_0} \\ \langle(\mathrm{x} - \theta_{0,1})(\mathrm{y} - \theta_{0,2})\rangle_{\vartheta_0} & \left\|\mathrm{y} - \theta_{0,2}\right\|_{\vartheta_0}^2 \end{pmatrix}.$$

Here the vectors are $\mathrm{x} = (x_1, \ldots, x_K)$, $\mathrm{y} = (y_1, \ldots, y_K)$ and $\theta_{0,1} = (\theta_{0,1}, \ldots, \theta_{0,1})$.

As there are at least three detectors not on the same line, this matrix is nondegenerate and the One-step MLE-process can be written in the standard form

$$\vartheta_{t,T}^\star = \vartheta_{\tau_T}^* + \mathbf{I}(\vartheta_{\tau_T}^*)^{-1} \int_{\tau_T}^{t} \frac{\dot{\mathrm{M}}_k(\vartheta_{\tau_T}^*, s)}{(t-\tau_T)\,\sigma_k^2} \left[\mathrm{d}X_k(s) - M_k(\vartheta_{\tau_T}^*, s)\mathrm{d}s\right], \qquad \tau_T < t \le T.$$

If we put $t = vT, v \in [\epsilon_T, 1]$, where $\epsilon_T = T^{-1+\delta} \to 0$, then the following asymptotic normality

$$\sqrt{t}\left(\vartheta_{t,T}^\star - \vartheta_0\right) \Longrightarrow \mathcal{N}\left(0, \mathbf{I}(\vartheta_0)^{-1}\right).$$

can be verified.

7.2 Indirect Observation of a Signal

7.2.1 Model of Observations

Khasminskii and Lazareva [94] considered the following partially observed system

$$\mathrm{d}X_t = Y_t\,\mathrm{d}t + \varepsilon\,\sigma\,\mathrm{d}W_t, \qquad X_0 = 0, \qquad 0 \le t \le T,$$

$$\mathrm{d}Y_t = S(\vartheta, t)\,\mathrm{d}t + \varepsilon\,\mathrm{d}V_t, \qquad Y_0 = y_0.$$

Here $W_t, 0 \le t \le T$ and $V_t, 0 \le t \le T$ are independent Wiener processes and $S(\vartheta, t), 0 \le t \le T$ is some deterministic signal depending on the unknown parameter $\vartheta \in \Theta \subset \mathcal{R}^d$. The parameters ε, σ and the function $S(\cdot, \cdot)$ are supposed to be known. The main problem considered in [94] was to estimate y_0, ϑ and to describe the properties of estimators in the asymptotics $\varepsilon \to 0$. It was shown that under regularity conditions the MLE is consistent, asymptotically normal, and asymptotically efficient. The case of both limits $\varepsilon \to 0$ and $\sigma \to 0$ was studied too.

In this section we study the properties of MDE, MLE, and One-step MLE for the model

$$\mathrm{d}X_t = f\,Y_t\,\mathrm{d}t + \varepsilon\,\sigma\,\mathrm{d}W_t, \qquad X_0 = 0, \qquad 0 \le t \le T, \tag{7.46}$$

$$\mathrm{d}Y_t = S(\vartheta, t)\,\mathrm{d}t + \varepsilon\,b\,\mathrm{d}V_t, \qquad Y_0 = \theta_1. \tag{7.47}$$

Here $\theta_1 = y_0$ and $\vartheta = (\theta_1, \theta_2, \ldots, \theta_d) \in \Theta$, but $S(\vartheta, t) = S(\theta_2, \ldots, \theta_d, t)$. The set $\Theta \subset \mathcal{R}^d, d > 1$ is open, convex, and bounded.

We suppose that the parameters $f, \sigma, \varepsilon, b$ are known and positive and the function $S(\vartheta, t), \vartheta \in \bar{\Theta}, t \in [0, T]$ is bounded.

The Kalman-Bucy filter for the process $m(\vartheta, t) = \mathbf{E}_\vartheta\left(Y_t|\mathfrak{F}_t^X\right)$ is given by the equations

$$\mathrm{d}m(\vartheta, t) = S(\vartheta, t)\,\mathrm{d}t + \frac{f\,\gamma(t)}{\varepsilon^2\sigma^2}\left[\mathrm{d}X_t - f m(\vartheta, t)\,\mathrm{d}t\right], \qquad m(\vartheta, 0) = \theta_1,$$

$$\frac{\mathrm{d}\gamma(t)}{\mathrm{d}t} = -\frac{f^2\gamma(t)^2}{\varepsilon^2\sigma^2} + \varepsilon^2 b^2, \qquad \gamma(0) = 0.$$

Here $m(\vartheta, t) = m(\theta_1, \ldots, \theta_d, t)$. If we denote $\gamma_*(t) = \varepsilon^{-2}\gamma(t)$, then the equation for $\gamma_*(\cdot)$ does not depend on ε

$$\frac{\mathrm{d}\gamma_*(t)}{\mathrm{d}t} = -\frac{f^2\gamma_*(t)^2}{\sigma^2} + b^2, \qquad \gamma_*(0) = 0.$$

Recall that the solution of this Riccati equation has an explicit expression (1.23)

$$\gamma_*(t) = \frac{b\sigma}{f}\tanh\left(\frac{fbt}{\sigma}\right), \qquad 0 \le t \le T.$$

Thus the equation for $m(\vartheta, t)$ can be written as follows:

$$\begin{aligned}
\mathrm{d}m(\vartheta, t) &= S(\vartheta, t)\,\mathrm{d}t + \frac{b}{\sigma}\tanh\left(\frac{fbt}{\sigma}\right)\left[\mathrm{d}X_t - f m(\vartheta, t)\,\mathrm{d}t\right] \\
&= S(\vartheta, t)\,\mathrm{d}t + \frac{fb}{\sigma}\tanh\left(\frac{fbt}{\sigma}\right)\left[m(\vartheta_0, t) - m(\vartheta, t)\right]\mathrm{d}t \\
&\quad + \varepsilon b\tanh\left(\frac{fbt}{\sigma}\right)\mathrm{d}\bar{W}_t, \qquad m(\vartheta, 0) = \theta_1.
\end{aligned}$$

The limit ($\varepsilon \to 0$) functions are

$$Y_t \longrightarrow y_t(\vartheta_0) = \theta_{0,1} + \int_0^t S(\vartheta_0, s)\,\mathrm{d}s, \qquad \varepsilon^{-1}\left(Y_t - y_t(\vartheta_0)\right) = b\,V_t,$$

$$X_t \longrightarrow x_t(\vartheta_0) = f \int_0^t y_s(\vartheta_0)\,\mathrm{d}s, \qquad \varepsilon^{-1}(X_t - x_t(\vartheta_0)) = \sigma W_t,$$

$$m(\vartheta, t) \longrightarrow y_t(\vartheta, \vartheta_0) = \theta_1 \phi(0, t) + \int_0^t \phi(s, t)\left[S(\vartheta, s) + \tanh\left(\frac{fbs}{\sigma}\right)\frac{fby_s(\vartheta_0)}{\sigma}\right]\mathrm{d}s.$$

Here $\vartheta_0 = (\theta_{0,1}\ldots, \theta_{0,d})$ is the true value and

$$\phi(s, t) = \exp\left(-\sigma^{-2}\int_s^t \gamma_*(v)\,\mathrm{d}v\right) = \left(\frac{\cosh\left(\frac{bf}{\sigma}s\right)}{\cosh\left(\frac{bf}{\sigma}t\right)}\right)^f.$$

7.2.2 Parameter Estimation

MDE

The system (7.46)–(7.47) can be written as one equation

$$\begin{aligned} X_t &= f\int_0^t Y_s\,\mathrm{d}s + \varepsilon\sigma W_t \\ &= \theta_{0,1} ft + f\int_0^t\int_0^s S(\vartheta_0, v)\,\mathrm{d}v\,\mathrm{d}s + \varepsilon f b\int_0^t V_s\,\mathrm{d}s + \varepsilon\sigma W_t \\ &= \theta_{0,1} ft + f\int_0^t S(\vartheta_0, v)(t - v)\,\mathrm{d}v + \varepsilon f b\int_0^t (t - s)\,\mathrm{d}V_s + \varepsilon\sigma W_t \\ &= S_*(\vartheta_0, t) + \varepsilon n(t), \qquad 0 \le t \le T, \end{aligned}$$

where

$$S_*(\vartheta, t) = \theta_1 ft + f\int_0^t S(\vartheta, v)(t - v)\,\mathrm{d}v, \quad n(t) = fb\int_0^t (t - s)\,\mathrm{d}V_s + \sigma W_t.$$

Thus this model of observations is

$$X_t = S_*(\vartheta_0, t) + \varepsilon n(t), \qquad 0 \le t \le T,$$

and we have a deterministic signal observed in the presence of small colored Gaussian noise $\varepsilon n(\cdot)$. The noise $n(\cdot)$ has the mean zero and covariance function

$$R(t, s) = \frac{f^2b^2}{2}|t - s|(s \wedge t)^2 + \frac{f^2b^2}{3}(s \wedge t)^3 + \sigma^2(s \wedge t).$$

The MDE $\check\vartheta_\varepsilon$ we define by the equation

$$\int_0^T \left[X_t - S_*\left(\check\vartheta_\varepsilon, t\right)\right]^2 \mathrm{d}t = \inf_{\vartheta\in\Theta}\int_0^T \left[X_t - S_*\left(\vartheta, t\right)\right]^2 \mathrm{d}t.$$

Introduce the matrices

$$\check{\mathbf{T}}\left(\vartheta\right) = \int_0^T \dot S_*\left(\vartheta, t\right)\dot S_*\left(\vartheta, t\right)^\top \,\mathrm{d}t,$$

$$\check{\mathbf{D}}\left(\vartheta\right) = \check{\mathbf{T}}\left(\vartheta\right)^{-1}\int_0^T\int_0^T R\left(t,s\right)\dot S_*\left(\vartheta, t\right)\dot S_*\left(\vartheta, s\right)^\top \,\mathrm{d}t\,\mathrm{d}s\,\check{\mathbf{T}}\left(\vartheta\right)^{-1},$$

and the function

$$\check g\left(\vartheta_0,\nu\right)^2 = \inf_{\|\vartheta-\vartheta_0\|>\nu}\int_0^T \left[S_*\left(\vartheta, t\right) - S_*\left(\vartheta_0, t\right)\right]^2 \mathrm{d}t$$

Conditions $\check{\mathscr{O}}$.

$\check{\mathscr{O}}_1$. *The function* $S(\vartheta,t)$ *is twice continuously differentiable on* $\bar\Theta$.
$\check{\mathscr{O}}_2$. *The matrix* $\check{\mathbf{T}}\left(\vartheta_0\right)$ *is uniformly nondegenerate*

$$\inf_{\vartheta_0\in\Theta}\ \inf_{\|\mathrm{e}\|=1,\mathrm{e}\in\mathcal{R}^d} \mathrm{e}^\top\check{\mathbf{T}}\left(\vartheta_0\right)\mathrm{e} > 0.$$

$\check{\mathscr{O}}_3$. *For all* $\nu>0$

$$\inf_{\vartheta_0\in\Theta}\check g\left(\vartheta_0,\nu\right) > 0.$$

Theorem 7.5 *Suppose that the conditions* $\mathscr{A}_*$, $\check{\mathscr{O}}$ *are fulfilled. Then the MDE* $\check\vartheta_\varepsilon$ *is uniformly on compacts* $\mathbb{K}\subset\Theta$ *consistent, asymptotically normal*

$$\frac{\check\vartheta_\varepsilon - \vartheta_0}{\varepsilon} \Longrightarrow \check\zeta\left(\vartheta_0\right) \sim \mathcal{N}\left(0, \check{\mathbf{D}}\left(\vartheta_0\right)\right), \tag{7.48}$$

and the moments converge: for any $p\geq 2$

$$\varepsilon^{-p}\mathbf{E}_{\vartheta_0}\|\check\vartheta_\varepsilon - \vartheta_0\|^p \rightarrow \mathbf{E}_{\vartheta_0}\left\|\check\zeta\left(\vartheta_0\right)\right\|^p. \tag{7.49}$$

The proof of this theorem is the same as Theorem 2.2.

MLE

There are several possibilities to write the likelihood ratio function in the case of observations $X_t = S(t) + n(t), 0 \leq t \leq T$, where $S(\cdot)$ is a deterministic function and $n(t), 0 \leq t \leq T$ is zero mean Gaussian process with continuous correlation function $R(t,s), t, s \in [0, T]$. The measures induced in the space $(\mathcal{C}[0,T], \mathfrak{B})$ by the random processes $X_t, 0 \leq t \leq T$ and $n(t), 0 \leq t \leq T$ respectively are denoted $\mathbf{P}_S$ and $\mathbf{P}_n$. The authors in [94] used the approach that is known as the *whitening of colored noise* [155]. The properties of the MLE are described in the asymptotics $\varepsilon \to 0$ and $\sigma \to 0$. We propose below to use the Kalman-Bucy filtering formalism for the solution of such problems.

The LR function $L(\vartheta, X^T)$ and MLE $\hat{\vartheta}_\varepsilon$ are defined by the relations

$$L\left(\vartheta, X^T\right) = \exp\left(\int_0^T \frac{f\,m(\vartheta,t)}{\varepsilon^2\sigma^2} \mathrm{d}X_t - \int_0^T \frac{f^2 m(\vartheta,t)^2}{2\varepsilon^2\sigma^2}\mathrm{d}t\right), \qquad \vartheta \in \Theta$$

and

$$L(\hat{\vartheta}_\varepsilon, X^T) = \sup_{\vartheta\in\Theta} L\left(\vartheta, X^T\right).$$

The calculation of the MLE can be slightly simplified if we consider the equation

$$L(\hat{\vartheta}_\varepsilon, \vartheta^*, X^T) = \sup_{\vartheta\in\Theta} L\left(\vartheta, \vartheta^*, X^T\right)$$

where $L\left(\vartheta, \vartheta_*, X^T\right) = L\left(\vartheta, X^T\right)/L\left(\vartheta^*, X^T\right)$ and $\vartheta^* \in \Theta$ is some fixed value. Indeed in this case

$$L\left(\vartheta, \vartheta^*, X^T\right) = \exp\left(\frac{f}{\varepsilon^2\sigma^2}\int_0^T \delta_m(\vartheta,t)\mathrm{d}X_t - \frac{f^2}{2\varepsilon^2\sigma^2}\int_0^T \delta_m(\vartheta,t)\,M(\vartheta,t)\right)\mathrm{d}t.$$

Here we denoted $\delta_m(\vartheta,t) = m(\vartheta,t) - m(\vartheta^*,t)$ and $M(\vartheta,t) = m(\vartheta,t) + m(\vartheta^*,t)$. The deterministic function $\delta_m(\vartheta,\cdot), \vartheta \in \Theta$ is solution of the ODE

$$\frac{\mathrm{d}\delta_m(\vartheta,\cdot)}{\mathrm{d}t} = S(\vartheta,t) - S\left(\vartheta^*,t\right) - \frac{fb}{\sigma}\tanh\left(\frac{fbt}{\sigma}\right)\delta_m(\vartheta,t),$$
$$\delta_m(\vartheta,0) = \theta_{0,1} - \theta^*_{0,1}.$$

and $M(\vartheta,t) = m(\vartheta,t) - m(\vartheta^*,t)$. Remark that the calculation of the stochastic integral in this LR function is more simple.

To calculate the Fisher information matrix $\mathbf{I}(\vartheta_0)$ we need the derivatives

$$\frac{\partial y_t(\vartheta,\vartheta_0)}{\partial\theta_1}=\phi(0,t)=\left(\cosh\left(fbt/\sigma\right)\right)^{-f},$$

$$\frac{\partial y_t(\vartheta,\vartheta_0)}{\partial\theta_k}=\int_0^t\left(\frac{\cosh\left(fbs/\sigma\right)}{\cosh\left(fbt/\sigma\right)}\right)^f\dot S_k(\vartheta,s)\,\mathrm{d}s,\qquad k=2,\ldots,d.$$

Here $\dot S_k(\vartheta,s)=\partial S(\vartheta,s)/\partial\theta_k$. Denote the vector $\dot{\mathrm{y}}_t(\vartheta,\vartheta_0)=\partial y_t(\vartheta,\vartheta_0)/\partial\vartheta$ and introduce the Fisher information matrix

$$\mathbf{I}(\vartheta_0)=\sigma^{-2}\int_0^T\dot{\mathrm{y}}_t(\vartheta_0,\vartheta_0)\,\dot{\mathrm{y}}_t(\vartheta_0,\vartheta_0)^\top\mathrm{d}t.$$

The elements of this matrix are

$$\mathrm{I}_{1,1}=\sigma^{-2}\int_0^T\left(\cosh\left(fbt/\sigma\right)\right)^{-2f}\mathrm{d}t,$$

$$\mathrm{I}_{1,k}=\mathrm{I}_{k,1}=\frac{1}{\sigma^2}\int_0^T\phi(0,t)\int_0^t\phi(r,t)\,\dot S_k(\vartheta,r)\,\mathrm{d}r\,\mathrm{d}t,\qquad k=2,\ldots,d,$$

$$\mathrm{I}_{m,k}=\frac{1}{\sigma^2}\int_0^T\phi(s,t)\int_0^t\phi(r,t)\,\dot S_k(\vartheta,r)\,\mathrm{d}r\int_0^t\phi(v,t)\,\dot S_k(\vartheta,v)\,\mathrm{d}v\mathrm{d}t,$$

$k,m=2,\ldots,d,$

Define the function

$$\breve g(\vartheta_0,\nu)^2=\inf_{\|\vartheta-\vartheta_0\|>\nu}\int_0^t\Big[\left(\theta_1-\theta_{0,1}\right)\phi(0,t)+\int_0^t\phi(s,t)\left[S(\vartheta,s)-S(\vartheta_0,s)\right]\mathrm{d}s\Big]^2\mathrm{d}t.$$

Conditions $\hat{\mathscr{O}}$.

$\hat{\mathscr{O}}_1$. *The function* $S(\vartheta,t)$ *has two continuous bounded derivatives w.r.t.* ϑ on $\bar\Theta$.

$\hat{\mathscr{O}}_2$. *The matrix* $\mathbf{I}(\vartheta_0)$ *is uniformly nondegenerate*

$$\inf_{\vartheta_0\in\Theta}\ \inf_{\|\mathrm{e}\|=1,\mathrm{e}\in\mathcal{R}^d}\mathrm{e}^\top\mathbf{I}(\vartheta_0)\,\mathrm{e}>0.$$

$\hat{\mathscr{O}}_3$. *For all* $\nu>0$

$$\inf_{\vartheta_0\in\Theta}\hat g(\vartheta_0,\nu)>0.$$

Theorem 7.6 *Let the conditions $\hat{\mathscr{O}}$ hold. Then the MLE $\hat{\vartheta}_\varepsilon$ is uniformly on compacts $\mathbb{K} \subset \Theta$ consistent, asymptotically normal*

$$\frac{\hat{\vartheta}_\varepsilon - \vartheta_0}{\varepsilon} \Longrightarrow \hat{\zeta}(\vartheta_0) \sim \mathcal{N}(0, \mathbf{I}(\vartheta_0)),$$

and the moments converge: for any $p \geq 2$

$$\varepsilon^{-p}\mathbf{E}_{\vartheta_0}\|\hat{\vartheta}_\varepsilon - \vartheta_0\|^p \to \mathbf{E}_{\vartheta_0}\left\|\hat{\zeta}(\vartheta_0)\right\|^p.$$

Proof We check the conditions of Theorem 1.10. Denote $\vartheta_u = \vartheta_0 + \varepsilon u$, where $u \in \mathbb{U}_\varepsilon = \{u : \vartheta_0 + \varepsilon u \in \Theta\}$. Remark that the difference $\delta_m(t, u) = m(\vartheta_u, t) - m(\vartheta_0, t)$ is a deterministic function because it is solution of the ODE

$$\frac{\mathrm{d}\delta_m(t, u)}{\mathrm{d}t} = S(\vartheta, t) - S(\vartheta_0, t) - \frac{fb}{\sigma}\tanh\left(\frac{fb}{\sigma}t\right)\delta_m(t, u), \qquad \delta_m(0, u) = u_1$$

and its solution is

$$\delta_m(t, u) = u_1\phi(0, t) + \int_0^t \phi(s, t)\left[S(\vartheta_u, s) - S(\vartheta_0, s)\right]\mathrm{d}s.$$

Therefore the normalized LR in this problem is

$$\begin{aligned} &Z_\varepsilon(\vartheta_0, u) \\ &= \exp\left(\int_0^T \frac{[m(\vartheta_u, t) - m(\vartheta_0, t)]}{\varepsilon\sigma}\mathrm{d}\bar{W}_t - \int_0^T \frac{[m(\vartheta_u, t) - m(\vartheta_0, t)]^2}{2\varepsilon^2\sigma^2}\mathrm{d}t\right) \\ &= \exp\left(u^\top \int_0^T \frac{\dot{y}_t(\vartheta_0, \vartheta_0)}{\sigma}\mathrm{d}\bar{W}_t - u^\top \int_0^T \frac{\dot{y}_t(\vartheta_0, \vartheta_0)\dot{y}_t(\vartheta_0, \vartheta_0)^\top}{2\sigma^2}\mathrm{d}t\, u \right.\\ &\qquad \left. + \|u\|\, O(\varepsilon)\right) \\ &= \exp\left(u^\top\Delta(\vartheta_0) - \frac{1}{2}u^\top\mathbf{I}(\vartheta_0)\,u + \|u\|\; O(\varepsilon)\right). \end{aligned}$$

Here

$$\Delta(\vartheta_0) = \int_0^T \frac{\dot{y}_t(\vartheta_0, \vartheta_0)}{\sigma}\mathrm{d}\bar{W}_t \sim \mathcal{N}(0, \mathbf{I}(\vartheta_0)).$$

The term $O(\varepsilon)$ has all polynomial moments.

Thus the family of measures is uniformly on compacts LAN on Θ.

By Lemma 1.1 for any $R > 0$ and any $m > d/2$ for the values $\|u_1\| + \|u_2\| \le R$ we have

$$\begin{aligned}\mathbf{E}_{\vartheta_0}&\left|Z_\varepsilon\left(\vartheta_0, u_2\right)^{1/2m} - Z_\varepsilon\left(\vartheta_0, u_1\right)^{1/2m}\right|^{2m}\\ &\le C_{1,m}T^{2m-1}\sigma^{-4m}\int_0^T \left[m\left(\vartheta_{u_2}, t\right) - m\left(\vartheta_{u_1}, t\right)\right]^{4m} \mathrm{d}t\\ &\quad + C_{2,m}T^{m-1/2}\sigma^{-2m}\int_0^T \left[m\left(\vartheta_{u_2}, t\right) - m\left(\vartheta_{u_1}, t\right)\right]^{2m} \mathrm{d}t\\ &\le C\,T^{m-1/2}\left(1 + R^{2m}\right)\|u_2 - u_1\|^{2m}. \qquad (7.50)\end{aligned}$$

As the difference $\delta_m(u, t)$ is a deterministic function,

$$\mathbf{E}_{\vartheta_0} Z_\varepsilon\left(\vartheta_0, u\right)^{1/2} = \exp\left(-\frac{1}{8\varepsilon^2\sigma^2}\int_0^T \delta_m\left(u, t\right)^2 \mathrm{d}t\right).$$

From the conditions $\hat{\mathscr{O}}_1$, $\hat{\mathscr{O}}_2$ it follows that for some small $\nu > 0$ and $\|\varepsilon u\| \le \nu$ we have

$$\frac{1}{\varepsilon^2\sigma^2}\int_0^T \delta_m\left(u, t\right)^2 \mathrm{d}t = u^\top \mathbf{I}\left(\vartheta_0\right) u\left(1 + o\left(\nu\right)\right) \ge \frac{1}{2}u^\top \mathbf{I}\left(\vartheta_0\right) u \ge \kappa_1\|u\|^2.$$

For $\|\varepsilon u\| > \nu$ by the condition $\hat{\mathscr{O}}_3$

$$\frac{1}{\sigma^2\varepsilon^2}\int_0^T \delta_m\left(u, t\right)^2 \mathrm{d}t \ge \frac{g\left(\vartheta_0, \nu\right)^2}{\varepsilon^2\sigma^2} \ge \frac{g\left(\vartheta_0, \nu\right)^2}{\sigma^2}\frac{\|u\|^2}{\left(\mathrm{Dim}\ \Theta\right)^2} \ge \kappa_2\|u\|^2.$$

Here $\mathrm{Dim}\ \Theta = \sup_{\vartheta,\vartheta' \in\Theta}\left\|\vartheta - \vartheta'\right\|$.

Thus if we put $\kappa = 8\left(\kappa_1 \wedge \kappa_2\right)$, then we can write

$$\mathbf{E}_{\vartheta_0} Z_\varepsilon\left(\vartheta_0, u\right)^{1/2} \le e^{-\kappa\|u\|^2}. \qquad (7.51)$$

The properties (7.50) and (7.51) of $Z_\varepsilon\left(\vartheta_0, \cdot\right)$ and the uniform LAN of the family of measures according to Theorem 1.10 prove Theorem 7.6. □

The same asymptotic properties, of course, have the BE $\tilde{\vartheta}_\varepsilon$.

7.2.3 One-Step MLE-Process

Let us fix the learning interval $[0, \tau]$, where $\tau < T$. The MDE is

$$\check{\vartheta}_{\tau,\varepsilon} = \arg\inf_{\vartheta\in\Theta}\int_0^{\tau}\left[X_t - S_*\left(\vartheta,t\right)\right]^2 \mathrm{d}t.$$

Suppose that the conditions $\check{\mathscr{O}}$ with T replaced by τ are fulfilled. Thus the MDE is consistent, asymptotically normal (7.48) and the moments converge (7.49). Remark that the derivatives $\dot{y}_t\left(\vartheta,\vartheta_0\right)$ do not depend on the true value ϑ_0 and can be written as $\dot{y}_t\left(\vartheta\right)$.

Introduce the Fisher information matrix-valued function

$$\mathbf{I}_\tau^t\left(\vartheta\right) = \sigma^{-2}\int_\tau^t \dot{y}_s\left(\vartheta\right)\dot{y}_s\left(\vartheta\right)^\top \mathrm{d}s, \qquad \tau < t \leq T$$

and the random process

$$\mathrm{d}m(\check{\vartheta}_{\tau,\varepsilon},t) = S(\check{\vartheta}_{\tau,\varepsilon},t)\mathrm{d}t + \frac{fb}{\sigma}\tanh\left(\frac{bt}{\sigma}\right)\left[\mathrm{d}X_t - m(\check{\vartheta}_{\tau,\varepsilon},t)\mathrm{d}t\right], \qquad \tau < t \leq T.$$

Remark that for this model there is no problem with the definition of the stochastic integral.

The One-step MLE-process is defined by the equality

$$\begin{aligned}\vartheta_{t,\varepsilon}^\star = \check{\vartheta}_{\tau,\varepsilon} + f\sigma^{-2}\mathbf{I}_0^t\left(\check{\vartheta}_{\tau,\varepsilon}\right)^{-1}&\Big\{\dot{y}_\tau(\check{\vartheta}_{\tau,\varepsilon})X_\tau \\ &-\int_0^\tau\left[X_s\,\dot{y}_s'\left(\check{\vartheta}_{\tau,\varepsilon}\right) - f\dot{y}_s\left(\check{\vartheta}_{\tau,\varepsilon}\right)m\left(\check{\vartheta}_{\tau,\varepsilon},s\right)\right]\mathrm{d}s \\ &+\int_\tau^t \dot{y}_s(\check{\vartheta}_{\tau,\varepsilon})\left[\mathrm{d}X_s - fm(\check{\vartheta}_{\tau,\varepsilon},s)\mathrm{d}s\right]\Big\}, \qquad \tau < t \leq T. \end{aligned} \tag{7.52}$$

Here we replaced the stochastic integral on the time interval $[0,\tau]$ by the ordinary integral following Lemma 1.3.

We add the *Conditions* $\mathscr{O}^\star$.

$\mathscr{O}_1^\star$. *The function* $S(\vartheta,t),\vartheta\in\Theta,t\in[0,T]$ *has three continuous bounded derivatives w.r.t.* ϑ *on* $\bar{\Theta}$.

$\mathscr{O}_2^\star$. *The matrix-function* $\mathbf{I}_0^t\left(\vartheta_0\right), t\in(0,T]$ *is uniformly nondegenerate*

$$\inf_{\vartheta_0\in\Theta}\ \inf_{\|\mathrm{e}\|=1,\mathrm{e}\in\mathcal{R}^d} \mathrm{e}^\top\mathbf{I}_0^t\left(\vartheta_0\right)\mathrm{e} > 0.$$

Below it is shown that the family of measures is LAN and thus the Hájek's lower bound is valid: for any estimator $\bar{\vartheta}_{t,\varepsilon}$ and any $t\in(\tau,T]$

$$\lim_{\nu\to0}\varliminf_{\varepsilon\to0}\sup_{\|\vartheta-\vartheta_0\|\leq\nu}\varepsilon^{-2}\mathbf{E}_\vartheta\left\|\bar{\vartheta}_{t,\varepsilon}-\vartheta\right\|^2 \geq \mathbf{E}_{\vartheta_0}\left\|\zeta_t\left(\vartheta_0\right)\right\|^2,$$

$$\zeta_t(\vartheta_0) \sim \mathcal{N}\left(0, \mathbf{I}_0^t(\vartheta_0)^{-1}\right).$$

The estimator $\vartheta_{t,\varepsilon}^*$ is called asymptotically efficient if

$$\lim_{\nu\to 0}\lim_{\varepsilon\to 0}\sup_{\|\vartheta-\vartheta_0\|\leq\nu} \varepsilon^{-2}\mathbf{E}_\vartheta\left\|\vartheta_{t,\varepsilon}^*-\vartheta\right\|^2 = \mathbf{E}_{\vartheta_0}\left\|\zeta_t(\vartheta_0)\right\|^2,$$

for all $\vartheta_0 \in \Theta$.

Theorem 7.7 *Let the conditions $\breve{\mathcal{O}}_{2-3}$ and $\mathcal{O}^\star$ hold. Then the One-step MLE-process $\vartheta_{t,\varepsilon}^\star$, $\tau < t \leq T$ is consistent and asymptotically normal*

$$\varepsilon^{-1}\left(\vartheta_{t,\varepsilon}^\star - \vartheta_0\right) \Longrightarrow \mathcal{N}\left(0, \mathbf{I}_0^t(\vartheta_0)^{-1}\right),$$

the polynomial moments converge and it is asymptotically efficient.

Proof The Taylor expansion at the point ϑ_0 allows us to write

$$\begin{aligned}
&\dot{\mathbf{y}}_\tau(\breve{\vartheta}_{\tau,\varepsilon})X_\tau - \int_0^\tau \left[X_s\,\dot{\mathbf{y}}_s'\left(\breve{\vartheta}_{\tau,\varepsilon}\right) - f\dot{\mathbf{y}}_s\left(\breve{\vartheta}_{\tau,\varepsilon}\right)m\left(\breve{\vartheta}_{\tau,\varepsilon},s\right)\right]\mathrm{d}s\\
&\quad= \dot{\mathbf{y}}_\tau(\vartheta_0)X_\tau - \int_0^\tau \left[X_s\,\dot{\mathbf{y}}_s'(\vartheta_0) - f\dot{\mathbf{y}}_s(\vartheta_0)\,m\left(\breve{\vartheta}_{\tau,\varepsilon},s\right)\right]\mathrm{d}s\\
&\qquad\times\left(1+O\left(\left\|\breve{\vartheta}_{\tau,\varepsilon}-\vartheta_0\right\|\right)\right)\\
&\quad= \int_0^\tau \dot{\mathbf{y}}_s(\vartheta_0)\left[\mathrm{d}X_s - fm\left(\breve{\vartheta}_{\tau,\varepsilon},s\right)\mathrm{d}s\right]\left(1+O\left(\left\|\breve{\vartheta}_{\tau,\varepsilon}-\vartheta_0\right\|\right)\right)\\
&\quad= f\int_0^\tau \dot{\mathbf{y}}_s(\vartheta_0)\left[m(\vartheta_0,s) - m\left(\breve{\vartheta}_{\tau,\varepsilon},s\right)\right]\mathrm{d}s\left(1+O\left(\left\|\breve{\vartheta}_{\tau,\varepsilon}-\vartheta_0\right\|\right)\right)\\
&\qquad+\quad \varepsilon\sigma\int_0^\tau \dot{\mathbf{y}}_s(\vartheta_0)\mathrm{d}\bar{W}_s\left(1+O\left(\left\|\breve{\vartheta}_{\tau,\varepsilon}-\vartheta_0\right\|\right)\right)\\
&\quad= f\int_0^\tau \dot{\mathbf{y}}_s(\vartheta_0)\dot{\mathbf{y}}_s(\vartheta_0)^\top\mathrm{d}s\left(\vartheta_0-\breve{\vartheta}_{\tau,\varepsilon}\right)\left(1+O\left(\left\|\breve{\vartheta}_{\tau,\varepsilon}-\vartheta_0\right\|\right)\right)\\
&\qquad+\quad \varepsilon\sigma\int_0^\tau \dot{\mathbf{y}}_s(\vartheta_0)\mathrm{d}\bar{W}_s\left(1+O\left(\left\|\breve{\vartheta}_{\tau,\varepsilon}-\vartheta_0\right\|\right)\right).
\end{aligned}$$

Recall that $\left\|\breve{\vartheta}_{\tau,\varepsilon}-\vartheta_0\right\| = O(\varepsilon)$. The Taylor expansions at the point ϑ_0 of the other terms in (7.52) lead to the following expressions

$$\int_{\tau}^{t} \frac{f^2 \dot{\mathbf{y}}_s(\check\vartheta_{\tau,\varepsilon})}{\varepsilon\sigma^2}\left[m\left(\vartheta_0,s\right)-m(\check\vartheta_{\tau,\varepsilon},s)\right]\mathrm{d}s$$

$$=\int_{\tau}^{t} \frac{f^2 \dot{\mathbf{y}}_s(\check\vartheta_{\tau,\varepsilon})\dot{\mathbf{y}}_s(\check\vartheta_{\tau,\varepsilon})^\top}{\sigma^2}\mathrm{d}s\,\frac{(\vartheta_0-\check\vartheta_{\tau,\varepsilon})}{\varepsilon}+\frac{\|\vartheta_0-\check\vartheta_{\tau,\varepsilon}\|^2}{\varepsilon}\,O\left(1\right)$$

$$=\mathbf{I}_\tau^t\left(\vartheta_0\right)\frac{(\vartheta_0-\check\vartheta_{\tau,\varepsilon})}{\varepsilon}+\,O\left(\varepsilon\right)$$

and

$$\int_{\tau}^{t} \frac{\dot{\mathbf{y}}_s(\check\vartheta_{\tau,\varepsilon})}{\sigma}\mathrm{d}\bar W_s=\int_{\tau}^{t} \frac{\dot{\mathbf{y}}_s(\vartheta_0)}{\sigma}\mathrm{d}\bar W_s\left(1+O\left(\varepsilon\right)\right).$$

Therefore

$$\begin{aligned}
\frac{\vartheta^\star_{t,\varepsilon}-\vartheta_0}{\varepsilon}&=\frac{\check\vartheta_{\tau,\varepsilon}-\vartheta_0}{\varepsilon}+\mathbf{I}_0^t\left(\vartheta_0\right)^{-1}\left[\int_0^\tau \frac{f\,\dot{\mathbf{y}}_s(\vartheta_0)}{\sigma}\,\mathrm{d}\bar W_s\right.\\
&\quad\left.+\int_\tau^t \frac{f\,\dot{\mathbf{y}}_s(\vartheta_0)}{\sigma}\,\mathrm{d}\bar W_s\right]\left(1+O\left(\varepsilon\right)\right)\\
&\quad+\mathbf{I}_0^t\left(\vartheta_0\right)^{-1}\left[\int_0^\tau \frac{f^2\,\dot{\mathbf{y}}_s(\vartheta_0)\dot{\mathbf{y}}_s(\vartheta_0)^\top}{\sigma^2}\,\mathrm{d}s\right.\\
&\quad\left.+\int_\tau^t \frac{f^2\,\dot{\mathbf{y}}_s(\vartheta_0)\dot{\mathbf{y}}_s(\vartheta_0)^\top}{\sigma^2}\,\mathrm{d}s\right]\frac{\left(\vartheta_0-\vartheta^\star_{\tau,\varepsilon}\right)}{\varepsilon}\left(1+O\left(\varepsilon\right)\right)\\
&=\frac{\check\vartheta_{\tau,\varepsilon}-\vartheta_0}{\varepsilon}+\mathbf{I}_0^t\left(\vartheta_0\right)^{-1}\int_0^t \frac{f\,\dot{\mathbf{y}}_s(\vartheta_0)}{\sigma}\,\mathrm{d}\bar W_s\left(1+O\left(\varepsilon\right)\right)\\
&\quad+\mathbf{I}_0^t\left(\vartheta_0\right)^{-1}\int_0^t \frac{f^2\,\dot{\mathbf{y}}_s(\vartheta_0)\dot{\mathbf{y}}_s(\vartheta_0)^\top}{\sigma^2}\,\mathrm{d}s\frac{\left(\vartheta_0-\vartheta^\star_{\tau,\varepsilon}\right)}{\varepsilon}\left(1+O\left(\varepsilon\right)\right)\\
&=\mathbf{I}_0^t\left(\vartheta_0\right)^{-1}\int_0^t \frac{f\,\dot{\mathbf{y}}_s(\vartheta_0)}{\sigma}\,\mathrm{d}\bar W_s\left(1+O\left(\varepsilon\right)\right)\\
&\Longrightarrow\mathbf{I}_0^t\left(\vartheta_0\right)^{-1}\int_0^t \frac{f\,\dot{\mathbf{y}}_s(\vartheta_0)}{\sigma}\,\mathrm{d}w\left(s\right)\equiv\eta_t^\star\left(\vartheta_0\right).
\end{aligned}$$

Note that the asymptotic efficiency follows from the uniform convergence of moments and the continuity of $\mathbf{I}_0^t\left(\vartheta\right)$. □

On-Line Estimation of the Signal

The estimator $\vartheta^\star_{t,\varepsilon}, \tau < t \leq T$ allows to construct on-line estimation of the signal $S(\vartheta, t), \tau < t \leq T$ by the obvious substitution

$$S^\star(t) = S\left(\vartheta^\star_{t,\varepsilon}, t\right), \qquad \tau < t \leq T.$$

It is easy to see that

$$\begin{aligned}\varepsilon^{-1}\left(S^\star(t) - S(\vartheta_0, t)\right) &= \dot{\mathbb{S}}(\vartheta_0, t)^\top \varepsilon^{-1}\left(\vartheta^\star_{t,\varepsilon} - \vartheta_0\right)(1 + O(\varepsilon))\\ &\Longrightarrow \dot{\mathbb{S}}(\vartheta_0, t)^\top \eta^\star_t(\vartheta_0).\end{aligned}$$

It is possible to verify the asymptotic efficiency of this estimator too.

7.3 Security Prices Process

7.3.1 Model of Observations

Let us consider one model, taken from financial mathematics, which is a slight generalization of the partially observable system discussed in Chap. 4. This model was studied in the work of Frydman and Lakner [57], where the authors constructed estimators of the parameters of this model using Expectation-Maximization approach (E-M algorithm). The observed process is a *single security* with the price process $S_t, 0 \leq t \leq T$ which follows the equation

$$\mathrm{d}S_t = \mu_t S_t \mathrm{d}t + \sigma S_t \mathrm{d}W_t, \qquad S_0, \tag{7.53}$$

where $\sigma > 0$ is a known constant, but $\mu_t, 0 \leq t \leq T$ (hidden) is a mean-reverting process with the dynamics

$$\mathrm{d}\mu_t = \alpha(\upsilon - \mu_t)\,\mathrm{d}t + \beta \mathrm{d}V_t, \qquad \mu_0. \tag{7.54}$$

Here $\alpha > 0, \beta > 0$ and υ are unknown parameters, and $W_t, t \geq 0$ and $V_t, t \geq 0$ are independent Wiener processes. The initial value S_0 is an observable constant whereas the initial value μ_0 is an unknown constant. This model has been found to be very flexible in modeling of security prices and its versions have been used extensively in the area of dynamic asset management, see the following references in [57]: Bielecki and Pliska [13], Lakner [130], and Kim and Omberg [96]. After the transformations

$$Y_t = \frac{1}{\beta}\mu_t - \frac{1}{\beta}\mu_0,$$

$$X_t = \frac{1}{\sigma}\ln\left(\frac{S_t}{S_0}\right) + \frac{\sigma}{2}t$$

proposed in [57] the systems (7.53) and (7.54) become

$$\mathrm{d}X_t = (f\,Y_t + q)\mathrm{d}t + \mathrm{d}W_t, \qquad X_0 = 0, \tag{7.55}$$

$$\mathrm{d}Y_t = -(a\,Y_t + h)\mathrm{d}t + \mathrm{d}V_t, \qquad Y_0 = 0, \tag{7.56}$$

where

$$f = \frac{\beta}{\sigma}, \qquad q = \frac{\mu_0}{\sigma}, \qquad a = \alpha, \qquad h = \frac{\alpha\,(\mu_0 - v)}{\beta}.$$

Let $\theta = (f, q, a, h)$. This parameter is estimated in [57] by observations $X^T = (X_t, 0 \le t \le T)$ with the help of E-M approach. In particular a computationally explicit multisteps E-M algorithm of the calculation of approximation of MLE is proposed.

Let us study the possibility of estimation of the same parameters using One-step process approach. First note that the solution of (7.56) can be written as follows:

$$\begin{aligned} Y_t &= -h\int_0^t e^{-a(t-s)}\mathrm{d}s + \int_0^t e^{-a(t-s)}\mathrm{d}V_s \\ &= -\frac{h}{a}\left[1 - e^{-at}\right] + \int_0^t e^{-a(t-s)}\mathrm{d}V_s = -\frac{h}{a} + \int_0^t e^{-a(t-s)}\mathrm{d}V_s + O\left(e^{-at}\right). \end{aligned}$$

Substitution of Y_t in (7.55) implies

$$\begin{aligned} \mathrm{d}X_t &= \left(q - \frac{fh}{a}\right)\mathrm{d}t + f\int_0^t e^{-a(t-s)}\mathrm{d}V_s\,\mathrm{d}t + \mathrm{d}W_t \\ &= g\,\mathrm{d}t + f\int_0^t e^{-a(t-s)}\mathrm{d}V_s\,\mathrm{d}t + \mathrm{d}W_t, \qquad X_0 = 0, \end{aligned} \tag{7.57}$$

where we omitted the term $O\left(e^{-at}\right)$ and denoted

$$g = q - \frac{fh}{a} = \frac{\mu_0}{\sigma} - \frac{\beta\alpha\,(\mu_0 - v)}{\sigma\beta\alpha} = \frac{v}{\sigma}.$$

Therefore the observed process depends on three parameters only: a, f, g. The link between a, f, g and $\alpha, \beta, \sigma, v, \mu_0$ is

$$\alpha = a, \qquad \beta = \sigma f, \qquad v = \sigma g.$$

Remark that the consistent ($T \to \infty$) estimation of μ_0 is impossible. To see it formally we write the system as

$$\mathrm{d}X_t = \sigma^{-1}\mu_t \,\mathrm{d}t + \mathrm{d}W_t, \qquad X_0 = 0, \quad 0 \le t \le T,$$

where $\mu_t, t \geq 0$ is solution of equation (7.54). Suppose that all other parameters are known. Then the equation for $m(\mu, t) = \mathbf{E}_\mu\left(\mu_t | \mathfrak{F}_t^X\right)$ is

$$\mathrm{d}m(\mu, t) = \alpha v \,\mathrm{d}t - \alpha m(\mu, t)\,\mathrm{d}t + \sigma^{-1}\gamma_t\left[\mathrm{d}X_t - \sigma^{-1}m(\mu, t)\,\mathrm{d}t\right], m(\mu, 0) = \mu.$$

Therefore the equation for the derivative $\dot{m}(\vartheta, t) = \partial m(\mu, t)/\partial\vartheta$ is

$$\frac{\mathrm{d}\dot{m}(\vartheta, t)}{\mathrm{d}t} = -\left[\alpha + \sigma^{-2}\gamma_t\right]\dot{m}(\vartheta, t), \qquad \dot{m}(\vartheta, 0) = 1.$$

Recall that the solution of the Riccati equation γ_t with constant coefficients converges to steady-state solution $\gamma_t \to \gamma_* = \sqrt{\sigma^2 + \alpha^2\sigma^4} - \alpha\sigma^2$. Therefore the Fisher information can be written as follows:

$$\mathrm{I}_0^T(\mu) = \int_0^T \sigma^{-2}\mathbf{E}_\mu \dot{m}(\mu, t)^2 \,\mathrm{d}t = \frac{1}{2\sigma\sqrt{1 + \alpha^2\sigma^2}}(1 + o(1)).$$

By the Cramer-Rao inequality for any estimator $\bar{\mu}_T$ we have

$$\mathbf{E}_\mu(\bar{\mu}_T - \mu)^2 \geq \frac{\left[1 + \dot{b}(\mu)\right]^2}{\mathrm{I}_0^T(\mu)} + b(\mu)^2$$

and as $\mathrm{I}_0^T(\mu) \not\to \infty$ the consistent estimation is impossible.

The observations (7.57) correspond to the partially observed system

$$\mathrm{d}X_t = (f\,Y_t + g)\mathrm{d}t + \mathrm{d}W_t, \qquad X_0 = 0, \qquad 0 \le t \le T, \tag{7.58}$$

$$\mathrm{d}Y_t = -a\,Y_t\,\mathrm{d}t + \mathrm{d}V_t, \qquad Y_0 = 0, \tag{7.59}$$

and the unknown parameter can be $\vartheta = (a, f, g)$. Equations (7.58) and (7.59) are the model of observations studied in this section. This model is close to that considered in Chap. 4, and the proofs are, of course, similar to the proofs presented there.

7.3.2 *Estimators*

MME

The true value is denoted $\vartheta_0 = (a_0, f_0, g_0)$. Introduce three statistics

$$R_{0,\tau_T} = \frac{X_{\tau_T}}{\tau_T}, \qquad R_{1,\tau_T} = \frac{1}{\tau_T}\sum_{k=1}^{\tau_T}[X_k - X_{k-1}]^2,$$

$$R_{2,\tau_T} = \frac{1}{\tau_T}\sum_{k=2}^{\tau_T}[X_k - X_{k-1}][X_{k-1} - X_{k-2}].$$

Here $\tau_T = \left[T^\delta\right], \delta \in (1/2, 1)$ and $[A]$ means the integer part of A.

The limits are

$$R_{0,\tau_T} = \frac{f}{\tau_T}\int_0^{\tau_T} Y_s\,\mathrm{d}s + g_0 + \frac{W_{\tau_T}}{\tau_T} \longrightarrow g_0,$$

$$R_{1,\tau_T} = \frac{1}{\tau_T}\sum_{k=1}^{\tau_T}\left[f_0\int_{k-1}^{k} Y_s\mathrm{d}s + g_0 + W_k - W_{k-1}\right]^2$$

$$= \frac{f_0^2}{\tau_T}\sum_{k=1}^{\tau_T}\eta_k^2 + g_0^2 + \frac{1}{\tau_T}\sum_{k=1}^{\tau_T}\zeta_k^2 + \frac{2f_0g_0}{\tau_T}\sum_{k=1}^{\tau_T}\eta_k + \frac{2f_0}{\tau_T}\sum_{k=1}^{\tau_T}\eta_k\zeta_k + \frac{2g_0}{\tau_T}\sum_{k=1}^{\tau_T}\zeta_k$$

$$\longrightarrow \frac{f_0^2}{a_0^3}\left[e^{-a_0} - 1 + a_0\right] + 1 + g_0^2 \equiv \Phi_1(\vartheta_0) + g_0^2.$$

Here we used the notation and limits from Sect. 4.2.1 (see Proposition 4.1), i.e.,

$$\eta_k = \int_{k-1}^{k} Y_s\mathrm{d}s, \qquad \zeta_k = W_k - W_{k-1}, \qquad \Phi_1(\vartheta) = \frac{f^2}{a^3}\left[e^{-a} - 1 + a\right] + 1.$$

Further

$$R_{2,\tau_T} = \frac{1}{\tau_T}\sum_{k=2}^{\tau_T}[f_0\eta_k + g_0 + \zeta_k][f_0\eta_{k-1} + g_0 + \zeta_{k-1}]$$

$$= \frac{f_0^2}{\tau_T}\sum_{k=1}^{\tau_T}\eta_k\eta_{k-1} + g_0^2 + \frac{1}{\tau_T}\sum_{k=1}^{\tau_T}\zeta_k\zeta_{k-1} + \frac{2f_0g_0}{\tau_T}\sum_{k=1}^{\tau_T}\eta_k$$

$$+ \frac{f_0}{\tau_T}\sum_{k=1}^{\tau_T}\eta_k\zeta_{k-1} + \frac{g_0}{\tau_T}\sum_{k=1}^{\tau_T}\zeta_{k-1}$$

$$\longrightarrow \frac{f_0^2}{2a_0^3}\left[1-e^{-a_0}\right]^2+g_0^2\equiv\Phi_2\left(\vartheta_0\right)+g_0^2.$$

The preliminary estimators can be chosen as follows:

1. The MME $g^*_{\tau_T}$ of g_0 is

$$g^*_{\tau_T}=\frac{X_{\tau_T}}{\tau_T}. \tag{7.60}$$

We have $\sqrt{\tau_T}(g^*_{\tau_T}-g_0)\Rightarrow\mathcal{N}\left(0,D\right)$.

2. The MME $a^*_{\tau_T}$ of a_0 is solution of the equation

$$\frac{e^{-a^*_{\tau_T}}-1+a^*_{\tau_T}}{\left(1-e^{-a^*_{\tau_T}}\right)^2}=\frac{R_{1,\tau_T}-1-g^{*2}_{\tau_T}}{2R_{2,\tau_T}-2g^{*2}_{\tau_T}}. \tag{7.61}$$

3. The MME $f^*_{\tau_T}$ of f_0 is

$$f^*_{\tau_T}=\frac{2(a^*_{\tau_T})^{3/2}\left(R_{2,\tau_T}-g^{*2}_{\tau_T}\right)^{1/2}_+}{(1-e^{-a^*_{\tau_T}})^2}. \tag{7.62}$$

Remark that in Sect. 4.2.1 we already verified that the function

$$h\left(x\right)=\frac{e^{-x}-1+x}{\left(1-e^{-x}\right)^2},\qquad x>0$$

is strictly increasing on $[0,+\infty]$, and therefore Eq. (7.61) has a unique solution.

As it follows from the direct generalizations of Propositions 4.1 and 4.3 we have the following proposition.

Proposition 7.3 *The MME* $\vartheta_{\tau_T}=\left(a^*_{\tau_T},f^*_{\tau_T},g^*_{\tau_T}\right)$ *is consistent, asymptotically normal*

$$\sqrt{\tau_T}(\vartheta^*_{\tau_T}-\vartheta_0)\Longrightarrow\mathcal{N}\left(0,\mathbf{R}\left(\vartheta_0\right)\right),$$

the polynomial moments converge and for any $p\geq 2$ *the estimate*

$$T^{p/2}\mathbf{E}_{\vartheta_0}\left\|\vartheta^*_{\tau_T}-\vartheta_0\right\|^p\leq C$$

holds.

Here $\mathbf{R}(\vartheta_0)$ is some bounded matrix and the constant $C > 0$ does not depend on T and ϑ_0.

MLE and BE

The Kalman-Bucy filter in the case of partially observed system (7.58) and (7.59) is

$$\mathrm{d}m(\vartheta,t) = -a\,m(\vartheta,t) + f\gamma(\vartheta,t)\left[\mathrm{d}X_t - (g + f m(\vartheta,t))\,\mathrm{d}t\right], \quad m(\vartheta,0) = 0, \tag{7.63}$$

$$\frac{\mathrm{d}\gamma(\vartheta,t)}{\mathrm{d}t} = -2a\gamma(\vartheta,t) - f^2\gamma(\vartheta,t)^2 + 1, \quad \gamma(\vartheta,0) = 0. \tag{7.64}$$

Recall that the Riccati equation with constant coefficients has an explicit solution

$$\gamma(\vartheta,t) = e^{-2r(\vartheta)t}\left[\frac{f^2}{2r(\vartheta)\sigma^2}\left(1 - e^{-2r(\vartheta)t}\right) - \frac{1}{\gamma_*(\vartheta)}\right]^{-1} + \gamma_*(\vartheta), \quad 0 \le t \le T,$$

where

$$r(\vartheta) = \left(a^2 + f^2\right)^{1/2}, \qquad \gamma_*(\vartheta) = \frac{r(\vartheta) - a}{f^2}.$$

It can be written as follows:

$$\gamma(\vartheta,t) = \frac{r(\vartheta) - a}{f^2} - \frac{r(\vartheta)}{f^2}\left[\frac{1 + a/r(\vartheta)}{1 - a/r(\vartheta)} + \frac{1}{2}e^{-2r(\vartheta)t}\right]^{-1} e^{-2r(\vartheta)t}.$$

It is easy to see that

$$\left|\gamma(\vartheta,t) - \gamma_*(\vartheta)\right| \le C\,e^{-2r(\vartheta)t}.$$

To simplify the calculations we will study system (7.63) and (7.64) in stationary regime, i.e., below we put $\gamma(\vartheta,t) \equiv \gamma_*(\vartheta)$. This means that the initial value for the Eq. (7.63) is $m(\vartheta,0) \sim \mathcal{N}(0, \gamma_*(\vartheta))$ and the equation is

$$\mathrm{d}m(\vartheta,t) = -r(\vartheta)\,m(\vartheta,t)\,\mathrm{d}t + f\gamma_*(\vartheta)\left[\mathrm{d}X_t - g\,\mathrm{d}t\right], \qquad m(\vartheta,0). \tag{7.65}$$

Let us write similar equation for the process $M(\vartheta,t) = f m(\vartheta,t)$

$$\begin{aligned}\mathrm{d}M(\vartheta,t) &= -r(\vartheta)\,M(\vartheta,t)\,\mathrm{d}t + \Gamma(\vartheta)\left[\mathrm{d}X_t - g\,\mathrm{d}t\right] \\ &= -r(\vartheta)\,M(\vartheta,t)\,\mathrm{d}t + \Gamma(\vartheta)\,\mathrm{d}\bar{W}_t \\ &\quad + \Gamma(\vartheta)\left[M(\vartheta_0,t) + g_0 - g\right]\mathrm{d}t, \qquad M(\vartheta,0),\end{aligned} \tag{7.66}$$

where $\Gamma(\vartheta) = r(\vartheta) - a$ and $M(\vartheta, 0) \sim \mathcal{N}(0, \Gamma(\vartheta))$. We denote $\bar{W}_t, t \geq 0$ the *innovation* Wiener process.

Therefore

$$M(\vartheta, t) = M(\vartheta, 0)\, e^{-r(\vartheta)t} + \Gamma(\vartheta) \int_0^t e^{-r(\vartheta)(t-s)} \mathrm{d}\bar{W}_s$$
$$+ \Gamma(\vartheta) \int_0^t e^{-r(\vartheta)(t-s)} \left[M(\vartheta_0, s) + g_0 - g\right] \mathrm{d}s. \qquad (7.67)$$

For the calculation of expectations it will be useful to write these integrals on the interval $(-\infty, t]$. Introduce an independent Wiener process $W_s^*, s \geq 0$ and put $\bar{W}_s = W_{-s}^*$ for $s < 0$. Then the stationary solution can be written as the sum of two integrals

$$M(\vartheta, t) = \Gamma(\vartheta) \int_{-\infty}^t e^{-r(\vartheta)(t-s)} \mathrm{d}\bar{W}_s$$
$$+ \Gamma(\vartheta) \int_{-\infty}^t e^{-r(\vartheta)(t-s)} \left[M(\vartheta_0, s) + g_0 - g\right] \mathrm{d}s$$

Denote vector $\dot{\mathrm{M}}^*(\vartheta, t) = \partial M^*(\vartheta, t)/\partial\vartheta = \left(\dot{M}_a(\vartheta, t), \dot{M}_f(\vartheta, t), \dot{M}_g(\vartheta, t) + 1\right)$ and remind that the Fisher information matrix is

$$\begin{aligned}
\mathbf{I}(\vartheta_0) &= \frac{1}{T}\mathbf{E}_\vartheta \left(\frac{\partial \ln L(\vartheta, X^T)}{\partial\vartheta} \frac{\partial \ln L(\vartheta, X^T)}{\partial\vartheta}^\top \right)\Bigg|_{\vartheta=\vartheta_0} \\
&= \frac{1}{T}\mathbf{E}_{\vartheta_0} \int_0^T \dot{\mathrm{M}}^*(\vartheta_0, t)\, \mathrm{d}\bar{W}_t \int_0^T \dot{\mathrm{M}}^*(\vartheta_0, t)^\top\, \mathrm{d}\bar{W}_t \\
&= \mathbf{E}_{\vartheta_0} \left(\frac{1}{T} \int_0^T \dot{\mathrm{M}}^*(\vartheta_0, t)\, \dot{\mathrm{M}}^*(\vartheta_0, t)^\top\, \mathrm{d}t \right) \\
&= \mathbf{E}_{\vartheta_0} \mathbf{I}_T(\vartheta_0),
\end{aligned}$$

where of course

$$\mathbf{I}_T(\vartheta_0) = \frac{1}{T} \int_0^T \dot{\mathrm{M}}^*(\vartheta_0, t)\, \dot{\mathrm{M}}^*(\vartheta_0, t)^\top\, \mathrm{d}t.$$

The equations for derivatives $\dot{M}_f(\vartheta, t)$, $\dot{M}_a(\vartheta, t)$, and $\dot{M}_g(\vartheta, t)$ we obtain by the formal derivation of Eq. (7.67).

$$\mathrm{d}\dot{M}_a(\vartheta, t) = -r(\vartheta)\, \dot{M}_a(\vartheta, t)\, \mathrm{d}t - \dot{r}_a(\vartheta)\, M(\vartheta, t)\, \mathrm{d}t + \dot{\Gamma}_a(\vartheta) \left[\mathrm{d}X_t - g\, \mathrm{d}t\right], \qquad (7.68)$$

$$\mathrm{d}\dot{M}_f(\vartheta,t) = -r(\vartheta)\,\dot{M}_f(\vartheta,t)\,\mathrm{d}t - \dot{r}_f(\vartheta)\,M(\vartheta,t)\,\mathrm{d}t + \dot{\Gamma}_f(\vartheta)\left[\mathrm{d}X_t - g\,\mathrm{d}t\right], \tag{7.69}$$

$$\frac{\mathrm{d}\dot{M}_g(\vartheta,t)}{\mathrm{d}t} = -r(\vartheta)\,\dot{M}_g(\vartheta,t) - \Gamma(\vartheta), \tag{7.70}$$

Remark that $\dot{r}_g(\vartheta) = 0$ and $\dot{\gamma}_{*g}(\vartheta) = 0$. Hence

$$\dot{M}_g(\vartheta,t) \longrightarrow \dot{M}_g(\vartheta) \equiv -\frac{\Gamma(\vartheta)}{r(\vartheta)} = \frac{a}{r(\vartheta)} - 1, \qquad \dot{M}_g(\vartheta) + 1 = \frac{a}{r(\vartheta)}.$$

Note that these derivatives are stationary processes too. Moreover it is possible to write the equation for the matrix $\ddot{\mathbf{M}}^*(\vartheta,t)$ and verify that it is also a stationary process.

In this three-dimensional case the Fisher information $\mathrm{I}(\vartheta)_{3\times 3}$ can be easily calculated with the help of the Fisher information in two-dimensional case $\vartheta = (a,f)$ studied in Sect. 4.2.2. We have

$$\mathbf{I}(\vartheta)_{3\times 3} = \begin{pmatrix} \dfrac{\Gamma(\vartheta)^2\left[r(\vartheta)a + (r(\vartheta)+a)^2\right]}{2ar(\vartheta)^3\left[r(\vartheta)+a\right]}, & \dfrac{f\left(a^2 + r(\vartheta)a - 2r(\vartheta)\right)}{2r(\vartheta)^3\left(r(\vartheta)+a\right)}, & 0 \\ \dfrac{f\left(a^2 + r(\vartheta)a - 2r(\vartheta)\right)}{2r(\vartheta)^3\left(r(\vartheta)+a\right)}, & \dfrac{f^2}{2r(\vartheta)^3}, & 0 \\ 0, & 0, & \dfrac{a^2}{r(\vartheta)^2} \end{pmatrix} \tag{7.71}$$

because

$$\mathrm{I}_{1,3}(\vartheta) = \mathbf{E}_\vartheta \int_0^T \left[\dot{M}_g(\vartheta,t) + 1\right]\dot{M}_a(\vartheta,t)\,\mathrm{d}t = 0,$$

$$\mathrm{I}_{2,3}(\vartheta) = \mathbf{E}_\vartheta \int_0^T \left[\dot{M}_g(\vartheta,t) + 1\right]\dot{M}_f(\vartheta,t)\,\mathrm{d}t = 0$$

and

$$\frac{1}{T}\int_0^T \left[\dot{M}_g(\vartheta,t) + 1\right]^2 \mathrm{d}t \longrightarrow \mathrm{I}_{3,3}(\vartheta) = \frac{a^2}{r(\vartheta)^2}.$$

Therefore we obtained the Fisher information matrix $\mathbf{I}(\vartheta)_{3\times 3} = \mathbf{I}(\vartheta)$. Remark that this matrix is nondegenerate (see (4.32) in a similar situation). This means that

$$\inf_{\vartheta\in\Theta}\inf_{e\in\mathcal{R}^3,\|e\|=1}\mathbf{I}(\vartheta)=\kappa_*>0.$$

Theorem 7.8 *The MLE $\hat\vartheta_T$ and BE $\tilde\vartheta_T$ are consistent, asymptotically normal*

$$\sqrt{T}(\hat\vartheta_T-\vartheta_0)\Longrightarrow\mathcal{N}\left(0,\mathbf{I}(\vartheta_0)^{-1}\right),\qquad \sqrt{T}(\tilde\vartheta_T-\vartheta_0)\Longrightarrow\mathcal{N}\left(0,\mathbf{I}(\vartheta_0)^{-1}\right),$$

and the polynomial moments converge.

Proof Introduce the normalized likelihood ratio (LR) process ($\varphi_T=T^{-1/2}$)

$$Z_T(\vartheta_0,u)=\frac{L\left(\vartheta_0+\varphi_T u,X^T\right)}{L\left(\vartheta_0,X^T\right)},\qquad u\in\mathbb{U}_T=\left(u\ :\ \vartheta_0+\varphi_T u\in\Theta\right).$$

The proof of this theorem consists in the verification of the conditions $\mathscr{I}\mathscr{K}_3$ of Theorems 1.10 and 1.12. Recall that these conditions are given in terms of normalized LR as follows:

1. The random function $Z_T(\vartheta_0,\cdot)$ admits the representation

$$Z_T(\vartheta_0,u)=\exp\left(u^\top\Delta_T\left(\vartheta_0,X^T\right)-\frac{1}{2}u^\top\mathbf{I}(\vartheta_0)\,u+r_T\right),\qquad u\in\mathbb{U}_T,\tag{7.72}$$

where $r_T\to 0$ and

$$\Delta_T\left(\vartheta_0,X^T\right)\Longrightarrow\mathcal{N}\left(0,\mathbf{I}(\vartheta_0)\right).\tag{7.73}$$

2. There exists a constant $C_1>0$ such that for any $R>0$ and $\|u_1\|\leq R$, $\|u_2\|\leq R$ the estimate

$$\mathbf{E}_{\vartheta_0}\left|Z_T(\vartheta_0,u_2)^{1/4}-Z_T(\vartheta_0,u_1)^{1/4}\right|^4\leq C_1\left(1+R^4\right)\|u_2-u_1\|^4\tag{7.74}$$

holds.
3. For any $n>2$ there exist constants $c_*>0$ and $C_2=C_2(n)>0$ such that

$$\mathbf{P}_{\vartheta_0}\left(Z_T(\vartheta_0,u)>e^{-c_*\|u\|^2}\right)\leq\frac{C_2}{\|u\|^n}.\tag{7.75}$$

Denote

$$\Delta_T\left(\vartheta_0,X^T\right)=\varphi_T\int_0^T\dot{\mathbf{M}}^*(\vartheta_0,t)\left[\mathrm{d}X_t-M(\vartheta_0,t)\,\mathrm{d}t\right]$$

and consider the difference (below $\vartheta_u = \vartheta_0 + \varphi_T u$)

$$\begin{aligned}
\mathbf{E}_{\vartheta_0} &\left(\int_0^T [M(\vartheta_u, t) - M(\vartheta_0, t)]\,\mathrm{d}\bar{W}_t - \Delta_T\left(\vartheta_0, X^T\right)\right)^2 \\
&= \int_0^T \mathbf{E}_{\vartheta_0}\left[M(\vartheta_u, t) - M(\vartheta_0, t) - \varphi_T\, u^\top \dot{\mathbf{M}}^*(\vartheta_0, t)\right]^2 \mathrm{d}t \\
&= \varphi_T^2 \int_0^T \mathbf{E}_{\vartheta_0}\left(\int_0^1 u^\top\left[\dot{\mathbf{M}}^*(\vartheta_0 + su, t) - \dot{\mathbf{M}}^*(\vartheta_0, t)\right]\mathrm{d}s\right)^2 \mathrm{d}t \\
&\le \varphi_T^2\, \|u\|^2 \int_0^1 \int_0^T \mathbf{E}_{\vartheta_0}\left[\dot{\mathbf{M}}^*(\vartheta_0 + su, t) - \dot{\mathbf{M}}^*(\vartheta_0, t)\right]^2 \mathrm{d}t\, \mathrm{d}s \\
&\le T^{-2}\, \|u\|^4 \sup_{s\in[0,1]} \int_0^T \mathbf{E}_{\vartheta_0} \|\ddot{\mathbf{M}}^*(\vartheta_0 + su, t)\|_m^2 \mathrm{d}t \longrightarrow 0.
\end{aligned}$$

Here the norm of matrix

$$\|\mathbf{N}\|_m = \sup_{e\in\mathcal{R}^3, \|e\|=1} e^\top \mathbf{N} e.$$

Similar arguments allow to verify the convergence

$$\begin{aligned}
\int_0^T & [M(\vartheta_u, t) - M(\vartheta_0, t)]^2\, \mathrm{d}t \\
&= u^\top \frac{1}{T} \int_0^T \dot{\mathbf{M}}^*(\vartheta_0, t)\, \dot{\mathbf{M}}^*(\vartheta_0, t)^\top \mathrm{d}t\, u\ \ (1 + o(1)) \\
&\longrightarrow u^\top \mathbf{I}(\vartheta_0)\, u. \qquad (7.76)
\end{aligned}$$

Therefore (7.72) holds and the convergence (7.73) follows from (7.76) and the CLT for stochastic integrals.

To verify (7.74) we recall the estimate (see Corollary 1.1)

$$\begin{aligned}
\mathbf{E}_{\vartheta_0} &\left| Z_T(\vartheta_0, u_2)^{1/4} - Z_T(\vartheta_0, u_1)^{1/4} \right|^4 \\
&\le C\, T^3 \int_0^T \mathbf{E}_{\vartheta_{u_2}} \delta_M(u_1, u_2, t)^8\, \mathrm{d}t + C\, T \int_0^T \mathbf{E}_{\vartheta_{u_1}} \delta_M(u_1, u_2, t)^4\, \mathrm{d}t
\end{aligned}$$

where we denoted $\delta_M(u_1, u_2, t) = M\left(\vartheta_{u_1}, t\right) - M\left(\vartheta_{u_2}, t\right)$. The stationary Gaussian process $\dot{\mathbf{M}}^*(\vartheta, t)$ has bounded moments and therefore

$$\mathbf{E}_{\vartheta_0}\left|Z_T(\vartheta_0,u_2)^{1/4}-Z_T(\vartheta_0,u_1)^{1/4}\right|^4\leq C\,\|u_2-u_1\|^8+C\,\|u_2-u_1\|^4$$
$$\leq C_1\left(1+R^4\right)\|u_2-u_1\|^4.$$

To study the limit of the integral

$$J_T(\vartheta,\vartheta_0)=\frac{1}{T}\int_0^T[M(\vartheta,t)+g-M(\vartheta_0,t)-g_0]^2\,\mathrm{d}t$$

we need the representation of the random process $\delta_M(t)=M(\vartheta,t)-M(\vartheta_0,t)$. The equation for $\delta_M(t)$ we obtain as the difference of the solutions of (7.66)

$$\mathrm{d}\delta_M(t)=r(\vartheta)\,\delta_M(t)\,\mathrm{d}t+[a_0-a]\,M(\vartheta_0,t)\,\mathrm{d}t+\Gamma(\vartheta)\,[g_0-g]\,\mathrm{d}t$$
$$+[\Gamma(\vartheta)-\Gamma(\vartheta_0)]\,\mathrm{d}\bar W_t.$$

Hence

$$\delta_M(t)=[a_0-a]\int_{-\infty}^t e^{-r(\vartheta)(t-s)}M(\vartheta_0,s)\,\mathrm{d}s+\Gamma(\vartheta)\,[g_0-g]\int_{-\infty}^t e^{-r(\vartheta)(t-s)}\mathrm{d}s$$
$$+[\Gamma(\vartheta)-\Gamma(\vartheta_0)]\int_{-\infty}^t e^{-r(\vartheta)(t-s)}\mathrm{d}\bar W_s.$$

Here

$$\Gamma(\vartheta)\,[g_0-g]\int_{-\infty}^t e^{-r(\vartheta)(t-s)}\mathrm{d}s=\frac{\Gamma(\vartheta)\,[g_0-g]}{r(\vartheta)}$$

and

$$M(\vartheta_0,s)=[r(\vartheta_0)-a_0]\int_{-\infty}^s e^{-a_0(s-q)}\mathrm{d}\bar W_q.$$

Further following the proof of Lemma 4.3 we obtain the representation

$$M(\vartheta,t)-M(\vartheta_0,t)=\frac{\Gamma(\vartheta)}{r(\vartheta)}[g_0-g]+\Gamma(\vartheta_0)\frac{(a_0-a)}{r(\vartheta)-a_0}\int_{-\infty}^t e^{-a_0(t-s)}\mathrm{d}\bar W_s$$
$$+\Gamma(\vartheta)\frac{(r(\vartheta)-r(\vartheta_0))}{r(\vartheta)-a_0}\int_{-\infty}^t e^{-r(\vartheta)(t-s)}\mathrm{d}\bar W_s$$
$$=\frac{\Gamma(\vartheta)}{r(\vartheta)}[g_0-g]+\Gamma(\vartheta_0)\frac{(a_0-a)}{r(\vartheta)-a_0}\int_{-\infty}^0 e^{-a_0q}\mathrm{d}\tilde W_q$$

$$+\Gamma(\vartheta)\frac{(r(\vartheta)-r(\vartheta_0))}{r(\vartheta)-a_0}\int_{-\infty}^{0}e^{-r(\vartheta)q}\mathrm{d}\tilde{W}_q$$

$$=\frac{\Gamma(\vartheta)}{r(\vartheta)}[g_0-g]+\int_{-\infty}^{0}H(\vartheta,\vartheta_0,q)\,\mathrm{d}\tilde{W}_q.$$

Therefore

$$G(\vartheta,\vartheta_0)\equiv\mathbf{E}_{\vartheta_0}[M(\vartheta,t)-M(\vartheta_0,t)]^2=\frac{\Gamma(\vartheta)^2}{r(\vartheta)^2}[g_0-g]^2$$
$$+\int_{-\infty}^{0}H(\vartheta,\vartheta_0,q)^2\,\mathrm{d}q. \qquad (7.77)$$

Let us verify the condition: for any $\nu>0$

$$\inf_{\|\vartheta-\vartheta_0\|>\nu}G(\vartheta,\vartheta_0)>0.$$

If for some $\nu>0$, $\vartheta_*\in\Theta$ and $\|\vartheta_*-\vartheta_0\|>\nu$ we have $G(\vartheta_*,\vartheta_0)=0$, then we obtain equalities $g_*=g_0$ and $H(\vartheta_*,\vartheta_0,q)^2=0$. But this implies the equalities $a_*=a_0$ and $f_*=f_0$.

Thus we have

$$\mathbf{E}_{\vartheta_0}J_T(\vartheta,\vartheta_0)=\frac{1}{T}\int_0^T\mathbf{E}_{\vartheta_0}[\delta_M(t)+g-g_0]^2\,\mathrm{d}t=G(\vartheta,\vartheta_0).$$

By the law of large numbers

$$J_T(\vartheta,\vartheta_0)\longrightarrow G(\vartheta,\vartheta_0)$$

and there exists a constant $C>0$ such that for any $n\geq 2$

$$\mathbf{E}_{\vartheta_0}\left|\sqrt{T}\,[J_T(\vartheta,\vartheta_0)-G(\vartheta,\vartheta_0)]\right|^n\leq C. \qquad (7.78)$$

For the proof of (7.78) see Lemma 1.6.

To prove (7.75) we denote $\delta_{M^*}(\vartheta,t)=M(\vartheta,t)+g-M(\vartheta_0,t)-g_0$ write

$$\mathbf{P}_{\vartheta_0}\left(Z_T(\vartheta_0,u)^{1/2}>e^{-\frac{c_*}{2}\|u\|^2}\right)$$
$$=\mathbf{P}_{\vartheta_0}\left(\int_0^T\frac{\delta_{M^*}(\vartheta_u,t)}{2}\mathrm{d}\bar{W}_t-\int_0^T\frac{\delta_{M^*}(\vartheta_u,t)^2}{8}\mathrm{d}t-\int_0^T\frac{\delta_{M^*}(\vartheta_u,t)^2}{8}\mathrm{d}t>\right.$$
$$\left.-\frac{c_*}{2}\|u\|^2\right)$$

$$
\begin{aligned}
&\leq \mathbf{P}_{\vartheta_0}\left(\int_0^T \frac{\delta_{M^*}(\vartheta_u,t)}{2}\mathrm{d}\bar{W}_t-\int_0^T \frac{\delta_{M^*}(\vartheta_u,t)^2}{8}\mathrm{d}t>c_1\left\|u\right\|^2\right)\\
&\qquad+\mathbf{P}_{\vartheta_0}\left(-\int_0^T \delta_{M^*}(\vartheta_u,t)^2\mathrm{d}t>-\left(4c_*-8c_1\right)\left\|u\right\|^2\right)\\
&\leq e^{-c_1\left\|u\right\|^2}+\mathbf{P}_{\vartheta_0}\left(-\int_0^T \delta_{M^*}(\vartheta_u,t)^2\,\mathrm{d}t>-\left(4c_*-8c_1\right)\left\|u\right\|^2\right).
\end{aligned}
$$

Here we used the Markov inequality and the equality

$$
\mathbf{E}_{\vartheta_0}\exp\left(\int_0^T \frac{\delta_{M^*}(\vartheta_u,t)}{2}\mathrm{d}\bar{W}_t-\int_0^T \frac{\delta_{M^*}(\vartheta_u,t)^2}{8}\mathrm{d}t\right)=1.
$$

The last probability we estimate for small values of $\nu>0$ on the following two sets: $\mathbb{A}_\nu=\left(u:\left\|T^{-1/2}u\right\|\leq\nu\right)$ and $\mathbb{A}_\nu^c=\left(u:\left\|T^{-1/2}u\right\|>\nu,u\in\mathbb{U}_T\right)$ separately. Let $u\in\mathbb{A}_\nu$, then

$$
\begin{aligned}
TG\left(\vartheta_u,\vartheta_0\right)&=u^\top\mathbf{I}\left(\vartheta_0\right)u\left(1+O\left(\nu\right)\right)\geq\left\|u\right\|^2\inf_{e\in R^3,\left\|e\right\|=1}e^\top\mathbf{I}\left(\vartheta_0\right)e\ \left(1+O\left(\nu\right)\right)\\
&\geq\frac{\kappa_*}{2}\left\|u\right\|^2.
\end{aligned}
$$

Remark that the estimate like (7.78)

$$
T^{n/2}\mathbf{E}_{\vartheta_0}\left\|\mathbf{I}\left(\vartheta_0\right)-\mathbf{I}_T\left(\vartheta_0\right)\right\|_m^n\leq C \tag{7.79}
$$

holds too.

Thus

$$
\begin{aligned}
&\mathbf{P}_{\vartheta_0}\left(u^\top\mathbf{I}\left(\vartheta_0\right)u-\int_0^T \delta_{M^*}(\vartheta_u,t)^2\,\mathrm{d}t>u^\top\mathbf{I}\left(\vartheta_0\right)u-\left(4c_*-8c_1\right)\left\|u\right\|^2\right)\\
&\qquad\leq\mathbf{P}_{\vartheta_0}\left(u^\top\left[\mathbf{I}\left(\vartheta_0\right)-\mathbf{I}_T\left(\vartheta_0\right)\right]u\left(1+O\left(\nu\right)\right)>\left[\frac{\kappa_*}{2}-\left(4c_*-8c_1\right)\right]\left\|u\right\|^2\right)\\
&\qquad=\mathbf{P}_{\vartheta_0}\left(u^\top\left[\mathbf{I}\left(\vartheta_0\right)-\mathbf{I}_T\left(\vartheta_0\right)\right]u\left(1+O\left(\nu\right)\right)>c_2\left\|u\right\|^2\right)\\
&\qquad\leq CT^{-n/2}\ T^{n/2}\mathbf{E}_{\vartheta_0}\left\|\mathbf{I}\left(\vartheta_0\right)-\mathbf{I}_T\left(\vartheta_0\right)\right\|_m^n\leq CT^{-n/2}\leq\frac{C}{\left\|u\right\|^n}
\end{aligned}
$$

where we put $c_2=\kappa_*/2-\left(4c_*-8c_1\right)$ and chose c_*,c_1 such that $c_2>0$, used estimates (7.79) and $T^{-n/2}\leq D\left(\vartheta_0\right)^n\left\|u\right\|$. Remind that $\vartheta_0+T^{-1/2}u\in\Theta$ and therefore

$$
T^{-1/2}\left\|u\right\|\leq\sup_{\vartheta\in\Theta}\left\|\vartheta-\vartheta_0\right\|\equiv D\left(\vartheta_0\right)
$$

Consider now the case $u \in \mathbb{A}_\nu^c$. Note that for $\|\vartheta_u - \vartheta_0\| > \nu$

$$G(\vartheta_u, \vartheta_0) \geq \inf_{\|\vartheta - \vartheta_0\| > \nu} G(\vartheta, \vartheta_0) \equiv G_\nu(\vartheta_0) \geq G_\nu(\vartheta_0) \frac{\|\vartheta_u - \vartheta_0\|^2}{D(\vartheta_0)^2} \geq \frac{c_3 \|u\|^2}{T}.$$

Here we used the estimate (7.77). We have

$$\begin{aligned}
&\mathbf{P}_{\vartheta_0}\left(-\int_0^T \delta_{M^*}(\vartheta_u, t)^2 \,\mathrm{d}t > -(4c_* - 8c_1)\|u\|^2\right)\\
&\leq \mathbf{P}_{\vartheta_0}\left(\int_0^T \left[G(\vartheta_u, \vartheta_0) - \delta_{M^*}(\vartheta_u, t)^2\right]\mathrm{d}t > TG(\vartheta_u, \vartheta_0) - (4c_* - 8c_1)\|u\|^2\right)\\
&\leq \mathbf{P}_{\vartheta_0}\left(T\left|G(\vartheta_u, \vartheta_0) - J_T(\vartheta_u, \vartheta_0)\right| > (c_3 - 4c_* + 8c_1)\|u\|^2\right)\\
&\leq C\,\frac{T^{n/2}}{\|u\|^{2n}} \leq \frac{C}{\|u\|^n}.
\end{aligned}$$

Here we used the estimates (7.78) and $T^{n/2} \leq \nu^{-n}\|u\|^n$. Of course we chose the constants c_* and c_1 such that $c_3 - 4c * +8c_1 > 0$.

Thus, we have verified conditions (7.72)–(7.75) and according to the indicated Theorems 1.10 and 1.12 the MLE and BE have the properties indicated in Theorem 7.8. □

Remark 7.3 The condition (7.77) has to be fulfilled always by the following reason. It is the limit of the Kulback-Leibner divergence of measures. If for some $\vartheta_* \neq \vartheta_0$ we have $G(\vartheta_*, \vartheta_0) = 0$, then the consistent estimation of ϑ_0 is impossible, but we already have the consistent MME $\vartheta^*_{\tau_T}$.

7.3.3 One-Step MLE-Process

We have the preliminary consistent MME $\vartheta^*_{\tau_T} = (a^*_{\tau_T}, f^*_{\tau_T}, g^*_{\tau_T})^\top$ and nondegenerate Fisher information matrix. Hence the One-step MLE-process $\vartheta^\star_{t,T}, \tau_T < t \leq T$ can be

$$\vartheta^\star_{t,T} = \vartheta^*_{\tau_T} + \mathbf{I}(\vartheta^*_{\tau_T})^{-1} t^{-1} \int_{\tau_T}^t \dot{\mathrm{M}}(\vartheta^*_{\tau_T}, s)\left[\mathrm{d}X_s - M(\vartheta^*_{\tau_T}, s)\mathrm{d}s - g^*_{\tau_T}\mathrm{d}s\right]. \tag{7.80}$$

For calculation of the processes $M(\vartheta^*_{\tau_T}, s), \tau_T \leq T$ and $\dot{\mathrm{M}}(\vartheta^*_{\tau_T}, s), \tau_T \leq T$ we solve Eqs. (7.66)–(7.70), where ϑ is replaced by $\vartheta^*_{\tau_T}$ and the initial values $M(\vartheta^*_{\tau_T}, \tau_T)$, $\dot{\mathrm{M}}(\vartheta^*_{\tau_T}, \tau_T)$ are obtained similar to (4.38)–(4.39).

Let us put $t = vT, v \in [\epsilon_0, 1]$ where $\epsilon \in (0, 1]$ and study the normalized difference $\sqrt{t}\left(\vartheta^\star_{t,T} - \vartheta_0\right)$ as $T \to \infty$.

Proposition 7.4 *The One-step MLE-process is asymptotically normal*

$$\sqrt{t}\left(\vartheta^\star_{t,T} - \vartheta_0\right) \Longrightarrow \mathcal{N}\left(0, \mathbf{I}(\vartheta_0)^{-1}\right). \tag{7.81}$$

Proof

$$\begin{aligned}\sqrt{t}\left(\vartheta^\star_{t,T} - \vartheta_0\right) &= \sqrt{t}\left(\vartheta^*_{\tau_T} - \vartheta_0\right)\\ &\quad + \frac{\mathbf{I}(\vartheta^*_{\tau_T})^{-1}}{\sqrt{t}}\int_{\tau_T}^{t} \dot{\mathbf{M}}(\vartheta^*_{\tau_T}, s)\left[\mathrm{d}X_s - M(\vartheta^*_{\tau_T}, s)\mathrm{d}s - g^*_{\tau_T}\mathrm{d}s\right]\\ &= \sqrt{t}\left(\vartheta^*_{\tau_T} - \vartheta_0\right) + \mathbf{I}(\vartheta^*_{\tau_T})^{-1}\frac{1}{\sqrt{t}}\int_{\tau_T}^{t} \dot{\mathbf{M}}(\vartheta^*_{\tau_T}, s)\mathrm{d}\bar{W}_s\\ &\quad + \mathbf{I}(\vartheta^*_{\tau_T})^{-1}\frac{1}{\sqrt{t}}\int_{\tau_T}^{t} \dot{\mathbf{M}}(\vartheta^*_{\tau_T}, s)\left[M(\vartheta_0, s) - M(\vartheta^*_{\tau_T}, s) + g_0 - g^*_{\tau_T}\right]\mathrm{d}s.\end{aligned} \tag{7.82}$$

Using the consistency of $\vartheta^*_{\tau_T}$ and the CLT we obtain the limit in distribution

$$\frac{1}{\sqrt{t}}\int_{\tau_T}^{t} \dot{\mathbf{M}}(\vartheta^*_{\tau_T}, s)\mathrm{d}\bar{W}_s \Longrightarrow \Delta(\vartheta_0), \qquad \Delta(\vartheta_0) \sim \mathcal{N}(0, \mathbf{I}(\vartheta_0)).$$

We have the relations

$$\begin{aligned}&\vartheta^*_{\tau_T} - \vartheta_0 = O(T^{-\delta/2}),\\ &M(\vartheta_0, t) - M(\vartheta^*_{\tau_T}, t) + g_0 - g^*_{\tau_T} = -\dot{\mathbf{M}}(\vartheta_0, t)^\top(\vartheta^*_{\tau_T} - \vartheta_0)(1 + O(T^{-\delta/2})),\\ &\mathbf{I}(\vartheta^*_{\tau_T})^{-1} = \mathbf{I}(\vartheta_0)^{-1}(1 + O(T^{-\delta/2})),\\ &\mathbf{J}_t = \frac{1}{t-\tau_T}\int_{\tau_T}^{t} \dot{\mathbf{M}}(\vartheta_0, s)\dot{\mathbf{M}}(\vartheta_0, s)^\top\mathrm{d}s \longrightarrow \mathbf{I}(\vartheta_0), \qquad \mathbf{J}_t - \mathbf{I}(\vartheta_0) = O\left(T^{-1/2}\right).\end{aligned}$$

Therefore

$$\begin{aligned}&\mathbf{I}(\vartheta^*_{\tau_T})^{-1}\frac{1}{\sqrt{t}}\int_{\tau_T}^{t} \dot{\mathbf{M}}(\vartheta^*_{\tau_T}, s)\left[M(\vartheta_0, s) - M(\vartheta^*_{\tau_T}, s) + g_0 - g^*_{\tau_T}\right]\mathrm{d}s\\ &\qquad = -\mathbf{I}(\vartheta_0)^{-1}\frac{1}{t}\int_{\tau_T}^{t} \dot{\mathbf{M}}(\vartheta_0, s)\dot{\mathbf{M}}(\vartheta_0, s)^\top\mathrm{d}s\ \sqrt{t}(\vartheta^*_{\tau_T} - \vartheta_0)(1 + O(T^{-\delta/2}))\end{aligned}$$

$$= -\sqrt{t}(\vartheta^*_{\tau_T} - \vartheta_0)(1 + O(T^{-\delta/2})).$$

Substitution of these relations in (7.82) gives us the following expression

$$\begin{aligned}\sqrt{t}(\vartheta^\star_{\tau_T} - \vartheta_0) &= \sqrt{t}(\vartheta^*_{\tau_T} - \vartheta_0) + \mathbf{I}(\vartheta^*_{\tau_T})^{-1}\frac{1}{\sqrt{t}}\int_{\tau_T}^t \dot{\mathbf{M}}(\vartheta^*_{\tau_T}, s)\mathrm{d}\bar{W}_s \\ &\quad - \sqrt{t}(\vartheta^*_{\tau_T} - \vartheta_0)(1 + O(T^{-\delta/2})) \\ &= \mathbf{I}(\vartheta^*_{\tau_T})^{-1}\frac{1}{\sqrt{t}}\int_{\tau_T}^t \dot{\mathbf{M}}(\vartheta^*_{\tau_T}, s)\mathrm{d}\bar{W}_s + \sqrt{t}(\vartheta^*_{\tau_T} - \vartheta_0)O(T^{-\delta/2}) \\ &\Longrightarrow \mathbf{I}(\vartheta_0)^{-1}\Delta\,(\vartheta_0) \quad \sim \quad \mathcal{N}\left(0, \mathbf{I}(\vartheta_0)^{-1}\right)\end{aligned}$$

because

$$\sqrt{t}(\vartheta^*_{\tau_T} - \vartheta_0)O(T^{-\delta/2}) = O(T^{-\delta+1/2}).$$

Recall that $\delta \in (1/2, 1]$. Therefore the convergence (7.82) was proved. □

7.3.4 Adaptive Filter and Extrapolation

Consider the problem of adaptive filtering of the process $\mu_t, 0 \leq t \leq T$ and extrapolation of the single security S_t and μ_t for the values $t > T$. It will be convenient to write first the equations for approximation of $m\,(\vartheta, t) = \mathbf{E}_\vartheta\left(Y_t|\mathfrak{F}^X_T\right)$ for $t \leq T$ and then we write equations for $m\,(\vartheta, T, t) = \mathbf{E}_\vartheta\left(Y_t|\mathfrak{F}^X_T\right)$ and $n\,(\vartheta, T, t) = \mathbf{E}_\vartheta\left(X_t|\mathfrak{F}^X_T\right)$, where $t > T$. Then this will allow us to define the adaptive extrapolation of S_t and μ_t.

Adaptive Filter for $\mu_t, 0 \leq t \leq T$

The solution we present in several steps.

- *Fix some $\delta \in (1/2, 1)$ and estimate the parameters a, f, g by observations $X^{\tau_T} = \left(X_s, 0 \leq s \leq \tau_T\right)$ by the method of moments. The MME is $\vartheta^*_{\tau_T} = (a^*_{\tau_T}, f^*_{\tau_T}, g^*_{\tau_T})^\top$ (see (7.60)–(7.62)).*
- *Calculate the One-step MLE-process $\vartheta^\star_{s,T} = (a^\star_{s,T}, f^\star_{s,T}, g^\star_{s,T})^\top, \tau_T < s \leq T$ by the formula (7.80). This requires the solution of equations (7.67)–(7.70), where ϑ is replaced by $\vartheta^*_{\tau_T}$ and calculation of the inverse Fisher information $\mathbf{I}(\vartheta^*_{\tau_T})^{-1}$.*
- *The adaptive filter for Y_s is given by the equation*

$$\mathrm{d}m^{\star}_{s,T} = -r\left(\vartheta^{\star}_{s,T}\right) m^{\star}_{s,T}\mathrm{d}s + f^{\star}_{s,T}\gamma_{*}\left(\vartheta^{\star}_{s,T}\right)\left[\mathrm{d}X_s - g^{\star}_{s,T}\mathrm{d}s\right], \qquad \tau_T \le s \le T, \tag{7.83}$$

where the initial value $m^{\star}_{\tau_T}$ is obtained similar to (4.38).

We can return to the initial notations, but as we mentioned above it is impossible to have consistent estimation of the four parameters. Suppose that μ_0 is known and we have to estimate α, β and υ. Then having the estimators $f^{\star}_{s,T}$, $a^{\star}_{s,T}$, and $g^{\star}_{s,T}$ we can write

$$\alpha^{\star}_{s,T} = a^{\star}_{s,T}, \qquad \beta^{\star}_{s,T} = \sigma\, f^{\star}_{s,T}, \qquad \upsilon^{\star}_{s,T} = \sigma\, g^{\star}_{s,T}$$

with the corresponding MME

$$\alpha^{*}_{\tau_T} = a^{*}_{\tau_T}, \qquad \beta^{*}_{\tau_T} = \sigma\, f^{*}_{\tau_T}, \qquad \upsilon^{*}_{\tau_T} = \sigma\, g^{*}_{\tau_T}.$$

Of course, the estimator $\theta^{\star}_{s,T} = \left(\alpha^{\star}_{s,T}, \beta^{\star}_{s,T}, \upsilon^{\star}_{s,T}\right)$ is consistent (converges to $\theta_0 = (\alpha_0, \beta_0, \upsilon_0)$) and asymptotically normal.

Recall that $\mu_s = \mu_0 + \beta Y_s$ and $\gamma_t \to \gamma_* = \sqrt{\sigma^2 + \alpha^2\sigma^4} - \alpha\sigma^2$. Hence $\mu\left(\vartheta, s\right) = \mu_0 + \beta m\left(\vartheta, s\right)$, $\sigma^{-1}\gamma_*\left(\vartheta\right) = \sqrt{1 + \alpha^2\sigma^2} - \alpha\sigma \equiv \gamma_{**}\left(\alpha\right)$ and the equation for $\mu\left(\vartheta, s\right)$ is

$$\mathrm{d}\mu\left(\theta, s\right) = -r\left(\theta\right)\beta^{-1}\mu\left(\theta, s\right)\mathrm{d}s + \gamma_{**}\left(\alpha\right)\left[\mathrm{d}X_s - \sigma^{-1}\upsilon\,\mathrm{d}s\right], \qquad \mu_0, \quad 0 \le s \le T.$$

The adaptive filter for μ_s, $\tau_T \le s \le T$ is

$$\mathrm{d}\mu^{\star}_{s,T} = -\frac{r\left(\theta^{\star}_{s,T}\right)}{\beta^{\star}_{s,T}}\mu^{\star}_{s,T}\,\mathrm{d}s + \gamma_{**}\left(\alpha^{\star}_{s,T}\right)\left[\mathrm{d}X_s - \sigma^{-1}\upsilon^{\star}_{s,T}\,\mathrm{d}s\right],$$

$$\mu^{\star}_{\tau_T,T} = \mu(\theta^{*}_{\tau_T}, \tau_T).$$

It is possible to describe the asymptotic of the error $\delta_\mu\left(s\right) = \mu^{\star}_{s,T} - \mu\left(\vartheta_0, s\right)$ following the same approach as in Sect. 4.3.

Extrapolation of Security Price Process $S_t, t > T$

The equation for extrapolation of $m\left(\vartheta, T, t\right)$ and $n\left(\vartheta, T, t\right)$ is (see Sect. 1.1.3 or Theorem 12.13 in [142])

$$\frac{m\left(\vartheta, T, t\right)}{\mathrm{d}t} = -a\, m\left(\vartheta, T, t\right), \qquad m\left(\vartheta, T, T\right) = m\left(\vartheta, T\right), \qquad t \ge T,$$

$$\frac{n(\vartheta, T, t)}{\mathrm{d}t} = g + f\, m(\vartheta, T, t), \qquad n(\vartheta, T, T) = X_T, \qquad t \geq T.$$

Therefore

$$\begin{aligned} m(\vartheta, T, t) &= m(\vartheta, T)\; e^{-a(t-T)}, \qquad t > T, \\ n(\vartheta, T, t) &= X_T + g(t-T) + f \int_T^t m(\vartheta, T, s)\,\mathrm{d}s \\ &= g(t-T) + f m(\vartheta, T) \int_T^t e^{-a(s-T)}\mathrm{d}s \\ &= X_T + g(t-T) + \frac{f}{a}\, m(\vartheta, T)\left[1 - e^{-a(t-T)}\right]. \end{aligned}$$

These equations allow us to write the adaptive extrapolations of $Y_t, t > T$ and $X_t, t > T$ as follows:

$$\begin{aligned} m^\star_{t,T} &= m^\star_{T,T}\; e^{-a^\star_{T,T}(t-T)}, \\ n^\star_{t,T} &= X_T + g^\star_{T,T}(t-T) + \frac{f^\star_{T,T}}{a^\star_{T,T}}\, m^\star_{T,T}\left[1 - e^{-a^\star_{T,T}(t-T)}\right]. \end{aligned}$$

Here $m^\star_{T,T}$ is solution of equation (7.83) at the moment $s = T$.

Having the estimation of the differences $\vartheta^\star_{T,T} - \vartheta_0$ and $m^\star_{T,T} - m(\vartheta_0, T)$, it is possible to describe the asymptotic of the errors $m^\star_{t,T} - m(\vartheta_0, T, t)$ and $n^\star_{t,T} - n(\vartheta_0, t, T)$ too.

The extrapolation $S^\star_{t,T}$ of the security price process $S_t, t > T$ can be

$$\begin{aligned} S^\star_{t,T} &= S_0 \exp\left(\sigma\, n^\star_{t,T} - \frac{\sigma^2}{2} t\right) \\ &= S_0 \exp\left(\sigma\, X_T + \sigma\, g^\star_{T,T}(t-T) + \frac{\sigma\, f^\star_{T,T}\, m^\star_{T,T}}{a^\star_{T,T}}\left[1 - e^{-a^\star_{T,T}(t-T)}\right] - \frac{\sigma^2}{2} t\right). \end{aligned}$$

7.4 Change Point Problems

We consider once more the partially observed model

$$\begin{aligned} \mathrm{d}X_t &= f(\vartheta, t)\, Y_t\, \mathrm{d}t + \varepsilon\sigma(t)\, \mathrm{d}W_t, \quad X_0 = 0, \quad 0 \leq t \leq T, \\ \mathrm{d}Y_t &= a(\vartheta, t)\, Y_t\, \mathrm{d}t + \varepsilon b(t)\, \mathrm{d}V_t, \qquad\qquad Y_0 = y_0 \neq 0, \end{aligned}$$

where $\vartheta \in \Theta = (\alpha, \beta)$, $\alpha > 0$, $\beta < T$, but in three different situations:

- *Change point ϑ in the observation equation*

$$f(\vartheta,t)=\left[g_t\,\mathbb{I}_{\{t<\vartheta\}}+h_t\,\mathbb{I}_{\{t\geq\vartheta\}}\right],\qquad a(\vartheta,t)=a(t),$$

- *Change point ϑ in the state equation*

$$a(\vartheta,t)=\left[g_t\,\mathbb{I}_{\{t<\vartheta\}}+h_t\,\mathbb{I}_{\{t\geq\vartheta\}}\right],\qquad f(\vartheta,t)=f(t),$$

- *Change points $\vartheta=(\theta_1,\theta_2)$ in both equations*

$$f(\vartheta,t)=\left[q_t\,\mathbb{I}_{\{t<\theta_1\}}+r_t\,\mathbb{I}_{\{t\geq\theta_1\}}\right],\quad a(\vartheta,t)=\left[g_t\,\mathbb{I}_{\{t<\theta_2\}}+h_t\,\mathbb{I}_{\{t\geq\theta_2\}}\right].$$

All functions are known and the functions $g_t\neq h_t$ and $q_t\neq r_t$. For all models we describe the behavior of the MLE and BE and discuss *off-line* adaptive filtering equations.

7.4.1 Change Point in the Observation Equation

Model of Observation

Consider the partially observed system

$$\begin{aligned}
\mathrm{d}X_t&=\left[g_t\,\mathbb{I}_{\{t<\vartheta\}}+h_t\,\mathbb{I}_{\{t\geq\vartheta\}}\right]Y_t\,\mathrm{d}t+\varepsilon\sigma(t)\,\mathrm{d}W_t, &X_0=0,\quad 0\leq t\leq T,\\
\mathrm{d}Y_t&=a(t)Y_t\,\mathrm{d}t+\varepsilon b(t)\,\mathrm{d}V_t, &Y_0=y_0\neq 0.
\end{aligned}$$

Here $W_t, 0\leq t\leq T$ and $V_t, 0\leq t\leq T$ are independent Wiener processes, and the functions $g_t,h_t,\sigma(t),a(t),b(t),t\in[0,T]$ and the initial value y_0 are known. The observations are $X^T=(X_t,0\leq t\leq T)$, and the main problem is to estimate the (unknown) parameter ϑ. Then we propose an approximation of the unobserved component $Y^T=(Y_t,0\leq t\leq T)$. As the best estimator of Y_t is its conditional expectation $m(\vartheta,t)=\mathbf{E}_\vartheta\left(Y_t|\mathfrak{F}_t^X\right)$ the problem is reduced to the estimation of the random function $m(\vartheta,t),0<t\leq T$.

Conditions $\mathscr{Q}$.

$\mathscr{Q}_1$. *Parameter $\vartheta\in\Theta=(\alpha,\beta),0<\alpha<\beta<T$.*
$\mathscr{Q}_2$. *The functions $g_t,h_t,\sigma(t),a(t),b(t),0\leq t\leq T$ are bounded.*
$\mathscr{Q}_3$. *The function $\sigma(t)^2>c_\sigma>0$.*
$\mathscr{Q}_4$. *The functions g_t,h_t are continuous and* $\inf_{\alpha<t<\beta}|g_t-h_t|>0$.

Let us denote $f(\vartheta,t)=g_t\,\mathbb{I}_{\{t<\vartheta\}}+h_t\,\mathbb{I}_{\{t\geq\vartheta\}}$. Then the Kalman-Bucy filter has the standard form

$$\mathrm{d}m(\vartheta,t)=a(t)m(\vartheta,t)\,\mathrm{d}t+\frac{f(\vartheta,t)\,\gamma_*(\vartheta,t)}{\sigma(t)^2}\left[\mathrm{d}X_t-f(\vartheta,t)\,m(\vartheta,t)\,\mathrm{d}t\right],\tag{7.84}$$

$$\frac{\mathrm{d}\gamma_*(\vartheta,t)}{\mathrm{d}t}=2a(t)\gamma_*(\vartheta,t)-\frac{f(\vartheta,t)^2\,\gamma_*(\vartheta,t)^2}{\sigma(t)^2}+b(t)^2,\tag{7.85}$$

subject to the initial values $m(\vartheta,0)=y_0$, $\gamma_*(\vartheta,0)=0$. Recall that $\gamma_*(\vartheta,t)=\varepsilon^{-2}\mathbf{E}_\vartheta\left(Y_t-m(\vartheta,t)\right)^2$.

The system (7.84) and (7.85) can be written as two different systems: the first one on the time interval $[0,\vartheta]$ is

$$\mathrm{d}m(t)=a(t)\,m(t)\,\mathrm{d}t+\frac{g_t\,\gamma_*(t)}{\sigma(t)^2}\left[\mathrm{d}X_t-g_t\,m(t)\,\mathrm{d}t\right],\qquad m(0)=y_0,\tag{7.86}$$

$$\frac{\mathrm{d}\gamma_*(t)}{\mathrm{d}t}=2a(t)\,\gamma_*(t)-\frac{g_t^2\,\gamma_*(t)^2}{\sigma(t)^2}+b(t)^2,\qquad \gamma_*(0)=0,$$

and on the time interval $[\vartheta,T]$ we have the system

$$\mathrm{d}m(\vartheta,t)=a(t)\,m(\vartheta,t)\,\mathrm{d}t+\frac{h_t\,\gamma_*(\vartheta,t)}{\sigma(t)^2}\left[\mathrm{d}X_t-h_t\,m(\vartheta,t)\,\mathrm{d}t\right],\tag{7.87}$$

$$\frac{\mathrm{d}\gamma_*(\vartheta,t)}{\mathrm{d}t}=2a(t)\,\gamma_*(\vartheta,t)-\frac{h_t^2\,\gamma_*(\vartheta,t)^2}{\sigma(t)^2}+b(t)^2.$$

The initial values are $m(\vartheta,\vartheta)=m(\vartheta)$ and $\gamma_*(\vartheta,\vartheta)=\gamma_*(\vartheta)$. Therefore the conditional expectation $m(\vartheta,t)=\mathbf{E}_\vartheta\left(Y_t|\mathfrak{F}_t^X\right)$ admits the representation

$$\begin{aligned}m(\vartheta,t)&=m(t)\;\mathbb{1}_{\{0\le t\le\vartheta\}}+m(\vartheta)\,\Phi(\vartheta,\vartheta,t)\\&\quad+\int_\vartheta^t\Phi(\vartheta,s,t)\,\frac{h_s\,\gamma_*(\vartheta,s)}{\sigma(s)^2}\mathrm{d}X_s\;\mathbb{1}_{\{\vartheta<t\le T\}},\end{aligned}\tag{7.88}$$

where

$$\Phi(\vartheta,s,t)=\exp\left(\int_s^t\left[a(r)-\frac{h_r^2\,\gamma_*(\vartheta,r)}{\sigma(r)^2}\right]\mathrm{d}r\right).$$

The true value is denoted ϑ_0. The limit ($\varepsilon=0$) system is

$$\frac{\mathrm{d}x_t(\vartheta_0)}{\mathrm{d}t}=\left[g_t\,\mathbb{1}_{\{t<\vartheta_0\}}+h_t\,\mathbb{1}_{\{t\ge\vartheta_0\}}\right]y_t,\qquad x_0(\vartheta)=0,$$

$$y_t=y_0\exp\left(\int_0^t a(s)\,\mathrm{d}s\right)$$

and

$$\frac{\mathrm{d}y_t\left(\vartheta,\vartheta_0\right)}{\mathrm{d}t}=A\left(\vartheta,t\right)y_t\left(\vartheta,\vartheta_0\right)+\frac{f\left(\vartheta,t\right)f\left(\vartheta_0,t\right)\gamma_*\left(\vartheta,t\right)}{\sigma\left(t\right)^2}y_t,\qquad y_0\left(\vartheta,\vartheta_0\right)=y_0,$$

where $y_t\left(\vartheta,\vartheta_0\right)=m\left(\vartheta,t\right)|_{\varepsilon=0}$, $A\left(\vartheta,t\right)=a\left(t\right)-f\left(\vartheta,t\right)^2\gamma_*\left(\vartheta,t\right)\sigma\left(t\right)^{-2}$ and we extend the notation $\gamma_*\left(\vartheta,t\right)$ on the time interval $[0,\vartheta]$ assuming that $\gamma_*\left(\vartheta,t\right)=\gamma_*\left(t\right),0\le t\le\vartheta$.

We have

$$\begin{aligned}x_t\left(\vartheta_0\right)&=\int_0^t g_s\,y_s\,\mathrm{d}s\,\mathbb{1}_{\{t<\vartheta_0\}}+\int_0^{\vartheta_0} g_s\,y_s\,\mathrm{d}s\,\mathbb{1}_{\{t\ge\vartheta_0\}}\\&\quad+\int_{\vartheta_0}^t h_s\,y_s\,\mathrm{d}s\,\mathbb{1}_{\{t>\vartheta_0\}},\quad 0\le t\le T.\end{aligned}$$

The function $x_t\left(\vartheta\right)$ has a derivative on ϑ:

$$\begin{aligned}\dot x_t\left(\vartheta\right)&=0,&&0\le t<\vartheta,\\\dot x_t\left(\vartheta\right)&=\left[g_\vartheta-h_\vartheta\right]y_\theta,&&\vartheta\le t\le T,\end{aligned}$$

and the random process $m\left(\vartheta,t\right),0\le t\le T$ is continuous with probability 1 function of ϑ (see (7.88)).

Parameter Estimators

MDE

Introduce the MDE

$$\check\vartheta_\varepsilon=\arg\inf_{\vartheta\in\Theta}\int_\alpha^\beta\left[X_t-\int_0^t f\left(\vartheta,s\right)y_s\mathrm{d}s\right]^2\mathrm{d}t,$$

and the function

$$\begin{aligned}g\left(\vartheta,\vartheta_0\right)&=\int_\alpha^\beta\left[X_t-\int_0^t f\left(\vartheta,s\right)y_s\mathrm{d}s\right]^2\mathrm{d}t\bigg|_{\varepsilon=0}\\&=\int_\alpha^\beta\left[\int_\alpha^t\left[f\left(\vartheta_0,s\right)-f\left(\vartheta,s\right)\right]y_s\mathrm{d}s\right]^2\mathrm{d}t\\&=\left|\vartheta-\vartheta_0\right|\left(\int_{\vartheta_0}^\vartheta\left[h_s-g_s\right]y_s\,\mathrm{d}s\right)^2.\end{aligned}$$

Therefore the condition of identifiability: for any $\nu > 0$

$$\inf_{|\vartheta-\vartheta_0|>\nu} g\left(\vartheta,\vartheta_0\right)=\nu\left(\int_{\vartheta_0}^{\vartheta_0+\nu}\left[h_s-g_s\right]y_s\,\mathrm{d}s\right)^2>0$$

by the condition $\mathcal{E}_4$ is fulfilled and the MDE is consistent. The proof follows the usual steps (see (2.14)).

To study the asymptotic of the normalized difference $\varepsilon^{-1}\left(\check{\vartheta}_\varepsilon-\vartheta_0\right)$ we note that this estimator is a solution of the MDEq

$$\int_\vartheta^\beta\left[X_t-\int_0^\vartheta g_sy_s\,\mathrm{d}s-\int_\vartheta^t h_sy_s\,\mathrm{d}s\right]\mathrm{d}t=0,\qquad \vartheta\in\Theta.$$

Let us write this equation as

$$Q\left(\vartheta,\varepsilon\right)=Q\left(\vartheta_0,\varepsilon\right)+\left(\vartheta-\vartheta_0\right)\dot{Q}(\tilde{\vartheta},\varepsilon)=0.$$

Then

$$\frac{\check{\vartheta}_\varepsilon-\vartheta_0}{\varepsilon}=-\frac{Q\left(\vartheta_0,\varepsilon\right)}{\varepsilon\dot{Q}(\tilde{\vartheta},\varepsilon)}.$$

We have

$$\left.\frac{\mathrm{d}Q\left(\vartheta,\varepsilon\right)}{\mathrm{d}\vartheta}\right|_{\vartheta=\vartheta_0,\varepsilon=0}=\left(\beta-\vartheta_0\right)\left[h_{\vartheta_0}-g_{\vartheta_0}\right]y_{\vartheta_0},$$

and

$$\frac{Q\left(\vartheta_0,\varepsilon\right)}{\varepsilon}=\int_{\vartheta_0}^\beta\left[\int_0^t f\left(\vartheta_0,s\right)\int_0^s\exp\left(\int_r^s a\left(v\right)\mathrm{d}v\right)b\left(r\right)\mathrm{d}V_r\,\mathrm{d}s\right.$$
$$\left.+\int_0^t\sigma\left(s\right)\mathrm{d}W_s\right]\mathrm{d}t\equiv\check{\zeta}\left(\vartheta_0\right).$$

Let us denote

$$D\left(\vartheta_0\right)=\frac{1}{\left(\beta-\vartheta_0\right)^2\left[g_{\vartheta_0}-h_{\vartheta_0}\right]^2y_{\vartheta_0}^2}\left[\int_0^\beta\left(\beta-s\vee\vartheta_0\right)^2\sigma\left(s\right)^2\mathrm{d}s\right.$$
$$\left.+\int_0^\beta b\left(s\right)^2\left(\int_{\vartheta_0\vee s}^\beta\int_s^t f\left(\vartheta_0,r\right)\phi\left(s,r\right)\mathrm{d}r\,\mathrm{d}t\right)^2\mathrm{d}s\right].$$

Therefore we obtained the asymptotic normality of this estimator.

Proposition 7.5 *Let the conditions $\mathscr{Q}$ hold. Then the MDE is consistent and asymptotically normal*

$$\varepsilon^{-1}\left(\check\vartheta_\varepsilon-\vartheta_0\right)\Longrightarrow\frac{\check\zeta\left(\vartheta_0\right)}{\left(\beta-\vartheta_0\right)\left[g_{\vartheta_0}-h_{\vartheta_0}\right]y_{\vartheta_0}}\sim\mathcal{N}\left(0,D\left(\vartheta_0\right)\right).$$

MLE and BE

The likelihood ratio in this problem is

$$L\left(\vartheta,X^T\right)=\exp\left(\int_0^T\frac{f\left(\vartheta,t\right)m\left(\vartheta,t\right)}{\varepsilon^2\sigma\left(t\right)^2}\mathrm{d}X_t-\int_0^T\frac{f\left(\vartheta,t\right)^2m\left(\vartheta,t\right)^2}{2\varepsilon^2\sigma\left(t\right)^2}\mathrm{d}t\right),\quad\vartheta\in\Theta,$$

but we will use a slightly different form of it

$$L\left(\vartheta,X^\beta\right)=\exp\left(\int_\alpha^\beta\frac{f\left(\vartheta,t\right)m\left(\vartheta,t\right)}{\varepsilon^2\sigma\left(t\right)^2}\mathrm{d}X_t-\int_\alpha^\beta\frac{f\left(\vartheta,t\right)^2m\left(\vartheta,t\right)^2}{2\varepsilon^2\sigma\left(t\right)^2}\mathrm{d}t\right),\quad\vartheta\in\Theta,$$

because the observations on the intervals $[0,\alpha]$ and $[\beta,T]$ have asymptotically negligible contributions in the problem of estimation ϑ_0.

The MLE $\hat\vartheta_\varepsilon$ and BE $\tilde\vartheta_\varepsilon$ are defined by the usual relations (2.32).

To describe the asymptotic properties of these estimators we introduce the limit likelihood ratio

$$Z\left(\vartheta_0,u\right)=\exp\left(\rho\left(\vartheta_0\right)W\left(u\right)-\frac{|u|}{2}\rho\left(\vartheta_0\right)^2\right),\qquad u\in\mathcal{R},$$

where $\rho\left(\vartheta\right)=\left[g_\vartheta-h_\vartheta\right]y_\vartheta\sigma\left(\vartheta\right)^{-1}$ and $W\left(u\right),u\in\mathcal{R}$ is a two-sided Wiener process. It can be written as $W\left(u\right)=W_+\left(u\right)1\!\!1_{\{u\geq0\}}+W_-\left(-u\right)1\!\!1_{\{u\leq0\}}$, where $W_+\left(u\right),u\geq0$ and $W_-\left(u\right),u\geq0$ are independent Wiener processes. Define two random variables $\hat u\left(\vartheta_0\right)$ and $\tilde u\left(\vartheta_0\right)$ by the relations

$$Z\left(\vartheta_0,\hat u\left(\vartheta_0\right)\right)=\sup_{u\in\mathcal{R}}Z\left(\vartheta_0,u\right),\qquad\tilde u\left(\vartheta_0\right)=\frac{\int_{\mathcal{R}}u\,Z\left(\vartheta_0,u\right)\mathrm{d}u}{\int_{\mathcal{R}}Z\left(\vartheta_0,u\right)\mathrm{d}u}.\tag{7.89}$$

The random variable $\hat u=\hat u\left(\vartheta_0\right)$ is a unique solution of the corresponding equation in (7.89) (see Lemma VII.2.5 in [73]).

The following minimax lower bound is valid for all estimators $\bar\vartheta_\varepsilon$

$$\lim_{\nu\to 0}\varliminf_{\varepsilon\to 0}\sup_{|\vartheta-\vartheta_0|\leq\nu}\varepsilon^{-4}\mathbf{E}_\vartheta\left(\bar\vartheta_\varepsilon-\vartheta_0\right)^2\geq\mathbf{E}_{\vartheta_0}\tilde u\left(\vartheta_0\right)^2. \tag{7.90}$$

For the proof see (7.17)–(7.18).

We call the estimator $\bar\vartheta_\varepsilon$ asymptotically efficient if for this estimator there is equality in (7.90) for all $\vartheta_0\in\Theta$.

Remark that

$$\mathbf{E}_{\vartheta_0}\hat u\left(\vartheta_0\right)^2=\frac{26}{\rho\left(\vartheta_0\right)^4}\qquad\text{and}\qquad\mathbf{E}_{\vartheta_0}\tilde u\left(\vartheta_0\right)^2=\frac{16\,z\left(3\right)}{\rho\left(\vartheta_0\right)^4}\approx\frac{19,5}{\rho\left(\vartheta_0\right)^4}.$$

Here $z\left(\cdot\right)$ is Riemann's zeta function

$$z\left(x\right)=\sum_{k=1}^{\infty}\frac{1}{k^x}.$$

The first equality was obtained in [187] (see also Theorem VII.3.1 in [73]), and the second was proved in [171] (see also [158] for more simple proof).

Theorem 7.9 *Let conditions* $\mathscr{Q}$ *hold. Then the MLE* $\hat\vartheta_\varepsilon$ *and BE* $\tilde\vartheta_\varepsilon$ *are uniformly on compacts consistent and converge in distribution*

$$\varepsilon^{-2}(\hat\vartheta_\varepsilon-\vartheta_0)\Longrightarrow\hat u\left(\vartheta_0\right),\qquad\qquad\varepsilon^{-2}(\tilde\vartheta_\varepsilon-\vartheta_0)\Longrightarrow\tilde u\left(\vartheta_0\right),$$

the polynomial moments of these estimators uniformly on compacts converge and the BE are asymptotically efficient.

Proof We have to verify the conditions of Theorems 1.9 and 1.11. In particular, we consider the normalization $\vartheta_u=\vartheta_0+\varepsilon^2 u$ of the LR

$$Z_\varepsilon\left(\vartheta_0,u\right)=\frac{L\left(\vartheta_u,X^\beta\right)}{L\left(\vartheta_0,X^\beta\right)},\qquad u\in\mathbb{U}_\varepsilon=\left(\frac{\alpha-\vartheta_0}{\varepsilon^2},\frac{\beta-\vartheta_0}{\varepsilon^2}\right)$$

and will verify the following properties of $Z_\varepsilon\left(\cdot\right)$:

- *The marginal distributions of* $Z_\varepsilon\left(\vartheta_0,\cdot\right)$ *converge to the marginal distributions of* $Z\left(\vartheta_0,\cdot\right)$ *uniformly on compacts* $\mathbb{K}\subset\Theta$*: for any* $u_1,\ldots,u_k$

$$\Big(Z_\varepsilon\left(\vartheta_0,u_1\right),\ldots,Z_\varepsilon\left(\vartheta_0,u_k\right)\Big)\Longrightarrow\Big(Z\left(\vartheta_0,u_1\right),\ldots,Z\left(\vartheta_0,u_k\right)\Big). \tag{7.91}$$

- *There exists a constant* $C>0$ *such that for any* $R>0$ *and any* $|u_1|\leq R,|u_2|\leq R,u_1,u_2\in\mathbb{U}_\varepsilon$

$$\sup_{\vartheta_0\in\mathbb{K}} \mathbf{E}_{\vartheta_0} \left|Z_\varepsilon\left(\vartheta_0,u_1\right)^{1/4}-Z_\varepsilon\left(\vartheta_0,u_2\right)^{1/4}\right|^4 \leq C\left(1+R^2\right)\left|u_2-u_1\right|^2. \tag{7.92}$$

- *For any $N>0$ there exists a constant $C=C_N>0$ such that*

$$\sup_{\vartheta_0\in\mathbb{K}} \mathbf{E}_{\vartheta_0} Z_\varepsilon\left(\vartheta_0,u\right)^{1/2} \leq \frac{C}{|u|^N}. \tag{7.93}$$

□

Lemma 7.11 *Let conditions $\mathscr{Q}$ hold then there exist a constant $C>0$ such that for $t\in[\vartheta_u\wedge\vartheta_0,T]$ the following estimate holds*

$$\mathbf{E}_{\vartheta_0}\left[m\left(\vartheta_u,t\right)-m\left(\vartheta_0,t\right)\right]^2 \leq C\,\varepsilon^4\left(u^2+|u|\right). \tag{7.94}$$

Proof Let us write Eq. (7.87) for $m\left(\vartheta_u,t\right)$ and $m\left(\vartheta_0,t\right)$ and take the difference $\delta_m\left(t\right)=m\left(\vartheta_u,t\right)-m\left(\vartheta_0,t\right)$. Then on the interval $\vartheta_0\leq t\leq\vartheta_0+\varepsilon^2u$ (for $u>0$) we obtain the equation

$$\begin{aligned}\mathrm{d}\delta_m\left(t\right)=&\left[a\left(t\right)-\frac{g_t^2\gamma_*\left(t\right)}{\sigma\left(t\right)^2}\right]\delta_m\left(t\right)\mathrm{d}t+\frac{g_t\gamma_*\left(t\right)}{\sigma\left(t\right)^2}\left[h_t-g_t\right]m\left(\vartheta_0,t\right)\mathrm{d}t\\&+\varepsilon\frac{g_t\gamma_*\left(t\right)-h_t\gamma_*\left(\vartheta_0,t\right)}{\sigma\left(t\right)}\mathrm{d}\bar W_t.\end{aligned}$$

Remark that for $0\leq t\leq\vartheta_0$ we have $m\left(\vartheta_u,t\right)=m\left(\vartheta_0,t\right)=m\left(t\right)$, where $m\left(t\right)$ is solution of equation (7.86). Therefore $\delta_m\left(t\right)=0, 0\leq t\leq\vartheta_0$ and for $\vartheta_0\leq t\leq\vartheta_0+\varepsilon^2u$ we have

$$\begin{aligned}\delta_m\left(t\right)=&\int_{\vartheta_0}^t\Phi\left(s,t\right)\frac{g_s\gamma_*\left(s\right)}{\sigma\left(s\right)^2}\left[h_s-g_s\right]m\left(\vartheta_0,s\right)\mathrm{d}s\\&+\varepsilon\int_{\vartheta_0}^t\Phi\left(s,t\right)\frac{g_s\gamma_*\left(s\right)-h_s\gamma_*\left(\vartheta_0,s\right)}{\sigma\left(s\right)}\mathrm{d}\bar W_s,\end{aligned}$$

where

$$\Phi\left(s,t\right)=\exp\left(\int_s^t\left[a\left(v\right)-\frac{g_v^2\gamma_*\left(v\right)}{\sigma\left(v\right)^2}\right]\mathrm{d}v\right).$$

Now we can write

$$\mathbf{E}_{\vartheta_0}\delta_m\left(t\right)^2\leq 2\mathbf{E}_{\vartheta_0}\left(\int_{\vartheta_0}^t\Phi\left(s,t\right)\frac{g_s\gamma_*\left(s\right)}{\sigma\left(s\right)^2}\left[h_s-g_s\right]m\left(\vartheta_0,s\right)\mathrm{d}s\right)^2$$

$$
\begin{aligned}
&+2\varepsilon^2\int_{\vartheta_0}^{t}\Phi\left(s,t\right)^2\frac{\left[g_s\gamma_*\left(s\right)-h_s\gamma_*\left(\vartheta_0,s\right)\right]^2}{\sigma\left(s\right)^2}{\rm d}s\\
&\leq 2\left(t-\vartheta_0\right)\int_{\vartheta_0}^{\vartheta_0+\varepsilon^2u}\Phi\left(s,t\right)^2\frac{g_s^2\gamma_*\left(s\right)^2}{\sigma\left(s\right)^4}\left[h_s-g_s\right]^2{\bf E}_{\vartheta_0}m\left(\vartheta_0,s\right)^2{\rm d}s\\
&+2\varepsilon^2\int_{\vartheta_0}^{\vartheta_0+\varepsilon^2u}\Phi\left(s,t\right)^2\frac{\left[g_s\gamma_*\left(s\right)-h_s\gamma_*\left(\vartheta_0,s\right)\right]^2}{\sigma\left(s\right)^2}{\rm d}s\\
&\leq C_1\,\varepsilon^4\,u^2+C_2\,\varepsilon^4\,\left|u\right|\leq C\,\varepsilon^4\left(u^2+\left|u\right|\right). \qquad (7.95)
\end{aligned}
$$

Here we used the estimate $t-\vartheta_0\leq\varepsilon^2u$.

Similar inequality we have in the case of negative u and $t\in\left[\vartheta_0+\varepsilon^2u,\vartheta_0\right]$.

Let $u>0$ and $t>\vartheta_0+\varepsilon^2u$. Then for the difference $\delta_m\left(t\right)$ we obtain the equation

$$
{\rm d}\delta_m\left(t\right)=\left[a\left(t\right)-\frac{h_t^2\gamma_*\left(\vartheta_u,t\right)}{\sigma\left(t\right)^2}\right]\delta_m\left(t\right){\rm d}t+\varepsilon\frac{h_t\left[\gamma_*\left(\vartheta_u,t\right)-\gamma_*\left(\vartheta_0,t\right)\right]}{\sigma\left(t\right)}{\rm d}W_t
$$

with the initial value

$$
\delta_m\left(\vartheta_u\right)=m\left(\vartheta_u,\vartheta_u\right)-m\left(\vartheta_0,\vartheta_u\right).
$$

The solution is

$$
\begin{aligned}
\delta_m\left(t\right)&=\left[m\left(\vartheta_u,\vartheta_u\right)-m\left(\vartheta_0,\vartheta_u\right)\right]\Phi\left(\vartheta_u,\vartheta_u,t\right)\\
&\quad+\varepsilon\int_{\vartheta_u}^{t}\Phi\left(\vartheta_u,s,t\right)\frac{h_s\left[\gamma_*\left(\vartheta_u,t\right)-\gamma_*\left(\vartheta_0,s\right)\right]}{\sigma\left(s\right)}{\rm d}W_s
\end{aligned}
$$

with the corresponding function $\Phi\left(\vartheta,s,t\right)$.

We have

$$
\begin{aligned}
{\bf E}_{\vartheta_0}\delta_m\left(t\right)^2&\leq 2\Phi\left(\vartheta_u,\vartheta_u,t\right)^2{\bf E}_{\vartheta_0}\left[m\left(\vartheta_u,\vartheta_u\right)-m\left(\vartheta_0,\vartheta_u\right)\right]^2\\
&\quad+\varepsilon^2\int_{\vartheta_u}^{t}\Phi\left(\vartheta_u,s,t\right)^2\frac{h_s^2\left[\gamma_*\left(\vartheta_u,t\right)-\gamma_*\left(\vartheta_0,s\right)\right]^2}{\sigma\left(s\right)^2}{\rm d}s. \qquad (7.96)
\end{aligned}
$$

To estimate the difference $\delta_\gamma\left(t\right)=\gamma_*\left(\vartheta_u,t\right)-\gamma_*\left(\vartheta_0,s\right)$ we write the equations for $\gamma_*\left(\vartheta_u,t\right)$ and $\gamma_*\left(\vartheta_0,s\right)$ and study $\delta_\gamma\left(t\right),\ \vartheta_0\leq t\leq\vartheta_0+\varepsilon^2u$ and $\delta_\gamma\left(t\right),\ \vartheta_0+\varepsilon^2u\leq t\leq T$ separately. On the first interval we obtain the expression

$$
\delta_\gamma\left(t\right)=\int_{\vartheta_0}^{t}\tilde\Phi\left(\vartheta_u,s,t\right)\left[g_s\gamma_*\left(\vartheta_u,s\right)\gamma_*\left(\vartheta_0,s\right)+h_s\gamma_*\left(\vartheta_0,s\right)^2\right]\sigma\left(s\right)^{-2}\left[h_s-g_s\right]{\rm d}s
$$

with some bounded function $\tilde{\Phi}(\vartheta_u, s, t)$. Therefore on this interval we have

$$|\gamma_*(\vartheta_u, t) - \gamma_*(\vartheta_0, s)| \leq C\,\varepsilon^2\,|u|\,.$$

The equation for $\delta_\gamma(t)$ on the interval $\left[\vartheta_0 + \varepsilon^2 u, T\right]$ is

$$\frac{\mathrm{d}\delta_\gamma(t)}{\mathrm{d}t} = \left[2a(t) - h_t^2\left[\gamma_*(\vartheta_u, t) + \gamma_*(\vartheta_0, s)\right]\sigma(t)^{-2}\right]\delta_\gamma(t)\,,$$
$$\delta_\gamma(\vartheta_u) = \gamma_*(\vartheta_u, \vartheta_u) - \gamma_*(\vartheta_0, \vartheta_u)\,.$$

Therefore

$$\left|\delta_\gamma(t)\right| \leq C\,\varepsilon^2\,|u|\,.$$

Moreover, it can be shown that the function $\gamma_*(\vartheta, t)\,,\ \vartheta \leq t \leq T$ has derivative on ϑ. Indeed, the equation for $\dot{\gamma}_*(\vartheta, t)$ is

$$\frac{\mathrm{d}\dot{\gamma}_*(\vartheta, t)}{\mathrm{d}t} = 2a(t)\,\dot{\gamma}_*(\vartheta, t) - \frac{2h_t^2\gamma_*(\vartheta, t)}{\sigma(t)^2}\dot{\gamma}_*(\vartheta, t)\,, \qquad \dot{\gamma}_*(\vartheta, \vartheta) = \dot{\gamma}_*(\vartheta)\,.$$

Therefore

$$\dot{\gamma}_*(\vartheta, t) = \dot{\gamma}_*(\vartheta)\exp\left(2\int_\vartheta^t\left[a(s) - \frac{h_s^2\gamma_*(\vartheta, s)}{\sigma(s)^2}\right]\mathrm{d}s\right), \qquad \vartheta < t \leq T,$$

where

$$\dot{\gamma}_*(\vartheta) = \left.\frac{\mathrm{d}\gamma_*(t)}{\mathrm{d}t}\right|_{t=\vartheta} = 2a(\vartheta)\,\gamma_*(\vartheta) - \frac{g_\vartheta^2\gamma_*(\vartheta)^2}{\sigma(\vartheta)^2} + b(\vartheta)^2\,.$$

Let us return to (7.96) and write for the first term in the RHS by (7.95)

$$\mathbf{E}_{\vartheta_0}\left[m(\vartheta_u, \vartheta_u) - m(\vartheta_0, \vartheta_u)\right]^2 \leq C\varepsilon^4\left(u^2 + |u|\right)$$

and for the second term

$$\varepsilon^2\int_{\vartheta_u}^t \Phi(\vartheta_u, s, t)^2\,\frac{h_s^2\left[\gamma_*(\vartheta_u, t) - \gamma_*(\vartheta_0, s)\right]^2}{\sigma(s)^2}\mathrm{d}s \leq C\,\varepsilon^6\,u^2.$$

Finally

$$\sup_{\vartheta_u < t \leq T}\mathbf{E}_{\vartheta_0}\left[m(\vartheta_u, t) - m(\vartheta_0, t)\right]^2 \leq C\varepsilon^4\left(u^2 + |u|\right).$$

This estimate together with (7.96) proves (7.94) □

For one-dimensional distributions the convergence (7.91) follows from the representation (below $u > 0$)

$$
\begin{aligned}
&\ln Z_\varepsilon(\vartheta_0, u) \\
&= \int_\alpha^\beta \frac{[f(\vartheta_u, t)\, m(\vartheta_u, t) - f(\vartheta_0, t)\, m(\vartheta_0, t)]}{\varepsilon \sigma_t}\, \mathrm{d}\bar W_t \\
&\quad - \int_\alpha^\beta \frac{[f(\vartheta_u, t)\, m(\vartheta_u, t) - f(\vartheta_0, t)\, m(\vartheta_0, t)]^2}{2\varepsilon^2 \sigma(t)^2}\, \mathrm{d}t \\
&= \int_{\vartheta_0}^{\vartheta_u} \frac{g_t - h_t}{\varepsilon \sigma(t)} m(\vartheta_0, t)\, \mathrm{d}\bar W_t - \frac{1}{2} \int_{\vartheta_0}^{\vartheta_u} \left(\frac{g_t - h_t}{\varepsilon \sigma(t)}\right)^2 m(\vartheta_0, t)^2\, \mathrm{d}t + o(1) \\
&= \frac{\left[g_{\vartheta_0} - h_{\vartheta_0}\right] y_{\vartheta_0}}{\sigma(\vartheta_0)} \frac{\bar W_{\vartheta_0 + \varepsilon^2 u} - \bar W_{\vartheta_0}}{\varepsilon} - \frac{\left[g_{\vartheta_0} - h_{\vartheta_0}\right]^2 y_{\vartheta_0}^2}{2\sigma(\vartheta_0)^2} |u| + o(1) \\
&= \rho(\vartheta_0)\, W_\varepsilon(u) - \frac{\rho(\vartheta_0)^2}{2} |u| + o(1) \\
&\Longrightarrow \rho(\vartheta_0)\, W(u) - \frac{|u|}{2} \rho(\vartheta_0)^2 .
\end{aligned}
$$

Here we used the relation $m(\vartheta_u, t) - m(\vartheta_0, t) \approx \varepsilon^2$ which corresponds (7.94). Therefore

$$
Z_\varepsilon(\vartheta_0, u) \Longrightarrow Z(\vartheta_0, u) = \exp\left(\rho(\vartheta_0)\, W(u) - \frac{|u|}{2} \rho(\vartheta_0)^2\right), \qquad u \in \mathcal{R}.
$$

Here we used the smoothness of $m(\vartheta_u, t)$ w.r.t. ϑ and ε which allows us to replace asymptotically $m(\vartheta_u, t)$ by $y_t(\vartheta_0)$. The proof of the convergence of multidimensional distributions follows from similar representation: for any $\lambda_1, \ldots, \lambda_k$

$$
\sum_{j=1}^k \lambda_j \ln Z_\varepsilon\left(\vartheta_0, u_j\right) = \sum_{j=1}^k \lambda_j \left[\rho(\vartheta_0)\, W_\varepsilon\left(u_j\right) - \frac{\rho(\vartheta_0)^2}{2} \left|u_j\right| + o(1)\right].
$$

This proves the convergence (7.91).

By Corollary 1.1 we have the estimate

$$
\begin{aligned}
&\mathbf{E}_{\vartheta_0} \left|Z_\varepsilon(\vartheta_0, u_1)^{1/4} - Z_\varepsilon(\vartheta_0, u_2)^{1/4}\right|^4 \\
&\le C_{1,4} \mathbf{E}_{\vartheta_{u_2}} \left(\int_\alpha^\beta \left[\frac{f\left(\vartheta_{u_2}, t\right) m\left(\vartheta_{u_2}, t\right) - f\left(\vartheta_{u_1}, t\right) m\left(\vartheta_{u_1}, t\right)}{\varepsilon \sigma(t)}\right]^2 \mathrm{d}t\right)^4
\end{aligned}
$$

$$+ C_{3,4}\mathbf{E}_{\vartheta_{u_2}}\left(\int_\alpha^\beta\left[\frac{f\left(\vartheta_{u_2},t\right)m\left(\vartheta_{u_2},t\right)-f\left(\vartheta_{u_1},t\right)m\left(\vartheta_{u_1},t\right)}{\varepsilon\sigma(t)}\right]\mathrm{d}W_t\right)^4.$$

The estimate (1.1) allows us to write

$$\mathbf{E}_{\vartheta_{u_2}}\left(\int_\alpha^\beta\left[\frac{f\left(\vartheta_{u_2},t\right)m\left(\vartheta_{u_2},t\right)-f\left(\vartheta_{u_1},t\right)m\left(\vartheta_{u_1},t\right)}{\varepsilon\sigma(t)}\right]\mathrm{d}W_t\right)^4$$
$$\leq C\mathbf{E}_{\vartheta_{u_2}}\left(\int_\alpha^\beta\left[\frac{f\left(\vartheta_{u_2},t\right)m\left(\vartheta_{u_2},t\right)-f\left(\vartheta_{u_1},t\right)m\left(\vartheta_{u_1},t\right)}{\varepsilon\sigma(t)}\right]^2\mathrm{d}t\right)^2.$$

Suppose that $u_2 > u_1$, then we can write (below $\delta_m(u_1,u_2,t) = m\left(\vartheta_{u_2},t\right) - m\left(\vartheta_{u_1},t\right)$)

$$\begin{aligned}
&f\left(\vartheta_{u_2},t\right)m\left(\vartheta_{u_2},t\right)-f\left(\vartheta_{u_1},t\right)m\left(\vartheta_{u_1},t\right)\\
&\quad=\left[f\left(\vartheta_{u_2},t\right)-f\left(\vartheta_{u_1},t\right)\right]m\left(\vartheta_{u_2},t\right)+f\left(\vartheta_{u_1},t\right)\left[m\left(\vartheta_{u_2},t\right)-m\left(\vartheta_{u_1},t\right)\right]\\
&\quad=\left[g_t-h_t\right]m\left(\vartheta_{u_2},t\right)\,\mathbb{1}_{\left\{\vartheta_{u_1}<t<\vartheta_{u_2}\right\}}+f\left(\vartheta_{u_1},t\right)\delta_m(u_1,u_2,t).
\end{aligned}$$

Therefore

$$\begin{aligned}
&\int_\alpha^\beta\left[\frac{f\left(\vartheta_{u_2},t\right)m\left(\vartheta_{u_2},t\right)-f\left(\vartheta_{u_1},t\right)m\left(\vartheta_{u_1},t\right)}{\varepsilon\sigma(t)}\right]^2\mathrm{d}t\\
&\quad=\int_\alpha^\beta\left[\frac{\left[g_t-h_t\right]m\left(\vartheta_{u_2},t\right)\,\mathbb{1}_{\left\{\vartheta_{u_1}<t<\vartheta_{u_2}\right\}}+f\left(\vartheta_{u_1},t\right)\delta_m(u_1,u_2,t)}{\varepsilon\sigma(t)}\right]^2\mathrm{d}t\\
&\quad\leq 2\int_{\vartheta_0+\varepsilon^2u_1}^{\vartheta_0+\varepsilon^2u_2}\left[\frac{\left[g_t-h_t\right]m\left(\vartheta_{u_2},t\right)}{\varepsilon\sigma(t)}\right]^2\mathrm{d}t+2\int_\alpha^\beta\frac{f\left(\vartheta_{u_1},t\right)^2\delta_m(u_1,u_2,t)^2}{\sigma(t)^2}\mathrm{d}t.
\end{aligned}$$

Using these estimates we obtain the expression

$$\mathbf{E}_{\vartheta_{u_2}}\left(\int_\alpha^\beta\left[\frac{f\left(\vartheta_{u_2},t\right)m\left(\vartheta_{u_2},t\right)-f\left(\vartheta_{u_1},t\right)m\left(\vartheta_{u_1},t\right)}{\varepsilon\sigma(t)}\right]^2\mathrm{d}t\right)^2$$
$$\leq C\left(u_2-u_1\right)^2+C\varepsilon^4\left(u_2-u_1\right)^4\leq C\left(u_2-u_1\right)^2.$$

Here we used the estimate $\varepsilon^2\left|u_2-u_1\right|\leq\beta-\alpha$.

For the second integral using similar calculations we obtain the estimate

$$\mathbf{E}_{\vartheta_{u_2}}\left(\int_\alpha^\beta\left[\frac{f\left(\vartheta_{u_2},t\right)m\left(\vartheta_{u_2},t\right)-f\left(\vartheta_{u_1},t\right)m\left(\vartheta_{u_1},t\right)}{\varepsilon\sigma\left(t\right)}\right]^2{\rm d}t\right)^4$$

$$\leq \mathbf{E}_{\vartheta_{u_2}}\left(2\int_{\vartheta_0+\varepsilon^2u_1}^{\vartheta_0+\varepsilon^2u_2}\left[\frac{\left[g_t-h_t\right]m\left(\vartheta_{u_2},t\right)}{\varepsilon\sigma\left(t\right)}\right]^2{\rm d}t\right.$$

$$\left.+2\int_\alpha^\beta\frac{f\left(\vartheta_{u_1},t\right)^2\delta_m\left(u_1,u_2,t\right)^2}{\varepsilon^2\sigma\left(t\right)^2}{\rm d}t\right)^4$$

$$\leq C\,\left(u_2-u_1\right)^4+C\,\varepsilon^8\left(u_2-u_1\right)^8\leq C\,\left(u_2-u_1\right)^4\leq C\,R^2\left(u_2-u_1\right)^2.$$

Therefore the condition (7.92) is fulfilled too.

To verify (7.93) we follow the proof of Lemma 2.5. Denote

$$G_\varepsilon\left(u,t\right)=\frac{f\left(\vartheta_u,t\right)m(\vartheta_u,t)-f\left(\vartheta_0,t\right)m(\vartheta_0,t)}{\varepsilon\sigma(t)},$$

$$G\left(u,t\right)=\frac{f\left(\vartheta_u,t\right)y_t(\vartheta_u,\vartheta_0)-f\left(\vartheta_0,t\right)y_t}{\varepsilon\sigma(t)}.$$

Consider the proof of (7.93) for the values $\varepsilon^2\left|u\right|\leq\nu$ and $\varepsilon^2\left|u\right|>\nu, u\in\mathbb{U}_\varepsilon$ separately. Here $\nu>0$ is some (small) real. Let $\varepsilon^2\left|u\right|\leq\nu$ and $u>0$. Then

$$\int_\alpha^\beta G\left(u,t\right)^2{\rm d}t$$

$$=\int_\alpha^\beta\left(\frac{\left[g_t-h_t\right]\mathbb{I}_{\{\vartheta_0<t<\vartheta_0+\varepsilon^2u\}}y_t(\vartheta_u,\vartheta_0)+\varepsilon^2uf\left(\vartheta_0,t\right)\dot y_t(\tilde\vartheta,\vartheta_0)}{\varepsilon\sigma\left(t\right)}\right)^2{\rm d}t$$

$$=\frac{1}{\varepsilon^2}\int_{\vartheta_0}^{\vartheta_0+\varepsilon^2u}\frac{\left[g_t-h_t\right]^2}{\sigma\left(t\right)^2}y_t(\vartheta_u,\vartheta_0)^2{\rm d}t+\varepsilon^2u\;O\left(1\right)+\varepsilon^2u^2O\left(1\right)$$

$$\geq\frac{u}{c_\sigma}\inf_{\vartheta_0<t<\vartheta_0+\varepsilon^2u}\left[g_t-h_t\right]^2y_t(\vartheta_u,\vartheta_0)^2-\nu\left|O\left(1\right)\right|-\nu\,u\left|O\left(1\right)\right|$$

$$\geq 2\kappa_1u-\nu\left|O\left(1\right)\right|-\nu\,u\left|O\left(1\right)\right|\geq\kappa_1u$$

for sufficiently small ν. Here we denoted $2\kappa=\inf_{\alpha<t<\beta}\left[g_t-h_t\right]^2y_t(\vartheta_u,\vartheta_0)^2>0$.

Therefore

$$\inf_{\vartheta_0\in\mathbb{K}}\int_\alpha^\beta G\left(u,t\right)^2{\rm d}t\geq\kappa_1\left|u\right|.$$

For the values $\varepsilon^2 |u| > \nu$ let us put $\vartheta = \vartheta_0 + \varepsilon^2 u$ (we have $|\vartheta - \vartheta_0| > \nu$) and suppose that

$$\inf_{|\vartheta-\vartheta_0|>\nu} \int_\alpha^\beta \left[\frac{f(\vartheta, t)\, y_t(\vartheta, \vartheta_0) - f(\vartheta_0, t)\, y_t}{\sigma(t)}\right]^2 \mathrm{d}t = 0. \tag{7.97}$$

Then it means that $f(\vartheta, t)\, y_t(\vartheta, \vartheta_0) \equiv f(\vartheta_0, t)\, y_t$ for all $t \in [\alpha, \beta]$. But the functions $f(\vartheta, t)$ and $f(\vartheta_0, t)$ have jumps at different points and the functions $y_t(\vartheta, \vartheta_0)$ and y_t are continuous. Therefore the equality (7.97) is impossible and we have

$$\inf_{|\vartheta-\vartheta_0|>\nu} \int_0^T \left[\frac{f(\vartheta, t)\, y_t(\vartheta, \vartheta_0) - f(\vartheta_0, t)\, y_t}{\varepsilon\sigma(t)}\right]^2 \mathrm{d}t \geq \frac{\kappa_2}{\varepsilon^2}$$
$$= \frac{\kappa_2}{\vartheta - \vartheta_0} u \geq \frac{\kappa_2}{\beta - \alpha} u = \kappa_3\, u$$

with some $\kappa_2 > 0$. For the negative values u we have a similar proof with the corresponding constants $\kappa_4 > 0$ and $\kappa_5 > 0$.

We can put now $\kappa = \min(\kappa_1, \kappa_3, \kappa_4, \kappa_5)$ and obtain the estimate

$$\inf_{\vartheta_0 \in \mathbb{K}} \int_\alpha^\beta G(u, t)^2\, \mathrm{d}t \geq \kappa\, |u|.$$

By direct calculations it can be shown that

$$\mathbf{E}_{\vartheta_0} |G_\varepsilon(u, t) - G(u, t)|^N \leq C\varepsilon^{2N} |u|^N.$$

These two last estimates allow to apply the same proof as in Lemma 2.5 and to obtain (7.93).

Remark 7.4 Let us remind one property else of the BE $\tilde{\vartheta}_\varepsilon$. For any compact $\mathbb{K} \subset \Theta$, any $H > 0$, any $N > 0$ and sufficiently small ε there exists a constant $C_N > 0$ such that

$$\sup_{\vartheta_0 \in \mathbb{K}} \mathbf{P}_{\vartheta_0} \left(\varepsilon^{-2} |\tilde{\vartheta}_\varepsilon - \vartheta_0| > H\right) \leq \frac{C_N}{H^N}.$$

For the proof see Theorem 1.5.2 in [73].

Remark 7.5 It is possible to consider the model with K change points in the drift of the observed process

$$\mathrm{d}X_t = \sum_{k=1}^K \left[g_{k,t} \mathbb{I}_{\{t<\theta_k\}} + h_{k,t} \mathbb{I}_{\{t\geq\theta_k\}}\right] Y_t\, \mathrm{d}t + \varepsilon\sigma(t)\, \mathrm{d}W_t, \quad X_0 = 0, \quad 0 \leq t \leq T,$$

$$\mathrm{d}Y_t = a(t)Y_t\,\mathrm{d}t + \varepsilon b(t)\,\mathrm{d}V_t, \qquad\qquad Y_0 = y_0 \neq 0.$$

Here $\vartheta = (\theta_1, \ldots, \theta_K) \in \Theta = (\alpha_1, \beta_1) \times \ldots \times (\alpha_K, \beta_K)$, $\theta_k \in (\alpha_k, \beta_k)$, $0 < \alpha_1, \beta_k < \alpha_{k+1}, \beta_K < T, k = 1, \ldots, K$ and

$$\inf_{\alpha_k < t \leq \beta_k} \left|h_{k,t} - g_{k,t}\right| > 0, \qquad k = 1, \ldots, K.$$

Introduce notation:

$$\vartheta = (\theta_1, \ldots, \theta_K), \quad \vartheta_0 = \left(\theta_{0,1}, \ldots, \theta_{0,K}\right), \quad \rho_k\left(\theta_{0,k}\right) = \frac{\left[h_{k,\theta_{0,k}} - g_{k,\theta_{0,k}}\right] y_{\theta_{0,k}}}{\sigma\left(\theta_k\right)},$$

$$Z_k\left(\theta_k, u_k\right) = \exp\left(\rho_k\left(\theta_k\right) W_k\left(u_k\right) - \frac{|u_k|}{2}\rho_k\left(\theta_k\right)^2\right), \qquad u_k \in \mathcal{R},$$

$$Z_k\left(\theta_k, \hat{u}_k\left(\theta_k\right)\right) = \sup_{u_k \in \mathcal{R}} Z_k\left(\theta_k, u_k\right), \qquad \tilde{u}_k\left(\theta_k\right) = \frac{\int_{\mathcal{R}} u Z_k\left(\theta_k, u\right) \mathrm{d}u}{\int_{\mathcal{R}} Z_k\left(\theta_k, u\right) \mathrm{d}u},$$

$$\hat{\mathrm{u}}\left(\vartheta\right) = \left(\hat{u}_1\left(\theta_1\right), \ldots, \hat{u}_K\left(\theta_K\right)\right), \qquad \tilde{\mathrm{u}}\left(\vartheta\right) = \left(\tilde{u}_1\left(\theta_1\right), \ldots, \tilde{u}_K\left(\theta_K\right)\right).$$

Here $W_1(\cdot), \ldots, W_k(\cdot)$ are independent two-sided Wiener processes.

It can be shown that the MLE $\hat{\vartheta}_\varepsilon$ and BE $\tilde{\vartheta}_\varepsilon$ are consistent, converge in distribution to the limit laws

$$\varepsilon^{-2}\left(\hat{\vartheta}_\varepsilon - \vartheta_0\right) \Longrightarrow \hat{\mathrm{u}}\left(\vartheta_0\right), \qquad \varepsilon^{-2}\left(\tilde{\vartheta}_\varepsilon - \vartheta_0\right) \Longrightarrow \tilde{\mathrm{u}}\left(\vartheta_0\right),$$

and the moments converge. The proof is similar to the proof of Theorem 7.9. For example, the normalized likelihood ratio

$$Z_\varepsilon\left(\vartheta_0, \mathrm{u}\right) = \frac{L\left(\vartheta_0 + \varepsilon^2 \mathrm{u}, X^T\right)}{L\left(\vartheta_0, X^T\right)}, \quad \mathrm{u} = (u_1, \ldots, u_K) \in \mathbb{U}_\varepsilon = \left(\mathrm{u}\ :\ \vartheta_0 + \varepsilon^2 \mathrm{u} \in \Theta\right)$$

converges to the limit

$$Z_\varepsilon\left(\vartheta_0, \mathrm{u}\right) \Longrightarrow Z\left(\vartheta_0, \mathrm{u}\right) = \prod_{k=1}^{K} Z_k\left(\theta_{0,k}, u_k\right), \qquad \mathrm{u} \in \mathcal{R}^K.$$

Remark that the random variables $\hat{u}_1\left(\theta_{0,1}\right), \ldots, \hat{u}_K\left(\theta_{0,K}\right)$ are independent and the random variables $\tilde{u}_1\left(\theta_{0,1}\right), \ldots, \tilde{u}_K\left(\theta_{0,K}\right)$ are independent too.

Adaptive Filter

Recall that on the time interval $[0, \vartheta_0]$ the random process $m(\vartheta_0, t)$ does not depend on ϑ_0 (see (7.86)). Therefore the construction of the approximation of $m(\vartheta_0, t), 0 \leq t \leq T$ has to be done for the values $m(\vartheta_0, t), \vartheta_0 \leq t \leq T$ only. The adaptive filter is done in two steps. First we estimate the parameter ϑ by observations $X_t, \alpha \leq t \leq \beta$ using Bayesian approach and obtain the Bayesian estimator $\tilde{\vartheta}_\varepsilon$. Then we have to substitute this estimator in Eq. (7.87), but we cannot use this equation directly because the stochastic Itô integral

$$\int_{\tilde{\vartheta}_\varepsilon}^{t} \frac{h_s \gamma_*(\tilde{\vartheta}_\varepsilon, s)}{\sigma(s)^2} \mathrm{d}X_s$$

is not defined.

Remark that as it follows from the proof of Theorem 7.9, all information concerning the unknown parameter is localized at the ε^2-vicinity of the true value. Moreover the derivative of $f(\vartheta, t)\, m(\vartheta, t)$ in ϑ does not exist. Therefore the One-step construction is not appropriate and for adaptive filter we need a different approach.

Introduce the condition

$\mathscr{Q}_5$. *The functions $h_t, g_t, \sigma(t), \alpha \leq t \leq \beta$ have continuous derivatives $h'_t, g'_t, \sigma'_t(t), \alpha \leq t \leq \beta$.*

The off-line construction is studied under the conditions $\mathscr{Q}_{1-5}$.

Equations (7.86) and (7.87) we rewrite as follows:

$$\mathrm{d}m(t) = A(t)\, m(t)\, \mathrm{d}t + B(t)\, \mathrm{d}X_t, \quad m(0) = y_0, \quad 0 \leq t \leq \vartheta,$$

$$\mathrm{d}m(\vartheta, t) = A(\vartheta, t)\, m(\vartheta, t)\, \mathrm{d}t + B(\vartheta, t)\, \mathrm{d}X_t, \quad m(\vartheta, \vartheta) = m(\vartheta), \quad \vartheta \leq t \leq T,$$

where

$$A(t) = a(t) - \frac{g_t^2 \gamma_*(t)}{\sigma(t)^2}, \qquad B(t) = \frac{g_t \gamma_*(t)}{\sigma(t)^2},$$

$$A(\vartheta, t) = a(t) - \frac{h_t^2 \gamma_*(\vartheta, t)}{\sigma(t)^2}, \qquad B(\vartheta, t) = \frac{h_t \gamma_*(\vartheta, t)}{\sigma(t)^2}.$$

We have

$$m(t) = y_0 \Phi(0, t) + \int_0^t \Phi(s, t)\, B(s)\, \mathrm{d}X_s, \qquad 0 \leq t \leq \vartheta, \tag{7.98}$$

$$m(\vartheta, t) = m(\vartheta)\, \Phi(\vartheta, \vartheta, t) + \int_\vartheta^t \Phi(\vartheta, s, t)\, B(\vartheta, s)\, \mathrm{d}X_s$$

$$= m(\vartheta)\,\Phi(\vartheta,\vartheta,t) + \Phi(\vartheta,\vartheta,t)\int_{\vartheta}^{t} \Phi(\vartheta,\vartheta,s)^{-1} B(\vartheta,s)\,\mathrm{d}X_s, \quad \vartheta \le t \le T. \tag{7.99}$$

Here

$$\Phi(s,t) = \exp\left(\int_s^t A(r)\,\mathrm{d}r\right) = \exp\left(\int_s^t [a(r) - g_r B(r)]\,\mathrm{d}r\right),$$

$$\Phi(\vartheta,s,t) = \exp\left(\int_s^t A(\vartheta,r)\,\mathrm{d}r\right) = \exp\left(\int_s^t [a(r) - h_r B(\vartheta,r)]\,\mathrm{d}r\right).$$

For the stochastic integrals in (7.98) and (7.99) we have the representations (see Lemma 1.7): if $t_0 < t$

$$\int_{t_0}^{t} \Phi(t_0,s)^{-1} B(s)\,\mathrm{d}X_s = \int_{t_0}^{t} \Phi(t_0,s)^{-1} \frac{g_s\gamma_*(s)}{\sigma(s)^2}\,\mathrm{d}X_s = \int_{t_0}^{t} Q(s)\;\mathrm{d}X_s$$

$$= Q(t)\,X_t - Q(t_0)\,X_{t_0} - \int_{t_0}^{t} X_s Q_s'(s)\,\mathrm{d}s \equiv \Psi(t_0,t,X),$$

$$\int_{t_0}^{t} \Phi(\vartheta,t_0,s)^{-1} B(\vartheta,s)\,\mathrm{d}X_s = \int_{t_0}^{t} \Phi(\vartheta,t_0,s)^{-1} \frac{h_s\gamma_*(\vartheta,s)}{\sigma(s)^2}\,\mathrm{d}X_s$$

$$= \int_{t_0}^{t} Q(\vartheta,s)\;\mathrm{d}X_s$$

$$= Q(\vartheta,t)\,X_t - Q(\vartheta,t_0)\,X_{t_0} - \int_{t_0}^{t} X_s Q_s'(\vartheta,s)\,\mathrm{d}s \equiv \Psi(\vartheta,t_0,t,X).$$

Hence we have the following representations for $t_0 < \vartheta$ and $t_0 > \vartheta$ respectively

$$m(t) = m(t_0)\,\Phi(t_0,t) + X_t\frac{g_t\gamma_*(t)}{\sigma(t)^2} - X_{t_0}\frac{g_{t_0}\gamma_*(t_0)}{\sigma(t_0)^2} - \Phi(t_0,t)\int_{t_0}^{t} X_s Q(s)_s'\,\mathrm{d}s,$$

$$m(\vartheta,t) = m(\vartheta,t_0)\,\Phi(\vartheta,t_0,t) + X_t\frac{h_t\gamma_*(\vartheta,t)}{\sigma(t)^2} - X_{t_0}\frac{h_{t_0}\gamma_*(\vartheta,t_0)}{\sigma(t_0)^2} - \Phi(\vartheta,t_0,t)\int_{t_0}^{t} X_s\,Q_s'(\vartheta,s)\,\mathrm{d}s.$$

Remark that the functions

$$Q(s) = \Phi(t_0,s)^{-1}\frac{g_s\gamma_*(s)}{\sigma(s)^2}, \qquad Q(\vartheta,s) = \Phi(\vartheta,t_0,s)^{-1}\frac{h_s\gamma_*(\vartheta,s)}{\sigma(s)^2}$$

have bounded derivatives on s.

Introduce the adaptive filter $m_t^\star, 0 \le t \le T$ by the following equations.

1. *On the time interval* $[0, \tilde\vartheta_\varepsilon]$ *the equations are*

$$\mathrm{d}m_t^\star = a(t)\, m_t^\star \mathrm{d}t + \frac{g_t\, \gamma_*(t)}{\sigma(t)^2}\left[\mathrm{d}X_t - g_t\, m_t^\star \mathrm{d}t\right], \qquad m_0^\star = y_0,$$

$$\frac{\mathrm{d}\gamma_*(t)}{\mathrm{d}t} = 2a(t)\,\gamma_*(t) - \frac{g_t^2\,\gamma_*(t)^2}{\sigma(t)^2} + b(t)^2, \qquad \gamma_*(0) = 0.$$

2. *On the time interval* $[\tilde\vartheta_\varepsilon, \beta]$ *the adaptive filter is*

$$m_t^\star = m_{\tilde\vartheta_\varepsilon}^\star\, \Phi(\tilde\vartheta_\varepsilon, \tilde\vartheta_\varepsilon, t) + X_t\, \frac{h_t \gamma_*^\star(\tilde\vartheta_\varepsilon, t)}{\sigma(t)^2} - X_{\tilde\vartheta_\varepsilon}\, \frac{h_{\tilde\vartheta_\varepsilon}\gamma_*^\star(\tilde\vartheta_\varepsilon)}{\sigma(\tilde\vartheta_\varepsilon)^2}$$
$$- \Phi(\tilde\vartheta_\varepsilon, \tilde\vartheta_\varepsilon, t)\int_{\tilde\vartheta_\varepsilon}^t X_s\, Q_s'(\tilde\vartheta_\varepsilon, s)\,\mathrm{d}s.$$

3. *On the time interval* $[\beta, T]$ *the adaptive filter is*

$$\mathrm{d}m_t^\star = a(t)\, m_t^\star \mathrm{d}t + \frac{h_t\, \gamma_*(\tilde\vartheta_\varepsilon, t)}{\sigma(t)^2}\left[\mathrm{d}X_t - h_t\, m_t^\star \mathrm{d}t\right], \qquad m_\beta^\star,$$

$$\frac{\mathrm{d}\gamma_*(\tilde\vartheta_\varepsilon, t)}{\mathrm{d}t} = 2a(t)\,\gamma_*^\star(\tilde\vartheta_\varepsilon, t) - \frac{h_t^2\,\gamma_*(\tilde\vartheta_\varepsilon, t)^2}{\sigma(t)^2} + b(t)^2, \qquad \gamma_*(\tilde\vartheta_\varepsilon, \beta).$$

The Error of Approximation Consider the error of approximation $\delta_m(t) = m_t^\star - m(\vartheta_0, t)$.

- For $t \in [0, \alpha]$ we have obviously $\delta_m(t) = 0$.
- Introduce the interval $[\vartheta_0 \wedge \tilde\vartheta_\varepsilon, \vartheta_0 \vee \tilde\vartheta_\varepsilon]$. We have $\delta_m(t) = 0, \alpha \le t \le \vartheta_0 \wedge \tilde\vartheta_\varepsilon$. There are two possibilities: $\vartheta_0 \le \tilde\vartheta_\varepsilon$ and $\vartheta_0 \ge \tilde\vartheta_\varepsilon$. If $\vartheta_0 \le \tilde\vartheta_\varepsilon$, then $[\vartheta_0 \wedge \tilde\vartheta_\varepsilon, \vartheta_0 \vee \tilde\vartheta_\varepsilon] = [\vartheta_0, \tilde\vartheta_\varepsilon]$ and for $t \in \left[\vartheta_0, \tilde\vartheta_\varepsilon\right]$ we have

$$\delta_m(t) = m(\vartheta_0)\left[\Phi(\vartheta_0, t) - \Phi(\vartheta_0, \vartheta_0, t)\right] + X_t \frac{\left[g_t\gamma_*(t) - h_t\gamma_*(\vartheta_0, t)\right]}{\sigma(t)^2}$$
$$+ X_{\vartheta_0}\frac{\left[h_{\vartheta_0} - g_{\vartheta_0}\right]\gamma_*(\vartheta_0)}{\sigma(\vartheta_0)^2} - \Phi(\vartheta_0, \vartheta_0, t)\int_{\vartheta_0}^t X_s\, Q(\vartheta_0, s)_s'\,\mathrm{d}s,$$
$$+ \Phi(\vartheta_0, t)\int_{\vartheta_0}^t X_s\, Q_s'(s)\,\mathrm{d}s$$
$$= m(\vartheta_0)\left[\Phi(\vartheta_0, t) - \Phi(\vartheta_0, \vartheta_0, t)\right]$$

$$
\begin{aligned}
&+\left[X_t - X_{\vartheta_0}\right]\frac{\left[g_t\gamma_*\left(t\right)-h_t\gamma_*\left(\vartheta_0,t\right)\right]}{\sigma\left(t\right)^2}\\
&+X_{\vartheta_0}\left[\frac{h_{\vartheta_0}\gamma_*\left(\vartheta_0\right)}{\sigma\left(\vartheta_0\right)^2}-\frac{h_t\gamma_*\left(\vartheta_0,t\right)^2}{\sigma\left(t\right)}+\frac{g_t\gamma_*\left(t\right)}{\sigma\left(t\right)^2}-\frac{g_{\vartheta_0}\gamma_*\left(\vartheta_0\right)}{\sigma\left(\vartheta_0\right)^2}\right]\\
&+\Phi\left(\vartheta_0,t\right)\int_{\vartheta_0}^{t}X_s\,Q_s'\left(s\right)\,\mathrm{d}s-\Phi\left(\vartheta_0,\vartheta_0,t\right)\int_{\vartheta_0}^{t}X_s\,Q\left(\vartheta_0,s\right)_s'\,\mathrm{d}s\\
&=\delta_1\left(t\right)+\delta_2\left(t\right)+\delta_3\left(t\right)+\delta_4\left(t\right)
\end{aligned}
$$

with obvious notations. Remark that $\delta_1(t), \delta_3(t)$, and $\delta_4(t)$ are continuously differentiable on t. Therefore

$$
\mathbf{E}_{\vartheta_0}\sup_{\vartheta_0\leq t\leq\tilde\vartheta_\varepsilon}\delta_k\left(t\right)^2\mathbb{I}_{\left\{\vartheta_0<\tilde\vartheta_\varepsilon\right\}}\leq C\mathbf{E}_{\vartheta_0}\left|\tilde\vartheta_\varepsilon-\vartheta_0\right|^2\leq C\,\varepsilon^4,\qquad k=1,3,4. \tag{7.100}
$$

Further,

$$
\mathbf{E}_{\vartheta_0}\sup_{\vartheta_0\leq t\leq\tilde\vartheta_\varepsilon}\delta_2\left(t\right)^2\mathbb{I}_{\left\{\vartheta_0<\tilde\vartheta_\varepsilon\right\}}\leq C\,\mathbf{E}_{\vartheta_0}\sup_{\vartheta_0\leq t\leq\tilde\vartheta_\varepsilon}\left[X_t-X_{\vartheta_0}\right]^2\mathbb{I}_{\left\{\vartheta_0<\tilde\vartheta_\varepsilon\right\}}.
$$

We have

$$
\begin{aligned}
\mathbf{E}_{\vartheta_0}\left[X_t-X_{\vartheta_0}\right]^2&\leq 2\mathbf{E}_{\vartheta_0}\left(\int_{\vartheta_0}^{t}h_s m\left(s\right)\mathrm{d}s\right)^2+2\varepsilon^2\mathbf{E}_{\vartheta_0}\left(\int_{\vartheta_0}^{t}\sigma\left(s\right)\mathrm{d}\bar W_s\right)^2\\
&\leq C\left(t-\vartheta_0\right)^2+C\varepsilon^2\left(t-\vartheta_0\right).
\end{aligned} \tag{7.101}
$$

The next step is not justified, and therefore the proof given here needs further precision. Using (7.101) we write

$$
\begin{aligned}
\mathbf{E}_{\vartheta_0}\sup_{\vartheta_0\leq t\leq\tilde\vartheta_\varepsilon}\delta_2\left(t\right)^2\mathbb{I}_{\left\{\vartheta_0<\tilde\vartheta_\varepsilon\right\}}&\leq C\mathbf{E}_{\vartheta_0}(\tilde\vartheta_\varepsilon-\vartheta_0)^2+C\varepsilon^2\mathbf{E}_{\vartheta_0}|\tilde\vartheta_\varepsilon-\vartheta_0|\\
&\leq C\,\varepsilon^4.
\end{aligned} \tag{7.102}
$$

Now the rate of approximation $\mathbf{E}_{\vartheta_0}\sup_{\vartheta_0\leq t\leq\tilde\vartheta_\varepsilon}\delta_m\left(t\right)^2\mathbb{I}_{\left\{\vartheta_0<\tilde\vartheta_\varepsilon\right\}}\leq C\varepsilon^4$ follows from (7.100) and (7.102). The case $\tilde\vartheta_\varepsilon<\vartheta_0$ can be studied similarly, and the error is of the same order. This is probably the correct rate of approximation, but as we mentioned above the proof has to be completed.

- For the values $\vartheta_0\vee\tilde\vartheta_\varepsilon\leq t\leq\beta$ the equations for $m_t^\star$ and $m\left(\vartheta_0,t\right)$ are both "after the jump" and therefore are smooth functions of $\tilde\vartheta_\varepsilon$. The expansion at the vicinity of ϑ_0 and the properties of the BE $\tilde\vartheta_\varepsilon$ give us the same rate of approximation ε^4.

- For the values $t \in [\beta, T]$ there is no problem with the stochastic integral because the estimator $\tilde\vartheta_\varepsilon$ is $\mathfrak{F}^X_\beta$ measurable. There always is a problem with the estimation of the initial error $m^\star_\beta - m\left(\vartheta_0, \beta\right)$. The expansion at the vicinity of ϑ_0 and the properties of the BE $\tilde\vartheta_\varepsilon$ give us once more the rate ε^4.

7.4.2 Change Point in the State Equation

Model of Observation

We have the partially observed linear system

$$\begin{aligned} \mathrm{d}X_t &= f\left(t\right) Y_t \,\mathrm{d}t + \varepsilon\sigma\left(t\right) \mathrm{d}W_t, \qquad X_0 = 0, \qquad 0 \le t \le T, \\ \mathrm{d}Y_t &= \left[g_t \mathbb{I}_{\{t<\vartheta\}} + h_t \mathbb{I}_{\{t\ge\vartheta\}}\right] Y_t \,\mathrm{d}t + \varepsilon\, b(t) \,\mathrm{d}V_t, \qquad Y_0 = y_0 \ne 0. \end{aligned}$$

The conditions are almost the same as above.

Conditions $\tilde{\mathscr{Q}}$.

$\mathscr{Q}_1$. *Parameter $\vartheta \in \Theta = (\alpha, \beta)\,, 0 < \alpha < \beta < T$.*
$\tilde{\mathscr{Q}}_2$. *The functions $g_t, h_t, \sigma(t), f(t), b(t), 0 \le t \le T$ are bounded.*
$\tilde{\mathscr{Q}}_3$. The functions $\sigma\left(t\right)^2 > c_\sigma > 0$ and $f\left(t\right) > c_f > 0$.
$\mathscr{Q}_4$. *The functions g_t, h_t are continuous and* $\inf_{\alpha<t<\beta} \left|g_t - h_t\right| > 0$.

Let us denote $a\left(\vartheta, t\right) = g_t \mathbb{I}_{\{t<\vartheta\}} + h_t \mathbb{I}_{\{t\ge\vartheta\}}$. Then the Kalman-Bucy filter has two different forms: on the time interval $[0, \vartheta]$ the conditional expectation $m\left(\vartheta, t\right) = m\left(t\right)$, i.e., does not depend on ϑ and is solution of the equation

$$\mathrm{d}m\left(t\right) = g_t m\left(t\right) \mathrm{d}t + \frac{f(t)\gamma_*\left(t\right)}{\sigma\left(t\right)^2}\left[\mathrm{d}X_t - f(t) m\left(t\right) \mathrm{d}t\right], \quad m\left(0\right) = y_0, \tag{7.103}$$

$$\frac{\mathrm{d}\gamma_*\left(t\right)}{\mathrm{d}t} = 2g_t\gamma_*\left(t\right) - \frac{f(t)^2\gamma_*\left(t\right)^2}{\sigma\left(t\right)^2} + b(t)^2, \quad \gamma_*\left(0\right) = 0$$

and on the time interval $[\vartheta, T]$ the filter is

$$\mathrm{d}m\left(\vartheta, t\right) = h_t m\left(\vartheta, t\right) \mathrm{d}t + \frac{f(t)\gamma_*\left(\vartheta, t\right)}{\sigma\left(t\right)^2}\left[\mathrm{d}X_t - f(t) m\left(\vartheta, t\right) \mathrm{d}t\right], \tag{7.104}$$

$$\frac{\mathrm{d}\gamma_*\left(\vartheta, t\right)}{\mathrm{d}t} = 2h_t\gamma_*\left(\vartheta, t\right) - \frac{f(t)^2\gamma_*\left(\vartheta, t\right)^2}{\sigma\left(t\right)^2} + b(t)^2.$$

The initial values are $m\left(\vartheta, \vartheta\right) = m\left(\vartheta\right)$ and $\gamma_*\left(\vartheta, \vartheta\right) = \gamma_*\left(\vartheta\right)$.

The limit system is

$$
\begin{aligned}
x_t(\vartheta_0) &= \int_0^t f(s)\, y_s(\vartheta_0)\,\mathrm{d}s, \qquad y_t(\vartheta_0) = y_0 \exp\left(\int_0^t a(\vartheta_0, s)\,\mathrm{d}s\right)\\
&= y_0\, \phi(\vartheta_0, 0, t),\\
\frac{\mathrm{d}y_t(\vartheta,\vartheta_0)}{\mathrm{d}t} &= A(\vartheta,t)\, y_t(\vartheta,\vartheta_0) + B(\vartheta,t)\, y_t(\vartheta_0), \qquad y_0(\vartheta,\vartheta_0) = y_0,\\
A(\vartheta,t) &= \left[g(t) - \frac{f(t)^2\gamma_*(t)}{\sigma(t)^2}\right]\mathbb{1}_{\{0\le t<\vartheta\}} + \left[h(t) - \frac{f(t)^2\gamma_*(\vartheta,t)}{\sigma(t)^2}\right]\mathbb{1}_{\{\vartheta\le t\le T\}}\\
&= A_g(t)\,\mathbb{1}_{\{0\le t<\vartheta\}} + A_h(\vartheta,t)\,\mathbb{1}_{\{\vartheta\le t\le T\}},\\
B(\vartheta,t) &= \frac{f(t)^2}{\sigma(t)^2}\left[\gamma_*(t)\,\mathbb{1}_{\{0\le t<\vartheta\}} + \gamma_*(\vartheta,t)\,\mathbb{1}_{\{\vartheta\le t\le T\}}\right]\\
&= B_g(t)\,\mathbb{1}_{\{0\le t<\vartheta\}} + B_h(\vartheta,t)\,\mathbb{1}_{\{\vartheta\le t\le T\}}.
\end{aligned}
$$

Here $y_t(\vartheta,\vartheta_0) = m(\vartheta,t)|_{\varepsilon=0}$. Of course, $X_t \longrightarrow x_t(\vartheta_0)$. To describe $y_t(\vartheta,\vartheta_0)$ we need further notations

$$
\phi_1(s,t) = \exp\left(\int_s^t g_r\,\mathrm{d}r\right), \quad \Phi_1(s,t) = \exp\left(\int_s^t \left[g_r - \frac{f(r)^2\gamma_*(r)}{\sigma(r)^2}\right]\mathrm{d}r\right),
$$

$$
\phi_2(s,t) = \exp\left(\int_s^t h_r\,\mathrm{d}r\right), \quad \Phi_2(\vartheta,s,t) = \exp\left(\int_s^t \left[h_r - \frac{f(r)^2\gamma_*(\vartheta,r)}{\sigma(r)^2}\right]\mathrm{d}r\right).
$$

We have the representations

$$
y_t(\vartheta_0) = y_0\phi_1(0,t)\,\mathbb{1}_{\{0\le t<\vartheta_0\}} + y_0\phi_1(0,\vartheta_0)\,\phi_2(\vartheta_0,t)\,\mathbb{1}_{\{\vartheta_0<t\le T\}}, \tag{7.105}
$$

$$
\begin{aligned}
y_t(\vartheta,\vartheta_0) &= \left[y_0\Phi_1(0,t) + \int_0^t \Phi_1(s,t)\,\frac{f(s)^2\gamma_*(s)}{\sigma(s)^2}\,y_s(\vartheta_0)\,\mathrm{d}s\right]\mathbb{1}_{\{0\le t\le\vartheta\}}\\
&+\left[y_\vartheta(\vartheta,\vartheta_0)\,\Phi_2(\vartheta,\vartheta,t) + \int_\vartheta^t \Phi_2(\vartheta,s,t)\,\frac{f(s)^2\gamma_*(\vartheta,s)}{\sigma(s)^2}\,y_s(\vartheta_0)\,\mathrm{d}s\right]\mathbb{1}_{\{\vartheta_0<t\le T\}}.
\end{aligned} \tag{7.106}
$$

To write the derivative $\dot y_t(\vartheta,\vartheta_0) = \mathrm{d}y_t(\vartheta,\vartheta_0)/\mathrm{d}\vartheta$ we first note that $\dot y_t(\vartheta,\vartheta_0) = 0,\ 0\le t<\vartheta$ and for $\vartheta\le t\le T$ the function $y_t(\vartheta,\vartheta_0)$ has the representation

$$
\begin{aligned}
y_t(\vartheta,\vartheta_0) &= e^{\int_0^\vartheta A_g(r)\mathrm{d}r + \int_\vartheta^t A_h(\vartheta,r)\mathrm{d}r}\left[y_0 + \int_0^\vartheta e^{-\int_0^s A_g(r)\mathrm{d}r} B_g(s)\, y_s(\vartheta_0)\,\mathrm{d}s\right.\\
&\left. + e^{-\int_0^\vartheta A_g(r)\mathrm{d}r}\int_\vartheta^t e^{-\int_\vartheta^s A_h(\vartheta,r)\mathrm{d}r} B_h(\vartheta,s)\, y_s(\vartheta_0)\,\mathrm{d}s\right].
\end{aligned}
$$

The derivative of this expression is obtained by the direct derivation of this expression.

Parameter Estimators

MDE

Introduce the MDE

$$\check{\vartheta}_{\varepsilon}=\arg\inf_{\vartheta\in\Theta}\int_{\alpha}^{T}\left[X_t-x_t\left(\vartheta\right)\right]^2\mathrm{d}t.$$

Denote

$$\begin{aligned}\phi\left(\vartheta,s,t\right)&=\exp\left(\int_s^t\left[g_r\,1\!\!1_{\{r<\vartheta\}}+h_r\,1\!\!1_{\{r\geq\vartheta\}}\right]\mathrm{d}r\right),\\ T\left(\vartheta\right)&=\int_{\vartheta}^{T}\dot{x}_t\left(\vartheta\right)^2\mathrm{d}t=y_0^2\left[g_\vartheta-h_\vartheta\right]^2\int_{\vartheta}^{T}\left(\int_0^t f\left(s\right)\phi\left(\vartheta,0,s\right)\mathrm{d}s\right)^2\mathrm{d}t\\ &=\left[g_\vartheta-h_\vartheta\right]^2\int_{\vartheta}^{T}x_t\left(\vartheta\right)^2\mathrm{d}t,\\ x_t^{(1)}\left(\vartheta_0\right)&=y_0\int_0^t f\left(s\right)\int_0^s\phi\left(\vartheta_0,r,s\right)b\left(r\right)\mathrm{d}V_r\,\mathrm{d}s+\int_0^t\sigma\left(s\right)\mathrm{d}W_s\\ &=y_0\int_0^t b\left(r\right)\int_r^t\phi\left(\vartheta_0,r,s\right)f\left(s\right)\mathrm{d}s\,\mathrm{d}V_r+\int_0^t\sigma\left(s\right)\mathrm{d}W_s\end{aligned}$$

and

$$\check{\zeta}\left(\vartheta_0\right)=T\left(\vartheta_0\right)^{-1}\int_{\vartheta_0}^{T}x_t^{(1)}\left(\vartheta_0\right)\dot{x}_t\left(\vartheta_0\right)\mathrm{d}t\sim\mathcal{N}\left(0,\check{D}\left(\vartheta_0\right)\right)$$

with corresponding $D\left(\vartheta_0\right)$.

Proposition 7.6 *Let the conditions $\tilde{\mathscr{Q}}$ hold. Then the MDE is consistent, asymptotically normal*

$$\varepsilon^{-1}\left(\check{\vartheta}_{\varepsilon}-\vartheta_0\right)\Longrightarrow\check{\zeta}\left(\vartheta_0\right)\sim\mathcal{N}\left(0,\check{D}\left(\vartheta_0\right)\right)$$

and the moments converge: for any $p>0$

$$\lim_{\varepsilon\to0}\varepsilon^{-p}\mathbf{E}_{\vartheta_0}\left|\check{\vartheta}_{\varepsilon}-\vartheta_0\right|^p=\mathbf{E}_{\vartheta_0}\left|\check{\zeta}\left(\vartheta_0\right)\right|^p.$$

Proof First we check the condition of identifiability. Consider the integral

$$g(\vartheta_0, \nu) = \inf_{|\vartheta-\vartheta_0|\geq\nu} \int_\alpha^T [x_t(\vartheta_0) - x_t(\vartheta)]^2 \, dt.$$

Suppose that for some $\vartheta_1 \neq \vartheta_0$ we have

$$\int_\alpha^\tau [x_t(\vartheta_0) - x_t(\vartheta_1)]^2 \, dt = 0.$$

Then for all $t \in [\alpha, T]$

$$\int_\alpha^t f(s)\, y_s(\vartheta_0)\, ds \equiv \int_\alpha^t f(s)\, y_s(\vartheta_1)\, ds.$$

This means that $y_s(\vartheta_0) = y_s(\vartheta_1)$ for all $s \in [\alpha, T]$. Hence

$$g_t \mathbb{I}_{\{t<\vartheta_0\}} + h_t \mathbb{I}_{\{t\geq\vartheta_0\}} = g_t \mathbb{I}_{\{t<\vartheta_1\}} + h_t \mathbb{I}_{\{t\geq\vartheta_1\}}.$$

Of course this equality is possible iff $\vartheta_1 = \vartheta_0$. Therefore

$$g(\vartheta_0, \nu) > 0 \qquad \text{for} \quad \text{any} \qquad \nu > 0.$$

The proof of the consistency of MDE under condition (7.106) follows the proof presented in (2.10).

To prove the asymptotic normality we use the MDEq

$$\int_{\check\vartheta_{\tau,\varepsilon}}^T \left[X_t - x_t(\vartheta_0) - \left(\check\vartheta_{\tau,\varepsilon} - \vartheta_0\right) \dot x_t(\tilde\vartheta)\right] \dot x_t(\check\vartheta_\varepsilon) dt = 0.$$

For the Gaussian process $x_t^{(1)}(\vartheta_0) = \varepsilon^{-1}(X_t - x_t(\vartheta_0))$ we have the expression (7.105) and the derivative is

$$\dot x_t(\vartheta) = 0, \qquad\qquad t < \vartheta,$$

$$\dot x_t(\vartheta) = y_0 [g_\vartheta - h_\vartheta] \int_0^t f(s)\, \phi(\vartheta, 0, s)\, ds = [g_\vartheta - h_\vartheta]\, x_t(\vartheta), \qquad t \geq \vartheta.$$

Hence

$$\varepsilon^{-1}\left(\check\vartheta_\varepsilon - \vartheta_0\right) = \left(\int_{\check\vartheta_\varepsilon}^T \dot x_t(\tilde\vartheta) \dot x_t(\check\vartheta_\varepsilon) dt\right)^{-1} \int_{\check\vartheta_\varepsilon}^T \frac{X_t - x_t(\vartheta_0)}{\varepsilon} \dot x_t(\check\vartheta_\varepsilon) dt$$

$$\Longrightarrow T(\vartheta_0)^{-1} \int_{\vartheta_0}^T x_t^{(1)}(\vartheta_0)\, \dot x_t(\vartheta_0)\, dt.$$

The convergence of the moments is proved following the same calculations as it was done in the proof of Theorem 2.1. □

MLE and BE

The LR function is

$$L\left(\vartheta, X^T\right) = \exp\left(\int_\alpha^T \frac{f\,(t)\, m\,(\vartheta,t)}{\varepsilon^2\sigma\,(t)^2}\, \mathrm{d}X_t - \int_\alpha^T \frac{f\,(t)^2\, m\,(\vartheta,t)^2}{2\varepsilon^2\sigma\,(t)^2}\, \mathrm{d}t\right), \qquad \vartheta \in \Theta,$$

where $m\,(\vartheta,t)\,, 0 \le t \le T$ is given by (7.103) and (7.104). The MLE $\hat\vartheta_\varepsilon$ and BE $\tilde\vartheta_\varepsilon$ are defined by Eq. (2.33).

The Fisher information

$$\mathrm{I}\,(\vartheta_0) = \int_{\vartheta_0}^T \frac{f(t)^2\, \dot y_t\,(\vartheta_0,\vartheta_0)^2}{\sigma\,(t)^2}\, \mathrm{d}t$$

by conditions $\tilde{\mathscr{Q}}$ is positive, and the Hájek's lower bound on the mean squared risk of all estimators is

$$\lim_{\nu\to 0}\,\underline{\lim}_{\varepsilon\to 0}\,\sup_{|\vartheta-\vartheta_0|\le\nu} \varepsilon^{-2}\mathbf{E}_\vartheta\left(\bar\vartheta_\varepsilon - \vartheta\right)^2 \ge \mathrm{I}\,(\vartheta_0)^{-1}.$$

Theorem 7.10 *Let the conditions $\tilde{\mathscr{Q}}$ hold. Then the MLE $\hat\vartheta_\varepsilon$ and BE $\tilde\vartheta_\varepsilon$ are uniformly consistent, asymptotically normal*

$$\varepsilon^{-1}\left(\hat\vartheta_\varepsilon - \vartheta_0\right) \Longrightarrow \mathcal{N}\left(0, \mathrm{I}\,(\vartheta_0)^{-1}\right), \qquad \varepsilon^{-1}\left(\tilde\vartheta_\varepsilon - \vartheta_0\right) \Longrightarrow \mathcal{N}\left(0, \mathrm{I}\,(\vartheta_0)^{-1}\right),$$

the polynomial moments of these estimators uniformly on compacts converge and both estimators are asymptotically efficient.

Proof We check the conditions of Theorems 1.10 and 1.12. Introduce the normalized LR

$$Z_\varepsilon\,(\vartheta_0,u) = \frac{L\left(\vartheta_0+\varepsilon u, X^T\right)}{L\left(\vartheta_0, X^T\right)}, \qquad u \in \mathbb{U}_\varepsilon = \left(\frac{\alpha-\vartheta_0}{\varepsilon}, \frac{\beta-\vartheta_0}{\varepsilon}\right)$$

and study the difference $m\,(\vartheta_0+\varepsilon u,t) - m\,(\vartheta_0,t)$. From representations (7.104)–(7.106) we obtain for $u > 0$ the relation

$$m\,(\vartheta_0+\varepsilon u,t) - m\,(\vartheta_0,t) = \varepsilon u\, \dot y_t\,(\vartheta_0,\vartheta_0) + \varepsilon^{3/2}u^{1/2}O\,(1)\,. \tag{7.107}$$

Recall that (see (7.104))

$$\mathrm{d}m\left(\vartheta,t\right)=A\left(\vartheta,t\right)m\left(\vartheta,t\right)\mathrm{d}t+\frac{f(t)^{2}\gamma_{*}\left(\vartheta,t\right)}{\sigma(t)^{2}}m\left(\vartheta_{0},t\right)\mathrm{d}t+\varepsilon\frac{f(t)\gamma_{*}\left(\vartheta,t\right)}{\sigma(t)}\,\mathrm{d}\bar{W}_{t},$$

The term $\varepsilon^{3/2}u^{1/2}O\left(1\right)$ appears from the integral like

$$\varepsilon\int_{\vartheta_0}^{\vartheta_0+\varepsilon u}\Phi_2\left(\vartheta,s,\vartheta_0+\varepsilon u\right)\frac{f\left(s\right)\gamma_{*}\left(s\right)}{\sigma\left(s\right)}\mathrm{d}\bar{W}_s.$$

Therefore

$$\int_0^T f\left(s\right)\frac{\left[m\left(\vartheta_0+\varepsilon u,t\right)-m\left(\vartheta_0,t\right)\right]}{\varepsilon\sigma\left(t\right)}\,\mathrm{d}\bar{W}_t=u\Delta_{\varepsilon}\left(\vartheta_0,X^T\right)+\varepsilon^{1/2}u^{1/2}O\left(1\right)$$

where

$$\begin{aligned}\Delta_{\varepsilon}\left(\vartheta_0,X^T\right)&=\int_{\vartheta_0}^{T}\frac{f(t)\dot{y}_t\left(\vartheta_0,\vartheta_0\right)}{\varepsilon\sigma(t)}\,\mathrm{d}\bar{W}_t+o\left(1\right)\Longrightarrow\int_{\vartheta_0}^{T}\frac{f(t)\dot{y}_t\left(\vartheta_0,\vartheta_0\right)}{\sigma(t)}\,\mathrm{d}W_t\\&\equiv\Delta\left(\vartheta_0\right).\end{aligned}$$

This implies LAN

$$Z_{\varepsilon}\left(\vartheta_0,u\right)\Longrightarrow Z\left(\vartheta_0,u\right)=\exp\left(u\Delta\left(\vartheta_0\right)-\frac{u^2}{2}\mathrm{I}\left(\vartheta_0\right)\right),\qquad u\in\mathcal{R},$$

where $\Delta\left(\vartheta_0\right)\sim\mathcal{N}\left(0,\mathrm{I}\left(\vartheta_0\right)\right)$. Hence we obtain the convergence of the finite-dimensional distributions of $Z_{\varepsilon}\left(\vartheta_0,\cdot\right)$.

Lemma 1.1 and representation (7.107) allow us to prove the estimate

$$\mathbf{E}_{\vartheta_0}\left|Z_{\varepsilon}\left(\vartheta_0,u_2\right)^{1/4}-Z_{\varepsilon}\left(\vartheta_0,u_1\right)^{1/4}\right|^4\leq C\left|u_2-u_1\right|^8+C\varepsilon^2\left|u_2-u_1\right|^2.\tag{7.108}$$

Further, consider the integral

$$g\left(\vartheta_0,\nu\right)=\inf_{\left|\vartheta-\vartheta_0\right|\geq\nu}\int_{\vartheta_0}^{T}f\left(s\right)^2\frac{\left[y_t\left(\vartheta,\vartheta_0\right)-y_t\left(\vartheta_0\right)\right]^2}{\sigma\left(t\right)^2}\,\mathrm{d}t$$

and suppose that for some $\nu>0$ we have $g\left(\vartheta_0,\nu\right)=0$. This means that there exists a value $\vartheta_1\neq\vartheta_0$ such that $y_t\left(\vartheta_1,\vartheta_0\right)\equiv y_t\left(\vartheta_0\right)$ for all $t\in\left[\vartheta_0,T\right]$. The comparison of (7.105) and (7.106) shows that this identity is impossible. Therefore $g\left(\vartheta_0,\nu\right)>0$ and the proof of Lemma 2.5 of the estimate: for any $N>0$ there exists a constant $C=C_N$ such that

$$\mathbf{E}_{\vartheta_0} Z_\varepsilon (\vartheta_0, u)^{1/2} \leq \frac{C}{|u|^N} \tag{7.109}$$

holds.

Now having LAN and estimates (7.108) and (7.109) by Theorems 1.10 and 1.12 we obtain all mentioned in Theorem 7.10 properties of estimators. □

Remark 7.6 It is possible to consider the model with K change points in the drift of the hidden component

$$\mathrm{d}X_t = f(t)\, Y_t\, \mathrm{d}t + \varepsilon\sigma(t)\, \mathrm{d}W_t, \qquad X_0 = 0, \quad 0 \leq t \leq T,$$

$$\mathrm{d}Y_t = \sum_{k=1}^{K} \left[g_{k,t} \mathbb{I}_{\{t<\theta_k\}} + h_{k,t} \mathbb{I}_{\{t\geq\theta_k\}} \right] Y_t\, \mathrm{d}t + \varepsilon b(t)\, \mathrm{d}V_t, \qquad Y_0 = y_0 \neq 0.$$

Here $\theta_k \in (\alpha_k, \beta_k)$, $\quad 0 < \alpha_1, \beta_k < \alpha_{k+1}, \beta_K < T, k = 1, \ldots, K$ and

$$\inf_{\alpha_k < t \leq \beta_k} \left| h_{k,t} - g_{k,t} \right| > 0, \qquad k = 1, \ldots, K.$$

It can be shown in this case too that the MLE $\hat{\vartheta}_\varepsilon$ and BE $\tilde{\vartheta}_\varepsilon$ are consistent, asymptotically normal

$$\varepsilon^{-1}\left(\hat{\vartheta}_\varepsilon - \vartheta_0\right) \Longrightarrow \left(0, \mathbf{I}(\vartheta_0)^{-1}\right), \qquad \varepsilon^{-1}\left(\tilde{\vartheta}_\varepsilon - \vartheta_0\right) \Longrightarrow \left(0, \mathbf{I}(\vartheta_0)^{-1}\right)$$

and their moments converge. Here $\mathbf{I}(\vartheta_0)$ is the Fisher information matrix.

Adaptive Filter

The construction of the adaptive *off-line* filter is similar to that of the adaptive filter in the case of change point in the observation equation. Suppose that conditions $\tilde{\mathscr{D}}$ hold and we already have the MLE $\hat{\vartheta}_\varepsilon$, which by Theorem 7.10 is consistent and asymptotically normal. The adaptive filter is similar to the one proposed above in the case of change point in observation equation. On the interval $[0, \alpha]$ we put

$$\mathrm{d}m_t^\star = \left[g_t - \frac{f(t)^2 \gamma_*(t)}{\sigma(t)^2} \right] m_t^\star \mathrm{d}t + \frac{f(t)\, \gamma_*(t)}{\sigma(t)^2} \mathrm{d}X_t, \qquad m_0^\star = y_0, \tag{7.110}$$

$$\frac{\mathrm{d}\gamma_*(t)}{\mathrm{d}t} = 2g_t \gamma_*(t) - \frac{f(t)^2\, \gamma_*(t)^2}{\sigma(t)^2} + b(t)^2, \qquad \gamma_*(0) = 0. \tag{7.111}$$

Then using observations $X_s, \alpha \leq s \leq T$ we construct the MLE $\hat{\vartheta}_\varepsilon$ and on the time interval $[\alpha, \hat{\vartheta}_\varepsilon]$ we use (7.110) and (7.111) for calculation of $m_t^\star, \alpha < t \leq \hat{\vartheta}_\varepsilon$.

The solution of Eq. (7.104) can be written without stochastic integral as follows (see Lemma 1.7):

$$m(\vartheta, t) = m(\vartheta, \vartheta)\,\Phi_2(\vartheta, \vartheta, t) + \Phi_2(\vartheta, \vartheta, t)\int_\vartheta^t \Phi_2(\vartheta, \vartheta, s)^{-1}\frac{f(s)\,\gamma_*(\vartheta, s)}{\sigma(s)^2}\mathrm{d}X_s$$

$$= m(\vartheta, \vartheta)\,\Phi_2(\vartheta, \vartheta, t) + X_t\frac{f(t)\,\gamma_*(\vartheta, t)}{\sigma(t)^2} - X_\vartheta\frac{f(\vartheta)\,\gamma_*(\vartheta, \vartheta)}{\sigma(\vartheta)^2}$$

$$- \Phi_2(\vartheta, \vartheta, t)\int_\vartheta^t X_s Q_s'(\vartheta, s)\,\mathrm{d}s$$

where

$$Q(\vartheta, s) = \Phi_2(\vartheta, \vartheta, s)^{-1}\frac{f(s)\,\gamma_*(\vartheta, s)}{\sigma(s)^2}.$$

The adaptive filter on the time interval $[\hat{\vartheta}_\varepsilon, T]$ is given by the equations

$$\mathrm{d}m_t^\star = m_{\hat{\vartheta}_\varepsilon}^\star\,\Phi_2(\hat{\vartheta}_\varepsilon, \hat{\vartheta}_\varepsilon, t) + X_t\frac{f(t)\gamma_*(\hat{\vartheta}_\varepsilon, t)}{\sigma(t)^2} - X_{\hat{\vartheta}_\varepsilon}\frac{f(\hat{\vartheta}_\varepsilon)\gamma_*(\hat{\vartheta}_\varepsilon, \hat{\vartheta}_\varepsilon)}{\sigma(\hat{\vartheta}_\varepsilon)^2}$$

$$- \Phi_2(\hat{\vartheta}_\varepsilon, \hat{\vartheta}_\varepsilon, t)\int_{\hat{\vartheta}_\varepsilon}^t X_s Q_s'(\hat{\vartheta}_\varepsilon, s)\mathrm{d}s.$$

The Riccati equation on the time interval $[\hat{\vartheta}_\varepsilon, T]$ is

$$\frac{\mathrm{d}\gamma_*(\hat{\vartheta}_\varepsilon, t)}{\mathrm{d}t} = 2h_t\gamma_*(\hat{\vartheta}_\varepsilon, t) - \frac{f(t)^2\,\gamma_*(\hat{\vartheta}_\varepsilon, t)^2}{\sigma(t)^2} + b(t)^2, \qquad \gamma_*(\hat{\vartheta}_\varepsilon, \hat{\vartheta}_\varepsilon) = \gamma_*(\hat{\vartheta}_\varepsilon).$$

We suppose that error of approximation $\mathbf{E}_{\vartheta_0}\left(m_t^\star - m(\vartheta_0, t)\right)$ is of order ε^2, but the detailed proof has to be done.

7.4.3 Change Points in Both Equations

The partially observed system is

$$\mathrm{d}X_t = \left[q_t\,\mathbb{I}_{\{t<\theta_1\}} + r_t\,\mathbb{I}_{\{t\geq\theta_1\}}\right]Y_t\,\mathrm{d}t + \varepsilon\sigma(t)\,\mathrm{d}W_t, \qquad X_0 = 0, \qquad 0 \leq t \leq T,$$

$$\mathrm{d}Y_t = \left[g_t\,\mathbb{I}_{\{t<\theta_2\}} + h_t\,\mathbb{I}_{\{t\geq\theta_2\}}\right]Y_t\,\mathrm{d}t + \varepsilon\,b(t)\,\mathrm{d}V_t, \qquad Y_0 = y_0 \neq 0,$$

where the unknown parameter is $\vartheta = (\theta_1, \theta_2)$. We study this model under the following conditions.

Conditions $\hat{\mathscr{Q}}$.

$\hat{\mathscr{Q}}_1$. *The parameter is* $\vartheta \in \Theta = (\alpha_1, \beta_1) \times (\alpha_2, \beta_2)$, $0 < \alpha_1 < \beta_1 < T$, $0 < \alpha_2 < \beta_2 < T$.

$\tilde{\mathscr{Q}}_2$. *The functions* $q_t, r_t, g_t, h_t, \sigma(t), b(t), 0 \leq t \leq T$ *are continuous.*

$\hat{\mathscr{Q}}_3$. *The function* $\sigma(t)^2 > c_\sigma > 0$.

$\hat{\mathscr{Q}}_4$. *The functions* q_t, r_t, g_t, h_t are *positive and satisfy the condition*

$$\inf_{\alpha_1 \leq t \leq \beta_1} |r_t - q_t| > 0, \qquad \inf_{\alpha_2 \leq t \leq \beta_2} |h_t - g_t| > 0.$$

The main problem is the same: to construct an approximation of the unobserved component Y_t. As the conditional expectation $m(\vartheta, t) = \mathbf{E}_\vartheta \left(Y_t | \mathfrak{F}_t^X\right)$ is the optimal estimator of Y_t, the problem is reduced to the estimation of $m(\vartheta, t), 0 < t \leq T$.

The solution follows the same lines as in the preceding two change-point problems. Therefore we start with the estimation of the unknown parameter and then the adaptive filter is obtained by the substitution of these estimators in the equations of optimal filtration. The proofs are close to the given above.

Denote $f(\theta_1, t) = q_t \mathbb{I}_{\{t<\theta_1\}} + r_t \mathbb{I}_{\{t \geq \theta_1\}}$ and $a(\theta_2, t) = g_t \mathbb{I}_{\{t<\theta_2\}} + h_t \mathbb{I}_{\{t \geq \theta_2\}}$. If ϑ is known, then the function $m(\vartheta, t), 0 \leq t \leq T$ can be found as a solution of the equations of Kalman-Bucy filter

$$\mathrm{d}m(\vartheta, t) = a(\theta_2, t)\, m(\vartheta, t)\,\mathrm{d}t + \frac{f(\theta_1, t)\,\gamma_*(\vartheta, t)}{\sigma(t)^2}\left[\mathrm{d}X_t - f(\theta_1, t)\, m(\vartheta, t)\,\mathrm{d}t\right], \tag{7.112}$$

$$\frac{\mathrm{d}\gamma_*(\vartheta, t)}{\mathrm{d}t} = 2a(\theta_2, t)\,\gamma_*(\vartheta, t) - \frac{f(\theta_1, t)^2\,\gamma_*(\vartheta, t)^2}{\sigma(t)^2} + b(t)^2,$$

subject to initial conditions $m(\vartheta, 0) = y_0$ and $\gamma_*(\vartheta, 0) = 0$.

The limit system is

$$\frac{\mathrm{d}x_t(\vartheta)}{\mathrm{d}t} = f(\theta_1, t)\, y_t(\theta_2), \qquad x_0(\vartheta) = 0,$$

$$\frac{\mathrm{d}y_t(\theta_2)}{\mathrm{d}t} = a(\theta_2, t)\, y_t(\theta_2), \qquad y_0(\theta_2) = y_0.$$

The limit ($\varepsilon = 0$) solution of equation (7.112) is the function

$$y_t(\vartheta, \vartheta_0) = y_0 \Phi(\vartheta, 0, t) + \int_0^t \Phi(\vartheta, s, t)\, \frac{f(\theta_1, s)\, f\left(\theta_{1,0}, s\right) \gamma_*(\vartheta, s)}{\sigma(s)^2}\, y_s\left(\theta_{0,2}\right) \mathrm{d}s$$

where

$$\Phi\left(\vartheta,s,t\right)=\exp\left(\int_s^t\left[a\left(\theta_2,r\right)-\frac{f\left(\theta_1,r\right)^2\gamma_*\left(\vartheta,r\right)}{\sigma\left(r\right)^2}\right]\mathrm{d}r\right).$$

Estimators

MDE

The MDE $\check\vartheta_\varepsilon$ is defined by the same equation

$$\check\vartheta_\varepsilon=\arg\inf_{\vartheta\in\Theta}\int_{\alpha_1\wedge\alpha_2}^T\left[X_t-x_t\left(\vartheta\right)\right]^2\mathrm{d}t.$$

Let us verify that the following identifiability condition holds.

For any $\nu>0$ and any $\vartheta_0\in\Theta$ the function

$$g\left(\vartheta_0,\nu\right)=\inf_{\left\|\vartheta-\vartheta_0\right\|>\nu}\int_{\alpha_1\wedge\alpha_2}^T\left[x_t\left(\vartheta_0\right)-x_t\left(\vartheta\right)\right]^2\mathrm{d}t>0.$$

Suppose that for some $\nu>0$ we have $g\left(\vartheta_0,\nu\right)=0$. Then there exists $\vartheta_1=\left(\theta_{1,1},\theta_{1,2}\right)\neq\vartheta_0=\left(\theta_{0,1},\theta_{0,2}\right)$ such that $x_t\left(\vartheta_0\right)=x_t\left(\vartheta_1\right)$. This equality implies the relation $f\left(\theta_{0,1},t\right)y_t\left(\theta_{0,2}\right)=f\left(\theta_{1,1},t\right)y_t\left(\theta_{1,2}\right)$ for $t\in\left[\alpha_1\wedge\alpha_2,T\right]$. For discontinuous function $f\left(\cdot\right)$ and continuous function $y_\cdot\left(\cdot\right)$ it is possible iff $\theta_{0,1}=\theta_{1,1}$. In this case we obtain the equality $y_t\left(\theta_{0,2}\right)=y_t\left(\theta_{1,2}\right)$ and $a\left(\theta_{0,2},t\right)=a\left(\theta_{1,2},t\right)$. Hence $\theta_{0,2}=\theta_{1,2}$.

As the condition of identifiability holds the consistency and asymptotic normality of the MDE $\check\vartheta_\varepsilon$ can be proved as in two cases above. Therefore

$$\varepsilon^{-1}\left(\check\vartheta_\varepsilon-\vartheta_0\right)\Longrightarrow\mathcal{N}\left(0,\mathbf{D}\left(\vartheta_0\right)\right),$$

where the matrix $\mathbf{D}\left(\vartheta_0\right)$ can be written explicitly.

MLE and BE

Let us denote

$$\rho\left(\vartheta_0\right)=\frac{\left[q_{\theta_{0,1}}-r_{\theta_{0,1}}\right]}{\sigma\left(\theta_{0,1}\right)}y_{\theta_{0,1}}\left(\theta_{0,2}\right),\qquad \mathrm{I}\left(\vartheta_0\right)=\int_{\theta_{2,0}}^T\frac{f\left(\theta_{0,1},t\right)^2\dot y_t\left(\vartheta_0,\vartheta_0\right)^2}{\sigma\left(t\right)^2}\mathrm{d}t,$$

$$\dot y_t\left(\vartheta_0,\vartheta_0\right)=\left.\frac{\mathrm{d}y_t\left(\vartheta,\vartheta_0\right)}{\mathrm{d}\theta_2}\right|_{\vartheta=\vartheta_0},\qquad \Delta\left(\vartheta_0\right)=\int_{\theta_{0,2}}^T\frac{f\left(\theta_{0,1},t\right)\dot y_t\left(\vartheta_0,\vartheta_0\right)}{\sigma\left(t\right)}\mathrm{d}w\left(s\right),$$

$$y_{\theta_{0,1}}\left(\theta_{0,2}\right)=y_0\exp\left(\int_0^{\theta_{0,1}}a\left(\theta_{0,2},s\right)\mathrm{d}s\right),$$

where $w\left(\cdot\right)$ is some Wiener process. Introduce the limit LR random field

$$
\begin{aligned}
Z(\vartheta_0,u,v) &= Z_1(\vartheta_0,u)\,Z_2(\vartheta_0,v)\\
&= \exp\left(\rho(\vartheta_0)\,W(u)-\frac{|u|}{2}\rho(\vartheta_0)^2+v\Delta(\vartheta_0)-\frac{v^2}{2}\mathrm{I}(\vartheta_0)\right),\\
&\qquad (u,v)\in\mathcal{R}^2.
\end{aligned}
$$

Here $W(\cdot)$ is double-sided Wiener process, $\Delta(\vartheta_0)\sim\mathcal{N}(0,\mathrm{I}(\vartheta_0))$ and remark that the random variable $\Delta(\vartheta_0)$ and the process $W(\cdot)$ are independent.

Introduce the random variables $\hat u=\hat u(\vartheta_0)$, $\tilde u=\hat u(\vartheta_0)$ and $\hat v=\hat v(\vartheta_0)$, $\tilde v=\tilde v(\vartheta_0)$ by the relations

$$
Z_1\left(\vartheta_0,\hat u(\vartheta_0)\right)=\sup_{u\in\mathcal{R}} Z_1(\vartheta_0,u),\qquad \tilde u(\vartheta_0)=\frac{\int_{\mathcal{R}} u\,Z_1(\vartheta_0,u)\,\mathrm{d}u}{\int_{\mathcal{R}} Z_1(\vartheta_0,u)\,\mathrm{d}u},
$$

$$
\hat\zeta(\vartheta_0)=\left(\hat u(\vartheta_0),\hat v(\vartheta_0)\right)^\top,\qquad \hat v(\vartheta_0)=\tilde v(\vartheta_0)=\frac{\Delta(\vartheta_0)}{\mathrm{I}(\vartheta_0)}\sim\mathcal{N}\left(0,\mathrm{I}(\vartheta_0)^{-1}\right),
$$

$$
\tilde\zeta(\vartheta_0)=\left(\tilde u(\vartheta_0),\tilde v(\vartheta_0)\right)^\top,
$$

and the matrix

$$
\boldsymbol{\varphi}_\varepsilon=\begin{pmatrix}\varepsilon^2, & 0\\ 0, & \varepsilon\end{pmatrix}.
$$

The following minimax lower bound is valid

$$
\lim_{\nu\to0}\varliminf_{\varepsilon\to0}\sup_{\|\vartheta-\vartheta_0\|\le\nu}\mathbf{E}_\vartheta\|\boldsymbol{\varphi}_\varepsilon{}^{-1}(\bar\vartheta_\varepsilon-\vartheta)\|^2\ge\mathbf{E}_\vartheta\|\tilde\zeta(\vartheta_0)\|^2.
$$

For the proof see (7.17)–(7.18).

As usual, we call an estimator asymptotically efficient if for this estimator we have equality here for all $\vartheta_0\in\Theta$.

Theorem 7.11 *Let conditions $\hat{\mathscr{Q}}$ hold. Then the MLE $\hat\vartheta_\varepsilon$ and BE $\tilde\vartheta_\varepsilon$ are uniformly consistent, converge in distribution*

$$
\boldsymbol{\varphi}_\varepsilon{}^{-1}(\hat\vartheta_\varepsilon-\vartheta_0)\Longrightarrow\hat\zeta(\vartheta_0),\qquad \boldsymbol{\varphi}_\varepsilon{}^{-1}(\tilde\vartheta_\varepsilon-\vartheta_0)\Longrightarrow\tilde\zeta(\vartheta_0),
$$

polynomial moments converge uniformly on compacts ($p>0$)

$$
\mathbf{E}_{\vartheta_0}\|\boldsymbol{\varphi}_\varepsilon{}^{-1}(\hat\vartheta_\varepsilon-\vartheta_0)\|^p\longrightarrow\mathbf{E}_{\vartheta_0}\|\hat\zeta(\vartheta_0)\|^p,
$$

$$
\mathbf{E}_{\vartheta_0}\|\boldsymbol{\varphi}_\varepsilon{}^{-1}(\tilde\vartheta_\varepsilon-\vartheta_0)\|^p\longrightarrow\mathbf{E}_{\vartheta_0}\|\tilde\zeta(\vartheta_0)\|^p
$$

and BE are asymptotically efficient.

Proof Introduce the normalized LR (below $\theta_u = \theta_{0,1} + \varepsilon^2 u, \theta_v = \theta_{0,2} + \varepsilon v, \vartheta_{u,v} = (\theta_u, \theta_v)$)

$$
\begin{aligned}
\ln Z_\varepsilon(\vartheta_0, u, v) &= \ln \frac{L\left(\theta_{0,1} + \varepsilon^2 u, \theta_{0,2} + \varepsilon v, X^T\right)}{L\left(\theta_{0,1}, \theta_{2,0}, X^T\right)} \\
&= \int_{\alpha_1 \wedge \alpha_2}^T \frac{\left[f(\theta_u, t)\, m\left(\vartheta_{u,v}, t\right) - f\left(\theta_{1,0}, t\right) m\left(\vartheta_0, t\right)\right]}{\varepsilon^2 \sigma(t)^2} \mathrm{d}X_t \\
&\quad - \int_{\alpha_1 \wedge \alpha_2}^T \frac{\left[f(\theta_u, t)\, m\left(\vartheta_{u,v}, t\right)^2 - f\left(\theta_{1,0}, t\right) m\left(\vartheta_0, t\right)^2\right]}{2\varepsilon^2 \sigma(t)^2} \mathrm{d}t \\
&= \int_{\alpha_1 \wedge \alpha_2}^T \frac{\left[f(\theta_u, t)\, m\left(\vartheta_{u,v}, t\right) - f\left(\theta_{1,0}, t\right) m\left(\vartheta_0, t\right)\right]}{\varepsilon \sigma(t)} \mathrm{d}\bar W_t \\
&\quad - \int_{\alpha_1 \wedge \alpha_2}^T \frac{\left[f(\theta_u, t)\, m\left(\vartheta_{u,v}, t\right) - f\left(\theta_{1,0}, t\right) m\left(\vartheta_0, t\right)\right]^2}{2\varepsilon^2 \sigma(t)^2} \mathrm{d}t.
\end{aligned}
$$

Consider the stochastic integral

$$
\begin{aligned}
&\int_{\alpha_1 \wedge \alpha_2}^T \frac{\left[f(\theta_u, t)\, m\left(\vartheta_{u,v}, t\right) - f\left(\theta_{1,0}, t\right) m\left(\vartheta_0, t\right)\right]}{\varepsilon \sigma(t)} \mathrm{d}\bar W_t \\
&\quad = \int_{\alpha_1 \wedge \alpha_2}^T \frac{\left[f(\theta_u, t) - f\left(\theta_{0,1}, t\right)\right] m\left(\vartheta_{u,v}, t\right)}{\varepsilon \sigma(t)} \mathrm{d}\bar W_t \\
&\qquad + \int_{\alpha_1 \wedge \alpha_2}^T \frac{f\left(\theta_{1,0}, t\right)\left[m\left(\vartheta_{u,v}, t\right) - m\left(\vartheta_0, t\right)\right]}{\varepsilon \sigma(t)} \mathrm{d}\bar W_t.
\end{aligned}
$$

Suppose that $u > 0$, $v > 0$. Then we have

$$
\begin{aligned}
&\int_{\alpha_1 \wedge \alpha_2}^T \frac{\left[f(\theta_u, t) - f\left(\theta_{0,1}, t\right)\right] m\left(\vartheta_{u,v}, t\right)}{\varepsilon \sigma(t)} \mathrm{d}\bar W_t \\
&\quad = \int_{\theta_{0,1}}^{\theta_{0,1}+\varepsilon^2 u} \frac{\left[q_t - r_t\right] m\left(\vartheta_{u,v}, t\right)}{\varepsilon \sigma(t)} \mathrm{d}\bar W_t \\
&\quad = \frac{\left[q_{\theta_{0,1}} - r_{\theta_{0,1}}\right]}{\sigma\left(\theta_{0,1}\right)} m\left(\vartheta_{u,v}, \theta_{0,1}\right) \frac{\bar W_{\theta_{0,1}+\varepsilon^2 u} - \bar W_{\theta_{0,1}}}{\varepsilon} (1 + o(1)) \\
&\quad = \frac{\left[q_{\theta_{0,1}} - r_{\theta_{0,1}}\right]}{\sigma\left(\theta_{1,0}\right)} y_{\theta_{0,1}}\left(\theta_{0,2}\right)\; w_\varepsilon(u)\,(1 + o(1)) = \rho\left(\vartheta_0\right)\; w_\varepsilon(u)\,(1 + o(1)) \\
&\quad \Longrightarrow \rho\left(\vartheta_0\right)\; W(u).
\end{aligned}
$$

By direct expansions we obtain the relation

$$m\left(\vartheta_{u,v},t\right)=m\left(\vartheta_{0},t\right)+\varepsilon\, v\,\dot m_{\theta_{0,2}}\left(\vartheta_{0},t\right)+\varepsilon^{2}\,u\,\dot m_{\theta_{0,1}}\left(\vartheta_{0},t\right)+O\left(\varepsilon^{2}\right). \tag{7.113}$$

Here $\dot m_{\theta_{0,2}}(\cdot)=\partial m(\cdot)/\partial\theta_2$ and $\dot m_{\theta_{1,0}}(\cdot)=\partial m(\cdot)/\partial\theta_1$. Therefore for the second stochastic integral we can write

$$\begin{aligned}
&\int_{\alpha_1\wedge\alpha_2}^{T}\frac{f\left(\theta_{0,1},t\right)\left[m\left(\vartheta_{u,v},t\right)-m\left(\vartheta_{0},t\right)\right]}{\varepsilon\sigma(t)}\,\mathrm{d}\bar W_t\\
&\qquad=u\int_{\theta_{0,2}}^{T}\frac{f\left(\theta_{0,1},t\right)\dot m_{\theta_{0,2}}\left(\vartheta_{0},t\right)}{\sigma(t)}\,\mathrm{d}\bar W_t+o(1)\\
&\qquad=u\int_{\theta_{0,2}}^{T}\frac{f\left(\theta_{0,1},t\right)\dot y_t\left(\vartheta_{0},\vartheta_{0},t\right)}{\sigma(t)}\,\mathrm{d}\bar W_t+o(1)\\
&\qquad\Longrightarrow u\int_{\theta_{0,2}}^{T}\frac{f\left(\theta_{0,1},t\right)\dot y_t\left(\vartheta_{0},\vartheta_{0},t\right)}{\sigma(t)}\,\mathrm{d}w(t)=u\Delta\left(\vartheta_0\right)
\end{aligned}$$

For the ordinary integral similar calculation leads to the expressions

$$\begin{aligned}
&\int_{\alpha_1\wedge\alpha_2}^{T}\frac{\left[f\left(\theta_{u},t\right)m\left(\vartheta_{u,v},t\right)-f\left(\theta_{0,1},t\right)m\left(\vartheta_{0},t\right)\right]^2}{2\varepsilon^2\sigma(t)^2}\,\mathrm{d}t\\
&\quad=\int_{\alpha_1\wedge\alpha_2}^{T}\frac{\left[f\left(\theta_{u},t\right)-f\left(\theta_{0,1},t\right)\right]^2m\left(\vartheta_{u,v},t\right)^2}{2\varepsilon^2\sigma(t)^2}\,\mathrm{d}t\\
&\qquad+\int_{\alpha_1\wedge\alpha_2}^{T}\frac{f\left(\theta_{0,1},t\right)^2\left[m\left(\vartheta_{u,v},t\right)-m\left(\vartheta_{0},t\right)\right]^2}{2\varepsilon^2\sigma(t)^2}\,\mathrm{d}t+o(1)\\
&\quad=\int_{\theta_{0,1}}^{\theta_{0,1}+\varepsilon^2u}\frac{\left[r_t-q_t\right]^2m\left(\vartheta_{u,v},t\right)^2}{2\varepsilon^2\sigma(t)^2}\,\mathrm{d}t\\
&\qquad+v^2\int_{\alpha_1\wedge\alpha_2}^{T}\frac{f\left(\theta_{0,1},t\right)^2\dot m_{\vartheta_2}\left(\vartheta_{u,v},t\right)^2}{2\varepsilon^2\sigma(t)^2}\,\mathrm{d}t+o(1)\\
&\quad=\frac{u}{2}\rho\left(\vartheta_0\right)^2+\frac{v^2}{2}\mathrm{I}\left(\vartheta_0\right)+o(1).
\end{aligned}$$

Therefore the convergence of the one-dimensional marginal distributions

$$Z_\varepsilon\left(\vartheta_0,u,v\right)\Longrightarrow Z\left(\vartheta_0,u,v\right)$$

holds. The convergence of multidimensional distributions can be shown as well.

We need two estimates for the LR $Z_\varepsilon(\vartheta_0, u, v) = Z_\varepsilon(u, v)$: there exist constants $C > 0$ such that for any $R > 0$ and $|u_i| + |v_i| \le R, i = 1, 2$

$$\mathbf{E}_{\vartheta_0} \left| Z_\varepsilon(u_2, v_2)^{1/8} - Z_\varepsilon(u_1, v_1)^{1/8} \right|^8 \le C\left(1 + R^6\right)|u_2 - u_1|^4 + C|v_2 - v_1|^8 \tag{7.114}$$

and

$$\mathbf{E}_{\vartheta_0} Z_\varepsilon(\vartheta_0, u, v)^{1/2} \le \frac{C}{\left(|u| + |v|^2\right)^N} \tag{7.115}$$

Introduce further notation

$$M(\vartheta, t) = f(\theta_1, t)\, m(\vartheta, t), \qquad F_t(\vartheta_2, \vartheta_1, t) = \frac{M\left(\vartheta_{u_2,v_2}, t\right) - M\left(\vartheta_{u_1,v_1}, t\right)}{\varepsilon\sigma(t)},$$

$$P_{t,\varepsilon} = \left[\frac{L\left(\vartheta_{u_2,v_2}, X^t\right)}{L\left(\vartheta_{u_1,v_1}, X^t\right)}\right]^{1/8}.$$

By Corollary 1.1 there exist constants C_1, C_2 such that

$$\begin{aligned}
&\mathbf{E}_{\vartheta_0} \left| Z_\varepsilon(u_2, v_2)^{1/8} - Z_\varepsilon(u_1, v_1)^{1/8} \right|^8 \\
&\quad \le C_1 \mathbf{E}_{\vartheta_{u_1,v_1}} \left(\int_{\alpha_1 \wedge \alpha_2}^T F_t(\vartheta_2, \vartheta_1, t)^2 P_{t,\varepsilon} \mathrm{d}t \right)^8 \\
&\qquad + C_2 \mathbf{E}_{\vartheta_{u_1,v_1}} \left(\int_{\alpha_1 \wedge \alpha_2}^T F_t(\vartheta_2, \vartheta_1, t) P_{t,\varepsilon} \mathrm{d}\bar{W}_t \right)^8.
\end{aligned}$$

Consider first the stochastic integral which can be written as follows (see (7.113)):

$$\begin{aligned}
\int_{\alpha_1 \wedge \alpha_2}^T F_t(\vartheta_2, \vartheta_1, t) P_{t,\varepsilon} \mathrm{d}\bar{W}_t &= \int_{\alpha_1 \wedge \alpha_2}^T \frac{\left[M\left(\vartheta_{u_2,v_2}, t\right) - M\left(\vartheta_{u_1,v_1}, t\right)\right]}{\varepsilon\sigma(t)} P_{t,\varepsilon} \mathrm{d}\bar{W}_t \\
&= \int_{\theta_{0,1}+\varepsilon^2 u_1}^{\theta_{0,1}+\varepsilon^2 u_2} [r_t - q_t]\, m\left(\vartheta_{u_2,v_2}, t\right) \varepsilon^{-1} \sigma(t)^{-1} P_{t,\varepsilon} \mathrm{d}\bar{W}_t \\
&\quad + \int_{\alpha_1 \wedge \alpha_2}^T f\left(\theta_{u_1}, t\right) \left[m\left(\vartheta_{u_2,v_2}, t\right) - m\left(\vartheta_{u_1,v_1}, t\right)\right] \varepsilon^{-1} \sigma(t)^{-1} P_{t,\varepsilon} \mathrm{d}\bar{W}_t.
\end{aligned}$$

Hence, using the inequality (1.1) we write

$$
\begin{aligned}
&\mathbf{E}_{\vartheta_{u_1,v_1}}\left(\int_{\theta_{0,1}+\varepsilon^2 u_1}^{\theta_{0,1}+\varepsilon^2 u_2}[r_t-q_t]\,m\left(\vartheta_{u_2,v_2},t\right)\varepsilon^{-1}\sigma\left(t\right)^{-1}P_{t,\varepsilon}\mathrm{d}\bar W_t\right)^8\\
&\quad\leq 28\left(u_2-u_1\right)^3\int_{\theta_{0,1}+\varepsilon^2 u_1}^{\theta_{0,1}+\varepsilon^2 u_2}\frac{[r_t-q_t]^8}{\varepsilon^2\sigma\left(t\right)^8}\mathbf{E}_{\vartheta_{u_1,v_1}}P_{t,\varepsilon}^8 m\left(\vartheta_{u_2,v_2},t\right)^8\mathrm{d}t\\
&\quad= 28\left(u_2-u_1\right)^3\int_{\theta_{0,1}+\varepsilon^2 u_1}^{\theta_{0,1}+\varepsilon^2 u_2}\frac{[r_t-q_t]^8}{\varepsilon^2\sigma\left(t\right)^8}\mathbf{E}_{\vartheta_{u_2,v_2}}m\left(\vartheta_{u_2,v_2},t\right)^8\mathrm{d}t\\
&\quad\leq C\,\left(u_2-u_1\right)^4.
\end{aligned}
$$

For estimation of the second stochastic integral we use the smoothness (7.113) of the function $m\left(\vartheta,t\right)$

$$
\begin{aligned}
&\mathbf{E}_{\vartheta_{u_1,v_1}}\left(\int_{\alpha_1\wedge\alpha_2}^{T}f\left(\theta_{u_1},t\right)\left[m\left(\vartheta_{u_2,v_2},t\right)-m\left(\vartheta_{u_1,v_1},t\right)\right]\varepsilon^{-1}\sigma\left(t\right)^{-1}P_{t,\varepsilon}\mathrm{d}\bar W_t\right)^8\\
&\quad\leq C\,\left(u_2-u_1\right)^8\varepsilon^8+C\,\left(v_2-v_1\right)^8\leq C\,\left(u_2-u_1\right)^4+C\,\left(v_2-v_1\right)^8.
\end{aligned}
$$

Here we used the estimate $\left(u_2-u_1\right)\varepsilon^2\leq\beta_1-\alpha_1$.

Similar calculations in the case of the ordinary integrals give us the estimate

$$
\mathbf{E}_{\vartheta_{u_1,v_1}}\left(\int_{\alpha_1\wedge\alpha_2}^{T}F_t\left(\vartheta_2,\vartheta_1,t\right)^2P_{t,\varepsilon}\mathrm{d}t\right)^8\leq C\left(u_2-u_1\right)^8+C\left(v_2-v_1\right)^{16}
$$

Therefore we obtaine (7.114).

To prove (7.115) we note that for the small $\nu>0$ and $\left|\theta_1-\theta_{0,1}\right|+\left|\theta_2-\theta_{0,2}\right|^2\leq\nu$ we have

$$
\begin{aligned}
&\int_{\alpha_1\wedge\alpha_2}^{T}\frac{\left[f\left(\theta_1,t\right)y_t\left(\vartheta,\vartheta_0\right)-f\left(\theta_{0,1},t\right)y_t\left(\theta_{0,2}\right)\right]^2}{\sigma\left(t\right)^2}\mathrm{d}t\\
&\quad=\left|\theta_1-\theta_{0,1}\right|\left|r_{\theta_{0,1}}-q_{\theta_{0,1}}\right|^2y_{\theta_{0,1}}\left(\theta_{0,2}\right)^2\left(1+O\left(\nu\right)\right)\\
&\qquad+\left|\theta_2-\theta_{0,2}\right|^2\int_{\theta_{0,2}}^{T}f\left(\theta_{0,1},t\right)^2\dot y_t\left(\vartheta_0,\vartheta_0\right)^2\mathrm{d}t\left(1+O\left(\nu\right)\right)\\
&\quad\geq\frac{1}{2}\left|\theta_1-\theta_{0,1}\right|\left|r_{\theta_{0,1}}-q_{\theta_{0,1}}\right|^2y_{\theta_{0,1}}\left(\theta_{0,2}\right)^2\\
&\qquad+\frac{1}{2}\left|\theta_2-\theta_{0,2}\right|^2\int_{\theta_{0,2}}^{T}f\left(\theta_{0,1},t\right)^2\dot y_t\left(\vartheta_0,\vartheta_0\right)^2\mathrm{d}t\\
&\quad\geq c_1^*\left(\left|\theta_1-\theta_{0,1}\right|+\left|\theta_2-\theta_{0,2}\right|^2\right), \qquad (7.116)
\end{aligned}
$$

where the constant $c_1^* > 0$ can be chosen not depending on ϑ_0.

If we suppose that for some $\vartheta_1 = \left(\theta_{1,1}, \theta_{2,1}\right) \neq \vartheta_0$

$$\int_{\alpha_1 \wedge \alpha_2}^{T} \frac{\left[f\left(\theta_{1,1}, t\right) y_t\left(\vartheta_1, \vartheta_0\right) - f\left(\theta_{0,1}, t\right) y_t\left(\theta_{0,2}\right)\right]^2}{\sigma(t)^2} \mathrm{d}t = 0,$$

then it implies that

$$f\left(\theta_{1,1}, t\right) y_t\left(\vartheta_1, \vartheta_0\right) = f\left(\theta_{0,1}, t\right) y_t\left(\theta_{0,2}\right), \qquad \alpha_1 \wedge \alpha_2 \leq t \leq T.$$

As the function $f(\cdot)$ is discontinuous such equality is possible iff $\theta_{1,1} = \theta_{0,1}$. In this case we have the equality $y_t\left(\vartheta_1^*, \vartheta_0\right) = y_t\left(\theta_{0,2}\right)$, where $\vartheta_1^* = \left(\theta_{0,1}, \theta_2\right)$. For the difference $\delta_y(t) = y_t\left(\vartheta_1^*, \vartheta_0\right) - y_t\left(\vartheta_0\right)$ we obtain the equation

$$\frac{\mathrm{d}\delta_y(t)}{\mathrm{d}t} = A\left(\vartheta_1^*, t\right) \delta_y(t) + \left[a\left(\theta_{2,1}, t\right) - a\left(\theta_{0,2}, t\right)\right] y_t\left(\theta_{0,2}\right), \qquad \delta_y(0) = 0.$$

Therefore $\delta_y(t) = 0$ corresponds to the equality

$$\int_{\alpha_2}^{t} \Phi\left(\vartheta_1^*, s, t\right) \left[a\left(\theta_{2,1}, s\right) - a\left(\theta_{0,2}, s\right)\right] y_s\left(\theta_{0,2}\right) \mathrm{d}s = 0$$

and to the equality

$$\int_{\alpha_2}^{t} \Phi\left(\vartheta_1^*, 0, s\right)^{-1} \left[a\left(\theta_{2,1}, s\right) - a\left(\theta_{0,2}, s\right)\right] y_s\left(\theta_{0,2}\right) \mathrm{d}s = 0.$$

This equality for all $t \in [\alpha_2, T]$ implies $a\left(\theta_{2,1}, s\right) = a\left(\theta_{0,2}, s\right)$ and the last equality is possible iff $\theta_{0,2} = \theta_{2,1}$.

Hence the following condition of identifiability is fulfilled: for any $\nu > 0$ and any $\vartheta_0 \in \Theta$

$$g\left(\vartheta_0, \nu\right) = \inf_{\|\vartheta - \vartheta_0\|_* \geq \nu} \int_{\alpha_1 \wedge \alpha_2}^{T} \frac{\left[f\left(\theta_1, t\right) y_t\left(\vartheta, \vartheta_0\right) - f\left(\theta_{0,1}, t\right) y_t\left(\theta_{0,2}\right)\right]^2}{\sigma(t)^2} \mathrm{d}t > 0.$$

Here we denoted $\|\vartheta - \vartheta_0\|_* = \left|\theta_1 - \theta_{0,1}\right| + \left|\theta_2 - \theta_{0,2}\right|^2$. This estimate allows us to write for $\|\vartheta - \vartheta_0\|_* \geq \nu$

$$\int_{\alpha_1 \wedge \alpha_2}^{T} \frac{\left[f\left(\theta_1, t\right) y_t\left(\vartheta, \vartheta_0\right) - f\left(\theta_{0,1}, t\right) y_t\left(\theta_{0,2}\right)\right]^2}{\sigma(t)^2} \mathrm{d}t \geq g\left(\vartheta_0, \nu\right)$$

$$\geq g\left(\vartheta_0, \nu\right) \frac{\left|\theta_1 - \theta_{0,1}\right| + \left|\theta_2 - \theta_{0,2}\right|^2}{\left(\beta_1 - \alpha_1\right) + \left(\beta_2 - \alpha_2\right)^2} \geq c_2^* \left\|\vartheta - \vartheta_0\right\|_*. \tag{7.117}$$

Define the constant $c^* = c_1^* \wedge c_2^*$. Then from (7.116) to (7.117) we obtain the estimate

$$\int_{\alpha_1\wedge\alpha_2}^{T} \frac{\left[f\left(\theta_{0,1}+\varepsilon^2 u,t\right) y_t\left(\vartheta_{u,v},\vartheta_0\right) - f\left(\theta_{0,1},t\right) y_t\left(\theta_{0,2}\right)\right]^2}{\varepsilon^2\sigma\left(t\right)^2}\mathrm{d}t \geq c^*\left(|u|+v^2\right).$$

Now the proof of (7.115) follows from the proof of Lemma 2.5.

Therefore the conditions $\mathscr{I}\mathscr{K}_2$ are fulfilled, and by Theorems 1.9 and 1.11 the estimators have the properties specified in Theorem 7.11. □

Adaptive Filter

Recall that the BE $\tilde\vartheta_\varepsilon = (\tilde\theta_{1,\varepsilon}, \tilde\theta_{2,\varepsilon})$ where

$$\tilde\theta_{1,\varepsilon} = \frac{\int_\Theta \theta_1 p\left(\vartheta\right) L\left(\vartheta, X^T\right)\mathrm{d}\vartheta}{\int_\Theta p\left(\vartheta\right) L\left(\vartheta, X^T\right)\mathrm{d}\vartheta}, \qquad \tilde\theta_{2,\varepsilon} = \frac{\int_\Theta \theta_2 p\left(\vartheta\right) L\left(\vartheta, X^T\right)\mathrm{d}\vartheta}{\int_\Theta p\left(\vartheta\right) L\left(\vartheta, X^T\right)\mathrm{d}\vartheta}.$$

The solution of equation (7.112) we write as follows:

$$m\left(\vartheta,t\right) = y_0\Phi\left(\vartheta,0,t\right) + \int_0^t \Phi\left(\vartheta,s,t\right) B\left(\vartheta,s\right)\mathrm{d}X_s$$

where

$$\Phi\left(\vartheta,s,t\right) = \exp\left(\int_s^t \left[a\left(\theta_2,l\right) - \frac{f\left(\theta_1,l\right)^2\gamma_*\left(\vartheta,l\right)}{\sigma\left(l\right)^2}\right]\mathrm{d}l\right),$$

$$B\left(\vartheta,s\right) = \frac{f\left(\theta_1,s\right)\gamma_*\left(\vartheta,s\right)}{\sigma\left(s\right)^2}.$$

To substitute estimator we will use the representation of the function $m\left(\vartheta,t\right)$ which does not contain a stochastic integral (see Lemma 1.3).

Suppose that $\theta_1 < \theta_2$ and denote

$$\Phi_{rg}\left(\vartheta,s,t\right) = \exp\left(\int_s^t \left[g_l - \frac{r_l^2\gamma_*\left(\vartheta,l\right)}{\sigma\left(l\right)^2}\right]\mathrm{d}l\right),$$

$$\Phi_{rh}\left(\vartheta,s,t\right) = \exp\left(\int_s^t \left[h_l - \frac{r_l^2\gamma_*\left(\vartheta,l\right)}{\sigma\left(l\right)^2}\right]\mathrm{d}l\right)$$

Suppose that $\tilde\theta_{1,\varepsilon} < \tilde\theta_{2,\varepsilon}$. The adaptive filter we define on three different intervals by the following equations:

- *Let* $t \in [0, \tilde\theta_{1,\varepsilon}]$, *then*

$$\mathrm{d}m_t^\star = g_t m_t^\star \mathrm{d}t + \frac{q_t\gamma_*(t)}{\sigma(t)^2}\left[\mathrm{d}X_t - q_t m_t^\star \mathrm{d}t\right], \quad m_0^\star = y_0, \tag{7.118}$$

$$\frac{\mathrm{d}\gamma_*(t)}{\mathrm{d}t} = 2g_t\gamma_*(t) - \frac{q_t^2\gamma_*(t)^2}{\sigma(t)^2} + b(t)^2, \quad \gamma_*(0) = 0. \tag{7.119}$$

- *Let* $t \in [\tilde\theta_{1,\varepsilon}, \tilde\theta_{2,\varepsilon}]$, *then the Riccati equation is*

$$\frac{\mathrm{d}\gamma_*(\tilde\theta_{1,\varepsilon}, t)}{\mathrm{d}t} = 2g_t\gamma_*(\tilde\theta_{1,\varepsilon}, t) - \frac{r_t^2\gamma_*(\tilde\theta_{1,\varepsilon}, t)^2}{\sigma(t)^2} + b(t)^2, \tag{7.120}$$

subject to the initial value $\gamma_*(\tilde\theta_{1,\varepsilon}, \tilde\theta_{1,\varepsilon}) = \gamma_*(\tilde\theta_{1,\varepsilon})$. *Here* $\gamma_*(\theta_1)$ *is the solution of* (7.119) *at the point* $t = \tilde\theta_{1,\varepsilon}$. *The equation for adaptive filter is*

$$m_t^\star = \left[m_{\tilde\theta_{1,\varepsilon}}^\star - X_{\tilde\theta_{1,\varepsilon}} B(\tilde\theta_{1,\varepsilon}, \tilde\theta_{1,\varepsilon})\right]\Phi_{rg}(\tilde\theta_{1,\varepsilon}, \tilde\theta_{1,\varepsilon}, t) + X_t B(\tilde\theta_{1,\varepsilon}, t)$$
$$- \int_{\tilde\theta_{1,\varepsilon}}^t X_s \Phi_{rg}(\tilde\theta_{1,\varepsilon}, s, t)\left[\left(\frac{r_s^2\gamma_*(\tilde\theta_{1,\varepsilon}, s)}{\sigma(s)^2} - g_s\right) B(\tilde\theta_{1,\varepsilon}, s) + B_s'(\tilde\theta_{1,\varepsilon}, s)\right]\mathrm{d}s.$$

- *Let* $t \in [\tilde\theta_{2,\varepsilon}, T]$, *then the Riccati equation is*

$$\frac{\mathrm{d}\gamma_*(\tilde\vartheta_\varepsilon, t)}{\mathrm{d}t} = 2h_t\gamma_*(\tilde\vartheta_\varepsilon, t) - \frac{r_t^2\gamma_*(\tilde\vartheta_\varepsilon, t)^2}{\sigma(t)^2} + b(t)^2,$$

subject to the initial value $\gamma_*(\tilde\vartheta_\varepsilon, \tilde\theta_{2,\varepsilon}) = \gamma_*(\tilde\theta_{1,\varepsilon}, \tilde\theta_{2,\varepsilon})$. *Here* $\gamma_*(\tilde\theta_{1,\varepsilon}, \tilde\theta_{2,\varepsilon})$ *is the solution of* (7.120) *at the point* $t = \tilde\theta_{2,\varepsilon}$. *The corresponding equation of the filter is*

$$m_t^\star = \left[m_{\tilde\theta_{2,\varepsilon}}^\star - X_{\tilde\theta_{2,\varepsilon}} B(\tilde\vartheta_\varepsilon, \tilde\theta_{2,\varepsilon})\right]\Phi_{rh}(\tilde\vartheta_\varepsilon, \tilde\theta_{2,\varepsilon}, t) + X_t B(\tilde\theta_\varepsilon, t)$$
$$- \int_{\tilde\theta_{2,\varepsilon}}^t X_s \Phi_{rh}(\tilde\vartheta_\varepsilon, s, t)\left[\left(\frac{r_s^2\gamma_*(\tilde\vartheta_\varepsilon, s)}{\sigma(s)^2} - h_s\right) B(\tilde\vartheta_\varepsilon, s) + B_s'(\tilde\vartheta_\varepsilon, s)\right]\mathrm{d}s. \tag{7.121}$$

If $\tilde\theta_{2,\varepsilon} < \tilde\theta_{1,\varepsilon}$, then the construction is quite similar to the one given above by the Eqs. (7.118)–(7.121).

Of course the error

$$\mathbf{E}_{\vartheta_0}\left[m_t^\star - m\left(\vartheta_0, t\right)\right]^2 \mathbb{I}_{\{0\leq t\leq \theta_m\}} = 0.$$

Here $\theta_m = \min(\theta_{1,0}, \theta_{0,2}, \tilde{\theta}_{1,\varepsilon}, \tilde{\theta}_{2,\varepsilon})$. For the other values of t the description of the error has to be done.

Remark 7.7 Note that the results of this section can be extended on the model with many change points in both equations like

$$\mathrm{d}X_t = \sum_{k=1}^{K}\left[q_{k,t}\,\mathbb{I}_{\{t<\theta_k\}} + r_{k,t}\,\mathbb{I}_{\{t\geq\theta_k\}}\right] Y_t\,\mathrm{d}t + \varepsilon\sigma\left(t\right)\mathrm{d}W_t, \quad X_0 = 0, \quad 0\leq t\leq T,$$

$$\mathrm{d}Y_t = \sum_{l=1}^{L}\left[g_{l,t}\,\mathbb{I}_{\{t<\tau_l\}} + h_{l,t}\,\mathbb{I}_{\{t\geq\tau_l\}}\right] Y_t\,\mathrm{d}t + \varepsilon b\left(t\right)\mathrm{d}V_t, \quad Y_0 = y_0 \neq 0.$$

The unknown parameter is $\vartheta = (\theta_1, \ldots, \theta_K, \tau_1, \ldots, \tau_L) \in \Theta \subset (0, T)^{K+L}$. The limit system is, of course,

$$x_t\left(\vartheta\right) = \int_0^t \sum_{k=1}^{K}\left[q_{k,s}\,\mathbb{I}_{\{s<\theta_k\}} + r_{k,s}\,\mathbb{I}_{\{s\geq\theta_k\}}\right] y_s\left(\tau\right)\mathrm{d}s, \qquad 0\leq t\leq T,$$

$$y_t\left(\tau\right) = y_0 \exp\left(\int_0^t \sum_{l=1}^{L}\left[g_{l,s}\,\mathbb{I}_{\{s<\tau_l\}} + h_{l,s}\,\mathbb{I}_{\{s\geq\tau_l\}}\right]\mathrm{d}s\right).$$

Here $\tau = (\tau_1, \ldots, \tau_L)$ and the functions $q_{k,t}, r_{k,t}, g_{k,l}, h_{l,t}$ are such that the corresponding identifiability condition holds.

The properties of the MLE and BE can be described following the same lines as it was done in the proof of Theorem 7.11.

7.5 BSDE Approximation

7.5.1 BSDE

The backward stochastic differential equation (BSDE) was first introduced in the linear case by Bismuth [16]. The general (nonlinear) case was initiated by Pardoux and Peng [160, 161]. Since that time the BSDE attracts large attention and intensively developed due to its importance in financial mathematics (see, e.g., El Karoui et al. [49], Ma and Yong [147], Pardoux and Răşcanu [162], Zhang [201], and the references therein).

For example, BSDE is a good mathematical model for description of risk-based optimal investment problem of insurer [51], where a simplified continuous-time economy with two investment vehicles, namely, a fixed interest security and a share, is considered. The BSDE approach is used to solve the game problem of two players: the insurer and the market. In the work [52] a BSDE approach is used to evaluate convex risk measures for unhedged positions of derivative securities in a continuous-time economy.

Let us recall the construction of the BSDE in the Markovian case with observable forward equation following [49]. We consider the one-dimensional processes only. We are given a filtered probability space $\left(\Omega, (\mathfrak{F}_t)_{t\in[0,T]}, \mathbf{P}\right)$ with the filtration $(\mathfrak{F}_t)_{t\in[0,T]}$ satisfying the *usual conditions*. Define the stochastic differential equation (called *forward*)

$$\mathrm{d}X_t = f(t, X_t)\,\mathrm{d}t + \sigma(t, X_t)\,\mathrm{d}W_t, \qquad X_0 = x_0, \qquad 0 \le t \le T, \tag{7.122}$$

where $W_t, \mathfrak{F}_t, 0 \le t \le T$ is standard Wiener process, and X_0 is $\mathfrak{F}_0$ measurable initial value. The trend coefficient $f(t, x)$ and diffusion coefficient $\sigma(t, x)^2$ satisfy the Lipschitz and linear growth conditions

$$|f(t, x) - f(t, y)| + |\sigma(t, x) - \sigma(t, y)| \le L\,|x - y|, \tag{7.123}$$

$$|f(t, x)| + |\sigma(t, x)| \le C\,(1 + |x|), \tag{7.124}$$

where $L > 0$ and $C > 0$ are some constants. By these conditions the stochastic differential equation (7.122) has a unique strong solution (Theorem 1.1).

The main problem is the following: *we are given Eq.* (7.122) *and two functions* $\Phi(x)$, $F(t, x, z, \sigma)$ *and we have to construct two processes* $(Z_t, \Sigma_t; \mathfrak{F}_t, 0 \le t \le T)$ *such that the solution of the stochastic differential equation*

$$\mathrm{d}Z_t = -F(t, X_t, Z_t, \Sigma_t)\,\mathrm{d}t + \Sigma_t\,\mathrm{d}W_t, \qquad 0 \le t \le T, \tag{7.125}$$

(called backward) has the final value $Z_T = \Phi(X_T)$.

This equation is often written in integral form as follows:

$$Z_t = \Phi(X_T) + \int_t^T F(s, X_s, Z_s, \Sigma_s)\,\mathrm{d}s - \int_t^T \Sigma_s\,\mathrm{d}W_s, \qquad 0 \le t \le T.$$

We always assume that the functions $F(t, x, y, z)$ and $\Phi(x)$ satisfy the uniform Lipschitz condition

$$|F(t, x_1, z_1, \sigma_1) - F(t, x_2, z_2, \sigma_2)| \le L\,(|x_1 - x_2| + |z_1 - z_2| + |\sigma_1 - \sigma_2|), \tag{7.126}$$

$$|\Phi(x_1) - \Phi(x_2)| \le L\,|x_1 - x_2|. \tag{7.127}$$

This is so-called *Markovian case* of BSDE or Forward-Backward SDE (FBSDE).

Conditions $\mathscr{R}$:

$\mathscr{R}_1$. The functions $f(\cdot), \sigma(\cdot), F(\cdot), \Phi(\cdot)$ are uniformly Lipschitz in (x, z, Σ) (see (7.123), (7.126), and (7.127)).

$\mathscr{R}_2$. The functions $f(t, 0), \sigma(t, 0), F(t, 0, 0, 0)$, and $\Phi(0)$ are bounded.

$\mathscr{R}_3$. The functions $f(\cdot), \sigma(\cdot)$, and $F(\cdot)$ are uniformly Hölder- $(1/2)$ continuous in t.

For the existence and uniqueness of the solution of (7.122) and (7.125) under conditions $\mathscr{R}$ see [162, 201].

The processes $(Z_t, \Sigma_t; \mathfrak{F}_t, 0 \leq t \leq T)$ can be constructed as follows. Let us denote $u(t, x)$ the solution of the partial differential equation (PDE)

$$\frac{\partial u}{\partial t} + f(t, x)\frac{\partial u}{\partial x} + \frac{\sigma(t, x)^2}{2}\frac{\partial^2 u}{\partial x^2} = -F\left(t, x, u, \sigma(t, x)\frac{\partial u}{\partial x}\right),$$

with the terminal condition $u(T, x) = \Phi(x)$ and put $Z_t = u(t, X_t)$. Then we obtain by Itô's formula

$$\mathrm{d}Z_t = \left[\frac{\partial u(t, X_t)}{\partial t} + f(t, X_t)\frac{\partial u(t, X_t)}{\partial x} + \frac{\sigma(t, X_t)^2}{2}\frac{\partial^2 u(t, X_t)}{\partial x^2}\right]\mathrm{d}t$$
$$+\sigma(t, X_t)\frac{\partial u(t, X_t)}{\partial x}\mathrm{d}W_t, \qquad Z_0 = u(0, X_0).$$

Below we use the notation

$$\frac{\partial u(t, X_t)}{\partial x} = u'_x(t, X_t) = \left.\frac{\partial u(t, x)}{\partial x}\right|_{x=X_t}.$$

Hence if we denote $\Sigma_t = \sigma(t, X_t)u'_x(t, X_t)$, then this equation becomes

$$\mathrm{d}Z_t = -F(t, X_t, Z_t, \Sigma_t)\,\mathrm{d}t + \Sigma_t\,\mathrm{d}W_t, \qquad Z_0 = u(0, X_0).$$

The terminal value $Z_T = u(T, X_T) = \Phi(X_T)$. Therefore the problem is solved and we obtained Eq. (7.125) with the given terminal value.

We are interested by the construction of the solution of such BSDE in the situations, where the forward equations depend on some unknown finite-dimensional parameter. We consider two models of observations. In the first model the forward equation has small diffusion coefficient and is directly observable, and in the second model the solution of the forward equation is observed in the presence of small white Gaussian noise (hidden forward equation) .

7.5.2 Observable Forward Equation

Let us consider the SDE (forward equation)

$$\mathrm{d}X_t = f(\vartheta, t, X_t)\,\mathrm{d}t + \varepsilon\sigma(t, X_t)\,\mathrm{d}W_t, \qquad X_0, \quad 0 \le t \le T, \tag{7.128}$$

where $f(\vartheta, t, x), \sigma(t, x); (\vartheta, t, x) \in \Theta \times [0, T] \times \mathcal{R}$ are some known deterministic functions satisfying (7.123) and (7.124). The unknown parameter $\vartheta \in \Theta$ and the set Θ for simplicity of exposition we take one-dimensional $\Theta = (\alpha, \beta)$, where α and β are finite.

We suppose that the corresponding functions $F(t, x, z, \sigma)$ and $\Phi(x)$ are given and satisfy the conditions (7.126), (7.127), $\mathscr{R}$, and we have to find an approximation $\hat{Z}_t, \hat{\Sigma}_t, 0 < t \le T$ of the random functions $Z_t, \Sigma_t, 0 < t \le T$ satisfying the backward SDE

$$\mathrm{d}Z_t = -F(t, X_t, Z_t, \Sigma_t)\,\mathrm{d}t + \Sigma_t\,\mathrm{d}W_t, \qquad Z_T = \Phi(X_T). \tag{7.129}$$

Remark that we cannot use the solution $u = u(t, x, \vartheta)$ of the corresponding PDE

$$u'_t + f(\vartheta, t, x)\,u'_x + \frac{\varepsilon^2\sigma(t, x)^2}{2}u''_{xx} = -F\left(t, x, u, \varepsilon\sigma(t, x)\,u'_x\right),$$
$$u(T, x, \vartheta) = \Phi(x) \tag{7.130}$$

for the estimator $\bar{Z}_t = u(t, X_t, \vartheta)$ because this solution depends on unknown parameter $\vartheta = \vartheta_0$ (true value).

The "natural" approximations $(\hat{Z}_t, \hat{\Sigma}_t; \mathfrak{F}_t, 0 \le t \le T)$ can be constructed as follows. Suppose that we already have the family $u(t, x, \vartheta), \vartheta \in \Theta$ of solutions of the partial differential equation (7.130). One way to obtain approximation $(\hat{Z}_t, \hat{\Sigma}_t)$ of (Z_t, Σ_t) is to find first an estimator-process $\vartheta_t^*, \mathfrak{F}_t, 0 < t \le T$ and then to put

$$\hat{Z}_t = u(t, X_t, \vartheta_t^*), \qquad \hat{\Sigma}_t = u'_x(t, X_t, \vartheta_t^*)\,\varepsilon\,\sigma(\vartheta_t^*, t, X_t).$$

If the estimator has good properties, say $\vartheta_t^* - \vartheta_0$ is small in some sense, then the error

$$\hat{Z}_t - Z_t = u(t, X_t, \vartheta_t^*) - u(t, X_t, \vartheta_0) \approx \frac{\partial u(t, X_t, \vartheta)}{\partial \vartheta}\left(\vartheta_t^* - \vartheta\right)$$

can be small as well.

Note that we do not write the BSDE for $\hat{Z}_t$ because this will involve the stochastic differential of ϑ_t^*, and the expression for this BSDE will be cumbersome. Of course this can be done, and asymptotically the difference between this equation and true BSDE will converge to zero, but we decided to avoid the corresponding technical difficulties.

Here $\vartheta_t^*; \mathfrak{F}_t, 0 \le t \le T$ is some *good estimator-process* of ϑ_0. In this problem the *good estimator* means (as in the preceding chapters) the following:

- *The estimator ϑ_t^* depends on $X^t = (X_s, 0 \le s \le t)$.*

- *Easy to calculate for each* $t \in (0, T]$.
- *Provides the asymptotically efficient in some sense estimation of* Z_t, *i.e.,*

$$\mathbf{E}_{\vartheta_0}(\hat{Z}_t - Z_t)^2 \to \min .$$

Therefore the main problem is *how to find a good estimator-process*. Recall that in this work we already used many times the One-step MLE-process $\vartheta_t^\star, \mathfrak{F}_t, 0 < t \leq T$ which possesses the required properties and of course can be used in this construction too. Below we consider a partially observed linear system, where the forward equation is hidden and is observed in the presence of small Gaussian noise, but before we discuss this construction in the more simple case of observable forward equation (7.128).

Introduce the family of functions

$$\mathcal{U}(\Theta) = \left\{u(t, x, \vartheta), (t, x, \vartheta) \in [0, T] \times \mathcal{R} \times \Theta\right\}$$

satisfying for all $\vartheta \in \Theta$ the partial differential equation (7.130).

The solution of this equation with $\varepsilon = 0$ we denote as $u^\circ(t, x, \vartheta)$, i.e.,

$$\frac{\partial u^\circ}{\partial t} + f(\vartheta, t, x)\frac{\partial u^\circ}{\partial x} = -F(t, x, u^\circ, 0), \qquad u^\circ(T, x, \vartheta) = \Phi(x).$$

Note that the function $u(t, x, \vartheta)$ converges to $u^\circ(t, x, \vartheta)$ as $\varepsilon \to 0$ (see [56] for conditions providing this convergence).

Parameter Estimation

MDE

Define a family of deterministic functions

$$\mathcal{X}(\Theta) = \{(x_s(\vartheta), 0 \leq s \leq T), \vartheta \in \Theta\},$$

where $x_s = x_s(\vartheta)$ is solution of the ordinary differential equation

$$\frac{\mathrm{d}x_s}{\mathrm{d}s} = f(\vartheta, s, x_s), \qquad x_0, \quad 0 \leq s \leq T.$$

Below $h(x) \in \mathcal{C}^k_{x,b}$ means that the function $h(x), x \in \mathcal{R}$ has k continuous bounded derivatives w.r.t. x and $h(x) \in \mathcal{C}^k_{x,p}$ means that the function $h(x)$ has k continuous derivatives w.r.t. x and the derivatives have polynomial majorants. As before, $\dot{f}(\vartheta, t, x) = \partial f(\vartheta, t, x)/\partial\vartheta$.

Conditions $\mathscr{R}$:

$\breve{\mathscr{R}}_1$. *The functions* $f(\vartheta,t,x), \sigma(t,x); (\vartheta,t,x) \in \Theta \times [0,T] \times \mathcal{R}$ *satisfy the conditions* (7.123) *and,* (7.124), $\mathscr{R}$, *where the constants* $C > 0$ *and* $L > 0$ *do not depend on* ϑ.

$\breve{\mathscr{R}}_2$. *The function* $f(\vartheta,t,x) \in \mathcal{C}^2_{\vartheta,p}$.

$\breve{\mathscr{R}}_3$. *The functions* $f(\vartheta,t,x) \in \mathcal{C}^2_{x,p}, \dot f(\vartheta,t,x) \in \mathcal{C}^1_{x,p}$.

Recall that under conditions $\breve{\mathscr{R}}_1, \breve{\mathscr{R}}_3$ we have the convergences

$$\sup_{0\leq s\leq T} |X_s - x_s(\vartheta_0)| \longrightarrow 0,$$

$$\sup_{0\leq s\leq T} \left|\varepsilon^{-1}(X_s - x_s(\vartheta_0)) - x_s^{(1)}(\vartheta_0)\right| \longrightarrow 0,$$

where $x_s^{(1)}(\vartheta_0) = x_s^{(1)}$ is a Gaussian process satisfying the equation

$$\mathrm{d}x_s^{(1)} = f_x'(\vartheta_0,s,x_s)\,x_s^{(1)}\,\mathrm{d}s + \sigma(s,x_s)\,\mathrm{d}W_s, \quad x_0^{(1)} = 0, \quad 0\leq s\leq T.$$

The convergence $X_t \to x_t$ was shown in Theorem 1.6. The second convergence follows a similar proof and can be found e.g., in [108].

As before we construct the estimator $\vartheta^\star_{t,\varepsilon}$ in *two steps*. First we estimate ϑ by observations $X^\tau = (X_s, 0\leq s\leq \tau)$ on small interval $[0,\tau]$ and obtain the estimator $\bar\vartheta_{\tau,\varepsilon}$. Then we use this estimator to construct the One-step MLE-process $\vartheta^\star_{t,\varepsilon}, \tau < t \leq T$ which is asymptotically equivalent to the MLE and can be easily calculated for the values $t \in [\tau, T]$.

Fix some (small) $\tau > 0$ and introduce the (MDE) $\breve\vartheta_{\tau,\varepsilon}$:

$$\left\|X - x\left(\breve\vartheta_{\tau,\varepsilon}\right)\right\|_\tau^2 = \inf_{\vartheta\in\Theta} \|X - x(\vartheta)\|_\tau^2 = \inf_{\vartheta\in\Theta} \int_0^\tau [X_t - x_t(\vartheta)]^2\,\mathrm{d}t.$$

Introduce notations:

$$g(\vartheta_0,\nu) = \inf_{|\vartheta-\vartheta_0|>\nu} \|x(\vartheta) - x(\vartheta_0)\|_\tau,$$

$$\phi(\vartheta_0,s,t) = \exp\left(\int_s^t f_x'(\vartheta_0,v,x_v(\vartheta_0))\,\mathrm{d}v\right),$$

$$\dot x_t(\vartheta_0) = \int_0^t \phi(\vartheta_0,s,t)\,\dot f(\vartheta_0,s,x_s(\vartheta_0))\,\mathrm{d}s,$$

$$T_\tau(\vartheta_0) = \int_0^\tau \dot x_s(\vartheta_0)^2\,\mathrm{d}s,$$

$$D_\tau(\vartheta_0) = T_\tau(\vartheta_0)^{-2} \int_0^\tau \left(\int_s^\tau \phi(\vartheta_0,s,t)\,\dot x_t(\vartheta_0)\,\mathrm{d}t\right)^2 \sigma(s,x_s(\vartheta_0))^2\,\mathrm{d}s,$$

$$\check{\eta}_\tau(\vartheta_0)=T_\tau(\vartheta_0)^{-1}\int_0^\tau\int_s^\tau \phi(\vartheta_0,s,t)\,\dot x_t(\vartheta_0)\,\mathrm{d}t\;\sigma(s,x_s(\vartheta_0))\,\mathrm{d}W_s.$$

Further conditions $\check{\mathscr{R}}$:

$\check{\mathscr{R}}_4$. *For any* $\nu>0$ *we have* $\inf_{\vartheta_0\in\Theta} g(\vartheta_0,\nu)>0.$

$\check{\mathscr{R}}_5$. $\inf_{\vartheta_0\in\Theta} T_\tau(\vartheta_0)>0.$

Proposition 7.7 *Let the conditions* $\check{\mathscr{R}}_{1-5}$ *hold. Then the MDE* $\check{\vartheta}_{\tau,\varepsilon}$ *is uniformly consistent, uniformly on compacts* $\mathbb{K}\in\Theta$ *asymptotically normal*

$$\varepsilon^{-1}\left(\check{\vartheta}_{\tau,\varepsilon}-\vartheta_0\right)\Longrightarrow\check{\eta}_\tau(\vartheta_0)\sim\mathcal{N}(0,D_\tau(\vartheta_0))$$

and the polynomial moments converge.

Proof The proof is very close to the proof of Theorem 2.2. □

One-Step MLE-Process

The limit ($\varepsilon=0$) Fisher information for the model (7.122) is

$$\mathrm{I}(\vartheta_0)=\int_0^T\frac{\dot f(\vartheta_0,s,x_s(\vartheta_0))^2}{\sigma(s,x_s(\vartheta_0))^2}\,\mathrm{d}s.$$

Here

$$\dot f(\vartheta_0,s,x_s(\vartheta_0))=\frac{\partial}{\partial\vartheta}f(\vartheta,s,x_s(\vartheta_0))\Big|_{\vartheta=\vartheta_0}.$$

Remark that the MLE $\hat\vartheta_\varepsilon$ in the case of these observations (under regularity conditions) is asymptotically normal

$$\frac{\hat\vartheta_\varepsilon-\vartheta_0}{\varepsilon}\Longrightarrow\mathcal{N}\left(0,\mathrm{I}(\vartheta_0)^{-1}\right).$$

The proof can be found in [106] or in [108].

Denote

$$\mathrm{I}_\tau^t(\vartheta_0,X)=\int_\tau^t\frac{\dot f(\vartheta_0,s,X_s)^2}{\sigma(s,X_s)^2}\,\mathrm{d}s,\qquad \tau\leq t\leq T,$$

the *Fisher information function* and introduce the One-step MLE-process as follows:

$$\vartheta^{\star}_{t,\varepsilon} = \check\vartheta_{\tau,\varepsilon} + \mathrm{I}_\tau^t\left(\check\vartheta_{\tau,\varepsilon}, X\right)^{-1}\int_\tau^t \frac{\dot f\left(\check\vartheta_{\tau,\varepsilon}, s, X_s\right)}{\sigma\left(s, X_s\right)^2}\left[\mathrm{d}X_s - f\left(\check\vartheta_{\tau,\varepsilon}, s, X_s\right)\mathrm{d}s\right],$$

$t \in [\tau, T]$.

We have

$$\begin{aligned}
&\frac{\vartheta^{\star}_{t,\varepsilon} - \vartheta_0}{\varepsilon}\\
&= \frac{\check\vartheta_{\tau,\varepsilon} - \vartheta_0}{\varepsilon} + \mathrm{I}_\tau^t\left(\check\vartheta_{\tau,\varepsilon}, X\right)^{-1}\int_\tau^t \frac{\dot f\left(\check\vartheta_{\tau,\varepsilon}, s, X_s\right)}{\sigma\left(s, X_s\right)}\mathrm{d}W_s\\
&\quad + \mathrm{I}_\tau^t\left(\check\vartheta_{\tau,\varepsilon}, X\right)^{-1}\int_\tau^t \frac{\dot f\left(\check\vartheta_{\tau,\varepsilon}, s, X_s\right)}{\varepsilon\sigma\left(s, X_s\right)^2}\left[f\left(\vartheta_0, s, X_s\right) - f\left(\check\vartheta_{\tau,\varepsilon}, s, X_s\right)\right]\mathrm{d}s\\
&= \frac{\check\vartheta_{\tau,\varepsilon} - \vartheta_0}{\varepsilon} + \mathrm{I}_\tau^t\left(\vartheta_0, X\right)^{-1}\int_\tau^t \frac{\dot f\left(\vartheta_0, s, X_s\right)}{\sigma\left(s, X_s\right)}\mathrm{d}W_s\left(1 + o\left(1\right)\right)\\
&\quad - \frac{\check\vartheta_{\tau,\varepsilon} - \vartheta_0}{\varepsilon}\mathrm{I}_\tau^t\left(\check\vartheta_{\tau,\varepsilon}, X\right)^{-1}\int_\tau^t \frac{\dot f\left(\check\vartheta_{\tau,\varepsilon}, s, X_s\right)^2}{\sigma\left(s, X_s\right)^2}\mathrm{d}s + \frac{\left(\check\vartheta_{\tau,\varepsilon} - \vartheta_0\right)^2}{\varepsilon}O\left(1\right)\\
&= \mathrm{I}_\tau^t\left(\vartheta_0, x\right)^{-1}\int_\tau^t \frac{\dot f\left(\vartheta_0, s, x_s\left(\vartheta_0\right)\right)}{\sigma\left(s, x_s\left(\vartheta_0\right)\right)}\mathrm{d}W_s\left(1 + o\left(1\right)\right) + o\left(1\right).
\end{aligned}$$

Here we used the Taylor expansion at the point $\check\vartheta_{\tau,\varepsilon}$

$$\begin{aligned}
f\left(\vartheta_0, s, X_s\right) - f\left(\check\vartheta_{\tau,\varepsilon}, s, X_s\right) &= \left(\vartheta_0 - \check\vartheta_{\tau,\varepsilon}\right)\dot f\left(\check\vartheta_{\tau,\varepsilon}, s, X_s\right)\\
&\quad + \frac{1}{2}\left(\vartheta_0 - \check\vartheta_{\tau,\varepsilon}\right)^2\ddot f\left(\bar\vartheta_{\tau,\varepsilon}, s, X_s\right)
\end{aligned}$$

and Proposition 7.7.

Of course, we will need two more conditions.

$\mathscr{R}_1^\star$. *The Fisher information function* $\mathrm{I}_\tau^t\left(\vartheta, x\right)$, $\tau < t \leq T$ *is positive*

$$\inf_{\vartheta\in\Theta}\mathrm{I}_\tau^t\left(\vartheta, x\right) > 0, \qquad \tau < t \leq T.$$

$\mathscr{R}_2^\star$. *The function* $\sigma\left(t, x\right)^2 \geq c$, *where the constant* $c > 0$ *does not depend on* t *and* x.

If the monotone random function $\mathrm{I}_\tau^t\left(\check\vartheta_{\tau,\varepsilon}, X\right) = 0$ for some t (this means that $\mathrm{I}_\tau^s\left(\check\vartheta_{\tau,\varepsilon}, X\right) = 0$ for all $s \in [\tau, t]$, then we put $\mathrm{I}_\tau^s\left(\check\vartheta_{\tau,\varepsilon}, X\right)^{-1} = 0$.

Note that if $\left|\dot{f}(\vartheta,\tau,x_\tau(\vartheta))\right| \geq \kappa$, where $\kappa > 0$ does not depend on ϑ, then the condition $\mathscr{R}_1^\star$ is fulfilled.

We proved the following proposition.

Proposition 7.8 *Let the conditions* $\breve{\mathscr{R}}_{1-5}$ *and* $\mathscr{R}_{1-2}^\star$ *hold. Then the estimator* $\vartheta_{t,\varepsilon}^\star$, $\tau < t \leq T$ *is consistent, uniformly on compacts asymptotically normal*

$$\varepsilon^{-1}\left(\vartheta_{t,\varepsilon}^\star - \vartheta_0\right) \Longrightarrow \eta_\tau^t \sim \mathcal{N}\left(0, \mathrm{I}_\tau^t(\vartheta,x)^{-1}\right), \tau < t \leq T$$

and the moments converge.

Remark that this estimator-process can be improved up to asymptotically efficient with the help of Lemma 1.3 as follows. Introduce further notation: $B(\vartheta,t,x) = \dot{f}(\vartheta,t,x)\,\sigma(t,x)^{-2}$ and re-write

$$\Delta^\tau(\vartheta,X) = \int_0^\tau B(\vartheta,t,x_t(\vartheta))\left[\mathrm{d}X_t - f(\vartheta,t,X_t)\right] \tag{7.131}$$

without stochastic integral as follows:

$$\begin{aligned}\Delta^\tau(\vartheta,X) = B(\vartheta,\tau,x_\tau(\vartheta))\,X_\tau &- \int_0^\tau X_t B_t'(\vartheta,t,x_t(\vartheta))\,\mathrm{d}t\\ &- \int_0^\tau X_t B_x'(\vartheta,t,x_t(\vartheta))\,f(\vartheta,t,x_t(\vartheta))\,\mathrm{d}t\\ &- \int_0^\tau B(\vartheta,t,x_t(\vartheta))\,f(\vartheta,t,X_t)\,\mathrm{d}t.\end{aligned} \tag{7.132}$$

Now we can define the One-step MLE-process for $t \in (\tau, T]$ as follows:

$$\begin{aligned}\vartheta_{t,\varepsilon}^\star = \breve{\vartheta}_{\tau,\varepsilon} + \mathrm{I}_0^t\left(\breve{\vartheta}_{\tau,\varepsilon},X\right)^{-1}&\left[\Delta^\tau\left(\breve{\vartheta}_{\tau,\varepsilon},X\right)\right.\\ &\left.+\int_\tau^t \frac{\dot{f}\left(\breve{\vartheta}_{\tau,\varepsilon},s,X_s\right)}{\sigma(s,X_s)^2}\left[\mathrm{d}X_s - f\left(\breve{\vartheta}_{\tau,\varepsilon},s,X_s\right)\mathrm{d}s\right]\right].\end{aligned} \tag{7.133}$$

By the direct calculation it can be verified that

$$\frac{\vartheta_{t,\varepsilon}^\star - \vartheta_0}{\varepsilon} \Longrightarrow \eta_t^\star(\vartheta_0) \sim \mathcal{N}\left(0, \mathrm{I}_0^t(\vartheta_0,x)^{-1}\right),$$

where

$$\eta_t^\star(\vartheta_0) = \mathrm{I}_0^t(\vartheta_0,x)^{-1}\int_0^t \frac{\dot{f}(\vartheta_0,s,x_s(\vartheta_0))}{\sigma(s,x_s(\vartheta_0))}\,\mathrm{d}w(s).$$

Here $w(s), 0 \le s \le t$ is some Wiener process.

Let us clarify the expansion of $\Delta^\tau\left(\check\vartheta_{\tau,\varepsilon}, X\right)$. Denote and

$$
\begin{aligned}
Q(\vartheta, t, x_t(\vartheta)) &= \dot B(\vartheta, t, x_t(\vartheta)) + B'_x(\vartheta, t, x_t(\vartheta))\,\dot x_t(\vartheta),\\
P(\vartheta, t, x_t(\vartheta)) &= B'_x(\vartheta, t, x_t(\vartheta))\left[\dot f(\vartheta, t, x_t(\vartheta)) + f'_x(\vartheta, t, x_t(\vartheta))\,\dot x_t(\vartheta)\right]\\
&\qquad + Q'_x(\vartheta, t, x_t(\vartheta))\, f(\vartheta, t, x_t(\vartheta)).
\end{aligned}
$$

We have

$$
\begin{aligned}
B\left(\check\vartheta_{\tau,\varepsilon}, t, x_t\left(\check\vartheta_{\tau,\varepsilon}\right)\right) &= B(\vartheta_0, t, x_t(\vartheta_0)) + \left(\check\vartheta_{\tau,\varepsilon} - \vartheta_0\right) Q(\vartheta_0, t, x_t(\vartheta_0))\\
&\qquad + \left(\check\vartheta_{\tau,\varepsilon} - \vartheta_0\right)^2 O(1).
\end{aligned}
$$

Thus

$$
\begin{aligned}
\Delta^\tau\left(\check\vartheta_{\tau,\varepsilon}, X\right) &= B(\vartheta_0, t, x_t(\vartheta_0))\, X_\tau - \int_0^\tau X_t B'_t(\vartheta_0, t, x_t(\vartheta_0))\,\mathrm{d}t\\
&\quad - \int_0^\tau X_t B'_x(\vartheta_0, t, x_t(\vartheta_0))\, f(\vartheta_0, t, x_t(\vartheta_0))\,\mathrm{d}t\\
&\quad - \int_0^\tau X_t B(\vartheta_0, t, x_t(\vartheta_0))\, f\left(\check\vartheta_{\tau,\varepsilon}, t, X_t\right)\mathrm{d}t\\
&\quad + \left(\check\vartheta_{\tau,\varepsilon} - \vartheta_0\right) Q(\vartheta, t, x_t(\vartheta))\, X_\tau\\
&\quad - \left(\check\vartheta_{\tau,\varepsilon} - \vartheta_0\right)\int_0^\tau X_t Q'_t(\vartheta_0, t, x_t(\vartheta_0))\,\mathrm{d}t\\
&\quad - \left(\check\vartheta_{\tau,\varepsilon} - \vartheta_0\right)\int_0^\tau X_t P(\vartheta_0, t, x_t(\vartheta_0))\,\mathrm{d}t\\
&\quad - \left(\check\vartheta_{\tau,\varepsilon} - \vartheta_0\right)\int_0^\tau X_t Q(\vartheta_0, t, x_t(\vartheta_0))\, f\left(\check\vartheta_{\tau,\varepsilon}, t, X_t\right)\mathrm{d}t\\
&\quad + \left(\check\vartheta_{\tau,\varepsilon} - \vartheta_0\right)^2 O(1).
\end{aligned}
$$

Using once more the relation between (7.131) and (7.132) we can write

$$
\begin{aligned}
\Delta^\tau\left(\check\vartheta_{\tau,\varepsilon}, X\right) &= \int_0^\tau B(\vartheta_0, t, x_t(\vartheta_0))\left[\mathrm{d}X_t - f\left(\check\vartheta_{\tau,\varepsilon}, t, X_t\right)\mathrm{d}t\right]\\
&\quad + \left(\check\vartheta_{\tau,\varepsilon} - \vartheta_0\right)\int_0^\tau Q(\vartheta_0, t, x_t(\vartheta_0))\left[\mathrm{d}X_t - f\left(\check\vartheta_{\tau,\varepsilon}, t, X_t\right)\mathrm{d}t\right]
\end{aligned}
$$

$$
+\left(\check{\vartheta}_{\tau,\varepsilon}-\vartheta_0\right)^2 O\left(1\right).
$$

Therefore

$$
\begin{aligned}
\varepsilon^{-1}\Delta^\tau\left(\check{\vartheta}_{\tau,\varepsilon},X\right) &= \int_0^\tau B\left(\vartheta_0,t,x_t\left(\vartheta_0\right)\right)\sigma\left(t,X_t\right)\mathrm{d}W_t\\
&\quad-\frac{\left(\check{\vartheta}_{\tau,\varepsilon}-\vartheta_0\right)}{\varepsilon}\int_0^\tau B\left(\vartheta_0,t,x_t\left(\vartheta_0\right)\right)\dot f\left(\bar\vartheta,t,X_t\right)\mathrm{d}t\\
&\quad+\left(\check{\vartheta}_{\tau,\varepsilon}-\vartheta_0\right)\int_0^\tau Q\left(\vartheta_0,t,x_t\left(\vartheta_0\right)\right)\sigma\left(t,X_t\right)\mathrm{d}W_t\\
&\quad-\frac{\left(\check{\vartheta}_{\tau,\varepsilon}-\vartheta_0\right)^2}{\varepsilon}\int_0^\tau Q\left(\vartheta_0,t,x_t\left(\vartheta_0\right)\right)\dot f\left(\bar\vartheta,t,X_t\right)\mathrm{d}t\\
&\quad+\frac{\left(\check{\vartheta}_{\tau,\varepsilon}-\vartheta_0\right)^2}{\varepsilon}O\left(1\right).
\end{aligned}
$$

Finally we obtain

$$
\begin{aligned}
\frac{\Delta^\tau\left(\check{\vartheta}_{\tau,\varepsilon},X\right)}{\varepsilon} &= \int_0^\tau \frac{\dot f\left(\vartheta_0,t,x_t\left(\vartheta_0\right)\right)}{\sigma\left(t,x_t\left(\vartheta_0\right)\right)}\mathrm{d}W_t-\frac{\left(\check{\vartheta}_{\tau,\varepsilon}-\vartheta_0\right)}{\varepsilon}\int_0^\tau \frac{\dot f\left(\vartheta_0,t,x_t\left(\vartheta_0\right)\right)^2}{\sigma\left(t,x_t\left(\vartheta_0\right)\right)^2}\mathrm{d}t\\
&\quad+\frac{\left(\check{\vartheta}_{\tau,\varepsilon}-\vartheta_0\right)^2}{\varepsilon}O\left(1\right).
\end{aligned}
$$

Now the estimator (7.133) admits the representation

$$
\begin{aligned}
\frac{\vartheta^\star_{t,\varepsilon}-\vartheta_0}{\varepsilon} &= \mathrm{I}_0^t\left(\vartheta_0\right)^{-1}\left[\int_0^\tau \frac{\dot f\left(\vartheta_0,s,x_s\left(\vartheta_0\right)\right)}{\sigma\left(s,x_s\left(\vartheta_0\right)\right)}\mathrm{d}W_s\right.\\
&\quad\left.+\int_\tau^t \frac{\dot f\left(\vartheta_0,s,x_s\left(\vartheta_0\right)\right)}{\sigma\left(s,x_s\left(\vartheta_0\right)\right)}\mathrm{d}W_s\right]\left(1+o\left(1\right)\right)+\frac{\left(\check{\vartheta}_{\tau,\varepsilon}-\vartheta_0\right)^2}{\varepsilon}O\left(1\right)\\
&\longrightarrow \mathrm{I}_0^t\left(\vartheta_0\right)^{-1}\int_0^t \frac{\dot f\left(\vartheta_0,s,x_s\left(\vartheta_0\right)\right)}{\sigma\left(s,x_s\left(\vartheta_0\right)\right)}\mathrm{d}W_s\sim\mathcal{N}\left(0,\mathrm{I}_0^t\left(\vartheta_0\right)^{-1}\right).
\end{aligned}
$$

This convergence is uniform on compacts, and thus the One-step MLE-process is asymptotically efficient for all $t\in(\tau,T]$.

Approximation of the Solution of BSDE

Remind that the observed process (7.128) depends on the unknown parameter ϑ and we have to estimate the random functions $(Z_t, \Sigma_t, \tau < t \leq T)$ satisfying the Eq. (7.129).

Suppose that the function $u(t, x, \vartheta) \in \mathcal{U}(\Theta)$ and $\vartheta^\star_{t,\varepsilon}, \tau < t \leq T$ is the One-step MLE-process defined by the equality (7.131).

Define the approximation of BSDE

$$Z^\star_t = u\left(t, X_t, \vartheta^\star_{t,\varepsilon}\right), \qquad \Sigma^\star_t = \varepsilon\sigma(t, X_t)\, u'_x\left(t, X_t, \vartheta^\star_{t,\varepsilon}\right)$$

Assume

$\mathscr{R}^\star_3$. *The functions $u(t, x, \vartheta), u'_x(t, x, \vartheta) \in \mathcal{C}^2_{\vartheta,p}$*

$\mathscr{R}^\star_4$. *The following convergence takes place*

$$\dot u(t, X_t, \vartheta_0) \longrightarrow \dot u^\circ(t, x_t(\vartheta_0), \vartheta_0), \quad \dot u'_x(t, X_t, \vartheta_0) \longrightarrow \dot u'^\circ_x(t, x_t(\vartheta_0), \vartheta_0)$$

Theorem 7.12 *Let the conditions $\mathscr{R}, \check{\mathscr{R}}_{1-5}, \mathscr{R}^\star_{1-4}$ hold. Then*

$$\varepsilon^{-1}\left(Z^\star_t - Z_t\right) \Longrightarrow \dot u^\circ(t, x_t(\vartheta_0), \vartheta_0)\; \eta^\star_t(\vartheta_0), \tag{7.134}$$

$$\varepsilon^{-2}\left(\Sigma^\star_t - \Sigma_t\right) \Longrightarrow \sigma(t, x_t(\vartheta_0))\, \dot u'^\circ_x(t, x_t(\vartheta_0), \vartheta_0)\; \eta^\star_t(\vartheta_0).$$

Proof We have

$$Z^\star_t - Z_t = u\left(t, X_t, \vartheta^\star_{t,\varepsilon}\right) - u(t, X_t, \vartheta_0) = \dot u\left(t, X_t, \bar\vartheta\right)\left(\vartheta^\star_{t,\varepsilon} - \vartheta_0\right)$$

where $X_t \rightarrow x_t(\vartheta_0)$, $\left|\bar\vartheta - \vartheta_0\right| \leq \left|\vartheta^\star_{t,\varepsilon} - \vartheta_0\right| \rightarrow 0$ and $\dot u(t, X_t, \vartheta) \rightarrow \dot u^\circ(t, x_t(\vartheta_0), \vartheta_0)$. Thus we obtain (7.134).

Further

$$\begin{aligned}\varepsilon^{-2}\left(\Sigma^\star_t - \Sigma_t\right) &= \varepsilon^{-1}\sigma(t, X_t)\left[u'_x\left(t, X_t, \vartheta^\star_{t,\varepsilon}\right) - u'_x(t, X_t, \vartheta_0)\right]\\ &= \sigma(t, X_t)\, \dot u'_x\left(t, X_t, \bar\vartheta\right)\, \varepsilon^{-1}\left(\vartheta^\star_{t,\varepsilon} - \vartheta_0\right)\\ &\Longrightarrow \sigma(t, x_t(\vartheta_0))\, \dot u'^\circ_x(t, x_t(\vartheta_0), \vartheta_0)\; \eta^\star_t(\vartheta_0).\end{aligned}$$

□

Remark 7.8 The proof seems to be elementary because we already required the necessary convergences without presenting the detailed conditions providing the properties of the solution of PDE. This is merely an illustrative example of a problem and model where the one-step MLE process developed in this paper can be useful. More about this problem can be found in [128]. The example of Black-Sholes model as the forward equation and the function $F(t, x, z, \Sigma) = bz + \gamma x \Sigma$ are discussed there together with the results of numerical simulation.

The next model with hidden forward equation is more appropriate for this work.

7.5.3 Hidden Forward Equation with Unknown Volatility

Model of Observations

Suppose that the forward equation is

$$\mathrm{d}Y_t = a(t)\, Y_t\, \mathrm{d}t + b(\vartheta, t)\, \mathrm{d}V_t, \qquad Y_0 = y_0, \quad 0 \le t \le T, \tag{7.135}$$

but the solution $Y^T = (Y_t, 0 \le t \le T)$ of this equation cannot be observed directly and we have the observations

$$\mathrm{d}X_t = f(t)\, Y_t\, \mathrm{d}t + \varepsilon\sigma(t)\, \mathrm{d}W_t, \qquad X_0 = 0, \quad 0 \le t \le T \tag{7.136}$$

only. Here $V_t, 0 \le t \le T$ and $W_t, 0 \le t \le T$ are independent Wiener processes. The parameter $\vartheta \in \Theta = (\alpha, \beta)$, where $|\alpha| + |\beta| < \infty$. Here $a(\cdot), b(\cdot), f(\cdot)$, and $\sigma(\cdot)$ are known functions and $\varepsilon \in (0, 1]$. This is the particular case of the model studied in Sect. 3.1.

We are given two functions $F(t, y, u, s)$, $\Phi(y)$, and we have to construct an approximation $\left(\bar{Z}_t, \bar{s}_t\right)$ of the solution (Z_t, s_t) of the corresponding BSDE

$$\mathrm{d}Z_t = -F(t, Y_t, Z_t, s_t)\, \mathrm{d}t + s_t\, \mathrm{d}V_t, \qquad Z_T = \Phi(Y_T), \quad 0 \le t \le T.$$

We say "approximation" because the forward equation is *hidden*, and moreover its volatility $b(\vartheta, t)$ depends on some unknown parameter ϑ. Therefore, we have no access to the process Y^T, and even if we have Y^T, the solution $U = U(t, y, \vartheta)$ of the corresponding PDE

$$U_t' + a(t)\, y\, U_y' + \frac{b(\vartheta, t)^2}{2}\, U_{yy}'' = -F\left(t, y, U, b(\vartheta, t)\, U_y\right), \quad U(T, y, \vartheta) = \Phi(y) \tag{7.137}$$

depends on the unknown parameter ϑ. We cannot put $\tilde{Z}_t = U(t, Y_t, \vartheta)$ because we do not know neither Y_t nor ϑ.

Introduce the *Conditions* $\breve{\mathscr{S}}$.

$\breve{\mathscr{S}}_1$. *The functions* $a(t), b(\vartheta, t), f(t), \sigma(t); \vartheta \in \Theta, t \in [0, T]$ *have continuous derivatives w.r.t.* t.

$\breve{\mathscr{S}}_2$. *The functions* $b(\vartheta,t), f(t), \sigma(t); \vartheta \in \Theta, t \in [0,T]$ *are separated from zero by a constant, which does not depend on* ϑ *and* t.

As we have no observations of the forward equation we change the statement of the problem and propose BSDE based on the best in the mean squared estimator of these process. Introduce the conditional expectation $\hat{Y}^T = (\hat{Y}_t, 0 \le t \le T)$, where $\hat{Y}_t = \mathbf{E}_\vartheta\left(Y_t|\mathfrak{F}_t^X\right)$. Now the corresponding BSDE can be written as follows:

$$dZ_t = -F(t,\hat{Y}_t, Z_t, s(t))dt + s(t)\,d\bar{W}_t, \qquad Z_T = \Phi(\hat{Y}_T), \quad 0 \le t \le T. \tag{7.138}$$

The Wiener process $\bar{W}_t, 0 \le t \le T$ we describe below. To construct Eq. (7.138) we need the equations of Kalman-Bucy filter for $\hat{Y}_t$, which we remind here. It will be convenient to denote $\hat{Y}_t = m(\vartheta,t)$ to show the dependence on ϑ. Remind (see (3.14) and (3.15)) that the equation for $m(\vartheta,t)$ is

$$dm(\vartheta,t) = \left[a(t) - \frac{\gamma(\vartheta,t)f(t)^2}{\varepsilon^2\sigma(t)^2}\right]m(\vartheta,t)\,dt + \frac{\gamma(\vartheta,t)f(t)}{\varepsilon^2\sigma(t)^2}dX_t. \tag{7.139}$$

Here $m(\vartheta,0) = y_0$ and $\gamma(\vartheta,t) = \mathbf{E}_\vartheta(Y_t - m(\vartheta,t))^2$ is the solution of the Riccati equation

$$\frac{\partial\gamma(\vartheta,t)}{\partial t} = 2a(t)\gamma(\vartheta,t) - \frac{\gamma(\vartheta,t)^2 f(t)^2}{\varepsilon^2\sigma(t)^2} + b(\vartheta,t)^2, \quad \gamma(\vartheta,0) = 0. \tag{7.140}$$

Introduce further notations

$$\gamma_*(\vartheta,t) = \frac{\gamma(\vartheta,t)}{\varepsilon}, \quad \gamma_0(\vartheta,t) = \frac{b(\vartheta,t)\sigma(t)}{f(t)}, \quad A_\varepsilon(\vartheta,t) = \frac{\gamma_*(\vartheta,t)f(t)}{\sigma(t)^2},$$

$$A_0(\vartheta,t) = \frac{b(\vartheta,t)}{\sigma(t)}, \qquad q_\varepsilon(\vartheta,t) = -a(t) + \frac{A_\varepsilon(\vartheta,t)f(t)}{\varepsilon}.$$

The true value we denote as ϑ_0. Equation (7.139) for $m(\vartheta_0,t)$ and Riccati equation (7.140) can be re-written as follows:

$$dm(\vartheta_0,t) = a(t)m(\vartheta_0,t)\,dt + A_\varepsilon(\vartheta_0,t)\sigma(t)\,d\bar{W}_t,$$

$$\frac{\partial\gamma_*(\vartheta_0,t)}{\partial t} = 2a(t)\gamma_*(\vartheta_0,t) - \frac{A_\varepsilon(\vartheta_0,t)^2\sigma(t)^2}{\varepsilon} + \frac{b(\vartheta_0,t)^2}{\varepsilon},$$

with initial values $m(\vartheta_0,0) = y_0$ and $\gamma_*(\vartheta_0,0) = 0$, respectively. Here $\bar{W}_t, \mathfrak{F}_t^X$, $0 \le t \le T$ is *innovation* Wiener process defined by the relation

$$\mathrm{d}X_t = f(t)\, m(\vartheta_0, t)\, \mathrm{d}t + \varepsilon\sigma(t)\, \mathrm{d}\bar{W}_t, \qquad X_0 = 0$$

(see Lemma 1.2).

Lemma 7.12 *Let the conditions $\breve{\mathscr{S}}_{1-2}$ hold, then for any $t_0 \in (0, T]$ we have the convergence*

$$\sup_{t_0 \le t \le T} |\gamma_*(\vartheta, t) - \gamma_0(\vartheta, t)| \longrightarrow 0, \qquad \sup_{t_0 \le t \le T} |A_\varepsilon(\vartheta, t) - A_0(\vartheta, t)| \longrightarrow 0$$

as $\varepsilon \to 0$.

Proof See Lemma 3.1. □

This lemma allows us to verify the following obvious result.

Let the conditions $\breve{\mathscr{S}}_{1-2}$ hold then for any $t_0 \in (0, T]$ we have the convergence

$$\sup_{t_0 \le t \le T} \mathbf{E}_{\vartheta_0} |m(\vartheta_0, t) - Y_t|^2 \le C\varepsilon \to 0$$

as $\varepsilon \to 0$. Recall that $\mathbf{E}_{\vartheta_0} |m(\vartheta_0, t) - Y_t|^2 = \gamma(\vartheta_0, t)$ and

$$\gamma(\vartheta_0, t) = \gamma_*(\vartheta_0, t)\, \varepsilon = \gamma_0(\vartheta_0, t)\,(1 + o(1))\, \varepsilon \le C\, \varepsilon.$$

Therefore for small ε the random process $m(\vartheta_0, t)$ is a good approximation of the solution Y_t of forward equation.

Equations (7.139) and (7.140) by Lemma 7.12 can be simplified if we replace $\gamma_*(\vartheta, t)$ by its limit $\gamma_0(\vartheta, t)$. Then Eq. (7.139) on the time interval $[t_0, T]$ becomes

$$\begin{aligned}
\mathrm{d}m(\vartheta, t) &= \left[a(t) - \frac{b(\vartheta, t)\, f(t)}{\varepsilon\sigma(t)}\right] m(\vartheta, t)\, \mathrm{d}t + \frac{b(\vartheta, t)}{\varepsilon\sigma(t)} \mathrm{d}X_t \\
&= a(t)\, m(\vartheta, t)\, \mathrm{d}t + b(\vartheta, t)\, \mathrm{d}\bar{W}_t \\
&\quad - \frac{b(\vartheta, t)\, f(t)}{\varepsilon\sigma(t)} \left[m(\vartheta, t) - m(\vartheta_0, t)\right] \mathrm{d}t, \qquad m(\vartheta, t_0).
\end{aligned}$$

If ϑ_0 is known, then

$$\begin{aligned}
\mathrm{d}m(\vartheta_0, t) &= \left[a(t) - \frac{b(\vartheta_0, t)\, f(t)}{\varepsilon\sigma(t)}\right] m(\vartheta_0, t)\, \mathrm{d}t + \frac{b(\vartheta_0, t)}{\varepsilon\sigma(t)} \mathrm{d}X_t \\
&= a(t)\, m(\vartheta_0, t)\, \mathrm{d}t + b(\vartheta_0, t)\, \mathrm{d}\bar{W}_t, \qquad m(\vartheta_0, t_0), \qquad t_0 \le t \le T.
\end{aligned} \tag{7.141}$$

Remark that the initial values $m(\vartheta, t_0)$ and $m(\vartheta_0, t_0)$ for these equations have exponentially small contribution in the calculation of $m(\vartheta, t)$ and $m(\vartheta_0, t)$ for any $t > t_0$ and we can always put, say, $m(\vartheta, t_0) = y_0$ and $m(\vartheta_0, t_0) = y_0$.

Thus to construct (7.138) we need the solution of the PDE

$$u'_t + a(t)\, y u'_y + \frac{b(\vartheta,t)^2}{2} u''_{yy} = -F\left(t, y, u, b(\vartheta,t)\, u'_y\right), \quad u(T, y, \vartheta_0) = \Phi(y). \tag{7.142}$$

Note that this equation coincides with the PDE (7.137). Therefore we return to the solution of the problem with the observable forward equation.

Suppose that we have the solutions $u(t, y, \vartheta), \vartheta \in \Theta$ of (7.142). Recall our program: first we estimate the parameter ϑ with the help of One-step MLE-process $\vartheta^\star_{t,\varepsilon}$, calculate the estimate $m^\star_t$ of $m(\vartheta_0, t)$, and then we obtain the estimator $Z^\star_t$, $S^\star_t$ of the random processes Z_t, $s(t)$ satisfying Eq. (7.138) by the following substitution of $\vartheta^\star_{t,\varepsilon}$

$$Z^\star_t = u\left(t, m^\star_t\right), \qquad S^\star_t = \sigma(t)\, u'_y\left(t, m^\star_t\right).$$

The last step is to evaluate the errors $Z^\star_t - Z_t$ and $S^\star_t - s(t)$.

Preliminary Estimators

Fix some *learning interval* $[0, \tau]$ and by observations $X^\tau = (X_t, 0 \le t \le \tau)$ estimate the integral

$$\Psi_\tau(\vartheta) = \int_0^\tau f(t)^2\, b(\vartheta, t)^2\, dt.$$

The estimator of $\Psi_\tau(\vartheta)$ is

$$\Psi^*_{\tau,\varepsilon} = \sum_{i=0}^{N_{\tau,\varepsilon}-1} \left(\frac{X_{t_{i+1}+\varepsilon} - X_{t_{i+1}}}{\varepsilon} - \frac{X_{t_i+\varepsilon} - X_{t_i}}{\varepsilon}\right)^2.$$

Here $t_{i+1} - t_i = \varepsilon^{1/3}$ and $N_{\tau,\varepsilon} = \left[\tau\, \varepsilon^{-1/3}\right]$ ($[A]$ is integer part of A). Remind that if the condition $\breve{\mathscr{S}}_1$ holds, then for any $p > 0$

$$\mathbf{E}_{\vartheta_0} \left|\Psi^*_{\tau,\varepsilon} - \Psi_\tau(\vartheta_0)\right|^p \le C\, \tau^{p/2} \varepsilon^{p/3}.$$

For the proof see Proposition 3.1.

Introduce the MME $\vartheta^*_{\tau,\varepsilon}$ as the solution of the equation

$$\Psi_\tau\left(\vartheta^*_{\tau,\varepsilon}\right) = \Psi_{\tau,\varepsilon}.$$

Conditions $\mathscr{S}^*$.

$\mathscr{S}_1^*$. *The function $b(\vartheta, t)$, $\vartheta \in \Theta, t \in [0, T]$ has three continuous derivatives w.r.t. ϑ.*

$\mathscr{S}_2^*$. *Identifiability condition: the function $\Psi_\tau(\vartheta)$, $\vartheta \in \bar{\Theta}$ is strictly monotone.*

Suppose for simplicity of exposition that $\Psi_\tau(\vartheta)$, $\vartheta \in \bar{\Theta}$ is increasing function. If the condition $\mathscr{S}_2^*$ holds and $\Psi_\tau\left(\vartheta_{\tau,\varepsilon}^*\right) \notin \Psi(\Theta)$, where $\Psi(\Theta) = (\Psi_\tau(\vartheta), \vartheta \in \Theta)$, then we can put $\vartheta_{\tau,\varepsilon}^* = \alpha$ (in the case $\Psi_\tau\left(\vartheta_{\tau,\varepsilon}^*\right) \leq \Psi_\tau(\alpha)$), otherwise $\vartheta_{\tau,\varepsilon}^* = \beta$.

Proposition 7.9 *Let the conditions $\breve{\mathscr{S}}_1$ and $\mathscr{S}^*$ hold. Then the MME $\check{\vartheta}_{\tau,\varepsilon}$ is uniformly consistent and for any $p > 0$ there exists a constant $C = C(p) > 0$ such that*

$$\sup_{\vartheta_0 \in \Theta} \varepsilon^{-p/2} \mathbf{E}_{\vartheta_0} \left|\vartheta_{\tau,\varepsilon}^* - \vartheta_0\right|^p \leq C.$$

For the proof of this and more general results see Proposition 3.2. Note that by condition $\mathscr{S}_2^*$ the derivative $\dot{\Psi}_\tau(\vartheta) > 0$.

One-Step MLE-Process

Once more we remind that this is particular case of the problem studied in Sect. 3.1. To introduce the One-step MLE-process we need the random processes $m\left(\vartheta_{\tau,\varepsilon}^*, s\right)$, $\tau \leq s \leq T$ and $\dot{m}\left(\vartheta_{\tau,\varepsilon}^*, s\right)$, $\tau \leq s \leq T$, defined by the equations

$$\mathrm{d}m\left(\vartheta_{\tau,\varepsilon}^*, t\right) = \left[a(t) - \frac{b\left(\vartheta_{\tau,\varepsilon}^*, t\right) f(t)}{\varepsilon\sigma(t)}\right] m\left(\vartheta_{\tau,\varepsilon}^*, t\right) \mathrm{d}t + \frac{b\left(\vartheta_{\tau,\varepsilon}^*, t\right)}{\varepsilon\sigma(t)} \mathrm{d}X_t,$$

$$\mathrm{d}\dot{m}\left(\vartheta_{\tau,\varepsilon}^*, t\right) = \left[a(t) - \frac{b\left(\vartheta_{\tau,\varepsilon}^*, t\right) f(t)}{\varepsilon\sigma(t)}\right] \dot{m}\left(\vartheta_{\tau,\varepsilon}^*, t\right) \mathrm{d}t + \frac{\dot{b}\left(\vartheta_{\tau,\varepsilon}^*, t\right)}{\varepsilon\sigma(t)} \mathrm{d}X_t - \frac{\dot{b}\left(\vartheta_{\tau,\varepsilon}^*, t\right) f(t)}{\varepsilon\sigma(t)} m\left(\vartheta_{\tau,\varepsilon}^*, t\right) \mathrm{d}t$$

with the initial values $m\left(\vartheta_{\tau,\varepsilon}^*, \tau\right)$ and $\dot{m}\left(\vartheta_{\tau,\varepsilon}^*, \tau\right)$, respectively.

Therefore

$$m\left(\vartheta_{\tau,\varepsilon}^*, t\right) = m\left(\vartheta_{\tau,\varepsilon}^*, \tau\right) \Phi\left(\vartheta_{\tau,\varepsilon}^*, \tau, t\right) + \int_\tau^t \Phi\left(\vartheta_{\tau,\varepsilon}^*, s, t\right) \frac{b\left(\vartheta_{\tau,\varepsilon}^*, s\right)}{\varepsilon\sigma(s)} \mathrm{d}X_s,$$

$$\dot{m}\left(\vartheta_{\tau,\varepsilon}^*, t\right) = \dot{m}\left(\vartheta_{\tau,\varepsilon}^*, \tau\right) \Phi\left(\vartheta_{\tau,\varepsilon}^*, \tau, t\right) + \int_\tau^t \Phi\left(\vartheta_{\tau,\varepsilon}^*, s, t\right) \frac{\dot{b}\left(\vartheta_{\tau,\varepsilon}^*, s\right)}{\varepsilon\sigma(s)} \mathrm{d}X_s - \int_\tau^t \Phi\left(\vartheta_{\tau,\varepsilon}^*, s, t\right) \frac{\dot{b}\left(\vartheta_{\tau,\varepsilon}^*, s\right) f(s)}{\varepsilon\sigma(s)} m\left(\vartheta_{\tau,\varepsilon}^*, s\right) \mathrm{d}s,$$

where we denoted

$$\Phi(\vartheta, s, t) = \exp\left(\int_s^t \left[a(v) - \frac{b\left(\vartheta_{\tau,\varepsilon}^*, v\right) f(v)}{\varepsilon\sigma(v)}\right] \mathrm{d}v\right).$$

For the calculation of initial values we consider similar equations on the time interval $[0, \tau]$ and replace the stochastic integral by the ordinary one following the relation (see Lemma 1.3)

$$\int_0^t H(\vartheta, s)\, \mathrm{d}X_s = H(\vartheta, t)\, X_t - \int_0^t X_s H_s'(\vartheta, s)\, \mathrm{d}s.$$

Another possibility is to omit the initial terms because thanks to the function $\Phi(\vartheta, \tau, t)$ the contribution of these terms in the calculation of the values $m\left(\vartheta_{\tau,\varepsilon}^*, t\right)$ and $\dot{m}\left(\vartheta_{\tau,\varepsilon}^*, t\right)$ is exponentially small.

If $\vartheta - \vartheta_0 = o(1)$, then the following representation takes place

$$\begin{aligned}\frac{f(t)\left[m(\vartheta, t) - m(\vartheta_0, t)\right]}{\varepsilon\sigma(t)} &= \frac{1}{\sqrt{\varepsilon}} \frac{\sqrt{f(t)}\left[b(\vartheta, t) - b(\vartheta_0, t)\right]}{\sqrt{2b(\vartheta_0, t)\sigma(t)}} \xi_{t,\varepsilon}\left(1 + o(1)\right) \\ &= \frac{(\vartheta - \vartheta_0)}{\sqrt{\varepsilon}} \frac{\sqrt{f(t)}\,\dot{b}(\vartheta_0, t)}{\sqrt{2b(\vartheta_0, t)\sigma(t)}} \xi_{t,\varepsilon}\left(1 + o(1)\right),\end{aligned}$$

where $\xi_{t,\varepsilon} \Rightarrow \xi_t$ and $\xi_t \sim \mathcal{N}(0, 1)$, $\tau \leq t \leq T$ are independent r.v.'s (see details in Sect. 3.2.2). Therefore the Fisher information is

$$\mathrm{I}(\vartheta) = \int_0^T \frac{f(s)\,\dot{b}(\vartheta, s)^2}{2b(\vartheta, s)\,\sigma(s)} \mathrm{d}s$$

and the Fisher information function is

$$\mathrm{I}_\tau^t(\vartheta) = \int_\tau^t \frac{f(s)\,\dot{b}(\vartheta, s)^2}{2b(\vartheta, s)\,\sigma(s)} \mathrm{d}s, \qquad \tau \leq t \leq T.$$

Now the One-step MLE-process can be defined by the usual formula

$$\vartheta_{t,\varepsilon}^\star = \vartheta_{\tau,\varepsilon}^* + \frac{1}{\mathrm{I}_\tau^t\left(\vartheta_{\tau,\varepsilon}^*\right)} \int_\tau^t \frac{f(s)\dot{m}(\vartheta_{\tau,\varepsilon}^*, s)}{\varepsilon\sigma(s)^2} \left[\mathrm{d}X_s - f(s)m(\vartheta_{\tau,\varepsilon}^*, s)\mathrm{d}s\right],$$

where $\tau < t \leq T$.

Further, introduce the random processes

$$\eta_{t,\varepsilon}^\star = \frac{\vartheta_{t,\varepsilon}^\star - \vartheta_0}{\sqrt{\varepsilon}}, \qquad \tau \leq t \leq T,$$

$$\eta_t^\star(\vartheta_0)=\frac{1}{\mathrm{I}_\tau^t(\vartheta_0)}\int_\tau^t\frac{\dot b(\vartheta_0,s)\sqrt{f(s)}}{\sqrt{2b(\vartheta_0,s)\sigma(s)}}\,\mathrm{d}w(s),\qquad \tau\leq t\leq T,$$

where $w(s), 0\leq s\leq T$ is some standard Wiener process.

We add one more condition.

$\mathscr{S}^\star$. *For any* $t_0\in(\tau,T]$

$$\inf_{\vartheta\in\Theta}\mathrm{I}_\tau^{t_0}(\vartheta)>0.$$

Remark that if the condition $|\dot b(\vartheta,\tau)|>c$, where the constant $c>0$ does not depend on ϑ holds, then this condition is fulfilled.

Proposition 7.10 *Let the conditions* $\check{\mathscr{S}}_1$, $\mathscr{S}^*$ *and* $\mathscr{S}^\star$ *hold, then the One-step MLE-process* $\vartheta_{t,\varepsilon}^\star$, $\tau<t\leq T$ *is uniformly consistent: for any* $\nu>0$ *and any* $t_0\in(\tau,T]$

$$\mathbf{P}_{\vartheta_0}\left(\sup_{t_0\leq t\leq T}\left|\vartheta_{t,\varepsilon}^\star-\vartheta_0\right|>\nu\right)\longrightarrow 0, \tag{7.143}$$

the stochastic process $\eta_{t,\varepsilon}^\star$, $t_0\leq t\leq T$ *converges in distribution in the measurable space* $(\mathcal{C}[t_0,T],\mathfrak{B})$ *to the random process* $\eta_t^\star(\vartheta_0)$, $t_0\leq t\leq T$

$$\eta_{\cdot,\varepsilon}^\star\Longrightarrow\eta_\cdot^\star(\vartheta_0),\qquad \eta_t^\star(\vartheta_0)\sim\mathcal{N}\left(0,\mathrm{I}_\tau^t(\vartheta_0)^{-1}\right). \tag{7.144}$$

Proof This is a simplified version of Theorem 3.2 because now the function $f(t)$ does not depend on ϑ and the parameter ϑ is one-dimensional. This proof is short, and we present it here too.

Consider the normalized difference

$$\frac{\vartheta_{t,\varepsilon}^\star-\vartheta_0}{\sqrt{\varepsilon}}=\frac{\vartheta_{\tau,\varepsilon}^*-\vartheta_0}{\sqrt{\varepsilon}}+\frac{1}{\mathrm{I}_\tau^t(\vartheta_{\tau,\varepsilon}^*)}\int_\tau^t\frac{f(s)\dot m(\vartheta_{\tau,\varepsilon}^*,s)}{\sqrt{\varepsilon}\sigma(s)}\mathrm{d}\bar W_s$$
$$+\frac{1}{\mathrm{I}_\tau^t(\vartheta_{\tau,\varepsilon}^*)}\int_\tau^t\frac{f(s)^2\dot m(\vartheta_{\tau,\varepsilon}^*,s)}{\varepsilon^{3/2}\sigma(s)^2}\left[m(\vartheta_0,s)-m(\vartheta_{\tau,\varepsilon}^*,s)\right]\mathrm{d}s.$$

We have the relations (see Sect. 3.2.2 for details)

$$m(\vartheta_{\tau,\varepsilon}^*,s)-m(\vartheta_0,s)=(\vartheta_{\tau,\varepsilon}^*-\vartheta_0)\dot m(\tilde\vartheta_{\tau,\varepsilon},s),$$

$$m(\vartheta_{\tau,\varepsilon}^*,s)-m(\vartheta_0,s)=(\vartheta_{\tau,\varepsilon}^*-\vartheta_0)\dot m(\vartheta_0,s)+\frac{1}{2}(\vartheta_{\tau,\varepsilon}^*-\vartheta_0)^2\ddot m(\tilde\vartheta_{\tau,\varepsilon},s),$$

$$\dot m(\vartheta_0,s)=\sqrt{\frac{\varepsilon\sigma(s)}{2b(\vartheta_0,s)f(s)}}\,\dot b(\vartheta_0,s)\,\xi_{s,\varepsilon}+\varepsilon\,R_{t,\varepsilon},$$

$$\mathrm{I}_\tau^t(\vartheta_{\tau,\varepsilon}^*)^{-1} = \mathrm{I}_\tau^t(\vartheta_0)^{-1} + (\vartheta_{\tau,\varepsilon}^* - \vartheta_0)\, Q_{t,\varepsilon}.$$

Here the functions $R_{t,\varepsilon}$ and $Q_{t,\varepsilon}$ are defined by these equations.

For any $\nu > 0$ we can write

$$\begin{aligned}
\mathbf{P}_{\vartheta_0}\left(\sup_{t_0\leq t\leq T}\left|\vartheta_{t,\varepsilon}^\star - \vartheta_0\right| > \nu\right) &\leq \mathbf{P}_{\vartheta_0}\left(\left|\vartheta_{\tau,\varepsilon}^* - \vartheta_0\right| \geq \frac{\nu}{3}\right)\\
&+ \mathbf{P}_{\vartheta_0}\left(\mathrm{I}_\tau^{t_0}(\vartheta_{\tau,\varepsilon}^*)^{-1}\left|\int_\tau^T \frac{f(s)\,\dot m(\vartheta_{\tau,\varepsilon}^*, s)}{\sigma(s)}\mathrm{d}\bar W_s\right| \geq \frac{\nu}{3}\right)\\
&+ \mathbf{P}_{\vartheta_0}\left(\frac{\left|\vartheta_{\tau,\varepsilon}^* - \vartheta_0\right|}{\mathrm{I}_\tau^{t_0}(\vartheta_{\tau,\varepsilon}^*)}\int_\tau^T \frac{f(s)^2\left|\dot m(\vartheta_{\tau,\varepsilon}^*, s)\dot m(\tilde\vartheta_{\tau,\varepsilon}, s)\right|}{\sqrt{\varepsilon}\sigma(s)^2}\mathrm{d}s \geq \frac{\nu}{3}\right).
\end{aligned}$$

Now the convergence (7.143) follows from the consistency of $\vartheta_{\tau,\varepsilon}^*$ and the following estimate of the moments of $\dot m(\cdot,\cdot)$: for any $p > 0$

$$\sup_{\vartheta_0\in\Theta}\sup_{\tau\leq t\leq T}\mathbf{E}_{\vartheta_0}\left|\dot m(\vartheta_0, t)\right|^p \leq C\varepsilon^{p/2}.$$

The proof of this estimate follows from the direct calculations.

Moreover, we have the convergences

$$\int_\tau^t \frac{f(s)^2\,\dot m(\vartheta_{\tau,\varepsilon}^*, s)^2}{\varepsilon\sigma(s)^2}\mathrm{d}s = \int_\tau^t \frac{f(s)\,\dot b(\vartheta_0, s)^2\,\xi_{s,\varepsilon}^2}{2b(\vartheta_0, s)\,\sigma(s)}\mathrm{d}s\,(1 + o(1)) \longrightarrow \mathrm{I}_\tau^t(\vartheta_0),$$

$$\begin{aligned}
\int_\tau^t \frac{f(s)\,\dot m(\vartheta_{\tau,\varepsilon}^*, s)}{\sqrt{\varepsilon}\sigma(s)}\mathrm{d}\bar W_s &= \int_\tau^t \frac{\sqrt{f(s)}\,\dot b(\vartheta_0, s)\,\xi_{s,\varepsilon}}{\sqrt{2b(\vartheta_0, s)\sigma(s)}}\mathrm{d}\bar W_s\,(1 + o(1))\\
&\Longrightarrow \mathcal{N}\left(0, \mathrm{I}_\tau^t(\vartheta_0)\right).
\end{aligned}$$

The random processes $R_{t,\varepsilon}$, $Q_{t,\varepsilon}$ have bounded polynomial moments.

Hence we can write the representation

$$\begin{aligned}
\frac{\vartheta_{t,\varepsilon}^\star - \vartheta_0}{\sqrt{\varepsilon}} &= \frac{\vartheta_{\tau,\varepsilon}^* - \vartheta_0}{\sqrt{\varepsilon}} + \frac{1}{\mathrm{I}_\tau^t(\vartheta_0)}\int_\tau^t \frac{\dot b(\vartheta_0, s)\sqrt{f(s)}\,\xi_{s,\varepsilon}}{\sqrt{2b(\vartheta_0, s)\sigma(s)}}\mathrm{d}\bar W_s + o(1)\\
&\quad - \frac{(\vartheta_{\tau,\varepsilon}^* - \vartheta_0)}{\sqrt{\varepsilon}}\frac{1}{\mathrm{I}_\tau^t(\vartheta_0)}\int_\tau^t \frac{\dot b(\vartheta_0, s)^2\,f(s)\,\xi_{s,\varepsilon}^2}{2b(\vartheta_0, s)\,\sigma(s)}\mathrm{d}s + \frac{(\vartheta_{\tau,\varepsilon}^* - \vartheta_0)^2}{\sqrt{\varepsilon}}P_{t,\varepsilon}\\
&= \frac{1}{\mathrm{I}_\tau^t(\vartheta_0)}\int_\tau^t \frac{\dot b(\vartheta_0, s)\sqrt{f(s)}\,\xi_{s,\varepsilon}}{\sqrt{2b(\vartheta_0, s)\sigma(s)}}\mathrm{d}\bar W_s + o(1) + \left(\frac{\vartheta_{\tau,\varepsilon}^* - \vartheta_0}{\varepsilon^{1/3}}\right)^2 P_{t,\varepsilon}\varepsilon^{1/6}.
\end{aligned} \tag{7.145}$$

The random process $P_{t,\varepsilon}$ has bounded polynomial moments. From this representation it follows that the One-step MLE-process is asymptotically normal: for all $t \in (\tau, T]$

$$\frac{\vartheta^\star_{t,\varepsilon} - \vartheta_0}{\sqrt{\varepsilon}} \Longrightarrow \eta^\star_t(\vartheta_0) \sim \mathcal{N}\left(0, \mathrm{I}^t_\tau(\vartheta_0)^{-1}\right).$$

The representation (7.145) allows us to obtain the convergence of the finite-dimensional distributions too

$$\left(\eta^\star_{t_1,\varepsilon}, \ldots, \eta^\star_{t_k,\varepsilon}\right) \Longrightarrow \left(\eta^\star_{t_1}(\vartheta_0), \ldots, \eta^\star_{t_k}(\vartheta_0)\right) \tag{7.146}$$

for any $k \geq 2$ and any $t_0 \leq t_1 < \ldots < t_k \leq T$.

Let us verify the condition

$$\mathbf{E}_{\vartheta_0} \left|\eta^\star_{t_1,\varepsilon} - \eta^\star_{t_2,\varepsilon}\right|^4 \leq C\left|t_1 - t_2\right|^2 \tag{7.147}$$

which together with the convergence (7.146) provides the weak convergence (7.144) of the random process $\eta^\star_{t,\varepsilon}$, $t_0 \leq t \leq T$. Introduce notation

$$J_1(t) = \int_\tau^t \frac{f(s)\,\dot m(\vartheta^*_{\tau,\varepsilon}, s)}{\sqrt{\varepsilon}\sigma(s)} \mathrm{d}\bar W_s, \qquad K(t) = \mathrm{I}^t_\tau(\vartheta^*_{\tau,\varepsilon})^{-1},$$

$$J_2(t) = \int_\tau^t \frac{f(s)^2\,\dot m(\vartheta^*_{\tau,\varepsilon}, s)[m(\vartheta_0, s) - m(\vartheta^*_{\tau,\varepsilon}, s)]}{\varepsilon\sigma(s)^2} \mathrm{d}s$$

Then we can write

$$\begin{aligned}
\mathbf{E}_{\vartheta_0}\left|\eta^\star_{t_1,\varepsilon} - \eta^\star_{t_2,\varepsilon}\right|^4 &\leq C\mathbf{E}_{\vartheta_0}\left|K(t_1)\,J_1(t_1) - K(t_2)\,J_1(t_2)\right|^4 \\
&\quad + C\mathbf{E}_{\vartheta_0}\left|K(t_1)\,J_2(t_1) - K(t_2)\,J_2(t_2)\right|^4 \\
&\leq C\mathbf{E}_{\vartheta_0}\left|(K(t_1) - K(t_2))\,J_1(t_1)\right|^4 \\
&\quad + C\mathbf{E}_{\vartheta_0}\left|(J_1(t_1) - J_1(t_2))\,K(t_2)\right|^4 \\
&\quad + C\mathbf{E}_{\vartheta_0}\left|(K(t_1) - K(t_2))\,J_2(t_1)\right|^4 \\
&\quad + C\mathbf{E}_{\vartheta_0}\left|(J_2(t_1) - J_2(t_2))\,K(t_2)\right|^4.
\end{aligned}$$

Using once more the estimates (7.132) we obtain

$$\begin{aligned}
&\mathbf{E}_{\vartheta_0}\left|K(t_1) - K(t_2)\right|^8 \leq C\left|t_2 - t_1\right|^8, && \mathbf{E}_{\vartheta_0}\left|J_1(t_2)\right|^8 \leq C, \\
&\mathbf{E}_{\vartheta_0}\left|(J_1(t_1) - J_1(t_2))\right|^8 \leq C\left|t_2 - t_1\right|^4, && \mathbf{E}_{\vartheta_0}\left|K(t_2)\right|^8 \leq C, \\
&\mathbf{E}_{\vartheta_0}\left|(J_2(t_1) - J_2(t_2))\right|^8 \leq C\left|t_2 - t_1\right|^8.
\end{aligned}$$

These estimates and the Cauchy-Schwartz inequality allow us to verify (7.147) and therefore to obtain (7.144). □

Approximation

Recall that the model is given by Eqs. (7.135) and (7.136), and the equations of Kalman-Bucy filter are reduced to Eq. (7.141) for $m_t = m(\vartheta_0, t)$

$$\mathrm{d}m_t = a(t)\, m_t\, \mathrm{d}t + b(\vartheta_0, t)\, \mathrm{d}\bar{W}_t, \qquad m_{t_0}, \qquad t_0 \le t \le T \tag{7.148}$$

where we omitted terms of order $O(\varepsilon)$.

As we wrote above if the process $Y^T = (Y_t, 0 \le t \le T)$ is observable, the parameter ϑ_0 is known and we have the solution of PDE (7.142)

$$u'_t + a(t)\, y u'_y + \frac{1}{2} b(\vartheta_0, t)^2 u''_{yy} = -F\left(t, y, u, b(\vartheta_0, t)\, u'_y\right), \; u(T, y, \vartheta_0) = \Phi(y)$$

then the BSDE

$$\mathrm{d}Z_t = -F(t, Y_t, Z_t, S_t)\, \mathrm{d}t + S_t\, \mathrm{d}V_t, \qquad Z_T = \Phi(Y_T), \qquad 0 \le t \le T,$$

can be obtained as follows: we set

$$Z_t = u(t, Y_t, \vartheta_0), \qquad S_t = b(\vartheta_0, t)\, u'_y(t, Y_t, \vartheta_0)$$

and apply Itô formula to Z_t.

As Y^T is hidden, Eq. (7.135) is replaced by the Eq. (7.148) which almost coincides with (7.135) but with one essential difference: in (7.148) we have innovation Wiener process. There is no difference between PDEs constructed in these two cases, but the BSDE in the second case is obtained by setting $Z_t^\circ = u(t, m_t, \vartheta_0)$, $S_t^\circ = b(\vartheta_0, t)\, u'_y(t, m_t, \vartheta_0)$ and applying once more the Itô formula

$$\mathrm{d}Z_t^\circ = -F\left(t, m_t, Z_t^\circ, S_t^\circ\right) \mathrm{d}t + S_t^\circ\, \mathrm{d}\bar{W}_t, \qquad Z_T^\circ = \Phi(Y_T) \qquad t_0 \le t \le T. \tag{7.149}$$

Remind that m_t is the best mean squared approximation of the process Y_t, and our goal is first to construct an adaptive filter $m_t^\star$ and then to propose an estimator $Z_t^\star, S_t^\star$ of the processes Z_t°, S_t°.

Note that the processes Z_t, S_t are defined on the whole interval $[0, T]$ and the process Z_t°, S_t° on the interval $[t_0, T]$ with any $t_0 \in (0, T]$. We do not describe the behavior of m_t at the vicinity of $t = 0$ and propose to extend the equation for Z_t° on the whole interval $[0, T]$ with the initial value $m(\vartheta_0, 0) = y_0$. Recall that Eq. (7.148) was obtained with the help of the asymptotics of the Riccati equation (Lemma 7.12).

As we have no $m(\vartheta_0, t)$ we introduce the adaptive Kalman filter (see (3.37))

$$\mathrm{d}m^{\star}_{t,\varepsilon} = \left[a(t) - \frac{f(t)\, b\left(\vartheta^{\star}_{t,\varepsilon}, t\right)}{\varepsilon\sigma(t)}\right] m^{\star}_{t,\varepsilon}\mathrm{d}t + \frac{b\left(\vartheta^{\star}_{t,\varepsilon}, t\right)}{\varepsilon\sigma(t)}\mathrm{d}X_t,$$

$$m^{\star}_{t_0,\varepsilon} = y_0,\ t_0 \le t \le T.$$

Suppose that we have a family

$$\mathcal{U} = \Big\{ (u(t, y, \vartheta), t_0 \le t \le T, y \in \mathcal{R}), \vartheta \in \Theta \Big\}$$

of solutions $u(t, y, \vartheta)$ of the PDE (7.142) and set

$$Z^{\star}_t = u\left(t, m^{\star}_{t,\varepsilon}, \vartheta^{\star}_{t,\varepsilon}\right), \qquad S^{\star}_t = b\left(\vartheta^{\star}_{t,\varepsilon}, t\right) u'_y\left(t, m^{\star}_{t,\varepsilon}, \vartheta^{\star}_{t,\varepsilon}\right), \quad t_0 \le t \le T. \tag{7.150}$$

We have no representation of $Z^{\star}_t$ as solution of the equation like (7.149). The stochastic differential of $Z^{\star}_t$ involves the differentials of $m^{\star}_{t,\varepsilon}$ and $\vartheta^{\star}_{t,\varepsilon}$. It can be written but will not be interesting for us. Therefore the proposed approximation of the solution of BSDE (7.149) is $Z^{\star}_t, t_0 \le t \le T$, where the volatility $s(t)$ is approximated with the help of $S^{\star}_t$. The errors of approximations $Z^{\star}_t - Z^{\circ}_t$ and $S^{\star}_t - S^{\circ}_t$ can be studied directly.

Introduce the *Conditions* $\mathscr{S}^{\diamond}$.

$\mathscr{S}^{\diamond}_1$. *The function* $u(t, y, \vartheta); t, y, \vartheta \in [0, T] \times \mathcal{R} \times \Theta$ *has two continuous derivatives w.r.t.* y *and* ϑ *and the derivatives have polynomial in* $|y|$ *majorants.*

$\mathscr{S}^{\diamond}_2$. *The functions* $F(t, y, u, s)$ *and* $\Phi(y)$ *satisfy the conditions* (7.126) *and* (7.127).

$\mathscr{S}^{\diamond}_3$. *The derivatives* $u'_y(\cdot), \dot{u}(\cdot), u'_\varepsilon(\cdot)$ *have polynomial majorants in* y.

Remark that $Z_t = u(t, m(\vartheta_0, t), \vartheta_0)$ is solution of BSDE

$$\mathrm{d}Z_t = -F(t, m(\vartheta_0, t), Z_t, s(t))\,\mathrm{d}t + s(t)\,\mathrm{d}\bar{W}_t, \qquad Z_T = \Phi(m(\vartheta_0, T)),$$

where $s(t) = b(\vartheta_0, t)\, u'(t, m(\vartheta_0, t), \vartheta_0)$. As the Eqs. (7.135) and (7.148) (up to $O(\varepsilon)$) coincide, the equations for $u(\cdot)$ coincide too. The difference is in the Wiener processes.

The corresponding limit BSDE is

$$\mathrm{d}Z_t = -F(t, Y_t, Z_t, s_t)\,\mathrm{d}t + s_t \mathrm{d}V_t, \qquad Z_T = \Phi(Y_T),$$

where $Z_t = U(t, Y_t, \vartheta_0), s_t = b(\vartheta_0, t)\, U'_y(t, Y_t, \vartheta_0)$.

We do not put $m\left(\vartheta^{\star}_{t,\varepsilon}, t\right)$ and $Z^{\star}_t = u\left(t, m\left(\vartheta^{\star}_{t,\varepsilon}, t\right), \vartheta^{\star}_{t,\varepsilon}, \varepsilon\right)$ because computationally this can be a difficult problem.

Theorem 7.13 *Let the conditions of Proposition 7.10 and $\mathscr{S}^{\diamond}_{1-2}$ hold. Then the process $Z^{\star}_t, \tau < t \leq T$ defined in (7.150) has the following convergence*

$$\frac{Z^{\star}_t - Z_t}{\sqrt{\varepsilon}} \Longrightarrow u'_y(t, Y_t, \vartheta_0)\sqrt{\frac{b(\vartheta_0, t)\,\sigma(t)}{2f(t)}}\left[\xi^{(1)}_t - \xi^{(2)}_t\right]$$

$$+\frac{\dot u(t, Y_t, \vartheta_0)}{\mathrm{I}^t_\tau(\vartheta_0)}\int_\tau^t\sqrt{\frac{\dot b(\vartheta_0, s)^2 f(s)}{2b(\vartheta_0, s)\sigma(s)}}\,\mathrm{d}w(s).$$

Here $\xi^{(1)}_t \sim \mathcal{N}(0,1), \xi^{(2)}_t \sim \mathcal{N}(0,1)$, and $w(s), \tau \leq s \leq T$ are mutually independent random variables and Wiener process.

Proof We have the relation

$$\begin{aligned}
Z^{\star}_t - Z_t &= u\left(t, m^{\star}_{t,\varepsilon}, \vartheta^{\star}_{t,\varepsilon}\right) - u(t, Y_t, \vartheta_0)\\
&= u\left(t, m^{\star}_{t,\varepsilon}, \vartheta^{\star}_{t,\varepsilon}\right) - u\left(t, m(\vartheta_0, t), \vartheta^{\star}_{t,\varepsilon}\right)\\
&\quad + u\left(t, m(\vartheta_0, t), \vartheta^{\star}_{t,\varepsilon}\right) - u(t, m(\vartheta_0, t), \vartheta_0)\\
&\quad + u(t, m(\vartheta_0, t), \vartheta_0) - u(t, Y_t, \vartheta_0)\\
&= u'_y\left(t, \tilde m_t, \vartheta^{\star}_{t,\varepsilon}\right)\left(m^{\star}_{t,\varepsilon} - m(\vartheta_0, t)\right)\\
&\quad + \dot u_\vartheta(t, m(\vartheta_0, t), \tilde\vartheta_{t,\varepsilon})\left(\vartheta^{\star}_{t,\varepsilon} - \vartheta_0\right)\\
&\quad + u'_y(t, \bar m_t, \vartheta_0)(m(\vartheta_0, t) - Y_t)\\
&= u'_y(t, Y_t, \vartheta_0)\left(m^{\star}_{t,\varepsilon} - m(\vartheta_0, t)\right)\left(1 + O\left(\sqrt{\varepsilon}\right)\right)\\
&\quad + \dot u_\vartheta(t, Y_t, \vartheta_0)\left(\vartheta^{\star}_{t,\varepsilon} - \vartheta_0\right)\left(1 + O\left(\sqrt{\varepsilon}\right)\right)\\
&\quad + u'_y(t, Y_t, \vartheta_0)(m(\vartheta_0, t) - Y_t)\left(1 + O\left(\sqrt{\varepsilon}\right)\right).
\end{aligned}$$

Here $\tilde m_t, \bar m_t, \tilde\vartheta_{t,\varepsilon}$ are some intermediate points in the corresponding expansions. We have to describe the quantities $\delta_\vartheta(t) = \vartheta^{\star}_{t,\varepsilon} - \vartheta_0$, $\delta_Y(t) = m(\vartheta_0, t) - Y_t$ and $\delta_m(t) = m^{\star}_t - m(\vartheta_0, t)$. Recall the relations obtained before (see (7.145), (3.38), and (3.39))

$$\delta_\vartheta(t) = \frac{1}{\mathrm{I}^t_\tau(\vartheta_0)}\int_\tau^t\frac{\dot b(\vartheta_0, s)\sqrt{f(s)}\,\xi_{s,\varepsilon}}{\sqrt{2b(\vartheta_0, s)\sigma(s)}}\,\mathrm{d}\bar W_s\,\sqrt{\varepsilon}\,(1 + o(1)),$$

$$\delta_Y(t) = \sqrt{\frac{b(\vartheta_0, t)\,\sigma(t)}{2f(t)}}\left(\xi^{(2)}_{t,\varepsilon} - \xi^{(1)}_{t,\varepsilon}\right)\sqrt{\varepsilon}\,(1 + o(1)).$$

Therefore

$$\frac{Z_t^\star - Z_t}{\sqrt{\varepsilon}} \Longrightarrow u_y'(t, Y_t, \vartheta_0, 0)\sqrt{\frac{b(\vartheta_0, t)\,\sigma(t)}{2f(t)}}\left[\xi_t^{(2)} - \xi_t^{(1)}\right]$$

$$+\frac{\dot u(t, Y_t, \vartheta_0, 0)}{\mathrm{I}_\tau^t(\vartheta_0)}\int_\tau^t \sqrt{\frac{\dot b(\vartheta_0, s)^2 f(s)}{2b(\vartheta_0, s)\sigma(s)}}\,\mathrm{d}w(s).$$

Here $\xi_t^{(1)} \sim \mathcal{N}(0,1), \xi_t^{(1)} \sim \mathcal{N}(0,1)$ and $w(s), 0 \le s \le T$ are mutually independent random variables and Wiener process. □

Similar limit can be written for the process $s_t^\star, \tau < t \le T$ too.
Let us introduce the random process

$$z(t, \vartheta_0, Y_t) = \frac{\dot u(t, Y_t, \vartheta_0, 0)}{\mathrm{I}_\tau^t(\vartheta_0)}\int_\tau^t \sqrt{\frac{\dot b(\vartheta_0, s)^2 f(s)}{2b(\vartheta_0, s)\sigma(s)}}\,\mathrm{d}w(s)$$

and note that the Gaussian process $Y_t, 0 \le t \le T$ and Wiener process $w(t), 0 \le t \le T$ are independent.

Corollary 7.1 *Let the conditions of Theorem 7.13 and $\mathscr{S}_3^\diamond$ hold. Then for any continuous functions $h(\cdot)$ we have the relation*

$$\varepsilon^{-1/2}\int_\tau^T h(t)\left[Z_t^\star - Z_t\right]\mathrm{d}t \Longrightarrow \int_\tau^T h(t)\,z(t, \vartheta_0, Y_t)\,\mathrm{d}t.$$

Proof The proof follows from the limits

$$\int_\tau^T r(Y_t, t)\,\xi_{t,\varepsilon}^{(1)}\mathrm{d}t \longrightarrow 0, \qquad \int_\tau^T g(Y_t, t)\,\hat\xi_{t,\varepsilon}^{(2)}\mathrm{d}t \longrightarrow 0$$

for any continuous functions $r(\cdot), g(\cdot)$ with finite moments (see (3.23)). □

Corollary 7.2 *Let the conditions of Theorem 7.13 and $\mathscr{S}_3^\diamond$ hold. Then for any $t \in (\tau, T]$*

$$\varepsilon^{-1}\mathbf{E}_{\vartheta_0}\left(\hat Z_t - Z_t\right)^2 \longrightarrow \frac{b(\vartheta_0, t)\,\sigma(t)}{f(t)}\mathbf{E}_{\vartheta_0}u_y'(t, Y_t, \vartheta_0)^2 + \frac{\mathbf{E}_{\vartheta_0}\dot u(t, Y_t, \vartheta_0)^2}{\mathrm{I}_\tau^t(\vartheta_0)}.$$

Proof The proof follows from the same arguments as in Corollary 7.1 and (3.25). □

The contribution of the approximation of the conditional expectation $m(\vartheta_0, t)$ by the value $m_t^\star$ of solution of recurrent equation (adaptive filtration) is negligible. This means that for this model of observations (7.122) and (7.123) with unknown parameter of volatility function Eq. (7.125) proposes the approximation $\hat m_t$ of $m(\vartheta_0, t)$ has error of order ε. This result can be applied in filtration theory.

Bibliographical Notes

The bibliography presented in this work is in no sense exhaustive. We mention mainly the publications that are closely related with the exposed results or were used to prove these results.

Chapter 1

Section 1.1 It is supposed that the reader is already familiar with the basic notions of the stochastic calculus including the stochastic integral, Itô formula, stochastic differential equation, likelihood ratio formula etc., which can be found in many books devoted to stochastic calculus. See, e.g., [47, 86, 141, 142, 159] or any other. The contents in the subsection on Kalman–Bucy filter (proposed in [84]) (Theorem on normal correlation, Innovation lemma, and the proof of the Theorem 1.3) follow exposition of these results in Liptser and Shiryaev [141, 142]. The expression (1.26) is taken from Arato [3]. More about solutions of Riccati equations can be foud in [14]. The central limit theorems for stochastic integrals can be found, e.g., in [105, 106] or in Jacod and Shiryaev [75].

Section 1.2 The presented methods of estimation (MME, MDE, TFE) are well-known and given here just as illustration of such estimators in the case of partially observed systems. Note that for such systems, the estimators MDE and TFE were introduced in [107]. For homogeneous diffusion processes, another TFE was proposed by Kasonga [87]. Theorems 1.9–1.12 were published in [73] and are given here for the convenience of citations. They are used to describe the properties of the MLE and BE in all considered in this work models. One-step MLE process is the key estimator of this work in the construction of adaptive filter. It was observed by Fisher [55], Le Cam [135] that one iteration of Gauss–Newton and a $\sqrt{n}$-consistent preliminary estimator allow to obtain the asymptotically efficient estimator called One-step MLE. See as well section 7.3 in Höpfner [69] for the details of Le

Y. A. Kutoyants, *Hidden Markov Processes and Adaptive Filtering*, Springer Series in Statistics, https://doi.org/10.1007/978-3-032-00052-1

Cam's construction of One-step MLE for LAN and LAMN (locally asymptotically mixing normal) families. Moreover, later it was understood as well that the rate of the preliminary estimator can be even worse, e.g., n^{δ}, where $\delta \in (1/4, 1/2]$ and neveretheless the One-step MLE will be asymptotically efficient (see, e.g., Theorem 5.5.4 in Zacks [200], Dzhaparidze [44]). For continuous-time stochastic processes, this effect can be seen, for example, in the works of Golubev [62] or Skorokhod and Khasminskii [177]. The One-step MLE process was introduced in [110, 128] and then applied to different models of observations [112–117, 119–122, 124, 126, 129]. My interest in multistep estimators [112] was attracted by the work of Kamatani and Uchida [85], where such estimators were proposed in the case of discrete-time observations of diffusion processes.

Section 1.3 The contents here are presented for the illustration of the proposed in this work construction of adaptive filters. The chosen homogeneous partially observed system allows to easily understand the computational advantage of recurrent estimation of the unknown parameter and adaptive filter.

Section 1.4 Cramér–Rao bound is a well-known inequality, which can be found in many books of mathematical statistics. For different versions of Van Trees [190] bound, see Gill and Levit [59]. The notion of LAN was introduced by Le Cam [134] and for Theorem 1.13 see Hájek [64]. Note that it was proved for the wide class of loss functions (see, e.g., [73, Theorem 2.12.1]) but for the simplicity of exposition we present it here in the case of quadratic loss funcion. The Bayesian estimators are described for the quadratic loss functions only by the same reason.

Chapter 2

Section 2.1 The MDE for this model of observations was introduced in [107] and studied in the works [48, 177]. MDE and TFE are discussed in [108, p. 271]. The details can be found in [108, Section 7.5]. For the general theory of optimality of MDE, see [153, 154]. Theorem 2.3 is new. The asymptotic normality of the MLE and BE was described in [108], and the asymptotic expansion by the powers of ε was obtained in [127]. The One-step MLE process, adaptive filter and its asymptotic efficiency were introduced in [129].

Section 2.2 The problem of initial value estimation for a slightly different partially observed system was discussed in [94]. The results of this section are published in [124]. We return to the problem of initial value and parameter estimation later in Sect. 7.2 for a different model. Remark that for nonlinear systems, such problem is much more complicate, see, e.g., Bobrovsky et al. [18] and the references therein.

Section 2.3 The equations of filtering for conditionally Gaussian partially observed systems were introduced by Liptser [139, 140] see also Chapters 11 and 12 in [142]. The results of this section are new but are direct generalization of the results of Sect. 2.1. Filtering theory for conditionally Gaussian processes (no unknown parameters) was developed by Bensoussan, Section 6.3 in [10].

Section 2.4 For nonlinear continuous-time systems due to difficulties related to the calculation of the conditional expectation, there are different approaches to the estimation of unknown parameters, see, e.g., Hijab [68], Campillo and Le Gland [22]. Le Gland and Wang [136] proved LAN for such models. James and Le Gland [76] proposed conditions of the consistency of MLE. Skorokhod and Khasminskii [177] verified LAN and showed the consistency and asymptotic efficiency of One-step estimator for slightly different nonlinear model (see Remark 2.15). Le Breton and Roubaud [133] proposed approximate filters for semilinear and nonlinear systems with colored noise. See as well Krener [100, 101]. The presented here results are new.

Section 2.5 For the linear systems with $\delta \in (0, 1/3]$ the properties of MLE were described in [114]. The results of this section published in [123]. Picard [163, 164] studied approximation of conditional expectation for nonlinear models with small noises of different orders in both equations (no unknown parameter). An exhaustive study of perturbation methods in nonlinear filtering problems of such models is given by Bensoussan, [10, Chapter 5]. Remark that the problem of parameter estimation for nonlinear models was studied by Konecny [99]. Theorems 7.9 and 7.10 are from [108] and Theorem 7.11 is new.

Chapter 3

Section 3.1 The results of this section were established in [113] and [119].

Section 3.2 The results of this section are new. For the properties of LAMN families, see, e.g., Höpfner [69].

Section 3.3 The results of this section were established in [122]. The properties of MDE in the case of n independent observations of one nonlinear partially observed system were studied in [125] in the asymptotics $n \to \infty$. The problem of approximation of conditional expectation for nonilinear partially observed systems (no unknown parameter) was studied by many authors. For example, Bobrovsky and Zakai [17] proposed asymptotic lower and upper bounds on the error of approximation. Then Katzur et al. [88, 89] used Kushner's [104] and Zakai's [199] equations and described the asymptotic expansions of the optimal filter. The EKF for nonlinear sytems was studied by Picard [165]. See as well analytical approach in nonlinear filtering problems for such models in Bensoussan [9, 10].

Chapter 4

Section 4.1 This section contains the well-known results (see [84] for the equations (4.3) and (4.4) and [3] for the asymptotic (4.5)).

Section 4.2 The construction of MME follows the same procedure as in the work [92] (see Sect. 5.2 of the next chapter). The asymptotic of TFE for the partially observed multidimensional homogeneous system was studied in the work [11]. The consistency and asymptotic normality of the MLE for homogeneous partially observed systems were established by many authors, see, e.g., Bagchi [5], Bagchi and Borkar [6], Le Breton [131, 132], Wei [192, 193], and some others. The one-dimensional case of Theorem 4.3 was published in [106, 109] and two-dimensional parameter case was studied by Kallianpur and Selukar [82]. Presented proof here is new. The properties of estimators including One-step MLE process discussed in this section were published in [115]. The conditions of identifiability and the equations for the derivatives of the conditional expectations are described in the work of Surace and Pfister [185] where numerical simulations of EM estimation method were realized too.

Section 4.3 These results were published in [121]. Another approach for construction of consistent recursive least squares estimators of unknown parameters and adaptive filter was proposed by van Schuppen [189]. The properties of the MLE in the case of multidimentional observation and hidden processes and discrete time observations for partially observed linear system were described by Kurisaki [103].

Note that the approximation of the filter in the case of nonlinear partially observed systems (with and without unknown parameters) extensively studied for discrete-time models and large samples asymptotics, but without strict mathematical justification. The authors proposed some reasonable algorithms, which then were justified by numerical simulations. See, e.g., the work by Haykin [67] and references (about 900) therein.

Chapter 5

Section 5.1 Equation (5.7) was obtained by Wonham [197]. For the invariant density see, e.g., [93].

Section 5.2 Theorems 5.1 and 5.4 were published in [92]. The properties of MLE in the case $\lambda = \mu = \vartheta$ (Theorem 5.3 but without condition $\mathscr{L}_1^\star$) are due to Chigansky [36].

Section 5.3 The results of this sections are new, but they are close to the results obtained in [93] (no unknown parameter).

Chapter 6

Section 6.1 This section contains the description of the model of observations. For the equations (6.3) and (6.4)), see Kalman [83].

Section 6.2 The problems of parameter estimation for the partially observed disceret-time systems and in particular the MME and MLE were studied by many authors (see,e.g. "Notes ..." given by van Handel [188], p.112–113). We can mention here as well Bellanger [8], Liung [144, 145], Chui and Chen [39], Hazewinkel and Willems (Ed's) [179], Haykin [67]. A new method of filtering for time series was proposed by Kunimoto and Sato [102]. Note that in [67] the recursive minimum mean-square estimation procedures were proposed too. For the recursive MLE methods, see Campillo and Mevel [23]. See as well the works of Mehra [151, 152]. The convergence of moments of the MLE and BE (Theorem 6.2) is new. The results of this section and Sect. 6.3 were published in [120]. See as well [126].

Chapter 7

Section 7.1 The results of this section concerning the fixed source were published in [117]. This is a *Gaussian version* of the similar problems studied in the case of Poisson observations and published in [34] (smooth case), [32] (cusp type singularity), [54] (change-point type singularity). See as well [118] Sect. 5.3. There is a vide engineering literature on these and related problems, see, e.g., Handbook [65] and references therein. The discussion of other problems (source tracking, one signal, two sources, large samples) is given without strict proofs because these proofs can be realized with the help of the presented before results for similar model of observations. For the problem of tracking moving source, see the poissonian version in [33].

Section 7.2 The properties of the MLE (Theorem 7.6) were described in the work Lazareva and Khasminskii [94]. The presented different proof here allows to give an explicite expression for the Fisher information matrix.

Section 7.3 The statement of the problem follows Frydman and Lakner [57], where the authors proposed EM algorithm for estimation of the unknown parameters. For the numerical realization of the EM algorithm for partially obsevred system see, e.g., Avido and Milheiro-Oliveira [4]. The results of this section are due to the Author [123].

Section 7.4 Theorems 7.9 and 7.10 were published in [108, Section 6.3]. Note that for the model with change point in the observations, the MLE can be consistent even in the case of misspecification [24, 108]. The problem of parameter estimation for the model with double disorder (in observation and state equations) we started with Le Gland and Campillo in 1992, but that time it was not finished. This publication (Theorem 7.11) can be considered as the finalization of it. The change-point probem for a different partially observed system was studied by Campillo [20, 21]. Similar problems for discrete-time systems with some applications were discussed in Gustaffson [63]. A survey of change-point problems for partially observed systems before 1976 can be found in the work of Willsky [195]. For another survey of the works published in 1954–1980, see Kligene and Telksnis [97].

Section 7.5 The first part of this section concerning the observable forward equation is taken from [128]. There are two rewiews of the works of this type problems [110] and [111]. The second part (hidden forward equation) was published in [35].

Bibliography

1. Abolmasoumi, A.H., Mohammadian, M. and Mili, L. (2022) Robust Kalman filter state estimation for gene regulatory networks. *IEEE/ACM Transactions on Computational Biology and Bioinformatics,* 20, 2, 1395–1405.
2. Almagbile, A., Wang, J. and Ding, W. (2010) Evaluating the performances of adaptive Kalman filter methods in GPS/INS integration. *Journal of Global Positioning Systems,* 9, 1, 33–40.
3. Arato, M. (1983) *Linear Stochastic Systems with Constant Coefficients. A Statistical Approach*, Lecture Notes in Control and Inform.Sciences, vol.45, Springer, New York.
4. Avido, N.O.B. and Milheiro-Oliveira, P. (2025) Parameter estimation of a partially observed hypoelliptic stochastic linear system. *Mathematics,* 13, 529.
5. Bagchi, A. (1981) A note on asymptotically efficient estimates of parameters in continuous-time dynamical systems. *SIAM. J. Control Optim.*, **19**, 2, 169–176.
6. Bagchi, A., Borkar,V. (1984) Parameter identification in infinite dimensional linear systems. *Stochastics*, **12**, 201–213.
7. Balakrishnan, A.V. (1973) *Stochastic Differential Systems. Filtering and Control: A Function Space Approach.* Springer, Berlin.
8. Bellanger, M.G. (2001) *Adaptive Digital Filters and Signal Analysis,* 2nd ed. Marcel Dekker, New York.
9. Bensoussan, A. (1988) On some approximation techniques in nonlinearfiltering, in *Proc. Stochastic Differential Systems, Stochastic Control and Applications*, Minneapolis, MN, Springer, Berlin,
10. Bensoussan, A., (1992) *Stochastic Control of Partially Observable Systems,* Cambridge University Press.
11. Bertrand P., Kutoyants Y.A. (1996) A minimum distance estimator for partially observed linear stochastic systems. *Statistics & Decisions*, 14, 323–342.
12. Bickel, P.J., Ritov, Y. and Rydén, T. (1988) Asymptotic normality of the maximum likelihood estimator for general hidden Markov models, *Ann. Statist.* 26, 1614–1635.
13. Bielecki, T.R. and Pliska, S.R. (1999). Risk sensitive dynamic asset management. *J. Appl.Math. Optim.* 39, 337–360.
14. Bittani, S., Laub, A.J. and Willems, J.C. (Eds) (1991) *The Riccati Equation.* Springer, Berlin.
15. Billingsley P. (1999) *Convergence of Probability Measures*, 2nd Ed., J.Wiley & Sons, N.Y.
16. Bismut, J.M. (1973) Conjugate convex functions in optimal stochastic control. *J. Math. Anal. Appl.*, 44, 384–404.
17. Bobrovsky, B.Z., Zakai, M.M. (1982) Asymptotic a priori estimates for the error in the nonlinear filtering problem. *IEEE Trans. Inform. Theory,* IT-28, 2, 371–376.

Y. A. Kutoyants, *Hidden Markov Processes and Adaptive Filtering*, Springer Series in Statistics, https://doi.org/10.1007/978-3-032-00052-1

18. Bobrovsky, B.Z., Zakai, M.M. and Zeitouni, O. (1988) Error bounds for the nonlinear filtering of signals with small diffusion coefficients. *IEEE Trans. Inform. Theory,* 34, 4, 710–721.
19. Burkholder, D.L., Davis, B. Gundy, R.F. (1972) Integral inequalities for convex functions of operators on martingales , *Proc. 6th Berkeley Symp. Math. Statistics and Probability* , 2, 223–240.
20. Campillo, F., (1986) La méthode d'approximation de Gauss-Galerkin en filtrage non linéaire. *RAIRO Modélisation Mathématique et Analyse Numérique (M2AN)*, 20, 2, 203–223.
21. Campillo, F., (2010) *Quelques applications des processus de diffusion: filtrage/statistique—controle - homogénéisation.* Mathématiques. Université Rennes 1, https://theses.hal.science/tel-00485401.
22. Campillo, F., Le Gland F. (1989) MLE for partially observed diffusions: direct maximization vs. the EM algorithm. *Stochastic Proces. App.*, 33, 2, 245–274.
23. Campillo, F., Mevel, L. (2005) Recursive maximum likelihood estimation for structural health monitoring: tangent filter implementations. in *Proceedings of the 44th IEEE Conference on Decision and Control,and the European Control Conference*, Dec. 2005, Seville, Spain, 5923–5928.
24. Campillo F., Le Gland F., Kutoyants Y. (2000) Small noise asymptotics of the glr test for off-line change detection in misspecified diffusion processes. *Stochastics and Stochastic Reports*, 70, 1, 109–129.
25. Capinski, M. and Polanski, A. (2015) Parameter estimation in systems biology models by using extended Kalman filter. In: Gruca, A., Brachman, A., Kozielski, S., Czachórski, T. (eds) *Man-Machine Interactions 4. Advances in Intelligent Systems and Computing*, Springer, Cham, 195–204.
26. Cappé, O., Moulines, E. and Rydén, T. (2005) *Inference in Hidden Markov Models*, Springer, New York.
27. Chan, K.S. (1993) Consistency and limiting distribution of the LSE of a TAR, *Ann. Stat.*, 21, 520–533.
28. Chan, K. S. (2010) Exploration of a Nonlinear World: An Appreciation of Howell Tong's Contributions to Statistics, World Scientific, Singapore.
29. Chan, N.H. and Kutoyants, Y.A. (2010) Recent developments of threshold estimation for nonlinear time series. *J. Japan Statist. Soc.*, 40, 2, 277–303.
30. Chan, N.H. and Kutoyants, Y.A. (2012) On parameter estimation of threshold autoregressive models. *Statist. Inference Stoch. Processes.*, 15, 81–104.
31. Chernoyarov, O.V., Dachian, S., Farinetto, C. and Kutoyants, Y.A. (2024) Localization of two radioactive sources on the plane. *Statist. Inf. Stoch. Proc.* 27, 1, 1–23.
32. Chernoyarov, O.V., Dachian S. and Kutoyants, Y.A. (2020) Poisson source localization on the plane. Cusp case. *Ann. Inst. Statist. Math.*, 72,5, 1137–1157.
33. Chernoyarov, O.V., Dachian, S.Y., Kutoyants, Y.A. (2025) Localization of moving Poisson source on the plane., to appear in *Ann. Inst. Stat. Math.*
34. Chernoyarov, O. V. and Kutoyants, Y.A. (2020) Poisson source localization on the plane. Smooth case. *Metrika*, 83, 4, 411–435.
35. Chernoyarov, O.V. and Kutoyants, Y.A. (2025) Approximation of BSDE with hidden forward equation and unknown volatility. *Econometrics and Statistics*, 36, 119–132.
36. Chigansky, P. (2009) Maximum likelihood estimation for hidden Markov models in continuous time, *Statist. Inf. Stoch. Proc.* 12, 139–163.
37. Chigansky, P. and van Handel, R. (2007) Model robustness of finite state nonlinear filtering over the infinite time horizon. *Ann. Appl. Probab.*, 17, 2, 688–715.
38. Chigansky, P. and Kutoyants, Y.A. (2012) On nonlinear TAR processes and threshold estimation. *Math. Methods Statist.*, 21, 2, 142–152.
39. Chui, C.K. and Chen, G. (2009) *Kalman Filtering with Real-time Applications.* 4th Ed., Springer, Berlin.

40. Davies, R.B. (1985) Asymptotic inference when the amount of information is random. in: L. Le Cam and R. Olsden, (Eds), *Proc. of the Berkley Symp. in Honor of J. Neymann and J. Kiefer,* vol. II, Wadsworth, 1985.
41. Dedecker, J., Doukhan, P., Lang, G., Leon, J.R., Louhichi, S., Prieur, C. (2007) *Weak Dependence: With Examples and Applications.* Springer, N.Y.
42. Dembo, A. and Zeitouni, O. (1986) Parameter estimation for partially observed continuous time processes via the EM algorithm. *Stoch. Proces. Applic.* 23, 91–113.
43. Diniz, P.S.R. (2020) *Adaptive Filtering. Algorithms and Practical Implementation* 5th ed., Springer, Cham.
44. Dzhaparidze, K.O. (1975) On simplified estimators of unknown parameters with good asymptotic properties. *Theory Probab. Appl.*, 19, 2, 347–358.
45. Douc, R., Moulines, E., Olsson, J. and van Handel, R. (2011) Consistency of the maximum likelihood estimator for general hidden Markov models. *Ann. Statist.,* 39, 1, 474–513.
46. Duflo, M. (2010) *Random Iterative Models.* Springer, New York.
47. Durett, R. (1996) *Stochastic Calculus: A Practical Introduction.* Boca Raton: CRC Press.
48. Ebady A.M. and Kutoyants Y.A. (1992) On minimum distance estimator of parameter of Kalman filter,(in Russian) *Utchen. Zap. Erev. Univ.*, 2, 89–94.
49. El Karoui N., Peng S. and Quenez M. (1997) Backward stochastic differential equations in finance. *Math. Fin.*, 7, 1–71.
50. Elliott, R.J., Aggoun, L. and Moor, J.B. (1995) *Hidden Markov Models*, Springer, New York.
51. Elliott, R.J. and Siu, T.K. (2011) A BSDE approach to a risk-based optimal investment of an insurer. *Automatica*, 47, 2, 253–261.
52. Elliott, R.J. and Siu, T.K. (2012) A BSDE approach to convex measures for derivative securities. *Stocha. Anal. Appl.*, 30, 6, 1083–1101.
53. Ephraim, Y. and Mehrav, N. (2002) Hidden Markov processes, *IEEE Trans. Inform. Theory*, 48, 1518–1569.
54. Farinetto, C., Kutoyants, Y.A. and Top, A. (2020) Radioactive source localization. Change-point case. *Ann. Inst. Statist. Math.*, 72, 3, 675–698.
55. Fisher, R.A. (1925) Theory of statistical estimation. *Proc. Cambridge Phylosophical Society.*, 22, 700–725.
56. Freidlin M.I. and Wentsell A.D. (1984) *Random Perturbations of Dynamical Systems*, Springer, New York.
57. Frydman, H. and Lakner, P. (2003) Maximum likelihood estimation of hidden Markov processes. *Ann. Appl. Probab.*, 13, 4, 1296–1312.
58. Genon-Catalot V. and Laredo C. (1986) *Non-parametric estimation for partially observed transient diffusion processes*, Stochastics, 18, 169–196.
59. Gill, R.D. and Levit, B.Ya. (1995) Applications of the van Trees inequality: A Bayessian Cramér–Rao bound. *Bernoulli*, 1/2, 59–79.
60. Gloter, A. and Jacod, J. (2001) Diffusion with measurement errors. I. Local asymptotic normality. *ESAIM: Probab. Stat.*, 5, 225–242.
61. Gloter, A. and Jacod, J. (2001) Diffusion with measurement errors. II. Optimal estimators. *ESAIM: Probab. Stat.*, 5, 243–260.
62. Golubev, G.K. (1984) Fisher's method of scoring in the problem of frequency estimation. *J. Soviet Mathematics*, 25, 1125–1139.
63. Gustaffson, F. (2000) *Adaptive Filtering and Change Detection*, J. Wiley & Sons, N.Y.
64. Hajek J. (1972) Local asymptotic minimax and admissibility in estimation. *Proc.6-th Berkeley Symp.on Math. Statist. and Probab.*, 1, 175–194.
65. *Handbook of Position Location: Theory, Practice and Advances*, (2019) Zekavat, S.A.R. and Buehrer, R.M., Ed's, 2 nd ed., John Wiley and Sons, Hoboken.
66. Harvey, A. (1987) Applications of the Kalman filter in econometrics. In *Advances in Econometrics: Fifth World Congress*, Bewley, Truman (ed's), Cambridge University Press, N.Y., 1, 285–313.
67. Haykin, S. (2014) *Adaptive Filter Theory.* Fifth Ed., Pearson Education, Boston.

68. Hidjab, O. (1981) Deterministic estimation and asymptotic stochastic estimation. In *Stochastic Systems: The Mathematics of Filtering and Identification and Applications.* M. Hazewinkel and J. C. Willems (eds.), D. Reidel, Dordrecht,
69. Höpfner, R. (2014) *Asymptotic Statistics, with a View to Stochastic Processes.* De Gruyter, Berlin.
70. Hörmander, L. (1967) Hypoelliptic second order differential equations. *Acta Math.* 119, 147–171.
71. Hu, C., Chen, W., Chen, Y. and Liu, D. (2003) Adaptive Kalman filtering for vehicule navigation. *J. Global Positioning systems*, 2, 1, 42–47.
72. Hutchinson, C. E.(1984) The Kalman filter applied to aerospace and electronic systems. *IEEE Transactions on Aerospace and Electronic Systems*, 20, 4, 500–504.
73. Ibragimov,I.A. and Khasminskii, R.Z. (1981) *Statistical Estimation: Asymptotic Theory*, Springer, New York.
74. Jacod, J., Li, Y., Mykland, P.A., Podolskij, M. and Vetter, M. (2009) Microstructure noise in the continuous case: The pre-averaging approach. *Stoch. Processes Appl.*, 119, 2249–2276.
75. Jacod, J. and Shiryaev, A.N. (2003) *Limit Theorems for Stochastic Processes*, Springer, Berlin.
76. James M., Le Gland F. (1995) Consistent parameter estimation for partially observed diffusions with small noise. *Appl. Math. Optim.*, 32, 47–72.
77. Jazwinski, A.H. (1970) *Stochastic Processes and Filrering Theory.* Academic Press, N.Y.
78. Jeganathan, P. (1982) On the asymptotic theory of estimation when the limit of the log-likelihood is mixed normal. Sankhya, Series A, 44, 173–212.
79. Kailath, T. (1974) A View of three decades of linear filtering theory. *IEEE Trans. Inform. Theory,* 20, 2, 146–181.
80. Kailath, T., Sayed, A.H. and Hassibi, B. (2000) *Linear Estimation.* Prentice Hall, New Jersey.
81. Kallianpur G. (1980) *Stochastic Filtering Theory,* Springer, New York.
82. Kallianpur G., Selukar R.S. (1991) Parameter estimation in linear filtering, *J. Multiv. Analysis*, 39, 284–304.
83. Kalman, R.E. (1960) A new approach to linear filtering and prediction problems. *ASME Journal of Basic Engineering,* 82, 35–45.
84. Kalman, R.E, and Bucy, R.S. (1961) New results in linear filtering and prediction theory. *Trans. ASME*, 83D, 95–100.
85. Kamatani, K. and Uchida, M. (2015) Hybrid multi-step estimators for stochastic differential equations based on sampled data. *Stat. Inference Stoch. Processes.* 18, 2, 177–204.
86. Karatzas, I., Shreve, S.E. (1991) *Brownian Motion and Stochastic Calculus.* New York: Springer-Verlag.
87. Kasonga, R.A. (1990) Parameter estimation by deterministic approximation of a stochastic differential equation, *Commun. Statist.-Stochastic Models,* 6, 1, 59–67.
88. Katzur, R., Bobrovsky, B.Z. and Schuss, Z. (1984) Asymptotic analysis of the optimal filtering problem for one-dimensional diffusions measured in a low noise channel, Part I, *SIAM J. Appl. Math.,* 44, 591–604.
89. Katzur, R., Bobrovsky, B.Z. and Schuss, Z. (1984) Asymptotic analysis of the optimal filtering problem for one-dimensional diffusions measured in a low noise channel, part II, *SIAM J. Appl. Math.,* 44, 1176–1191.
90. Khasminskii,R.Z. (1966) On stochastic processes defined by differential equations with small parameter, *Theory Probab. Appl.* 11, 240–259.
91. Khasminskii, R. (2012) *Stochastic Stability of Differential Equations.* 2nd Ed., Springer, Berlin.
92. Khasminskii, R.Z. and Kutoyants, Y.A. (2018) On parameter estimation of hidden telegraph process, *Bernoulli,* 24, 2064–2090.
93. Khasminskii, R.Z. and Lazareva, B.V. (1992) On some filtration procedure for jump Markov process observed in white Gaussian noise. *Ann. Statist.,* 20, 4, 2153–2160.
94. Khasminskii, R.Z. and Lazareva, B.V. (1996) Parameter estimation of a signal from linear indirect observations. *Math. Methods Statist.,* 5, 1, 61–76.

95. Khasminskii, R. and Zeitouni, O. (1996) Asymptotic filtering for finite state Markov chains. *Stochastic processes and their applications*, 63, 1, 1–10.
96. Kim, T. and Omberg, E. (1996) Dynamic nonmyopic portfolio behavior. *Rev. Financial Studies*, 9, 141–161.
97. Kligene and Telksnys, L. (1983) Methods of detecting instants of change of random process properties. *Autom. Remote Control,* 44, 10, 1241–1283.
98. Kolesnik, A.D. and Ratanov, N. (2013) *Telegraph Processes and Option Pricing,* Springer, Heidelberg.
99. Konecny F. (1990) Maximum likelihood estimation of a drift parameter from a partially observed diffusion in a case of small measurement noise, *Statistics and Decisions*, 8, 115–130.
100. Krener, A.J. (1986) The asymptotic approximation of nonlinear filters by linear filters. Byrnes, C.I. and Lindquist, A. (ed's) *Theory and Applications of Nonlinear Control Systems.* Elsevier North-Holland, Amsterdam.
101. Krener, A.J. (2003). The convergence of the extended Kalman filter. In: Rantzer, A., Byrnes, C.I. (eds)*Directions in Mathematical Systems Theory and Optimization.* Lecture Notes in Control and Information Sciences, vol 286, 3–13. Springer, Berlin, Heidelberg.
102. Kunitomo, N. and Sato, S. (2025) *The SIML Filtering Method for Noisy Non-stationary Economic Time Series.* Springer, Singapore.
103. Kurisaki, M. (2023) Parameter estimation for ergodic linear SDEs from partial and discrete observations, *Statist. Inference Stoch. Processes.*, 26, 2, 279–330.
104. Kushner, H.J. (1962) On the differential equations satisfied by conditional probablitity densities of Markov processes, with applications. *J. SIAM Control,* ser. A, 2, 1, 106–119.
105. Kutoyants, Y.A. (1975) On a problem of testing hypotheses and asymptotic normality of stochastic integrals.*Theory Probab. Appl.,* 20, 376–384.
106. Kutoyants, Y.A. (1984) *Parameter Estimation for Stochastic Processes*, Heldermann, Berlin.
107. Kutoyants, Y.A (1991) Minimum distance parameter estimation for diffusion type observations. *C.R. Acad. Sci. Paris,* 312, Serie I, 637–642.
108. Kutoyants, Y.A (1994) *Identification of Dynamical Systems with Small Noise*, Springer, Dordrecht.
109. Kutoyants, Y.A. (2004) *Statistical Inference for Ergodic Diffusion Processes*, Springer, London.
110. Kutoyants, Y.A. (2014) Approximation of the solution of the backward stochastic differential equation. Small noise, large samples and high frequency cases. *Proceedings of the Steklov Institute of Mathematics*, 287, 133–154.
111. Kutoyants, Y.A. (2016) On approximation of BSDE and Multi-step MLE-processes. *Probab., Uncertainty, and Quantitative Risk*, 1, 1, 1–22.
112. Kutoyants, Y.A. (2017) On the multi-step MLE-process for ergodic diffusion, *Stoch. Process. Appl.*, 127, 2243–2261.
113. Kutoyants, Y.A. (2019) On parameter estimation of hidden Ornstein-Uhlenbeck process. *J. Multivariate Anal.*, 169, 1, 248–263.
114. Kutoyants, Y.A. (2020) Hidden Markov model where higher noise makes smaller errors. ArXiv: 2010.07596.
115. Kutoyants, Y.A. (2019) On parameter estimation of hidden ergodic Ornstein-Uhlenbeck process. *Electron. J. Statist.*, 13, 4508–4526.
116. Kutoyants, Y.A. (2020) Parameter estimation for continuous time hidden Markov processes. *Autom. Remote Control*, 81, 3, 446–469.
117. Kutoyants, Y.A. (2021) On localization of source by hidden Gaussian processes with small noise, *Ann. Inst. Stat. Math.*, 73, 4, 671–702.
118. Kutoyants, Y.A. (2023) *Introduction to the Statistics of Poisson Processes.* Springer, Cham.
119. Kutoyants, Y.A. (2024) Volatility estimation of hidden Markov process and adaptive filtration, *Stoch. Processes Appl.*, July, 173, 104381.
120. Kutoyants, Y.A. (2024) Hidden AR process and adaptive Kalman filter. *Ann. Inst. Stat. Math.*, 77, 1, 61–103.

121. Kutoyants, Y.A. (2025) Hidden ergodic O-U process and adaptive Kalman filter. *Statist. Inference Stoch. Processes.*, 28, 1, https://doi.org/10.1007/s11203-024-09321-6
122. Kutoyants, Y.A. (2025) Extended adaptive Kalman filter with low noise observations. To appear in *Probability, Uncertainty and Quantitative Risk.*
123. Kutoyants, Y.A. (2025) Identification of the security prices process . Submitted.
124. Kutoyants, Y.A. (2025) Adaptive Kalman filter for systems with unknown initial values in the state equations. To appear in *Communications in Statistics - Theory and Methods.*
125. Kutoyants Y.A., Le Gland F., Rakotozafy R. (1993) *Identification d'un système non-lineaire partiellement observé par la méthode de la distance minimale*, INRIA, Rapports de Recherche N 2115, inria-00074547.
126. Kutoyants Y.A., Motrunich, A. (2016) On multi-step MLE-process for Markov sequences. *Metrika*, 79, 705–724.
127. Kutoyants Y.A., Pohlmann H. (1994) Parameter estimation for Kalman-Bucy filter with small noise,*Math. Operationsforsch. u. Statist., ser.Statist.*, 25, 307–323.
128. Kutoyants, Y.A. and Zhou, L. (2014) On approximation of the backward stochastic differential equation. *J. Stat. Plann. Infer.* 150, 111–123.
129. Kutoyants, Y.A. and Zhou, L. (2021) On parameter estimation of the hidden Gaussian process in perturbed SDE. *Electron. J. Statist.* 15, 1, 211–234.
130. Lakner, P. (1998) Optimal trading strategy for an investor: The case of partial information. *Stoch. Process. Appl.*, 76, 77–97.
131. Le Breton, A. (1976) *Sur l'Estimation de Paramètres dans les Modèles Différentielles Stochastiques Multidimensionnels.* Thèse d'Etat, Université Scientifique et Mdical de Grenoble.
132. Le Breton, A. (1977) Identifiability of a continuous time system with unknown noise covariance. *Recent Developments in Statistics.* J.R. Barra *et al.*, editors. North-Holland, Amsterdam, 515–523.
133. Le Breton, A. and Roubaud, M.-C. (2000) Asymptotic optimality of approximate filters in stochastic systems with colored noises. *SIAM J. Control Optim.*, 39, 3, 917–927.
134. Le Cam, L. (1960) Locally asymptotically normal families of distributions, *Univ. Calif.Publ. Statist.*, 3, 27–98.
135. Le Cam, L. (1974) *Notes on Asymptotic Methods in Statistical Decision Theory.* Centre de Recherches Mathématiques, University of Montrea, Montreal.
136. Le Gland, F., Wang, B. (2002) Asymptotic normality in partially observed diffusions with small noise: Application to FDI. In *Stochastic Theory and Control*, LNCIS 280, Springer, Berlin, 267–282.
137. Levanoy, D., Shwartz, A., Zeitouni, O. (1994) Recursive identification in continuous-time stochastic processes. *Stochastic Process. Appl.*, **49**, 245–275.
138. Leroux, B.G. (1992) Maximum-likelihood estimation for hidden Markov models. *Stoch. Process. Appl.*, 40, 127–143.
139. Liptser, R.S.(1967) On filtration and prediction of components in diffusion Markov processes. *Theory Probab. Appl.*, 12, 4, 697–698.
140. Liptser, R.S.(1974) Conditionally Gaussian random processes. *Problems Inform. Transmission,* 10, 2, 151–167.
141. Liptser, R.S. and Shiryayev, A.N. (2001) *Statistics of Random Processes, I: General Theory*, 2nd Ed., Springer, New York.
142. Liptser, R.S. and Shiryayev, A.N. (2001) *Statistics of Random Processes, II: Applications*, 2nd Ed., Springer, New York.
143. Liu, W., Principe, J.C. and Haykin, S. (2010) *Kernel Adaptive Filtering,* Wiley, New Jersey.
144. Ljung, L. (1979) Asymptotic behavior of the extended Kalman filter as a parameter estimator for linear systems. *IEEE Trans. Autom. Control* , 24(1), 36–50.
145. Ljung, L., (1987) *System Identification: Theory for the User*, Prentice Hall, New Jersey.
146. Luo, X. (2013) *GPS Stochastic Modeling*. Springer, New York.
147. Ma, J. and Yong, J. (1999) *Forward-Backward Stochastic Differential Equations and their Applications.* Lecture Notes in Mathematics. Springer, Berlin.

148. Mamon, R.S. and Elliott, R.J. (ed's) (2007) *Hidden Markov Models in Finance.* Springer, N.Y.
149. Mamon, R.S. and Elliott, R.J. (ed's) (2014) *Hidden Markov Models in Finance. Further Developments and Applications.* Springer, N.Y.
150. Manolakis, D.G., Ingle, V.K., Kogon, S.M. (2000) *Statistical and Adaptive Signal Processing,* McGraw Hill, New York.
151. Mehra, R.K. (1970) On the identification of variances and adaptive Kalman filtering. *IEEE Trans. Automat. Control,* AC-15, 175–184,
152. Mehra, R.K. (1971) On-Line identification of linear dynamic systems with applications to Kalman Filtering. *IEEE Trans. Automat. Control,* AC-16, 1, 12–21.
153. Millar P.W. (1983) The minimax principle in asymptotic statistical theory. *Lecture Notes in Math.*, 976, 76–265.
154. Millar P.W. (1984) A general approach to the optimality of minimum distance estimators, *Trans. Amer. Math. Soc.*, 286, 377–418.
155. Mohany, N. (1987) *Signals, Fitering and Detection*, Van Nostrand, N.Y.
156. Nakakita, S.H., Kaino, Y. and Uchida, M. (2021) Quasi-likelihood analysis and Bayes-type estimators of an ergodic diffusion plus noise. *Ann. Inst. Stat. Math.*, 73, 177–225.
157. Nevelson, M.B. and Khasminskii, R.Z. (1998) *Stochastic Approximation and Recursive Estimation.* American Mathematical Society, Providence, Rhode Island.
158. Novikov, A., Kordzakhia, N., Ling, T. (2014) On moments of Pitman estimators: the case of fractional Brownian motion. *Theory Prob. Appl.* 58, 4, 601–614.
159. Oksendal, B. (1995) *Stochastic Differential Equations.* 4th Ed., Springer, Berlin.
160. Pardoux, E. and Peng, S.(1990) Adapted solution of a backward stochastic differential equation. *SystemsControl Lett.*, 14, 1, 55–61.
161. Pardoux, E. and Peng, S.(1992) Backward stochastic differential equation and quasilinear parabolic differential equations. In *Stochastic Partial Differential Equations and Their Applications,* Lecture Notes in Control and Information Sciences, 176, 200–217.
162. Pardoux, E. and Răşcanu, A. (2014) *Stochastic Differential Equations, Backward SDEs, Partial Differential Equations.* Springer, Cham.
163. Picard, J. (1986) Nonlinear filtering of one-dimensional diffusions in the case of a high signal-to-noise ratio. *SIAM J. Appl. Math.* 46, 6, 1098–1125.
164. Picard, J. (1991) Efficiency of the extended Kalman filter for nonlinear systems with small noise. *SIAM J. Appl. Math.* 51, 843–885.
165. Picard, J. (1993) Estimation of the quadratic variation of nearly observed semimartingales with application to filtering, *SIAM J. Control Optim.*, 31 , pp. 494 517.
166. Pflug, G.C. (1982) A statistically important Gaussian process. *Stochastic Process. Appl.*, 13, 45–47.
167. Podolskij, M. and Vetter, M. (2009) Estimation of volatility functionals in the simultaneous presence of microstructure noise and jumps. Bernoulli, 634–658.
168. Reif, K., Günter, S., Yaz, E. and Unhbehauen, R. (2000) Stochastic stability of the continuous-time extended Kalman filter. *IEE Proc.-Control Theory Appl.*, 147, 1, 45–52.
169. Reiss, M. (2011) Asymptotic equivalence for inference on the volatility from noisy observations. *Ann. Statist.*, 39, 2, 772–802.
170. Rigatos, G.G. (2015) *Nonlinear Control and Filtering Using Differential Flatness Approaches.* Springer, Cham.
171. Rubin, H. and Song, K. (1995) Exact computation of the asymptotic efficiency of maximum likelihood estimators of a discontinuous signal in a Gaussian white noise. *Ann. Statist.*, 23, 3, 732–739.
172. Rutan, S.C. (1991) Adaptive Kalman filtering. *Anal. Chem.*, 63 (22), 1103A-1109A.
173. Särkkä, S. and Svensson, L. (2023) *Bayesian Filtering and Smoothing.* 2nd ed., Cambridge University Press, Cambridge.
174. Sayed, A.H. (2008) *Adaptive Filters.* John Wiley & Sons, New Jersey.
175. Schneider, W. (1988) Analytical uses of Kalman filtering in econometricsa survey. *Statistical Papers,* 29, 3–33.

176. Shumway, R.H. and Stoffer, D.S. (2017) *Time Series Analysis and Its Applications. With R Examples.* 4th Ed., Springer, Cham.
177. Skorokhod, A.V. and Khasminskii, R.Z. (1996) On parameter estimation by indirect observations. *Probl. Inf. Transm.*, 32, 1, 58–68.
178. Stepanov, O. A. (2011). Kalman filtering: Past and present. An outlook from Russia. (On the occasion of the 80th birthday of Rudolf Emil Kalman)". *Gyroscopy and Navigation.* 2 (2), 105.
179. *Stochastic Systems: The Mathematics of Filtering and Identification and Applications.* Hazewinkel, M. and Willems, J.C. (Ed's), D.Reidel, Dordrecht.
180. Storvik, G. (2002) Particle filters for state space models with the presence of unknown static parameters. *IEEE Trans. Signal Process.,* 50, 2, 281–289.
181. Stratonovich, R.L. (1959). Optimum nonlinear systems which bring about a separation of a signal with constant parameters from noise. *Radiofizika,* 2,6, 892–901.
182. Stratonovich, R.L. (1959). On the theory of optimal non-linear filtering of random functions. *Theory of Probab. Appl.*, 4, 223–225.
183. Stratonovich, R.L. (1960) Application of the Markov processes theory to optimal filtering. *Radio Engineering and Electronic Physics*, 5, 11, 1–19.
184. Stratonovich, R.L. (1960). Conditional Markov Processes. *Theory Probab. Appl.*, 5, 156–178.
185. Surace, S.C. and Pfister, J.-P. (2019) Online maximum-likelihood estimation of the parameters of partially observed diffusion processes. *IEEE Trans. Automat. Control*, 64, 7, 2814–2829.
186. Tamura, T., Watanabe, Y. (2011) Hypoellipticity and ergodicity of thewonham filter as a diffusion process. *Appl. Math. Optim.*, 64, 13–36.
187. Terent'yev A.S. (1968) Probability distribution of a time location of an absolute maximum at the output of a sinchronized filter, *Radioengineering and Electronics*, 13, 4, 652–657.
188. van Handel, R. (2008) *Hidden Markov Models*, Lecture Notes, Princeton (https://web.math.princeton.edu/~rvan/orf557/hmm080728.pdf).
189. van Schuppen, J.H. (1983) Convergence results for continuous-time adaptive stochastic filtering algorithms. *J. Math. Anal. App.,* 96, 209–225.
190. Van Trees, H.L. (1968) *Detection, Estimation and Modulation Theory.* Part I. Wiley, New York.
191. Wang, W., Liu, Z.-Y., Xie, R.-R. (2006) Quadratic extended Kalman filter approach for GPS/INS integration. *Aerospace Science and Technology,* 10, 8, 709–713.
192. Wei, C. (2018) Estimation for parameters in partially observed linear stochastic system. *IAENG International Journal of Applied Mathematics.* 48, 2, 118–122.
193. Wei, C. (2019) Estimation for incomplete information stochastic systems from discrete observations, *Adv. Difference Equ.,* 2019:2027, 1–16.
194. Widrow, B and Stearns, S.D. (1985) *Adaptive Signal Processing.* Prentice Hall, Englewood Cliffs.
195. Willsky, A.S. (1976) A survey of design methods for failure detection in dynamic systems. *Automatica, Journal of IFAC*, 12, 601–611.
196. Wells, C. (2010) *The Kalman Filter in Finance*. Springer, New York.
197. Wonham, W. M. (1965) Some applications of stochastic differential equations to optimal nonlinear filtering. *SIAM J. Contr.*, Ser. A, 2, 347–369.
198. Yu, X. and Li, J. (2021) Adaptive Kalman filter for linear systems with additive and multiplicative noises. *J. Latex Class Files.* 14, 8, 1–11.
199. Zakai, M. (1969) On the optimal filtering of diffusion processes. *Z. Wahrsch. Verw. Geb.*, 11, 230–243.
200. Zacks, S. (1971) *The Theory of Statistical Inference.* John Wiley & Sons, New York.
201. Zhang, J. (2017) *Backward Stochastic Differential Equations*. Springer, New York.

Index

A
Adaptive filter, 53, 118, 122, 131, 139, 149, 160, 231, 234, 235, 282, 388, 431, 613, 636
  EKF, 341
  EKF-1, 177, 309
  EKF-2, 187
Adaptive Kalman filter, 480, 483
Asymptotic derivatives, 328

B
Backward stochastic differential equation (BSDE)
  approximation, 625, 636
  Markovian case, 615

C
Central limit theorem, 27

E
EKF-1, 165
Empirical quadratic variation, 328
Equation
  backward, 615
  forward, 615, 616, 626
  Forward-Backward SDE, 615
  hidden forward, 616
  minimum distance, 79
  Riccati, 10, 191, 252, 388, 492, 547, 550, 613, 627
  Wonham, 405, 434
Estimator
  asymptotically efficient, 63, 67, 125, 254, 291, 363, 371, 421, 505, 514
  Bayesian, 45, 93, 157, 199, 201, 217, 252, 298, 362, 363, 420, 495, 536
  Cramér-Rao efficient, 57
  Hájek efficient, 63
  least squares, 520, 526, 529, 535
  locally asymptotically minimax, 85
  maximum likelihood, 40, 93, 157, 199, 201, 217, 252, 298, 362, 363, 420, 495, 536, 553
  method of moments, 36, 90, 143, 239, 297, 336, 348, 358, 396, 412, 426, 448, 479, 548, 564, 629
  method of substitution, 239
  minimum distance, 37, 78, 101, 133, 154, 168, 194, 249, 552, 580, 598, 605, 619
  One-step MDE process, 86
  One-step MLE, 50, 101, 110
  One-step MLE-process, 51, 101, 110, 117, 122, 136, 158, 160, 221, 224, 226, 227, 234, 274, 309, 377, 386, 427, 431, 465, 471, 480, 533, 546, 557, 573, 620, 622, 631
  One-step pseudo-MLE-process, 175, 186
  pseudo-MLE, 170, 183
  recurrent One-step MDE-process, 87
  recursive One-step MLE-process, 383, 386, 387, 396, 482, 483
  trajectory fitting, 38, 72
  Two-step MLE-process, 52
  Van Trees efficient, 60
Extended Kalman filter, 310
Extrapolation, 23

Y. A. Kutoyants, *Hidden Markov Processes and Adaptive Filtering*, Springer Series in Statistics, https://doi.org/10.1007/978-3-032-00052-1

F

Family of measures
  locally asymptotically mixing normal, 150
  locally asymptotically normal, 92
Fisher information, 92, 363, 372, 385, 452, 600, 620, 631
Fisher information function, 620, 631
  empirical, 308
  limit, 308
Fisher information matrix, 385, 421, 458, 462, 497, 533, 537, 538, 546, 549, 553
  empirical, 431

G

Grönwall's lemma, 26, 140, 155, 165, 167, 233, 285, 295, 498, 506

H

Hidden TAR process, 462

I

Inequality
  Burkholder-Davis-Gundy, 3
Infinitesimal rate matrix, 411
Initial value
  random, 142
  unknown, 127
Invariant density, 405

K

Kalman-Bucy filter, 10, 17, 23, 33, 48, 55, 70, 144, 191, 344, 543, 550, 578, 596, 604, 627
Kalman filter, 440, 442
  extended, 165
Kulback–Leibner divergency, 85

L

Likelihood function, 443
Likelihood ratio, 5, 405, 420, 422, 495, 513, 536, 553, 582
  limit, 504, 513, 582, 605
  normalized, 505
Local asymptotic mixing normal, 300
Local asymptotic normality, 61, 158, 208, 268, 497, 601
Lower bound
  for adaptive filter, 64, 124, 289, 398, 485
  Cramér–Rao, 56
  Hájek, 92, 254, 363, 421, 472, 497, 557, 600
  minimax, 504, 514, 582, 606
  Van Trees, 59

M

Maximum likelihood equation, 44
Minimum distance equation, 44

N

Normalized likelihood ratio, 40

O

Observation equation, 8

P

Popcorn noise, 403
Process
  autoregressive, 439
  conditionally Gaussian, 152, 293
  innovation Wiener, 12, 129, 405, 492, 627
  mean-reverting, 560
  Ornstein-Uhlenbeck, 8, 33, 343

R

Random telegraph signal, 403
Riemann's zeta function, 583

S

Single security, 560
Source detection
  change-point case, 513
  cusp-type case, 503
  smooth case, 496
Source tracking, 535
State equation, 8

T

Telegraph process, 403
Two sources, 542

U

Uniform local asymptotic normality, 94, 364, 422, 473

9783032000514